P9-API-387

Physics of
Semiconductor Devices

Physics of Semiconductor Devices

SECOND EDITION

S. M. Sze

Bell Laboratories, Incorporated
Murray Hill, New Jersey

A WILEY-INTERSCIENCE PUBLICATION

JOHN WILEY & SONS

New York • Chichester • Brisbane • Toronto • Singapore

Library of Congress Cataloging in Publication Data:

Sze, S. M., 1936-
Physics of semiconductor devices.

"A Wiley-Interscience publication."
Includes index.
1. Semiconductors. I. Title.

TK7871.85.S988 1981 537.6'22 81-213
ISBN 0-471-05661-8 AACR2

Printed in the United States of America

20 19 18 17 16 15 14 13 12

To My Wife

Preface

Since the invention of the bipolar transistor in 1947, the semiconductor device field has grown rapidly. Coincident with this growth, semiconductor-device literature has also burgeoned and diversified. For access to this massive amount of information, there is a need for a book giving a comprehensive introductory account of device physics and operational principles, with references. In 1969, the first edition of *Physics of Semiconductor Devices* was published with the intention of meeting such a need. It is perhaps somewhat surprising that the book has so long held its place as one of the main textbooks for advanced undergraduate and graduate courses, and as a major reference for scientists in semiconductor-device research and development.

In the last decade, more than 40,000 papers on semiconductor devices have been published, with numerous breakthroughs in device concepts and performance. The book clearly needed substantial revision if it were to continue to serve its purpose. In the second edition of *Physics of Semiconductor Devices*, 80 percent of the material has been revised or updated, and the material has been totally reorganized. About 1,000 references have been cited, of which 70 percent were published in the last decade, and over 600 technical illustrations are included, of which 65 percent are new.

Most of the important semiconductor devices are included, and they are divided into four groups: bipolar, unipolar, microwave, and photonic devices. A brief historical review is given in the introduction to each chapter. Subsequent sections present the physics and mathematical formulations of the devices. The sections are arranged in a logical sequence without heavy reliance on the original papers. Each chapter is more or less independent of the other chapters, so readers can use the book as a reference and instructors can rearrange the device chapters or select ones appropriate for their classes.

In the course of the writing many people have assisted me and offered their support. I would like, first, to express my appreciation to the management of Bell Laboratories for providing the environment in which I worked on the book; without their support this book could not have been written. I have benefited significantly from the suggestions of the reviewers: Drs. J. M. Andrews, D. E. Aspnes, W. E. Beadle, J. R. Brews, H. J. Boll, C. C. Chen, W. Fichtner, H. Fukui, H. K. Gummel, D. Kahng, T. P. Lee, M. P. Lepselter, E. H. Nicollian, W. C. Niehaus, P. T. Panousis, T. Paoli, R. M. Ryder, M. Shoji, G. E. Smith, K. K. Thornber, and S. H. Wemple of Bell Laboratories; Professors H. C. Casey, Duke University, C. R. Crowell, University of Southern California, W. S. Feng, National Taiwan University, S. K. Ghandhi, Rensselaer Polytechnic Institute, H. Kroemer, University of California, M. Lampert, Princeton University, H. Melchior, Swiss Federal Institute of Technology, R. H. Rediker, Massachusetts Institute of Technology, and H. W. Thim, Technical University of Vienna; and Drs. L. L. Chang, IBM Corporation, G. Gibbons, Plessey Research Limited, and R. N. Hall, General Electric Company.

I am further indebted to Ms. D. McGrew, Ms. J. Chee, Ms. K. R. Funk, and Mr. M. Lynch for technical editing of the entire manuscript, to Messrs. E. Labate, B. A. Stevens, and H. H. Teitelbaum for their literature searches, and to Ms. A. W. Talcott for providing more than five thousand technical papers on semiconductor devices cataloged at the Murray Hill Library of Bell Laboratories. Thanks are also due Mr. W. H. Shafer of the Center for Information and Numerical Data Analysis and Synthesis (CINDAS), Purdue University, for providing up-to-date references on semiconductor properties. I wish to thank Ms. J. T. McCarthy and Ms. V. J. Maye, who typed various sections of the book in its revision stage, and Mr. G. Holmfelt and members of the Drafting Department of Bell Laboratories, who furnished hundreds of technical illustrations used in this book. In each case where an illustration was used from another published source, I have applied for and received permission from the copyright holder even though all illustrations were then adapted and redrawn. I appreciate being granted these permissions. At my publishers, John Wiley and Sons, I want to acknowledge Mr. G. Novotny who encouraged me to undertake this new edition, and Ms. V. Aldzeris, Ms. R. Farkas and Mr. R. Fletcher who handled the production of this book. Finally, I wish especially to thank my wife Therese Ling-yi, my son Raymond, and my daughter Julia for their assistance in many ways, including typing the first draft and preparing the final manuscript.

S. M. Sze

Murray Hill, New Jersey
May 1981

Contents

Introduction

The book is divided into five parts:

Part 1: résumé of physics and properties of semiconductors
Part 2: bipolar devices
Part 3: unipolar devices
Part 4: microwave devices
Part 5: photonic devices

Part 1, Chapter 1, is intended as a summary of materials properties, to be used throughout the book as a basis for understanding and calculating device characteristics. Energy-band, carrier distribution, and transport properties are briefly surveyed, with emphasis on the three most important semiconductors: germanium (Ge), silicon (Si), and gallium arsenide (GaAs). A compilation of the recommended or the most accurate values for these semiconductors is given in the illustrations of Chapter 1 and in the appendixes for convenient reference.

Part 2, Chapters 2 through 4, treats bipolar devices in which both electrons and holes are involved in the transport processes. The basic device technology and *p-n* junction characteristics are considered in Chapter 2. Because the *p-n* junction is the building block of most semiconductor devices, the *p-n* junction theory serves as the foundation of the physics of semiconductor devices. Chapter 3 treats the bipolar transistor, that is, the interaction between two closely coupled *p-n* junctions. The bipolar transistor is one of the most important semiconductor devices. The invention of the bipolar transistor in 1947 was the beginning of the modern electronics era. The thyristor, which is basically three closely coupled *p-n* junctions in the form of a *p-n-p-n* structure, is discussed in Chapter 4. Thyristors have a wide range of power-handling capability; they can handle currents from a few milliamperes to thousands of amperes and voltages extending above 5000 V.

Part 3, Chapters 5 through 8, deals with unipolar devices in which only one type of carrier predominantly participates in the conduction mechanism. The metal–semiconductor contact, in Chapter 5, is electrically similar to a one-sided abrupt *p-n* junction, yet it can be operated as a majority-carrier device with inherent fast response. The metal–semiconductor contact on heavily doped semiconductors constitutes the most important form of ohmic contacts. The junction field-effect transistor (JFET) and metal–semiconductor field-effect transistor (MESFET), described in Chapter 6, are generically similar devices. Both utilize the electric field to control a current flow that is parallel to the junction rather than perpendicular to it. The surface physics and metal–oxide–semiconductor (MOS) devices are considered in Chapters 7 and 8. A knowledge of the 'interface traps" associated with these devices is important not only because of the devices themselves but also because of their relevance to the stability and reliability of all other semiconductor devices. The charge-coupled device (CCD), which is especially useful for signal processing and image sensing, is formed using closely coupled MOS capacitors. The metal–oxide–semiconductor field-effect transistor (MOSFET), discussed in Chapter 8, is the most important device for very-large-scale integrated (VLSI) circuits. MOSFETs are used extensively in semiconductor memories and microprocessors having thousands of individual components per chip.

Part 4, Chapters 9 through 11, considers some special microwave devices. When a *p-n* junction is doped so heavily on both sides that the field becomes sufficiently high for quantum-mechanical tunneling, the interesting features of tunnel diode behavior are seen (Chapter 9). When a *p-n* junction or a metal–semiconductor contact is operated in avalanche breakdown, under proper conditions we have an IMPATT diode that can generate microwave radiation. The operational characteristics of IMPATT diodes and some related devices are presented in Chapter 10. Microwave oscillation can be generated by the mechanism of electron transfer from a high-mobility lower-energy valley in the conduction band to a low-mobility higher-energy valley. The transferred-electron device is considered in Chapter 11.

Part 5, Chapters 12 through 14, deals with photonic devices that can detect, generate, and convert optical energy to electric energy, or vice versa. The light-emitting diode (LED) and semiconductor laser are discussed in Chapter 12. Both devices are important sources for optical-fiber communication systems. Various photodetectors with high quantum efficiency and high response speed are discussed in Chapter 13. As the worldwide energy demand increases, there is a need to develop alternative energy sources. The solar cell is considered a major candidate because it can convert sunlight directly to electricity with high efficiency. Various configurations of solar cells and their operational characteristics are considered in Chapter 14.

A remark on notation: To keep the notation simple, it is necessary to use the simple symbols more than once, with different meanings for different devices. For example, the symbol α is used as the common-base current gain for a bipolar transistor, as the optical absorption coefficient for a photodetector, and as the impact ionization coefficient for an IMPATT diode. This usage is considered preferable to the alternative, which would be to use alpha only once and then be forced to find more complicated symbols for the other uses. Within each chapter, however, each symbol is used with only one meaning and is defined the first time it appears. Many symbols do have the same or similar meanings consistently throughout this book; they are summarized in Appendix A for convenient reference.

At present, the electronics field in general and the semiconductor-device field in particular are so dynamic and so fast-changing that today's concepts may be obsolete tomorrow. It is therefore important for us to understand the fundamental physical processes and to equip ourselves with sufficient background in physics and mathematics to digest, appreciate, and meet the challenge of these dynamic fields. It is important to point out that many of the devices, especially the unipolar devices and photonic devices, are still under intensive investigation.[1] Their ultimate performance is by no means fully understood at the present time. The material presented in this book is intended to serve as a foundation. The references listed at the end of each chapter can supply more information.

REFERENCE

1 S. M. Sze, "Semiconductor Device Development in the 1970s and 1980s—A Perspective," Proc. IEEE, *69*, 1121 (1981).

PART I

SEMICONDUCTOR PHYSICS

1

Physics and Properties of Semiconductors—A Résumé

1.1 INTRODUCTION

The physics of semiconductor devices is naturally dependent on the physics of semiconductors themselves. This chapter presents a summary of the physics and properties of semiconductors. It represents only a small cross section of the vast literature on semiconductors; only those subjects pertinent to device operations are included here.

For detailed consideration of semiconductor physics, the reader should consult the standard textbooks or reference works by Dunlap,[1] Madelung,[2] Moll,[3] Moss,[4] and Smith.[5]

To condense a large amount of information into a single chapter, three tables and over 30 illustrations drawn from experimental data are compiled and presented here. This chapter emphasizes the three most important semiconductors: germanium (Ge), silicon (Si), and gallium arsenide (GaAs).

Germanium and silicon have been studied extensively. Gallium arsenide has been intensively investigated in recent years. It has different properties than germanium and silicon; particular properties studied are its direct bandgap for photonic application and its intervalley-carrier transport and high mobility for generation of microwaves.

1.2 CRYSTAL STRUCTURE

Three primitive basis vectors, **a**, **b**, and **c**, describe a crystalline solid such that the crystal structure remains invariant under translation through any vector that is the sum of integral multiples of these basis vectors. In other words, the direct lattice sites can be defined by the set[6]

$$\mathbf{R} = m\mathbf{a} + n\mathbf{b} + p\mathbf{c} \tag{1}$$

where m, n, and p are integers.

Figure 1 shows some important unit cells (direct lattices). A great many important semiconductors have diamond or zincblende lattice structures which belong to the tetrahedral phases; that is, each atom is surrounded by four equidistant nearest neighbors which lie at the corners of a tetrahedron.

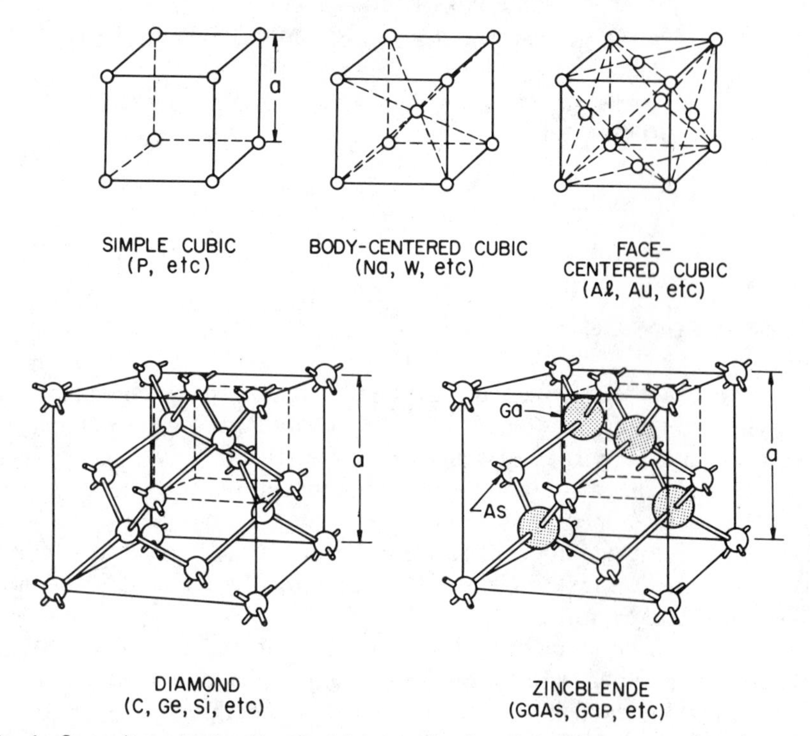

Fig. 1 Some important unit cells (direct lattices) and their representative elements or compounds; a is the lattice constant.

The bond between two nearest neighbors is formed by two electrons with opposite spins. The diamond and the zincblende lattices can be considered as two interpenetrating face-centered cubic lattices. For the diamond lattice, such as silicon, all the atoms are silicon; in a zincblende lattice, such as gallium arsenide, one sublattice is gallium and the other is arsenic. Gallium arsenide is a III–V compound, since it is formed from elements of groups III and V of the periodic table. Most III–V compounds crystallize in the zincblende structure;[2,7] however, many semiconductors (including some III–V compounds) crystallize in the wurtzite or rock-salt structures.

Figure 2*a* shows the wurtzite lattice, which can be considered as two

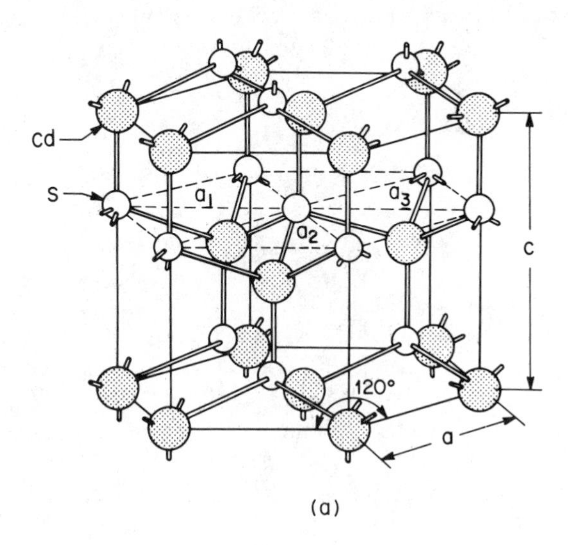

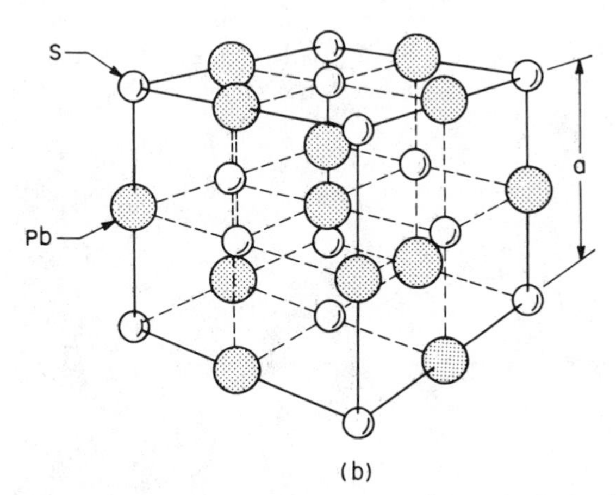

Fig. 2 Two unit cells of compound semiconductors. (*a*) Wurtzite lattice (CdS, ZnS, etc.). (*b*) Rock-salt lattice (PbS, PbTe, etc.).

interpenetrating hexagonal close-packed lattices (e.g., the sublattices of cadmium and sulfur). The wurtzite structure has a tetrahedral arrangement of four equidistant nearest neighbors, similar to a zincblende structure. Figure 2*b* shows the rock-salt lattice, which can be considered as two interpenetrating face-centered cubic lattices. In the rock-salt structure, each atom has six nearest neighbors.

Appendix F gives a summary of the lattice constants of important semiconductors, together with their crystal structures.[8,9] Note that some compounds, such as zinc sulfide and cadmium sulfide, can crystallize in both zincblende and wurtzite structures.

For a given set of the direct basis vectors, a set of reciprocal lattice basis vectors $\mathbf{a}^*$, $\mathbf{b}^*$, $\mathbf{c}^*$ can be defined such that

$$\mathbf{a}^* \equiv 2\pi \frac{\mathbf{b}\times\mathbf{c}}{\mathbf{a}\cdot\mathbf{b}\times\mathbf{c}}, \qquad \mathbf{b}^* \equiv 2\pi \frac{\mathbf{c}\times\mathbf{a}}{\mathbf{a}\cdot\mathbf{b}\times\mathbf{c}}, \qquad \mathbf{c}^* \equiv 2\pi \frac{\mathbf{a}\times\mathbf{b}}{\mathbf{a}\cdot\mathbf{b}\times\mathbf{c}} \tag{2}$$

so that $\mathbf{a}\cdot\mathbf{a}^* = 2\pi$; $\mathbf{a}\cdot\mathbf{b}^* = 0$, and so on; and the general reciprocal lattice vector is given by

$$\mathbf{G} = h\mathbf{a}^* + k\mathbf{b}^* + l\mathbf{c}^* \tag{3}$$

where h, k, and l are integers.

It follows that the product $\mathbf{G}\cdot\mathbf{R} = 2\pi \times$ integer, and therefore that each vector of the reciprocal lattice is normal to a set of planes in the direct lattice, and that the volume V_c^* of a unit cell of the reciprocal lattice is inversely proportional to the volume V_c of a unit cell of the direct lattice; that is, $V_c^* = (2\pi)^3/V_c$, where $V_c \equiv \mathbf{a}\cdot\mathbf{b}\times\mathbf{c}$.

A convenient method of defining the various planes in a crystal is to use Miller indices, which are determined by first finding the intercepts of the plane with the three basis axes in terms of the lattice constants, and then taking the reciprocals of these numbers and reducing them to the smallest three integers having the same ratio. The result is enclosed in parentheses (hkl) as the Miller indices for a single plane or a set of parallel planes. Figure 3 shows the Miller indices of important planes in a cubic crystal.

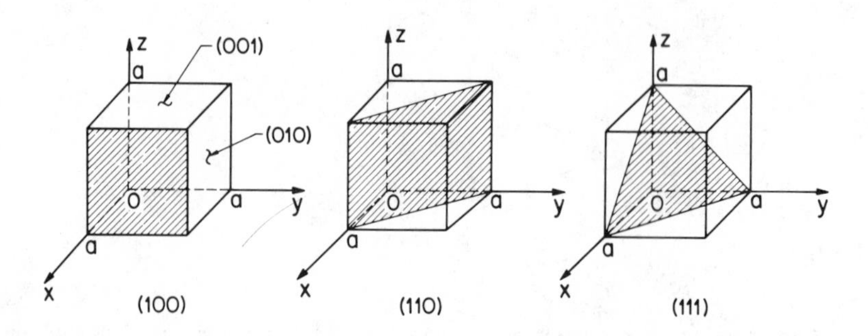

Fig. 3 Miller indices of some important planes in a cubic crystal.

Some other conventions are given as follows:

- $(\bar{h}kl)$: for a plane that intercepts the x axis on the negative side of the origin.
- $\{hkl\}$: for planes of equivalent symmetry such as {100} for (100), (010), (001), $(\bar{1}00)$, $(0\bar{1}0)$, and $(00\bar{1})$ in cubic symmetry.
- $[hkl]$: for the direction of a crystal such as [100] for the x axis.
- $\langle hkl \rangle$: for a full set of equivalent directions.
- $[a_1a_2a_3c]$: for a hexagonal lattice. Here it is customary to use four axes (Fig. 2*a*) with the c axis as the [0001] direction.

For the two semiconductor elements, germanium and silicon, the easiest breakage, or cleavage, planes are the {111} planes. In contrast, gallium arsenide, which has a similar lattice structure but also has a slight ionic component in the bonds, cleaves on {110} planes.

The unit cell of a reciprocal lattice can be represented by a Wigner–Seitz cell. The Wigner–Seitz cell is constructed by drawing perpendicular bisector planes in the reciprocal lattice from the chosen center to the nearest equivalent reciprocal lattice sites. Figure 4*a* shows a typical example for a face-centered cubic structure.[10] If one first draws lines from the center point (Γ) to the eight corners of the cube, then forms the bisector planes, the result is the truncated octahedron within the cube—a Wigner–Seitz cell. It can be shown that[11] a face-centered cubic (fcc) direct lattice with lattice constant a has a body-centered cubic (bcc) reciprocal lattice with spacing $4\pi/a$. Thus the Wigner–Seitz cell shown in Fig. 4*a* is the unit cell of the reciprocal lattice of an fcc direct lattice. The Wigner–Seitz cell for a

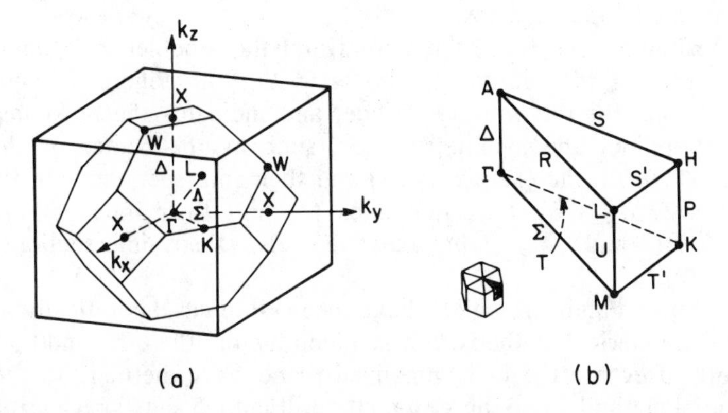

Fig. 4 (*a*) Brillouin zone for diamond and zincblende lattices. (*b*) Brillouin zone for wurtzite lattice. The most important symmetry points and symmetry lines, are also indicated: Γ: $2\pi/a(0, 0, 0)$, zone center; *L*: $2\pi/a(1/2, 1/2, 1/2)$, zone edge along ⟨111⟩ axes (Λ); *X*: $2\pi/a(0, 0, 1)$, zone edge along ⟨100⟩ axes Δ; *K*: $2\pi/a(3/4, 3/4, 0)$, zone edge along ⟨110⟩ axes (Σ). (After Brillouin, Ref. 10; Cohen, Ref. 12.)

hexagonal structure can be similarly constructed;[12] Fig. 4*b* shows the result. The symbols used in Fig. 4 are adopted from group theory and some will be used in Section 1.3.

1.3 ENERGY BANDS

The band structure of a crystalline solid, that is, the energy–momentum (E–k) relationship, is usually obtained by solving the Schrödinger equation of an approximate one-electron problem. The Bloch theorem, one of the most important theorems basic to band structure, states that if a potential energy $V(\mathbf{r})$ is periodic with the periodicity of the lattice, then the solutions $\phi_{\mathbf{k}}(\mathbf{r})$ of the Schrödinger equation[11, 13]

$$\left[-\frac{\hbar^2}{2m}\nabla^2 + V(\mathbf{r})\right]\phi_{\mathbf{k}}(\mathbf{r}) = E_{\mathbf{k}}\phi_{\mathbf{k}}(\mathbf{r}) \tag{4}$$

are of the form

$$\phi_{\mathbf{k}}(\mathbf{r}) = e^{j\mathbf{k}\cdot\mathbf{r}}U_n(\mathbf{k}, \mathbf{r}) = \text{Bloch function} \tag{5}$$

where $U_n(\mathbf{k}, \mathbf{r})$ is periodic in $\mathbf{r}$ with the periodicity of the direct lattice and n is the band index. From the Bloch theorem one can show that the energy $E_{\mathbf{k}}$ is periodic in the reciprocal lattice, that is, $E_{\mathbf{k}} = E_{\mathbf{k}+\mathbf{G}}$, where $\mathbf{G}$ is given by Eq. 3. For a given band index, to label the energy uniquely, it is sufficient to use only $\mathbf{k}$'s in a primitive cell of the reciprocal lattice. The standard convention is to use the Wigner–Seitz cell in the reciprocal lattice (Fig. 4). This cell is called the Brillouin zone or the first Brillouin zone.[10] It is thus evident that we can reduce any momentum $\mathbf{k}$ in the reciprocal space to a point in the Brillouin zone, where any energy state can be given a label in the reduced zone schemes.

The Brillouin zone for the diamond and the zincblende lattices is the same as that of the fcc and is shown in Fig. 4*a*. Figure 4*b* shows the Brillouin zone for the wurtzite lattice, and indicates the most important symmetry points and symmetry lines, such as the center of the zone $[\Gamma = 2\pi/a(0, 0, 0)]$, the $\langle 111\rangle$ axes (Λ) and their intersections with the zone edge $[L = 2\pi/a(\frac{1}{2}, \frac{1}{2}, \frac{1}{2})]$, the $\langle 100\rangle$ axes (Δ) and their intersections $[X = 2\pi/a(0, 0, 1)]$, and the $\langle 110\rangle$ axes (Σ) and their intersections $[K = 2\pi/a(\frac{3}{4}, \frac{3}{4}, 0)]$.

The energy bands of solids have been studied theoretically using a variety of numerical methods. For semiconductors the three methods most frequently used are the orthogonalized plane-wave method,[14, 15] the pseudopotential method[16], and the $\mathbf{k}\cdot\mathbf{p}$ method.[5] Figure 5 shows recent results of studies of the energy-band structures of Ge, Si, and GaAs.[17] Notice that for any semiconductor there is a forbidden energy region in which allowed states cannot exist. Energy regions or energy bands are permitted above and below this energy gap. The upper bands are called the conduction

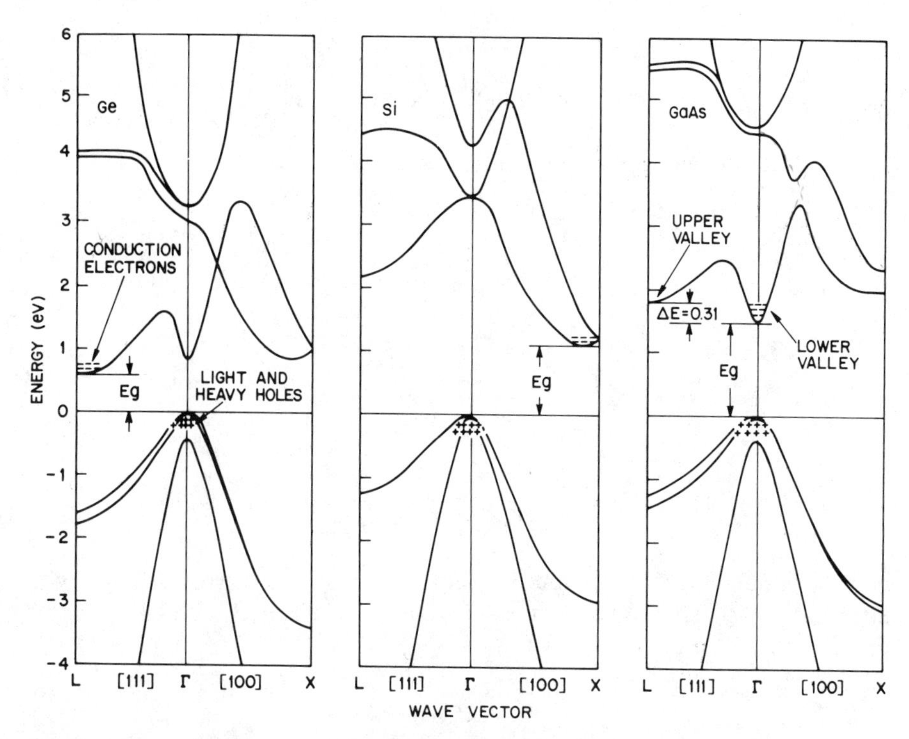

Fig. 5 Energy-band structures of Ge, Si, and GaAs, where E_g is the energy bandgap. Plus (+) signs indicate holes in the valence bands and minus (−) signs indicate electrons in the conduction bands. (After Chelikowsky and Cohen, Ref. 17.)

bands; the lower bands, the valence bands. The separation between the energy of the lowest conduction band and that of the highest valence band is called the bandgap E_g, which is the most important parameter in semiconductor physics. Before we discuss the details of the band structure, we consider the simplified band picture shown in Fig. 6. In this figure the bottom of the conduction band is designated E_C, and the top of the valence band E_V. The electron energy is conventionally defined to be positive when measured upward, and the hole energy is positive when measured downward. The bandgaps of some important semiconductors are listed[9, 18] in Appendix G.

The valence band in the zincblende structure consists of four subbands when spin is neglected in the Schrödinger equation, and each band is doubled when spin is taken into account. Three of the four bands are degenerate at $k = 0$ (Γ point) and form the upper edge of the band, and the fourth band forms the bottom. Furthermore, the spin-orbit interaction causes a splitting of the band at $k = 0$. As shown in Fig. 5 along a given direction the two top valence bands can be approximately fitted by two parabolic bands with different curvatures: the heavy-hole band (the wider

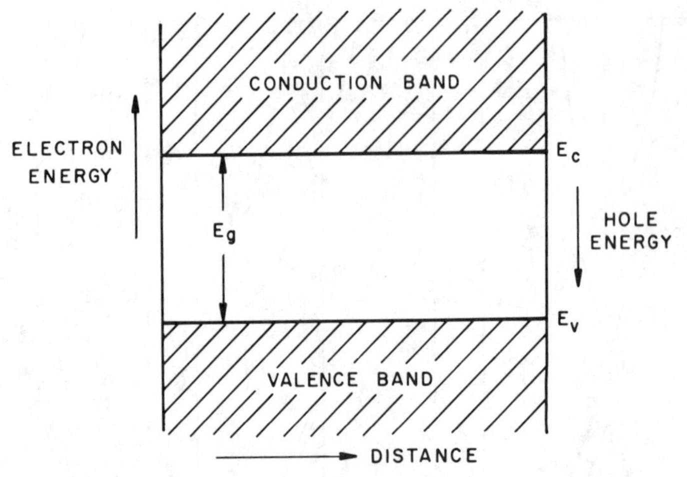

Fig. 6 Simplified band diagram of a semiconductor.

band with smaller $\partial^2 E/\partial k^2$) and the light-hole band (the narrower band with larger $\partial^2 E/\partial k^2$). The effective mass in general is tensorial with components m^*_{ij} defined as

$$\frac{1}{m^*_{ij}} \equiv \frac{1}{\hbar^2} \frac{\partial^2 E(\mathbf{k})}{\partial k_i \, \partial k_j}. \tag{6}$$

The effective mass is listed in Appendix G for important semiconductors.

The conduction band consists of a number of subbands (Fig. 5). The bottom of the conduction band can appear along the ⟨111⟩ axes (Λ or L), along the ⟨100⟩ axes (Δ or X), or at $k = 0$ (Γ). Symmetry considerations alone do not determine the location of the bottom of the conduction band. Experimental results show, however, that in Ge it is along the ⟨111⟩ axes, in Si along the ⟨100⟩ axes, and in GaAs at $k = 0$. Figure 7 shows the shapes of

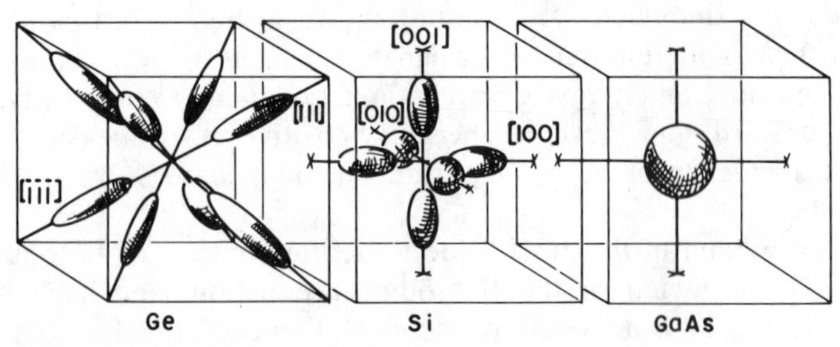

Fig. 7 Shapes of constant energy surfaces in Ge, Si, and GaAs. For Ge there are eight half-ellipsoids of revolution along the ⟨111⟩ axes, and the Brillouin zone boundaries are at the middle of the ellipsoids. For Si there are six ellipsoids along the ⟨100⟩ axes with the centers of the ellipsoids located at about three-fourths of the distance from the Brillouin zone center. For GaAs the constant energy surface is a sphere at zone center. (After Ziman, Ref. 19.)

the constant-energy surfaces.[19] For Ge there are eight half-ellipsoids of revolution along the ⟨111⟩ axes. The Brillouin zone boundaries are at the middle of the ellipsoids, and the constant-energy surfaces are centered about the L points, making four full ellipsoids. For Si there are six ellipsoids along the ⟨100⟩ axes, with the centers of the ellipsoids located at about three-fourths of the distance from the Brillouin zone center. For GaAs the constant energy surface is a sphere at the zone center. By fitting experimental results to parabolic bands, we obtain the electron effective masses; one for GaAs, two for Ge, and two for Si: m_l^* along the symmetry axes and m_t^* transverse to the symmetry axes. Appendix G also gives these values.

At room temperature and under normal atmosphere, the values of the bandgap are 0.66 eV for Ge, 1.12 eV for Si, and 1.42 eV for GaAs. These values are for high-purity materials. For highly doped materials the bandgaps become smaller. Experimental results show that the bandgaps of most semiconductors decrease with increasing temperature. Figure 8 shows

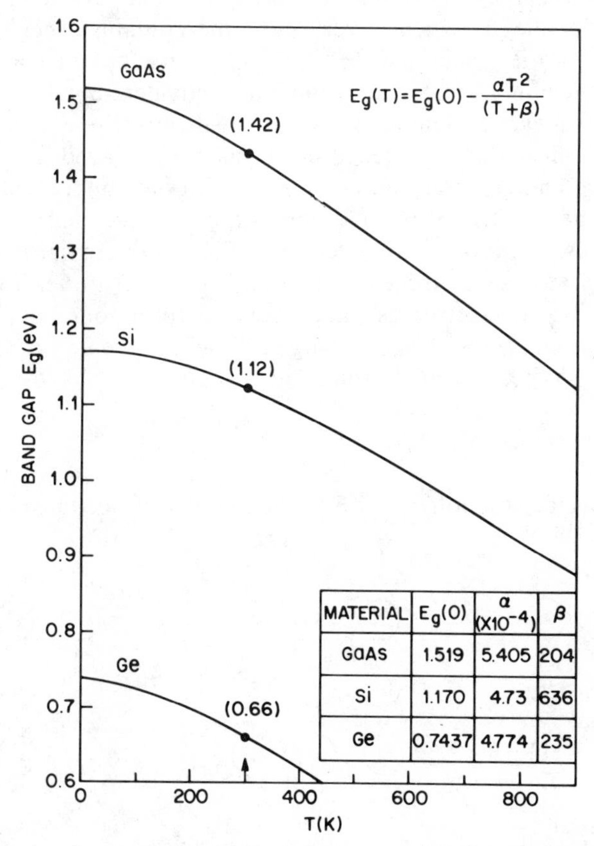

MATERIAL	$E_g(0)$	α ($\times 10^{-4}$)	β
GaAs	1.519	5.405	204
Si	1.170	4.73	636
Ge	0.7437	4.774	235

Fig. 8 Energy bandgaps of Ge, Si, and GaAs as a function of temperature. (After Thurmond, Ref. 20.)

variations of bandgaps as a function of temperature for Ge, Si, and GaAs.[20] The bandgap approaches 0.743, 1.17, and 1.519 eV, respectively, for the three semiconductors at 0 K. The variation of bandgaps with temperature can be expressed approximately by a universal function $E_g(T) = E_g(0) - \alpha T^2/(T + \beta)$, where $E_g(0)$, α, and β are given in Fig. 8. The temperature coefficient dE_g/dT is negative for the aforementioned three semiconductors. Some semiconductors have positive dE_g/dT; for example, the bandgap of PbS (Appendix G) increases from 0.286 eV at 0 K to 0.41 eV at 300 K. Near room temperature, the bandgaps of Ge and GaAs increase with pressure,[21] $dE_g/dP = 5 \times 10^{-6}$ eV/(kg/cm^2) for Ge, and about 12.6×10^{-6} eV/(kg/cm^2) for GaAs, and the Si bandgap decreases with pressure, $dE_g/dP = -2.4 \times 10^{-6}$ eV/(kg/cm^2).

1.4 CARRIER CONCENTRATION AT THERMAL EQUILIBRIUM

Figure 9 shows three basic bond pictures of a semiconductor. Figure 9*a* shows intrinsic silicon, which is very pure and contains a negligibly small amount of impurities; each silicon atom shares its four valence electrons with the four neighboring atoms, forming four covalent bonds (also see Fig. 1). Figure 9*b* shows schematically an *n*-type silicon, where a substitutional phosphorus atom with five valence electrons has replaced a silicon atom, and a negative-charged electron is "donated" to the conduction band. The silicon is *n* type because of the addition of the negative charge carrier, and the phosphorus atom is called a "donor." Figure 9*c* similarly shows that when a boron atom with three valence electrons substitutes for a silicon atom, an additional electron is "accepted" to form four covalent bonds around the boron, and a positive-charged "hole" is created in the valence band. This is *p* type, and the boron is an "acceptor."

1.4.1 Intrinsic Semiconductor

We now consider the intrinsic case. The number of occupied conduction-

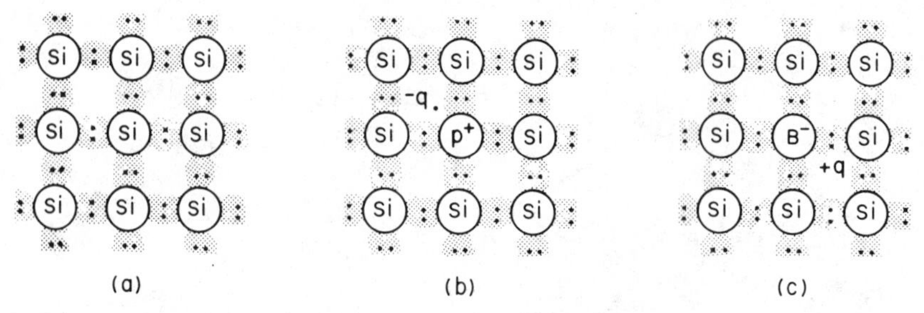

Fig. 9 Three basic bond pictures of a semiconductor. (*a*) Intrinsic Si with negligible impurities. (*b*) *n*-type Si with donor (phosphorus). (*c*) *p*-type Si with acceptor (boron).

band levels is given by

$$n = \int_{E_C}^{E_{top}} N(E)F(E)\,dE \tag{7}$$

where E_C is the energy at the bottom of the conduction band and E_{top} is the energy at the top. The density of states $N(E)$ can be approximated by the density near the bottom of the conduction band for low-enough carrier densities and temperatures:

$$N(E) = M_c \frac{\sqrt{2}}{\pi^2} \frac{(E - E_C)^{1/2}}{\hbar^3} (m_{de})^{3/2} \tag{8}$$

where M_c is the number of equivalent minima in the conduction band and m_{de} is the density-of-state effective mass for electrons:[5]

$$m_{de} = (m_1^* m_2^* m_3^*)^{1/3} \tag{9}$$

where m_1^*, m_2^*, m_3^* are the effective masses along the principal axes of the ellipsoidal energy surface, for example, in silicon $m_{de} = (m_l^* m_t^{*2})^{1/3}$. The Fermi–Dirac distribution function $F(E)$ is given by

$$F(E) = \frac{1}{1 + \exp\left(\frac{E - E_F}{kT}\right)} \tag{10}$$

where k is Boltzmann's constant, T the absolute temperature, and E_F the Fermi energy, which can be determined from the charge neutrality condition (see Section 1.4.3).

The integral, Eq. 7, can be evaluated to be

$$n = N_C \frac{2}{\sqrt{\pi}} F_{1/2}\left(\frac{E_F - F_C}{kT}\right) \tag{11}$$

where N_C is the effective density of states in the conduction band and is given by

$$N_C \equiv 2\left(\frac{2\pi m_{de} kT}{h^2}\right)^{3/2} M_C \tag{12}$$

and $F_{1/2}(\eta_f)$ is the Fermi–Dirac integral (Fig. 10).[22] For the Boltzmann statistics case, that is, for the Fermi level several kT below E_C in nondegenerate semiconductors, the integral approaches $\sqrt{\pi}\, e^{\eta_f}/2$ and Eq. 11 becomes

$$n = N_C \exp\left(-\frac{E_C - E_F}{kT}\right). \tag{13}$$

Similarly, we can obtain the hole density near the top of the valence band:

$$p = N_V \frac{2}{\sqrt{\pi}} F_{1/2}\left(\frac{E_V - E_F}{kT}\right) \tag{14}$$

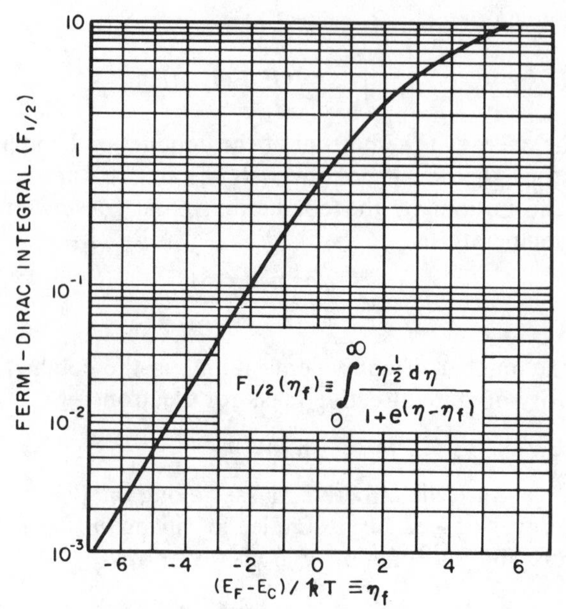

Fig. 10 Fermi–Dirac integral $F_{1/2}$ as a function of Fermi energy. (After Blackmore, Ref. 22.)

where N_V is the effective density of states in the valence band and is given by

$$N_V \equiv 2\left(\frac{2\pi m_{dh}kT}{h^2}\right)^{3/2} \tag{15}$$

where m_{dh} is the density-of-state effective mass of the valence band:[5]

$$m_{dh} = (m_{lh}^{*\,3/2} + m_{hh}^{*\,3/2})^{2/3} \tag{16}$$

where the subscripts refer to "light" and "heavy" hole masses previously discussed, Eq. 6. Again under nondegenerate conditions

$$p = N_V \exp\left(-\frac{E_F - E_V}{kT}\right). \tag{17}$$

For intrinsic semiconductors at finite temperatures continuous thermal agitation exists, which results in excitation of electrons from the valence band to the conduction band and leaves an equal number of holes in the valence band, that is, $n = p = n_i$, where n_i is the intrinsic carrier density. This process is balanced by recombination of the electrons in the conduction band with holes in the valence band.

The Fermi level for an intrinsic semiconductor (which by definition is nondegenerate) is obtained by equating Eqs. 13 and 17:

$$E_F = E_i = \frac{E_C + E_V}{2} + \frac{kT}{2}\ln\left(\frac{N_V}{N_C}\right) = \frac{E_C + E_V}{2} + \frac{3kT}{4}\ln\left(\frac{m_{dh}}{m_{de}M_c^{2/3}}\right). \tag{18}$$

Hence the Fermi level E_i of an intrinsic semiconductor generally lies very close to the middle of the bandgap.

The intrinsic carrier density is obtained from Eqs. 13, 17, and 18:

$$np = n_i^2 = N_C N_V \exp(-E_g/kT) \tag{19}$$

or

$$n_i = \sqrt{N_C N_V}\, e^{-E_g/2kT}$$
$$= 4.9 \times 10^{15} \left(\frac{m_{de} m_{dh}}{m_0^2}\right)^{3/4} M_c^{1/2} T^{3/2} e^{-E_g/2kT} \tag{19a}$$

where $E_g \equiv (E_C - E_V)$, and m_0 is the free-electron mass. Figure 11 shows

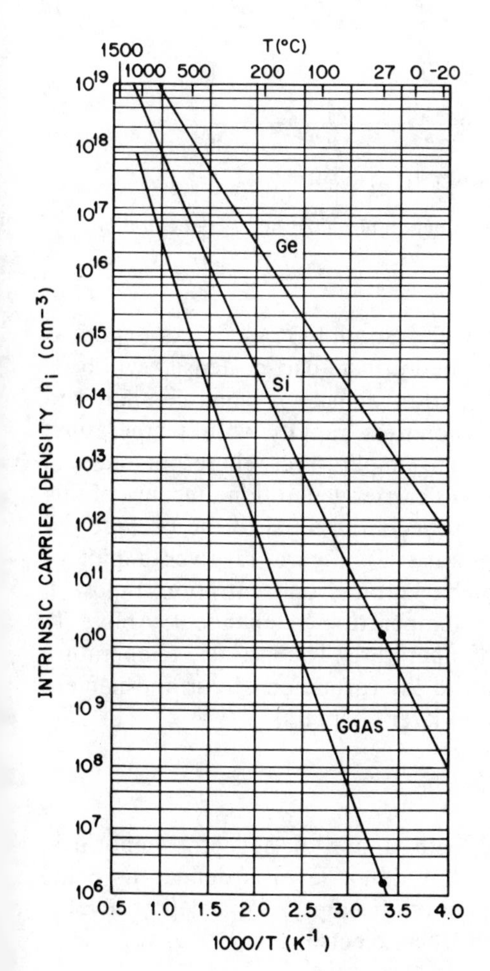

Fig. 11 Intrinsic carrier densities of Ge, Si, and GaAs as a function of reciprocal temperature. (After Thurmond, Ref. 20.)

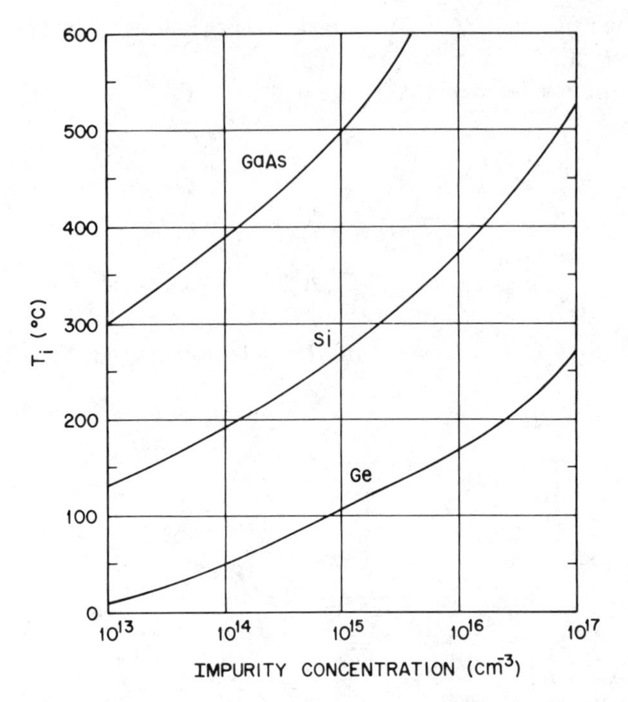

Fig. 12 Intrinsic temperature as a function of background concentration.

the temperature dependence of n_i for Ge, Si, and GaAs.[20,23] As expected, the larger the bandgap is, the smaller the intrinsic carrier density will be.

At room temperature, the intrinsic carrier density is small compared to device doping levels. However, n_i increases rapidly with temperature, doubling every 11°C for silicon. At high temperatures, therefore, thermal generation can be the dominant process of carrier generation. Because of this thermal effect the carrier concentration becomes equal to the background concentration at the intrinsic temperature T_i. Figure 12 gives a plot of intrinsic temperature as a function of background concentration. Below T_i the carrier concentration is relatively temperature independent. Above T_i, however, it rises exponentially with temperature. The intrinsic temperature is an important parameter; its relation to the formation of current filament and second breakdown is considered in later chapters.

1.4.2 Donors and Acceptors

When a semiconductor is doped with donor or acceptor impurities, impurity energy levels are introduced. A donor level is defined as being neutral if filled by an electron, and positive if empty. An acceptor level is neutral if empty, and negative if filled by an electron.

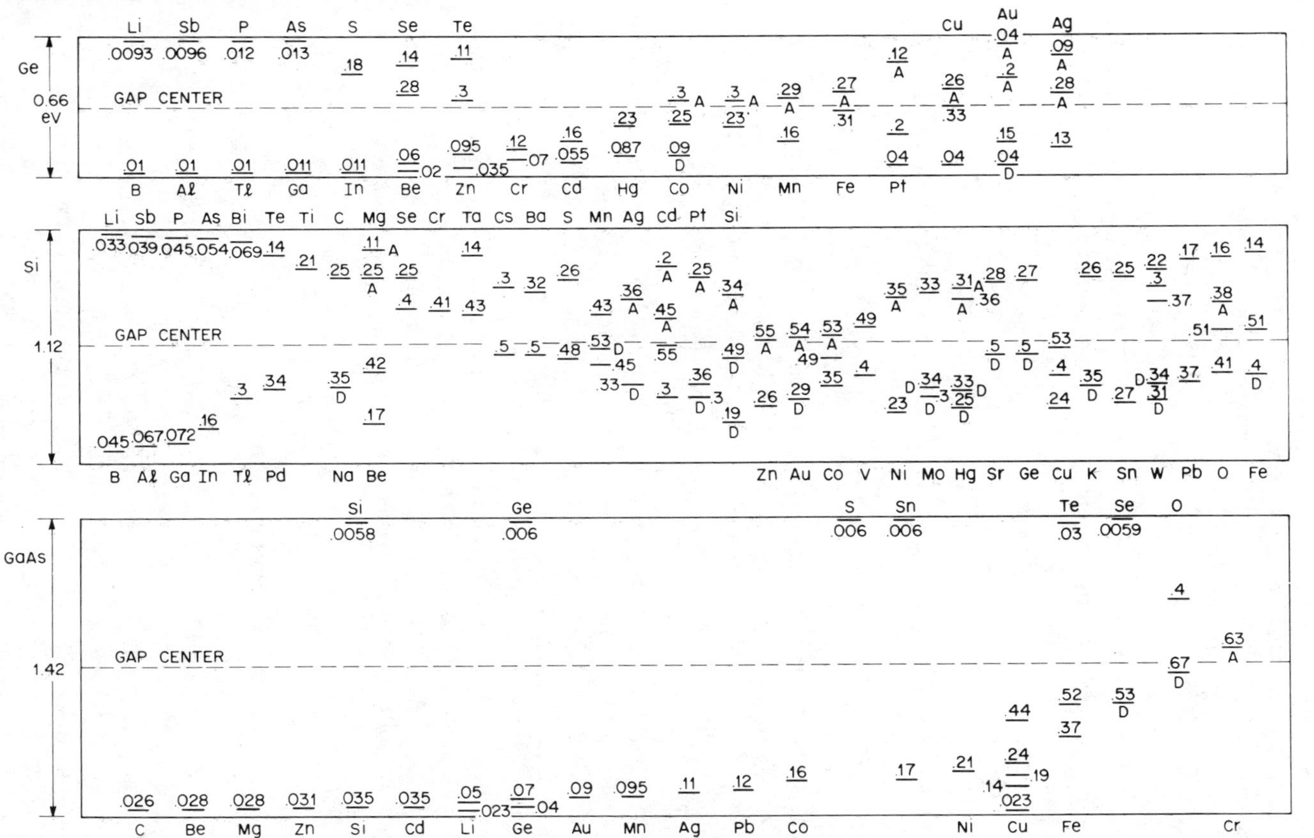

Fig. 13 Measured ionization energies for various impurities in Ge, Si, and GaAs. The levels below the gap centers are measured from the top of the valence band and are acceptor levels unless indicated by D for donor level. The levels above the gap centers are measured from the bottom of the conduction-band level and are donor levels unless indicated by A for acceptor level. The bandgaps at 300 K are 0.66, 1.12, and 1.42 eV for Ge, Si, and GaAs, respectively. (After Conwell, Ref. 27; Sze and Irvin, Ref. 28; Milnes, Ref. 24.)

The simplest calculation of impurity energy levels is based on the hydrogen-atom model. The ionization energy for the hydrogen atom is

$$E_H = \frac{m_0 q^4}{32\pi^2 \epsilon_0^2 \hbar^2} = 13.6\,\text{eV} \tag{20}$$

where ϵ_0 is the free-space permittivity. The ionization energy for the donor E_d, can be obtained by replacing m_0 by the conductivity effective mass[5] of electrons

$$m_{ce} = 3\left(\frac{1}{m_1^*} + \frac{1}{m_2^*} + \frac{1}{m_3^*}\right)^{-1} \tag{21}$$

and by replacing ϵ_0 by the permittivity of the semiconductor ϵ_s in Eq. 20:

$$E_d = \left(\frac{\epsilon_0}{\epsilon_s}\right)^2 \left(\frac{m_{ce}}{m_0}\right) E_H. \tag{22}$$

The ionization energy for donors as calculated from Eq. 22 is 0.006 eV for Ge, 0.025 eV for Si, and 0.007 eV for GaAs. The hydrogen-atom calculation for the ionization level for the acceptors is similar to that for the donors. We consider the unfilled valence band as a filled band plus an imaginary hole in the central force field of a negatively charged acceptor. The calculated acceptor ionization energy (measured from the valence-band edge) is 0.015 eV for Ge, 0.05 eV for Si, and about 0.05 eV for GaAs.

The simple hydrogen-atom model given above certainly cannot account for the details of ionization energy, particularly the deep levels in semiconductors.[24–26] However, the calculated values do predict the correct order of magnitude of the true ionization energies for shallow impurities. Figure 13 shows the measured ionization energies for various impurities in Ge, Si, and GaAs.[24–28] Note that it is possible for a single atom to have many levels; for example, gold in Ge has three acceptor levels and one donor level in the forbidden energy gap.[29]

1.4.3 Calculation of Fermi Level

The Fermi level for the intrinsic semiconductor (Eq. 18) lies very close to the middle of the bandgap. Figure 14*a* depicts this situation, showing schematically from left to right the simplified band diagram, the density of states $N(E)$, the Fermi–Dirac distribution function $F(E)$, and the carrier concentrations. The shaded areas in the conduction band and the valence band are the same; indicating that $n = p = n_i$ for the intrinsic case.

When impurity atoms are introduced, the Fermi level must adjust itself to preserve charge neutrality (Fig. 14*b* and *c*). Consider the case shown in Fig. 14*b*, where donor impurities with the concentration $N_D(\text{cm}^{-3})$ are added to the crystal. To preserve electrical neutrality the total negative charges (electrons and ionized acceptors) must equal the total positive

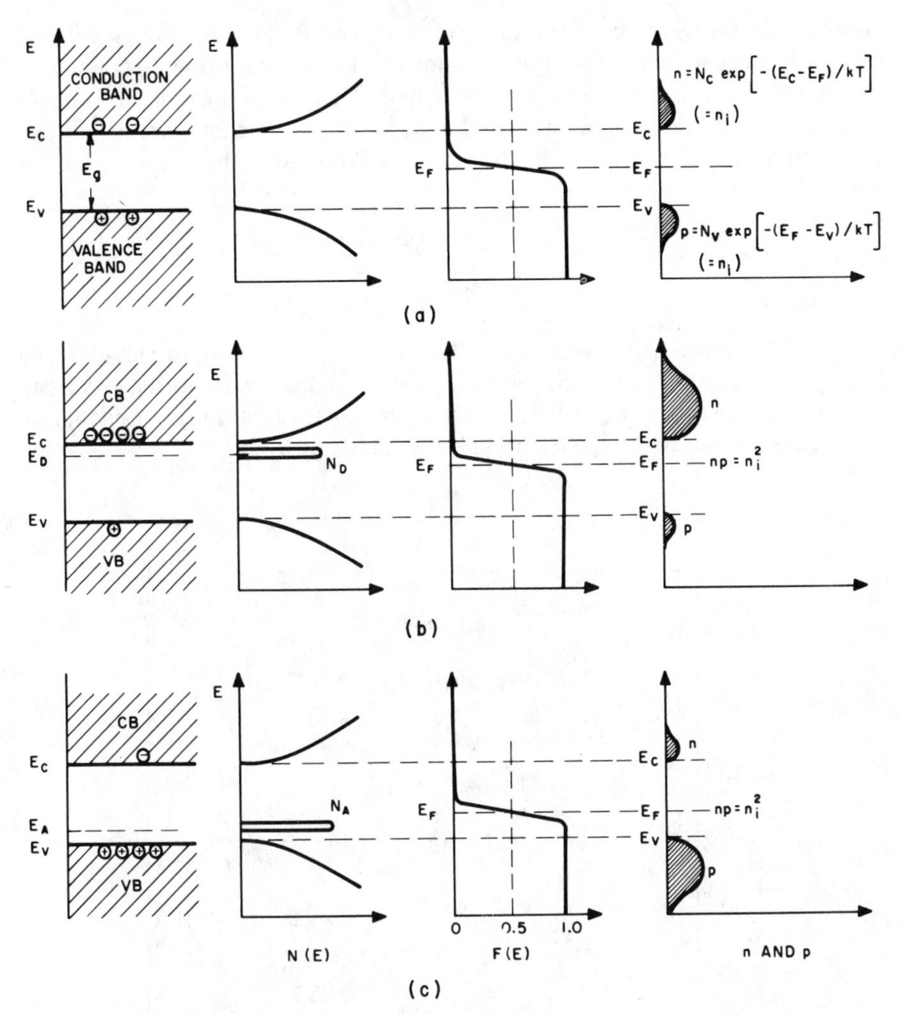

Fig. 14 Schematic band diagram, density of states, Fermi–Dirac distribution, and the carrier concentrations for (*a*) intrinsic, (*b*) *n*-type, and (*c*) *p*-type semiconductors at thermal equilibrium. Note that $pn = n_i^2$ for all three cases.

charges (holes and ionized donors), or for the present case

$$n = N_D^+ + p \tag{23}$$

where n is the electron density in the conduction band, p the hole density in the valence band, and N_D^+ the number of ionized donors, given by[31]

$$N_D^+ = N_D \left[1 - \frac{1}{1 + \frac{1}{g} \exp\left(\frac{E_D - E_F}{kT}\right)} \right] \tag{24}$$

where g is the ground-state degeneracy of the donor impurity level and equals 2 because of the fact that a donor level can accept one electron with either spin or can have no electron. When acceptor impurities of concentration N_A are added to a semiconductor crystal, a similar expression can be written for the charge neutrality condition, and the expression for ionized acceptors is

$$N_A^- = \frac{N_A}{1 + g \exp\left(\frac{E_A - E_F}{kT}\right)} \tag{25}$$

where the ground-state degeneracy factor g is 4 for acceptor levels. The value is 4 because in Ge, Si, and GaAs each acceptor impurity level can accept one hole of either spin and the impurity level is doubly degenerate as a result of the two degenerate valence bands at $\mathbf{k} = 0$.

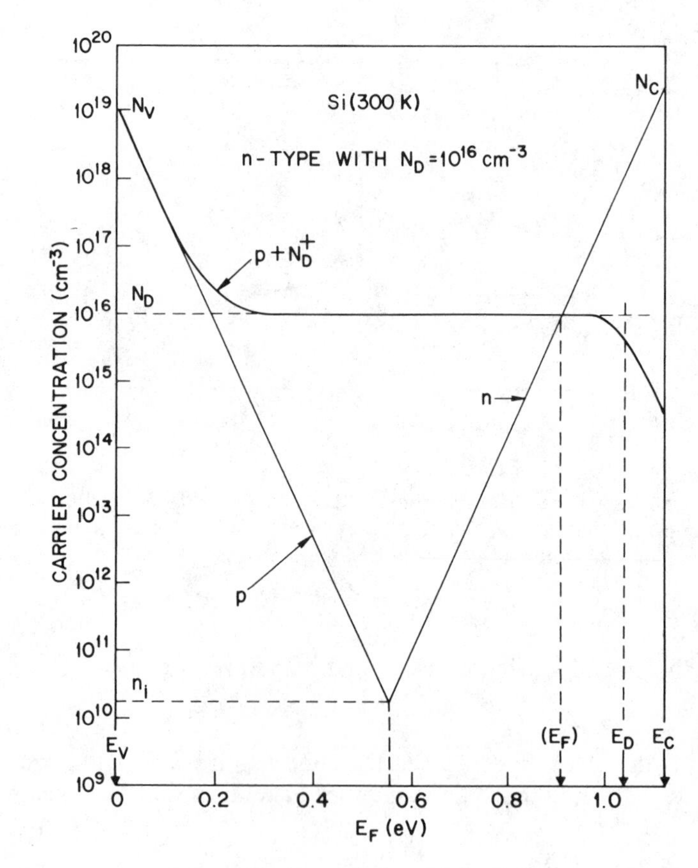

Fig. 15 Graphical method to determine the Fermi energy level E_F. (After Shockley, Ref. 31.)

Rewriting the neutrality condition Eq. 23, we obtain

$$N_C \exp\left(-\frac{E_C - E_F}{kT}\right) = N_D \frac{1}{1 + 2\exp\left(\frac{E_F - E_D}{kT}\right)} + N_V \exp\left(\frac{E_V - E_F}{kT}\right). \quad (26)$$

For a set of given N_C, N_D, N_V, E_C, E_D, E_V, and T, the Fermi level E_F can be uniquely determined from Eq. 26. Figure 15 illustrates an elegant graphical method[31] to determine E_F. For this particular solution (with $N_D = 10^{16}\,\text{cm}^{-3}$, $T = 300\,\text{K}$) the Fermi level is close to the conduction-band edge and adjusts itself so that almost all the donors are ionized. For another temperature, one can first evaluate the values of N_C and N_V which are proportional to $T^{3/2}$ and then obtain from Fig. 11 the value of $n_i(T)$ that determines the intercept of the lines $n(E_F)$ and $p(E_F)$; a new Fermi level is thus obtained. As the temperature is lowered sufficiently, the Fermi level rises toward the donor level (for n-type semiconductors) and the donor level is partially filled with electrons. The approximate expression for the electron density is then[5]

$$n \simeq \left(\frac{N_D - N_A}{2N_A}\right) N_C \exp(-E_d/kT) \quad (27)$$

for a partially compensated semiconductor and for

$$N_A \gg \tfrac{1}{2} N_C \exp(-E_d/kT)$$

where $E_d \equiv (E_C - E_D)$, or

$$n \simeq \frac{1}{\sqrt{2}} (N_D N_C)^{1/2} \exp(-E_d/2kT) \quad (28)$$

for $N_D \gg \frac{1}{2} N_C \exp(-E_d/kT) \gg N_A$. Figure 16 shows a typical example, where n is plotted as a function of the reciprocal temperature. At high temperatures we have the intrinsic range since $n \approx p \gg N_D$. At very low temperatures most impurities are frozen out and the slope is given by Eq. 27 or Eq. 28, depending on the compensation conditions. The electron density, however, remains essentially constant over a wide range of temperatures ($\sim$100 to 500 K in Fig. 16).

Figure 17 shows the Fermi level for silicon as a function of temperature and impurity concentration[32] and the dependence of the bandgap on temperature (see Fig. 8).

When impurity atoms are added, the np product is still given by Eq. 19, which is called the mass-action law, and the product is independent of the added impurities. At relatively elevated temperatures, most donors and acceptors are ionized, so the neutrality condition can be approximated by

$$n + N_A = p + N_D. \quad (29)$$

Equations 19 and 29 can be combined to give the concentration of elec-

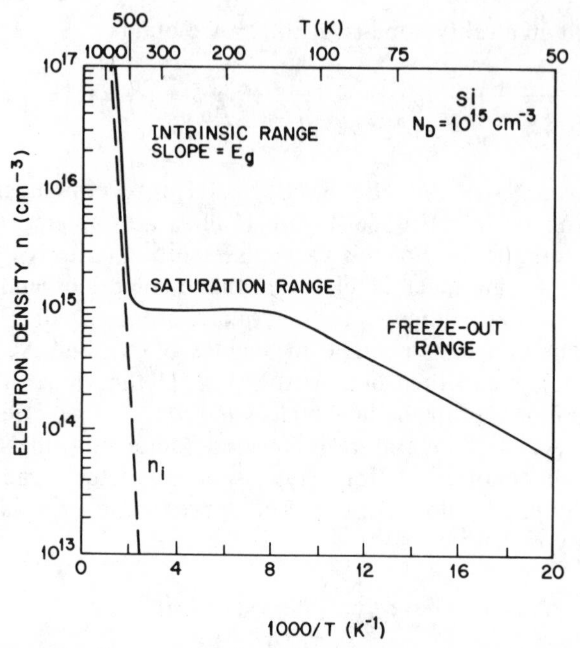

Fig. 16 Electron density as a function of temperature for a Si sample with donor impurity concentration of 10^{15} cm^{-3}. (After Smith, Ref. 5.)

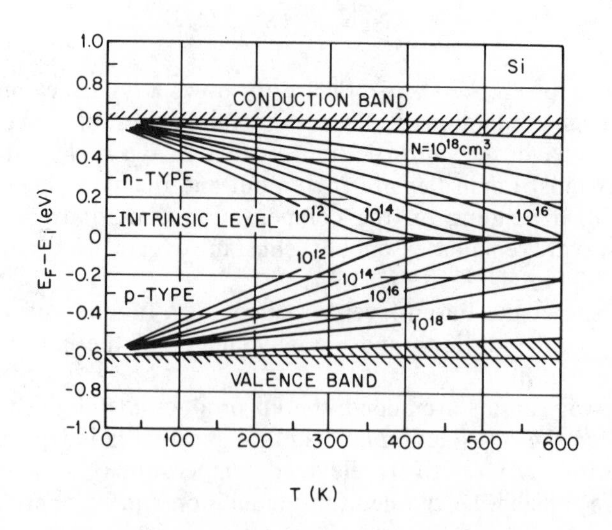

Fig. 17 Fermi level for Si as a function of temperature and impurity concentration. The dependence of the bandgap on temperature is also incorporated in the figure. (After Grove, Ref. 32.)

trons and holes in an n-type semiconductor:

$$n_{no} = \tfrac{1}{2}\left[(N_D - N_A) + \sqrt{(N_D - N_A)^2 + 4n_i^2}\right]$$
$$\approx N_D \quad \text{if} \quad |N_D - N_A| \gg n_i \quad \text{and} \quad N_D \gg N_A \tag{30}$$

$$p_{no} = n_i^2/n_{no} \simeq n_i^2/N_D \tag{31}$$

and

$$E_C - E_F = kT \ln\left(\frac{N_C}{N_D}\right), \tag{32}$$

or from Eq. 18,

$$E_F - E_i = kT \ln\left(\frac{n_{no}}{n_i}\right). \tag{33}$$

The concentration of holes and electrons in a p-type semiconductor is given by

$$p_{po} = \tfrac{1}{2}\left[(N_A - N_D) + \sqrt{(N_A - N_D)^2 + 4n_i^2}\right]$$
$$\approx N_A \quad \text{if} \quad |N_A - N_D| \gg n_i \quad \text{and} \quad N_A \gg N_D \tag{34}$$

$$n_{po} = n_i^2/p_{po} \simeq n_i^2/N_A \tag{35}$$

and

$$E_F - E_V = kT \ln\left(\frac{N_V}{N_A}\right) \tag{36}$$

or

$$E_i - E_F = kT \ln\left(\frac{p_{po}}{n_i}\right). \tag{37}$$

In the formulas above the subscripts n and p refer to the type of semiconductors, and the subscripts o refer to the thermal equilibrium condition. For n-type semiconductors the electron is referred to as the majority carrier and the hole as the minority carrier, since the electron concentration is the larger of the two. The roles are reversed for p-type semiconductors.

1.5 CARRIER TRANSPORT PHENOMENA

1.5.1 Mobility

At low electric field the drift velocity v_d is proportional to the electric field strength $\mathscr{E}$, and the proportionality constant is defined as the mobility μ in cm^2/V-s, or

$$v_d = \mu\mathscr{E}. \tag{38}$$

For nonpolar semiconductors, such as Ge and Si, the presence of acoustic phonons and ionized impurities results in carrier scattering which significantly affects the mobility. The mobility from acoustic phonon interaction, μ_l, is given by[33]

$$\mu_l = \frac{\sqrt{8\pi}\, q\hbar^4 C_{11}}{3E_{ds} m^{*5/2} (kT)^{3/2}} \sim (m^*)^{-5/2} T^{-3/2} \tag{39}$$

where C_{11} is the average longitudinal elastic constant of the semiconductor, E_{ds} the displacement of the edge of the band per unit dilation of the lattice, and m^* the conductivity effect mass. From Eq. 39 mobility decreases with the temperature and with the effective mass.

The mobility from ionized impurities μ_i can be described by[34]

$$\mu_i = \frac{64\sqrt{\pi}\,\epsilon_s^2 (2kT)^{3/2}}{N_I q^3 m^{*1/2}} \left\{ \ln\left[1 + \left(\frac{12\pi\epsilon_s kT}{q^2 N_I^{1/3}}\right)^2\right]\right\}^{-1}$$
$$\sim (m^*)^{-1/2} N_I^{-1} T^{3/2} \tag{40}$$

where N_I is the ionized impurity density and ϵ_s the permittivity. The mobility is expected to decrease with the effective mass but to increase with the temperature. The combined mobility, which includes the two mechanisms above, is

$$\mu = \left(\frac{1}{\mu_l} + \frac{1}{\mu_i}\right)^{-1}. \tag{41}$$

For polar semiconductors such as GaAs optical-phonon scattering is significant. The combined mobility can be approximated by[35]

$$\mu \sim (m^*)^{-3/2} T^{1/2}. \tag{42}$$

In addition to the scattering mechanisms discussed above, other mechanisms also affect the actual mobility. For example, (1) the intravalley scattering in which an electron is scattered within an energy ellipsoid (Fig. 7) and only long-wavelength phonons are involved; and (2) the intervalley scattering in which an electron is scattered from the vicinity of one minimum to another minimum and an energetic phonon is involved.

Figure 18 shows the measured mobilities of Ge, Si, and GaAs versus impurity concentrations at room temperature.[9,36–38] As the impurity concentration increases (at room temperature most impurities are ionized) the mobility decreases, as predicted by Eq. 40. Also as m^* increases, μ decreases; thus for a given impurity concentration the electron mobilities for these semiconductors are larger than the hole mobilities (Appendix G lists the effect masses).

Figure 19 shows the temperature effect on mobility for n-type and p-type silicon samples.[37] For lower impurity concentrations the mobility decreases with temperature as predicted by Eq. 39. The measured slopes, however, are different from $(-\frac{3}{2})$ because of other scattering mechanisms. For pure materials near room temperature the mobility varies as $T^{-1.66}$ and $T^{-2.33}$ for

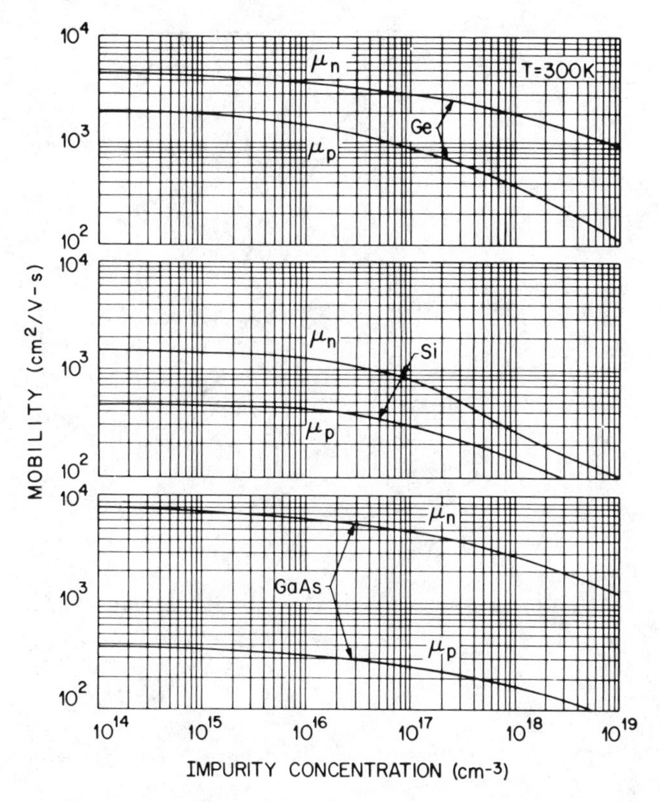

Fig. 18 Drift mobility of Ge, Si, and GaAs at 300 K versus impurity concentration. (After Casey and Panish, Ref. 9; Prince, Ref. 36; Beadle, Plummer, and Tsai, Ref. 38.)

n- and *p*-type Ge, respectively; as $T^{-2.42}$ and $T^{-2.20}$ for *n*- and *p*-type Si, respectively; and as $T^{-1.0}$ and $T^{-2.1}$ for *n*- and *p*-type GaAs, respectively.

The carrier diffusion coefficient (D_n for electrons and D_p for holes, is another important parameter associated with mobility. In thermal equilibrium the relationship between D_n and μ_n (or D_p and μ_p) is given by[5]

$$D_n = 2\left(\frac{kT}{q}\mu_n\right)F_{1/2}\left(\frac{E_F - E_C}{kT}\right)\bigg/F_{-1/2}\left(\frac{E_F - E_C}{kT}\right) \tag{43}$$

where $F_{1/2}$ and $F_{-1/2}$ are Fermi–Dirac integrals. Equation 43 can be expressed as[74]

$$D_n = \frac{\mu_n kT}{q}\left[1 + 0.35355\left(\frac{n}{N_C}\right) - 9.9\times10^{-3}\left(\frac{n}{N_C}\right)^2 + 4.45\times10^{-4}\left(\frac{n}{N_C}\right)^3 + \cdots\right] \tag{43a}$$

where n is the electron concentration and N_C is the effective density of states in the conduction band. For most cases, the first and second terms including n/N_C give sufficient accuracy. The extension to holes is obvious.

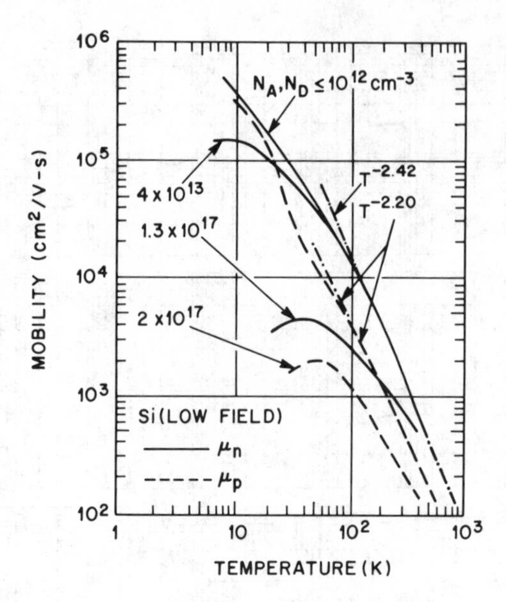

Fig. 19 Mobility of electrons and holes in Si as a function of temperature. (After Jacoboni et al., Ref. 37.)

For nondegenerate semiconductors, that is, where n is much smaller than N_C, Eq. 43a reduces to

$$D_n = \left(\frac{kT}{q}\right)\mu_n \tag{44a}$$

and similarly

$$D_p = \left(\frac{kT}{q}\right)\mu_p. \tag{44b}$$

Equations 44a and 44b are known as the Einstein relationship. At 300 K $kT/q = 0.0259$ V, and values of D are readily obtainable from the mobility results shown in Fig. 18. The mobilities discussed above are the conductivity mobilities, which have been shown to be equal to the drift mobilities.[27] They are, however, different from the Hall mobilities considered in the next section.

1.5.2 Resistivity and Hall Effect

The resistivity ρ is defined as the proportionality constant between the electric field and the current density J:

$$\mathscr{E} = \rho J. \tag{45}$$

Its reciprocal value is the conductivity, that is, $\sigma = 1/\rho$, and

$$J = \sigma\mathscr{E}. \tag{46}$$

For semiconductors with both electrons and holes as carriers, we obtain

$$\rho = \frac{1}{\sigma} = \frac{1}{q(\mu_n n + \mu_p p)}. \tag{47}$$

If $n \gg p$, as in n-type semiconductors,

$$\rho \simeq \frac{1}{q\mu_n n} \tag{48}$$

or

$$\sigma \simeq q\mu_n n. \tag{48a}$$

The most common method for measuring resistivity is the four-point probe method (insert, Fig. 20).[38,39] A small current from a constant-current source is passed through the outer two probes and the voltage is measured between the inner two probes. For a thin wafer with thickness W much smaller than either a or d, the sheet resistance R_s is given by

$$R_s = \frac{V}{I} \cdot \text{CF} \qquad \Omega/\text{square} \tag{49}$$

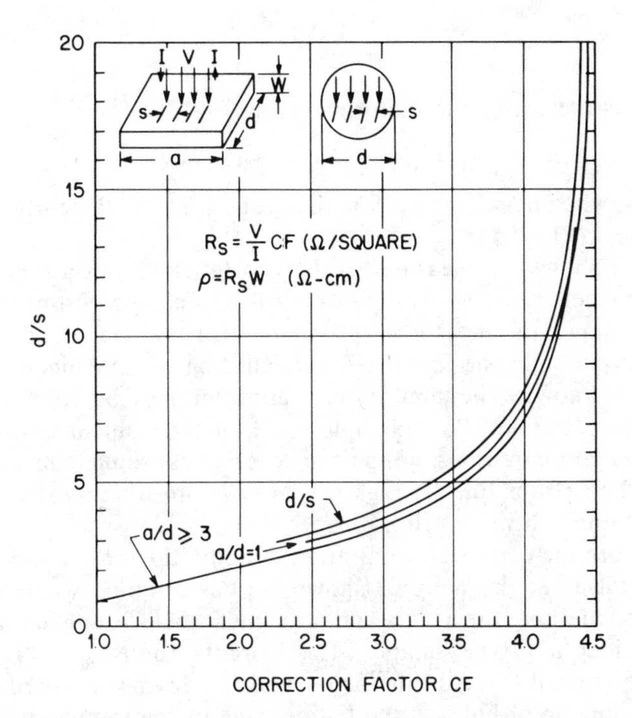

Fig. 20 Correction factor for measurement of resistivity using a four-point probe. (After Beadle, Plummer, and Tsai, Ref. 38.)

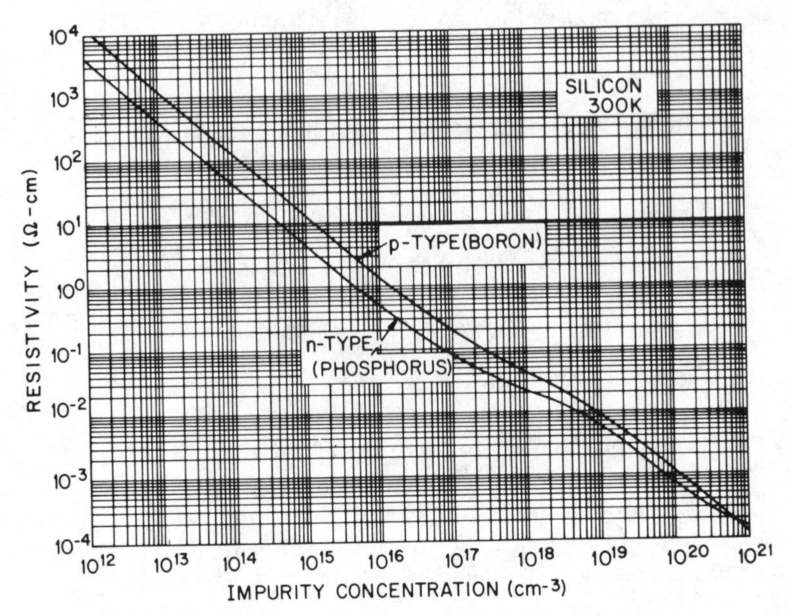

Fig. 21 Resistivity versus impurity concentration for silicon at 300 K. (After Beadle, Plummer, and Tsai, Ref. 38.)

where CF is the correction factor shown in Fig. 20. The resistivity is then

$$\rho = R_s W \qquad \Omega\text{-cm}. \tag{49a}$$

In the limit when $d \gg S$, where S is the probe spacing, the correction factor becomes $(\pi/\ln 2) = 4.54$.

Figure 21 shows the measured resistivity (at 300 K) as a function of the impurity concentration (n-type phosphorus and p-type boron) for silicon.[38] Figure 22 shows the measured resistivities for Ge, GaAs, and GaP.[28,38,40] Thus we can obtain the impurity concentration of a semiconductor if its resistivity is known. The impurity concentration may be different from the carrier concentration. For example, in a p-type silicon with 10^{17} cm^{-3} gallium acceptor impurities, un-ionized acceptors at room temperature make up about 23% (from Eq. 25, Figs. 13 and 17); in other words, the carrier concentration is only 7.7×10^{16} cm^{-3}.

To measure the carrier concentration directly, the most common method uses the Hall effect.[41] Figure 23 shows the basic setup[42] where an electric field is applied along the x axis and a magnetic field is applied along the z axis. Consider a p-type sample. The Lorentz force $qv_x \times \mathscr{B}_z$ exerts an average downward force on the holes, and the downward-directed current causes a piling up of holes at the bottom side of the sample, which in turn gives rise to an electric field $\mathscr{E}_y$. Since there is no net current along the y

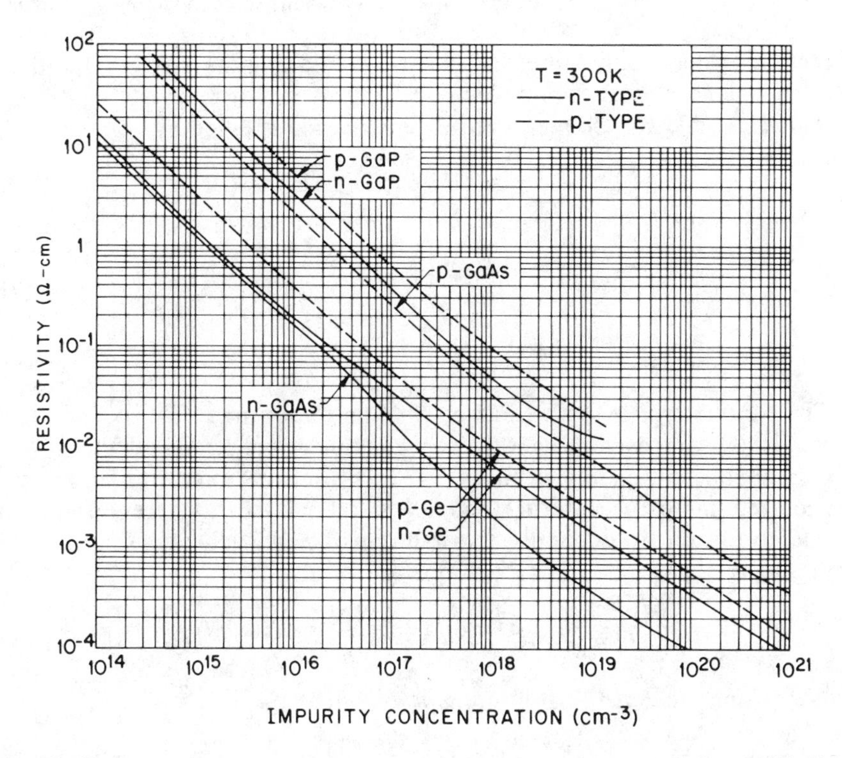

Fig. 22 Resistivity versus impurity concentration for Ge, GaAs, and GaP at 300 K. (After Sze and Irvin, Ref. 28; Beadle, Plummer, and Tsai, Ref. 38.)

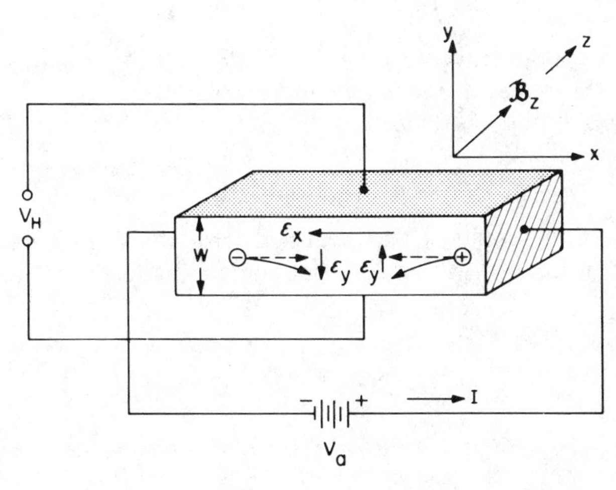

Fig. 23 Basic setup to measure carrier concentration using the Hall effect.

direction in the steady state, the electric field along the y axis (Hall field) exactly balances the Lorentz force.

This Hall field can be measured externally and is given by

$$\mathscr{E}_y = (V_y/W) = R_H J_x \mathscr{B}_z \tag{50}$$

where R_H is the Hall coefficient and is given by[5]

$$R_H = r\frac{1}{q}\frac{p - b^2 n}{(p + bn)^2}, \qquad b \equiv \mu_n/\mu_p \tag{51}$$

$$r \equiv \langle \tau^2 \rangle / \langle \tau \rangle^2. \tag{52}$$

The parameter τ is the mean free time between carrier collisions, which depends on the carrier energy, for example, for semiconductors with spherical constant-energy surfaces, $\tau \sim E^{-1/2}$ for phonon scattering, $\tau \sim E^{3/2}$ for ionized impurity scattering, and in general, $\tau = aE^{-s}$, where a and s are constants. From Boltzmann's distribution for nondegenerate semiconductors, the average value of the mth power of τ is

$$\langle \tau^m \rangle = \int_0^\infty \tau^m E^{3/2} \exp(-E/kT)\, dE \Big/ \int_0^\infty E^{3/2} \exp(-E/kT)\, dE \tag{53}$$

so that using the general form of τ, we obtain

$$\langle \tau^2 \rangle = a^2 (kT)^{-2s} \Gamma(\tfrac{5}{2} - 2s)/\Gamma(\tfrac{5}{2}) \tag{54a}$$

and

$$\langle \tau \rangle^2 = [a(kT)^{-s} \Gamma(\tfrac{5}{2} - s)/\Gamma(\tfrac{5}{2})]^2 \tag{54b}$$

where $\Gamma(n)$ is the gamma function defined as

$$\Gamma(n) \equiv \int_0^\infty x^{n-1} e^{-x}\, dx, \qquad \Gamma(\tfrac{1}{2}) = \sqrt{\pi}.$$

From the expression above we obtain $r = 3\pi/8 = 1.18$ for phonon scattering and $r = 315\pi/512 = 1.93$ for ionized impurity scattering.

The Hall mobility μ_H is defined as the product of the Hall coefficient and conductivity:

$$\mu_H = |R_H \sigma|. \tag{55}$$

The Hall mobility should be distinguished from the drift mobility μ_n (or μ_p), as given in Eq. 48a, which does not contain the factor r. From Eq. 51, if $n \gg p$,

$$R_H = r\left(\frac{-1}{qn}\right) \tag{56a}$$

and if $p \gg n$,

$$R_H = r\left(\frac{+1}{qp}\right). \tag{56b}$$

Thus the carrier concentration and carrier type (electron or hole) can be obtained directly from the Hall measurement provided that one type of carrier dominates.

In the preceding discussion the applied magnetic field was assumed to be small enough that there is no change in the resistivity of the sample. However, under strong magnetic fields, a significant increase in the resistivity is observed: the so-called magneto-resistance effect. For spherical-energy surfaces the ratio of the incremental resistivity to the bulk resistivity at zero magnetic field is given by[5]

$$\frac{\Delta\rho}{\rho_0} = \left\{\left[\frac{\Gamma^2(\frac{5}{2})\Gamma(\frac{5}{2}-3s)}{\Gamma^3(\frac{5}{2}-s)}\right]\left(\frac{\mu_n^3 n + \mu_p^3 p}{\mu_n n + \mu_p p}\right) - \left[\frac{\Gamma(\frac{5}{2})\Gamma(\frac{5}{2}-2s)}{\Gamma^2(\frac{5}{2}-s)}\right]^2\left(\frac{\mu_n^2 n - \mu_p^2 p}{\mu_n n + \mu_p p}\right)^2\right\}\mathcal{B}_z^2. \tag{57}$$

The ratio is proportional to the square of the magnetic field component perpendicular to the direction of the current flow. For $n \gg p$, $(\Delta\rho/\rho_0) \sim \mu_n^2\mathcal{B}_z^2$. A similar result can be obtained for the case $p \gg n$.

1.5.3 Recombination Processes

Whenever the thermal-equilibrium condition of a physical system is distributed (i.e., $pn \neq n_i^2$), processes exist to restore the system to equilibrium (i.e., $pn = n_i^2$). Figure 24 shows the basic recombination processes. Figure 24*a* illustrates the band-to-band recombination where an electron–hole pair recombines. This transition of an electron from the conduction band to the valence band is made possible by emission of a photon (radiative process) or by transfer of the energy to another free electron or hole (Auger process). The latter process is the inverse of impact ionization, and the former process is the inverse of direct optical transitions, which are important for most III–V compounds with direct energy gaps.

Figure 24*b* shows single-level recombination in which only one trapping energy level is present in the bandgap, and Fig. 24*c* shows multiple-level recombination in which more than one trapping level is present in the bandgap.

The single-level recombination can be described by four processes: electron capture, electron emission, hole capture, and hole emission. The recombination rate U (in units of cm^{-3}/s) is given by[43–45]

$$U = \frac{\sigma_p\sigma_n v_{th}(pn - n_i^2)N_t}{\sigma_n\left[n + n_i\exp\left(\frac{E_t - E_i}{kT}\right)\right] + \sigma_p\left[p + n_i\exp\left(-\frac{E_t - E_i}{kT}\right)\right]} \tag{58}$$

where σ_p and σ_n are the hole and electron capture cross sections, respectively, v_{th} the carrier thermal velocity equal to $\sqrt{3kT/m^*}$, N_t the trap density, E_t the trap energy level, E_i the intrinsic Fermi level, and n_i the intrinsic carrier density. Obviously in thermal equilibrium, $pn = n_i^2$ and

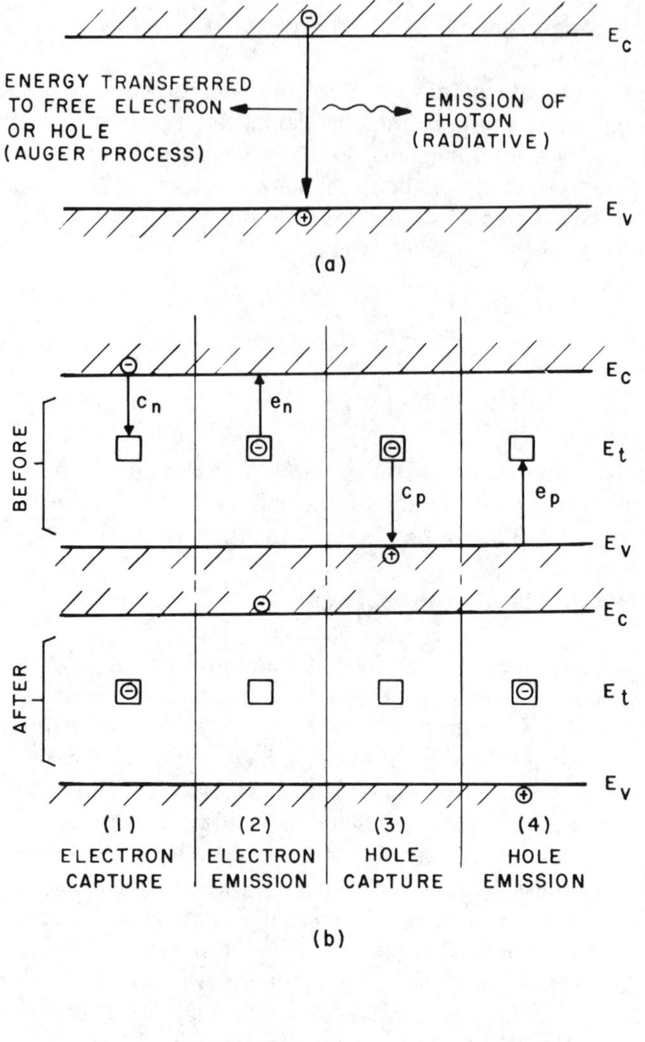

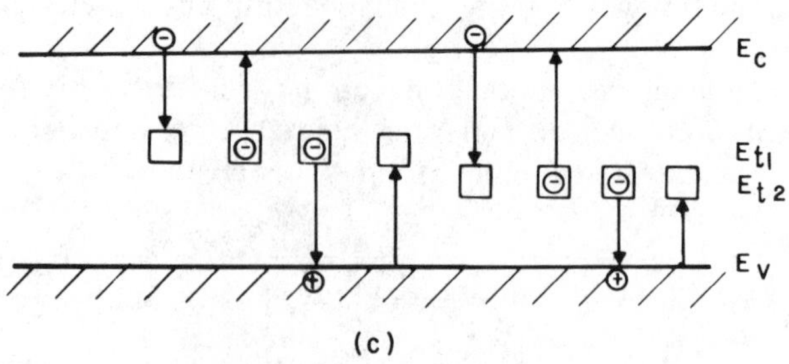

Fig. 24 Recombination processes. (*a*) Band-to-band recombination (radiative or Auger process). (*b*) Single-level recombination. (*c*) Multiple-level recombination. (After Sah, Noyce, and Shockley, Ref. 43.)

$U = 0$. Furthermore, under the simplified condition $\sigma_n = \sigma_p = \sigma$, Eq. 58 reduces to

$$U = \sigma v_{th} N_t \frac{pn - n_i^2}{n + p + 2n_i \cosh\left(\dfrac{E_t - E_i}{kT}\right)}. \tag{59}$$

The recombination rate approaches a maximum as the energy level of the recombination center approaches midgap (i.e., $E_t \approx E_i$). Thus the most effective recombination centers are those located near the middle of the bandgap.

Under low injection conditions, that is, when the injected carriers ($\Delta n = \Delta p$) are much fewer in number than the majority carriers, the recombination process may be characterized by the expression

$$U = \frac{p_n - p_{no}}{\tau_p} \tag{60}$$

where p_{no} is the equilibrium minority-carrier concentration, $p_n = \Delta p + p_{no}$ and τ_p is the minority-carrier lifetime. In an n-type semiconductor, where $n \approx n_{no}$ = the equilibrium majority-carrier concentration, and $n \gg n_i$ and p_n, Eq. 58 becomes

$$U = \sigma_p v_{th} N_t (p_n - p_{no}). \tag{61}$$

Comparing Eqs. 60 and 61 yields the minority-carrier lifetime (hole lifetime) in an n-type semiconductor, and

$$\tau_p = \frac{1}{\sigma_p v_{th} N_t}. \tag{62}$$

Similarly for a p-type semiconductor, the electron lifetime

$$\tau_n = \frac{1}{\sigma_n v_{th} N_t}. \tag{63}$$

For multiple-level traps the recombination processes have gross qualitative features that are similar to those of the single-level case. The behavioral details are, however, different, particularly in the high-injection-level condition (i.e., where $\Delta n = \Delta p \approx$ majority-carrier concentration), where the asymptotic lifetime is an average of the lifetimes associated with all the positively charged, negatively charged, and neutral trapping levels.

Equations 62 and 63 have been verified experimentally by using solid-state diffusion and high-energy radiation. Many impurities have energy levels close to the middle of the bandgap (Fig. 13). These impurities are efficient recombination centers. A typical example is gold in silicon;[29] the minority-carrier lifetime decreases linearly with the gold concentration over the range 10^{14} to 10^{17} cm^{-3}, where τ decreases from about 2×10^{-7} s to 2×10^{-10} s. This effect is important in some switching-device applications when a short lifetime is a desirable feature. Another method of changing the minority-carrier lifetime is high-energy-particle irradiation, which causes displacement of host atoms and damage to lattices. These, in turn, introduce energy levels in the bandgap. For example,[18] electron irradiation in Si gives rise to an acceptor level at 0.4 eV above the valence band and a donor level at 0.36 eV below the conduction band; neutron irradiation creates an acceptor level at 0.56 eV; deuteron irradiation gives rise to an interstitial state with an energy level 0.25 eV above the valence band. Similar results are obtained for Ge, GaAs, and other semiconductors. Unlike the solid-state diffusion, the radiation-induced trapping centers may be annealed out at relatively low temperatures.

The minority-carrier lifetime has generally been measured using the photoconduction effect[46] (PC) or the photoelectromagnetic effect[47] (PEM). The basic equation for the PC effect is given by

$$J_{PC} = q(\mu_n + \mu_p)\,\Delta n\mathscr{E} \tag{64}$$

where J_{PC} is the incremental current density as a result of illumination and $\mathscr{E}$ is the applied electric field along the sample. The quantity Δn is the incremental carrier density or the number of electron–hole pairs per volume created by the illumination, which equals the product of the generation rate of electron–hole pairs resulting from photon G and the lifetime τ, or $\Delta n = \tau G$. The lifetime is thus given by

$$\tau = \frac{\Delta n}{G} = \frac{J_{PC}}{G\mathscr{E}q(\mu_n + \mu_p)} \sim J_{PC}. \tag{65}$$

A setup to measure τ will be discussed in Section 1.7. For the PEM effect we measure the short-circuit current, which appears when a constant magnetic field $\mathscr{B}_z$ is applied perpendicular to the direction of incoming radiation. The current density is given by

$$J_{PEM} = q(\mu_n + \mu_p)\mathscr{B}_z \frac{D}{L}(\tau G) \tag{66}$$

where $L \equiv \sqrt{D\tau}$ is the diffusion length. The lifetime is given by

$$\tau = \left[\frac{J_{PEM}}{\mathscr{B}_z\sqrt{D}Gq(\mu_n + \mu_p)}\right]^2 \sim (J_{PEM}/\mathscr{B}_z)^2. \tag{67}$$

1.6 PHONON SPECTRA AND OPTICAL, THERMAL, AND HIGH-FIELD PROPERTIES OF SEMICONDUCTORS

In the preceding section we considered the effect of low to moderately high electric fields on the transport of carriers in semiconductors. In this section we briefly consider other effects and properties of semiconductors that are important to the operation of semiconductor devices.

1.6.1 Phonon Spectra

It is well known that for a one-dimensional lattice with only nearest-neighbor coupling and two different masses, m_1 and m_2, placed alternately, the frequencies of oscillation are given by[3]

$$\nu_\pm = \left[\alpha_f\left(\frac{1}{m_1}+\frac{1}{m_2}\right) \pm \alpha_f\sqrt{\left(\frac{1}{m_1}+\frac{1}{m_2}\right)^2 - 4\sin^2(qa)/m_1m_2}\right]^{1/2} \tag{68}$$

where α_f is the force constant, q the wave number, and a the lattice spacing. The frequency ν_- tends to be proportional to q near $q = 0$. This branch is the acoustic branch, because it is the analog of a long-wavelength vibration of the lattice, and the frequency corresponds to frequency of sound in such a medium. The frequency ν_+ tends to

$$\left[2\alpha_f\left(\frac{1}{m_1}+\frac{1}{m_2}\right)\right]^{1/2}$$

as q approaches zero. This branch, separated considerably from the acoustic mode, is the optical branch, because the frequency ν_+ is generally in the optical range. For the acoustic mode the two sublattices of the atoms with different masses move in the same direction, whereas for the optical mode they move in opposite directions.

For a three-dimensional lattice with one atom per unit cell, such as a simple cubic, body-centered, or face-centered cubic lattice, only three acoustic modes exist. For a three-dimensional lattice with two atoms per unit cell, such as Ge, Si, and GaAs, three acoustic modes and three optical modes exist. Longitudinally polarized modes are modes with the displacement vectors of each atom along the direction of the wave vector; thus we have one longitudinal acoustic mode (LA) and one longitudinal optical mode (LO). Modes with atoms moving in the planes normal to the wave vector are called transversely polarized modes. We have two transverse acoustic modes (TA) and two transverse optical modes (TO).

Figure 25 shows the measured results for Ge, Si, and GaAs.[48–50] Note that at small q's, with LA and TA modes, the energies are proportional to q. The first-order Raman phonon energy is the longitudinal optical phonon energy at $q = 0$. Their values are 0.037 eV for Ge, 0.063 eV for Si, and 0.035 eV for GaAs. Appendix H lists these results, together with other important properties of Ge, Si, and GaAs.

1.6.2 Optical Property

Optical measurement constitutes the most important means of determining the band structures of semiconductors. Photon-induced electronic transitions can occur between different bands, which lead to the determination of the energy bandgap, or within a single band such as the

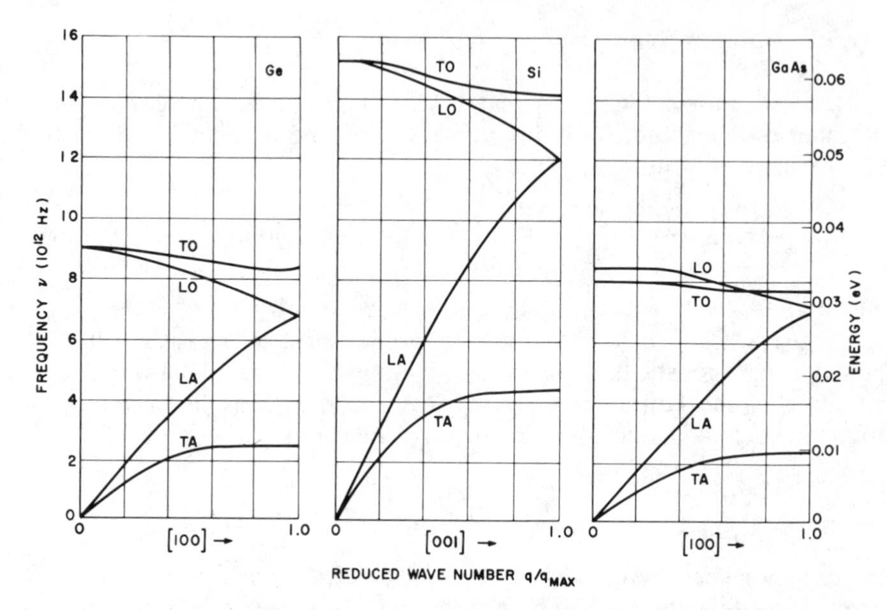

Fig. 25 Measured phonon spectra in Ge, Si, and GaAs, where TO stands for transverse optical modes, LO for longitudinal optical modes, TA for transverse acoustic modes, and LA for longitudinal acoustic modes. (After Brockhouse and Iyengar, Ref. 48; Brockhouse, Ref. 49; Waugh and Dolling, Ref. 50.)

free-carrier absorption. Optical measurements can also be used to study lattice vibrations.

The transmission coefficient T and the reflection coefficient R are the two important quantities generally measured. For normal incidence they are given by

$$T = \frac{(1 - R^2)\exp(-4\pi x/\lambda)}{1 - R^2 \exp(-8\pi x/\lambda)} \tag{69}$$

$$R = \frac{(1-\bar{n})^2 + k^2}{(1+\bar{n})^2 + k^2} \tag{70}$$

where λ is the wavelength, $\bar{n}$ the refractive index, k the absorption constant, and x the thickness of the sample. The absorption coefficient per unit length α is given by

$$\alpha \equiv \frac{4\pi k}{\lambda}. \tag{71}$$

By analyzing the $T - \lambda$ or $R - \lambda$ data at normal incidence, or by making observations of R or T for different angles of incidence, both $\bar{n}$ and k can be obtained and related to transition energy between bands.

Near the absorption edge the absorption coefficient can be expressed as[5]

$$\alpha \sim (h\nu - E_g)^\gamma \tag{72}$$

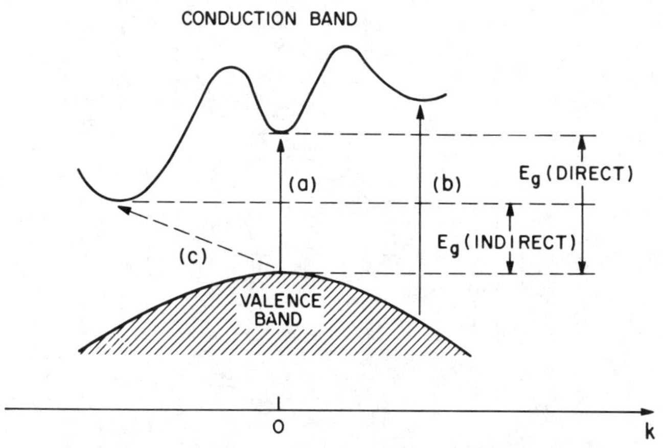

Fig. 26 Optical transitions: (*a*) and (*b*) direct transitions; (*c*) indirect transition involving phonons.

where $h\nu$ is the photon energy, E_g is the bandgap, and γ is a constant. In the one-electron approximation γ equals $\frac{1}{2}$ and $\frac{3}{2}$ for allowed direct transitions and forbidden direct transitions, respectively [with $k_{\min} = k_{\max}$ as transitions (*a*) and (*b*) shown in Fig. 26]; the constant γ equals 2 for indirect transitions [transition (*c*) shown in Fig. 26], where phonons are involved. In addition, γ equals $\frac{1}{2}$ for allowed indirect transitions to exciton states, where an exciton is a bound electron–hole pair with energy levels in the bandgap and moves through the crystal lattice as a unit.

Near the absorption edge, where the values of $(h\nu - E_g)$ become comparable with the binding energy of an exciton, the Coulomb interaction between the free hole and electron must be taken into account. For $h\nu \lesssim E_g$ the absorption merges continuously into the absorption caused by the higher excited states of the exciton. When $h\nu \gg E_g$, higher energy bands participate in the transition processes, and complicated band structures are reflected in the absorption coefficient.

Figure 27 plots the experimental absorption coefficients near and above the fundamental absorption edge (band-to-band transition) for Ge, Si, and GaAs.[51–54] The shift of the curves toward higher photon energies at lower temperature is obviously associated with the temperature dependence of the bandgap (Fig. 8).

1.6.3 Thermal Property

When a temperature gradient is applied to a semiconductor in addition to an applied electric field, the total current density (in one dimension) is[5]

$$J = \sigma \left(\frac{1}{q} \frac{\partial E_F}{\partial x} - \mathscr{P} \frac{\partial T}{\partial x} \right) \tag{73}$$

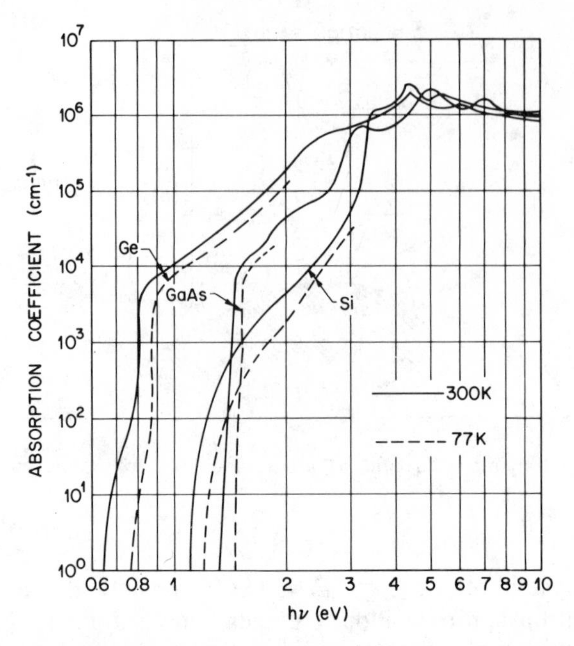

Fig. 27 Measured absorption coefficients near and above the fundamental absorption edge for pure Ge, Si, and GaAs. (After Dash and Newman, Ref. 51; Philipp and Taft, Ref. 52; Hill, Ref. 53; Casey, Sell, and Wecht, Ref. 54.)

where σ is the conductivity, E_F the Fermi energy, and $\mathcal{P}$ the differential thermoelectric power. For a nondegenerate semiconductor with a mean free time between collisions $\tau \sim E^{-s}$ as discussed previously, the thermoelectric power is given by

$$\mathcal{P} = -\frac{k}{q}\left\{\frac{[\frac{5}{2} - s + \ln(N_C/n)]n\mu_n - [\frac{5}{2} - s - \ln(N_V/p)]p\mu_p}{n\mu_n + p\mu_p}\right\} \tag{74}$$

where k is the Boltzmann constant and N_C and N_V are the effective density of states in the conduction and valence bands, respectively.

Equation 74 indicates that the thermoelectric power is negative for n-type semiconductors and positive for p-type semiconductors, a fact often used to determine the conduction type of a semiconductor. The thermoelectric power can also be used to determine the position of the Fermi level relative to the band edges. At room temperature the thermoelectric power $\mathcal{P}$ of p-type silicon increases with resistivity: 1 mV/K for a 0.1 Ω-cm sample and 1.7 mV/K for a 100 Ω-cm sample. Similar results (except a change of the sign for $\mathcal{P}$) can be obtained for n-type silicon samples.

Another important quantity in thermal effect is the thermal conductivity κ, which, if $\tau \sim E^{-s}$ for both electrons and holes, is given by

$$\kappa = \kappa_L + \frac{(\frac{5}{2} - s)k^2\sigma T}{q^2} + \frac{k^2\sigma T}{q^2}\frac{(5 - 2s + E_g/kT)^2 np\mu_n\mu_p}{(n\mu_n + p\mu_p)^2}. \tag{75}$$

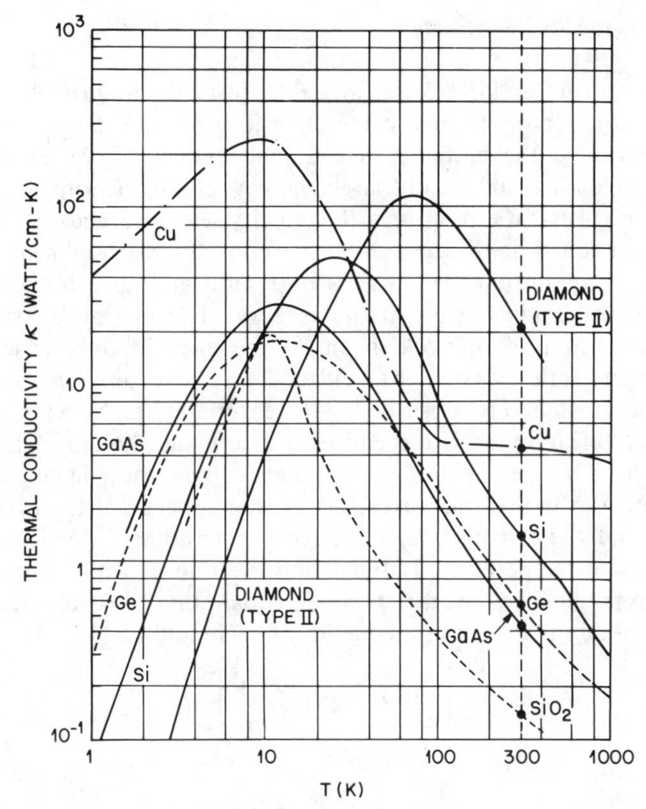

Fig. 28 Measured thermal conductivity versus temperature for pure Ge, Si, GaAs, Cu, diamond type II, and SiO_2. (After Ho, Powell, and Liley, Ref. 55; Holland, Ref. 56; Armstrong, Ref. 57.)

The first, second, and third terms on the right-hand side of Eq. 75 represent the lattice contribution, electronic contribution, and contributions due to mixed conduction, respectively. The contributions of conduction carriers to the thermal conductivity are in general quite small. The third term, however, may be quite large when $E_g \gg kT$. The thermal conductivity first increases with T at low temperatures and then decreases with temperature at higher temperatures.

Figure 28 shows the measured thermal conductivity as a function of lattice temperature for Ge, Si, and GaAs.[55,56] Appendix H lists their room-temperature values. Figure 28 also shows the thermal conductivities[55] for Cu, diamond (type II) and SiO_2.[57] Copper is the most commonly used metal for thermal conduction in p-n junction devices; diamond (type II) has the highest room-temperature thermal conductivity known to date and is useful as the thermal sink for junction lasers and IMPATT oscillators (to be discussed later).

1.6.4 High-Field Property

As discussed in Section 1.5.1 at low electric fields the drift velocity in a semiconductor is proportional to the electric field, and the proportionality constant is called the mobility that is independent of the electric field. When the fields are sufficiently large, however, nonlinearities in mobility and in some cases saturation of drift velocity are observed. At still larger fields, impact ionization occurs. First we consider the nonlinear mobility.

At thermal equilibrium the carriers both emit and absorb phonons, and the net rate of exchange of energy is zero. The energy distribution at thermal equilibrium is Maxwellian. In the presence of an electric field the carriers acquire energy from the field and lose it to phonons by emitting more phonons than are absorbed. At reasonably high fields the most frequent scattering event is the emission of optical phonons. The carriers thus, on the average, acquire more energy than they have at thermal equilibrium. As the field increases, the average energy of the carriers also increases, and they acquire an effective temperature T_e which is higher than the lattice temperature T. Balancing the rate at which energy transferred from the field to the carriers by an equal rate of loss of energy to the lattice, we obtain from the rate equation, for Ge and Si:[3]

$$\frac{T_e}{T} = \frac{1}{2}\left\{1 + \left[1 + \frac{3\pi}{8}\left(\frac{\mu_0 \mathscr{E}}{C_s}\right)^2\right]^{1/2}\right\} \tag{76}$$

and

$$v_d = \mu_0 \mathscr{E} \sqrt{\frac{T}{T_e}} \tag{77}$$

where μ_0 is the low-field mobility, $\mathscr{E}$ the electric field, and C_s the velocity of sound.

When $\mu_0 \mathscr{E}$ is comparable to C_s the mobility deviates from constant value at low fields, and Eqs. 76 and 77 reduce to

$$T_e \simeq T\left[1 + \frac{3\pi}{32}\left(\frac{\mu_0 \mathscr{E}}{C_s}\right)^2\right] \tag{78}$$

and

$$v_d \simeq \mu_0 \mathscr{E}\left[1 - \frac{3\pi}{64}\left(\frac{\mu_0 \mathscr{E}}{C_s}\right)^2\right]. \tag{79}$$

When the field increases until $\mu_0 \mathscr{E} \simeq 8C_s/3$, the carrier temperature doubles over the crystal temperature and the mobility drops by 30%. Finally at sufficiently high fields, the drift velocities for Ge and Si approach a saturation velocity:

$$v_s = \sqrt{\frac{8E_p}{3\pi m_0}} \sim 10^7 \quad \text{cm/s} \tag{80}$$

where E_p is the optical-phonon energy (listed in Appendix H).

The velocity–field relationship is more complicated for GaAs. We must consider the band structure of GaAs (Fig. 5). A high-mobility valley ($\mu \approx 4000$ to $8000\,\text{cm}^2/\text{V-s}$) is located at the Brillouin zone center, and a low-mobility satellite valley ($\mu \approx 100\,\text{cm}^2/\text{V-s}$) along the $\langle 111 \rangle$ axes,[58] about 0.3 eV higher in energy. The effective mass of the electrons is $0.068m_0$ in the lower valley and about $1.2m_0$ in the upper valley; thus the density of states of the upper valley is about 70 times that of the lower valley, from Eq. 12. As the field increases, the electrons in the lower valley can be field-excited to the normally unoccupied upper valley, resulting in a differential negative resistance in GaAs. The intervalley transfer mechanism and the velocity–field relationship are considered in more detail in Chapter 11.

Figure 29*a* shows the measured room-temperature drift velocities versus electric field for high-purity Ge, Si, and GaAs.[37, 59, 60] For high-impurity dopings, the drift velocity at low fields decreases. However, the velocity at high fields is essentially independent of impurity dopings, and approaches a saturation value.[75] For Ge, the saturation velocities v_s for electrons and holes are about 6×10^6 cm/s; in Si, v_s is 1×10^7 cm/s. For GaAs a wide range of differential negative mobility for fields above 3×10^3 V/cm exists, and the high-field velocity approaches 6×10^6 cm/s. Figure 29*b* shows the temperature dependence of electron saturation velocity.[37, 61, 62] As the temperature increases, the saturation velocity for Si and GaAs decreases.

We next consider impact ionization. When the electric field in a semiconductor is increased above a certain value, the carriers gain enough energy so that they can excite electron–hole pairs by impact ionization. The electron–hole pair generation rate G from impact ionization is given by

$$G = \alpha_n n v_n + \alpha_p p v_p \tag{81}$$

where α_n is the electron ionization rate defined as the number of electron–hole pairs generated by an electron per unit distance traveled. Similarly, α_p is the analogously defined ionization rate for holes. Both α_n and α_p are strongly dependent on the electric field.

A physical expression for the ionization rate is given by[76]

$$\alpha(\mathscr{E}) = (q\mathscr{E}/E_I)\exp\{-\mathscr{E}_I/[\mathscr{E}(1+\mathscr{E}/\mathscr{E}_p)+\mathscr{E}_{kT}]\} \tag{82}$$

where E_I is the high-field, effective ionization threshold energy, and $\mathscr{E}_{kT}$, $\mathscr{E}_p$, and $\mathscr{E}_I$ are threshold fields for carriers to overcome the decelerating effects of thermal, optical-phonon, and ionization scattering, respectively. For Si, the value of E_I is found to be 3.6 eV for electrons and 5.0 eV for holes. Over a limited field range, Equation 82 can be reduced to

$$\alpha(\mathscr{E}) = (q\mathscr{E}/E_I)\exp(-\mathscr{E}_I/\mathscr{E}), \quad \text{if} \quad \mathscr{E}_p > \mathscr{E} > \mathscr{E}_{kT} \tag{83a}$$

or

$$\alpha(\mathscr{E}) = (q\mathscr{E}/E_I)\exp(-\mathscr{E}_I\mathscr{E}_p/\mathscr{E}^2), \quad \text{if} \quad \mathscr{E} > \mathscr{E}_p \quad \text{and} \quad \mathscr{E} > \sqrt{\mathscr{E}_p\mathscr{E}_{kT}}. \tag{83b}$$

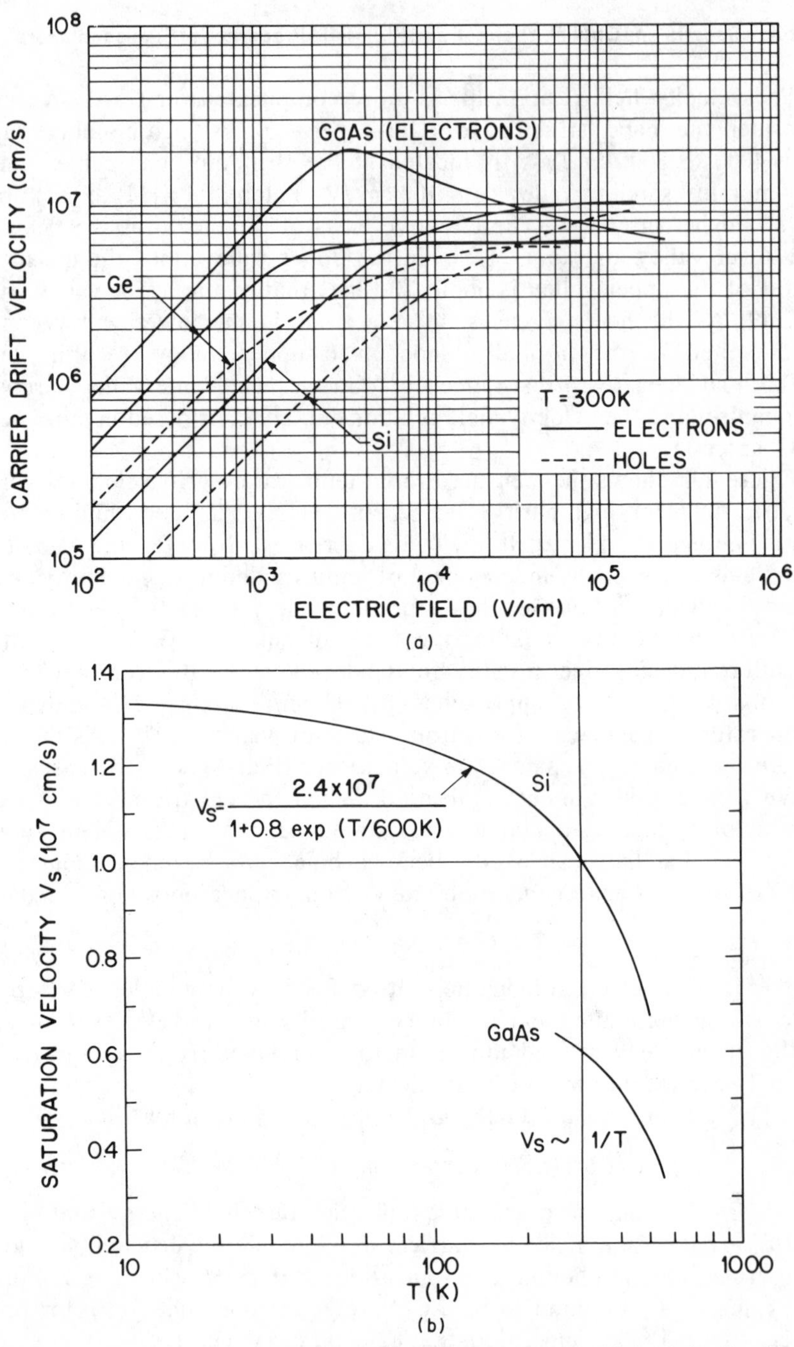

Fig. 29 (*a*) Measured carrier velocity versus electric field for high purity Ge, Si, and GaAs. For highly doped samples, the initial lines are lower than indicated here. In a high-field region, however, the velocity is essentially independent of dopings. (After Jacoboni et al., Ref. 37; Smith, Inoue, and Frey, Ref. 59; Ruch and Kino, Ref. 60.) (*b*) Saturated electron velocity versus temperature in Si and GaAs. (After Jacoboni et al., Ref. 37; Okamoto and Ikeda, Ref. 61; Kramer and Mircea, Ref. 62.)

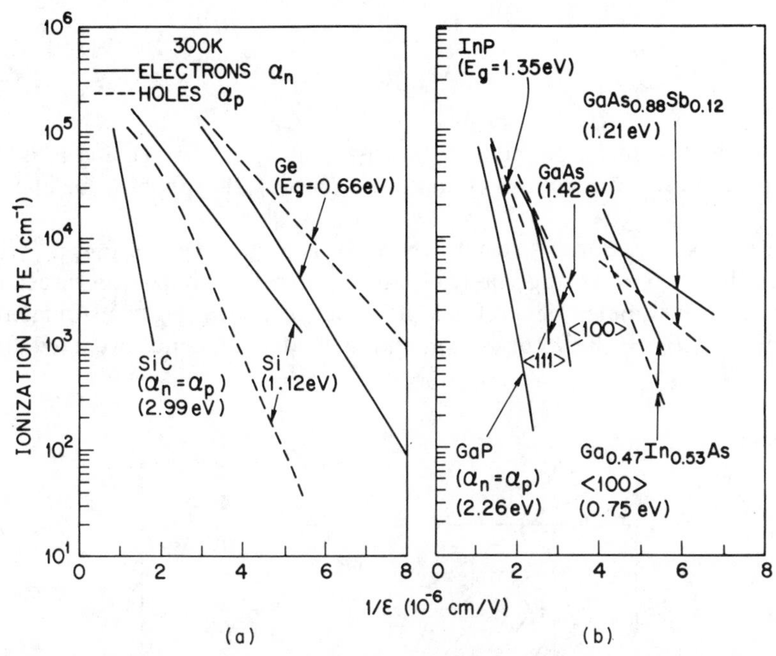

Fig. 30 Ionization rates at 300 K versus reciprocal electric field for Ge, Si, GaAs, and some IV–IV and III–V compound semiconductors. (After Logan and Sze, Ref. 63; Grant, Ref. 64; Glover, Ref. 65; Pearsall et al., Ref. 66; Umebu, Choudhury, and Robson, Ref. 67; Logan and White, Ref. 68; Pearsall, Ref. 69; Pearsall, Nahory, and Pollack, Ref. 70.)

Figure 30*a* shows the experimental results of the ionization rates[63–65] for Ge, Si, and SiC. Figure 30*b* shows the measured ionization rates of GaAs and a few other III–V binary and ternary compounds.[66–70] These results are obtained by using photomultiplication measurements on *p-n* junctions. Note that for certain semiconductors, such as GaAs, the ionization rate is a function of crystal orientation. At a given ionization rate (e.g., $10^4\ \text{cm}^{-1}$) the reciprocal electric field generally increases with decreasing bandgap. Note that Eq. 83a is applicable to most semiconductors shown in Fig. 30, except GaAs and GaP, for which Eq. 83b is applicable.

The temperature dependence of the ionization rate can be expressed by modifying Baraff's three-parameter theory.[71,72] The parameters are: E_I, the ionization threshold energy; λ, the optical-phonon mean free path; and $\langle E_p \rangle$, the average energy loss per phonon scattering. The values λ and $\langle E_p \rangle$ are given by[69]

$$\lambda = \lambda_0 \tanh\left(\frac{E_p}{2kT}\right) \tag{84}$$

$$\langle E_p \rangle = E_p \tanh\left(\frac{E_p}{2kT}\right) \tag{85}$$

and

$$\frac{\lambda}{\lambda_0} = \frac{\langle E_p \rangle}{E_p} \tag{86}$$

where E_p is the optical-phonon energy (listed in Appendix H), and λ_0 is the high-energy low-temperature asymptotic value of the phonon mean free path.

Figure 31 shows Baraff's result, where $\alpha\lambda$ is plotted versus $E_I/q\mathscr{E}\lambda$, with $\langle E_p \rangle/E_I$, the ratio of average optical-phonon energy to ionization threshold energy, as a parameter. Since for a given set of ionization measurements, the values of $E_I(\approx 1.0\,\text{eV}$ for Ge), α, and its field dependence are fixed, one

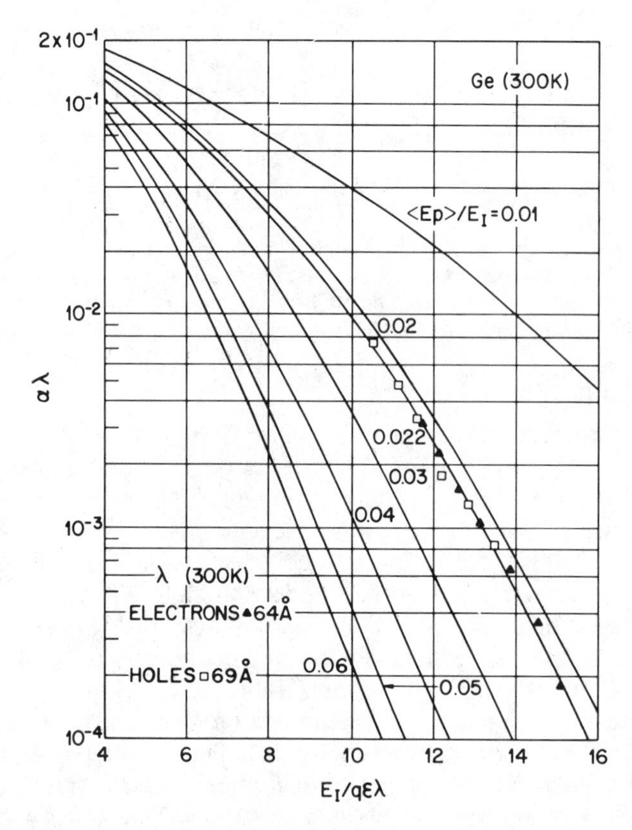

Fig. 31 Baraff's plot-product of ionization rate and optical phonon mean-free path ($\alpha\lambda$) versus $E_I/q\mathscr{E}\lambda$, where $E_I = 1.0\,\text{eV}$ and $\mathscr{E}$ is the electric field. The running parameter is the ratio $\langle E_p \rangle/E_I$, where $\langle E_p \rangle$ is the average optical-phonon energy. The solid curves are theoretical results. (After Baraff, Ref. 71.) The experimental data are obtained from Ge p-n junctions with $\lambda = 64$ Å for electrons and $\lambda = 69$ Å for holes. (After Logan and Sze Ref. 63.)

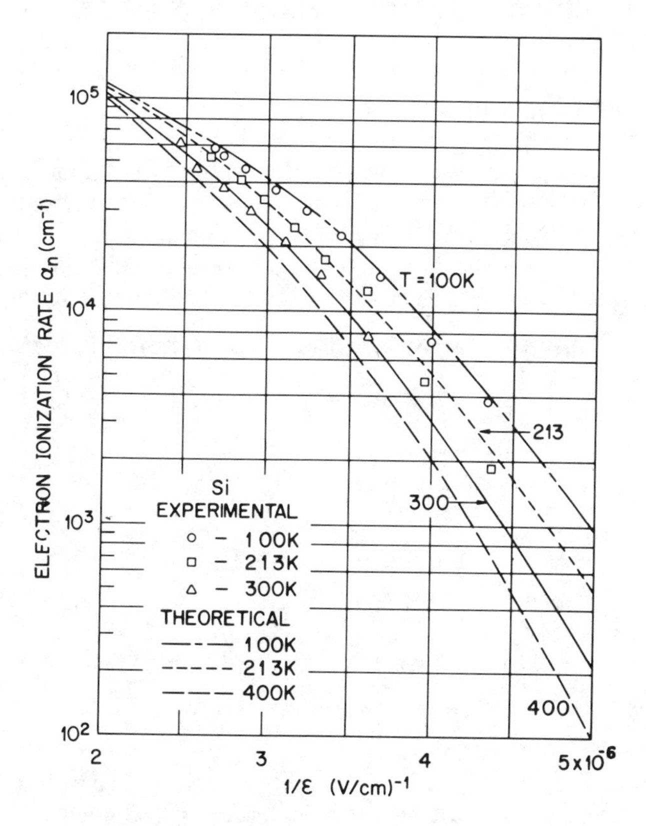

Fig. 32 Electron ionization rate versus reciprocal electric field in Si for four temperatures. (After Crowell and Sze, Ref. 72.)

can thus obtain the optical-phonon mean free path λ by fitting the ionization data to the Baraff plot. A typical result is shown in Fig. 31 for Ge at 300 K. The value of $\langle E_p \rangle / E_I$ is 0.022. We obtain 64 Å for the room-temperature electron–phonon mean free path, and 69 Å for the hole–phonon mean free path. Similar results have been obtained for Si, GaAs, and GaP. From the room-temperature data one can obtain the value of λ_0 from Eq. 84. Appendix H lists the average values of λ_0. Once the value λ_0 is known, the values of λ at various temperatures can be predicted, and from the temperature dependence of $\langle E_p \rangle$, Eq. 84, the correct Baraff plot can be chosen. Figure 32 shows the theoretical predicted electron ionization rates in silicon as obtained from the foregoing approach, together with the experimental results at three different temperatures. The agreement is satisfactory. Note that at a given electric field, the ionization rate decreases with increasing temperature.

1.7 BASIC EQUATIONS FOR SEMICONDUCTOR-DEVICE OPERATION

1.7.1 Basic Equations[31]

The basic equations for semiconductor-device operation describe the static and dynamic behavior of carriers in semiconductors under the influence of external fields that cause deviation from the thermal-equilibrium conditions. The basic equations can be classified in three groups: Maxwell equations, current-density equations, and continuity equations.

Maxwell Equations for Homogeneous and Isotropic Materials

$$\nabla \times \mathscr{E} = -\frac{\partial \mathscr{B}}{\partial t} \tag{87}$$

$$\nabla \times \mathscr{H} = \frac{\partial \mathscr{D}}{\partial t} + \mathbf{J}_{\text{cond}} = \mathbf{J}_{\text{tot}} \tag{88}$$

$$\nabla \cdot \mathscr{D} = \rho(x, y, z) \tag{89}$$

$$\nabla \cdot \mathscr{B} = 0 \tag{90}$$

$$\mathscr{B} = \mu_0 \mathscr{H} \tag{91}$$

$$\mathscr{D}(\mathbf{r}, t) = \int_{-\infty}^{t} \epsilon_s(t - t')\mathscr{E}(\mathbf{r}, t')\, dt' \tag{92}$$

where Eq. 92 reduces to $\mathscr{D} = \epsilon_s \mathscr{E}$ under static or very low frequency conditions; $\mathscr{E}$ and $\mathscr{D}$ are the electric field and displacement vector, respectively; $\mathscr{H}$ and $\mathscr{B}$ are the magnetic field and induction vector, respectively; ϵ_s and μ_0 are the permittivity and permeability, respectively; $\rho(x, y, z)$ is the total electric charge density; $\mathbf{J}_{\text{cond}}$ the conduction current density; and $\mathbf{J}_{\text{tot}}$ the total current density (including both conduction and convection current components and $\nabla \cdot \mathbf{J}_{\text{tot}} = 0$). Among the six equations above the most important is the Poisson equation, Eq. 89, which determines the properties of the *p-n* junction depletion layer (discussed in Chapter 2).

Current-Density Equations

$$\mathbf{J}_n = q\mu_n n\mathscr{E} + qD_n \nabla n \tag{93}$$

$$\mathbf{J}_p = q\mu_p p\mathscr{E} - qD_p \nabla p \tag{94}$$

$$\mathbf{J}_{\text{cond}} = \mathbf{J}_n + \mathbf{J}_p. \tag{95}$$

Here $\mathbf{J}_n$ and $\mathbf{J}_p$ are the electron current density and the hole current density, respectively. They consist of the drift component caused by the field and the diffusion component caused by the carrier concentration gradient. The values of the electron and hole mobilities (μ_n and μ_p) have been given in Section 1.5.1. For nondegenerate semiconductors the carrier diffusion constants (D_n and D_p) and the mobilities are given by the Einstein relationship [$D_n = (kT/q)\mu_n$, etc.].

For a one-dimensional case, Eqs. 93 and 94 reduce to

$$J_n = q\mu_n n\mathscr{E} + qD_n \frac{\partial n}{\partial x} = q\mu_n \left(n\mathscr{E} + \frac{kT}{q}\frac{\partial n}{\partial x}\right) \tag{93a}$$

$$J_p = q\mu_p p\mathscr{E} - qD_p \frac{\partial p}{\partial x} = q\mu_p \left(p\mathscr{E} - \frac{kT}{q}\frac{\partial p}{\partial x}\right) \tag{94a}$$

which are valid for low electric fields. At sufficiently high fields the term $\mu_n\mathscr{E}$ or $\mu_p\mathscr{E}$ should be replaced by the saturation velocity v_s. These equations do not include the effect from an externally applied magnetic field. With an applied magnetic field, another current of $\mathbf{J}_{n\perp} \tan\theta_n$ and $\mathbf{J}_{p\perp} \tan\theta_p$ should be added to Eqs. 93 and 94, respectively, where $\mathbf{J}_{n\perp}$ is the current component of $\mathbf{J}_n$ perpendicular to the magnetic field and $\tan\theta_n \equiv q\mu_n n R_H |\mathscr{H}|$ (which has negative value because the Hall coefficient R_H is negative for electrons); similar results are obtained for the hole current.

Continuity Equations

$$\frac{\partial n}{\partial t} = G_n - U_n + \frac{1}{q}\nabla\cdot\mathbf{J}_n \tag{96}$$

$$\frac{\partial p}{\partial t} = G_p - U_p - \frac{1}{q}\nabla\cdot\mathbf{J}_p \tag{97}$$

where G_n and G_p are the electron and hole generation rate, respectively (cm^{-3}/s), caused by external influence such as the optical excitation with high-energy photons or impact ionization under large electric fields. The electron recombination rate in p-type semiconductors is U_n. Under low injection conditions (i.e., when the injected carrier density is much less than the equilibrium majority carrier density), U_n can be approximated by the expressions $(n_p - n_{po})/\tau_n$, where n_p is the minority-carrier density, n_{po} the thermal-equilibrium minority-carrier density, and τ_n the electron (minority) lifetime. There is a similar expression for the hole recombination rate with lifetime τ_p. If the electrons and holes are generated and recombined in pairs with no trapping or other effects, $\tau_n = \tau_p$.

For the one-dimensional case under a low-injection condition, Eqs. 96 and 97 reduce to

$$\frac{\partial n_p}{\partial t} = G_n - \frac{n_p - n_{po}}{\tau_n} + n_p\mu_n\frac{\partial\mathscr{E}}{\partial x} + \mu_n\mathscr{E}\frac{\partial n_p}{\partial x} + D_n\frac{\partial^2 n_p}{\partial x^2} \tag{96a}$$

$$\frac{\partial p_n}{\partial t} = G_p - \frac{p_n - p_{no}}{\tau_p} - p_n\mu_p\frac{\partial\mathscr{E}}{\partial x} - \mu_p\mathscr{E}\frac{\partial p_n}{\partial x} + D_p\frac{\partial^2 p_n}{\partial x^2}. \tag{97a}$$

1.7.2 Simple Examples

Decay of Photoexcited Carriers Consider an n-type sample, as shown in Fig. 33a, that is illuminated with light and where the electron–hole pairs are generated uniformly throughout the sample with a generation rate G. The boundary conditions are $\mathscr{E} = 0$ and $\partial p_n/\partial x = 0$. We have from Eq. 97a:

$$\frac{\partial p_n}{\partial t} = G - \frac{p_n - p_{no}}{\tau_p}. \tag{98}$$

At steady state, $\partial p_n/\partial t = 0$ and

$$p_n = p_{no} + \tau_p G = \text{constant}. \tag{99}$$

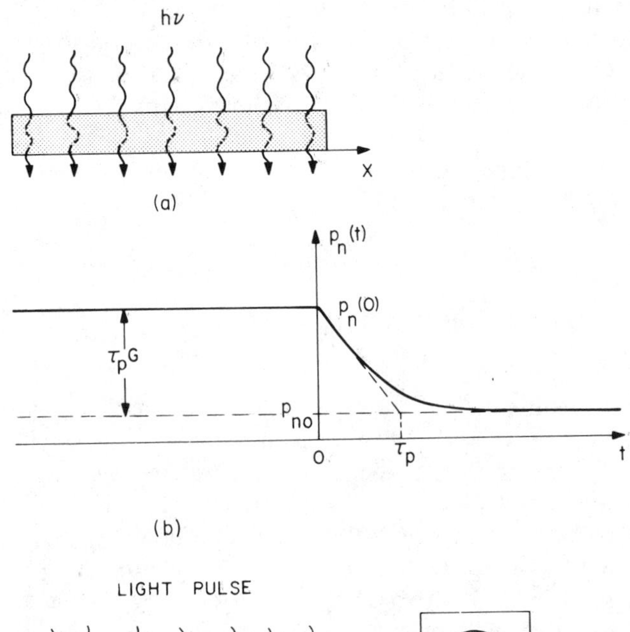

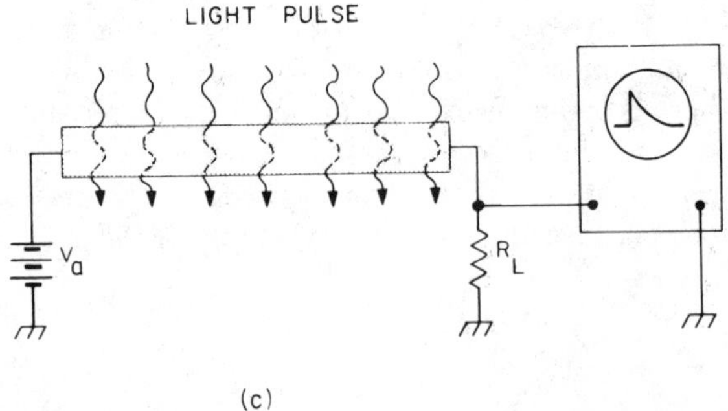

Fig. 33 Decay of photoexcited carriers. (a) n-type sample under constant illumination. (b) Decay of minority carriers (holes) with time. (c) Schematic experimental setup to measure minority carrier lifetime. (After Stevenson and Keyes, Ref. 46.)

If at an arbitrary time, say $t = 0$, the light is suddenly turned off, the boundary conditions are $p_n(0) = p_{no} + \tau_p G$, as given in Eq. 99, and $p_n(t \to \infty) = p_{no}$. The differential equation is now

$$\frac{\partial p_n}{\partial t} = -\frac{p_n - p_{no}}{\tau_p} \tag{100}$$

and the solution is

$$p_n(t) = p_{no} + \tau_p G e^{-t/\tau_p}. \tag{101}$$

Figure 33*b* shows the variation of p_n with time.

The example above presents the main idea of the Stevenson–Keyes method for measuring minority-carrier lifetime.[46] Figure 33*c* shows a schematic setup. The excess carriers generated uniformly throughout the sample by the light pulses cause a momentary increase in the conductivity. The increase manifests itself by a drop in voltage across the sample when a constant current is passed through it. The decay of this photoconductivity can be observed on an oscilloscope and is a measure of the lifetime. (The pulse width must be much less than the lifetime.)

Steady-State Injection from One Side Figure 34*a* shows another simple example where excess carriers are injected from one side (e.g., by high-energy photons that create electron–hole pairs at the surface only). Referring to Fig. 27, note that for $h\nu = 3.5$ eV, the absorption coefficient is about 10^6 cm^{-1}, in other words, the light intensity decreases by $1/e$ in a distance of 100 Å.

At steady state there is a concentration gradient near the surface. The differential equation is, from Eq. 97*a*,

$$\frac{\partial p_n}{\partial t} = 0 = -\frac{p_n - p_{no}}{\tau_p} + D_p \frac{\partial^2 p_n}{\partial x^2}. \tag{102}$$

The boundary conditions are $p_n(x = 0) = p_n(0) =$ constant value and $p_n(x \to \infty) = p_{no}$. The solution of $p_n(x)$ is

$$p_n(x) = p_{no} + [p_n(0) - p_{no}]e^{-x/L_p} \tag{103}$$

where the diffusion length is $L_p \equiv \sqrt{D_p \tau_p}$ (Fig. 34*a*). The maximum values of L_p and $L_n (\equiv \sqrt{D_n \tau_n})$ are of the order of 1 cm in germanium and silicon, but only of the order of 10^{-2} cm in gallium arsenide.

If the second boundary condition is changed so that all excess carriers at $x = W$ are extracted or $p_n(W) = p_{no}$, then we obtain from Eq. 102 a new solution,

$$p_n(x) = p_{no} + [p_n(0) - p_{no}] \left[\frac{\sinh\left(\frac{W - x}{L_p}\right)}{\sinh(W/L_p)} \right]. \tag{104}$$

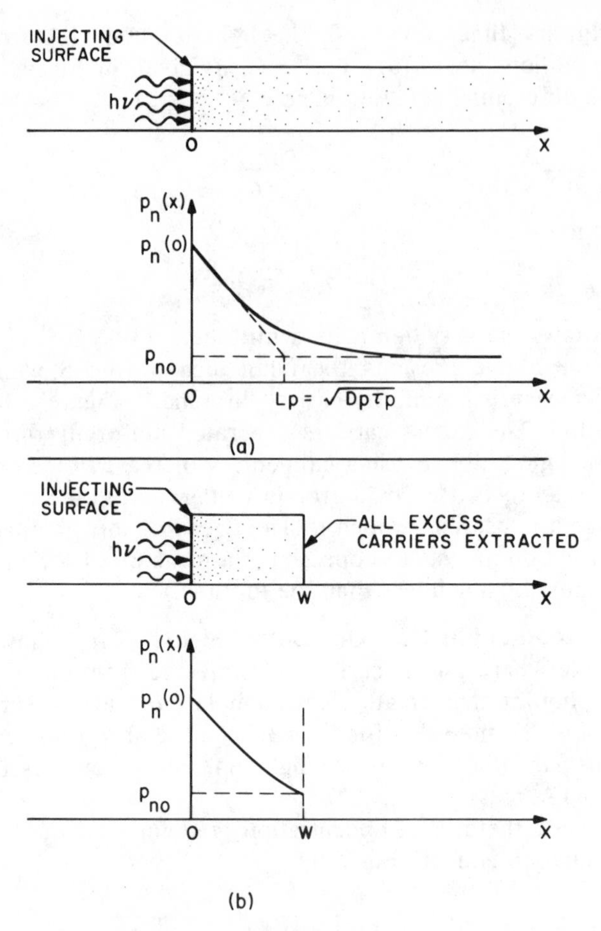

Fig. 34 Steady-state carrier injection from one side. (*a*) Semiinfinite sample. (*b*) Sample with length *W*.

This result is shown in Fig. 34*b*. The current density at $x = W$ is given by Eq. 94a:

$$J_p = -qD_p \left.\frac{\partial p}{\partial x}\right|_W = q[p_n(0) - p_{no}]\frac{D_p}{L_p}\frac{1}{\sinh(W/L_p)}. \tag{105}$$

It will be shown later that Eq. 105 is related to the current gain in junction transistors (Chapter 3).

Transient and Steady-State Diffusion When localized light pulses generate excess carriers in a semiconductor (Fig. 35*a*), the transport equation after the pulse is given by Eq. 97a by setting $G = 0$ and $d\mathscr{E}/dx = 0$:

$$\frac{\partial p_n}{\partial t} = -\frac{p_n - p_{no}}{\tau_p} - \mu_p \mathscr{E}\frac{\partial p_n}{\partial x} + D_p\frac{\partial^2 p_n}{\partial x^2}. \tag{106}$$

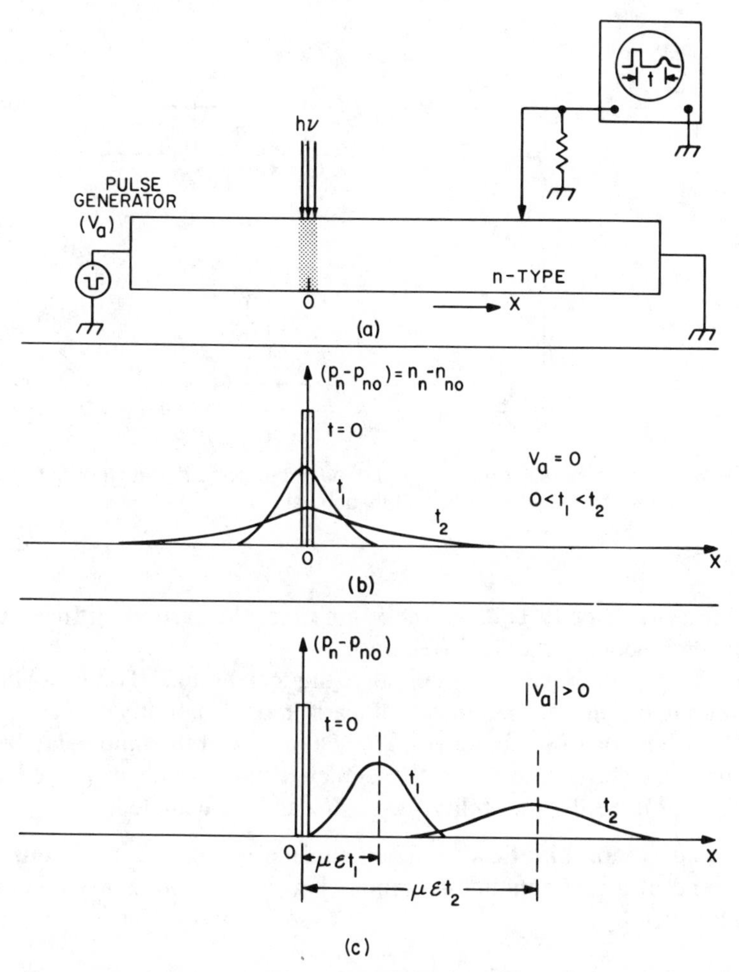

Fig. 35 Transient and steady-state carrier diffusion. (*a*) Experimental setup. (*b*) Without applied field. (*c*) With applied field. (After Haynes and Shockley, Ref. 73.)

If no field is applied along the sample $\mathscr{E} = 0$, the solution is given by

$$p_n(x, t) = \frac{N}{\sqrt{4\pi D_p t}} \exp\left(-\frac{x^2}{4D_p t} - \frac{t}{\tau_p}\right) + p_{no} \tag{107}$$

where N is the number of electrons or holes generated per unit area. Figure 35*b* shows this solution as the carriers diffuse away from the point of injection, and also they recombine.

If an electric field is applied along the sample, the solution is in the form of Eq. 107, but with x replaced by $(x - \mu_p\mathscr{E}t)$ (Fig. 35*c*); thus the whole "package" of excess carrier moves toward the negative end of the sample

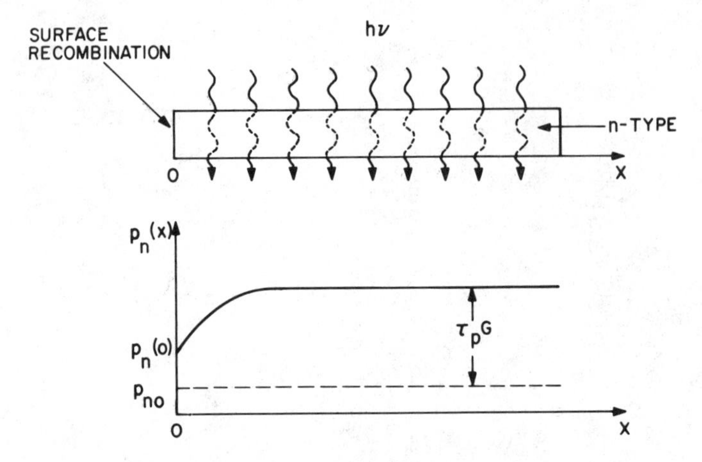

Fig. 36 Surface recombination at $x = 0$. The minority-carrier distribution near the surface is affected by the surface recombination velocity.

with the drift velocity $\mu_p \mathscr{E}$. At the same time, the carriers diffuse outward and recombine as in the field-free case.

The example above is essentially the celebrated Haynes–Shockley experiment for the measurement of carrier drift mobility in semiconductors.[73] With known sample length, applied field, and the time delay between the applied electric pulse and the detected pulse (both displayed on the oscilloscope), the drift mobility $\mu = x/\mathscr{E}t$ can be calculated.

Surface Recombination[32] When surface recombination is introduced at one end of a semiconductor sample (Fig. 36), the boundary condition at $x = 0$ is given by

$$qD_n \left.\frac{\partial p_n}{\partial x}\right|_{x=0} = qS_p[p_n(0) - p_{no}] \tag{108}$$

which states that the minority carriers that reach the surface recombine there; the constant S_p with units cm/s is defined as the surface recombination velocity. The boundary condition at $x = \infty$ is given by Eq. 99. The differential equation is

$$\frac{\partial p_n}{\partial t} = G - \frac{p_n - p_{no}}{\tau_p} + D_p \frac{\partial^2 p_n}{\partial x^2}. \tag{109}$$

The solution of the equation subject to the boundary conditions above is

$$p_n(x) = p_{no} + \tau_p G \left[1 - \frac{\tau_p S_p \exp(-x/L_p)}{L_p + \tau_p S_p}\right] \tag{110}$$

which is plotted in Fig. 36 for a finite S_p. When $S_p \to 0$, then $p_n(x) \to p_{no} + \tau_p G$, which was obtained previously (Eq. 99); when $S_p \to \infty$, then

$p_n(x) \rightarrow p_{no} + \tau_p G[1 - \exp(-x/L_p)]$, and the minority carrier density at the surface approaches its thermal equilibrium value p_{no}. Analogous to the low-injection bulk-recombination process, in which the reciprocal of the minority carrier lifetime $(1/\tau)$ is equal to $\sigma_p v_{th} N_t$, the surface recombination velocity is given by

$$S_p = \sigma_p v_{th} N_{st} \qquad (111)$$

where N_{st} is the number of surface trapping centers per unit area at the boundary region.

REFERENCES

1 W. C. Dunlap, *An Introduction to Semiconductors*, Wiley, New York, 1957.

2 O. Madelung, *Physics of III–V Compounds*, Wiley, New York, 1964.

3 J. L. Moll, *Physics of Semiconductors*, McGraw-Hill, New York, 1964.

4 T. S. Moss, Ed., *Handbook on Semiconductors*, Vols. 1–4, North-Holland, Amsterdam, 1980.

5 R. A. Smith, *Semiconductors*, 2nd ed., Cambridge University Press, London, 1979.

6 See, for example, C. Kittel, *Introduction to Solid State Physics*, Wiley, New York, 1976.

7 R. K. Willardson and A. C. Beer, Eds., *Semiconductors and Semimetals*, Vol. 2, *Physics of III–V Compounds*, Academic, New York, 1966.

8 W. B. Pearson, *Handbook of Lattice Spacings and Structure of Metals and Alloys*, Pergamon, New York, 1967.

9 H. C. Casey, Jr., and M. B. Panish, *Heterostructure Lasers*, Academic, New York, 1978.

10 L. Brillouin, *Wave Propagation in Periodic Structures*, 2nd ed., Dover, New York, 1963.

11 J. M. Ziman, *Principles of the Theory of Solids*, Cambridge University Press, London, 1964.

12 M. L. Cohen, "Pseudopotential Calculations for II–VI Compounds," D. G. Thomas, Ed., *II–VI Semiconducting Compounds*, W. A. Benjamin, New York, 1967, p. 462.

13 C. Kittel, *Quantum Theory of Solids*, Wiley, New York, 1963.

14 L. C. Allen, "Interpolation Scheme for Energy Bands in Solids," *Phys. Rev.* **98**, 993 (1955).

15 F. Herman, "The Electronic Energy Band Structure of Silicon and Germanium," *Proc. IRE*, **43**, 1703 (1955).

16 J. C. Phillips, "Energy-Band Interpolation Scheme Based on a Pseudopotential," *Phys. Rev.*, **112**, 685 (1958).

17 J. R. Chelikowsky and M. L. Cohen, "Nonlocal Pseudopotential Calculations for the Electronic Structure of Eleven Diamond and Zinc-Blende Semiconductors," *Phys. Rev.*, **B14**, 556 (1976).

18 M. Neuberger, *Germanium Data Sheets*, DS-143 (Feb. 1965, Oct. 1960); *Silicon Data Sheets*, DS-137 (May 1964, July 1968); *Gallium Arsenide Data Sheets*, DS-144 (Apr. 1965, Sept. 1967). Compiled from Data Sheets of Electronic Properties Information Center (EPIC), Hughes Aircraft Co., Culver City, Calif.
(a) R. Dalven, "A Review of the Semiconductor Properties of PbTe, PbSe, PbS and PbO," *Infrared Phys.*, **9**, 141 (1969).
(b) I. Strzalkowski, S. Joshi, and C. R. Crowell, "Dielectric Constant and Its Temperature Dependence for GaAs, CdTe and ZnSe," *Appl. Phys. Lett.*, **28**, 350 (1976).

(c) G. H. Jensen, "Temperature Dependence of Bandgap in ZnO from Reflection Data," *Phys. Status Solidi*, **64**, K51 (1974).
(d) F. P. Kesamanly, "GaN: Band Structure, Properties and Potential Applications," *Sov. Phys. Semicond.* **8**, 147 (1974).

19 J. M. Ziman, *Electrons and Phonons*, Clarendon, Oxford, 1960.

20 C. D. Thurmond, "The Standard Thermodynamic Function of the Formation of Electrons and Holes in Ge, Si, GaAs and GaP," *J. Electrochem. Soc.*, **122**, 1133 (1975).

21 W. Paul and D. M. Warschauer, Eds., *Solids under Pressure*, McGraw-Hill, New York, 1963.

22 J. S. Blackmore, "Carrier Concentrations and Fermi Levels in Semiconductors," *Electron. Commun.*, **29**, 131 (1952).

23 R. N. Hall and J. H. Racette, "Diffusion and Solubility of Copper in Extrinsic and Intrinsic Germanium, Silicon, and Gallium Arsenide," *J. Appl. Phys.*, **35**, 379 (1964).

24 A. G. Milnes, *Deep Impurities in Semiconductors*, Wiley, New York, 1973.

25 J. Hermanson and J. C. Phillips, "Pseudopotential Theory of Exciton and Impurity States," *Phys. Rev.*, **150**, 652 (1966).

26 J. Callaway and A. J. Hughes, "Localized Defects in Semiconductors," *Phys. Rev.*, **156**, 860 (1967).

27 E. M. Conwell, "Properties of Silicon and Germanium, Part II," *Proc. IRE*, **46**, 1281 (1958).

28 S. M. Sze and J. C. Irvin, "Resistivity, Mobility, and Impurity Levels in GaAs, Ge, and Si at 300 K," *Solid State Electron.*, **11**, 599 (1968).

29 W. M. Bullis, "Properties of Gold in Silicon," *Solid State Electron.*, **9**, 143 (1966).

30 K. B. Wolfstirn, "Holes and Electron Mobilities in Doped Silicon from Radio Chemical and Conductivity Measurements," *J. Phys. Chem. Solids*, **16**, 279 (1960).

31 W. Shockley, *Electrons and Holes in Semiconductors*, D. Van Nostrand, Princeton, N.J., 1950.

32 A. S. Grove, *Physics and Technology of Semiconductor Devices*, Wiley, New York, 1967.

33 J. Bardeen and W. Shockley, "Deformation Potentials and Mobilities in Nonpolar Crystals," *Phys. Rev.*, **80**, 72 (1950).

34 E. Conwell and V. F. Weisskopf, "Theory of Impurity Scattering in Semiconductors," *Phys. Rev.*, **77**, 388 (1950).

35 H. Ehrenreich, "Band Structure and Electron Transport in GaAs," *Phys. Rev.*, **120**, 1951 (1960).

36 M. B. Prince, "Drift Mobility in Semiconductors I, Germanium," *Phys. Rev.*, **92**, 681 (1953).

37 C. Jacoboni, C. Canali, G. Ottaviani, and A. A. Quaranta, "A Review of Some Charge Transport Properties of Silicon," *Solid State Electron.*, **20**, 77 (1977).

38 W. F. Beadle, R. D. Plummer, and J. C. C. Tsai, *Quick Reference Manual for Semiconductor Engineers*, to be published.

39 F. M. Smits, "Measurement of Sheet Resistivities with the Four-Point Probe," *Bell Syst. Tech. J.*, **37**, 711 (1958).

40 J. C. Irvin, "Resistivity of Bulk Silicon and of Diffused Layers in Silicon," *Bell Syst. Tech. J.*, **41**, 387 (1962).

41 E. H. Hall, "On a New Action of the Magnet on Electric Currents," *Am. J. Math.*, **2**, 287 (1879).

42 L. J. Van der Pauw, "A Method of Measuring Specific Resistivity and Hall Effect of Disc or Arbitrary Shape," *Philips Res. Rep.*, **13**, 1 (Feb. 1958).

43 C. T. Sah, R. N. Noyce, and W. Shockley, "Carrier Generation and Recombination in *p-n* Junction and *p-n* Junction Characteristics," *Proc. IRE*, **45**, 1228 (1957).

44 R. N. Hall, "Electron–Hole Recombination in Germanium," *Phys. Rev.*, **87**, 387 (1952).

45 W. Shockley and W. T. Read, "Statistics of the Recombination of Holes and Electrons," *Phys. Rev.*, **87**, 835 (1952).

46 D. T. Stevenson and R. J. Keyes, "Measurement of Carrier Lifetime in Germanium and Silicon," *J. Appl. Phys.*, **26**, 190 (1955).

47 W. W. Gartner, "Spectral Distribution of the Photomagnetic Electric Effect," *Phys. Rev.*, **105**, 823 (1957).

48 B. N. Brockhouse and P. K. Iyengar, "Normal Modes of Germanium by Neutron Spectrometry," *Phys. Rev.*, **111**, 747 (1958).

49 B. N. Brockhouse, "Lattice Vibrations in Silicon and Germanium," *Phys. Rev. Lett.*, **2**, 256 (1959).

50 J. L. T. Waugh and G. Dolling, "Crystal Dynamics of Gallium Arsenide," *Phys. Rev.*, **132**, 2410 (1963).

51 W. C. Dash and R. Newman, "Intrinsic Optical Absorption in Single-Crystal Germanium and Silicon at 77°K and 300°K," *Phys. Rev.*, **99**, 1151 (1955).

52 H. R. Philipp and E. A. Taft, "Optical Constants of Germanium in the Region 1 to 10 eV," *Phys. Rev.*, **113**, 1002 (1959); "Optical Constants of Silicon in the Region 1 to 10 eV," *Phys. Rev. Lett.*, **8**, 13 (1962).

53 D. E. Hill, "Infrared Transmission and Fluorescence of Doped Gallium Arsenide," *Phys. Rev.*, **133**, A866 (1964).

54 H. C. Casey, Jr., D. D. Sell, and K. W. Wecht, "Concentration Dependence of the Absorption Coefficient for *n*- and *p*-type GaAs between 1.3 and 1.6 eV," *J. Appl. Phys.*, **46**, 250 (1975).

55 C. Y. Ho, R. W. Powell, and P. E. Liley, *Thermal Conductivity of the Elements—A Comprehensive Review*, American Chemical Society and American Institute of Physics, New York, 1975.

56 M. G. Holland, "Phonon Scattering in Semiconductors from Thermal Conductivity Studies," *Phys. Rev.*, **134**, A471 (1964).

57 B. H. Armstrong, "Thermal Conductivity in SiO_2", in S. T. Pantelides, Ed., *The Physic of SiO_2 and Its Interfaces*, Pergamon, New York, 1978.

58 D. E. Aspnes, "GaAs Lower Conduction-Band Minima: Ordering and Properties," *Phys. Rev.*, **B14**, 5331 (1976).

59 P. Smith, M. Inoue, and J. Frey, "Electron Velocity in Si and GaAs at Very High Electric Fields," *Appl. Phys. Lett.*, **37**, 797 (1980).

60 J. G. Ruch and G. S. Kino, "Measurement of the Velocity–Field Characteristics of Gallium Arsenide," *Appl. Phys. Lett.*, **10**, 40 (1967).

61 H. Okamoto and M. Ikeda, "Measurement of the Electron Drift Velocity in Avalanching GaAs Diodes," *IEEE Trans. Electron. Devices*, **ED-23**, 372 (1976).

62 B. Kramer and A. Mircea, "Determination of Saturated Electron Velocity in GaAs," *Appl. Phys. Lett.*, **26**, 623 (1975).

63 R. A. Logan and S. M. Sze, "Avalanche Multiplication in Ge and GaAs *p-n* Junctions," Proc. Int. Conf. Phys. Semicond., Kyoto, and *J. Phys. Soc. Jpn. Suppl.*, **21**, 434 (1966).

64 W. N. Grant, "Electron and Hole Ionization Rates in Epitaxial Silicon at High Electric Fields," *Solid State Electron.*, **16**, 1189 (1973).

65 G. H. Glover, "Charge Multiplication in Au–SiC (6H) Schottky Junction," *J. Appl. Phys.*, **46**, 4842 (1975).

66 T. P. Pearsall, F. Capasso, R. E. Nahory, M. A. Pollack, and J. R. Chelikowsky, "The Band Structure Dependence of Impact Ionization by Hot Carriers in Semiconductors GaAs," *Solid State Electron.*, **21**, 297 (1978).

67 I. Umebu, A. N. M. M. Choudhury, and P. N. Robson, "Ionization Coefficients Measured in Abrupt InP Junction," *Appl. Phys. Lett.*, **36**, 302 (1980).

68 R. A. Logan and H. G. White, "Charge Multiplication in GaP *p-n* Junctions," *J. Appl. Phys.*, **36**, 3945 (1965).

69 T. P. Pearsall, "Impact Ionization Rates for Electrons and Holes in $Ga_{0.47}In_{0.53}As$," *Appl. Phys. Lett.*, **36**, 218 (1980).

70 T. P. Pearsall, R. E. Nahory, and M. A. Pollack, "Impact Ionization Rates for Electrons and Holes in $GaAs_{1-x}Sb_x$ Alloys," *Appl. Phys. Lett.*, **28**, 403 (1976).

71 G. A. Baraff, "Distribution Junctions and Ionization Rates for Hot Electrons in Semiconductors," *Phys. Rev.*, **128**, 2507 (1962).

72 C. R. Crowell and S. M. Sze, "Temperature Dependence of Avalanche Multiplication in Semiconductors," *Appl. Phys. Lett.*, **9**, 242 (1966).

73 J. R. Haynes and W. Shockley, "The Mobility and Life of Injected Holes and Electrons in Germanium," *Phys. Rev.*, **81**, 835 (1951).

74 H. Kroemer, "The Einstein Relation for Degenerate Carrier Concentration," *IEEE Trans. Elec. Dev.* **ED-25**, 850 (1978).

75 K. K. Thornber, "Relation of Drift Velocity to Low-Field Mobility and High Field Saturation Velocity," *J. Appl. Phys.*, **51**, 2127 (1980).

76 K. K. Thornber, "Applications of Scaling to Problems in High-Field Electronic Transport," *J. Appl. Phys.* **52**, 279 (1981).

PART II

BIPOLAR DEVICES

2

p-n Junction Diode

- INTRODUCTION
- BASIC DEVICE TECHNOLOGY
- DEPLETION REGION AND DEPLETION CAPACITANCE
- CURRENT–VOLTAGE CHARACTERISTICS
- JUNCTION BREAKDOWN
- TRANSIENT BEHAVIOR AND NOISE
- TERMINAL FUNCTIONS
- HETEROJUNCTION

2.1 INTRODUCTION

The *p-n* junctions are of great importance both in modern electronic applications and in understanding other semiconductor devices. The *p-n* junction theory serves as the foundation of the physics of semiconductor devices. The basic theory of current–voltage characteristics of *p-n* junctions was established by Shockley.[1] This theory was then extended by Sah, Noyce, and Shockley[2], and by Moll.[3]

In this chapter we first briefly discuss the basic device technology which is pertinent not only to *p-n* junctions but also to most semiconductor devices. Then the basic equations presented in Chapter 1 are used to develop the ideal static and dynamic characteristics of *p-n* junctions. Departures from the ideal characteristics due to generation and recombination in the depletion layer, to high injection, and to series resistance effects are then discussed. Junction breakdown (especially that due to avalanche multiplication) is considered in detail, after which transient behaviors and noise performance in *p-n* junction are presented.

A *p-n* junction is a two-terminal device. Depending on doping profile, device geometry, and biasing condition, a *p-n* junction can perform various terminal functions, which are considered briefly in Section 2.7. The chapter closes with attention to an important group of devices, the heterojunctions, which are junctions formed between dissimilar semiconductors (e.g., *n*-type Ge on *p*-type GaAs).

2.2 BASIC DEVICE TECHNOLOGY

In this chapter we are concerned primarily with silicon technology,[4] because its development is more advanced than that of any other semiconductor. Some important device fabrication methods are shown in Fig. 1.

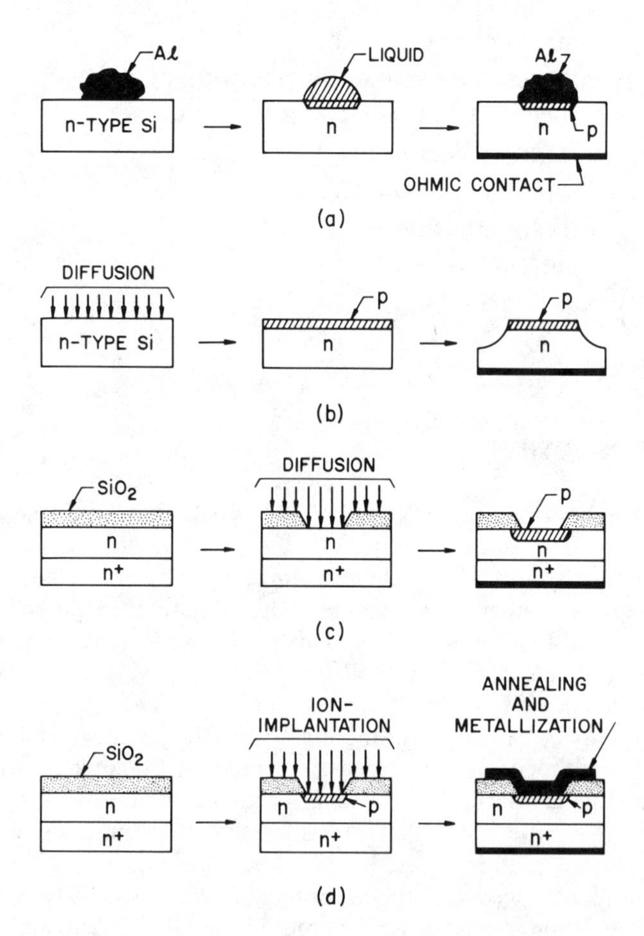

Fig. 1 Some device fabrication methods. (*a*) Alloyed junction. (*b*) Diffused mesa junction. (*c*) Diffused planar junction on epitaxial substrate. (*d*) Ion implantation.

In the alloy method,[5] Fig. 1*a*, a small pellet of aluminum is placed on an *n*-type ⟨111⟩-oriented silicon wafer. The system is then heated to a temperature slightly higher than the eutectic temperature (~580°C for the Al–Si system) so that a small puddle of molten Al–Si mixture is formed. The temperature is then lowered and the puddle begins to solidify. A recrystallized portion, which is saturated with the acceptor impurities and with the same crystal orientation, forms the heavily doped *p*-type region (p^+) on the *n*-type substrate. The aluminum button on top can be used as an ohmic contact for the *p*-type region. For the ohmic contact on the *n*-type wafer, a Au–Sb alloy (with ~0.1% Sb) can be evaporated onto the wafer and alloyed at about 400°C to form a heavily doped *n*-type region (n^+). On a *p*-type wafer, the roles of the aluminum and the Au–Sb alloy can be interchanged to form an n^+ junction on top and p^+ ohmic contact on the bottom of the wafer. The junction location obtained by the alloy method depends critically on the temperature–time alloying cycle and is difficult to control precisely.

The solid-state diffusion method was developed in 1956 to give more precise control of the impurity profile.[6] A diffused mesa junction method, where *p*-type impurities (e.g., boron in the form of BBr_3) are diffused into the *n*-type substrate, is shown in Fig. 1*b*. After diffusion, portions of the surface are protected (e.g., by wax or metal contacts), and the rest are etched out to form the mesa structures.

A new degree of control over the lateral geometry of the diffused junction is achieved by using an insulating layer that can prevent most donor and acceptor impurities from diffusing through it.[7] A typical example is shown in Fig. 1*c*. A thin layer of silicon dioxide (~1 μm) is thermally grown on silicon. With the help of lithographic techniques (e.g., photolithography, x-ray, or electron-beam lithography), portions of the oxide can be removed and windows (or patterns) cut in the oxide. The impurities will diffuse only through the exposed silicon surface, and *p-n* junctions will form in the oxide windows. This process, the planar process,[8] has become the principal method of fabricating semiconductor devices and integrated circuits.*

Also shown in Fig. 1*c* is the epitaxial substrate, for example, an *n* layer on n^+ substrate.[9] Such a substrate is generally used in the planar process to reduce series resistance. Epitaxy, derived from the Greek words *epi*, meaning "on", and *taxis*, meaning "arrangement", describes a technique of crystal growth by chemical reaction used to form, on the surface of crystal, thin layers of semiconductor materials with lattice structures identical to

*The planar process uses techniques which were previously known, such as oxide masking against diffusion. The distinguishing feature is their use in combination, permitting an unprecedented fineness in controlling the sizes and shapes of electrodes and diffused regions. The name "planar" comes from the requirement that the wafer surface must be approximately flat. If the surface is rough, the liquid photoresist does not coat it evenly, and imperfections result.

those of the crystal. In this method, lightly doped high-resistivity epitaxial layers are grown on and supported by a heavily doped low-resistivity substrate thus ensuring both the desired electrical properties and mechanical strength.

Figure 1*d* shows a *p-n* junction formed by ion implantation.[10] To date, this method gives the most precise control of an impurity profile. Ion implantation can be done at room temperature, and the implantation-induced lattice damages can be removed by annealing at about 700°C or less. Therefore, ion implantation is a relatively low-temperature process compared to diffusion, which is generally done at 1000°C or higher.

We shall now briefly discuss the four main processes of planar technology: epitaxial growth, oxidation, diffusion of impurities, and ion implantation.

Epitaxial layers can be formed by the vapor-phase growth technique.[4] The basic reaction that results in the growth of silicon layers is $SiCl_4 + 2H_2 \rightarrow Si(solid) + 4HCl(gas)$. Typically, the silicon layer is grown at a rate of about 1 μm/min at 1200°C or higher temperature with a mole fraction of $SiCl_4$ at 0.01% (ratio of $SiCl_4$ molecules to the total number of molecules in the gas). Epitaxial layers can also be formed by liquid-phase growth,[11] which has been extensively used for compound semiconductors, or by molecular-beam epitaxy,[12, 13] which can give precise control of semiconductor compositions down to atomic dimensions.

The most frequently used method to form silicon dioxide films is by thermal oxidation[14] of silicon through the chemical reaction: $Si(solid) + O_2(dry\ oxygen) \rightarrow SiO_2(solid)$ or $Si(solid) + 2H_2O(steam) \rightarrow SiO_2(solid) + 2H_2$. It can be shown that for short reaction times the oxide thickness increases linearly with time, and for prolonged oxidation the thickness varies as the square root of time—the so-called parabolic relationship.[15] When a silicon dioxide film of thickness d is formed, a layer of silicon of thickness $0.45\,d$ is consumed. Figure 2 shows the experimental results of the oxide thickness as a function of reaction time and temperature for both dry oxygen growth and steam growth.

At a given temperature, the oxidation rate in steam is about 5 to 10 times higher than for dry oxygen. Also, at lower temperatures, the oxidation rates show pronounced dependence on crystal orientation.[16]

The simple one-dimensional diffusion process can be given by the Fick equation,[17]

$$\frac{\partial C(x,t)}{\partial t} = D\frac{\partial^2 C(x,t)}{\partial x^2} \qquad (1)$$

where C is the impurity concentration and D the diffusion coefficient. This expression is similar to that given in Eq. 96a of Chapter 1, without generation, recombination, or electric field. For a "limited source" condition, with the total amount of impurities S, the solution of Eq. 1 is given

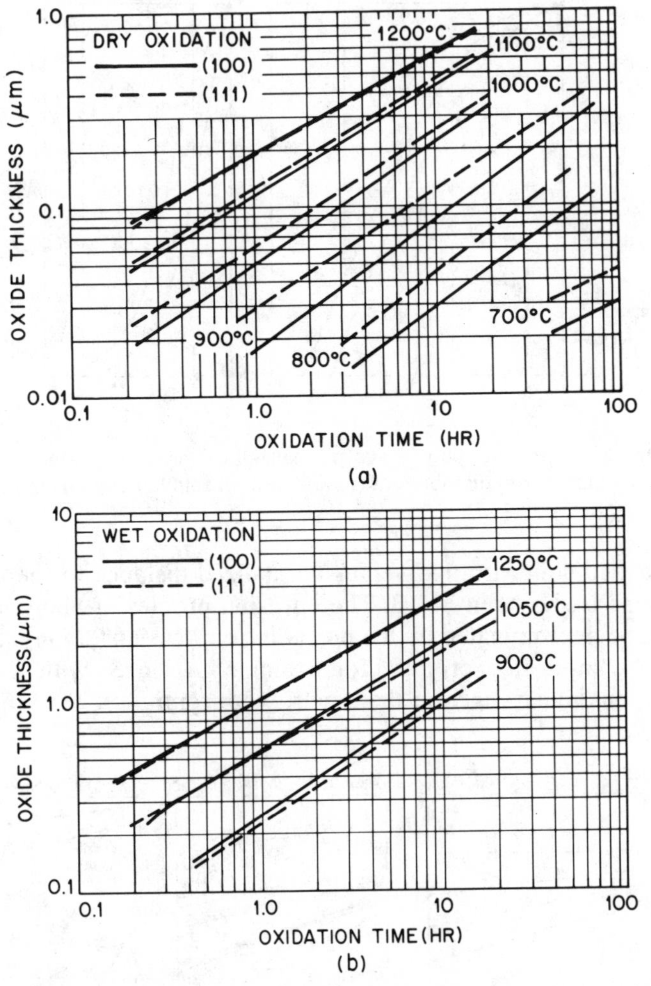

Fig. 2 Experimental results of silicon dioxide thickness as a function of the reaction time and temperature for two substrate orientations. (*a*) Dry oxygen growth. (*b*) Steam growth. (After Meindl et al., Ref. 16.)

by the Gaussian function

$$C(x, t) = \frac{S}{\sqrt{\pi D t}} \exp\left(-\frac{x^2}{4Dt}\right). \tag{2a}$$

For the "constant surface concentration" condition with a surface concentration C_s, the solution of Eq. 1 is given by the error function complement

$$C(x, t) = C_s \operatorname{erfc}\left(\frac{x}{2\sqrt{Dt}}\right). \tag{2b}$$

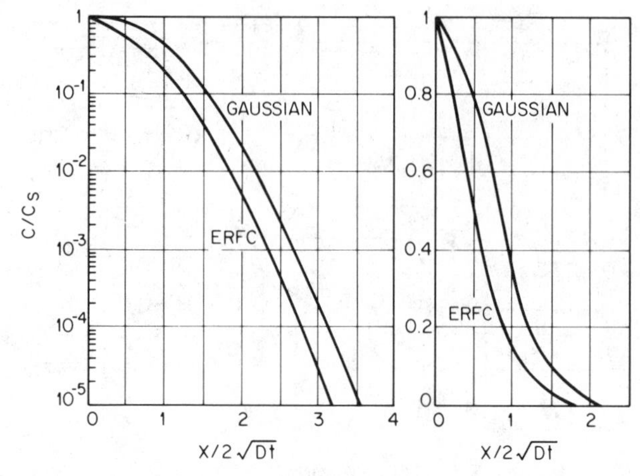

Fig. 3 Normalized concentration versus normalized distance for Gaussian and error function complement (erfc) distributions plotted in both semilog and linear scales. (After Carslaw and Jaeger, Ref. 17.)

The normalized concentration versus normalized distance to the foregoing two solutions is shown in Fig. 3. The diffusion profiles of many impurities can indeed be approximated by the preceding expressions. Many, however, have more complicated profiles, for example, As in Si with a diffusion process depending strongly on the impurity concentration.[18]

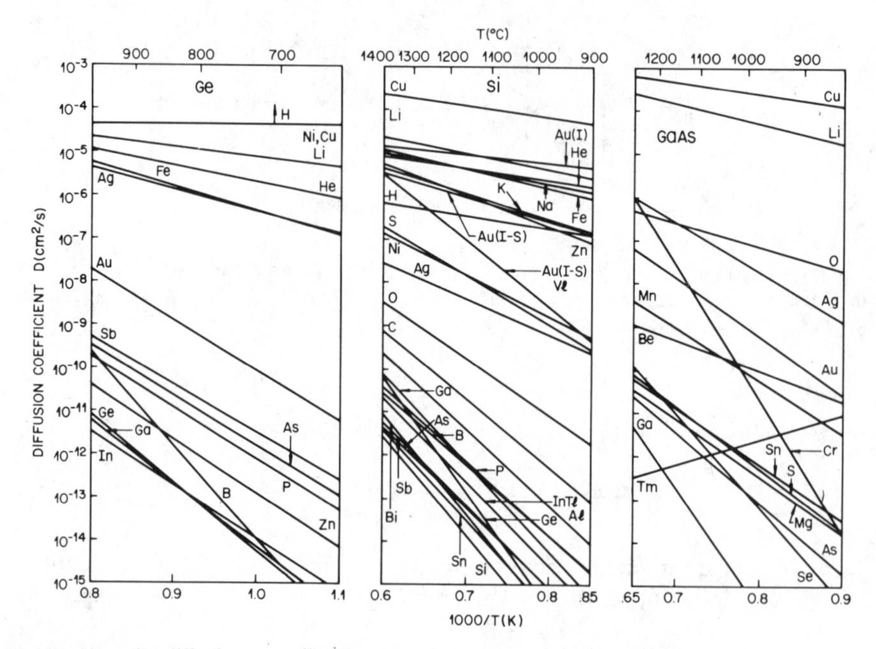

Fig. 4 Impurity diffusions coefficients as a function of temperature for Ge, Si, and GaAs. (After Burger and Donovan, Ref. 19; Kendall and DeVries, Ref. 20.)

The diffusion coefficient D depends on temperature and impurity concentration. Under low-concentration conditions, D becomes independent of impurity concentration. (In low-concentration conditions, the impurity concentration is less than the intrinsic carrier concentration at the diffusion temperature, for example, at 1100°C, $n_i \approx 10^{19}\ \text{cm}^{-3}$, as shown in Fig. 11 of Chapter 1.) In a limited temperature range and under low-concentration conditions, the diffusion coefficient can be described by

$$D(T) = D_0 \exp(-\Delta E/kT) \tag{3}$$

where D_0 is the diffusion coefficient extrapolated to infinite temperature, and ΔE is the activation energy of diffusion. Values of $D(T)$ for Ge, Si, and GaAs are plotted in Fig. 4 for various impurities.[19,20] These values are for low-impurity-concentration cases; as impurity concentration increases, $D(T)$ becomes increasingly concentration-dependent.

The impurity diffusion coefficient is related to the solid solubility of the

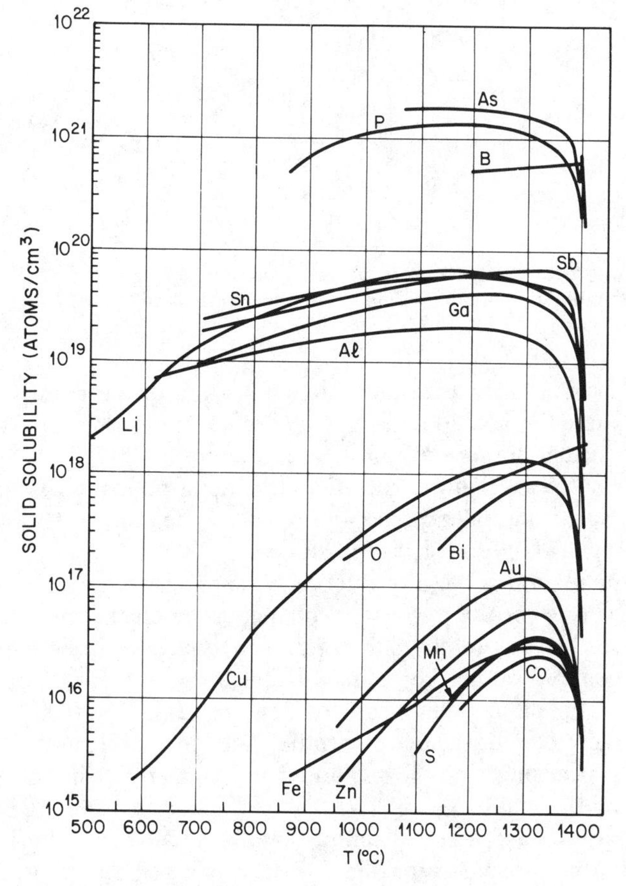

Fig. 5 Solid solubility of various elements in Si as a function of temperature. (After Trumbore, Ref. 21.)

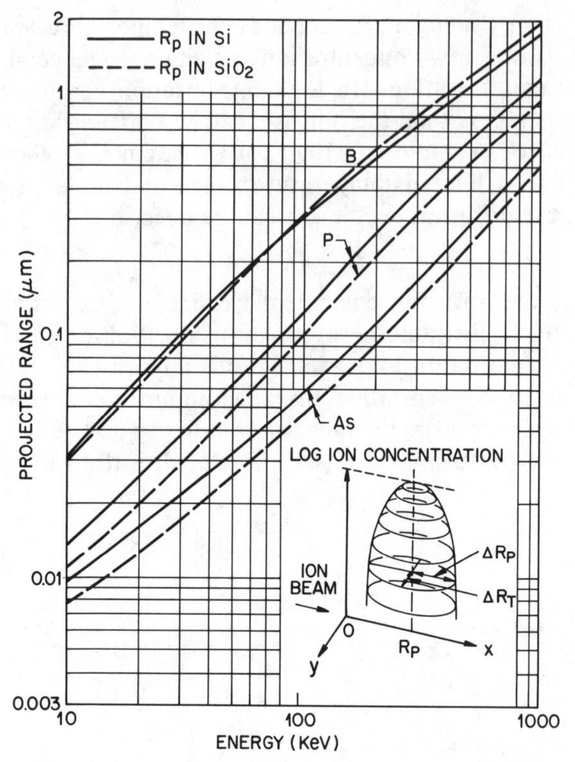

Fig. 6 Projected range of boron, phosphorus, and arsenic ions in Si and SiO_2 as a function of implantation energy. The insert shows the distribution of the implanted ions. (After Pickar, Ref. 10.)

impurity, which is the maximum concentration of an impurity that can be accommodated in a solid at any given temperature. The solid solubilities of some important impurities in Si are plotted in Fig. 5 as a function of temperature.[21] This figure shows that arsenic or phosphorus should be used as the impurity in making heavily doped *n*-type silicon, and boron should be used for heavily doped *p*-type silicon.

Ion implantation is the introduction of energetic charged atomic particles into a substrate for the purpose of changing the electrical, metallurgical, or chemical properties of the substrate. The typical ion energies considered are between 10 and 400 keV, and typical ion doses vary from 10^{11} to 10^{16} ions/cm^2. The main advantages of ion implantation include (1) precise control over total dose, depth profile, and area uniformity; (2) low-temperature processing; and (3) implanted junctions that can be self-aligned to the edge of the mask.

For an ion beam with an infinitesimally small beam diameter, the ion distribution in the substrate (Fig. 6, insert) is given by[10]

$$n(x, y) = \frac{S}{(2\pi)^{3/2}\,\Delta R_p\,\Delta R_T^2} \exp\left[-\left(\frac{x - R_p}{\sqrt{2}\Delta R_p}\right)^2\right] \exp\left[-\left(\frac{y}{\sqrt{2}\Delta R_T}\right)^2\right] \quad (4)$$

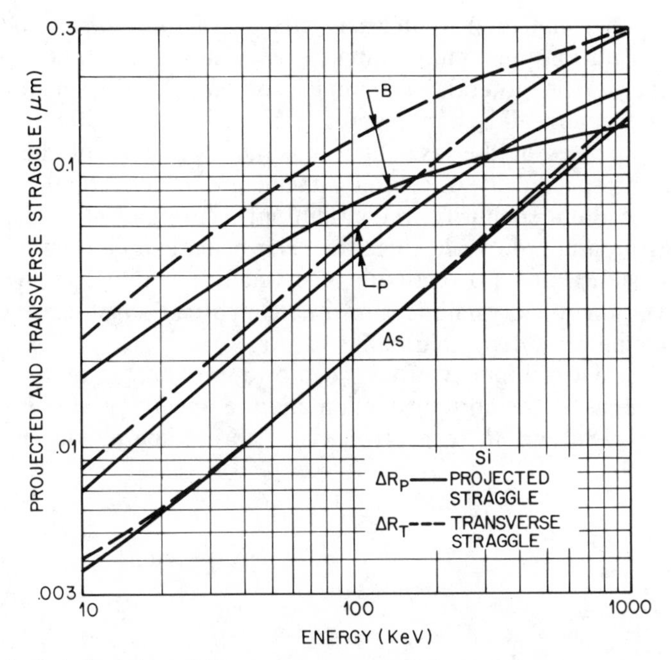

Fig. 7 Projected straggle and transverse straggle of boron, phosphorus, and arsenic in silicon as a function of implantation energy. (After Pickar, Ref. 10.)

where s is the incident rate of ions per second, R_p the projected range, ΔR_p the projected straggle, and ΔR_T the transverse straggle. Figure 6 shows values of R_p in Si and SiO_2 for boron, phosphorus, and arsenic ions as a function of implantation energy. The projected range increases approximately linearly with the energy. For boron, the projected ranges in Si and SiO_2 are quite close; however, for phosphorus and arsenic, the projected range in SiO_2 is about 20% lower than that in Si. The projected straggle and transverse straggle in Si for these ions are shown in Fig. 7. The straggles also increase with increasing energy. The ratio of ΔR_p to R_p over the energy range is about 0.2 to 0.5.

For an implantation from an infinitesimal beam that is scanned uniformly across the substrate surface, the y dependence of the doping density drops out in regions several ΔR_T from the edge of the scanned area. The doping distribution then becomes

$$n(x) = \frac{\phi}{\sqrt{2\pi}\,\Delta R_p} \exp\left[-\left(\frac{x - R_p}{\sqrt{2}\,\Delta R_p}\right)^2\right] \tag{4a}$$

which describes the Gaussian distribution in doping density of the implanted ions, where ϕ is the total number of ions per unit area, and the quantity $\phi/(\sqrt{2\pi}\Delta R_p)$ is the peak doping concentration at $x = R_p$.

Since 1974, laser processing of semiconductors has been extensively studied.[22,23] High-intensity laser radiations [such as the pulsed ruby laser and continuous-wave (cw) argon laser] offer possibilities to remove damages associated with ion implantation and to recrystallize amorphous semiconductor layers. The potential advantages of laser processing include (1) control of the annealing depth and the impurity profile through the absorption properties of the laser light and the dwell time of the pulsed or swept beam; specifically, the laser annealing can activate the implanted impurities without impurity redistribution; (2) highly localized lateral processing on a micron scale since the laser beam can be focused down to such dimensions; and (3) regrowth of crystalline material from an amorphous layer on the crystalline substrate or formation of large-grain-size polycrystalline films deposited on insulators.

In practice, most impurity profiles can be approximated by the following two limiting cases: the abrupt junction and the linearly graded junction, as shown in Fig. 8*a* and 8*b* respectively. The abrupt approximation provides

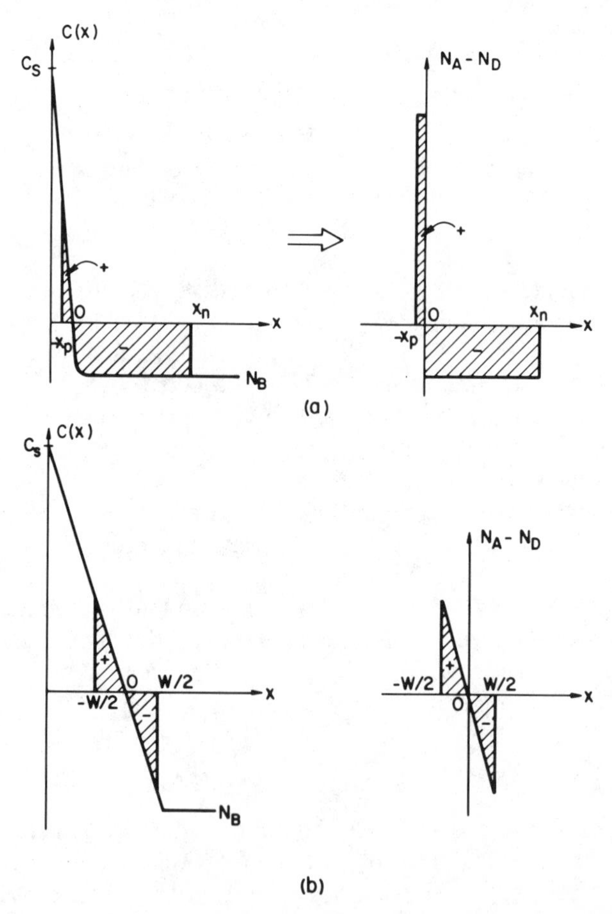

Fig. 8 Approximate doping profiles. (*a*) Abrupt junction. (*b*) Linearly graded junction.

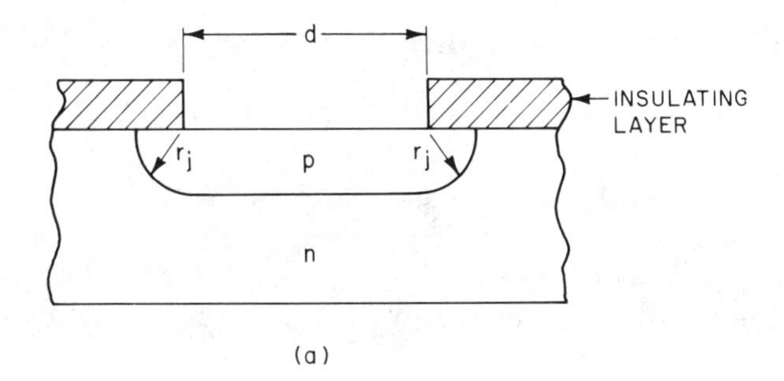

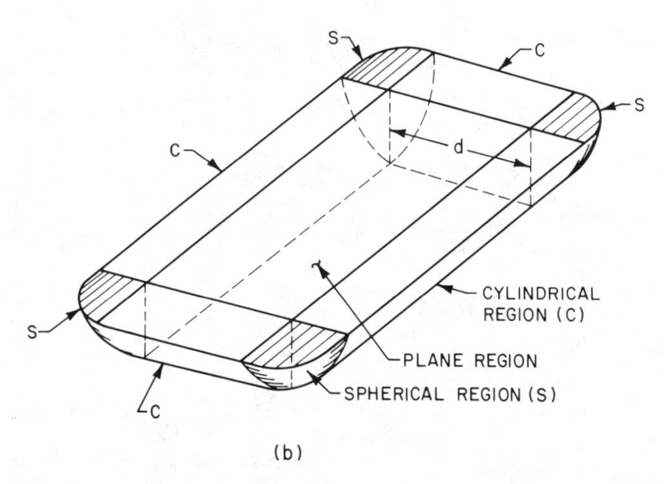

Fig. 9 (*a*) Planar diffusion process which forms junction curvature near the edges of the diffusion mask. r_j is the radius of curvature. (*b*) The formation of approximately cylindrical and spherical regions by diffusion through a rectangular mask. (After Lee and Sze, Ref. 24.)

an adequate description for alloyed junctions, shallowly diffused junctions, and ion-implanted junctions. The linearly graded approximation is reasonable for deeply diffused junctions.

Another important effect results from the planar processes. When a *p-n* junction is formed by diffusion into a bulk semiconductor through a window in an insulating layer, the impurities will diffuse downward and also sideways. Hence the junction consists of a plane (or flat) region with approximately cylindrical edges, as shown[24] in Fig. 9*a*. In addition, if the diffusion mask contains sharp corners, the junction near the corner will be roughly spherical in shape (Fig. 9*b*). These spherical and cylindrical regions have profound effects on the junction, especially for the avalanche multiplication process,[25] discussed in Section 2.5.

2.3 DEPLETION REGION AND DEPLETION CAPACITANCE

2.3.1 Abrupt Junction

Diffusion Potential and Depletion-Layer Width When the impurity concentration in a semiconductor changes abruptly from acceptor impurities N_A to donor impurities N_D, as shown in Fig. 10*a*, one obtains an abrupt junction. In particular, if $N_A \gg N_D$, one obtains a one-sided abrupt junction or p^+n junction.

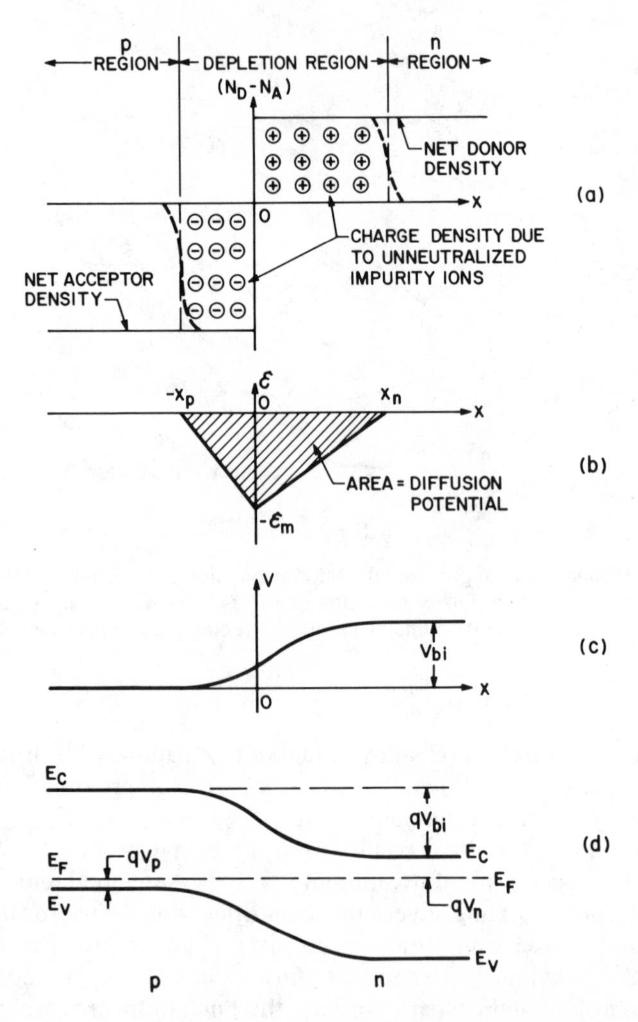

Fig. 10 Abrupt *p-n* junction in thermal equilibrium. (*a*) Space-charge distribution. The dashed lines indicate the majority-carrier distribution tails. (*b*) Electric field distribution. (*c*) Potential variation with distance where V_{bi} is the built-in potential. (*c*) Energy-band diagram.

We first consider the thermal equilibrium condition, that is, one with no applied voltage and no current flow. From Eqs. 33 and 93a in Chapter 1,

$$J_n = 0 = q\mu_n\left(n\mathscr{E} + \frac{kT}{q}\frac{\partial n}{\partial x}\right) = \mu_n n \frac{\partial E_F}{\partial x} \tag{5}$$

or

$$\frac{\partial E_F}{\partial x} = 0. \tag{5a}$$

Similarly,

$$J_p = 0 = \mu_p p \frac{\partial E_F}{\partial x}. \tag{6}$$

Thus the condition of zero net electron and hole currents requires that the Fermi level must be constant throughout the sample. The diffusion potential, or built-in potential V_{bi}, as shown in Fig. 10*b*, *c*, and *d*, is equal to

$$\begin{aligned} qV_{bi} &= E_g - (qV_n + qV_p) \\ &= kT \ln\left(\frac{N_C N_V}{n_i^2}\right) - \left[kT \ln\left(\frac{N_C}{n_{no}}\right) + kT \ln\left(\frac{N_V}{p_{po}}\right)\right] \\ &= kT \ln\left(\frac{n_{no} p_{po}}{n_i^2}\right) \simeq kT \ln\left(\frac{N_A N_D}{n_i^2}\right). \end{aligned} \tag{7}$$

Since at equilibrium $n_{no}p_{no} = n_{po}p_{po} = n_i^2$,

$$V_{bi} = \frac{kT}{q}\ln\left(\frac{p_{po}}{p_{no}}\right) = \frac{kT}{q}\ln\left(\frac{n_{no}}{n_{po}}\right). \tag{7a}$$

Equation 7a gives the relationship between the hole and electron densities on either side of the junction:

$$p_{no} = p_{po}\exp\left(-\frac{qV_{bi}}{kT}\right) \tag{8a}$$

$$n_{po} = n_{no}\exp\left(-\frac{qV_{bi}}{kT}\right). \tag{8b}$$

The approximate values of V_{bi} for one-sided abrupt *p-n* junctions in Ge, Si, and GaAs are shown in Fig. 11.

Since in thermal equilibrium the electric field in the neutral regions (far from the junction at either side) of the semiconductor must be zero, the total negative charge per unit area in the *p* side must be precisely equal to the total positive charge per unit area in the *n* side:

$$N_A x_p = N_D x_n. \tag{9}$$

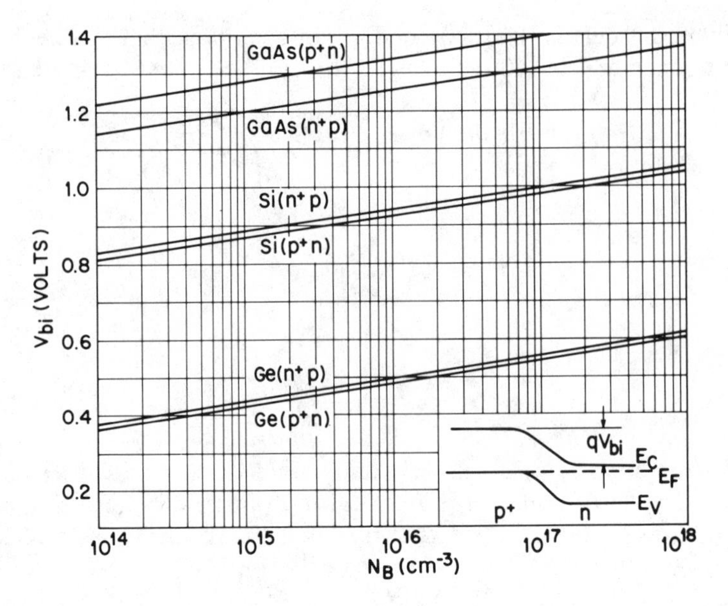

Fig. 11 Built-in potential for one-sided abrupt junctions in Ge, Si, and GaAs, where p^+ is for the heavily doped *p* side and n^+ is for the heavily doped *n* side. The background doping N_B is for the impurity concentration of the lightly doped side.

From Poisson's equation we obtain (for abrupt approximation)

$$-\frac{\partial^2 V}{\partial x^2} \equiv \frac{\partial \mathscr{E}}{\partial x} = \frac{\rho(x)}{\epsilon_s} = \frac{q}{\epsilon_s}[p(x) - n(x) + N_D^+(x) - N_A^-(x)] \tag{10}$$

or

$$-\frac{\partial^2 V}{\partial x^2} \approx \frac{q}{\epsilon_s} N_D \qquad \text{for} \quad 0 < x \le x_n \tag{10a}$$

$$-\frac{\partial^2 V}{\partial x^2} \approx \frac{-q}{\epsilon_s} N_A \qquad \text{for} \quad -x_p \le x < 0. \tag{10b}$$

The electric field is then obtained by integrating Eqs. 10a and 10b as shown in Fig. 10*b*:

$$\mathscr{E}(x) = -\frac{qN_A(x + x_p)}{\epsilon_s} \qquad \text{for} \quad -x_p \le x < 0 \tag{11a}$$

and

$$\mathscr{E}(x) = -\mathscr{E}_m + \frac{qN_D x}{\epsilon_s}$$

$$= \frac{qN_D}{\epsilon_s}(x - x_n) \qquad \text{for} \quad 0 < x \le x_n \tag{11b}$$

where $\mathscr{E}_m$ is the maximum field that exists at $x = 0$ and is given by

$$|\mathscr{E}_m| = \frac{qN_D x_n}{\epsilon_s} = \frac{qN_A x_p}{\epsilon_s}. \tag{12}$$

Integrating Eq. 10 once again, Fig. 10*c*, gives the potential distribution $V(x)$ and the built-in potential V_{bi}:

$$V(x) = \mathscr{E}_m\left(x - \frac{x^2}{2W}\right) \tag{13}$$

$$V_{bi} = \tfrac{1}{2}\mathscr{E}_m W \equiv \tfrac{1}{2}\mathscr{E}_m (x_n + x_p) \tag{14}$$

where W is the total depletion width. Eliminating $\mathscr{E}_m$ from Eqs. 12 and 14 yields

$$W = \sqrt{\frac{2\epsilon_s}{q}\left(\frac{N_A + N_D}{N_A N_D}\right) V_{bi}} \tag{15}$$

for a two-sided abrupt junction. For a one-sided abrupt junction, Eq. 15 reduces to

$$W = \sqrt{\frac{2\epsilon_s V_{bi}}{qN_B}} \tag{15a}$$

where $N_B = N_D$ or N_A depending on whether $N_A \gg N_D$ or vice versa.

A more accurate result for the depletion-layer width can be obtained from Eq. 10 by considering the majority-carrier contribution in addition to the impurity concentration, that is, $\rho \approx -q[N_A - p(x)]$ on the p side and $\rho \approx q[N_D - n(x)]$ on the n side. The depletion width is essentially the same as given by Eq. 15, except that V_{bi} is replaced by $(V_{bi} - 2kT/q)$. The correction factor $2kT/q$ comes about because of the two majority-carrier distribution tails[26] (electrons in n side and holes in p side, as shown by the dashed lines in Fig. 10*a*). Each contributes a correction factor kT/q. The correction is simply the dipole moment of the "error" distribution—the true carrier distribution minus the abrupt distribution. The depletion-layer width at thermal equilibrium for a one-sided abrupt junction becomes

$$\begin{aligned} W &= \sqrt{\frac{2\epsilon_s}{qN_B}(V_{bi} - 2kT/q)} \\ &= L_D\sqrt{2(\beta V_{bi} - 2)} \end{aligned} \tag{16}$$

where $\beta = q/kT$ and L_D is the Debye length, which is a characteristic length for semiconductors. The Debye length is defined as

$$L_D \equiv \sqrt{\frac{\epsilon_s kT}{q^2 N_B}} = \sqrt{\frac{\epsilon_s}{qN_B\beta}}. \tag{17}$$

At thermal equilibrium the depletion-layer widths of abrupt junctions are about $6L_D$ for Ge, $8L_D$ for Si, and $10L_D$ for GaAs. The Debye length as a function of doping density is shown in Fig. 12 for silicon at room tem-

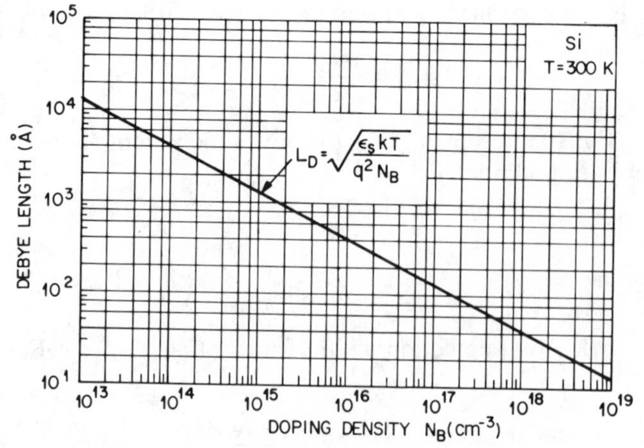

Fig. 12 Debye length in Si as a function of doping density.

perature. For a doping density of 10^{16} cm^{-3}, the Debye length is 400 Å; for other dopings, L_D will vary as $1/\sqrt{N_B}$, that is, a reduction by a factor of 3.16 per decade.

The values of W as a function of the doping concentration for one-sided abrupt junctions in silicon are shown in Fig. 13 (dashed line for zero bias). When a voltage V is applied to the junction, the total electrostatic potential

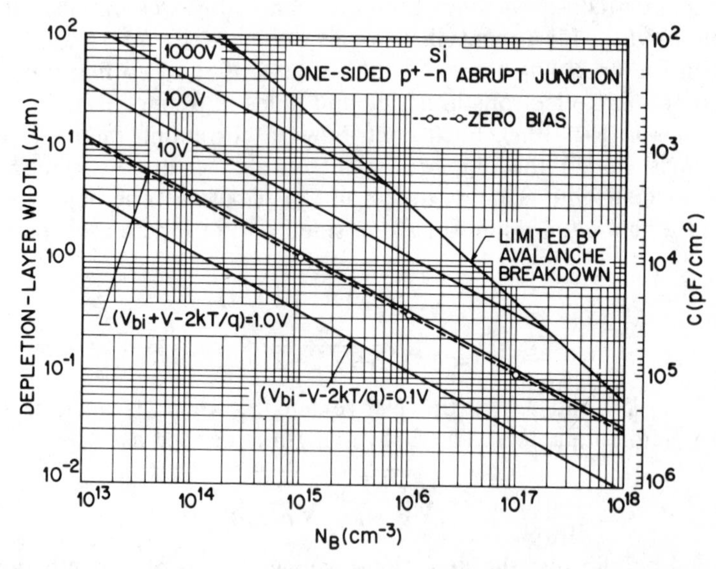

Fig. 13 Depletion-layer width and depletion-layer capacitance per unit area as a function of doping for one-sided abrupt junction in Si. The dashed line is for the case of zero-bias voltage.

variation across the junction is given by $(V_{bi} + V)$ for reverse bias (positive voltage on n region with respect to p region) and by $(V_{bi} - V)$ for forward bias. Substituting these voltage values in Eq. 16 yields the depletion-layer width as a function of the applied voltage. The results for one-sided abrupt junctions in silicon are shown in Fig. 13. The values below the zero-bias line (dashed line) are for the forward-biased condition; and above, for the reverse-biased condition.

These results can also be used for GaAs since both Si and GaAs have approximately the same static dielectric constants. To obtain the depletion-layer width for Ge, one must multiply the results of Si by the factor

$$\sqrt{\epsilon_s(\text{Ge})/\epsilon_s(\text{Si})} = 1.16.$$

The simple model above can give adequate predictions for most abrupt p-n junctions. However, for strongly asymmetrical junctions or for devices with ultrashallow junction depths, numerical analysis may be needed for accurate results.[27] The electric field region near the junction cannot be confined to the shaded region as shown in the left diagram of Fig. 8a, because any doping gradient creates a field (refer to Eq. 5). The strong doping gradient clearly extends well outside the shaded region. Figure 14 shows an example of a diffused junction 0.25 μm deep with $C_s = 2 \times 10^{20}$ cm^{-3}, an erfc profile, and $N_A = 5 \times 10^{15}$ cm^{-3}. The equilibrium energy-band diagram is shown in Fig. 14a. Note that the conduction band goes below the Fermi level near the surface due to high surface concentration. The electric field profile is shown in Fig. 14b. There are several important differences in the field profiles between the simple model and the numerical results. First, the actual field region on the diffused side is five times wider than it is in the simple model. Second, the electric field is always larger than 10^4 V/cm throughout the surface region, so that transport processes can be strongly affected. The space-charge distribution (assuming negligible surface-charge effects) is shown in Fig. 14c. Apparently, the space charge on the diffused side is considerably widened beyond the depletion-layer width $-x_p$ in the simple model.

Depletion-Layer Capacitance The depletion-layer capacitance per unit area is defined as $C \equiv dQ_c/dV$, where dQ_c is the incremental increase in charge per unit area upon an incremental change of the applied voltage dV.

For one-sided abrupt junctions, the capacitance per unit area is given by

$$C \equiv \frac{dQ_c}{dV} = \frac{d(qN_BW)}{d[(qN_B/2\epsilon_s)W^2]} = \frac{\epsilon_s}{W} = \sqrt{\frac{q\epsilon_s N_B}{2}}\,(V_{bi} \pm V - 2kT/q)^{-1/2}$$

$$= \frac{\epsilon_s}{\sqrt{2}L_D}(\beta V_{bi} \pm \beta V - 2)^{-1/2} \qquad \text{F/cm}^2 \tag{18}$$

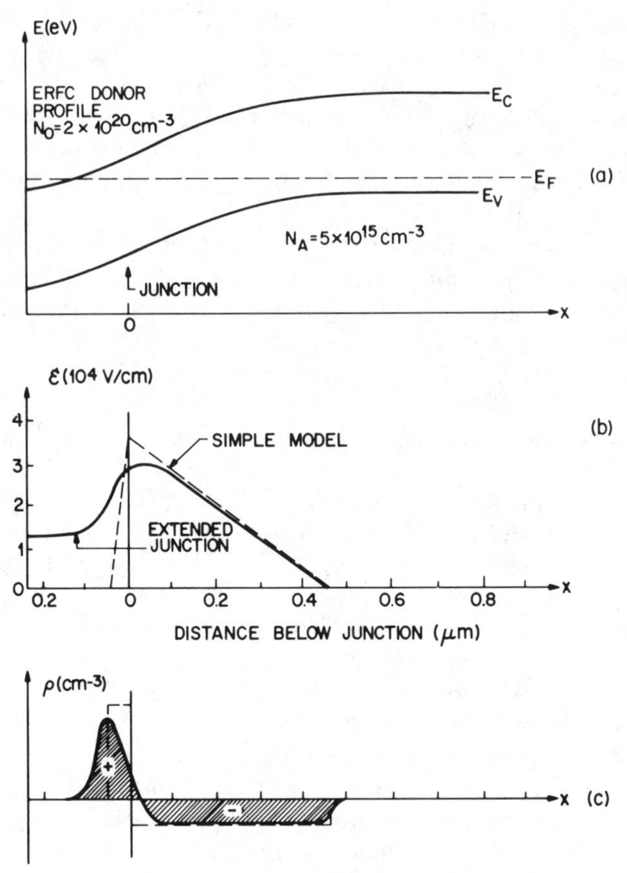

Fig. 14 Strongly asymmetrical junction. (*a*) Energy-band diagram. (*b*) Electric field distribution. (*c*) Space-charge distribution. (After Redfield, Ref. 27.)

or

$$\frac{1}{C^2} = \frac{2L_D^2}{\epsilon_s^2}(\beta V_{bi} \pm \beta V - 2) \tag{18a}$$

$$\frac{d(1/C^2)}{dV} \cong \frac{2L_D^2\beta}{\epsilon_s^2} = \frac{2}{q\epsilon_s N_B} \tag{18b}$$

where the $\pm$ signs are for the reverse- and forward-bias conditions, respectively. It is clear from Eq. 18a that by plotting $1/C^2$ versus V, a straight line should result for a one-sided abrupt junction. The slope gives the impurity concentration of the substrate (N_B), and the intercept (at $1/C^2 = 0$) gives ($V_{bi} - 2kT/q$). The results of the capacitance are also shown in Fig. 13. Note that, for the forward bias, a diffusion capacitance exists in addition to the depletion capacitance mentioned previously. The diffusion capacitance is discussed in Section 3.4.

Equation 18b holds for more general distributions than just for the abrupt p^+-n junction. For a general distribution we have

$$\frac{d(1/C^2)}{dV} = \frac{2}{q\epsilon_s N(W)} \tag{18c}$$

and

$$W = \frac{\epsilon_s}{C(V)} \tag{18d}$$

where $N(W)$ is the doping density at $x = W$.

Note also that the capacitance–voltage data are insensitive to changes in the doping profiles that occur in a distance less than a Debye length of the highly doped side, and so the doping profiles determined by the C–V method should be expected to provide a spatial resolution of only the order of a Debye length.[28]

2.3.2 Linearly Graded Junction

Consider the thermal equilibrium case first. The impurity distribution for linearly graded junctions is shown in Fig. 15*a*. The Poisson equation for this case is

$$-\frac{\partial^2 V}{\partial x^2} \equiv \frac{\partial \mathscr{E}}{\partial x} = \frac{\rho(x)}{\epsilon_s} = \frac{q}{\epsilon_s}(p - n + ax) \approx \frac{q}{\epsilon_s} ax \qquad -\frac{W}{2} \le x \le \frac{W}{2} \tag{19}$$

where a is the impurity gradient in cm^{-4}. By integrating Eq. 19 once, we obtain the field distribution shown in Fig. 15*b*:

$$\mathscr{E}(x) = -\frac{qa}{\epsilon_s} \frac{(W/2)^2 - x^2}{2} \tag{20}$$

with the maximum field $\mathscr{E}_m$ at $x = 0$,

$$|\mathscr{E}_m| = \frac{qaW^2}{8\epsilon_s}. \tag{20a}$$

Integrating Eq. 19 once again gives the built-in potential shown in Fig. 15*c*:

$$V_{bi} = \frac{qaW^3}{12\epsilon_s} \tag{21}$$

or

$$W = \left(\frac{12\epsilon_s V_{bi}}{qa}\right)^{1/3}. \tag{21a}$$

Since the values of the impurity concentrations at the edges of the depletion region ($-W/2$ and $W/2$) are equal to $aW/2$, the built-in potential for linearly graded junctions can be approximated by an expression similar

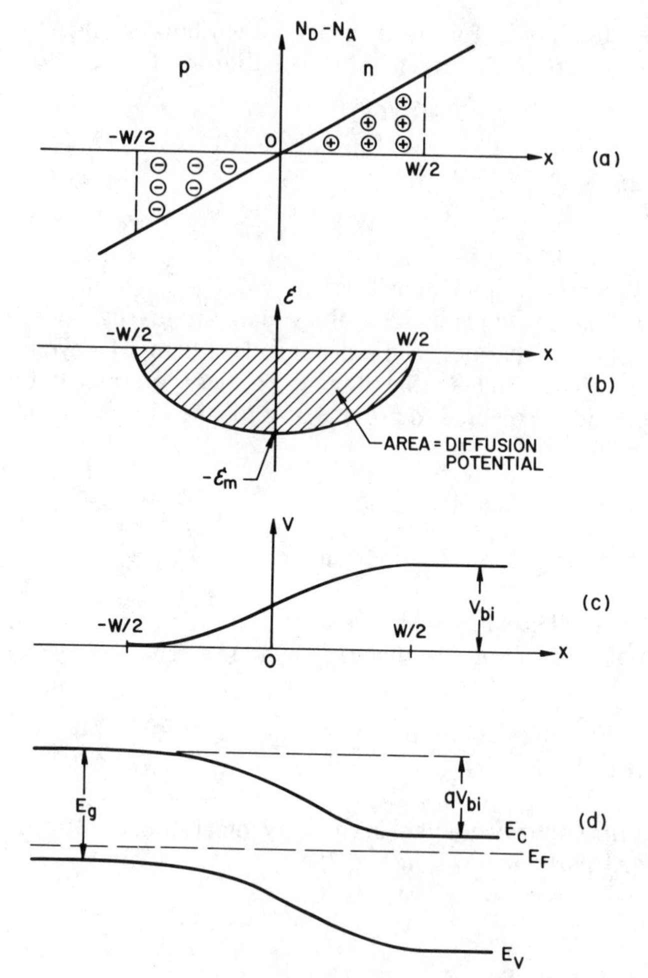

Fig. 15 Linearly graded junction in thermal equilibrium. (*a*) Space-charge distribution. (*b*) Electric field distribution. (*c*) Potential variation with distance. (*d*) Energy-band diagram.

to Eq. 7:

$$V_{bi} \simeq \frac{kT}{q} \ln \left[\frac{(aW/2)(aW/2)}{n_i^2} \right] = \frac{kT}{q} \ln \left(\frac{aW}{2n_i} \right)^2. \tag{22}$$

The depletion-layer capacitance for a linearly graded junction is given by

$$C \equiv \frac{dQ_c}{dV} = \frac{d(qaW^2/8)}{d(qaW^3/12\epsilon_s)} = \frac{\epsilon_s}{W} = \left[\frac{qa\epsilon_s^2}{12(V_{bi} \pm V)} \right]^{1/3} \quad \text{F/cm}^2 \tag{23}$$

where the signs + and − are for the reverse and forward bias, respectively.
Based on an accurate numerical technique,[29] the depletion-layer capaci-

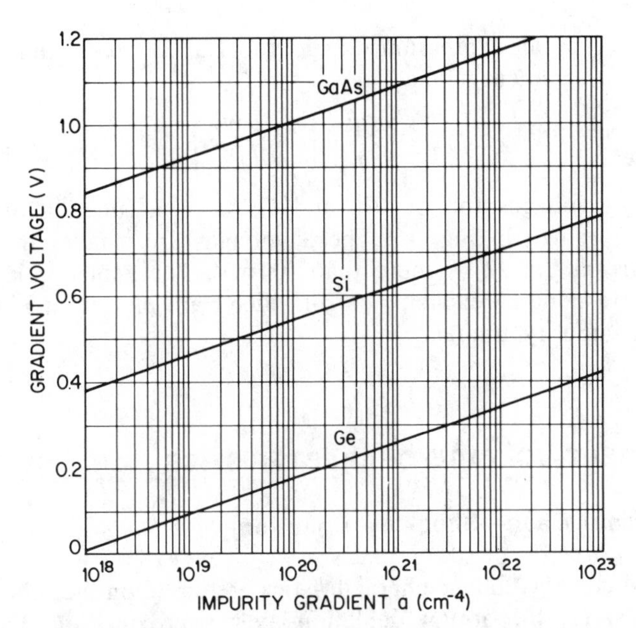

Fig. 16 Gradient voltage for linearly graded junctions in Ge, Si, and GaAs.

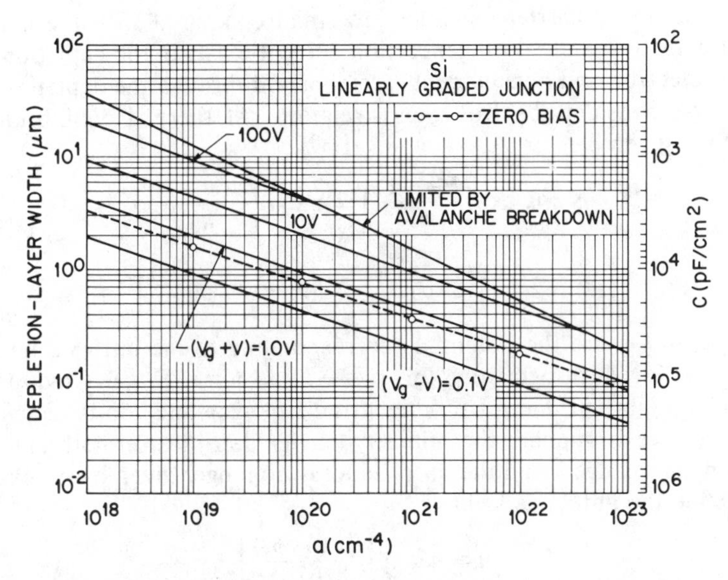

Fig. 17 Depletion-layer width and depletion-layer capacitance per unit area as a function of impurity gradient for linearly graded junctions in Si. The dashed line is for the case of zero-bias voltage.

tance is given by an expression identical to Eq. 23 except that the V_{bi} is replaced by a "gradient voltage" V_g:

$$V_g = \frac{2}{3}\frac{kT}{q}\ln\left[\frac{a^2\epsilon_s kT/q}{8qn_i^3}\right]. \tag{24}$$

The gradient voltages for Ge, Si, and GaAs as a function of impurity gradient are shown in Fig. 16. These voltages are smaller than the V_{bi} calculated from Eq. 22 by more than 100 mV. The depletion-layer width and the corresponding capacitance for silicon are plotted in Fig. 17 as a function of impurity gradient.

2.4 CURRENT–VOLTAGE CHARACTERISTICS

2.4.1 Ideal Case—Shockley Equation[1]

The ideal current–voltage characteristics are based on the following four assumptions: (1) the abrupt depletion-layer approximation; that is, the built-in potential and applied voltages, are supported by a dipole layer with abrupt boundaries, and outside the boundaries the semiconductor is assumed to be neutral; (2) the Boltzmann approximation; that is, throughout the depletion layer, the Boltzmann relations similar to Eqs. 33 and 37 of Chapter 1 are valid; (3) the low injection assumption; that is, the injected minority carrier densities are small compared with the majority-carrier densities; and (4) no generation current exists in the depletion layer, and the electron and hole currents are constant through the depletion layer.

We first consider the Boltzmann relation. At thermal equilibrium this relation is given by

$$n = n_i \exp\left(\frac{E_F - E_i}{kT}\right) \equiv n_i \exp\left[\frac{q(\psi - \phi)}{kT}\right] \tag{25a}$$

$$p = n_i \exp\left(\frac{E_i - E_F}{kT}\right) \equiv n_i \exp\left[\frac{q(\phi - \psi)}{kT}\right] \tag{25b}$$

where ψ and ϕ are the potentials corresponding to the intrinsic level and the Fermi level, respectively (or $\psi \equiv -E_i/q$, $\phi \equiv -E_F/q$). Obviously, at thermal equilibrium, the pn product from Eqs. 25a and 25b is equal to n_i^2. When voltage is applied, the minority-carrier densities on both sides of a junction are changed, and the pn product is no longer given by n_i^2. We shall now define the imrefs as follows:

$$n \equiv n_i \exp\left[\frac{q(\psi - \phi_n)}{kT}\right] \tag{26a}$$

$$p = n_i \exp\left[\frac{q(\phi_p - \psi)}{kT}\right] \tag{26b}$$

where ϕ_n and ϕ_p are the imrefs or quasi-Fermi levels for electrons and holes, respectively. From Eqs. 26a and 26b we obtain

$$\phi_n \equiv \psi - \frac{kT}{q} \ln\left(\frac{n}{n_i}\right) \tag{27a}$$

$$\phi_p \equiv \psi + \frac{kT}{q} \ln\left(\frac{p}{n_i}\right). \tag{27b}$$

The pn product becomes

$$pn = n_i^2 \exp\left[\frac{q(\phi_p - \phi_n)}{kT}\right]. \tag{28}$$

For a forward bias, $(\phi_p - \phi_n) > 0$ and $pn > n_i^2$; on the other hand, for a reversed bias, $(\phi_p - \phi_n) < 0$ and $pn < n_i^2$.

From Eq. 93 of Chapter 1, Eq. 26a, and from the fact that $\mathscr{E} \equiv -\nabla\psi$, we obtain

$$\begin{aligned}\mathbf{J}_n &= q\mu_n\left(n\mathscr{E} + \frac{kT}{q}\nabla n\right) = q\mu_n n(-\nabla\psi) + q\mu_n\frac{kT}{q}\left[\frac{qn}{kT}(\nabla\psi - \nabla\phi_n)\right]\\ &= -q\mu_n n\nabla\phi_n.\end{aligned} \tag{29}$$

Similarly, we obtain

$$\mathbf{J}_p = -q\mu_p p\nabla\phi_p. \tag{30}$$

Thus the electron and hole current densities are proportional to the gradients of the electron and hole imref, respectively. If $\phi_n = \phi_p = \phi =$ constant (at thermal equilibrium), then $\mathbf{J}_n = \mathbf{J}_p = 0$.

The idealized potential distributions and the carrier concentrations in a p-n junction under forward-bias and reverse-bias conditions are shown in Fig. 18. The variations of ϕ_n and ϕ_p with distance are related to the carrier concentrations as given in Eq. 27. Since the electron density n varies in the junction from the n side to the p side by many orders of magnitude, while the electron current J_n is almost constant, it follows that ϕ_n must also be almost constant over the depletion layer. The electrostatic potential difference across the junction is given by

$$V = \phi_p - \phi_n. \tag{31}$$

Equations 28 and 31 can be combined to give the electron density at the boundary of the depletion-layer region on the p side ($x = -x_p$):

$$n_p = \frac{n_i^2}{p_p}\exp\left(\frac{qV}{kT}\right) = n_{po}\exp\left(\frac{qV}{kT}\right) \tag{32}$$

where n_{po} is the equilibrium electron density on the p side. Similarly,

$$p_n = p_{no}\exp\left(\frac{qV}{kT}\right) \tag{33}$$

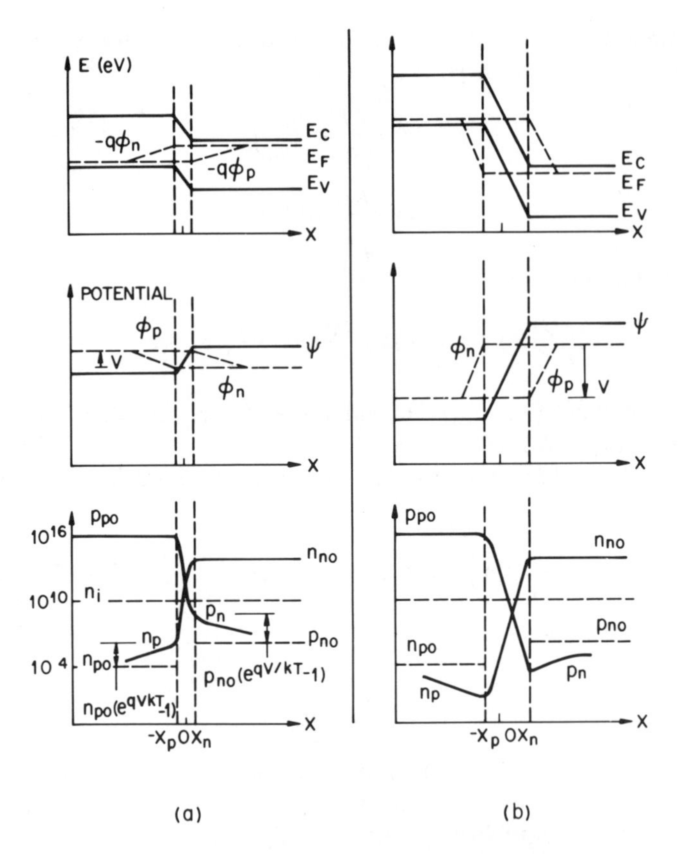

Fig. 18 Energy-band diagram; intrinsic Fermi level (ψ); quasi-Fermi level, also referred to as imref (ϕ_n for electrons, ϕ_p for holes); and carrier distributions under (*a*) forward-biased conditions and (*b*) reverse-biased conditions. (After Shockley, Ref. 1.)

at $x = x_n$ for the n-type boundary. The preceding equations are the most important boundary conditions for the ideal current–voltage equation.

From the continuity equations we obtain for the steady state:

$$-U + \mu_n \mathscr{E} \frac{\partial n_n}{\partial x} + \mu_n n_n \frac{\partial \mathscr{E}}{\partial x} + D_n \frac{\partial^2 n_n}{\partial x^2} = 0 \tag{34a}$$

$$-U - \mu_p \mathscr{E} \frac{\partial p_n}{\partial x} - \mu_p p_n \frac{\partial \mathscr{E}}{\partial x} + D_p \frac{\partial^2 p_n}{\partial x^2} = 0. \tag{34b}$$

In these equations, U is the net recombination rate. The charge neutrality holds approximately, so that $(n_n - n_{no}) \approx (p_n - p_{no})$. Multiplying Eq. 34a by $\mu_p p_n$ and Eq. 34b by $\mu_n n_n$, and combining with the Einstein relation, $D = (kT/q)\mu$, we have

$$-\frac{p_n - p_{no}}{\tau_a} + D_a \frac{\partial^2 p_n}{\partial x^2} - \frac{n_n - p_n}{n_n/\mu_p + p_n/\mu_n} \frac{\mathscr{E} \partial p_n}{\partial x} = 0 \tag{35}$$

where

$$D_a = \frac{n_n + p_n}{n_n/D_p + p_n/D_n} = \text{ambipolar diffusion coefficient} \tag{36}$$

$$\tau_a = \frac{p_n - p_{no}}{U} = \frac{n_n - n_{no}}{U} = \text{ambipolar lifetime.} \tag{37}$$

From the low-injection assumption (e.g., $p_n \ll n_n \approx n_{no}$ in the n-type semiconductor), Eq. 35 reduces to

$$-\frac{p_n - p_{no}}{\tau_p} - \mu_p \mathscr{E} \frac{\partial p_n}{\partial x} + D_p \frac{\partial^2 p_n}{\partial x^2} = 0 \tag{38}$$

which is Eq. 34b except that the term $\mu_p p_n \partial \mathscr{E}/\partial x$ is missing; under the low-injection assumption, this term is of the same order as the neglected terms.

In the neutral region where there is no electric field, Eq. 38 reduces further to

$$\frac{\partial^2 p_n}{\partial x^2} - \frac{p_n - p_{no}}{D_p \tau_p} = 0. \tag{39}$$

The solution of Eq. 39 with the boundary condition Eq. 33 and $p_n(x = \infty) = p_{no}$ gives

$$p_n - p_{no} = p_{no}(e^{qV/kT} - 1)e^{-(x-x_n)/L_p} \tag{40}$$

where

$$L_p \equiv \sqrt{D_p \tau_p}. \tag{41}$$

And at $x = x_n$,

$$J_p = -qD_p \left. \frac{\partial p_n}{\partial x} \right|_{x_n} = \frac{qD_p p_{no}}{L_p} (e^{qV/kT} - 1). \tag{42}$$

Similarly, we obtain for the p side

$$J_n = qD_n \left. \frac{\partial n_p}{\partial x} \right|_{-x_p} = \frac{qD_n n_{po}}{L_n} (e^{qV/kT} - 1). \tag{43}$$

The minority-carrier densities and the current densities for the forward-bias and reverse-bias condition are shown in Fig. 19.

The total current is given by the sum of Eqs. 42 and 43:

$$J = J_p + J_n = J_s(e^{qV/kT} - 1), \tag{44}$$

$$J_s \equiv \frac{qD_p p_{no}}{L_p} + \frac{qD_n n_{po}}{L_n}. \tag{45}$$

Equation 45 is the celebrated Shockley equation,[1] which is the ideal diode law. The ideal current–voltage relation is shown in Fig. 20a and b in the linear and semilog plots, respectively. In the forward direction (positive

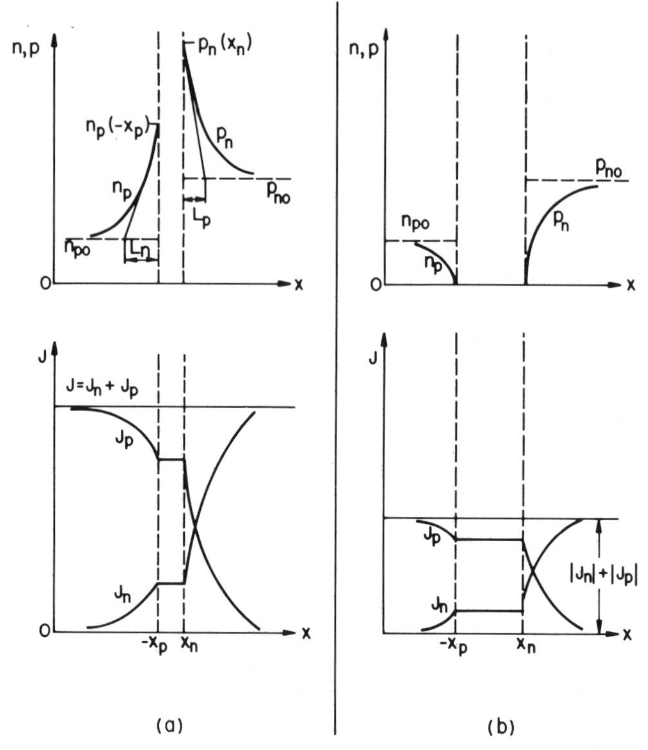

Fig. 19 Carrier distributions and current densities (both linear plots) for (*a*) forward-biased conditions and (*b*) reverse-biased conditions. (After Shockley, Ref. 1.)

bias on p) for $V > 3kT/q$, the rate of rise is constant, Fig. 20*b*; at 300 K for every decade change of current, the voltage changes by 59.5 mV ($= 2.3kT/q$). In the reverse direction, the current density saturates at $-J_s$.

We shall now briefly consider the temperature effect on the saturation current density J_s. We shall consider only the first term in Eq. 45, since the second term will behave similarly to the first one. For the one-sided p^+n abrupt junction (with donor concentration N_D), $p_{no} \gg n_{po}$, the second term also can be neglected. The quantities D_p, p_{no}, and L_p ($\equiv \sqrt{D_p \tau_p}$) are all temperature-dependent. If D_p/τ_p is proportional to T^γ, where γ is a constant, then

$$J_s \simeq \frac{qD_p p_{no}}{L_p} \simeq q\sqrt{\frac{D_p}{\tau_p}}\frac{n_i^2}{N_D}$$
$$\sim \left[T^3 \exp\left(-\frac{E_g}{kT}\right)\right] T^{\gamma/2} = T^{(3+\gamma/2)} \exp\left(-\frac{E_g}{kT}\right). \tag{46}$$

The temperature dependence of the term $T^{(3+\gamma/2)}$ is not important compared with the exponential term. The slope of a plot J_s versus $1/T$ is determined by the energy gap E_g. It is expected that in the reverse direction, where

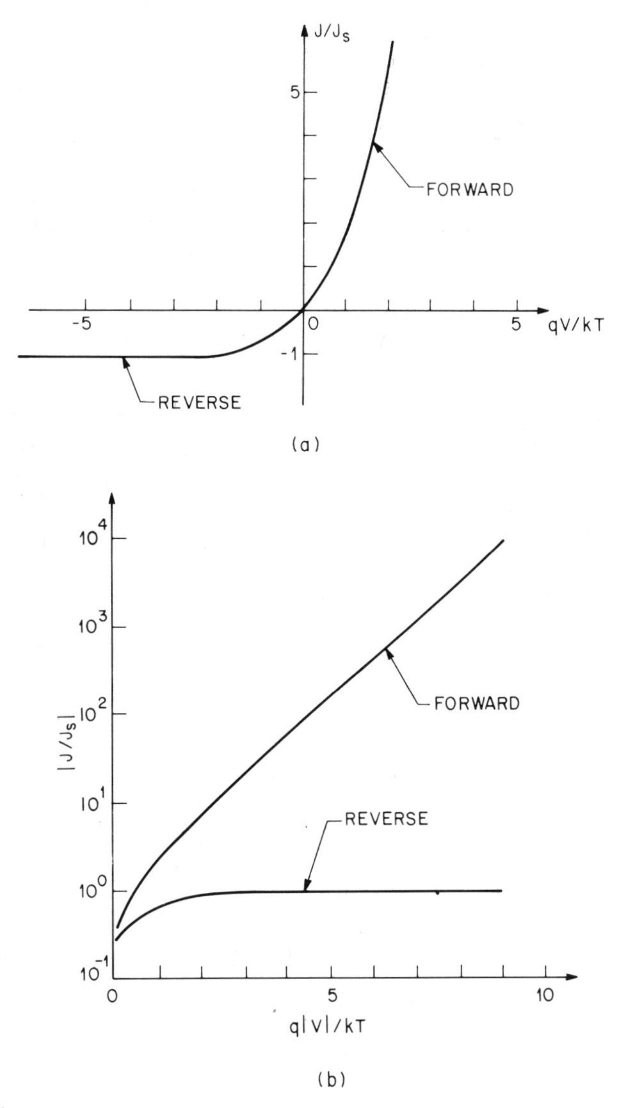

Fig. 20 Ideal current–voltage characteristics. (*a*) Linear plot. (*b*) Semilog plot.

$|J_R| \sim J_s$, the current will increase approximately as $e^{-E_g/kT}$ with temperature; and in the forward direction, where $J_F \sim J_s e^{qV/kT}$, the current will increase approximately as $\exp[-(E_g - qV)/kT]$.

2.4.2 Generation–Recombination Process[2]

The Shockley equation adequately predicts the current–voltage characteristics of germanium *p-n* junctions at low current densities. For Si and

GaAs p-n junctions, however, the ideal equation can give only qualitative agreement. The departures from the ideal are mainly due to: (1) the surface effect, (2) the generation and recombination of carriers in the depletion layer, (3) the tunneling of carriers between states in the bandgap, (4) the high-injection condition that may occur even at relatively small forward bias, and (5) the series resistance effect. In addition, under sufficiently larger field in the reverse direction, the junction will break down (as a result, for example, of avalanche multiplication). The junction breakdown will be discussed in Section 2.5.

The surface effects on p-n junctions are due primarily to ionic charges on or outside the semiconductor surface that induce image charges in the semiconductor and thereby cause the formation of the so-called surface channels or surface depletion-layer regions. Once a channel is formed, it modifies the junction depletion region and gives rise to surface leakage current. The details of the surface effect are discussed in Chapters 7 and 8. For Si planar p-n junctions, the surface leakage current is generally much smaller than the generation current in the depletion region.

Consider first the generation current under the reverse-bias condition. Because of the reduction in carrier concentration under reverse bias ($pn \ll n_i^2$), the dominant recombination–generation processes as discussed in Chapter 1 are those of emission. The rate of generation of electron–hole pairs can be obtained from Eq. 58 of Chapter 1 with the condition $p < n_i$ and $n < n_i$:

$$U = -\left[\frac{\sigma_p \sigma_n v_{th} N_t}{\sigma_n \exp\left(\dfrac{E_t - E_i}{kT}\right) + \sigma_p \exp\left(\dfrac{E_i - E_t}{kT}\right)}\right] n_i \equiv -\frac{n_i}{\tau_e} \tag{47}$$

where τ_e is the effective lifetime and is defined as the reciprocal of the expression in brackets. The current due to the generation in the depletion region is thus given by

$$J_{\text{gen}} = \int_0^W q|U|dx \simeq q|U|W = \frac{qn_i W}{\tau_e} \tag{48}$$

where W is the depletion-layer width. If the effective lifetime is a slowly varying function of temperature, the generation current will then have the same temperature dependence as n_i. At a given temperature, J_{gen} is proportional to the depletion-layer width, which, in turn, is dependent on the applied reverse bias. It is thus expected that

$$J_{\text{gen}} \sim (V_{bi} + V)^{1/2} \tag{49a}$$

for abrupt junctions, and

$$J_{\text{gen}} \sim (V_{bi} + V)^{1/3} \tag{49b}$$

for linearly graded junctions.

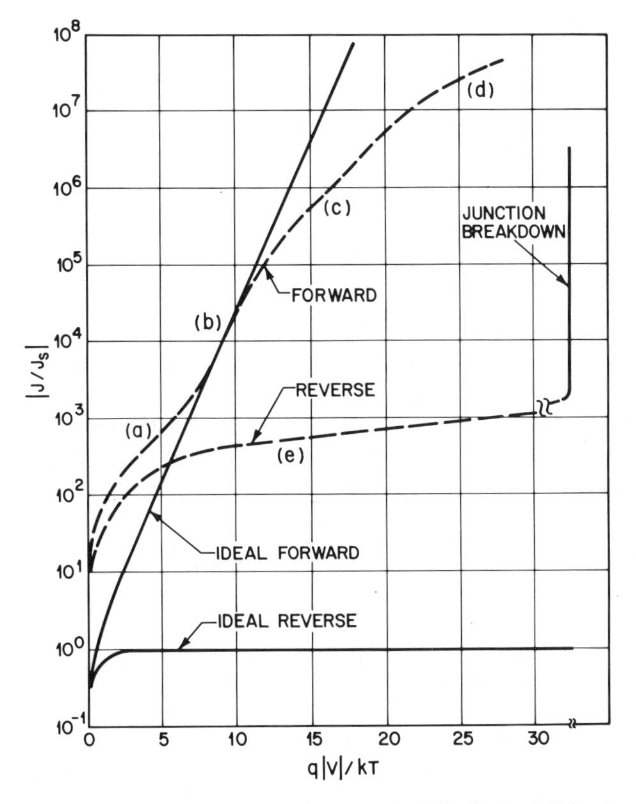

Fig. 21 Current–voltage characteristics of a practical Si diode. (*a*) Generation–recombination current region. (*b*) Diffusion current region. (*c*) High-injection region. (*d*) Series resistance effect. (*e*) Reverse leakage current due to generation–recombination and surface effects. (After Moll, Ref. 3.)

The total reverse current (for $p_{no} \gg n_{po}$ and $|V| > 3kT/q$) can be approximately given by the sum of the diffusion components in the neutral region and the generation current in the depletion region:

$$J_R = q\sqrt{\frac{D_p}{\tau_p}}\frac{n_i^2}{N_D} + \frac{qn_iW}{\tau_e}. \tag{50}$$

For semiconductors with large values of n_i (such as Ge), the diffusion component will dominate at room temperature and the reverse current will follow the Shockley equation; but if n_i is small (such as for Si), the generation current may dominate. A typical result[3] for Si is shown in Fig. 21, curve (e). At sufficiently high temperatures, however, the diffusion current will dominate.

At forward bias, where the major recombination–generation processes in the depletion region are the capture processes, we have a recombination current J_{rec} in addition to the diffusion current. Substituting Eq. 28 in Eq. 58

of Chapter 1 yields

$$U = \frac{\sigma_p \sigma_n v_{th} N_t n_i^2 (e^{qV/kT} - 1)}{\sigma_n \left[n + n_i \exp\left(\frac{E_t - E_i}{kT} \right) \right] + \sigma_p \left[p + n_i \exp\left(\frac{E_i - E_t}{kT} \right) \right]}. \tag{51}$$

Under the assumptions that $E_i = E_t$ and $\sigma_n = \sigma_p = \sigma$, Eq. 51 reduces to

$$U = \frac{\sigma v_{th} N_t n_i^2 (e^{qV/kT} - 1)}{n + p + 2n_i} \tag{52}$$

$$= \frac{\sigma v_{th} N_t n_i^2 (e^{qV/kT} - 1)}{n_i \left\{ \exp\left[\frac{q(\psi - \phi_n)}{kT} \right] + \exp\left[\frac{q(\phi_p - \psi)}{kT} \right] + 2 \right\}}. \tag{52a}$$

The maximum value of U exists in the depletion region where ψ is halfway between ϕ_p and ϕ_n, or $\psi = (\phi_n + \phi_p)/2$, and so the denominator of Eq. 52a becomes $2n_i[\exp(qV/2kT) + 1]$. We obtain for $V > kT/q$,

$$U \simeq \frac{1}{2} \sigma v_{th} N_t n_i \exp\left(\frac{qV}{2kT} \right) \tag{53}$$

and

$$J_{\text{rec}} = \int_0^W qU\, dx \approx \frac{qW}{2} \sigma v_{th} N_t n_i \exp\left(\frac{qV}{2kT} \right) \sim n_i N_t. \tag{54}$$

Similar to the generation current in reverse bias, the recombination current in forward bias is also proportional to n_i. The total forward current can be approximated by the sum of Eqs. 44 and 54 for $p_{no} \gg n_{po}$ and $V > kT/q$:

$$J_F = q \sqrt{\frac{D_p}{\tau_p}} \frac{n_i^2}{N_D} \exp\left(\frac{qV}{kT} \right) + \frac{qW}{2} \sigma v_{th} N_t n_i \exp\left(\frac{qV}{2kT} \right). \tag{55}$$

The experimental results in general can be represented by the empirical form,

$$J_F \sim \exp\left(\frac{qV}{nkT} \right) \tag{56}$$

where the factor n equals 2 when the recombination current dominates [Fig. 21, curve (a)] and n equals 1 when the diffusion current dominates [Fig. 21, curve (b)]. When both currents are comparable, n has a value between 1 and 2.

2.4.3 High-Injection Condition

At high current densities (under the forward-bias condition) such that the injected minority-carrier density is comparable with the majority concentration, both drift and diffusion current components must be considered. The individual conduction current densities can always be given by Eqs. 29

and 30 and are repeated here:

$$J_p = -q\mu_p p \nabla \phi_p$$

$$J_n = -q\mu_n n \nabla \phi_n.$$

Since J_p, q, μ_p, and p are positive, the hole imref decreases monotonically to the right as shown in Fig. 18*a*. Similarly, the electron imref increases monotonically to the left. Thus everywhere the separation of the imrefs must be less than or equal to the applied voltage, and therefore[30]

$$pn \le n_i^2 \exp\left(\frac{qV}{kT}\right) \tag{57}$$

even under the high-injection condition. Note also that the foregoing argument does not depend on recombination in the depletion region. As long as recombination takes place somewhere, currents will flow.

To illustrate the high-injection case, we present in Fig. 22 plots of numerical results for intrinsic Fermi level (ψ), imrefs (ϕ_n and ϕ_p), and carrier concentrations for a silicon *p-n* step junction with the following parameters: $N_A = 10^{18}\,\text{cm}^{-3}$, $N_D = 10^{16}\,\text{cm}^{-3}$, $\tau_n = 3\times10^{-10}$ s, and $\tau_p = 8.4\times10^{-10}$ s. The current densities in Fig. 22*a*, *b*, and *c* are 10, 10^3, and 10^4 A/cm^2. At 10 A/cm^2 the diode is in the low-injection regime. Almost all of the potential drop occurs across the junction. The hole concentration on the *n* side is small compared to the electron concentration. At 10^3 A/cm^2 the electron concentration near the junction exceeds the donor concentration appreciably. An ohmic potential drop appears on the *n* side. At

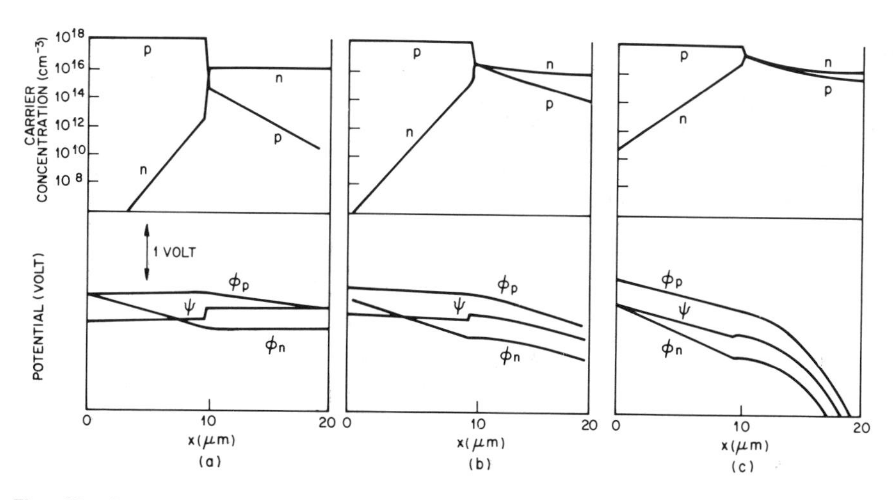

Fig. 22 Carrier concentrations, intrinsic Fermi level (ψ) and imrefs for a Si *p-n* junction operated at different current densities. (*a*) 10 A/cm^2. (*b*) 10^3 A/cm^2. (*c*) 10^4 A/cm^2. (After Gummel, Ref. 30.)

10^4 A/cm^2 we have very high injection; the potential drop across the junction is insignificant compared to ohmic drops on both sides. Even though only the center region of diode is shown in Fig. 22, it is apparent that the separation of the imrefs at the junction is less than or equal to the difference in the hole imref to the left of the junction and the electron imref to the right of the junction for all forward-bias levels.

From Fig. 22*b* and *c*, the carrier densities at the *n* side of the junction are comparable ($n \approx p$). Substituting this condition in Eq. 57, we obtain $p_n(x = x_n) \approx n_i \exp(qV/2kT)$. The current then becomes roughly proportional to $\exp(qV/2kT)$, as shown in Fig. 21, curve (c).

At high-injection levels we should consider another effect associated with the finite resistivity in the quasi-neutral regions of the junction. This resistance absorbs an appreciable amount of the voltage drop between the diode terminals. This effect is shown in Fig. 21, curve (d). The series resistance effect can be substantially reduced by the use of epitaxial materials.

2.4.4 Diffusion Capacitance

The depletion-layer capacitance considered previously accounts for most of the junction capacitance when the junction is reverse-biased. When forward-biased, there is in addition a significant contribution to junction capacitance from the rearrangement of minority carrier density, the so-called diffusion capacitance. When a small ac signal is applied to a junction that is forward-biased to a voltage V_0 and current density J_0, the total voltage and current are defined by

$$V(t) = V_0 + V_1 e^{j\omega t}$$

$$J(t) = J_0 + J_1 e^{j\omega t} \tag{58}$$

where V_1 and J_1 are the small-signal amplitude of the voltage and current density, respectively. The electron and hole densities at the depletion region boundaries can be obtained from Eqs. 32 and 33 by using $(V_0 + V_1 e^{j\omega t})$ instead of V. The small-signal ac component of the hole density is given by

$$\tilde{p}_n(x, t) = p_{n1}(x) e^{j\omega t} \tag{59}$$

We obtain for $V_1 \ll V_0$,

$$p_n = p_{no} \exp\left[\frac{q(V_0 + V_1 e^{j\omega t})}{kT}\right]$$

$$\simeq p_{no} \exp\left(\frac{qV_0}{kT}\right) + \frac{p_{no} q V_1}{kT} \exp\left(\frac{qV_0}{kT}\right) e^{j\omega t}. \tag{60}$$

A similar expression is obtained for the electron density. The first term in Eq. 60 is the dc component, and the second term is the small-signal ac component at the depletion-layer boundary $[p_{n1}(x_n)e^{j\omega t}]$. Substituting p_n into the continuity equation (Eq. 97 of Chapter 1 with $G_p = 0$) yields

$$j\omega\tilde{p}_n = -\frac{\tilde{p}_n}{\tau_p} + D_p \frac{\partial^2 \tilde{p}_n}{\partial x^2}$$

or

$$\frac{\partial^2 \tilde{p}_n}{\partial x^2} - \frac{\tilde{p}_n}{D_p\tau_p/(1+j\omega\tau_p)} = 0. \tag{61}$$

Equation 61 is identical to Eq. 39 if the carrier lifetime is expressed as

$$\tau_p^* = \frac{\tau_p}{1+j\omega\tau_p}. \tag{62}$$

We can then obtain the alternating current density from Eq. 44 by making the appropriate substitutions:

$$J_1 = \frac{qV_1}{kT}\left[\frac{qD_p p_{no}}{L_p/\sqrt{1+j\omega\tau_p}} + \frac{qD_n n_{po}}{L_n/\sqrt{1+j\omega\tau_n}}\right]\exp\left(\frac{qV_0}{kT}\right). \tag{63}$$

Equation 63 leads directly to the ac admittance:

$$Y \equiv \frac{J_1}{V_1} = G_d + j\omega C_d. \tag{64}$$

For relatively low frequencies ($\omega\tau_p, \omega\tau_n \ll 1$), the diffusion conductance G_{d0} is given by

$$G_{d0} = \frac{q}{kT}\left(\frac{qD_p p_{no}}{L_p} + \frac{qD_n n_{po}}{L_n}\right)e^{qV_0/kT} \qquad \text{mho/cm}^2 \tag{65}$$

which has exactly the same value obtained by differentiating Eq. 44. The

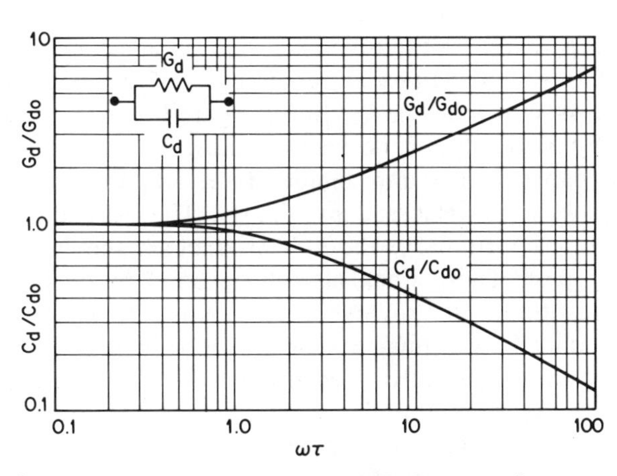

Fig. 23 Normalized diffusion conductance and diffusion capacitance versus $\omega\tau$. Insert shows the equivalent circuit of a p-n junction under forward bias.

low-frequency diffusion capacitance C_{d0} is given by

$$C_{d0} = \frac{q}{kT}\left(\frac{qL_p p_{no}}{2} + \frac{qL_n n_{po}}{2}\right)e^{qV_0/kT} \qquad \text{F/cm}^2. \tag{66}$$

The frequency dependence of the conductance and capacitance is shown in Fig. 23 as a function of the normalized frequency $\omega\tau$ where only one term in Eq. 63 is considered (e.g., the term contains p_{no} if $p_{no} \gg n_{po}$). The insert shows the equivalent circuit of the ac admittance. It is clear from Fig. 23 that the diffusion capacitance decreases with increasing frequency. For large frequencies, C_d is approximately $(\omega)^{-1/2}$. The diffusion capacitance, however, increases with the direct current level ($\sim e^{qV_0/kT}$). For this reason, C_d is especially important at low frequencies and under forward-bias conditions.

2.5 JUNCTION BREAKDOWN[31]

When a sufficiently high field is applied to a p-n junction, the junction "breaks down" and conducts a very large current. There are basically three breakdown mechanisms: thermal instability, tunneling effect, and avalanche multiplication. We consider the first two mechanisms briefly, and discuss avalanche multiplication in detail.

2.5.1 Thermal Instability

Breakdown due to thermal instability is responsible for the maximum dielectric strength in most insulators at room temperature, and is also a major effect in semiconductors with relatively small bandgaps (e.g., Ge). Because of the heat dissipation caused by the reverse current at high reverse voltage, the junction temperature increases. This temperature increase, in turn, increases the reverse current in comparison with its value at lower voltages. The temperature effect[32] on reverse current–voltage characteristics is shown in Fig. 24. In this figure the reverse currents J_s are represented by a family of horizontal lines. Each line represents the current at a constant junction temperature, and the current varies as $T^{3+\gamma/2}\exp(-E_g/kT)$, as discussed previously. The heat dissipation hyperbolas which are proportional to the I–V product are shown as straight lines in the log-log plot. These lines also correspond to curves of constant junction temperature. The reverse current–voltage characteristic of the junction is obtained by joining the intersection points of the curves of constant junction temperature. Because of the heat dissipation at high reverse voltage, the characteristic shows a negative differential resistance. In this case the diode will be destroyed unless some special measure such as a large series-limiting resistor is used. This effect is called thermal instability. The voltage V_U is called the turnover voltage. For p-n junctions

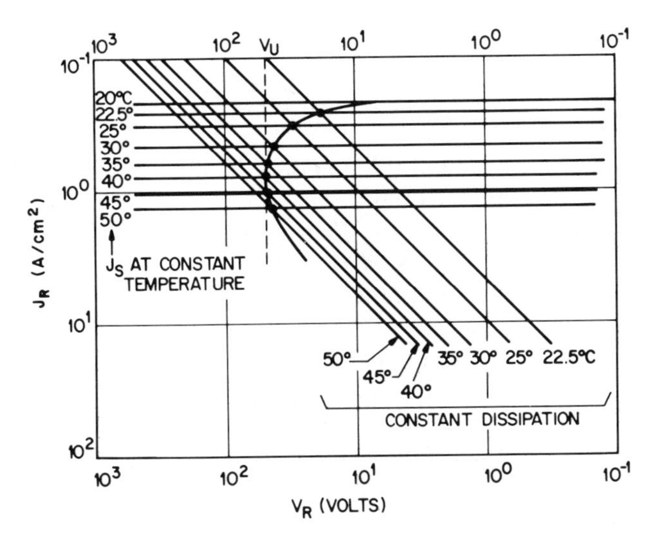

Fig. 24 Reverse current–voltage characteristics of thermal breakdown, where V_U is the turnover voltage. (*Note*: Direction of coordinate increases are opposite to usual conventions.) (After Strutt, Ref. 32.)

with relatively large saturation currents (e.g., Ge), the thermal instability is important at room temperature. At very low temperatures, however, thermal instability becomes less important compared with other mechanisms.

2.5.2 Tunneling Effect

We next consider the tunneling effect. It is well known that for a one-dimensional square energy barrier with barrier height E_0 and thickness W, the quantum-mechanical transmission probability T_t is given by[33]

$$T_t = \left[1 + \frac{E_0^2 \sinh^2 \kappa W}{4E(E_0 - E)}\right]^{-1} \tag{67}$$

with

$$\kappa \equiv \sqrt{\frac{2m(E_0 - E)}{\hbar^2}}$$

where E is the energy of the carrier. The probability decreases monotonically with decreasing E. When $\kappa W \gg 1$, the probability becomes

$$T_t \approx \frac{16E(E_0 - E)}{E_0^2} \exp(-2\kappa W). \tag{67a}$$

A similar expression has been obtained for p-n junctions. The detailed mathematical treatment is given in Chapter 9. The tunneling current density is given by[31]

$$J_t = \frac{\sqrt{2m^*}q^3\mathscr{E}V}{4\pi^2\hbar^2E_g^{1/2}}\exp\left(-\frac{4\sqrt{2m^*}E_g^{3/2}}{3q\mathscr{E}\hbar}\right) \tag{68}$$

where $\mathscr{E}$ is the electric field at the junction, E_g the bandgap, V the applied voltage, and m^* the effective mass.

When the field approaches 10^6 V/cm in Ge and Si, significant current begins to flow by means of the band-to-band tunneling process. To obtain such a high field, the junction must have relatively high impurity concentrations on both the p and n sides. The mechanism of breakdown for Si and Ge junctions with breakdown voltages less than about $4E_g/q$ is found to be due to the tunneling effect. For junctions with breakdown voltages in excess of $6E_g/q$, the mechanism is caused by the avalanche multiplication. At voltages between 4 and 6 E_g/q, the breakdown is due to a mixture of both avalanche and tunneling. Since the energy bandgaps E_g in Ge, Si, and GaAs decrease with increasing temperature (refer to Chapter 1), the breakdown voltage in these semiconductors due to the tunneling effect has a negative temperature coefficient; that is, the voltage decreases with increasing temperature. This is because a given breakdown current J_t can be reached at smaller reverse voltages (or fields) at higher temperatures, Eq. 68. A typical example is shown in Fig. 25. This temperature effect is generally used to distinguish the tunneling mechanism from the avalanche mechanism, which has a positive temperature coefficient; that is, the breakdown voltage increases with increasing temperature.

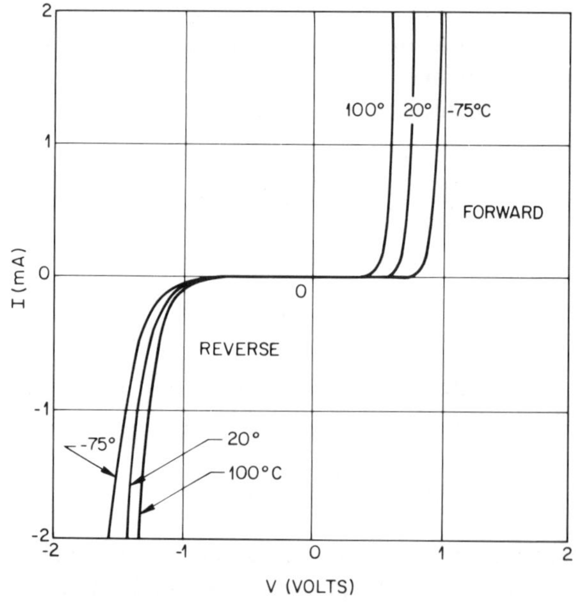

Fig. 25 Current–voltage characteristics of tunneling breakdown. (After Strutt, Ref. 32.)

2.5.3 Avalanche Multiplication

Avalanche multiplication (or impact ionization) is the most important mechanism in junction breakdown, since the avalanche breakdown voltage imposes an upper limit on the reverse bias for most diodes, on the collector voltage of bipolar transistors (Chapter 3) and on the drain voltages of MESFETs (Chapter 6) and MOSFETs (Chapter 8). In addition, the impact ionization mechanism can be used to generate microwave power, as in IMPATT devices (Chapter 10), and to detect optical signals, as in avalanche photodetectors (Chapter 13).

We first derive the basic ionization integral which determines the breakdown condition. Assume that a current I_{po} is incident at the left-hand side of the depletion region with width W. If the electric field in the depletion region is high enough that electron–hole pairs are generated by the impact ionization process, the hole current I_p will increase with distance through the depletion region and reaches a value M_pI_{po} at W. Similarly, the electron current I_n will increase from $x = W$ to $x = 0$. The total current $I(= I_p + I_n)$ is constant at steady state. The incremental hole current at x equals the number of electron–hole pairs generated per second in the distance dx,

$$d(I_p/q) = (I_p/q)(\alpha_p\, dx) + (I_n/q)(\alpha_n\, dx) \tag{69}$$

or

$$dI_p/dx - (\alpha_p - \alpha_n)I_p = \alpha_n I. \tag{70}$$

The electron and hole ionization rates (α_n and α_p) have been considered in Chapter 1.

The solution* of Eq. 70 with boundary condition that $I = I_p(W) = M_pI_{po}$ is given by

$$I_p(x) = I\left\{\frac{1}{M_p} + \int_0^x \alpha_n \exp\left[-\int_0^x (\alpha_p - \alpha_n)\, dx'\right] dx\right\}\Big/\exp\left[-\int_0^x (\alpha_p - \alpha_n)\, dx'\right] \tag{71}$$

where M_p is the multiplication factor of holes and is defined as

$$M_p \equiv \frac{I_p(W)}{I_p(0)}. \tag{72}$$

Equation 71 can be written as

$$1 - \frac{1}{M_p} = \int_0^W \alpha_p \exp\left[-\int_0^x (\alpha_p - \alpha_n)\, dx'\right] dx. \tag{73}$$

*Equation 70 has the form $y' + Py = Q$, where $y \equiv I_p$. The standard solution is

$$y = \left[\int_0^x Qe^{\int_0^x P\,dx'}\, dx + C\right]\Big/ e^{\int_0^x P\,dx'}$$

where C is the constant of integration.

The avalanche breakdown voltage is defined as the voltage where M_p approaches infinity. Hence the breakdown condition is given by the ionization integral

$$\int_0^W \alpha_p \exp\left[-\int_0^x (\alpha_p - \alpha_n)\, dx'\right] dx = 1. \tag{74}$$

If the avalanche process is initiated by electrons instead of holes, the ionization integral is given by

$$\int_0^W \alpha_n \exp\left[-\int_x^W (\alpha_n - \alpha_p)\, dx'\right] dx = 1. \tag{75}$$

Equations 74 and 75 are equivalent;[34] that is, the breakdown condition depends only on what is happening within the depletion region and not on the carriers (or primary current) that initiate the avalanche process. The situation does not change when a mixed primary current initiates the breakdown, so either Eq. 74 or Eq. 75 gives the breakdown condition.

For semiconductors with equal ionization rates ($\alpha_n = \alpha_p = \alpha$) such as GaP, Eqs. 74 or 75 reduce to the simple expression

$$\int_0^W \alpha\, dx = 1. \tag{76}$$

From the breakdown conditions described above and the field dependence of the ionization rates, the breakdown voltages, maximum electric field, and depletion-layer width can be calculated. As discussed previously, the electric field and potential in the depletion layer are determined from the solutions of Poisson's equation. Depletion-layer boundaries that satisfy Eq. 74 can be obtained numerically using an iteration method. With known boundaries we obtain

$$V_B(\text{breakdown voltage}) = \frac{\mathscr{E}_m W}{2} = \frac{\epsilon_s \mathscr{E}_m^2}{2q} (N_B)^{-1} \tag{77a}$$

for one-sided abrupt junctions, and

$$V_B = \frac{2\mathscr{E}_m W}{3} = \frac{4\mathscr{E}_m^{3/2}}{3} \left(\frac{2\epsilon_s}{q}\right)^{1/2} (a)^{-1/2} \tag{77b}$$

for linearly graded junctions, where N_B is the ionized background impurity concentration of the lightly doped side, ϵ_s the semiconductor permittivity, a the impurity gradient, and $\mathscr{E}_m$ the maximum field.

Figure 26 shows the calculated breakdown voltage[35] as a function of N_B for abrupt junctions in Ge, Si, ⟨100⟩-oriented GaAs, and GaP. The experimental results are generally in good agreement with the calculated values.[36] The dashed lines in the figure indicate the upper limit of N_B for which the avalanche breakdown calculation is valid. This limitation is based on the criterion $6E_g/q$. Above these values the tunneling mechanism will also contribute to the breakdown process and eventually dominates.

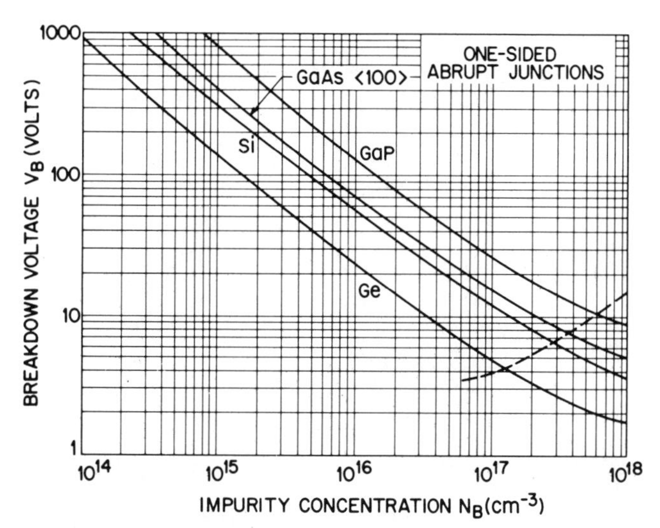

Fig. 26 Avalanche breakdown voltage versus impurity concentration for one-sided abrupt junctions in Ge, Si, ⟨100⟩-oriented GaAs, and GaP. The dashed line indicates the maximum doping beyond which the tunneling mechanism will dominate the voltage breakdown characteristics. (After Sze and Gibbons, Ref. 35.)

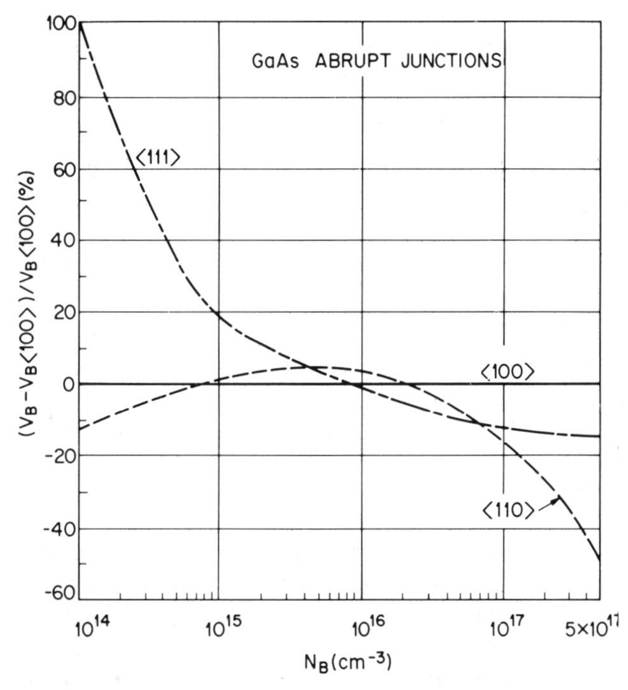

Fig. 27 Orientation dependence of avalanche breakdown voltage in one-sided abrupt GaAs junctions. (After Lee and Sze, Ref. 37.)

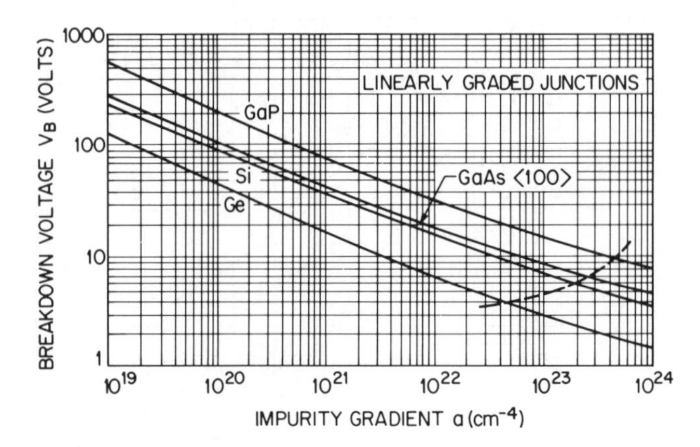

Fig. 28 Avalanche breakdown voltage versus impurity gradient for linearly graded junctions in Ge, Si, ⟨100⟩-oriented GaAs, and GaP. The dashed line indicates the maximum gradient beyond which the tunneling mechanism will set in. (After Sze and Gibbons, Ref. 35.)

In GaAs, the ionization rates depend on crystal orientations (refer to Chapter 1). Figure 27 shows a comparison[37] of V_B in ⟨111⟩ and ⟨110⟩ orientations with respect to that in ⟨100⟩. Note that at around 10^{16} cm^{-3}, the breakdown voltages are essentially independent of orientations. At lower dopings, V_B in ⟨111⟩ becomes the largest; whereas at higher dopings, V_B in ⟨100⟩ is the largest.

Figure 28 shows the calculated breakdown voltage versus the impurity gradient for linearly graded junctions in these semiconductors. The dashed line indicates the upper limit of a for which the avalanche breakdown calculation is valid.

The calculated values of the maximum field $\mathscr{E}_m$ and the depletion-layer width at breakdown for the four semiconductors above are shown[35] in Fig. 29 for the abrupt junctions and in Fig. 30 for the linearly graded junctions. For the Si junctions, the maximum field can be expressed as[38]

$$\mathscr{E}_m = \frac{4 \times 10^5}{1 - \frac{1}{3}\log_{10}(N_B/10^{16})} \qquad \text{V/cm} \tag{78}$$

where N_B is in cm^{-3}.

Because of the strong dependence of the ionization rates on the field, the maximum field varies very slowly with either N_B or a. Thus as a first approximation we can assume that, for a given semiconductor, $\mathscr{E}_m$ has a fixed value. Then from Eq. 77 we obtain $V_B \sim N_B^{-1.0}$ for abrupt junctions and $V_B \sim a^{-0.5}$ for linearly graded junctions. Figures 26 and 28 show that the foregoing patterns are generally followed. Also as expected, for a given N_B or a, the breakdown voltage increases with the energy bandgap, since the avalanche process requires band-to-band excitations.

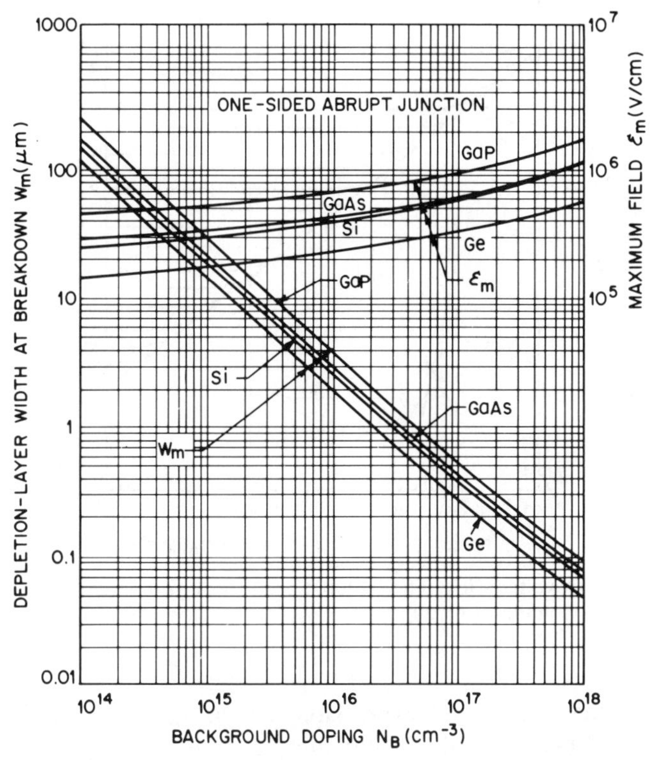

Fig. 29 Depletion-layer width and maximum field at breakdown for one-sided abrupt junctions in Ge, Si, ⟨100⟩-oriented GaAs, and GaP. (After Sze and Gibbons, Ref. 35.)

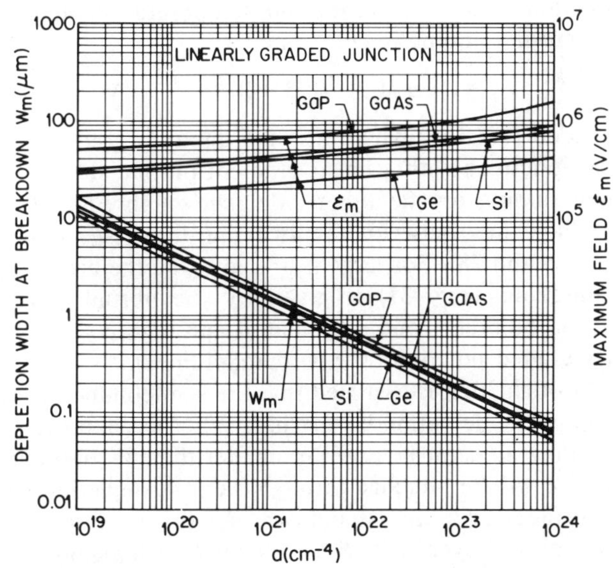

Fig. 30 Depletion-layer width and maximum field at breakdown for linearly graded junctions in Ge, Si, ⟨100⟩-oriented GaAs, and GaP. (After Sze and Gibbons, Ref. 35.)

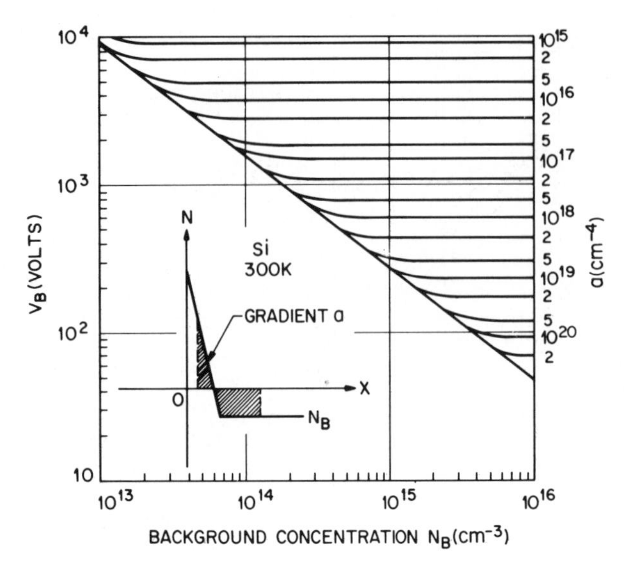

Fig. 31 Breakdown voltage for diffused junctions. The insert shows the space-charge distribution. (After Ghandhi, Ref. 39.)

An approximate universal expression can be given as follows for the results above comprising all semiconductors studied:

$$V_B \cong 60(E_g/1.1)^{3/2}(N_B/10^{16})^{-3/4} \qquad \text{V} \tag{79a}$$

for abrupt junctions where E_g is the room-temperature bandgap in eV, and N_B is the background doping in cm^{-3}; and

$$V_B \cong 60(E_g/1.1)^{6/5}(a/3 \times 10^{20})^{-2/5} \qquad \text{V} \tag{79b}$$

for linearly graded junctions where a is the impurity gradient in cm^{-4}.

For diffused junctions with a linear gradient on one side of the junction and a constant doping on the other side (shown in Fig. 31, insert), the breakdown voltage lies between the two limiting cases considered previously[39] (Figs. 26 and 28). For large a and low N_B, the breakdown voltage of diffused junctions (Fig. 31) is given by the abrupt junction results (bottom line); on the other hand, for small a and high N_B, V_B will be given by the linearly graded junction results (parallel lines).

In Figs. 26 through 30, it is assumed that the semiconductor layer is thick enough to support the depletion-layer width W_m at breakdown (Fig. 29). If, however, the semiconductor layer W is smaller than W_m (shown in Fig. 32, insert), the device will be punched through (i.e., the depletion layer reaches the $n - n^+$ interface) prior to breakdown. As the reverse bias increases further, the device will eventually break down. The maximum electric field $\mathscr{E}_m$ is essentially the same as for the non-punched-through diode. Therefore, the breakdown voltage V_{PT} for the punched-through diode can be

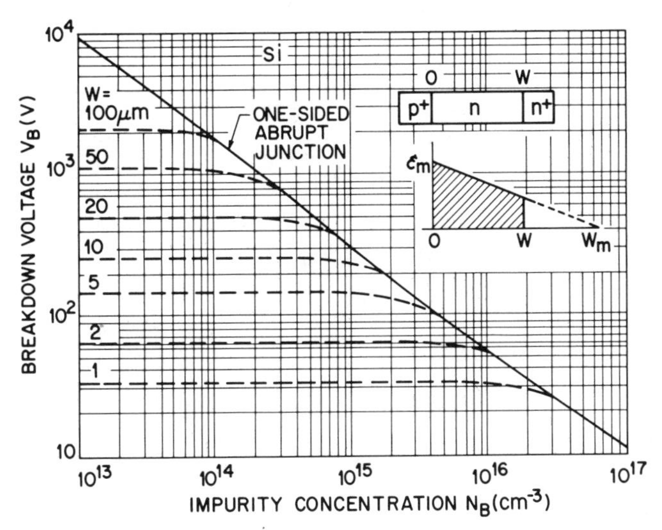

Fig. 32 Breakdown voltage for p^+-π-n^+ and p^+-ν-n^+ junctions, where π is for lightly doped p-type and ν for lightly doped n-type. W is the thickness of the π or ν region.

given by

$$\frac{V_{PT}}{V_B} = \frac{\text{shaded area in Fig. 32 insert}}{(\mathscr{E}_m W_m)/2}$$

$$= \left(\frac{W}{W_m}\right)\left(2 - \frac{W}{W_m}\right). \tag{80}$$

The punch-through usually occurs when the doping concentration N_B becomes sufficiently low as in a p^+-π-n^+ or an p^+-ν-n^+ diode, where π stands for a lightly doped p-type and ν for a lightly doped n-type semiconductor. The breakdown voltages for such diodes as calculated from Eq. 80 are shown in Fig. 32 as a function of the background doping for Si one-sided abrupt junction formed on epitaxial substrates (e.g., ν on n^+ with the epitaxial-layer thickness W as a parameter). For a given thickness, the breakdown voltage approaches a constant value as the doping decreases, corresponding to the punch-through of the epitaxial layer.

The results in Figs. 26 through 32 are for avalanche breakdowns at room temperature. At higher temperatures the breakdown voltage increases. A simple explanation of this increase is that hot carriers passing through the depletion layer under a high field lose part of their energy to optical phonons after traveling each electron–phonon mean free path λ. The value of λ decreases with increasing temperature, (Eq. 84 of Chapter 1). Therefore, the carriers lose more energy to the crystal lattice along a given distance at constant field. Hence the carriers must pass through a greater potential difference (or higher voltage) before they can acquire sufficient energy to generate an electron–hole pair. The detailed calculations have

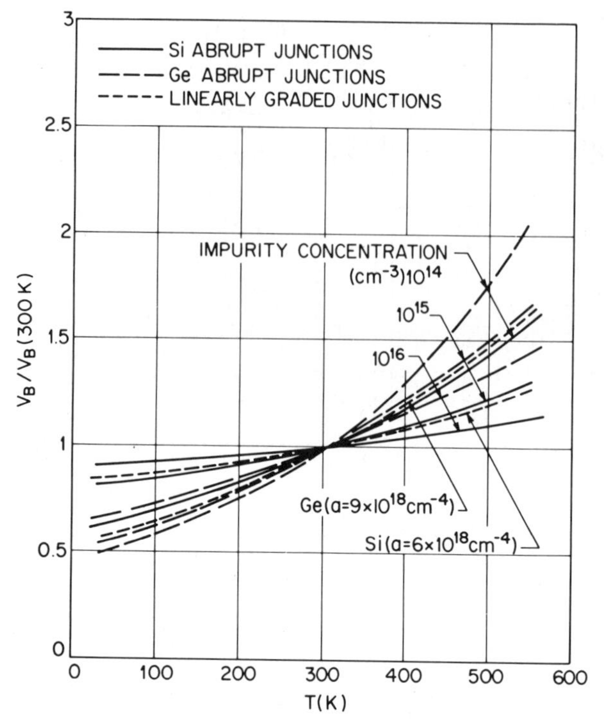

Fig. 33 Normalized avalanche breakdown voltage versus lattice temperature. The breakdown voltage increases with temperature. (After Crowell and Sze, Ref. 40.)

been done by the use of a modification of Baraff's theory[40] as discussed in Chapter 1. The predicted values of V_B normalized to the room-temperature value are shown in Fig. 33 for Ge and Si. For the same doping profile, the predicted percentage change on V_B with temperature is about the same for GaAs as it is for Ge and for GaP as it is for Si junctions. Note that there are substantial increases of the breakdown voltage, especially for lower dopings (or small gradient) at higher temperatures.[41] Figure 34 shows the measured results,[42] which agree quite well with this theory.

For planar junctions, the very important junction curvature effect should be considered. A schematic diagram of a planar junction has been shown in Fig. 9*b*. Since the cylindrical and/or spherical regions of the junction have a higher field intensity, the avalanche breakdown voltage is determined by these regions. The potential $V(r)$ and the electric field $\mathscr{E}(r)$ in a cylindrical or spherical *p-n* junction can be calculated from Poissons's equation:

$$\frac{1}{r^n}\frac{d}{dr}[r^n\mathscr{E}(r)] = \frac{\rho(r)}{\epsilon_s} \tag{81}$$

where n equals 1 for the cylindrical junction, and 2 for the spherical

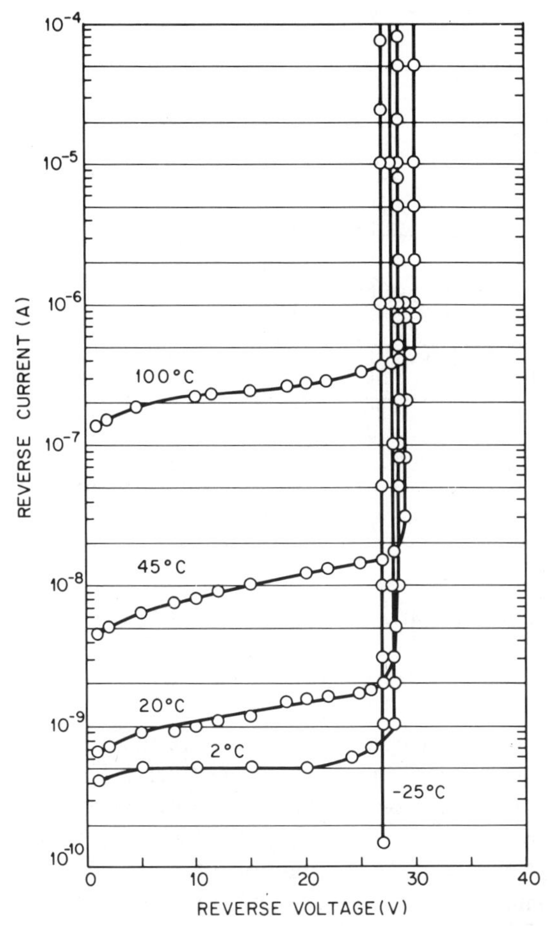

Fig. 34 Temperature dependence of reverse *I–V* characteristics of a microplasma-free n^+-p Si diode with $N_B = 2.5 \times 10^{16}\ \text{cm}^{-3}$ and an *n*-type guard ring. The temperature coefficient is 0.024 V/°C. (After Goetzberger et al., Ref. 42.)

junction. The solution for $\mathscr{E}(r)$ can be obtained from Eq. 81 and is given by

$$\mathscr{E}(r) = \frac{1}{\epsilon_s r^n} \int_{r_j}^{r} r^n \rho(r) dr + \frac{\text{constant}}{r^n} \tag{82}$$

where r_j is the radius of curvature of the metallurgical junction, and the constant must be adjusted so that the breakdown condition Eq. 74 or 75 is satisfied.

The calculated results for Si one-sided abrupt junctions at 300 K can be expressed by simple analytical equations:[39]

$$\frac{V_{CY}}{V_B} = \left[\frac{1}{2}(\eta^2 + 2\eta^{6/7}) \ln(1 + 2\eta^{-8/7}) - \eta^{6/7}\right] \tag{83}$$

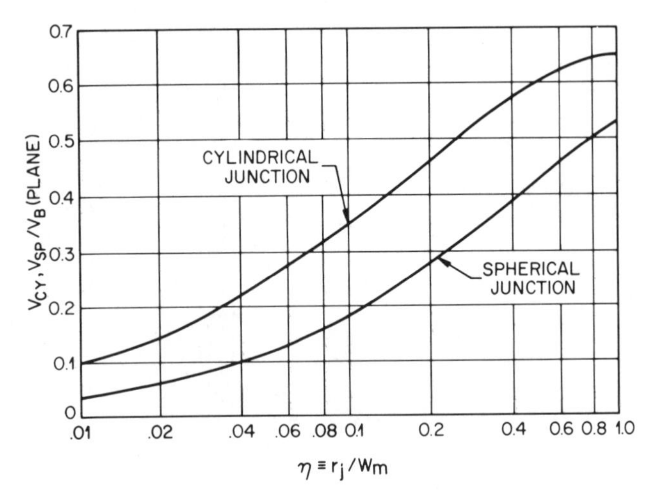

Fig. 35 Normalized breakdown voltage of cylindrical and spherical junction as a function of the normalized radius of curvature. (After Ghandhi, Ref. 39.)

for cylindrical junctions, and

$$\frac{V_{SP}}{V_B} = [\eta^2 + 2.14\eta^{6/7} - (\eta^3 + 3\eta^{13/7})^{2/3}] \tag{84}$$

for spherical junctions, where V_{CY} and V_{SP} are the breakdown voltages of cylindrical and spherical junctions, respectively, V_B is the breakdown voltage of a plane junction having the same background doping, and $\eta \equiv r_j/W_m$. Figure 35 illustrates the breakdown voltages for cylindrical and spherical abrupt junctions as a function of η. Clearly, as the radius of curvature becomes smaller, so does the breakdown voltage. For linearly graded cylindrical or spherical junctions, the calculated results show that the breakdown voltage is relatively independent of its radius of curvature.[25]

2.6 TRANSIENT BEHAVIOR AND NOISE

2.6.1 Transient Behavior

For switching applications the transition from forward bias to reverse bias must be nearly abrupt and the transient time short. In Fig. 36*a* a simple circuit is shown where a forward current I_F is flowing in the *p-n* junction; at time $t = 0$, the switch S is suddenly thrown to the right, and initial reverse current $I_R \simeq V/R$ flows. The transient time is defined as the time in which the current reaches 10% of the initial current I_R, and is equal to the sum of t_1 and t_2 as shown in Fig. 36*b*, where t_1 and t_2 are the time intervals for the constant-current phase and the decay phase, respectively. Consider

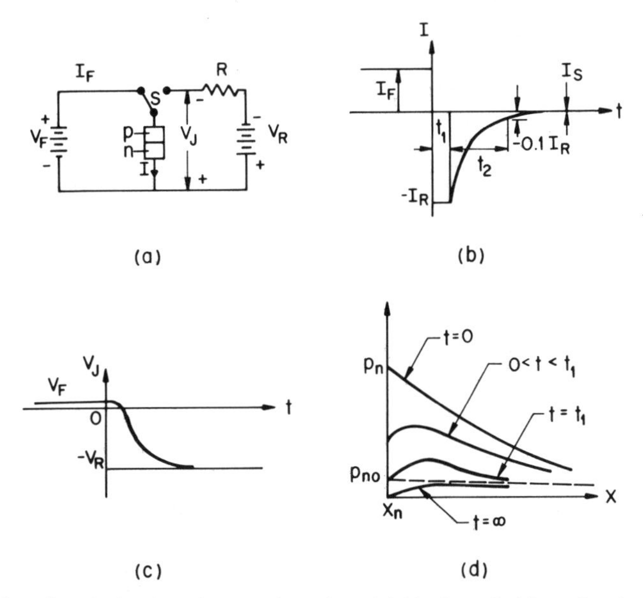

Fig. 36 Transient behavior of a *p-n* junction. (*a*) Basic switching circuit. (*b*) Transient response. (*c*) Junction voltage as a function of time. (*d*) Minority-carrier distribution for various time intervals. (After Kingston, Ref. 43.)

the constant-current phase (also called storage phase) first. The continuity equation as given in Chapter 1 can be written for the *n*-type side (assume that $p_{po} \ll n_{no}$) as

$$\frac{\partial p_n(x, t)}{\partial t} = D_p \frac{\partial^2 p_n(x, t)}{\partial x^2} - \frac{p_n(x, t) - p_{no}}{\tau_p} \tag{85}$$

where τ_p is the minority-carrier lifetime. The boundary conditions are that at $t = 0$ the initial distribution of holes is a steady-state solution to the diffusion equation, and that the voltage across the junction is given from Eq. 33 as

$$V_j = \frac{kT}{q} \ln\left[\frac{p_n(0, t)}{p_{no}}\right]. \tag{86}$$

The distribution of the minority-carrier density p_n with time is shown[43] in Fig. 36*d*. From Eq. 86 it can be calculated that, as long as $p_n(0, t)$ is greater than p_{no} (in the interval $0 < t < t_1$), the junction voltage V_J remains of the order of kT/q, as shown in Fig. 36*c*, and the current I_R is approximately given by V/R = constant. Hence in this time interval the reverse current is constant and we have the constant-current phase. However, at or near t_1, the hole density approaches zero, the junction voltage tends to minus infinity, and a new boundary condition now holds. This phase is the decay phase with the boundary condition $p(0, t) = p_{no}$ = constant. The solutions

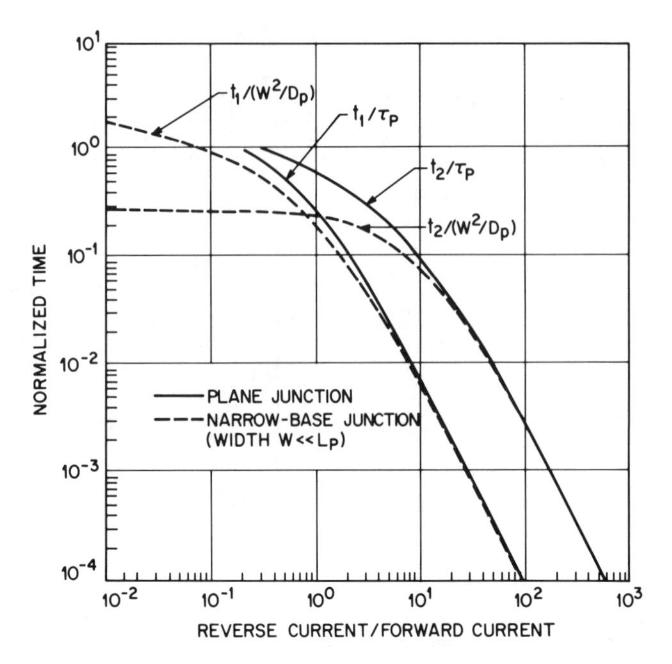

Fig. 37 Normalized time versus the ratio of reverse current to forward current. (After Kingston, Ref. 43.)

have been given by Kingston,[43] and the times t_1 and t_2 are given by the transcendental equations

$$\operatorname{erf}\sqrt{\frac{t_1}{\tau_p}} = \frac{1}{1 + I_R/I_F} \tag{87}$$

$$\operatorname{erf}\sqrt{\frac{t_2}{\tau_p}} + \frac{\exp(-t_2/\tau_p)}{\sqrt{\pi(t_2/\tau_p)}} = 1 + 0.1\left(\frac{I_R}{I_F}\right). \tag{88}$$

The results are shown in Fig. 37 where the solid lines are for the plane junction with the length of the n-type material W much greater than the diffusion length ($W \gg L_p$), and the dashed lines are for the narrow-base junction with $W \ll L_p$. For a large ratio I_R/I_F, the transient time can be approximated by

$$t_1 + t_2 \simeq \frac{\tau_p}{2}\left(\frac{I_R}{I_F}\right)^{-1} \tag{89a}$$

for $W \gg L_p$, or

$$t_1 + t_2 \simeq \frac{W^2}{2D_p}\left(\frac{I_R}{I_F}\right)^{-1} \tag{89b}$$

for $W \ll L_p$. If one switches a junction ($W \gg L_p$) from forward 10 mA to reverse 10 mA ($I_R/I_F = 1$), the time for the constant-current phase is $0.3\tau_p$, and that for the decay phase is about $0.6\tau_p$. Total transient time is then

$0.9\tau_p$. For a fast switch, one thus requires that τ_p be small. The lifetime τ_p can be substantially reduced by introducing impurities with deep levels in the forbidden gap (such as gold in silicon).

2.6.2 Noise

The term "noise" refers to spontaneous fluctuations in the current passing through, or the voltage developed across, semiconductor bulk materials or devices. Since the devices are mainly used to measure small physical quantities or to amplify small signals, spontaneous fluctuations in current or voltage set a lower limit to the quantities to be measured or the signals to be amplified. It is important to know the factors contributing to these limits, to use this knowledge to optimize operating conditions, and to find new methods and new technologies to reduce noise.

Observed noise is generally classified into thermal noise, flicker noise, and shot noise. Thermal noise occurs in any conductor or semiconductor and is caused by the random motion of the current carriers. The open-circuit mean-square voltage $\langle V_n^2 \rangle$ of thermal noise is given by[44,45]

$$\langle V_n^2 \rangle = 4kTBR \tag{90}$$

where k is the Boltzmann constant, T the absolute temperature in K, B the bandwidth in Hz, and R the real part of the impedance between terminals. At room temperature, for a semiconductor material with 1 kΩ resistance, the root-mean-square voltage $\sqrt{\langle V_n^2 \rangle}$ measured with a 1-Hz bandwidth is only about 4 nV (1 nV = 10^{-9} V).

Flicker noise is distinguished by its peculiar spectral distribution which is proportional to $1/f^{\alpha}$ with α generally close to unity (the so-called $1/f$ noise). Flicker noise is important at lower frequencies. For most semiconductor devices, the origin of flicker noise is due to the surface effect. The $1/f$ noise-power spectrum has been correlated both qualitatively and quantitatively with the lossy part of the metal–insulator–semiconductor (MIS) gate impedance due to carrier recombination at the interface traps.

Shot noise constitutes the major noise in most semiconductor devices. It is independent of frequency (white spectrum) at low and intermediate frequencies. At higher frequencies the shot-noise spectrum also becomes frequency-dependent. The mean-square noise current of shot noise for a p-n junction is given by

$$\langle i_n^2 \rangle = 2qBI \tag{91}$$

where I is the current, which is positive in the forward and negative in the reverse direction. For low injection the total mean-square noise current (neglecting $1/f$ noise) is given by

$$\langle i_n^2 \rangle = 4kTBG - 2qBI. \tag{92}$$

From the Shockley equation we obtain

$$G = \frac{\partial I}{\partial V} = \frac{\partial}{\partial V}\left[I_s(e^{qV/kT} - 1)\right] = \frac{qI_s}{kT}\, e^{qV/kT}. \tag{93}$$

Substituting Eq. 93 into Eq. 92 yields for the forward-bias condition,

$$\langle i_n^2 \rangle = 2qI_sBe^{qV/kT} + 2qBI_s. \tag{94}$$

Experimental measurements indeed confirm that the mean-square noise current is proportional to the saturation current I_s, which can be varied by irradiation.

2.7 TERMINAL FUNCTIONS

A p-n junction is a two-terminal device that can perform various terminal functions, depending upon its biasing conditions as well as its doping profile and device geometry. In this section we discuss briefly some interesting device performances based on current–voltage, capacitance–voltage, and breakdown characteristics discussed in previous sections. Many other related two-terminal devices will be considered in subsequent chapters (e.g., tunnel diode in Chapter 9 and IMPATT diode in Chapter 10).

2.7.1 Rectifier

A rectifier is a p-n junction diode that is specifically designed to rectify alternating current, that is, to give a very low resistance to current flow in one direction and a very high resistance in the other direction. The forward and reverse resistances of a rectifier can be easily derived from the current–voltage relationship of a practical diode,

$$I = I_s(e^{qV/nkT} - 1) \tag{95}$$

where I_s is the saturation current and the factor n generally has a value between 1 (for diffusion current) and 2 (for recombination current). The forward dc (or static) resistance R_F and small-signal (or dynamic) resistance r_F are obtainable from Eq. 95:

$$R_F \equiv \frac{V_F}{I_F}\left(\simeq \frac{V_F}{I_s}\, e^{-qV_F/nkT} \qquad \text{for} \quad V \geq 3kT/q\right) \tag{96a}$$

$$r_F = \frac{dV_F}{dI_F} = \frac{nkT}{qI_F}. \tag{96b}$$

The reverse dc resistance R_R and small-signal resistance r_R are given by

$$R_R \equiv \frac{V_R}{I_R} \simeq \left(\frac{V_R}{I_s} \qquad \text{for} \quad |V_R| \geq 3kT/q\right) \tag{97a}$$

$$r_R \equiv \frac{\partial V_R}{\partial I_R} = \frac{nkT}{qI_s}\, e^{q|V_R|/kT}. \tag{97b}$$

Comparing Eqs. 96 and 97 shows that the dc rectification ratio R_R/R_F varies with $\exp(qV_F/nkT)$, while the ac rectification ratio r_R/r_F varies with $I_F/[I_s \exp(-q|V_R|/kT)]$.

Rectifiers generally have slow switching speeds; that is, a significant time delay is necessary to obtain high impedance after switching from the forward-conduction state to the reverse-blocking state. This time delay (proportional to the minority-carrier lifetime as shown in Fig. 37) is of little consequence in rectifying 60-Hz currents. For high-frequency applications, the lifetime should be sufficiently reduced to maintain rectification efficiency. The majority of rectifiers has power-dissipation capabilities from 0.1 to 10 W, reverse breakdown voltages from 50 to 2500 V (for a high-voltage rectifier two or more *p-n* junctions are connected in series), and switching times from 50 ns for low-power diodes to about 500 ns for high-power diodes.

2.7.2 Voltage Regulator

A voltage regulator is a *p-n* junction diode operated in the reverse direction up to its breakdown voltage. Prior to breakdown, the diode has a very high resistance; after breakdown the diode has a very small dynamic resistance. The voltage is thus limited (or regulated) by the breakdown voltage.

Most voltage regulators are made of Si, because of the low saturation current in Si diodes and the advanced Si technology. As discussed in Section 3.5, for breakdown voltage V_B larger than $6E_g/q$ (~8 V for Si), the breakdown mechanism is mainly avalanche multiplication, and the temperature coefficient of V_B is positive. For $V_B < 4E_g/q$ (~5 V for Si), the breakdown mechanism is band-to-band tunneling, and the temperature coefficient of V_B is negative. For $4E_g/q < V_B < 6E_g/q$, the breakdown is due to a combination of these two mechanisms. One can connect, for example, a negative-temperature-coefficient diode in series with a positive-temperature-coefficient diode to produce a low-temperature-coefficient regulator (with a temperature coefficient of the order of 0.002% per °C), which is suitable as a voltage reference.

2.7.3 Varistor

A varistor (*vari*able resis*tor*) is a two-terminal device that shows nonohmic behavior.[46] Equations 96 and 97 have shown the nonohmic characteristics of a *p-n* junction diode. Similar nonohmic characteristics are obtainable from metal–semiconductor diodes, considered in Chapter 5. An interesting application of varistors is their use as symmetrical fractional-voltage (~0.5 V) limiter by connecting two diodes in parallel, oppositely poled. The two-diode unit will exhibit the forward $I-V$ characteristics in either direction.

2.7.4 Varactor

The term "varactor" comes from the words *var*iable re*actor* and means a device whose reactance can be varied in a controlled manner with a bias voltage. Varactor diodes are widely used in parametric amplification, harmonic generation, mixing, detection, and voltage-variable tuning.

The basic capacitance–voltage relationships have already been derived in Section 2.3. We shall now extend the previous derivations of abrupt and linearly graded doping distributions to a more general case. The one-dimensional Poisson equation is given as

$$\frac{\partial^2 V}{\partial x^2} = -\frac{qN}{\epsilon_s} \tag{98}$$

where N is the generalized doping distribution as shown in Fig. 38*a* (assuming one side is very heavily doped):

$$N = Bx^m \qquad \text{for} \quad x \geq 0. \tag{99}$$

For $m = 0$ we have $B = N_B$ corresponding to the uniformly doped (or one-sided abrupt junction) case. For $m = 1$, the doping profile corresponds to a one-sided linearly graded case. For $m < 0$, the device is called a "hyper-abrupt" junction. The hyper-abrupt doping profile can be achieved by an epitaxial process or by ion implantation. The boundary conditions are $V(x = 0) = 0$ and $V(x = W) = V + V_{bi}$, where V is the applied voltage and V_{bi} is the built-in voltage.

Integrating Poisson's equation with the boundary conditions, we obtain

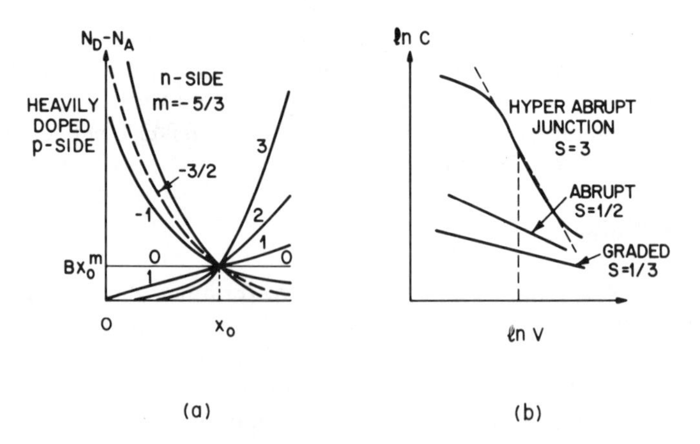

Fig. 38 (a) Various impurity distributions for varactors. (*b*) Log-log plot of depletion-layer capacitance versus reverse-biased voltage. (After Norwood and Shatz, Ref. 47; Moline and Foxhall, Ref. 48.)

for the depletion-layer width and the differential capacitance per unit area[47]

$$W = \left[\frac{\epsilon_s(m+2)(V+V_{bi})}{qB}\right]^{1/(m+2)} \tag{100}$$

$$C \equiv \frac{\partial Q_c}{\partial V} = \left[\frac{qB(\epsilon_s)^{m+1}}{(m+2)(V+V_{bi})}\right]^{1/(m+2)} \sim (V+V_{bi})^{-s} \tag{101}$$

$$s \equiv \frac{1}{m+2}$$

where Q_c, the charge per unit area, is equal to the product of ϵ_s and the maximum electric field (at $x = 0$).

One important parameter in characterizing the varactor is the sensitivity $s(V)$ defined by[48]

$$s \equiv -\frac{dC}{C}\frac{V}{dV} = \frac{-d(\log C)}{d(\log V)} = \frac{1}{m+2}\,. \tag{102}$$

The larger the s, the larger will be the capacitance variation with biasing voltage. For linearly graded junctions, $m = 1$ and $s = \frac{1}{3}$; for abrupt junctions, $m = 0$ and $s = \frac{1}{2}$; for hyper-abrupt junctions with $m = -1$, $-\frac{3}{2}$, or $-\frac{5}{3}$, the value of s is 1, 2, or 3, respectively. The capacitance–voltage relationships for these junction diodes are shown in Fig. 38*b*. The hyper-abrupt junction, as expected, has the highest sensitivity and gives rise to the largest capacitance variation.

The simplified equivalent circuit of a varactor is shown[47] in the Fig. 39 insert, where C_J is the junction capacitance, R_S is the series resistance, and

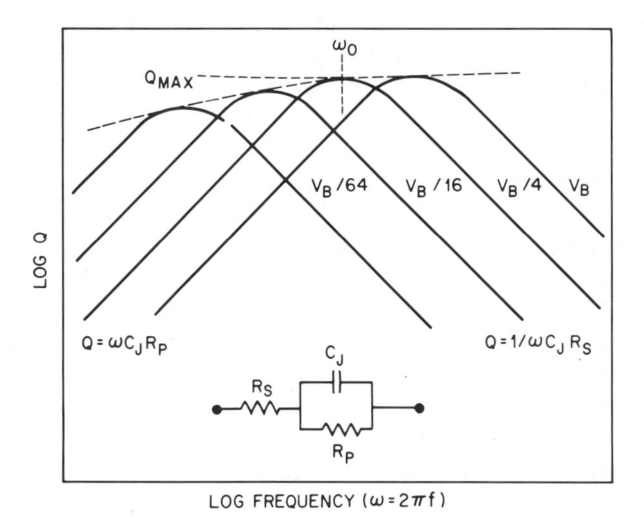

Fig. 39 Quality factor Q of a varactor versus frequency for various bias voltages. The insert shows the equivalent circuit. (After Norwood and Shatz, Ref. 47.)

R_P is the parallel equivalent resistance of generation–recombination current, diffusion current, and surface leakage current. Both C_J and R_S decrease with the reverse-bias voltage, while R_P generally increases with voltage. The efficiency of a varactor is expressed as a quality factor Q, which is the ratio of energy stored to energy dissipated:

$$Q \simeq \frac{\omega C_J R_P}{1 + \omega^2 C_J^2 R_P R_S}. \tag{103}$$

This expression can be differentiated to obtain the angular frequency ω_0 for maximum Q, and the value of this maximum, Q_{max}. These expressions are shown as

$$\omega_0 \simeq \frac{1}{C_J (R_P R_S)^{1/2}} \tag{104}$$

$$Q_{max} \approx \left(\frac{R_P}{4R_S}\right)^{1/2}. \tag{105}$$

Figure 39 shows a qualitative graph of the relationship between Q, frequency, and bias voltage. For a given bias, Q varies as $\omega C_J R_P$ at low frequencies and as $1/\omega C_J R_S$ at high frequencies. The maximum bias voltage is limited by the breakdown voltage V_B.

2.7.5 Fast-Recovery Diode

Fast-recovery diodes are designed to give ultrahigh switching speed. The devices can be classified into two types: diffused *p-n* junction diodes and metal–semiconductor diodes. The equivalent circuit of both types can be represented by the varactor diode (Fig. 39, insert). The general switching behavior of both types can be described by Fig. 36*b*.

The total recovery time $(t_1 + t_2)$ for a *p-n* junction diode can be substantially reduced by introducing recombination centers, such as Au in Si. Although the recovery time is directly proportional to the lifetime τ, as shown in Fig. 37, it is not possible, unfortunately, to reduce recovery times to zero by introducing an extremely large number of recombination centers N_t, because the reverse generation current of a *p-n* junction is proportional to N_t (Eqs. 47 and 48). For direct bandgap semiconductors, such as GaAs, the minority-carrier lifetimes are generally much smaller than that of Si. This results in ultra-high-speed GaAs *p-n* junction diodes with recovery times of the order of 0.1 ns or less. For Si the practical recovery time is in the range of 1 to 5 ns.

The metal–semiconductor diode (Schottky diode) also exhibits ultra-high-speed characteristics, because most Schottky diodes are majority-carrier devices and the minority-carrier storage effect is negligible. We discuss metal–semiconductor contacts in detail in Chapter 5.

2.7.6 Charge-Storage Diode

In contrast to fast-recovery diodes, a charge-storage diode is designed to store charge while conducting in the forward direction and upon switching to conduct for a short period in the reverse direction. A particularly interesting charge-storage diode is the step-recovery diode (also called the snapback diode) which conducts in the reverse direction for a short period then abruptly cuts off the current as the stored charges have been dispelled. This cutoff occurs in the range of picoseconds and results in a fast-rising wavefront rich in harmonics. Because of these characteristics, step-recovery diodes are used as harmonic generators and pulse formers. Most charge-storage diodes are made from Si with relatively long minority-carrier lifetimes ranging from 0.5 to 5 μs. Note that the lifetimes are about 1000 times longer than for fast-recovery diodes.

2.7.7 *p-i-n* Diode

A *p-i-n* diode is a *p-n* junction with a doping profile tailored so that an intrinsic layer, the "*i* region," is sandwiched between a *p* layer and an *n* layer, Fig. 40*a*. In practice, however, the idealized *i* region is approximated by either a high-resistivity *p* layer (referred to as π layer) or a high-resistivity *n* layer (ν layer). The impurity distribution, space-charge density, and field distribution in *p-i-n* and *p*-π-*n* diodes are shown[49] in Fig. 40*b*, *c*, and *d*, respectively. Because of low doping in the *i* region, most of the potential will drop across this region. For a practical *p-i-n* diode the impurity distribution in the *p* and *n* layers varies more gradually than that shown in Fig. 40. It can be fabricated, for example, using (1) the epitaxial process, (2) the diffusion of *p* and *n* regions into a high-resistivity semiconductor substrate, and (3) the ion-drift (e.g., lithium) method to introduce the highly compensated intrinsic region.[50]

The *p-i-n* diode has found wide applications in microwave circuits. It can be used as a microwave switch with essentially constant depletion-layer capacitance and high power-handling capability. The switching speed[51] is approximately given by $W/2v_s$, where v_s is the saturation velocity across the *i* region. In addition, a *p-i-n* diode can be used as a variolosser (variable attenuator) by controlling the device resistance which varies approximately linearly with the forward current. It can also modulate signals up to the GHz range. Furthermore, the forward characteristics of a thyristor (refer to Chapter 4) in its ON state closely resemble those of a *p-i-n* diode.

Figure 32 gives the breakdown voltage of a *p-i-n* diode under reverse biases. Because the maximum field $\mathscr{E}_m$ in Si at lower dopings is about 2.5×10^5 V/cm, the breakdown voltage is then

$$V_B \approx \mathscr{E}_m W \approx 25 W \quad (\text{in } \mu\text{m}) \qquad \text{V}. \tag{106}$$

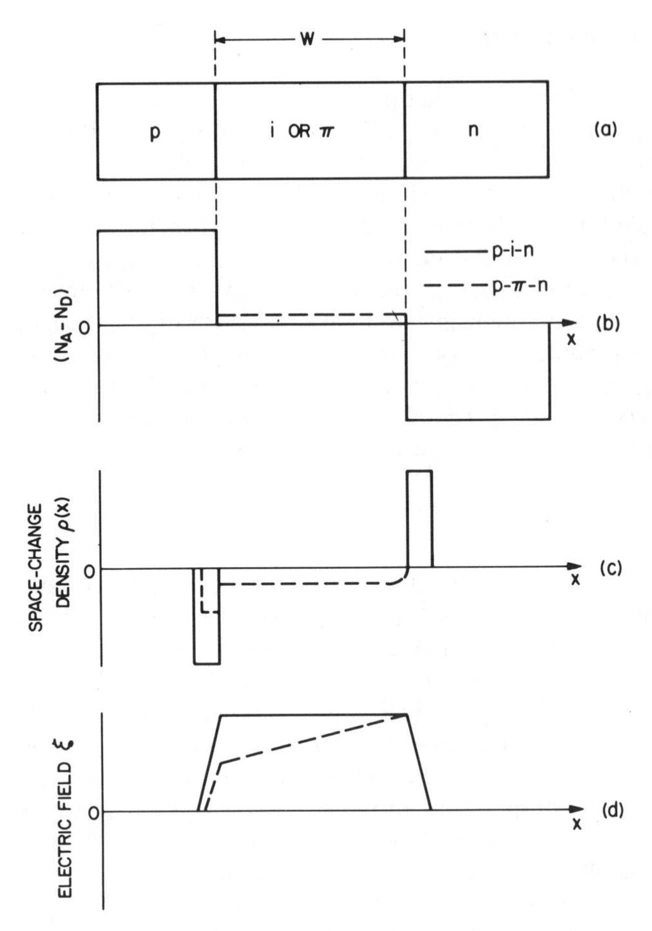

Fig. 40 Impurity distribution, space-charge density, and field distribution in *p-i-n* and *p-π-n* junctions. (After Veloric and Prince, Ref. 49.)

The measured reverse-biased capacitance and series resistance of a *p-i-n* diode designed for low-power switching applications are shown[52] in Fig. 41*a*. Note that the capacitance becomes $\epsilon_s A/W$ (A is the area) at about 5 V, which is far from the breakdown voltage of approximately 75 V. Beyond 10 V, the capacitance decreases only 3% further. The series resistance R_s consists of two components:

$$R_s = R_i + R_c \tag{107}$$

where R_i is the resistance of the *i* region and R_c the contact resistance. As the reverse bias increases, R_i approaches zero and the series resistance decreases rapidly toward an asymmetric value corresponding to the contact resistance.

Under forward conditions, holes are injected from the p^+-*i* contact and

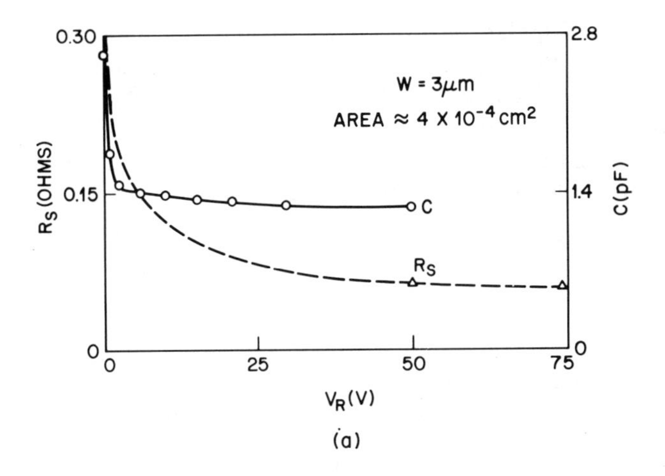

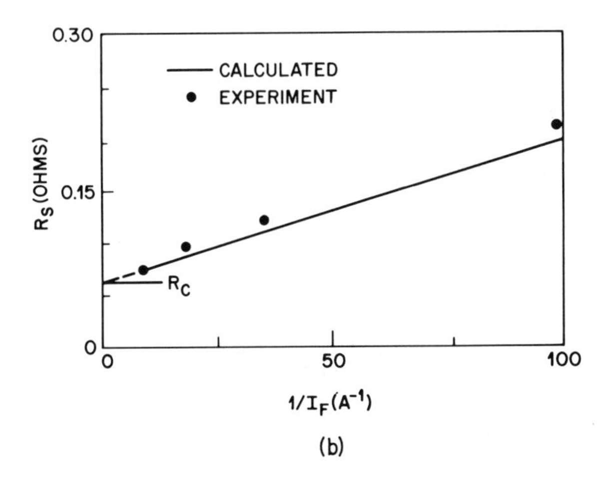

Fig. 41 (*a*) Depletion-layer capacitance and series resistance versus reverse voltage. (*b*) Series resistance versus reciprocal of forward current. (After Chiang and Denlinger, Ref. 52.)

electrons are injected from the i-n^+ contact. We first consider the current flow due to electron–hole recombinations in the i region. The current density is given by[39, 53]

$$J = \int_0^W qU\,dx \tag{108}$$

where U, the recombination rate, is equal to $n(x)/\tau$. If n' is the average injected electron concentration in the i region, then

$$J = \frac{qn'W}{\tau_a} \tag{109}$$

where τ_a is the ambipolar lifetime. If we assume that the carrier concentration throughout the i region is approximately constant, the diffusion current can be neglected. As the injected carrier density is much higher than the i-region doping concentration, the p-i-n diode is generally operated in the high-injection condition, that is, $n' \approx p' \gg n_i$, where p' is the average injected hole density in the i region. The total drift current is then

$$J = q\mu_n n'\mathscr{E}' + q\mu_p p'\mathscr{E}' = q(\mu_n + \mu_p)n'\mathscr{E}'$$

$$= \frac{q}{kT}\left(1 + \frac{1}{b}\right)qD_n n'\mathscr{E}' \tag{110}$$

where $b \equiv \mu_n/\mu_p$ and $\mathscr{E}'$ is the average electric field in the i region. The ambipolar diffusion coefficient given in Eq. 36 becomes

$$D_a = \frac{2D_n}{1+b}. \tag{111}$$

Substituting Eq. 111 into Eq. 110 gives

$$J = \frac{q}{kT}\frac{(b+1)^2}{2b}\, qD_a n'\mathscr{E}'. \tag{112}$$

The voltage drop across the i region, V_i, is given by

$$V_i = \mathscr{E}'W. \tag{113}$$

From Eqs. 109 through 113 and noting that L_a ($\equiv \sqrt{D_a\tau_a}$) is the ambipolar diffusion length:

$$V_i = \frac{kT}{q}\frac{2b}{(1+b)^2}\left(\frac{W}{L_a}\right)^2. \tag{114}$$

For silicon, where $b \approx 3$,

$$V_i \approx \frac{3kT}{8q}\left(\frac{W}{L_a}\right)^2. \tag{115}$$

The resistance R_i is given by

$$R_i \equiv \frac{V_i}{I_F} = \left(\frac{3kTW^2}{8qD_a\tau_a}\right)\frac{1}{I_F}. \tag{116}$$

An example of the series resistance in forward bias is shown in Fig. 41b. Note that the resistance varies as $1/I_F$ and the extrapolated value at $1/I_F \to 0$ gives the contact resistance R_c.

A more accurate result for V_i can be obtained by solving Eq. 35 with appropriate boundary conditions (Fig. 42). Note that for $W/L_a \le 2$, V_i is very close to the value given by Eq. 115. For $W/L_a > 2$, however, V_i increases rapidly and can be approximated by[39]

$$V_i \simeq \frac{3\pi kT}{8q}\exp\left(\frac{W}{2L_a}\right). \tag{117}$$

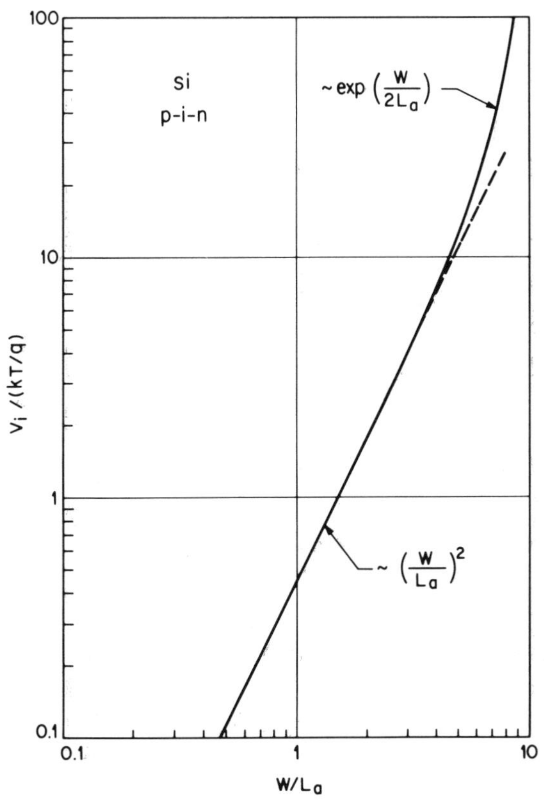

Fig. 42 Voltage drop in the intrinsic region of a *p-i-n* junction versus W/L_a, where W is the width of the *i*-region and L_a is the ambipolar diffusion length. (After Ghandhi, Ref. 39.)

The condition $W = 2L_a$ thus marks the transition between a "short" and a "long" *p-i-n* diode. For the short structure, where $W \leq 2L_a$, the spatial variation in carrier concentration over the *i* region has little effect on the voltage drop (less than 0.05 V at room temperature) and can be ignored. For the long structure, however, the voltage drop across the *i* region can be quite high. In a typical device design, the *i* region thickness is determined by the required breakdown voltage. To maintain a short-structure characteristic, L_a must be large. Under high-injection conditions, the ambipolar diffusion coefficient D_a decreases with increasing injected carrier concentration because of carrier–carrier scattering effects. The minority-carrier lifetime will also be reduced because of Auger recombination processes at high carrier concentration. Therefore, L_a will decrease as current density increases, causing the ratio W/L_a to increase.

Figure 43 shows the results of calculations[54] on high-voltage devices with $W = 600\ \mu\text{m}$ and $\tau_a = 30\ \mu\text{s}$. Curve (a) considers only recombination in the

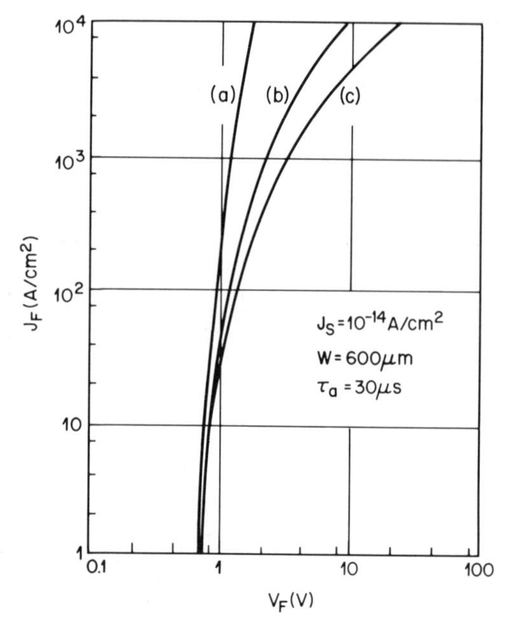

Fig. 43 Forward current–voltage characteristics. (*a*) Recombination only. (*b*) Including carrier–carrier scattering. (*c*) Including Auger recombination as well. (After Burtscher, Dannhauser, and Krausse, Ref. 54.)

i region, curve (b) includes the effect of carrier–carrier scattering, and curve (c) includes Auger recombination as well. The steady worsening of device characteristics is revealed in this progression. Clearly, the Auger recombination and carrier–carrier scattering set a limit on device performance.

2.8 HETEROJUNCTION

A heterojunction is a junction formed between two dissimilar semiconductors. When the two semiconductors have the same type of conductivity, the junction is called an isotype heterojunction. When the conductivity types differ, the junction is called an anisotype heterojunction. In 1951, Shockley proposed the abrupt heterojunction to be used as an efficient emitter–base junction in a bipolar transistor.[55] In the same year, Gubanov published theoretical papers on heterojunctions.[56] Kroemer later analyzed a similar, although graded, heterojunction as a wide-gap emitter.[57] Since then, heterojunctions have been extensively studied, and many important applications have been made, among them the room-temperature injection laser, light-emitter diode, photodetector, and solar cell. In addition, by forming periodic layered heterojunctions with layer thickness of the order of 100 Å, we have the so-called superlattice structures. The heterojunctions have been reviewed by Milnes and Feucht,[58] Sharma and Purohit,[59] and Casey and Panish.[11]

2.8.1 Basic Device Model

The energy-band model of an ideal abrupt heterojunction without interface traps was proposed by Anderson[60] based on the previous work of Shockley. We consider this model next, since it can adequately explain most transport processes, and only slight modification of the model is needed to account for nonideal cases such as interface traps. Figure 44*a* shows the

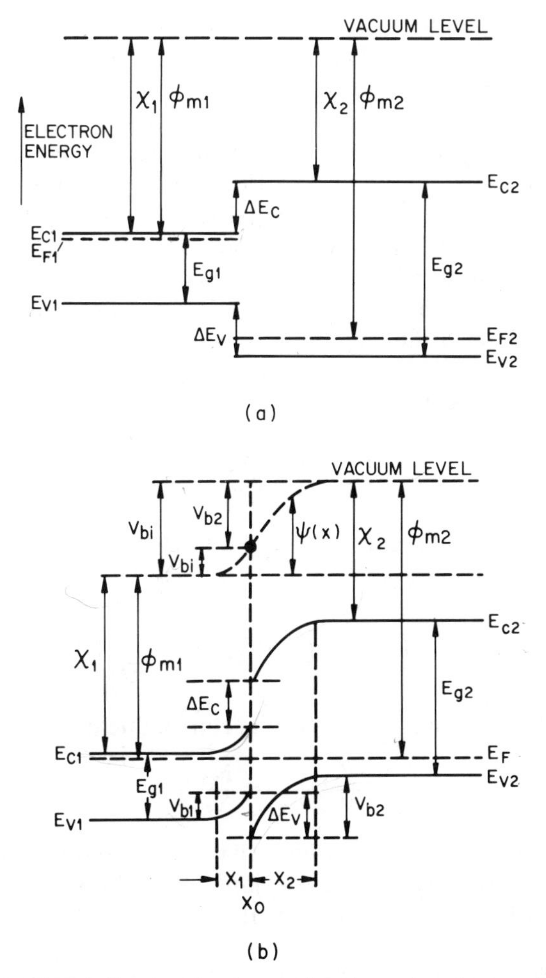

Fig. 44 (*a*) Energy-band diagram for two isolated semiconductors in which space-charge neutrality is assumed to exist in each region. (*b*) Energy-band diagram of an ideal *n-p* anisotype heterojunction at thermal equilibrium. (After Anderson, Ref. 60.)

energy-band diagram of two isolated pieces of semiconductors. The two semiconductors were assumed to have different bandgaps E_g, different permittivities ϵ, different work functions ϕ_m, and different electron affinities χ. Work function and electron affinity are defined as that energy required to remove an electron from the Fermi level E_F and from the bottom of the conduction band E_C, respectively, to a position just outside the material (vacuum level). The difference in energy of the conduction-band edges in the two semiconductors is represented by ΔE_C and that in the valence-band edges by ΔE_V. Figure 44*a* shows that $\Delta E_C = (\chi_1 - \chi_2)$. This electron affinity rule $\Delta E_C = \Delta\chi$ may not be a valid assumption. However, by choosing ΔE_C as an empirical quantity, the Anderson model remains unaltered.[60a]

When a junction is formed between these semiconductors, the energy-band profile at equilibrium is as shown in Fig. 44*b* for an *n*-*p* anisotype heterojunction. Since the Fermi level must coincide on both sides in equilibrium and the vacuum level is everywhere parallel to the band edges and is continuous, the discontinuity in conduction-band edges (ΔE_C) and valence-band edges (ΔE_V) is invariant with doping in those cases where E_g and χ are not functions of doping (i.e., nondegenerate semiconductors). The total built-in potential V_{bi} is equal to the sum of the partial built-in voltage $(V_{b1} + V_{b2})$, where V_{b1} and V_{b2} are the electrostatic potential supported at equilibrium by semiconductors 1 and 2, respectively.

The depletion widths and capacitance can be obtained by solving Poisson's equation for the step junction on either side of the interface. One boundary condition is the continuity of electric displacement, that is, $\epsilon_1 E_1 = \epsilon_2 E_2$ at the interface. We obtain

$$x_1 = \left[\frac{2N_{A2}\epsilon_1\epsilon_2(V_{bi} - V)}{qN_{D1}(\epsilon_1 N_{D1} + \epsilon_2 N_{A2})}\right]^{1/2} \tag{118}$$

$$x_2 = \left[\frac{2N_{D1}\epsilon_1\epsilon_2(V_{bi} - V)}{qN_{A2}(\epsilon_1 N_{D1} + \epsilon_2 N_{A2})}\right]^{1/2} \tag{119}$$

and

$$C = \left[\frac{qN_{D1}N_{A2}\epsilon_1\epsilon_2}{2(\epsilon_1 N_{D1} + \epsilon_2 N_{A2})(V_{bi} - V)}\right]^{1/2}. \tag{120}$$

The relative voltage supported in each semiconductor is

$$\frac{V_{b1} - V_1}{V_{b2} - V_2} = \frac{N_{A2}\epsilon_2}{N_{D1}\epsilon_1} \tag{121}$$

where $V = V_1 + V_2$. It is apparent that the foregoing expressions will reduce to the expression for the *p*-*n* junction (homojunction) discussed in Section 2.3, where both sides of the heterojunction have the same materials.

The case of an *n*-*n* isotype heterojunction of the two semiconductors is somewhat different. Since the work function of the wide-gap semiconductor is smaller, the energy bands will be bent oppositely to those for the *n*-*p* case* (see Fig. 45*a*).[61] The relation between $V_{b1}-V_1$ and $V_{b2}-V_2$ can be found from the boundary condition of continuity of electric displacement at the interface. For an accumulation in region 1 governed by Boltzmann statistics (for detailed derivation, see Section 7.2), the electric displacement $\mathscr{D}_1$ at x_0 is given by

$$\mathscr{D}_1 = \epsilon_1 \mathscr{E}_1(x_0) = \left\{2\epsilon_1 q N_{D1}\left[\frac{kT}{q}\left(\exp\frac{q(V_{b1}-V_1)}{kT}-1\right)-(V_{b1}-V_1)\right]\right\}^{1/2}. \tag{122}$$

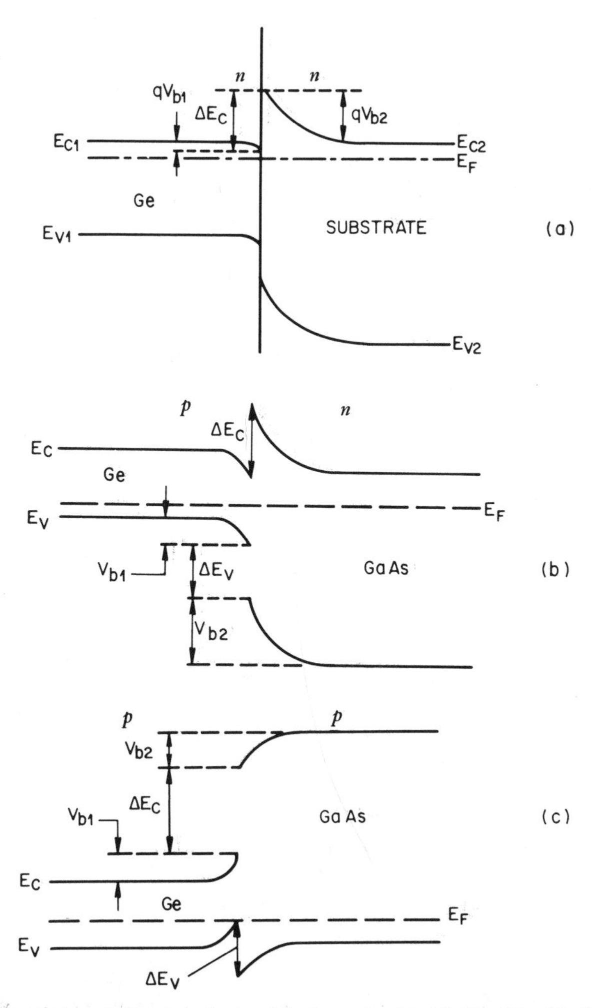

Fig. 45 (*a*) Energy-band diagram for an ideal *n-n* isotype heterojunction. (After Chang, Ref. 61.) (*b*) and (*c*) Energy-band diagrams for ideal *p-n* and *p-p* heterojunctions, respectively. (After Anderson, Ref. 60.)

*The convention is to list the material with the smaller bandgap as the first symbol.

The electric displacement at the interface for a depletion in region 2 is given by

$$\mathcal{D}_2 = \epsilon_2 \mathcal{E}_2(x_0) = [2\epsilon_2 q N_{D2}(V_{b2} - V_2)]^{1/2}. \tag{123}$$

Equating Eqs. 122 and 123 gives a relation between $(V_{b1} - V_1)$ and $(V_{b2} - V_2)$ that is quite complicated. However, if the ratio $\epsilon_1 N_{D1}/\epsilon_2 N_{D2}$ is of the order of unity and $V_{bi}(\equiv V_{b1} + V_{b2}) \gg kT/q$, we obtain[61]

$$\exp\left[\frac{q(V_{b1} - V_1)}{kT}\right] \simeq \frac{q}{kT}(V_{bi} - V) \tag{124}$$

where V is the total applied voltage and is equal to $(V_1 + V_2)$. Also shown in Fig. 45 are the idealized equilibrium energy-band diagrams for p-n (narrow-gap p-type and wide-gap n-type) and p-p heterojunctions.

For the current–voltage characteristics we shall consider an interesting case shown in Fig. 45a. The conduction mechanism is governed by thermionic emission (refer to Chapter 5 for details) and the current density is given by[61]

$$J = A^*T^2 \exp\left(-\frac{qV_{b2}}{kT}\right)\left[\exp\left(\frac{qV_2}{kT}\right) - \exp\left(\frac{qV_1}{kT}\right)\right] \tag{125}$$

where A^* is the effective Richardson constant. Substituting Eq. 124 into Eq. 125 yields the current–voltage relationship:

$$J = J_0\left(1 - \frac{V}{V_{bi}}\right)\left[\exp\left(\frac{qV}{kT}\right) - 1\right] \tag{126}$$

where

$$J_0 \equiv \frac{qA^*TV_{bi}}{k} \exp\left(-\frac{qV_{bi}}{kT}\right).$$

This expression is somewhat different from that for metal–semiconductor contact. The value of J_0 is different and so is its temperature dependence. The reverse current never saturates but increases linearly with voltage at large V. In the forward, the dependence of J on qV/kT can be approximated by an exponential function or $J \sim \exp(qV/nkT)$.

2.8.2 Heterojunction Devices

The successful application of heterojunctions to various devices is due mainly to the epitaxial technology to grow lattice-matched isotype or anisotype heterojunctions with virtually no interface traps.[62] Heterojunctions have been used in bipolar devices as wide-gap emitters and in unipolar devices for MESFET applications. We consider them in Chapters 3 and 6, respectively. The most important applications of heterojunctions are in photonic devices, including semiconductor lasers, photodetectors, and solar cells. We shall consider their characteristics in detail in Chapters 12 through 14.

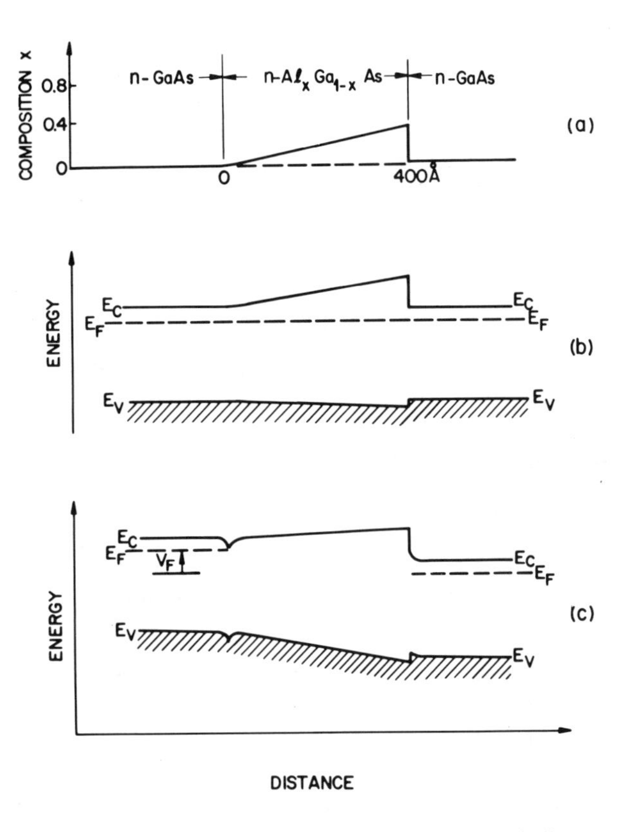

Fig. 46 (*a*) Composition variation. (*b*) Equilibrium energy-band diagram. (*c*) Energy-band diagram under forward bias for a sawtooth-shaped composition grading structure. (After Allyn, Gossard, and Weigmann, Ref. 63.)

In this section we consider briefly a few novel heterojunction configurations that may have potential applications. Figure 46*a* shows[63] a unipolar rectifying structure having a sawtooth-shaped composition grading of ternary compound $Al_xGa_{1-x}As$ sandwiched between layers of *n*-type GaAs. As the composition x increases from 0 to 0.4, the bandgap of $Al_xGa_{1-x}As$ increases linearly from 1.42 eV to 1.92 eV, which gives rise to the equilibrium energy-band diagram shown in Fig. 46*b*. Under forward bias, the voltage drop occurs across the graded layer, reducing the slope of the potential barrier and allowing increased thermionic emission over the barrier (Fig. 46*c*). In the reverse bias, the electrons will be inhibited from passing through the abrupt potential discontinuity at the sharp edge of the sawtooth. Rectification characteristics have been observed for this device operated at 77 and 300 K.

The superlattice structures include (1) a multilayered heterojunction arrangement with typical layer thickness of the order of 80 to 100 Å, and (2) a periodic alternation of the doping of only one semiconductor to form a

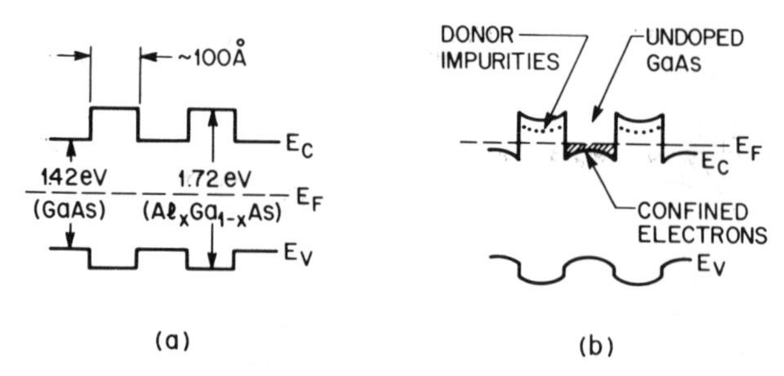

Fig. 47 (*a*) Energy-band diagram for undoped superlattice structure of GaAs (1.42 eV) and $Al_{0.3}Ga_{0.7}As$. (*b*) Energy-band diagram for modulation-doped superlattice structure. (After Dingle et al., Ref. 64.)

series of homojunctions. Molecular-beam epitaxy is known to produce atomically smooth layers and to allow very precise control over grown layer thicknesses. The schematic energy-band diagram of GaAs-$Al_xGa_{1-x}As$ superlattice structure is shown[64] in Fig. 47*a*. The structure is undoped. Therefore, the Fermi level lies near the middle of the bandgap.

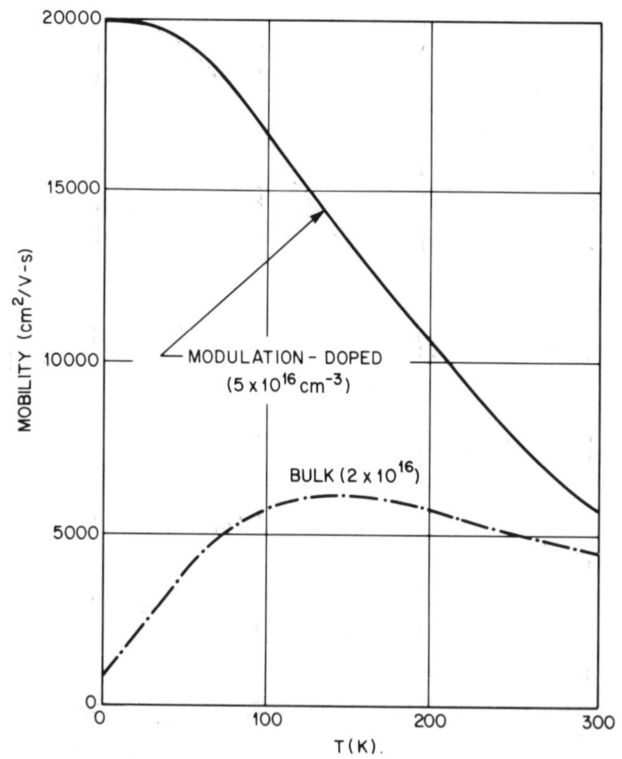

Fig. 48 Mobility versus temperature for bulk GaAs and modulation-doped superlattice structure. (After Dingle et al., Ref. 64.)

For composition $x = 0.3$, the bandgap difference is about 300 meV. We can modulate the doping by synchronizing the deposition of Al and Si (a donor to $Al_xGa_{1-x}As$) so that only the $Al_xGa_{1-x}As$ layers are doped with Si impurities. The energy-band diagram for the modulation-doped superlattice is shown in Fig. 47*b*. The Fermi level now moves closer to the conduction-band edge. Since the GaAs conduction band edge lies lower in energy than the $Al_xGa_{1-x}As$ donor states, electrons from the donors will move into the GaAs regions. Now all mobile carriers are confined to the GaAs layers, and their parent donor impurities (in the $Al_xGa_{1-x}As$ layers) are spatially separated from each other. Thus the electron density in the GaAs channel may greatly exceed the density of the neutral and ionized impurity scattering centers in the channel, leading to considerable change in mobility behavior in the temperature and carrier density regime, where impurity scatterings are important. Figure 48 shows the measured mobility parallel to the multilayer as a function of temperature. Note that the modulation-doped superlattice structure has a substantially higher mobility than that of the bulk material. If a voltage is applied perpendicular to the multilayers, resonant tunneling may occur, giving enhanced tunneling current at voltages near the quasi-stationary states of the superlattice potential well. These properties may lead to many useful device possibilities.[64–66]

REFERENCES

1 W. Shockley, "The Theory of *p-n* Junctions in Semiconductors and *p-n* Junction Transistors," *Bell Syst. Tech. J.*, **28**, 435 (1949); *Electrons and Holes in Semiconductors*, D. Van Nostrand, Princeton, N. J., 1950.

2 C. T. Sah, R. N. Noyce, and W. Shockley, "Carrier Generation and Recombination in *p-n* Junction and *p-n* Junction Characteristics," *Proc. IRE*, **45**, 1228 (1957).

3 J. L. Moll, "The Evolution of the Theory of the Current–Voltage Characteristics of *p-n* Junctions," *Proc. IRE*, **46**, 1076 (1958).

4 For example, see A. G. Grove, *Physics and Technology of Semiconductor Devices*, Wiley, New York, 1967.

5 R. N. Hall and W. C. Dunlap, "*p-n* Junctions Prepared by Impurity Diffusion," *Phys. Rev.*, **80**, 467 (1950).

6 M. Tanenbaum and D. E. Thomas, "Diffused Emitter and Base Silicon Transistors," *Bell Syst. Tech. J.*, **35**, 1 (1956).

7 C. J. Frosch and L. Derrick, "Surface Protection and Selective Masking during Diffusion in Silicon," *J. Electrochem. Soc.*, **104**, 547 (1957).

8 J. A. Hoerni, "Planar Silicon Transistor and Diodes," IRE Electron Devices Meet., Washington, D.C., 1960.

9 H. C. Theuerer, J. J. Kleimack, H. H. Loar, and H. Christenson, "Epitaxial Diffused Transistors," *Proc. IRE*, **48**, 1642 (1960).

10 For a review, see, for example, K. A. Pickar, "Ion Implantation in Silicon-Physics, Processing and Microelectronic Devices," in R. Wolfe, Ed., *Applied Solid State Science*, Vol. 5, Academic, New York, 1975.

11 H. C. Casey, Jr., and M. B. Panish, *Heterostructure Lasers*, Academic, New York, 1978.

12 A. Y. Cho, "Recent Developments in Molecular Beam Epitaxy," *J. Vac. Sci. Technol.*, **16**, 275 (1979).

13 J. C. Bean, "Growth of Doped Silicon Layers by Molecular Beam Epitaxy," in F. F. Y. Wang, Ed., *Impurity Doping Processes in Silicon*, North-Holland, Amsterdam, 1981.

14 M. M. Atalla, "Semiconductor Surfaces and Films; the Silicon–Silicon Dioxide System," in H. Gatos, Ed., *Properties of Elemental and Compound Semiconductors*, Vol. 5, Interscience, New York, 1960, pp. 163–181.

15 B. E. Deal and A. S. Grove, "General Relationship for the Thermal Oxidation of Silicon," *J. Appl. Phys.*, **36**, 3770 (1965).

16 J. P. Meindl, R. W. Dutton, K. C. Saraswat, J. D. Plummer, T. I. Kamins, and B. E. Deal, "Silicon Epitaxy and Oxidation," in F. Van de Wiele, W. L. Engl, and P. O. Jespers, Eds., *Process and Device Modeling for Integrated Circuit Design*, Noordhoff, Leyden, 1977.

17 For a general reference, see H. S. Carslaw and J. C. Jaeger, *Conduction Heat in Solids*, 2nd ed., Oxford University Press, London, 1959.

18 R. B. Fair, "Concentration Profiles of Diffused Dopants in Silicon," in F. F. Y. Wang, Ed., *Impurity Doping Processes in Silicon*, North-Holland, Amsterdam, 1981.

19 R. M. Burger and R. P. Donovan, Eds., *Fundamentals of Silicon Integrated Device Technology*, Vol. 1, Prentice-Hall, Englewood Cliffs, N.J., 1967.

20 D. L. Kendall and D. B. DeVries, "Diffusion in Silicon," in R. R. Haberecht and E. L. Kern, Eds., *Semiconductor Silicon*, Electrochemical Society, New York, 1969, p. 358.

21 F. A. Trumbore, "Solid Solubilities of Impurity Elements in Germanium and Silicon," *Bell Syst. Tech. J.*, **39**, 205 (1960).

22 I. B. Khaibullin et al., VINITI dep. N2661 (1974).

23 S. D. Ferris, H. J. Leamy, and J. M. Poate, Eds., *Laser–Solid Interactions and Laser Processing*, American Institute of Physics, New York, 1979.

24 T. P. Lee and S. M. Sze, "Depletion Layer Capacitance of Cylindrical and Spherical *p-n* Junctions," *Solid State Electron.*, **10**, 1105 (1967).

25 S. M. Sze and G. Gibbons, "Effect of Junction Curvature on Breakdown Voltages in Semiconductors," *Solid State Electron.*, **9**, 831 (1966).

26 C. G. B. Garrett and W. H. Brattain, "Physical Theory of Semiconductor Surfaces," *Phys. Rev.*, **99**, 376 (1955); C. Kittel and H. Kroemer, *Thermal Physics*, 2nd ed., W. H. Freeman and Co., San Francisco, 1980.

27 D. Redfield, "Revised Model of Asymmetric *p-n* Junctions," *Appl. Phys. Lett.*, **35**, 182 (1979).

28 W. C. Johnson and P. T. Panousis, "The Influence of Debye Length on the *C–V* Measurement of Doping Profiles," *IEEE Trans. Electron. Devices*, **ED-18**, 965 (1971).

29 B. R. Chawla and H. K. Gummel, "Transition Region Capacitance of Diffused *p-n* Junctions," *IEEE Trans. Electron Devices*, **ED-18**, 178 (1971).

30 H. K. Gummel, "Hole–Electron Product of *p-n* Junctions," *Solid State Electron.*, **10**, 209 (1967).

31 For a general discussion, see J. L. Moll, *Physics of Semiconductors*, McGraw-Hill, New York, 1964.

32 M. J. O. Strutt, *Semiconductor Devices*, Vol. 1, *Semiconductor and Semiconductor Diodes*, Academic, New York, 1966, Chap. 2.

33 L. J. Schiff, *Quantum Mechanics*, 2nd ed., McGraw-Hill, New York, 1955.

34 P. J. Lundberg, private communication.

35 S. M. Sze and G. Gibbons, "Avalanche Breakdown Voltages of Abrupt and Linearly Graded *p-n* Junctions in Ge, Si, GaAs, and GaP," *Appl. Phys. Lett.*, **8**, 111 (1966).

36 R. M. Warner, Jr., "Avalanche Breakdown in Silicon Diffused Junctions," *Solid State Electron.*, **15**, 1303 (1972).

37 M. H. Lee and S. M. Sze, "Orientation Dependence of Breakdown Voltage in GaAs," *Solid State Electron.*, **23**, 1007 (1980).

38 F. Waldhauser, private communication.

39 S. K. Ghandhi, *Semiconductor Power Devices*, Wiley, New York, 1977.

40 C. R. Crowell and S. M. Sze, "Temperature Dependence of Avalanche Multiplication in Semiconductors," *Appl. Phys. Lett.*, **9**, 242 (1966).

41 C. Y. Chang, S. S. Chiu, and L. P. Hsu, "Temperature Dependence of Breakdown Voltage in Silicon Abrupt *p-n* Junctions," *IEEE Trans. Electron Devices*, **ED-18**, 391 (1971).

42 A. Goetzberger, B. McDonald, R. H. Haitz, and R. M. Scarlet, "Avalanche Effects in Silicon *p-n* Junction. II. Structurally Perfect Junctions," *J. Appl. Phys.*, **34**, 1591 (1963).

43 R. H. Kingston, "Switching Time in Junction Diodes and Junction Transistors," *Proc. IRE*, **42**, 829 (1954).

44 A. Van der Ziel, *Noise in Measurements*, Wiley, New York, 1976.

45 A. Van der Ziel and C. H. Chenette, "Noise in Solid State Devices," in *Advances in Electronics and Electron Physics*, Vol. 46, Academic, New York, 1978.

46 J. P. Levin, "Theory of Varistor Electronic Properties", *Crit. Rev. Solid State Sci.*, **5**, 597 (1975).

47 For a review, see M. H. Norwood and E. Shatz, "Voltage Variable Capacitor Tuning—A Review," *Proc. IEEE*, **56**, 788 (1968).

48 R. A. Moline and G. F. Foxhall, "Ion-Implanted Hyperabrupt Junction Voltage Variable Capacitors," *IEEE Trans. Electron Devices*, **ED-19**, 267 (1972).

49 H. S. Veloric and M. B. Prince, "High Voltage Conductivity-Modulated Silicon Rectifier," *Bell Syst. Tech. J.*, **36**, 975 (1957).

50 E. M. Pell, "Ion Drift in an *n-p* Junction," *J. Appl. Phys.*, **31**, 291 (1960); also J. W. Mager, "Characteristics of *p-i-n* Junction Produced by Ion-Drift Techniques in Silicon," *J. Appl. Phys.*, **33**, 2894 (1962).

51 G. Lucovsky, R. F. Schwarz, and R. B. Emmons, "Transit-Time Considerations in *p-i-n* Diodes," *J. Appl. Phys.*, **35**, 622 (1964).

52 Y. S. Chiang and E. J. Denlinger, "Low-Resistance All-Epitaxial *pin* Diode for Ultra-High-Frequency Applications," *RCA Rev.*, **38**, 390 (1977).

53 R. N. Hall, "Power Rectifiers and Transistors," *Proc. IRE*, **40**, 1512 (1952).

54 J. Burtscher, F. Dannhauser, and J. Krausse, "Recombination in Thyristor and Rectifier in Silicon," *Solid State Electron.*, **18**, 35 (1975).

55 W. Shockley, U.S. Patent 2,569,347 (1951).

56 A. I. Gubanov, *Zh. Tekh. Fiz.*, **21**, 304 (1951); *Zh. Eksp. Teor. Fiz.*, **21**, 721 (1951).

57 H. Kroemer, "Theory of a Wide-Gap Emitter for Transistors," *Proc. IRE*, **45**, 1535 (1957).

58 A. G. Milnes and D. L. Feucht, *Heterojunctions and Metal–Semiconductor Junctions*, Academic, New York, 1972.

59 B. L. Sharma and R. K. Purohit, *Semiconductor Heterojunctions*, Pergamon, London, 1974.

60 R. L. Anderson, "Experiments on Ge–GaAs Heterojunctions," *Solid State Electron.*, **5**, 341 (1962).

60a W. R. Frensley and H. Kroemer, "Theory of the Energy-Band Lineup at an Abrupt Semiconductor Heterojunction," *Phys. Rev. B.*, **16**, 2642 (1977).

61 L. L. Chang, "The Conduction Properties of Ge–$GaAs_{1-x}P_x$ *n-n* Heterojunctions," *Solid State Electron.*, **8**, 721 (1965).

62 D. V. Lang and R. A. Logan, "A Search for Interface States in an LPE GaAs/$Al_xGa_{1-x}As$ Heterojunction," *Appl. Phys. Lett.*, **31**, 683 (1977).

63 C. L. Allyn, A. C. Gossard, and W. Wiegmann, "New Rectifying Semiconductor Structure by Molecular Beam Epitaxy," *Appl. Phys. Lett.*, **36**, 373 (1980).

64 R. Dingle, H. L. Stormer, A. C. Gossard, and W. Wiegmann, "Electron Mobilities in Modulation-Doped Semiconductor Heterojunction Superlattices," *Appl. Phys. Lett.*, **33**, 665 (1978).

65 L. L. Chang, L. Esaki, and R. Tsu, "Resonant Tunneling in Semiconductor Double Barriers," *Appl. Phys. Lett.*, **24**, 593 (1974).

66 K. Hess, H. Morkoc, H. Shichijo, and B. G. Streetman, "Negative Differential Resistance through Real-Space Electron Transfer," *Appl. Phys. Lett.*, **35**, 469 (1979).

3

Bipolar Transistor

3.1 INTRODUCTION

The bipolar transistor (*tran*sfer re*sistor*), one of the most important semiconductor devices, was invented by a research team at Bell Laboratories in 1947. It has had an unprecedented impact on the electronic industry in general and on solid-state research in particular. Prior to 1947 semiconductors were only used as thermistors, photodiodes, and rectifiers. In 1948 John Bardeen and Walter Brattain announced the development of the point-contact transistor.[1] In the following year William Shockley's classic paper on junction diodes and transistors was published.[2]

Since then the transistor theory has been extended to include high-frequency, high-power, and switching behaviors. Many breakthroughs have been made in transistor technology, particularly in the alloy-junction[3] and grown-junction techniques,[4] and in zone-refining,[5] diffusion,[6] epitaxial,[7] planar,[8] beam-lead,[9] ion implantation,[10] lithography, and dry etching[11] technologies. These breakthroughs have helped increase the power and frequency capabilities as well as the reliability of transistors. In addition, application of semiconductor physics, transistor theory, and transistor technology has broadened our knowledge and improved other semiconductor devices as well.

Bipolar transistors are now key elements, for example, in high-speed computers, in vehicles and satellites, and in all modern communication and power systems. Many books have been written on bipolar transistor physics, design, and application. Among them are standard texts by Phillips,[12] and Gartner,[13] and a series of books by the Semiconductor Electronics Education Committee,[14] Pritchard,[15] Ghandhi,[16] and Muller and Kamins.[17]

3.2 STATIC CHARACTERISTICS

3.2.1 Basic Current–Voltage Relationship

In this section we consider the basic dc characteristics of *p-n-p* and *n-p-n* bipolar transistors. Figure 1 shows the symbols and nomenclatures for *p-n-p* and *n-p-n* transistors. The arrow indicates the direction of current flow under normal operating conditions, that is, forward-biased emitter junction and reverse-biased collector junction. A bipolar transistor can be connected in three circuit configurations, depending on which lead is common to the input and output circuits. Figure 2 shows the common-base, common-emitter, and common-collector configurations for a *p-n-p* transistor. The current and voltage conventions are given for normal operations. All signs and polarities should be inverted for an *n-p-n* transistor. In the following discussion we consider *p-n-p* transistors; the results are applicable to the *n-p-n* transistor with an appropriate change of polarities.

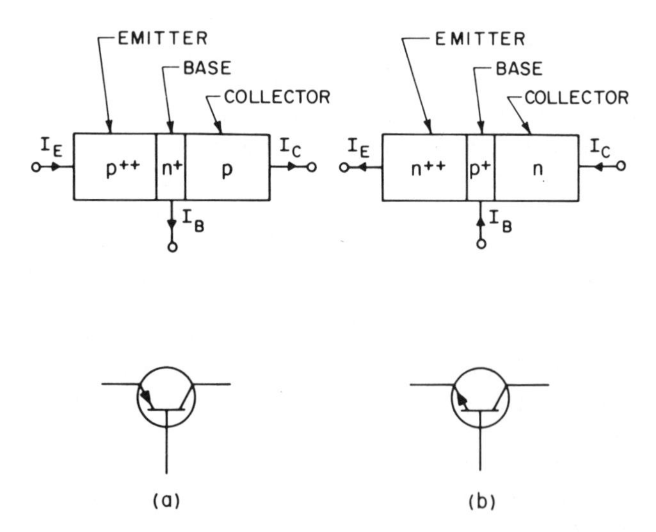

Fig. 1 Symbols and nomenclatures of (*a*) *p-n-p* transistors and (*b*) *n-p-n* transistors.

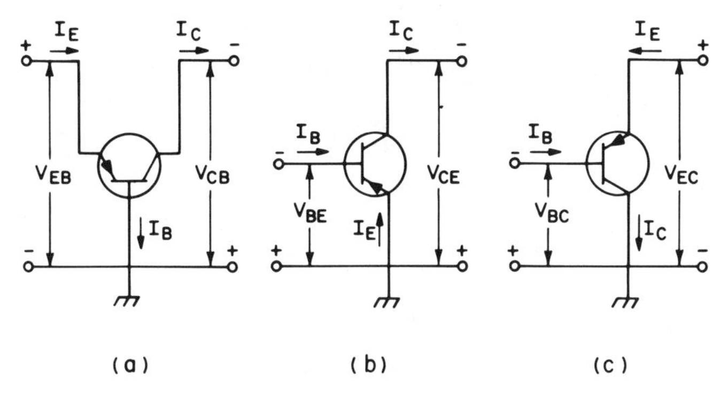

Fig. 2 Three configurations of *p-n-p* transistors: (*a*) common-base, (*b*) common-emitter, and (*c*) common-collector.

Figure 3*a* is a schematic of a *p-n-p* transistor connected as an amplifier with common-base configuration. Figure 3*b* shows a schematic doping profile for the transistor with regions of uniform impurity density, and Fig. 3*c* shows the corresponding band diagram under normal operating conditions.

The static characteristics can be readily derived from the *p-n* junction theory discussed in Chapter 2. To illustrate the major properties of a transistor, we assume that the current–voltage relationship of the emitter and collector junctions is given by the ideal diode equation,[2] that is, the effects due to surface recombination–generation, series resistance, and high-level injection are neglected. These effects will be considered later.

As in Fig. 3*b*, where all the potential drops occur across the junction depletion region, the continuity and current density equations govern the steady-state characteristics. For the neutral base region, these equations are given by

$$0 = -\frac{p - p_B}{\tau_B} + D_B \frac{\partial^2 p}{\partial x^2} \tag{1}$$

$$J_p = -qD_B \frac{\partial p}{\partial x} \tag{2a}$$

$$J_n = J_{\text{tot}} + qD_B \frac{\partial p}{\partial x} \tag{2b}$$

where p_B is the equilibrium minority-carrier density in the base, J_{tot} is the total conduction current density, and τ_B and D_B are the minority-carrier lifetime and diffusion coefficients, respectively. The conditions at the emitter depletion-layer edges for the excess carrier concentrations are

$$\begin{aligned} p'(0) &\equiv p(0) - p_B = p_B \left[\exp\left(\frac{qV_{EB}}{kT}\right) - 1\right] \\ n'(-x_E) &= n(-x_E) - n_E = n_E \left[\exp\left(\frac{qV_{EB}}{kT}\right) - 1\right] \end{aligned} \tag{3}$$

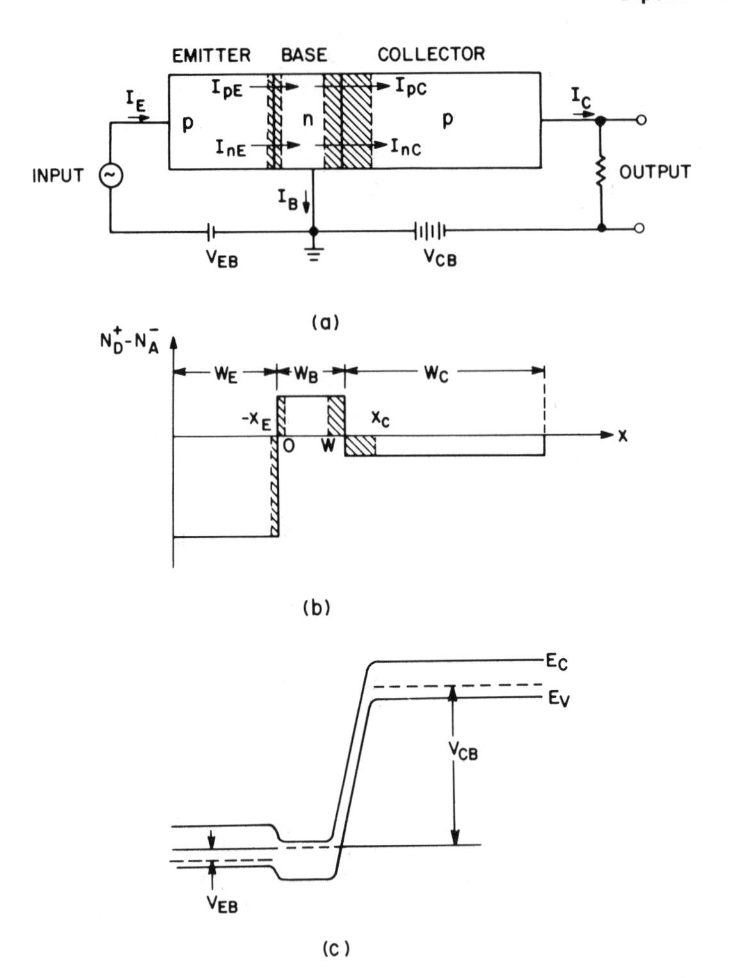

Fig. 3 (a) A *p-n-p* transistor connected in common-base configuration for amplifier application. (*b*) Doping profiles of a transistor with abrupt impurity distributions. (c) Energy-band diagram under normal operating conditions.

where n_E is the equilibrium minority-carrier density (electrons) in the emitter. A similar set of equations can be written for the collector junction:

$$
\begin{aligned}
p'(W) &= p(W) - p_B = p_B \left[\exp\left(\frac{qV_{CB}}{kT}\right) - 1\right] \\
n'(x_C) &= n(x_C) - n_C = n_C \left[\exp\left(\frac{qV_{CB}}{kT}\right) - 1\right].
\end{aligned}
\tag{4}
$$

The solutions for the minority-carrier distributions, that is, the hole distribution in the base from Eq. 1 and electron distributions in the emitter and collector, are given by

$$p(x) = p_B + \left[\frac{p'(W) - p'(0)e^{-W/L_B}}{2\sinh(W/L_B)}\right] e^{x/L_B} - \left[\frac{p'(W) - p'(0)e^{W/L_B}}{2\sinh(W/L_B)}\right] e^{-x/L_B} \tag{5}$$

$$n(x) = n_E + n'(-x_E)\exp\left[\frac{(x + x_E)}{L_E}\right], \qquad x < -x_E \tag{6}$$

$$n(x) = n_C + n'(x_C)\exp\left[-\frac{(x - x_C)}{L_C}\right], \qquad x > x_C \tag{7}$$

where $L_B = \sqrt{\tau_B D_B}$ is the diffusion length of holes in the base, and L_E and L_C are the diffusion lengths in the emitter and collector, respectively. Equation 5 is important because it correlates the base width W to the minority-carrier distribution. If $W \to \infty$ or $W/L_B \gg 1$, Eq. 5 reduces to

$$p(x) = p_B + p(0)e^{-x/L_B} \tag{8}$$

which is identical to the case of a p-n junction. In this case, there is no communication between the emitter and collector currents, which are determined by the density gradient at $x = 0$ and $x = W$, respectively. The "transistor" action is thus lost. From Eqs. 2 and 3 we can obtain the total dc emitter current as a function of the applied voltages:

$$\begin{aligned}
I_E &= AJ_p(x = 0) + AJ_n(x = -x_E) \\
&= A\left(-qD_B \left.\frac{\partial p}{\partial x}\right|_{x=0}\right) + A\left(-qD_E \left.\frac{\partial n}{\partial x}\right|_{x=-x_E}\right) \\
&= Aq\frac{D_B p_B}{L_B}\coth\left(\frac{W}{L_B}\right)\left[(e^{qV_{EB}/kT} - 1) - \frac{1}{\cosh(W/L_B)}(e^{qV_{CB}/kT} - 1)\right] \\
&\quad + Aq\frac{D_E n_E}{L_E}(e^{qV_{EB}/kT} - 1)
\end{aligned} \tag{9}$$

and for the total dc collector current

$$\begin{aligned}
I_C &= AJ_p(x = W) + AJ_n(x = x_C) \\
&= A\left(-qD_B \left.\frac{\partial p}{\partial x}\right|_{x=W}\right) + A\left(-qD_C \left.\frac{\partial n}{\partial x}\right|_{x=x_C}\right) \\
&= Aq\frac{D_B p_B}{L_B}\frac{1}{\sinh(W/L_B)}\left[(e^{qV_{EB}/kT} - 1) - \cosh\left(\frac{W}{L_B}\right)(e^{qV_{CB}/kT} - 1)\right] \\
&\quad - Aq\frac{D_C n_C}{L_C}(e^{qV_{CB}/kT} - 1)
\end{aligned} \tag{10}$$

where A is the cross-sectional area of the transistor. The difference between these two currents is small and appears as the base current:

$$I_B = I_E - I_C. \tag{11}$$

We shall now modify the doping distribution in the base layer of Fig. 3*b*

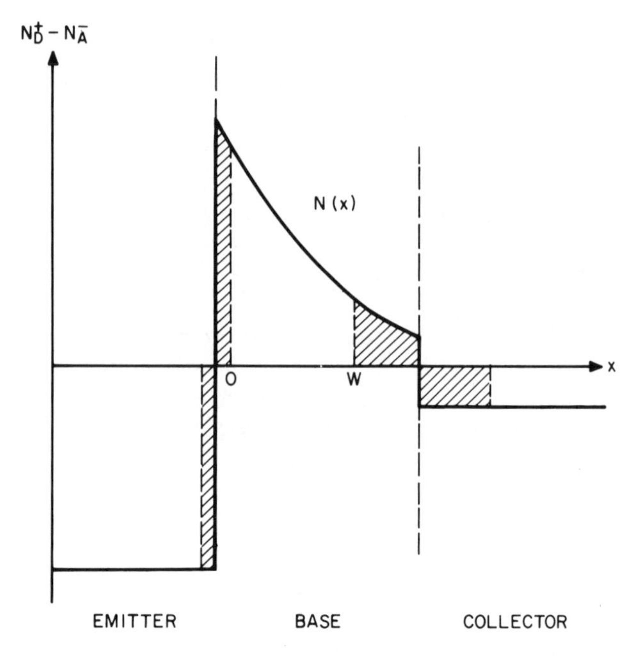

Fig. 4 Transistor doping profile with an impurity gradient in the base region. (After Moll and Ross, Ref. 19.)

and consider a more general base impurity distribution,[19] as shown in Fig. 4. A transistor with such doping distribution is called a drift transistor, since a built-in electric field enhances the hole drift in the base. The donor density N and the electron density in the base for $N \gg n_i$ are given by

$$n \approx N = n_i \exp\left[\frac{q(\psi - \phi)}{kT}\right] \tag{12}$$

where n_i is the intrinsic carrier concentration, ϕ the Fermi potential, and ψ the intrinsic Fermi potential. From Eq. 12 we obtain for the built-in field

$$\mathscr{E} \equiv -\frac{d\psi}{dx} = -\frac{kT}{q}\frac{1}{N}\frac{dN}{dx}. \tag{13}$$

The hole current density is given by

$$J_p = q\mu_B p\mathscr{E} - qD_B\frac{dp}{dx}. \tag{14}$$

Substituting Eq. 13 into 14 yields

$$J_p = -qD_B\left(\frac{p}{N}\frac{dN}{dx} + \frac{dp}{dx}\right). \tag{15}$$

The steady-state solution to Eq. 15 with the boundary condition $p = 0$ at

$x = W$ is

$$p = \frac{J_p}{qD_B} \frac{1}{N(x)} \int_x^W N(x)\, dx. \tag{16}$$

The hole concentration at $x = 0$ is given by

$$p(x=0) = \frac{J_p}{qD_B} \frac{1}{n_{BO}} \int_0^W N(x)\, dx \simeq p_{BO} \exp\left(\frac{qV_{EB}}{kT}\right) \tag{17}$$

where n_{BO} is defined as the donor concentration at $x = 0$ and p_{BO} is the equilibrium hole concentration at $x = 0$ (so that $n_{BO}p_{BO} = n_i^2$). The current $I_p = AJ_p$, where A is the area, is given by

$$I_p = \frac{qAD_B n_i^2}{\int_0^W N(x)\, dx} \exp\left(\frac{qV_{EB}}{kT}\right) = I_1 \exp\left(\frac{qV_{EB}}{kT}\right). \tag{18}$$

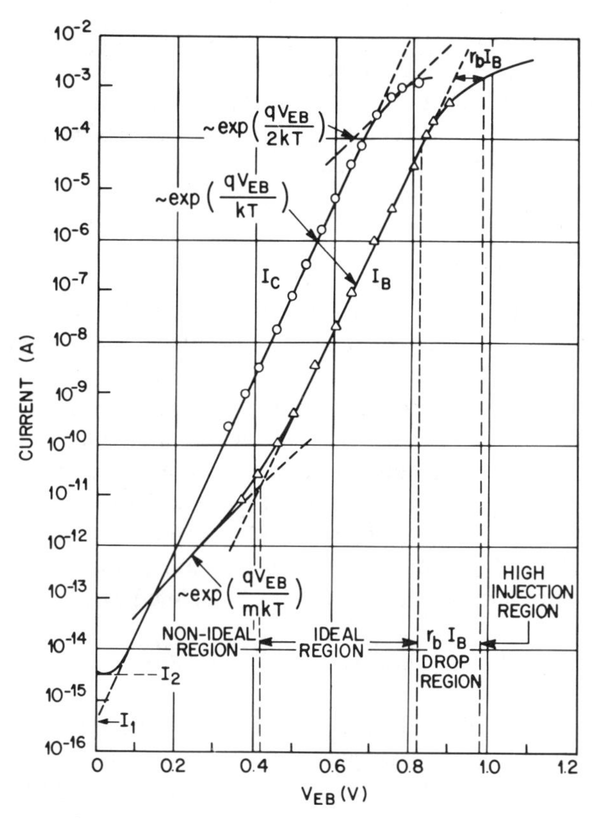

Fig. 5 Collector and base current as functions of emitter–base voltage. (After Jespers, Ref. 20.)

The total collector current is given by

$$I_C = I_1 \exp\left(\frac{qV_{EB}}{kT}\right) + I_2 \tag{19}$$

where I_2 is the saturation current. Figure 5 shows a typical experimental result.[20] Note that the exponential law of Eq. 19 is very closely obeyed over most of the current range, except at very high current densities where the injected-carrier density becomes comparable or larger than the collector doping concentration. The constant I_1 can be obtained by extrapolating the current to $V_{EB} = 0$. The number of impurities per unit area in the base (also called the Gummel number)[21] can be obtained from Eq. 18:

$$Q_b \equiv \int_0^W N(x)\,dx = \frac{q}{I_1} AD_B n_i^2. \tag{20}$$

For silicon bipolar transistors the Gummel number is about 10^{12} to $10^{13}\ \text{cm}^{-2}$.

Figure 5 also shows a typical base current characteristic. Four regions are observed: (1) the low-current nonideal region, in which the base current varies as $\exp(qV_{EB}/mkT)$ with $m \sim 2$; (2) the ideal region; (3) the moderate-injection region, characterized by significant voltage drop through the base resistance; and (4) the high-injection region. To improve the current characteristic in the low-current region, the trap densities in the depletion region and at the semiconductor surface must be reduced. Base doping profile and device configuration can be modified to minimize base resistance and high-injection effects.

3.2.2 Current Gain

When a p-n-p transistor is biased into its active region, as shown in Fig. 3a, the emitter current consists of two components—the hole component $I_{pE} = AJ_p(x = 0)$, injected into the base, and the electron component $I_{nE} = AJ_n(x = x_E)$, injected from the base into the emitter region. The collector current also consists of two components—the hole component $I_{pC} = AJ_p(x = W)$, and the electron component $I_{nC} = AJ_n(x = x_C)$. These current components are given in Eqs. 9 and 10.

The common-base current gain α_0, also referred to as h_{FB} from the four-terminal hybrid parameters (where the subscripts F and B refer to forward and common base, respectively), is defined as

$$\alpha_0 \equiv h_{FB} = \frac{\partial I_C}{\partial I_E} = \frac{\partial I_{pE}}{\partial I_E}\frac{\partial I_{pC}}{\partial I_{pE}}\frac{\partial I_C}{\partial I_{pC}}. \tag{21}$$

The first of these terms, $\partial I_{pE}/\partial I_E$, is defined as the emitter efficiency γ; $\partial I_{pC}/\partial I_{pE}$ is called the base transport factor α_T; and $\partial I_C/\partial I_{pC}$ is the collector multiplication factor M. The static common-base current gain is thus given by

$$\alpha_0 = \gamma\alpha_T M \simeq \gamma\alpha_T \tag{22}$$

since the transistor is normally operated at a collector–base bias well below the avalanche breakdown voltage. The static common-emitter current gain β_0, also referred to as h_{FE}, is defined as

$$\beta_0 \equiv h_{FE} = \frac{\partial I_C}{\partial I_B}. \tag{23}$$

From Eq. 11 note that α_0 and β_0 are related to each other by

$$\beta_0 = \frac{\alpha_0}{1-\alpha_0}. \tag{24}$$

Because the value of α_0 in well-designed bipolar transistors is close to unity, β_0 is generally much larger than 1. For example, if α_0 is 0.99, β_0 is 99; and if α_0 is 0.998, β_0 is 499.

Under normal operation of a *p-n-p* transistor, $V_{EB} > 0$ and $V_{CB} \ll 0$, so that the terms in Eqs. 9 and 10 associated with V_{CB} can be neglected. The current gain can be obtained from Eqs. 9 and 10 as

$$\gamma(\text{emitter efficiency}) = \frac{\text{incremental hole current from the emitter}}{\text{incremental total emitter current}}$$

$$= \frac{\partial AJ_p(x=0)}{\partial I_E} = \left[1 + \frac{n_E}{p_B}\frac{D_E}{D_B}\frac{L_B}{L_E}\tanh\left(\frac{W}{L_B}\right)\right]^{-1} \tag{25}$$

and

$$\alpha_T \text{ (base transport factor)}$$

$$= \frac{\text{incremental hole current reaching collector}}{\text{incremental hole current from the emitter}}$$

$$= \frac{J_p(x=W)}{J_p(x=0)} = \frac{1}{\cosh(W/L_B)} \approx 1 - \frac{W^2}{2L_B^2}. \tag{26}$$

Note that both γ and α_T are less than unity; the extent to which they depart from unity represents an electron current that must be supplied from the base contact. For bipolar transistors with base width less than one-tenth of the diffusion length, $\alpha_T > 0.995$; and the current gain is given almost entirely by the emitter efficiency. Under the condition $\alpha_T \sim 1$,

$$h_{FE} = \frac{\gamma}{1-\gamma} = \frac{p_B D_B L_E}{n_E D_E L_B}\coth\left(\frac{W}{L_B}\right)$$

$$\sim \frac{p_B}{n_E}\left(\frac{1}{W}\right) \sim \frac{N_E}{N_B W} \sim \frac{N_E}{Q_b} \tag{27}$$

where N_E and N_B are the emitter and base doping, respectively, and Q_b is the Gummel number defined in Eq. 20. Therefore, for a given N_E, the static common-emitter current gain is inversely proportional to Q_b. Figure 6 shows this relationship for ion-implanted transistors having the same

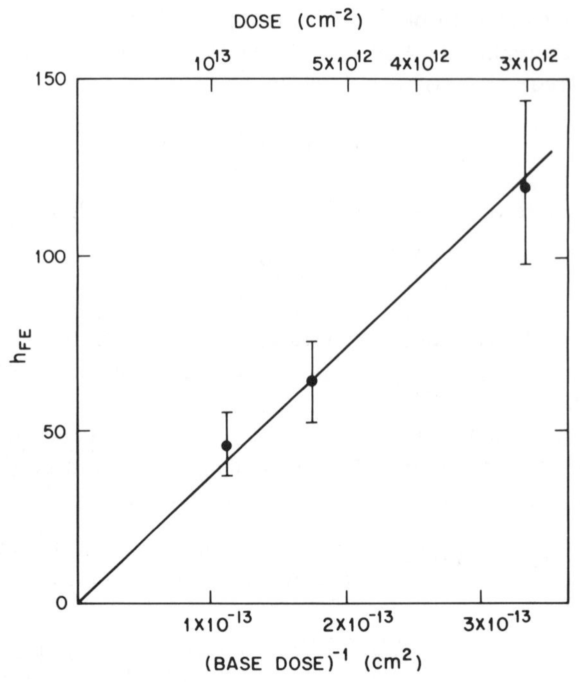

Fig. 6 Common-emitter current gain versus the inverse of the base implantation dose for 5-GHz all-implanted transistors. (After Payne et al., Ref. 22.)

emitter doping.[22] The base ion dose is directly proportional to Q_b; and we see that as the dose decreases, h_{FE} increases.

The current gain generally varies with collector current. A representative plot is shown in Fig. 7, which is obtained from Fig. 5 using Eq. 23. At very low collector current, the contribution of the recombination–generation current (also called the Sah–Noyce–Shockley current)[23] in the emitter

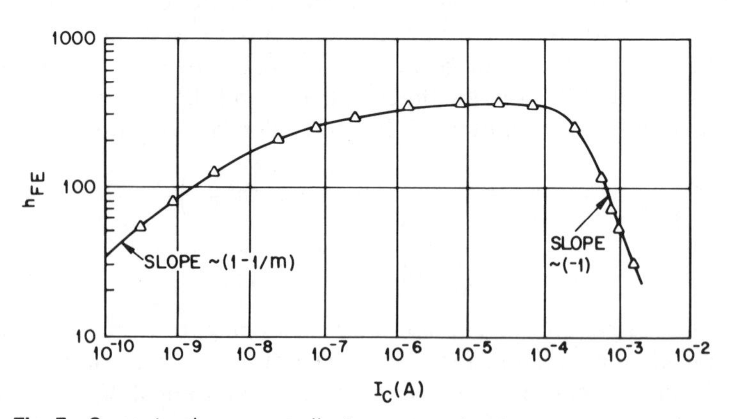

Fig. 7 Current gain versus collector current for the transistor in Fig. 5.

depletion region, and the surface leakage current may be large compared with the useful diffusion current of minority carriers across the base, so that the efficiency is low. The current gain h_{FE} increases with the collector current as follows:

$$h_{FE} = \frac{\partial I_C}{\partial I_B} \sim \frac{e^{qV_{EB}/kT}}{e^{qV_{EB}/mkT}} = \exp\left[\frac{qV_{EB}}{kT}\left(1 - \frac{1}{m}\right)\right] \sim (I_C)^{1-1/m}. \tag{28}$$

By minimizing the bulk and surface traps, h_{FE} can be improved at low-current levels.[24] As the base current reaches the ideal region, h_{FE} increases to a high plateau. For still higher collector current, the injected minority-carrier density in the base approaches the majority-carrier density there (the high-level injection condition), and the injected carriers effectively increase the base doping, which, in turn, causes the emitter efficiency to decrease. The detailed analysis can be obtained by solving the continuity equation and current equations with both diffusion and drift components. The decrease of current gain with increasing I_C is referred to as the Webster effect.[25] As shown in Fig. 7, at high-level injection h_{FE} varies as $(I_C)^{-1}$:

$$h_{FE} = \frac{\partial I_C}{\partial I_B} \sim \frac{e^{qV_{EB}/2kT}}{e^{qV_{EB}/kT}} = e^{-qV_{EB}/2kT} \sim (I_C)^{-1}. \tag{29}$$

In Eq. 27, there is another dominant factor besides the Gummel number—the emitter doping concentration N_E. To improve h_{FE}, the emitter should be much more heavily doped than the base, that is, $N_E/N_B \gg 1$. However, as the emitter doping becomes very high, we have to consider the bandgap narrowing effect and the Auger effect; both cause reductions of h_{FE}.

The bandgap narrowing in heavily doped silicon has been studied based on the stored electrostatic energy of majority–minority carrier pairs, and the bandgap reduction ΔE_g is given by[26]

$$\Delta E_g = \frac{3q^2}{16\pi\epsilon_s}\left(\frac{q^2 N_E}{\epsilon_s kT}\right)^{1/2}. \tag{30}$$

At room temperature, the bandgap narrowing follows the relationship

$$\Delta E_g = 22.5(N_E/10^{18})^{1/2} \quad \text{meV} \tag{31}$$

where the emitter doping is in cm^{-3}. Figure 8 shows the experimental data, which are in good agreement with Eq. 31.

The intrinsic carrier density in the emitter is now

$$n_{iE}^2 = N_C N_V \exp[-(E_g - \Delta E_g)/kT] = n_i^2 \exp(\Delta E_g/kT) \tag{32}$$

where N_C and N_V are the densities of states in the conduction and valence band, respectively, and n_i is the intrinsic carrier density without the bandgap narrowing effect. The minority carrier concentrations in Eq. 27

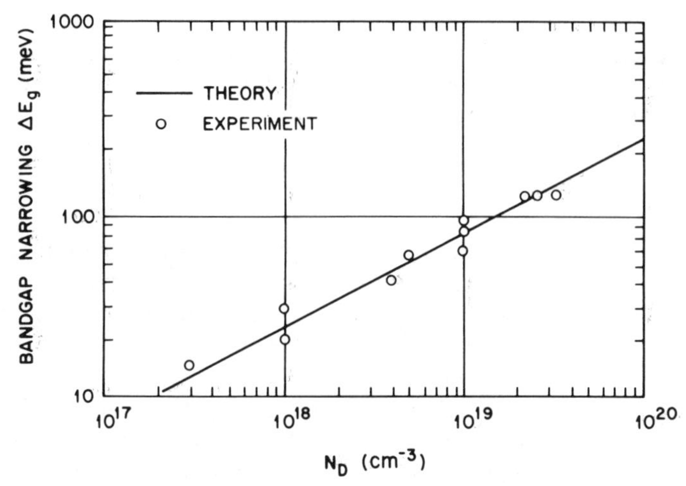

Fig. 8 Comparison between experimental data and theory for bandgap narrowing in silicon. (After Lanyon and Tuft, Ref. 26.)

can be written as

$$p_B = \frac{n_i^2}{N_B} \tag{33a}$$

and

$$n_E = \frac{n_{iE}^2}{N_E} = \frac{n_i^2}{N_E}\exp(\Delta E_g/kT). \tag{33b}$$

Therefore,

$$h_{FE} \sim \frac{p_B}{n_E} \sim \exp(-\Delta E_g/kT). \tag{34}$$

As E_g increases, the current gain decreases.

The Auger recombination is the direct recombination between an electron and a hole, accompanied by the transfer of energy to another free hole.[16] Such a process, involving two holes and one electron, occurs when electrons are injected into a heavily doped p^+ region, as in the emitter of a p^+-n-p transistor. Auger recombination is the inverse process of avalanche multiplication. The Auger lifetime τ_A is given by $1/G_p p^2$, where p is the majority carrier concentration, and G_p is the recombination rate ($1 \sim 2 \times 10^{-31}$ cm^6/s for Si at room temperature). In like manner, recombination in a heavily doped n^+ region can occur by involving two electrons and one hole with $\tau_A = 1/G_n n^2$. The electron (minority) lifetime τ in a p-type emitter is given by

$$\frac{1}{\tau} = \frac{1}{\tau_p} + \frac{1}{\tau_A} \tag{35}$$

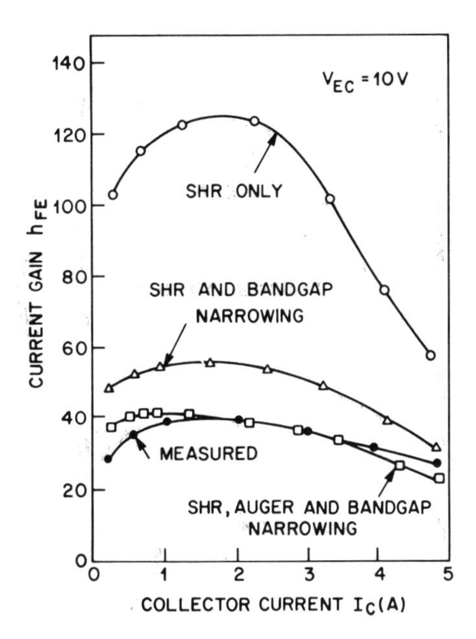

Fig. 9 Comparison of calculated current gain with measured current gain versus collector current. (After McGrath and Navon, Ref. 27.)

where τ_p is the lifetime of the Sah–Noyce–Shockley type of recombination. As the carrier concentration increases, the Auger recombination becomes dominant, causing a reduction of the emitter minority lifetime, which, in turn, reduces the emitter diffusion length L_E, causing degradation of the emitter efficiency, Eq. 25.

Figure 9 shows a two-dimensional analysis of the current gain versus the collector current. The measured results are also shown.[27] Device characteristics are generated for (1) the Shockley–Hall–Read (SHR) process only, (2) SHR and bandgap narrowing, and (3) SHR and Auger recombination and bandgap narrowing. Figure 9 clearly shows that both bandgap narrowing and Auger recombination must be considered for accurate current gain prediction. The relative importance of the foregoing three effects is a function of emitter depth and injection level.

In modern bipolar transistors with a lightly doped epitaxial collection region, the current gain is affected by the relocation of the high-field region from point A to point B under high-current condition (Fig. 10).[28] The effective base width increases from W_B to $(W_B + W_C)$. This high-field-relocation phenomenon is referred to as the Kirk effect,[29] which increases the effective base Gummel number Q_b, causing a reduction of h_{FE}. It is important to point out that under a high-injection condition where the currents are large enough to produce substantial fields in the collector region, the classic concept of well-defined transition regions at emitter–base and

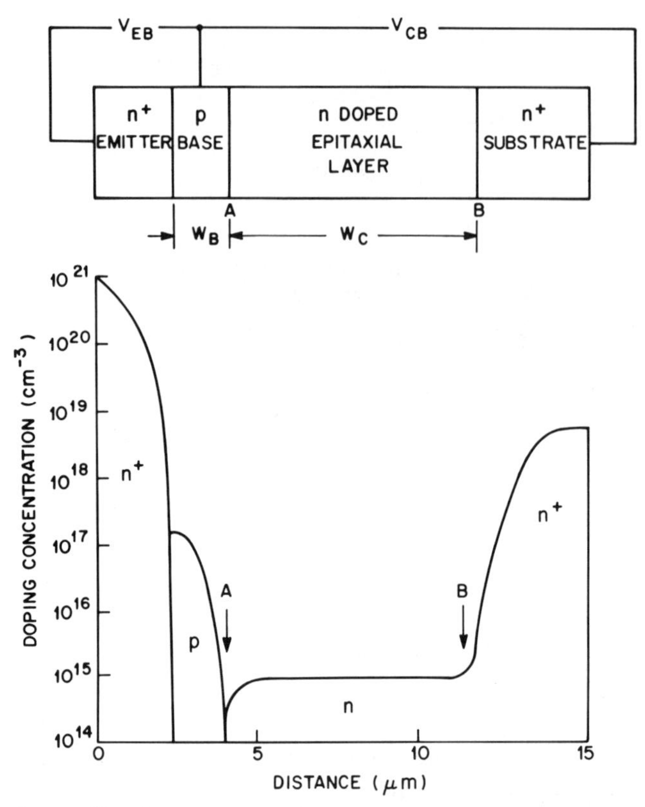

Fig. 10 Doping profile of an *n-p-n* transistor with collector epitaxial layer. (After Poon, Gummel, and Scharfetter, Ref. 28.)

base–collector junctions is no longer valid. One must solve the basic differential equations (current density, continuity, and Poisson's equations) numerically with boundary conditions applied only at the electric terminals. Figure 11 shows the computed results of the electric field distributions for $|V_{CB}| = 2$ V and various collector current densities for the doping profile of Fig. 10. Note that as the current increases, the peak electric field moves from point A to point B.

As indicated in Fig. 11, the current-induced base width W_{CIB} depends on the collector doping concentration and the collector current density. The current-induced base width is given by[16]

$$W_{CIB} = W_C \left[1 - \left(\frac{J_1 - qv_s N_C}{J_C - qv_s N_C}\right)^{1/2}\right] \tag{36}$$

and

$$J_1 \equiv qv_s \left(N_C + \frac{2\epsilon_s V_{CB}}{qW_C^2}\right)$$

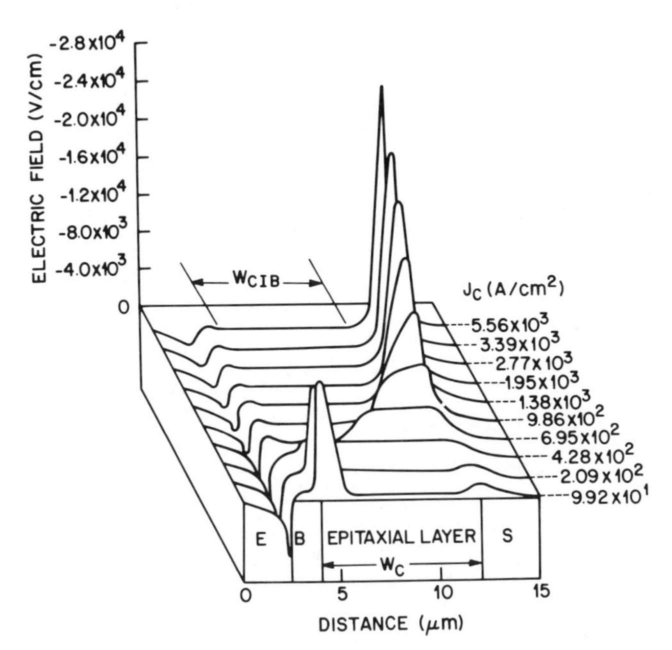

Fig. 11 Electric field distributions as a function of distance for various collector current densities. The doping profile is shown in Fig. 10. (After Poon, Gummel, and Scharfetter, Ref. 28.)

where v_s is the saturation velocity (10^7 cm/s in silicon at 300 K), N_C the epitaxial-layer doping, and V_{CB} the applied collector–base voltage. As J_C becomes larger than J_1, W_{CIB} increases; and when J_C becomes much larger than J_1, W_{CIB} approaches W_C.

3.2.3 Output Characteristics

In Section 3.2.2 we saw that the currents in the three terminals of a transistor are related by the minority-carrier distribution in the base region. For a transistor with high-emitter efficiency, the expressions for the dc emitter and collector currents, Eqs. 9 and 10, reduce to terms proportional to the minority-carrier gradient ($\partial p/\partial x$) at $x = 0$ and $x = W$, respectively. We can thus summarize the fundamental relationships of a transistor as follows:

1. The applied voltages control the boundary densities through the terms $\exp(qV/kT)$.
2. The emitter and collector currents are given by the minority (hole) density gradients at the junction boundaries, that is, $x = 0$ and $x = W$.
3. The base current is the difference between the emitter and collector currents.

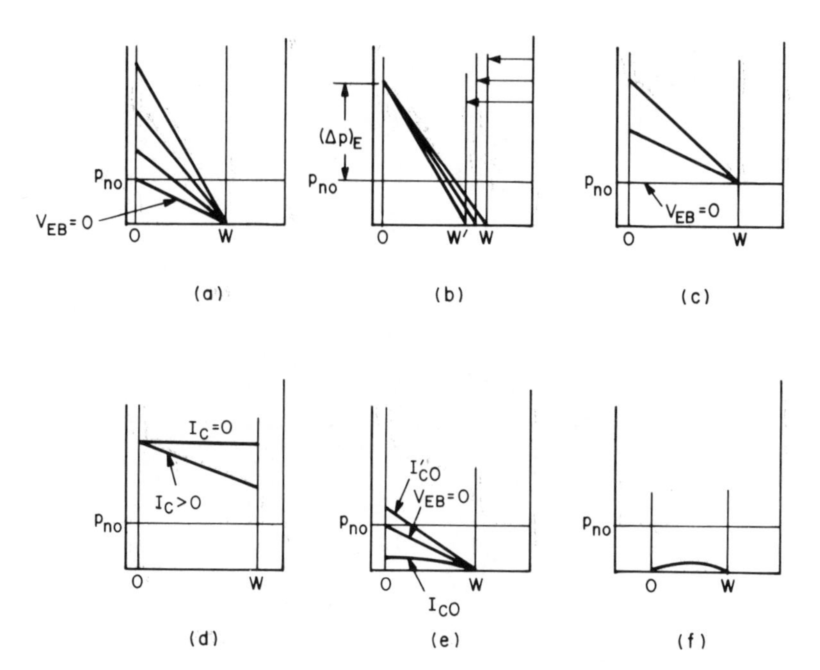

Fig. 12 Hole density in the base region of a *p-n-p* transistor for various applied voltages. (*a*) Normal polarities: V_{CB} = const., V_{EB} varying. (*b*) Normal polarities: V_{EB} = const., V_{CB} varying. (*c*) V_{EB} positive, $V_{CB} = 0$. (*d*) Both junctions are forward-biased. (e) Conditions with currents I_{CO} and I'_{CO}. (*f*) Both junctions are reverse-biased. (After Morant, Ref. 30.)

Figure 12 shows the hole distribution in the base region of a *p-n-p* transistor for various applied voltages.[30] The dc characteristics can be interpreted by means of these diagrams.

For a given transistor, the emitter current I_E and the collector current I_C are functions of the applied voltages V_{EB} and V_{CB}, that is, from Eqs. 9 and 10, $I_E = f_1(V_{EB}, V_{CB})$ and $I_C = f_2(V_{EB}, V_{CB})$.

Figure 13 shows a representative set of output characteristics for common-base and common-emitter configuration. For the common-base configuration (Fig. 13*a*), the collector current is practically equal to the emitter current ($\alpha_0 \approx 1$) and virtually independent of V_{CB}. The collector current remains practically constant, even down to zero voltage where the excess holes are still extracted by the collector, as indicated by the hole profile shown in Fig. 12*c*. To reduce the collector current to zero, a small forward voltage (~ 1 V for Si) must be applied to the collector, which sufficiently increases the hole density at W to make it equal to that of the emitter at $x = 0$ (Fig. 12*d*).

The collector saturation current I_{CO} (also denoted by I_{CBO}) is measured with the emitter open circuit. This current is considerably smaller than the ordinary reverse current of a *p-n* junction, because the presence of the

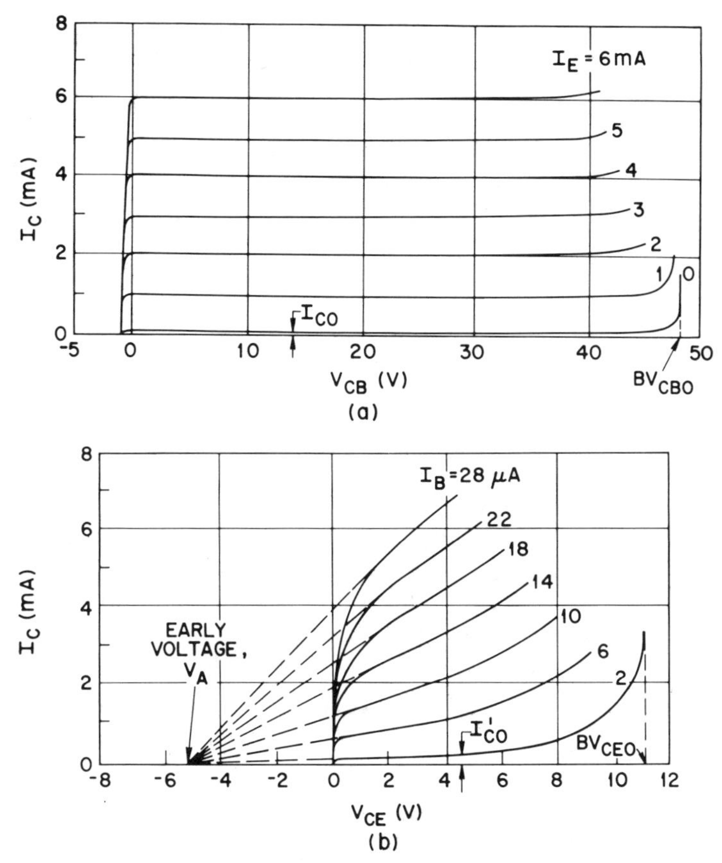

Fig. 13 Output characteristics for a *p-n-p* transistor in (*a*) common-base configuration, (*b*) common-emitter configuration. (After Morant, Ref. 30; Gummel and Poon, Ref. 35.)

emitter junction with a zero hole gradient at $x = 0$ (corresponding to zero emitter current) reduces the hole gradient at $x = W$ (Fig. 12*e*). The current I_{CO} is therefore smaller than when the emitter junction is short-circuited ($V_{EB} = 0$).

As V_{CB} increases to the value BV_{CBO}, the collector current starts to increase rapidly. Generally, this increase is due to the avalanche breakdown of the collector–base junction, and the breakdown voltage is similar to that considered in Chapter 2 for *p-n* junctions. For a very narrow base width or a base with relatively low doping, the breakdown may also be caused by the punch-through effect, that is, the neutral base width is reduced to zero at a sufficient V_{CB} and the collector depletion region is in direct contact with the emitter depletion region. At this point, the collector is effectively short-circuited to the emitter, and a large current can flow.

We now consider the output characteristics of the common-emitter configuration. Figure 13*b* shows the output (I_C versus V_{CE}) characteristics of a typical *p-n-p* transistor. Note that the current gain ($h_{FE} = \partial I_C/\partial I_B$) is considerable and the current increases with increasing V_{CE}. The saturation current I'_{CO}, which is the collector current with zero base current (base open-circuited), is much larger than I_{CO}, because from Eq. 11

$$I_B = I_E - I_C = I_E - (I_{CO} + \alpha_0 I_E). \tag{37}$$

Therefore,

$$I'_E(I_B = 0) = \frac{I_{CO}}{1-\alpha_0}. \tag{38}$$

Since the emitter and collector currents are equal in this condition (Fig. 12*e*), $I'_{CO} = I'_E$ and

$$I_{CEO} = I'_{CO} = \frac{I_{CO}}{1-\alpha_0} \approx \beta_0 I_{CO} = \beta_0 I_{CBO}. \tag{39}$$

As V_{CE} increases, the base width W decreases, causing an increase in β_0 (Fig. 12*b*). The lack of saturation in the common-emitter output characteristic is due to the large increase of β_0 with V_{CE} and is referred to as the Early effect.[31] The voltage V_A at which the extrapolated output curves meet is called the Early voltage. For a transistor with base width W_B much larger than the depletion region in the base, the Early voltage is given by

$$V_A \simeq \frac{qN_B W_B^2}{\epsilon_s}. \tag{40}$$

For small collector–emitter voltages, the collector current falls rapidly to zero. The voltage V_{CE} is divided between the two junctions to give the emitter a smaller forward bias, and the collector a larger reverse bias. To maintain a constant base current, the potential across the emitter junction must remain essentially constant. Thus when V_{CE} is reduced below a certain value ($\sim$1 V for the silicon transistor), the collector junction will reach zero bias (Fig. 12*c*). With further reduction in V_{CE} the collector is actually forward-biased (Fig. 12*d*), and the collector current falls rapidly because of the rapid decrease of the hole gradient at $x = W$.

The breakdown voltage under the open-base condition can be obtained as follows. Let M be the multiplication factor at the collector junction and be approximated by

$$M = \frac{1}{1-(V/BV_{CBO})^n} \tag{41}$$

where BV_{CBO} is the common-base breakdown voltage, and n is a constant. When the base is open-circuited, we have $I_E = I_C = I$. The currents I_{CO} and $\alpha_0 I_E$ are multiplied by M when they flow across the collector junction (Fig. 14). We have

$$M(I_{CO} + \alpha_0 I) = I \tag{42}$$

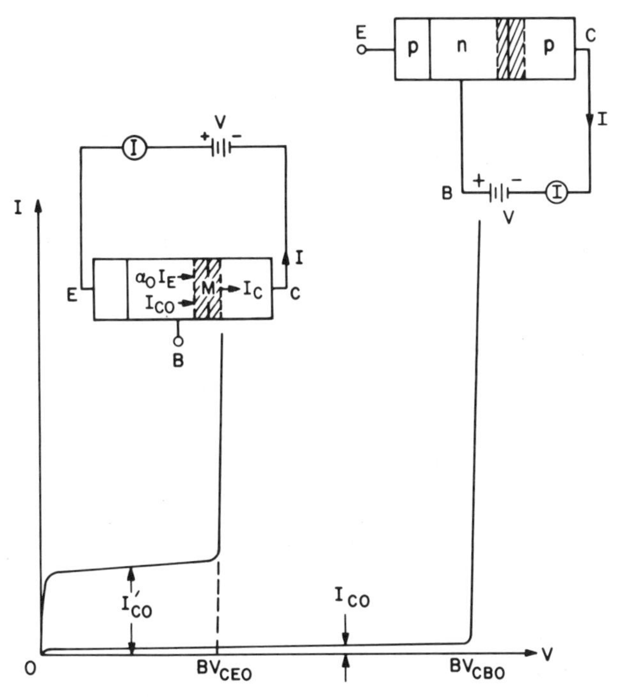

Fig. 14 Breakdown voltage BV_{CBO} and saturation current I_{CO} for common-base configuration, and corresponding qualities BV_{CEO} and I'_{CO} for common-emitter configuration. (After Gartner, Ref. 13.)

or

$$I = \frac{MI_{CO}}{1-\alpha_0 M}. \tag{43}$$

Current I will be limited only by external resistances when $\alpha_0 M = 1$. From the condition $\alpha_0 M = 1$ and Eq. 41, the breakdown voltage BV_{CEO} for the common-emitter configuration is given by

$$BV_{CEO} = BV_{CBO}(1-\alpha_0)^{1/n}. \tag{44}$$

For $\alpha_0 \approx 1$, the value of BV_{CEO} is much smaller than BV_{CBO}.

3.2.4 Device Modeling

Ebers–Moll Model[32] Device modeling aims at relating physical device parameters to device terminal characteristics. Device modeling is especially important for integrated circuits, since simple and accurate device models are needed to predict the performance of a circuit. Generally, by making a model more accurate we make it more complex. Therefore, there is a trade-off between accuracy and complexity.[33,34]

The basic model for the bipolar transistor is the Ebers–Moll model,

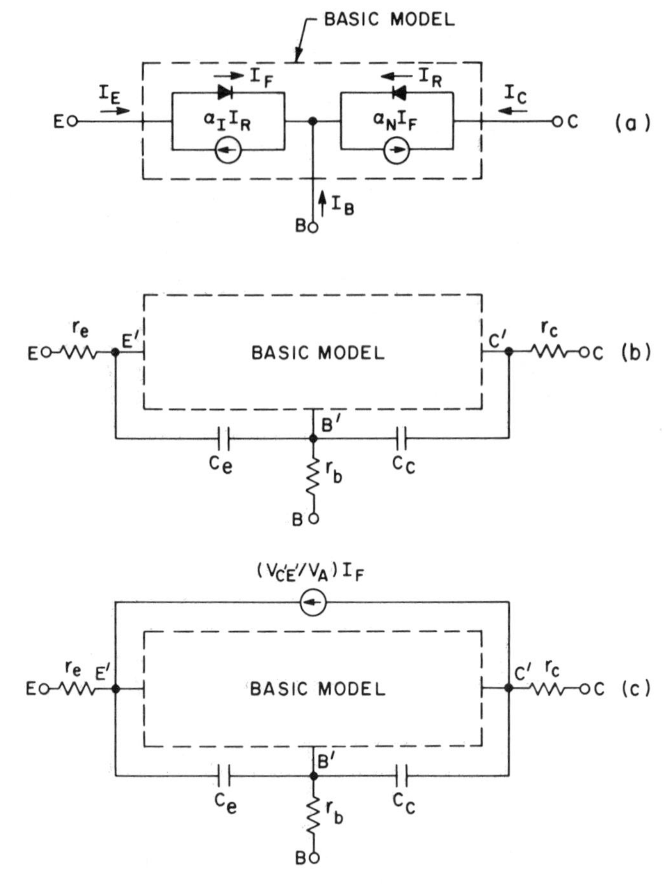

Fig. 15 Circuit diagram of Ebers–Moll model. (*a*) Basic model. (*b*) Model with additional series resistances and depletion capacitances. (*c*) Model with additional current source for Early effect. (After Ebers and Moll, Ref. 32.)

which has two diodes connected back to back and two current sources (Fig. 15*a*). The current sources are driven by the diode currents, which are assumed to have ideal characteristics,

$$I_F = I_{FO}(e^{qV_{EB}/kT} - 1) \tag{45a}$$

$$I_R = I_{RO}(e^{qV_{CB}/kT} - 1) \tag{45b}$$

where I_{FO} and I_{RO} are the saturation currents of the normally forward- and reverse-biased diodes, respectively. The terminal currents are

$$I_E = I_F - \alpha_I I_R \tag{46a}$$

$$I_C = I_R - \alpha_N I_F \tag{46b}$$

$$I_B = -(1-\alpha_N)I_F - (1-\alpha_I)I_R \tag{46c}$$

where α_N and α_I are the forward and reverse common-base current gains, respectively. The equations above give relations between the terminal currents I_E and I_C, and the terminal voltages V_{EB} and V_{CB}. The basic model has four parameters: I_{RO}, I_{FO}, α_N, and α_I.

Referring to Eqs. 9 and 10 derived previously, we can write the following general expressions for the emitter and collector currents:

$$I_E = a_{11}(e^{qV_{EB}/kT} - 1) + a_{12}(e^{qV_{CB}/kT} - 1) \tag{47a}$$

$$I_C = a_{21}(e^{qV_{EB}/kT} - 1) + a_{22}(e^{qV_{CB}/kT} - 1)\,. \tag{47b}$$

Comparing Eq. 47 to 46 gives

$$\begin{aligned} a_{11} &= I_{FO} \\ a_{12} &= -\alpha_I I_{RO} \\ a_{21} &= -\alpha_N I_{FO} \\ a_{22} &= I_{RO}. \end{aligned} \tag{48}$$

From the reciprocity characteristic of the two-port device, $a_{12} = a_{21}$, so that $\alpha_I I_{RO} = \alpha_N I_{FO}$. Therefore, only three parameters are required for the basic model.

To improve the accuracy of the model, series resistances and depletion capacitances are added to the basic model[33] (Fig. 15*b*). Note that the diodes are controlled by the internal junction voltages $V_{E'B'}$ and $V_{C'B'}$, and no longer by the externally applied voltages. To include the Early effect in the model, an extra current source between the internal emitter and collector terminals can be added (see Fig. 15*c*, where V_A is the Early voltage). By now the model parameters have increased from 3 to 9. Additional parameters can be added to the basic model to account for variations of α_N and α_I with current density and operating frequency, and a diode can be added to the base lead to account for the two-dimensional current crowding effect along the base–emitter junction (this effect will be considered in Section 3.4). As can be seen, to make a device model more accurate, we have to use more parameters and the device model becomes more complex.

Gummel–Poon Model[35] The Gummel–Poon model is based on an integral charge relation that relates electrical terminal characteristics to the base charge. This model is very accurate, taking many physical effects into account, but many parameters are required for its characterization; up to 25 parameters are needed to cover a wide range of operations. Simplified versions of the Gummel–Poon model have been derived, which finally lead back to the basic Ebers–Moll model.

To obtain the integral charge relation, first consider the current equations derived in Chapter 2:

$$J_n = q\mu_n n \frac{\partial \phi_n}{\partial x} \tag{49a}$$

$$J_p = -q\mu_p p \frac{\partial \phi_p}{\partial x} \tag{49b}$$

where

$$n = n_i \exp[q(\psi - \phi_n)/kT] \tag{50a}$$

$$p = n_i \exp[q(\phi_p - \psi)/kT]. \tag{50b}$$

The space derivative of the pn product can be written as

$$\frac{d}{dx}(pn) = \frac{q(pn)}{kT}\left(\frac{\partial \phi_p}{\partial x} - \frac{\partial \phi_n}{\partial x}\right). \tag{51}$$

Using Eq. 49 and integrating from $x = 0$ to $x = W$ (shown in Fig. 3), and assuming negligible recombination, we obtain from Eq. 51,

$$(pn)_{x=0} - (pn)_{x=W} = \frac{J_{CC}}{kT}\int_0^W \frac{n(x)}{\mu_p}\, dx \tag{52}$$

where J_{CC} is the current that would flow from emitter to collector if the transistor had unity gain. Substituting Eq. 50 into Eq. 52 yields

$$\exp[q(\phi_p - \phi_n)/kT]|_{x=0} - \exp[q(\phi_p - \phi_n)/kT]|_{x=W} = \frac{J_{CC}}{n_i^2 kT}\int_0^W \frac{n(x)}{\mu_p}\, dx. \tag{53}$$

We shall assume that the electron imref ϕ_n is constant in the base. Therefore,

$$\begin{aligned} V_{EB} &= \phi_p(0) - \phi_n(0) \\ V_{CB} &= \phi_p(W) - \phi_n(W). \end{aligned} \tag{54}$$

These voltages differ from terminal voltages by ohmic drops. Equation 53 becomes

$$I_{CC} = AJ_{CC} = (qn_iA)^2 D_B \frac{e^{qV_{EB}/kT} - e^{qV_{CB}/kT}}{qA\int_0^W n(x)\, dx} \tag{55}$$

where A is the active area. The Gummel–Poon model is based on Eq. 53, which links junction voltages, collector current, and base charge. The modeling problem reduces to modeling the base charge

$$Q_B = qA\int_0^W n(x)\, dx \tag{56}$$

which consists of five components:

$$Q_B = Q_{BO} + Q_{jE} + Q_{jC} + Q_{dE} + Q_{dC} \tag{56a}$$

where Q_{BO} is the zero-bias charge, Q_{jE} and Q_{jC} are charges associated with emitter and collector depletion capacitance, and Q_{dE} and Q_{dC} are minority-carrier charges associated with emitter and collector diffusion capacitances. As the injection level increases, the diffusion capacitances increase, thereby giving rise to the high-injection gain degradation.

We can rewrite Eq. 55 in the form

$$J_{CC} = I_F - I_R \tag{57}$$

where

$$I_F = I_S Q_{BO} \frac{e^{qV_{EB}/kT} - 1}{Q_B} \tag{58a}$$

$$I_R = I_S Q_{BO} \frac{e^{qV_{CB}/kT} - 1}{Q_B}. \tag{58b}$$

Note that Eqs. 58a and 58b resemble Eqs. 45a and 45b in the Ebers–Moll model. The changes Q_{dE} in Eq. 56 can be expressed as $B\tau_F I_F$, where τ_F is the lifetime associated with minority carriers in forward current and B is a factor that is usually equal to unity, but may become larger than unity from the Kirk effect. The charge Q_{dC} can be expressed as $\tau_R I_R$, where τ_R is the lifetime of minority carriers in reverse current.

Substituting Eqs. 58a and 58b into Eq. 56a, we obtain a quadratic equation for Q_B, with the solution

$$Q_B = \frac{Q_{BO} + Q_{jE} + Q_{jC}}{2} + \left\{ \left(\frac{Q_{BO} + Q_{jE} + Q_{jC}}{2} \right)^2 + I_S Q_{BO} [B\tau_F (e^{qV_{EB}/kT} - 1) + \tau_R (e^{qV_{CB}/kT} - 1)] \right\}^{1/2}. \tag{59}$$

The base current is given by

$$I_B = dQ_B/dt + I_{\text{rec}} \tag{60}$$

where the base recombination current can be separated into two parts,

$$I_{\text{rec}} = I_{EB} + I_{CB} \tag{61}$$

and

$$I_{EB} = I_1(e^{qV_{EB}/kT} - 1) + I_2(e^{qV_{EB}/m_e kT} - 1) \tag{61a}$$

$$I_{CB} = I_3(e^{qV_{EB}/m_c kT} - 1). \tag{61b}$$

In these equations, m_e and m_c are the emitter and collector ideality factors. For ideal currents, m_e and m_c are both equal to 1; for depletion recombination–generation currents, m_e and m_c are both equal to 2. The total emitter and collector currents are now given by

$$I_E = I_{CC} + I_{BB} + \tau_F (dI_F/dt) + C_{jE}(dV_{EB}/dt) \tag{62a}$$

$$I_C = I_{CC} - I_{CB} - \tau_R (dI_R/dt) + C_{jC}(dV_{EC}/dt). \tag{62b}$$

Figure 16 shows the circuit diagram of the Gummel–Poon model, complete with series resistances. Since Q_B is voltage-dependent, the effect of high injection in the base ($\tau_F I_F$ becoming larger than Q_{BO}) is included. The current-induced base push-out (Kirk effect) is represented by the factor B,

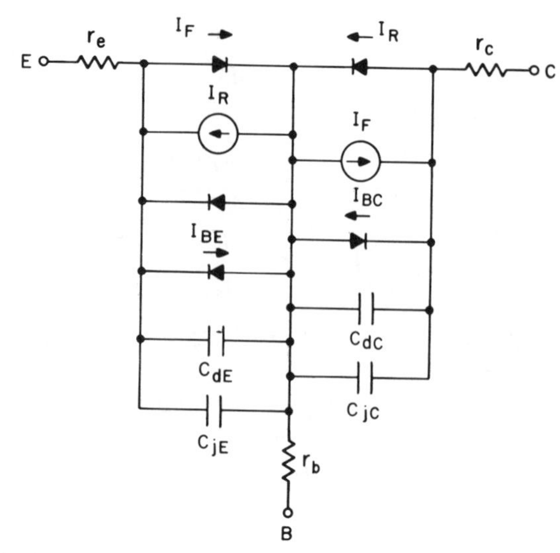

Fig. 16 Circuit diagram of Gummel–Poon model. (After Gummel and Poon, Ref. 35.)

which is a function of I_C and V_{CB}. The emitter part I_{BE} of the base current is modeled by two diodes in parallel, one ideal and one with an ideality factor $m_e > 1$. This makes the current gain at low current levels bias-dependent. The voltage dependence of Q_{jC} ($= C_{jC}V_{CB}$) models the Early effect. The Gummel–Poon model is very accurate; many physical effects have been taken into account through the bias-dependent Q_B. However, it requires as many as 25 parameters for its characterization. For a specific circuit analysis, a compromise between accuracy and model complexity must be made.

3.3 MICROWAVE TRANSISTOR

In this section we consider bipolar transistors operated in the microwave region (above 1 GHz). In the insert in Fig. 17, we show a typical microwave silicon transistor. Since the electron mobility in silicon is higher than the hole mobility, all microwave silicon transistors are of the *n-p-n* type. An epitaxial wafer of n on n^+ is used as the substrate to reduce the collector series resistance. An insulating layer (e.g., thermally grown SiO_2) is formed on the surface. The base and emitter layers are formed by diffusion or ion implantation. Typically, arsenic is used as an emitter dopant to reduce the emitter-push effect and to improve emitter efficiency.[36] Variations in horizontal geometry (such as varying the number of emitter and base stripes) yield the different current capabilities of the transistor, while changes in doping profiles result in frequency and breakdown voltage differences.

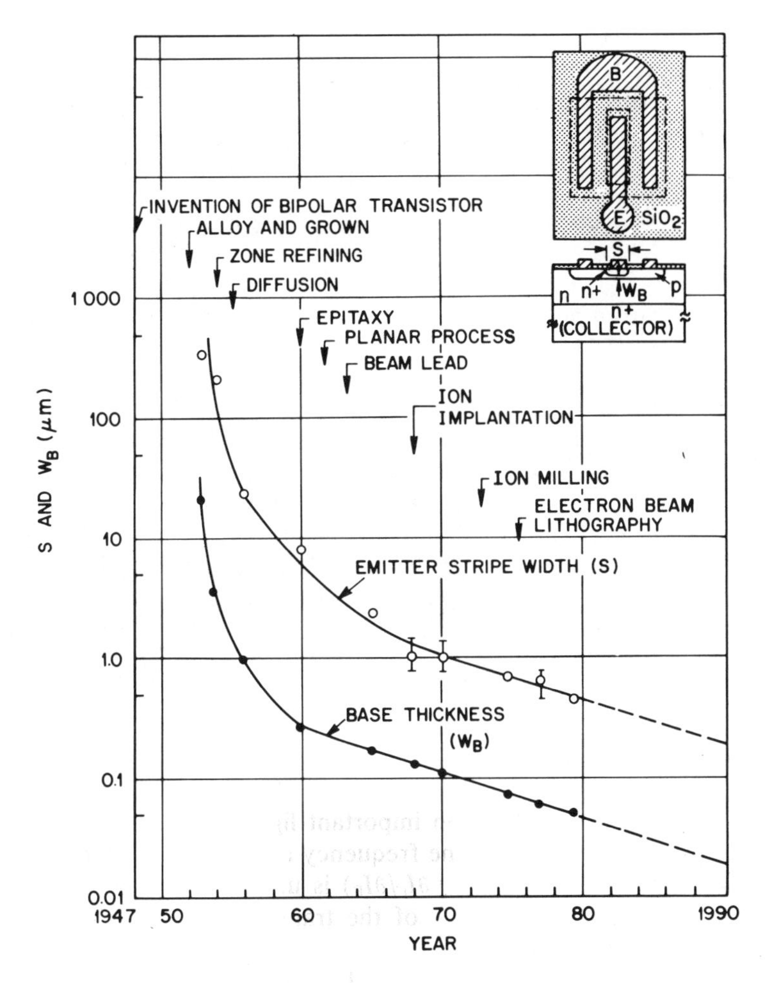

Fig. 17 Reduction of the two critical dimensions, emitter stripe width and base-layer thickness for discrete transistors, since 1952. Also indicated are major events associated with transistor development. Insert shows a stripe-geometry transistor. For integrated circuits, the critical dimensions are about a factor of 10 larger to avoid emitter–collector shorts. (After Edwards, Ref. 37; Labuda and Clemens, Ref. 11.)

A low-frequency transistor differs from a microwave transistor in the dimensions of the active areas, and in the control of the wafer and package parasitics. To achieve microwave capability, the dimensions of the active areas and the parasitics should be considerably reduced. For microwave applications, the two critical dimensions are the emitter stripe width S and the base thickness W_B. Figure 17 shows the reduction of these dimensions since 1952, and also indicates the major events associated with transistor development.[11,37] The developments of the diffusion process and the ion implantation technology are mainly responsible for reducing the vertical

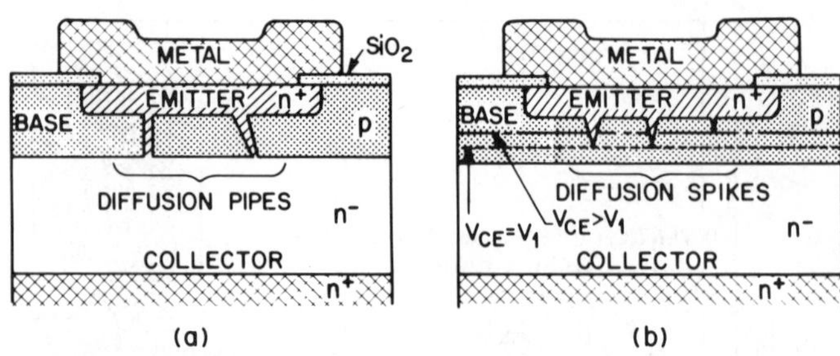

Fig. 18 (*a*) Diffusion pipes and (*b*) diffusion spikes through the base along the dislocations. (After Wang and Kakihana, Ref. 38.)

geometry, while advancement in lithographic technology helps to reduce the horizontal dimension. At the present time the emitter stripe width is in the submicron region, and the base thickness can be as small as a few hundred angstroms. As the base width decreases, it is of paramount importance to eliminate the emitter–collector shorts caused by diffusion pipes or diffusion spikes through the base along dislocations (Fig. 18).[38] One must employ processes that eliminate oxidation-induced stacking faults, epitaxial-growth-induced slip dislocations, and other process-induced defects.[39]

3.3.1 Cutoff Frequency

The cutoff frequency f_T is an important figure of merit for microwave transistors and is defined[40] as the frequency at which the common emitter, short-circuit current gain h_{fe} ($\equiv \partial I_C/\partial I_B$) is unity. The cutoff frequency is related to the physical structure of the transistor through the emitter-to-collector delay time τ_{ec} by

$$f_T = \frac{1}{2\pi\tau_{ec}}. \tag{63}$$

Delay time τ_{ec} represents the sum of four delays encountered sequentially by the carriers as they flow from the emitter to the collector:

$$\tau_{ec} = \tau_E + \tau_B + \tau_C + \tau'_C \tag{64}$$

where τ_E is the emitter depletion-layer charging time:

$$\tau_E = r_e(C_e + C_c + C_p) \approx \frac{kT}{qI_E}(C_e + C_c + C_p) \tag{65}$$

where r_e is the emitter resistance, C_e the emitter capacitance, C_c the collector capacitance, C_p any other parasitic capacitance connected to the base lead, and I_E the emitter current which is essentially equal to the collector current I_C. The expression for r_e is obtained by differentiating Eq. 9 with respect to the emitter voltage.

The second delay in Eq. 64 is the base-layer charging time, given by

$$\tau_B = \frac{W^2}{\eta D_B} \tag{66}$$

where $\eta = 2$ for the uniformly doped base layer. Equation 66 can be obtained by substituting $L_B = \sqrt{D_B\tau_B/(1 + j\omega\tau_B)}$ in Eqs. 25 and 26 to give the small-signal common-base current gain:[41]

$$\alpha \approx \frac{1}{\cosh(W\sqrt{(1 + j\omega\tau_B)/D_B\tau_B})} \approx \frac{1}{1 + jW^2\omega/2D_B}. \tag{67}$$

The charging time τ_B is defined as $1/2\pi f_\alpha$ where f_α is generally called the alpha cutoff frequency at which the gain has fallen to $1/\sqrt{2}$ of its low-frequency value. In Eq. 67 the contribution of the emitter efficiency γ to the charging time is small and is neglected. For a nonuniformly doped base, for example, the drift transistor shown in Fig. 4, the factor η in Eq. 66 should be replaced by a larger number. If the built-in field $\mathscr{E}_{bi}$ is a constant the factor η is given by[42]

$$\eta \approx 2\left[1 + \left(\frac{\mathscr{E}_{bi}}{\mathscr{E}_0}\right)^{3/2}\right] \tag{68}$$

where $\mathscr{E}_0 = 2D_B/\mu_B W$. For $\mathscr{E}_{bi}/\mathscr{E}_0 = 10$, η is about 60; thus considerable reduction in τ_B can be achieved by a large built-in field. This built-in field can be obtained automatically in a practical transistor using the base-diffusion process. A typical example is shown in Fig. 10 for a high-frequency double-diffused epitaxial n-p-n transistor.

The third delay is the collector depletion-layer transit time (Fig. 3):

$$\tau_C = \frac{x_C - W}{2v_s} \tag{69}$$

where v_s is the saturation velocity in the collector.

The fourth delay is the collector charging time:

$$\tau'_C = r_c C_c \tag{70}$$

where r_c is the collector series resistance and C_c the collector capacitance. For epitaxial transistors, r_c can be substantially reduced and the charging time τ'_C can be neglected as compared to other delay times.

The cutoff frequency f_T is given by

$$f_T = \frac{1}{2\pi\tau_{ec}} = \left\{2\pi\left[\frac{kT(C_e + C_c + C_p)}{qI_C} + \frac{W^2}{\eta D_B} + \frac{x_C - W}{2v_s}\right]\right\}^{-1}. \tag{71}$$

Clearly, from Eq. 71, to increase the cutoff frequency, the transistor should have a very narrow base thickness (one of the critical dimensions shown in Fig. 17), a narrow collector region, and should be operated at a high-current level. As the collector width decreases, however, there is a corresponding decrease in breakdown voltage. Therefore, compromises must be made for high-frequency and high-voltage operation.

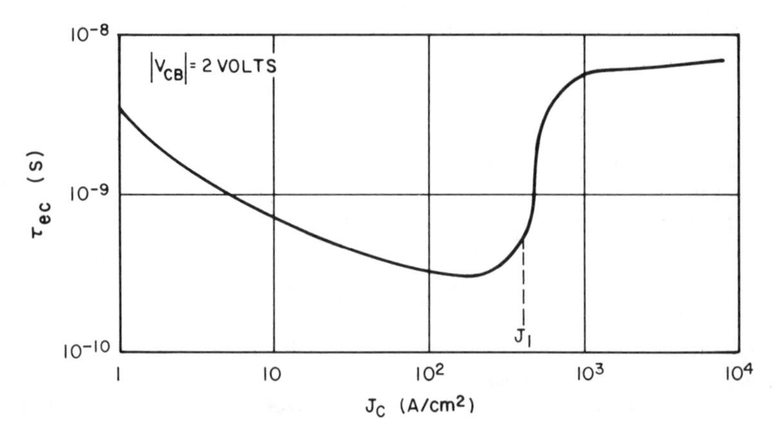

Fig. 19 Emitter–collector delay time as a function of collector current density for the device shown in Fig. 10. (After Poon, Gummel and Scharfetter, Ref. 28.)

As the operating current increases, the cutoff frequency increases because the emitter charging time τ_E is inversely proportional to the current. However, as the current becomes sufficiently high, the injected minority-carrier density is comparable to or larger than the base doping and the effective base thickness increases from W_B to $(W_B + W_C)$ as discussed in Section 3.2. Figure 19 shows the calculated emitter-to-collector delay time τ_{ec} for the transistor shown[28] in Fig. 10. At low current densities, τ_{ec} decreases with J_c as predicted by Eq. 71, and the collector current J_C is carried mainly by the drift component, so that

$$J_C \approx q\mu_C N_C \mathscr{E}_C \tag{72}$$

where μ_C, N_C, and $\mathscr{E}_C$ are the mobility, impurity doping, and electric field, respectively, in the collector epitaxial layer. As the current increases, delay time τ_{ec} reaches a minimum and increases rapidly around J_1, where J_1 is the current at which the largest uniform electric field $\mathscr{E}_C = (V_{CO} + |V_{CB}|)/W_C$ can exist where V_{CO} is the collector built-in potential and V_{CB} is the applied collector–base voltage. Beyond this point, the current cannot be carried totally by the drift component throughout the collector epitaxial region. The current J_1 is given from Eq. 72 as

$$J_1 = q\mu_C N_C (V_{CO} + |V_{CB}|)/W_C. \tag{73}$$

Because of the previously mentioned Kirk effect, there is an optimum collector current that gives the maximum cutoff frequency. It should be pointed out that as V_{CB} increases, the corresponding value of J_1 also increases.

3.3.2 Microwave Characterization

To characterize the microwave performance, scattering parameters (s parameters) are extensively used because they are easier to measure at

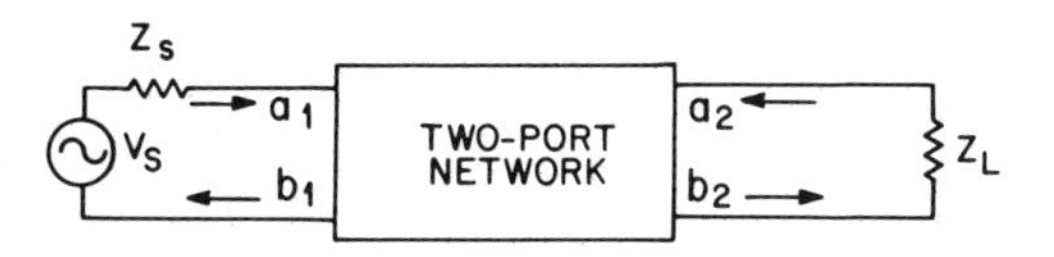

Fig. 20 Two-port network showing incident wave (a_1, a_2) and reflected waves (b_1, b_2) used in s-parameter definitions.

high frequencies than other kinds of parameters.[43,44] Figure 20 shows a general two-port network with incident (a_1, a_2) and reflected (b_1, b_2) waves to be used in s-parameter definitions. The linear equations describing the two-port network are

$$\begin{bmatrix} b_1 \\ b_2 \end{bmatrix} = \begin{bmatrix} s_{11} & s_{12} \\ s_{21} & s_{22} \end{bmatrix} \begin{bmatrix} a_1 \\ a_2 \end{bmatrix} \tag{74}$$

where the s parameters s_{11}, s_{22}, s_{12}, and s_{21} are

$s_{11} = \left.\dfrac{b_1}{a_1}\right|_{a_2=0}$ = input reflection coefficient with output terminated by a matched load ($Z_L = Z_0$ sets $a_2 = 0$, where Z_0 is the reference impedance)

$s_{22} = \left.\dfrac{b_2}{a_2}\right|_{a_1=0}$ = output reflection coefficient with input terminated by a matched load ($Z_S = Z_0$ sets $a_1 = 0$)

$s_{21} = \left.\dfrac{b_2}{a_1}\right|_{a_2=0}$ = forward-transmission gain with output terminated in a matched load

$s_{12} = \left.\dfrac{b_1}{a_2}\right|_{a_1=0}$ = reverse-transmission gain with input terminated in a matched load.

We shall define several figures of merit for microwave transistors using the s parameters. The power gain G_p is the ratio of power delivered to load to power input to the network:[44]

$$G_p(\text{power gain}) = \frac{|s_{21}|^2(1-\Gamma_L^2)}{(1-|s_{11}|^2)+\Gamma_L^2(|s_{22}|^2-D^2)-2\,\mathrm{Re}(\Gamma_L N)} \tag{75}$$

where

$$\Gamma_L = (Z_L - Z_0)/(Z_L + Z_0)$$

and

$$D = s_{11}s_{22} - s_{12}s_{21}$$

$$N = s_{22} - Ds_{11}^*.$$

In Eq. 75 Re means the real part, and the asterisk denotes the complex conjugate.

The stability factor K indicates if a transistor will oscillate upon applying a combination of passive load and source impedance with no external feedback. The factor is given by

$$K = \frac{1 + |D|^2 - |s_{11}|^2 - |s_{22}|^2}{2|s_{12}s_{21}|}. \tag{76}$$

If K is larger than 1, the device is unconditionally stable, that is, in the absence of external feedback, a passive load or source impedance will not cause oscillation. If K is less than 1, the device is potentially unstable, that is, applying a certain combination of passive load and source impedance can induce oscillation.

The maximum available gain $G_{a\max}$ is the maximum power gain that can be realized by a particular transistor without external feedback. The maximum available gain is given by the forward power gain of the transistor when the input and output are simultaneously and conjugately matched, and is defined only for an unconditionally stable transistor ($K > 1$):

$$G_{a\max} = \left|\frac{s_{21}}{s_{12}}\left(K + \sqrt{K^2 - 1}\right)\right|. \tag{77}$$

It is obvious from Eq. 77 that, when $K < 1$, the terms in parentheses become a complex number and $G_{a\max}$ is not defined.

The unilateral gain is the forward power gain in a feedback amplifier with its reverse power gain set to zero by adjusting a lossless reciprocal feedback network around the transistor. Unilateral gain is independent of header reactances and common-lead configuration. This gain is defined as

$$U = \frac{|s_{11}s_{22}s_{12}s_{21}|}{(1 - |s_{11}|^2)(1 - |s_{22}|^2)}. \tag{78}$$

We shall now combine the above two-port analysis with device internal parameters. Figure 21 shows the simplified equivalent circuits for a high-frequency bipolar transistor. The device parameters have been defined previously. The small-signal common-base current gain is α. We shall consider the stripe-base geometry shown in Fig. 17 with emitter stripe width S, length L, and spaced S from the base stripes on either side. The collector capacitance can be approximated by $C_c = C_0SL$, where C_0 is the collector capacitance per unit area. The base resistance for this geometry is approximately $r_b = r_0S/L$ with $r_0 \simeq \rho_B/W$ where ρ_B is the average resistivity of the base layer.

The small-signal common-base current gain α is defined as

$$\alpha \equiv h_{fb} = \frac{dI_C}{dI_E}. \tag{79}$$

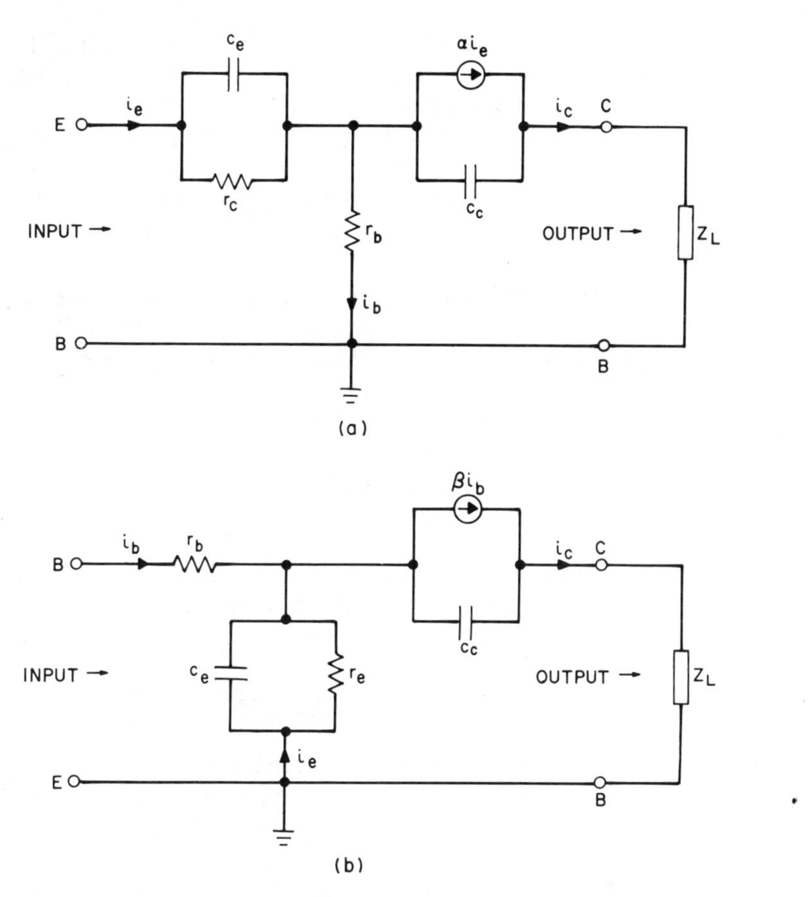

Fig. 21 Simplified microwave equivalent circuit for (*a*) common-base and (*b*) common-emitter configurations.

Similarly, the small-signal common-emitter current gain β is defined as

$$\beta = h_{fe} = \frac{dI_C}{dI_B}. \tag{80}$$

From Eqs. 21, 22, 79, and 80 we obtain

$$\begin{aligned} \alpha &= \alpha_0 + I_E \frac{\partial \alpha_0}{\partial I_E} \\ \beta &= \beta_0 + I_B \frac{\partial \beta_0}{\partial I_B} \end{aligned} \tag{81}$$

and

$$\beta = \frac{\alpha}{1-\alpha}.$$

At low-current levels, both α_0 and β_0 increase with current (Fig. 7), and α and β are larger than their corresponding static values. At high-current levels, however, the opposite is true.

For the equivalent circuit shown in Fig. 21*a*, the gain is given by[68]

$$U \equiv \frac{|\alpha(f)|^2}{8\pi f r_b C_c \left\{-\text{Im}[\alpha(f)] + \dfrac{2\pi f r_e C_c}{1 + 4\pi^2 f^2 r_e^2 C_e^2}\right\}} \tag{82}$$

where Im[$\alpha(f)$] is the imaginary part of the common-base current gain. If $\alpha(f)$ can be expressed as $\alpha_0/(1 + jf/f_T)$, and if $f < f_T$, Im[$\alpha(f)$] can be approximated by $-\alpha_0 f/f_T$ or $-\alpha_0 \omega \tau_{ec}$. The gain is then given by

$$\begin{aligned} U &\simeq \frac{\alpha_0}{16\pi^2 r_b C_c f^2 \left(\tau_{ec} + r_e C_c/\alpha_0\right)} \\ &= \frac{\alpha_0/f^2}{16\pi^2 S^2 r_0 C_0 \tau_{ec}^*} \end{aligned} \tag{83}$$

where the relationships $r_b = r_0 S/L$ and $C_c = C_0 SL$ have been used, and τ_{ec}^* is the sum of τ_{ec} and $r_e C_c/\alpha_0$. If $\alpha_0 \approx 1$ and $\tau_{ec} > r_e C_c$, Eq. 83 reduces to the simplified form

$$U = \frac{f_T}{8\pi f^2 r_b C_c} = \frac{f_T/f^2}{8\pi S^2 r_0 C_0}. \tag{84}$$

Another important figure of merit is the maximum oscillation frequency f_{max}, which is the frequency at which unilateral gain becomes unity. From Eqs. 83 and 84, the extrapolated value of f_{max} is given by

$$f_{max} \simeq \frac{1}{4\pi S}\left(\frac{\alpha_0}{r_0 C_0 \tau_{ec}^*}\right)^{1/2} \tag{85a}$$

or

$$f_{max} \simeq \frac{1}{2S}\left(\frac{f_T}{2\pi r_0 C_0}\right)^{1/2}. \tag{85b}$$

Note that both unilateral gain and maximum oscillation frequency will increase with decreasing S, which is why the emitter stripe width S is a critical dimension for microwave application.

Another important figure of merit is the noise figure, which is the ratio of total mean-square noise voltage at the output of the transistor to mean-square noise voltage at the output resulting from thermal noise in source resistance R_S. At lower frequencies the dominant noise source in a transistor is due to the surface effect that gives rise to the $1/f$ noise spectrum. At medium and high frequencies, the noise figure is given by[45]

$$NF = 1 + \frac{r_b}{R_S} + \frac{r_e}{2R_S} + \frac{(1-\alpha_0)[1 + (1-\alpha_0)^{-1}(f/f_\alpha)^2](R_S + r_b + r_e)^2}{2\alpha_0 r_e R_S}. \tag{86}$$

From Eq. 86 it can be shown that at medium frequencies where $f \ll f_\alpha$, the noise figure is essentially a constant determined by r_b, r_e, $(1-\alpha_0)$, and R_S. The optimum termination R_S can be calculated from the condition $d(NF)/dR_S = 0$. The corresponding noise figure is referred to as $NF_{\min}$. For low-noise design, a low value of $(1-\alpha_0)$, that is, a high α_0, is very important. At high frequencies beyond the "corner" frequency $f = \sqrt{1-\alpha_0} f_\alpha$, the noise figure increases approximately as f^2.

3.3.3 Device Geometry and Performance

Figure 17 shows the basic stripe-base bipolar transistor. At present all microwave bipolar transistors are planar and almost all are a silicon *n-p-n* type. The geometry falls into three general configurations (Fig. 22):[46] (1) interdigitated, (2) overlay, and (3) mesh structures. As discussed previously, because of the voltage drop along the base–emitter junction, the emitter current tends to flow near the emitter periphery. Therefore, the current-carrying capability of a bipolar transistor is proportional to the emitter periphery. The structures in Fig. 22 all have large ratios of emitter periphery to emitter area.

For all transistor structures, a final metallization process is used to make ohmic contacts to the emitter, base, and collector.

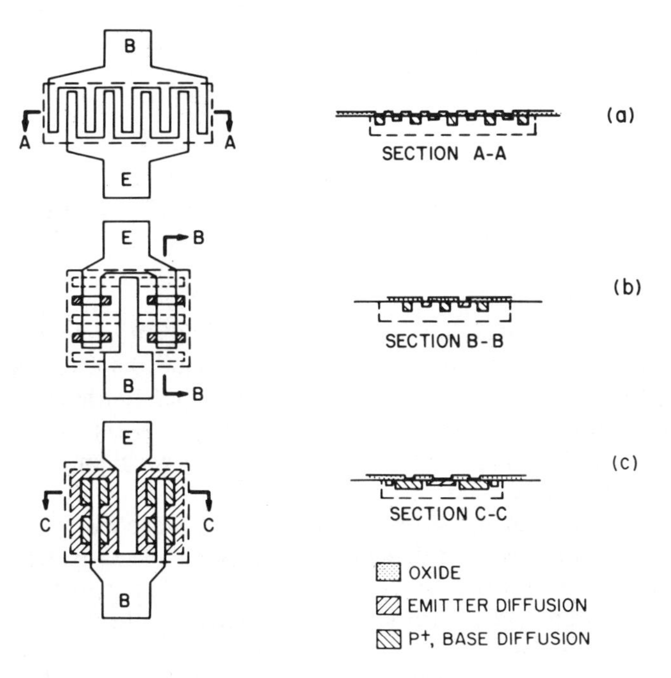

Fig. 22 Three types of microwave transistor geometry: (*a*) interdigitated, (*b*) overlay, and (*c*) mesh. (After Cooke, Ref. 46.)

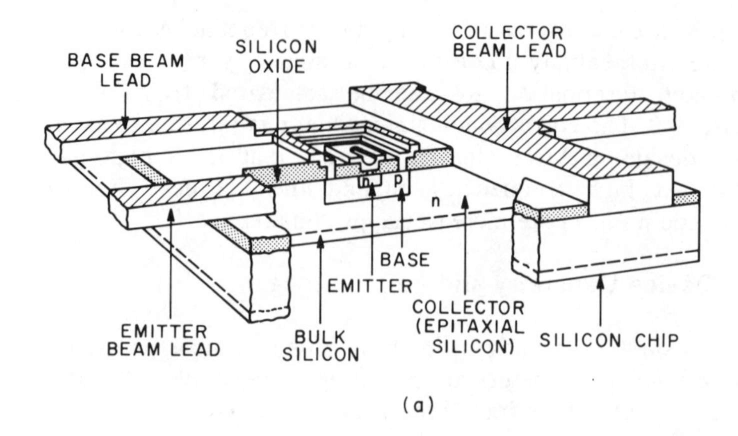

(a)

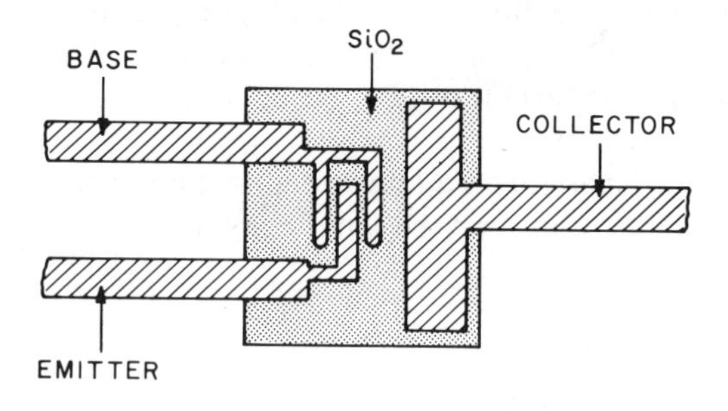

(b)

Fig. 23 Beam-lead transistor structure (*a*) schematic and (*b*) top view of an actual transistor. (After Lepselter, Ref. 9.)

An important metallization process is the use of beam-lead technology.[9,47] Figure 23 shows a silicon high-frequency beam-lead transistor with stripe-base geometry as shown in Fig. 17. Metal leads about 10 μm thick are used for structural support of the silicon chip as well as for electrical contacts. With beam-lead technology, the devices made have excellent reliability and electrical performance.

To reduce the spacing between emitter–base electrodes, a stepped-electrode transistor has been developed (Fig. 24*a*).[48] The inverse trapezoid shapes are formed by selective etches in polycrystalline silicon. The edges of the inverse trapezoids form shadows to separate the evaporated base and emitter electrodes. The spacing between the emitter diffused layer and the base contact can be made as small as 0.4 μm or less, and cutoff frequency of 8.4 GHz has been obtained.

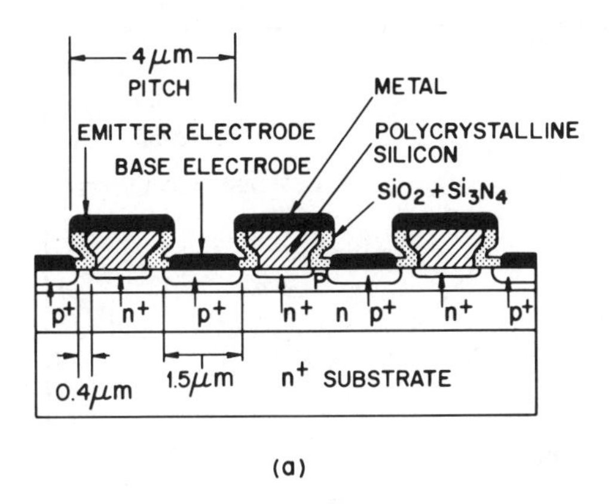

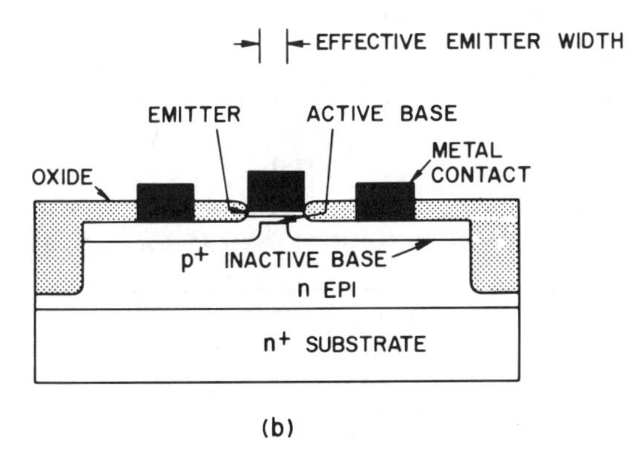

Fig. 24 (*a*) Stepped-electrode transistor structure. (*b*) Implanted-base transistor structure. (After Sakai et al., Ref. 48; Archer, Ref. 49.)

Figure 24*b* shows another improved microwave transistor structure where the active and inactive base regions are formed by separate, independently optimized processes.[49] Because the effective emitter stripe width and the base resistance are reduced, high cutoff frequency and low noise figure are obtained. Figure 25 shows the power gain and noise performance of such a transistor. These results are obtained from the s-parameter measurements. The unilateral gain U varies as f^{-2}, in agreement with Eq. 84. The extrapolated value at $U = 1$ (0 dB) gives the f_{max}, which is about 25 GHz for this microwave transistor. The maximum available gain $G_{a\max}$ and the common-emitter current gain h_{fe} also vary as f^{-2}. The cutoff frequency f_T is obtainable at $h_{fe} = 1$ (0 dB) to be above 5 GHz. The minimum noise figure increases from about 1 dB at 1 GHz to 4 dB at 8 GHz.

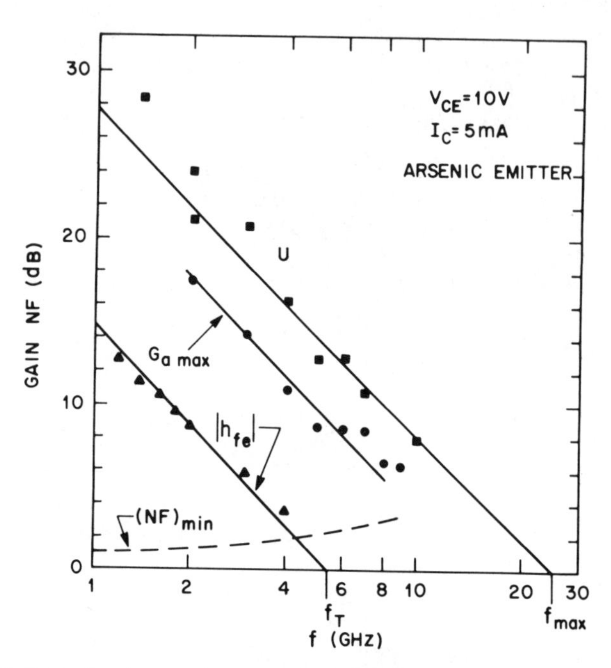

Fig. 25 Unilateral gain, maximum available gain, common-emitter current gain, and noise figure of an implanted bipolar transistor. (After Archer, Ref. 49.)

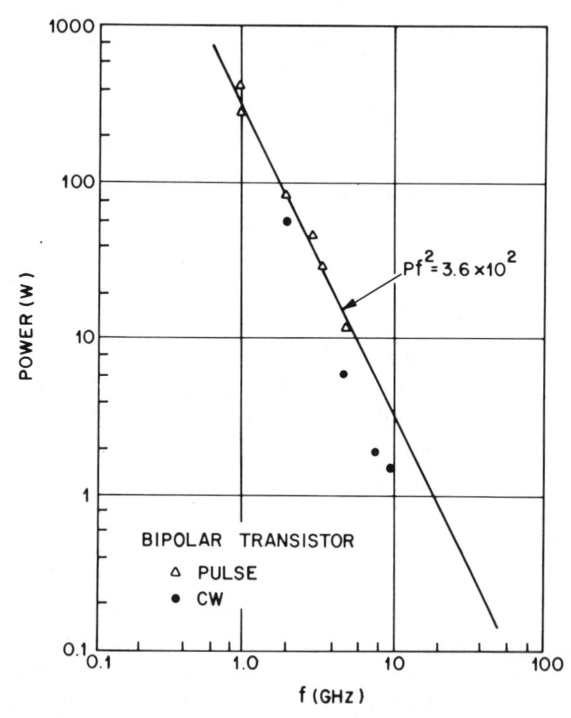

Fig. 26 Output power versus frequency for state-of-the-art microwave bipolar transistors. (After Allison, Ref. 50.)

Figure 26 shows the microwave power output as a function of frequency for the state-of-the-art bipolar transistors.[50] The power output varies as $1/f^2$ as a result of the limitations of the avalanche breakdown field and carrier saturation velocity[18] (refer to Chapter 10 on device power-frequency limitations). Under pulse condition, about 500 W can be obtained at 1 GHz. For cw operation, 60 W at 2 GHz, 6 W at 5 GHz, and 1.5 W at 10 GHz have been achieved. With the development of new fabrication and processing technologies, we can expect a threefold increase over the presently realized performance.

3.4 POWER TRANSISTOR

3.4.1 Temperature Distribution and Emitter Ballasting Resistor

Power transistors are designed for power amplification and for handling high voltages and large currents. For power transistors the main concern is with the absolute values of power and the limitation of operation imposed by second breakdown. For microwave transistors discussed in Section 3.3, the emphasis was on cutoff frequency and power gain. There is, however, no clear-cut boundary between power and microwave transistors, because the power–frequency product is mainly limited by material parameters.[18]

In a power transistor as the power increases, the junction temperature T_j increases. The maximum T_j is limited by the temperature at which the base region becomes intrinsic. Above T_j the transistor action ceases, since by then the collector is effectively short-circuited to the emitter. To improve transistor performance, the encapsulation must be improved to provide an adequate heat sink for efficient thermal dissipation.

To handle a large amount of power, the stripe width S and the base thickness W_B should be appropriately adjusted. The interdigitated and overlay structures (Fig. 22) have also been used for power transistors to handle large currents and to distribute the current more uniformly.

Figure 27*a* and *b* show the current and temperature distribution, respectively, along half the transistor emitter–base junction (emitter stripe width is 250 μm), under four different biasing conditions.[51] The total input power levels are the same (60 W). Note that at higher voltages and lower currents (curves A and B), the current is concentrated at the center of the stripe and the temperature rise at the stripe center can be very high. At lower voltages and higher currents (curve D), the current is crowded near the emitter periphery—the emitter crowding effect—but the temperature rise at the stripe center is much lower. Therefore, the high-voltage low-current condition imposes a more severe thermal problem.

A useful technique to distribute evenly the current in an interdigitated or overlay transistor is to add a distributed emitter resistance R_E so that any undesired increase in the current through a particular emitter will be

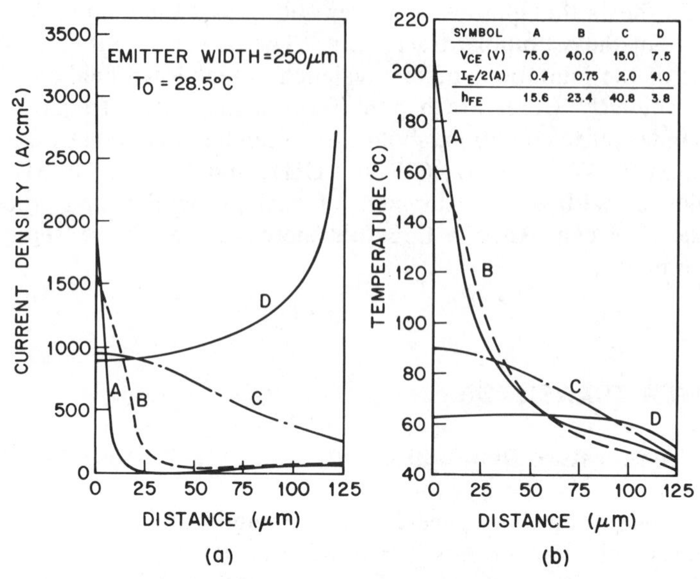

Fig. 27 (*a*) Current-density distribution and (*b*) temperature distribution along half the transistor center finger base–emitter junction. (After Gaur, Navon, and Teerlinck, Ref. 51.)

limited by the resistor. Such series resistors are referred to as stabilizing resistors or emitter ballasting resistors. The Fig. 28 insert shows a circuit configuration of the ballasting resistor, where the resistor R_E is in series with the emitter resistor r_e. The collector current is given by[52]

$$I_C = I_{SO} \exp\left\{\frac{qV_{in} - qI_C[(R_E + r_e)/\alpha + r_b(1/\alpha - 1) - E_g(T_j)]}{k(T_0 + R_{th} I_C V_C)}\right\} \tag{87}$$

where I_{SO} is a constant, independent of temperature; E_g is the bandgap at junction temperature T_j; T_0 is the ambient temperature; and R_{th} is the thermal resistance. Figure 28 shows the measured data and the calculated curves based on Eq. 87. The agreement is very good; when the ballasting resistor is small (less than 0.74 Ω), the differential resistance dV_{IN}/dI becomes negative at high current levels and current runaway occurs. For higher values of R_E, the transistor becomes unconditionally stable since the differential resistance is always positive.

3.4.2 Second Breakdown

The use of power transistors and other semiconductor devices is often limited by a phenomenon called "second breakdown" which is marked by an abrupt decrease in device voltage with a simultaneous internal constriction of current. The second breakdown phenomenon was first reported

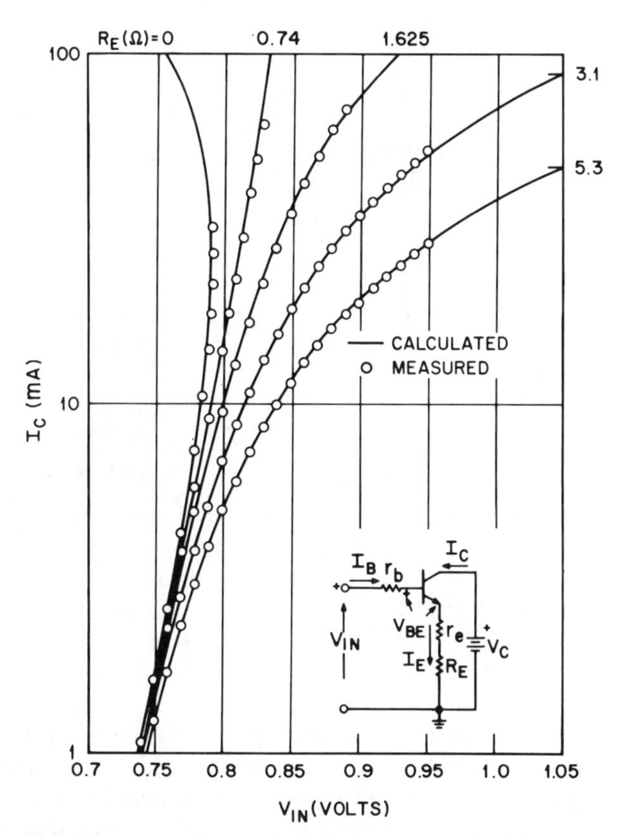

Fig. 28 Calculated and measured results for the collector current versus voltage for different emitter ballasting resistors R_E. The insert shows the circuit configuration with R_E. (After Arnold and Zoroglu, Ref. 52.)

by Thornton and Simmons,[53] and has since been under extensive study in high-power semiconductor devices.[54,55] For high-power transistors, the device must be operated within a certain safe region so that permanent damage caused by the second breakdown can be avoided.

Figure 29 shows the general features of the I_C versus V_{CE} characteristics of a transistor under second breakdown conditions.[56] The avalanche breakdown (first breakdown) occurs when the applied emitter–collector voltage reaches a value given by Eq. 44. As the voltage increases further, second breakdown occurs. The experimental results can generally be treated as consisting of four stages: the first stage leads to instability I at the breakover voltage; the second to switching from the high- to low-voltage region; the third to the low-voltage high-current range; the fourth stage to destruction (marked by D in Fig. 29).

The initiation of instability is mainly caused by the temperature effect. When a pulse with given power $P = I_C \cdot BV_{CEO}$ is applied to a transistor, a

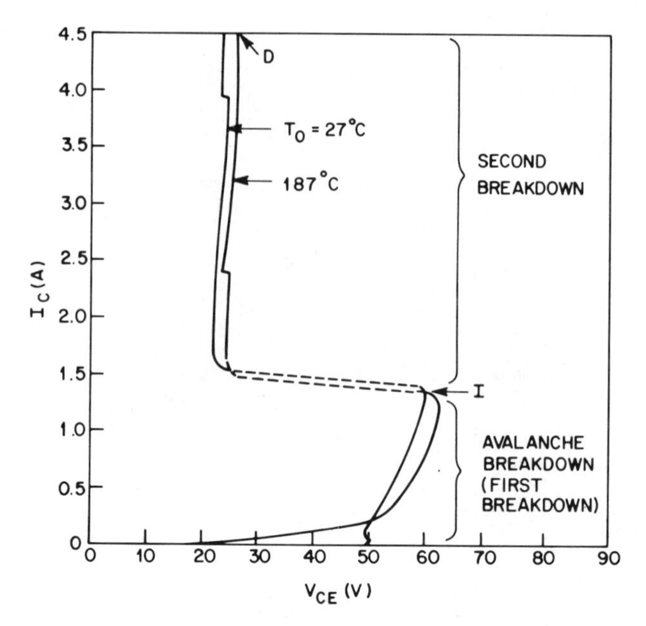

Fig. 29 Current–voltage characteristics of second breakdown at two ambient temperatures. (After Dunn and Nuttall, Ref. 56.)

time delay follows. The device is then triggered into the second breakdown condition. This time is called the triggering time. Figure 30 shows a typical plot[57] of the triggering time versus applied pulse power for various ambient temperatures. For the same triggering time τ, the triggering temperature T_{tr}, which is the temperature at the "hot" spot prior to second breakdown, is approximately related to the pulse power P at different ambient temperatures T_0 by the thermal relation

$$T_{tr} - T_0 = C_1 P \tag{88}$$

where C_1 is a constant. From Fig. 30 note that for a given ambient temperature the relationship between the pulse power and the triggering time is approximately

$$\tau \sim \exp(-C_2 P) \tag{89}$$

where C_2 is a constant.

Substituting Eq. 88 into Eq. 89 yields

$$\tau \sim \exp\left[-\frac{C_2}{C_1}(T_{tr} - T_0)\right]. \tag{90}$$

The triggering temperature T_{tr} depends on various device parameters and geometries. For most silicon diodes and transistors, T_{tr} is the temperature at which the intrinsic concentration n_i equals the collector doping concen-

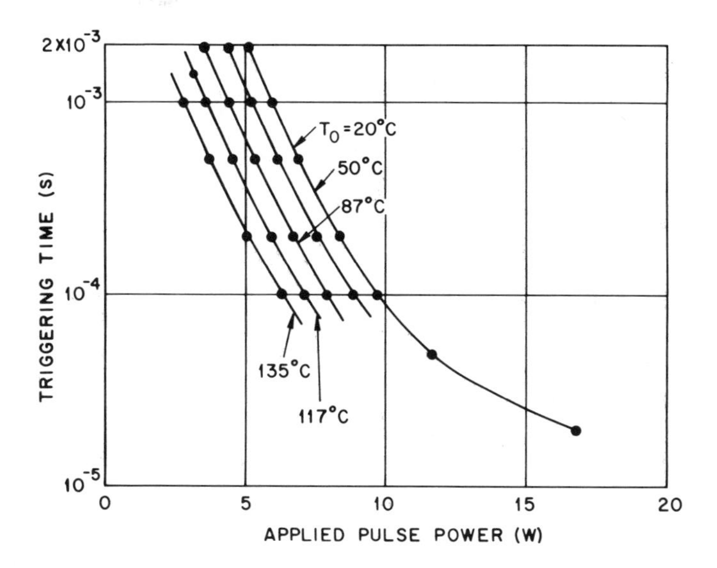

Fig. 30 Second breakdown triggering time versus applied pulse power for various ambient temperatures. (After Melchior and Strutt, Ref. 57.)

tration (see Fig. 12 in Chapter 1). The hot spot is usually located near the center of the device. For different doping concentrations the values of T_{tr} vary, and for different device geometries the value of C_2/C_1 varies resulting in a large variation of the triggering time as a result of its exponential dependence on the parameters in Eq. 90.

After instability the voltage collapses across the junction. During this second stage of the breakdown process, the resistance of the breakdown spot becomes drastically reduced. In the third low-voltage stage, the semiconductor is at high temperature and is intrinsic in the vicinity of the breakdown spot. As the current continues to increase, the breakdown spot melts, resulting in the fourth stage of destruction.

To safeguard a transistor from permanent damage the safe operating area (SOA) must be specified. Figure 31 shows a typical example for a silicon power transistor operated in the common-emitter configuration. The collector load lines for specific circuits must fall below the limits indicated by the applicable curve. The data are based upon a peak junction temperature T_j of 150°C. The dc thermal limit of the SOA is determined from the thermal resistance of the device. The thermal resistance R_{th} is given by[58]

$$R_{th} = \frac{T_j - T_0}{P} \tag{91}$$

where P is power dissipated. Therefore, the thermal limit defines the limit

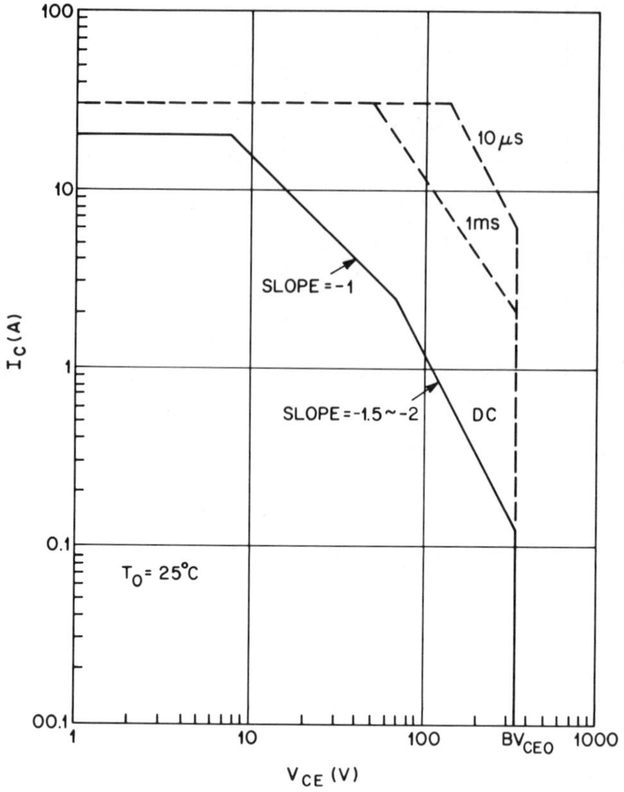

Fig. 31 Safety operating area (SOA) for transistor operation. (After Oettinger, Blackburn, and Rubin, Ref. 58.)

for the maximum allowed junction temperature:

$$R_{th}(\text{peak}) = \frac{T_{j(\max)} - T_0}{(I_C \times V_{CE})_{\text{limit}}}. \tag{92}$$

If $T_{j(\max)} = 150°\text{C}$ and R_{th}(peak) is assumed constant, then

$$(I_C \times V_{CE})_{\text{limit}} = \frac{T_{j(\max)} - T_0}{R_{th}(\text{peak})} = \text{constant}. \tag{93}$$

Thus a straight-line relation with a slope of -1 exists between $\ln I_C$ and $\ln V_{CE}$ (Fig. 31). As shown in Fig. 27, at higher voltages and lower currents the temperature rise at the stripe center can be substantial. This temperature rise is responsible for the second breakdown, and the slope is generally between -1.5 and -2. The device is eventually limited by the first breakdown voltage BV_{CEO} in the SOA, as indicated by the vertical line. For pulse operations, the SOA can be extended to higher current values. At higher ambient temperatures, the thermal limitation reduces the power that can be handled by the device, Eq. 93, and the SOA is reduced.

3.5 SWITCHING TRANSISTOR

A switching transistor is a transistor, designed to function as a switch, that can change its state, say from the high-voltage low-current (off) condition to the low-voltage high-current (on) condition, in a very short time. The basic operating conditions of switching transistors are different from those of microwave transistors, because switching is a large-signal transient process while microwave transistors are generally concerned with small-signal amplification. The basic device geometries, however, are similar to those for microwave transistors (Fig. 17).

The most important parameters for a switching transistor are current gain and switching time. To improve the current gain, usually lower doping in the base region is used. The transistor can be doped with gold to introduce midgap recombination centers and thus reduce the switching time.

A switching transistor can be operated in various switching modes. Figure 32*a* shows the two basic modes and their corresponding load lines. The modes are classified as saturated and current modes, which are determined by the portion of the transistor output characteristic curve

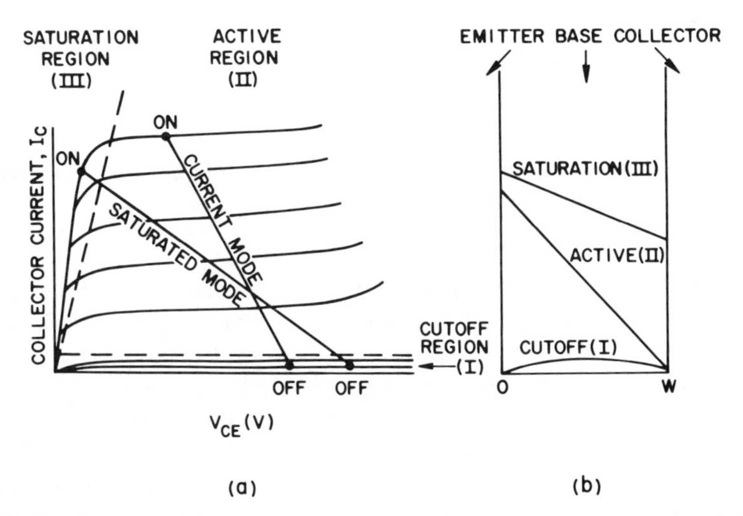

Fig. 32 (*a*) Operation regions and switching modes of a switching transistor, (*b*) Minority-carrier densities in the base for cutoff, active, and saturation regions. (After Moll, Ref. 59.)

utilized. The output characteristic curve can be divided into three regions:

Region I: cutoff region, collector current off, emitter and collector junctions reverse-biased.

Region II: active region, emitter forward-biased and collector reverse-biased.

Region III: saturation region, emitter and collector both forward-biased.

Figure 32*b* shows the corresponding minority-carrier distributions in the base for the cutoff, active, and saturation regions.[59]

For all switching modes the switch-off condition is characterized by an excursion of the load line into the cutoff region of the transistor. Therefore, the operating mode is determined primarily by the direct current level in the switch-on condition and by the location of the operating points. The most common mode of operation is the saturated mode, which most nearly duplicates the function of an ideal switch. The transistor is virtually open-circuited between the emitter and collector terminal in the off condition and short-circuited in the on condition. The current-mode operation is useful for high-speed switching, since the storage delay time associated with the excursion of the transistor into the saturation region is eliminated.

We shall now consider the switching behavior of a transistor based on the Ebers–Moll model. Referring to Eqs. 47a and 47b derived previously, the coefficient a's can be determined from the following four quantities which can be directly measured:

I_{EO}: the reverse saturation current of the emitter junction with the collector open-circuited,

$$e^{qV_{EB}/kT} \ll 1, \qquad I_C = 0.$$

I_{CO}: the reverse saturation current of the collector junction with the emitter open-circuited,

$$e^{qV_{CB}/kT} \ll 1, \qquad I_E = 0.$$

α_N: the normal current gain under the normal operating conditions where the emitter is forward-biased and the collector is reverse-biased. The collector current is given by $I_C = -\alpha_N I_E + I_{CO}$.

α_I: the inverse current gain under inverted operating conditions; that is, the emitter is reverse-biased and the collector is forward-biased. The emitter current is given by $I_E = -\alpha_I I_C + I_{EO}$. For most transistors, α_N is greater than α_I. Because the emitter area is usually smaller than the collector area, the latter is much more effective in collecting the carriers, which diffuse away from the emitter, than vice versa.

From the quantities above and from Eqs. 47a and 47b, we obtain the coefficients:

$$
\begin{aligned}
a_{11} &= -\frac{I_{EO}}{1-\alpha_N\alpha_I} \\
a_{12} &= \frac{\alpha_I I_{CO}}{1-\alpha_N\alpha_I} \\
a_{21} &= \frac{\alpha_N I_{EO}}{1-\alpha_N\alpha_I} \\
a_{22} &= -\frac{I_{CO}}{1-\alpha_N\alpha_I}.
\end{aligned}
\tag{94}
$$

In regions I and II the collector junction is reverse-biased. Equations 47a and 47b reduce to

$$I_E = -\frac{I_{EO}}{1-\alpha_N\alpha_I} e^{qV_{EB}/kT} + \frac{(1-\alpha_N)I_{EO}}{1-\alpha_N\alpha_I} \tag{95a}$$

$$I_C = \frac{\alpha_N I_{EO}}{1-\alpha_N\alpha_I} e^{qV_{EB}/kT} + \frac{(1-\alpha_I)I_{CO}}{1-\alpha_N\alpha_I}. \tag{95b}$$

The equivalent circuit corresponding to Eq. 95 is shown in Fig. 33*a*. In this circuit, base resistance r_b, emitter resistance r_e, and collector resistance r_c

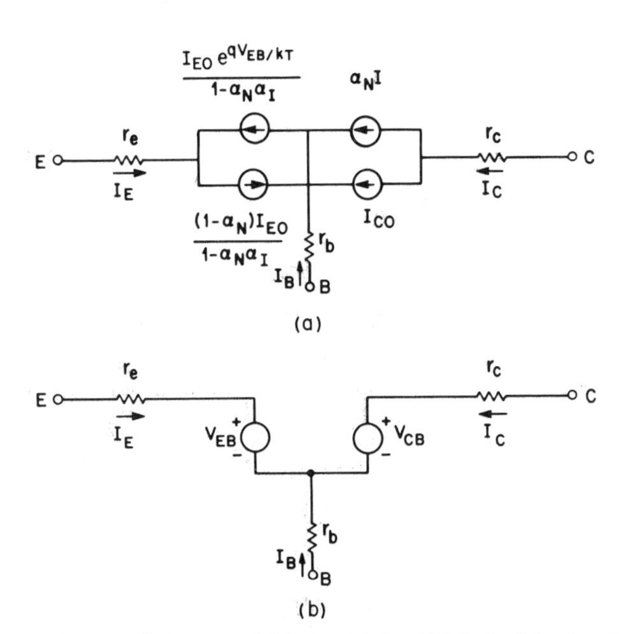

Fig. 33 Equivalent currents for switching transistors (*a*) in regions I and II and (*b*) in region III. (After Ebers and Moll, Ref. 32.)

have been added to account for the finite resistivity of the semiconductor. In region III it is most convenient to consider the currents as independent variables. From the equations above we obtain

$$V_{EB} = \frac{kT}{q} \ln\left(-\frac{I_E + \alpha_I I_C}{I_{EO}} + 1\right) \tag{96a}$$

$$V_{CB} = \frac{kT}{q} \ln\left(-\frac{I_C + \alpha_N I_E}{I_{CO}} + 1\right). \tag{96b}$$

Figure 33*b* shows the equivalent circuit for this region. Equations 95 and 96 form the basis on which a nonlinear large-signal switching problem can be analyzed.

To characterize a switching transistor, we must consider the following five quantities: current-carrying capability, maximum open-circuit voltage, off impedance, on impedance, and switching time. The current-carrying capability is determined by the allowable power dissipation and is related to the thermal limitation as it would be to a power transistor. The maximum open-circuit voltage is determined by the breakdown or punch-through voltage discussed previously. The impedance at the off or on condition can be obtained from Eqs. 95 and 96 using appropriate boundary conditions. For example, for a common-base configuration, the off and on impedances are given by

$$V_C/I_C(\text{off, region I}) = \frac{V_C(1-\alpha_N\alpha_I)}{I_{CO} - \alpha_N I_{EO}} \tag{97}$$

and

$$V_C/I_C(\text{on, region III}) = \frac{kT}{qI_C} \ln\left(-\frac{I_C + \alpha_N I_E}{I_{CO}}\right). \tag{98}$$

It is apparent from Eq. 97 that the off impedance will be high for small reverse saturation currents I_{CO} and I_{EO} of the junctions. The on impedance, Eq. 98 is approximately inversely proportional to the collector current I_C, and is very small when I_C is large. In practice the ohmic resistances Fig. 33*b* contribute to the total impedance of the transistor and must be added.

We shall now consider the switching time, which is the time required for a transistor to switch from the off to the on condition, or vice versa; where in general the turn-on time is different from the turn-off time.[59] Figure 34*a* shows a switching circuit for a transistor connected in the common-base configuration. When a pulse is applied to the emitter terminal (Fig. 34*b*), from time $t = 0$ to t_1 (Fig. 34*c*) the transistor is being "turned on" and the transient is determined by the active region parameters (region II) of the transistor. At time t_1, the operating point enters the current saturation region (region III). The period of time required for the current to reach 90% of its current saturation value ($= V_{CC}/R_L$) is called the turn-on time τ_0. At time t_2, the emitter current is reduced to zero, and the turn-off transient

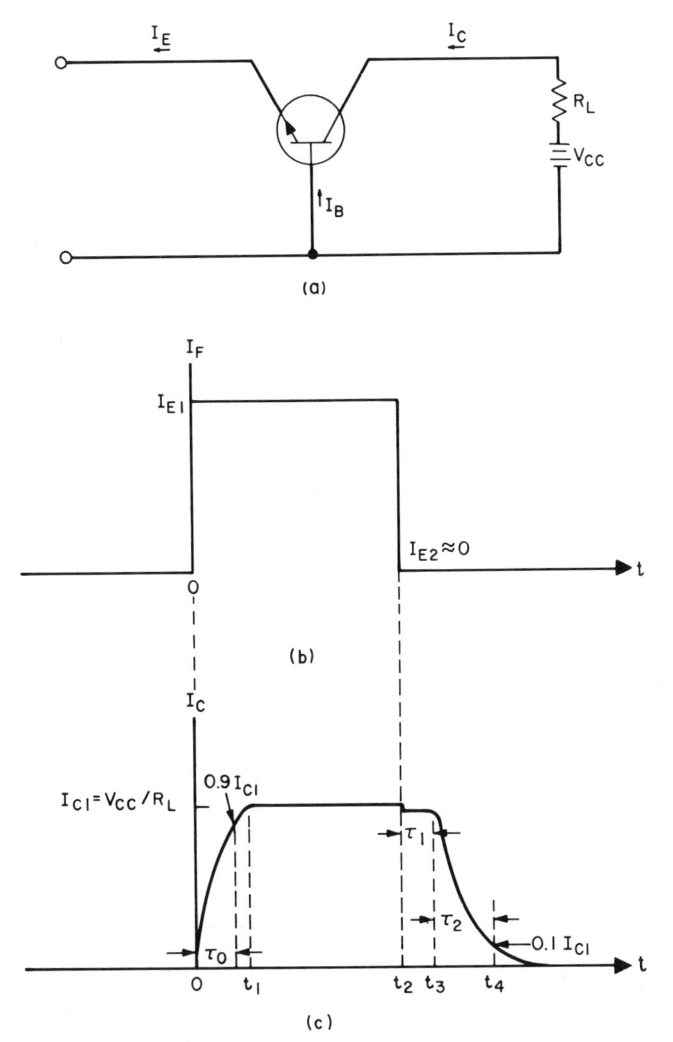

Fig. 34 (a) Switching circuit using an *n-p-n* transistor. (*b*) Input emitter current pulse. (*c*) Corresponding collector current response, where τ_0 is the turn-on time, τ_1 the storage time, and τ_2 the decay time. (After Moll, Ref. 59.)

begins. From t_2 to t_3 the minority-carrier density in the base layer is large, corresponding to the operation in region III (Fig. 32*b*), but decaying toward zero. During the time τ_1 the collector has a low impedance, and the collector current is determined by the external circuit. At t_3 the carrier density near the collector junction is nearly zero. At this point the collector junction impedance increases rapidly, and the transistor begins to operate in active region II. The time interval τ_1 is the carrier storage time. After time t_3 transient behavior is calculated from the active region parameters.

At time t_4 the collector current has decayed to 10% of its maximum value. The interval of time τ_2, from t_3 to t_4, is called the decay time.

The turn-on time τ_0 can be obtained from the transient response in the active region. For a step input function I_{E1} the Laplace transform is given by I_{E1}/s. If the common-base current gain α is expressed as $\alpha_N/(1+j\omega/\omega_N)$ where ω_N is the alpha cutoff frequency at which $\alpha/\alpha_N = 1/\sqrt{2}$, the Laplace transform of the current gain is $\alpha_N/(1+s/\omega_N)$. Thus the collector current in the Laplace transform notation is given by

$$I_C(s) = \frac{\alpha_N}{1+s/\omega_N}\frac{I_{E1}}{s}. \tag{99}$$

The inverse transform of this equation is given by

$$I_C = I_{E1}\alpha_N(1-e^{-\omega_N t}). \tag{100}$$

If we denote $I_{C1} \approx V_{CC}/R_L$ as the saturation value of the collector current, τ_0 is given from Eq. 100 by setting $I_C = 0.9I_{C1}$:

$$\tau_0 = \frac{1}{\omega_N}\ln\left(\frac{I_{E1}}{I_{E1} - 0.9I_{C1}/\alpha_N}\right). \tag{101}$$

Based on a similar approach as outlined above, the storage time and decay time for common-base configurations are given as follows.[59]

$$\tau_1 = \frac{\omega_N + \omega_I}{\omega_N\omega_I(1-\alpha_N\alpha_I)}\ln\left[\frac{I_{E1} - I_{E2}}{(I_{C1}/\alpha_N) - I_{E2}}\right] \tag{102}$$

$$\tau_2 = \frac{1}{\omega_N}\ln\left(\frac{I_{C1} - \alpha_N I_{E2}}{0.1I_{C1} - \alpha_N I_{E2}}\right) \tag{103}$$

where ω_I is the inverted alpha cutoff frequency, and I_{E1} and I_{E2} are indicated in Fig. 34. From Eq. 102 note that the storage time τ_1 becomes equal to zero if the transistor does not enter saturation region III (as in current mode), because in this case $I_{C1} = \alpha_N I_{E1}$. For the common-emitter configuration, the equations above can be used with appropriate changes of quantities: for τ_0 and τ_2, ω_N is replaced by its corresponding beta cutoff frequency $\omega_N(1-\alpha_N)$, I_{E1} and I_{E2} are replaced by I_{B1} and I_{B2}, respectively, and α_N is replaced by $\alpha_N/(1-\alpha_N)$; for τ_1 the latter two operations apply: I_{E1} and I_{E2} are replaced by I_{B1} and I_{B2}, respectively, and α_N is replaced by $\alpha_N/(1-\alpha_N)$.

It is apparent from the above equations that the switching times, that is, the turn-on time τ_0 and the turn-off time $(\tau_1+\tau_2)$, are inversely proportional to the cutoff frequencies. To increase the switching speed, one must increase the cutoff frequencies. It is important to note that the cutoff frequencies of most switching transistors are limited by the collector storage capacitance which should be reduced to increase the cutoff frequency.

3.6 RELATED DEVICE STRUCTURES

3.6.1 Scaled Device

For integrated-circuit (IC) applications, especially for VLSI (very large scale integration), bipolar transistors must be reduced in size to meet the high-speed high-density requirement. Figure 35 illustrates the reduction in bipolar transistor sizes in the past decade.[60] Note that the main difference of a transistor in IC as compared to a discrete transistor is that all electrode contacts are located on the top surface of the wafer, and each transistor must be electrically isolated to prevent interactions between devices. Prior to 1970, junction isolation was used (Fig. 35*a*), and the *p*-isolation region was always reverse-biased with respect to the *n*-type collector. In 1971, thermal oxide was used for device isolation, resulting in substantial reduction in device size (Fig. 35*b*). In 1975, the emitter was extended to the walls of the oxide for an additional 50% reduction in area (Fig. 35*c*). Presently, all the linear dimensions have been scaled down, devices have reduced by a factor of 2, resulting in a reduction of the device area by a factor of 4 and the base width by a factor of 2 (Fig. 35*d*).

Figure 36 shows a three-dimensional view of a transistor with oxide isolation. The isolation region can be merged directly into the device structure, reducing not only the silicon area but also significantly improving device performance because parasitic capacitance is reduced. With the

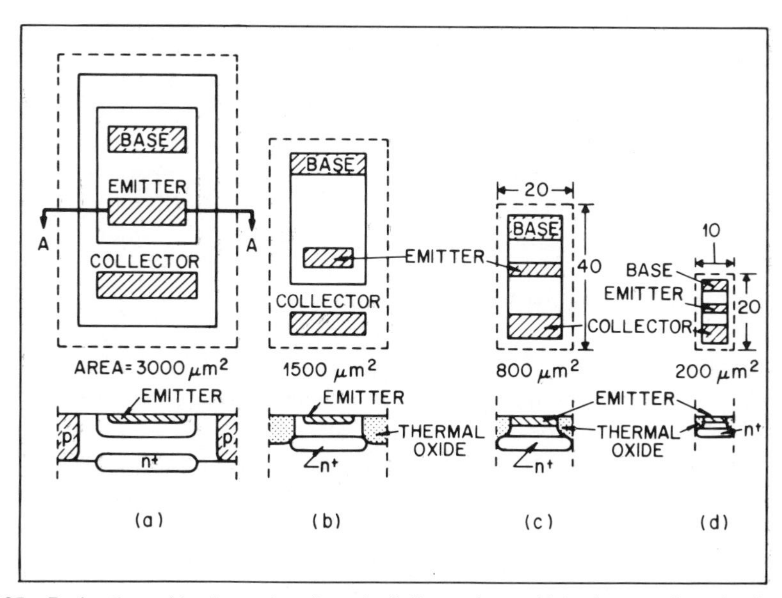

Fig. 35 Reduction of horizontal and vertical dimensions of bipolar transistor in the past decade. (*a*) Junction isolation; (*b*) Oxide isolation; (*c*) and (*d*) Scaled oxide isolation. (After Rice, Ref. 60.)

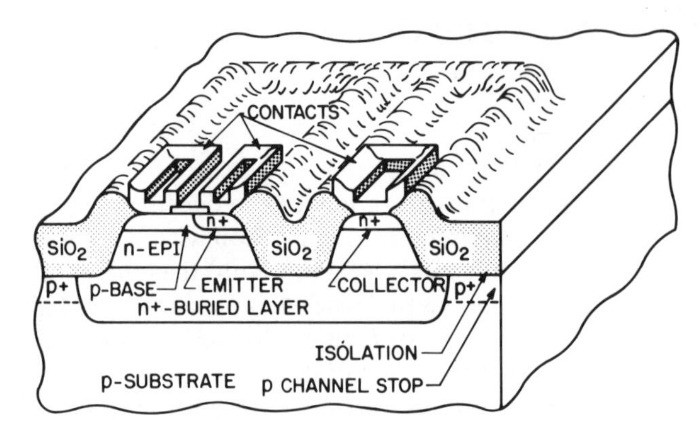

Fig. 36 Three-dimensional view of a transistor with oxide isolation. (After Labuda and Clemens, Ref. 11.)

development of ion implantation, lithographic, and dry etching technologies, we can expect that bipolar transistors with submicron dimensions and power-delay products less than 1 pJ will exist in the near future.

3.6.2 Integrated Injection Logic[61,62]

Since its introduction in 1972, integrated injection logic (I^2L) has been extensively used in IC logic and memory designs. Attractive features of I^2L include compatibility with bipolar transistor processing, ease of layout, and high packing density. Figure 37 shows the electrical schematic diagram and a cross section of the I^2L. The I^2L has a *p-n-p* lateral transistor (Q_1) and an inverted vertical *n-p-n* transistor Q_2 with multiple collector contacts. Current is injected from Q_1 into the base of Q_2. Since I^2L does not require isolation regions or resistors, its circuit density can be very high. The collectors can be replaced by Schottky diodes (refer to Chapter 5) to further improve the speed performance.[63]

3.6.3 Heterojunction Transistor[64]

Figure 38 shows the band diagram of a heterojunction transistor with a wide-gap emitter. The device has an *n*-type $Al_xGa_{1-x}As$ emitter, *p*-type GaAs base, and *n*-type GaAs collector. The potential advantages of heterojunction transistors include (1) higher emitter efficiency, since holes (minority carriers for the emitter) flowing from the base to emitter are blocked by the high barrier in the valence band; (2) decreased base resistance, since the base can be heavily doped without sacrificing emitter efficiency; (3) less emitter current crowding, because of a low voltage drop along the emitter–base junction; (4) improved frequency response because of higher current gain and lower base resistance; and (5) wider

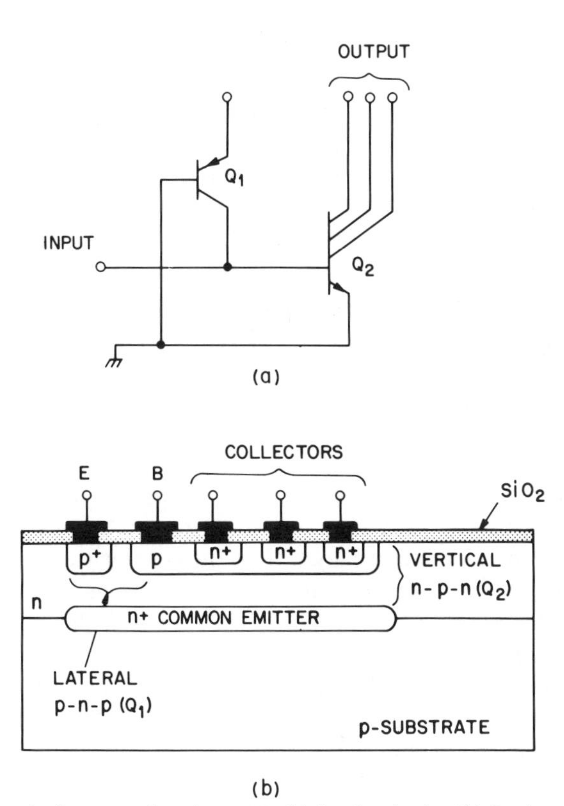

Fig. 37 (*a*) Circuit diagram of an integrated injection logic. (*b*) Device cross section of an integrated injection logic.

temperature range of operation; a heterojunction transistor can be operated at a higher temperature (~350°C) because of its larger bandgap and can be operated down to liquid-helium temperature (4 K) because of its shallow impurity levels. To date, current gain $\beta_0 \approx 350$ has been obtained.[65] However, because of technological problems, the cutoff frequency is limited to about 1 GHz.

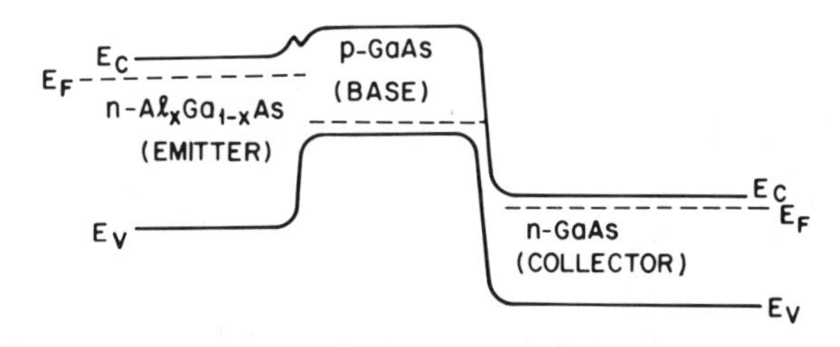

Fig. 38 Energy-band diagram of an *n-p-n* heterojunction transistor.

3.6.4 Hot-Electron Transistors

A hot electron is an electron with energy more than a few kT above the Fermi energy, where k and T are Boltzmann's constant and lattice temperature, respectively. Thus the electron is not in thermal equilibrium with the lattice. Many bipolar transistor-like three-terminal structures having hot-electron transport from emitter to collector have been proposed. The first of these structures, a metal–insulator–metal–insulator–metal structure (MIMIM) in which current flows through the insulator layer by tunneling (Fig. 39*a*) was proposed[66] in 1960. The device performance can be improved by replacing the collector insulator by a Schottky barrier (Fig. 39*b*). Using a Schottky emitter structure would improve the device even more (Fig. 39*c*). A space-charge-limited emitter structure was also proposed (Fig. 39*d*).

The main difference in these transistors is the method used to inject electrons into the base.[67] Only the metal-base transistor has the potential to give a better microwave performance than the bipolar transistor. However, experimental α_0 at room temperature is low, of the order of 0.3, which is obtained from a Si–Au–Ge transistor with 90-Å gold film.[68] By using advanced technology, such as molecular-beam epitaxy, to grow single-crystal metal film on semiconductors,[69] and operating the device at low temperatures, one can expect to obtain good microwave-frequency performance with higher current gain.

3.6.5 Permeable-Base Transistor[70]

Figure 40 shows a three-dimensional drawing of a permeable-base transistor. The device consists of a four-layer sandwich: the n^+ GaAs sub-

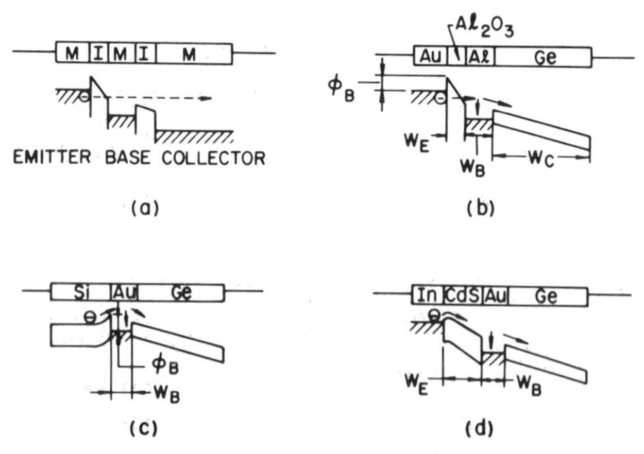

Fig. 39 Four types of hot-electron transistors. (*a*) MIMIM structure. (*b*) Tunnel transistor. (*c*) Metal-base transistor. (*d*) Space-charge-limited transistor. (After Moll, Ref. 67.)

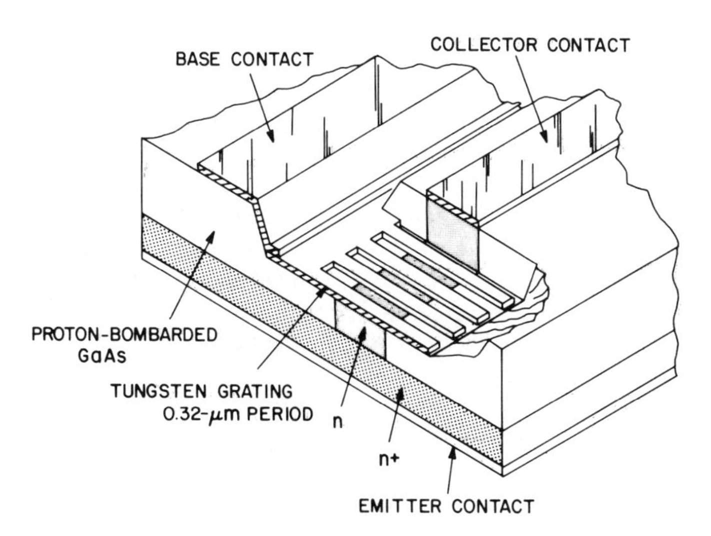

Fig. 40 Schematic diagram of a permeable-base transistor. (After Bozler et al., Ref. 70.)

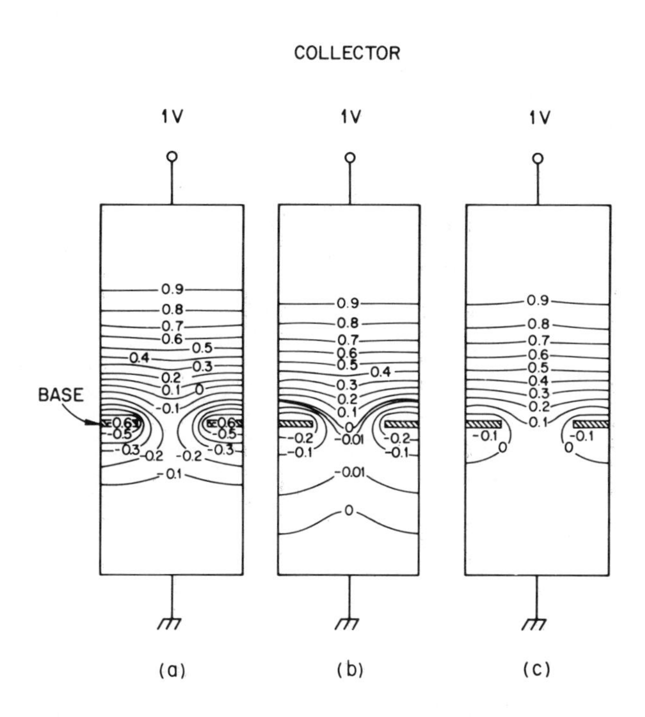

Fig. 41 Calculated equipotential lines in the unit cell of a permeable-base transistor shown in cross section for three different base-bias conditions. (*a*) $V_{BE} = 0$ V. (*b*) $V_{BE} = 0.3$ V. (*c*) $V_{BE} = 0.5$ V. (After Bozler et al., Ref. 70.)

strate, the *n*-type GaAs emitter layer, the patterned metal film (tungsten with a thickness of 200 Å and a Schottky-barrier height of 0.8 V), and the *n*-type collector layer. The slit width and the metal line width are both 1600 Å. A permeable-base transistor is made by using x-ray lithography and epitaxial overgrowth of metal structures. Figure 41 shows the calculated equipotential lines in a unit cell shown in cross section for three different bias conditions. In Fig. 41*a* one volt is applied to the collector with both emitter and base at 0 V. As an electron travels from emitter to collector, it must move into a negative potential region, which means that it must pass over an energy barrier. The barrier is lowest (0.25 V) in the center of the opening between metal regions and highest (0.6 V) at the metal–semiconductor interface. The large barrier causes the current density at the collector contact to be small ($\sim 5\ A/cm^2$). If, however, the base is forward-biased to 0.3 V (Fig. 41*b*), the barrier is lowered and a current of $900\ A/cm^2$ flows at the collector contact. With even larger base bias, the carrier accumulation develops with the current beginning to be space-charge-limited (Fig. 41*c*). The common-base current gain remains greater than 0.99 for all base bias levels below 0.4 V, because of the Schottky-barrier potential. When the base voltage reaches 0.5 V, α_0 is reduced to 0.96 and for higher base voltages, α_0 decreases rapidly. With such a fine base grating and an appropriate carrier concentration, barrier-limited current flow can exist at high current densities, resulting in a large transconductance and a large f_{max}. The barrier-limited current flow (similar to a bipolar transistor) and the similarity in structure to a metal-base transistor leads to the name "permeable-base transistor." Experimental results[70] on initial devices show an f_{max} of 17 GHz, a gain of 13 dB at 4 GHz, and a noise figure of 3.5 dB at 4 GHz. Numerical studies indicate that much higher f_{max} may be obtained by appropriate scaling of the dimensions.

REFERENCES

1 J. Bardeen and W. H. Brattain, "The Transistor, A Semiconductor Triode," *Phys. Rev.*, **74**, 230 (1948).

2 W. Shockley, "The Theory of *p-n* Junctions in Semiconductors and *p-n* Junction Transistors," *Bell Syst. Tech. J.*, **28**, 435 (1949).

3 R. N. Hall and W. C. Dunlap, "*p-n* Junctions Prepared by Impurity Diffusion," *Phys. Rev.*, **80**, 467 (1950).

4 G. K. Teal, M. Sparks, and E. Buehlor, "Growth of Germanium Single Crystals Containing *p-n* Junctions," *Phys. Rev.*, **81**, 637 (1951).

5 W. H. Pfann, "Principles of Zone-Refining," *Trans. AIME*, **194**, 747 (1952).

6 M. Tanenbaum and D. E. Thomas, "Diffused Emitter and Base Silicon Transistor," *Bell Syst. Tech. J.*, **35**, 1 (1956).

7 H. C. Theuerer, J. J. Kleimack, H. H. Loar, and H. Christenson, "Epitaxial Diffused Transistors," *Proc. IRE*, **48**, 1642 (1960).

8 J. A. Hoerni, "Planar Silicon Transistor and Diodes," IRE Electron Devices Meet., Washington, D.C., 1960.

9 M. P. Lepselter, "Beam-Lead Technology," *Bell Syst. Tech. J.*, **45**, 233 (1966).

10 W. Shockley, U.S. Patent 2,787,564 (1954). For a view on ion implantation, see J. F. Gibbons, "Ion Implantation in Semiconductors—Part I, Range Distribution Theory and Experiments," *Proc. IEEE*, **56**, 295 (1968).

11 E. F. Labuda and J. T. Clemens, "Integrated Circuit Technology," in R. E. Kirk and D. F. Othmer, Eds., *Encyclopedia of Chemical Technology*, Wiley, New York, 1980.

12 A. B. Phillips, *Transistor Engineering*, McGraw-Hill, New York, 1962.

13 W. W. Gartner, *Transistors, Principle, Design and Application*, D. Van Nostrand, Princeton, N.J., 1960.

14 SEEC (Semiconductor Electronics Education Committee), 4 vols.

(1) R. B. Adler, A. C. Smith, and R. L. Longini, *Introduction to Semiconductor Physics, SEEC*, Vol. 1, Wiley, New York, 1966.

(2) P. E. Gray, D. DeWitt, A. R. Boothroyd, and J. F. Gibbons, *Physical Electronics and Circuit Models of Transistors*, SEEC, Vol. 2, Wiley, New York, 1966.

(3) C. L. Searle, A. R. Boothroyd, E. J. Angelo, P. E. Gray, and D. O. Pederson, *Elementary Circuit Properties of Transistors*, SEEC, Vol. 3, Wiley, New York, 1966.

(4) R. D. Thornton, D. DeWitt, E. R. Chenette, and P. E. Gray, *Characteristics and Limitations of Transistors*, SEEC, Vol. 4, Wiley, New York, 1966.

15 R. L. Pritchard, *Electrical Characteristics of Transistors*, McGraw-Hill, New York, 1967.

16 S. K. Ghandhi, *Semiconductor Power Devices*, Wiley, New York, 1977.

17 R. S. Muller and T. I. Kamins, *Device Electronics for Integrated Circuits*, Wiley, New York, 1977.

18 E. O. Johnson, "Physical Limitations on Frequency and Power Parameters of Transistors," *IEEE Int. Conv. Rec.*, Pt. 5, p. 27 (1965).

19 J. L. Moll and I. M. Ross, "The Dependence of Transistor Parameters on the Distribution of Base Layer Resistivity," *Proc. IRE*, **44**, 72 (1956).

20 P. G. A. Jespers, "Measurements for Bipolar Devices," in F. Van de Wiele, W. L. Engl, and P. G. Jespers, Eds., *Process and Device Modeling for Integrated Circuit Design*, Noordhoff, Leyden, 1977.

21 H. K. Gummel, "Measurement of the Number of Impurities in the Base Layer of a Transistor," *Proc. IRE*, **49**, 834 (1961).

22 R. S. Payne, R. J. Scavuzzo, K. H. Olson, J. M. Nacci, and R. A. Moline, "Fully Ion-Implanted Bipolar Transistors," *IEEE Trans. Electron Devices*, **ED-21**, 273 (1974).

23 C. T. Sah, R. N. Noyce, and W. Shockley, "Carrier Generation and Recombination in *p-n* Junction and *p-n* Junction Characteristics," *Proc. IRE*, **45**, 1228 (1957).

24 W. M. Werner, "The Influence of Fixed Interface Charges on Current Gain Fallout of Planar *n-p-n* Transistors," *J. Electrochem. Soc.*, **123**, 540 (1976).

25 W. M. Webster, "On the Variation of Junction-Transistor Current Amplification Factor with Emitter Current," *Proc. IRE*, **42**, 914 (1954).

26 H. P. D. Lanyon and R. A. Tuft, "Bandgap Narrowing in Heavily Doped Silicon," *IEEE Tech. Dig.*, Int. Electron Device Meet., 1978, p. 316.

27 E. J. McGrath and D. H. Navon, "Factors Limiting Current Gain in Power Transistors," *IEEE Trans. Electron Devices*, **ED-24**, 1255 (1977).

28 H. C. Poon, H. K. Gummel, and D. L. Scharfetter, "High Injection in Epitaxial Transsistors," *IEEE Trans. Electron Devices*, **ED-16**, 455 (1969).

29 C. T. Kirk, "A Theory of Transistor Cutoff Frequency (f_T) Fall-Off at High Current Density," *IEEE Trans. Electron Devices*, **ED-9**, 164 (1962).

30 M. J. Morant, *Introduction to Semiconductor Devices*, Addison-Wesley, Reading, Mass., 1964.

31 J. M. Early, "Effects of Space-Charge Layer Widening in Junction Transistors," *Proc. IRE*, **40**, 1401 (1952).

32 J. J. Ebers and J. L. Moll, "Large-Signal Behavior of Junction Transistors," *Proc. IRE*, **42**, 1761 (1954).

33 H. C. deGraaff, "Review of Models for Bipolar Transistors," in Ref. 20.

34 I. E. Getreu, *Modeling the Bipolar Transistor*, Elsevier, New York, 1978.

35 H. K. Gummel and H. C. Poon, "An Integral Charge Control Model of Bipolar Transistors," *Bell Syst. Tech. J.*, **49**, 827 (1970).

36 K. Tsukamato, Y. Akasaka, Y. Watari, Y. Kusano, Y. Hirose, and G. Nakamura, "Arsenic-Implanted Emitter and Its Application to UHF Power Transistors," *Jpn. J. Appl. Phys.*, **17**, *Suppl.*, **17-1**, 187 (1977).

37 R. Edwards, "Fabrication Control Is Key to Microwave Performance," *Electronics*, **41**, 109 (Feb. 1968).

38 A. C. M. Wang and S. Kakihana, "Leakage and h_{FE} Degradation in Microwave Bipolar Transistors," *IEEE Trans. Electron Devices*, **ED-21**, 667 (1974).

39 L. C. Parrillo, R. S. Payne, T. F. Seidel, McD. Robinson, G. W. Reutlinger, D. E. Post, and R. L. Field, "The Reduction of Emitter–Collector Shorts in a High-Speed, All Implanted, Bipolar Technology," *IEEE Tech. Dig.*, Int. Electron Device Meet., 1979, p. 348.

40 R. L. Pritchard, J. B. Angell, R. B. Adler, J. M. Early, and W. M. Webster, "Transistor Internal Parameters for Small-Signal Representation," *Proc. IRE*, **49**, 725 (1961).

41 J. L. Moll, *Physics of Semiconductors*, McGraw-Hill, New York, 1964.

42 A. N. Daw, R. N. Mitra, and N. K. D. Choudhury, "Cutoff Frequency of a Drift Transistor," *Solid State Electron.*, **10**, 359 (1967).

43 K. Kurokawa, "Power Waves and the Scattering Matrix," *IEEE Trans. Microwave Theory Tech.*, **MTT-13**, 194 (1965).

44 "*S*-Parameter Techniques for Faster, More Accurate Network Design," Application Note 95-1, *Hewlett-Packard J.*, **18**, No. 6 (Feb. 1967).

45 E. G. Nielson, "Behavior of Noise Figure in Junction Transistors," *Proc. IRE*, **45**, 957 (1957).

46 H. F. Cooke, "Microwave Transistors: Theory and Design," *Proc. IEEE*, **59**, 1163 (1971).

47 L. D. Yau and T. N. Tsai, "Fabrication of a Low-Noise Beam-Leaded Microwave Bipolar Transistor by Electron and Photolithography," *IEEE Trans. Electron Devices*, **ED-25**, 413 (1978).

48 T. Sakai, Y. Sunohara, Y. Sakakibara, and J. Murota, "Stepped Electrode Transistor," *Jpn. J. Appl. Phys.*, **16**, *Suppl.*, **16-1**, 43 (1977).

49 J. A. Archer, "Low-Noise Implanted-Base Microwave Transistors," *Solid State Electron.*, **17**, 387 (1974).

50 R. Allison, "Silicon Bipolar Microwave Power Transistors," *IEEE Trans. Microwave Theory Tech.*, **MTT-27**, 415 (1979).

51 S. P. Gaur, D. H. Navon, and R. W. Teerlinck, "Transistor Design and Thermal Stability," *IEEE Trans. Electron Devices*, **ED-20**, 527 (1973).

52 R. P. Arnold and D. S. Zoroglu, "A Quantitative Study of Emitter Ballasting," *IEEE Trans. Electron Devices*, **ED-21**, 385 (1974).

53 C. G. Thornton and C. D. Simmons, "A New High Current Mode of Transistor Operation," *IRE Trans. Electron Devices*, **ED-5**, 6 (1958).

54 H. A. Schafft, "Second-Breakdown—A Comprehensive Review," *Proc. IEEE*, **55**, 1272 (1967).

55 N. Klein, "Electrical Breakdown in Solids," in L. Marton, Ed., *Advances in Electronics and Electron Physics*, Academic, New York, 1968.

56 L. Dunn and K. I. Nuttall, "An Investigation of the Voltage Sustained by Epitaxial Bipolar Transistors in Current Mode Second Breakdown," *Int. J. Electron.*, **45**, 353 (1978).

57 H. Melchior and M. J. O. Strutt, "Secondary Breakdown in Transistors," *Proc. IEEE*, **52**, 439 (1964).

58 F. F. Oettinger, D. L. Blackburn, and S. Rubin, "Thermal Characterization of Power Transistors," *IEEE Trans. Electron Devices*, **ED-23**, 831 (1976).

59 J. L. Moll, "Large-Signal Transient Response of Junction Transistors," *Proc. IRE*, **42**, 1773 (1954).

60 D. Rice, "Isoplanar-S Scales Down for New Heights in Performance," *Electronics*, **52**, 137 (1979).

61 K. Hart and A. Slob, "Integrated Injection Logic—A New Approach to LSI", *ISSCC Dig. Tech. Pap.*, p. 92 (1972); *IEEE J. Solid State Circuits*, **SC-7**, 346 (1972).

62 H. H. Berger and S. K. Widemann, "Merged Transistor Logic—A Low-Cost Bipolar Logic Concept," *ISSCC Dig. Tech. Pap.*, p. 90 (1972); also *IEEE J. Solid State Circuits*, **SC-7**, 340 (1972).

63 H. H. Berger and S. K. Wiedmann, "Advanced Merged Transistor Logic by Using Schottky Junctions," *Microelectronics*, **7**, 35 (1976).

64 W. Shockley, U.S. Patent 2,569,347 (1951).

65 M. Konagai and K. Takahashi, "GaAlAs–GaAs Heterojunction Transistors with High Injection Efficiency," *J. Appl. Phys.*, **46**, 2120 (1975).

66 C. A. Mead, "Tunnel-Emission Amplifiers," *Proc. IRE*, **48**, 359 (1960).

67 J. L. Moll, "Comparison of Hot Electrons and Related Amplifiers," *IEEE Trans. Electron Devices*, **ED-10**, 299 (1963).

68 S. M. Sze and H. K. Gummel, "Appraisal of Semiconductor–Metal–Semiconductor Transistors," *Solid State Electron.*, **9**, 751 (1966).

69 A. Y. Cho and P. D. Dernier, "Single-Crystal-Aluminum Schottky-Barrier Diodes Prepared by Molecular-Beam Epitaxy (MBE) on GaAs," *J. Appl. Phys.*, **49**, 3328 (1978).

70 C. O. Bozler, G. D. Alley, R. A. Murphy, D. C. Flanders, and W. T. Lindley, "Fabrication and Microwave Performance of the Permeable Base Transistor," *IEEE Tech. Dig.*, Int. Electron Device Meet., 1979, p. 384.

4

Thyristors

4.1 INTRODUCTION

The name "thyristor" applies to a general family of semiconductor devices that exhibit bistable characteristics and can be switched between a high-impedance, low-current OFF state and a low-impedance, high-current ON state. The operations of thyristors are intimately related to the bipolar transistor, in which both electrons and holes are involved in the transport processes. The name "thyristor" is derived from "gas thyratron," since the electrical characteristics of both devices are similar in many respects.

Following Shockley's concept of the "hook" collector[1] in 1950, Ebers developed a two-transistor analogue to explain the characteristics of a basic thyristor, a multilayered *p-n-p-n* device.[2] The detailed device principles and the first working *p-n-p-n* devices were reported by Moll et al.[3] This work has since served as the basis for all work in the study of thyristors. Because of their two stable states (ON and OFF) and low power dissipations in these states, thyristors have found unique usefulness in applications ranging from speed control in home appliances to switching and power inversion in high-voltage transmission lines. Thyristors are now available with current ratings from a few milliamperes to over 5000 A and voltage ratings extending above 10,000 V.[4] The device principles and applications have been written by Gentry et al.[5] and Blicher.[6] A comprehensive treatment on the operation and fabrication technology of thyristors can be found in *Semiconductor Power Devices* by Ghandhi.[7]

4.2 BASIC CHARACTERISTICS

The basic thyristor structure is shown in Fig. 1. It is a four-layer *p-n-p-n* device having three *p-n* junctions, J1, J2, and J3, in series. The contact

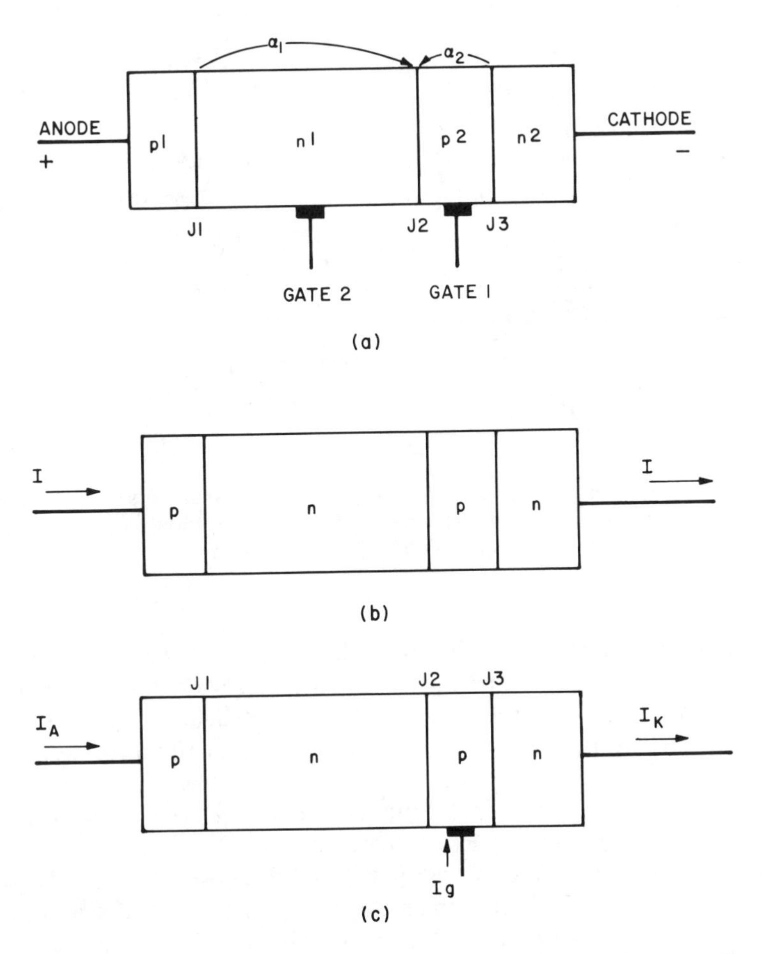

Fig. 1 (*a*) General thyristor with anode, cathode, and two gate electrodes. There are three *p-n* junctions in series J1, J2, and J3. Current gain α_1 is for the *p-n-p* transistor and α_2 for the *n-p-n*. Under forward blocking condition, as shown, the center junction J2 is reverse-biased and serves as a common collector for the *p-n-p* and *n-p-n* transistors. (*b*) Two-terminal *p-n-p-n* diode (Shockley diode). (*c*) Three-terminal thyristor.

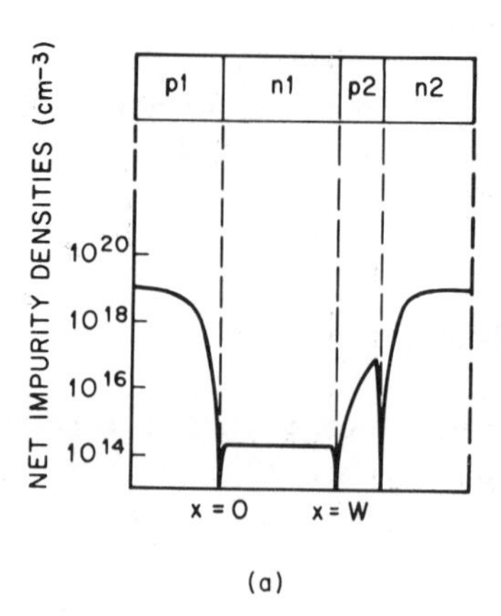

(a)

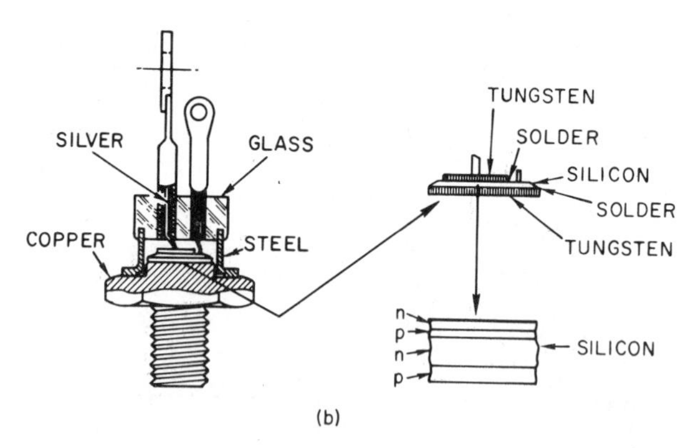

(b)

Fig. 2 (a) Typical doping profile. The most important parameters are the doping concentration and the width of the $n1$ region. (b) A cross-sectional view of a medium-current thyristor. (After Gentry et al., Ref. 5.)

electrode to the outer p layer is called the anode and that to the outer n layer is called the cathode. For a general p-n-p-n device, there may be two gate electrodes (also referred to as base) connected to the inner n and p layers. If there is no gate electrode, the device is operated as a two-terminal p-n-p-n, or Shockley, diode (Fig. 1*b*). With one gate electrode, the device then has three terminals and is commonly called the semiconductor-controlled rectifier (SCR) or thyristor (Fig. 1*c*).

A typical doping profile of an alloy-diffused p-n-p-n device is shown in Fig. 2*a*. An n-type wafer is chosen as the starting material. Then a diffusion step is used to form the $p1$ and $p2$ layer simultaneously. Finally an n-type layer is alloyed (or diffused) into one side of the wafer to form the $n2$ layer. A cross section of a medium-current thyristor is shown[5] in Fig. 2*b*. The copper block serves as the heat sink.

The basic current–voltage characteristic of a thyristor (with or without

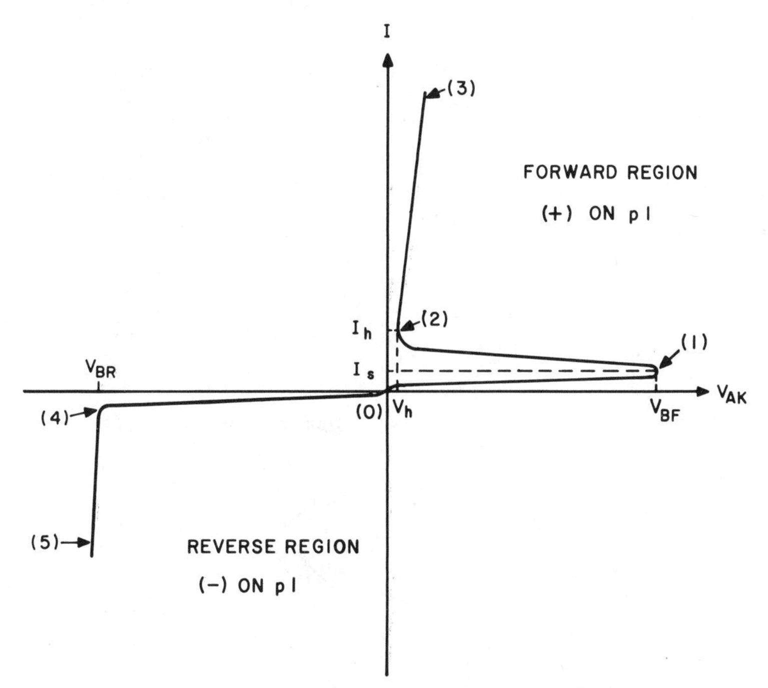

Fig. 3 Current–voltage characteristics of a thyristor.

any gate electrodes) that has a number of complex regions is shown in Fig. 3. In region 0–1 the device is in the forward blocking or OFF state with very high impedance. Forward breakover (or switching) occurs where $dV/dI = 0$, and we define a forward breakover voltage V_{BF}, and a switching current I_s (also called the turn-on current) shown in Fig. 3. Region 1–2 is the negative resistance region, and region 2–3 is the forward conducting or ON state. At point 2, where again $dV/dI = 0$, we define the holding current I_h and holding voltage V_h. Region 0–4 is in the reverse blocking state, and region 4–5 is the reverse breakdown region.

A thyristor operated in the forward region is thus a bistable device that can switch from a high-impedance, low-current, OFF state to a low-impedance, high-current, ON state, or vice versa. In this section, we consider the basic thyristor characteristics shown in Fig. 3, that is, the reverse blocking, forward blocking, and forward conduction modes.

4.2.1 Reverse Blocking

Breakdown Voltage Two basic factors limit the reverse breakdown voltage and the forward breakover voltage: avalanche breakdown and depletion-layer punch-through.

In the reverse blocking mode, the applied anode voltage is negative with

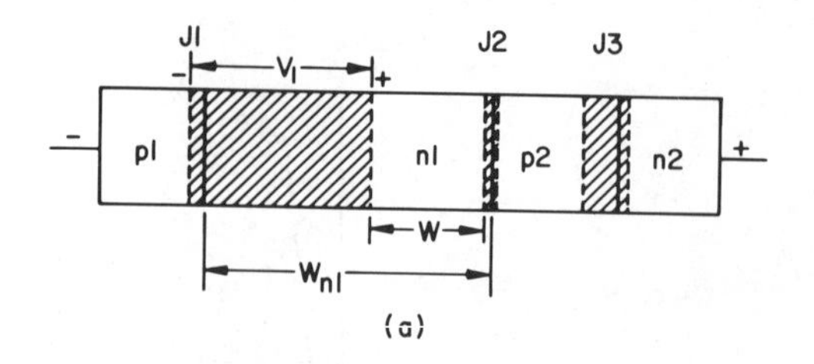

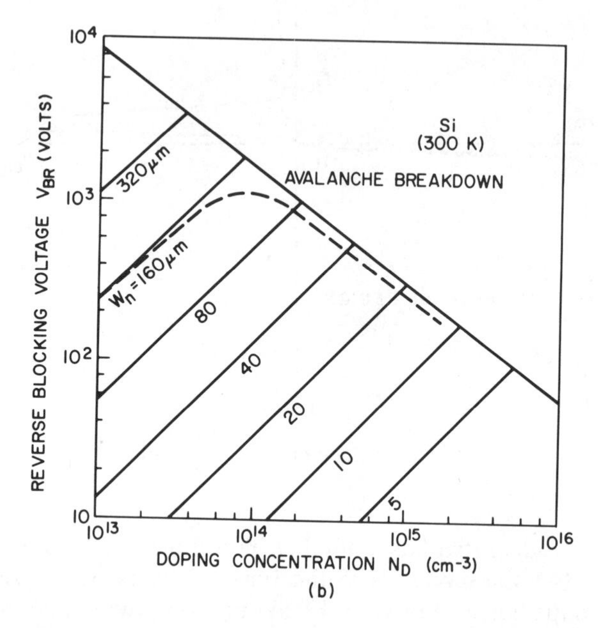

Fig. 4 Reverse blocking capability of a thyristor. The avalanche breakdown line indicates the maximum voltage attainable in the $n1$ layer with doping concentration as a parameter. The parallel lines indicate the punch-through of the $n1$ layer for various layer widths. (After Herlet, Ref. 9.)

respect to the cathode; junctions J1 and J3 are reverse-biased and J2 is forward-biased (Fig. 4*a*). For the doping profile (Fig. 2*a*), most of the applied reverse voltage will drop across J1. Depending on the thickness of the $n1$ layer W_{n1}, the breakdown will be caused by avalanche multiplication if the depletion-layer width at breakdown is less than W_{n1}, or caused by punch-through if the whole width W_{n1} is filled out first by the depletion layer at which the junction J1 is effectively shorted to J2.

For a one-sided abrupt silicon p^+-n junction with a heavily doped $p1$ region, the avalanche breakdown voltage at room temperature is given by[7,8] (refer to Chapter 2)

$$V_B = 5.34 \times 10^{13}(N_{n1})^{-0.75} \quad \text{V} \tag{1}$$

where N_{n1} is the doping concentration of the $n1$ region. For a linearly graded junction the avalanche breakdown voltage is given by[7,8]

$$V_B = 9.17 \times 10^9 a^{-0.4} \qquad \text{V} \tag{2}$$

where a is the impurity gradient in cm^{-4}. The punch-through voltage for the one-sided abrupt junction is given by

$$V_{PT} \simeq \frac{qN_{n1}W_{n1}^2}{2\epsilon_s}. \tag{3}$$

Figure 4*b* shows the fundamental limit of the reverse blocking capability of silicon thyristors.[9] For example, for $W_{n1} = 160\ \mu\text{m}$ the maximum breakdown voltage is limited to below 2000 V, which occurs at $N_{n1} = 8.5 \times 10^{13}\ \text{cm}^{-3}$; for lower dopings the breakdown voltage is limited by punch-through and for higher dopings by avalanche multiplication.

The actual reverse blocking voltage lies below these limits, because the junction J1 is coupled with its adjacent junction J2 as in an open-base common-emitter *p-n-p* transistor, and the transistor action will lower the breakdown voltage. The reverse breakdown condition corresponds to that for the common-emitter configuration, which is $M = 1/\alpha_1$, and the breakdown voltage is given by (refer to Chapter 3)

$$V_{BR} = V_B(1 - \alpha_1)^{1/n} \tag{4}$$

where M is the avalanche multiplication factor, α_1 the common-base current gain, V_B the avalanche breakdown voltage of the $p1$–$n1$ junction, and n a constant (~6 for p^+-n diodes and 4 for n^+-p diodes). Since $(1 - \alpha_1)^{1/n}$ is less than unity, the reverse breakdown voltage of a thyristor will be less than V_B. We can estimate the influence of α_1 on V_{BR}. The injection efficiency γ is close to unity for most practical situations, since the $p2$ region is heavily doped. The current gain is therefore equal to the transport factor α_T:

$$\alpha_1 = \gamma\alpha_T \approx \alpha_T = \text{sech}(W/L_{n1}) \tag{5}$$

where L_{n1} is the hole diffusion length in the $n1$ region and

$$W = W_{n1}[1 - (V/V_{pT})^{1/2}]. \tag{6}$$

For a given W_{n1} and L_{n1}, the ratio W/L_{n1} will decrease as the reverse voltage increases. Therefore, the base transport factor becomes more important as the reverse voltage approaches the punch-through limit. Figure 4*b* shows an example for the reverse blocking voltage with $W_{n1} = 160\ \mu\text{m}$ and $L_{n1} = 150\ \mu\text{m}$ (dashed curve). Note that V_{BR} approached the V_{PT} for low dopings in the $n1$ region. As the doping increases, V_{BR} always lies slightly below V_B, because of the finite value of W/L_{n1}.

Neutron Doping[10] For high-power, high-voltage thyristors, large areas are used, frequently an entire wafer (up to 100 mm or larger in diameter) for a single device. This size imposes stringent requirements on the

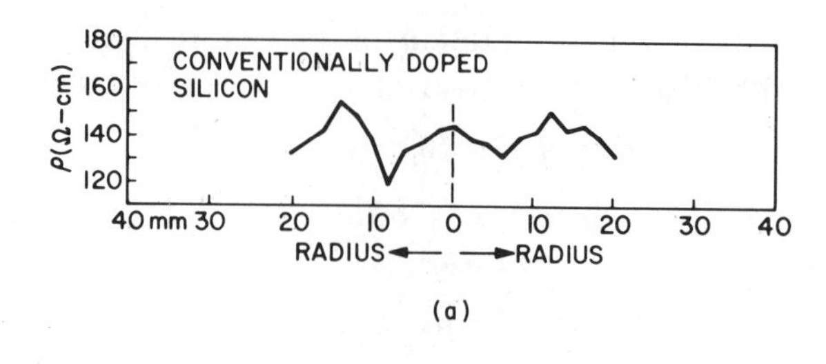

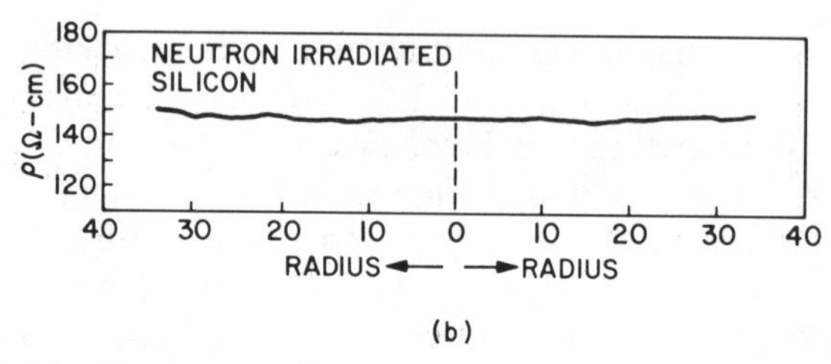

Fig. 5 Typical lateral macroscopic resistivity distributions in conventionally doped silicon (a) and in silicon doped by neutron irradiation (*b*). (After Haas and Schnoller, Ref. 11.)

uniformity of the starting material. To obtain tight tolerance for the resistivity and homogeneous distribution of impurity dopant, the neutron irradiation method is employed. Usually, the float-zone silicon slice having an average resistivity well in excess of that required is used. The slice is then irradiated with thermal neutrons. This process gives rise to fractional transmutation of silicon into phosphorus and dopes the silicon *n*-type:

$$\mathrm{Si}_{14}^{30} + \text{neutron} \rightarrow \mathrm{Si}_{14}^{31} + \gamma \text{ ray} \xrightarrow[2.62\text{ hr}]{} \mathrm{P}_{15}^{31} + \beta \text{ ray}.$$

The half-life is 2.62 h. Since the penetration range of neutrons in silicon is about 100 cm, doping is very uniform throughout the slice. Figure 5 compares the lateral macroscopic resistivity distributions in conventionally doped silicon and in silicon doped by neutron irradiation using spreading resistance measurements.[11] The resistivity variations are about ±15% for the conventionally doped silicon and about ±1% for neutron doping.

Beveled Structures To maximize the breakdown voltage in a thyristor, usually planar junctions formed by alloy or diffusion are used, since cylindrical or spherical junctions have lower breakdown voltages (refer to Chapter 2). Even for plane junctions, premature breakdown can still occur at the surface, leading to a concentration of current at the edge and reducing the surge current capability of the thyristor. By using beveled structures, the surface field can be lowered significantly compared to the bulk field, ensuring that the breakdown will occur uniformly in the bulk.

Figure 6 shows the beveled structures. A positive bevel angle results in a junction of decreasing cross-sectional area when going from the heavily

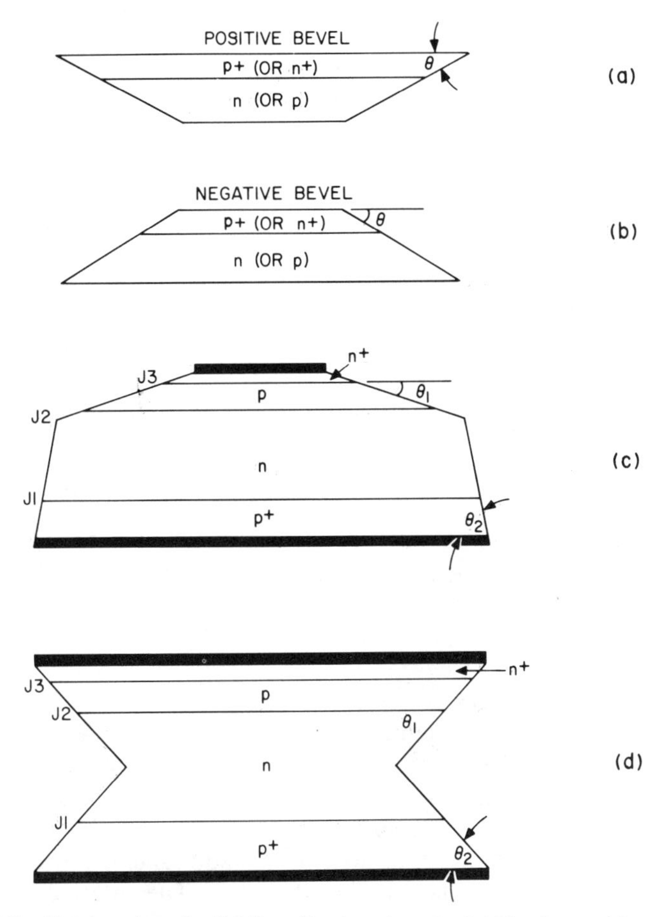

Fig. 6 (*a*) Positive bevel angle. (*b*) Negative bevel angle. (*c*) Thyristor with one negative bevel angle and one positive bevel angle. (*d*) Thyristor with two positive bevel angles.

doped side to the lightly doped side (Fig. 6*a*). The negative bevel angle, on the other hand, has an increasing area going in the same direction (Fig. 6*b*). Two beveled thyristor structures are shown in Fig. 6*c* and *d*. Figure 6*c* has a negative bevel angle for junctions J2 and J3, and a positive bevel angle for J1. Figure 6*d* has positive bevel angles for all three junctions.[12]

For the positive-beveled junction, the surface field, as a first-order approximation, is reduced by the factor sin θ. Figure 7 shows the calculated values[13] of the electric field from a two-dimensional Poisson equation for a p^+-n junction under a reverse bias of 600 V. The internal electric field in the bulk is also shown. Note that the peak field on the surface is always less than that in the bulk; as the bevel angle is reduced, so is the peak electric field; and the position of the peak field shifts into the lightly doped

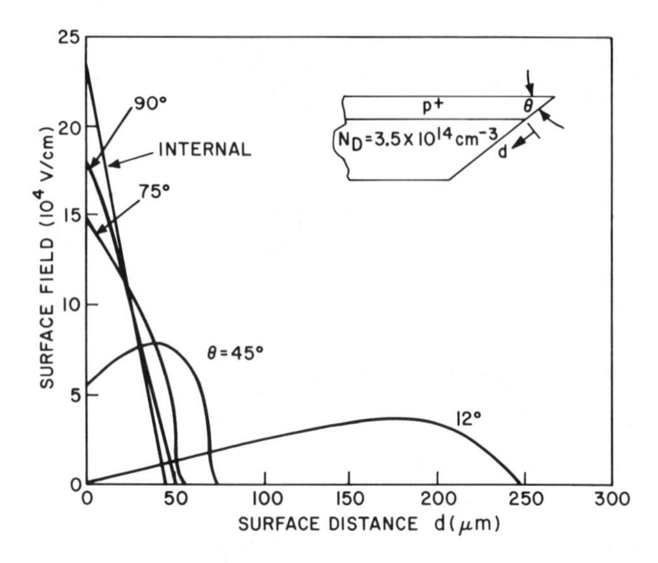

Fig. 7 Surface fields of positive beveled device shown in the insert as a function of bevel angle. (After Davies and Gentry, Ref. 13.)

region as the bevel angle is reduced. The breakdown voltage for the positive-beveled junction is identical to that of plane junction.

For the negative-beveled junction to have internal breakdown, the *p-n* junction must not be highly asymmetric and must have very small negative bevel angles. Calculated results show that for small negative bevel angles, the peak surface field is less than that in the bulk and is located in the *p* region (where *p*1 has a higher doping than *n*1). The internal breakdown occurs at a voltage below that given for the plane junction. Figure 8 shows[14] the bulk breakdown voltage normalized to the plane junction breakdown voltage as a function of the effective bevel angle:

$$\theta_{\text{eff}} = (0.04)\theta\left(\frac{W}{d}\right)^2 \tag{7}$$

where W and d are the respective plane junction depletion-layer widths shown in the insert. To approach the plane junction breakdown voltage, low values of bevel angles should be used.

4.2.2 Forward Blocking

Variation of Alphas For forward blocking, the anode voltage is positive with respect to the cathode, and only the center junction J2 is reverse-biased. Junctions J1 and J2 are forward-biased. Most of the applied voltage will drop across J2. To understand the forward blocking characteristics, we shall use the method of the two-transistor analog.[2] Figure 1

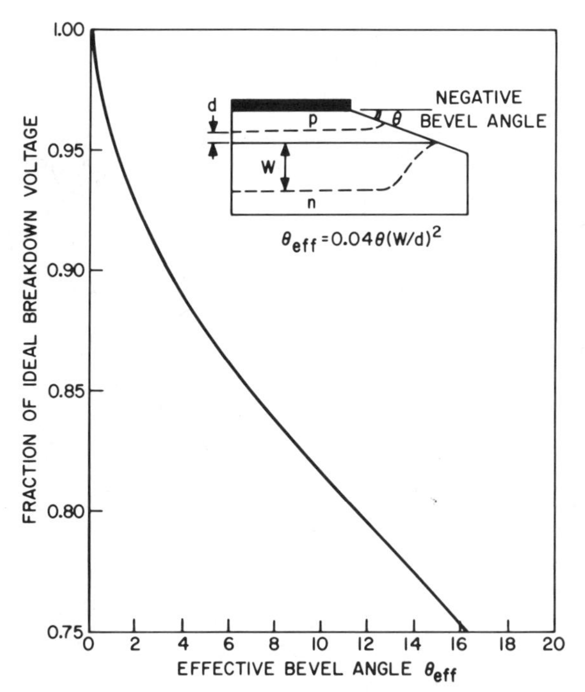

Fig. 8 Normalized breakdown voltage versus bevel parameter for negative beveled structures shown in the insert. (After Adler and Temple, Ref. 14.)

shows that the thyristor can be considered as a *p-n-p* transistor and an *n-p-n* transistor connected with the collector of one transistor attached to the base of the other, and vice versa, as shown in Fig. 9*a* and *b* for a three-terminal thyristor. The center junction acts as the collector of holes from J1 and of electrons from J3.

The relationship between emitter, collector, and base currents (I_E, I_C, and I_B, respectively) and the dc common-base current gain α_1 for a *p-n-p* transistor is shown in Fig. 9*c*, where I_{CO} is the collector–base reverse saturation current. Similar relationships can be obtained for an *n-p-n* transistor, except that the currents are reversed. From Fig. 9*b* it is evident that the collector current of the *n-p-n* transistor provides the base drive for the *p-n-p* transistor. Also, the collector current of the *p-n-p* transistor along with gate current I_g supplies the base drive for the *n-p-n* transistor. Thus a regeneration situation results when the total loop gain exceeds unity.

The base current of the *p-n-p* transistor is

$$I_{B1} = (1 - \alpha_1)I_A - I_{CO1} \tag{8}$$

which is supplied by the collector of the *n-p-n* transistor. The collector current of the *n-p-n* transistor with a dc common-base current gain α_2 is

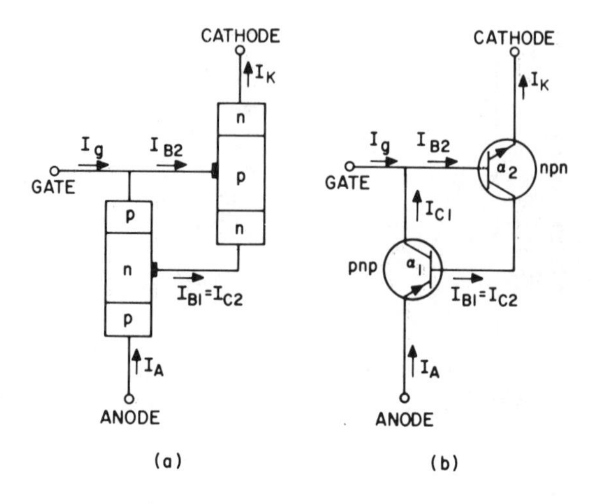

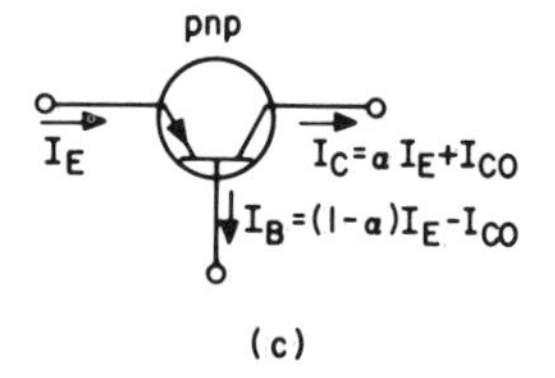

Fig. 9 (*a*) Two-transistor approximation of a three-terminal thyristor. (*b*) Same as (*a*) using transistor notations. (*c*) Current relationships in a *p-n-p* transistor. (After Ebers, Ref. 2.)

given by

$$I_{C2} = \alpha_2 I_K + I_{CO2}. \tag{9}$$

By equating I_{B1} and I_{C2}, we obtain

$$(1-\alpha_1)I_A - I_{CO1} = \alpha_2 I_K + I_{CO2}.$$

Since $I_K = I_A + I_g$, from Eq. 9 we obtain

$$I_A = \frac{\alpha_2 I_g + I_{CO1} + I_{CO2}}{1-(\alpha_1+\alpha_2)}. \tag{10}$$

It will be shown later that both α_1 and α_2 are functions of the current I_A and generally increase with increasing current. Equation 10 gives the static characteristic of the device up to the breakover voltage. Beyond this point the device acts as a *p-i-n* diode. Note that all the current components in the numerator of Eq. 10 are small, hence I_A is small unless $(\alpha_1+\alpha_2)$ approaches unity. At this point the denominator of the equation approaches zero and forward breakover or switching will occur. It is worth-

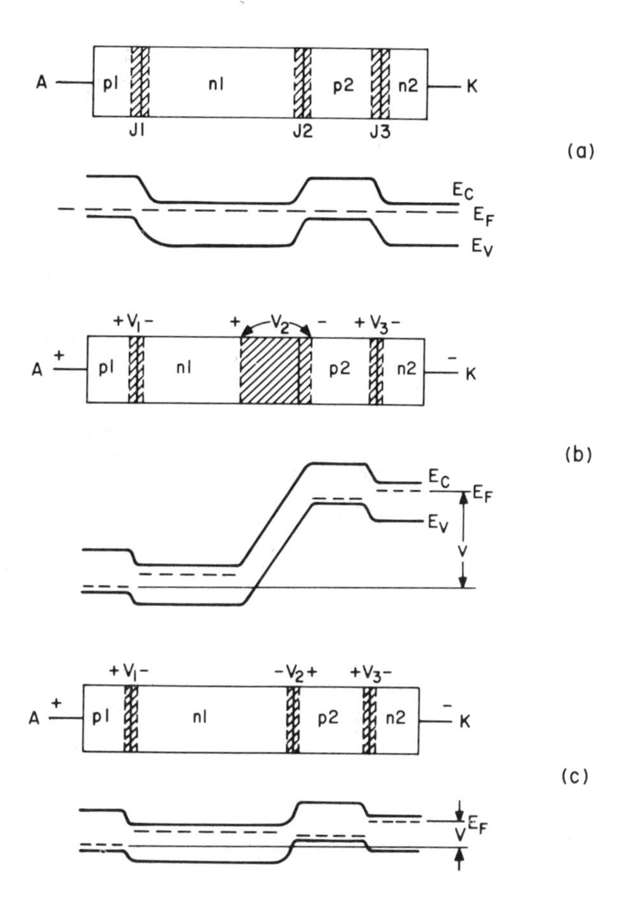

Fig. 10 Energy-band diagram for forward regions. (*a*) Equilibrium condition. (*b*) Forward OFF state, where most of the voltage drops across the center junction J2. (*c*) Forward ON state, where all three junctions are forward-biased.

while to point out that if the polarities of the anode and cathode are reversed, junctions J1 and J3 are reverse-biased while J2 is forward-biased. Under this condition there is no switching action, since only the center junction acts as an emitter, and no regenerative process can take place.

The depletion-layer widths and the corresponding energy-band diagrams for the equilibrium, forward OFF state, and forward ON state are shown in Fig. 10*a*, *b*, and *c*, respectively. In equilibrium there is at each junction a depletion region with a built-in potential that is determined by the impurity doping profile. When a positive voltage is applied to the anode, junction J2 will tend to become reverse-biased, while J1 and J3 will be forward-biased. The anode-to-cathode voltage drop is approximately equal to the algebraic sum of the junction drops:

$$V_{AK} = V_1 + V_2 + V_3. \tag{11}$$

As the voltage increases, the current will increase, causing α_1 and α_2 to increase. Because of the regenerative nature of these processes, the device is eventually switched to its ON state. Upon switching, the current through the device must be limited by an external load resistance; otherwise, the device would destroy itself if the supply voltage were sufficiently high. In this ON state, J2 is also forward-biased, as shown in Fig. 10*c*, and the voltage drop V_{AK} is given by $(V_1 - |V_2| + V_3)$ which is approximately equal to the voltage drop across one forward-biased *p-n* junction plus a saturated transistor.

Switching of a thyristor occurs when $dV_{AK}/dI_A = 0$. This condition is generally reached before $(\alpha_1 + \alpha_2) = 1$. We shall now show that the switching will begin when the sum of the small-signal alphas reaches unity.[15] Let us consider the situation that results when the gate current I_g is increased by a small amount ΔI_g. As a consequence of this increase, the anode current will increase by an amount ΔI_A, so the incremental cathode current is given by

$$\Delta I_K = \Delta I_A + \Delta I_g. \tag{12}$$

The small-signal alphas are defined as

$$\tilde{\alpha}_1 \equiv \frac{dI_C}{dI_A} = \lim_{\Delta I_A \to 0} \frac{\Delta I_C}{\Delta I_A} \tag{13a}$$

$$\tilde{\alpha}_2 \equiv \frac{dI_C}{dI_K} = \lim_{\Delta I_K \to 0} \frac{\Delta I_C}{\Delta I_K}. \tag{13b}$$

The hole current collected by J2 will be $\tilde{\alpha}_1 \Delta I_A$ and the electron current will be $\tilde{\alpha}_2 \Delta I_K$. Equating the change in anode current to the change in current across J2, we obtain

$$\Delta I_A = \tilde{\alpha}_1 \Delta I_A + \tilde{\alpha}_2 \Delta I_K. \tag{14}$$

Substituting Eq. (14) into (12) yields

$$\frac{\Delta I_A}{\Delta I_g} = \frac{\tilde{\alpha}_2}{1 - (\tilde{\alpha}_1 + \tilde{\alpha}_2)}. \tag{15}$$

When $(\tilde{\alpha}_1 + \tilde{\alpha}_2)$ becomes unity, any small increase in I_g will cause the device to become unstable, since from Eq. 15 a small increase in I_g will cause an infinite increase in I_A. Although gate current was used in the analysis, the same effect can be obtained with a slight increase in temperature or voltage.

The dc common-base current gain α_1 of a transistor is given by

$$\alpha_1 = \alpha_T \gamma \tag{16}$$

where α_T is the transport factor defined as the ratio of the injected current reaching the collector junction to the injected current and γ is the injection efficiency defined as the ratio of the injected current to the total emitter current. From Fig. 9*c* we have the relationship

$$I_C = \alpha_1 I_E + I_{CO}. \tag{17}$$

By differentiating Eq. 17 with respect to emitter current, we obtain the small-signal alpha:

$$\tilde{\alpha}_1 \equiv \frac{dI_C}{dI_E} = \alpha_1 + I_E \frac{\partial \alpha_1}{\partial I_E}. \tag{18}$$

Substituting Eq. 16 into Eq. 18 yields

$$\tilde{\alpha}_1 = \gamma\left(\alpha_T + I_E \frac{\partial \alpha_T}{\partial I_E}\right) + \alpha_T I_E \frac{\partial \gamma}{\partial I_E}. \tag{19}$$

The simplest approximations for α_T and γ are given by

$$\alpha_T = \frac{1}{\cosh(W/\sqrt{D\tau})} \simeq 1 - \frac{W^2}{2D\tau}, \tag{20a}$$

$$\gamma \simeq \frac{1}{1 + N_B W / N_E L_E} \tag{20b}$$

where W is the base width, D and τ are, respectively, the diffusion coefficient and lifetime of minority carriers in the base, N_B and N_E are the base and emitter concentrations, respectively, and L_E is the diffusion length in the emitter. To obtain large values of alpha, one must use small values of $W/\sqrt{D\tau}$ and N_B/N_E.

To investigate the dependence of dc alphas and small-signal alphas on current, we must use the more detailed calculation, considering both diffusion and drift current components. The hole currents at junctions J1 and J2 can be calculated from the equation

$$I_p(x) = qA_s\left(p_n \mu_p \mathscr{E} - D_p \frac{\partial p_n}{\partial x}\right) \tag{21}$$

where A_s is the area of the junction. The continuity equation for the *n*1 region as shown in Fig. 2*a* is given by

$$\frac{\partial p_n}{\partial t} = -\frac{p_n - p_{no}}{\tau_p} - \mu_p \mathscr{E} \frac{\partial p_n}{\partial x} + D_p \frac{\partial^2 p_n}{\partial x^2}. \tag{22}$$

And the boundary conditions are $p_n(x=0) = p_{no} \exp(\beta V)$ where $\beta \equiv q/kT$, and $p_n(x = W) = 0$. The steady-state solution of Eq. 22 subject to the boundary conditions is

$$\begin{aligned} p_n(x) = {} & p_{no} \exp(\beta V) \exp[(C_1 + C_2)x] - p_{no}[\exp(\beta V)\exp(C_2 W) \\ & + \exp(-C_1 W)] \exp(C_1 x) \operatorname{csch}(C_2 W) \sinh(C_2 x) \\ & + p_{no} \exp(C_1 x) \operatorname{csch}(C_2 W) \sinh(x - W)C_2 + p_{no} \end{aligned} \tag{23}$$

where

$$C_1 \pm C_2 = \frac{\mu_p \mathscr{E}}{2D_p} \pm \left[\left(\frac{\mu_p \mathscr{E}}{2D_p}\right)^2 + \frac{1}{D_p \tau_p}\right]^{1/2}.$$

From Eqs. 21, 22, and 23 we obtain for the transport factor

$$\alpha_T = \frac{C_2 \exp(C_1 W)}{C_1 \sinh(C_2 W) + C_2 \cosh(C_2 W)}. \tag{24}$$

The injection efficiency is given by

$$\gamma \equiv \frac{I_p}{I_p + I_n + I_r} \simeq \frac{I_p}{I_p + I_r} = \frac{I_{po} \exp(\beta V)}{I_{po} \exp(\beta V) + I_R \exp(\beta V/n)} \tag{25}$$

where I_p and I_n are the injected current flowing into the base and emitter regions, respectively, I_r is the space-charge recombination current given by $I_R \exp(\beta V/n)$, where I_R and n are constants (generally $1 < n < 2$), and $I_{po} = qD_pA_sp_n[C_1 + C_2 \coth(C_2W)]$. For the doping profile of Fig. 2*a*, $p_{po}(p1) \gg n_{no}(n1)$, the current I_n can be neglected in Eq. 25.

We can now calculate α_1 from Eqs. 24 and 25 as a function of the emitter current and the base layer width (W). In addition, we can combine Eqs. 19, 24, and 25 to give the small-signal alpha. The results are shown in Fig. 11 for the doping profile shown in Fig. 2*a* and for some typical parameters of silicon.[16] Note that for the current range shown, the small-signal alpha is always greater than the dc alpha. The ratio of the base width to diffusion

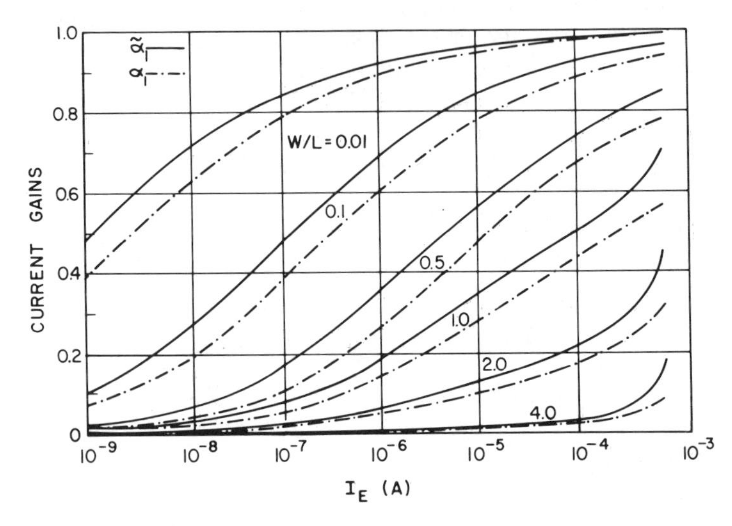

Fig. 11 Small-signal alpha and dc alpha as functions of current and base width for a transistor with the following parameters, $n_{no} = 3 \times 10^{14}$ cm^{-3}, $p_{no} = 7.5 \times 10^5$ cm^{-3}, A_s = 0.16 mm^2, $\mu_n = 1400$ cm^2/V-s, $\mu_p = 500$ cm^2/V-s, $D_p = 13$ cm^2/s, $\tau_p = 0.5$ μs, $L_p = 25.5$ μm, $I_R = 2.5 \times 10^{-10}$ A and $n = 1.5$. (After Yang and Voulgaris, Ref. 16.)

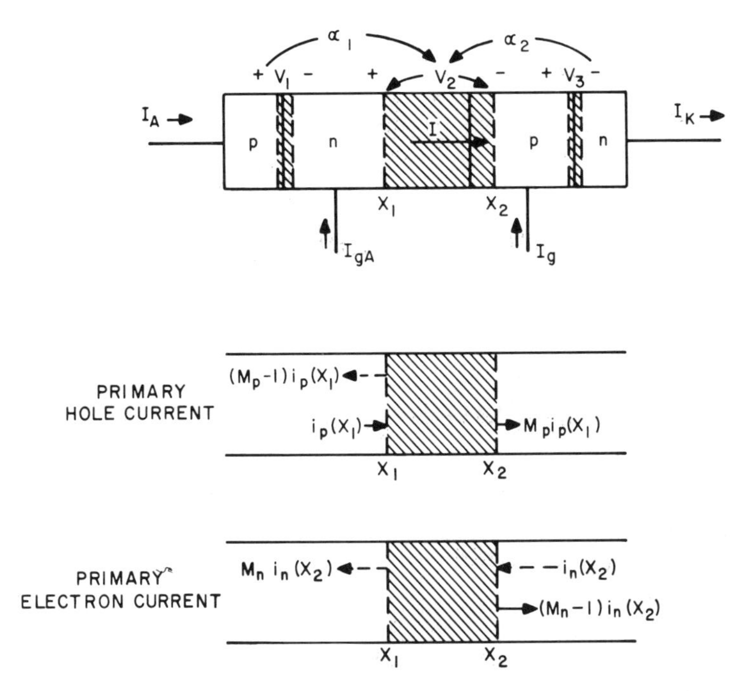

Fig. 12 Generalized thyristor. The current flowing through the center junction is I. Electron i_n and hole i_p primary currents generate $M_n i_n$ and $M_p i_p$, respectively, under avalanche multiplication conditions.

length W/L is an important device parameter in determining the variation of gain with current. For small values of W/L, the transport factor is independent of current, and the gain varies with current only through the injection efficiency. This condition applies to the narrow base-width sections of the devices (n-p-n section). For larger values of W/L, both transport factor and injection efficiency are functions of current (p-n-p section). Thus the value of the gain can, in principle, be tailored to the desired range by choosing the proper diffusion length and doping profile.

Forward Breakover Voltage To obtain the forward breakover voltage V_{BF}, we shall consider a general thyristor with leads connected to all four layers as shown in Fig. 12. Reference directions for voltages and currents are shown in the figure. We assume that the center junction of the device remains reverse-biased. We also assume that the voltage drop V_2 across this junction is sufficient to produce avalanche multiplication of carriers as they travel across the depletion region. We denote the multiplication factor for electrons by M_n and that for holes by M_p; both are functions of V_2. Because of the multiplication, a steady hole current $I_p(x_1)$ entering the depletion region at x_1 becomes $M_p I_p(x_1)$ at $x = x_2$. A similar result will be obtained for an electron current $I_n(x_2)$ entering the depletion layer at x_2.

The total current I is given by

$$I = M_p I_p(x_1) + M_n I_n(x_2). \tag{26}$$

Since $I_p(x_1)$ is actually the collector current of the p-n-p transistor, we can express $I_p(x_1)$ as Fig. 9c,

$$I_p(x) = \alpha_1(I_A)I_A + I_{CO1}. \tag{27}$$

Similarly, we can express the primary electron current $I_n(x_2)$ as

$$I_n(x) = \alpha_2(I_K)I_K + I_{CO2}. \tag{28}$$

Substituting Eqs. 27 and 28 into Eq. 26 yields

$$I = M_p[\alpha_1(I_A)I_A + I_{CO1}] + M_n[\alpha_2(I_K)I_K + I_{CO2}]. \tag{29}$$

If we assume that $M_p = M_n = M$, Eq. 29 reduces to

$$\frac{1}{M(V_2)} = \frac{\alpha_1(I_A)I_A}{I} + \frac{\alpha_2(I_K)I_K}{I} + \frac{I_0}{I} \tag{30}$$

where $I_0 = I_{CO1} + I_{CO2}$.

If $I_g = I_{gA} = 0$ (where I_{gA} is the current in the anode gate) and $I = I_A = I_K$, Eq. 30 reduces to

$$\frac{1}{M(V_2)} = \alpha_1(I) + \alpha_2(I) + I_0/I. \tag{31}$$

The multiplication M can be expressed as

$$M(V_2) = \frac{1}{1 - (V_2/V_B)^n} \tag{32}$$

where V_B is the breakdown voltage considered in Section 4.2.1, and n is a constant. The forward breakover voltage can now be obtained from Eqs. 31 and 32 under the condition that $I \gg I_0$, and we have

$$M(V_2) = \frac{1}{\alpha_1 + \alpha_2} = \frac{1}{1 - (V_{BF}/V_B)^n} \tag{33}$$

or

$$V_{BF} = V_B(1 - \alpha_1 - \alpha_2)^{1/n}. \tag{34}$$

Comparison with the reverse breakdown voltage $V_{BR} = V_B(1 - \alpha_1)^{1/n}$ shows that V_{BF} is always less than V_{BR}. For small values of $(\alpha_1 + \alpha_2)$, V_{BF} will be essentially the same as the reverse breakdown voltage shown in Fig. 4. For values of $(\alpha_1 + \alpha_2)$ close to 1, the breakover voltages can be substantially less than V_{BR}.

4.2.3 Forward Conduction

When the thyristor is in the ON state, all three junctions are forward-biased. Holes are injected from the $p1$ region and electrons from the $n2$

region. The $n1$-$p2$-$n2$ device acts like a saturated transistor and provides a remote contact to the $n1$ region. Therefore, the device behaves like a $p1$-i-$n2$ (p^+-i-n^+) diode.

For a p^+-i-n^+ diode with i-region width of W, the forward current density is accounted for by the rate at which holes and electrons recombine within the i region. The current density is thus given by

$$J = \int_0^W qR\,dx \tag{35}$$

where R is the recombination rate that can be expressed as[17]

$$R = G(n^2p + p^2n) + \frac{np - n_i^2}{\tau_{po}(n + n_i) + \tau_{no}(p + n_i)} \tag{36}$$

where the first term is due to Auger processes and the Auger coefficient G is found to be $1 \sim 2 \times 10^{-31}\,\text{cm}^6/\text{s}$ for silicon; the second term is due to midgap recombination traps, and τ_{po} and τ_{no} are the hole and electron lifetimes, respectively. In the limit that $n = p \gg n_i$, Eq. 36 reduces to

$$\tau_{\text{eff}} = \frac{n}{R} = \left(2Gn^2 + \frac{1}{\tau_{po} + \tau_{no}}\right)^{-1}. \tag{37}$$

If the carrier concentration is approximately constant throughout the i region, the current density from Eqs. 35 and 37 is

$$J = qnW/\tau_{\text{eff}}. \tag{38}$$

We can also write the current density as

$$\begin{aligned} J &= q(\mu_n + \mu_p)n\bar{\mathscr{E}} \\ &= \frac{q}{kT}\frac{(b+1)^2}{2b}\,qD_a n\bar{\mathscr{E}} \end{aligned} \tag{39}$$

where $\bar{\mathscr{E}}$ is the average electric field, b is the ratio μ_n/μ_p, and D_a is the ambipolar diffusion coefficient.

The voltage drop across the i region V_m is given by

$$V_m = 2d\bar{\mathscr{E}}. \tag{40}$$

Combining Eqs. 38 and 39 yields

$$V_m = \frac{kT}{q}\frac{8b}{(1+b)^2}\left(\frac{d^2}{D_a\tau_{\text{eff}}}\right) \tag{41}$$

or

$$V_m \sim (\tau_{\text{eff}})^{-1}. \tag{42}$$

Because V_m is inversely proportional to the effective lifetime, longer τ_{eff} should be used to reduce V_m. Calculated values[18] of τ_{eff} are shown in Fig. 13 for a number of the ambipolar lifetime $\tau_a = (\tau_{po} + \tau_{no})$. At low carrier concentrations, the effective lifetime is equal to the ambipolar lifetime;

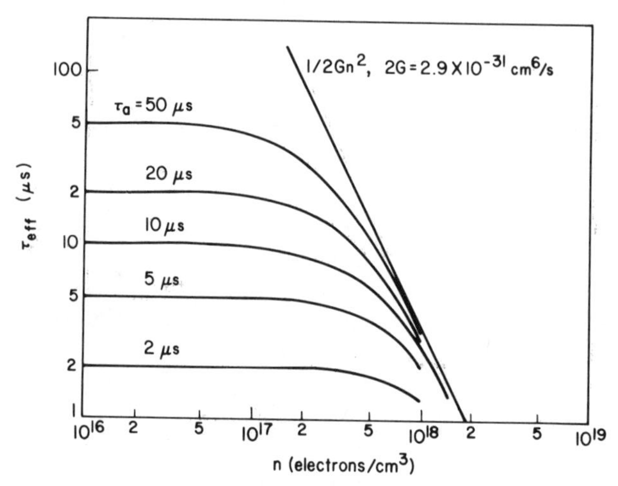

Fig. 13 Effective lifetime under high-injection condition. τ_a is the ambipolar lifetime and G is the Auger coefficient. (After Krausse, Ref. 18.)

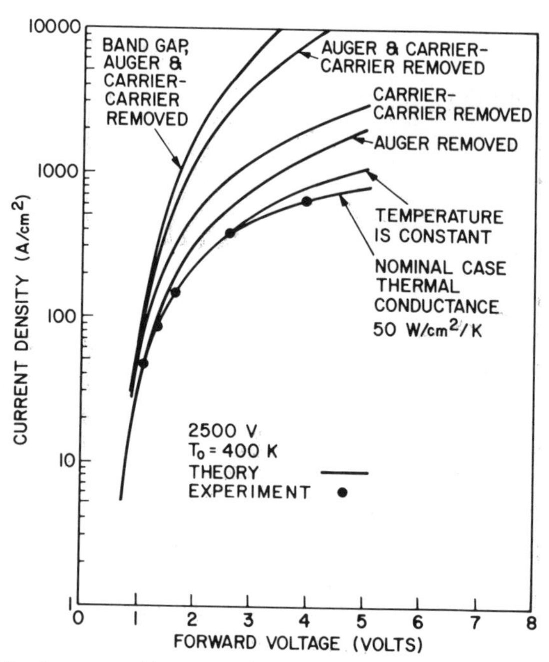

Fig. 14 Theoretical curves illustrate the relative importance of various physical mechanisms, including heat flow on the current–voltage characteristics of a 2500-V thyristor. Also shown are the measured results. (After Adler, Ref. 17.)

however, at carrier concentrations above 10^{17} cm^{-3}, the effective lifetime falls rapidly as n^{-2} due to Auger processes.

Numerical analysis for the forward conduction has been done incorporating various physical mechanisms. A series of calculated I–V curves for a 2500-V thyristor are shown[17] in Fig. 14 for a heat-sink temperature of 400 K. The inscription of each curve indicates the physical mechanisms that were removed. For example, "carrier–carrier removed" means the removal of carrier–carrier scattering in the numerical analysis. The 1000-A/cm^2 level is associated with maximum surge operation while 100-A/cm^2 level is associated with maximum steady-state operation. As can be seen, carrier–carrier scattering and Auger recombination are important limiting mechanisms both at the surge operation levels and at the 100-A/cm^2 level. The bandgap narrowing has virtually no effect until the current density is above 1000 A/cm^2. The midgap trap recombination becomes the limiting factor at levels below 100 A/cm^2, and is also important at the surge level. The junction temperature effect becomes important when the current density is larger than 500 A/cm^2. The bottom curve is the "nominal case" incorporating all the mechanisms described above. Also shown are the experimental results, which are in excellent agreement with the "nominal case."

4.3 SHOCKLEY DIODE AND THREE-TERMINAL THYRISTOR

4.3.1 Static *I–V* Characteristics

As mentioned previously, the Shockley diode is a two-terminal *p-n-p-n* device.[19a, b] From the general equation, Eq. 30, we can develop a graphical method to analyze the I–V characteristics. Since $I_g = I_{gA} = 0$ and $I_A = I_K = I$ in a Shockley diode, Eq. 30 reduces to

$$\frac{1}{M(V_2)} = \alpha_1(I) + \alpha_2(I) + I_0/I = f(I) \tag{43}$$

where the multiplication $M(V_2)$ is given by Eq. 32. We shall assume that I_0 is some known constant, and α_1 and α_2 are known functions of current similar to those shown in Fig. 11. The graphical solution of Eq. 43 is illustrated[20] in Fig. 15. We first obtain the function $f(I)$ by adding the three curves shown in Fig. 15*a*. Figure 15*b* is a plot of $1/M$ versus M with the vertical scale identical to that in Fig. 15*a* and the horizontal scale identical to that in Fig. 15*c* which is a plot of Eq. 32. We now choose a value I in Fig. 15*d* at which we want to know the voltage drop. We project vertically upward to find the corresponding value of $f(I)$, point (1). Then we project horizontally to the left and locate a point on the $1/M$ versus M plot, point (2). With known M we can project vertically downward to find the required value (V_2/V_B) in Fig. 15*c*, point (3), and horizontally back to point (4). Point (4) gives the normalized voltage drop required to sustain the given current

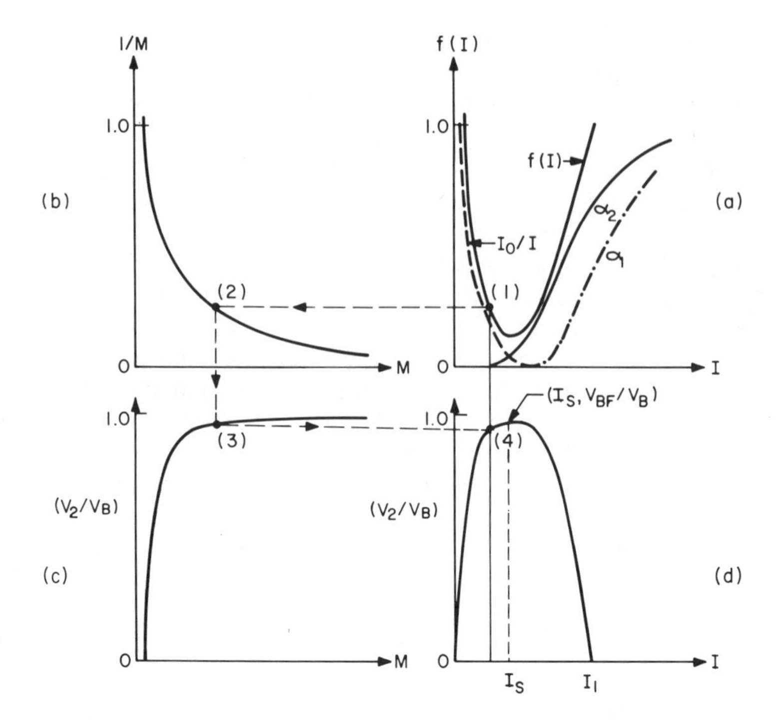

Fig. 15 Graphical solution of current–voltage characteristics of a Shockley diode. (After Gibbons, Ref. 20.)

I. The entire I–V characteristic can thus be obtained by repeating this geometrical construction process. The result is shown in Fig. 15*d*.

Note from the figure that the switching point (I_s, V_{BF}) occurs at the location where the function $f(I)$ reaches its minimum. The holding point is defined as the low-voltage, high-current point at which $dV/dI = 0$. This analysis does not enable us to find this point. However, the holding point can be approximately related to the coordinates (I_1, 0) at which $f(I) = 1$. At $f(I) = 1$, $M(V_2) = 1$, this means that voltage V_2 is zero. If $V_2 = 0$, the saturation current I_0 of the center junction goes to zero. Then from Eq. 43 we have $\alpha_1(I) + \alpha_2(I) = 1$. For known α_1 and α_2 the current at which the center junction reaches zero bias can be determined. The voltage across the entire device at this point will simply be the sum of forward voltage drops across the two outer junctions (about 1.2 to 1.4 V for silicon devices.)

For current larger than I_1, the entire junction becomes forward-biased. (The analysis above cannot be extended to the forward case, since we have assumed that the junction J2 remains reverse-biased.) As the current increases beyond I_1, the voltage drop across the entire device continues to decrease and the device continues to exhibit negative dynamic resistance

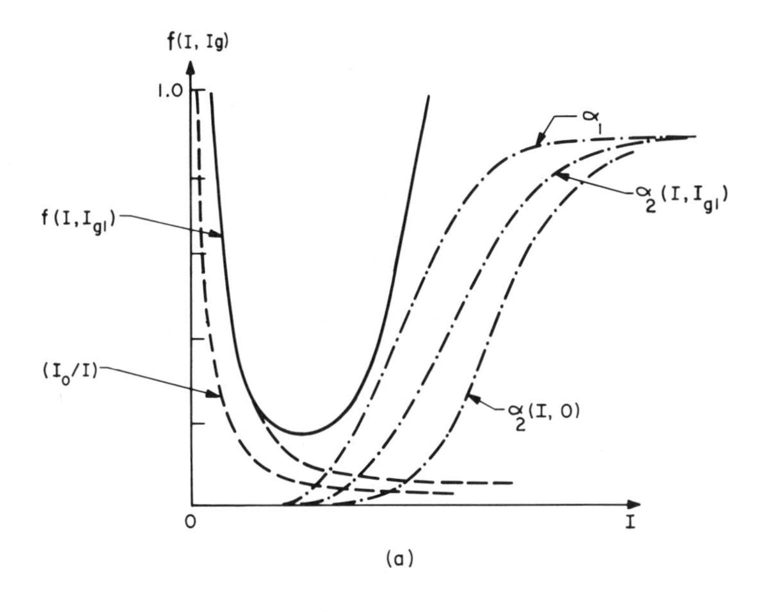

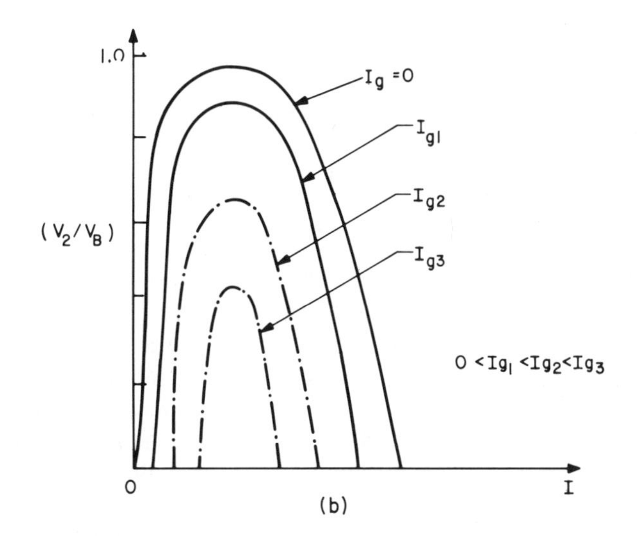

Fig. 16 Graphical solution of current–voltage characteristics of a gate-controlled thyristor. (After Gibbons, Ref. 20.)

up to current I_h. Above this current, the center junction voltage drop is comparable to the emitter junction voltage, and the dynamic resistance of the entire device again becomes positive.[21] Beyond the point (I_h, V_h), the device is in forward conduction as discussed in Section 4.2.3.

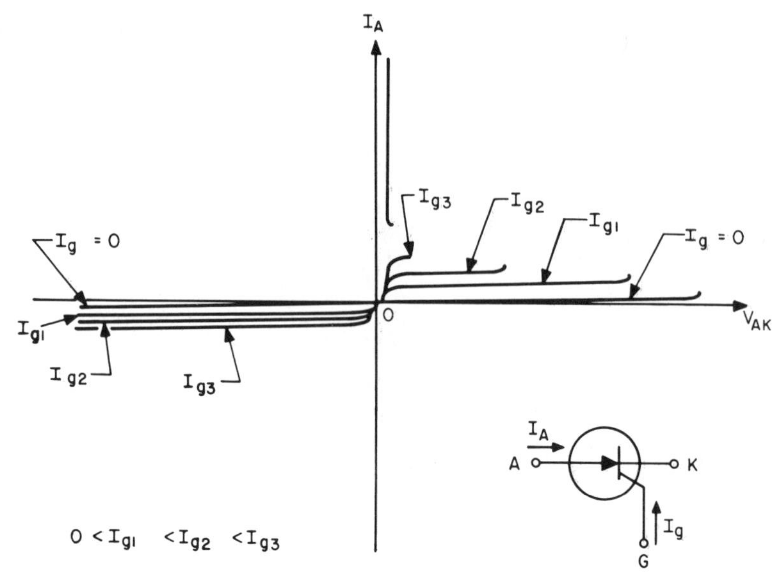

Fig. 17 Effect of gate current on current–voltage characteristics of a thyristor. (After Gentry et al., Ref. 5.)

For the thyristor with one gate electrode, Eq. 30 can be expressed as

$$\begin{aligned}\frac{1}{M(V_2)} &= \alpha_1(I) + \alpha_2(I + I_g) + \frac{\alpha_2(I + I_g)}{I} I_g + \frac{I_0}{I} \\ &= f(I, I_g).\end{aligned} \tag{44}$$

In Eq. 44 the current I_K is replaced by $I + I_g$ and $I_{gA} = 0$. This equation would be identical to Eq. 43 if I_g were equal to zero. The $f(I, I_g)$ curve and I–V characteristics of the structure for $I_g = 0$ are shown in Fig. 16. The I–V characteristics for various values of I_g are obtained by replotting $\alpha_2(I + I_g)$ for each value of I_g and including the term $\alpha_2(I + I_g)/I$ in $f(I, I_g)$. This generates a set of $f(I, I_g)$ curves. We note that as I_g increases, the switching voltage decreases. This gives rise to the gate turn-on property of the structure.

The complete I–V characteristics for the gate-triggered thyristor are shown in Fig. 17 for a family of different gate currents. In the forward blocking state, the curves are similar to those shown in Fig. 16*b* except for a change of coordinates.

4.3.2 Turn-On and Turn-Off Times

To switch a thyristor from OFF to ON requires that the current be raised to a level high enough to satisfy the condition $\tilde{\alpha}_1 + \tilde{\alpha}_2 = 1$. A number of methods can be used to trigger thyristors from the OFF to the ON state.

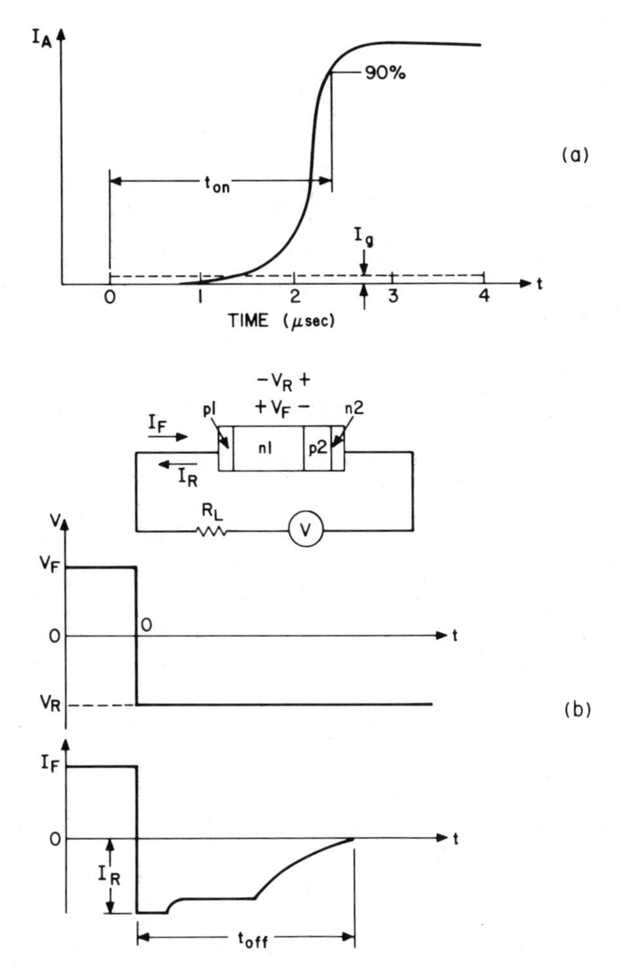

Fig. 18 (a) Turn-on characteristics when a current step I_g is applied to a thyristor. (*b*) Turn-off characteristics where the voltage suddenly changes polarity. (After Gentry et al., Ref. 5.)

The voltage triggering is the only method of switching a Shockley diode. Voltage triggering can be accomplished in two ways: by slowly raising the forward voltage until the breakover voltage is reached, or by applying the anode voltage rapidly, referred to as dV/dt triggering, considered in Section 4.3.3.

Current triggering is the most important method of switching a three-terminal thyristor. When a triggering current (e.g., gate current) is applied, the anode current through a thyristor does not respond immediately. The anode current can be characterized by a turn-on time as shown in Fig. 18*a*. Because of the regenerative nature of a thyristor, the turn-on time is approximately the geometric mean value of the diffusion times in the $n1$ and $p2$ regions, or

$$t_{on} = \sqrt{t_1 t_2} \tag{45}$$

where $t_1 \equiv W_{n1}^2/2D_p$, $t_2 \equiv W_{p2}^2/2D_n$, W_{n1} and W_{p2} are the layer widths of the $n1$ and $p2$ regions, respectively, and D_p and D_n are the hole and electron diffusion coefficients, respectively.

The result above can be derived from Fig. 9*b* with the help of the charge-control approach. We shall assume that the stored charges in the *p-n-p* and *n-p-n* transistors are Q_1 and Q_2, respectively. The collector currents in the transistors are then given by $I_{c2} \simeq Q_1/t_1$ and $I_{c1} \simeq Q_2/t_2$, respectively. Under the ideal condition that $dQ_1/dt = I_{c2}$ and $dQ_2/dt = I_g + I_{c1}$, we obtain the following equation:

$$\frac{d^2Q_1}{dt^2} - \frac{Q_1}{t_1 t_2} = \frac{I_g}{t_2}. \tag{46}$$

The solution of Eq. 46 is of the form $\exp(-t/t_{on})$ with the time constant t_{on} given by Eq. 45. To reduce the turn-on time, one must employ devices with narrow $n1$ and $p2$ layer widths. This requirement, however, is in contrast to that for large breakdown voltage, and is the reason that high-power, high-voltage thyristors have long turn-on times.

When a thyristor is in the ON state, all three junctions are forward-biased. Consequently, in the device, excess minority and majority carriers exist and increase with forward current. To switch back to the blocking state, these excess carriers must be swept out by an electric field or must decay by recombination.[22,23] A typical turn-off current waveform is shown in Fig. 18*b*. The major time delay is due to the recombination time in layer $n1$. Since the hole current through the structure is proportional to the excess charge in $n1$, we can write

$$I = I_F \exp\left(-\frac{t}{\tau_p}\right) \tag{47}$$

where $I = I_F$ at $t = 0$, and τ_p is the minority-carrier lifetime. This current must drop below the holding current I_h to permit the device to block forward voltage. Thus the turn-off time is

$$t_{off} = \tau_p \ln\left(\frac{I_F}{I_h}\right). \tag{48}$$

To obtain a small turn-off time, we must reduce the lifetime in layer $n1$. This reduction can be achieved by introducing recombination centers, such as gold and platinum, in silicon during the diffusion process, or using electron and gamma-ray irradiation.[24–26] Gold has an acceptor level near the midgap of silicon to serve as an efficient generation–recombination center. The leakage current increases with increasing gold doping. As a result, the forward breakover voltage decreases with gold doping. This decrease

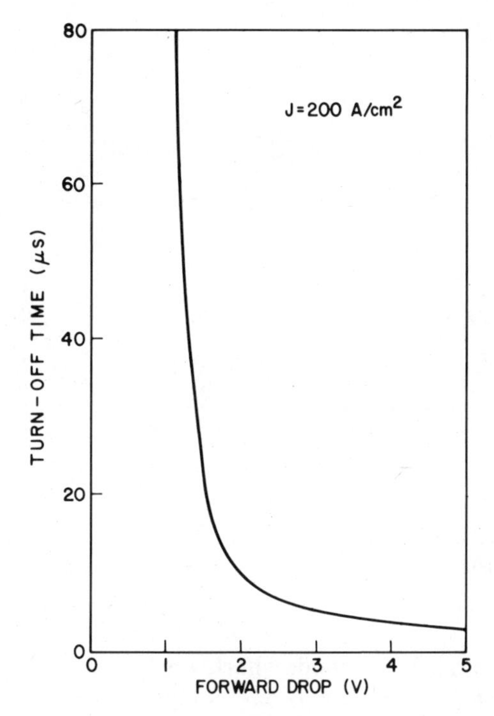

Fig. 19 Typical trade-off relationship between forward drop and turn-off time of power thyristors. (After Schlegel, Ref. 27.)

does not happen with platinum doping or electron irradiation. The reduction in lifetime will cause the forward voltage drop in ON state, Eq. 42, to increase. A typical trade-off relationship[27] between the forward drop and turn-off time for power thyristors is shown in Fig. 19. A compromise would be a turn-off time of 10 μs and a forward voltage drop of about 2 V.

To shorten the turn-off time, a common circuit practice is to apply a reverse bias between the gate and the cathode during the turn-off phase. This method is called gate-assisted turn-off.[28,29] The improvement comes about because the reverse-biased gate can divert most of the forward recovery current which would otherwise flow through the cathode during the reapplication of the forward anode voltage.

4.3.3 Cathode Short and *dV/dt* Effect

In modern Shockley diode and thyristor designs, cathode shorts are used to improve device performance.[6,7] A schematic diagram of a thyristor with cathode shorts is shown in Fig. 20*a*. A two-transistor equivalent circuit is shown in Fig. 20*b*, where the total cathode current $I_{K'}$ is the sum of I_K and the shunt current I_{shunt}. If the resistance R_{shunt} is so small that most anode

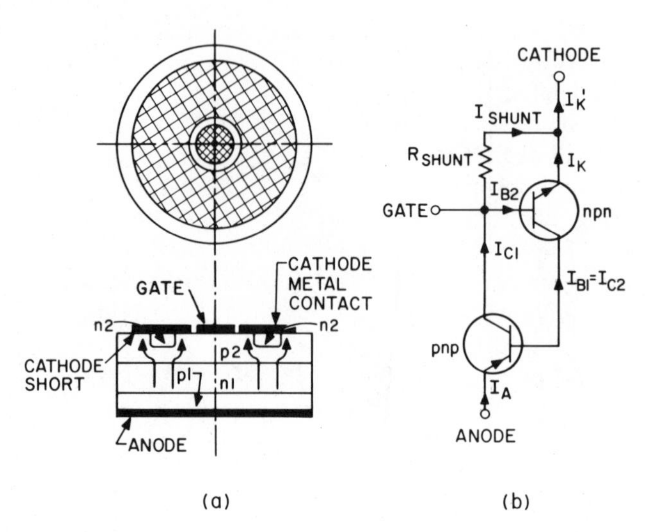

Fig. 20 (*a*) Shorted-cathode thyristor. Some current will flow through the cathode shorts. (*b*) Two transistor analogs of a thyristor with cathode short.

current flows through the cathode short, then I_K is much smaller than I_A. From Eq. 30 for $I_{gA} = 0$ and $I = I_A$, we obtain

$$I_A = \frac{MI_0}{1 - M\alpha_1 - M\alpha_2(I_K/I_A)}. \tag{49}$$

With no cathode shorts, I_K equals I_A, so that

$$I_A = \frac{MI_0}{1 - M\alpha_1 - M\alpha_2}. \tag{50}$$

Thus V_{BF} occurs when $M = 1/(\alpha_1 + \alpha_2)$ as given in Eq. 33. For the case $I_K \ll I_A$ when cathode shorts are present, Eq. 49 becomes

$$I_A = \frac{MI_0}{1 - M\alpha_1}. \tag{51}$$

Under this condition, the forward breakover voltage becomes equal to the reverse blocking voltage as given by Eq. 4.

Under transient condition, a blocking thyristor can switch to its forward ON state at voltages well below the breakover voltage. The reduction in breakover voltage depends on the magnitude of the anode voltage and its rate of increase. This phenomenon is called the dV/dt effect. This effect, which can be used to turn on a thyristor, is called dV/dt triggering. The dV/dt effect is due to the rapidly varying anode voltage giving rise to a displacement current $d(CV)/dt$, where C is the capacitance of the junction J2. This current, in turn, can cause $(\tilde{\alpha}_1 + \tilde{\alpha}_2)$ to approach 1; then switching occurs. In power thyristors, dV/dt ratings must be high so that large V_{BF}

can be maintained. To improve the dV/dt rating, one can reverse-bias the gate–cathode circuit so that the displacement current is drawn from the gate and will not affect the current gains. The lifetimes in the $n1$ and $p2$ regions can also be reduced to reduce the alphas at any current levels; this reduction of α will degrade the forward conduction mode.

An effective method to improve dV/dt rating is to use cathode shorts.[30] As shown in Fig. 20a, the displacement current in the $p2$ region will flow into the shorts. The variable α_2 of the n-p-n transistor is not affected by the displacement current. The cathode shorts can substantially improve the dV/dt capability. Typically, a 20-V/μs rating is obtainable in thyristors without cathode shorts. For shorted devices, dV/dt can be increased by a factor of 10 to 100 or more.

4.3.4 The *dI/dt* Limitation[7]

Initially, in the thyristor turn-on process, only a small area of the cathode region near the gate contact begins to conduct. This highly conducting region supplies the necessary forward current to turn on adjacent regions until the conduction process spreads over the entire cross section of the cathode. The spreading of the conduction process is characterized by a spreading velocity v_{sp}. We assume a concentric structure as shown in Fig. 20a with a centrally located gate of radius r_0, and that the anode current and anode-to-cathode voltage are linear functions of time. During the turn-on process,

$$i_A = \frac{dI_A}{dt}\, t \tag{52}$$

$$v_{AK} = V_{AK}\left(1 - \frac{t}{t_0}\right) \tag{53}$$

$$r = r_0 + v_{sp} t \tag{54}$$

where V_{AK} is the steady-state anode-to-cathode voltage; t_0 the switching OFF time; and i_A, v_{AK}, and r are the time-varying values of current, voltage, and spreading radius of the ON state, respectively.

The instantaneous power dissipated and the turned-on area are

$$i_A v_{AK} = V_{AK} \frac{dI_A}{dt}\left(1 - \frac{t}{t_0}\right) t \tag{55}$$

$$\text{area} = \pi[(r_0 + v_{sp} t)^2 - r_0^2]. \tag{56}$$

So the power density is given by

$$P = \frac{i_A v_{AK}}{\text{area}} = \frac{V_{AK}(dI_A/dt)(1 - t/t_0)t}{\pi[(r_0 + v_{sp} t)^2 - r_0^2]} \tag{57}$$

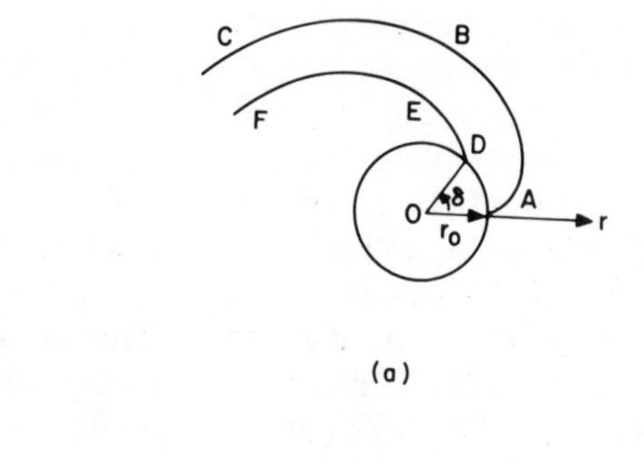

(a)

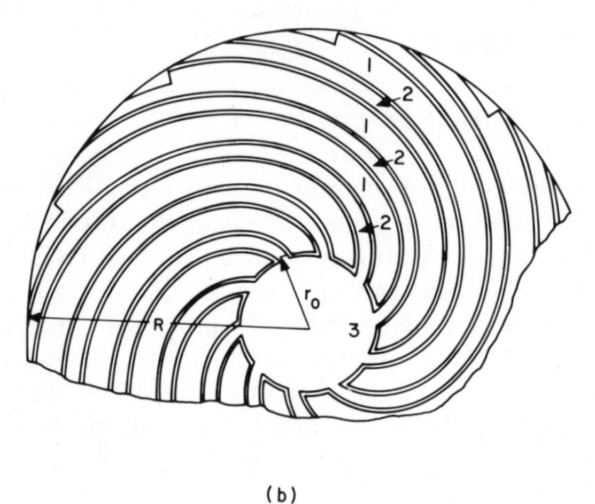

(b)

Fig. 21 The involute structure. (After Storm and St. Clair, Ref. 31.)

and the temperature rise of the hot spot is then

$$\Delta T = \frac{1}{\rho C} \int_0^\infty P\,dt \sim (dI_A/dt) \tag{58}$$

where ρ and C are the density and specific heat of silicon, respectively. Equation 58 shows that for a fixed value of V_{AK}, the temperature rise is proportional to dI_A/dt. To avoid overheating and permanent damage to the device, the allowable dI_A/dt is thus an important rating. To increase the dI/dt capability, one can overdrive the gate to enlarge the initial turn-on area or to decrease the W/L ratio of the $n1$ region so that the spreading velocity will be increased.

A number of interdigitated designs have been developed so that no part of the cathode is greater than a certain maximum allowed distance from the gate electrode. An elegant example is the involute pattern[31] shown in Fig. 21. The polar equation for the involute ABC in Fig. 21a is

$$r = r_0(1 + \theta^2)^{1/2} \tag{59}$$

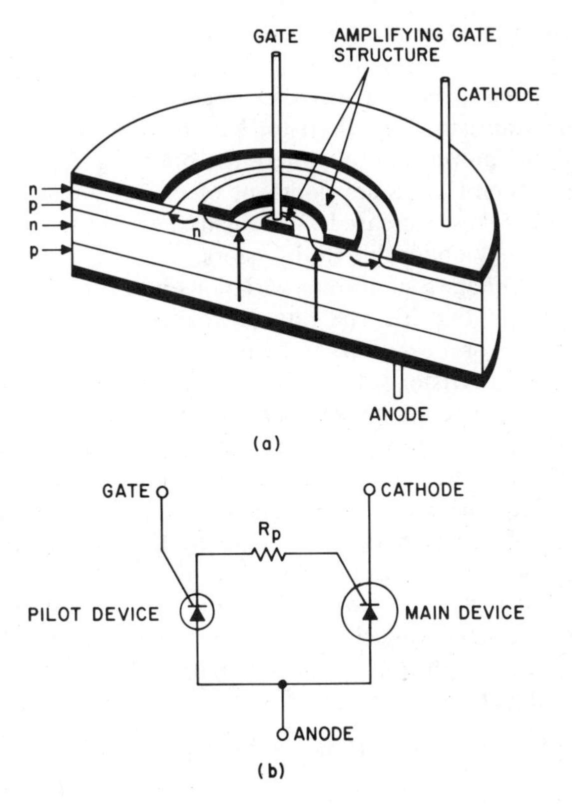

Fig. 22 (*a*) Amplifying gate arrangement. (*b*) Equivalent circuit. (After Gentry and Moyson, Ref. 32.)

and for involute *DEF* is

$$r = r_0[1 + (\theta - \delta)^2]^{1/2}. \tag{60}$$

These two involutes are always equidistant from each other by a fixed amount equal to $r_0\delta$. In the detailed gate–cathode structure shown in Fig. 21*b*, the cathode contacts are mesas marked by 1 and the gate metallizations are marked by 2. The latter are connected to the gate contact 3.

Another method to enlarge the initial turn-on area is the use of an amplifying gate[32] (Fig. 22*a*). The equivalent circuit of a thyristor with an amplifying gate is shown in Fig. 22*b*. When a small triggering current is applied to the central gate, the amplifying gate structure, which serves as a pilot device, will turn on rapidly because of its small lateral dimensions. The pilot current is much larger than the original triggering current, and it provides a strong driving current to the main device. The larger the driving current is, the larger is the initial turn-on area for the main thyristor.

4.3.5 Maximum Operating Frequency

At low switching speeds, the thyristor is generally a more efficient switch than a bipolar transistor. The thyristor has thus essentially dominated the field of industrial power control, where the operating frequency is usually 50 or 60 Hz. Recently, the development of circuit application for higher switching speeds has increased. We shall now consider the maximum operating frequency obtainable in thyristors.

The terminal voltage and current variations in a thyristor during switching[33] are shown in Fig. 23. The rate of change of current in the device during switch on or off, dI/dt, is a major factor affecting turn-on and turn-off times of thyristors. The dI/dt (OFF) is determined primarily by external circuitry, and in maximizing this parameter it is necessary to ensure that dI/dt (ON) ratings, as discussed in the preceding section, are not exceeded. The turn-off time t_{off} has been considered in Section 4.3.2. The rate at which the forward voltage is reapplied to the thyristor dV/dt after a reverse recovery period is limited by the capacitive displacement current. This current may cause the alpha of the *n-p-n* transistor to rise sufficiently to switch on the thyristor before the full forward voltage has been applied and before any signal has been applied to the gate. This effect can be substantially reduced by cathode shorts. The forward recovery time is the sum of the above three components:

$$t_{fr} = I_F/(dI/dt) + t_{off} + V_{BF}/(dV/dt) \tag{61}$$

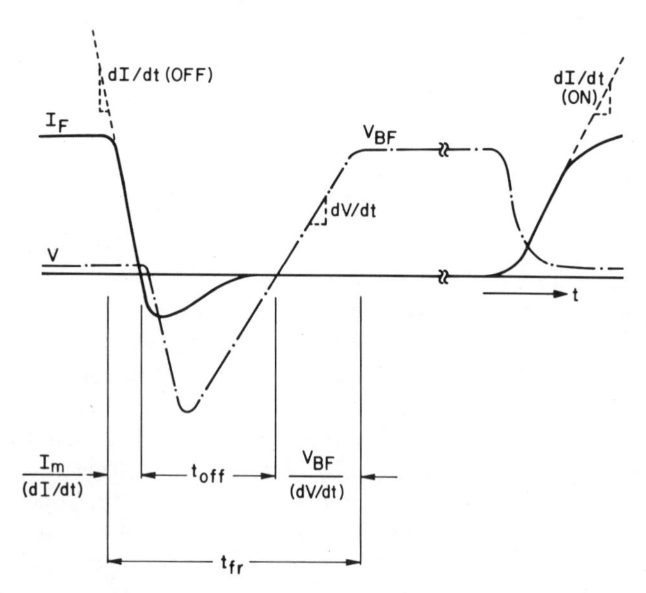

Fig. 23 Terminal voltage and current variations in a thyristor during switching. (After Roberts and Wilkinson, Ref. 33.)

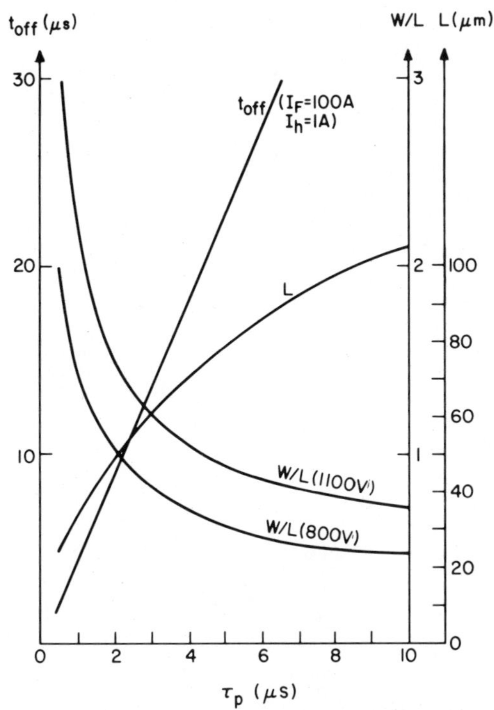

Fig. 24 Relationship between turn-off time, diffusion length, and $n1$ base width as a function of minority lifetime in $n1$ region. (After Roberts and Wilkinson, Ref. 33.)

where I_F is the peak forward current and V_{BF} is the forward breakover voltage. The maximum operating frequency is given by

$$f_m = \frac{1}{2t_{fr}}. \tag{62}$$

The relationship between the turn-off time t_{off}, the diffusion length L, and the $n1$ base width as a function of minority lifetime in $n1$ region are shown in Fig. 24 for a 800-V and a 1100-V thyristor. The figure shows that t_{off} increases linearly with τ_p in accordance with Eq. 48. The forward blocking voltage varies inversely as frequency for any given value of the ratio W/L. For a given blocking voltage, W is generally fixed and the ratio W/L varies inversely as $\sqrt{\tau_p}$.

The maximum operating frequency of conventionally fabricated thyristors blocking 800 V and 1100 V are shown in Fig. 25 for two values each of current, dI/dt, and dV/dt. For example, a 800-V, 100-A thyristor with $W/L = 0.75$ ($t_{off} = 18$ μs in Fig. 24) will have a f_m of about 20 kHz given a dV/dt capability of 100 V/μs and a dI/dt of 100 A/μs. To improve f_m, either dV/dt or the dI/dt rating would have to be increased or the

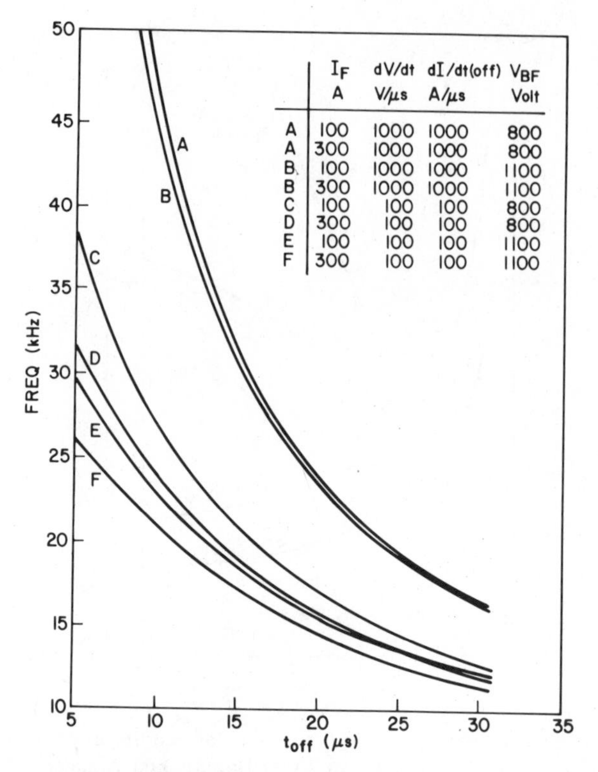

Fig. 25 Maximum operating frequencies of various thyristors with operating parameters given in the figure. (After Roberts and Wilkinson, Ref. 33.)

complexity of the device structure increased, or both, so that for the same blocking voltage and forward current, a shorter lifetime would be feasible and a shorter turn-off time result.

4.4 RELATED POWER THYRISTORS

4.4.1 Reverse Conducting Thyristors

A reverse conducting thyristor (RCT) is a three-terminal, multilayer thyristor, which behaves similarly to a conventional thyristor in the forward direction, but conducts large current in the reverse direction. Its uses include applications in electroluminescent drivers and ac bidirectional switching circuits. The unique device feature of the RCT is that it has both cathode short and anode short. A cross-sectional diagram[34] of a RCT is shown in Fig. 26*a*. Note that only a positive bevel for the J2 junction is needed. An n^+ region is introduced between the p and n regions near the anode. Hence in the reverse direction, when the anode is negatively biased

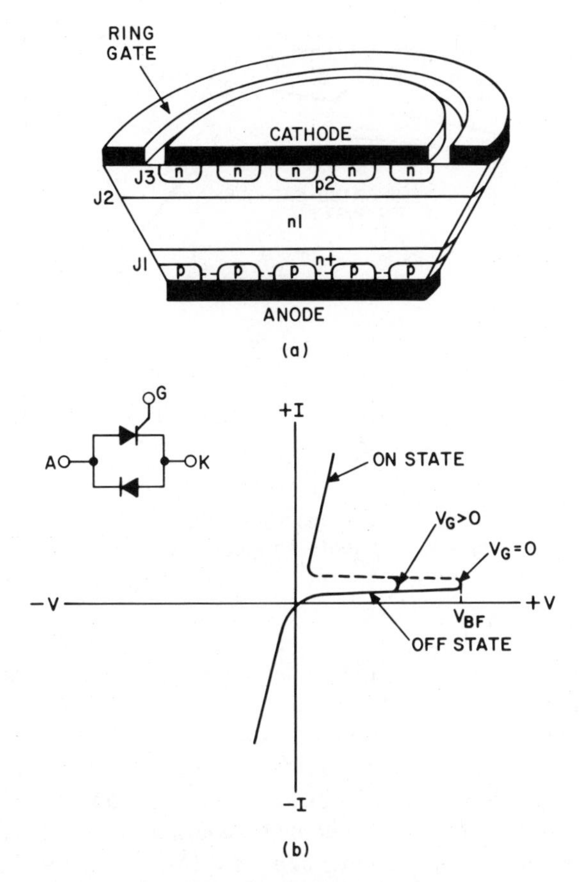

Fig. 26 (*a*) Cross section of a high-voltage, high-temperature reverse-conducting thyristor (RCT). (*b*) Current–voltage characteristics and device symbol of a RCT. (After Kokosa and Tuft, Ref. 34.)

with respect to the cathode, the junction J1 which usually supports the reverse bias is now shorted to the anode. A large current will conduct in the reverse direction. The forward $I-V$ characteristics are the same as a conventional thyristor. The $I-V$ curves and the device symbol are shown in Fig. 26*b*.

The power RCT devices can be operated at junction temperatures above 150°C as compared to 125°C for the conventional thyristors. Thus a RCT can handle higher currents in the ON state.

Figure 27 compares the forward blocking voltage and junction temperature for a RCT and a shorted-cathode thyristor. As discussed before, shorted-cathode thyristor has V_{BF} equal to the reverse blocking voltage, which is higher than that for a thyristor without cathode shorts. The RCT in the forward blocking state is independent of the influence of the current

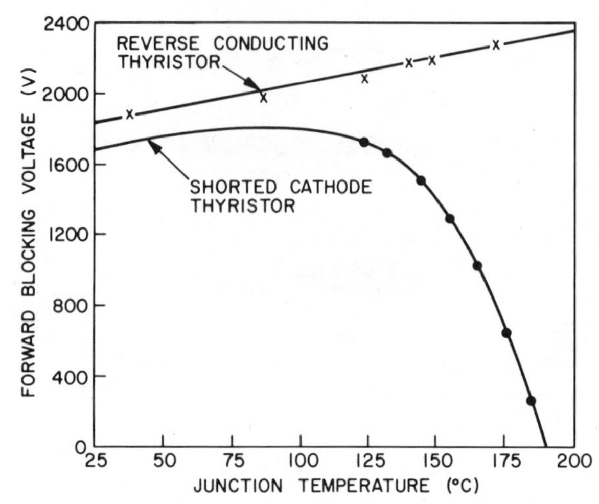

Fig. 27 Temperature dependence of the forward breakover voltage for a reverse-conducting thyristor compared to that of a shorted-cathode thyristor. (After Kokosa and Tuft, Ref. 34.)

gains, and the V_{BF} is equal to the avalanche breakdown voltage of junction J2. Since avalanche breakdown has a positive temperature coefficient, V_{BF} of the RCT will increase with temperature (Fig. 27). Note that above 125°C, V_{BF} for shorted-cathode thyristor decreases rapidly with temperature, whereas V_{BF} for the RCT continues to increase beyond 150°C.

The RCT can also have fast recovery characteristics, because the forward blocking capability is independent of transistor action. Consequently, a RCT can be more heavily doped with gold or platinum than can a conventional thyristor without deteriorating the forward blocking capability.

4.4.2 Light-Activated Thyristor

The light-activated thyristor, also called the light-activated switch (LAS), is a three-terminal four-layer reverse blocking thyristor which can be turned on by exceeding its light threshold level. The LAS enables perfect electrical isolation between power and trigger circuits through the use of fiber-optical transmission of the trigger energy. Its applications include photoelectric control, position monitoring, light coupled, and triggering circuits.

A simplified LAS device structure and its symbol are shown[35] in Fig. 28. The device has a cathode short to improve its dV/dt capability and its temperature stability of the forward breakover voltage. The cathode area is irradiated homogeneously from a light source through an optical fiber up to a radius r_l, resulting in a uniform generation of electron–hole pairs within

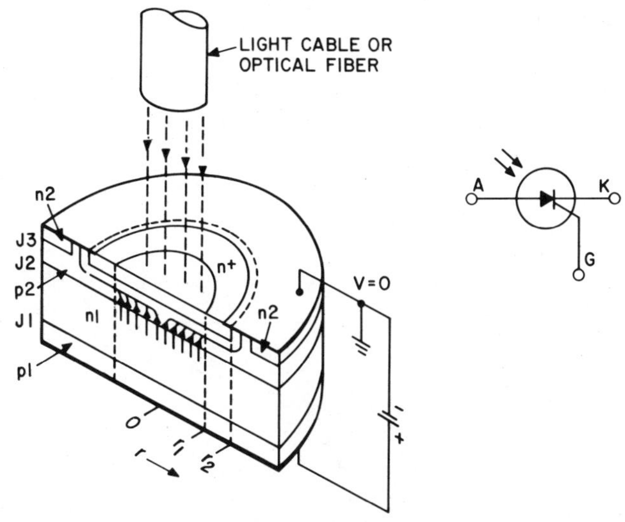

Fig. 28 Schematic diagram and symbol of a light-activated thyristor having a shorted cathode. (After Gerlach, Ref. 35.)

the irradiated area. The inner cathode region is annularly shorted at $r = r_2$.

The most important contribution to the external current is due to electron–hole pairs generated inside the space-charge region of the reverse-biased junction J2. These electron–hole pairs will be separated by the electric field within a time interval of the order of 1 ns, or instantaneously, compared to the turn-on time of a typical thyristor (of the order of 1 μs). Holes are transported to the $p2$ region and electrons to the $n1$ region. Both regions are supplied with equal amounts of majority carriers without any time delay to the carrier generation, so that at the moment of light turn-on the anode current increases abruptly by the amount of photocurrent as shown in Fig. 29.

The photocurrent is then amplified in the two-transistor p-n-p-n structure with regenerative action. The anode current continues to increase after some delay caused by the transit time of the injected minority carriers through the base regions. If the sum of alphas remains smaller than 1 (I_{ph1} to reach I_{A1}), the current approaches a stationary value asymptotically in a time interval t_m equal to the average minority lifetimes in $n1$ and $p2$ regions. Let I_A^* be the stationary current for which the sum of alphas equals 1, and I_{ph2} a photocurrent for which the anode current I_{A2} approaches a stationary value higher than I_A^*. Then, a short time after the anode current has exceeded the value of I_A^*, the regenerative action of the feedback current will start and lead to a rapid increase of the anode current. The thyristor will switch to its ON state, as Fig. 29 illustrates. The turn-on will shift to shorter time delays with increasing photocurrent, and therefore the turn-on time becomes shorter with enhanced light intensity.

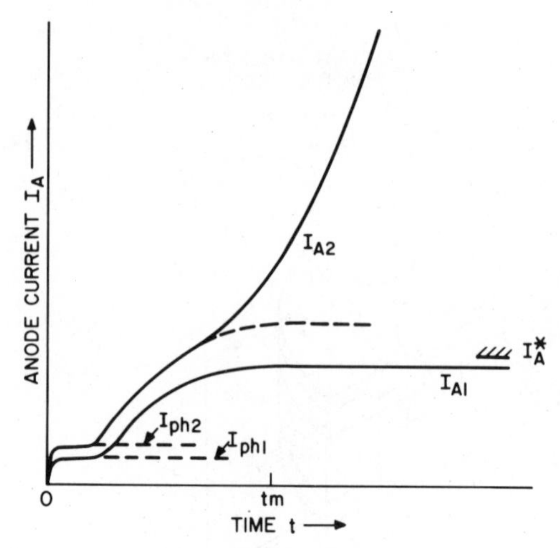

Fig. 29 Turn-on characteristics of a light-activated thyristor for different photocurrents. (After Gerlach, Ref. 35.)

The experimental turn-on characteristics of a high-voltage thyristor are shown in Fig. 30. The dashed curve represents the turn-on with gate current. Curve a shows turn-on at low light energy (25 mW-s) and curve b at higher drive (50 mW-s). At higher drive, the turn-on delay is very short.

The amount of photocurrent to trigger a thyristor depends on the wavelength λ of the light. For silicon the peak spectral response occurs at $\lambda = 0.85$ to $1.0\ \mu m$. The effective light sources with λ in this range include GaAs–GaAlAs double-heterostructure lasers, GaAs lasers, and GaAs light-emitting diodes (refer to Chapter 12).

A power thyristor can be turned on with very small light power

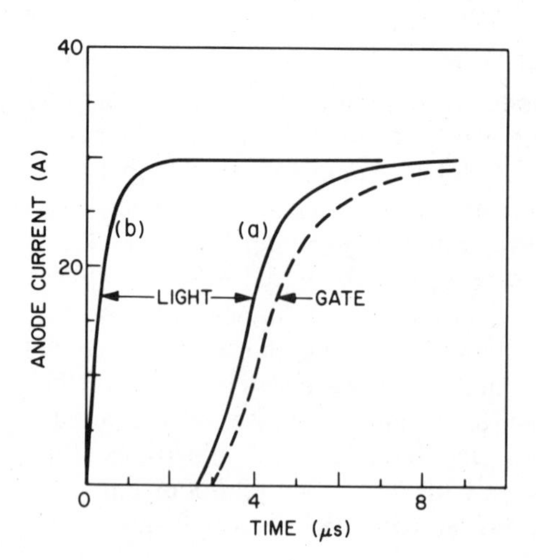

Fig. 30 Experimental turn-on characteristics of a high-power, light-activated thyristor by a neodymium laser. Energy of the laser pulse: (a) 2.5×10^{-5} W-s, (b) 5×10^{-4} W-s. (After Gerlach, Ref. 35.)

(~0.2 mW for a 3-kV thyristor), because the light power can be focused onto a very small area. For example, for a single glass fiber with 100 μm diameter, the initial turn-on area can be less than 10^{-2} mm^2. Therefore, power density in the initial turn-on area can be very high. For the shorted-cathode LAS, (Fig. 28), the required minimum light power varies approximately inversely as r_2/r_1. Hence a larger r_2/r_1 ratio can reduce the light power. However, even at $r_2/r_1 = 5$, a light power of about 5 mW is necessary for firing, which is an order of magnitude higher than an unshorted-cathode LAS. Therefore, there is a trade-off between the light power and the dV/dt capability.

Once the light power turns on the initial area, the regenerative action of the device will enlarge the turned-on area, and eventually the full cathode will be on. After triggering has been brought into action and when the anode current prevails over the photocurrent, the light power can be turned off without any change of the anode current.

4.4.3 Gate Turn-Off Thyristor

A gate turn-off thyristor (GTO) is a four-layer reverse blocking thyristor that can be turned on with a positive gate voltage and turned off with a negative gate voltage. A conventional thyristor is turned off generally by reducing the anode current to below the holding current or by reversing the anode current. A GTO can be used for applications in inverter, pulse generator, chopper, and dc switching circuits. The GTO is often used in preference to transistor in high-speed, high-power applications because of its ability to withstand higher voltage in its OFF state.

A schematic GTO circuit is shown[36] in Fig. 31*a*. In a one-dimensional description of the turn-off process, one can consider a GTO having a negative gate current of magnitude I_g. By neglecting all leakage currents, the base drive required to sustain the *n-p-n* transistor in its ON state is equal to $(1-\alpha_2)I_K$. The actual base current is $(\alpha_1 I_A - I_g)$. Therefore, the

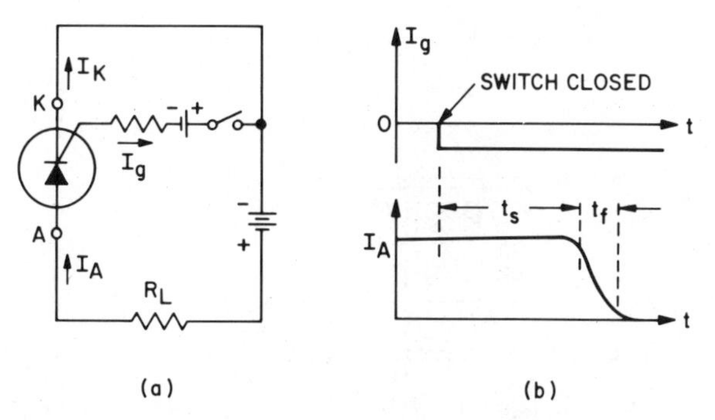

Fig. 31 (*a*) Circuit diagram for gate turn-off thyristor, GTO. (*b*) Turn-off characteristics of a GTO. (After Wolley, Ref. 36.)

turn-off condition is

$$\alpha_1 I_A - I_g < (1-\alpha_2) I_K. \tag{63}$$

Since $I_A = I_K + I_g$, the required I_g is given from Eq. 63:

$$I_g > \left(\frac{\alpha_1 + \alpha_2 - 1}{\alpha_2}\right) I_A. \tag{64}$$

We shall define the ratio I_A/I_g as the turn-off gain β_{off}:

$$\beta_{off} \equiv \frac{I_A}{I_g} < \frac{\alpha_2}{\alpha_1 + \alpha_2 - 1}. \tag{65}$$

A high β_{off} can be obtained by making α_2 of the n-p-n transistor as close to unity as possible, and at the same time making α_1 of the p-n-p transistor small.

In an actual thyristor, the turn-off is a two-dimensional process. Prior to the applied negative I_g, both transistors are heavily saturated in their ON state. The removal of excess stored charges is an important part of the turn-off process. This removal of stored charges results in a storage time delay t_s, followed by a fall time t_f (Fig. 31*b*), after which the thyristor is in its OFF state.

As soon as a negative bias is applied to the gate, stored charges will be removed from $p2$ by the gate current. This removal is an inverse of the current spreading during the turn-on process. Because of the voltage drop due to the lateral current in $p2$, the junction J3 becomes less positively biased as we proceed from the center of the device toward the gate contact (Fig. 32). Eventually, the portion of J3 that is closest to the gate contact becomes reverse-biased. At this point, all the forward current will be squeezed into the remaining portion of J3 that is still forward-biased. The forward current will be progressively squeezed into a smaller and smaller

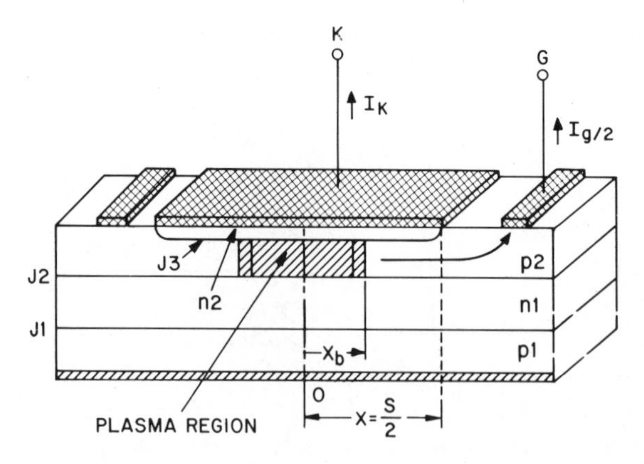

Fig. 32 Plasma focusing in the *p* base of a gate turn-off thyristor. (After Wolley, Ref. 36.)

region until some limiting dimension is reached. At that limit, the remaining excess charge in $p2$ is removed, and storage phase is over. The storage time is given by the expression[36]

$$t_s = t_{p2}(\beta_{\text{off}} - 1)\ln\left(\frac{Sx_0/W_{p2}^2 + 2x_0^2/W_{p2}^2 - \beta_{\text{off}} + 1}{4x_0^2/W_{p2}^2 - \beta_{\text{off}} + 1}\right) \tag{66}$$

where t_{p2} is the transit time through the $p2$, S the length of the cathode electrode, x_0 the decay constant of electron concentration for $x > x_b$ and is of the order of an electron diffusion length, and W_{p2} is the width of $p2$ region. The storage time will increase with increasing turn-off gain β_{off}. There is a trade-off between the storage time and the turn-off gain. To reduce the storage time, low β_{off} (corresponding to large gate current) will be used.

The fall time in Fig .31*b* corresponds to the time required to expand the depletion layer across J2 and to remove the hole charges in this region. The total charge per unit area in $n1$ region is

$$Q \simeq qp^*W(V_A) \simeq J_A t_f \tag{67}$$

where p^* is the average hole concentration in the $n1$ region, W the depletion-layer width for a given anode voltage V_A, and J_A the anode current density. From Eq. 67 the fall time is given by

$$t_f \simeq \frac{qp^*W(V_A)}{J_A}. \tag{68}$$

The fall time decreases with increasing anode current density, and increases with $\sqrt{V_A}$ because W varies approximately as $\sqrt{V_A}$.

Reliable operation of GTO can be obtained when the final area of the squeezed plasma is large enough to prevent excessive current density. This requirement has resulted in the use of interdigitated designs, such as the involute structure (Fig. 21). The use of an amplifying gate is also desirable to achieve fast turn-on. The main difference between GTO and the previously discussed gate-assisted turn-off thyristor is that the former can be turned off by the application of a negative bias on the gate, while the anode is kept positive with respect to the cathode. On the other hand, the latter requires commutation of the supply voltage to turn it off, and a reverse gate bias is used to reduce the turn-off time.

4.5 DIAC AND TRIAC

The diac (*di*ode *ac* switch) and triac (*tri*ode *ac* switch) are bidirectional thyristors.[37,38] They have ON and OFF states for positive or negative anode voltages and are therefore useful in ac applications.

The two diac structures are the ac trigger diode and the bidirectional *p-n-p-n* diode switch. The former is simply a three-layer device similar in

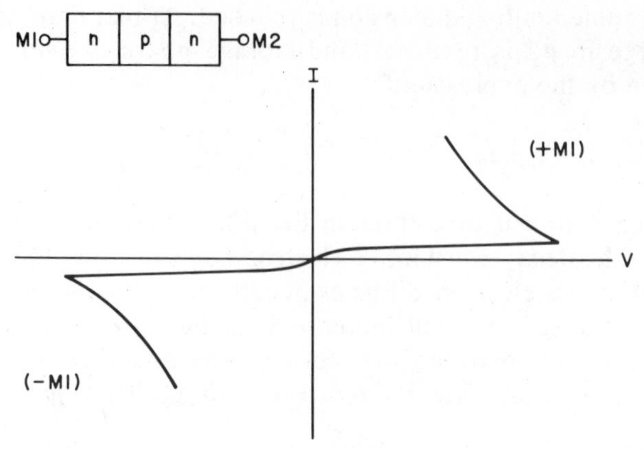

Fig. 33 Typical characteristics for an ac trigger diode. The insert shows the device structure.

construction to a bipolar transistor, except that doping concentrations at the two junctions are approximately the same and no contact is made to the base region. The equal doping levels result in a symmetrical, bidirectional characteristic as shown in Fig. 33, where a schematic cross-sectional diagram of the device is shown in the insert. When a voltage of either polarity is applied to a diac, one junction is forward-biased and the other is reverse-biased. The current is limited by the leakage current of the reverse-biased junction. When the applied voltage is sufficiently large, breakdown occurs at $BV_{CBO}(1-\alpha)^{1/n}$, where BV_{CBO} is the avalanche breakdown voltage of the *p-n* junction, α the common-base current gain, and n is a constant. This expression is the same for the breakdown voltage of an open base *n-p-n* transistor (refer to Chapter 3). As the current increases after breakdown, α will increase, causing a reduction of the terminal voltage. This reduction gives rise to the negative region.

The bidirectional *p-n-p-n* diode switch behaves like two conventional Shockley diodes connected in antiparallel to permit the accommodation to voltage signals of two polarities, as in Fig. 34*a*, where M1 stands for main terminal 1, and M2 for main terminal 2. Using the shorted-cathode principle, we can integrate this arrangement into a single two-terminal diac as shown in Fig. 34*b*. The symmetry of this structure results in identical performance for either polarity of applied voltage. The symmetrical $I-V$ characteristics and the device symbol are shown in Fig. 34*c*. Similar to the Shockley diode, the diac can be triggered into conduction by exceeding the breakover voltage or by dV/dt triggering. Because of its regenerative action, the bidirection *p-n-p-n* diode switch has a larger negative resistance and smaller forward drop than that of an ac trigger diode.

The triac can switch the current in either direction by applying low-voltage, low-current pulse of either polarity between the gate and one of

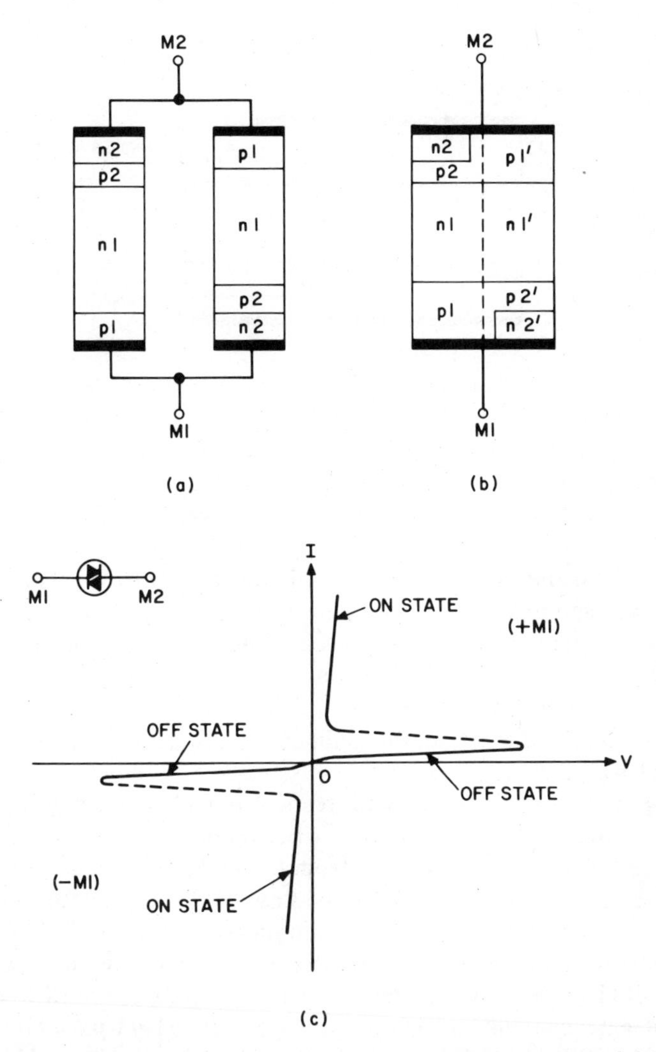

Fig. 34 (*a*) Two Shockley diodes connected in antiparallel. (*b*) Integration of the diodes into a single two-terminal device. (*c*) Current–voltage characteristics and device symbol for a diac.

the two main terminals, M1 and M2 (Fig. 35). The triac is very useful in light dimming, motor speed control, temperature control, and other applications. The triac structure is considerably more complicated than a conventional thyristor. In addition to the $p1$-$n1$-$p2$-$n2$ basic four layers, there are a junction gate $n3$ region and a $n4$ region in contact with M1. Note also that $p1$ is shorted to $n4$, $p2$ is shorted to $n2$, and $n3$ is shorted to $p2$ by M1, M2, and the gate electrode, respectively.

The device operations under various biasing conditions are illustrated in

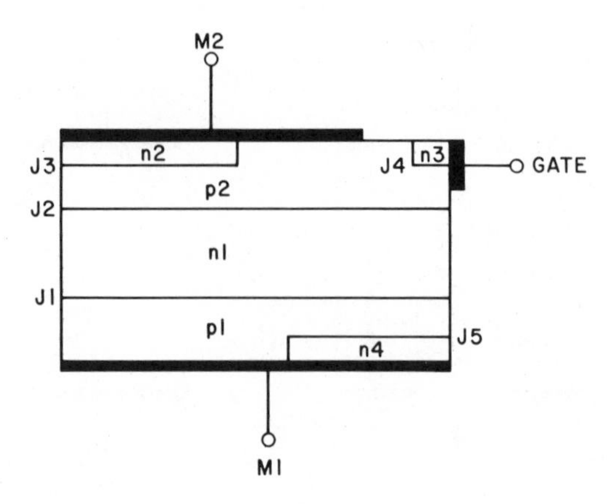

Fig. 35 Cross section view of a triac, a six-layer structure having five *p-n* junctions and three electrode shorts.

Fig. 36. When the main terminal M1 is positive with respect to M2 and a positive voltage is applied to the gate (also with respect to M2), the device behavior is identical to that of a conventional thyristor (Fig. 36*a*). The junction J4 is reverse-biased and is inactive; the gate current is supplied through the gate short near the *n*3 region. Since junction J5 is also reverse-biased and inactive, the main current is carried through the left side of *p*1-*n*1-*p*2-*n*2 section.

In Fig. 36*b*, M1 is positive with respect to M2, but a negative voltage is applied to the gate. The junction J4 between *n*3 and *p*2 is now forward-biased, and electrons are injected from the *n*3 to *p*2. The auxiliary thyristor *p*1-*n*1-*p*2-*n*3 will be turned on by the flow of the lateral base current in *p*2 toward the *n*3 gate because of the gain increase in the transistor *n*3-*p*2-*n*1. Full conduction of this auxiliary thyristor results in the current flowing out of this device and toward the *n*2 region. This current will provide the required gate current and trigger the left-side *p*1-*n*1-*p*2-*n*2 thyristor into conduction. When M1 is negatively biased with respect to M2, and V_G is positively biased, the junction J3 becomes forward-biased between M2 and the shorted gate (Fig. 36*c*). Electrons are injected from *n*2 to *p*2 and diffuse to *n*1, resulting in an increase of the forward bias of J2. By the regenerative action, eventually full current is carried through the short at M2. The gate junction J4 is reverse-biased and is inactive. The full device current is carried through the right-side *p*2-*n*1-*p*1-*n*4 thyristor.

Figure 36*d* shows the condition for M1 negative with respect to M2 and V_G is also negative. In this condition, the junction J4 is forward-biased, and triggering is initiated by injection of electrons from *n*3 to *n*1 region. This action lowers the potential at *n*1, causing holes to be injected from *p*2 into *n*1 region. These holes provide the base current drive for the *p*2-*n*1-*p*1 transistor, and the right side *p*2-*n*1-*p*1-*n*4 thyristor is eventually turned on.

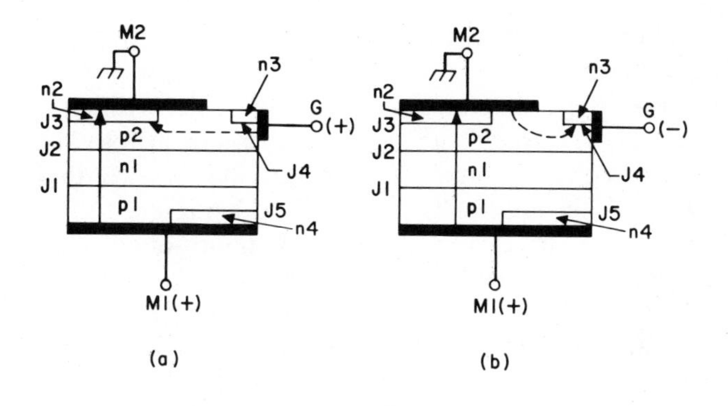

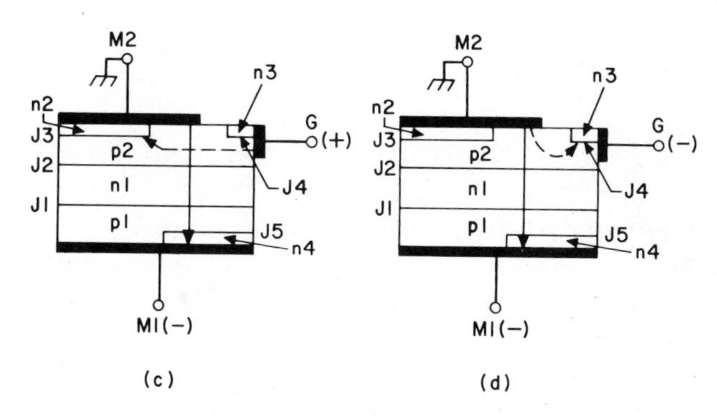

Fig. 36 Current flow in four different triggering modes of a triac. (After Gentry, Scace, and Flowers, Ref. 37.)

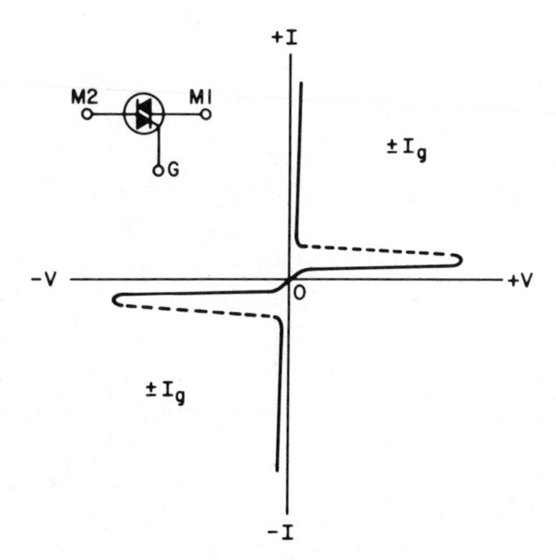

Fig. 37 Current–voltage characteristics and device symbol of a triac.

Since J3 is reverse-biased, the main current is carried from the short at M2 through the $n4$ region.

The current–voltage characteristics and the device symbol of a triac are shown in Fig. 37. The triac is a symmetrical triode switch that can control loads supplied with ac power. The integration of two thyristors on a single chip results in only half the structure being used at any one time (Fig. 36). Therefore, triac area utilization is poor—about equal to that of two independently connected thyristors. The main advantages of the device are its perfect matching of output characteristics, and the elimination of one package and additional external connections. However, their input characteristics are grossly mismatched. The diac is essential to drive these devices to compensate for the mismatched input. Triacs now have encompassed a wide range of operating voltages (up to 1600 V) and currents (over 300 A).

4.6 UNIJUNCTION TRANSISTOR AND TRIGGER THYRISTORS

In this section we consider a few low-power thyristors. They are useful for applications involving low voltages and low currents. One of the most important applications is to serve as a triggering device for the power thyristors.

4.6.1 Unijunction Transistor

The unijunction transistor (UJT) belongs to the thyristor family. The UJT has a high-impedance OFF state and a low-impedance ON state similar to a thyristor. Switching from the OFF state to the ON state, however, is caused by conductivity modulation and not by regenerative bipolar transistor action. The UJT is a three-terminal device having one emitter junction and two base ohmic contacts. The device has evolved from the alloyed germanium bar structure originally discussed by Shockley, Pearson, and Haynes.[39] At that time the structure was called a filamentary transistor. As the device developed through the cube UJT, the diffused planar structure, and the epitaxial planar structure, the terms "double-base diode" and finally "unijunction transistor" were coined for the device.[40–44] A schematic diagram of a UJT is shown in Fig. 38*a*. The two ohmic contacts are called base 1 (B1) and base 2 (B2). The *p-n* junction located between B1 and B2 is called the emitter junction.

The equivalent circuit of a UJT is shown in Fig. 38*b*. In the normal operating condition, the base 1 terminal is grounded and a positive bias voltage V_{BB} is applied at base 2 (Fig. 38*a*). The resistance between B1 and B2 is designated by R_{BB}, that between B2 and A by R_{B2}, and that between A and B1 by R_{B1} ($R_{BB} = R_{B2} + R_{B1}$). The G_p and G_n are the excess hole and electron conductance between the emitter and base 1. The applied voltage

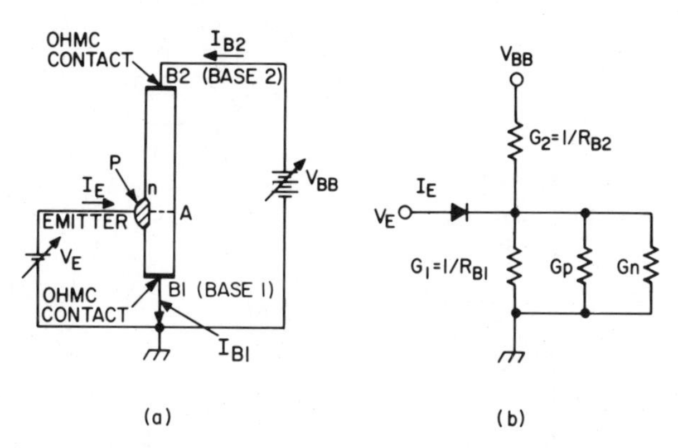

Fig. 38 (*a*) Schematic diagram of unijunction transistor (UJT). (*b*) Equivalent circuit of UJT.

V_{BB} establishes a current and an electric field along the semiconductor bar and produces a voltage on the *n* side of the emitter junction, which is a fraction η of the applied voltage V_{BB}. The fraction η is called the intrinsic stand-off ratio and is given by

$$\eta \equiv \frac{R_{B1}}{R_{B1}+R_{B2}} = \frac{R_{B1}}{R_{BB}}. \tag{69}$$

When the emitter voltage V_E is less than ηV_{BB}, the emitter junction is reverse-biased and only a small reverse saturation current flows in the emitter circuit. If the voltage V_E exceeds ηV_{BB} by an amount equal to the forward voltage drop of the emitter junction, holes will be injected into the bar. Because of the electric field within the semiconductor bar, these holes will move toward base 1 and increase the conductivity of the bar in the region between the emitter and base 1. As I_E is increased, the emitter voltage will decrease because of the increased conductivity, and the device will exhibit a negative resistance characteristic.

The device characteristic and the symbol for UJT are shown in Fig. 39. The two important points on the curve are the peak point and the valley point. At these two points the slope $dV_E/dI_E = 0$. The region with current less than I_P is called the cutoff region (OFF state). The region between the peak and valley part is called the negative resistance region; here the conductivity modulation is important. The region with current larger than I_V is called the saturation region (ON state). The switching time from the peak to the valley point depends on the device geometry and the biasing condition. The time is proportional to the distance between the emitter and base 1 contact.[45]

A UJT can be used in a triggering circuit as shown[46] in Fig. 40*a*. The capacitor C_1 charges through R_1 until the emitter voltage reaches the V_P, whereupon the UJT switches to the ON state and discharges C_1 through

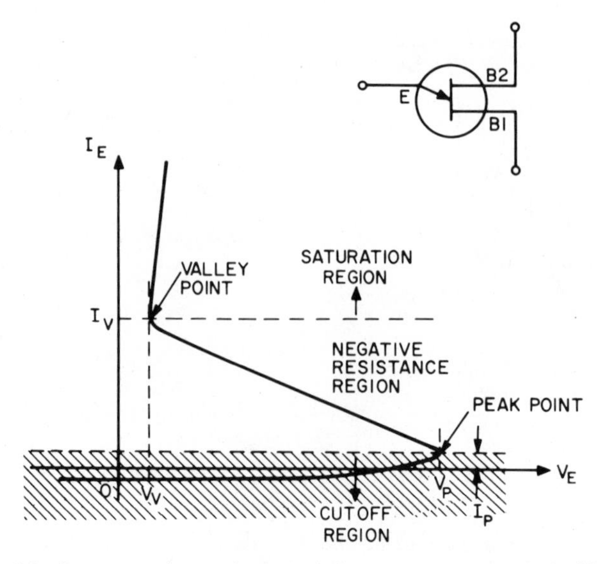

Fig. 39 Current–voltage characteristics and device symbol of UJT.

R_{B1}. When the emitter voltage reaches approximately 2 V, the UJT switches to the OFF state and the cycle is repeated. The waveforms of V_E and the output voltage are shown in Fig. 40*b*. The period of oscillation is given by

$$T \simeq R_1 C_1 \ln\left(\frac{1}{1-\eta}\right). \tag{70}$$

The output voltage can be coupled to a thyristor gate for triggering applications.

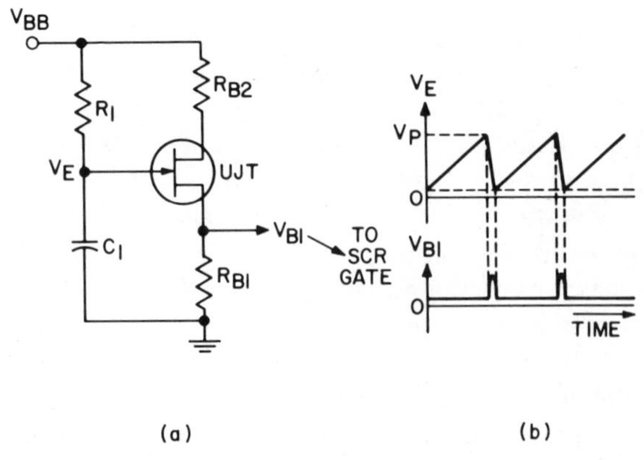

Fig. 40 (*a*) Basic UJT relaxation oscillator trigger circuit. (*b*) Output waveforms. (After *SCR Manual*, Ref. 46.)

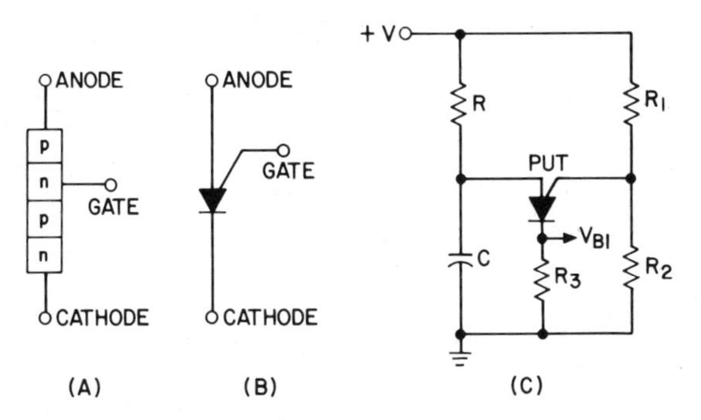

Fig. 41 Programmable unijunction transistor. (*a*) Structure. (*b*) Symbol. (*c*) Typical circuit. (After Blicher, Ref. 6.)

4.6.2 Programmable Unijunction Transistor

The programmable unijunction transistor (PUT) is basically a *p-n-p-n* device with an anode gate (gate contact made to the *n*-base instead of to the *p*-base as in a conventional thyristor). Figure 41 shows the device structure and symbol for PUT together with a typical PUT circuit. The waveforms observed are similar to that for UJT, as shown in Fig. 40*b*. The period of oscillation is also given by Eq. 70 except that $\eta = R_2/(R_1 + R_2)$, where R_1 and R_2 are the resistances of the voltage divider. For a given RC product, the oscillation frequency can be adjusted by varying the circuit elements R_1 and/or R_2. Therefore, PUT is more flexible in frequency adjustment than UJT.

4.6.3 Silicon Unilateral Switch

The silicon unilateral switch (SUS) is a *p-n-p-n* device with an anode gate, similar to a PUT. It is a small thyristor with a built-in avalanche diode between the anode gate and cathode. When the anode gate voltage exceeds that of the avalanche diode, the SUS turns on. The I–V characteristics of SUS are similar to that of a thyristor. The SUS can be used in a relaxation trigger circuit similar to that of Fig. 40*a*. Since the switching voltage is fixed by the breakdown voltage and the turn-on current is high, the lower and upper limits of the time delays obtainable from SUS are more restricted than UJT. Because the breakdown voltage of the built-in avalanche diode is essentially independent of temperature, the V_{BF} is also temperature-independent. Therefore, the SUS has the advantage for stable, low-voltage triggering.

4.6.4 Silicon Bilateral Switch

The silicon bilateral switch (SBS) consists of two SUSs connected back to back and integrated on the same silicon substrate. The I–V characteristics are similar to those of a triac. The SBS operates as a switch under positive and negative polarities of the applied voltage. The SBS is especially useful for triggering the triac with alternate $\pm V_G$ pulses.

4.7 FIELD-CONTROLLED THYRISTOR

A field-controlled thyristor (FCT) is a power-switching device consisting of a p-ν-n diode with multiple grids as shown[47] in Fig. 42*a*. The grid structure is similar to that of power field-effect transistors.[48]

When the anode and cathode junctions are forward-biased and the grid contacts are open, electrons and holes are injected into the ν base region, lowering its resistivity and resulting in the low voltage drop. This is the ON state. When a reverse bias is applied to the grids with respect to the cathode, the current that has been going from anode to cathode is diverted to the grids, because it is now an efficient hole collector (Fig. 42*b*). If the applied grid bias is large enough, the depletion regions meet under the cathode contact and a potential barrier is established as illustrated in Fig. 42*c*, which shows the equipotential lines within the depletion region. The potential goes from a positive value associated with the anode, through zero to some negative potential, then back to zero at the grounded cathode. The potential well thus established represents a barrier to electrons and prevents them from being injected at the cathode. Without a source of electrons, holes cannot be injected at the anode, and hence the device is in the forward blocking or OFF state.

The current–voltage characteristics of a FCT are shown[47,49] in Fig. 43. For the ON state, typical forward voltage across the device is around 1 V. For the OFF state, the maximum forward blocking voltage V_{AK} (voltage between anode and cathode), increases with increasing negative bias on the grid V_{GK}. A forward blocking gain is defined as

$$\mu \equiv -\frac{\Delta V_{AK}}{\Delta V_{GK}}. \tag{71}$$

Figure 44*a* shows[50] the forward blocking voltage as a function of the grid bias and grid depth for FCTs fabricated on 100 Ω-cm phosphorus-doped silicon wafers. At a given grid bias, the forward blocking voltage increases with grid depth. The forward blocking gain is shown in Fig. 44*b*. The gain increases exponentially with the grid depth. It has also been found that the turn-off decreases by a factor of 2 to 4 when the grid depth increases from 16 to 36 μm.

The FCT can exhibit faster turn on and turn off than the conventional thyristor, because the device can be turned on by simply removing the grid bias and turned off by removing minority carriers; no regenerative cycle needs to be interrupted. Since FCTs are nonregenerative, they have immunity to static dV/dt just as the transistor has and can operate at higher temperature than the conventional thyristor.

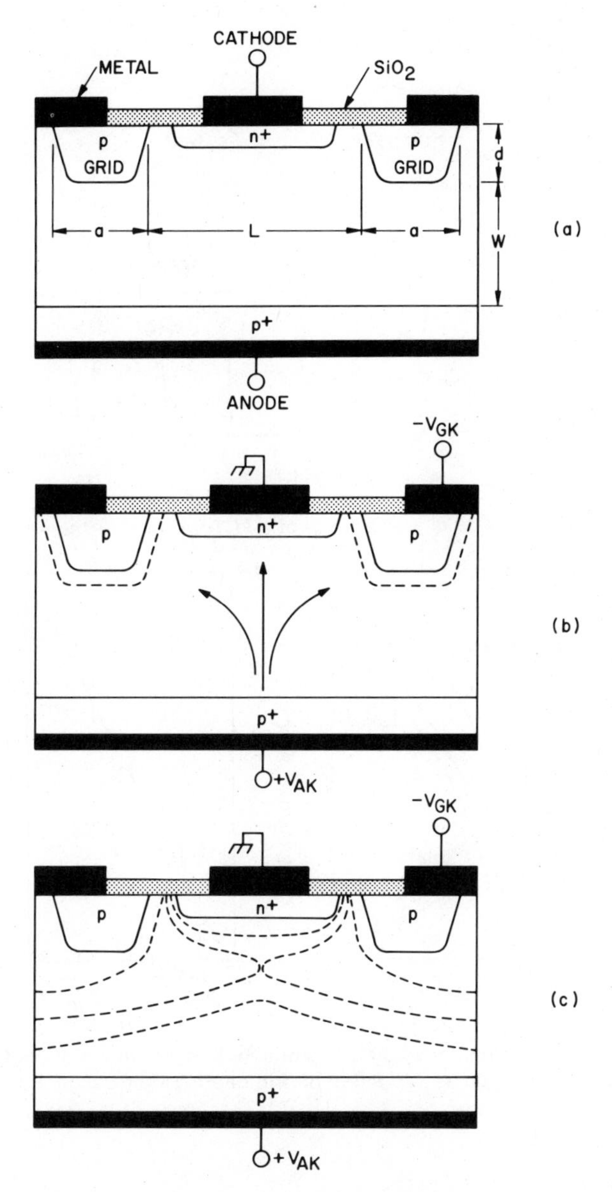

Fig. 42 (*a*) Cross section of planar field-controlled thyristor (FCT). (*b*) Cathode current is diverted to the reverse-biased grid. (*c*) Equipotentials in the depletion region under forward blocking condition. (After Houston et al., Ref. 47.)

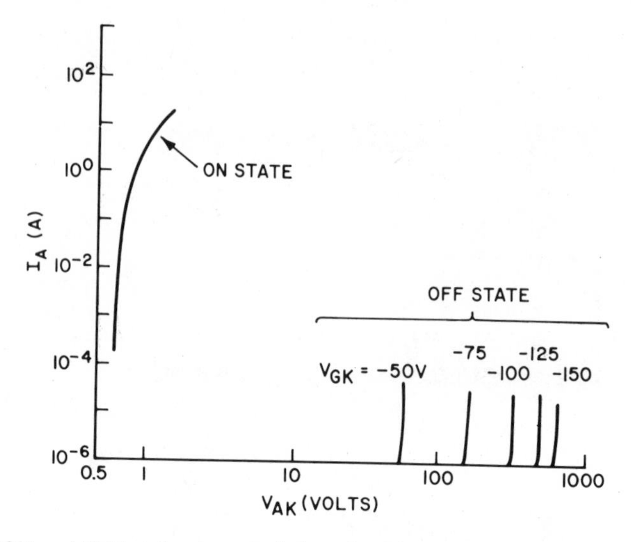

Fig. 43 ON and OFF state characteristics of a FCT. (After Houston et al., Ref. 47.)

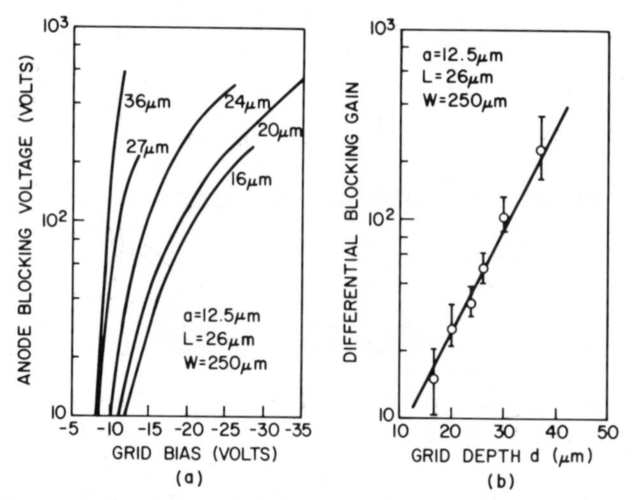

Fig. 44 (*a*) Forward blocking characteristics of a FCT as a function of the grid bias and grid depth. (*b*) Blocking gain as a function of grid depth. (After Baliga, Ref. 50.)

REFERENCES

1 W. Shockley, *Electrons and Holes in Semiconductors*, D. Van Nostrand, Princeton, N.J., 1950, p. 112.

2 J. J. Ebers, "Four-Terminal *p-n-p-n* Transistors," *Proc. IRE*, **40**, 1361 (1952).

3 J. L. Moll, M. Tanenbaum, J. M. Goldey, and N. Holonyak, "*p-n-p-n* Transistor Switches," *Proc. IRE*, **44**, 1174 (1956).

4 *Thyristor DATA Book*, DATA Inc., Pine Brook, N.J., 1979.

5 F. E. Gentry, F. W. Gutzwieler, N. H. Holonyak, and E. E. Von Zastrow, *Semiconductor Controlled Rectifiers*, Prentice-Hall, Englewood Cliffs, N.J., 1964.

6 A. Blicher, *Thyristor Physics*, Springer, New York, 1976.

7 S. K. Ghandhi, *Semiconductor Power Devices*, Wiley, New York, 1977.

8 S. M. Sze and G. Gibbons, "Avalanche Breakdown Voltages of Abrupt and Linearly Graded *p-n* Junctions in Ge, Si, GaAs, and GaP," *Appl. Phys. Lett.*, **8**, 111 (1966).

9 A. Herlet, "The Maximum Blocking Capability of Silicon Thyristors," *Solid State Electron.*, **8**, 655 (1965).

10 Special Issue on High-Power Semiconductor Devices, *IEEE Trans. Electron Devices*, **ED-23** (1976).

11 E. W. Haas and M. S. Schnoller, "Phosphorus Doping of Silicon by Means of Neutron Irradiation," *IEEE Trans. Electron Devices*, **ED-23**, 803 (1976).

12 J. Cornu, S. Schweitzer, and O. Kuhn, "Double Positive Beveling: A Better Edge Contour for High Voltage Devices," *IEEE Trans. Electron Devices*, **ED-21**, 181 (1974).

13 R. L. Davies and F. E. Gentry, "Control of Electric Field at the Surface of *p-n* Junctions," *IEEE Trans. Electron Devices*, **ED-11**, 313 (1964).

14 M. S. Adler and V. A. K. Temple, "A General Method for Predicting the Avalanche Breakdown Voltage of Negative Bevelled Devices," *IEEE Trans. Electron Devices*, **ED-23**, 956 (1976).

15 F. E. Gentry, "Turn-on Criterion for *p-n-p-n* Devices," *IEEE Trans. Electron Devices*, **ED-11**, 74 (1964).

16 E. S. Yang and N. C. Voulgaris, "On the Variation of Small-Signal Alphas of a *p-n-p-n* Device with Current," *Solid State Electron.*, **10**, 641 (1967).

17 M. S. Adler, "Accurate Calculation of the Forward Drop and Power Dissipation in Thyristors," *IEEE Trans. Electron Devices*, **ED-25**, 16 (1978).

18 J. Krausse, "Auger Recombination in Forward Biased Silicon Rectifiers and Thyristors," *Solid State Electron.*, **17**, 427 (1974).

19a W. H. Schroen, "Characteristics of a High-Current, High-Voltage Shockley Diode," *IEEE Trans. Electron Devices*, **ED-17**, 694 (1970).

19b C. K. Chu, J. E. Johnson, and J. B. Brewster, "1200 V and 5000 A Peak Reverse Blocking Diode Thyristor," *Jpn. J. Appl. Phys.*, **16**, *Suppl.*, **16-1**, 537 (1977).

20 J. F. Gibbons, "Graphical Analysis of the *I–V* Characteristics of Generalized *p-n-p-n* Devices," *Proc. IEEE*, **55**, 1366 (1967).

21 J. F. Gibbons, "A Critique of the Theory of *p-n-p-n* Devices," *IEEE Trans. Electron Devices*, **ED-11**, 406 (1964).

22 E. S. Yang, "Turn-off Characteristics of *p-n-p-n* Devices," *Solid State Electron.*, **10**, 927 (1967).

23 T. S. Sundresh, "Reverse Transient in *p-n-p-n* Triodes," *IEEE Trans. Electron Devices*, **ED-14**, 400 (1967).

24 B. J. Baliga and E. Sun, "Comparison of Gold, Platinum, and Electron Irradiation for Controlling Lifetime in Power Rectifiers," *IEEE Trans. Electron Devices*, **ED-24**, 685 (1977).

25 I. Dudeck and R. Kassing, "Gold as an Optimum Recombination Center for Power Rectifiers and Thyristors," *Solid State Electron.*, **20**, 1033 (1977).

26 B. J. Baliga and S. Krishna, "Optimization of Recombination Levels and Their Capture Cross Section in Power Rectifiers and Thyristors," *Solid State Electron.*, **20**, 225 (1977).

27 E. S. Schlegel, "A Technique for Optimizing the Design of Power Semiconductor Devices," *IEEE Trans. Electron Devices*, **ED-23**, 925 (1976).

28 J. Shimizu, H. Oka, S. Funakawa, H. Gamo, T. Ilda, and A. Kawakami, "High-Voltage High-Power Gate-Assisted Turn-Off Thyristor for High-Frequency Use," *IEEE Trans. Electron Devices*, **ED-23**, 883 (1976).

29 E. Schlegel, "Gate Assisted Turn-off Thyristors," *IEEE Trans. Electron Devices*, **ED-23**, 888 (1976).

30 A. Munoz-Yague and P. Leturcq, "Optimum Design of Thyristor Gate-Emitter Geometry," *IEEE Trans. Electron Devices*, **ED-23**, 917 (1976).

31 H. F. Storm and J. G. St. Clair, "An Involute Gate–Emitter Configuration for Thyristors," *IEEE Trans. Electron Devices*, **ED-21**, 520 (1974).

32 F. E. Gentry and J. Moyson, "The Amplifying Gate Thyristor," Paper No. 19.1, IEEE Meet. Prof. Group Electron Devices, Washington, D.C., 1968.

33 F. M. Roberts and E. L. G. Wilkinson, "The Relative Merits of Thyristors and Power Transistors for Fast Power-Switching Application," *Int. J. Electron.*, **33**, 319 (1972).

34 R. A. Kokosa and B. R. Tuft, "A High-Voltage High-Temperature Reverse Conducting Thyristor," *IEEE Trans. Electron Devices*, **ED-17**, 667 (1970).

35 W. Gerlach, "Light Activated Power Thyristors," *Inst. Phys. Conf. Ser.*, **32**, 111 (1977).

36 E. D. Wolley, "Gate Turn-Off in *p-n-p-n* Devices," *IEEE Trans. Electron Devices*, **ED-13**, 590 (1966).

37 F. E. Gentry, R. I. Scace, and J. K. Flowers, "Bidirectional Triode *p-n-p-n* Switches," *Proc. IEEE*, **53**, 355 (1965).

38 J. F. Essom, "Bidirectional Triode Thyristor Applied Voltage Rate Effect Following Conduction," *Proc. IEEE*, **55**, 1312 (1967).

39 W. Shockley, G. L. Pearson, and J. R. Haynes, "Hole Injection in Germanium-Quantitative Studies and Filamentary Transistors," *Bell Syst. Tech. J.*, **28**, 344 (1949).

40 V. A. Bluhm and T. P. Sylvan, "A High Performance Unijunction Transistor Using Conductivity Modulation of Spreading Resistance," *Solid State Des.*, **5**, 26 (1964).

41 L. S. Senhouse, "A Unique Filamentary-Transistor Structure," Paper No. 23.6, IEEE Electron Device Meet., Washington, D.C., Oct. 1967.

42 I. A. Lesk and V. P. Mathis, "The Double-Base Diode—A New Semiconductor Device," *IRE Conv. Rec.*, Pt. 6, p. 2 (1953).

43 F. N. Trofimenkoff and G. J. Huff, "DC Theory of the Unijunction Transistor," *Int. J. Electron.*, **20**, 217 (1966).

44 L. E. Clark, "Now, New Unijunction Geometries," *Electronics*, **38**, 93 (1965).

45 D. L. Scharfetter and A. G. Jordan, "Reactive Effects in Semiconductor Filaments Due to Conductivity Modulation and an Extension of the Theory of the Double-Base Diode," *IRE Trans. Electron Devices*, **ED-9**, 461 (1962).

46 *SCR Manual*, 5th ed., General Electric, Syracuse, N.Y., 1972.

47 D. E. Houston, S. Krishna, D. E. Piccone, R. J. Einke, and Y. S. Sun, "A Field Terminated Diode," *IEEE Trans. Electron Devices*, **ED-23**, 905 (1976).

48 R. Zuleeg, "Multi-Channel Field-Effect Transistor Theory and Experiment," *Solid State Electron.*, **10**, 559 (1967).

49 R. Barandon and P. Laurenceau, "Power Bipolar Gridistor," *Electron. Lett.*, **12**, 486 (1976).

50 B. J. Baliga, "Grid Depth Dependence of the Characteristics of Vertical Channel Field Controlled Thyristors," *Solid State Electron.*, **22**, 237 (1979).

PART III

UNIPOLAR DEVICES

5

Metal–Semiconductor Contacts

- INTRODUCTION
- ENERGY-BAND RELATION
- SCHOTTKY EFFECT
- CURRENT TRANSPORT PROCESSES
- CHARACTERIZATION OF BARRIER HEIGHT
- DEVICE STRUCTURES
- OHMIC CONTACT

5.1 INTRODUCTION

The earliest systematic investigation on metal–semiconductor rectifying systems is generally attributed to Braun, who in 1874 noted the dependence of the total resistance on the polarity of the applied voltage and on the detailed surface conditions.[1] The point-contact rectifier in various forms found practical applications[2] beginning in 1904. In 1931, Wilson formulated the transport theory of semiconductors based on the band theory of solids.[3] This theory was then applied to the metal–semiconductor contacts. In 1938, Schottky suggested that the potential barrier could arise from stable space charges in the semiconductor alone without the presence of a chemical layer.[4] The model arising from this consideration is known as the Schottky barrier. In 1938, Mott also devised an appropriate theoretical model for swept-out metal–semiconductor contacts that is known as the Mott barrier.[5] The basic theory and the historical development of rectifying metal–semiconductor contacts have been summarized by Henisch in *Rectifying Semiconductor Contacts*.[6]

Because of their importance in direct current and microwave applications and as tools in the analysis of other fundamental physical parameters, metal–semiconductor contacts have been studied extensively. The transport processes and device technology have been reviewed by Rhoderick and by Rideout.[7,8] Subsequent chapters will consider applications of these contacts. Specifically, metal–semiconductor contacts have been used as the gate electrodes of a field-effect transistor (MESFETs in Chapter 6), the drain and source contacts in MOSFETs (Chapter 8), the electrodes for high-power IMPATT oscillators (Chapter 10), the third terminal in a transferred-electron device (Chapter 11), and photodetectors and solar cells (Chapters 13 and 14).

5.2 ENERGY-BAND RELATION

When metal is making contact with a semiconductor, a barrier will be formed at the metal–semiconductor interface. In this section we consider the basic energy-band diagrams. We also show that the depletion layer of a metal–semiconductor contact is similar to that of the one-sided abrupt (e.g., p^+-n) junction.

5.2.1 Ideal Condition and Surface States

When a metal is making intimate contact with a semiconductor, the Fermi levels in the two materials must be coincident at thermal equilibrium. We will first consider two limiting cases;[6] a more general result will be considered in Section 5.5. The two limiting cases are shown in Fig. 1. Figure 1*a* shows the electronic energy relations at an ideal contact between a metal and an *n*-type semiconductor in the absence of surface states. At far left, the metal and semiconductor are not in contact, and the system is not in thermal equilibrium. If a wire is connected between the semiconductor and the metal so that charge will flow from the semiconductor to the metal and thermal equilibrium is established, the Fermi levels on both sides line up. Relative to the Fermi level in the metal, the Fermi level in the semiconductor is lowered by an amount equal to the difference between the two work functions.

The work function is the energy difference between the vacuum level and the Fermi level. This quantity is denoted by $q\phi_m$ (ϕ_m in volts) for the metal, and equal to $q(\chi + V_n)$ in the semiconductor, where $q\chi$ is the electron affinity measured from the bottom of the conduction band E_C to the vacuum level, and qV_n is the energy difference between E_C and the Fermi level. The potential difference $q\phi_m - q(\chi + V_n)$ is called the contact potential. As the distance δ decreases, an increasing negative charge is built up at the metal surface. An equal and opposite charge (positive) must exist in the semiconductor. Because of the relatively low carrier concen-

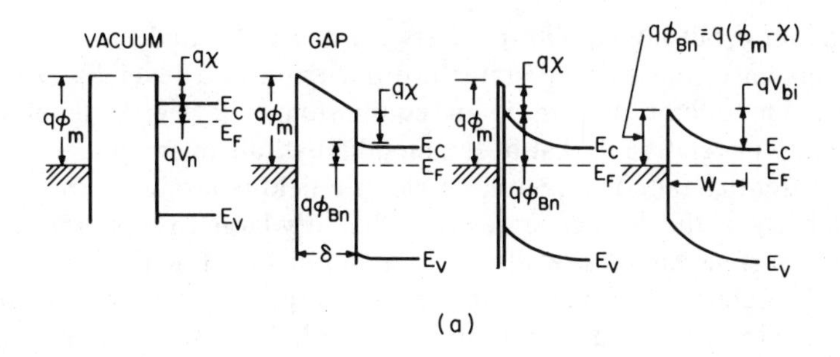

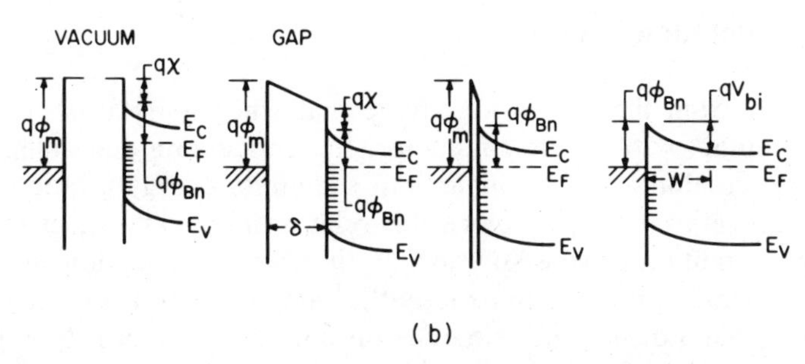

Fig. 1 Energy-band diagrams of metal–semiconductor contacts. (After Henisch, Ref. 6.)

tration, this positive charge is distributed over a barrier region near the semiconductor surface. When δ is small enough to be comparable with interatomic distances, the gap becomes transparent to electrons, and we obtain the limiting case, as shown on the far right (Fig. 1*a*). It is clear that the limiting value of the barrier height $q\phi_{Bn}$ is given by

$$q\phi_{Bn} = q(\phi_m - \chi). \tag{1}$$

The barrier height is simply the difference between the metal work function and the electron affinity of the semiconductor. For an ideal contact between a metal and a p-type semiconductor, the barrier height $q\phi_{Bp}$ is given by

$$q\phi_{Bp} = E_g - q(\phi_m - \chi). \tag{2}$$

For a given semiconductor and for any metals, the sum of the barrier heights on n-type and p-type substrates is thus expected to be equal to the bandgap, or

$$q(\phi_{Bn} + \phi_{Bp}) = E_g. \tag{3}$$

The second limiting case is shown in Fig. 1*b*, where a large density of surface states is present on the semiconductor surface. At far left, the figure shows equilibrium between the surface states and the bulk of the semiconductor but nonequilibrium between the metal and the semiconductor. In this case, the surface states are occupied to a level E_F. When the metal–semiconductor system is in equilibrium, the Fermi level of the semiconductor relative to that of the metal must fall an amount equal to the contact potential and, as a result, an electric field is produced in the gap δ. If the density of the surface states is sufficiently large to accommodate any additional surface charges resulting from diminishing δ without appreciably altering the occupation level E_F, the space charge in the semiconductor will remain unaffected. As a result, the barrier height is determined by the property of the semiconductor surface and is independent of the metal work function.

5.2.2 Depletion Layer

It is clear from the discussion above that when a metal is brought into intimate contact with a semiconductor, the conduction and valence bands of the semiconductor are brought into a definite energy relationship with the Fermi level in the metal. Once this relationship is known, it serves as a boundary condition on the solution of the Poisson equation in the semiconductor, which proceeds in exactly the same manner as in p-n junctions. The energy-band diagrams for metals on both n-type and p-type materials are shown, under different biasing conditions, in Fig. 2.

Under the abrupt approximation that $\rho \simeq qN_D$ for $x < W$, and $\rho \simeq 0$ and $dV/dx \simeq 0$ for $x > W$, where W is the depletion width, the results for the metal–semiconductor barrier are similar to those of the one-sided abrupt p^+-n junction and we obtain

$$W(\text{depletion width}) = \sqrt{\frac{2\epsilon_s}{qN_D}\left(V_{bi} - V - \frac{kT}{q}\right)} \tag{4}$$

$$|\mathscr{E}(x)| = \frac{qN_D}{\epsilon_s}(W - x) = \mathscr{E}_m - \frac{qN_D}{\epsilon_s}x \tag{5}$$

$$V(x) = \frac{qN_D}{\epsilon_s}\left(Wx - \frac{1}{2}x^2\right) - \phi_{Bn} \tag{6}$$

where the term kT/q arises from the contribution of the majority-carrier distribution tail (electrons in n side) and $\mathscr{E}_m$ is the maximum field strength which occurs at $x = 0$:

$$\mathscr{E}_m = \mathscr{E}(x=0) = \sqrt{\frac{2qN_D}{\epsilon_s}\left(V_{bi} - V - \frac{kT}{q}\right)} = \frac{2(V_{bi} - V - kT/q)}{W}. \tag{7}$$

The space charge Q_{sc} per unit area of the semiconductor and the depletion-layer capacitance C per unit area are given by

$$Q_{sc} = qN_DW = \sqrt{2q\epsilon_sN_D\left(V_{bi} - V - \frac{kT}{q}\right)} \quad \text{C/cm}^2 \tag{8}$$

$$C \equiv \frac{|\partial Q_{sc}|}{\partial V} = \sqrt{\frac{q\,\epsilon_sN_D}{2(V_{bi} - V - kT/q)}} = \frac{\epsilon_s}{W} \quad \text{F/cm}^2. \tag{9}$$

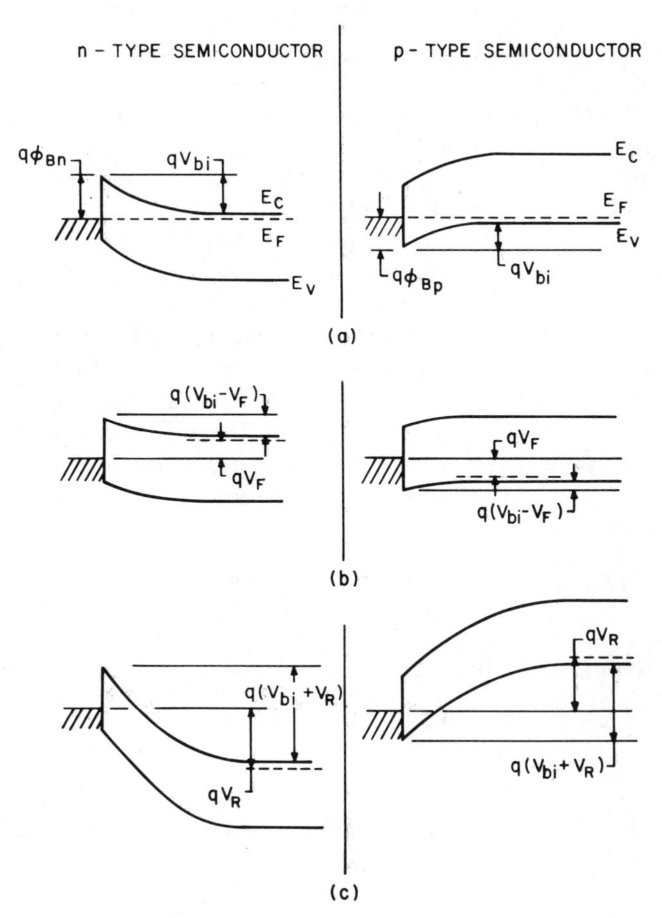

Fig. 2 Energy-band diagram of metal n-type and metal p-type semiconductors under different biasing conditions (a) Thermal equilibrium. (b) Forward bias. (c) Reverse bias.

Equation (9) can be written in the form

$$\frac{1}{C^2} = \frac{2(V_{bi} - V - kT/q)}{q\epsilon_s N_D} \tag{10a}$$

or

$$-\frac{d(1/C^2)}{dV} = \frac{2}{q\,\epsilon_s N_D} \tag{10b}$$

$$N_D = \frac{2}{q\epsilon_s}\left[-\frac{1}{d(1/C^2)/dV}\right]. \tag{10c}$$

If N_D is constant throughout the depletion region, one should obtain a straight line by plotting $1/C^2$ versus V. If N_D is not a constant, the differential capacitance method can be used to determine the doping profile from Eq. 10c.

5.3 SCHOTTKY EFFECT

The Schottky effect is the image-force-induced lowering of the potential energy for charge carrier emission when an electric field is applied. Consider a metal–vacuum system first. The minimum energy necessary for an electron to escape into vacuum from an initial energy at the Fermi level is defined as the work function $q\phi_m$ (ϕ_m in volts) as shown in Fig. 3. For metals, $q\phi_m$ is of the order of a few electron volts and varies from 2 to 6 eV. The values of $q\phi_m$ are generally very sensitive to surface contamination. The most reliable values for clean surfaces are given[9] in Fig. 4.

When an electron is at a distance x from the metal, a positive charge will be induced on the metal surface. The force of attraction between the electron and the induced positive charge is equivalent to the force that would exist between the electron and an equal positive charge located at $-x$. This positive charge is referred to as the image charge. The attractive force, called the image force, is given by

$$F = \frac{-q^2}{4\pi(2x)^2\epsilon_0} = \frac{-q^2}{16\pi\epsilon_0 x^2} \tag{11}$$

where ϵ_0 is the permittivity of free space. The work done by an electron in

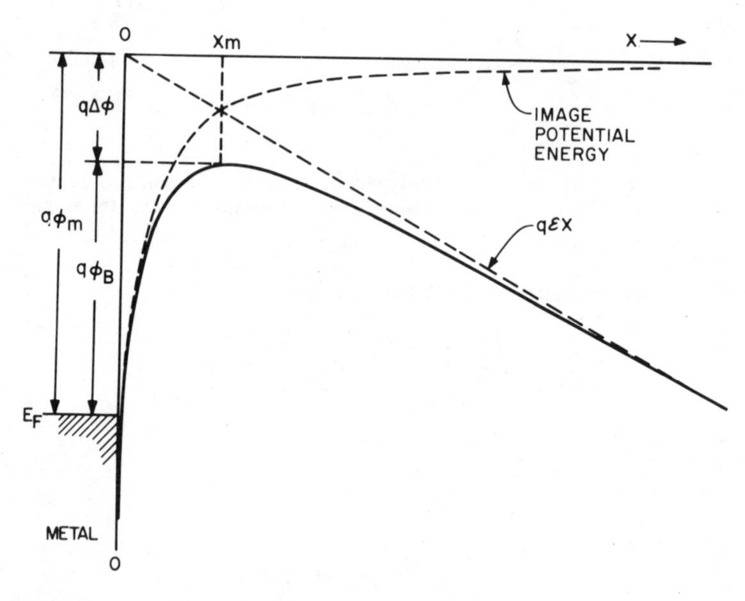

Fig. 3 Energy-band diagram between a metal surface and a vacuum. The metal work function is $q\phi_m$. The effective work function (or barrier) is lowered when an electric field is applied to the surface. The lowering is due to the combined effects of the field and the image force.

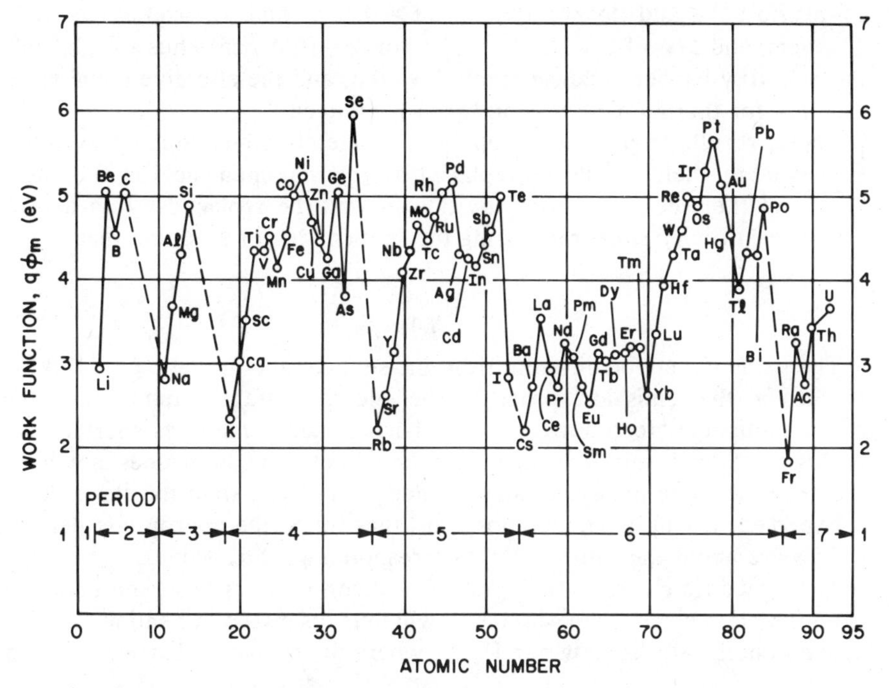

Fig. 4 Metal work function for a clean metal surface in a vacuum versus atomic number. Note the periodic nature of the increase and decrease of the work functions within each group. (After Michaelson, Ref. 9.)

the course of its transfer from infinity to the point x is given by

$$E(x) = \int_{\infty}^{x} F\,dx = \frac{q^2}{16\pi\epsilon_0 x}. \tag{12}$$

The energy above corresponds to the potential energy of an electron at a distance x from the metal surface, shown in Fig. 3, and is measured downwards from the x axis.

When an external field $\mathscr{E}$ is applied, the total potential energy PE as a function of distance (measured downward from the x axis) is given by the sum

$$PE(x) = \frac{q^2}{16\pi\epsilon_0 x} + q\mathscr{E}x \qquad \text{eV}. \tag{13}$$

The Schottky barrier lowering (also referred to as image force lowering) $\Delta\phi$ and the location of the lowering x_m (as shown in Fig. 3), are given by the condition $d[PE(x)]/dx = 0$, or

$$x_m = \sqrt{\frac{q}{16\pi\epsilon_0\mathscr{E}}} \qquad \text{cm} \tag{14}$$

$$\Delta\phi = \sqrt{\frac{q\mathscr{E}}{4\pi\epsilon_0}} = 2\mathscr{E}x_m \qquad \text{V}. \tag{15}$$

From Eqs. 14 and 15 we obtain $\Delta\phi = 0.12$ V and $x_m \simeq 60$ Å for $\mathscr{E} = 10^5$ V/cm; and $\Delta\phi = 1.2$ V and $x_m \simeq 10$ Å for $\mathscr{E} = 10^7$ V/cm. Thus at high fields the Schottky barrier is considerably lowered, and the effective metal work function for thermionic emission ($q\phi_B$) is reduced.

These results can also be applied to metal–semiconductor systems. However, the field should be replaced by the maximum field at the interface, and the free-space permittivity ϵ_0 should be replaced by an appropriate permittivity ϵ_s characterizing the semiconductor medium, that is,

$$\Delta\phi = \sqrt{\frac{q\mathscr{E}}{4\pi\epsilon_s}}. \tag{15a}$$

The value ϵ_s may be different from the semiconductor static permittivity. If during the emission process, the electron transit time from the metal–semiconductor interface to the barrier maximum x_m is shorter than the dielectric relaxation time, the semiconductor medium does not have enough time to be polarized, and smaller permittivity than the static value is expected. It will be shown, however, that for Si the appropriate permittivities are about the same as their corresponding static values.

The dielectric constant (ϵ_s/ϵ_0) in gold–silicon barriers has been obtained from the photoelectric measurement, which is discussed in Section 5.5. The experimental result is shown in Fig. 5, where the measured barrier lowering

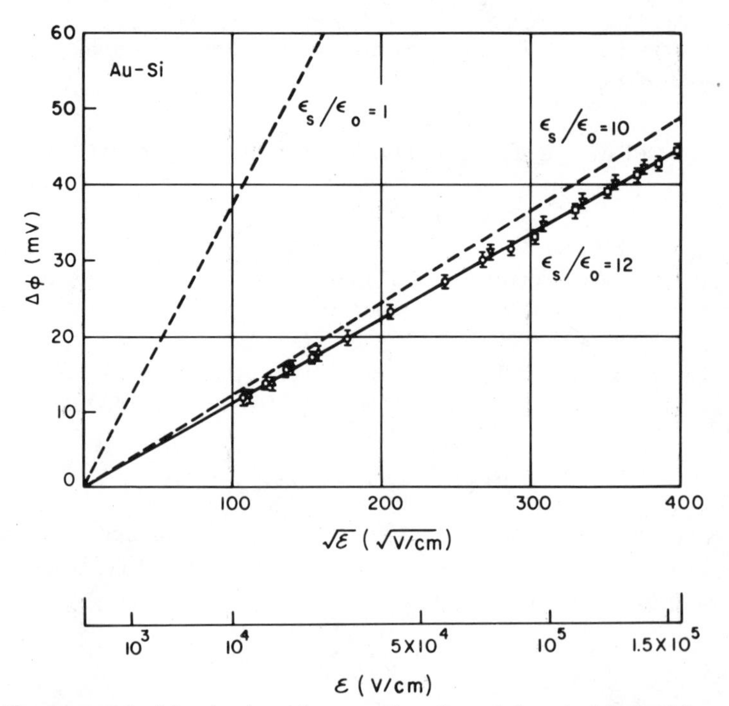

Fig. 5 Measurement of barrier lowering as a function of the electric field in a Au–Si diode. (After Sze, Crowell, and Kahng, Ref. 10.)

is plotted as a function of the square root of the electric field.[10] From Eq. 15a the image-force dielectric constant is determined to be 12 ± 0.5. For $\epsilon_s/\epsilon_0 = 12$, the distance x_m varies between 10 and 50 Å, as in the field range shown in Fig. 5. Assuming a carrier velocity of the order of 10^7 cm/s, the transit time for these distances should be between 1×10^{-14} and 5×10^{-14} s. The image-force dielectric constant should thus be comparable to the dielectric constant of approximately 12 for electromagnetic radiation of roughly these periods (wavelengths between 3 and 15 μm).[11] The dielectric constant of silicon is essentially constant (11.7) from dc to $\lambda = 1\ \mu$m, therefore, the lattice has time to polarize while the electron is traversing the depletion layer. The photoelectric measurements and data deduced from the optical constants are thus in excellent agreement. For Ge and GaAs the dependence of the optical dielectric constant on wavelength is similar to that of Si. The image-force permittivities of these semiconductors in the foregoing field range are thus expected to be approximately the same as the corresponding static values.

Figure 6 shows the energy diagram incorporating the Schottky effect for a metal n-type semiconductor under different biasing conditions. Note that for forward bias ($V > 0$), the barrier height $q\phi_{BO} - q\Delta\phi_F$ is slightly larger than the barrier height at zero bias, $q\phi_{Bn}$. For reverse bias ($V < 0$), the barrier height $q\phi_{BO} - q\Delta\phi_R$ is slightly smaller. Because of the larger values of ϵ_s in a metal–semiconductor system, the barrier lowerings above are

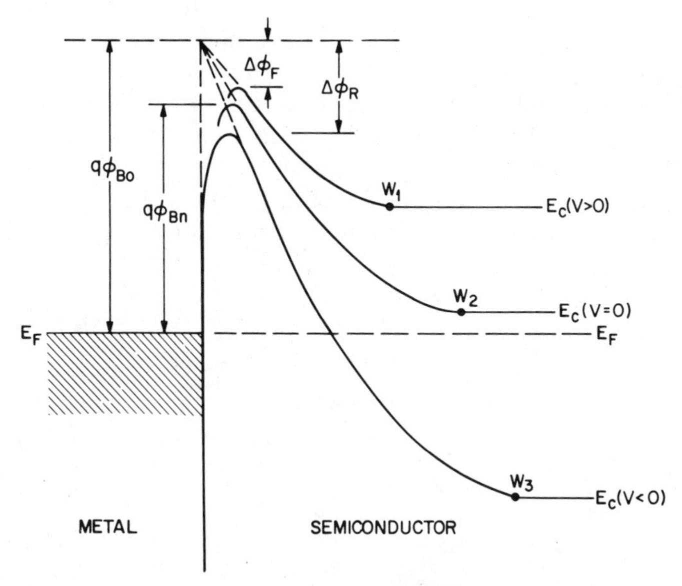

Fig. 6 Energy-band diagram incorporating Schottky effect for a metal n-type semiconductor under different biasing conditions. The intrinsic barrier height is $q\phi_{BO}$. The barrier height at thermal equilibrium is $q\phi_{Bn}$. The barrier lowerings under forward and reverse bias are $\Delta\phi_F$ and $\Delta\phi_R$, respectively. (After Rideout, Ref. 8.)

smaller than those for a corresponding metal–vacuum system. For example, for $\epsilon_s = 12\epsilon_0$, $\Delta\phi$ as obtained from Eq. 15a is only 0.035 V for $\mathscr{E} = 10^5$ V/cm and even smaller for smaller fields. Although the barrier lowering is small, it does have a profound effect on current transport processes in metal–semiconductor systems. These are considered in Section 5.4.

5.4 CURRENT TRANSPORT PROCESSES

The current transport in metal–semiconductor contacts is mainly due to majority carriers, in contrast to *p-n* junctions, where the minority carriers are responsible. Figure 7 shows four basic transport processes under forward bias (the inverse processes occur under reverse bias).[7]

The four processes are (1) transport of electrons from the semiconductor over the potential barrier into the metal [the dominant process for Schottky diodes with moderately doped semiconductors (e.g., Si with $N_D \leq 10^{17}$ cm^{-3}) operated at moderate temperatures (e.g., 300 K)], (2) quantum-mechanical tunneling of electrons through the barrier (important for heavily doped semiconductors and responsible for most ohmic contacts), (3) recombination in the space-charge region [identical to the recombination process in a *p-n* junction (refer to chapter 2)] and (4) hole injection from the metal to the semiconductor (equivalent to recombination in the neutral region). In addition, we may have edge leakage current due to a high electric field at the contact periphery or interface current due to traps at the metal–semiconductor interface. Various methods have been used to improve the interface quality, and many device structures have been proposed to reduce or eliminate the edge leakage currents (see Section 5.6).

We shall first consider the transport of electrons over the potential barrier. For high-mobility semiconductors (e.g., Si) the transport can be adequately described by the thermionic emission theory. We shall also

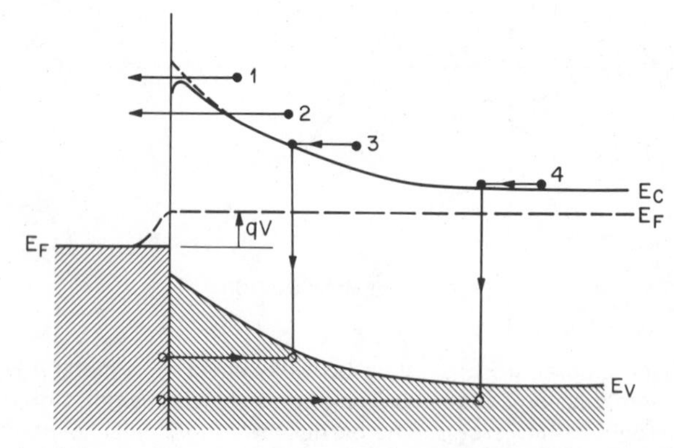

Fig. 7 Four basic transport processes under forward bias. (After Rhoderick, Ref. 7.)

consider the diffusion theory applicable to low-mobility semiconductors and a generalized thermionic-diffusion theory that is a synthesis of the preceding two theories.

5.4.1 Thermionic Emission Theory

The thermionic emission theory by Bethe[12] is derived from the assumptions that (1) the barrier height $q\phi_{Bn}$ is much larger than kT, (2) thermal equilibrium is established at the plane that determines emission, and (3) the existence of a net current flow does not affect this equilibrium, so that one can superimpose two current fluxes—one from metal to semiconductor, the other from semiconductor to metal, each with a different imref. Because of these assumptions, the shape of the barrier profile is immaterial and the current flow depends solely on the barrier height. The current density $J_{s\to m}$ from the semiconductor to the metal is then given by the concentration of electrons with energies sufficient to overcome the potential barrier and traversing in the x direction:

$$J_{s\to m} = \int_{E_F+q\phi_B}^{\infty} qv_x \, dn \tag{16}$$

where $E_F + q\phi_B$ is the minimum energy required for thermionic emission into the metal, and v_x is the carrier velocity in the direction of transport. The electron density in an incremental energy range is given by

$$\begin{aligned} dn &= N(E)F(E)\,dE \\ &= \frac{4\pi(2m^*)^{3/2}}{h^3}\sqrt{E-E_C}\exp[-(E-E_C+qV_n)/kT]\,dE \end{aligned} \tag{17}$$

where $N(E)$ and $F(E)$ are the density of states and the distribution function, respectively; m^* is the effective mass of the semiconductor; and qV_n is $(E_C - E_F)$.

If we postulate that all the energy of electrons in the conduction band is kinetic energy, then

$$E - E_C = \tfrac{1}{2}m^*v^2 \tag{18a}$$

$$dE = m^*v\,dv \tag{18b}$$

$$\sqrt{E-E_C} = v\sqrt{m^*/2}. \tag{18c}$$

Substituting Eq. 18 into Eq. 17 gives

$$dn = 2\left(\frac{m^*}{h}\right)^3 \exp\left(-\frac{qV_n}{kT}\right)\exp\left(-\frac{m^*v^2}{2kT}\right)(4\pi v^2 dv). \tag{19}$$

Equation 19 gives the number of electrons per unit volume that have speeds between v and $v + dv$ distributed over all directions. If the speed is resolved into its components along the axes with the x axis parallel to the transport direction, we have

$$v^2 = v_x^2 + v_y^2 + v_z^2. \tag{20}$$

With the transformation $4\pi v^2\,dv = dv_x\,dv_y\,dv_z$, we obtain from Eqs. 16, 19, and 20

$$\begin{aligned} J_{s\to m} &= 2q\left(\frac{m^*}{h}\right)^3 \exp(-qV_n/kT) \\ &\quad \int_{v_{ox}}^{\infty} v_x \exp(-m^*v_x^2/2kT)\,dv_x \int_{-\infty}^{\infty} \exp(-m^*v_y^2/2kT)\,dv_y \\ &\quad \times \int_{-\infty}^{\infty} \exp(-m^*v_z^2/2kT)\,dv_z \\ &= \left(\frac{4\pi q m^* k^2}{h^3}\right) T^2 \exp(-qV_n/kT) \exp\left(-\frac{m^* v_{ox}^2}{2kT}\right). \end{aligned} \tag{21}$$

The velocity v_{ox} is the minimum velocity required in the x direction to surmount the barrier and is given by

$$\tfrac{1}{2} m^* v_{ox}^2 = q(V_{bi} - V) \tag{22}$$

where V_{bi} is the built-in potential at zero bias (Fig. 1*a*). Substituting Eq. 22 into Eq. 21 yields

$$\begin{aligned} J_{s\to m} &= \left(\frac{4\pi q m^* k^2}{h^3}\right) T^2 \exp\left[-\frac{q(V_n + V_{bi})}{kT}\right] \exp\left(\frac{qV}{kT}\right) \\ &= A^* T^2 \exp\left(-\frac{q\phi_B}{kT}\right) \exp\left(\frac{qV}{kT}\right) \end{aligned} \tag{23}$$

where ϕ_B is the barrier height and equals the sum of V_n and V_{bi}, and

$$A^* = \frac{4\pi q m^* k^2}{h^3} \tag{24}$$

is the effective Richardson constant for thermionic emission, neglecting the effects of optical phonon scattering and quantum mechanical reflection (see Section 5.4.3). For free electrons the Richardson constant A is 120 A/cm^2/K^2. When the image force lowering is considered, the barrier height ϕ_B in Eq. 23 is reduced by $\Delta\phi$.

For semiconductors with isotropic effective mass in the lowest minimum of the conduction band such as n-type GaAs, $A^*/A = m^*/m_0$ where m^* and m_0 are the effective mass and the free-electron mass, respectively. For multiple-valley semiconductors the appropriate Richardson constant A^* associated with a single energy minimum is given by[13]

$$\frac{A_1^*}{A} = \frac{1}{m_0}\left(l_1^2 m_y^* m_z^* + l_2^2 m_z^* m_x^* + l_3^2 m_x^* m_y^*\right)^{1/2} \tag{25}$$

where l_1, l_2, and l_3 are the direction cosines of the normal to the emitting plane relative to the principal axes of the ellipsoid, and m_x^*, m_y^*, and m_z^* are the components of the effective mass tensor. For Ge the emission in the conduction band arises from minima at the edge of the Brillouin zone in the ⟨111⟩ direction. These minima are equivalent to four ellipsoids with longitudinal mass $m_l^* = 1.64m_0$ and transverse mass $m_t^* = 0.082m_0$. The sum of all the A_1^* values has a minimum in the ⟨111⟩ direction:

$$\left(\frac{A^*}{A}\right)_{n\text{-Ge}\langle 111\rangle} = m_t^*/m_0 + [(m_t^*)^2 + 8m_l^* m_t^*]^{1/2}/m_0 = 1.11. \tag{26}$$

The maximum A^* occurs for the ⟨100⟩ direction:

$$\left(\frac{A^*}{A}\right)_{n\text{-Ge}\langle 100\rangle} = \frac{4}{m_0}\left[\frac{(m_t^*)^2 + 2m_t^* m_l^*}{3}\right]^{1/2} = 1.19. \tag{27}$$

For Si the conduction band minima occur in the ⟨100⟩ directions and $m_l^* = 0.98m_0$, $m_t^* = 0.19m_0$. All minima contribute equally to the current in the ⟨111⟩ direction, yielding the maximum A^*:

$$\left(\frac{A^*}{A}\right)_{n\text{-Si}\langle 111\rangle} = \frac{6}{m_0}\left[\frac{(m_t^*)^2 + 2m_t^* m_l^*}{3}\right]^{1/2} = 2.2. \tag{28}$$

The minimum value of A^* occurs for the ⟨100⟩ direction:

$$\left(\frac{A^*}{A}\right)_{n\text{-Si}\langle 100\rangle} = 2m_t^*/m_0 + 4(m_l^* m_t^*)^{1/2}/m_0 = 2.1. \tag{29}$$

For holes in Ge, Si, and GaAs the two energy maxima at $\mathbf{k} = 0$ give rise to approximately isotropic current flow from both the light and heavy holes. Adding the currents due to these carriers, we obtain

$$\left(\frac{A^*}{A}\right)_{p\text{-type}} = (m_{lh}^* + m_{hh}^*)/m_0. \tag{30}$$

Table 1 gives a summary of the values[13] of (A^*/A).

Since the barrier height for electrons moving from the metal into the semiconductor remains the same, the current flowing into the semiconductor is thus unaffected by the applied voltage. It must therefore be equal to the current flowing from the semiconductor into the metal when thermal equilibrium prevails (i.e., when $V = 0$). The corresponding current density

Table 1 Values of A^*/A

Semiconductor	Ge	Si	GaAs	
p-type	0.34	0.66	0.62	
n-type ⟨111⟩	1.11	2.2	0.068 (low field)	1.2 (high field)
n-type ⟨100⟩	1.19	2.1	0.068 (low field)	1.2 (high field)

is obtained from Eq. 23 by setting $V = 0$,

$$J_{m\to s} = -A^*T^2 \exp\left(-\frac{q\phi_{Bn}}{kT}\right). \tag{31}$$

The total current density is given by the sum of Eqs. 23 and 31.

$$\begin{aligned} J_n &= \left[A^*T^2 \exp\left(-\frac{q\phi_{Bn}}{kT}\right)\right]\left[\exp\left(\frac{qV}{kT}\right) - 1\right] \\ &= J_{ST}\left[\exp\left(\frac{qV}{kT}\right) - 1\right] \end{aligned} \tag{32}$$

where

$$J_{ST} \equiv A^*T^2 \exp\left(-\frac{q\phi_{Bn}}{kT}\right). \tag{33}$$

Equation 32 is similar to the transport equation for p-n junctions. However, the expressions for the saturation current densities are quite different.

5.4.2 Diffusion Theory

The diffusion theory by Schottky[4] is derived from the assumptions that (1) the barrier height is much larger than kT, (2) the effect of electron collisions within the depletion region is included, (3) the carrier concentrations at $x = 0$ and $x = W$ are unaffected by the current flow (i.e., they have their equilibrium values), and (4) the impurity concentration of the semiconductor is nondegenerate.

Since the current in the depletion region depends on the local field and the concentration gradient, we must use the current density equation:

$$\begin{aligned} J_x = J_n &= q\left[n(x)\mu\mathscr{E} + D_n \frac{\partial n}{\partial x}\right] \\ &= qD_n\left[-\frac{qn(x)}{kT}\frac{\partial V(x)}{\partial x} + \frac{\partial n}{\partial x}\right]. \end{aligned} \tag{34}$$

Under the steady-state condition, the current density is independent of x, and Eq. 34 can be integrated using $\exp[-qV(x)/kT]$ as an integrating factor. We then have

$$J_n \int_0^W \exp\left[-\frac{qV(x)}{kT}\right] dx = qD_n\left\{n(x)\exp\left[-\frac{qV(x)}{kT}\right]\right\}_0^W \tag{35}$$

and the boundary conditions

$$\begin{aligned} qV(0) &= -q(V_n + V_{bi}) = -q\phi_{Bn} \\ qV(W) &= -qV_n - qV \\ n(0) &= N_C \exp\left[-\frac{E_C(0) - E_F}{kT}\right] = N_C \exp\left(-\frac{q\phi_{Bn}}{kT}\right) \\ n(W) &= n = N_C \exp\left(-\frac{qV_n}{kT}\right). \end{aligned} \tag{36}$$

Substituting Eq. 36 into Eq. 35 yields

$$J_n = qN_C D_n \left[\exp\left(\frac{qV}{kT}\right) - 1\right] \Big/ \int_0^W \exp\left[-\frac{qV(x)}{kT}\right] dx. \tag{37}$$

For Schottky barriers, neglecting image-force effect, the potential distribution is given by Eq. 6 or

$$qV(x) = \frac{q^2 N_D}{\epsilon_s}\left(Wx - \frac{x^2}{2}\right) - q\phi_{Bn}.$$

Substituting into Eq. 37 and expressing W in terms of $V_{bi} + V$, leads to

$$\begin{aligned} J_n &\cong \left\{\frac{q^2 D_n N_C}{kT}\left[\frac{q(V_{bi} - V)2N_D}{\epsilon_s}\right]^{1/2} \exp\left(-\frac{q\phi_{Bn}}{kT}\right)\right\}\left[\exp\left(\frac{qV}{kT}\right) - 1\right] \\ &= J_{SD}\left[\exp\left(\frac{qV}{kT}\right) - 1\right]. \end{aligned} \tag{38}$$

The current density expressions of the diffusion and thermionic emission theories, Eqs. 32 and 38, are basically very similar. However, the "saturation current density" J_{SD} for the diffusion theory varies more rapidly with the voltage but is less sensitive to temperature compared with the "saturation current density" J_{ST} of the thermionic emission theory.

5.4.3 Thermionic Emission-Diffusion Theory

A synthesis of the thermionic emission and diffusion approaches described above has been proposed by Crowell and Sze.[14] This approach is derived from the boundary condition of a thermionic recombination velocity v_R near the metal-semiconductor interface.

Since the diffusion of carriers is strongly affected by the potential configuration in the region through which the diffusion occurs, we consider the electron potential energy $q\psi(x)$ versus distance incorporating the Schottky lowering effect as shown in Fig. 8 for a metal–semiconductor barrier. We consider the case where the barrier height is large enough that the charge density between the metal surface and $x = W$ is essentially that of the ionized donors (i.e., W is the edge of the electron depletion layer). The rounding of $q\psi$ near the metal–semiconductor interface is due to the superimposed effects of the electric field associated with the ionized donors (shown by the dashed extrapolation of ψ) and the attractive image force experienced by an electron when it approaches the metal. As drawn, the applied voltage V between the metal and the semiconductor bulk would give rise to a flow of electrons into the metal. The imref $(-q\phi)$ associated with the electron current density J in the barrier is also shown schematically as a function of distance in Fig. 8. Throughout the region between x_m and W,

$$J = -q\mu n \frac{d\phi_n}{dx} \tag{39}$$

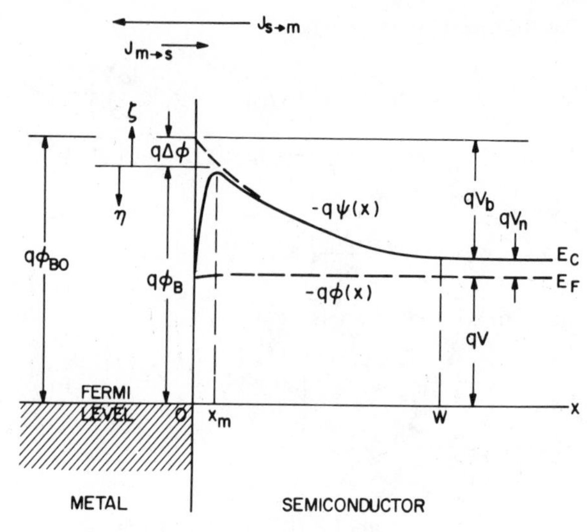

Fig. 8 Energy-band diagram incorporating the Schottky effect. The electron potential energy is $q\psi(x)$, and the quasi-Fermi level is $q\phi(x)$.

where the electron density at the point x is given by

$$n = N_C e^{-q(\phi_n - \psi)/kT} \tag{40}$$

where N_C is the effective density of states in the conduction band, and T is the electron temperature. We will assume that the region between x_m and W is isothermal and that the electron temperature is equal to the lattice temperature.

If the portion of the barrier between x_m and the interface ($x = 0$) acts as a sink for electrons, we can describe the current flow in terms of an effective recombination velocity v_R at the potential energy maximum:

$$J = q(n_m - n_0)v_R \tag{41}$$

where n_m is the electron density at x_m when the current is flowing. n_0 is a quasi-equilibrium electron density at x_m, the density that would occur if it were possible to reach equilibrium without altering the magnitude or position of the potential energy maximum. Both ϕ and ψ can be conveniently measured with respect to the Fermi level in the metal. Then

$$\phi(W) = -V$$

$$n_0 = N_C e^{-q\phi_{Bn}/kT}$$

and

$$n_m = N_C \exp\left[\frac{-q\phi(x_m) - q\phi_{Bn}}{kT}\right] \tag{42}$$

where $q\phi_{Bn}$ is the barrier height, and $q\phi(x_m)$ is the imref at x_m.

If n is eliminated from Eqs. 39 and 40 and the resulting expression for ϕ is integrated between x_m and W,

$$\exp\left[\frac{q\phi(x_m)}{kT}\right] - \exp\left(\frac{qV}{kT}\right) = -\frac{J}{\mu N_C kT}\int_{x_m}^{W} \exp\left(\frac{-q\psi}{kT}\right) dx. \tag{43}$$

Then from Eqs. 41, 42, and 43,

$$J = \frac{qN_C v_R}{1 + v_R/v_D} \exp\left(-\frac{q\phi_{Bn}}{kT}\right)\left[\exp\left(\frac{-qV}{kT}\right) - 1\right] \tag{44}$$

where

$$v_D \equiv \left[\int_{x_m}^{W} \frac{q}{\mu kT} \exp\left[-\frac{q}{kT}(\phi_{Bn} + \psi)\right] dx\right]^{-1} \tag{45}$$

is an effective diffusion velocity associated with the transport of electrons from the edge of the depletion layer at W to the potential energy maximum. If the electron distribution is Maxwellian for $x \geq x_m$, and if no electrons return from the metal other than those associated with the current density qn_0v_R, the semiconductor acts as a thermionic emitter. Then v_R is the thermal velocity given by

$$\begin{aligned} v_R &= \int_0^{\infty} v_x \exp(-m^* v_x^2/2kT)\, dv_x \Big/ \int_{-\infty}^{\infty} \exp(-m^* v_x^2/2kT)\, dv_x \\ &= (kT/2m^*\pi)^{1/2} = A^*T^2/qN_C \end{aligned} \tag{46}$$

where A^* is the effective Richardson constant, as shown in Table 1. At 300 K, v_R is 7.0×10^6, 5.2×10^6, and 1.0×10^7 cm/s for $\langle 111 \rangle$ oriented n-type Ge, $\langle 111 \rangle$ n-type Si, and n-type GaAs, respectively. If $v_D \gg v_R$, the pre-exponential term in Eq. 44 is dominated by v_R and the thermionic emission theory most nearly applies. If, however, $v_D \ll v_R$, the diffusion process is dominant. If we were to neglect image-force effects, and if the electron mobility were independent of the electric field, v_D would be equal to $\mu\mathscr{E}$, where $\mathscr{E}$ is the electric field in the semiconductor near the boundary. The standard diffusion result as given by Eq. 38 would then be obtained, and

$$J \simeq qN_C\mu\mathscr{E} \exp\left(-\frac{q\phi_{Bn}}{kT}\right)\left[\exp\left(\frac{qV}{kT}\right) - 1\right]. \tag{47}$$

To include image-force effects on the calculation of v_D, the appropriate expression for ψ in Eq. 45 is

$$\psi = \phi_{Bn} + \Delta\phi - \mathscr{E}x - \frac{q}{16\pi\epsilon_s x} \tag{48}$$

where $\Delta\phi$ is the barrier lowering as given by Eq. 15a (assuming that the electric field is constant for $x < x_m$). Substitution of Eq. 48 into Eq. 45 yields the results that $v_D \simeq \mu\mathscr{E}$ for $\Delta\phi < kT/q$, and v_D reduces to 0.3 $\mu\mathscr{E}$ when $\Delta\phi$ increases to 20 kT/q.

In summary, Eq. 44 gives a result which is a synthesis of Schottky's diffusion theory and Bethe's thermionic emission theory, and which predicts currents in essential agreement with the thermionic emission theory if $\mu\mathscr{E}(x_m) > v_R$. The latter criterion is more rigorous than Bethe's condition $\mathscr{E}(x_m) > kT/q\lambda$, where λ is the carrier mean free path.

In the preceding section a recombination velocity v_R associated with thermionic emission was introduced as a boundary condition to describe the collecting action of the metal in a Schottky barrier. In many cases an appreciable probability exists that an electron which crosses the potential energy maximum will be backscattered by electron optical-phonon scattering.[15,16] As a first approximation the probability of electron emission over the potential maximum can be given by $f_p = \exp(-x_m/\lambda)$. In addition, the electron energy distribution can be further distorted from a Maxwellian distribution because of quantum-mechanical reflection of electrons by the Schottky barrier and also because of tunneling of electrons through the barrier.[17,18] The ratio f_Q of the total current flow, considering the quantum-mechanical tunneling and reflection, to the current flow neglecting these effects depends strongly on the electric field and the electron energy measured from the potential maximum.

The complete expression of the J–V characteristics taking into account f_p and f_Q is thus

$$J = J_S(e^{qV/kT} - 1) \tag{49}$$

$$J_S = A^{**}T^2 \exp\left(-\frac{q\phi_{Bn}}{kT}\right) \tag{50}$$

where

$$A^{**} = \frac{f_p f_Q A^*}{1 + f_p f_Q v_R/v_D}. \tag{50a}$$

Figure 9 shows the calculated room-temperature values[19] of the effective Richardson constant, A^{**}, for metal–Si systems with an impurity concentration of 10^{16} cm^{-3}. We note that for electrons (n-type Si), A^{**} in the field range 10^4 to 2×10^5 V/cm remains essentially at a constant value of about 110 A/cm^2/K^2. For holes (p-type Si), A^{**} in this field range also remains essentially constant but at a considerably lower value (~ 30 A/cm^2/K^2).

We conclude from the foregoing discussions that at room temperature in the electric field range of 10^4 to about 10^5 V/cm, the current transport mechanism in most Ge, Si, and GaAs Schottky barrier diodes is due to thermionic emission of majority carriers.

The spatial dependence of the imref near the metal–semiconductor interface has been studied by substituting Eqs. 6 and 40 into Eq. 39 and evaluating the imref difference, $\phi(W) - \phi(0)$. The imref as shown in Fig. 8 is essentially flat throughout the depletion region.[20] The imref difference, $\phi(W) - \phi(0)$ for a Au–Si diode with $N_D = 1.2\times10^{15}$ cm^{-3}, is only 8 mV for a forward bias of 0.2 V at 300 K. At higher doping levels the difference is

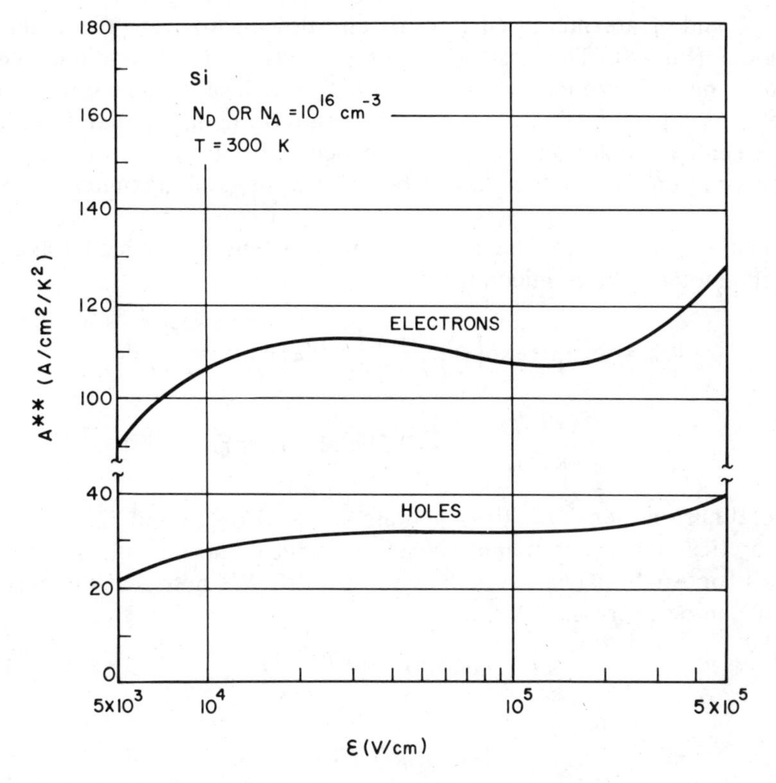

Fig. 9 Calculated effective Richardson constant A^{**} versus electric field for metal-silicon barriers. (After Andrews and Lepselter, Ref. 19.)

even smaller. These results further confirm that for high-mobility semiconductors with moderate dopings, the thermionic emission theory is applicable.

5.4.4 Tunneling Current

For a heavily doped semiconductor or for operation at low temperatures, the tunneling current may become the dominant transport process. The expression for $J_{s\to m}$ as shown in Eq. 16 will now be modified to include both the thermionic emission and tunneling components.[18] The current $J_{s\to m}$ is proportional to the quantum transmission coefficient multiplied by the occupation probability in the semiconductor and the unoccupied probability in the metal, that is,

$$J_{s\to m} = \frac{A^*T}{k}\int_0^\infty T(\zeta)\exp\left[\frac{-q(V_b+V_n+\zeta-\Delta\phi)}{kT}\right]d\zeta + \frac{A^*T}{k}\int_0^{q(V_b-\Delta\phi)} F_s(V)T(\eta)(1-F_m)\,d\eta \tag{51}$$

where ζ and η are measured upward and downward from the potential maximum (Fig. 8). The first term corresponds to the thermionic components and will reduce to Eq. 23 if $T(\zeta) = 1$. The second term is the tunneling component. F_s and F_m are the Fermi–Dirac distribution functions for the semiconductor and the metal, respectively. $T(\zeta)$ and $T(\eta)$ are the transmission coefficients above and below the potential maximum, respectively.

Similar expression can be given for the current $J_{m\to s}$ which traverses from the metal to the semiconductor:

$$J_{m\to s} = -\frac{A^*T}{k}\exp\left(\frac{-q\phi_{Bn}}{kT}\right)\int_0^\infty T(\zeta)\exp\left(-\frac{\zeta}{kT}\right)d\zeta$$

$$-\frac{A^*T}{k}\int_0^{q(V_b-\Delta\phi)} F_m T(\eta)(1-F_s)\,d\eta. \tag{52}$$

The total current density is the algebraic sum of Eqs. 51 and 52.

Theoretical and experimental values[18] of typical current–voltage characteristics for Au–Si barriers are shown in Fig. 10. We note that the current density can be expressed as

$$J = J_S[\exp(qV/nkT) - 1] \tag{53}$$

or

$$J \simeq J_S \exp(qV/nkT) \qquad \text{for} \quad V \gg kT/q \tag{53a}$$

where J_S is the saturation current density obtained by extrapolating the current density from the log-linear region to $V = 0$ and n is the ideality factor defined as

$$n \equiv \frac{q}{kT}\frac{\partial V}{\partial(\ln J)}. \tag{53b}$$

The saturation current density and n are plotted in Fig. 11 for Au–Si diodes as a function of doping concentration with temperature as a parameter.[18] It is interesting to note that J_S is essentially a constant for low dopings but begins to increase rapidly when $N_D > 10^{17}\ \text{cm}^{-3}$. The ideality factor, n, is very close to unity at low dopings and high temperatures. However, it can depart substantially from unity when the doping is increased or the temperature is lowered.

If the tunneling component dominates the current flow (e.g., for high dopings and low temperatures), the transmission coefficient has the form[21]

$$T(\eta) \sim \exp(-q\phi_{Bn}/E_{00}) \tag{54}$$

where

$$E_{00} \equiv \frac{q\hbar}{2}\sqrt{\frac{N_D}{\epsilon_s m^*}}$$

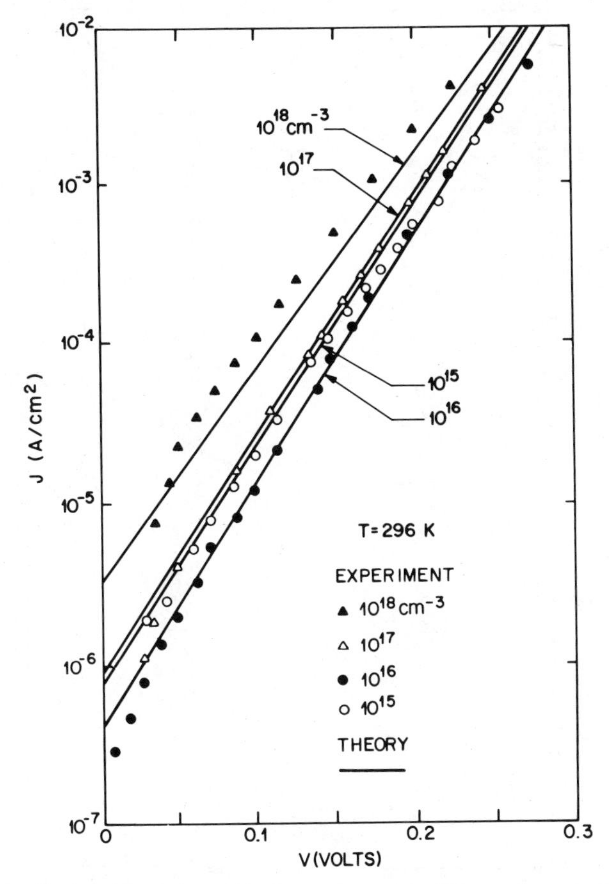

Fig. 10 Theoretical and experimental values of current–voltage characteristics for Au–Si barriers. (After Chang and Sze, Ref. 18.)

and the tunneling current has a similar expression:

$$J_t \sim \exp(-q\phi_{Bn}/E_{00}). \tag{54a}$$

This equation indicates that the tunneling current will increase exponentially with $\sqrt{N_D}$.

Figure 12 shows the ratio of the tunneling current to the thermionic current of a Au–Si barrier.[18] Note that for $N_D \leq 10^{17}\ \text{cm}^{-3}$ and $T \geqslant 300\ K$, the ratio is much less than unity and the tunneling component can be neglected. However, for higher dopings and lower temperatures, the ratio can become much larger than unity, indicating that the tunneling current becomes dominant.

5.4.5 Minority-Carrier Injection Ratio

The Schottky barrier diode is a majority-carrier device under low-injection conditions. At sufficiently large forward bias, the minority-carrier

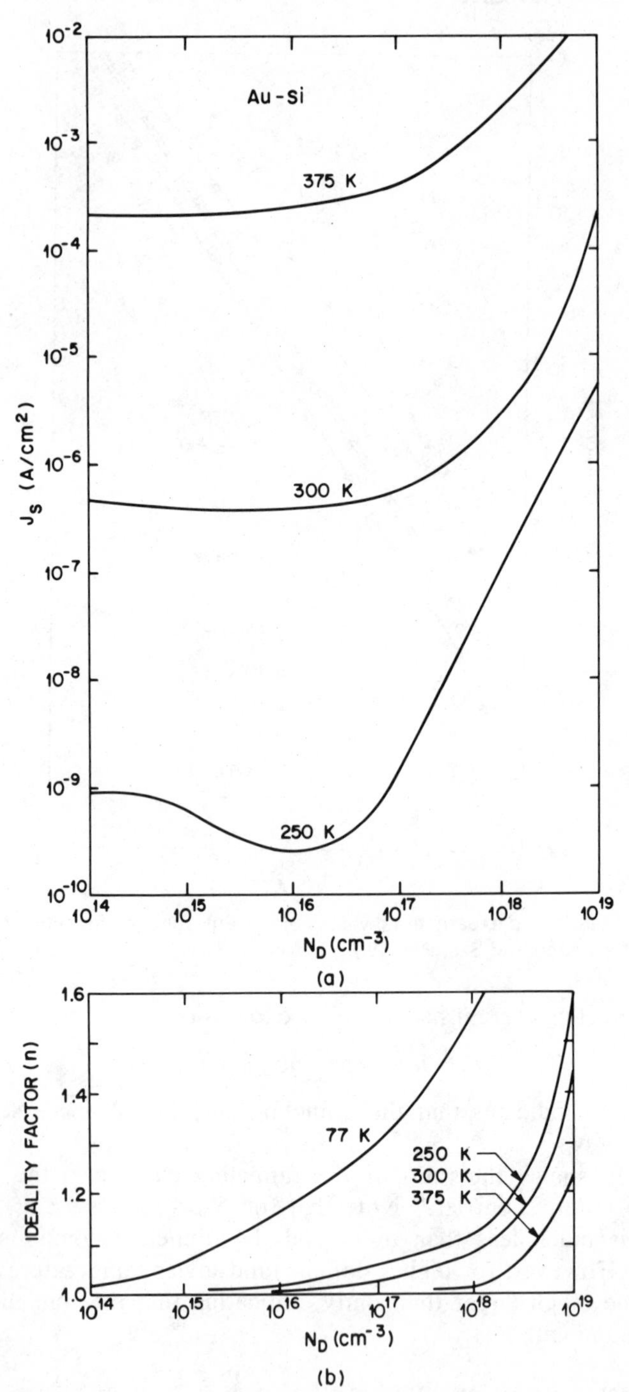

Fig. 11 (*a*) Saturation current density versus doping concentration for Au–Si barriers at three temperatures. (*b*) Ideality factor *n* versus doping concentration at different temperatures. (After Chang and Sze, Ref. 18.)

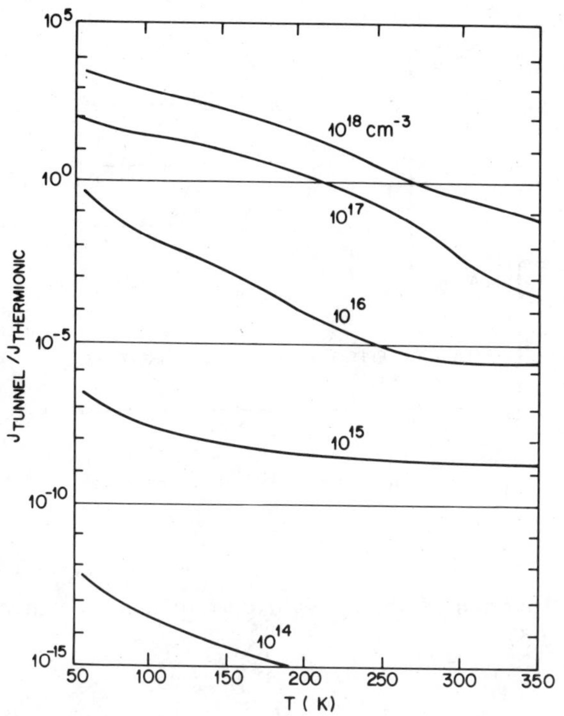

Fig. 12 Ratio of tunneling current component to the thermionic current component of a Au–Si barrier. The tunneling current will dominate ($J_t/J_{th} > 1$) at higher dopings and lower temperatures. (After Chang and Sze, Ref. 18.)

injection ratio γ (ratio of minority-carrier current to total current) increases with current due to the enhancement of the drift-field component, which becomes much larger than the diffusion current.

At steady state, the one-dimensional continuity and current density equations for the minority carriers are given by

$$0 = -\frac{p_n - p_{no}}{\tau_p} - \frac{1}{q}\frac{\partial J_p}{\partial x} \tag{55}$$

$$J_p = q\mu_p p_n \mathscr{E} - qD_p \frac{\partial p_n}{\partial x}. \tag{56}$$

We consider the energy-band diagram as shown in Fig. 13 where x_1 is the boundary of the depletion layer, and x_2 occurs at the interface between the n-type epitaxial layer and the n^+ substrate. From the rectifying theory as discussed in Chapter 2, the minority-carrier density at x_1 is

$$p_n(x_1) = p_{no}\left[\exp\left(\frac{qV}{kT}\right) - 1\right] \simeq \frac{n_i^2}{N_D}\left[\exp\left(\frac{qV}{kT}\right) - 1\right] \tag{57}$$

where N_D is the n-type donor concentration. The quantity $p_n(x)$ at $x = x_1$

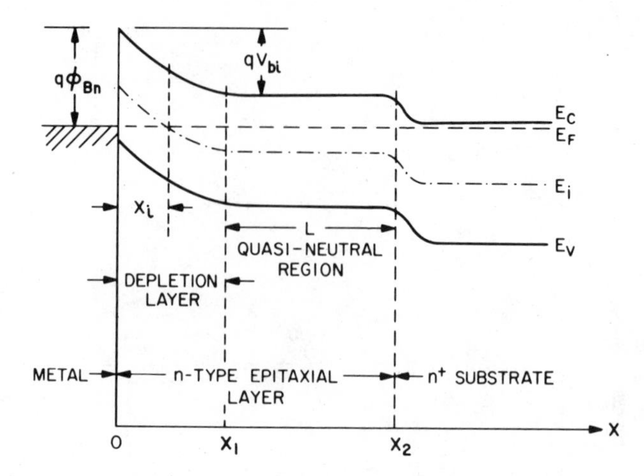

Fig. 13 Energy-band diagram of an epitaxial Schottky barrier. (After Scharfetter, Ref. 23.)

expressed as a function of the forward current density can be obtained from Eqs. 49 and 57:

$$p_n(x_1) = \frac{n_i^2}{N_D}\frac{J}{J_S}. \tag{58}$$

The boundary condition on $p_n(x)$ at $x = x_2$ can be stated in terms of a transport velocity, $v_T = D_p/L_p$, for the minority-carriers

$$J_p(x_2) = qv_Tp_n = q\left(\frac{D_p}{L_p}\right)p_{no}\left[\exp\left(\frac{qV}{kT}\right) - 1\right] \quad \text{for} \quad L \ll L_p \tag{59}$$

where D_p and L_p are the minority-carrier diffusion constant and diffusion length, respectively, and L is the distance of the quasi-neutral region.

For the low-injection conditions, the minority carrier drift component in Eq. 56 is negligible in comparison with the diffusion term, and the injection ratio γ is given by

$$\gamma \equiv \frac{J_p}{J_p + J_n} \simeq \frac{J_p}{J_n} = \frac{qn_i^2 D_p}{N_D L_p A^{**} T^2 \exp(-q\phi_{Bn}/kT)}. \tag{60}$$

For Au–Si diodes with $N_D = 10^{16}\ \text{cm}^{-3}$, the injection ratio has been measured[22] to be very small, of the order of 5×10^{-5}, in agreement with Eq. 60.

For sufficiently large forward bias, however, the electric field causes a significant carrier-drift current component that eventually dominates the minority-carrier current. From Eqs. 56 and 58 we obtain for the high-current-limiting condition

$$\gamma \sim q\mu_p p_n \mathscr{E} \sim \frac{J}{J_S} \tag{61}$$

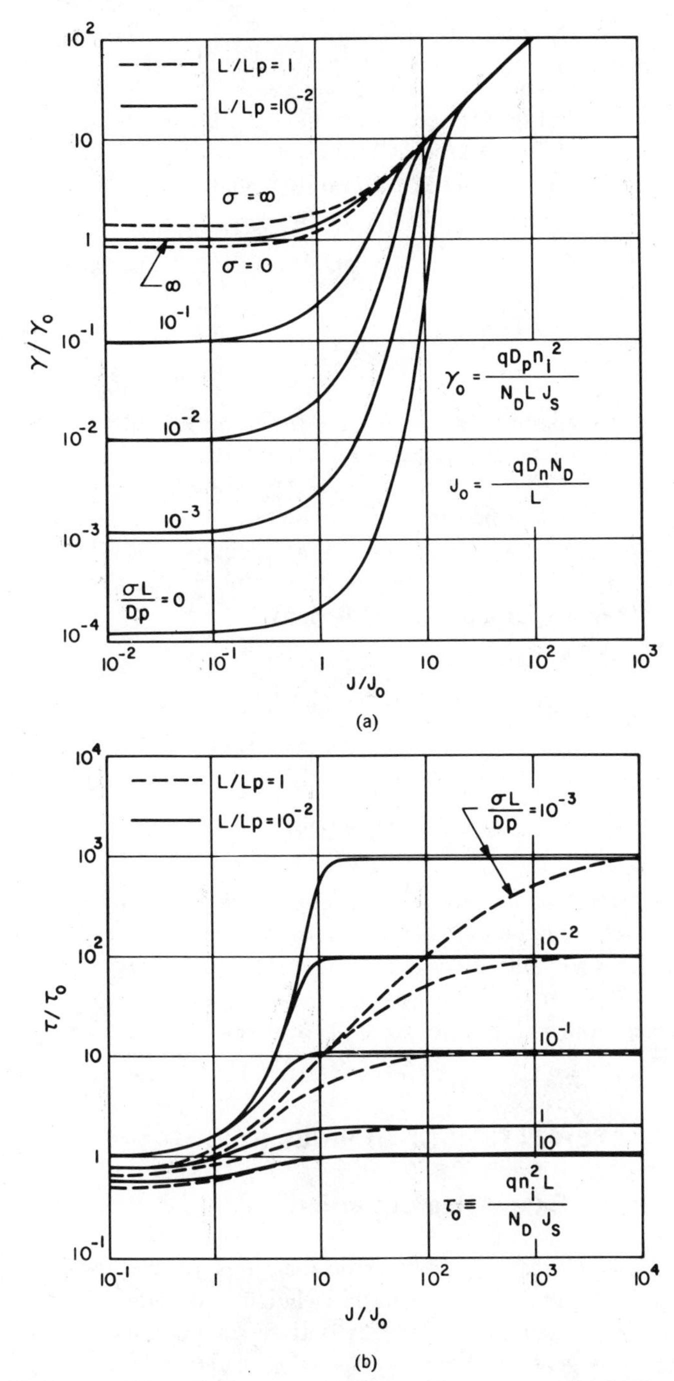

Fig. 14 (*a*) Normalized minority-carrier injection ratio versus normalized current density. (*b*) Normalized minority-carrier storage time versus normalized current density. (After Scharfetter, Ref. 23.)

and the injection ratio increases linearly with the current density. For example, a gold-n-type silicon diode with $N_D = 10^{15}\ \text{cm}^{-3}$ and $J_S = 5 \times 10^{-7}\ \text{A/cm}^2$ would be expected to have an injection ratio of about 5% at a current density of $350\ \text{A/cm}^2$. The intermediate cases have been considered by Scharfetter,[23] and the computed results are shown in Fig. 14a, where the normalization factors are given by

$$J_0 \equiv \frac{qD_n N_D}{L} \tag{62a}$$

$$\gamma_0 \equiv \frac{qD_p n_i^2}{N_D L J_S}. \tag{62b}$$

Figure 14*a* clearly shows that to reduce the minority-carrier injection ratio one must use a metal–semiconductor system with large N_D (corresponding to low resistivity material), large J_S (corresponding to small barrier height), and small n_i (corresponding to large bandgap).

Another quantity associated with the injection ratio is the minority storage time τ_s, which is defined as the minority carrier stored in the quasi-neutral region per unit current density:

$$\tau_s \equiv \frac{\int_{x_1}^{x_2} qp(x)\,dx}{J}. \tag{63}$$

For high current limit, τ_s is given by

$$\tau_s \simeq \frac{qn_i^2 L_p}{N_D J_S}. \tag{64}$$

The results for τ_s versus the current density are shown in Fig. 14*b*, where similar parameters as in Fig. 14*a* are used. For example, in a Au–Si diode with $N_D = 1.5 \times 10^{14}\ \text{cm}^{-3}$, $L = 7\ \mu\text{m}$, and $D_p/L_p = 2000\ \text{cm/s}$, the storage time for $J = 10\ \text{A/cm}^2$ is about 1 ns. If N_D is increased to $1.5 \times 10^{16}\ \text{cm}^{-3}$, the storage time would decrease to 0.01 ns, even at a current density of $1000\ \text{A/cm}^2$.

5.5 CHARACTERIZATION OF BARRIER HEIGHT

5.5.1 General Expression of Barrier Height

We have considered the basic band diagrams of metal–semiconductor contacts in Section 5.2. The barrier heights of metal–semiconductor systems are, in general, determined by both the metal work function and the surface states. A general expression[24] of the barrier height can be obtained on the basis of the following two assumptions: (1) with intimate contact between the metal and the semiconductor and with an interfacial layer of atomic dimensions, this layer will be transparent to electrons and can

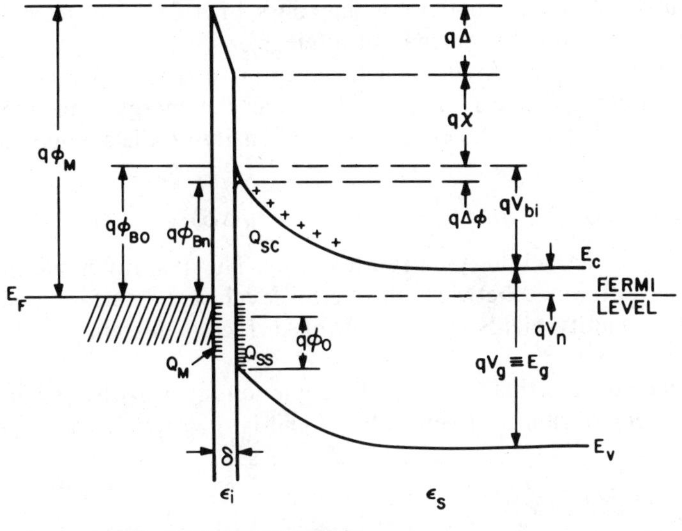

ϕ_M = WORK FUNCTION OF METAL
ϕ_{Bn} = BARRIER HEIGHT OF METAL-SEMICONDUCTOR BARRIER
ϕ_{BO} = ASYMPTOTIC VALUE OF ϕ_{Bn} AT ZERO ELECTRIC FIELD
ϕ_0 = ENERGY LEVEL AT SURFACE
$\Delta\phi$ = IMAGE FORCE BARRIER LOWERING
Δ = POTENTIAL ACROSS INTERFACIAL LAYER
χ = ELECTRON AFFINITY OF SEMICONDUCTOR
V_{bi} = BUILT-IN POTENTIAL
ϵ_s = PERMITTIVITY OF SEMICONDUCTOR
ϵ_i = PERMITTIVITY OF INTERFACIAL LAYER
δ = THICKNESS OF INTERFACIAL LAYER
Q_{sc} = SPACE-CHARGE DENSITY IN SEMICONDUCTOR
Q_{ss} = SURFACE-STATE DENSITY ON SEMICONDUCTOR
Q_M = SURFACE-CHARGE DENSITY ON METAL

Fig. 15 Detailed energy-band diagram of a metal *n*-type semiconductor contact with an interfacial layer of the order of atomic distance. (After Cowley and Sze, Ref. 24.)

withstand potential across it, and (2) the surface states per unit area per electron volt at the interface are a property of the semiconductor surface and are independent of the metal.

A more detailed energy-band diagram of a metal *n*-type semiconductor contact is shown in Fig. 15. The various quantities used in the derivation that follows are defined in this figure. The first quantity of interest is the energy level $q\phi_0$ at the surface. The energy level coincided with the Fermi level before the metal–semiconductor contact was formed. It specified the level below which all surface states must have been filled for charge neutrality at the surface.[25] The second quantity is $q\phi_{Bn}$, the barrier height of the metal–semiconductor contact; it is this barrier that must be surmounted by electrons flowing from the metal into the semiconductor. The interfacial layer will be assumed to have a thickness of a few angstroms and will therefore be essentially transparent to electrons.

We consider a semiconductor with acceptor surface states whose density is D_s states/cm^2/eV, and D_s is a constant over the energy range from $q\phi_0$ to the Fermi level. The surface-state charge density on the semiconductor Q_{ss} is given by

$$Q_{ss} = -qD_s(E_g - q\phi_0 - q\phi_{Bn} - q\,\Delta\phi) \qquad \text{C/cm}^2 \tag{65}$$

where $q\,\Delta\phi$ is the Schottky-barrier lowering. The quantity in parentheses is simply the difference between the Fermi level at the surface and $q\phi_0$. D_s times this quantity yields the number of surface states above $q\phi_0$ that are full.

The space charge that forms in the depletion layer of the semiconductor at thermal equilibrium is given in Eq. 8 and is rewritten as

$$Q_{sc} = \sqrt{2q\epsilon_s N_D\left(\phi_{Bn} - V_n + \Delta\phi - \frac{kT}{q}\right)} \qquad \text{C/cm}^2. \tag{66}$$

The total equivalent surface charge density on the semiconductor surface is given by the sum of Eqs. 65 and 66. In the absence of any space-charge effects in the interfacial layer, an exactly equal and opposite charge, Q_M (C/cm^2), develops on the metal surface. For thin interfacial layers such effects are negligible, and Q_M can be written as

$$Q_M = -(Q_{ss} + Q_{sc}). \tag{67}$$

The potential Δ across the interfacial layer can be obtained by the application of Gauss's law to the surface charge on the metal and semiconductor:

$$\Delta = -\delta\frac{Q_M}{\epsilon_i} \tag{68}$$

where ϵ_i is the permittivity of the interfacial layer and δ its thickness. Another relation for Δ can be obtained by inspection of the energy-band diagram of Fig. 15:

$$\Delta = \phi_m - (\chi + \phi_{Bn} + \Delta\phi). \tag{69}$$

This relation results from the fact that the Fermi level must be constant throughout this system at thermal equilibrium.

If Δ is eliminated from Eqs. 68 and 69, and Eq. 67 is used to substitute for Q_M, we obtain

$$(\phi_m - \chi) - (\phi_{Bn} + \Delta\phi) = \sqrt{\frac{2q\epsilon_s N_D\delta^2}{\epsilon_i^2}\left(\phi_{Bn} + \Delta\phi - V_n - \frac{kT}{q}\right)} - \frac{qD_s\delta}{\epsilon_i}(E_g - q\phi_0 - q\phi_{Bn} - q\,\Delta\phi). \tag{70}$$

Equation 70 can now be solved for ϕ_{Bn}. Introducing the quantities

$$c_1 \equiv \frac{2q\epsilon_s N_D \delta^2}{\epsilon_i^2} \tag{71a}$$

$$c_2 \equiv \frac{\epsilon_i}{\epsilon_i + q^2 \delta D_s} \tag{71b}$$

we can write the solution to Eq. 70 as

$$\phi_{Bn} = \left[c_2(\phi_m - \chi) + (1 - c_2)\left(\frac{E_g}{q} - \phi_0\right) - \Delta\phi \right] + \left\{ \frac{c_2^2 c_1}{2} - c_2^{3/2}\left[c_1(\phi_m - \chi) + (1 - c_2)\left(\frac{E_g}{q} - \phi_0\right)\frac{c_1}{c_2} - \frac{c_1}{c_2}\left(V_n + \frac{kT}{q}\right) + \frac{c_2 c_1^2}{4} \right]^{1/2} \right\}. \tag{72}$$

Equation 71a can be used to calculate c_1 if values of δ and ϵ_i are estimated; for vacuum-cleaved or well-cleaned semiconductor substrates the interfacial layer will have a thickness of atomic dimensions (i.e., 4 or 5 Å). The permittivity of such a thin layer can be well approximated by the free-space value, and since this approximation represents a lower limit for ϵ_i, it leads to an overestimation of c_2. For $\epsilon_s \approx 10\epsilon_0$, $\epsilon_i = \epsilon_0$, and $N_D < 10^{18}$ cm^{-3}, c_1 is small, of the order of 0.01 V, and the $\{\cdot\}$ term in Eq. 72 is estimated to be less than 0.04 V. Neglecting the $\{\cdot\}$ term in Eq. 72 reduces the equation to

$$\phi_{Bn} = c_2(\phi_m - \chi) + (1 - c_2)\left(\frac{E_g}{q} - \phi_0\right) - \Delta\phi \equiv c_2\phi_m + c_3. \tag{73}$$

If c_2 and c_3 can be determined experimentally and if χ is known, then

$$\phi_0 = \frac{E_g}{q} - \frac{c_2\chi + c_3 + \Delta\phi}{1 - c_2} \tag{74}$$

and from Eq. 71b

$$D_s = \frac{(1 - c_2)\epsilon_i}{c_2 \delta q^2}. \tag{75}$$

Using the previous assumptions for δ and ϵ_i, we obtain

$$D_s \simeq 1.1 \times 10^{13}(1 - c_2)/c_2 \qquad \text{states/cm}^2\text{/eV}. \tag{75a}$$

The two limiting cases considered previously can be obtained directly from Eq. 73:

1 When $D_s \to \infty$, then $c_2 \to 0$ and

$$q\phi_{Bn} = (E_g - q\phi_0) - q\Delta\phi. \tag{76a}$$

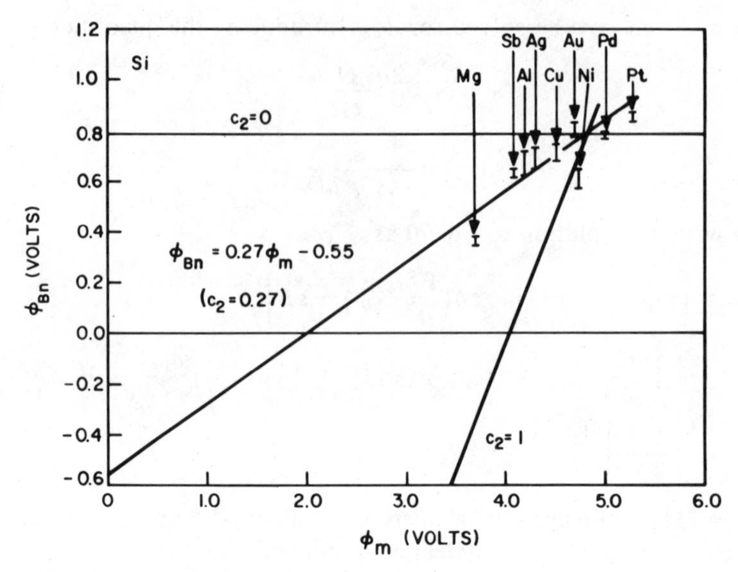

Fig. 16 Experimental results of barrier heights for metal *n*-type silicon contacts. (After Cowley and Sze, Ref. 24.)

In this case the Fermi level at the interface is "pinned" by the surface states at the value $q\phi_0$ above the valence band. The barrier height is independent of the metal work function and is determined entirely by the doping and surface properties of the semiconductor.

2 When $D_s \rightarrow 0$, then $c_2 \rightarrow 1$ and

$$q\phi_{Bn} = q(\phi_m - \chi) - q\Delta\phi. \tag{76b}$$

This equation for the barrier height of an ideal Schottky barrier where surface-state effects are neglected, is identical to Eq. 1 (except for the Schottky lowering term).

The experimental results of the metal *n*-type silicon system are shown in Fig. 16. A least-squares straight line fit to the data yields

$$\phi_{Bn} = 0.27\phi_m - 0.55.$$

Comparing this expression with Eq. 73 and using Eqs. 74 and 75a we obtain $c_2 = 0.27$, $q\phi_0 = 0.33$ eV, and $D_s = 4 \times 10^{13}$ states/cm^2/eV. Similar results are obtained for GaAs, GaP, and CdS, which are shown in Fig. 17 and listed in Table 2.

We note that the values of $q\phi_0$, for Si, GaAs, and GaP are very close to one-third of the bandgap. Similar results are obtained for other semiconductors.[26] This fact indicates that most covalent semiconductor surfaces

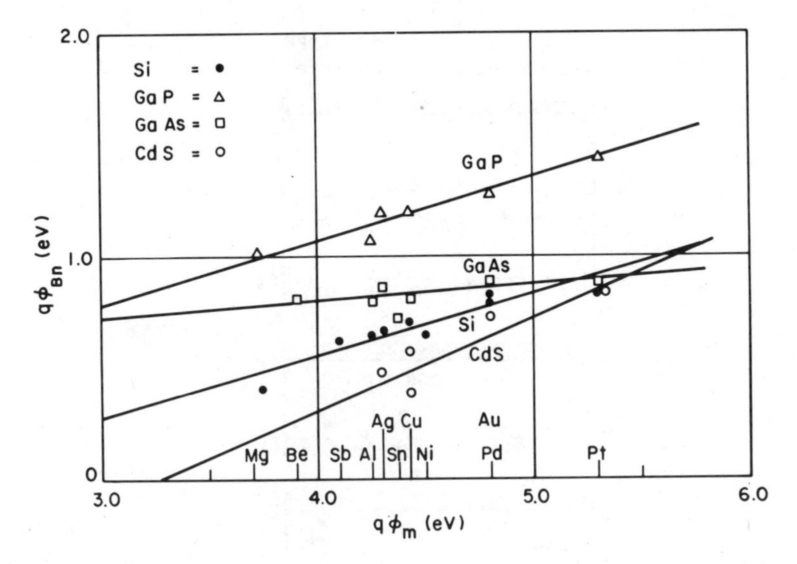

Fig. 17 Similar results for other metal–semiconductor contacts. (After Cowley and Sze, Ref. 24.)

have a high peak density of surface states or defects near one-third of the gap from the valence-band edge. The theoretical calculation by Pugh[27] for ⟨111⟩ diamond indeed gives a narrow band of surface states slightly below the center of the forbidden gap. It is thus expected that a similar situation may exist for other semiconductors.

For III–V compounds, extensive measurements using photoemission spectroscopy indicate that the Schottky-barrier formation is due to defects formed near the interface by deposition of the metal.[72] Figure 18 shows[73] the surface Fermi-level positions obtained for a number of metals on GaAs, GaSb, and InP. Note that the surface Fermi level for a given III–V compound is pinned at a position independent of the metal. (It is also interesting to note that the surface Fermi level for oxygen atoms is also

Table 2 Summary of Barrier Height Data and Calculations for Si, GaP, GaAs, and CdS

Semi-conductor	c_2	c_3 (V)	χ (V)	$D_s \times 10^{-13}$ (eV^{-1}/cm^2)	$q\phi_0$ (eV)	$q\phi_0/E_g$
Si	0.27 ± 0.05	-0.55 ± 0.22	4.05	2.7 ± 0.7	0.30 ± 0.36	0.27
GaP	0.27 ± 0.03	-0.01 ± 0.13	4.0	2.7 ± 0.4	0.66 ± 0.2	0.294
GaAs	0.07 ± 0.05	$+0.49 \pm 0.24$	4.07	12.5 ± 10.0	0.53 ± 0.33	0.38
CdS	0.38 ± 0.16	-1.20 ± 0.77	4.8	1.6 ± 1.1	1.5 ± 1.5	0.6

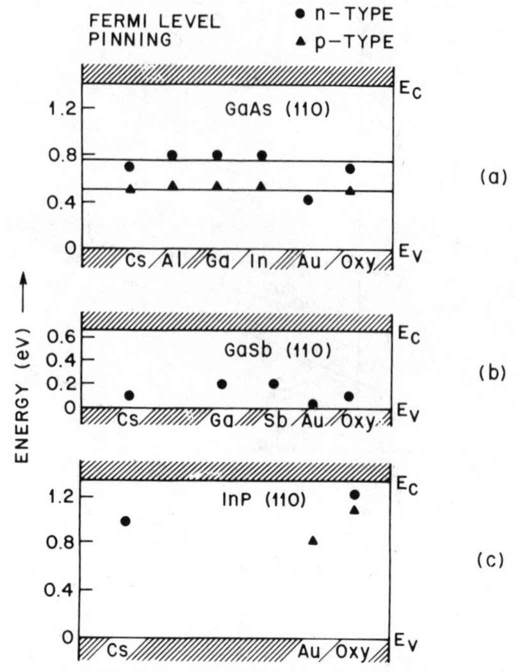

Fig. 18 Surface Fermi-level position for a number of metals and oxygen on GaAs, GaSb, and InP. Note that there is little dependence on the chemical nature of the metals and oxygen. (After Spicer et al., Ref. 73.)

pinned at the same position.) This pinning of surface Fermi level can explain the fact that for most III–V compounds, the barrier height is essentially independent of metal work function.

For ionic semiconductors, such as CdS and ZnS, the barrier height generally depends strongly on the metal; and a correlation has been found between interface behavior and the electronegativity. The electronegativity X_M is defined as the power of an atom in a molecule to attract electrons to itself. Figure 19 shows Pauling's electronegativity scale.[28] Note that the periodicity is similar to that for the work function (Fig. 4).

Figure 20*a* shows a plot[29] of the barrier height versus the electronegativity of metals deposited on Si, GaSe, and SiO_2. We define the slope of the plot as an index of interface behavior:

$$S \equiv \frac{d\phi_{Bn}}{dX_M}. \tag{77}$$

We can plot the index S as a function of the electronegativity difference of the semiconductors shown in Fig. 20*b*. Note a sharp transition from the covalent semiconductors (such as GaAs with $\Delta X = 0.4$) to ionic semiconductors (such as AlN with $\Delta X = 1.5$). For semiconductors with $\Delta X < 1$,

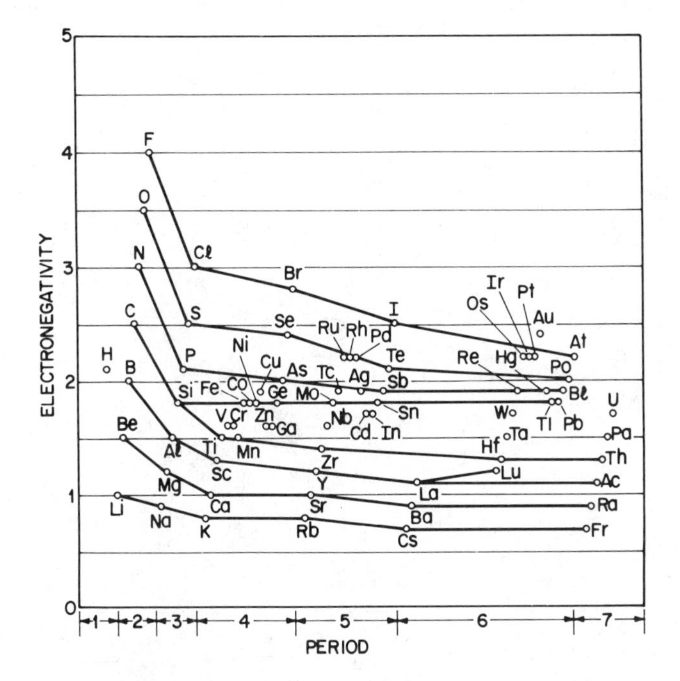

Fig. 19 Pauling's electronegativity scale. Note the trend of increasing electronegativity within each group. (After Pauling, Ref. 28.)

the index S is small, indicating that the barrier height is only weakly dependent on metal electronegativity (or the work function). On the other hand, for $\Delta X > 1$, the index S approaches 1, and the barrier height is dependent on the metal electronegativity (or the work function).

For technological applications in silicon integrated circuits, an important class of Schottky barrier contacts has been developed in which a chemical reaction between the metal and the underlying silicon is induced. The formation of metal silicides by solid–solid metallurgical reaction provides reliable and reproducible Schottky barriers, because the interface chemical reactions are well defined and can be maintained under good control. Figure 21 shows empirical values for the barrier heights on n-type silicon of 12 transition-metal silicides plotted against the heat of formation of the compounds.[71] With the exception of PtSi, the data can be expressed as

$$\phi_{Bn} = 0.81 - 0.17(\Delta H) \tag{78}$$

where ΔH is the heat of formation in eV. At zero heat of formation, the value of 0.81 eV for the barrier height is related to the properties of the free silicon surface. The slope of the curve can be understood by recognizing that the barrier height between two samples of n-type silicon must be zero. Thus the straight line in Fig. 21 must intersect the abscissa at the

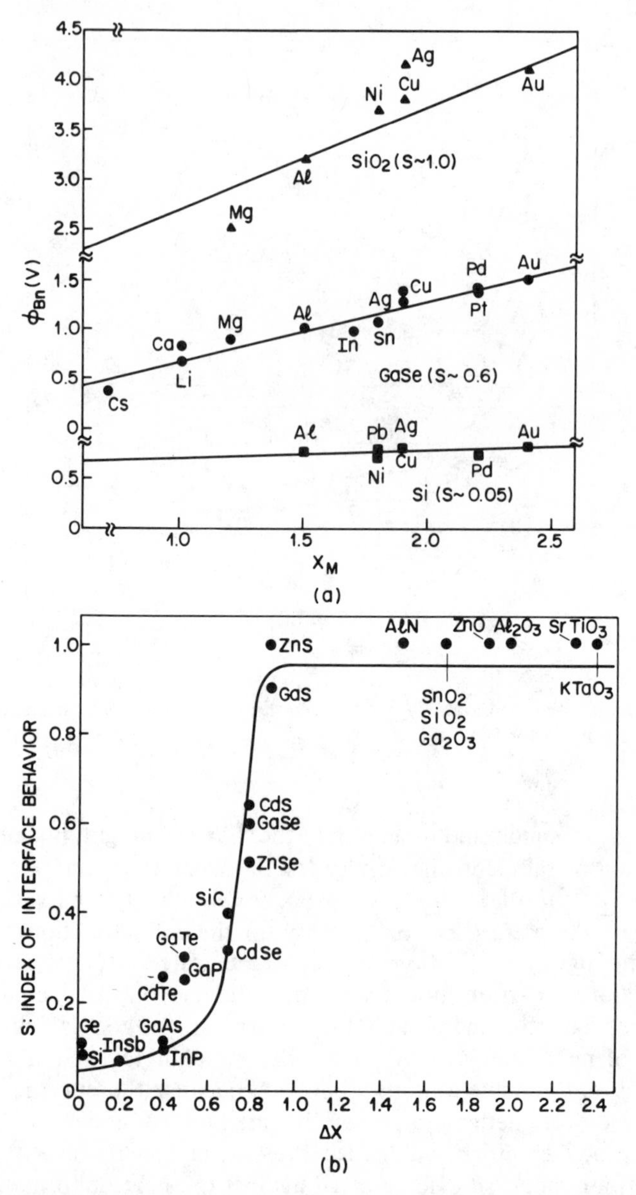

Fig. 20 (*a*) Barrier height versus electronegativity of metals deposited on Si, GaSe, and SiO_2. (*b*) Index of interface behavior S as a function of the electronegativity difference of the semiconductors. (After Kurtin, McGill, and Mead, Ref. 29.)

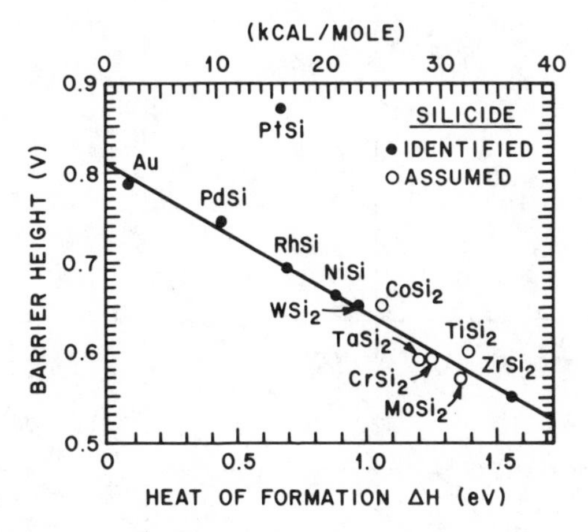

Fig. 21 Barrier height of transition metal silicides versus heat of formation. (After Andrews, Ref. 71.)

binding energy of silicon (4.67 eV). This value is very close to the value ΔH obtained from Eq. 78 by setting $\phi_{Bn} = 0$.

5.5.2 Measurement of Barrier Height

Basically, four methods are used to measure the barrier height of a metal–semiconductor contact: the current–voltage, activation energy, capacitance–voltage, and photoelectric methods.[30]

Current–Voltage Measurement For moderately doped semiconductors, the I–V characteristics in the forward direction with $V > 3kT/q$ is given by Eq. 49:

$$J = A^{**}T^2 \exp\left(-\frac{q\phi_{B0}}{kT}\right) \exp\left[\frac{q(\Delta\phi + V)}{kT}\right] \tag{79}$$

where ϕ_{B0} is the zero-field asymptotic barrier height as shown in Fig. 15, A^{**} is the effective Richardson constant, and $\Delta\phi$ is the Schottky barrier lowering. Since both A^{**} and $\Delta\phi$ are functions of the applied voltage, the forward J–V characteristic (for $V > 3kT/q$) is represented by $J \sim \exp(qV/nkT)$, as given previously in Eq. 53, where n is the ideality factor:

$$n \equiv \frac{q}{kT}\frac{\partial V}{\partial(\ln J)}$$
$$= \left[1 + \frac{\partial\,\Delta\phi}{\partial V} + \frac{kT}{q}\frac{\partial(\ln A^{**})}{\partial V}\right]^{-1}. \tag{80}$$

Typical examples are shown in Fig. 22, where $n = 1.02$ for the W–Si diode and $n = 1.04$ for the W–GaAs diode.[31] The extrapolated value of current

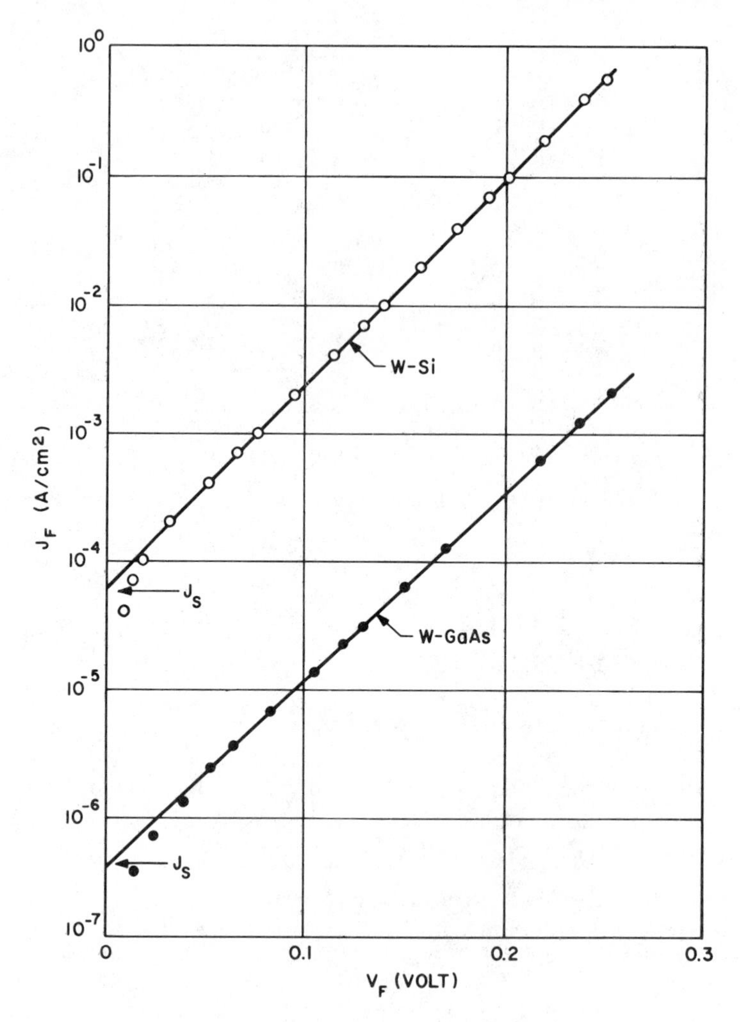

Fig. 22 Forward current density versus applied voltage of W-Si and W-GaAs diodes. (After Crowell, Sarace, and Sze, Ref. 31.)

density at zero voltage is the saturation current J_S, and the barrier height can be obtained from the equation

$$\phi_{Bn} = \frac{kT}{q} \ln \left(\frac{A^{**}T^2}{J_S} \right). \tag{81}$$

The value of ϕ_{Bn} is not very sensitive to the choice of A^{**}, since at room temperature, a 100% increase in A^{**} will cause an increase of only 0.018 V in ϕ_{Bn}. The theoretical relationship between J_S and ϕ_{Bn} (or ϕ_{Bp}) at room temperature is plotted in Fig. 23 for $A^{**} = 120$ A/cm²/K². For other values of A^{**}, parallel lines can be drawn on this plot to obtain the proper

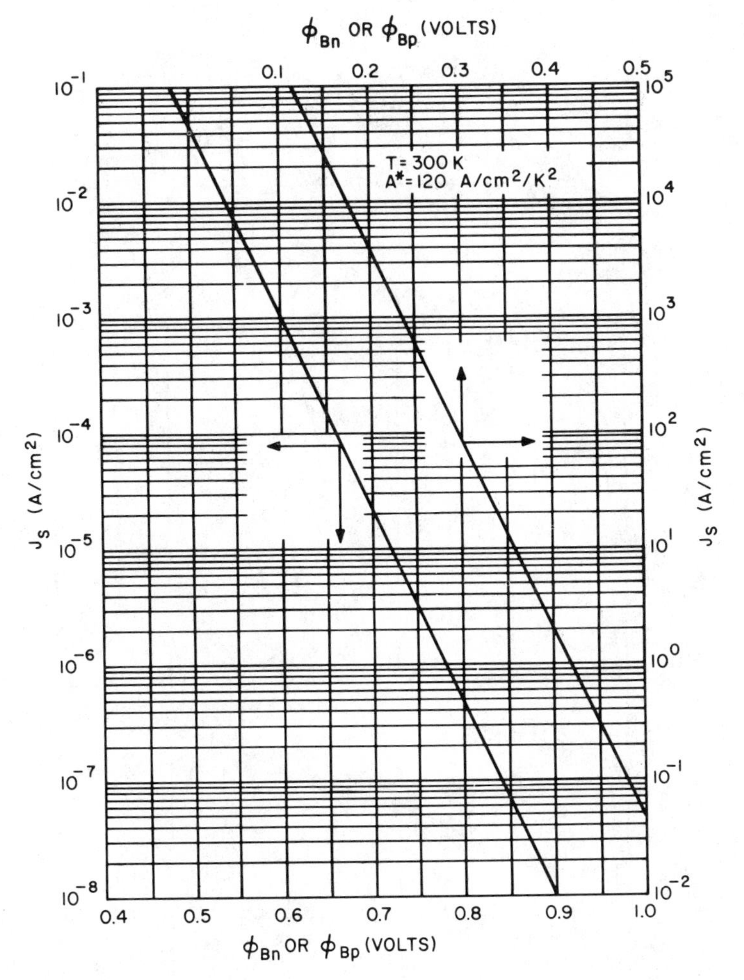

Fig. 23 Theoretical saturation current density at 300 K versus barrier height for a Richardson constant of 120 $A/cm^2/K^2$.

relationship. In the reverse direction, the dominant effect is due to the Schottky-barrier lowering, or

$$
\begin{aligned}
J_R &\simeq J_S \quad (\text{for } V_R > 3kT/q) \\
&= A^{**}T^2 \exp\left(-\frac{q\phi_{B0}}{kT}\right) \exp\left(+\frac{q\sqrt{q\mathscr{E}/4\pi\epsilon_s}}{kT}\right)
\end{aligned} \tag{82}
$$

where

$$
\mathscr{E} = \sqrt{\frac{2qN_D}{\epsilon_s}\left(V + V_{bi} - \frac{kT}{q}\right)}.
$$

If the barrier height $q\phi_{Bn}$ is sufficiently smaller than the bandgap so that the

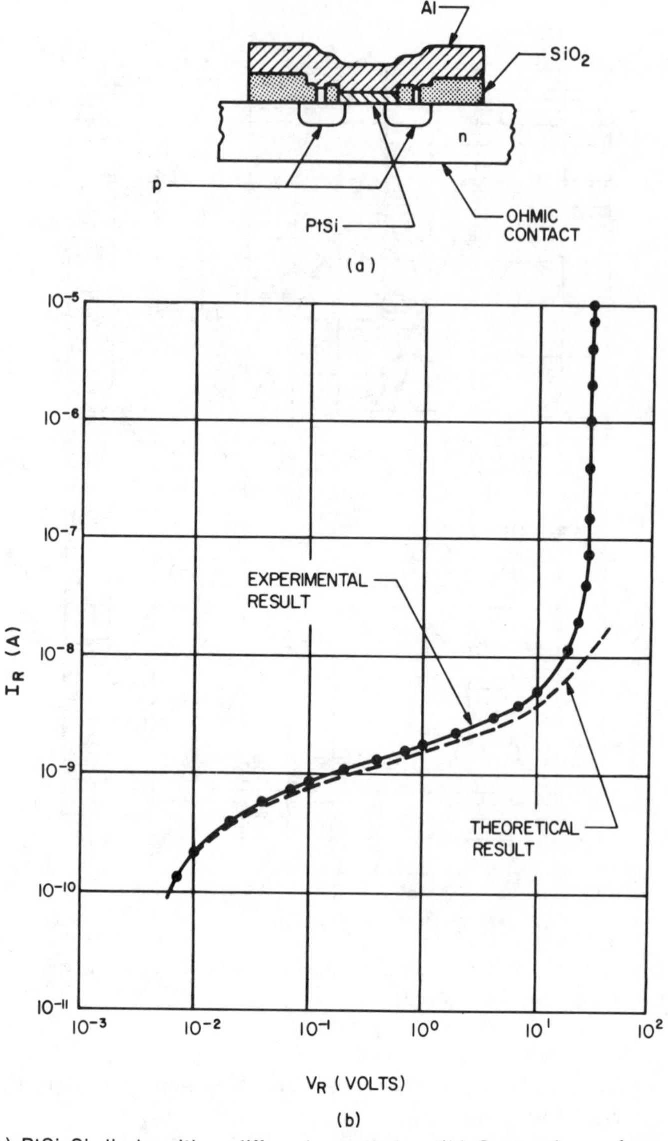

Fig. 24 (*a*) PtSi–Si diode with a diffused guard ring. (*b*) Comparison of experimental with theoretical prediction of Eq. 82 for a PtSi–Si diode. (After Lepselter and Sze, Ref. 32.)

depletion-layer generation–recombination current is small in comparison with the Schottky emission current, then the reverse current will increase gradually with the reverse bias as given by Eq. 82.

For most of the practical Schottky diodes, however, the dominant reverse current component is the edge-leakage current, which is caused by the sharp edge around the periphery of the metal plate. This sharp-edge effect is similar to the junction curvature effect (with $r_j \rightarrow 0$) as discussed in Chapter 2. To eliminate this effect, metal–semiconductor diodes have been fabricated with a diffused guard ring as shown[32] in Fig. 24*a*. The guard ring is a deep *p*-type diffusion, and the doping profile is tailored to give the *p-n* junction a higher breakdown voltage than that of the metal–semiconductor contact. Because of the elimination of the sharp-edge effect, near-ideal forward and reverse *I–V* characteristics have been obtained. Figure 24*b* shows a comparison between experimental measurement from a PtSi–Si diode with guard ring and theoretical calculation based on Eq. 82. The agreement is excellent. The sharp increase of current near 30 V is due to avalanche breakdown and is expected for the diode with a donor concentration of $2.5 \times 10^{16}\ \text{cm}^{-3}$.

The efficacy of guard ring structures in preventing premature breakdown and surface leakage can be ascertained by studying reverse leakage current as a function of diode diameter at constant reverse bias. For this purpose, arrays of Schottky diodes with different diameters can be formed on the semiconductor. Figure 25 shows the measured reverse leakage currents as

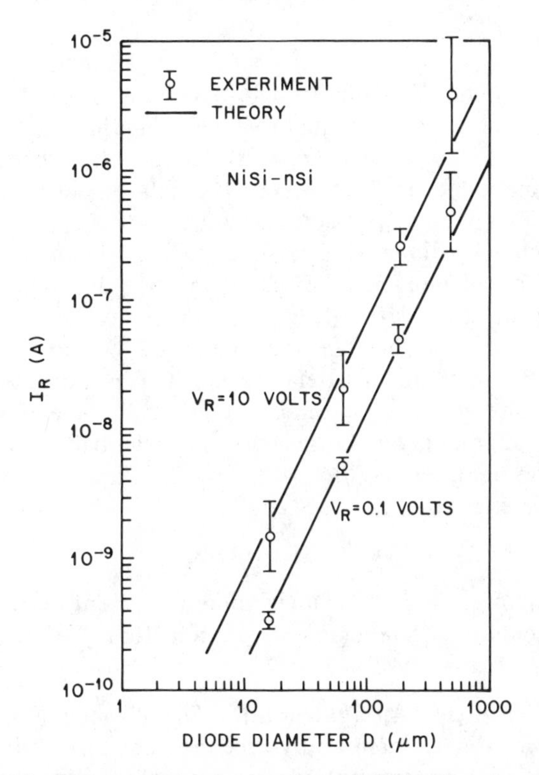

Fig. 25 Reverse leakage current as a function of diode diameter for NiSi–Si diodes formed on *n*-type silicon with $N_D = 6 \times 10^{15}\ \text{cm}^{-3}$. (After Andrews and Koch, Ref. 33.)

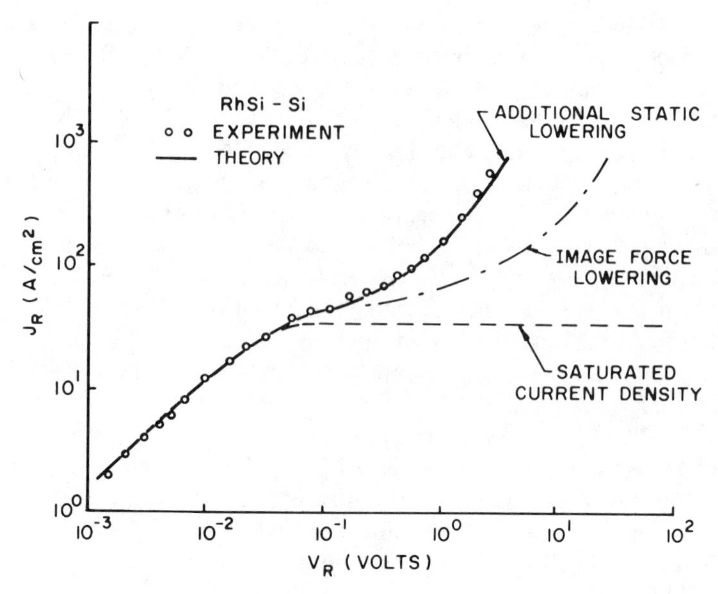

Fig. 26 Theory and experiment of reverse characteristics for a RhSi–Si diode. (After Andrews and Lepselter, Ref. 19.)

a function of diode diameter.[33] The solid lines drawn through the experimental data have slopes equal to 2, showing that the leakage currents are proportional to the device area. If, on the other hand, the leakage currents are dominated by edge effects, the data would be expected to lie along straight lines with slopes equal to unity.

For some Schottky diodes, the reverse current has an additional voltage dependence. This dependence arises from the fact that if the metal–semiconductor interface is free from intervening layers of oxide and other contaminants, the electrons in the metal have wave functions that penetrate into the semiconductor energy gap. This is a quantum-mechanical effect that results in a static dipole layer at the metal–semiconductor interface. The dipole layer causes the intrinsic barrier height to vary slightly with the field, so $\partial\phi_{BO}/\partial\mathscr{E}_m \neq 0$. To a first approximation the static lowering can be expressed as

$$(\Delta\phi)_{\text{static}} \simeq \alpha\mathscr{E}_m \tag{83}$$

where $\alpha \equiv \partial\phi_{BO}/\partial\mathscr{E}_m$. Figure 26 shows good agreement between the theory and measurements of the reverse current in a RhSi–Si diode, based on an empirical value of $\alpha = 17$ Å.

Activation Energy Measurement The principal advantage of Schottky-barrier determination by means of an activation energy measurement is that no assumption of electrically active area is required. This feature is particularly important in the investigation of novel or unusual metal–semiconductor interfaces, because often the true value of the contacting area is not known. In the case of poorly cleaned or incompletely reacted surfaces, the electrically active area may be only a small fraction of the geometric area. On the other hand, a strong metallurgical reaction could result in rough nonplanar metal–semiconductor interface with an electrically active area that is larger than the apparent geometric area.

If Eq. 49 is multiplied by A_e, the electrically active area, we obtain

$$\ln(I_F/T^2) = \ln(A_e A^{**}) - q(\phi_{Bn} - V_F)/kT \tag{84}$$

where $q(\phi_{Bn} - V_F)$ is the activation energy. Over a limited range of temperature (e.g., $273\text{ K} < T < 373\text{ K}$), the value of A^{**} and ϕ_{Bn} are essentially temperature-independent. Thus for a given forward bias V_F, the slope of a plot of $\ln(I_F/T^2)$ versus $1/T$ yields the barrier height ϕ_{Bn}, and the ordinate intercept at $1/T = 0$ yields the product of the electrically active area A_e and the effective Richardson constant A^{**}.

To illustrate the importance of the activation energy method in the investigation of interfacial metallurgical reactions, Figure 27 shows the

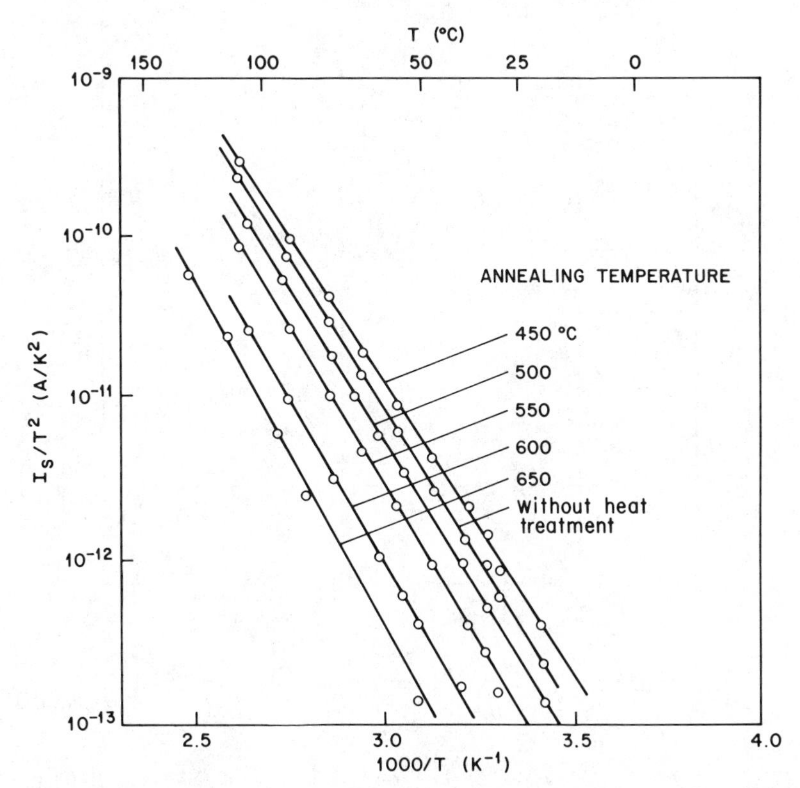

Fig. 27 Activation energy plots for determination of barrier height. (After Chino, Ref. 47.)

activation energy plots of the forward current in Al–n-type Si contacts annealed at various temperatures.[47] The slopes of these plots indicate a nearly linear increase of effective Schottky barrier height from 0.71 to 0.81 V for annealing temperatures between 450 and 650°C. These observations were also confirmed with I–V and C–V measurements.

Obviously, when the Al–Si eutectic temperature (~580°) is reached, the true metallurgical nature of the metal–semiconductor must be considerably modified. Determination of the ordinate intercepts from the plots shown in Fig. 27 indicates that the electrically active area increases by a factor of 2 when the annealing temperature exceeds the Al–Si eutectic temperature.

Capacitance–Voltage Measurement The barrier height can also be determined by the capacitance measurement. When a small ac voltage is superimposed upon a dc bias, charges of one sign are induced on the metal surface and charges of the opposite sign in the semiconductor. The relationship between C and V is given by Eq. 9. Figure 28 shows some typical results where $1/C^2$ is plotted against the applied voltage. From the

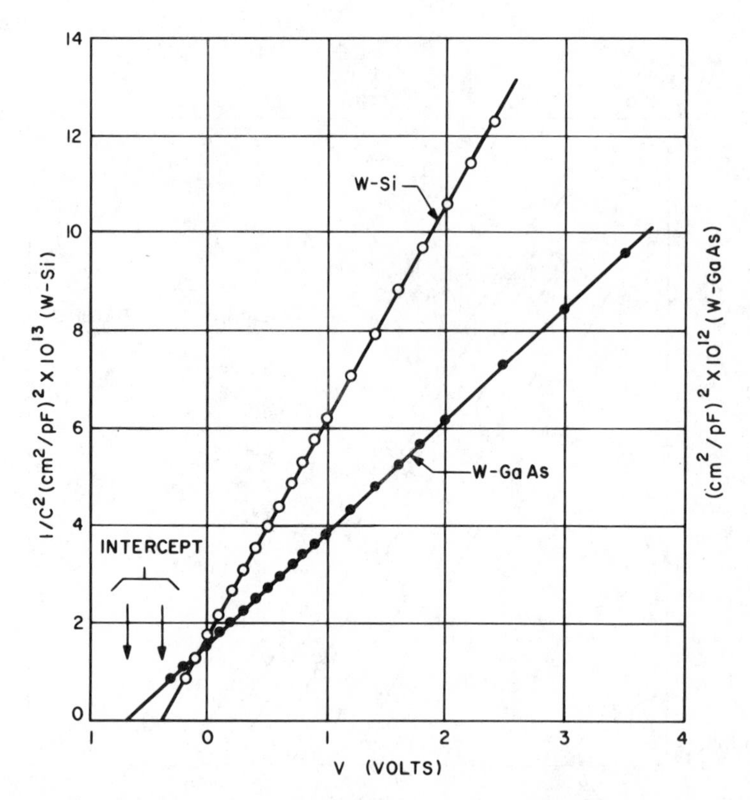

Fig. 28 $1/C^2$ versus applied voltage for W–Si and W–GaAs diodes. (After Crowell, Sarace, and Sze, Ref. 31.)

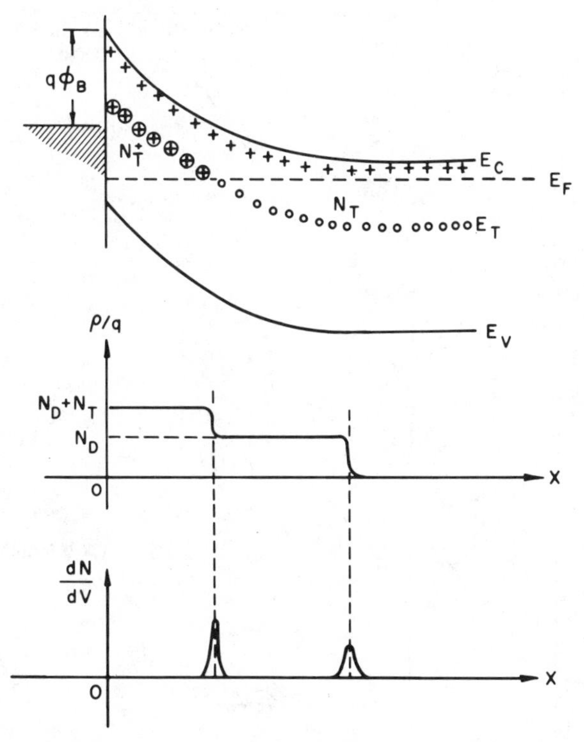

Fig. 29 Semiconductor with one shallow donor level and one deep donor level. N_D and N_T are the shallow donor and deep donor concentration, respectively. (After Roberts and Crowell, Ref. 35.)

intercept on the voltage axis, the barrier height can be determined:[31,34]

$$\phi_{Bn} = V_i + V_n + \frac{kT}{q} - \Delta\phi \tag{85}$$

where V_i is the voltage intercept, and V_n the depth of the Fermi level below the conduction band, which can be computed if the doping concentration is known. From the slope the carrier density can be determined (Eq. 10c). (This method can also be used to measure the doping variation in an epitaxial layer.)

The C–V measurement can also be used to study deep impurity levels. Figure 29 shows a semiconductor with one shallow doping level and one deep donor level.[35] Under bias, all the donors above the Fermi level will be ionized, giving a higher doping concentration near the interface. When a small ac signal is superimposed on the dc bias and when the deep level can follow the signal, there will be an additional contribution of dN/dV to the capacitance. Figure 30 shows $1/C^2$ versus V for various frequencies. The low-frequency curves can reveal the properties of the deep impurities. To obtain the barrier height of semiconductor with one shallow level and one

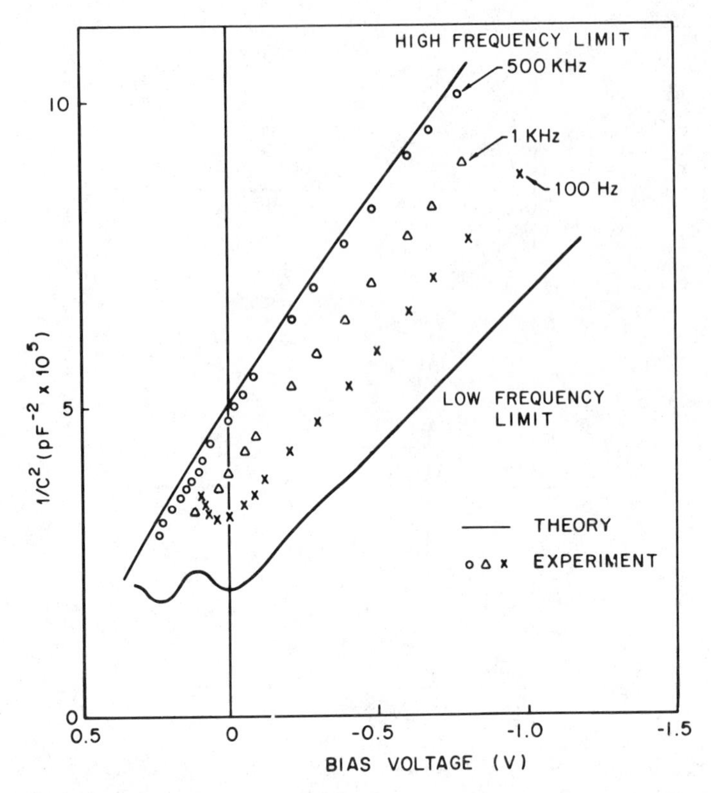

Fig. 30 $1/C^2$ versus voltage for various frequencies. (After Roberts and Crowell, Ref. 35.)

deep level impurities (Fig. 29), we need to measure the C–V curves at two different temperatures.[39]

Photoelectric Measurement The photoelectric measurement is an accurate and direct method of determining the barrier height.[36] When a monochromatic light is incident upon a metal surface, photocurrent may be generated. The basic setup is shown in Fig. 31. For the front illumination the light can generate excited electrons in the metal, process (1), if $h\nu > q\phi_{Bn}$, and also can generate electron–hole pairs in the semiconductor, process (2), if the metal film is thin enough and $h\nu > E_g$. For the back illumination, photoelectrons can be generated, process (1), if $h\nu > q\phi_{Bn}$; however, when $h\nu > E_g$, the light will be strongly absorbed at the back semiconductor surface, and the photoexcited electron–hole pairs have very small probability of reaching the metal–semiconductor interface.

The photocurrent per absorbed photon R as a function of the photon

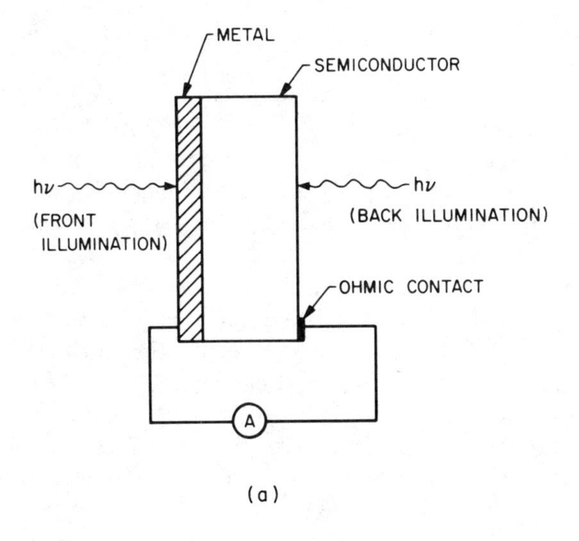

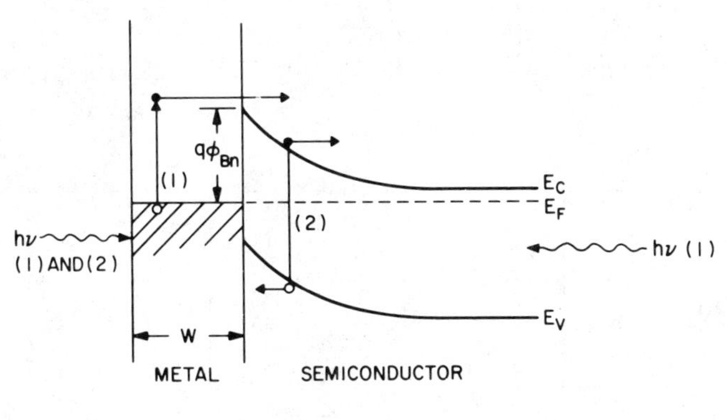

Fig. 31 (*a*) Schematic setup for photoelectric measurement. (*b*) Energy-band diagram for photoexcitation processes.

energy, $h\nu$, is given by the Fowler theory:[37]

$$R \sim \frac{T^2}{\sqrt{E_s - h\nu}} \left[\frac{x^2}{2} + \frac{\pi^2}{6} - \left(e^{-x} - \frac{e^{-2x}}{4} + \frac{e^{-3x}}{9} - \cdots \right) \right] \qquad \text{for} \quad x \geq 0 \tag{86}$$

where $h\nu_0$ is the barrier height ($q\phi_{Bn}$), E_s the sum of $h\nu_0$ and the Fermi energy measured from the bottom of the metal conduction band, and $x \equiv h(\nu - \nu_0)/kT$. Under the condition that $E_s \gg h\nu$, and $x > 3$, Eq. 86 reduces to

$$R \sim (h\nu - h\nu_0)^2 \qquad \text{for} \quad h(\nu - \nu_0) > 3kT \tag{86a}$$

or

$$\sqrt{R} \sim h(\nu - \nu_0). \tag{86b}$$

When the square root of the photoresponse is plotted as a function of photon energy, a straight line should be obtained, and the extrapolated value on the energy axis should give directly the barrier height. Figure 32 shows the photoresponse of W–Si and W–GaAs diodes, the barrier heights of 0.65 and 0.80 eV, respectively.

The photoelectric measurement can be used to study other device and material parameters. It has been used to determine the image-force dielectric constant of Au–Si diodes.[10] By measuring the shift of the photothreshold under different reverse biases, one can determine the image-force lowering $\Delta\phi$. From a plot of $\Delta\phi$ versus $\sqrt{\mathscr{E}}$, the image for dielectric constant (ϵ_s/ϵ_0) can be determined, as shown previously in Fig. 5. The photoelectric measurement has been used to study the temperature dependence of the barrier height.[38] The photothreshold is measured as a function of temperature of Au–Si diodes. The shift of photothreshold correlates reasonably well with the temperature dependence of the silicon-bandgap. This result implies that the Fermi level at the Au–Si interface is

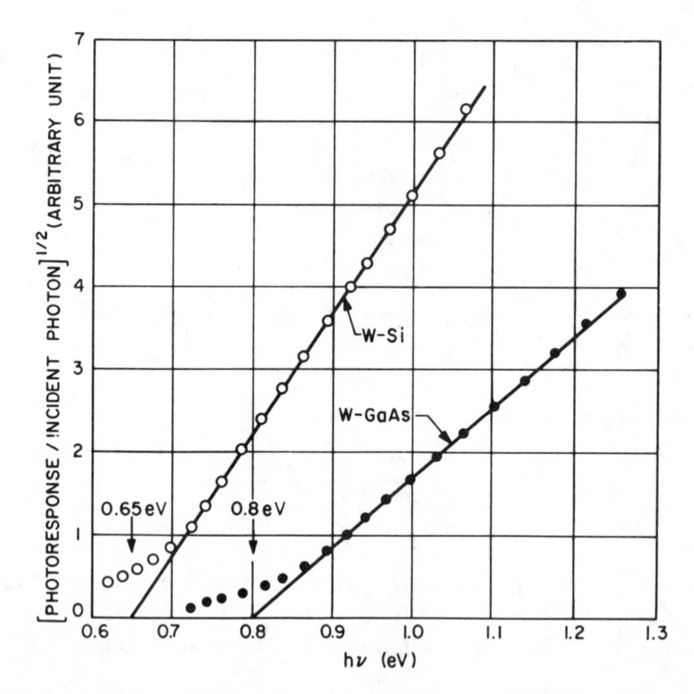

Fig. 32 Square root of the photoresponse per incident photon versus photon energy for W–Si and W–GaAs diodes. The extrapolated values are the corresponding barrier height $q\phi_{Bn}$. (After Crowell, Sarace, and Sze, Ref. 31.)

Table 3 Measured Schottky-Barrier Heights (Volts at 300 K)

Semi-conductor	Type	E_g (eV)	Ag	Al	Au	Cr	Cu	Hf	In	Mg	Mo	Ni	Pb	Pd	Pt	Ta	Ti	W
Diamond	*p*	5.47			1.71													
Ge	*n*	0.66	0.54	0.48	0.59		0.52		0.64			0.49	0.38					0.48
Ge	*p*		0.50		0.30				0.55									
Si	*n*	1.12	0.78	0.72	0.80	0.61	0.58	0.58		0.40	0.68	0.61		0.81	0.90		0.50	0.67
Si	*p*		0.54	0.58	0.34	0.50	0.46				0.42	0.51	0.55				0.61	0.45
SiC	*n*	3.00		2.00	1.95													
AlAs	*n*	2.16			1.20										1.00			
AlSb	*p*	1.63			0.55													
BN	*p*	7.50			3.10													
BP	*p*	6.00			0.87													
GaSb	*n*	0.67			0.60													
GaAs	*n*	1.42	0.88	0.80	0.90		0.82	0.72							0.84	0.85		0.80
GaAs	*p*		0.63		0.42			0.68										
GaP	*n*	2.24	1.20	1.07	1.30	1.06	1.20	1.84		1.04	1.13	1.27			1.45		1.12	
GaP	*p*				0.72													
InSb	*n*	0.16	0.18[a]		0.17[a]													
InAs	*p*	0.33			0.47[a]													
InP	*n*	1.29	0.54		0.52													
InP	*p*				0.76													
CdS	*n*	2.43	0.56	Ohmic	0.78		0.50					0.45	0.59	0.62	1.10		0.84	
CdSe	*n*	1.70	0.43		0.49		0.33								0.37			
CdTe	*n*		0.81	0.76	0.71										0.76			
ZnO	*n*	3.20		0.68	0.65		0.45		0.30					0.68	0.75	0.30		
ZnS	*n*	3.60	1.65	0.80	2.00		1.75		1.50	0.82				1.87	1.84	1.10		
ZnSe	*n*		1.21	0.76	1.36		1.10		0.91				1.16		1.40			
PbO	*n*		0.95						0.93			0.96	0.95					

[a] 77 K.

Table 4 Barrier Height of Metal Silicide on *n*-Type Si

Metal Silicide	ϕ_B (V)	Structure	Forming Temperature (°C)	Melting Temperature (°C)
CoSi	0.68	Cubic	400	1460
$CoSi_2$	0.64	Cubic	450	1326
$CrSi_2$	0.57	Hexagonal	450	1475
HfSi	0.53	Orthorhombic	550	2200
IrSi	0.93	—	300	—
MnSi	0.76	Cubic	400	1275
$Mn_{11}Si_{19}$	0.72	Tetragonal	800[a]	1145
$MoSi_2$	0.55	Tetragonal	1000[a]	1980
Ni_2Si	0.7–0.75	Orthorhombic	200	1318
NiSi	0.66–0.75	Orthorhombic	400	992
$NiSi_2$	0.7	Cubic	800[a]	993
Pd_2Si	0.72–0.75	Hexagonal	200	1330
PtSi	0.84	Orthorhombic	300	1229
RhSi	0.69	Cubic	300	—
$TaSi_2$	0.59	Hexagonal	750[a]	2200
$TiSi_2$	0.60	Orthorhombic	650	1540
WSi_2	0.65	Tetragonal	650	2150
$ZrSi_2$	0.55	Orthorhombic	600	1520

[a]Can be $\leq 700°C$ under clean interface condition.

pinned in relation to the valence-band edge, and this is in agreement with our discussion in Section 5.5.1.

Measured Barrier Heights The *I–V*, *C–V*, activation energy, and photoelectric methods have been used to measure the barrier heights. For intimate contacts with clean interface, these methods generally yield consistent barrier heights within ± 0.02 V. A large discrepancy between different methods may result from such causes as contamination in the interface, intervening insulating layer, edge leakage current, or deep impurity levels.

The measured Schottky barrier heights are listed[39–41] in Table 3 for some elemental and compound semiconductors. The barrier heights are representative values for metal–semiconductor contacts made by deposition of high-purity metals in a good vacuum system onto cleaved or chemically cleaned semiconductor surfaces. As expected, silicon and GaAs metal–semiconductor contacts are most extensively studied. Among the metals, gold, aluminum, and platinum are most commonly used. The barrier heights of metal silicides on *n*-type silicon are listed[42–43a] in Table 4.

It should be pointed out that the barrier height is generally sensitive to

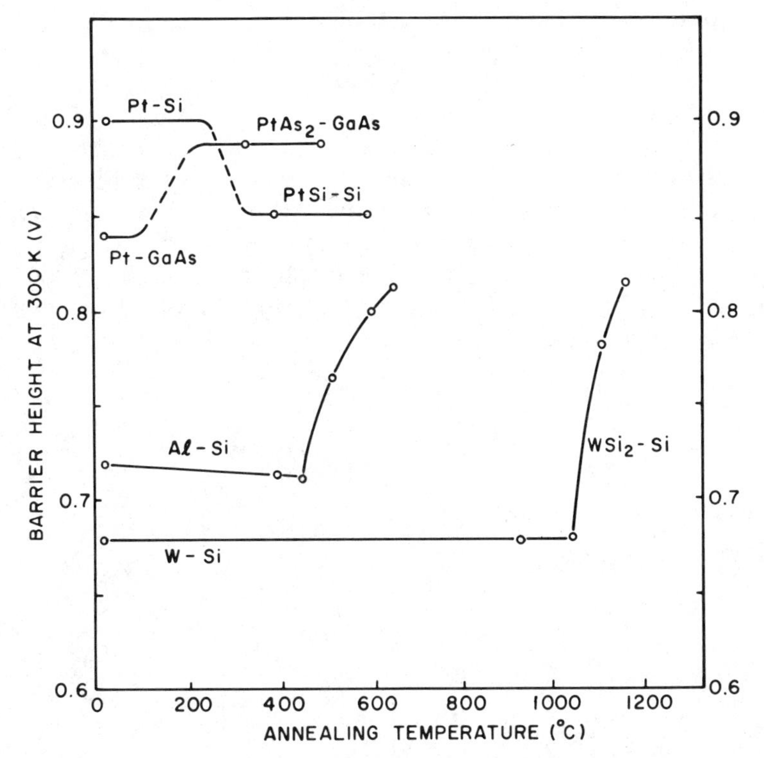

Fig. 33 Barrier heights on *n*-type Si and GaAs measured at room temperature after annealing at various temperatures.

pre-deposition and post-deposition heat treatments.[44] Figure 33 shows the barrier heights on *n*-type Si and GaAs measured at room temperature after annealing at various temperatures. The barrier height of a Pt–Si diode is 0.9 V. After annealing at 300°C or higher temperatures, PtSi is formed at the interface[45] and ϕ_{Bn} decreases to 0.85 V. For Pt–GaAs contact the barrier height increases from 0.84 V to 0.87 V when $PtAs_2$ is formed at the interface.[46] When an Al–Si diode is annealed above 450°C, the barrier height begins to increase[47] presumably due to diffusion of Si in Al (also see Fig. 27). For a W–Si diode the barrier height remains constant until the annealing temperature is above 1000°C, where WSi_2 is formed.[48]

5.5.3 Barrier Height Adjustment

For a standard Schottky barrier, the barrier height is determined primarily by the character of the metal and the metal–semiconductor interface property and is nearly independent of the doping. Usual Schottky barriers on a given semiconductor (e.g., *n*- or *p*-type Si) therefore give a finite number of choices for barrier height (Tables 3 and 4).

By introducing a thin layer (~ 100 Å or less) of semiconductor with a controllable number of dopants on a semiconductor surface (e.g., by ion implantation), the effective barrier height for a given metal–semiconductor contact can be varied.[49–51] This approach is particularly useful when a metal having the most desirable metallurgical properties required for reliable device operation can be selected, and the effective barrier height between this metal and the semiconductor can be adjusted in a controlled manner.

Figure 34*a* shows the idealized controlled barrier contacts with a thin n^+ layer or a thin p^+ layer on an n-type substrate for barrier reduction or barrier raising, respectively. Consider the reduction of barrier first. The field distribution, Fig. 34*b* is given by

$$\mathscr{E} = -\mathscr{E}_m + qn_1x/\epsilon_s \qquad \text{for} \quad 0 < x < a$$

$$= -\frac{qn_2}{\epsilon_s}(W - x) \qquad \text{for} \quad a < x < W \tag{87}$$

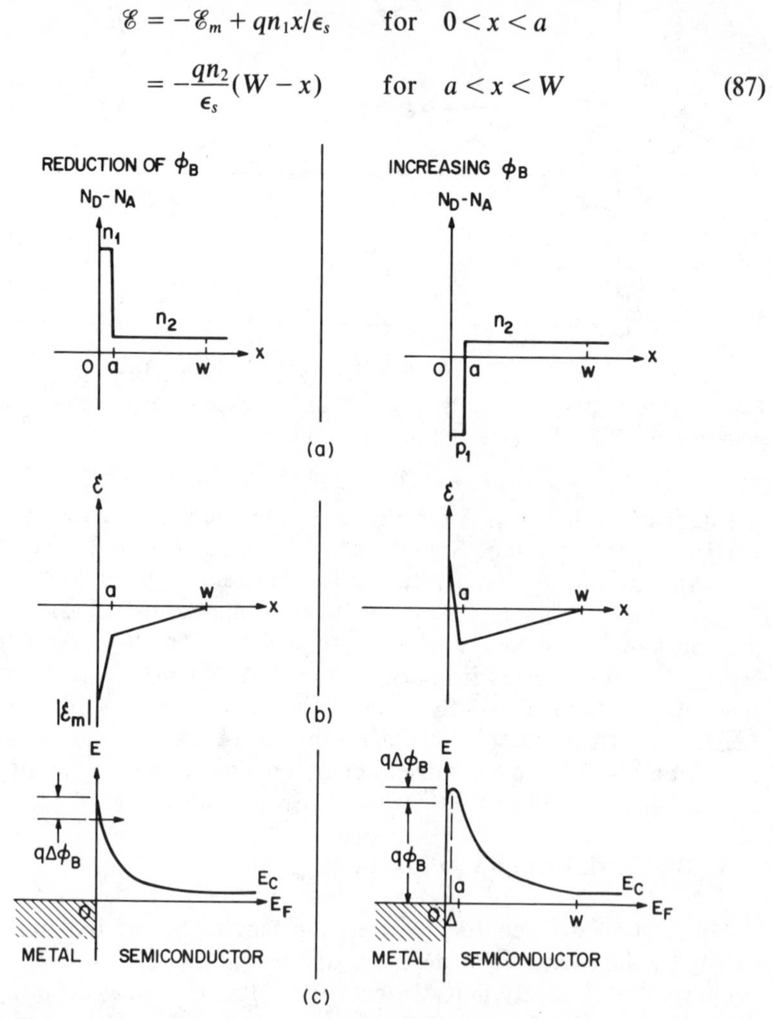

Fig. 34 Idealized controlled barrier contacts with a thin n^+ layer or a thin p^+ layer on an n-type substrate for barrier reduction or barrier raising, respectively.

where $\mathscr{E}_m$ is the maximum electric field at the metal–semiconductor interface, and is given by

$$\mathscr{E}_m = \frac{q}{\epsilon_s}[n_1 a + n_2(W - a)]. \tag{88}$$

The image-force lowering due to $\mathscr{E}_m$ is given by

$$\Delta\phi = \sqrt{q\mathscr{E}_m/4\pi\epsilon_s}. \tag{89}$$

For Si and GaAs Schottky barriers with n_2 of the order of $10^{16}/\text{cm}^3$ or less, the zero-bias value of $n_2(W - a)$ is about $10^{11}\ \text{cm}^{-2}$. Therefore, if $n_1 a$ is made sufficiently larger than $10^{11}\ \text{cm}^{-2}$, Eqs. 88 and 89 can be reduced to

$$\mathscr{E}_m \simeq qn_1 a/\epsilon_s \tag{90}$$

$$\Delta\phi \simeq \frac{q}{\epsilon_s}\sqrt{\frac{n_1 a}{4\pi}}. \tag{91}$$

For $n_1 a = 10^{12}$ and $10^{13}\ \text{cm}^{-2}$, the corresponding lowerings are 0.045 and 0.14 V, respectively.

Although the image-force lowering contributes to the barrier reduction, generally the tunneling effect is more significant. For $n_1 a = 10^{13}\ \text{cm}^{-2}$, the maximum field from Eq. 90 is 1.6×10^6 V/cm, which is the zero-bias field of a Au–Si Schottky diode with a doping of $10^{19}\ \text{cm}^{-3}$. From Fig. 11*a*, the saturation current density for such a diode is about $10^{-3}\ \text{A/cm}^2$ corresponding to an effective barrier height of 0.6 V (see Fig. 23), a reduction of 0.2 V from the 0.8 V barrier of the Au–Si diode. The calculated effective barrier height as a function of $\mathscr{E}_m$ is shown in Fig. 35 for Si and GaAs barriers.[52] By increasing the maximum field from 10^5 V/cm to 10^6 V/cm, one generally can reduce the effective barrier by 0.2 V in Si and over 0.3 V in GaAs.

For a given application, the parameters n_1 and "a" should be properly chosen that in the forward direction the large Schottky barrier lowering and the added tunneling current will not substantially degrade the ideality factor n. In the reverse direction, they will not cause large leakage current in the required bias range.

If opposite doping is formed in the thin semiconductor layer at the interface, the effective barrier can be increased. As in Fig. 34*a*, if the n^+ region is replaced by p^+, it can be shown that the energy-band profile will be $q\phi_B$ at $x = 0$ and reach a maximum at $x = \Delta$, where

$$\Delta = \frac{1}{p_1}[ap_1 - (W - a)n_2]. \tag{92}$$

The effective barrier height occurs at $x = \Delta$ and is given by

$$\phi_B' = \phi_B + \mathscr{E}_m\Delta - \frac{qp_1\Delta^2}{2\epsilon_s}. \tag{93}$$

Equation 93 approaches $(\phi_B + qp_1 a^2/2\epsilon_s)$ if $p_1 \gg n_2$ and $ap_1 \gg n_2 W$. There-

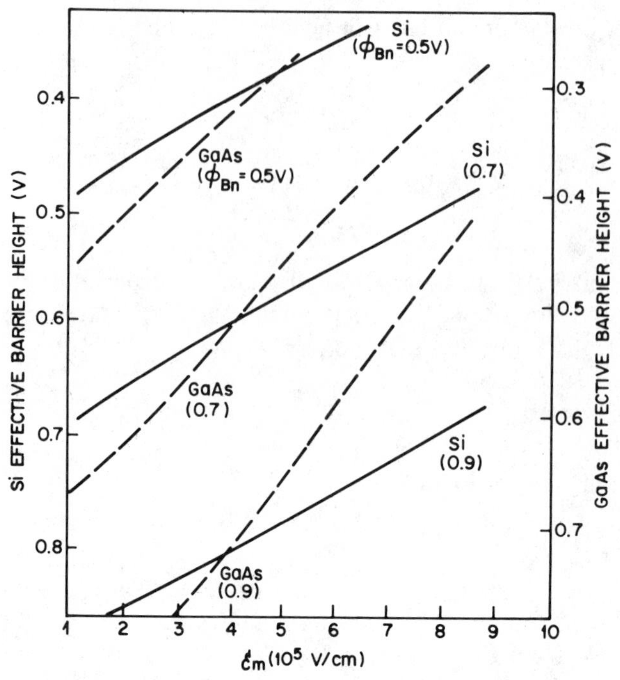

Fig. 35 Calculated effective barrier height for Si and GaAs metal–semiconductor contacts. (After Shannon, Ref. 52.)

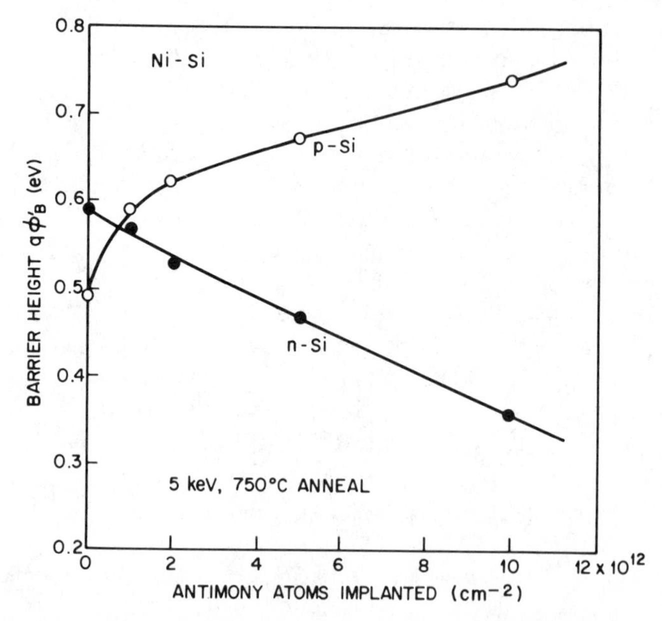

Fig. 36 Effective barrier height for holes in *p*-type substrates and electrons in *n*-type substrates as a function of the implanted antimony dose. (After Shannon, Ref. 52.)

fore, as the product ($p_1 a$) increases, the effective barrier height will increase accordingly.

Figure 36 shows the measured results[52] of Ni–Si diodes with shallow antimony implantation on the surface. As the implant dose increases, the effective barrier height decreases for n-type substrates and increases for p-type substrates.

5.6 DEVICE STRUCTURES

As mentioned previously, Schottky diode behavior is electrically similar to a one-sided abrupt p-n junction, and yet the Schottky diode can be operated as a majority-carrier device with inherent fast response. Thus the terminal functions of a p-n junction diode can also be performed by a Schottky diode with one exception for the charge-storage diode. This is because the charge-storage time in a majority-carrier device is extremely small. In this section we consider various device structures and point out their related applications.

The earliest device structure is the point-contact rectifier using a small metal wire with a sharp point making contact with a semiconductor. The contact may be just a simple mechanical contact or it may be formed by electrical discharge processes that may result in a small alloyed p-n junction.

A point-contact rectifier usually has poor forward and reverse I–V characteristics compared with a planar Schottky diode. Its characteristics are also difficult to predict from theory, since the rectifiers are subject to wide variations such as the whisker pressure, contact area, crystal structure, surface treatment, whisker composition, and heat or forming processes. The advantage of a point-contact rectifier is its small area, which can give very small capacitance, a desirable feature for microwave application. The disadvantages are its large spreading resistance ($R_S = \rho/2\pi r_0$, where r_0 is the radius of the hemispheric point contact); large leakage current due mainly to the surface effect, which gives rise to poor rectification ratio; and soft reverse breakdown characteristics due to a large concentration of field beneath the metal point.

Most modern metal–semiconductor diodes are made by a planar process, and the metal–semiconductor contacts are formed by various methods, including thermal evaporation, chemical decomposition, electron-gun bombardment, sputtering, or plating of metals onto chemically etched, mechanically polished, vacuum-cleaved, back-sputtered, heat-treated, or ion-bombarded semiconductor surfaces. Since most metal–semiconductor contacts are formed in a vacuum system,[53] an important parameter concerning vacuum deposition of metals is the vapor pressure, which is defined as the pressure exerted when a solid or liquid is in equilibrium with its own vapor. Vapor pressure[54] versus temperature for the more common elements is shown in Fig. 37.

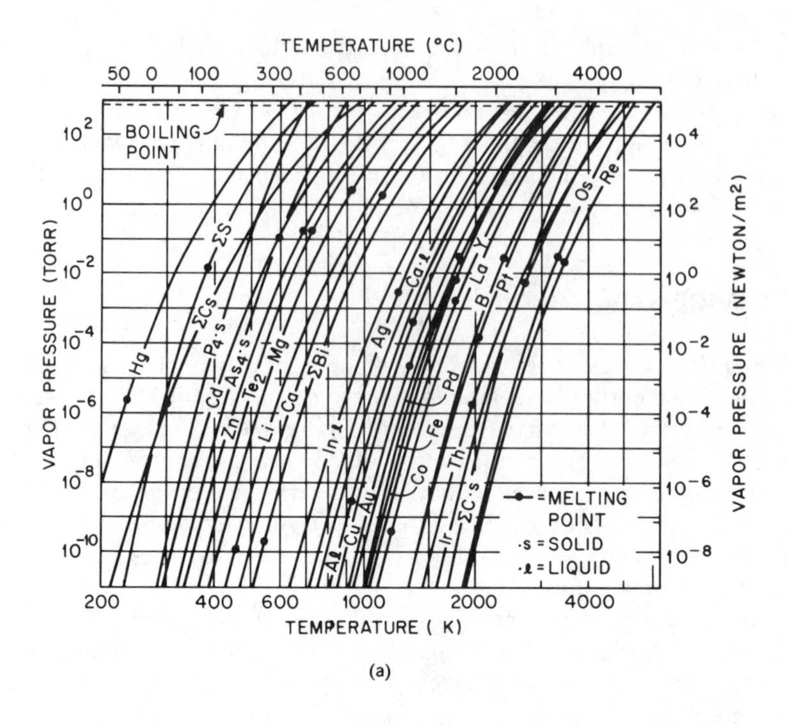

(a)

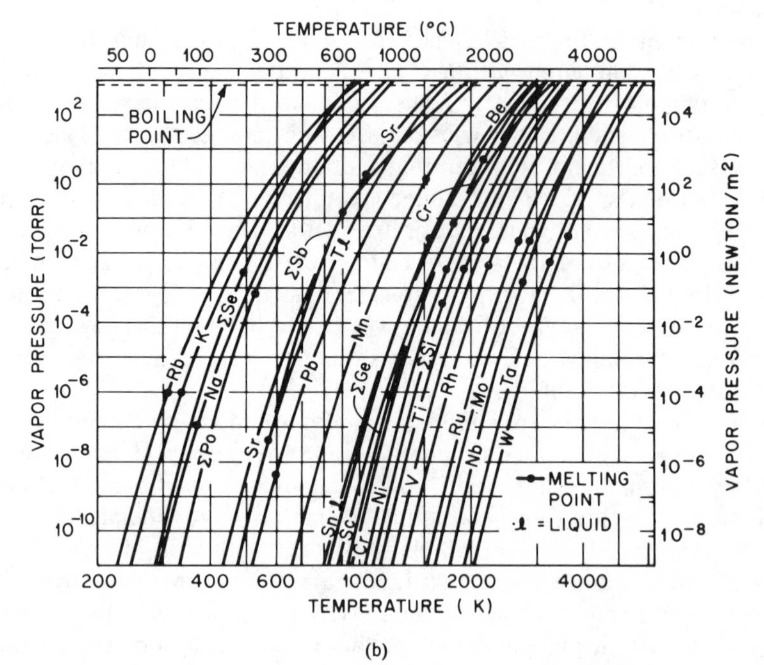

(b)

Fig. 37 Vapor pressure versus temperature for solid and liquid elements. (After Honig, Ref. 54.)

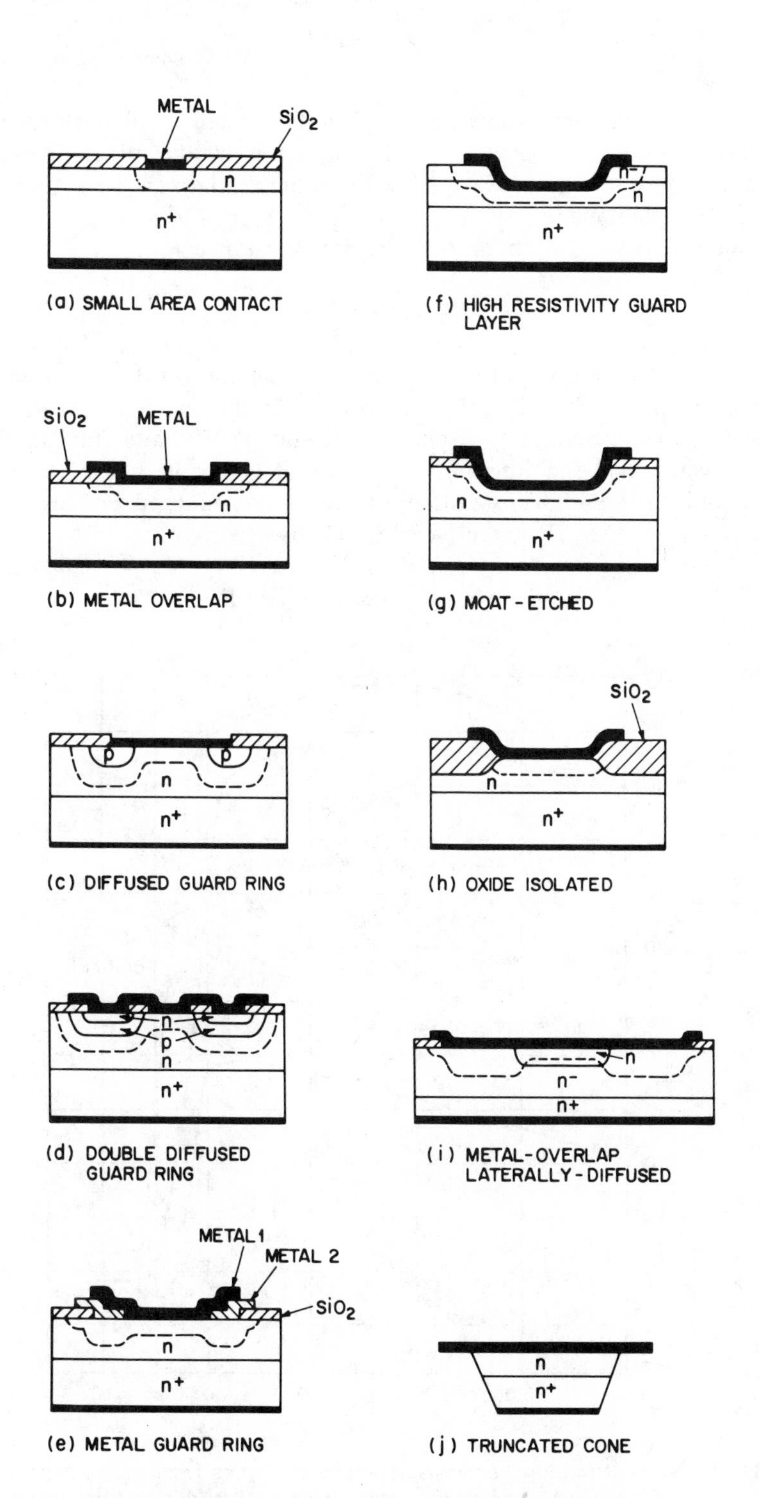

Fig. 38 Various metal–semiconductor device structures.

The small-area contact device, Fig. 38*a*, fabricated by planar process on epitaxial n on n^+ substrate, is useful as a microwave mixer diode.[55] To achieve good performance, we have to minimize the series resistance and the diode capacitance.

From Eq. 53 we can obtain the junction resistance R_j:

$$R_j \equiv \frac{\partial V}{\partial I} = \frac{nkT}{qJA_j} \tag{94}$$

where A_j is the junction area. Typical experimental results of R_j versus I are shown in Fig. 39 for Au–Si and Au–GaAs diodes. Also shown is the result for Si point contact, discussed previously. We note for sufficiently high forward bias the junction resistance does not decrease to zero as predicted by Eq. 94 but instead approaches a constant value. This value is the series resistance R_S given (see Fig. 13) by

$$R_S = \frac{1}{A_j}\int_{x_1}^{x_2} \rho(x)\,dx + \frac{\rho_B}{4r} + R_c \tag{95}$$

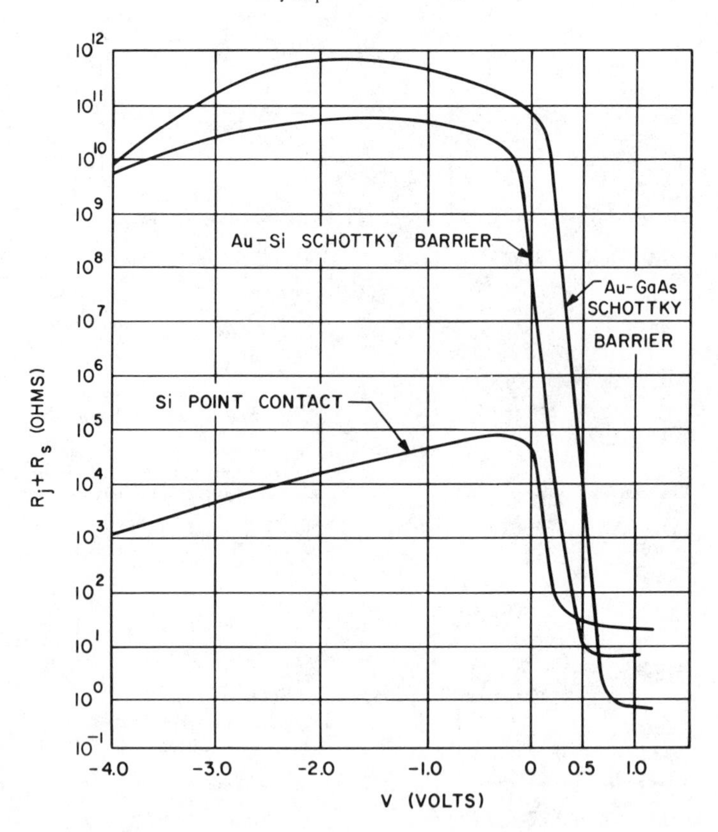

Fig. 39 The sum of the junction resistance and the series resistance versus applied voltage for Au–Si, Au–GaAs, and point-contact diodes. (After Irvin and Vanderwall, Ref. 56.)

where the first term on the right-hand side is the series resistance of the quasi-neutral region, and x_1 and x_2 are the depletion-layer edge and the epitaxial layer–substrate boundary, respectively. The second term is the spreading resistance of the metal–semiconductor barrier substrate with a resistivity ρ_B and a circular area of radius $r(A_j = \pi r^2)$. The last term R_c is the resistance due to the ohmic contact with the substrate.

An important figure of merit for microwave application of the Schottky diodes is the forward bias cutoff frequency f_{c0}, which is defined as

$$f_{c0} \equiv \frac{1}{2\pi R_F C_F} \tag{96}$$

where R_F and C_F are the resistance and capacitance, respectively, at a forward bias of 0.1 V to the flat-band condition.[55a] The value of f_{c0} is considerably smaller than the corresponding cutoff frequency using zero-bias capacitance, and can be used as a lower limit for practical consideration. A typical result[56] is shown in Fig. 40. Note that for a given doping and junction diameter (e.g., 10 μm), the Schottky diode of n-type GaAs gives the highest cutoff frequency, mainly because the electron mobility is considerably higher in GaAs.

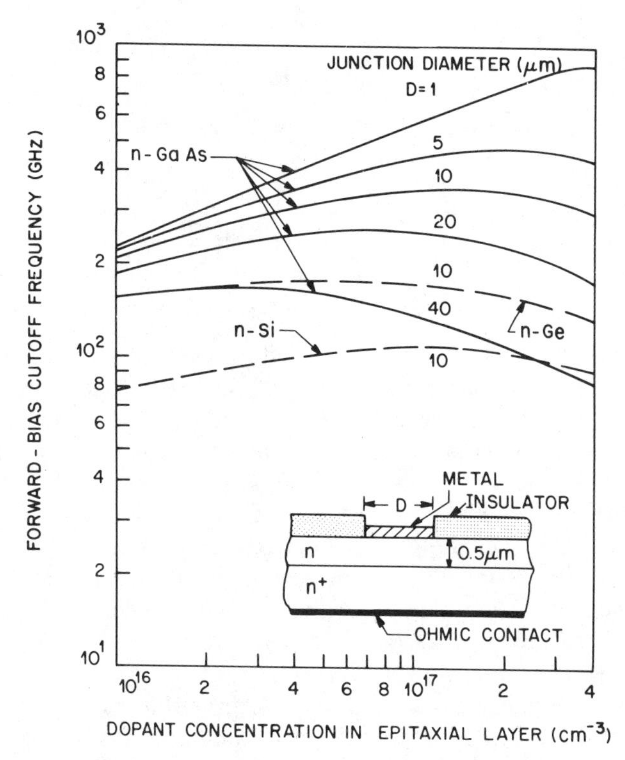

Fig. 40 Forward-bias cutoff frequency versus doping concentration in the epitaxial layer for a 0.5-μm epitaxial layer and with various junction diameters.

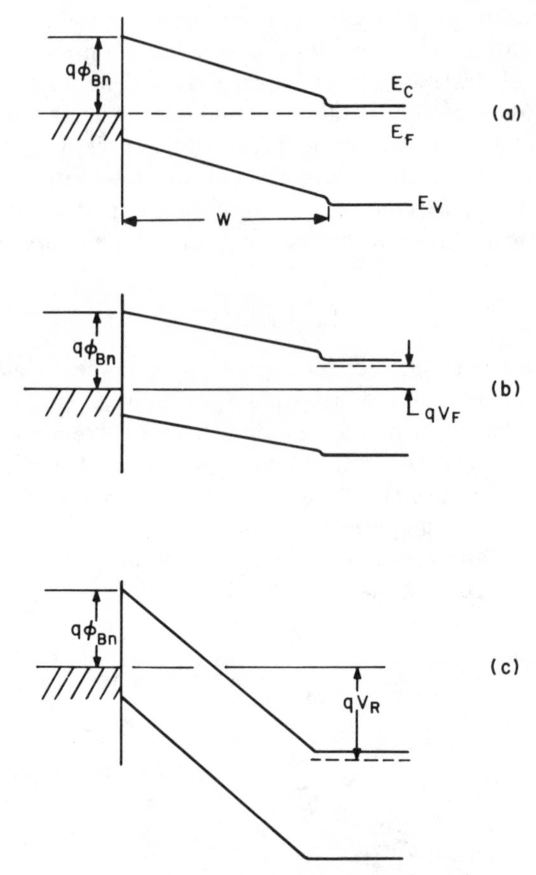

Fig. 41 Band diagram for Mott barrier at various biasing conditions. (After Mott, Ref. 5.)

To improve reliability and stability, devices that have larger contact diameters without increased diode capacitances are desirable. It has been shown that the Mott barrier can meet the requirement. A Mott barrier is a metal–semiconductor contact in which the epitaxial layer is very lightly doped and much narrower than the required depletion layer width; the layer is swept out even under forward bias. Figure 41 shows the band diagram of a Mott barrier. The depletion capacitance per unit area is a constant and is given by ϵ_s/W. Since for a given cutoff frequency, the depletion width W can be made much wider than that of a standard Schottky diode, the Mott diode diameter can be made much larger.[57] The current transport in a Mott barrier is dominated by diffusion, given by Eq. 38.

The metal overlap structure,[58] Fig. 38*b*, gives near-ideal forward I–V characteristics and low leakage current at moderate reverse bias, but the electrode sharp-edge effect will increase the reverse current when a large reverse bias is applied. This structure is extensively used in integrated

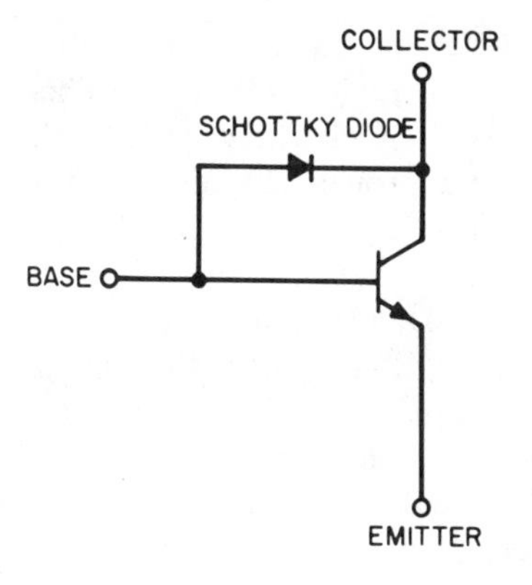

Fig. 42 Composite transistor with a Schottky diode connected between the base and collector terminals of a bipolar transistor. (After Tada and Laraya, Ref. 60.)

circuits since it can be formed as an integral part of the metallization. An important example is the clamped transistor[59,60] (Fig. 42). A Schottky diode can be incorporated into the base-collector terminal to form a clamped (composite) transistor with a very short saturation time constant. Fabrication is simply achieved by allowing the base contact to straddle the base diffusion with the surrounding undiffused epitaxial layer in the standard buried-collector technology.[41] In the saturation region the collector junction of the original transistor is slightly forward-biased instead of reverse-biased. If the forward voltage drop in the Schottky diode is much lower than the base–collector ON voltage of the original transistor, most of the excess base current flows through the diode in which minority carriers are not stored. Thus the saturation time is reduced markedly compared with that of the original transistor.

To eliminate the electrode sharp-edge effects, many device structures have been proposed. Figure 38*c* uses a diffused guard ring[32] to give near-ideal forward and reverse characteristics, as shown previously in Fig. 24. This structure is useful as a tool to study static characteristics; however, it suffers from long recovery time and large parasitic capacitance due to the adjacent *p-n* junction. Figure 38*d* uses a double-diffused guard ring[61] to reduce the recovery time, but the process is relatively complicated. The guard-ring structure, Fig. 38*e*, has been proposed to use two metals with different barrier heights. However, large variations in barrier height are generally difficult to obtain for covalent semiconductors. Another guard-ring structure with a high resistivity layer on top of the active layer[62] is shown in Fig. 38*f*. Since the dielectric constant of the high resistivity is higher than that of an insulator, the parasitic capacitance is generally higher than the structure shown in Fig. 38*b*. In Fig. 38*g*, the diode is surrounded by a void or moat.[63] In this case, reliability problems can result from burying contaminants in the moat. Another approach is to use oxide isolation,[64] to reduce the edge field, Fig. 38*h*. This approach requires a special planar process to incorporate a local oxidation step.

The metal-overlap laterally diffused structure[65] is basically a double (parallel) Schottky diode that does not involve a p-n junction, Fig. 38*i*. This structure gives nearly ideal forward and reverse I–V characteristics with very short reverse recovery time. However, the process involves extra oxidation and diffusion steps, and the outer n^- ring may increase the device capacitance.

For certain microwave power generators (e.g., IMPATT diodes) one uses the truncated-cone structure,[66] Fig. 38*j*. The angle between the metal overhang and the semiconductor cone must be larger than 90° so that the electric field at the contact periphery is always smaller than that in the center. This angle ensures that the avalanche breakdown will occur uniformly inside the metal–semiconductor contact.

5.7 OHMIC CONTACT

An ohmic contact is defined as a metal–semiconductor contact that has a negligible contact resistance relative to the bulk or spreading resistance of the semiconductor. A satisfactory ohmic contact should not significantly perturb device performance, and it can supply the required current with a voltage drop that is sufficiently small compared with the drop across the active region of the device.

We shall first consider specific contact resistance that is the reciprocal of the derivative of current density with respect to voltage. When evaluated at zero bias, this specific contact resistance R_c is an important figure of merit for ohmic contacts:[67]

$$R_c \equiv \left(\frac{\partial J}{\partial V}\right)^{-1}_{V=0}. \tag{97}$$

For metal–semiconductor contacts with lower doping concentrations, the thermionic–emission current dominates the current transport, as given in Eq. 49. Therefore,

$$R_c = \frac{k}{qA^*T}\exp\left(\frac{q\phi_{Bn}}{kT}\right). \tag{98}$$

In deriving this equation, we have neglected the small voltage dependence of the barrier height. Equation 98 shows that low barrier height should be used to obtain small R_c.

For contacts with higher dopings, the tunneling process will dominate, and the current is given by Eq. 54a. Then

$$R_c \sim \exp\left(\frac{q\phi_{Bn}}{E_{00}}\right) = \exp\left[\frac{2\sqrt{\epsilon_s m^*}}{\hbar}\left(\frac{\phi_{Bn}}{\sqrt{N_D}}\right)\right]. \tag{99}$$

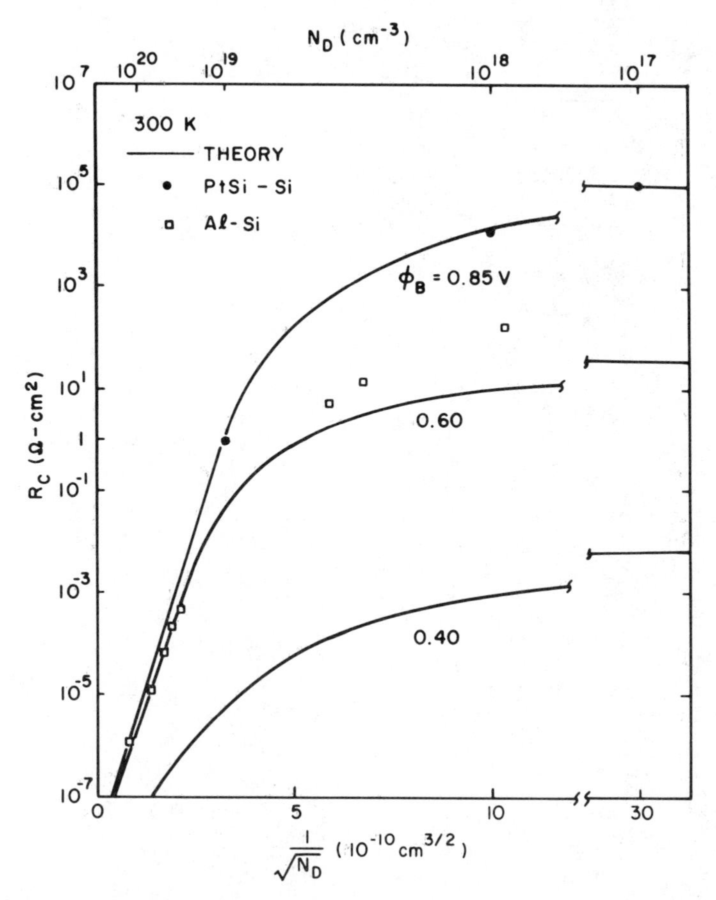

Fig. 43 Theoretical and experimental values of specific contact resistance. (After Chang, Fang, and Sze, Ref. 67; Yu, Ref. 68.)

Equation 99 shows that in the tunneling range the specific contact resistance depends strongly on doping concentration and varies exponentially with the factor $(\phi_{Bn}/\sqrt{N_D})$.

The calculated values of R_c are plotted[67,68] in Fig. 43 as a function of $1/\sqrt{N_D}$. For $N_D \geq 10^{19}$ cm^{-3}, R_c is dominated by the tunneling process and decreases rapidly with increased doping. On the other hand, for $N_D \leq 10^{17}$ cm^{-3}, the current is due to thermionic emission, and R_c is essentially independent of doping. Also shown in Fig. 43 are experimental data for PtSi–Si and Al–Si diodes. They are in good agreement with the calculated values.

Figure 43 shows that high doping concentration, low barrier height, or both must be used to obtain low values of R_c. And these are exactly the approaches used for all ohmic contacts (Fig. 44).

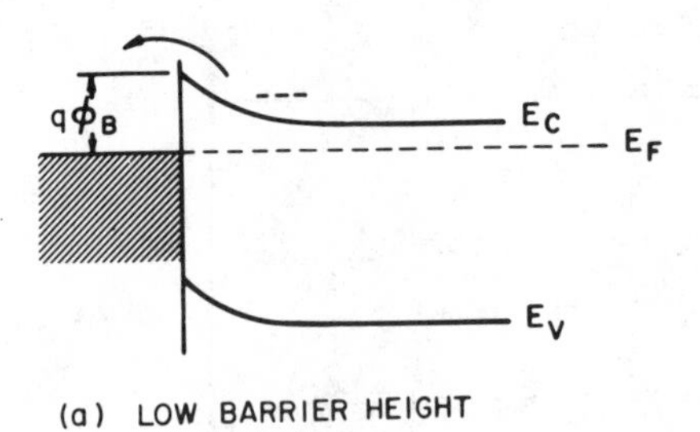

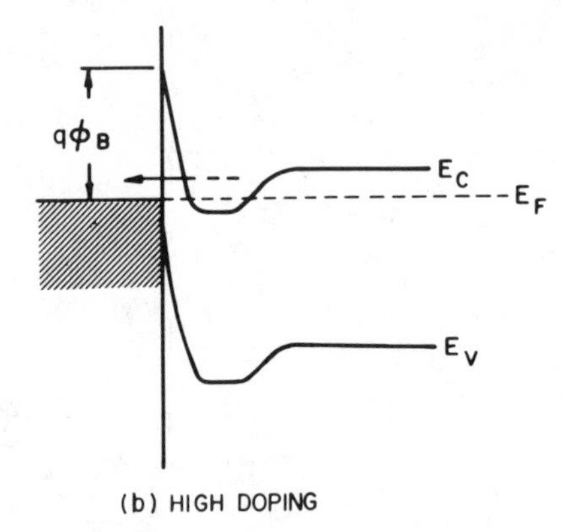

Fig. 44 Low barrier height and/or high doping for ohmic contacts.

It is difficult to make ohmic contacts on wide-gap semiconductors. A metal does not generally exist with a low-enough work function to yield a low barrier. In such cases the general technique for making an ohmic contact involves the establishment of a heavily doped surface layer such as metal–n^+-n or metal–p^+-p contact by various methods, such as shallow diffusion, alloy regrowth, in-diffusion of a dopant contained in the contact material, double epitaxy, and ion implantation. For ohmic contacts on Ge and Si, Au–Sb alloy (with 0.1% Sb) is first evaporated onto n-type semiconductors. These contacts are then alloyed at the corresponding eutectic temperature into the semiconductors under an inert gas (such as argon or nitrogen).[69] For GaAs and other III–V compound semiconductors, various technologies have been developed for the ohmic contacts.[70] A summary is listed in Table 5.

Table 5 Ohmic Contact Technology for III–V and Mixed III–V Compound Semiconductors

III–V	E_g (eV)	Type	Contact Material	Technique	Alloy Temperature (°C)
AlN	5.9	Semi-i	Si	Preform	
		Semi-i	Al, Al–In	Preform	1500–1800
		Semi-i	Mo, W	Sputter	1000
AlP	2.45	n	Ga–Ag	Preform	500–1000
AlAs	2.16	n, p	In–Te	Preform	150
		$n. p$	Au	Preform	160
		n, p	Au–Ge	Preform	700
		n	Au–Sn	Preform	
GaN	3.36	Semi-i	Al–In	Preform	
GaP	2.26	p	Au–Zn(99:1)	Preform, evap.	700
		p	Au–Ge	Preform	
		n	Au–Sn(62:38)	Preform	360
		n	Au–Si(98:2)	Evap.	700
GaAs	1.42	p	Au–Zn(99:1)	Electroless, evap.	600
		p	In–Au(80:20)	Preform	
		n	Au–Ge(88:12)	Evap.	
		n	In–Au(90:10)	Evap.	350–450
		n	Au–Si(94:6)	Evap.	550
		n	Au–Sn(90:10)	Evap.	300
		n	Au–Te(98:2)	Evap.	350–700
GaSb	0.72	p	In	Preform	500
		n	In	Preform	
InP	1.35	p	In	Preform	
		n	In, In–Te	Preform	350–600
		n	Ag–Sn	Preform, evap.	350–600 600
InAs	0.36	n	In	Preform	
			Sn–Te(99:1)	Preform	
InSb	0.17	n	In	Preform	
		n	Sn–Te(99:1)	Preform	
$GaAs_{1-x}P_x$	1.42–2.31	p	Au–Zn	Evap.	500
		p	Al	Evap.	500
		n	Au–Ge–Ni	Evap.	450
		n	Au–Sn	Evap.	450
$Al_xGa_{1-x}As$	1.42–2.16	p	Au–In	Electroplate	400–450
		p	Au–Zn	Evap.	
		p	Al	Evap.	500
		n	Au–Ge–Ni	Evap.	500
		n	Au–Sn	Evap., electroless	450–485 450
		n	Au–Si	Evap.	
$Ga_{1-x}In_xSb$	0.70–0.17	n	Sn–Te	Evap.	
$Al_xGa_{1-x}P$	2.31–2.45	n	Sn	Preform	
$Ga_{1-x}In_xAs$	1.47–0.35	n	Sn	Preform	
$InAs_xSb_{1-x}$	0.17–0.35	n	In–Te	Preform	

REFERENCES

1 F. Braun, "Über die Stromleitung durch Schwefelmetalle," *Ann. Phy. Chem.*, **153**, 556 (1874).

2 J. C. Bose, U.S. Patent 775,840 (1904).

3 A. H. Wilson, "The Theory of Electronic Semiconductors," *Proc. R. Soc. Lond. Ser. A*, **133**, 458 (1931).

4 W. Schottky, "Halbleitertheorie der Sperrschicht," *Naturwissenschaften*, **26**, 843 (1938).

5 N. F. Mott, "Note on the Contact between a Metal and an Insulator or Semiconductor," *Proc. Cambr. Philos. Soc.*, **34**, 568 (1938).

6 H. K. Henisch, *Rectifying Semiconductor Contacts*, Clarendon, Oxford, 1957.

7 E. H. Rhoderick, *Metal-Semiconductor Contacts*, Clarendon, Oxford, 1978; "Transport Processes in Schottky Diodes," in K. M. Pepper, Ed, *Inst. Phys. Conf. Ser., No.* 22, Institute of Physics, Manchester, England, 1974, p. 3.

8 V. L. Rideout, "A Review of the Theory, Technology and Applications of Metal–Semiconductor Rectifiers," *Thin Solid Films*, **48**, 261 (1978).

9 H. B. Michaelson, "Relation between an Atomic Electronegativity Scale and the Work Function," *IBM J. Res. Dev.*, **22**, 72 (1978).

10 S. M. Sze, C. R. Crowell, and D. Kahng, "Photoelectric Determination of the Image Force Dielectric Constant for Hot Electrons in Schottky Barriers," *J. Appl. Phys.*, **35**, 2534 (1964).

11 C. D. Salzberg and G. G. Villa, "Infrared Refractive Indexes of Silicon Germanium and Modified Selenium Glass," *J. Opt. Soc. Am.*, **47**, 244 (1957).

12 H. A. Bethe, "Theory of the Boundary Layer of Crystal Rectifiers," *MIT Radiat. Lab. Rep. 43–12* (1942).

13 C. R. Crowell, "The Richardson Constant for Thermionic Emission in Schottky Barrier Diodes," *Solid State Electron.*, **8**, 395 (1965).

14 C. R. Crowell and S. M. Sze, "Current Transport in Metal–Semiconductor Barriers," *Solid State Electron*, **9**, 1035 (1966).

15 C. R. Crowell and S. M. Sze, "Electron–Optical–Phonon Scattering in the Emitter and Collector Barriers of Semiconductor–Metal–Semiconductor Structures," *Solid State Electron.*, **8**, 979 (1965).

16 C. W. Kao, L. Anderson, and C. R. Crowell, "Photoelectron Injection at Metal–Semiconductor Interface," *Surface Sci.*, **95**, 321 (1980).

17 C. R. Crowell and S. M. Sze, "Quantum-Mechanical Reflection of Electrons at Metal–Semiconductor Barriers: Electron Transport in Semiconductor–Metal–Semiconductor Structures," *J. Appl. Phys.*, **37**, 2685 (1966).

18 C. Y. Chang and S. M. Sze, "Carrier Transport across Metal–Semiconductor Barriers," *Solid State Electron.*, **13**, 727 (1970).

19 J. M. Andrews and M. P. Lepselter, "Reverse Current–Voltage Characteristics of Metal-Silicide Schottky Diodes," *Solid State Electron.*, **13**, 1011 (1970).

20 C. R. Crowell and M. Beguwala, "Recombination Velocity Effects on Current Diffusion and Imref in Schottky Barriers," *Solid State Electron.*, **14**, 1149 (1971).

21 F. A. Padovani and R. Stratton, "Field and Thermionic-Field Emission in Schottky Barriers," *Solid State Electron.*, **9**, 695 (1966).

22 A. Y. C. Yu and E. H. Snow, "Minority Carrier Injection of Metal–Silicon Contacts," *Solid State Electron.*, **12**, 155 (1969).

23 D. L. Scharfetter, "Minority Carrier Injection and Charge Storage in Epitaxial Schottky Barrier Diodes," *Solid State Electron.*, **8**, 299 (1965).

24 A. M. Cowley and S. M. Sze, "Surface States and Barrier Height of Metal–Semiconductor Systems," *J. Appl. Phys.*, **36**, 3212 (1965).

25 J. Bardeen, "Surface States and Rectification at a Metal Semiconductor Contact," *Phys. Rev.*, **71**, 717 (1947).

26 C. A. Mead and W. G. Spitzer, "Fermi-Level Position at Metal-Semiconductor Interfaces," *Phys. Rev.*, **134**, A713 (1964).

27 D. Pugh, "Surface States on the ⟨111⟩ Surface of Diamond," *Phys. Rev. Lett.*, **12**, 390 (1964).

28 L. Pauling, *The Nature of The Chemical Bond*, 3rd ed., Cornell University Press, Ithaca, N.Y., 1960.

29 S. Kurtin, T. C. McGill, and C. A. Mead, "Fundamental Transition in Electronic Nature of Solids," *Phys. Rev. Lett.* **22**, 1433 (1969).

30 C. A. Mead, "Metal-Semiconductor Surface Barriers," *Solid State Electron.*, **9**, 1023 (1966).

31 C. R. Crowell, J. C. Sarace, and S. M. Sze, "Tungsten–Semiconductor Schottky-Barrier Diodes," *Trans. Met. Soc. AIME*, **233**, 478 (1965).

32 M. P. Lepselter and S. M. Sze, "Silicon Schottky Barrier Diode with Near-Ideal I–V Characteristics," *Bell Syst. Tech. J.*, **47**, 195 (1968).

33 J. M. Andrews and F. B. Koch, "Formation of NiSi and Current Transport across the NiSi–Si Interface," *Solid State Electron.*, **14**, 901 (1971).

34 A. M. Goodman, "Metal–Semiconductor Barrier Height Measurement by the Differential Capacitance Method—One Carrier System," *J. Appl. Phys.*, **34**, 329 (1963).

35 G. I. Roberts and C. R. Crowell, "Capacitive Effects of Au and Cu Impurity Levels in Pt n-type Si Schottky Barriers," *Solid State Electron.*, **16**, 29 (1973).

36 C. R. Crowell, W. G. Spitzer, L. E. Howarth, and E. Labate, "Attenuation Length Measurements of Hot Electrons in Metal Films," *Phys. Rev.*, **127**, 2006 (1962).

37 R. H. Fowler, "The Analysis of Photoelectric Sensitivity Curves for Clean Metals at Various Temperatures," *Phys. Rev.*, **38**, 45 (1931).

38 C. R. Crowell, S. M. Sze, and W. G. Spitzer, "Equality of the Temperature Dependence of the Gold–Silicon Surface Barrier and the Silicon Energy Gap in Au n-type Si Diodes," *Appl. Phys. Lett.*, **4**, 91 (1964).

39 M. Beguwala and C. R. Crowell, "Characterization of Multiple Deep Level Systems in Semiconductor Junctions by Admittance Measurements," *Solid State Electron.*, **17**, 203 (1974).

40 J. O. McCaldin, T. C. McGill, and C. A. Mead, "Schottky Barriers on Compound Semiconductors: The Role of the Anion," *J. Vac. Sci. Technol.*, **13**, 802 (1976).

41 J. M. Andrews, "The Role of the Metal–Semiconductor Interface in Silicon Integrated Circuit Technology," *J. Vac. Sci. Technol.*, **11**, 972 (1974).

42 J. M. Andrews and J. C. Phillips, "Chemical Bonding and Structure of Metal–Semiconductor Interfaces," *Phys. Rev. Lett.*, **35**, 56 (1975).

43 G. J. van Gurp, "The Growth of Metal Silicide Layers on Silicon," in H. R. Huff and E. Sirtl, Eds., *Semiconductor Silicon 1977*, Electrochemical Society, Princeton, N.J., 1977, p. 342.

43a I. Ohdomari, K. N. Tu, F. M. d'Heurle, T. S. Kuan, and S. Petersson, "Schottky-Barrier Height of Iridium Silicide," *Appl. Phys. Lett.*, **33**, 1028 (1978).

44 J. L. Saltich and L. E. Terry, "Effects of Pre- and Post-Annealing Treatments on Silicon Schottky Barrier Diodes," *Proc. IEEE*, **58**, 492 (1970).

45 A. K. Sinha, "Electrical Characteristics and Thermal Stability of Platinum Silicide-to-Silicon Ohmic Contacts Metalized with Tungsten," *J. Electrochem. Soc.*, **120**, 1767 (1973).

46 A. K. Sinha, T. E. Smith, M. H. Read, and J. M. Poate, "*n*-GaAs Schottky Diodes Metallized with Ti and Pt/Ti," *Solid State Electron.*, **19**, 489 (1976).

47 K. Chino, "Behavior of Al–Si Schottky Barrier Diodes under Heat Treatment," *Solid State Electron.*, **16**, 119 (1973).

48 Y. Itoh and N. Hashimoto, "Reaction-Process Dependence of Barrier Height between Tungsten Silicide and *n*-Type Silicon," *J. Appl. Phys.*, **40**, 425 (1969).

49 J. M. Shannon, "Reducing the Effective Height of a Schottky Barrier Using Low-Energy Ion Implantation," *Appl. Phys. Lett.*, **24**, 369 (1974).

50 J. M. Shannon, "Increasing the Effective Height of a Schottky Barrier Using Low-Energy Ion Implantation," *Appl. Phys. Lett.*, **25**, 75 (1974).

51 J. M. Andrews, R. M. Ryder, and S. M. Sze, "Schottky Barrier Diode Contacts," U.S. Patent 3,964,084 (1976).

52 J. M. Shannon, "Control of Schottky Barrier Height Using Highly Doped Surface Layers," *Solid State Electron.*, **19**, 537 (1976).

53 For general references on vacuum deposition, see L. Holland, *Vacuum Deposition of Thin Films*, Chapman & Hall, London, 1966; A. Roth, *Vacuum Technology*, North-Holland, Amsterdam, 1976.

54 R. E. Honig, "Vapor Pressure Data for the Solid and Liquid Elements," *RCA Rev.*, **23**, 567 (1962).

55 D. T. Young and J. C. Irvin, "Millimeter Frequency Conversion Using Au-*n*-type GaAs Schottky Barrier Epitaxy Diode with a Novel Contacting Technique," *Proc. IEEE.*, **53**, 2130 (1965); D. Kahng and R. M. Ryder, "Small Area Semiconductor Devices," U.S. Patent 3,360,851 (1968).

55a N. C. Vanderwal, "A Microwave Schottky-Barrier Varistor Using GaAs for Low Series Resistance," IEEE Int. Electron Device Meet., Washington, D.C., Oct. 18–20, 1967.

56 J. C. Irvin and N. C. Vanderwal, "Schottky-Barrier Devices," in H. A. Watson, Ed., *Microwave Semiconductor Devices and Their Circuit Applications*, McGraw-Hill, New York, 1968.

57 M. McColl and M. F. Millea, "Advantages of Mott Barrier Mixer Diodes," *Proc. IEEE*, **61**, 499 (1973).

58 A. Y. C. Yu and C. A. Mead, "Characteristics of Al–Si Schottky Barrier Diode," *Solid State Electron.*, **13**, 97 (1970).

59 R. H. Baker, "Maximum Efficiency Switching Circuit," *MIT Lincoln Lab. Rep. TR*-110, Lexington, Mass. (1956).

60 K. Tada and J. L. R. Laraya, "Reduction of the Storage Time of a Transistor Using a Schottky-Barrier Diode," *Proc. IEEE*, **55**, 2064 (1967).

61 J. L. Saltich and L. E. Clark, "Use of a Double Diffused Guard Ring to Obtain Near Ideal *I–V* Characteristics in Schottky-Barrier Diodes," *Solid State Electron.*, **13**, 857 (1970).

62 K. J. Linden, "GaAs Schottky Mixer Diode with Integral Guard Layer Structure," *IEEE Trans. Electron Devices*, **ED–23**, 363 (1976).

63 C. Rhee, J. L. Saltich, and R. Zwernemann, "Moat-Etched Schottky Barrier Diode Displaying Near Ideal *I–V* Characteristics," *Solid State Electron.*, **15**, 1181 (1972).

64 N. G. Anantha and K. G. Ashar, *IBM. J. Res. Dev.*, **15**, 442 (1971).

65 A. Rusu, C. Bulucea, and C. Postolache, "The Metal-Overlap-Laterally-Diffused (MOLD) Schottky Diode," *Solid State Electron.*, **20**, 499 (1977).

66 D. J. Coleman Jr., J. C. Irvin, and S. M. Sze, "GaAs Schottky Diodes with Near-Ideal Characteristics," *Proc. IEEE*, **59**, 1121 (1971).

67 C. Y. Chang, Y. K. Fang, and S. M. Sze, "Specific Contact Resistance of Metal–Semiconductor Barriers," *Solid State Electron.*, **14**, 541 (1971).

68 A. Y. C. Yu, "Electron Tunneling and Contact Resistance of Metal–Silicon Contact Barriers," *Solid State Electron.*, **13**, 239 (1970).

69 M. P. Lepselter and J. M. Andrews, "Ohmic Contacts to Silicon," in B. Schwartz, Ed., *Ohmic Contacts to Semiconductors*, The Electrochemical Society Symposium Series, New York 1969, p. 159.

70 V. L. Rideout, "A Review of the Theory and Technology for Ohmic Contacts to Group III–V Compound Semiconductors," *Solid State Electron.*, **18**, 541 (1975).

71 J. M. Andrews, *Extended Abstracts*, Electrochem. Soc. Spring Meet. Abstr. 191, 1975, p. 452.

72 W. E. Spicer, P. W. Chye, C. M. Garner, I. Lindau, and P. Pianetta, "The Surface Electronic Structure of III–V Compounds and the Mechanism of Fermi Level Pinning by Oxygen (Passivation) and Metals (Schottky Barriers)," *Surface Sci.*, **86**, 763 (1979).

73 W. E. Spicer, I. Lindau, P. Skeath, C. Y. Su, and P. Chye, "Unified Mechanism for Schottky-Barrier Formation and III-V Oxide Interface States," *Phys. Rev. Lett.*, **44,** 420 (1980).

6

JFET and MESFET

6.1 INTRODUCTION

The junction field-effect transistor (JFET), first analyzed[1] by Shockley in 1952, is basically a voltage-controlled resistor. Because its conduction process involves predominantly one kind of carrier, the JFET is called a "unipolar" transistor to distinguish it from the bipolar junction transistor, in which both types of carriers are involved. Based on Shockley's theoretical treatment, the first working JFET was reported by Dacey and Ross, who later also considered the effect of field dependent mobility.[2,3]

The metal–semiconductor field-effect transistor (MESFET) was proposed by Mead in 1966 and subsequently fabricated by Hooper and Lehrer using a GaAs epitaxial layer on semiinsulating GaAs substrate.[4,5] The operation of a MESFET is identical to that of a JFET. In the MESFET, however, a metal–semiconductor rectifying contact instead of a *p-n* junction is used for the gate electrode. The MESFET offers certain processing and performance advantages, such as low-temperature formation of the metal–semiconductor barrier (as compared to a *p-n* junction made by diffusion or grown processes), low resistance and low *IR* drop along the channel width, and good heat dissipation for power devices (the rectifying contact can also serve as an efficient thermal contact to heat sink). On the other hand, the JFET can have various gate configurations, such as a

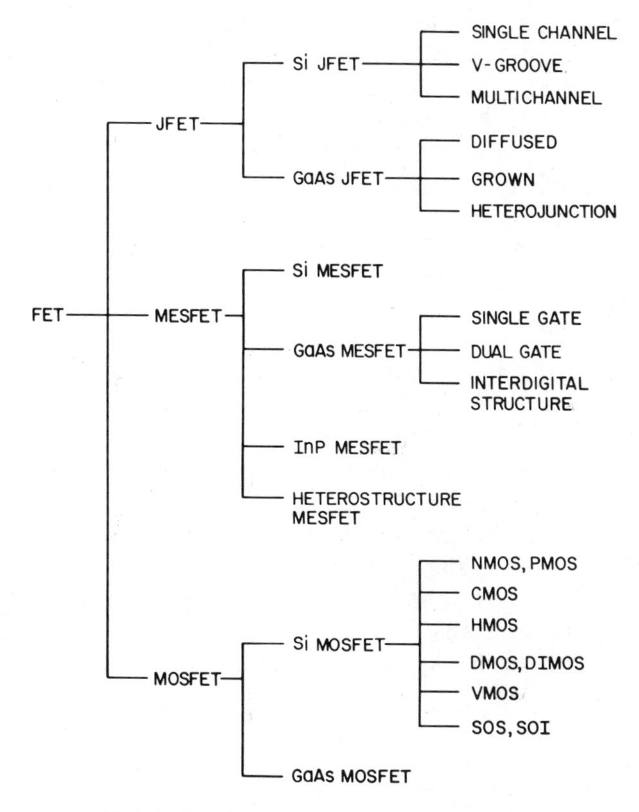

Fig. 1 Family tree of field-effect transistors.

heterojunction or a buffered layer gate, that improve high-frequency performances.

A family tree of field-effect transistors is shown in Fig. 1. In addition to the JFET and MESFET, we have the MOSFET (metal–oxide–semiconductor field-effect transistor), considered in Chapter 8. Extensive studies have been made of the three classes of field-effect transistors using various semiconductor materials and device structures. The basic device characteristics, noise behavior, and microwave performance of the JFET and MESFET have been reviewed by Hauser,[6] Pucel, Haus, and Statz[7] and Liechti,[8] respectively.

Field-effect transistors offer many attractive features for applications in analog switching, high-input-impedance amplifiers, microwave amplifications, and integrated circuits. The FETs have considerably higher input impedance than bipolar transistors, which allows the input of a FET to be more readily matched to the standard microwave system. The FET has a negative temperature coefficient at high current levels; that is, the current decreases as temperature increases. This characteristic leads to a more uniform temperature distribution over the device area and prevents the FET from thermal runaway or second breakdown, which occurs in the bipolar transistor. The device is thermally stable, even when the active area is large or when many devices are connected in parallel. Because FETs are unipolar devices, they do not suffer from minority-carrier storage effects, and consequently, have higher switching speeds and higher cutoff frequencies. In addition, the devices are basically square-law or linear devices; intermodulation and cross-modulation products are much smaller than those of bipolar transistors.

6.2 BASIC DEVICE CHARACTERISTICS

A schematic diagram of a JFET is shown in Fig. 2. The JFET consists of a conductive channel provided with two ohmic contacts, one acting as the source and the other as the drain. When a positive voltage V_D is applied to the drain with respect to the source, electrons flow from source to drain. Hence the source acts as the origin of the carriers and the drain acts as the sink. The third electrode, the gate, forms a rectifying junction with the channel. The device is basically a voltage-controlled resistor, and its resistance can be varied by varying the width of the depletion layer extending into the channel region.

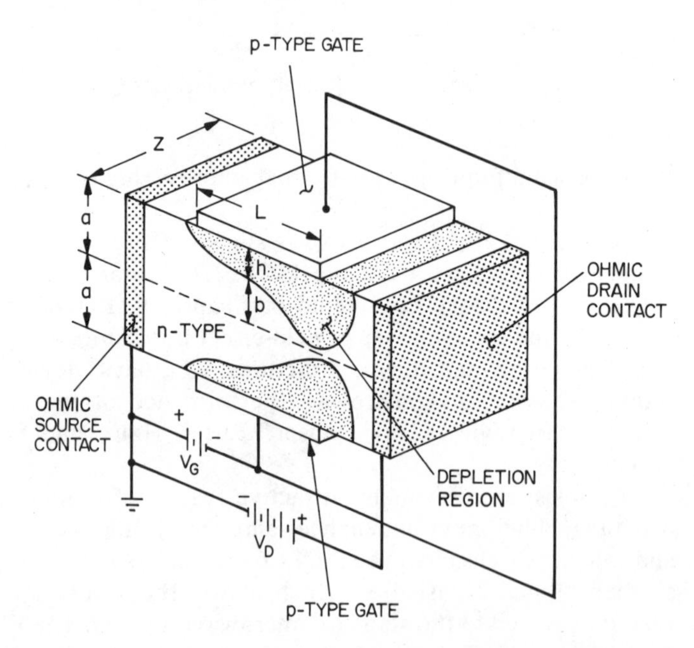

Fig. 2 Shockley's model of the junction field-effect transistor. (After Dacey and Ross, Ref. 2.)

In Fig. 2 the basic device dimensions are the channel length L (also called the gate length), channel width Z, channel depth a, depletion-layer width h, and channel opening b. The voltage polarities shown are for an n-channel JFET. The polarities are inverted for a p-channel JFET. The source electrode is generally grounded, and the gate and drain voltages are measured with respect to the source. When $V_G = V_D = 0$, the device is in equilibrium and no current flows. For a given V_G (zero or negative values), the channel current increases as the drain voltage increases. Eventually, for sufficiently large V_D, the current will saturate at a value I_{Dsat}.

The basic current–voltage characteristics of a JFET are shown in Fig. 3, where the drain current is plotted against the drain voltage for various gate voltages. We can divide the characteristic into three regions: the linear region, where the drain voltage is small and I_D is proportional to V_D; the saturation region, where the current remains essentially constant and is independent of V_D; and the breakdown region, where the drain current increases rapidly with a slight increase of V_D. As the reverse gate bias increases, both the saturation current I_{Dsat} and the corresponding saturation voltage V_{Dsat} decrease. These decreases occur because of the reduced initial channel width, which results in larger initial resistance.

We shall now derive the current–voltage characteristics for long-channel FETs (i.e., channel length $L \gg$ channel depth a) based on the follow-

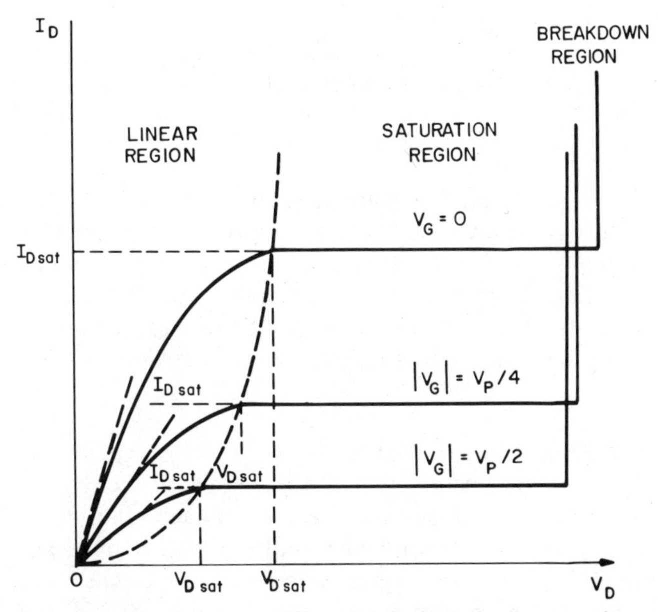

Fig. 3 Basic I–V characteristics of a JFET, which include the linear region, saturation region, and breakdown region. V_P is the pinch-off voltage. For a given voltage V_G, the current and voltage at the point where saturation occurs are designed by I_{Dsat} and V_{Dsat}, respectively.

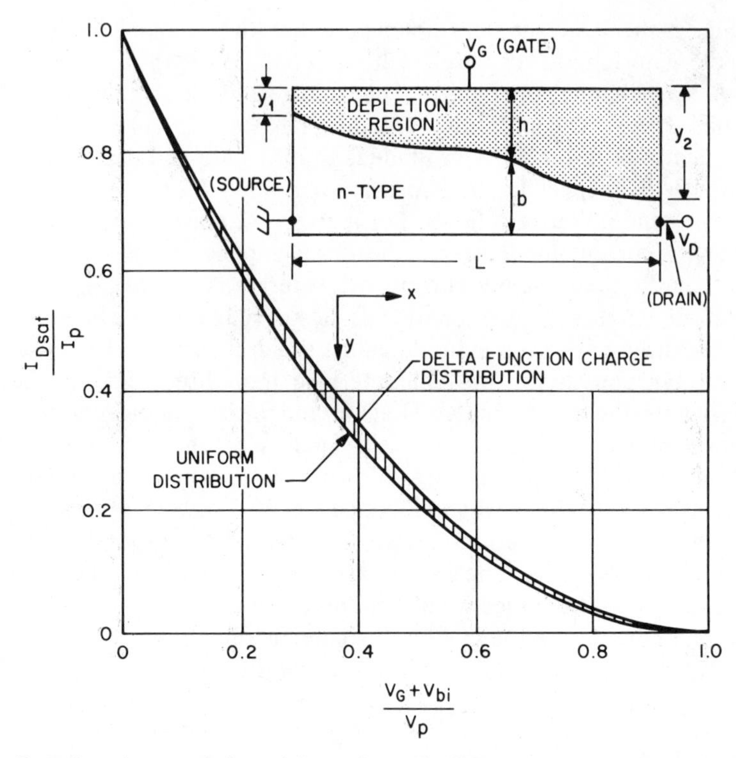

Fig. 4 Transfer characteristics of long-channel JFETs for two specific charge distributions. Insert shows a cross-sectional view of the upper half of a JFET. The depletion widths are y_1 and y_2 at the source and drain ends. (After Bockemuehl, Ref. 9; Middlebrook and Richer, Ref. 10.)

ing assumptions: (1) gradual channel approximation, (2) abrupt depletion layer, and (3) constant mobility. The short-channel and two-dimensional effects are considered in Section 6.3. Because of the symmetry shown in Fig. 2, we need only consider the upper half of the device, as shown in the insert in Fig. 4, where h is the depletion-layer width at an arbitrary point, and y_1 and y_2 are the widths at the source and drain electrodes, respectively.

6.2.1 Uniform Charge Distribution

For a uniformly doped n region, under the gradual channel approximation, the depletion-layer width h varies only gradually along the channel (x direction) and one can solve the one-dimensional Poisson equation in the y direction:

$$-\frac{d^2V}{dy^2} = \frac{d\mathscr{E}_y}{dy} = \frac{\rho(y)}{\epsilon_s} \tag{1}$$

or

$$-\frac{d^2V}{dy^2}=\frac{qN_D}{\epsilon_s} \tag{1a}$$

where $\mathscr{E}_y$ is the electric field in the y direction. The depletion-layer width at a distance x from the source is given by the abrupt junction expression

$$h=\{2\epsilon_s[V(x)+V_G+V_{bi}]/qN_D\}^{1/2} \tag{2}$$

where V_{bi} is the built-in potential between the p^+n junction and is given by $(kT/q)\ln(N_D/n_i)$. The voltage $V(x)$ is the applied drain voltage at x with respect to the source. For n-channel devices, the gate voltage is negative with respective to the source, and we shall use the absolute value of V_G in Eq. 2 and subsequent equations. The depletion widths at the source and drain ends are

$$y_1=[2\epsilon_s(V_G+V_{bi})/qN_D]^{1/2} \qquad \text{at} \quad x=0 \tag{3}$$

$$y_2=[2\epsilon_s(V_D+V_G+V_{bi})/qN_D]^{1/2} \qquad \text{at} \quad x=L.$$

The maximum value of y_2 is equal to a, and the corresponding voltage is called the pinch-off voltage, defined as

$$V_P=V(y_2=a)\equiv qN_Da^2/2\epsilon_s. \tag{4}$$

The current density in the x direction (along the channel) is given by the ohmic-law equation:

$$J_x=\sigma(x)\mathscr{E}_x \tag{5}$$

or

$$J_x=qN_D\mu\mathscr{E}_x \tag{5a}$$

where J_x is the current density, $\sigma(x)$ the conductivity, $\mathscr{E}_x$ the electrical field along the x direction ($\mathscr{E}_x=-dV/dx$), and μ the mobility, which is assumed to be field-independent. The drain (or channel) current for the upper half-channel region is then given by

$$I_D=qN_D\mu\left(\frac{dV}{dx}\right)(a-h)Z \tag{6}$$

or

$$I_D\,dx=Z\mu qN_D(a-h)\,dV. \tag{6a}$$

The differentiation of the drain voltage dV can be obtained from Eq. 2:

$$dV=\frac{qN_D}{\epsilon_s}h\,dh. \tag{7}$$

Substituting dV into Eq. 6 and integrating from $x = 0$ to $x = L$ yields

$$I_D = \frac{1}{L}\int_{y_1}^{y_2} Z\mu q N_D(a-h)\frac{qN_D}{\epsilon_s} h\, dh$$

$$= \frac{Z\mu q^2 N_D^2 a^3}{6\epsilon_s L}\left[\frac{3}{a^2}(y_2^2 - y_1^2) - \frac{2}{a^3}(y_2^3 - y_1^3)\right]. \tag{8}$$

We shall introduce I_P as the pinch-off current and three normalized depletion widths:

$$I_P \equiv Z\mu q^2 N_D^2 a^3/6\epsilon_s L \tag{9}$$

$$u \equiv h/a = [(V + V_G + V_{bi})/V_P]^{1/2}$$

$$u_1 \equiv y_1/a = [(V_G + V_{bi})/V_P]^{1/2} \tag{10}$$

$$u_2 \equiv y_2/a = [(V_D + V_G + V_{bi})/V_P]^{1/2}.$$

Equation 8 now becomes

$$I_D = I_P[3(u_2^2 - u_1^2) - 2(u_2^3 - u_1^3)] \tag{11}$$

or

$$I_D = I_P\{3V_D/V_P - 2[(V_D + V_G + V_{bi})^{3/2} - (V_G + V_{bi})^{3/2}]/V_P^{3/2}\}. \tag{11a}$$

For a given V_G, the maximum current I_{Dsat} occurs at the point where the channel is pinched off. The current is obtained from Eq. 11 by setting $u_2 = 1$ (or $y_2 = a$):

$$I_{Dsat} = I_P(1 - 3u_1^2 + 2u_1^3)$$

$$= I_P\left[1 - 3\left(\frac{V_G + V_{bi}}{V_P}\right) + 2\left(\frac{V_G + V_{bi}}{V_P}\right)^{3/2}\right]. \tag{12}$$

The current–voltage characteristics calculated from Eq. 11a are shown in Fig. 3, where the saturation current I_{Dsat} is given by Eq. 12 and the saturation voltage is given by

$$V_{Dsat} = V_P - V_G - V_{bi} = \frac{qN_D a^2}{2\epsilon_s} - V_G - \frac{kT}{q}\ln\left(\frac{N_D Na}{n_i{}^2}\right). \tag{13}$$

For drain voltages beyond V_{Dsat}, the drain current is assumed to remain essentially the same as the saturation current. As the drain voltage increases further, avalanche breakdown of the gate-to-channel diode will eventually occur, and the drain current will suddenly increase. The breakdown occurs at the drain end of the channel, where the reverse voltage is highest:

$$V_B(\text{breakdown voltage}) = V_D + V_G \tag{14}$$

or

$$V_D = V_B - V_G.$$

As mentioned previously, the absolute value of V_G is used. At $V_G = 0$, breakdown occurs at $V_D = V_B$. For higher V_G, the breakdown voltage V_B remains the same, but the drain voltage at breakdown reduces to $(V_B - V_G)$.

From Eq. 11 we can derive two important device parameters: the transconductance g_m and the channel conductance (also called the drain conductance) g_D:

$$g_m \equiv \frac{\partial I_D}{\partial V_G} = \frac{2Z\mu q N_D}{L}(y_2 - y_1) \tag{15a}$$

$$g_D \equiv \frac{\partial I_D}{\partial V_D} = \frac{2Z\mu q N_D}{L}(a - y_2). \tag{15b}$$

In the linear region ($V_D \to 0$), the drain conductance is given by Eq. 15b:

$$g_{DO}(V_D \to 0) = g_{\max}\left(1 - \sqrt{\frac{V_G + V_{bi}}{V_P}}\right) \tag{16}$$

where $g_{\max} \equiv qN_D a\mu Z/L$. The transconductance in the saturation region is

$$g_m = \frac{\partial I_{D\text{sat}}}{\partial V_G} = g_{\max}(1 - u_1) = g_{\max}\left(1 - \sqrt{\frac{V_G + V_{bi}}{V_P}}\right) \tag{17}$$

which is identical to g_D in the linear region.

6.2.2 Arbitrary Charge Distribution[9]

For an arbitrary doping distribution in the channel region, we shall define an integral form of the charge density $Q(Y)$ as

$$Q(Y) \equiv \int_0^Y \rho(y)\,dy \qquad \text{C/cm}^2 \tag{18}$$

or

$$Q(h) \equiv \int_0^h \rho(y)\,dy \qquad \text{C/cm}^2 \tag{18a}$$

where $\rho(y)$ is the charge density in C/cm^3. The dependence of the reverse bias voltage $V(h)$ on h and $\rho(y)$ can be derived from Poisson's equation, Eq. 1. Integrating the equation from $y = 0$ to $y = h$ yields

$$\mathscr{E}_y \equiv -\frac{\partial V}{\partial y} = \frac{1}{\epsilon_s}\int_0^y \rho(y)\,dy + \text{constant.} \tag{19}$$

The integration constant can be determined from the boundary condition that $\mathscr{E}_y = 0$ at $y = h$ for an abrupt depletion layer and is obtainable from Eq. 19, which becomes $-(1/\epsilon_s)\int_0^h \rho(y)\,dy$. We thus have from Eqs. 18 and 19:

$$\frac{\partial V}{\partial y} = \frac{1}{\epsilon_s}\left[\int_0^h \rho(y)\,dy - \int_0^y \rho(y)\,dy\right]$$
$$= \frac{1}{\epsilon_s}[Q(h) - Q(y)]. \tag{20}$$

Integrating once more from $y = 0$ to $y = h$, we obtain the voltage $V(h)$ across the depletion layer, which includes both the applied and built-in voltages:

$$V(h) = \frac{1}{\epsilon_s}\left[Q(h)\int_0^h dy - \int_0^h Q(y)\,dy\right]$$

$$= \frac{1}{\epsilon_s}\left[hQ(h) - \int_0^h Q(y)\,dy\right] \tag{21}$$

or

$$V(h) = \frac{1}{\epsilon_s}\int_0^h y\rho(y)\,dy. \tag{22}$$

Using integration by parts, the right-hand expressions in Eqs. 21 and 22 can be readily shown to be identical.

The maximum value for the upper limit of the integration occurs at $h = a$, and the corresponding voltage is called the pinch-off voltage as defined previously. Beyond this point the channel current remains essentially constant. The pinch-off voltage is given by

$$V_P(\text{pinch-off voltage}) = V(h = a) = \frac{1}{\epsilon_s}\int_0^a y\rho(y)\,dy. \tag{23}$$

Differentiating Eq. 22 yields

$$dV/dh = h\rho(h)/\epsilon_s \tag{24}$$

which shows that the voltage change required to move the depletion boundary a given distance, increases with the value h and is proportional to the space-charge density at that boundary. The junction capacitance per unit area is given by

$$C \equiv dQ(h)/dV = \left(\frac{dQ}{dh}\right)\left(\frac{dh}{dV}\right) \qquad \text{F/cm}^2. \tag{25}$$

The depletion layer thus acts as a plane capacitor with plate distance h, and the capacitance is independent of the charge distribution profile.

We next consider the current–voltage characteristics and the transconductance. The drain current density was given previously in Eq. 5:

$$I_D = Z\mu\frac{dV}{dx}\int_h^a \rho(y)\,dy \tag{26}$$

or

$$I_D\,dx = Z\mu(dV/dh)\,dh\int_h^a \rho(y)\,dy. \tag{26a}$$

Substituting Eq. 24 and integrating with the boundary conditions $h = y_1$ at

$x = 0$ and $h = y_2$ at $x = L$, we obtain

$$\int_0^L I_D dx = I_D \cdot L = \frac{2Z\mu}{\epsilon_s} \int_0^L h\rho(h)\, dh \int_h^a \rho(y)\, dy \tag{27}$$

or

$$I_D = \frac{2Z\mu}{\epsilon_s L} \int_{y_1}^{y_2} [Q(a) - Q(h)] h\rho(h)\, dh. \tag{28}$$

Equation 28 is the basic equation of long-channel JFET.

From Eq. 28 we can derive g_m:

$$g_m \equiv \frac{\partial I_D}{\partial V_G} = \frac{\partial I_D}{\partial y_1}\frac{\partial y_1}{\partial V_G} + \frac{\partial I_D}{\partial y_2}\frac{\partial y_2}{\partial V_G}. \tag{29}$$

The partial derivatives are obtained from Eqs. 24 and 28:

$$g_m = \frac{2Z\mu}{L}[Q(y_2) - Q(y_1)] \tag{29a}$$

which shows that g_m is equal to the conductance of the rectangular section of the semiconductor extending from $y = y_1$ to $y = y_2$. The channel conductance g_D can be obtained from Eqs. 24 and 28 in a similar manner:

$$g_D = \frac{\partial I_D}{\partial V_D} = \frac{2Z\mu}{L}[Q(a) - Q(y_2)]. \tag{30}$$

This value approaches zero as $y_2 \approx a$ at which $V_D + V_G = V_P - V_{bi}$. It is interesting to compare Eq. 29a with Eq. 30. For $V_D \to 0$, we have $y_2 \to y_1$ and g_D is proportional to $[Q(a) - Q(y_1)]$. On the other hand, when $V_D + V_G \geq V_P$ and $y_2 \to a$, then g_m is also proportional to the same charge difference. Hence we obtain the following useful expression for any arbitrary charge distribution:

$$g_{DO}(V_D \to 0) = g_m(V_D \gg V_P) = \frac{2Z\mu}{L}[Q(a) - Q(y_1)]$$

$$\equiv g_{\max}\left[1 - \frac{Q(y_1)}{Q(a)}\right] \tag{31}$$

where

$$g_{\max} \equiv \frac{2Z\mu}{L} Q(a).$$

The uniformly doped charge distribution is a special case of the foregoing results. The current–voltage characteristics of other charge distribution profiles can be calculated from Eq. 28. Table 1 summarizes the results for three cases.[9] Case B represents the uniform charge distribution discussed previously. Cases A and C represent two limiting cases in which the charge distributions are delta functions located at $y = 2a$ and $y = 0$, respectively. We note that the dimensionless ratio $g_{\max} V_P / I_P$ depends only on the charge

Table 1 Field-Effect Equations for Specific Charge Distributions in a Rectangular Structure for Reflected-Type JFET (Total Channel Depth 2*a*)

Parameter	Common Factor	Multiplying Factors for Specific Distributions		
		A: All Charge at $y = 2a$	B: Uniform	C: All Charge at $y = 0$
g_{max}	$\frac{2Z\mu\rho a}{L}$	1	1	1
V_P	$\frac{4\rho a^2}{\epsilon_s}$	$\frac{1}{4}$	$\frac{1}{8}$	0
I_P	$\frac{8Z\mu\rho^2 a^3}{\epsilon_s L}$	$\frac{1}{8}$	$\frac{1}{24}$	0
$\frac{g_{max} V_P}{I_P}$	1	2	3	4

distribution parameters. However, this ratio is limited to the range 2 to 4. For case A, the current–voltage relationship is given by

$$I_{Dsat} = I_P \left[1 - \left(\frac{V_G + V_{bi}}{V_P} \right) \right]^2 . \tag{32}$$

Figure 4 compares Eqs. 32 and 12. Note that the range of possible transfer characteristics, as indicated by the shaded region, is surprisingly narrow. The shaded region can be represented by

$$\left[1 - \left(\frac{V_G + V_{bi}}{V_P} \right) \right]^n$$

where n varies[10] between 2 and 2.25.

6.2.3 Normally-Off FET

The symbols for JFET and MESFET are shown in Fig. 5. In the previous discussions we considered only the n-type normally-on (or depletion mode) device, that is, the device with an n-type conductive channel at $V_G = 0$.

For high-speed low-power applications, the n-type normally-off (or enhancement-mode) device is very attractive. A normally-off device is one that does not have a conductive channel at $V_G = 0$; that is, the built-in potential V_{bi} of the gate junction is sufficient to totally deplete the channel region. When a positive gate bias larger than a threshold voltage is applied, the channel current will begin to flow. The threshold voltage V_T is approximately given by

$$V_T \simeq V_{bi} - V_P \tag{33}$$

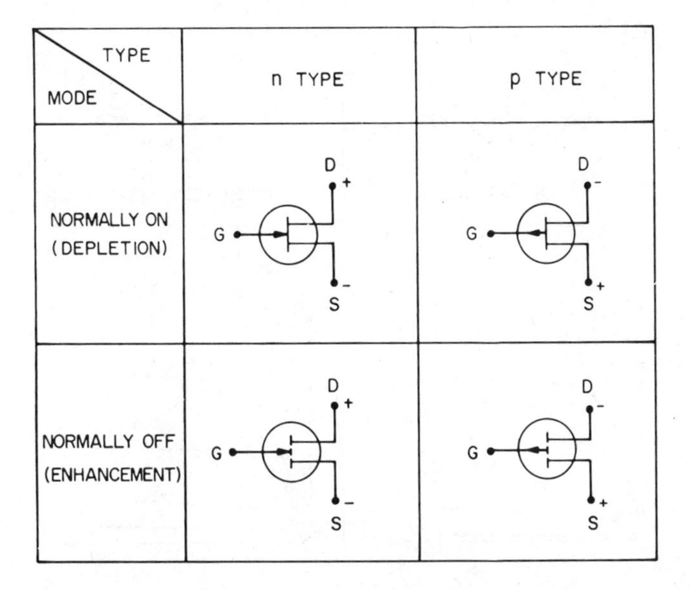

Fig. 5 Symbols for *n*-type and *p*-type normally-on and normally-off JFET and MESFET.

or

$$V_{bi} \approx V_T + V_P \tag{33a}$$

where V_P is the pinch-off voltage defined in Eqs. 4 and 23. Near the threshold, the drain current in the saturation region can be obtained by substituting V_{bi} in Eq. 33a to Eq. 12, and using the Taylor series expansion around the point $V_G = V_T$. The result is[11]

$$I_D = \frac{Z\mu\epsilon_s}{2aL}(V_G - V_T)^2. \tag{34}$$

This relation is similar to the enhancement-mode MOSFET equation (Chapter 8) with the channel depth $2a$ replacing the insulator thickness.

The basic current–voltage characteristics of normally-on and normally-off devices are similar. Figure 6 compares these two modes of operation. The main difference is the shift of threshold voltage along the V_G axis. The normally-off device has no current conduction at $V_G = 0$, and the current varies as in Eq. 34 when $V_G > V_T$. Since the built-in potential of the gate is about 1 V or less, the forward bias on gate is limited to about 0.5 V to avoid excessive gate current. In subsequent discussions, we shall concentrate on the normally-on devices. The results are applicable to normally-off devices with appropriate change of gate voltages.

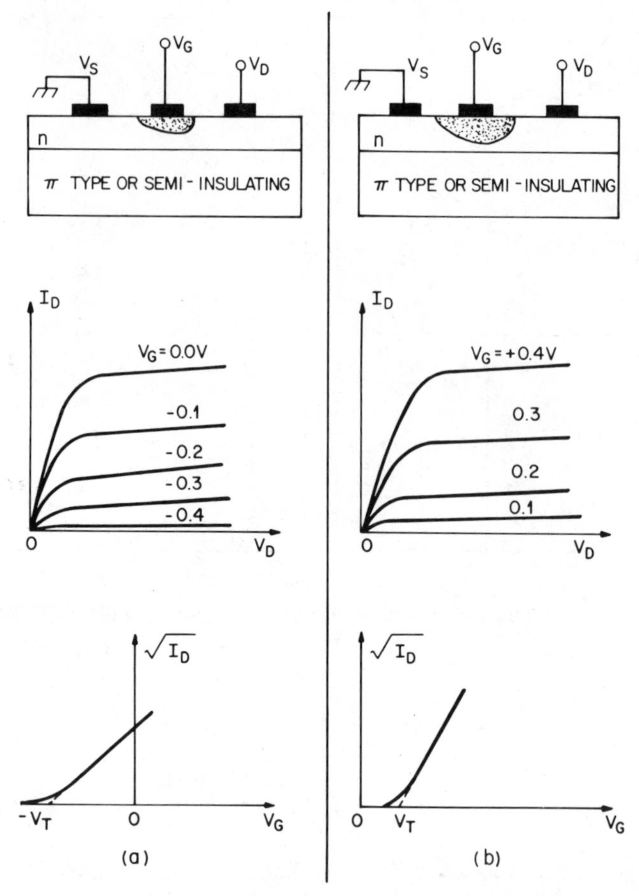

Fig. 6 Comparison of *I–V* characteristics. (*a*) Normally-on MESFET. (*b*) Normally-off MESFET.

6.3 GENERAL CHARACTERISTICS

6.3.1 Field-Dependent Mobility

For long-channel devices (i.e., $L \gg a$), the assumptions in Section 6.2 are generally valid, and Eqs. 12 and 28 can describe the device behavior reasonably well. For FETs with short channels (and small channel length/height ratios), significant discrepancies are encountered between experiment and basic theory. One main reason for the discrepancies is the field-dependent mobility, leading to high-field velocity saturation. Figure 7 shows the dependence of the drift velocity versus electric field for Si, GaAs, and InP.[12,13] At low fields the velocity increases linearly with the

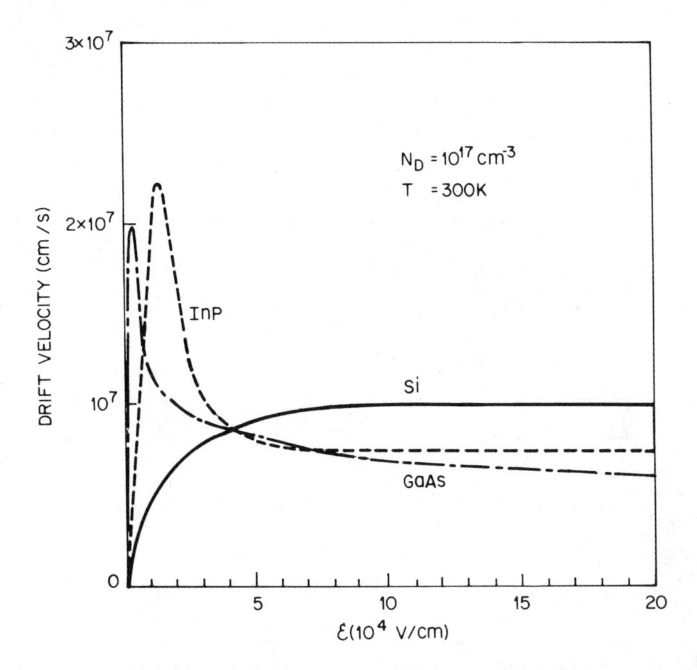

Fig. 7 Drift velocity versus electric field for Si, GaAs, and InP. (After Jacoboni et al., Ref. 12; Smith, Inoue, and Frey, Ref. 13.)

field, and the slope corresponds to a constant mobility ($\mu \equiv dv/d\mathscr{E}$). For Si the drift velocity reaches its saturation value of 10^7 cm/s at fields above 5×10^4 V/cm. For GaAs and InP, the drift velocity first reaches a peak value and then decreases toward a velocity of about $6 \sim 8 \times 10^6$ cm/s.

We shall assume a simple analytic expression for the drift velocity in Si as shown in Fig. 8:

$$v = \frac{\mu \mathscr{E}_x}{1 + \mu \mathscr{E}_x / v_s} \tag{35}$$

where μ is the low-field mobility, v_s the saturation velocity, and $\mathscr{E}_x = dV/dx$ is the longitudinal field in the channel. The channel current Eq. 6 is modified as

$$I_D = qN_D \frac{\mu \mathscr{E}_x}{1 + \mu \mathscr{E}_x / v_s} (a - h)Z. \tag{36}$$

Using the normalized depletion-layer widths, Eq. 10, the drain current becomes[14]

$$I_D = 6I_P(1 - u)uu'/(1 + 2uu'z) \tag{37}$$

where

$$u' = \partial u / \partial (x/L) = (L/V_P)\mathscr{E}_x(2u)^{-1}$$

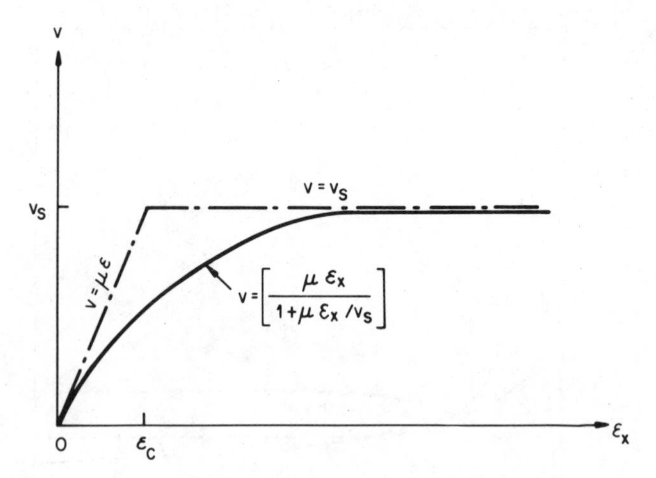

Fig. 8 Approximations for the velocity-field curves.

and

$$z \equiv \mu V_P / v_x L \tag{38}$$

which is the ratio between the small-field velocity extrapolated linearly to the field V_P/L and the saturation velocity. For the constant-mobility case, z equals zero. Integrating Eq. 37 between source $x = 0$ and a position x along the channel provides

$$x/L = \frac{I_P}{I_D}\left[3\left(1 - \frac{z}{3}\frac{I_D}{I_P}\right)(u^2 - u_1^2) - 2(u^3 - u_1^3)\right] \tag{39}$$

where the boundary conditions at $x = 0$ are $V_D = 0$ and

$$u^2\big|_{x=0} \equiv u_1^2 = \frac{V_G + V_{bi}}{V_P}.$$

At $x = L$, $u = u_2$ and Eq. 39 becomes[14]

$$I_D = \frac{I_P[3(u_2^2 - u_1^2) - 2(u_2^3 - u_1^3)]}{1 + \mu V_D / v_s L}. \tag{40}$$

Comparing Eqs. 40 and 11 shows that for given I_P, u_1, and u_2, the drain current is reduced by a factor of $(1 + \mu V_D / v_s L)$ due to the field-dependent mobility.

The normalized I_D versus V_D characteristics according to Eq. 40 are plotted in Fig. 9*a* for the constant-mobility case with $z = 0$, and in Fig. 9*b* for $z = 3$ with the normalized gate bias as parameter. Note that the drain current is substantially reduced because of the field-dependent mobility effect. Equation 40 has a maximum at $u_2 = u_m$, given by

$$u_m^3 + 3u_m\left(\frac{1}{z} - u_1^2\right) + 2u_1^3 - \frac{3}{z} = 0. \tag{41}$$

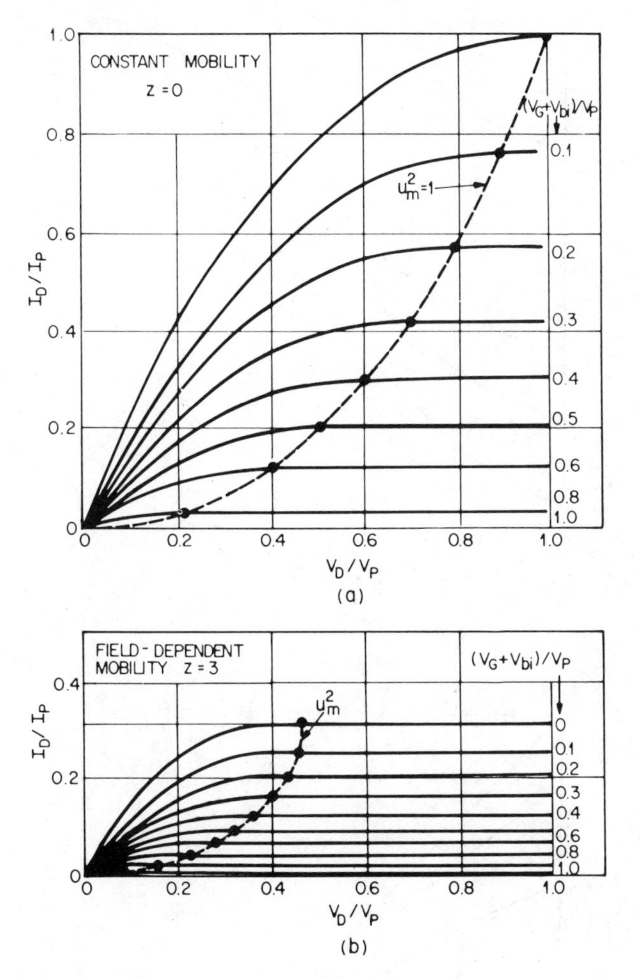

Fig. 9 Plots of normalized drain current versus drain voltage with gate voltage as parameter for (*a*) constant mobility case $z = 0$ and (*b*) field-dependent mobility $z = 3$. (After Lehovec and Zuleeg, Ref. 14.)

By substituting the value from Eq. 41 for u_1^3 in Eq. 40, the saturation drain current can be obtained:

$$I_{D\text{sat}} = \frac{3I_P(1-u_m)}{z}. \tag{42}$$

Solutions of u_m from Eq. 41 for various values of z are plotted in Fig. 10*a*. Also shown is $u_m = 1$, which refers to the boundary condition $(V_D + V_G + V_{bi}) = V_P$ for the constant-mobility case. Figure 10*b* is a plot of $V_{D\text{sat}}/V_P$ versus $(V_G + V_{bi})/V_P$, that is, the drain voltage versus gate voltage at onset of saturation for various values of z. Note that with increasing z, the saturation of drain current is reached at smaller values of drain voltage.

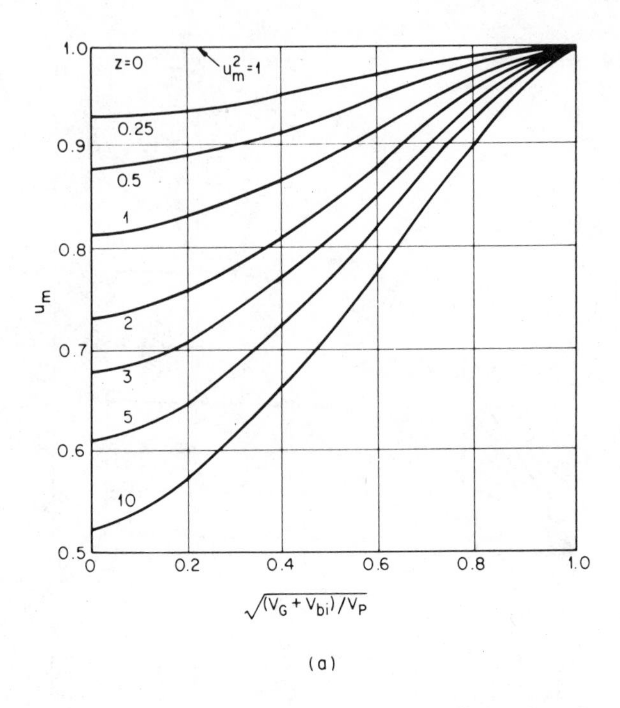

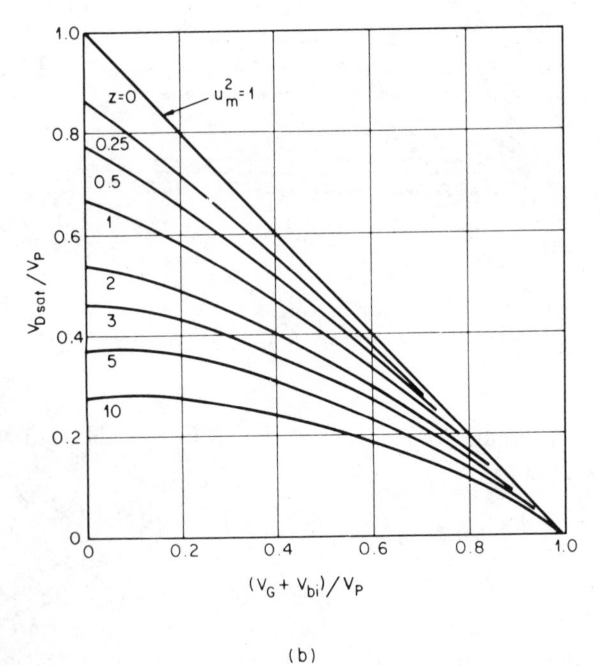

Fig. 10 (*a*) Normalized saturated drain voltage u_m versus $(V_G + V_{bi})/V_P$, where the absolute value of V_G is used. (*b*) Normalized V_{Dsat} versus $(V_G + V_{bi})/V_P$. (After Lehovec and Zuleeg, Ref. 14.)

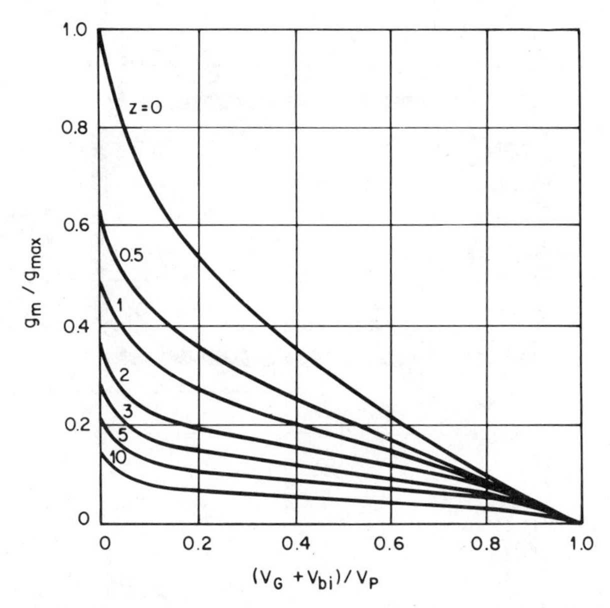

Fig. 11 Normalized transconductance versus $(V_G + V_{bi})/V_P$. (Ref. 14.)

The transconductance in the saturation region can be obtained from Eq. 42:

$$g_m = \frac{\partial I_{Dsat}}{\partial V_G} = \frac{g_{max}(u_m - u_1)}{1 + z(u_m^2 - u_1^2)} . \tag{43}$$

This expression reduces to Eq. 17 for $z = 0$ and $u_m = 1$. Figure 11 shows curves of g_m/g_{max} versus gate bias. For large values of z, the transconductance is almost independent of the gate bias.

6.3.2 Two-Region Model

The field-dependent mobility approaches discussed above give reasonable predictions for short-channel JFET and MESFET in Si. For GaAs, the velocity-field characteristic is more complicated, and the velocity saturation occurs at a much lower field. A different approximation has been proposed[7] for the velocity-field curve, as shown in Fig. 8. At low fields, the mobility is assumed constant, and the velocity varies linearly with $\mathscr{E}_x$ up to a critical field $\mathscr{E}_c$. For fields above $\mathscr{E}_c$, the velocity is assumed constant. For operation above the pinch-off, the JFET is divided into two regions (Fig. 12). Region I, near the source, is the constant-mobility region, and the gradual-channel approximation described in Section 6.2 is applicable. Region II, near the drain, is the velocity-saturation region; in this region, a conductive channel of finite width is postulated to account for current continuity. The plane corresponding to the onset of velocity saturation is

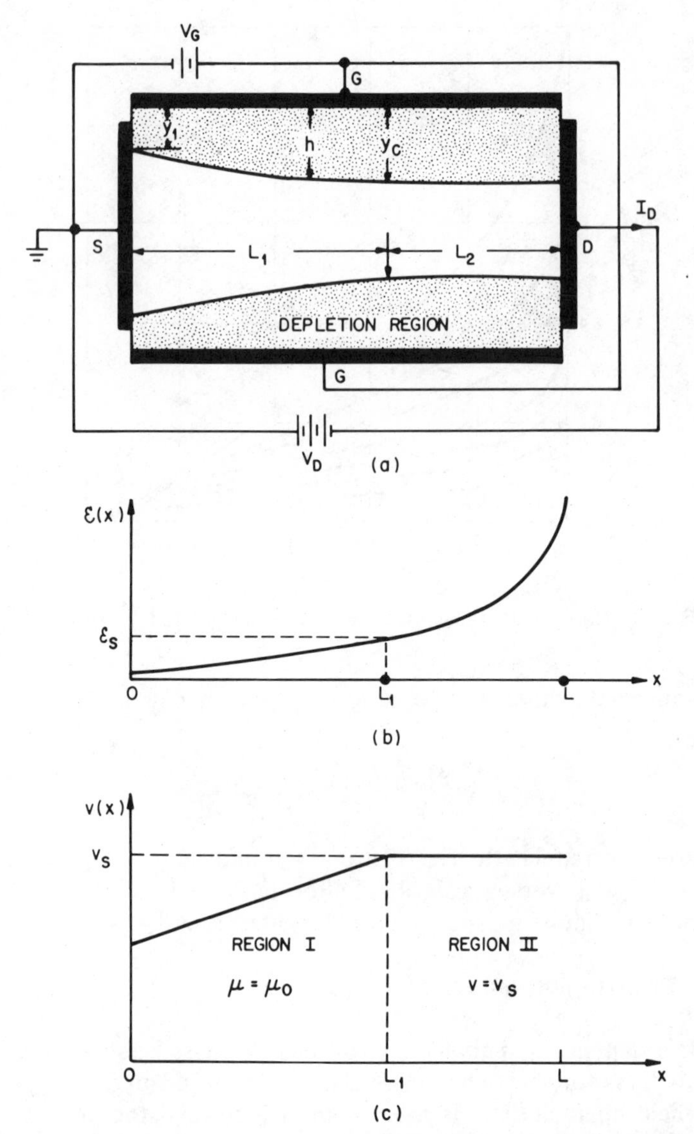

Fig. 12 Two-region model. Region I has constant mobility and region II has saturated velocity. The velocity begins to saturate at the depletion width y_c. (After Pucel, Haus, and Statz, Ref. 7).

not "pinned" at the drain end of the gate as required in the field-dependent mobility approximation. The plane is allowed to move into the channel with bias voltage variations as the field distribution demands. Its position is determined by the location at which the longitudinal channel field equals the critical value $\mathscr{E}_c$. The two-region model above is therefore applicable to operating conditions for all I–V characteristics from linear to saturation regime.

As shown in Fig. 12a, the depletion width at the joining point is y_c. We shall define a normalized width

$$u_c = y_c/a.$$

Consider region I first. Integrating Eq. 6 from $x = 0$ to $x = L_1$ (instead of to $x = L$), one obtains an expression identical to Eq. 11 except that u_2 is replaced by u_c and I_P is replaced by I_1:

$$I_D = I_1[3(u_c^2 - u_1^2) - 2(u_c^3 - u_1^3)] \tag{44}$$

and

$$I_1 \equiv Z\mu q^2 N_D^2 a^3 / 6\epsilon_s L_1.$$

We can determine L_1 by utilizing the current continuity between regions I and II. In region II the carriers are assumed to travel at their saturation velocity $v_s = \mu\mathscr{E}_c$. Thus in region II,

$$I_D = qN_D v_s(a - y_c)Z = I_S(1 - u_c) \tag{45}$$

where $I_S = qN_D v_s aZ$ is the open-channel saturation current corresponding to the maximum drain current that could exist if the channel were totally undepleted and the carriers traveled at their saturation velocity.

Equating Eqs. 44 and 45, we obtain the following relation:

$$L_1 = zL\frac{(u_c^2 - u_1^2) - \frac{2}{3}(u_c^3 - u_1^3)}{1 - u_c} \tag{46}$$

where z is defined in Eq. 38. The larger z is, the greater the role of velocity saturation. For microwave FETs, z is of the order of 2 to 20, with the higher values corresponding to the devices operated in the 10-GHz range.

Once the normalized depletion widths u_c and u_1 are known, the length L_1 is specified and the current I_D is determined. Conversely, given the normalized width u_1 and the current I_D, u_c can be determined from Eq. 45 and hence L_1 is specified for Eq. 46.

For a given channel current I_D, the potential drop from source to drain can be obtained by integrating the longitudinal electric field from $x = 0$ to $x = L$. In region I, the potential drop is simply

$$V(\text{region I}) = V_P - V_S = V_P(u_c^2 - u_1^2). \tag{47}$$

In region II, the longitudinal field is determined by free charges on the drain electrode. Using Laplace's equation and keeping the lowest space harmonic, we obtain

$$V(\text{region II}) \simeq \frac{2a}{\pi}\mathscr{E}_c \cos\frac{\pi y}{2a}\sinh\left[\frac{\pi(L - L_1)}{2a}\right]. \tag{48}$$

The source–drain potential is the sum of Eqs. 47 and 48:

$$V_D = V_P\left\{(u_c^2 - u_1^2) + \frac{2}{\pi}\frac{a}{zL}\sinh\left[\frac{\pi(L - L_1)}{2a}\right]\right\}. \tag{49}$$

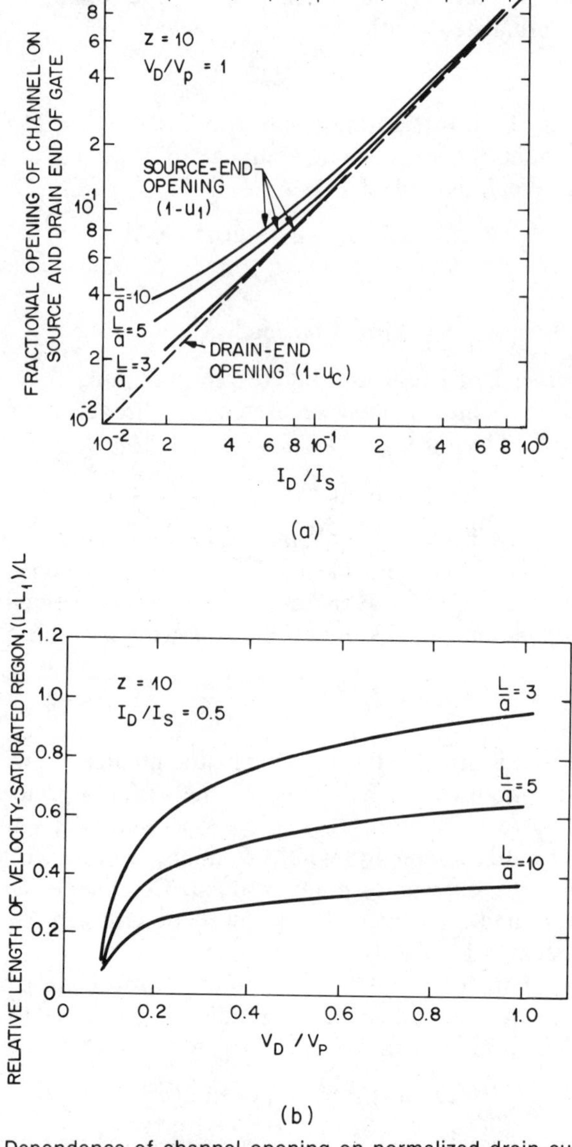

Fig. 13 (*a*) Dependence of channel opening on normalized drain current for $z = 10$ and $V_D/V_P = 1$. (*b*) Relative length of the velocity-saturated region as a function of the normalized drain voltage and L/a. (After Pucel, Haus, and Statz, Ref. 7.)

Equations 49 and 46 form a pair that allows one to eliminate L_1 and solve for u_c with known V_G and V_D. Once u_c is obtained, Eq. 45 then yields I_D.

Figure 13*a* shows[7] the normalized channel openings at the source and drain ends of the gate, $(1 - u_1)$ and $(1 - u_c)$, as a function of I_D/I_S for various values of z and the ratio L/a. In the usual operating range of current, I_D/I_S is of the order of 0.5 or less. Note that the percentage variation of the channel width from source to drain is about 10 to 20%, except at very small currents. In other words, the channel boundary is nearly parallel to the gate electrode plane. Figure 13*b* shows the strong penetration of the velocity-saturated region into the gate region. For example, for $L/a = 3$, region II encompasses about 95% of the region under the gate for $V_D = V_P$.

6.3.3 Saturated-Velocity Model

Another limiting case is the saturated-velocity model,[15] which is expected to be valid in the limit of very short gates in which full current saturation is assumed under the gate. This saturated current is directly modulated by the difference between the depletion width h and the channel depth a. For uniformly doped material, this model yields

$$I = qv_s Z(a - h)N_D. \tag{50}$$

Equation 50 agrees quite well with experimental results of short-gate ($L \leqslant 2\ \mu m$) GaAs MESFETs; and the corresponding effective saturation velocity v_s at 300 K is 1.2×10^7 cm/s.

For other doping profiles $\rho(y)$, the model gives

$$I = v_s Z \int_h^a \rho(y)\, dy. \tag{51}$$

The electric field and the voltage across the depletion layer have been obtained previously in Eqs. 19 and 22, respectively. Equations 22 and 51 imply

$$dI/dh = v_s Z \rho(h) \tag{52a}$$

and

$$dV/dh = h\rho(h)/\epsilon_s \tag{52b}$$

Thus

$$g_m = \frac{dI}{dV} = v_s Z \epsilon_s / h(V_G). \tag{53}$$

The linearity of the transfer characteristics (constant g_m) is approached by those profiles in which the depletion depth $h(V_G)$ changes very little as a function of the gate voltage. The transfer characteristics for various doping profiles[15] are shown in Fig. 14. Since the source–drain potential does not appear in this model, the voltage V is the gate potential that includes the built-in potential V_{bi}. Note that all three types of nonuniform dopings achieve linearity (i.e., $g_m \equiv dI/dV =$ constant) as the appropriate variable parameter is taken to its limit. The results above are quite different from the constant-mobility case, in which the doping profile has negligible effect

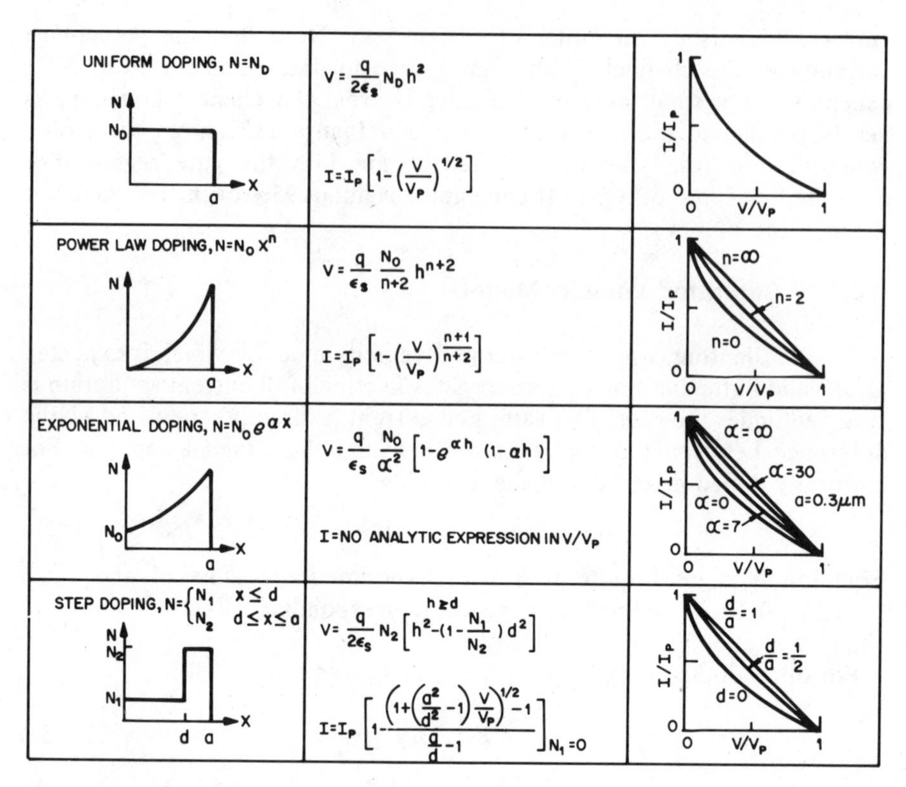

Fig. 14 Transfer characteristics of various charge distributions. (After Williams and Shaw, Ref. 15.)

on the transfer characteristics. Although Eq. 53 implies a reduction of g_m as a result of increased gate voltage, the important quantity g_m/C_{GS} remains unaffected where C_{GS} is the gate-to-source capacitance. This is because $C_{GS} = \epsilon_s/h$, and Eq. 53 gives

$$g_m/C_{GS} = v_s Z = \text{constant}. \tag{54}$$

Experimental results have shown that FETs with graded channel doping[16] or step dopings[17] have improved linearity. A typical example is shown in Fig. 15. The graded-channel device has a constant $g_m =$ 9.5 mmhos over the entire range $0 < V_G < 12$ V.

6.3.4 Two-Dimensional Considerations

In the discussions above, we derived the current–voltage characteristics of JFET and MESFET in closed or analytical forms based on various approximations. As the gate length becomes shorter and the drain voltage becomes larger, two-dimensional effects will dominate the device opera-

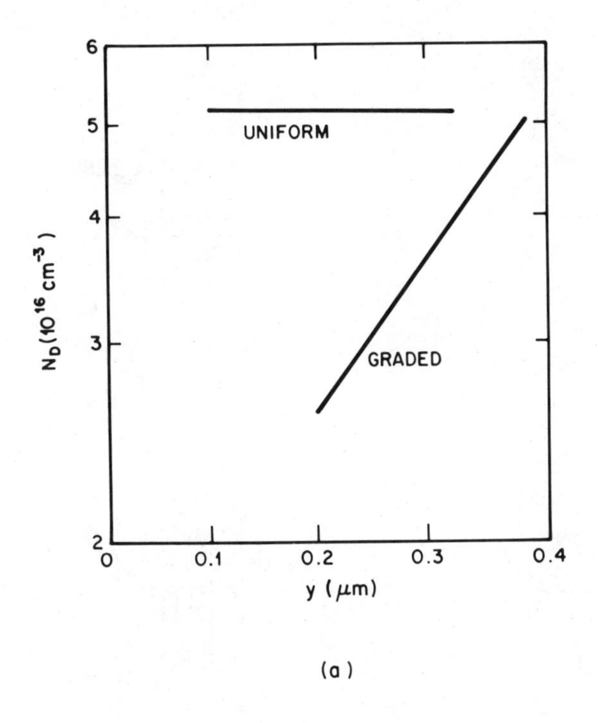

(a)

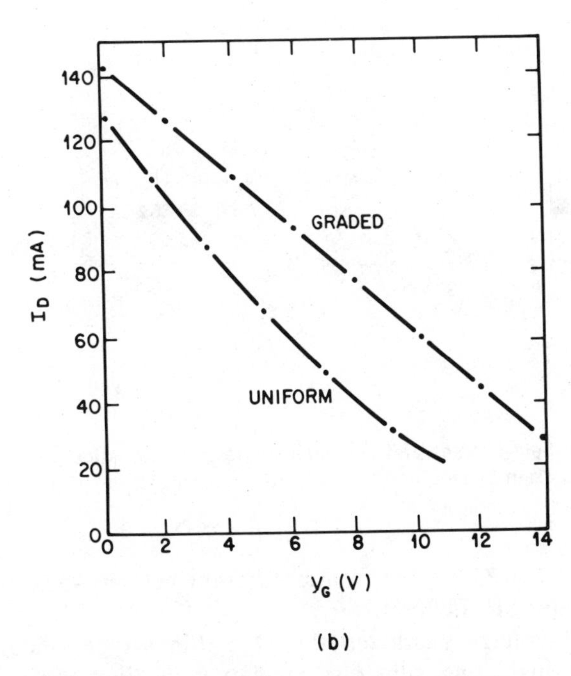

(b)

Fig. 15 (*a*) Measured doping profiles and (*b*) transfer characteristics of graded and uniformly doped FETs. (After Williams and Shaw, Ref. 16.)

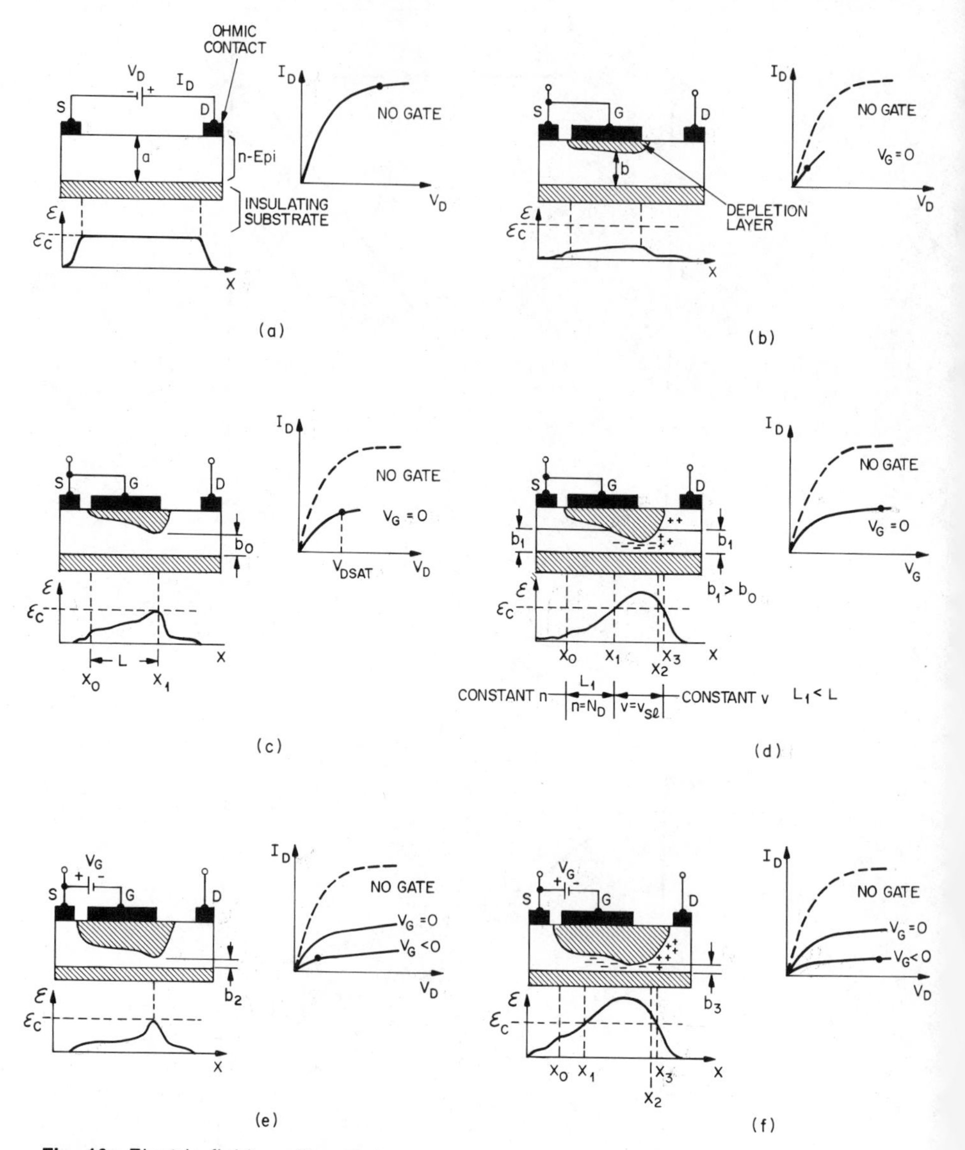

Fig. 16 Electric field profiles and current–voltage characteristics of Si MESFET under various gate and drain biasing conditions. (After Liechti, Ref. 8.)

tion.[18] Since Si and GaAs have quite different velocity–field relationships, we shall consider Si FET first.

The current–voltage characteristic[8] of a thin n-type silicon layer supported by an insulating substrate is shown in Fig. 16*a*. As discussed previously, at small drain voltages the silicon layer behaves like a linear resistor. For larger voltages, the electron drift velocity does not increase at the same rate as the field. As a result, the current falls below the initial resistor line. As V_D is further increased, the electrons reach the saturation velocity. At this drain voltage, the current starts to saturate. In Fig. 16*b* a *p-n* junction or a metal–semiconductor rectifying contact is added between the source and the drain. This contact creates a depletion layer that acts like an insulating region and constricts the channel opening available for current flow in the *n* layer. The width of the depletion layer depends on the applied voltages. Under the condition that the gate is shorted to the source and a small drain voltage is applied, the depletion layer has a finite width and the conductive channel beneath has a smaller cross section b than in Fig. 16*a*. Consequently, the resistance between the source and drain is larger, as shown in Fig. 16*b*.

The drain current is given by

$$I_D = Zqn(x)v(x)b(x). \tag{55}$$

The electron density $n(x)$ is equal to the doping density N_D as long as the field does not exceed the critical value $\mathscr{E}_c$. The voltage along the channel increases from zero at the source to V_D at the drain. Thus the gate contact becomes increasingly reverse-biased, and the depletion width becomes wider as we proceed from source to drain. The resulting decrease in channel opening b must be compensated by an increase in electric field and electron velocity to maintain a constant current through the channel. As V_D is increased further, the electrons reach the saturation velocity at the drain end of the gate (Fig. 16*c*). The channel is constricted to the smallest cross section b_0 under the gate, the electric field reaches the critical value at this point, and I_D starts to saturate.

If the drain voltage is increased beyond V_{Dsat}, the depletion region widens toward the drain. The point x_1, where the electrons reach the saturation velocity, moves slightly toward the source (Fig. 16*d*). As x_1 moves closer to the source, the voltage at x_1 decreases. Consequently, the channel opening b_1 widens, and more current is injected into the velocity saturation region. This action results in a positive slope of the I_D–V_D curve and a finite drain resistance beyond current saturation.[19]

Proceeding from x_1 toward the drain, the channel potential increases, the depletion layer widens, and the channel opening becomes narrower than b_1. Since the velocity is saturated, the change in channel width must be compensated by a change in carrier density to maintain constant current. According to Eq. 55, an electron accumulation layer forms between x_1 and x_2, where the channel opening is smaller than b_1. At x_2 the channel opening is again b_1, and the negative space charge changes to a positive space charge to preserve constant current. This positive space charge is caused by partial electron depletion. The electron velocity remains saturated between x_2 and x_3 due to the field added by the negative space charge. Therefore, the drain voltage applied in excess of V_{Dsat} forms a dipole layer in a channel that extends beyond the drain end of the gate.

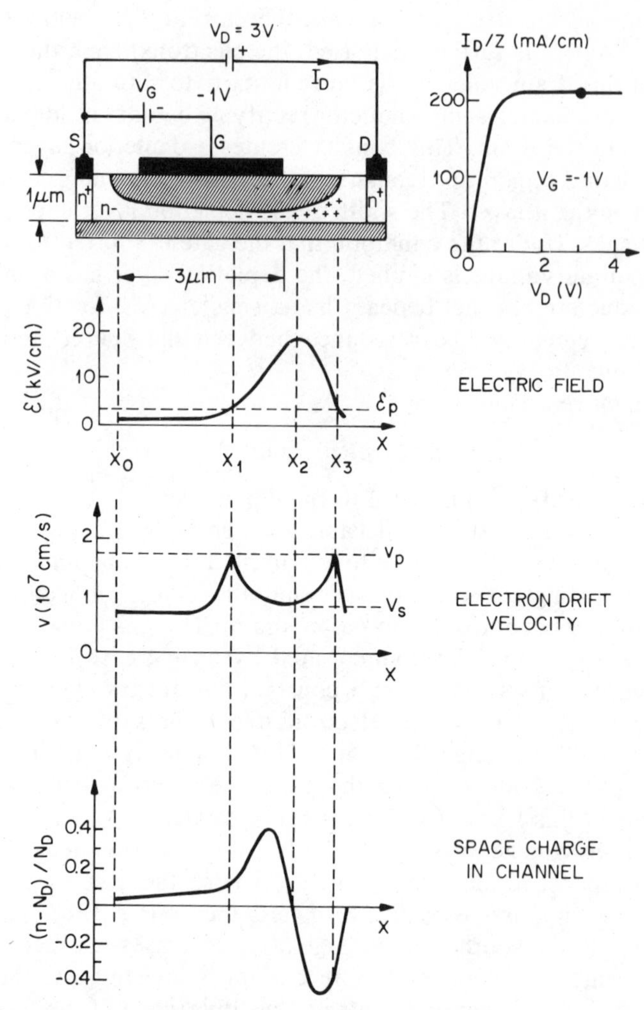

Fig. 17 Channel cross section, electric field, drift velocity, and space-charge distribution in the channel of a GaAs MESFET operated in current-saturation region. (After Liechti, Ref. 8.)

When a negative voltage is applied to the gate, Fig. 16*e*, the depletion width becomes larger. For small V_D, the channel acts as a linear resistor with larger resistance due to narrower channel opening. As V_D increases, the critical field is reached at a drain current lower than in the $V_G = 0$ case due to the larger channel resistance. For a further increase in V_D, the current remains saturated. At even larger V_D, a dipole layer will be formed at the drain end in order to maintain current continuity across the channel. (Fig. 16*f*).

In GaAs the velocity–field curve is more complicated. Figure 17 shows

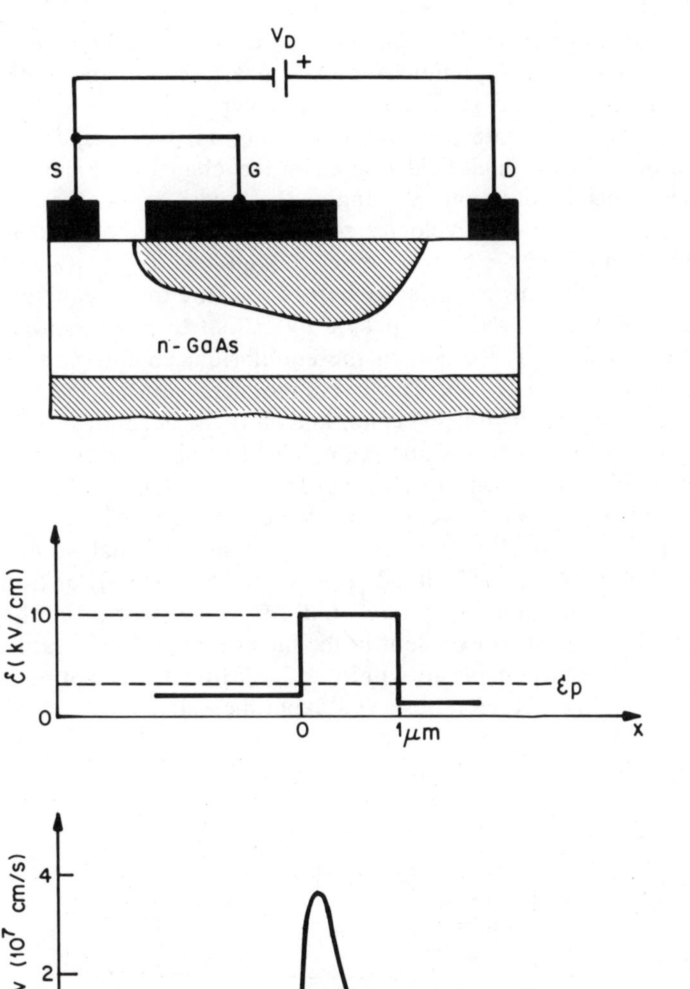

Fig. 18 Velocity overshoot of electrons as they enter the high-field region ($\mathscr{E} > \mathscr{E}_p$) under the gate. (After Liechti, Ref. 8.)

the key features of a GaAs MESFET operated in the saturation region.[8, 20] The narrowest channel opening is under the drain end of the gate. The drift velocity rises to a peak at x_1 and falls to the low saturated value under the gate edge. To preserve current continuity, heavy electron accumulation has to form in this region, because the channel opening is narrowing, and, in addition, the electrons are moving progressively slower with increasing x. Exactly the opposite occurs between x_2 and x_3. The channel widens and the electrons move faster, causing a strong depletion layer. The charges in the accumulation and depletion layers are nearly equal, and most of the drain voltage drops in this stationary dipole layer.

For very short gates, the electrons may not reach equilibrium transport conditions in the high-field region of the channel.[21] Figure 18 shows the nonequilibrium situation. As long as the field is below the threshold value $\mathscr{E}_p$ at which the drift velocity reaches its peak, the electrons remain in equilibrium. If the electrons enter a high-field region ($\mathscr{E} > \mathscr{E}_p$), they are accelerated to a higher velocity before relaxing to the equilibrium velocity. Figure 18 also shows the expected overshoot to more than twice the peak velocity and the relaxation to the equilibrium condition, after traveling a distance of about 1 μm. The overshoot will shorten the electron transit time through the high-field region and shift the accumulation layer into the gap between the gate and the drain. This overshoot is expected to improve high-frequency response, especially for GaAs FET.

Figure 19 shows the current–voltage characteristics for Si and GaAs MESFETs obtained by two-dimensional numerical analysis (solid curves).[22] Also shown are the calculated results (dashed curves) for Si based on the field-dependent mobility model and for GaAs based on the two-region model. Agreement is excellent in the linear region. In the saturation region, agreement is reasonable to within ±15%. However, the dips in I-V curves for GaAs have not been observed experimentally.

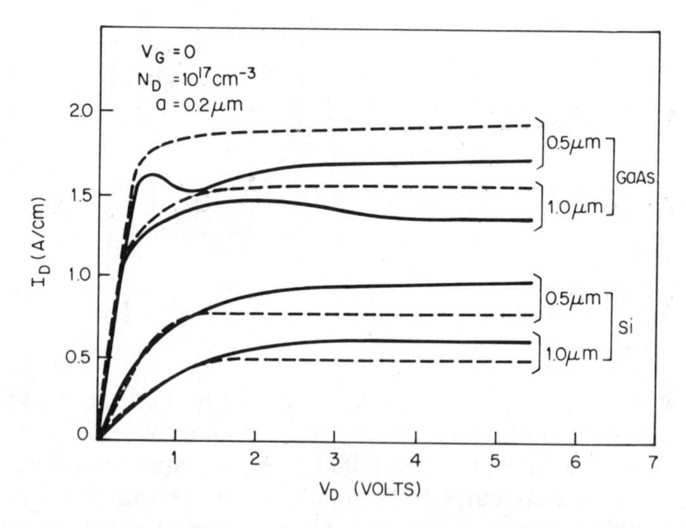

Fig. 19 Current–voltage characteristics for Si and GaAs 1-μm and 0.5-μm MESFETs. Solid lines are from two-dimensional computer simulation and dashed curves from analytical formulas. (After Wada and Frey, Ref. 22.)

6.4 MICROWAVE PERFORMANCE

6.4.1 Small-Signal Equivalent Circuit

Field-effect transistors, especially GaAs MESFETs, are useful for low-noise amplification, high-efficiency power generation, and high-speed logic application. We shall first consider the small-signal equivalent circuit of an MESFET. A lumped-element circuit for operation in the saturation region in common-source configuration is shown[8] in Fig. 20*a*. The location of the element in the FET structure is illustrated in Fig. 20*b*. In the intrinsic FET, the elements $(C_{DG} + C_{GS})$ are the total gate-to-channel capacitance; the input resistances R_i and R_{DS} under the gate show the effects of the channel resistance. The extrinsic (parasitic) elements include the source resistance R_S, the drain resistance R_D, and the substrate capacitance C_{DS}.

The current in the reverse-biased gate-to-channel junction can be expressed as

$$I_G = I_S[\exp(qV_G/nkT) - 1] \tag{56}$$

where n is the diode ideality factor $(1 < n < 2)$ and I_S is the saturation current. The input resistance is given by

$$R_i \equiv (\partial I_G/\partial V_G)^{-1} = nkT/q(I_G - I_S). \tag{57}$$

At $I_G \to 0$, the input resistance at room temperature is about 250 MΩ for $I_S = 10^{-10}$ A. The FET obviously has a very high input resistance.

The source and drain series resistances, which cannot be modulated by the gate voltages, will introduce *IR* drop between the gate and the source and drain contacts. They will reduce the drain conductance as well as the transconductance. The voltages V_D and V_G in Eq. 11 should then be replaced by $[V_D - I_D(R_S + R_D)]$ and $(V_G - I_D R_S)$, respectively. In the linear region, the resistances R_S, $1/g_{DO}$, and R_D are in series, and the measured drain conductance is given by $g_{DO}/[1 + (R_S + R_D)g_{DO}]$. A similar result can be obtained for the measured transconductance. In the saturation region, however, the transconductance is affected only by the source resistance. The drain resistance R_D will cause an increase of the drain voltage at which current saturation occurs. Beyond that voltage, $V_D > V_{Dsat}$, the magnitude of V_D has no effect on the drain current. Thus R_D has no further effect on g_m, so the transconductance measured is equal to $g_m/(1 + R_S g_m)$.

Under high-frequency operations, two factors limit the frequency response of a FET: the transit time and the *RC* time constant. The transit-time effect is the result of a finite time being required for carriers to travel from source to drain. For the constant mobility case and the saturated velocity case, the transit time is given by

$$\tau = \frac{L}{\mu \mathscr{E}_x} \approx \frac{L^2}{\mu V_D} \tag{58}$$

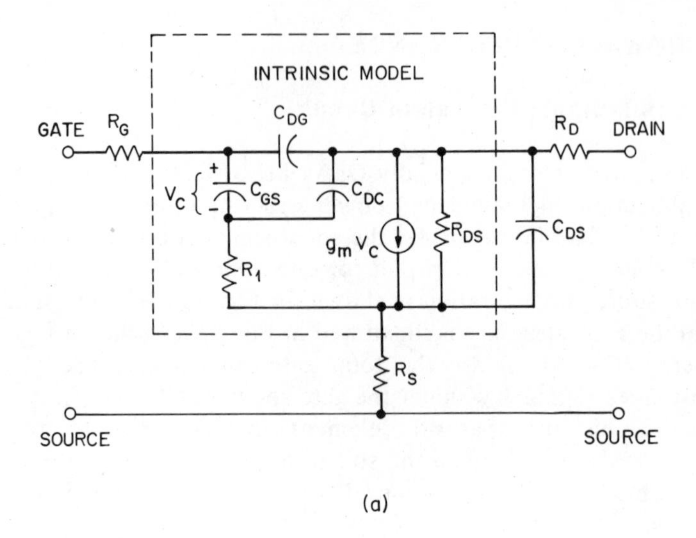

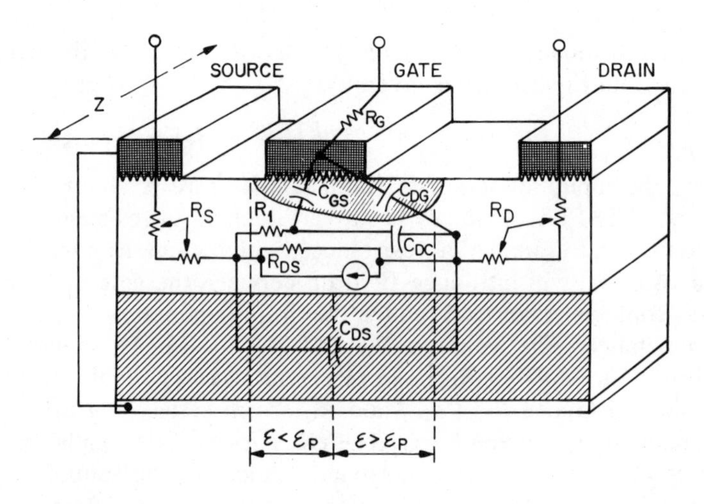

Fig. 20 (*a*) Equivalent circuit of a MESFET and (*b*) physical origin of the circuit elements. (After Liechti, Ref. 8.)

and

$$\tau = L/v_s \tag{59}$$

respectively. For a 1-μm gate length, the transit time in a GaAs FET is of the order of 10 ps (10^{-11} s). This transit time is usually small compared to the *RC* time constant resulting from the input gate capacitance and the transconductance.

A frequency f_T is defined as the frequency at which the current through C_{GS} is equal to that of the current generator $g_m V_C$ in the intrinsic FET:

$$f_T = \frac{g_m}{2\pi C_{GS}} \left(= \frac{1}{2\pi\tau} = \frac{v_s}{2\pi L} \right). \tag{60}$$

The maximum frequency of oscillation f_{max} is given by

$$f_{max} \simeq \frac{f_T}{2\sqrt{r_1 + f_T \tau_3}} \tag{61}$$

where r_1 is the input-to-output resistance ratio,

$$r_1 = (R_G + R_i + R_S)/R_{DS}$$

and τ_3 is a time constant,

$$\tau_3 \equiv 2\pi R_G C_{DG}.$$

The unilateral gain is given by

$$U \simeq (f_{max}/f)^2. \tag{62}$$

The unilateral gain will decrease at 6 dB/octave as the frequency increases. At f_{max}, unity power gain is reached. To maximize f_{max}, the frequency f_T and the resistance ratio R_i/R_{DS} must be optimized in the intrinsic MESFET. In addition, the extrinsic resistances R_G and R_S and the feedback capacitance C_{DG} must also be minimized.

6.4.2 Power-Frequency Limitations

The frequency limitations of FETs are dependent on device geometry and material parameters. In Si and GaAs, electrons have a higher mobility than holes have. Therefore, only n-channel FETs are used in microwave applications. Since the low-field mobility in GaAs is about five times higher than that of silicon, the frequency f_T is expected to be higher in GaAs.

In the device geometry, the most important parameter is the gate length L. Decreasing L will decrease the capacitance C_{GS} and increase the transconductance; consequently, f_T improves.

The theoretical f_T as a function of gate length is shown in Fig. 21 for Si, GaAs, and InP.[23,24] Note that InP is expected to have higher f_T than GaAs because of its high peak velocity.[25] For Si and GaAs MESFET with gate length 0.5 μm or less, the devices will have f_T in the millimeter-wave region ($\geq$30 GHz).

For a gate electrode to have adequate control of the current transport across the channel, the gate length must be larger than the channel depth,[24] that is $L/a > 1$. To reduce L and keep $L/a > 1$, the channel depth has to be reduced, which implies a higher doping level. In practical Si and GaAs FETs, the highest doping level is about 5×10^{17} cm^{-3} because of breakdown

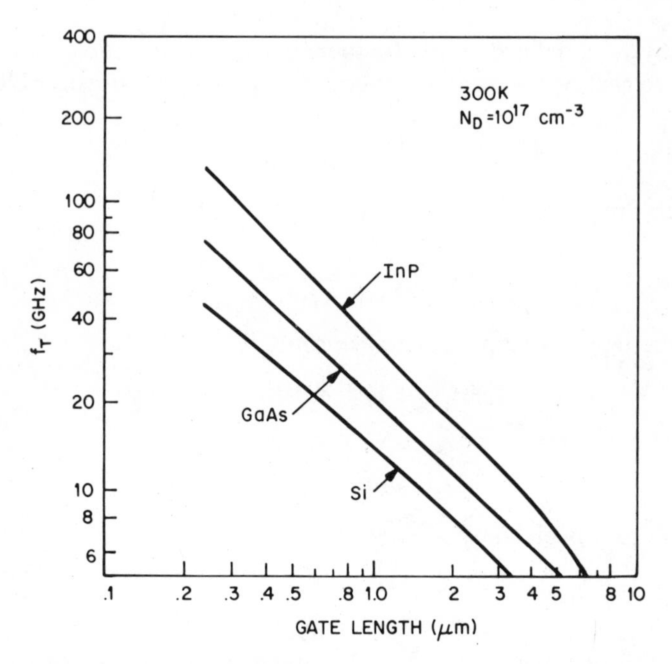

Fig. 21 Theoretical cutoff frequency as a function of gate length for Si, GaAs, and InP. (After Reiser and Wolf, Ref. 23; Maloney and Frey, Ref. 24.)

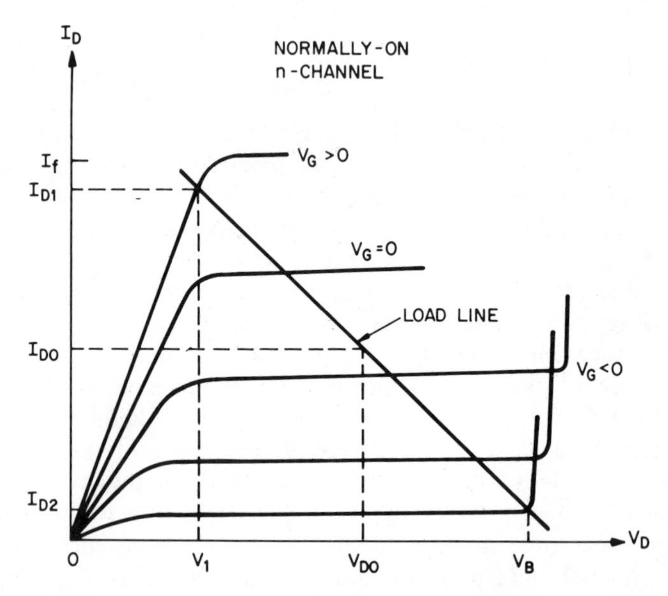

Fig. 22 *I–V* characteristic of a power MESFET; I_f is the drain current under forward gate bias, V_B is the breakdown voltage, and (I_{DO}, V_{DO}) is the operating biasing point. (After DiLorenzo and Wisseman, Ref. 48.)

phenomena. This doping level limits the minimum gate length to about 0.1 μm with a corresponding maximum f_T of the order of 100 GHz.

Figure 22 shows the I–V characteristics of a power MESFET. The maximum output voltage and current swings give an output power for a sinusoidal waveform:

$$P_{\text{out}} = \frac{(I_{D1} - I_{D2})(V_B - V_1)}{8} \tag{63}$$

where I_{D1} is close to I_f corresponding to the maximum channel current under the maximum allowed forward gate bias, and V_B is the avalanche breakdown voltage. To maximize the power, both I_f and V_B should be maximized. The maximum channel current[26] can be obtained by equating Eqs. 44 and 45 and setting $u_1 = 0$:

$$u_2^3 - 1.5u_2^2 - 1.5(u_2 - 1)/z = 0. \tag{64}$$

The solution to Eq. 64 gives the minimum channel depletion width at the drain $u_{\min}$. The maximum channel current is then given by Eq. 45:

$$I_m = I_S(1 - u_{\min}). \tag{65}$$

The calculated channel depth versus N_D is shown[26] in Fig. 23 for various L and I_m/Z. For a given L and I_m/Z, the channel depth varies approximately as $(N_D)^{-1}$. For example, for $L = 1\ \mu$m, $N_D = 10^{17}$, and $a = 0.2\ \mu$m, the maximum channel current per unit width is 3 A/cm.

To maximize V_B, one should choose low channel doping. However, for a given L and a, as N_D decreases, the maximum channel current also decreases. Therefore, a trade-off exists between high channel current and high breakdown voltage. For very small channel depths such that the charge per unit area in the channel Q_c (equal to $N_D a$) is less than $\sim 2 \times 10^{12}$ cm^{-2}, one must consider the two-dimensional effect of electric fields at the drain. This effect leads to a breakdown voltage given by[47]

$$V_B = 5 \times 10^{13}/Q_C(\text{cm}^{-2}) \qquad \text{V}. \tag{66}$$

This breakdown voltage can be substantially larger than the one-dimensional breakdown voltage of a p^+-n junction with doping N_D (refer to Chapter 2).

For high-power operation, the device temperature increases. This increase causes a reduction of the mobility and saturation velocity, since the mobility varies[27] as $[T(\text{K})]^{-2}$ and velocity as $[T(\text{K})]^{-1}$. Therefore, the FET has negative temperature coefficient and will be thermally stable under high-power operation. The state-of-the-art power-frequency performance[28] of GaAs MESFETs is shown in Fig. 24. The power varies as f^{-2}, similar to the electronic limitation discussed in Chapter 10.

With further miniaturation to submicron dimensions and with reduction of parasitics, FETs of higher powers operated at higher frequencies can be made.

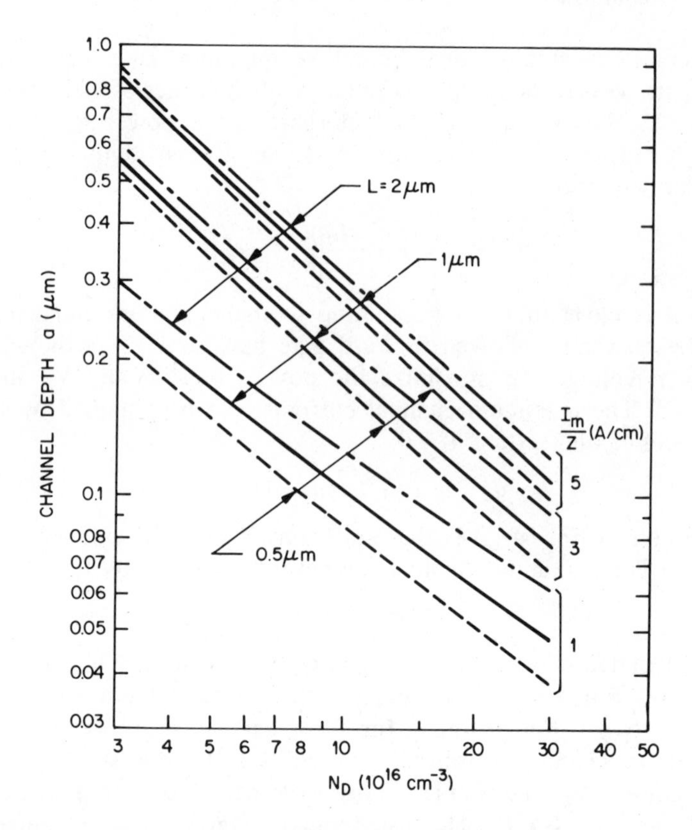

Fig. 23 Calculated channel depth as a function of channel doping for a GaAs MESFET with various gate lengths and maximum channel currents per unit gate width. (After Fukui, Ref. 26.)

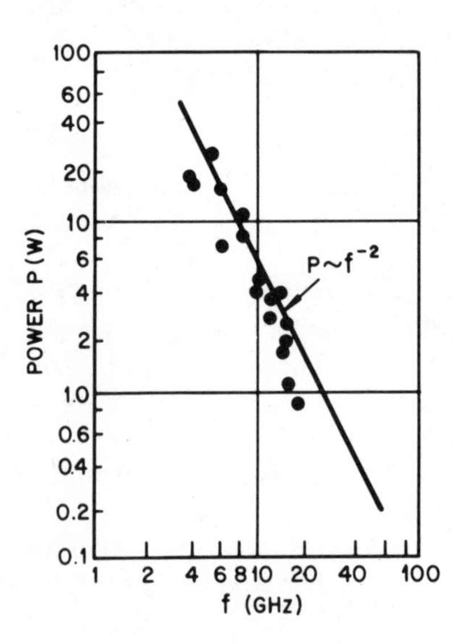

Fig. 24 State-of-the-art power output versus frequency for GaAs MESFET. (After Ikoma, Ref. 28.)

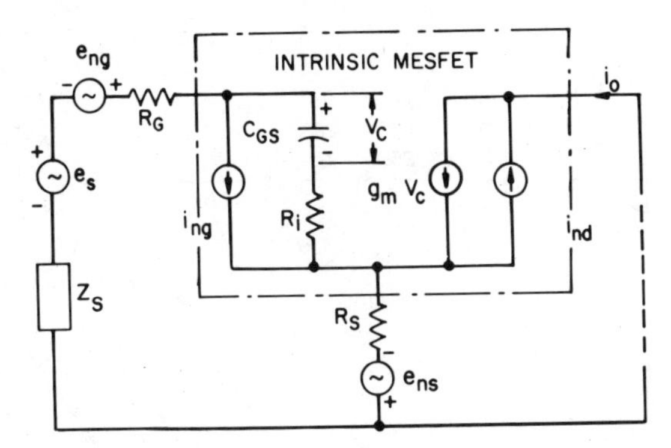

Fig. 25 Equivalent circuit for noise analysis. (After Pucel, Haus, and Statz, Ref. 7.)

6.4.3 Noise Behavior

The JFET and MESFET are low-noise devices, because only majority carriers participate in their operations. However, in practical devices, extrinsic resistances are unavoidable, and the parasitic resistances are mainly responsible for the noise behavior.

The equivalent circuit used for noise analysis is shown[7] in Fig. 25. Noise sources i_{ng}, i_{nd}, e_{ng}, and e_{ns} represent the induced gate noise, drain circuit noise, thermal noise of the gate metallization resistance R_G, and thermal noise of the source series resistance R_S, respectively. The e_s and Z_s are the signal source voltage and source impedance. The circuit within the semi-dashed lines corresponds to the intrinsic MESFET. The optimum noise figure for GaAs MESFET has been obtained[29] from the equivalent circuit:

$$F_0 \simeq 1 + fL\sqrt{g_m(R_S + R_G)}/4. \tag{67}$$

Clearly for low-noise performances, one should reduce the gate length and minimize the parasitic source and gate resistances. Calculated and measured[29,30] noise figures are shown in Fig. 26. At a given frequency, the noise decreases with decreasing gate length. The noise also decreases with decreasing channel width, because of the reduction of the gate resistance for narrower gate.

The graded channel FET[15] (Fig. 15) has been found to yield less noise than uniformly doped devices having the same geometry (1 to 3 dB lower in noise). This difference in noise is because the noise figure is related to g_m. The reduction of g_m (but not g_m/C_{GS}) for graded-channel FET gives superior noise performance.

6.4.4 Device Structures

The schematic diagrams of various power microwave MESFETs are shown[48] in Fig. 27. All devices have a semiinsulating layer as the substrate.

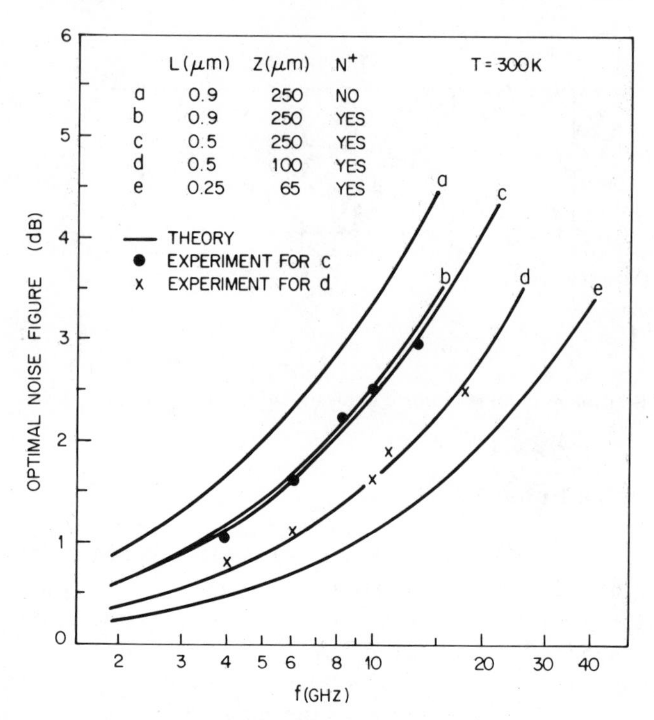

Fig. 26 Theoretical and experimental optional noise figure versus frequency for GaAs MESFETs with different channel lengths and channel widths. (After Fukui, Ref. 29; Hewitt et al., Ref. 30.)

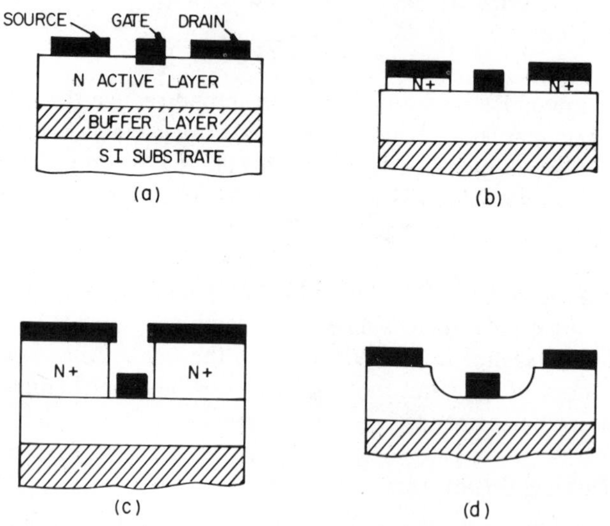

Fig. 27 Various source and drain contacts of GaAs MESFETs. (After DiLorenzo and Wisseman, Ref. 48.)

A buffer layer is epitaxially grown and followed by an active layer. The buffer layer serves to eliminate defects in the semiinsulating layer. Figure 27*a* shows the planar alloyed contacts. Since the ohmic contacts to the source and drain are important to the power performance and device reliability,[31] various approaches have been considered to reduce the contact resistance and increase the drain breakdown characteristic.[49] In Fig. 27*b*, an epitaxy n^+ layer is grown over the active n channel, and then the n^+ layer is selectively removed in the region between the source and the drain to form the gate. An extension of this method is shown in Fig. 27*c*, where a very thick n^+ layer is used under the source and drain, and a fabrication sequence utilizes an overhanging source and drain electrode to produce a self-aligned gate. A recessed contact structure is shown in Fig. 27*d*. By reducing the electric field at the drain contact, high breakdown voltages are obtained. Figure 28 shows a power

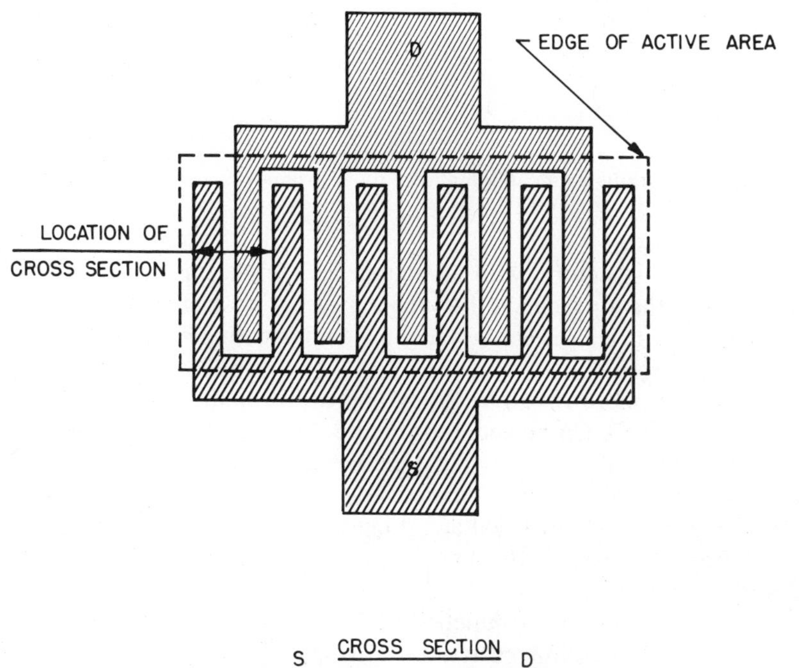

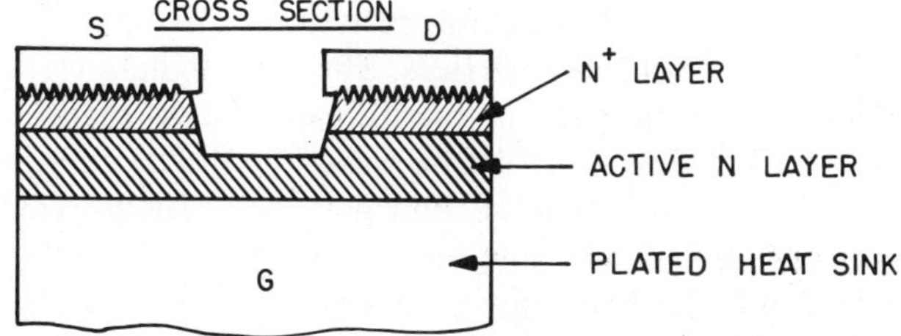

Fig. 28 Power MESFET with plated heat sink and interdigital source and drain fingers. (After Blocker, Macksey, and Adams, Ref. 32.)

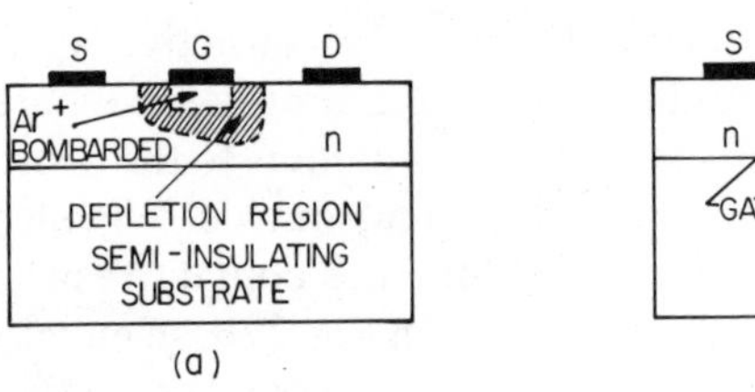

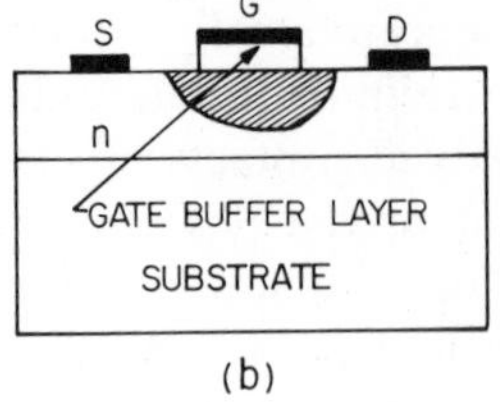

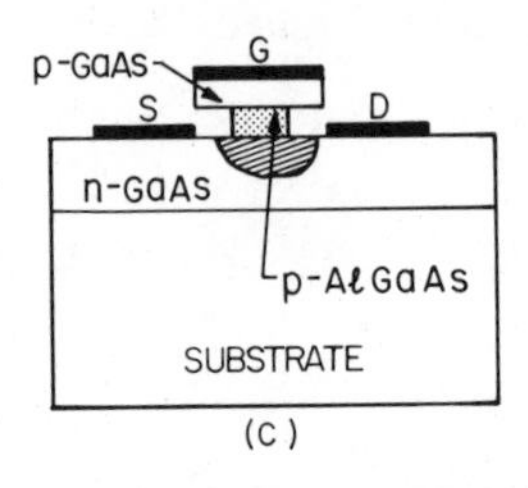

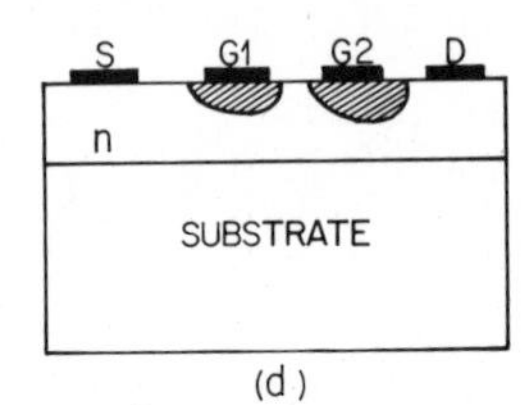

Fig. 29 Various gate configurations to improve device performances.

GaAs MESFET structure with a plated heat sink on the gate.[32] The metallization on the top side has interdigital source and drain fingers. Outside the active area, the GaAs is converted into semiinsulating material by proton bombardment.

To extend the microwave frequency and to lower the noise, many gate configurations have been considered. Figure 29*a* shows a structure with a semiinsulated gate fabricated by Ar^+ bombardment in the gate region.[33] The device can reduce the gate capacitance, lower the gate leakage current, and increase the gate breakdown voltage. Figure 29*b* shows a similar structure with a buffer-layer gate;[34] the buffer layer is inserted between the gate metal and the active layer.

Figure 29*c* shows a heterojunction JFET.[35] The leakage currents associated with heterojunction structures are appreciably lower than those obtained by Schottky barriers. The self-aligning technique has been used to achieve devices with submicron gate length. Figure 29*d* shows a dual-gate MESFET.[36] The second gate (near the drain) can improve the forward-to-reverse isolation. The second gate can also prevent the high-field region between the second gate and the drain from reaching back under the first gate. Both the heterojunction and gate structures are candidates for low-noise applications.

Double heterostructure MESFETs have been studied recently. Figure 30*a* shows a cross-sectional view of such a device with III–V ternary

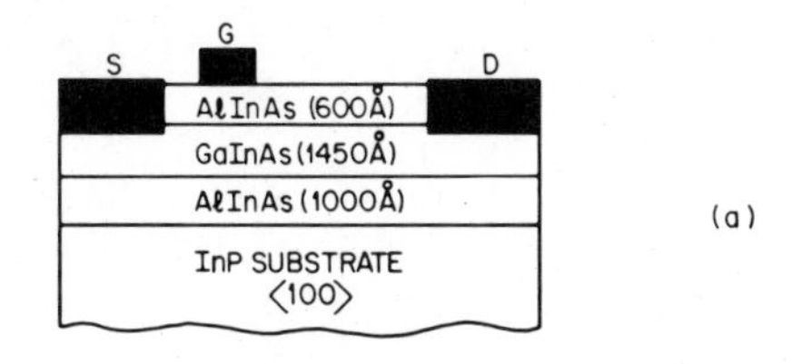

(a)

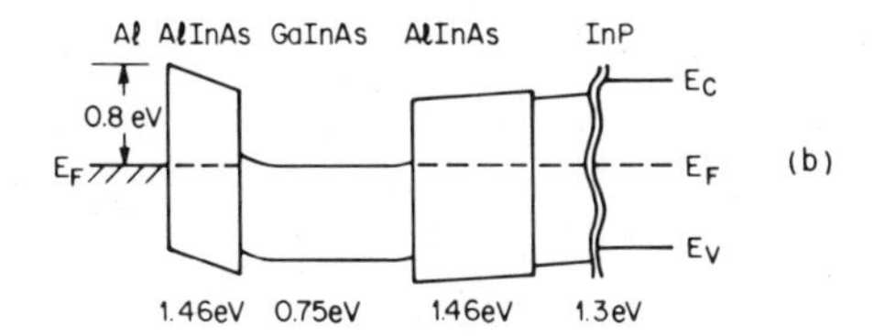

(b)

Fig. 30 (*a*) Cross-sectional view of a double heterostructure MESFET. (*b*) Energy-band diagram. (After Barnard et al., Ref. 50.)

compound $Ga_{0.47}In_{0.53}As$ as the active channel layer. The metal and semiconductor layers are grown successively on ⟨100⟩ InP semiinsulating substrates using molecular-beam epitaxy technique.[50] The semiconductor layers have good lattice match to the InP substrate, implying a low density of interface traps. Figure 30*b* shows the energy-band diagram at equilibrium. The top $Al_{0.48}In_{0.52}As$ layer forms a Schottky barrier with the aluminum gate ($\phi_{Bn} = 0.8$ V), so that electrons in the channel are confined in the active $Ga_{0.47}In_{0.53}As$ layer. By using materials such as $Ga_{0.47}In_{0.53}As$ with higher low-field mobility and higher peak velocity than GaAs, higher transconductances have been obtained. These MESFETs are potentially useful for high-speed applications.

6.5 RELATED FIELD-EFFECT DEVICES

6.5.1 Current Limiter

We shall consider two classes of current-regulator diodes that are two-terminal field-effect devices: the field-effect diode and the saturated-velocity diode.

The operation principle of a field-effect diode (FED) is the same as that of a shorted-gate-to-source junction field-effect transistor.[37,38] Its I–V characteristic is shown in Fig. 31, which is similar to the curve shown in Fig. 3 for $V_G = 0$. We shall consider four important parameters pertinent to the operation of current limiters: the limiting current I_l, the saturation voltage V_{sat}, the slope g_l in the limiting range, and the breakdown voltage. For an FED, these parameters have already been discussed in connection with the JFET. To reduce V_{sat}, one can use devices with a small channel depth a and a low channel-doping concentration (Eq. 4). To reduce I_l, one can reduce N_D and a or reduce the ratio Z/L (Eq. 9). To decrease g_l, one

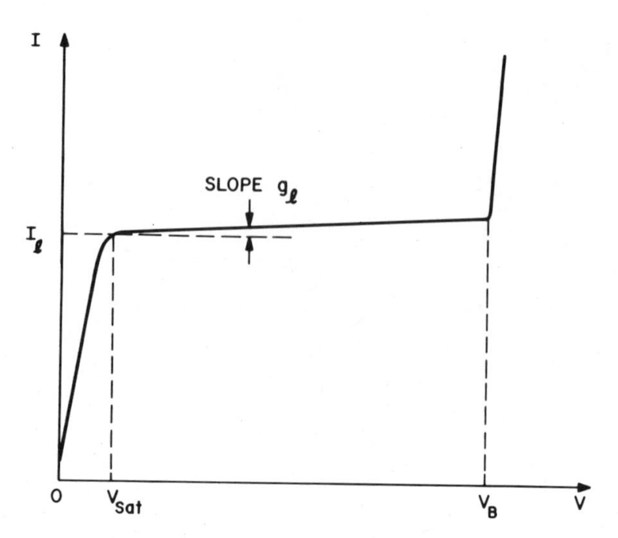

Fig. 31 Basic characteristics of a current limiter (field-effect diode).

must increase L. To increase V_B, a low channel-doping concentration should be used.

For the saturated-velocity diode, the current-limiting characteristic is obtained by employing the effect that at high electric fields the carrier drift velocity saturates.[39] A schematic geometry is shown in Fig. 32*a* where a *p*-type Ge substrate is used. The high-field region is a shallow *n*-type diffused layer about 0.5 μm deep and 3 μm long. The reason for choosing Ge for this diode is that its velocity saturation characteristics occur at low field (~4 kV/cm) in contrast to those in Si (~30 kV/cm).

The four important parameters as shown in Fig. 31 will now be discussed. The limiting current is given by

$$I_l = qN_D v_s A + I_S \tag{68}$$

where v_s is the saturation velocity, A the area, and I_S the *p-n* junction reverse-biased saturation current. The current I_S increases with increasing temperature while the velocity v_s decreases with temperature. A minimum in I_l is thus expected when these two competing mechanisms cancel.

The saturation voltage V_{sat} is given by

$$V_{sat} = \mathscr{E}_s L + I_l R_c \tag{69}$$

where $\mathscr{E}_s$ is the electric field at the onset of velocity saturation, and R_c is the residual resistance associated with the contacts. In an ideal limiter, V_{sat} is zero; in a practical limiter, V_{sat} should be as small as possible.

The slope g_l results from two effects: first, the electron drift velocity in the saturation range is not completely field independent, and second, there

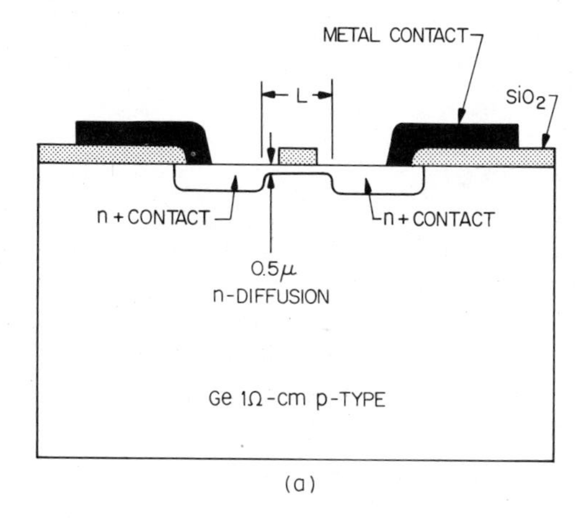

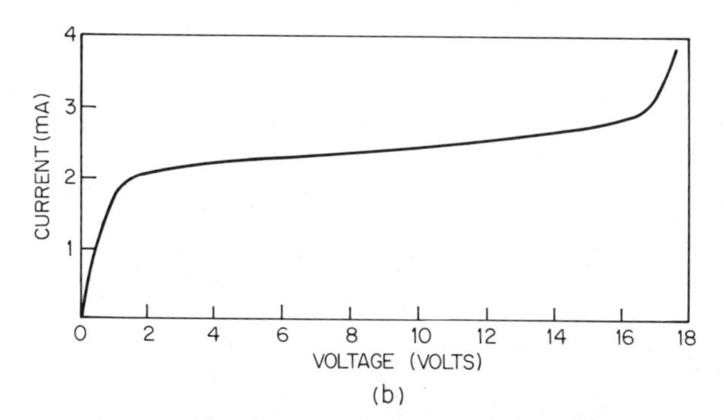

Fig. 32 (*a*) Schematic geometry of a current limiter (saturated-velocity diode). (*b*) Experimental I–V characteristics. (After Boll, Iwersen, and Perry, Ref. 39.)

is a space-charge-limited resistance due to injected carriers. This resistance is given by $(L^2/2\epsilon_s v_s A)$ as considered in Chapter 10.

The breakdown voltage depends on the impact ionization in the conduction channel. The electric field will increase linearly approximately from the negative contact to a value $\mathscr{E}_m$ at the positive contact. The breakdown voltage is given by

$$V_B \approx \mathscr{E}_m L/2 \tag{70}$$

where $\mathscr{E}_m/2$ is the average field in the channel. For Ge, $\mathscr{E}_m$ is about 1.5×10^5 V/cm at breakdown. The breakdown voltage is then expected to be about 20 V for a channel length of 3 μm.

The experimental results are shown in Fig. 32*b*. The saturated-velocity diode can be operated at higher speed than the field-effect diode, because for the same current level I_l, the depletion-layer width of the former can be made much greater than that of the corresponding FED so that the shunt-capacitance of a saturated-velocity diode is smaller than the input capacitance of the FED.

6.5.2 V-Groove FET

In the V-groove FET, a nonplanar structure, the nonplanar feature is used to optimize device characteristics.[40,41] The device can exhibit a higher transconductance and a lower turn-on resistance than the conventional planar FET. A cross section of a V-groove FET is shown in Fig. 33*a*. The device has three sets of V grooves that form the gate, drain and source, and isolation regions. The starting material is a ⟨100⟩-oriented *n*-type silicon epitaxial layer on *p*-type substrates. An anisotropic etching process is used to etch V-grooves in silicon. The etching process is self-stopping and the walls of the groove slope down from the horizontal at 54.7°. Another version of the V-groove FET is shown in Fig. 33*b* where a channel groove

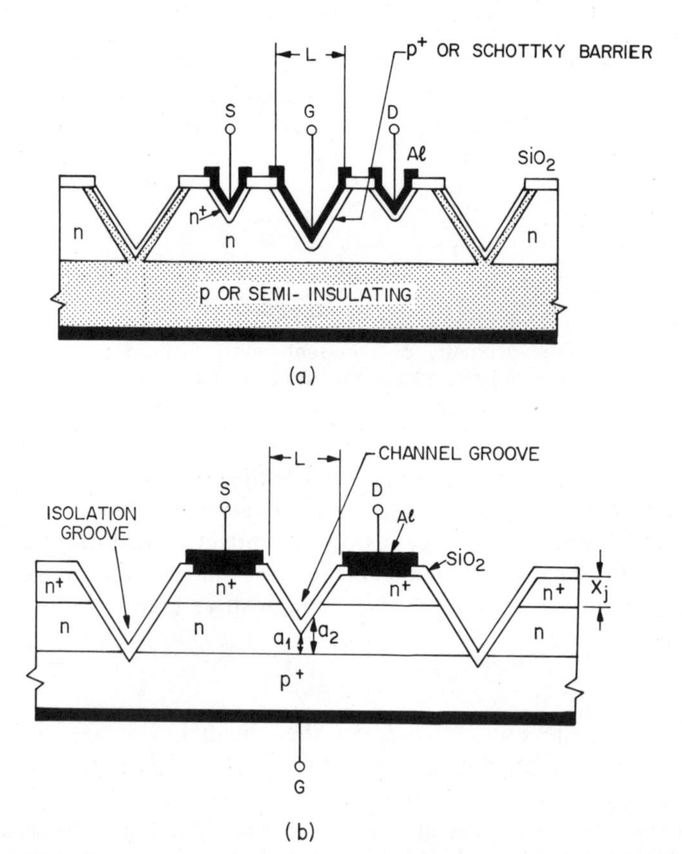

Fig. 33 (*a*) V-groove gate FET and (*b*) V-groove channel FET. (After Mok and Salama, Ref. 41.)

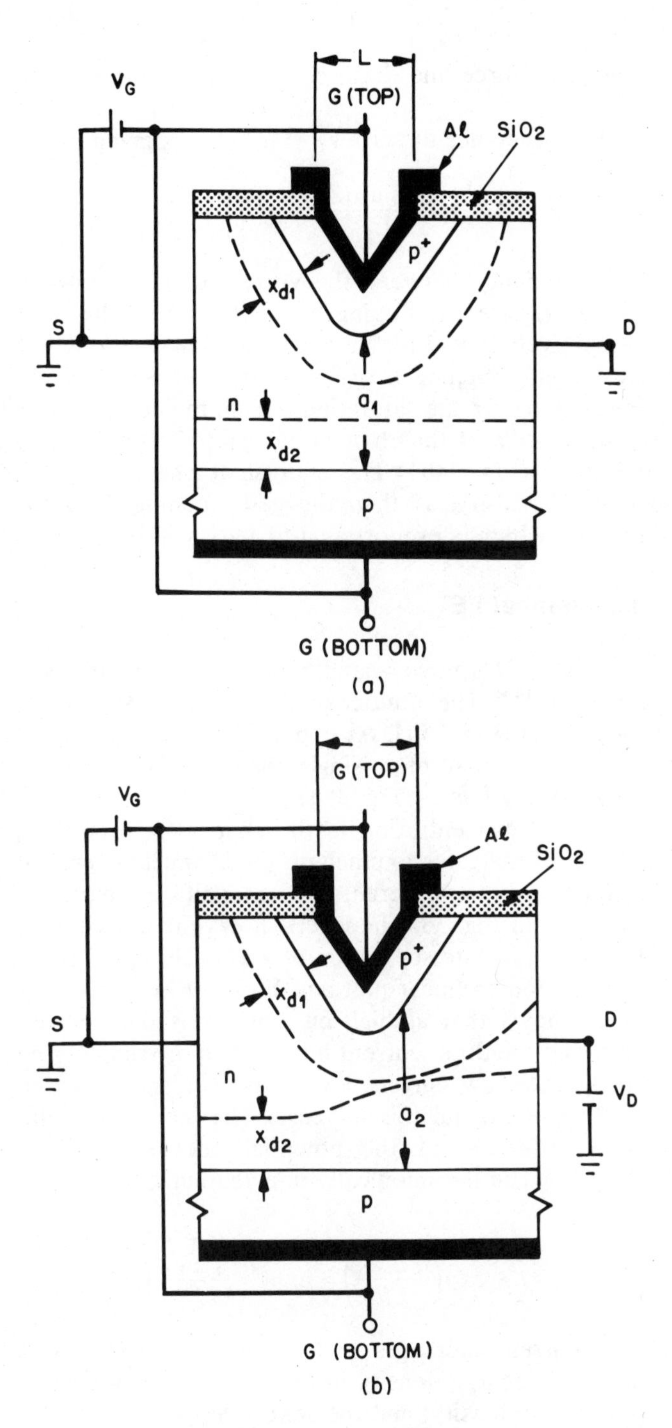

Fig. 34 Operation of V-groove FET for (*a*) $V_D = 0$ and (*b*) V_D in saturation region. (After Mok and Salama, Ref. 41.)

is etched, and the source and drain contacts are made to an n^+ epitaxial layer.

The channel conductance at zero V_D (Fig. 34a) is given by

$$g_D = \frac{qN_D\mu a_1 Z}{L_1}\left[1+\left(\frac{V_G}{V_P}\right)^{1/2}\right] \tag{71}$$

where a_1 is the distance between the vortex of the V-groove and the p region; N_A and N_D are the doping concentration in the n layer and p substrate, respectively, and L_1 is an equivalent channel length. Since L_1 is smaller than the gate opening L, g_D is larger than a planar FET with the same device geometry. In the saturation region, the device is pinched off at a point near the middle of the channel, instead of at the drain end as in a conventional FET. (Fig. 34b). The equivalent channel length L_2 in the saturation region is also less than the gate opening. Consequently, the transconductance, which is proportional to $1/L_2$, will be larger.

6.5.3 Multichannel FET

To improve the FET's power-handling capability, multichannel devices have been studied.[42–44] The multichannel FET is basically many JFETs connected in parallel (Fig. 35a). As expected, the current–voltage characteristic is similar[45] to that of a single-channel JFET, except the drain current is substantially higher (Fig. 35b).

When the doping concentration in the channel is low enough for the built-in potentials of the gates to pinch off the channel at zero gate bias, the device will then have a different current–voltage characteristic.[44] An example is shown[46] in Fig. 36. The insert shows the device with a channel doping of 10^{14} cm^{-3} and gate separation of 4 μm. The I_D–V_D characteristics are shown in Fig. 36a in linear plot and Fig. 36b in semilogarithmic plot. The linear plot shows that at high current levels the device has triode characteristics, and the drain current increases approximately linearly with the drain voltage. At low current levels, the current increases exponentially. Figure 37 can explain this increase. The incremental drain voltage ΔV_D causes a reduction of the potential between the gates, where $\Delta V_G \simeq (L_1/L_2)\Delta V_D$. The thermionically injected current from the source is given by

$$I \sim \exp\left(\frac{q\,\Delta V_G}{kT}\right) \simeq \exp\left(\frac{qL_1\,\Delta V_D}{kTL_2}\right). \tag{72}$$

Therefore, the current increases exponentially at first. As the current increases, the space-charge density in the channel regions becomes larger than the fixed-charge density, and the space-charge limited effect, causing the drain current to vary approximately linearly with drain voltage.

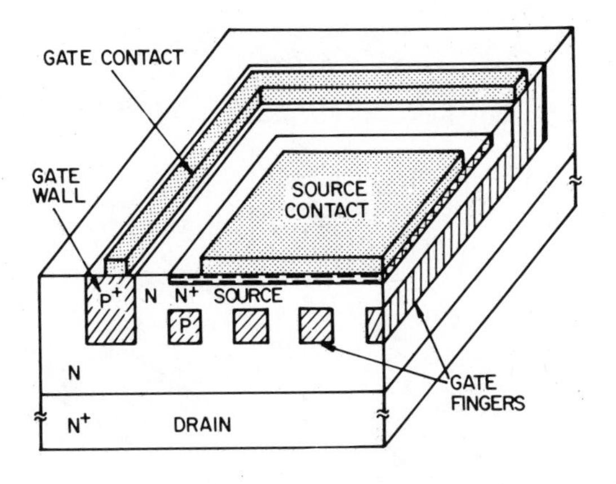

(a)

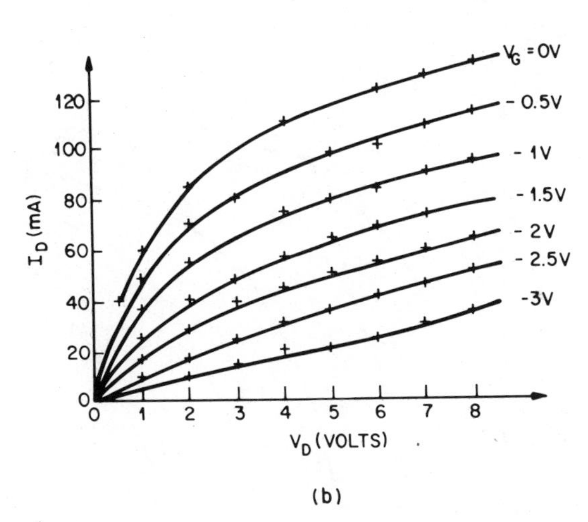

(b)

Fig. 35 (a) Isometric view showing plan and cross section of a multichannel FET and (b) experimental I–V characteristics for a device with $N_D = 10^{16}\ \text{cm}^{-3}$. (After Lecrosnier and Pelous, Ref. 45.)

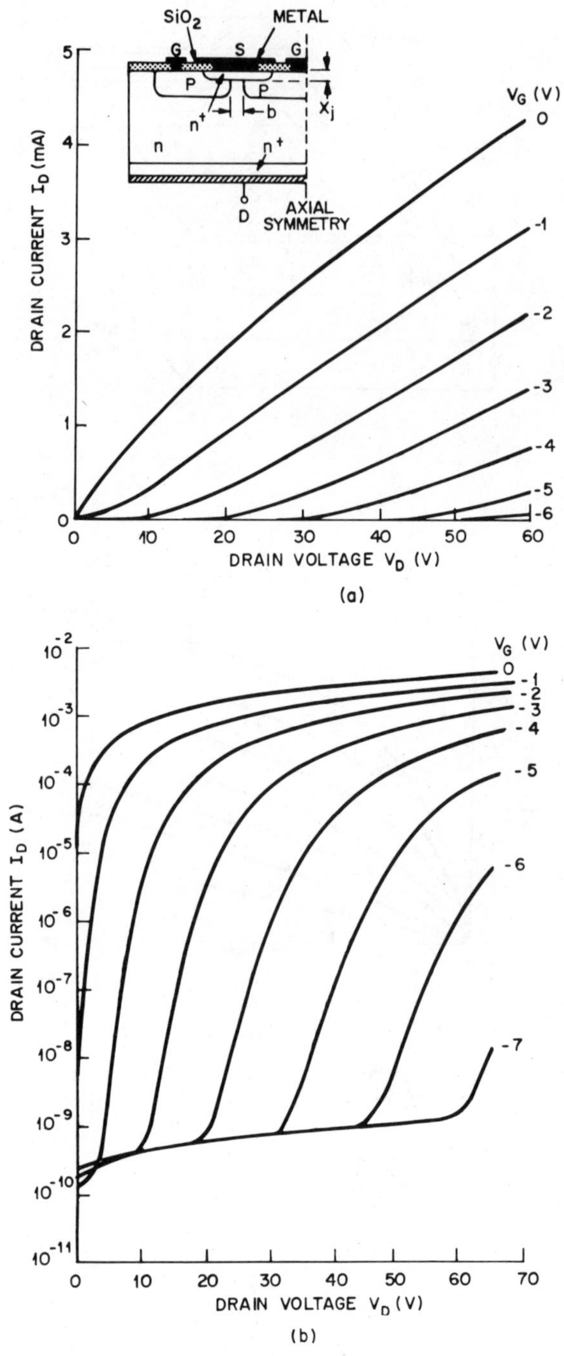

Fig. 36 Measured *I–V* characteristic of a multichannel FET with $N_D = 10^{14}\ \text{cm}^{-3}$: (*a*) linear plot and (*b*) semilogarithmic plot. The insert shows the device cross section. (After Morenza and Esteve, Ref. 46.)

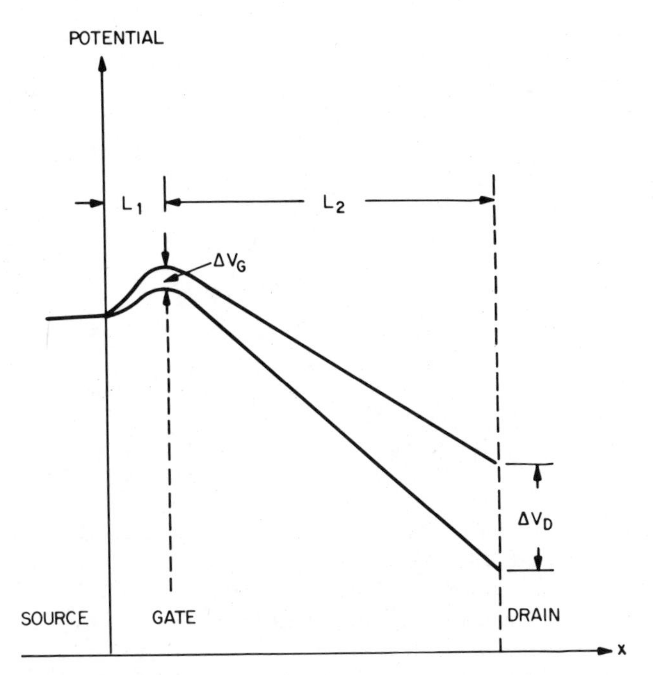

Fig. 37 Potential between the source and drain of a multichannel FET. The channel region is pinched off at $V_G = 0$.

REFERENCES

1 W. Shockley, "A Unipolar Field-Effect Transistor," *Proc. IRE*, **40**, 1365 (1952).

2 G. C. Dacey and I. M. Ross, "Unipolar Field-Effect Transistor," *Proc. IRE*, **41**, 970 (1953).

3 G. C. Dacey and I. M. Ross, "The Field-Effect Transistor," *Bell Syst. Tech. J.*, **34**, 1149 (1955).

4 C. A. Mead, "Schottky Barrier Gate Field-Effect Transistor," *Proc. IEEE*, **54**, 307 (1966).

5 W. W. Hooper and W. I. Lehrer, "An Epitaxial GaAs Field-Effect Transistor," *Proc. IEEE*, **55**, 1237 (1967).

6 J. R. Hauser, "Junction Field Effect Transistors," in R. M. Burger and R. P. Donovan, Eds., *Fundamental of Silicon Integrated Device Technology*, Vol. 2, Prentice-Hall, Englewood Cliffs, N.J., 1968, Chap. 3.

7 R. A. Pucel, H. A. Haus, and H. Statz, "Signal and Noise Properties of GaAs Microwave Field-Effect Transistors," in L. Martin, Ed., *Advances in Electronics and Electron Physics*, Vol. 38, Academic, New York, 1975, p. 195.

8 C. A. Liechti, "Microwave Field-Effect Transistors—1976," *IEEE Trans. Microwave Theory Tech.*, **MTT-24**, 279 (1976).

9 R. R. Bockemuehl, "Analysis of Field-Effect Transistors with Arbitrary Charge Distribution," *IEEE Trans. Electron Devices*, **ED-10**, 31 (1963).

10 R. D. Middlebrook and I. Richer, "Limits on the Power-Law Exponent for Field-Effect Transistor Transfer Characteristics," *Solid State Electron.*, **6**, 542 (1963).

11 R. Zuleeg, J. K. Notthoff, and K. Lehovec, "Femtojoule High-Speed Planar GaAs E-JFET Logic," *IEEE Trans. Electron Devices*, **ED-25**, 628 (1978).

12 C. Jacoboni, C. Canali, G. Ottaviani, and A. A. Quaranta, "A Review of Some Charge Transport Properties of Silicon," *Solid State Electron.*, **20**, 77 (1977).

13 P. Smith, M. Inoue, and J. Frey, "Electron Velocity in Si and GaAs at Very High Electric Fields," *Appl. Phys. Lett.*, **37**, 797 (1980).

14 K. Lehovec and R. Zuleeg, "Voltage–Current Characteristics of GaAs JFETs in the Hot Electron Range," *Solid State Electron.*, **13**, 1415 (1970).

15 R. E. Williams and D. W. Shaw, "Graded Channel FET's Improved Linearity and Noise Figure," *IEEE Trans. Electron Devices*, **ED-25**, 600 (1978).

16 R. E. Williams and D. W. Shaw, "GaAs FETs with Graded Channel Doping Profiles," *Electron. Lett.*, **13**, 408 (1977).

17 R. A. Pucel, "Profile Design for Distortion Reduction in Microwave Field-Effect Transistors," *Electron. Lett.* **14**, 204 (1978).

18 D. P. Kennedy and R. R. O'Brien, "Computer Aided Two-Dimensional Analysis of the Junction Field-Effect Transistor," *IBM J. Res. Dev.* **14**, 95 (1970).

19 K. Lehovec and R. Miller, "Field Distribution in Junction Field Effect Transistors at Large Drain Voltages," *IEEE Trans. Electron Devices*, **ED-22**, 273 (1975).

20 B. Himsworth, "A Two-Dimensional Analysis of GaAs Junction Field-Effect Transistors with Long and Short Channels," *Solid State Electron.*, **15**, 1353 (1972).

21 J. Ruch, "Electron Dynamics in Short Channel Field Effect Transistors," *IEEE Trans. Electron Devices*, **ED-19**, 652 (1972).

22 T. Wada and J. Frey, "Physical Basis of Short-Channel MESFET Operation," *IEEE Trans. Electron Devices*, **ED-26**, 476 (1979).

23 M. Reiser and P. Wolf, "Computer Study of Submicrometer FETs," *Electron. Lett.* **8**, 254 (1972).

24 T. J. Maloney and J. Frey, "Frequency Limits of GaAs and InP Field-Effect Transistors at 300 K and 77 K with Typical Active Layer Doping," *IEEE Trans. Electron Devices*, **ED-23**, 519 (1976).

25 J. S. Barrera and R. J. Archer, "InP Schottky-Gate Field-Effect Transistors," *IEEE Trans. Electron Devices*, **ED-22**, 1023 (1975).

26 H. Fukui, "Drain Current Limitations and Temperature Effects in GaAs MESFET." *IEEE Tech. Dig.*, Int. Electron Device Meet. 1978, pp. 140–143.

27 L. J. Sevin, *Field Effect Transistors*, McGraw-Hill, New York, 1965.

28 T. Ikoma, "Status of Microwave and High Speed Devices," Proc. 7th Bienn. Cornell Electr. Eng. Conf. Active Microwave Semicond. Devices and Circuits, Cornell University, Ithaca, N.Y., p. 7 (1979).

29 H. Fukui, "Optimal Noise Figure of Microwave GaAs MESFETs," *IEEE Trans. Electron Devices*, **ED-26**, 1032 (1979).

30 B. S. Hewitt, H. M. Cox, H. Fukui, J. V. DiLorenzo, W. O. Schlosser, and D. E. Iglesias, "Low Noise GaAs MESFETS," *Electron. Lett.*, **12**, 309 (1976).

31 K. Mizuishi, H. Kurono, H. Sato, and H. Kodera, "Degradation Mechanics of GaAs MESFETs," *IEEE Trans. Electron Devices*, **ED-26**, 1008 (1979).

32 T. Blocker, H. Macksey, and R. Adams, "X-Band RF Power Performance of GaAs FET's," *IEEE Tech. Dig.*, Int. Electron Device Meet., 1974, pp. 288–291.

33 H. M. Macksey, D. W. Shaw, and W. R. Wisseman, "GaAs Power FETs with Semi-Insulating Gates," *Electron. Lett.*, **12**, 192 (1976).

34 A. Nagashima, S. Umebachi, and G. Kano, "Calculation of Microwave Performance of Buffer Layer Gate GaAs MESFETs," *IEEE Trans. Electron Devices*, **ED-25**, 537 (1978).

35 S. Umebachi, K. Ashahi, M. Inoue, and G. Kano, "A New Heterojunction Gate GaAs FET," *IEEE Trans. Electron Devices*, **ED-12**, 613 (1975).

36 J. A. Turner, A. J. Waller, E. Kelly, and D. Parker, "Dual-Gate GaAs Microwave Field-Effect Transistor," *Electron. Lett.*, **7**, 661 (1971).

37 R. M. Warner, W. H. Jackson, E. I. Doucette, and H. A. Sone, "A Semiconductor Current Limiter," *Proc. IRE*, **47**, 45 (1959).

38 H. Lawrence, "A Diffused Field-Effect Current Limiter," *IRE Trans. Electron Devices*, **ED-9**, 82 (1962).

39 H. J. Boll, J. E. Iwersen, and E. W. Perry, "High-Speed Current Limiters," *IEEE Trans. Electron Devices*, **ED-13**, 904 (1966).

40 C. A. Salama and J. G. Oakes, "Nonplanar Power Field-Effect Transistors," *IEEE Trans. Electron Devices*, **ED-25**, 1222 (1978).

41 T. K. Mok and C. A. T. Salama, "The Characteristics and Applications of a V-Shaped Notched Channel Field-Effect Transistor (VFET)," *Solid State Electron.*, **19**, 159 (1976).

42 S. Teszner and R. Gicquel, "Gridistor—A New Field Effect Device," *Proc. IEEE*, **52**, 1502 (1964).

43 R. Zuleeg, "Multichannel Field-Effect Transistor Theory and Experiment," *Solid State Electron.*, **10**, 559 (1967).

44 J. I. Nishizawa, T. Terasaki, and J. Shibata, "Field Effect Transistor Versus Analog Transistor (Static Induction Transistor)," *IEEE Trans. Electron Devices*, **ED-22**, 185 (1975).

45 D. R. Lecrosnier and G. P. Pelous, "Ion Implanted FET for Power Applications," *IEEE Trans. Electron Devices*, **ED-21**, 113 (1974).

46 J. L. Morenza and D. Esteve, "Entirely Diffused Vertical Channel JFET: Theory and Experiment," *Solid State Electron.* **21**, 739 (1978).

47 S. H. Wemple, W. C. Niehaus, H. M. Cox, J. V. DiLorenzo, and W. O. Schlosser, "Control of Gate–Drain Avalanche in GaAs MESFETs," *IEEE Trans. Electron Devices*, **ED-27**, 1013 (1980).

48 J. V. DiLorenzo and W. R. Wisseman, "GaAs Power MESFET: Design, Fabrication and Performance," *IEEE Trans. Microwave Theory Tech.*, **MTT-27**, 367 (1979).

49 M. Fukuta, K. Syama, H. Suzuki, K. Nakayama, and H. Ishidawa, "Power GaAs MESFET with High Drain–Source Breakdown Voltage," *IEEE Trans. Microwave Theory Tech.*, **MTT-24**, 312 (1976).

50 J. Barnard, H. Ohno, C. E. C. Wood, and L. F. Eastman, "Double Heterostructure $Ga_{0.47}In_{0.53}As$ MESFETs with Submicron Gates," *IEEE Electron Devices, Lett.*, **EDL-1,** *174* (1980).

7

MIS Diode and Charge-Coupled Device

- INTRODUCTION
- IDEAL MIS DIODE
- $Si–SiO_2$ MOS DIODE
- CHARGE-COUPLED DEVICE

7.1 INTRODUCTION

The metal–insulator–semiconductor (MIS) diode is the most useful device in the study of semiconductor surfaces. Since the reliability and stability of all semiconductor devices are intimately related to their surface conditions, an understanding of the surface physics with the help of MIS diodes is of great importance to device operations. In this chapter we are concerned primarily with the metal–oxide–silicon (MOS) system. This system has been extensively studied because it is directly related to most planar devices and integrated circuits.

The MIS structure was first proposed as a voltage-variable capacitor in 1959 by Moll[1] and by Pfann and Garrett.[2] Its characteristics were then analyzed by Frankl[3] and Lindner.[4] The MIS diode was first employed in the study of a thermally oxidized silicon surface by Terman[5] and by Lehovec and Slobodskoy.[6] A comprehensive and in-depth treatment of the $Si–SiO_2$ MOS diode can be found in *MOS Physics and Technology* by Nicollian and Brews.[7]

The charge-coupling principle was first presented by Boyle and Smith[8] in 1970. Amelio, Tompsett, and Smith demonstrated the first charge-coupled device (CCD).[9] The CCD in its simplest form is an array of closely spaced MOS diodes. Under the application of a proper sequence of clock voltage pulses, a CCD can move quantities of electrical charge in a controlled

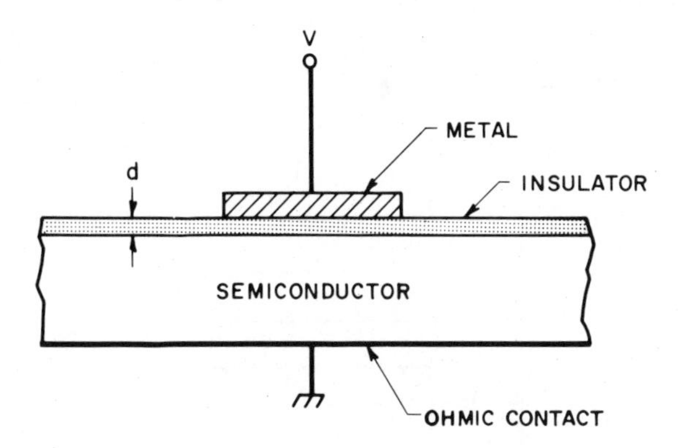

Fig. 1 Metal–insulator–semiconductor (MIS) diode.

manner across a semiconductor substrate. Using this basic mechanism, CCD can perform a wide range of electronic functions, including image sensing, data storage, signal processing, and logic operations. Sequin and Tompsett[10] and Kim[11] have given comprehensive reviews on CCD physics and the IEEE press has published a compilation of important CCD papers.[12]

7.2 IDEAL MIS DIODE

The metal–insulator–semiconductor (MIS) structure is shown in Fig. 1, where d is the thickness of the insulator and V is the applied voltage on the metal field plate. Throughout this chapter we use the convention that the voltage V is positive when the metal plate is positively biased with respect to the ohmic contact, and V is negative when the metal plate is negatively biased with respect to the ohmic contact.

The energy-band diagram of an ideal MIS structure for $V = 0$ is shown in Fig. 2, where Figs. 2a and b are for n-type and p-type semiconductors, respectively. An ideal MIS diode is defined as follows. (1) At zero applied bias, energy difference between the metal work function ϕ_m and the semiconductor work function is zero, or the work-function difference ϕ_{ms} is zero:

$$\phi_{ms} \equiv \phi_m - \left(\chi + \frac{E_g}{2q} - \psi_B\right) = 0 \qquad \text{for } n\text{-type} \tag{1a}$$

$$\phi_{ms} \equiv \phi_m - \left(\chi + \frac{E_g}{2q} + \psi_B\right) = 0 \qquad \text{for } p\text{-type} \tag{1b}$$

where ϕ_m is the metal work function, χ the semiconductor electron affinity, χ_i the insulator electron affinity, E_g the bandgap, ϕ_B the potential barrier

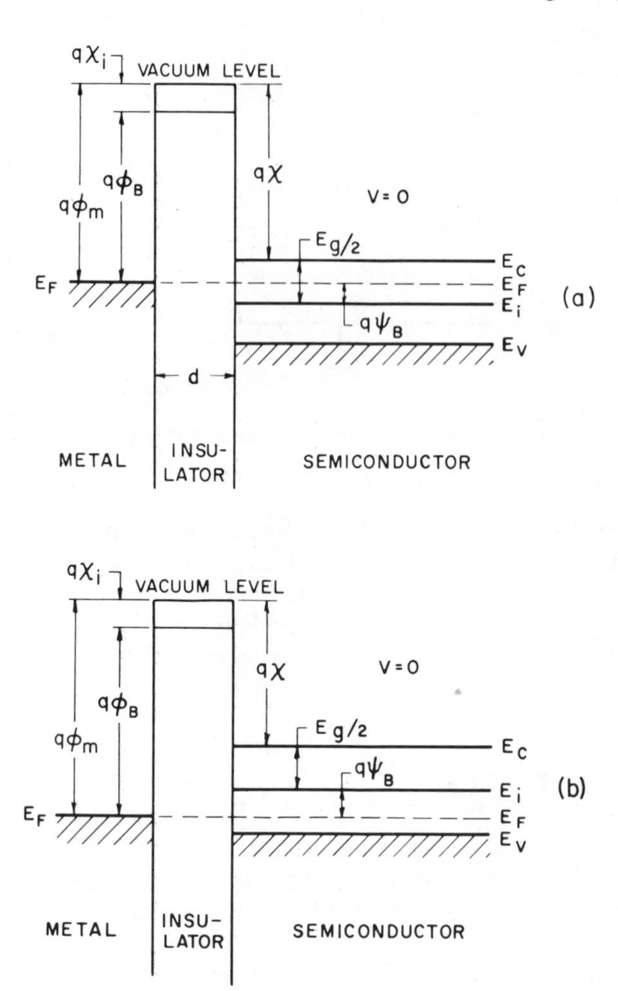

Fig. 2 Energy-band diagrams of ideal MIS diodes at $V = 0$. (*a*) *n*-type semiconductor. (*b*) *p*-type semiconductor.

between the metal and the insulator, and ψ_B the potential difference between the Fermi level E_F and the intrinsic Fermi level E_i. In other words, the band is flat (flat-band condition) when there is no applied voltage. (2) The only charges that can exist in the structure under any biasing conditions are those in the semiconductor and those with the equal but opposite sign on the metal surface adjacent to the insulator. (3) There is no carrier transport through the insulator under dc biasing conditions, or the resistivity of the insulator is infinity. The ideal MIS diode theory to be considered in this section serves as a foundation for understanding practical MIS structures and to exploring the physics of semiconductor surfaces.

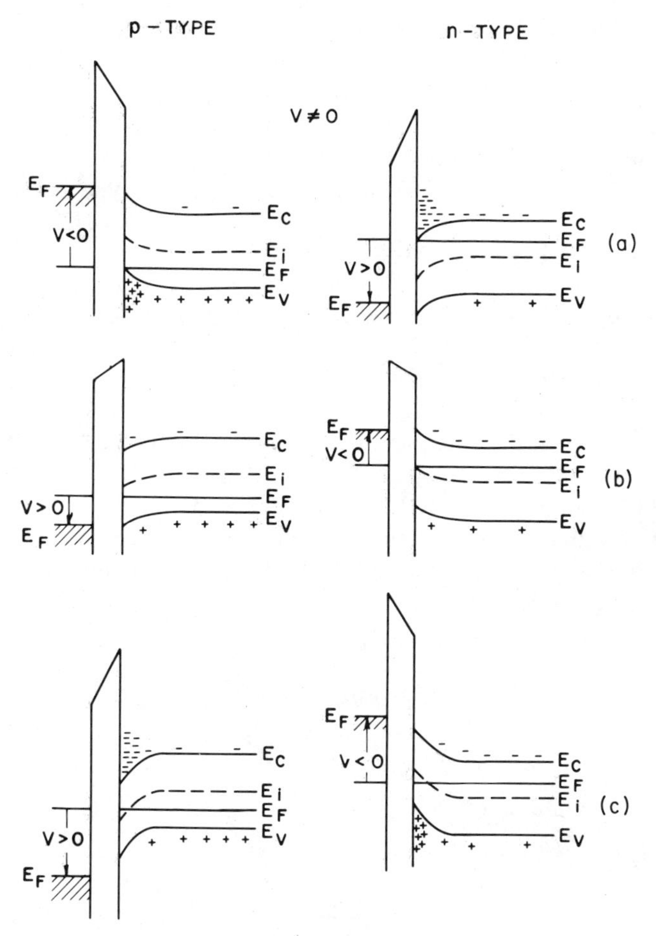

Fig. 3 Energy-band diagrams for ideal MIS diodes when $V \neq 0$, for the following cases: (*a*) accumulation; (*b*) depletion; (*c*) inversion.

When an ideal MIS diode is biased with positive or negative voltages, basically three cases may exist at the semiconductor surface (Fig. 3). Consider the *p*-type semiconductor first. When a negative voltage ($V < 0$) is applied to the metal plate, the top of the valence band bends upward and is closer to the Fermi level (Fig. 3*a*). For an ideal MIS diode, no current flows in the structure [or $d(\text{Imref})/dx = 0$], so the Fermi level remains constant in the semiconductor. Since the carrier density depends exponentially on the energy difference ($E_F - E_V$), this band bending causes an accumulation of majority carriers (holes) near the semiconductor surface. This is the "accumulation" case. When a small positive voltage ($V > 0$) is applied, the bands bend downward, and the majority carriers are depleted (Fig. 3*b*). This is the "depletion" case. When a larger positive voltage is applied, the bands bend even more downward so that the intrinsic level E_i at the surface crosses over the Fermi level E_F (Fig. 3*c*). At this point the number of electrons (minority carriers) at the surface is larger than that of the holes, the surface is thus inverted, and this is the "inversion" case. Similar results can be obtained for the *n*-type semiconductor. The polarity of the voltage, however, should be changed for the *n*-type semiconductor.

7.2.1 Surface Space-Charge Region

In this subsection we derive the relations between the surface potential, space charge, and electric field. These relations are then used to derive the capacitance–voltage characteristics of the ideal MIS structure in Section 7.2.2.

Figure 4 shows a more detailed band diagram at the surface of a *p*-type semiconductor. The potential ψ is defined as zero in the bulk of the semiconductor and is measured with respect to the intrinsic Fermi level E_i as shown. At the semiconductor surface, $\psi = \psi_s$, and ψ_s is called the surface potential. The electron and hole concentrations as a function of ψ are given by the following relations:

$$n_p = n_{po} \exp(q\psi/kT) = n_{po} \exp(\beta\psi) \tag{2}$$

$$p_p = p_{po} \exp(-q\psi/kT) = p_{po} \exp(-\beta\psi) \tag{3}$$

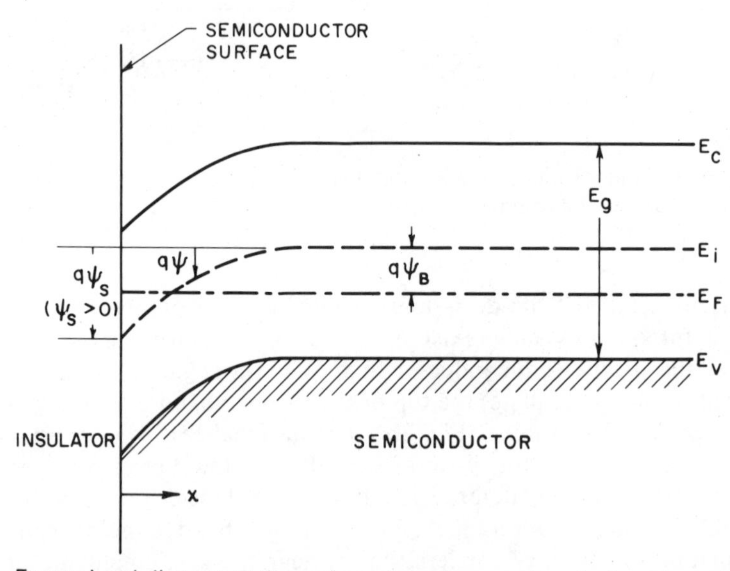

Fig. 4 Energy-band diagram at the surface of a *p*-type semiconductor. The potential ψ, defined as zero in the bulk, is measured with respect to the intrinsic Fermi level E_i. The surface potential ψ_s is positive as shown. (*a*) Accumulation occurs when $\psi_s < 0$. (*b*) Depletion occurs when $\psi_B > \psi_s > 0$. (*c*) Inversion occurs when $\psi_s > \psi_B$.

where ψ is positive when the band is bent downward (as shown in Fig. 4), n_{po} and p_{po} are the equilibrium densities of electrons and holes, respectively, in the bulk of the semiconductor, and $\beta \equiv q/kT$. At the surface the densities are

$$n_s = n_{po} \exp(\beta\psi_s)$$
$$p_s = p_{po} \exp(-\beta\psi_s). \tag{4}$$

From previous discussions and with the help of Eq. 4, the following regions of surface potential can be distinguished:

$\psi_s < 0$	accumulation of holes (bands bend upward)
$\psi_s = 0$	flat-band condition
$\psi_B > \psi_s > 0$	depletion of holes (bands bend downward)
$\psi_s = \psi_B$	midgap with $n_s = p_s = n_i$ (intrinsic concentration)
$\psi_s > \psi_B$	inversion (electron enhancement, bands bend downward)

The potential ψ as a function of distance can be obtained by using the one-dimensional Poisson equation

$$\frac{d^2\psi}{dx^2} = -\frac{\rho(x)}{\epsilon_s} \tag{5}$$

where ϵ_s is the permittivity of the semiconductor and $\rho(x)$ is the total space-charge density given by

$$\rho(x) = q(N_D^+ - N_A^- + p_p - n_p) \tag{6}$$

where N_D^+ and N_A^- are the densities of the ionized donors and acceptors, respectively. Now, in the bulk of the semiconductor, far from the surface, charge neutrality must exist. Therefore, $\rho(x) = 0$ and $\psi = 0$, and we have

$$N_D^+ - N_A^- = n_{po} - p_{po}. \tag{7}$$

In general, for any value of ψ we have from Eqs. 2 and 3

$$p_p - n_p = p_{po} \exp(-\beta\psi) - n_{po} \exp(\beta\psi). \tag{8}$$

The resultant Poisson's equation to be solved is therefore

$$\frac{\partial^2\psi}{\partial x^2} = -\frac{q}{\epsilon_s}[p_{po}(e^{-\beta\psi} - 1) - n_{po}(e^{\beta\psi} - 1)]. \tag{9}$$

Integrating Eq. 9 from the bulk toward the surface[13]

$$\int_0^{\partial\psi/\partial x} \left(\frac{\partial\psi}{\partial x}\right) d\left(\frac{\partial\psi}{\partial x}\right) = -\frac{q}{\epsilon_s}\int_0^{\psi} [p_{po}(e^{-\beta\psi} - 1) - n_{po}(e^{\beta\psi} - 1)]\, d\psi \tag{10}$$

gives the relation between the electric field ($\mathscr{E} \equiv -d\psi/dx$) and the potential ψ:

$$\mathscr{E}^2 = \left(\frac{2kT}{q}\right)^2 \left(\frac{qp_{po}\beta}{2\epsilon_s}\right)\left[(e^{-\beta\psi} + \beta\psi - 1) + \frac{n_{po}}{p_{po}}(e^{\beta\psi} - \beta\psi - 1)\right]. \tag{11}$$

We shall introduce the following abbreviations:

$$L_D \equiv \sqrt{\frac{kT\epsilon_s}{p_{po}q^2}} \equiv \sqrt{\frac{\epsilon_s}{qp_{po}\beta}} \tag{12}$$

and

$$F\left(\beta\psi, \frac{n_{po}}{p_{po}}\right) \equiv \left[(e^{-\beta\psi} + \beta\psi - 1) + \frac{n_{po}}{p_{po}}(e^{\beta\psi} - \beta\psi - 1)\right]^{1/2} \geq 0 \tag{13}$$

where L_D is called the extrinsic Debye length for holes. Thus the electric field becomes

$$\mathscr{E} = -\frac{\partial\psi}{\partial x} = \pm\frac{\sqrt{2}\,kT}{qL_D}\,F\left(\beta\psi, \frac{n_{po}}{p_{po}}\right) \tag{14}$$

with positive sign for $\psi > 0$ and negative sign for $\psi < 0$. To determine the electric field at the surface, we let $\psi = \psi_s$:

$$\mathscr{E}_s = \pm\frac{\sqrt{2}\,kT}{qL_D}\,F\left(\beta\psi_s, \frac{n_{po}}{p_{po}}\right). \tag{15}$$

Similarly, by Gauss's law the space charge per unit area required to produce this field is

$$Q_s = -\epsilon_s\mathscr{E}_s = \mp\frac{\sqrt{2}\epsilon_s kT}{qL_D}\,F\left(\beta\psi_s, \frac{n_{po}}{p_{po}}\right). \tag{16}$$

To determine the change in hole density, Δp, and electron density, Δn, per unit area when the ψ at the surface is shifted from zero to a final value ψ_s, it is necessary to evaluate the following expressions:[14]

$$\begin{aligned}\Delta p &= p_{po}\int_0^\infty (e^{-\beta\psi} - 1)\,dx \\ &= \frac{qp_{po}L_D}{\sqrt{2}\,kT}\int_{\psi_s}^0 \frac{e^{-\beta\psi} - 1}{F(\beta\psi, n_{po}/p_{po})}\,d\psi \qquad \text{cm}^{-2}\end{aligned} \tag{17}$$

$$\begin{aligned}\Delta n &= n_{po}\int_0^\infty (e^{\beta\psi} - 1)\,dx \\ &= \frac{qn_{po}L_D}{\sqrt{2}\,kT}\int_{\psi_s}^0 \frac{e^{\beta\psi} - 1}{F(\beta\psi, n_{po}/p_{po})}\,d\psi \qquad \text{cm}^{-2}.\end{aligned} \tag{18}$$

A typical variation of the space-charge density Q_s as a function of the surface potential ψ_s is shown in Fig. 5 for a p-type silicon with $N_A = 4\times10^{15}\ \text{cm}^{-3}$ at room temperature. Note that for negative ψ_s, Q_s is positive and corresponds to the accumulation region. The function F is dominated by the first term in Eq. 13, that is, $Q_s \sim \exp(q|\psi_s|/2kT)$. For $\psi_s = 0$, we have the flat-band condition and $Q_s = 0$. For $\psi_B > \psi_s > 0$, Q_s is negative and we have the depletion case. The function F is now dominated by the second term, that is, $Q_s \sim \sqrt{\psi_s}$. For $\psi_s \gg \psi_B$, we have the inversion case with the function F dominated by the fourth term, that is, $Q_s \sim -\exp(q\psi_s/2kT)$. Also

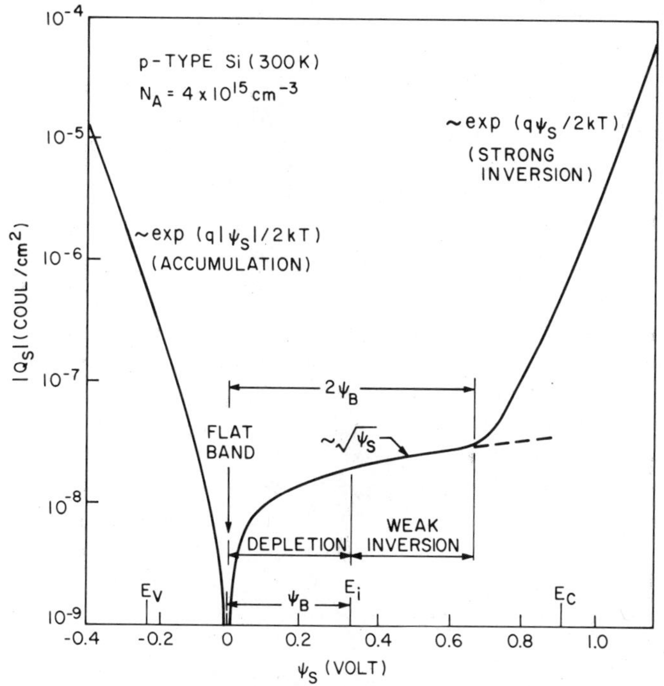

Fig. 5 Variation of space-charge density in the semiconductor as a function of the surface potential ψ_s for a p-type silicon with $N_A = 4 \times 10^{15}$ cm^{-3} at room temperature; ψ_B is the potential difference between the Fermi level and the intrinsic level of the bulk semiconductor. (After Garrett and Brattain, Ref. 13.)

note that the strong inversion begins at a surface potential,

$$\psi_s(\text{inv}) \simeq 2\psi_B = \frac{2kT}{q} \ln\left(\frac{N_A}{n_i}\right). \tag{19}$$

The differential capacitance of the semiconductor depletion layer is given by

$$C_D \equiv \frac{\partial Q_s}{\partial \psi_s} = \frac{\epsilon_s}{\sqrt{2}\, L_D} \frac{[1 - e^{-\beta\psi_s} + (n_{po}/p_{po})(e^{\beta\psi_s} - 1)]}{F(\beta\psi_s, n_{po}/p_{po})} \quad \text{F/cm}^2. \tag{20}$$

At flat-band conditions, that is, $\psi_s = 0$, C_D can be obtained by expanding the exponential terms into series, and we obtain

$$C_D(\text{flat-band}) = \epsilon_s / L_D \qquad \text{F/cm}^2. \tag{21}$$

7.2.2 Ideal MIS Curves

Figure 6*a* shows the band diagram of an ideal MIS structure with the band bending of the semiconductor identical to that shown in Fig. 4. The charge distribution is shown in Fig. 6*b*. For charge neutrality of the system,

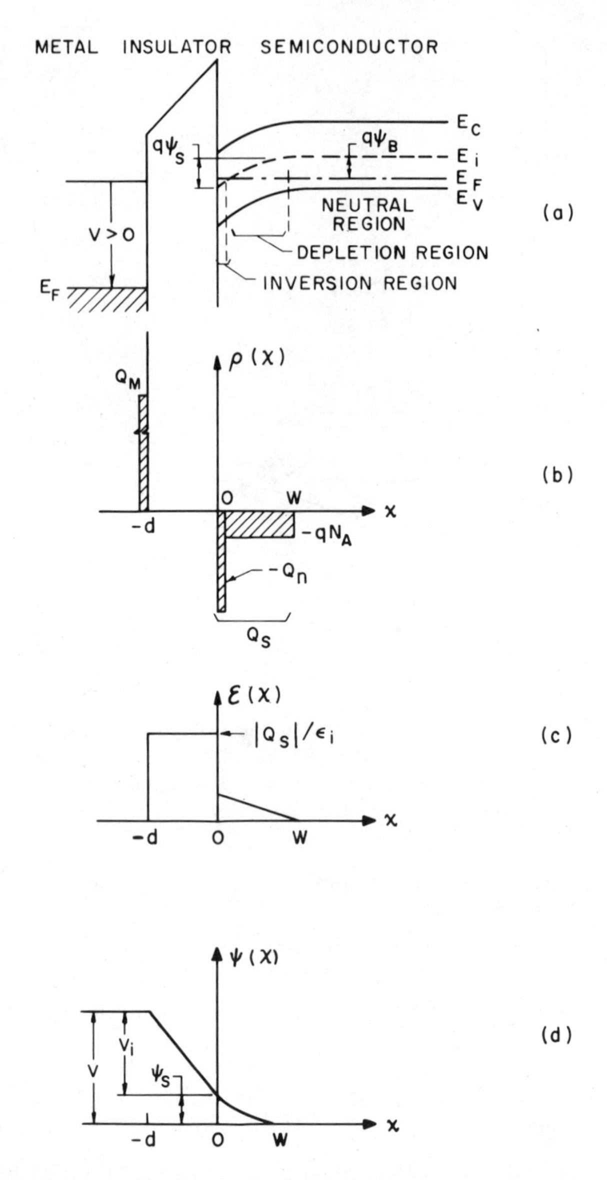

Fig. 6 (*a*) Band diagram of an ideal MIS diode. (*b*) Charge distribution under inversion condition. (*c*) Electric field distribution. (*d*) Potential distribution.

it is required that

$$Q_M = Q_n + qN_AW = Q_s \tag{22}$$

where Q_M is charges per unit area on the metal, Q_n is the electrons per unit area in the inversion region, qN_AW is the ionized acceptors per unit area in the space-charge region with space-charge width W, and Q_s is the total charges per unit area in the semiconductor. The electric field and the potential as obtained by first and second integrations of Poisson's equation are shown in Fig. 6*c* and *d*, respectively.

Clearly, in the absence of any work-function differences, the applied voltage will partly appear across the insulator and partly across the silicon. Thus

$$V = V_i + \psi_s \tag{23}$$

where V_i is the potential across the insulator and is given (Fig. 6*c*) by

$$V_i = \mathscr{E}_i d = \frac{|Q_s|d}{\epsilon_i} \left(\equiv \frac{|Q_s|}{C_i}\right). \tag{24}$$

The total capacitance C of the system is a series combination of the insulator capacitance C_i $(= \epsilon_i/d)$, and the semiconductor depletion-layer capacitance C_D:

$$C = \frac{C_i C_D}{C_i + C_D} \quad \text{F/cm}^2. \tag{25}$$

For a given insulator thickness d, the value of C_i is constant and corresponds to the maximum capacitance of the system. The capacitance C_D as given by Eq. 20 depends on the voltage. Combination of Eqs. 20, 23, 24, and 25 gives the complete description of the ideal MIS curve as shown in Fig. 7, curve (a). Of particular interest is the total capacitance at flat-band

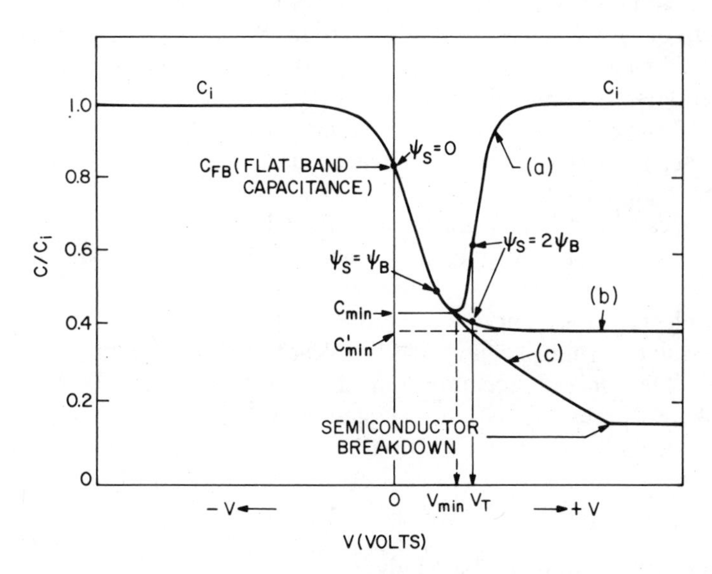

Fig. 7 MIS capacitance–voltage curves. (*a*) Low frequency. (*b*) High frequency. (*c*) Deep depletion. (After Grove et al., Ref. 16.)

condition, that is, $\psi_s = 0$. From Eqs. 21 and 25, we obtain

$$C_{FB}(\psi_s = 0) = \frac{\epsilon_i}{d + (\epsilon_i/\epsilon_s)L_D} = \frac{\epsilon_i}{d + (\epsilon_i/\epsilon_s)\sqrt{kT\epsilon_s/p_{po}q^2}} \tag{26}$$

where ϵ_i and ϵ_s are the permittivities of the insulator and the semiconductor, respectively, and L_D is the extrinsic Debye length given by Eq. 12.

In describing this curve we begin at the left side (negative voltage), where we have an accumulation of holes and therefore a high differential capacitance of the semiconductor. As a result the total capacitance is close to the insulator capacitance. As the negative voltage is reduced sufficiently, a depletion region which acts as a dielectric in series with the insulator is formed near the semiconductor surface, and the total capacitance decreases. The capacitance goes through a minimum and then increases again as the inversion layer of electrons forms at the surface. The minimum capacitance and the corresponding minimum voltage are designated $C_{\min}$ and $V_{\min}$, respectively (Fig. 7). Note that the increase of the capacitance depends on the ability of the electron concentration to follow the applied ac signal. This only happens at low frequencies where the recombination–generation rates of minority carriers (in our example, electrons) can keep up with the small-signal variation and lead to charge exchange with the inversion layer in step with the measurement signal. Experimentally, it is found that for the metal–SiO_2–Si system the frequency is between 5 and 100 Hz.[15,16] As a consequence, MIS curves measured at higher frequencies do not show the increase of capacitance on the right side, Fig. 7, curve (b). Figure 7, curve (c) shows the capacitance curve, under deep depletion conditions (pulse condition), which is directly related to the operation of a CCD to be considered in Section 7.4. At even higher voltages, impact ionization will occur at the semiconductor surface to be discussed later in connection with the avalanche effect in an MIS diode.

Figure 7 also shows the corresponding surface potentials. For an ideal MIS diode, the flat-band capacitance occurs at $V = 0$, where $\psi_s = 0$. The depletion region corresponds to a surface potential range from $\psi_s = 0$ to $\psi_s = \psi_B$. Weak inversion begins at $\psi_s = \psi_B$, which is slightly less than $V_{\min}$, and the onset of strong inversion occurs at $\psi_s = 2\psi_B$, as indicated in the figure.

The high-frequency curve can be obtained using an approach analogous to a one-sided abrupt p-n junction.[17,70] When the semiconductor surface is depleted, the ionized acceptors in the depletion region are given by $(-qN_AW)$, where W is the depletion width. Integrating Poisson's equation yields the potential distribution in the depletion region:

$$\psi = \psi_s\left(1 - \frac{x}{W}\right)^2 \tag{27}$$

where the surface potential ψ_s is given by

$$\psi_s = \frac{qN_AW^2}{2\epsilon_s}. \tag{27a}$$

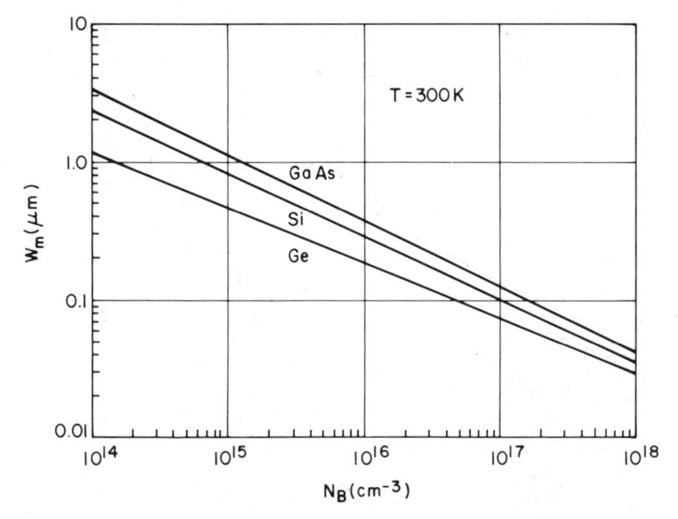

Fig. 8 Maximum depletion layer width versus impurity concentration of the semiconductors Ge, Si, and GaAs under heavy-inversion condition.

When the applied voltage increases, ψ_s and W increase. Eventually, strong inversion will occur. As shown in Fig. 5, strong inversion begins at $\psi_s(\text{inv}) \simeq 2\psi_B$. Once strong inversion occurs, the depletion-layer width reaches a maximum. When the bands are bent down far enough that $\psi_s = 2\psi_B$, the semiconductor is effectively shielded from further penetration of the electric field by the inversion layer and even a very small increase in band bending (corresponding to a very small increase in the depletion-layer width) results in a very large increase in the charge density within the inversion layer. Accordingly, the maximum width W_m of the surface depletion region under steady-state condition can be obtained from Eqs. 19 and 27a:

$$W_m \simeq \sqrt{\frac{2\epsilon_s \psi_s(\text{inv})}{qN_A}} = \sqrt{\frac{4\epsilon_s kT \ln(N_A/n_i)}{q^2 N_A}}. \tag{28}$$

The relationship between W_m and the impurity concentration is shown in Fig. 8 for Ge, Si, and GaAs, where N_B is equal to N_A for p-type and N_D for n-type semiconductors. Another quantity of interest is the so-called turn-on voltage, V_T, at which strong inversion occurs. From Eqs. 19 and 23, we obtain

$$V_T(\text{strong inversion}) = \frac{Q_s}{C_i} + 2\psi_B. \tag{29}$$

Because at the onset of strong inversion, $Q_s = qN_A W$ from Eq. 22, the turn-on voltage, also called the threshold voltage, is given by

$$V_T \simeq \frac{\sqrt{2\epsilon_s q N_A (2\psi_B)}}{C_i} + 2\psi_B. \tag{29a}$$

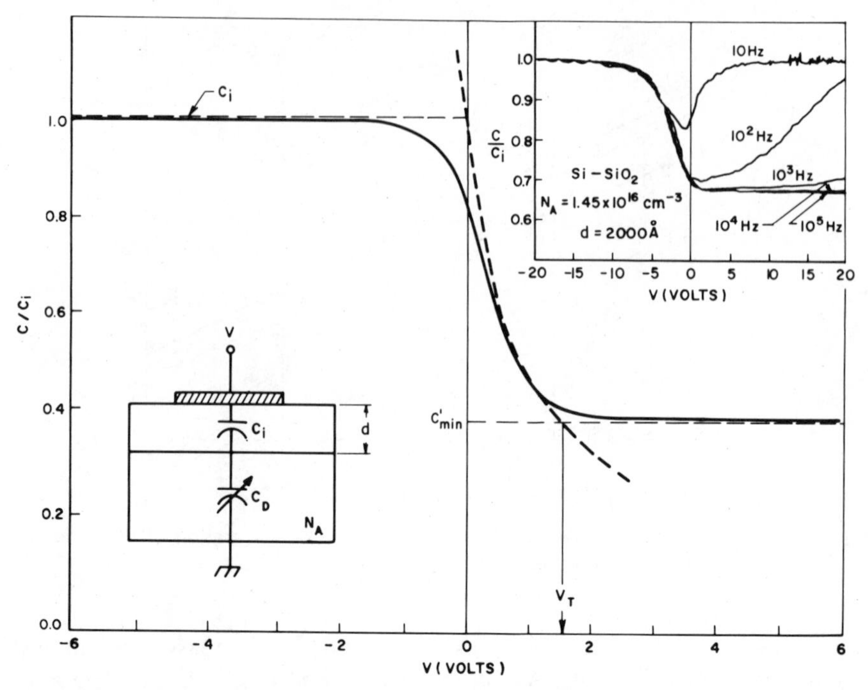

Fig. 9 High-frequency MIS capacitance–voltage curve showing its approximated segments (dashed lines). The insert shows the frequency effect. (After Grove et al., Ref. 16.)

The corresponding total capacitance is given by

$$C'_{\min} \simeq \frac{\epsilon_i}{d + (\epsilon_i/\epsilon_s)W_m}. \tag{30}$$

Figure 9 shows the high-frequency capacitance curve with its approximated segments (dashed curves). The insert shows the measured MIS curves at different frequencies.[16] Note that the onset of the low-frequency curves occurs at $f \lesssim 100$ Hz.

The ideal MIS curves of the metal–SiO_2–Si system have been computed for various oxide thicknesses and semiconductor doping densities.[18] Figure 10 shows typical examples for p-type silicon. Note that as the oxide film becomes thinner, larger variation of the capacitance is obtained. Figure 11 shows the dependence of ψ_s on applied voltage for the same systems as in Fig. 10. Figures 12 through 14 show the normalized flat-band capacitance (C_{FB}/C_i), the normalized minimum capacitances ($C_{\min}/C_i$) and ($C'_{\min}/C_i$), and the threshold voltage (V_T) and the minimum voltage ($V_{\min}$) versus oxide thickness with silicon doping concentration as the parameter, respectively. The conversion to n-type silicon is achieved simply by changing the sign of the voltage axes. Converting to other insulators requires scaling the oxide thickness with the ratio of the permittivities of SiO_2 and the other

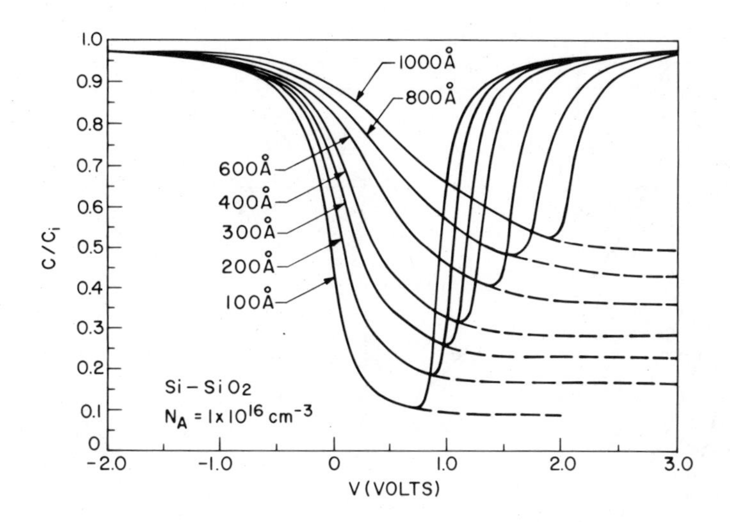

Fig. 10 Ideal MIS *C–V* curve. Solid lines for low frequencies. Dashed lines for high frequencies. (After Goetzberger, Ref. 18.)

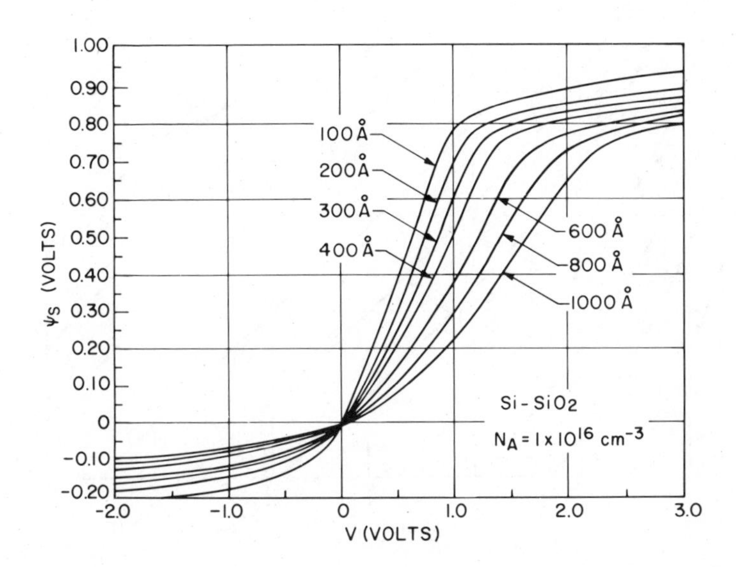

Fig. 11 Surface potential versus applied voltage for ideal MOS diodes. (After Goetzberger, Ref. 18.)

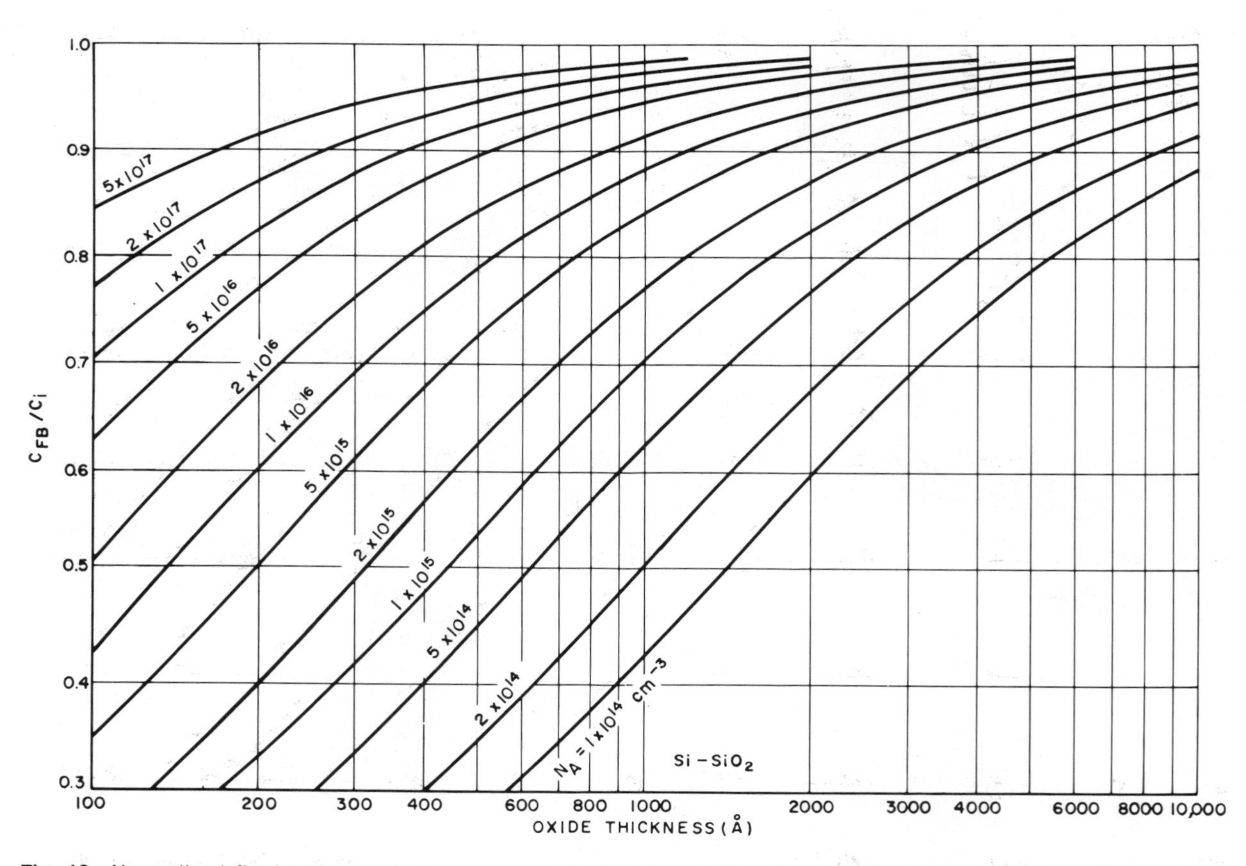

Fig. 12 Normalized flat-band capacitance versus oxide thickness with silicon doping as the parameter for ideal MIS diodes. (After Goetzberger, Ref. 18.)

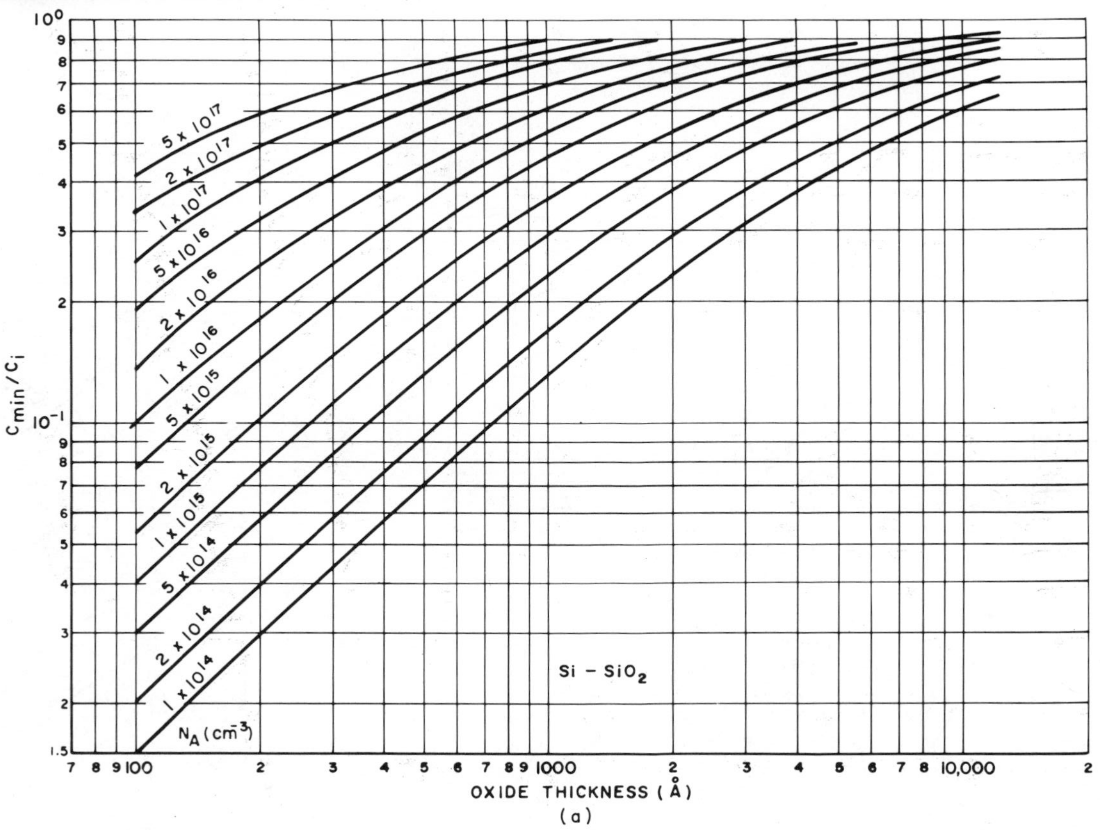

Fig. 13 (*a*) Normalized minimum capacitance versus oxide thickness with silicon doping as the parameter for ideal MIS diodes under low-frequency condition. (*b*) Normalized minimum capacitance versus oxide thickness with silicon doping as the parameter for ideal MIS diodes under high-frequency condition. (After Goetzberger, Ref. 18.)

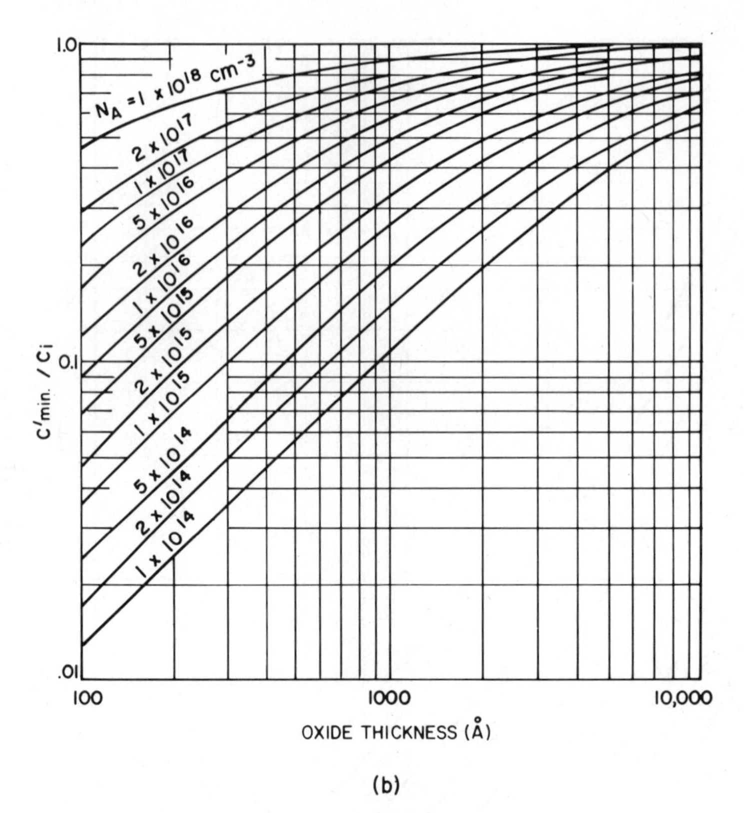

Fig. 13*b*.

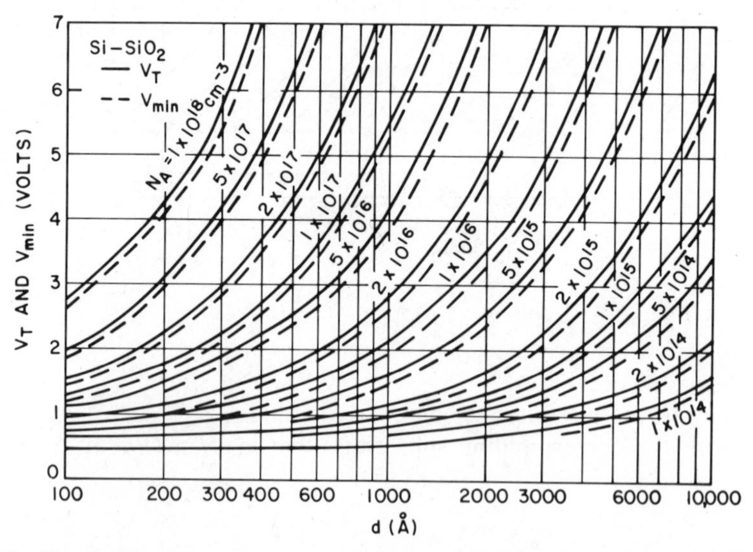

Fig. 14 Threshold voltage (V_T) and minimum voltage (V_{min}) versus oxide thickness with silicon doping as the parameter for ideal MIS diodes. (After Goetzberger, Ref. 18.)

insulator

$$d_c = d_i \frac{\epsilon_i(\mathrm{SiO_2})}{\epsilon_i(\mathrm{insulator})} \tag{31}$$

where d_c is the thickness to be used in these curves, d_i is the actual thickness of the insulator, ϵ_i(insulator) is the permittivity of the new insulator, and $\epsilon_i(SiO_2) = 3.5 \times 10^{-13}$ F/cm. For other semiconductors, the MIS curves similar to those in Fig. 9 can be constructed by using Eqs. 26 through 29.

The ideal MIS curves shown in Figs. 10 through 14 will be used in subsequent sections to compare with experimental results and to understand practical MIS systems.

7.3 $Si\text{–}SiO_2$ MOS DIODE

Of all the MIS diodes, the metal–SiO_2–Si (MOS) diode is by far the most important. The exact nature of the Si–SiO_2 interface is not yet fully understood. An appealing picture[7] of the interface is that the chemical composition of the interfacial region, as a consequence of thermal oxidation, is a single-crystal silicon followed by a monolayer of SiO_x, that is, incompletely oxidized silicon, then a strained region of SiO_2 roughly $10 \sim 40$ Å deep, and the remainder stoichiometric, strain-free, amorphous SiO_2 (the compound SiO_x is stoichiometric when $x = 2$ and nonstoichiometric when $2 > x > 1$). For a practical MOS diode, interface traps and oxide charges exist that will, in one way or another, affect the ideal MOS characteristics.

The basic classifications of these traps and charges are shown[19] in Fig. 15: (1) interface trapped charges Q_{it}, which are charges located at the Si–SiO_2 interface with energy states in the silicon-forbidden bandgap and which can exchange charges with silicon in a short time; Q_{it} can possibly be produced by excess silicon (trivalent silicon), excess oxygen, and impurities; (2) fixed oxide charges Q_f, which are located at or near the interface and are immobile under an applied electric field; (3) oxide trapped charges Q_{ot}, which can be created, for example, by x-ray radiation or hot-electron injection; these traps are distributed inside the oxide layer; and (4) mobile ionic charges Q_m, such as sodium ions, which are mobile within the oxide under bias-temperature aging conditions.

The foregoing Q's are the effective net charges per unit area (i.e., C/cm^2). To convert to the effective net number of charges per unit area, one can use N with the corresponding subscripts, that is, $N = Q/q$ in number of charges/cm^2. Because interface-trap levels are distributed across the silicon energy bandgap, we shall define an interface-trap density, D_{it}:

$$D_{it} = \frac{1}{q}\frac{dQ_{it}}{dE} \qquad \text{number of charges/cm}^2\text{-eV}. \tag{32}$$

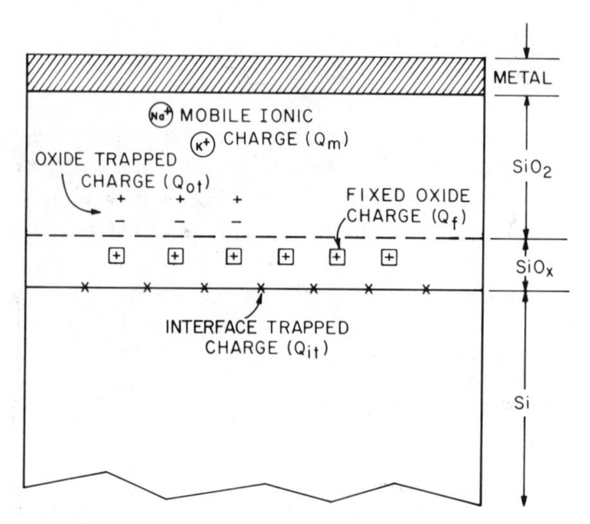

Fig. 15 Terminology for charges associated with thermally oxidized silicon. (After Deal, Ref. 19.)

7.3.1 Interface Trapped Charge

Tamm,[20] Shockley,[21] and others[7] have studied the interface-trapped charge Q_{it} (historically called the interface state, fast state, or surface state) and have shown that Q_{it} exists within the forbidden gap due to the interruption of the periodic lattice structure, at the surface of a crystal. Shockley and Pearson experimentally found the existence of Q_{it} in their surface conductance measurement.[22] Measurements on clean surfaces[23] in an ultra-high-vacuum system confirm that Q_{it} is very high—of the order of the density of surface atoms ($\sim 10^{15}$ atoms/cm^2). For the present MOS diodes having thermally grown SiO_2 on Si, most of the interface trapped charge can be neutralized by low-temperature (450°C) hydrogen annealing. The value of Q_{it} can be as low as 10^{10} cm^{-2}, which amounts to about one interface trapped charge per 10^5 surface atoms.

An interface trap is considered a donor if it can become neutral or positive by donating (giving up) an electron. An acceptor interface trap can become neutral or negative by accepting an electron. The distribution functions for the interface traps are similar to those for the bulk impurity levels as discussed in Chapter 1:

$$F_{SD}(E_t) = \left[1 - \frac{1}{1 + \frac{1}{g}\exp\left(\frac{E_t - E_F}{kT}\right)}\right] = \frac{1}{1 + g\exp\left(\frac{E_F - E_t}{kT}\right)} \tag{33a}$$

for donor interface traps and

$$F_{SA}(E_t) = \frac{1}{1 + \frac{1}{g}\exp\left(\frac{E_t - E_F}{kT}\right)} \tag{33b}$$

for acceptor interface traps, where E_t is the energy of the interface trap and g is the ground-state degeneracy, which is 2 for donor and 4 for acceptor.

When a voltage is applied, the interface-trap levels move up or down with the valence and conductance bands while the Fermi level remains fixed. A change of charge in the interface trap occurs when it crosses the Fermi level. This change of charge contributes to the MIS capacitance and alters the ideal MIS curve. The basic equivalent circuit[24] incorporating the interface-trap effect is shown in Fig. 16*a*. In the figure, C_i and C_D are the insulator capacitance and the semiconductor depletion-layer capacitance, respectively, and are identical to those shown in the insert of Fig. 9. C_s and R_s are the capacitance and resistance associated with the interface traps, and are functions of surface potential. The product C_sR_s is defined as the interface-trap lifetime, which determines the frequency behavior of the interface traps. The parallel branch of the equivalent circuit in Fig. 16*a* can be converted into a frequency-dependent capacitance C_p in parallel with a frequency-dependent conductance G_p, as shown in Fig. 16*b*, where

$$C_p = C_D + \frac{C_s}{1 + \omega^2\tau^2} \tag{34}$$

and

$$\frac{G_p}{\omega} = \frac{C_s\omega\tau}{1 + \omega^2\tau^2} \tag{35}$$

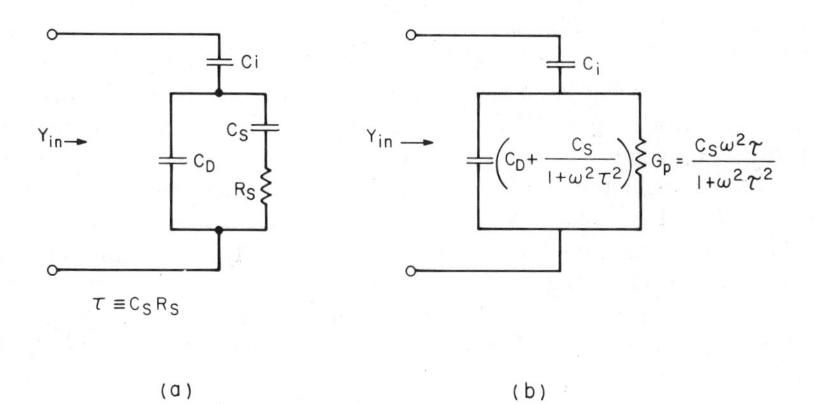

Fig. 16 Equivalent circuit including interface-trap effect, where C_s and R_s are associated with interface-trap density. (After Nicollian and Goetzberger, Ref. 24.)

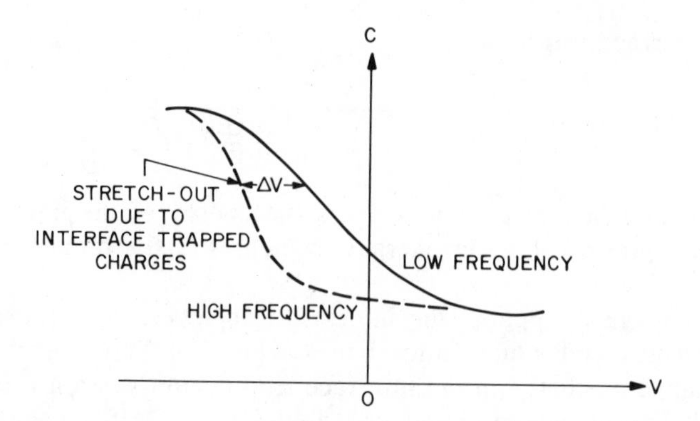

Fig. 17 Capacitance stretch-out due to interface trapped charges.

with $\tau \equiv C_s R_s$. The input admittance Y_{in} is given by

$$Y_{\text{in}} \equiv G_{\text{in}} + j\omega C_{\text{in}} \tag{36}$$

where

$$G_{\text{in}} = \frac{\omega^2 C_s \tau C_i^2}{(C_i + C_D + C_s)^2 + \omega^2 \tau^2 (C_i + C_D)^2} \tag{36a}$$

$$C_{\text{in}} = \frac{C_i}{C_i + C_D + C_s}\left[C_D + C_s \frac{(C_i + C_D + C_s)^2 + \omega^2 \tau^2 C_D (C_i + C_D)}{(C_i + C_D + C_s)^2 + \omega^2 \tau^2 (C_i + C_D)^2}\right]. \tag{36b}$$

Capacitance Method Either the capacitance measurement or the conductance measurement can be used to evaluate the interface-trapped charges, because in Eqs. 36a and 36b both the input conductance and the input capacitance contain similar information about the interface traps. It will be shown that the conductance technique can give more accurate results, especially for MOS diodes with relatively low interface-trap density ($\sim 10^{10}$ cm^{-2}/eV). The capacitance measurement, however, can give rapid evaluation of flat-band shift and the total interface trapped charge Q_{it}.

Figure 17 shows the C–V stretch-out due to interface trapped charges. At high frequencies ($\omega\tau \gg 1$), interface traps cannot follow the ac voltage swing, so the capacitance expression in Eq. 36b reduces to Eq. 25, which yields the high-frequency curve (dashed line), free of capacitance due to interface traps. The influence of interface traps on the biasing voltage, however, causes a shift of the ideal MOS curve along the voltage axis, because, when interface traps are present, more charges on the metal are necessary to create a given surface potential. The interface-trap density is given by[7]

$$D_{it} = \frac{C_i}{q}[(d\psi_s/dV)^{-1} - 1] - \frac{C_D}{q} \qquad \text{cm}^{-2}\,\text{eV}^{-1} \tag{37}$$

where $d\psi_s/dV$ is the slope of a ψ_s versus C curve, and C_D is calculated from the known doping density from Eq. 20. The integration method is employed to evaluate ψ_s versus V using low-frequency capacitance measurement.[25] When oxide trapped charges and mobile ionic charges can be neglected, we obtain, from Eqs. 23 and 25 and the fact that $dQ = C_i dV_i = CdV$,

$$\left(\frac{\partial \psi_s}{\partial V}\right) = 1 - \frac{C}{C_i} \tag{38}$$

and

$$\frac{d\psi_s}{dV_i} = \frac{C_i}{C} - 1. \tag{39}$$

Integrating Eq. 38 from V_1 to V_2 yields

$$\psi_s(V_1) - \psi_s(V_2) = \int_{V_2}^{V_1} \left[1 - \frac{C}{C_i}\right] dV. \tag{40}$$

Equation 40 indicates that the surface potential at any applied voltage can be determined by integrating a curve of $(1 - C/C_i)$. Note that Eq. 40 is valid only when the interface trapped charges are in equilibrium at all times during the measurement of $C(V)$; that is, the measurement frequency must be low enough so that all interface traps can follow both the gate bias and the ac signal. The requirement of charge neutrality in the MIS system gives us the following relation. Referring to Fig. 6*b*, in addition to the space-charge density in the semiconductor (Q_s), we now have $D_{it}(q\psi_s) = D_{it}^d(q\psi_s) + D_{it}^a(q\psi_s)$, where D_{it}^d and D_{it}^a are the donor and acceptor interface-trap densities, respectively. The charge neutrality requirement gives

$$\frac{\epsilon_i V_i}{d} = q \int_{E_V}^{E_C} [D_{it}^d F_{SD}(E_t) - D_{it}^a F_{SA}(E_t)] dE_t + Q_s. \tag{41}$$

Differentiating Eq. 41 with respect to ψ_s gives

$$\frac{\partial \psi_s}{\partial V_i} = \frac{\epsilon_i/d}{(dQ_s/d\psi_s) + qD_{it}(q\psi_s)}. \tag{42}$$

From Eqs. 39 and 40, a curve of $\partial\psi_s/\partial V_i$ versus ψ_s can be obtained directly using the low-frequency capacitance measurement of an MOS diode. If the doping density of the semiconductor and the temperature are known, comparing this curve to the one given by Eq. 42 accurately determines D_{it}.

Conductance Method Nicollian and Goetzberger give a detailed and comprehensive discussion of the conductance method.[26] Difficulty arises in the capacitance measurement, because the interface-trap capacitance must be extracted from the measured capacitance which consists of oxide capacitance, depletion-layer capacitance, and interface-trap capacitance. As previously mentioned, both the capacitance and conductance as functions of voltage and frequency contain identical information about interface

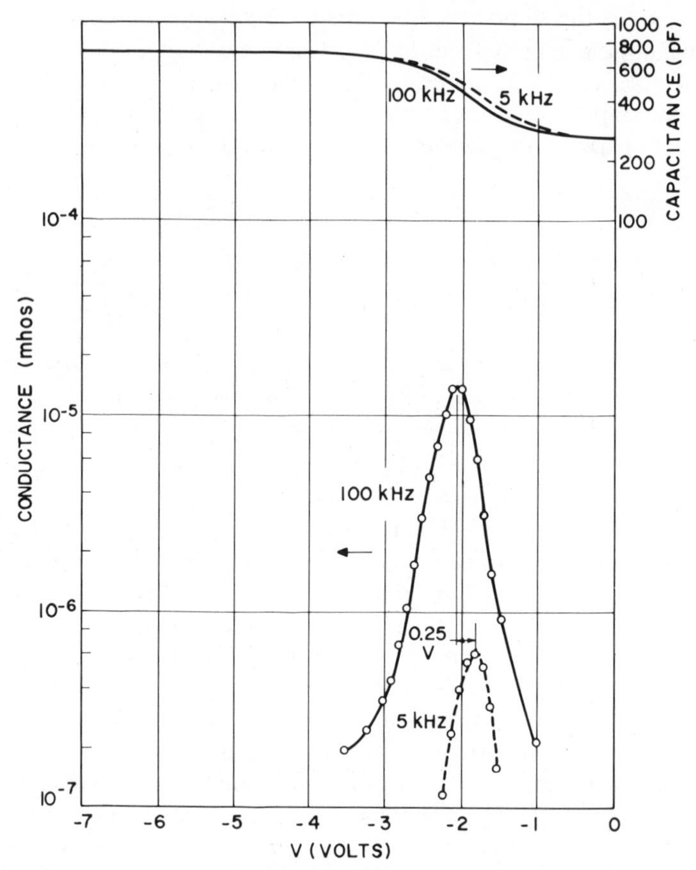

Fig. 18 Comparison of MIS capacitance measurement and conductance measurement at two frequencies. (After Nicollian and Goetzberger, Ref. 26.)

traps, greater inaccuracies arise in extracting this information from the measured capacitance, because the difference between two capacitances must be calculated. This difficulty does not apply to the measured conductance (which is directly related to the interface traps).

Thus conductance measurements yield more accurate and reliable results, particularly when D_{it} is low as in the thermally oxidized SiO_2–Si system, as illustrated in Fig. 18. This figure shows the measured capacitance and measured conductance at 5 and 100 kHz. The largest capacitance spread is only 14% while the magnitude of the conductance peak increases by over one order of magnitude in this frequency range.

The simplified equivalent circuit in Fig. 16 illustrates the principle of the MIS conductance technique. The admittance of the MIS diode is measured by a bridge across the diode terminals. The insulator capacitance is measured in the region of strong accumulation. The admittance of the circuit is then converted into an impedance. The reactance of the insulator capacitance is subtracted from this impedance and the resulting impedance converted back into an admittance. This leaves C_D in parallel with the series R_sC_s network of the interface traps. The capacitance and equivalent parallel conductance divided by ω are given by Eqs. 34 and 35. Equation 35 does not contain C_D and depends only on the interface trap branch of the equivalent circuit. At a given bias, G_p/ω can be measured as a function of frequency. A plot of G_p/ω versus $\omega\tau$ goes through a maximum when $\omega\tau = 1$, and gives τ directly. The value of G_p/ω at the maximum is $C_s/2$. Thus, equivalent parallel conductance corrected for C_i gives C_s and τ $(= R_sC_s)$ directly from the measured conductance. Once C_s is known, the interface-trap density is obtained by using the relation $D_{it} = C_s/qA$, where A is the metal plate area.

Figure 19 shows typical results of D_{it} in a $Si-SiO_2$ system.[27] Near the midgap D_{it} is relatively constant, but it increases toward the conduction

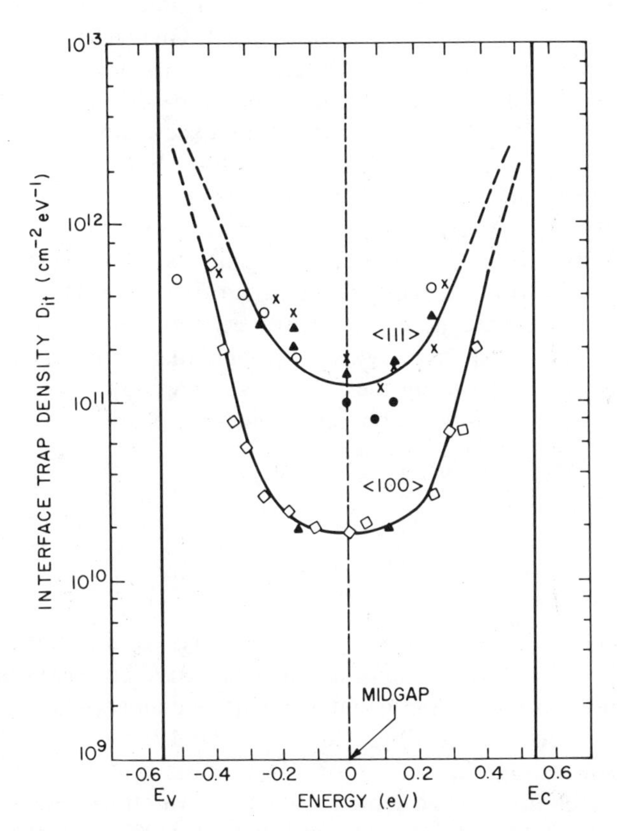

Fig. 19 Interface-trap density in thermally oxidized silicon. (After White and Cricchi, Ref. 27.)

Table 1 Properties of Silicon Crystal Planes

Orientation	Plane Area of Unit Cell (cm^2)	Atoms in Area	Available Bonds in Area	$Atoms/cm^2$	Available $Bonds/cm^2$
$\langle 111 \rangle$	$\sqrt{3}a^2/2$	2	3	7.85×10^{14}	11.8×10^{14}
$\langle 110 \rangle$	$\sqrt{2}a^2$	4	4	9.6×10^{14}	9.6×10^{14}
$\langle 100 \rangle$	a^2	2	2	6.8×10^{14}	6.8×10^{14}

and valence-band edges. Orientation dependence is particularly important. In $\langle 100 \rangle$ orientation D_{it} is about an order of magnitude smaller than that in $\langle 111 \rangle$. This result has been correlated with the available bonds per unit area on the silicon surface.[28,29] Table 1 shows the properties of silicon crystal planes oriented along $\langle 111 \rangle$, $\langle 110 \rangle$, and $\langle 100 \rangle$ directions. It is apparent that the (111) surface has the largest number of available bonds per centimeter squared, and the (100) surface has the smallest. One would expect that the (100) surface has the lowest oxidation rate. If we assume that the origin of interface traps is due to excess silicon in the oxide, then the lower the oxidation rate, the smaller the amount of the excess silicon; thus the (100) surface should have the smallest interface-trap density, as confirmed in Fig. 19. Therefore, all modern silicon MOSFETs are fabricated on $\langle 100 \rangle$-oriented substrates (refer to Chapter 8).

Figure 20 shows the variation of the time constant τ versus surface potential for MOS diodes with steam-grown oxides on $\langle 100 \rangle$ silicon substrates where ψ_B is the potential difference between intrinsic level and Fermi level, and $\bar{\psi}_s$ is the average surface potential (to be discussed later). These curves can be fitted by the following expressions and are similar to those for the generation–recombination processes as discussed in Chapter 1:

$$\tau = \frac{1}{\bar{v}\sigma_p n_i} \exp\left[-\frac{q(\psi_B - \bar{\psi}_s)}{kT}\right] \quad \text{for } p\text{-type}$$

$$\tau = \frac{1}{\bar{v}\sigma_n n_i} \exp\left[\frac{q(\psi_B - \bar{\psi}_s)}{kT}\right] \quad \text{for } n\text{-type} \tag{43}$$

where σ_p and σ_n are the capture cross sections of holes and electrons, respectively, and $\bar{v}$ is the average thermal velocity. These results indicate that the capture cross section is independent of energy. The capture cross sections obtained[26] from Fig. 20 are $\sigma_p = 4.3 \times 10^{-16}\,cm^2$ and $\sigma_n = 8.1 \times 10^{-16}\,cm^2$, where the values of $\bar{v} = 10^7\,cm/s$ and $n_i = 1.6 \times 10^{10}\,cm^{-3}$ have been used. For $\langle 111 \rangle$-oriented silicon the variation of time constant versus surface potential is similar to that of $\langle 100 \rangle$, and the measured capture cross sections are $\sigma_p = 2.2 \times 10^{-16}\,cm^2$ and $\sigma_n = 5.9 \times 10^{-16}\,cm^2$.

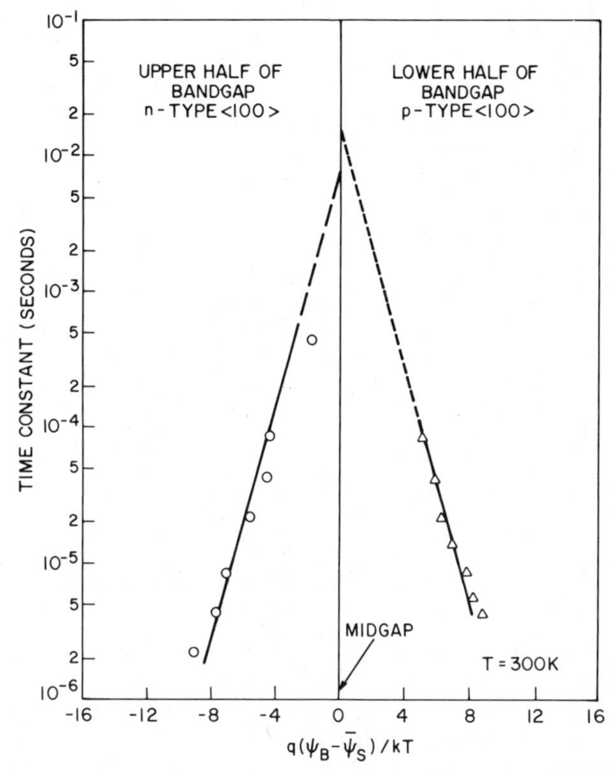

Fig. 20 Variation of time constant versus surface potential. (After Nicollian and Goetzberger, Ref. 26.)

Figure 19 shows that the interface-trap density in the Si–SiO_2 system is comprised of many levels. The levels are so closely spaced in energy that they cannot be distinguished as separate levels, and actually appear as a continuum over the bandgap of the semiconductor. The equivalent circuit for an MIS diode with a single-level time constant should therefore be modified (Fig. 16). We must also consider the statistical fluctuation of surface potential due to surface charges (this includes the fixed oxide charges Q_f and the interface trapped charges Q_{it}) because, from Eq. 43, a small fluctuation in $\bar{\psi}_s$ causes a large fluctuation in τ. Assuming that surface charges are randomly distributed in the plane of the interface, the electric field at the semiconductor surface will fluctuate over the plane of the interface. This is shown schematically in the insert in Fig. 21. Figure 21 shows calculated values of G_p/ω as a function of frequency for a Si–SiO_2 MOS diode biased in the depletion (solid broad curve), and in the weak inversion (solid narrow curve) based on the time-constant dispersion resulting from the interface-trap continuum and the statistical (Poisson) distribution of surface charges ($Q_{it} + Q_f$). Experimental results are also shown

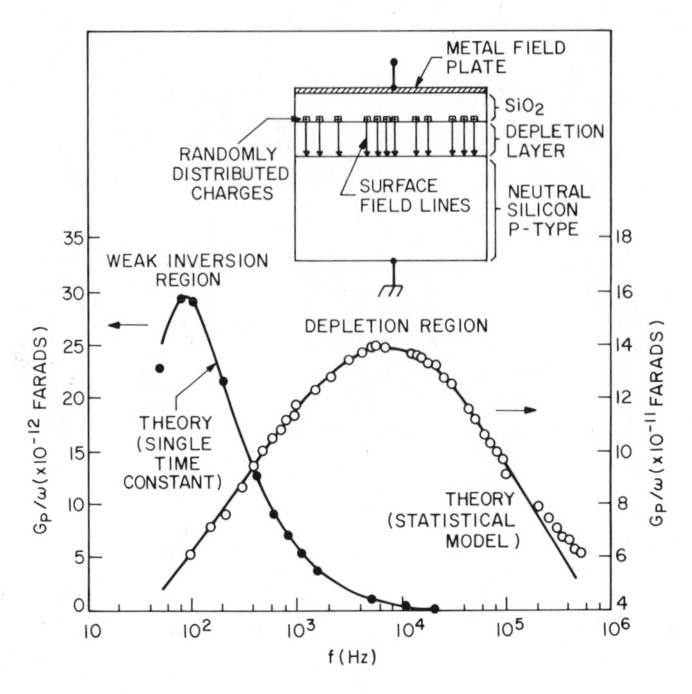

Fig. 21 G_p/ω versus frequency for a Si–SiO_2 MIS diode biased in the depletion region (broad curve) and in the weak inversion region (narrow curve). Circles are experimental results. Lines are the theoretical calculations. Insert shows the randomly distributed fixed oxide charge that causes time-constant dispersion. (After Nicollian and Goetzberger, Ref. 26.)

(open and solid circles); their excellent agreement with the statistical results indicates the importance of the statistical model.

The results above can be explained with the help of the modified equivalent circuits shown in Fig. 22. Figure 22*a* shows the time-constant dispersion caused primarily by statistical fluctuation of surface potential. Each subnetwork consisting of C_s and R_s in series represents a time constant of the continuum of interface traps in a characteristic area A_c that is proportional to the square of the depletion width. This circuit corresponds to the depletion case. Figure 22*b* shows the case of the midgap region, where $q\psi_s = q\psi_B \pm$ a few kT. R_{ns} and R_{ps} represent the capture resistances for the electrons and holes, respectively. We now have two resistances for each subnetwork; when the surface potential is equal to or greater than ψ_B, the minority-carrier density at the semiconductor surface will be equal to or greater than the magnitude of the majority carrier density at the surface. The minority-carrier density can thus no longer be ignored as in the depletion case, Fig. 22*a*. Figure 22*c* represents weak inversion ($2\psi_B > \psi_s > \psi_B$); whereas Fig. 22*b* is simplified by the fast response (small R_{ns}) of minority carriers such that all R_{ns}'s are shorted

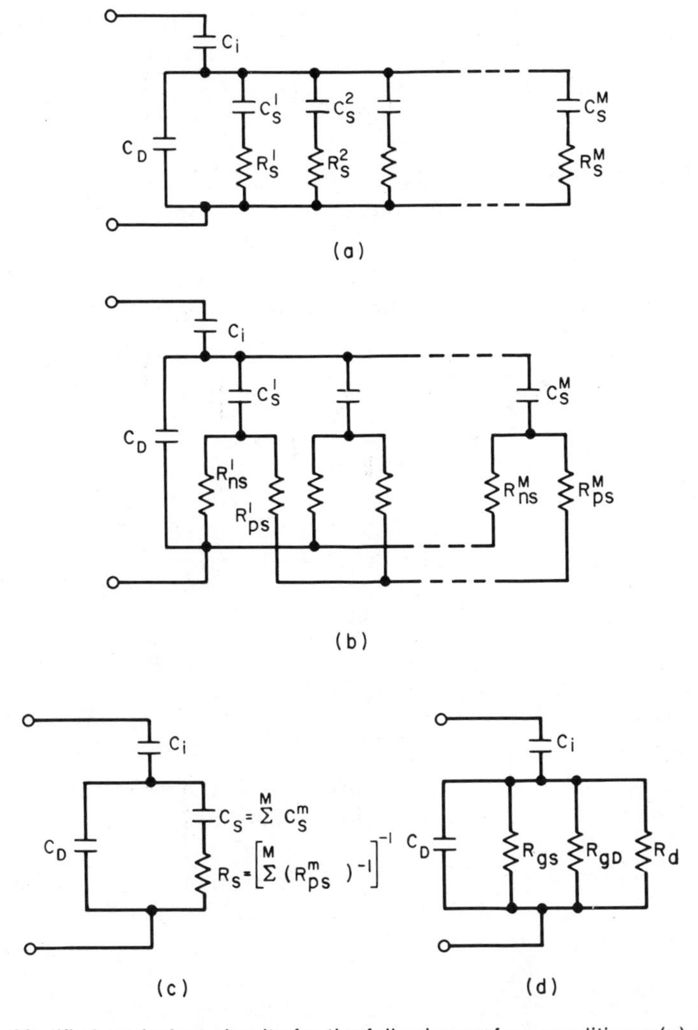

Fig. 22 Modified equivalent circuits for the following surface conditions: (*a*) depletion; (*b*) midgap; (*c*) weak inversion; (*d*) heavy inversion. (After Nicollian and Goetzberger, Ref. 26; Hofstein and Warfield, Ref. 15.)

together, the equivalent interface trap capacitance is given by the sum of each individual C_s, and the resistance by the sum of the capture resistances for the majority carriers. Therefore, in weak inversions good agreement is obtained between the experimental result and the single time constant theory. Figure 22*d* shows the circuit for heavy inversion[15] ($\psi_s > 2\psi_B$), where C_i and C_D are the insulator capacitance and the depletion capacitance, respectively; R_d is the resistance associated with the diffusion current of the minority carriers from the bulk to the edge of the depletion region and through the depletion region to the surface; R_{gs} is the resistance associated with the current flow of holes between the valence band and the interface traps, with the current flow of electrons between the conductance band and the interface traps, and with the flow of the majority carriers from the bulk to the surface; and R_{gD} is the resistance associated with the finite generation and recombination within the depletion layer, which is found to be the dominant factor in controlling the frequency response of the inversion layer. The interface trap capacitance C_s can be assumed to be essentially zero ($C_s \ll C_D$) under heavy inversion conditions. Figure 22*d* shows that, in the heavy inversion region, the interface traps have only a minor effect on the MIS characteristics.

7.3.2 Oxide Charges

Oxide charges include the oxide fixed charge Q_f, the oxide trapped charge Q_{ot}, and the mobile ionic charge Q_m, as shown in Fig. 15.

The fixed oxide charge Q_f has the following properties: It is fixed and cannot be charged or discharged over a wide variation of ψ_s; it is located within the order of 30 Å of the Si–SiO_2 interface;[7] its density is not greatly affected by the oxide thickness or by the type or concentration of impurities in the silicon; it is generally positive and depends on oxidation and annealing conditions, and on silicon orientation. It has been suggested that excess silicon (trivalent silicon) or the loss of an electron from excess oxygen centers (nonbridging oxygen) near the Si–SiO_2 interface is the origin of fixed oxide charge. In electrical measurements, Q_f can be regarded as a charge sheet located at the Si–SiO_2 interface.

Figure 23 shows the shift along the voltage axis of a high-frequency C–V curve when positive or negative Q_f is present at the interface.[7] The voltage shift is measured with respect to an ideal C–V curve where $Q_f = 0$. For both *n*-type and *p*-type substrates, positive Q_f causes the C–V curve to shift to more negative values of gate bias with respect to the ideal C–V curve, while negative Q_f causes the C–V curve to shift to more positive gate voltage.

The voltage shift of the C–V curve caused by Q_f can be explained with the help of Fig. 24, which shows a cross section through an MOS diode having positive Q_f and a negative gate bias. Charge neutrality requires every negative charge on the gate to be compensated by an equal and opposite charge in the oxide and the silicon. For the ideal case, $Q_f = 0$, the entire compensation charge comes from the ionized donors. However, for a practical MOS diode with a positive Q_f, part of the compensating charge must consist of Q_f and the rest of ionized donors. Figure 24 shows some field lines from Q_f terminating on negative charges on the gate and fewer field lines from ionized donors terminating on the gate than would exist if $Q_f = 0$. Because fewer ionized donors are required, the silicon depletion-layer width will be smaller than with $Q_f = 0$ at any given gate bias. Thus the

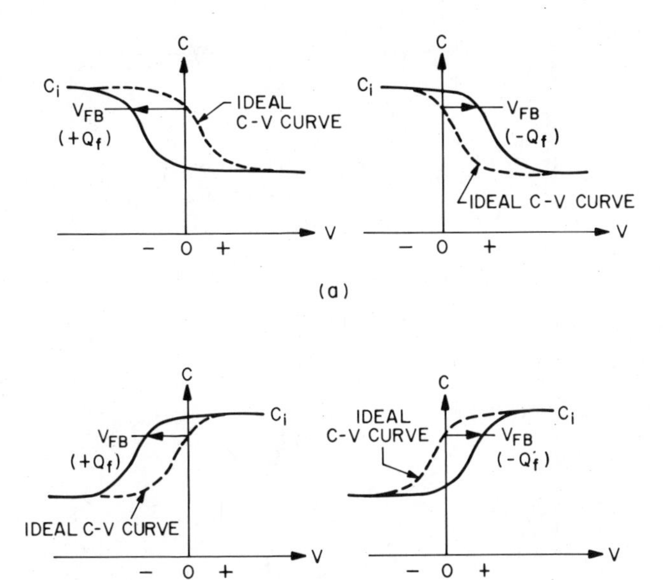

Fig. 23 *C–V* curve shift along the voltage axis due to positive or negative fixed oxide charge. (*a*) For *p*-type semiconductor. (*b*) For *n*-type semiconductor. (After Nicollian and Brews, Ref. 7.)

capacitance will be higher than for the ideal case for all values of gate bias in the depletion and weak inversion regions. The result is a shift of the *C–V* curve toward more negative gate bias for positive Q_f (left diagram in Fig. 23*b*). For negative Q_f the *C–V* curve shifts to the opposite direction (Fig. 23). The magnitude of the shift is given by

$$\Delta V_f = \frac{Q_f}{C_i}. \tag{44}$$

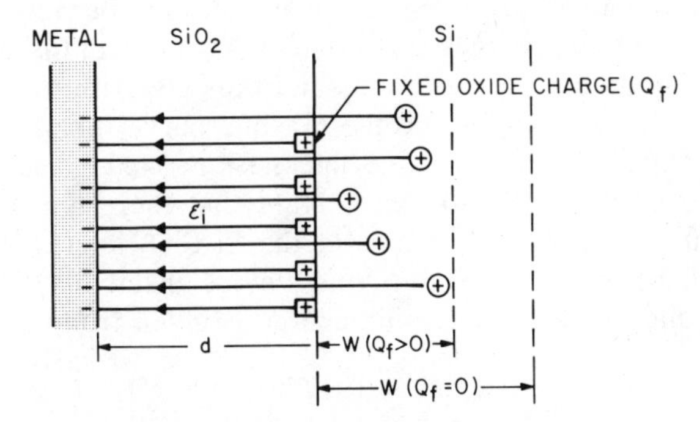

Fig. 24 Effect of fixed oxide charge on MIS diode. (After Nicollian and Goetzberger, Ref. 24.)

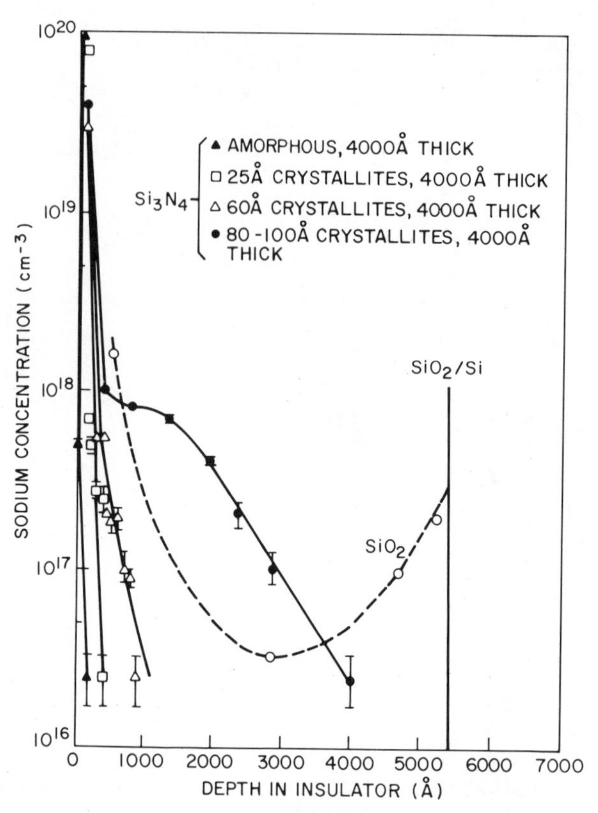

Fig. 25 Sodium concentration versus depth in silicon dioxide and silicon nitride with different crystallite sizes. (After Yon, Ko, and Kuper, Ref. 31; Dalton and Drobek, Ref. 32.)

It was first demonstrated by Snow et al.[30] that alkali ions, such as sodium, in thermally grown SiO_2 films are mainly responsible for the instability of the oxide-passivated devices. Reliability problems in semiconductor devices operated at high temperatures and voltages may be related to trace contamination by alkali metal ions, because under these conditions mobile ionic charges can move back and forth through the oxide layer, depending on biasing conditions and thus give rise to voltage shifts.

Figure 25 shows the sodium profile (dashed curve) obtained after bias-temperature drift.[31] The initial oxide thickness is 5400 Å and the acceptor density in the silicon is $5 \times 10^{14}\,\text{cm}^{-3}$. Note that there is a large concentration ($3 \times 10^{17}\,\text{cm}^{-3}$) of sodium ions at the Si–SiO_2 interface. These ions cause large flat-band voltage shift and device instability. The flat-band voltage shift due to the mobile ionic charge is given from Gauss's law by

$$\Delta V_m = \frac{Q_m}{C_i} = \frac{1}{C_i}\left[\frac{1}{d}\int_0^d x\rho_m(x)\,dx\right] \tag{45}$$

where Q_m is the effective net charge of mobile ions per unit area at the Si–SiO_2 interface and $\rho_m(x)$ is the volume charge density of the mobile ions (i.e., C/cm^3). Figure 25 also shows sodium diffusion results in silicon nitride films[32] having different crystallite sizes. For amorphous Si_3N_4, there is very little sodium penetration. To prevent mobile ionic charge contamination of the oxide during device life, one can protect it with a film impervious to mobile ions, such as using amorphous or small crystallite silicon nitride. Other sodium barrier layers include Al_2O_3 and phosphosilicate glass. The metal–(Al_2O_3–SiO_2)–Si composite insulator system[33] can give a voltage shift in the C–V curve toward negative voltage, which implies that a p-type silicon surface can be inverted at zero or negative bias voltage. The phosphosilicate glass,[34] which is SiO_2 rich in P_2O_5, can be formed on the outside of an SiO_2 layer during the phosphorus diffusion process and can substantially reduce the instability of a contaminated MOS diode, because the sodium has a much greater solubility in glass than in SiO_2.

Silicon nitride also can be used in a composite insulator system[35] of metal–(Si_3N_4-SiO_2)–Si in a nonvolatile memory application (refer to Chapter 8). This system has the desired features of low interface–trap density because of the clean Si–SiO_2 interface, and is immune to ion drift because of the outer Si_3N_4 film. In the visible range, Si_3N_4 films have a slightly larger dielectric constant than the SiO_2 films have. Table 2 compares the colors of SiO_2 and Si_3N_4 films at various ranges of film thickness.[36]

Table 2 Color Comparison of SiO_2 and Si_3N_4 Films

Order	Color	SiO_2 Thickness Range[a] (μm)	Si_3N_4 Thickness Range (μm)
	Silicon	0–0.027	0–0.020
	Brown	0.027–0.053	0.020–0.040
	Golden brown	0.053–0.073	0.040–0.055
	Red	0.073–0.097	0.055–0.073
	Deep blue	0.097–0.010	0.073–0.077
1st	Blue	0.10–0.12	0.077–0.093
	Pale blue	0.12–0.13	0.093–0.10
	Very pale blue	0.13–0.15	0.10–0.11
	Silicon	0.15–0.16	0.11–0.12
	Light yellow	0.16–0.17	0.12–0.13
	Yellow	0.17–0.20	0.13–0.15
	Orange red	0.20–0.24	0.15–0.18
1st	Red	0.24–0.25	0.18–0.19
	Dark red	0.25–0.28	0.19–0.21
2nd	Blue	0.28–0.31	0.21–0.23
	Blue-green	0.31–0.33	0.23–0.25
	Light green	0.33–0.37	0.25–0.28
	Orange yellow	0.37–0.40	0.28–0.30
2nd	Red	0.40–0.44	0.30–0.33

[a]The ratio of refractive index $= \dfrac{\bar{n}(Si_3N_4)}{\bar{n}(SiO_2)} = \dfrac{1.97}{1.48} = 1.33 = \dfrac{SiO_2 \text{ thickness}}{Si_3N_4 \text{ thickness}}$.

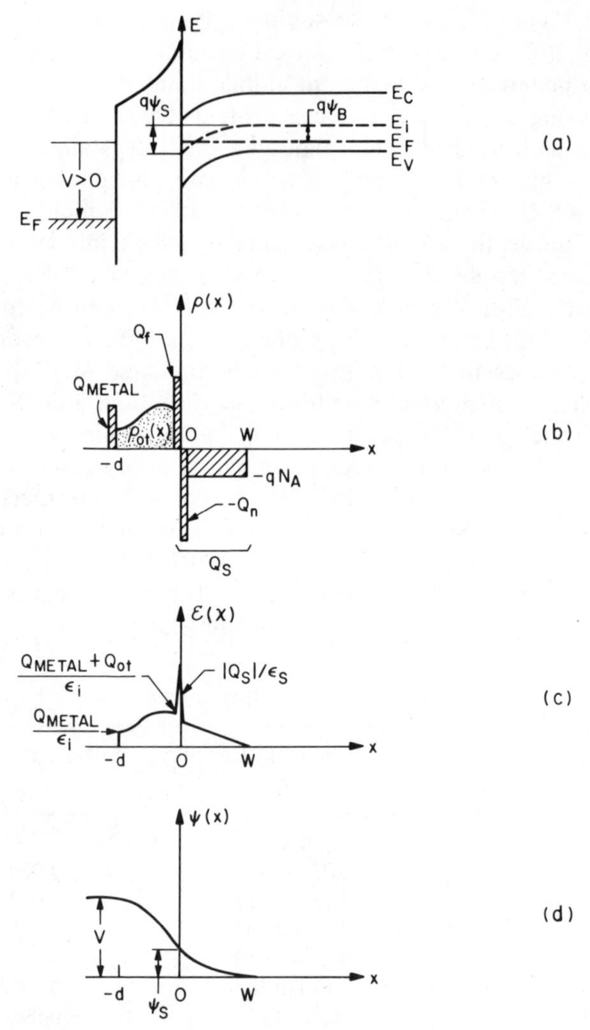

Fig. 26 An MIS diode with fixed oxide charge and oxide trapped charge. (*a*) Band diagram. (*b*) Charge distribution. (*c*) Field. (*d*) Potential.

Oxide trapped charge also can cause a voltage shift of the MOS $C-V$ curve. These oxide traps are associated with defects in SiO_2. The oxide traps are usually electrically neutral, and are charged by introducing electrons and holes into the oxide. Figure 26 shows the band diagram, the charge distribution, the electric field, and the potential for an MOS diode with both fixed oxide charge and oxide trapped charge. Comparing this figure with Fig. 6, note that for the same surface potential ψ_s, the applied voltage V is reduced, indicating a voltage shift of the $C-V$ curve toward negative voltage. The shift due to the oxide trapped charge is given by

$$\Delta V_{ot} = \frac{Q_{ot}}{C_i} = \frac{1}{C_i}\left[\frac{1}{d}\int_0^d x\rho_{ot}(x)\,dx\right] \tag{46}$$

where Q_{ot} is the effective net charge in bulk oxide traps per unit area at the Si–SiO_2 interface, and $\rho_{ot}(x)$ is the volume oxide-trap density. The total voltage shift due to all the oxide charges is given by

$$\Delta V = \Delta V_f + \Delta V_m + \Delta V_{ot} = \frac{Q_o}{C_i} \tag{47}$$

where $Q_o \equiv (Q_f + Q_m + Q_{ot})$ is the sum of the effective net oxide charge per unit area at the Si–SiO_2 interface.

7.3.3 Work-Function Difference and External Influences

Work-Function Difference For an ideal MIS diode, it has been assumed that the work-function difference (Fig. 2) for an *n*-type semiconductor,

$$\phi_{ms} \equiv \phi_m - \left(\chi + \frac{E_g}{2q} - \psi_B\right) \tag{48}$$

is zero. If the value of ϕ_{ms} is not zero, and if oxide charges Q_o exist, Eq. 47 (assuming negligible interface traps), the experimental capacitance–voltage curve, will be shifted from the ideal theoretical curve by an amount

$$V_{FB} = \phi_{ms} - \frac{Q_o}{C_i} = \phi_{ms} - \frac{Q_f + Q_m + Q_{ot}}{C_i} \tag{49}$$

where V_{FB} is called the flat-band voltage shift. If negligible mobile ions or oxide trapped charges exist, Eq. 49 reduces to

$$V_{FB} = \phi_{ms} - \frac{Q_f}{C_i}. \tag{49a}$$

The energy band for the Si–SiO_2 interface has been obtained from electron photoemission measurements;[37] SiO_2 bandgap is found to be about 9 eV, and the electron affinity ($q\chi_i$) is 0.9 eV. The metal work function on MOS diodes has been studied using the photoresponse measurement and the MOS capacitance–voltage measurement. Figure 27 shows the cube root of photoresponse versus photon energy for MOS diodes using various metals.[38] The intercept on the $h\nu$ axis corresponds to the metal–SiO_2 barrier energy $q\phi_B$. The metal work function is given by the sum of ϕ_B and χ_i, where χ_i is the electron affinity of the oxide (refer to Fig. 2). Similar results are obtained from the MOS capacitance measurements (Fig. 27, insert). From Eq. 49, if two different metals are deposited as field plates on the same oxidized silicon sample, the displacement between the two experimental MOS curves will represent the difference in metal work functions $(\phi_{m1} - \phi_{m2})$ or $(\phi_{B1} - \phi_{B2})$.

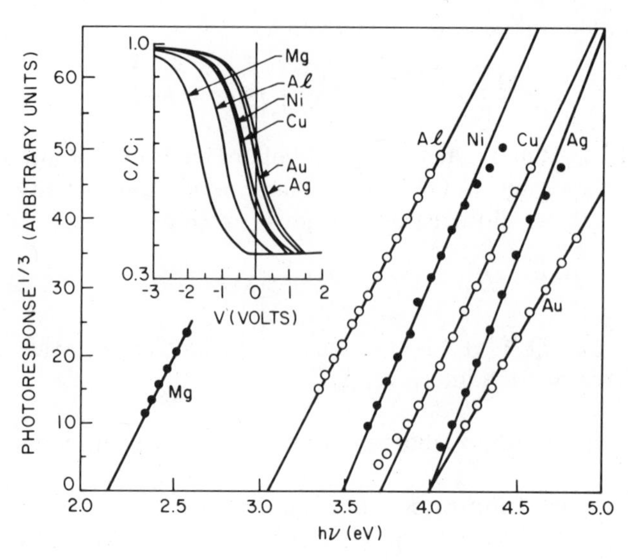

Fig. 27 Photoresponse versus photon energy for MIS diodes using various metals. Insert shows the corresponding *C–V* curves. (After Deal, Snow, and Mead, Ref. 38.)

Hence if the value of ϕ_m for any one metal is known, the ϕ_m values for other metals can be determined. The results are shown in Table 3; also shown are the vacuum work functions. The metal work functions as obtained from the photoresponse and the capacitance curve are in excellent agreement. These values are, however, different from that of the vacuum work function, which is no surprise, since the deposited metal films are polycrystalline and the oxide–metal interface is quite different from the vacuum–metal (single-crystal) interface as used in the vacuum work-function measurement. It has also been found using these methods that the silicon–silicon dioxide barrier is independent of silicon orientation to within 0.1 V.

Table 3 Metal Work Function (Volts)

Metal	ϕ_m (C–V)	ϕ_m (Photoresponse)	ϕ_m (Vacuum Work Function)
Mg	3.35	3.15	3.7
Al	4.1[a]	4.1	4.25
Ni	4.55	4.6	4.5
Cu	4.7	4.7	4.25
Au	5.0	5.0	4.8
Ag	5.1	5.05	4.3

[a]Value of ϕ_m for Al (4.1 V) is the sum of the barrier height (3.2 V) and the SiO_2 electron affinity (0.9 V).

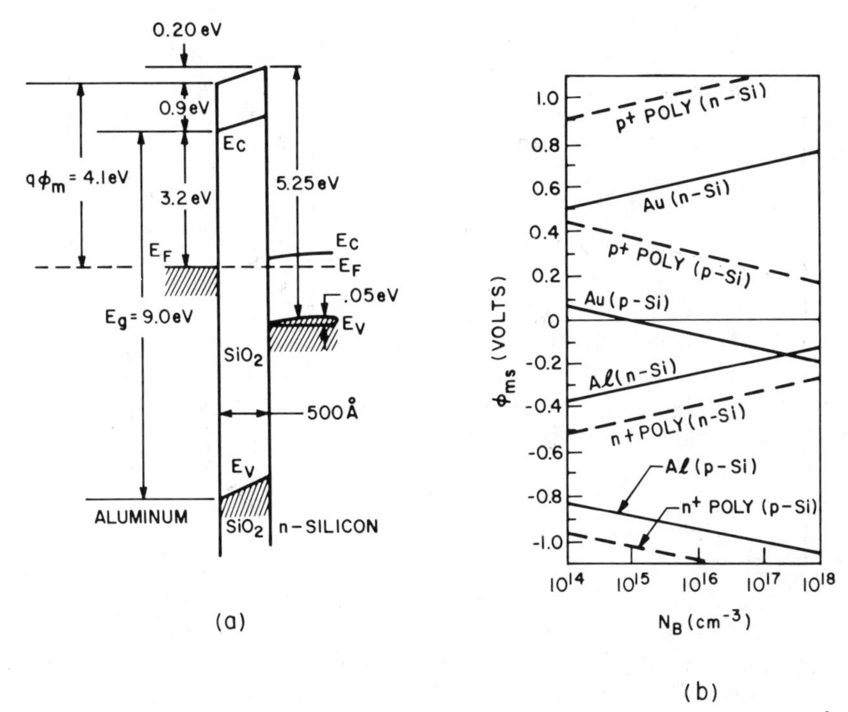

Fig. 28 (*a*) Energy-band diagram of Al–SiO_2–Si diode with oxide thickness of 500 Å and $N_A = 10^{16}$ cm^{-3}. (*b*) Work-function difference ϕ_{ms} versus doping for degenerate polysilicon, Al, and Au electrodes. (After Deal, Snow, and Mead, Ref. 38; Werner, Ref. 39.)

The foregoing results show that ϕ_{ms} can significantly affect the silicon surface potential, and that in evaluating the oxide fixed charge from C–V curves, the voltage shift must be corrected for ϕ_{ms}, Eq. 49. Figure 28*a* shows the band diagram of an MOS diode with an Al metal electrode and an oxide of 500 Å thermally grown on *n*-type Si (10^{16} cm^{-3}). Because for this case $\phi_m = 4.1$ V and the work function for Si ($\chi + E_g/2q - \psi_B$) is 4.35 V, the work-function difference ϕ_{ms} is -0.25 V (i.e., 4.1 to 4.35).

In modern integrated-circuit processing, heavily doped polysilicon has been used extensively to replace Al as the gate electrode. For an n^+ polysilicon gate, the Fermi level essentially coincides with the bottom of the conduction band, and the effective work function ϕ_m is equal to the Si electron affinity ($\chi_{Si} = 4.15$ V). For a p^+ polysilicon gate, the Fermi level coincides with the top of the valence band, and the effective work function ϕ_m is equal to the sum of χ_{Si} and E_g/q (5.25 V).

Figure 28*b* shows the work-function difference[38,39] as a function of Si doping concentration for Al, Au, p^+ polysilicon, and n^+ polysilicon gates. By an appropriate choice of electrode metal (or polysilicon), both *n*-type and *p*-type silicon surfaces can be varied from accumulation to inversion.

External Influences External influences such as temperature, illumination, ionization radiation, and hot-carrier injection can strongly affect the MOS diode behavior.

For an MOS diode, the charge in the inversion layer can communicate with the bulk under steady-state conditions only by means of generation–recombination processes. At room temperature the inversion-layer cutoff frequencies in $Si–SiO_2$ systems are normally below 100 Hz, sometimes below 1 Hz. At lower temperatures the buildup of the inversion charge is very slow. A forward bias of about 0.25 V must be reached across the space-charge region before a noticeable injection of the inversion charge occurs.[40] The true inversion capacitance–voltage curve can only be measured by allowing a long time for equilibration at each bias point. At elevated temperatures, however, the generation is more rapid, and the temperature effect on the MOS characteristics, especially the generation mechanisms, can be easily studied.

The appropriate equivalent circuit for the heavy inversion region was shown in Fig. 22*d*. Figure 29 shows[41] the total conductance G, $1/R_{gs} + 1/R_{gD} + 1/R_d$, for an *n*-type sample and plots it as a function of $1/T$. As discussed in Chapter 2, the space-charge recombination process is proportional to n_i with an activation energy of $E_g/2$, and the diffusion process is proportional to n_i^2 with an activation energy of E_g. Figure 29 shows that the space-charge recombination is the dominant effect ($1/R_{gD} \sim n_i$) up to about

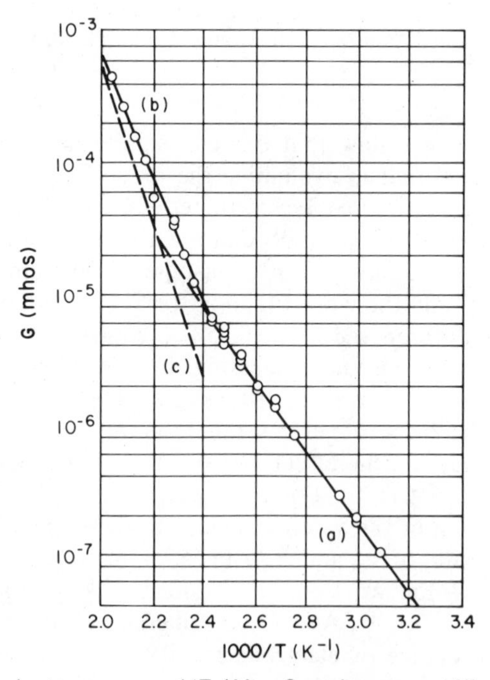

Fig. 29 Conductance versus $1/T$. (After Goetzberger and Nicollian, Ref. 41.)

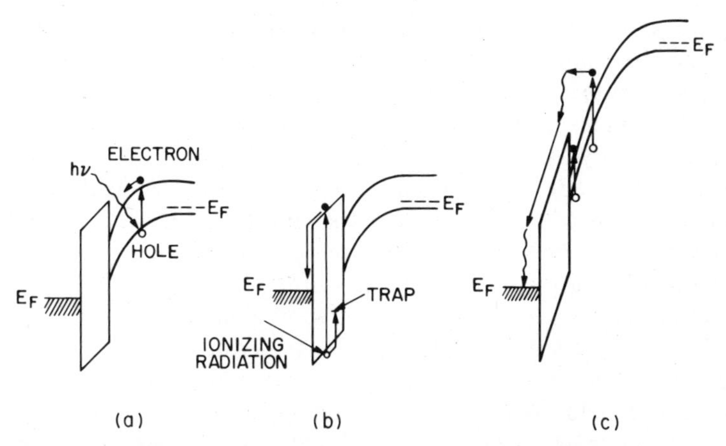

Fig. 30 Band diagrams of MIS diode under the following conditions: (*a*) illumination; (*b*) radiation; (*c*) avalanche injection. (After Nicollian and Brews, Ref. 7.)

140°C. In this range the experimental activation energy is 0.56 eV, line (a), in excellent agreement with the expected activation energy of n_i or $(E_g/2)$. Above 140°C a new process dominates, as shown by the break in the $1/T$ plot. The high-temperature region, line (c), is obtained by subtracting the influence of space-charge generation from the total conductance, line (b) – line (a). The activation energy is found to be 1.17 eV (E_g), corresponding to the expected result for the diffusion process ($1/R_d \sim n_i^2$). The foregoing results demonstrate the validity of the equivalent circuit shown in Fig. 22*d* for the heavy inversion region.[15]

The main effect of illumination on the MIS capacitance–voltage curves is that the capacitance in the heavy inversion region approaches the low-frequency value as the intensity of illumination is increased. Two basic mechanisms are responsible for this effect. The first is the decrease in the time constant τ_{inv} of minority-carrier generation in the inversion layer.[16] The second is the generation of electron–hole pairs by photons, which causes a decrease of the surface potential ψ_s under constant applied voltage [42] as shown in Fig. 30*a*. This decrease of ψ_s results in a reduction of the width of the space-charge layer, with a corresponding increase of the capacitance. The second mechanism is dominant when the measurement frequency is high.

Figure 30*b* shows the band diagram under ionization radiation such as x ray[43] or γ ray.[44] Ionization radiation passing through the oxide creates hole–electron pairs in the oxide by breaking Si–O bonds. The electric field applied across the oxide during radiation exposure drives the generated electrons toward the positive electrode (the gate) and drives the holes toward the Si–SiO$_2$ interface (see Fig. 30*b*). The electrons are considerably more mobile than the holes as they rapidly drift toward the positive electrode, where most flow out into the external circuit; the holes drift much more slowly toward the negative electrode and become trapped in the interfacial region near the silicon. The trapped holes at the Si–SiO_2 interface constitute the radiation-induced positive oxide charge observed. These trapped holes may also be responsible for the increased interface trap density usually associated with ionizing radiation.[7]

If the gate electrode is biased negatively with respect to the silicon, the same phenomenon will occur, except that the hole trapping will occur at the metal–SiO_2 interface. In this case, the effective oxide trapped charge, Eq. 46, is small, and the C–V curve has little shift along the voltage axis. Experimental results agree with such a trapping model.

In avalanche injection,[45] the MOS diode is operated in the deep depletion condition, Fig. 7, curve (c). Charge carriers are accelerated by a rapidly applied electric field. When the gate voltage is high enough to produce avalanche breakdown in the silicon, carriers generated in the surface-depletion layer will be accelerated to sufficient energy for impact ionization to occur, creating a plasma of energetic hole–electron pairs in the silicon surface, Fig. 30*c*. Some of the electrons or holes created will have enough energy to surmount the interfacial energy barrier and enter the SiO_2. The energy barrier for electron injection is 3.2 eV (i.e., $q\chi_{Si} - q\chi_i = 4.1–0.9$), whereas for hole injection it is 4.7 eV [i.e., $E_g(SiO_2) - E_g(Si) - q\chi_{Si} - q\chi_i$]. Generally, electrons have a higher injection probability because of the lower energy barrier.

The avalanche breakdown in the MOS diode under the deep depletion condition has been calculated[46] based on a two-dimensional model (see Fig. 31, insert). The electric field along the Si–SiO_2 interface shows a peak value $\mathscr{E}_m$ near the edge of the gate electrode. The breakdown voltage is defined as the gate voltage that makes the ionization integral equal to unity when integrated along the line from $\mathscr{E}_m$ to a point at the depletion-layer boundary. Figure 31 shows the result for various oxide thicknesses and doping concentrations. As can be seen, for a given oxide thickness, a doping exists which yields minimum breakdown voltage; edge breakdown (i.e., $\mathscr{E}_m > \mathscr{E}_1$ shown in the insert) dominates to the left of the minimum, and uniform breakdown (i.e., $\mathscr{E}_m = \mathscr{E}_1$) dominates to the right. Under the uniform breakdown condition, the surface breakdown field $\mathscr{E}_1$ in Si increases with doping (Chapter 2). Therefore, the oxide field $\mathscr{E}_1\epsilon_s/\epsilon_i$ will increase, resulting in higher voltages. To avoid edge breakdown, the ratio of d/W_{max} must be larger than 0.3, where W_{max} is the maximum depletion-layer width at breakdown.

Figure 32 shows the capacitance and equivalent parallel conductance as functions of voltage for a steam-grown oxide before and after avalanche injection of electrons. The hot-electron injection causes a flat-band shift of the C–V curve toward more positive voltage, indicating an increase of negative oxide charges. The increase of the equivalent parallel conductance also shows an increase of interface-trap density (from 1.2×10^{11} to 7.9×10^{11} cm^{-2}/eV). Hot-carrier or avalanche injection is closely related to many

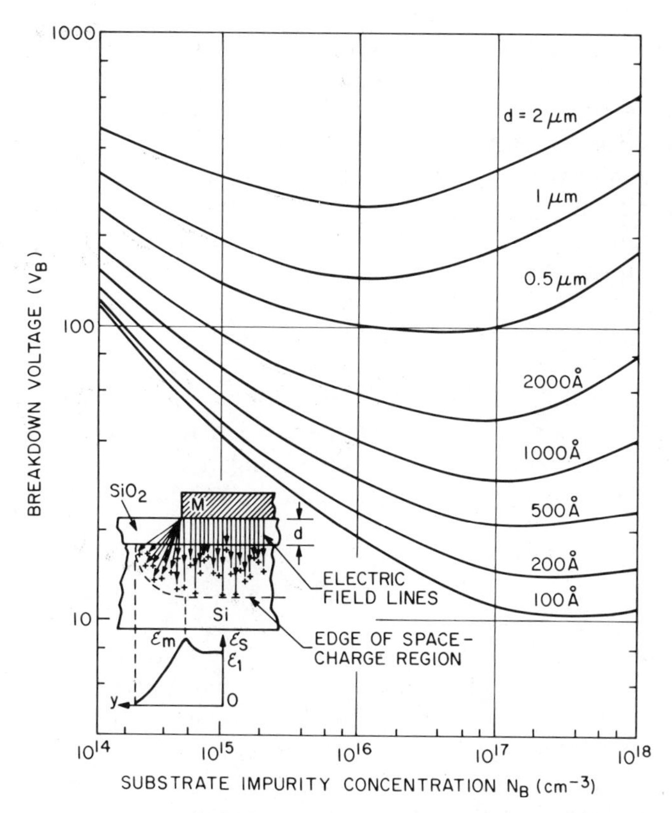

Fig. 31 Breakdown voltage of MIS diode in deep-depletion condition versus silicon doping concentration with oxide thickness as the parameter. The insert shows the electric field variation along the Si surface. (After Rusu and Bulucea, Ref. 46.)

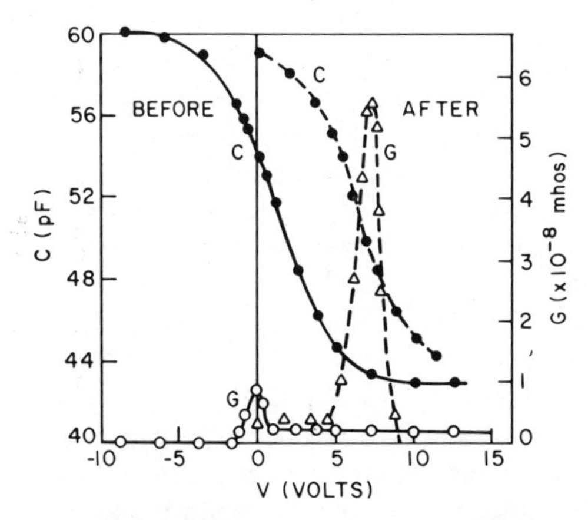

Fig. 32 Capacitance and conductance versus voltage for a MIS diode before and after avalanche injection of electrons. (After Nicollian and Brews, Ref. 7.)

MOS device operations. For example, in a MOSFET, channel carriers can be accelerated by the source-to-drain electric field to sufficient energy to surmount the Si–SiO_2 interfacial energy barrier. Once in the oxide, some carriers may get trapped, creating both fixed oxide charge and interface traps. These effects are undesirable because they cause a change in device characteristics during operation. On the other hand, avalanche injection from the drain junction of a MOSFET across an oxide layer will charge up a polysilicon gate. This idea has been used as a memory element.

7.3.4 Carrier Transport in Insulating Films

In an ideal MIS diode the conductance of the insulating film is assumed to be zero. Real insulators, however, show carrier conduction when the electric field or temperature is sufficiently high. To estimate the electric field in an insulator under biasing conditions, we obtain from Eqs. 16 and 24 that

$$\mathscr{E}_i = \mathscr{E}_s \left(\frac{\epsilon_s}{\epsilon_i}\right) \tag{50}$$

where $\mathscr{E}_i$ and $\mathscr{E}_s$ are the electric fields in the insulator and the semiconductor, respectively, and ϵ_i and ϵ_s are the corresponding permittivities. For the Si–SiO_2 system, the field for silicon at avalanche breakdown[47] is about 3×10^5 V/cm; the corresponding field in the oxide is then three times larger ($\epsilon_{Si}/\epsilon_{SiO_2} = 11.7/3.9$), that is, about 10^6 V/cm. At this field the electron and hole conduction in the SiO_2 are negligible even at elevated temperatures. However, mobile ions such as sodium can transport through the oxide and give rise to device instability and a hysteresis effect. For ultrathin SiO_2 or under a very high electric field, tunneling will occur. In either Si_3N_4 or Al_2O_3 the conductance is generally much higher than in SiO_2.

Table 4 summarizes the basic conduction processes in insulators. The Schottky emission process is similar to the process discussed in Chapter 5, where thermionic emission across the metal–insulator interface or the insulator–semiconductor interface are responsible for carrier transport. A plot of $\ln(J/T^2)$ versus $1/T$ yields a straight line with a slope determined by the permittivity ϵ_i of the insulator. The Frenkel–Poole emission,[48] shown in the insert of Fig. 33, is due to field-enhanced thermal excitation of trapped electrons into the conduction band. For trap states with coulomb potentials, the expression is virtually identical to that of the Schottky emission. The barrier height, however, is the depth of the trap potential well, and the quantity $\sqrt{q/\pi\epsilon_i}$ is larger than in the case of Schottky emission by a factor of 2, since the barrier lowering is twice as large due to the immobility of the positive charge. The tunnel emission is caused by field ionization of trapped electrons into the conduction band or by electrons tunneling from the metal Fermi energy into the insulator conduction band. The tunnel emission has the strongest dependence on the applied voltage but is essentially independent of the temperature. The space-charge-limited current results from a carrier

Table 4 Basic Conduction Processes in Insulators

Process	Expression[a]	Voltage and Temperature Dependence[b]
Schottky emission	$J = A^*T^2 \exp\left[\frac{-q(\phi_B - \sqrt{q\mathscr{E}/4\pi\epsilon_i})}{kT}\right]$	$\sim T^2 \exp(+a\sqrt{V}/T - q\phi_B/kT)$
Frenkel–Poole emission	$J \sim \mathscr{E} \exp\left[\frac{-q(\phi_B - \sqrt{q\mathscr{E}/\pi\epsilon_i})}{kT}\right]$	$\sim V \exp(+2a\sqrt{V}/T - q\phi_B/kT)$
Tunnel or field emission	$J \sim \mathscr{E}^2 \exp\left[-\frac{4\sqrt{2m^*}(q\phi_B)^{3/2}}{3q\hbar\mathscr{E}}\right]$	$\sim V^2 \exp(-b/V)$
Space-charge-limited	$J = \frac{8\epsilon_i \mu V^2}{9d^3}$	$\sim V^2$
Ohmic	$J \sim \mathscr{E} \exp(-\Delta E_{ae}/kT)$	$\sim V \exp(-c/T)$
Ionic conduction	$J \sim \frac{\mathscr{E}}{T} \exp(-\Delta E_{ai}/kT)$	$\sim \frac{V}{T} \exp(-d'/T)$

[a] A^* = effective Richardson constant, ϕ_B = barrier height, $\mathscr{E}$ = electric field, ϵ_i = insulator dynamic permittivity, m^* = effective mass, d = insulator thickness, ΔE_{ae} = activation energy of electrons, ΔE_{ai} = activation energy of ions, and $a \equiv \sqrt{q/(4\pi\epsilon_i d)}$.
[b] $V = \mathscr{E}d$. Positive constants independent of V or T are b, c, and d'.

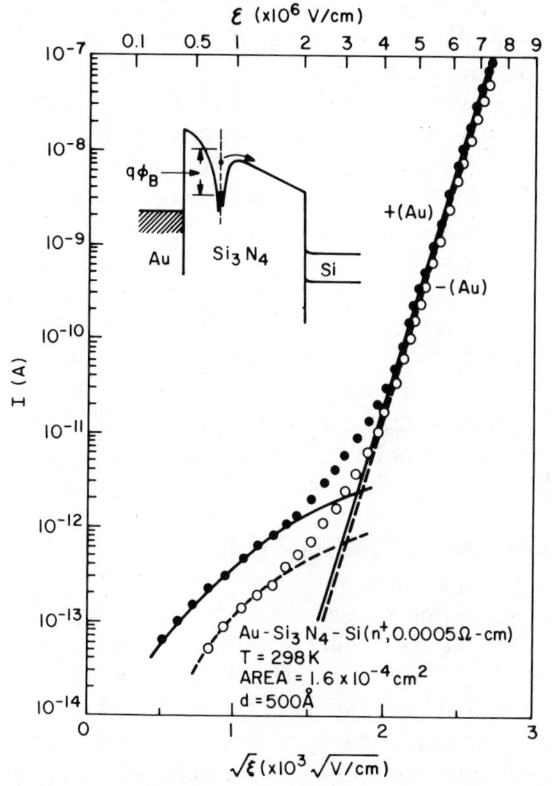

Fig. 33 Current–voltage characteristics of Au–Si_3N_4–Si diode at room temperature. Insert shows Frenkel–Poole emission from trapped electrons. (After Sze, Ref. 50.)

injected into the insulator, where no compensating charge is present. The current for the unipolar trap-free case is proportional to the square of the applied voltage. At low voltage and high temperature, current is carried by thermally excited electrons hopping from one isolated state to the next. This mechanism yields an ohmic characteristic exponentially dependent on temperature. The ionic conduction is similar to a diffusion process. Generally, the dc ionic conductivity decreases during the time the electric field is applied, because ions cannot be readily injected into or extracted from the insulator. After an initial current flow, positive and negative space charges will build up near the metal–insulator and the semiconductor–insulator interfaces, causing a distortion of the potential distribution. When the applied field is removed, large internal fields remain which cause some, but not all, ions to flow back toward their equilibrium position; hysteresis effects result.

For a given insulator, each conduction process may dominate in certain temperature and voltage ranges. The processes are also not exactly independent of one another and should be carefully examined. For example, for the large space-charge effect, the tunneling characteristic is found to be very similar to the Schottky-type emission.[49]

An example of the conduction processes for silicon nitride films[50] is shown in Fig. 33. The films are deposited on degenerate silicon substrate (0.0005 Ω-cm n-type) by the process of reaction of $SiCl_4$ and NH_3 at 1000°C. An MIS diode is made by evaporating Au onto the Si_3N_4 film. Figure 33 shows $\ln I$ versus $\sqrt{\mathscr{E}}$. The +(Au) curve is for the gold-electrode positive and the −(Au) is for the gold-electrode negative. Note that the two curves are virtually identical. The slight difference (especially at low fields) is believed to be mainly due to the difference in barrier heights at the gold–nitride and nitride–silicon interfaces. Also note that there are two distinct regions. In high electric fields the current varies exponentially with the square root of the field; at low fields, the characteristic is ohmic. It has been found that at room temperature for a given field, the characteristics of current density versus field are essentially independent of the film thickness, device area, electrode materials, and polarity of the electrodes. These results strongly suggest that the current is bulk-controlled rather than electrode-controlled as in Schottky-barrier diodes. At low temperatures the current–voltage characteristic becomes nearly independent of the temperature.

Figure 34 shows plots of current density versus $1/T$ for three different insulators. For a Si_3N_4 film with an applied field of 5.3 MV/cm, the conduction current can be separated into three components: J_1, J_2, and J_3. The current J_1 is due to Frenkel–Poole emission, which dominates at high temperatures and high fields. The dynamic dielectric constant, ϵ_i/ϵ_0, obtained from the slopes of Fig. 34 is found to be 5.5, in agreement with the optical measurement.[36] The current J_2 is due to tunnel emission of trapped electrons into the conduction band, which dominates at low temperatures and high fields. The current J_3 is the ohmic component, which

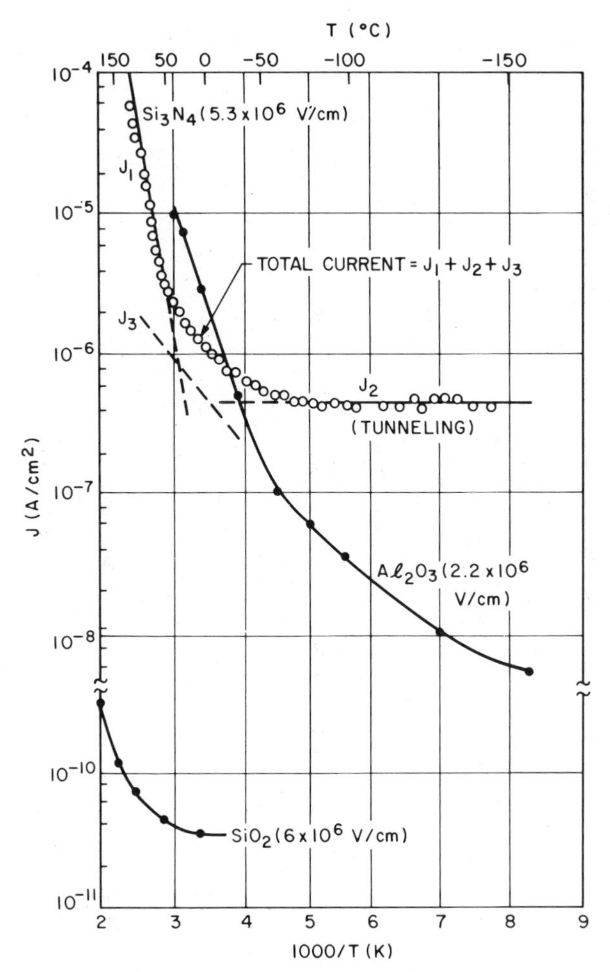

Fig. 34 Current density versus $1/T$ for Si_3N_4, Al_2O_3, and SiO_2 films. (After Sze, Ref. 50; Johnson, Ref. 51; Av-Ron et al., Ref. 52.)

contributes at low fields and moderate temperatures. The Al_2O_3 film shows similar behavior.[51] The SiO_2 film has the lowest conductance at a given field.[52] For example, at 6 MV/cm the room-temperature current density in SiO_2 is about 4×10^{-11} A/cm^2, which is many orders of magnitude lower than Si_3N_4 or Al_2O_3.

Another important parameter in the insulator is the maximum dielectric strength, also called the breakdown field. The essential breakdown processes in insulators are thermal and electric breakdowns.[53] These are competing processes and the process with the lower breakdown field dominates, causing the breakdown. At low temperatures, the breakdown is due to electric processes and is nearly independent of temperature; at high tem-

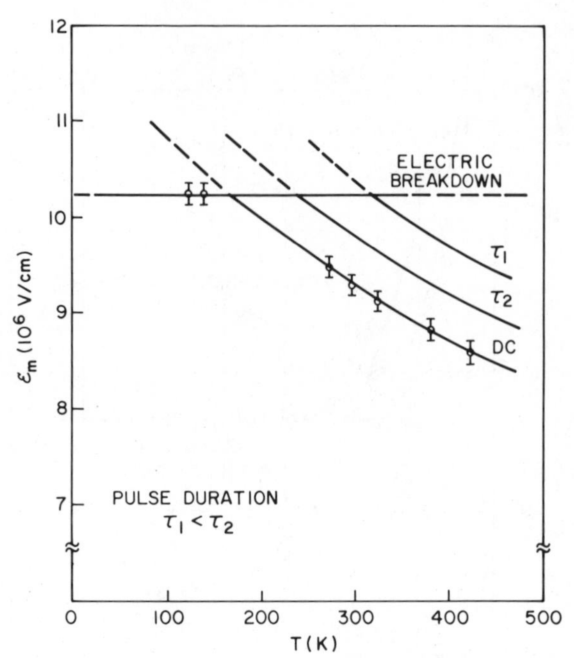

Fig. 35 Breakdown field versus temperature in Si_3N_4. (After Klein, Ref. 53; Sze, Ref. 50.)

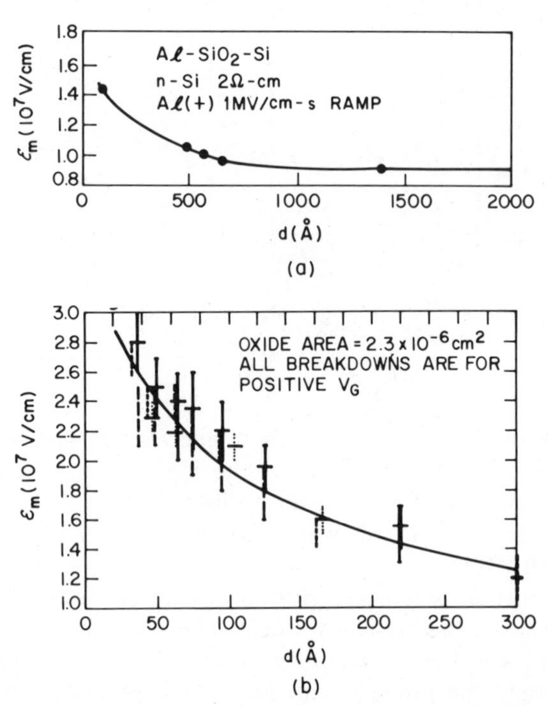

Fig. 36 Breakdown field versus oxide thickness in SiO_2. (*a*) From 100 to 2000 Å. (*b*) From 40 to 300 Å. (After Osburn and Ormond, Ref. 54; Harari, Ref. 55.)

peratures, thermal breakdown occurs. An example of the breakdown field strength versus temperature characteristics is shown in Fig. 35. Also shown are data points for Si_3N_4. The value of $\mathscr{E}_m$ for thermal breakdown can be obtained by equating the Joule heat with that lost by heat transfer and is given by[50]

$$\mathscr{E}_m \simeq \left(\frac{\pi \epsilon_i}{q}\right)(\phi_B - CT)^2 \tag{51}$$

where ϕ_B is the barrier height and C is a slowly varying function of T and is dependent on the pulse duration. Note from Fig. 35 that at low temperature, $\mathscr{E}_m$ approaches 10^7 V/cm in Si_3N_4 films; and at high temperature, $\mathscr{E}_m$ varies approximately as Eq. 51.

For thermally grown SiO_2 films, the $\mathscr{E}_m$ versus T behavior is similar to that of Si_3N_4. The maximum field strength in SiO_2 is found to be dependent on film thickness, and $\mathscr{E}_m$ increases with decreasing thickness as shown[54,55] in Fig. 36. For films with thicknesses above 1000 Å, $\mathscr{E}_m$ is essentially constant at 9×10^6 V/cm (Fig. 36*a*). For very thin SiO_2 films, $\mathscr{E}_m$ can approach 3×10^7 V/cm, which appears to be close to the maximum field strength of the Si–O bond (Fig. 36*b*).

7.4 CHARGE-COUPLED DEVICE

A charge-coupled device (CCD) is basically an array of closely spaced MOS diodes. In operation, the information is represented by a quantity of electric charge (called a charge packet), which is distinct from conventional devices, where current and voltage levels are generally used. Under the application of a proper sequence of clock voltage pulses, the MOS diode array is biased into the deep depletion, and the charge packet can be stored and transferred in a controlled manner across a semiconductor substrate.

The basic types of CCDs are surface channel (SCCD) and buried channel (BCCD). In the SCCD, charge is stored and transferred at the semiconductor surface, whereas in the BCCD the doping of the semiconductor substrate is modified so that storage and transfer of the charge packet takes place in the bulk semiconductor just beneath the semiconductor surface. Several different types of electrode configurations and clocking techniques are also used in making a practical CCD.

7.4.1 Charge Storage

An MOS diode operated in deep depletion is the basic element of a CCD (surface channel). Figure 37*a* shows the energy-band diagram for deep depletion with a zero signal-charge ($Q_{sig} = 0$) condition, where ψ_{so} is the surface potential, and $(V_G - V_{FB})$ is the effective applied gate voltage (the voltage applied to the metal-gate electrode will now be called the gate voltage

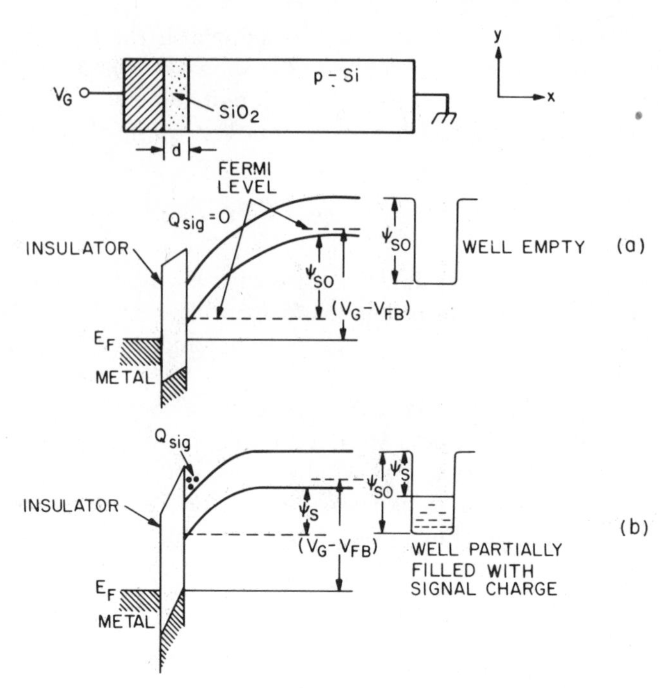

Fig. 37 Energy-band diagrams for a surface-channel MOS diode. (*a*) Band bending at deep depletion and the empty potential well. (*b*) Band bending at the Si–SiO_2 interface and the partially filled potential-well representation. (After Barbe, Ref. 56.)

V_G, and V_{FB} is the flat-band voltage shift).[56] The potential well formed by the potential minimum at the semiconductor surface is also shown. For $Q_{sig} = 0$, the well is empty.

When a signal charge packet is stored at the semiconductor surface, the surface potential decreases, corresponding to the filling of the potential well as shown in Fig. 37*b*.

For an empty well, the depth of the potential well, ψ_{so}, can be obtained from the previously derived equations. From Eqs. 23, 24, and 27a, the gate voltage and the surface potential ψ_s under deep depletion are given by

$$V_G - V_{FB} = V_i + \psi_s = \frac{qN_AW}{C_i} + \psi_s \tag{52}$$

$$\psi_s = qN_AW^2/2\epsilon_s . \tag{53}$$

The depletion width W can be larger than W_m, the maximum depletion width under steady state. Eliminating W from Eqs. 52 and 53 yields a relationship between the gate voltage and the surface potential,

$$V_G - V_{FB} = \psi_s + \frac{1}{C_i}\sqrt{2\epsilon_s qN_A\psi_s} . \tag{54}$$

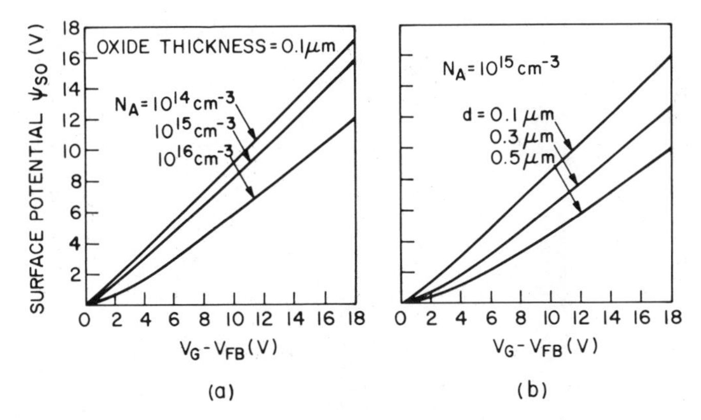

Fig. 38 (*a*) Surface potential versus gate voltage for 0.1-μm SiO_2. (*b*) Surface potential versus gate voltage for acceptor doping of 10^{15} cm^{-3}. (After Kim, Ref. 11.)

Figure 38*a* shows this relationship for various dopings[11] with $d = 0.1$ μm, and Fig. 38*b* shows it for various oxide thicknesses with $N_A = 10^{15}$ cm^{-3}. Note that for a given gate voltage, a different surface potential can be obtained by varying either d or N_A. For example, for $(V_G - V_{FB}) = 10$ V and $N_A = 10^{15}$ cm^{-3}, the surface potential decreases from 8.5 V to 4.2 V when the oxide thickness increases from 0.1 μm to 0.5 μm. This fact will be used later to construct the two-phase CCD and to discuss channel confinement.

When a signal charge packet is stored at the semiconductor surface, the surface field and the oxide field become

$$\mathscr{E}_s = (Q_{sig} + qN_A W)/\epsilon_s \tag{55a}$$

$$\mathscr{E}_i = (Q_{sig} + qN_A W)/\epsilon_i \tag{55b}$$

where Q_{sig} is the stored signal charge per unit surface area. The gate voltage is then given by

$$V_G - V_{FB} = \frac{Q_{sig}}{C_i} + \frac{qN_A W}{C_i} + \psi_s\,. \tag{56}$$

The surface potential is again given by Eq. 53. Therefore, the gate voltage becomes

$$V_G - V_{FB} = \frac{Q_{sig}}{C_i} + \frac{\sqrt{2\epsilon_s qN_A\psi_s}}{C_i} + \psi_s\,. \tag{57}$$

Equation 57 can be solved for ψ_s, resulting in

$$\psi_s = V'_G + V_0 - (2V'_G V_0 + V_0^2)^{1/2} \tag{58}$$

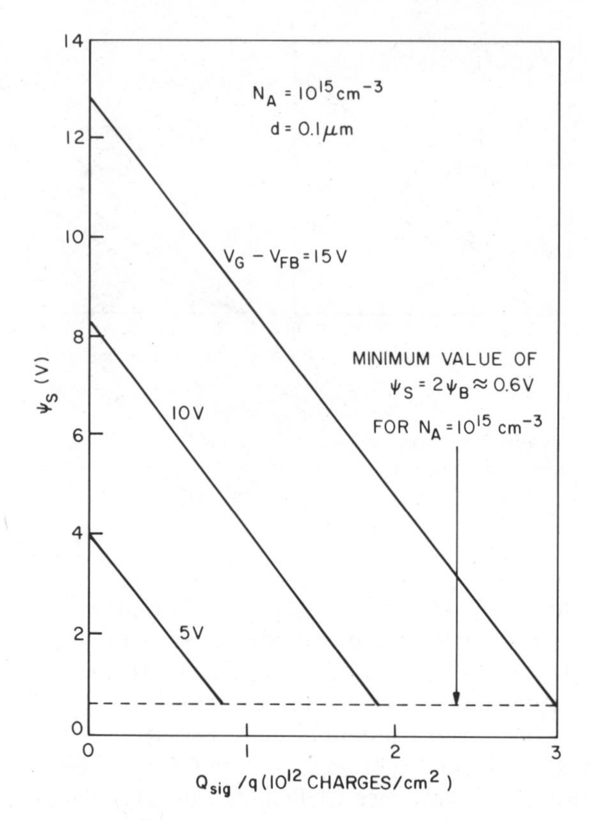

Fig. 39 Variation of surface potential versus signal charge density and gate voltage. (After Beynon, Ref. 57.)

with

$$V'_G \equiv V_G - V_{FB} - \frac{Q_{sig}}{C_i} \tag{58a}$$

$$V_0 \equiv qN_A\epsilon_s/C_i^2 . \tag{58b}$$

Figure 39 shows the surface potential as a function of stored signal charge for various gate voltages.[57] For a given gate voltage, ψ_s decreases essentially linearly as the stored charge increases. The potential-well model is justified, because this linear relationship between ψ_s and Q_{sig} suggests a simple hydraulic system for the charge-storage mechanism. Filling a potential well, that is, carriers are present in the well, causes the surface potential to decrease in a linearly related fashion (Fig. 37*b*).

To convert the signal charge per unit area Q_{sig} (C/cm^2) to surface volume density n_s (electrons/cm^3), where $n_s = n_{po} \exp(q\psi_s/kT)$, Eq. 4, we first obtain the depletion-layer width from Eqs. 53 and 56:

$$W = \frac{\epsilon_s}{C_i}\left(\sqrt{1+\frac{V_G'}{2V_0}}-1\right) \tag{59}$$

where V_0 is given by Eq. 58b. The surface potential for the deep depletion is given by Eqs. 13 and 15 with the second and fourth terms of Eq. 13 dominating:

$$\begin{aligned}\mathscr{E}_s^2 &\simeq 2\left(\frac{kT}{qL_D}\right)^2\left(\frac{q\psi_s}{kT}+\frac{n_{po}}{N_A}\,e^{q\psi_s/kT}\right)\\ &= \left(\frac{qN_AW}{\epsilon_s}\right)^2 + 2\left(\frac{kT}{qL_D}\right)^2\frac{n_s}{N_A}.\end{aligned} \tag{60}$$

Combining Eqs. 55a and 60 and solving for n_s, we obtain

$$n_s = \frac{WQ_{\text{sig}}}{qL_D^2} + \frac{1}{2N_A}\left(\frac{Q_{\text{sig}}}{qL_D}\right)^2. \tag{61}$$

The relationship[58] of n_s versus Q_{sig} as calculated from Eq. 61 is shown in Fig. 40. As can be seen, over the signal charge range of 4×10^{10} to 10^{12} cm^{-2}, the surface volume density n_s varies as $(Q_{\text{sig}})^{3/2}$.

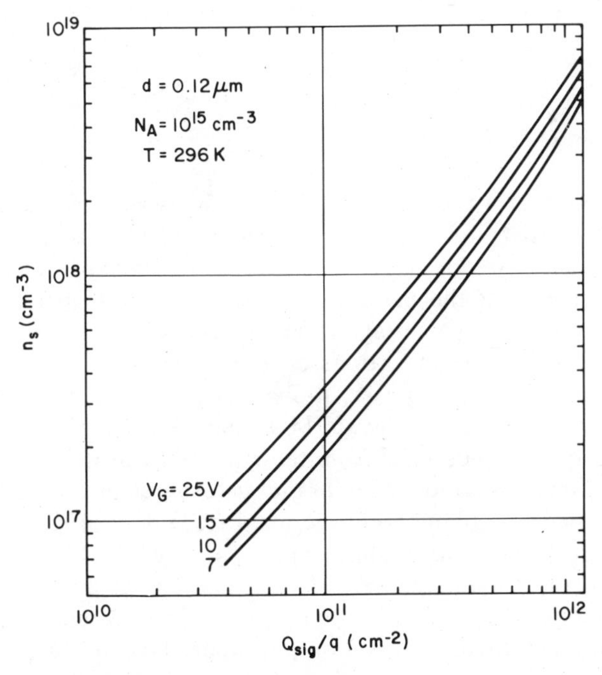

Fig. 40 Electron surface volume concentration as a function of surface-charge density and gate voltage. (After Ong and Pierret, Ref. 58.)

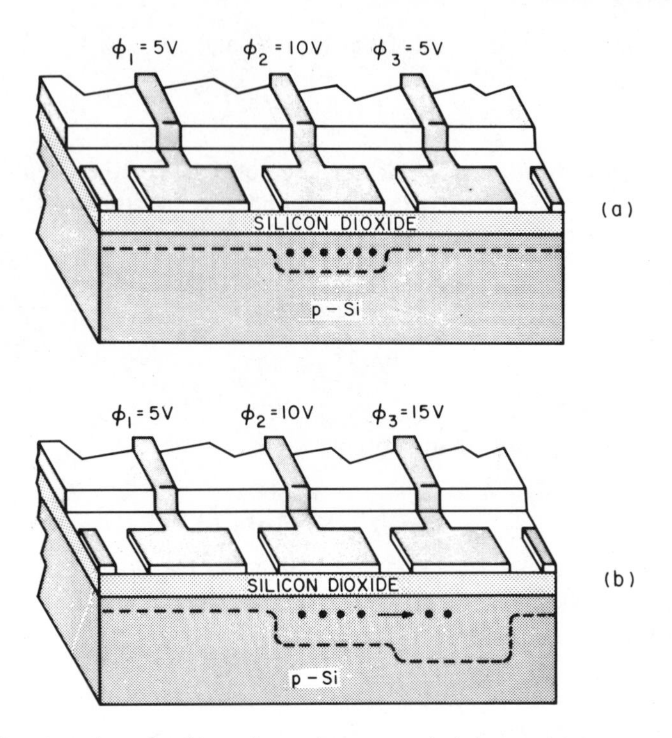

Fig. 41 Cross section of a three-phase charge-coupled device. (*a*) High voltage on ϕ_2. (*b*) ϕ_3 is pulsed to higher voltage for charge transfer. (After Boyle and Smith, Ref. 59.)

7.4.2 Basic CCD Structure

A cross section of a typical three-phase CCD is shown[59] in Fig. 41. The basic device consists of a closely spaced array of MOS diodes on a continuous insulator (oxide) layer that covers the semiconductor substrate. Figure 41*a* shows that the middle gate, which has a higher applied gate voltage, serves as a charge-storage element. As the right gate is pulsed to a higher voltage charges begin to transfer to it (Fig. 41*b*).

Figure 42 shows[11] a three-phase, *n*-channel CCD together with its basic input and output structures. The six MOS diodes or electrodes connected to the ϕ_1, ϕ_2, and ϕ_3 clock lines form the main body of the CCD; the input diode, input gate, output diode, and output gate are elements that inject and detect charge packets to and from the main CCD body. Figure 43*a* shows the clock waveforms and output signal for the CCD,[11] and Fig. 43*b* illustrates the corresponding potential wells and charge distributions.

At $t = t_1$, clock line ϕ_1 is at a high voltage and ϕ_2 and ϕ_3 are at low voltages. The input diode (ID) and output diode (OD) are biased with high positive voltages to prevent inversion of the surface under the input gate (IG) and output gate (OG). Thus the surfaces under IG and OG are in deep

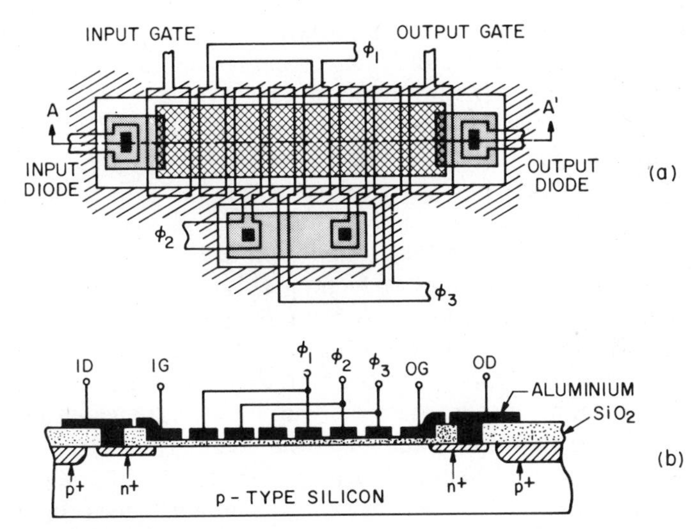

Fig. 42 An *n*-channel charge-coupled device. (a) Layout of the device. (*b*) Cross-sectional diagram along *AA'*. (After Kim, Ref. 11.)

depletion, and ID and OD cannot supply electrons into the main CCD array (i.e., all potential wells under the CCD array are empty). The potential well under ϕ_1 will be deeper than the potential wells under ϕ_2 and ϕ_3 because a higher voltage is applied to ϕ_1 at $t = t_1$. At $t = t_2$, the voltage of ID is lowered so that electrons can inject (flow) to the potential well under the first ϕ_1 electrode through the IG. At the end of injection, the surface

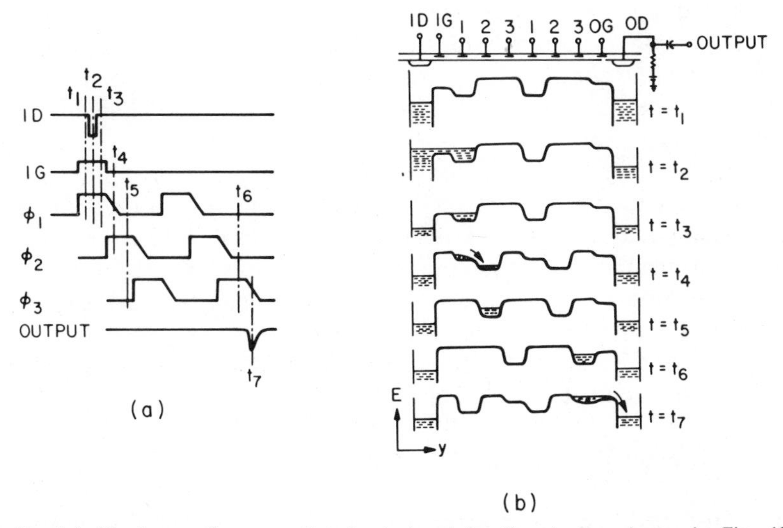

Fig. 43 (*a*) Clock waveforms and output signal for the device shown in Fig. 42. (*b*) Potential and charge distribution of the device shown in Fig. 42, with the clock shown in (*a*). (After Kim, Ref. 11.)

potentials under the IG and ϕ_1 electrode will be the same as the input diode potential. Electrons are now stored, under the IG and the first ϕ_1 electrode. At $t = t_3$, the voltage of ID is returned to a high value; electrons under the IG and the excess electrons under the first ϕ_1 electrode will be taken out of the device through the ID lead, creating a well-defined charge packet under the first ϕ_1 electrode. At $t = t_4$, the voltage applied to ϕ_1 is returning to the low value while the ϕ_2 electrodes have high voltage applied to them. The electrons stored under ϕ_1 are then transferred to the ϕ_2 electrode because the surface potential under ϕ_2 is higher. This process is called charge transfer. The voltage on ϕ_1 has a slowly falling edge, because the charge carriers require a finite time to transport across the width of the electrode. At $t = t_5$, the charge-transfer process is completed and the original charge packet is now stored under the first ϕ_2 electrode. This process is repeated and at $t = t_6$ the injected charge packet is stored under the second ϕ_3 electrode. At $t = t_7$, the voltage of the ϕ_3 electrodes is returning to the low value and pushing the electrons to the output diode, thereby giving an output signal proportional to the size of the charge packet at the output terminal (Fig. 43*a*).

For analog and memory devices, charge packets are introduced by applying suitable voltage to a *p-n* junction at the input of the CCD as described above. For optical imaging applications, the charge packets are formed as a result of electron–hole pair generation caused by light incident on the semiconductor substrate. The magnitude of the output signal will be proportional to the light intensity.

Many electrode structures and clocking schemes have been proposed and implemented. Two ways of fabricating three-phase electrode structures, shown[60] in Fig. 44*a* and *b*, avoid the need for making the small gaps (1–2 μm) required for the single-level structure shown in Fig. 42. Figure 44*c* and *d* show two structures with overlapping polysilicon electrodes which are clocked with four and two phases, respectively. A particular advantage of the structures Fig. 44*b* and *d* is that they are electrically immune to intralevel shorts.

To keep the signal charge confined in the lateral direction,[10] a lateral confinement (channel stop) of the transfer channel has to be defined (Fig. 45*a*). A combination of a thick field oxide outside the active transfer area with a light channel ion implantation is an attractive approach (Fig. 45*b*). Figure 45*c* shows a diffusion channel stop clamping the surface potential and defining the extent of the channel. One also can use a field shield, formed by an additional electrode placed under the transfer electrodes and biased into the accumulation (Fig. 45*d*).

7.4.3 Charge Transfer and Frequency Response

The three basic charge-transfer mechanisms are thermal diffusion, self-induced drift, and fringing field effect. For a small amount of signal charge, thermal diffusion is the dominant transfer mechanism. The total charge

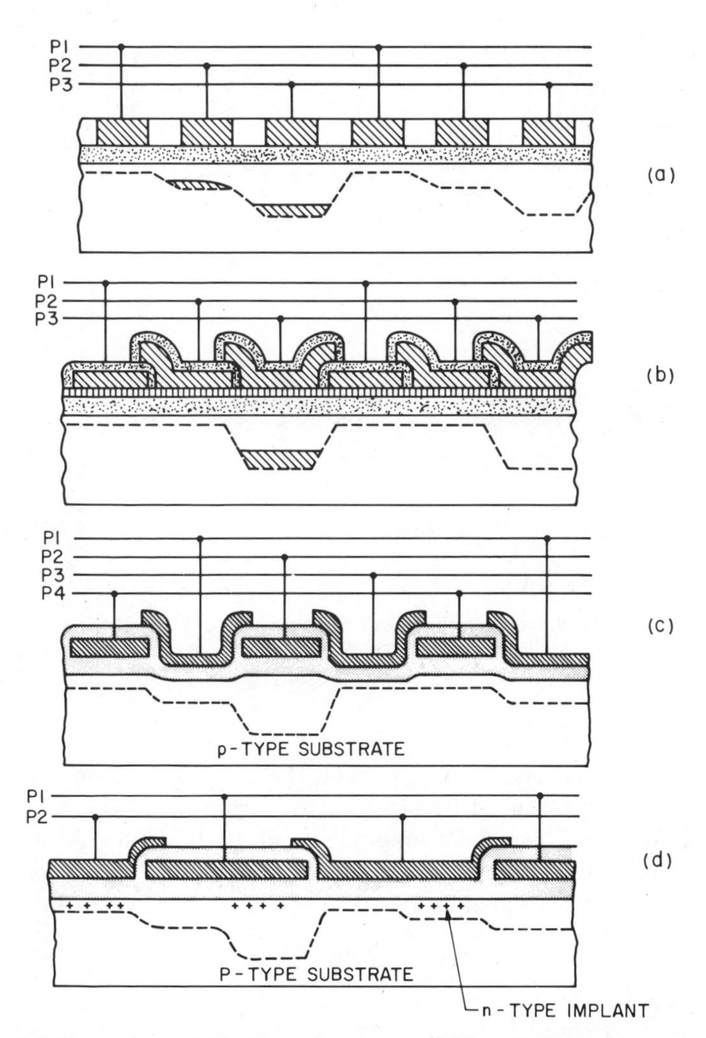

Fig. 44 (a) Three-phase CCD electrode structure with selectively doped regions in a layer of intrinsic polysilicon. (*b*) Three-phase CCD using three overlapping levels of oxidized silicon. (*c*) Basic four-phase, two-level polysilicon electrode structure. (*d*) Coplanar two-phase, two level polysilicon electrode structure. (After Tompsett, Ref. 60.)

under the storage electrode decreases exponentially with time, and the time constant is given by[11]

$$\tau_{th} = \frac{4L^2}{\pi^2 D_n} \tag{62}$$

where L is the length of the electrode and D_n is the minority carrier diffusion constant.

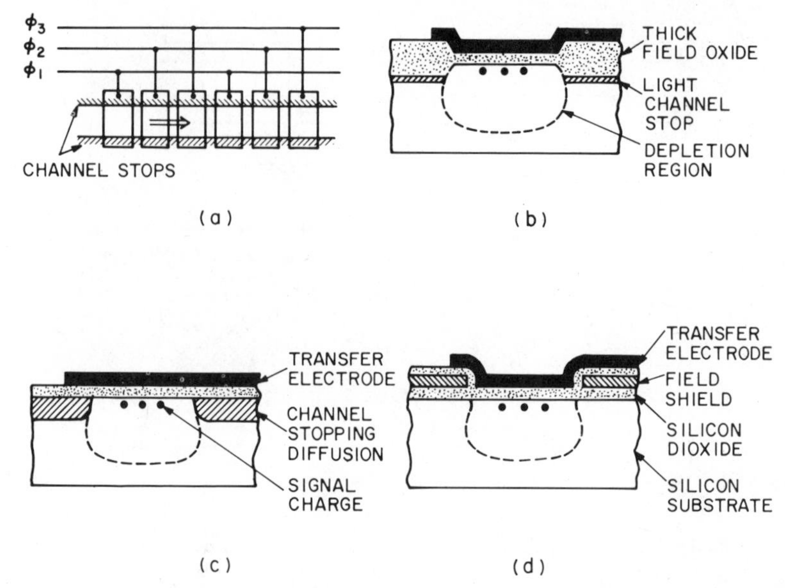

Fig. 45 Methods of channel confinement of the width of CCD channel. (*a*) Channel stops along the transfer electrodes. (*b*) Thick field oxide. (*c*) Diffusion channel stop. (*d*) Polysilicon field shield. (After Sequin and Tompsett, Ref. 10.)

For a reasonably large charge packet, the transfer is dominated by the self-induced drift produced by electrostatic repulsion of the carriers. The magnitude of the self-induced longitudinal electrical field $\mathscr{E}_{ys}$ can be estimated by taking the gradient of the surface potential (which is assumed to vary linearly with the signal carrier concentration n_s, as given by Eq. 61)

$$\mathscr{E}_{ys} = \frac{q}{C_i}\frac{\partial n_s(y, t)}{\partial y}. \tag{63}$$

The decay of the initial charge packet Q due to the self-induced field is given by[10]

$$\frac{Q(t)}{Q} = \frac{t_0}{t + t_0} \tag{64}$$

with

$$t_0 \equiv \pi L^3 W_e C_i / 2\mu_n Q$$

where W_e is the width of the electrode and μ_n is the mobility of the carriers.

The surface potential under the storage electrode is affected by the voltage applied to the adjacent electrodes due to two-dimensional coupling of the electrostatic potential. The applied voltage results in a surface electric field, even in the absence of signal charge at the interface. This

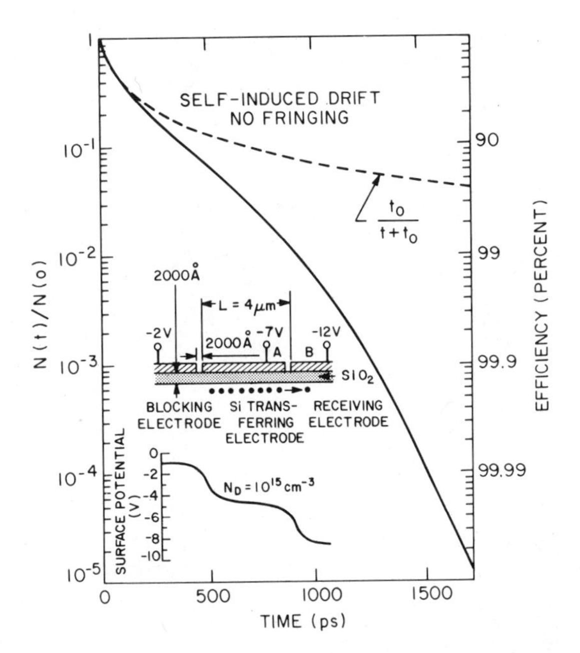

Fig. 46 Normalized total remaining charge versus time for 4-μm gate length and 10^{15}-cm^{-3} doping. The dashed line indicates how the charge transfer would proceed in the absence of a fringing field. The insert shows the surface potential variations. (After Carnes, Kosonocky, and Ramberg, Ref. 61.)

fringing field is a function of the oxide thickness, electrode length, substrate doping, and gate voltage. The insert in Fig. 46 gives an example.[61] The minimum fringing field shown is 2×10^3 V/cm. Because the fringing field is present even at very low charge concentration, the last bit of the signal charge will be transferred by the fringing field. Figure 46 shows the ratio of the total remaining charge normalized to the initial charge as a function of time (solid line). The dashed line indicates how the charge transfer would proceed in the absence of a fringing field.

We shall now define a transfer efficiency η, which is the ratio of charge transferred to electrode B to the initial charge stored under electrode A (insert in Fig. 46):

$$\eta = 1 - \frac{N(t=T)}{N(t=0)}. \tag{65}$$

A closely related concept is the transfer inefficiency ϵ, defined as

$$\epsilon \equiv 1 - \eta = N(t=T)/N(t=0). \tag{66}$$

Figure 46 shows that transfer efficiencies larger than 99.99% (or a transfer inefficiency less than 10^{-4}) can be obtained in the presence of the fringing field for clock frequencies of several tens of megabits per second. As

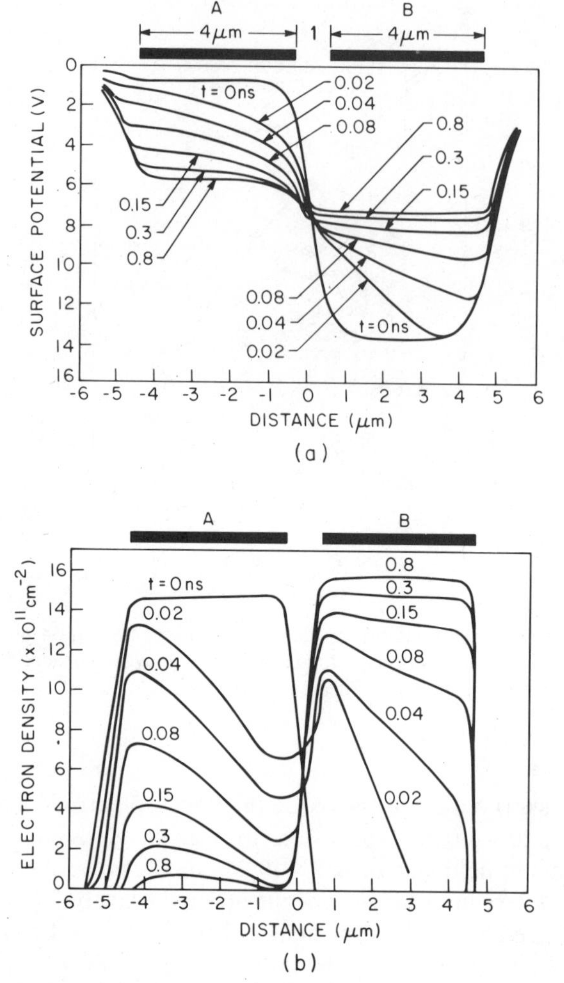

Fig. 47 (*a*) Time-dependent surface potential distribution under the storage and transfer gates for 4-μm gate lengths. (*b*) Transient charge distribution under the gates. (After Elsaid, Chamberlain, and Watt, Ref. 62.)

frequencies become higher, gate length must be reduced to increase the fringing field.

The time-dependent surface potential and the transient behavior of the charge distribution have been computed using a two-dimensional model based on the charge continuity and current transport equation. Figure 47 shows a representative result.[62] Figure 47*a* shows that at the beginning of the charge-transfer process, the speed of the charge transfer is high because of a strong self-induced drift and a fringing field that results in a high drift velocity. After 0.8 ns from the beginning of the charge transfer, the surface potential changes very little, indicating that the amount of charge left to be transferred is small. The potential difference between the two neighboring potential wells after 0.8 ns (which is very close to the final potential difference when all the charges are completely transferred) is about 1.5 V. When the two potential wells approach each other, the transfer slows down considerably. Figure 47*b* shows the transient behavior of the charge distribution. The electrons under the storage gate *A* are distributed more widely than the electrons under the transfer gate *B*, because the fringing field near the edges of the potential wells forces electrons to move to the center of the wells. The fringing field of the potential well under gate *B* is stronger than that of the potential well under gate *A*. Therefore, electrons under gate *B* are localized near the center of the gate. Also note from Fig. 47*b* that after 0.8 ns, about 99% of the electrons have been transferred.

In the discussion above, we considered only the free electrons in the conduction band. We have not considered the transition of electrons between the conduction band and any energy levels in the bandgap, such as interface traps. Thus the charge-transfer mechanism treated here is called the free charge-transfer model. The transfer efficiency at high frequencies can be described by the free charge-transfer model and is limited by the clock rates for a given device. The maximum operating frequency can be well above 10 MHz for CCD with gate lengths smaller than 10 μm.

At medium frequencies trapping the signal charge at the interface traps determines the transfer efficiency. Figure 48 shows trapping or release of carriers from interface traps under each electrode as charge packets are transferred along a three-phase CCD.[63] When charge packets come in contact with empty interface traps, these traps are filled instantaneously, but when the signal charge has moved on, the interface traps release carriers, with a whole spectrum of much slower time constants (Fig. 20). Some trapped charges are released so rapidly from interface traps that they can move into the correct charge packet, but others are released into trailing packets (Fig. 48), which results in a charge loss from the leading charge packet and in a "tail" behind the last packet of a sequence. The

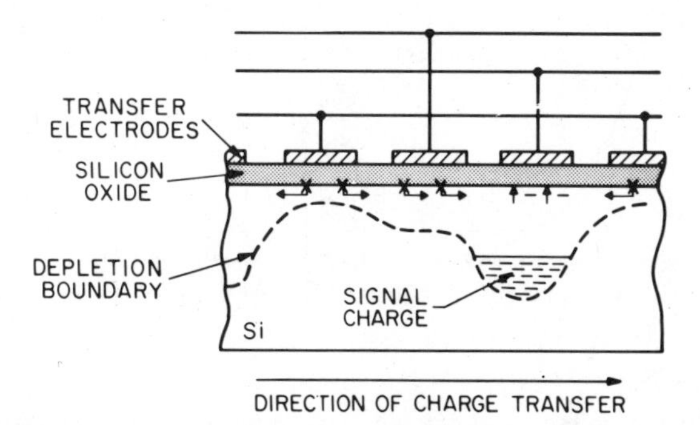

Fig. 48 Net trapping or release of carriers from interface traps under each electrode as charge packet transfers along a three-phase CCD. (After Tompsett, Ref. 63.)

transfer inefficiency due to interface traps is given by

$$\epsilon \simeq \frac{qkTD_{it}}{C_i V_s} \ln(p+1) \tag{67}$$

where V_s is the change in surface potential caused by the signal charge, D_{it} is the interface trap density, and p is the number of phases ($p = 3$ for Fig. 48). To reduce ϵ, the interface trap density must be low.

The effect of these interface traps can be substantially reduced by passing a constant background charge or "fat zero" through the MOS diode array. Then to first order, the interface traps remain permanently filled and as a consequence interact with the desired signal to a minimal extent. The fat-zero charges are, in practice, about 10 to 25% of the full potential-well capacity. The main disadvantage of a fat zero is the reduction of dynamic range.

At low clock frequencies, the frequency limit is determined by the dark current. The dark-current density per unit surface area J_{dark} can be expressed as[11]

$$J_{\text{dark}} = \frac{qn_iW}{2\tau} + \frac{qD_n}{L_n}\frac{n_i^2}{N_A} + \frac{qS_0n_i}{2} \tag{68}$$

where the first term on the right-hand side is the bulk generation inside the depletion region, the second term is the diffusion current at the edge of the depletion region, and the last term is the surface generation current (τ is the minority lifetime, D_n the diffusion constant, L_n the diffusion length, and S_0 the surface recombination velocity).

The effect of the dark current on the MOS capacitance behavior is shown[64] in Fig. 49. When a voltage pulse V_G is applied to the gate, the MOS

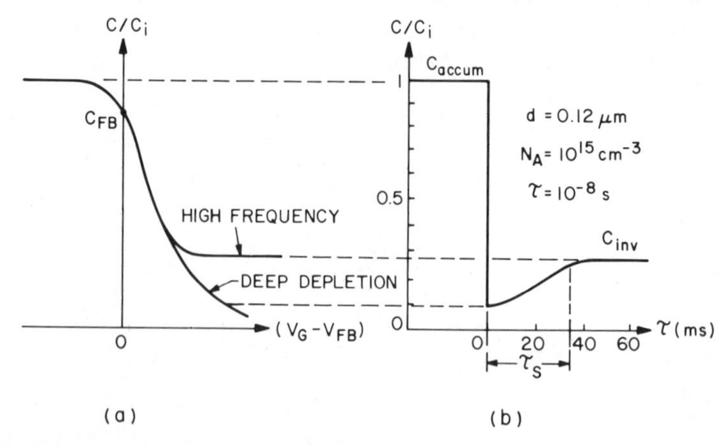

Fig. 49 Capacitance–voltage and capacitance–time characteristics of an MIS diode. τ_s is the storage time. (*a*) Normalized capacitance versus gate voltage. (*b*) Normalized capacitance versus time. (After Tao et al., Ref. 64.)

diode instantaneously changes from accumulation to deep depletion (Fig. 49*a*). If the gate voltage remains fixed at V_G, thermally generated minority carriers will gradually fill the surface potential well, and the MOS capacitance will increase toward its asymptotic inversion condition. The storage time τ_s depends on the rate and magnitude of the dark-current generation (Fig. 49*b*). Typical τ_s for the Si–SiO_2 system is of the order of 10^{-3} to 10^{-2} s. The clock frequency should be high enough so that the contribution of the dark current to the signal packet during charge transfer is negligible.

The low-frequency limit of a CCD can be estimated by comparing the charge accumulated from dark current with the saturation charge. If a CCD is clocked continuously at a constant frequency f, the output signal due to dark current is[11]

$$Q_{\text{dark}} = AJ_{\text{dark}}N/pf \tag{69}$$

where A is the area of the electrode, N the number of electrodes, and p the number of phases for a CCD (e.g., in Fig. 42, $N = 6$, $p = 3$). The maximum signal charge a CCD can handle is

$$Q_{\max} = AC_i\,\Delta\psi_s \tag{70}$$

where $\Delta\psi_s$ is the maximum surface potential change due to the maximum signal charge. Thus

$$Q_{\text{dark}}/Q_{\max} = J_{\text{dark}}N/(pfC_i\,\Delta\psi_s). \tag{71}$$

If this ratio is equal to 10^{-3}, for a 128-triplet-gate CCD ($N = 128$) with 0.1-μm gate oxide, $J_{\text{dark}} = 10^{-9}$ A/cm^2, and $\Delta\psi_s = 5$ V, the low-frequency limit due to dark current is about 1 kHz.

In digital applications, a given amount of signal charge is represented by "1" and an empty well is represented by "0". Figure 50 presents the frequency-response curve[65] for the normalized worst-case 1/0 difference, that is, $(1_{wc} - 0_{wc})/1_{ss}$. Because the 1/0 difference is directly related to the charge-transfer efficiency, the data imply that the transfer efficiency characterizes device operation over a wide range of clock frequencies.

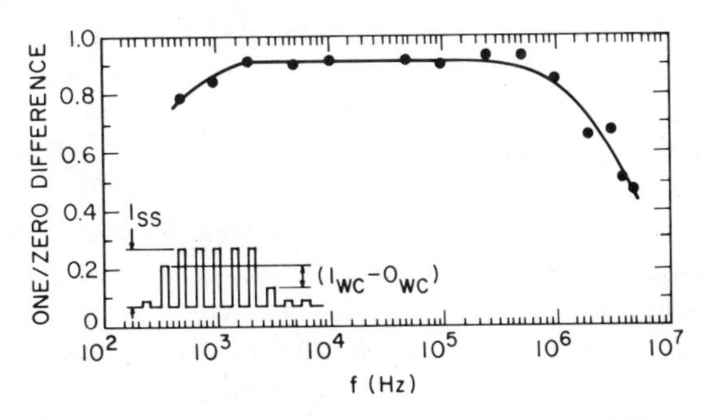

Fig. 50 Frequency response of an SCCD. (After Agusta and Harroun, Ref. 65.)

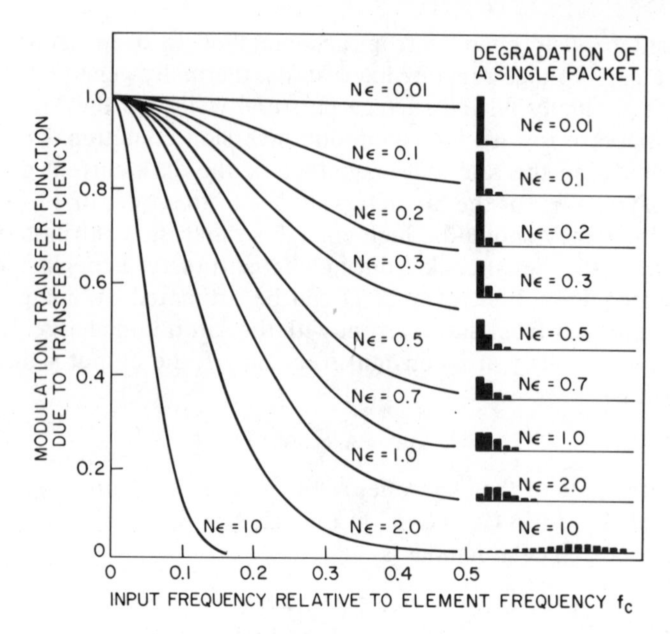

Fig. 51 Effects of transfer inefficiency product $N\epsilon$ on spatial frequency response. (After Tompsett, Ref. 60.)

The low-frequency degradation of frequency response is due to the buildup of dark current in the charge packets, which distort the size of the signal charge. At very high frequencies, transfer efficiency falls rapidly because there is not enough time to allow for complete charge transfer.

To improve the low-frequency response, one must reduce all the dark-current components in Eq. 68 by having a long minority lifetime, large diffusion length, and low surface recombination velocity. To extend high-frequency performance, one can reduce the gate length (L), maximize surface mobility (by using electrons instead of holes in the charge packet), and minimize electrode spacings. The higher electron mobility in GaAs makes it possible to design ultra-high-speed CCDs in GaAs. A GaAs BCCD has been operated up to 500 MHz clock frequency.[71]

Transfer inefficiency can introduce an extra phase delay. Figure 51 shows both frequency response and degradation of a single charge packet as a function of the $N\epsilon$ product, where N represents the number of transfers to move the signal packet to the output.[60] One can see the spreading of an individual charge packet into the tailing charge packets for different values of $N\epsilon$. The left-hand station in each frame represents the position where the original charge packet is expected to appear in an ideal CCD. Charge delayed by transfer inefficiency emerges in later time slots, shown toward the right. For $N\epsilon \geq 1$ the inadequacy of the transfer efficiency is clear, because the main amount of charge no longer appears in the leading station.

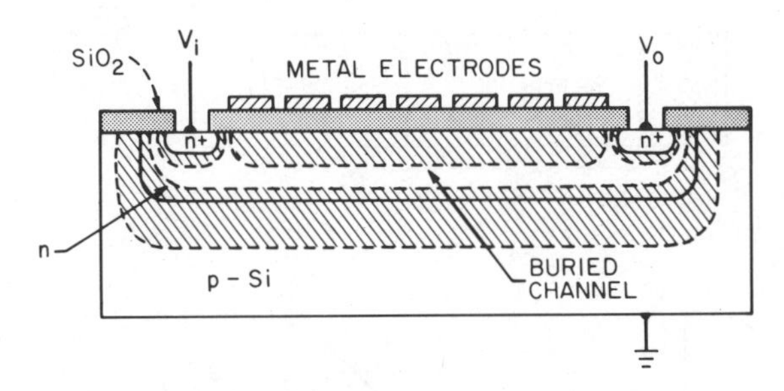

Fig. 52 Cross-sectional view of a BCCD. (After Walden et al., Ref. 67.)

7.4.4 Buried Channel CCD

In the surface channel CCD, minority-carrier charge packets are moved along the surface of a semiconductor with proper clock voltage pulses. One major limitation of this CCD is the effect of interface traps. To circumvent this problem and improve transfer efficiency, the buried-channel CCD (BCCD) was proposed[66] in which the charge packets do not flow at the semiconductor surface; instead, they are confined to a channel that lies beneath the surface. The BCCD has the potential of eliminating the interface trapping. A schematic cross-sectional view of a BCCD is shown[67] in Fig. 52. It consists of an n-type semiconductor layer on a p-type substrate with n^+ contacts at either end of the channel.

The energy-band diagrams for the BCCD are shown[68] in Fig. 53. When no signal charge is present, the narrow n-type region is fully depleted under a positive voltage pulse applied to the gate electrode (Fig. 53b). As signal charges are introduced to the buried channel, they will be stored in the channel. The energy-band diagram is shown in Fig. 53c.

The potential distribution of Fig. 53b can be obtained analytically using the depletion approximation for the case where the impurity concentrations are constant in the n-type and p-type regions. The Poisson equations for the potential are given by[11]

$$\frac{d^2\psi}{dx^2} = 0 \qquad -d < x < 0 \tag{72a}$$

$$\frac{d^2\psi}{dx^2} = -qN_D/\epsilon_s \qquad 0 < x < x_n \tag{72b}$$

$$\frac{d^2\psi}{dx^2} = qN_A/\epsilon_s \qquad x_n < x < x_n + x_p. \tag{72c}$$

The boundary conditions are (1) $\psi = (V_G - V_{FB})$ at $x = -d$, (2) $\psi = 0$ at $x = (x_n + x_p)$, and (3) the requirements of continuity of the potential and the electric displacement at $x = 0$ and $x = x_n$. Solving Eqs. 72 with these

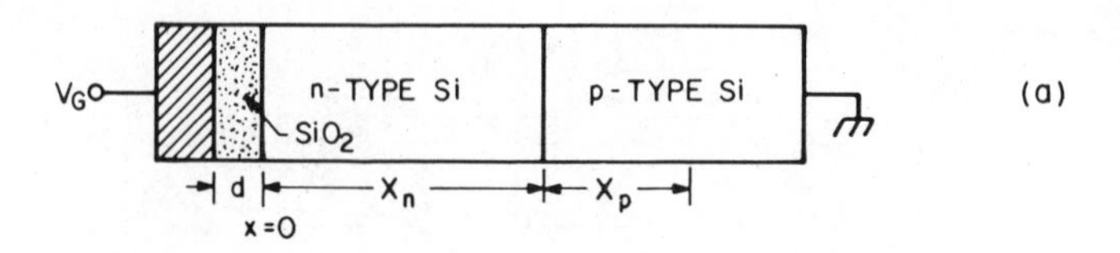

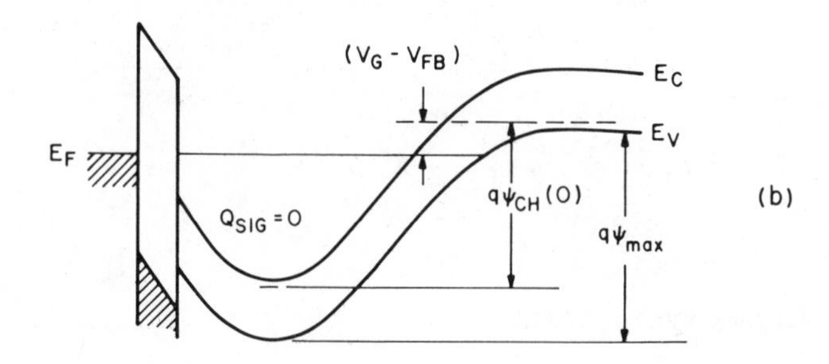

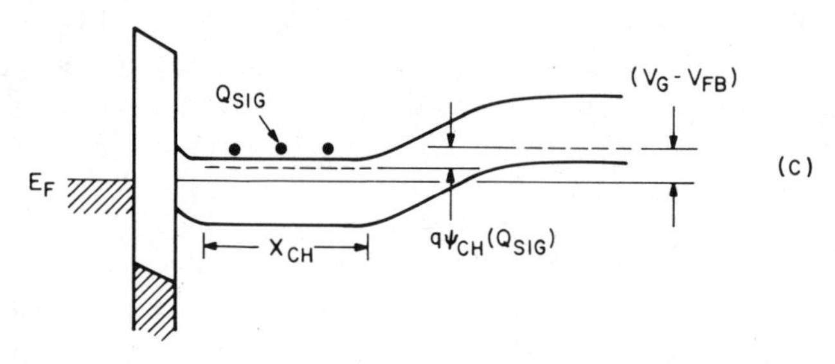

Fig. 53 (*a*) BCCD. (*b*) Energy band for an empty well. (*c*) Energy band when a signal packet is present. (After Burt, Ref. 68.)

boundary conditions yields the maximum potential depth ψ_{max} as shown in Fig. 53*b*,

$$\psi_{max} = \psi_J \left(1 + \frac{N_A}{N_D}\right) \tag{73}$$

and

$$V_G - V_{FB} + V_1 = \psi_J + \sqrt{V_{ox}\psi_J} \tag{74}$$

where

$$\psi_J \equiv \frac{qN_A}{2\epsilon_s} x_p^2$$

$$V_{ox} \equiv \frac{2qN_A x_n^2}{\epsilon_s} \left(1 + \frac{\epsilon_s}{\epsilon_i} \frac{d}{x_n}\right)^2$$

$$V_1 \equiv \frac{qN_D x_n^2}{2\epsilon_s} \left(1 + \frac{2\epsilon_s}{\epsilon_i} \frac{d}{x_n}\right).$$

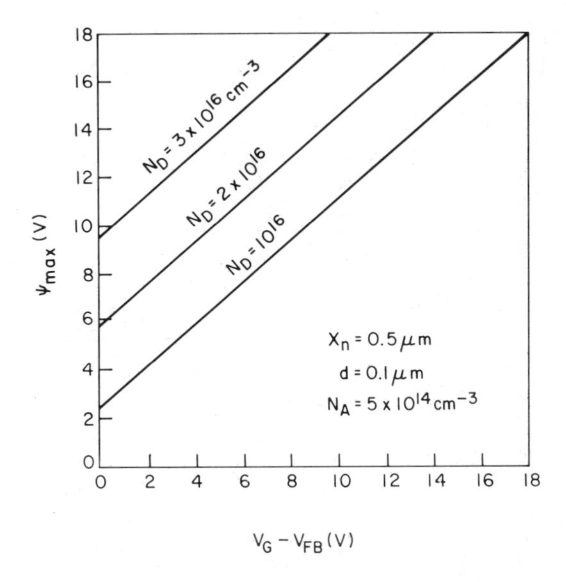

Fig. 54 Channel potential ψ_{max} versus gate voltage in a BCCD. (After Kim, Ref. 11.)

Figure 54 shows the relationship between the gate voltage and ψ_{max} for device parameters typical of a BCCD.

When a charge packet is present in the buried channel, the potential distribution can be obtained from Eqs. 72 with $[N_D - n(x)]$ replacing N_D in Eq. 72*b*, and $[-p(x)+n(x)+N_A]$ replacing N_A in Eq. 72c, where $n(x)$ and $p(x)$ are the free electron and free hole density, respectively. Figure 55*a* shows the calculated potential distribution for various sizes of charge packets.[69] The total amount of charge Q is normalized to $N_A L_D$, where L_D is the Debye length. For $N_A = 10^{14}$ cm^{-3}, the Debye length is 0.415 μm and $N_A L_D = 4.15 \times 10^9$ charges/cm^2. The corresponding charge distributions versus distance are shown in Fig. 55*b*. When the charge packet increases, the charge distribution becomes wider and moves toward the insulator–semiconductor interface.

Figure 56 shows a two-dimensional calculation of potential along the buried channel. For comparison, it also shows the potential plot for a surface device. Clearly, the BCCD has a greater potential gradient under the transferring electrode, which helps to speed up charge transfer. A transfer inefficiency to 10^{-4} to 10^{-5} is readily achieved in the BCCD, and is an order of magnitude smaller than a typical SCCD having the same device geometry.

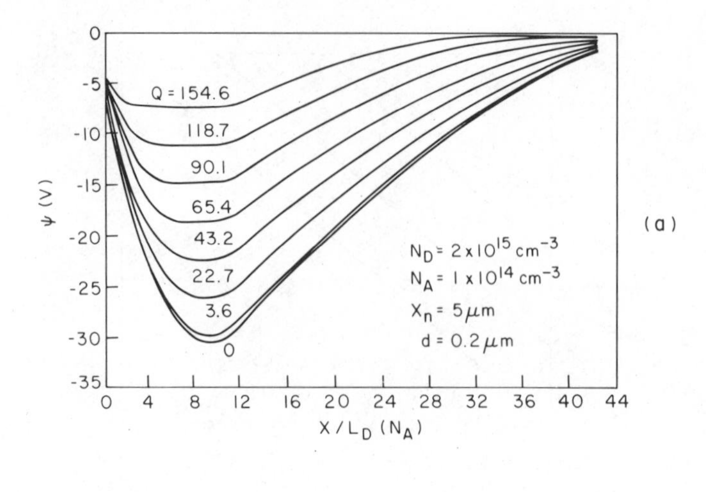

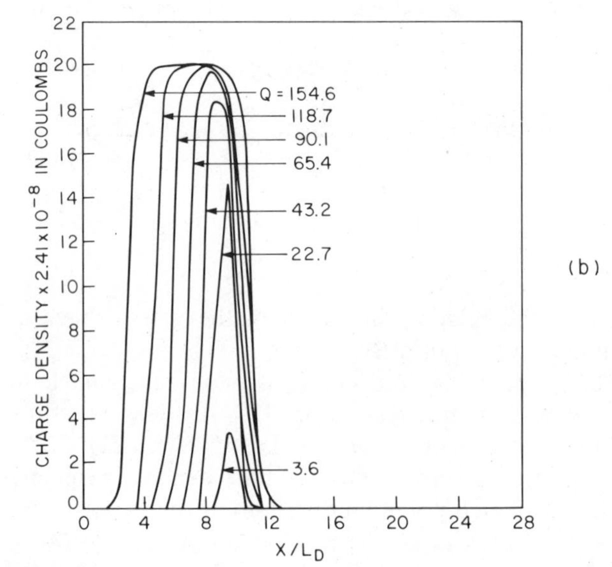

Fig. 55 (*a*) Channel potential versus distance for a BCCD with uniformly doped *n*-type and *p*-type regions and $V_G = 4$ V. (*b*) The corresponding charge density versus distance. (After Kent, Ref. 69.)

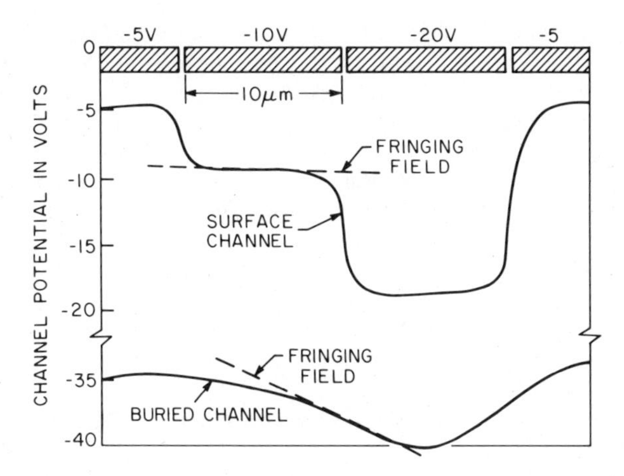

Fig. 56 Two-dimensional calculation of the potential along a BCCD. Shown for comparison is the potential plot for an SCCD. (After Walden et al., Ref. 67.)

REFERENCES

1. J. L. Moll, "Variable Capacitance with Large Capacity Change," *Wescon Conv. Rec.*, Pt. 3, p. 32 (1959).
2. W. G. Pfann and C. G. B. Garrett, "Semiconductor Varactor Using Space-Charge Layers," *Proc. IRE*, **47**, 2011 (1959).
3. D. R. Frankl, "Some Effects of Material Parameters on the Design of Surface Space-Charge Varactors," *Solid State Electron.*, **2**, 71 (1961).
4. R. Lindner, "Semiconductor Surface Varactor," *Bell Syst. Tech. J.*, **41**, 803 (1962).
5. L. M. Terman, "An Investigation of Surface States at a Silicon/Silicon Dioxide Interface Employing Metal–Oxide–Silicon Diodes," *Solid State Electron.*, **5**, 285 (1962).
6. K. Lehovec and A. Slobodskoy, "Field-Effect Capacitance Analysis of Surface States on Silicon," *Phys. Status Solidi*, **3**, 447 (1963).
7. E. H. Nicollian and J. R. Brews, *MOS Physics and Technology*, Wiley, New York, 1982.
8. W. S. Boyle and G. E. Smith, "Charge Coupled Semiconductor Devices," *Bell Syst. Tech. J.*, **49**, 587 (1970).
9. G. F. Amelio, M. F. Tompsett, and G. E. Smith, "Experimental Verification of the Charge Coupled Diode Concept," *Bell Syst. Tech. J.*, **49**, 593 (1970).
10. C. H. Sequin and M. F. Tompsett, *Charge Transfer Devices*, Academic, New York, 1975.
11. C. K. Kim, "The Physics of Charge-Coupled Devices," in M. J. Howes and D. V. Morgan, Eds., *Charge-Coupled Devices and Systems*, Wiley, New York, 1979, p. 1.
12. R. Melen and D. Buss, Eds., *Charge-Coupled Devices: Technology and Applications*, IEEE Press, New York, 1977.
13. C. G. B. Garrett and W. H. Brattain, "Physical Theory of Semiconductor Surfaces," *Phys. Rev.*, **99**, 376 (1955).
14. R. H. Kingston and S. F. Neustadter, "Calculation of the Space Charge, Electric Field, and Free Carrier Concentration at the Surface of a Semiconductor," *J. Appl. Phys.*, **26**, 718 (1955).

15 S. R. Hofstein and G. Warfield, "Physical Limitation on the Frequency Response of a Semiconductor Surface Inversion Layer," *Solid State Electron.*, **8**, 321 (1965).

16 A. S. Grove, B. E. Deal, E. H. Snow, and C. T. Sah, "Investigation of Thermally Oxidized Silicon Surfaces Using Metal–Oxide–Semiconductor Structures," *Solid State Electron.*, **8**, 145 (1965).

17 A. S. Grove, E. H. Snow, B. E. Deal, and C. T. Sah, "Simple Physical Model for the Space-Charge Capacitance of Metal–Oxide–Semiconductor Structures," *J. Appl. Phys.*, **33**, 2458 (1964).

18 A. Goetzberger, "Ideal MOS Curves for Silicon," *Bell Syst. Tech. J.*, **45**, 1097 (1966).

19 B. E. Deal, "Standardized Terminology for Oxide Charges Associated with Thermally Oxidized Silicon," *IEEE Trans. Electron Devices*, **ED-27**, 606 (1980).

20 I. Tamm, "Über eine mögliche Art der Elektronenbindung an Kristalloberflächen," *Phys. Z. Sowjetunion*, **1**, 733 (1933).

21 W. Shockley, "On the Surface States Associated with a Periodic Potential," *Phys. Rev.*, **56**, 317 (1939).

22 W. Shockley and G. L. Pearson, "Modulation of Conductance of Thin Films of Semiconductors by Surface Charges," *Phys. Rev.*, **74**, 232 (1948).

23 F. G. Allen and G. W. Gobeli, "Work Function, Photoelectric Threshold and Surface States of Atomically Clean Silicon," *Phys. Rev.*, **127**, 150 (1962).

24 E. H. Nicollian and A. Goetzberger, "MOS Conductance Technique for Measuring Surface State Parameters," *Appl. Phys. Lett.*, **7**, 216 (1965).

25 C. N. Berglund, "Surface States at Steam-Grown Silicon–Silicon Dioxide Interface," *IEEE Trans. Electron Devices*, **ED-13**, 701 (1966).

26 E. H. Nicollian and A. Goetzberger, "The Si–SiO_2 Interface–Electrical Properties as Determined by the MIS Conductance Technique," *Bell Syst. Tech. J.*, **46**, 1055 (1967).

27 M. H. White and J. R. Cricchi, "Characterization of Thin-Oxide MNOS Memory Transistors," *IEEE Trans. Electron Devices*, **ED-19**, 1280 (1972).

28 B. E. Deal, M. Sklar, A. S. Grove, and E. H. Snow, "Characteristics of the Surface-State Charge (Q_{ss}) of Thermally Oxidized Silicon," *J. Electrochem. Soc.*, **114**, 266 (1967).

29 J. R. Ligenza, "Effect of Crystal Orientation on Oxidation Rates of Silicon in High Pressure Steam," *J. Phys. Chem.*, **65**, 2011 (1961).

30 E. H. Snow, A. S. Grove, B. E. Deal, and C. T. Sah, "Ion Transport Phenomena in Insulating Films," *J. Appl. Phys.*, **36**, 1664 (1965).

31 E. Yon, W. H. Ko, and A. B. Kuper, "Sodium Distribution in Thermal Oxide on Silicon by Radiochemical and MOS Analysis," *IEEE Trans. Electron Devices*, **ED-13**, 276 (1966).

32 J. V. Dalton and J. Drobek, "Structure and Sodium Migration in Silicon Nitride Films," *J. Electrochem. Soc.*, **115**, 865 (1968).

33 G. T. Cheney, R. M. Jacobs, H. W. Korb, H. E. Nigh, and J. Stack, "Al_2O_3–SiO_2 IGFET Integrated Circuits," Paper No. 2.2, IEEE Device Meet., Washington, D.C., Oct. 18–21, 1967.

34 E. H. Snow and B. E. Deal, "Polarization Phenomena and Other Properties of Phosphosilicate Glass Films on Silicon," *J. Electrochem. Soc.*, **113**, 2631 (1966).

35 T. L. Chu, J. R. Szedon, and C. H. Lee, "The Preparation and *C–V* Characteristics of Si–Si_3N_4 and Si–SiO_2–Si_3N_4 Structure," *Solid State Electron.*, **10**, 897 (1967).

36 F. Reizman and W. Van Gelder, "Optical Thickness Measurement of SiO_2–Si_3N_4 Films on Silicon," *Solid State Electron.*, **10**, 625 (1967).

37 R. Williams, "Photoemission of Electrons from Silicon into Silicon Dioxide," *Phys. Rev.*, **140**, A569 (1965).

38 B. E. Deal, E. H. Snow, and C. A. Mead, "Barrier Energies in Metal–Silicon Dioxide–Silicon Structures," *J. Phys. Chem. Solids*, **27**, 1873 (1966).

39 W. M. Werner, "The Work Function Difference of the MOS-System with Aluminum Field Plates and Polycrystalline Silicon Field Plates," *Solid State Electron.*, **17**, 769 (1974).

40 A. Goetzberger, "Behavior of MOS Inversion Layers at Low Temperature," *IEEE Trans. Electron Devices*, **ED-14**, 787 (1967).

41 A Goetzberger and E. H. Nicollian, "Temperature Dependence of Inversion Layer Frequency Response in Silicon," *Bell Syst. Tech. J.*, **46**, 513 (1967).

42 J. Grosvalet and C. Jund, "Influence of Illumination on MIS Capacitance in the Strong Inversion Region," *IEEE Trans. Electron Devices*, **ED-14**, 777 (1967).

43 D. R. Collins and C. T. Sah, "Effects of X-Ray Irradiation on the Characteristics of MOS Structures," *Appl. Phys. Lett.*, **8**, 124 (1966).

44 E. H. Snow, A. S. Grove, and D. J. Fitzgerald, "Effect of Ionization Radiation on Oxidized Silicon Surfaces and Planar Devices," *Proc. IEEE*, **55**, 1168 (1967).

45 E. H. Nicollian, A. Goetzberger, and C. N. Berglund, "Avalanche Injection Currents and Charging Phenomena in Thermal SiO_2," *Appl. Phys. Lett.*, **15**, 174 (1969).

46 A. Rusu and C. Bulucea, "Deep-Depletion Breakdown Voltage of SiO_2/Si MOS Capacitors," *IEEE Trans. Electron Devices*, **ED-26**, 201 (1979).

47 S. M. Sze and G. Gibbons, "Effects of Junction Curvature on Breakdown Voltage in Semiconductors," *Solid State Electron.*, **9**, 831 (1966).

48 J. Frenkel, "On the Theory of Electric Breakdown of Dielectrics and Electronic Semiconductors," *Tech. Phys. USSR*, **5**, 685 (1938); "On Pre-Breakdown Phenomena in Insulators and Electronic Semiconductors," *Phys. Rev.*, **54**, 647 (1938).

49 J. J. O'Dwyer, *The Theory of Electrical Conduction and Breakdown in Solid Dielectrics*, Clarendon, Oxford, 1973.

50 S. M. Sze, "Current Transport and Maximum Dielectric Strength of Silicon Nitride Films," *J. Appl. Phys.*, **38**, 2951 (1967).

51 W. C. Johnson, "Study of Electronic Transport and Breakdown in Thin Insulating Films," *Tech. Rep. No. 7*, Princeton University, 1979.

52 M. Av-Ron, M. Shatzkes, T. H. DiStefano, and I. B. Cadoff, "The Nature of Electron Tunneling in SiO_2" in S. T. Pantelider, Ed., *The Physics of SiO_2 and Its Interfaces*, Pergamon, New York, 1978, p. 46.

53 N. Klein, "Electrical Breakdown in Solids," *Advances in Electronics and Electron Physics*, Vol. 26, Academic, New York, 1969.

54 C. M. Osburn and D. W. Ormond, "Dielectric Breakdown in Silicon Dioxide Films on Silicon," *J. Electrochem. Soc., Solid State Sci. Technol.*, **119**, 591 (1972).

55 E. Harari, "Dielectric Breakdown in Electrically Stressed Thin Films of Thermal SiO_2," *J. Appl. Phys.*, **49**, 2478 (1978).

56 D. F. Barbe, "Imaging Devices Using the Charge-Coupled Concept," *Proc. IEEE*, **63**, 38 (1975).

57 J. D. E. Beynon, "The Basic Principles of Charge-Coupled Devices," *Microelectronics*, **7**, 7 (1975).

58 D. G. Ong and R. F. Pierret, "Approximate Formula for Surface Carrier Concentration in Charge-Coupled Devices," *Electron. Lett.*, **10**, 6 (1974).

59 W. S. Boyle and G. E. Smith, "Charge-Coupled Devices—A New Approach to MIS Device Structures," *IEEE Spectrum*, **8**, 18 (1971).

60 M. F. Tompsett, "Video-Signal Generation," in T. P. McLean and P. Schagen, Eds., *Electronic Imaging*, Academic, New York, 1979, p. 55.

61 J. E. Carnes, W. F. Kosonocky, and E. G. Ramberg, "Free Charge Transfer in Charge-Coupled Devices," *IEEE Trans. Electron Devices*, **ED-19**, 798 (1972).

62 M. H. Elsaid, S. G. Chamberlain, and L. A. K. Watt, "Computer Model and Charge Transport Studies in Short Gate Charge-Coupled Devices," *Solid State Electron.*, **20**, 61 (1977).

63 M. F. Tompsett, "The Quantitative Effect of Interface States on the Performance of Charge-Coupled Devices," *IEEE Trans. Electron. Devices*, **ED-20**, 45 (1973).

64 T. F. Tao, J. R. Ellis, L. Kost, and A. Doshier, "Feasibility Study of PbTe and PbSnTe Infrared Charge Coupled Imager," *Proc. Appl. Conf. CCD*, Naval Electron. Lab. Center, San Diego, 1973, p. 259.

65 B. Agusta and T. Harroun, "Conceptual Design of an Eight Megabyte High Performance Charge-Coupled Storage Device," *Proc. Appl. Conf. CCD*, Naval Electron. Lab. Center, San Diego, 1973, p. 55.

66 W. S. Boyle and G. E. Smith, U.S. Patent 3,792,322 (1974).

67 R. H. Walden, R. H. Krambeck, R. J. Strain, J. McKenna, N. L. Schryer, and G. E. Smith, "The Buried Channel Charge Coupled Device," *Bell Syst. Tech. J.*, **51**, 1635 (1972).

68 D. J. Burt, "Basic Operation of the Charge Coupled Device," Int. Conf. Technol. Appl. CCD, University of Edinburgh, 1974, p. 1.

69 W. H. Kent, "Charge Distribution in Buried-Channel Charge-Coupled Devices," *Bell Syst. Tech. J.*, **52**, 1009 (1973).

70 J. R. Brews, "A Simplified High-Frequency MOS Capacitance Formula," *Solid State Electron.*, **20**, 607 (1977).

71 I. Deyhimy, R. C. Eden, R. J. Anderson, and I. S. Harris, Jr., "A 500-MHz GaAs Charge-Coupled Device," *Appl. Phys. Lett.*, **36**, 151 (1980).

8

MOSFET

8.1 INTRODUCTION

The metal–oxide–semiconductor field-effect transistor (MOSFET) is the most important device for very-large-scale integrated circuits such as microprocessors and semiconductor memories. MOSFET is also becoming an important power device. It has many acronyms including IGFET (insulated-gate field-effect transistor) MISFET (metal–insulator–semiconductor field-effect transistor) and MOST (metal–oxide–semiconductor transistor). The principle of the surface field-effect transistor was first proposed in the early 1930s by Lilienfeld[1] and Heil.[2] It was subsequently studied by Shockley and Pearson[3] in the late 1940s. In 1960, Kahng and Atalla[4] proposed and fabricated the first MOSFET using a thermally oxidized silicon structure. The basic device characteristics have been subsequently studied by Ihantola and Moll,[5,6] Sah,[7] and Hofstein and Heiman.[8] The technology, application, and device physics have been reviewed by Wallmark and Johnson,[9] Richman,[10] and Brews.[11]

Because the current in a MOSFET is transported predominantly by carriers of one polarity only (e.g., electrons in an *n*-channel device), the MOSFET is usually referred to as a unipolar device. The MOSFET is a member of the family of field-effect transistors. The other members, JFETs and MESFETs, have already been considered in Chapter 6. Although MOSFETs have been made with various semiconductors such as Ge,[12] Si, and GaAs,[13] and use various insulators such as SiO_2, Si_3N_4, and Al_2O_3, the most important system is the Si–SiO_2 combination. Hence most of the results in this chapter are obtained from the Si–SiO_2 system.

We first consider the basic device characteristics of the so-called long-channel MOSFET; that is, the channel length L is much longer than the sum of the source and drain depletion-layer widths $(W_S + W_D)$.* This serves as a foundation to understand short-channel, that is, $L \lesssim (W_S + W_D)$, and related MOSFET devices.

Figure 1 shows[14] the reduction of the minimum device dimension since the beginning of the integrated circuit era in 1959. Figure 1 also shows that the minimum dimension will shrink continuously; the 1-μm barrier for commercial devices may be overcome by 1990. The reduction of device dimensions is driven by the requirement that integrated circuits of high complexity be fabricated. The number of components per integrated-circuit

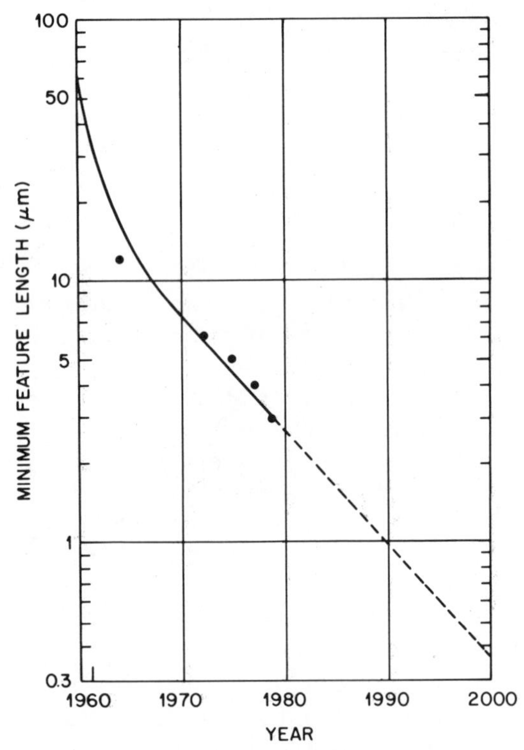

Fig. 1 The minimum device dimension in an integrated circuit as a function of the year for commercial devices. (After Ref. 14.)

*These terms will be defined in Section 8.2.

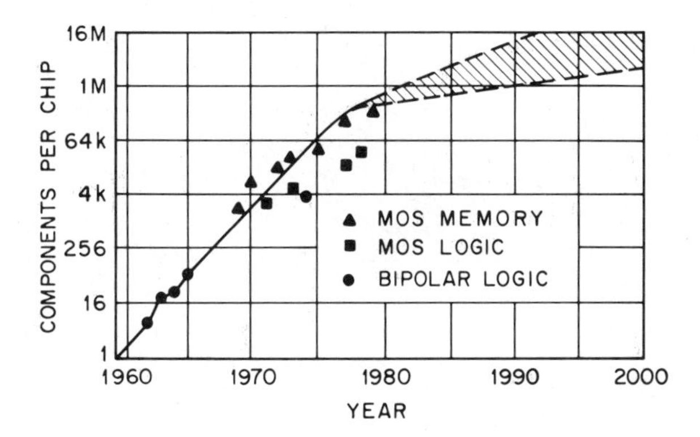

Fig. 2 Complexity of integrated circuits as a function of the year. (After Moore, Ref. 15.)

chip has grown exponentially[15] since 1959 (Fig. 2). The rate of growth is expected to slow down because of a lack of product definition and design. However, a complexity of 1 million or more devices per chip may be available around 1990 using 1-μm or submicron device geometries. As the channel length becomes shorter, one has to consider short-channel effects due to two-dimensional potential, high-field transport and oxide charging. Many device structures have been proposed to improve MOSFET performance. Some representative structures as well as the nonvolatile semiconductor memory, basically a MOSFET with a multilayer gate structure, will be discussed.

8.2 BASIC DEVICE CHARACTERISTICS

The basic structure of a metal–oxide–semiconductor field-effect transistor (MOSFET) is illustrated in Fig. 3. It is a four-terminal device and consists of a p-type semiconductor substrate into which two n^+ regions, the source and drain, are formed* (e.g., by ion implantation). The metal contact on the insulator is called gate; heavily doped polysilicon or a combination of silicide and polysilicon can also be used as the gate electrode. The basic device parameters are the channel length L, which is the distance between the two metallurgical n^+-p junctions; the channel width Z; the insulator thickness d; the junction depth r_j; and the substrate doping N_A. In a silicon integrated circuit, a MOSFET is surrounded by a thick oxide (called the field oxide to distinguish it from the gate oxide) to isolate it from adjacent devices.

The source contact will be used as the voltage reference throughout this

*This is an n-channel device; one may consider a p-channel device by exchanging p for n and reversing the polarity of the voltage.

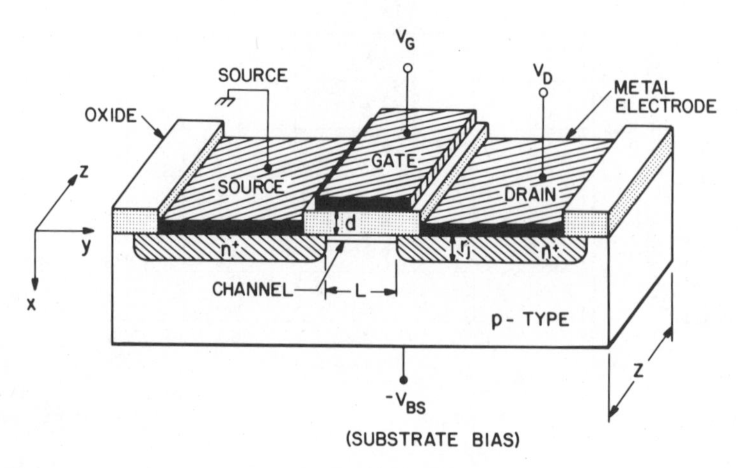

Fig. 3 Schematic diagram of a MOSFET. (After Kahng and Atalla, Ref. 4.)

chapter. When no voltage is applied to the gate, the source-to-drain electrodes correspond to two *p-n* junctions connected back to back. The only current that can flow from source to drain is the reverse leakage current.* When a sufficiently large positive bias is applied to the gate so that a surface inversion layer (or channel) is formed between the two n^+ regions, the source and the drain are then connected by a conducting-surface *n* channel through which a large current can flow. The conductance of this channel can be modulated by varying the gate voltage. The back-surface contact (or substrate contact) can have the reference voltage or be reverse-biased; the back-surface voltage will also affect the channel conductance.

8.2.1 Nonequilibrium Condition

When a voltage is applied across the source–drain contacts, the MOS structure is in a nonequilibrium condition; that is, the imref of the minority carriers (electrons, in the present case) is lowered from the equilibrium Fermi level. To show more clearly the band bending across the device, Fig. 4*a* shows[16] the MOSFET turned 90°. The two-dimensional, flat-band, zero-bias ($V_G = V_D = V_{BS} = 0$) equilibrium condition is shown in Fig. 4*b*. The equilibrium conditions under a gate bias that causes surface inversion are shown in Fig. 4*c*. The nonequilibrium condition with both drain and gate biases is shown in Fig. 4*d*, where we note the separation of the imrefs of electrons and holes; the hole imref E_{Fp} remains at the bulk Fermi level while the electron imref E_{Fn} (minority in the present case) is lowered

*This is the *n*-channel normally-off (enhancement-type) MOSFET. Other types will be discussed later.

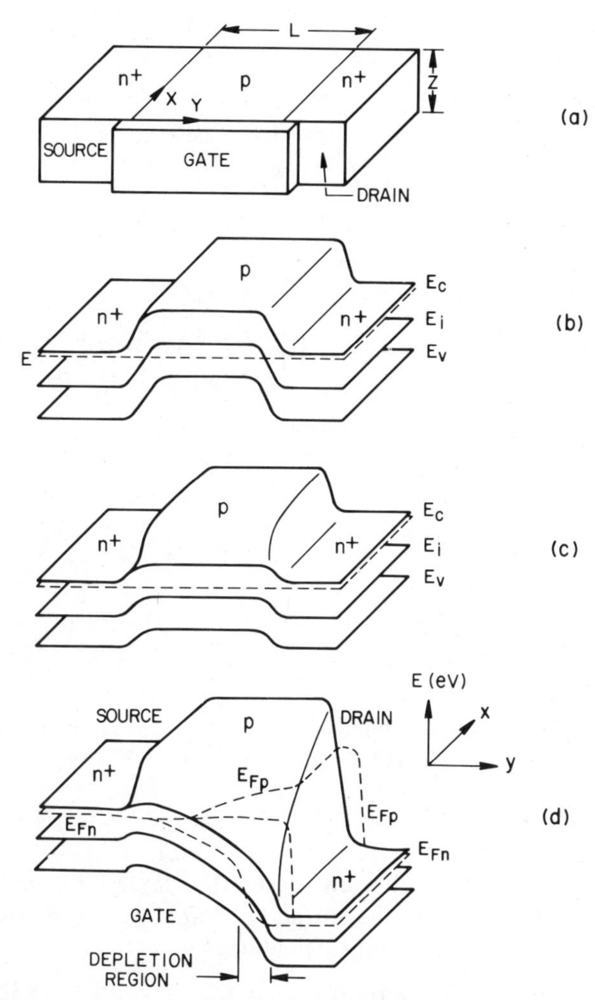

Fig. 4 Two-dimensional band diagram of an *n*-channel MOSFET. (*a*) Device configuration. (*b*) Flat-band zero-bias equilibrium condition. (*c*) Equilibrium condition under a gate bias. (*d*) Nonequilibrium condition under both gate and drain biases. (After Pao and Sah, Ref. 16.)

toward the drain contact. Figure 4*d* shows that the gate voltage required for inversion at the drain is larger than the equilibrium case in which $\psi_s(\text{inv}) \simeq 2\psi_B$. This is because the applied drain bias lowers the electron imref, and an inversion layer can be formed only when the potential at the surface crosses over the imref of the minority carrier.

Figure 5 shows a comparison[17] of the charge distribution and energy-band variation of an inverted *p* region for the equilibrium case and the nonequilibrium case at the drain. For the equilibrium case (discussed in Chapter 7), the surface depletion region reaches a maximum width W_m at

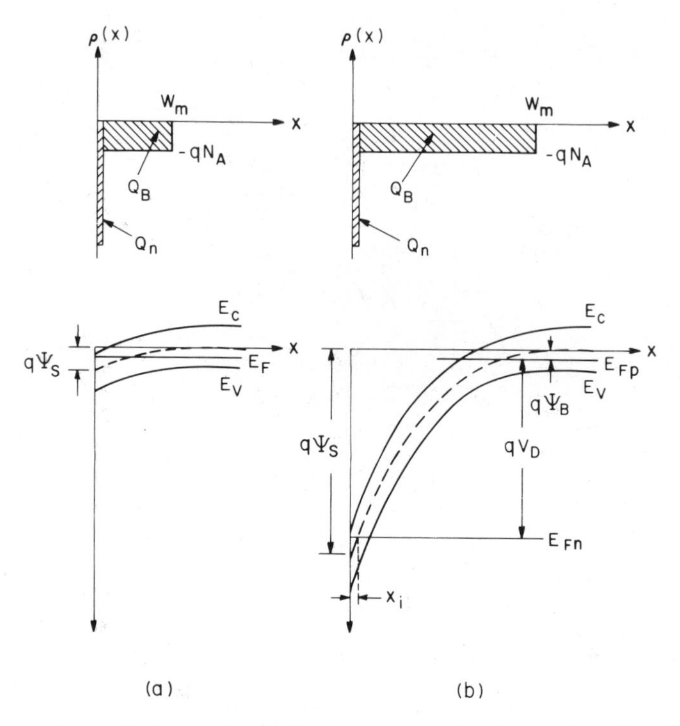

Fig. 5 Comparison of charge distribution and energy band variation of an inverted *p* region for (*a*) the equilibrium case and (*b*) the nonequilibrium case at the drain. (After Grove and Fitzgerald, Ref. 17.)

inversion. For the nonequilibrium case, the depletion-layer width is a function of the bias V_D, and the surface potential ψ_s at the onset of strong inversion is given, to a good approximation, by

$$\psi_s(\text{inv}) \simeq V_D + 2\psi_B . \tag{1}$$

The derivation for the characteristic of the surface-space charge under the nonequilibrium condition is similar to that in Chapter 7. The two assumptions are that (1) the imref for the majority carriers of the substrate does not vary with distance from the bulk to the surface, and (2) the imref for the minority carriers of the substrate is separated by the applied junction bias V_D from the imref for the majority carriers; that is, $E_{Fp} = E_{Fn} + qV_D$ for a p substrate. The first assumption introduces little error when the surface is inverted, because majority carriers are then only a negligible part of the surface space charge; the second assumption is correct under the inversion condition, because minority carriers are an important part of the surface-space-charge region when the surface is inverted.

Based on these assumptions, the one-dimensional Poisson equation for the surface-space-charge region at the drain is given by

$$\frac{\partial^2 \psi}{\partial x^2} = -\frac{q}{\epsilon_s}(N_D^+ - N_A^- + p - n) \tag{2}$$

where

$$\begin{aligned} N_D^+ - N_A^- &= n_{po} - p_{po}, \qquad p_{po} \simeq N_A \\ p &= p_{po} e^{-\beta\psi} \\ n &= n_{po} e^{\beta\psi - \beta V_D}, \qquad \beta \equiv q/kT. \end{aligned} \tag{3}$$

Following the same approach as in Chapter 7, we obtain

$$\mathscr{E} = -\frac{\partial \psi}{\partial x} = \pm \frac{\sqrt{2}kT}{qL_D} F\left(\beta\psi, V_D, \frac{n_{po}}{p_{po}}\right) \tag{4}$$

and

$$Q_s = -\epsilon_s \mathscr{E}_s = \mp \frac{\sqrt{2}\epsilon_s kT}{qL_D} F\left(\beta\psi_s, V_D, \frac{n_{po}}{p_{po}}\right) \tag{5}$$

where

$$F\left(\beta\psi, V_D, \frac{n_{po}}{p_{po}}\right) \equiv \left[e^{-\beta\psi} + \beta\psi - 1 + \frac{n_{po}}{p_{po}} e^{-\beta V_D}(e^{\beta\psi} - \beta\psi e^{\beta V_D} - 1)\right]^{1/2} \tag{6}$$

and

$$L_D \equiv \left(\frac{kT\epsilon_s}{p_{po}q^2}\right)^{1/2}. \tag{7}$$

The surface charge per unit area after strong inversion is given by

$$Q_s = Q_n + Q_B \tag{8}$$

where

$$Q_B = -qN_A W_m = -\sqrt{2qN_A\epsilon_s(V_D + 2\psi_B)} \tag{9}$$

and Q_n, the charge due to minority carriers within the inversion layer, is

$$|Q_n| \equiv q\int_0^{x_i} n(x)\,dx = q\int_{\psi_s}^{\psi_B} \frac{n(\psi)\,d\psi}{d\psi/dx} \tag{10}$$

or

$$|Q_n| = q\int_{\psi_s}^{\psi_B} \frac{n_{po}e^{(\beta\psi - \beta V_D)}\,d\psi}{(\sqrt{2}kT/qL_D)F(\beta\psi, V_D, n_{po}/p_{po})} \tag{11}$$

where x_i denotes the point at which the intrinsic Fermi level intersects the imref for electrons. For the practical doping ranges in silicon, the value of x_i is quite small, of the order of 30 to 300 Å. Equation 11 is the basic formula for long-channel MOSFET, and can be evaluated numerically.

Under strong inversion conditions, a simplified expression for Q_n can be obtained from a charge-sheet model[18] and is given by

$$|Q_n| = \sqrt{2}qN_AL_D\left\{\left[\beta\psi_s + \left(\frac{n_{po}}{p_{po}}\right)e^{(\beta\psi_s-\beta V_D)}\right]^{1/2} - (\beta\psi_s)^{1/2}\right\}. \tag{12}$$

This expression for Q_n is derived under the condition $V_{BS}=0$. When a substrate reverse bias is applied, the depletion width increase, and the term βV_D in Eq. 12 is replaced by $\beta(V_D+V_{BS})$.

8.2.2 Linear and Saturation Regions

We shall first present a qualitative discussion of device operation. Let us consider that a voltage is applied to the gate, causing an inversion at the semiconductor surface, Fig. 6*a*. If a small drain voltage is applied, a current will flow from the source to the drain through the conducting channel. Thus the channel acts as a resistance, and the drain current I_D is proportional to the drain voltage V_D. This is the linear region. As the drain voltage increases, it eventually reaches a point at which the channel depth x_i at $y=L$ is reduced to zero; this is called the pinch-off point, Fig. 6*b*. Beyond the pinch-off point the drain current remains essentially the same, because for $V_D > V_{D\,\text{sat}}$, the voltage at Y remains the same, $V_{D\,\text{sat}}$. Thus the number of carriers arriving at point Y from the source, and hence the current flowing from source to drain, remains the same apart from a decrease in L to the value L' (Fig. 6*c*). Carrier injection from Y into the drain-depletion region is quite similar to the case of carrier injection from an emitter–base junction to the base–collector depletion region of a bipolar transistor.

We shall now derive the basic MOSFET characteristics under the following idealized conditions: (1) the gate structure corresponds to an ideal MOS diode as defined in Chapter 7; that is, there are no interface traps, fixed oxide charge, or work-function difference, and so on; (2) only drift current will be considered; (3) carrier mobility in the inversion layer is constant; (4) doping in the channel is uniform; (5) reverse leakage current is negligibly small; and (6) the transverse field ($\mathscr{E}_x$ in the x direction) in the channel is much larger than the longitudinal field ($\mathscr{E}_y$ in the y direction). The last condition corresponds to the so-called gradual channel approximation.

Under such idealized conditions, the total charge induced in the semiconductor per unit area Q_s at a distance y from the source is given by

$$Q_s(y) = [-V_G + \psi_s(y)]C_i \tag{13}$$

where $C_i \equiv \epsilon_i/d$ is the capacitance per unit area. The charge in the inversion layer is given by

$$\begin{aligned} Q_n(y) &= Q_s(y) - Q_B(y) \\ &= -[V_G - \psi_s(y)]C_i - Q_B(y). \end{aligned} \tag{14}$$

The surface potential $\psi_s(y)$ at inversion can be approximated by $2\psi_B +$

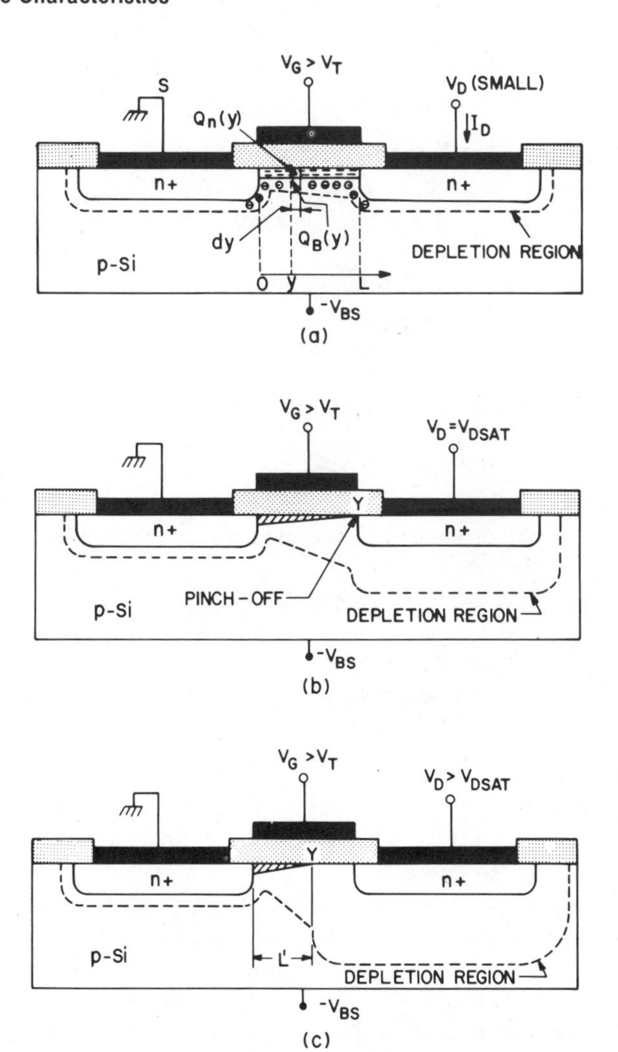

Fig. 6 (*a*) MOSFET operated in the linear region (low drain voltage). (*b*) MOSFET operated at onset of saturation. The point *Y* indicates the pinch-off point. (*c*) MOSFET operated beyond saturation and the effective channel length is reduced.

$V(y)$, where $V(y)$ is the reverse bias between point y and the source electrode (which is assumed to be grounded). The charge within the surface depletion region $Q_B(y)$ was given previously as

$$Q_B(y) = -qN_AW_m = -\sqrt{2\epsilon_s qN_A[V(y)+2\psi_B]}. \tag{15}$$

Substituting Eq. 15 into Eq. 14 yields

$$Q_n(y) = -[V_G - V(y) - 2\psi_B]C_i + \sqrt{2\epsilon_s qN_A[V(y)+2\psi_B]}. \tag{16}$$

The conductivity of the channel can be approximated by

$$\sigma(x) = qn(x)\mu_n(x). \tag{17}$$

The channel conductance is then given by

$$g = \frac{Z}{L}\int_0^{x_i} \sigma(x)\,dx. \tag{18}$$

For a constant mobility, the channel conductance becomes

$$g = \frac{qZ\mu_n}{L}\int_0^{x_i} n(x)\,dx = qZ\mu_n|Q_n|/L. \tag{19}$$

The channel resistance of an elemental section dy, Fig. 6*a*, is given by

$$dR = \frac{dy}{gL} = \frac{dy}{Z\mu_n|Q_n(y)|} \tag{20}$$

and the voltage drop across this elemental section is given by

$$dV = I_D\,dR = \frac{I_D\,dy}{Z\mu_n|Q_n(y)|} \tag{21}$$

where I_D is the drain current and is a constant independent of y. Substituting Eq. 16 into Eq. 21 and integrating from the source ($y = 0$, $V = 0$) to the drain ($y = L$, $V = V_D$) yields

$$I_D = \frac{Z}{L}\mu_n C_i\left\{\left(V_G - 2\psi_B - \frac{V_D}{2}\right)V_D - \frac{2}{3}\frac{\sqrt{2\epsilon_s qN_A}}{C_i}\left[(V_D + 2\psi_B)^{3/2} - (2\psi_B)^{3/2}\right]\right\} \tag{22}$$

for the present idealized case.

Equation 22 predicts that for a given V_G the drain current first increases linearly with drain voltage (the linear region), then gradually levels off, approaching a saturated value (the saturation region). The basic output characteristic of an idealized MOSFET is shown in Fig. 7. The dashed line indicates the locus of the drain voltage ($V_{D\,\mathrm{sat}}$) at which the current reaches a maximum value.

We shall now consider the above-mentioned two regions. For the case of small V_D, Eq. 22 reduces to

$$I_D \simeq \frac{Z}{L}\mu_n C_i\left[(V_G - V_T)V_D - \left(\frac{1}{2} + \frac{\sqrt{\epsilon_s qN_A/\psi_B}}{4C_i}\right)V_D^2\right] \tag{23}$$

or

$$I_D \simeq \left(\frac{Z}{L}\right)\mu_n C_i(V_G - V_T)V_D \qquad \text{for} \quad V_D \ll (V_G - V_T) \tag{23a}$$

where V_T (the threshold voltage) is given by

$$V_T = 2\psi_B + \frac{\sqrt{2\epsilon_s qN_A(2\psi_B)}}{C_i}. \tag{24}$$

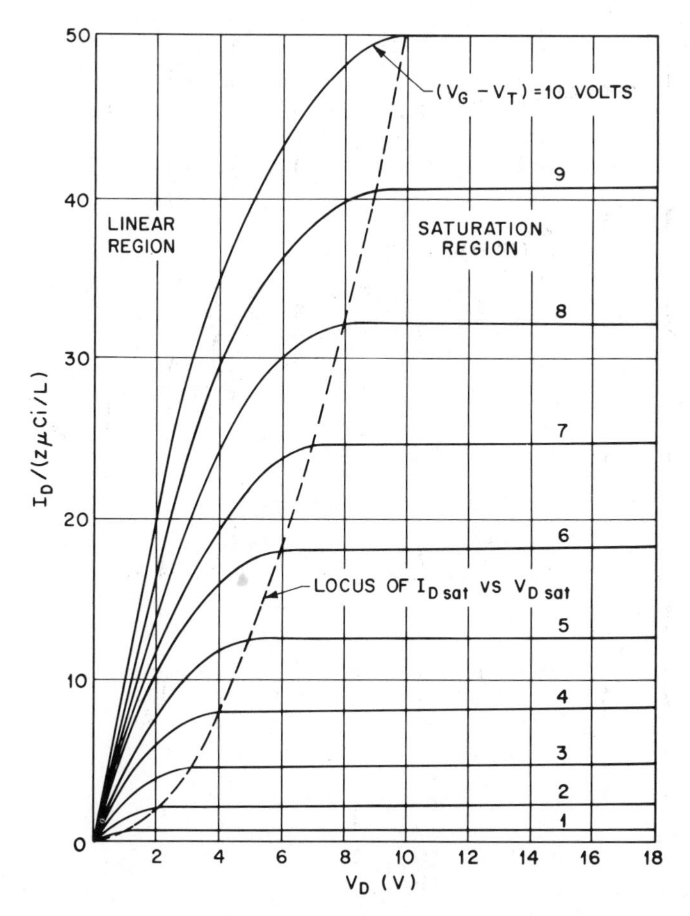

Fig. 7 Idealized drain characteristics (I_D versus V_D) of a MOSFET. The dashed line indicates the locus of the saturation drain voltage ($V_{D\,sat}$). For $V_D > V_{D\,sat}$, the drain current remains constant.

The calculated values of V_T as a function of semiconductor doping density and insulator thickness were shown in Chapter 7 for the Si–SiO_2 system. By plotting I_D versus V_G (for a given small V_D), the threshold voltage can be deduced from the linearly extrapolated value at the V_G axis. In the linear region, Eq. 23a, the channel conductance g_D and the transconductance g_m are given as

$$g_D \equiv \left.\frac{\partial I_D}{\partial V_D}\right|_{V_G=\text{const}} = \frac{Z}{L}\,\mu_n C_i (V_G - V_T) \tag{25}$$

$$g_m \equiv \left.\frac{\partial I_D}{\partial V_G}\right|_{V_D=\text{const}} = \frac{Z}{L}\,\mu_n C_i V_D\,. \tag{26}$$

When the drain voltage is increased to a point such that the charge in the inversion layer $Q(y)$ at $y = L$ becomes zero, the number of mobile electrons at the drain experiences a drastic fall-off. This point, called pinch-off, is analogous to the junction field-effect transistor. The drain voltage and the drain current at this point are designated as $V_{D\,sat}$ and $I_{D\,sat}$, respectively. Beyond the pinch-off point we have the saturation region. The value of $V_{D\,sat}$ is obtained from Eq. 16 under the condition $Q_n(L) = 0$:

$$V_{D\,sat} = V_G - 2\psi_B + K^2\left(1 - \sqrt{1 + 2V_G/K^2}\right) \tag{27}$$

where $K \equiv \sqrt{\epsilon_s q N_A}/C_i$. The saturation current $I_{D\,sat}$ can be obtained by substituting Eq. 27 into Eq. 22:

$$I_{D\,sat} \simeq \frac{mZ}{L} \mu_n C_i (V_G - V_T)^2. \tag{28}$$

where m is a function of doping concentration and approaches $\frac{1}{2}$ at low dopings.[11]

The threshold voltage V_T in the saturation region is the same as given by Eq. 24 for low substrate dopings and thin insulator layers. For higher dopings, V_T becomes V_G-dependent. The transconductance in the saturation region when Eq. 28 applies is given by

$$g_m = \left.\frac{\partial I_D}{\partial V_G}\right|_{V_D=\text{const.}} = \frac{2mZ}{L} \mu_n C_i (V_G - V_T). \tag{29}$$

In previous discussions, we made many assumptions to bring out the most important characteristics of the MOSFET. We shall now remove the first two assumptions and consider the effects due to a nonideal gate MOS and diffusion current. The main effect of the fixed oxide charges and the difference in work functions is to cause a voltage shift corresponding to the flat-band voltage V_{FB}. This in turn causes a change in the threshold voltage V_T; in the linear region V_T becomes

$$\begin{aligned} V_T &= V_{FB} + 2\psi_B + \frac{\sqrt{2\epsilon_s q N_A (2\psi_B)}}{C_i} \\ &= \left(\phi_{ms} - \frac{Q_f}{C_i}\right) + 2\psi_B + \frac{\sqrt{4\epsilon_s q N_A \psi_B}}{C_i}. \end{aligned} \tag{30}$$

When a substrate bias is applied, the threshold voltage becomes

$$V_T = V_{FB} + 2\psi_B + \sqrt{2\epsilon_s q N_A (2\psi_B + V_{BS})}/C_i \tag{31}$$

or

$$\begin{aligned} \Delta V_T &= V_T(V_{BS}) - V_T(V_{BS} = 0) \\ &= \frac{\sqrt{2\epsilon_s q N_A}}{C_i}\left(\sqrt{2\psi_B + V_{BS}} - \sqrt{2\psi_B}\right) \\ &= \frac{a}{\beta}\left(\sqrt{2\beta\psi_B + \beta V_{BS}} - \sqrt{2\beta\psi_B}\right) \end{aligned} \tag{32}$$

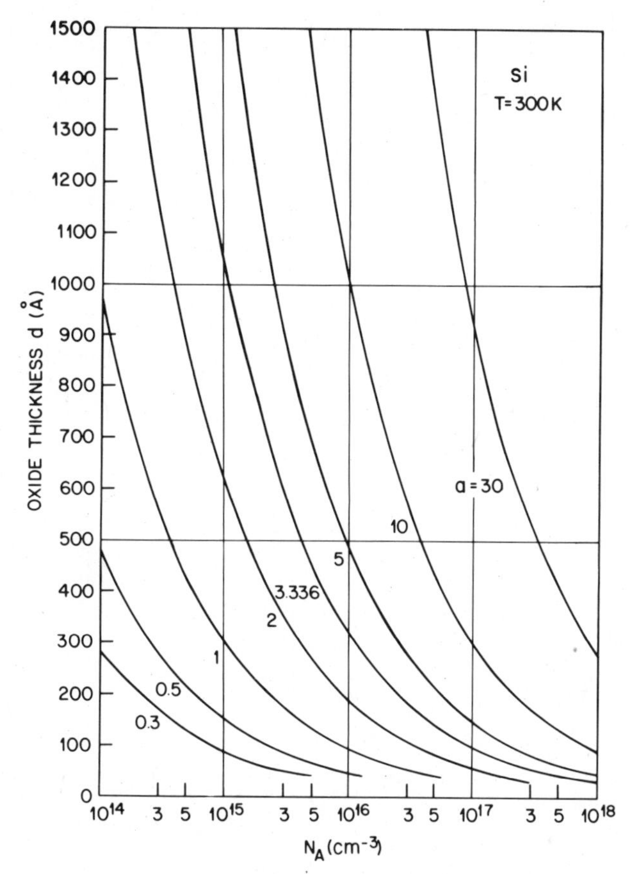

Fig. 8 Oxide thickness versus substrate doping for various *a* values. (After Brews, Ref. 19.)

where

$$a \equiv \sqrt{2}(\epsilon_s/L_D)/C_i = 2(\epsilon_s/\epsilon_i)(d/L_D). \tag{33}$$

In Fig. 8, oxide thickness versus substrate doping is plotted for given a values[19] using Eq. 33. The a values increase with increasing doping and oxide thickness.

Threshold voltage shift versus V_{BS} is plotted in Fig. 9 for various a values. As the a value increases, ΔV_T also increases. For a given a value, the resulting variation in ΔV_T is indicated by vertical bars for substrate dopings ranging from 10^{15} to 10^{17} cm^{-3} (Fig. 9). The primary influence upon ΔV_T is the choice of a itself; the influence of doping or oxide thickness upon ΔV_T, independent of a, is minor.

To consider the effect of the diffusion current component, we refer to Fig. 4 for the nonequilibrium condition. The drain current density including

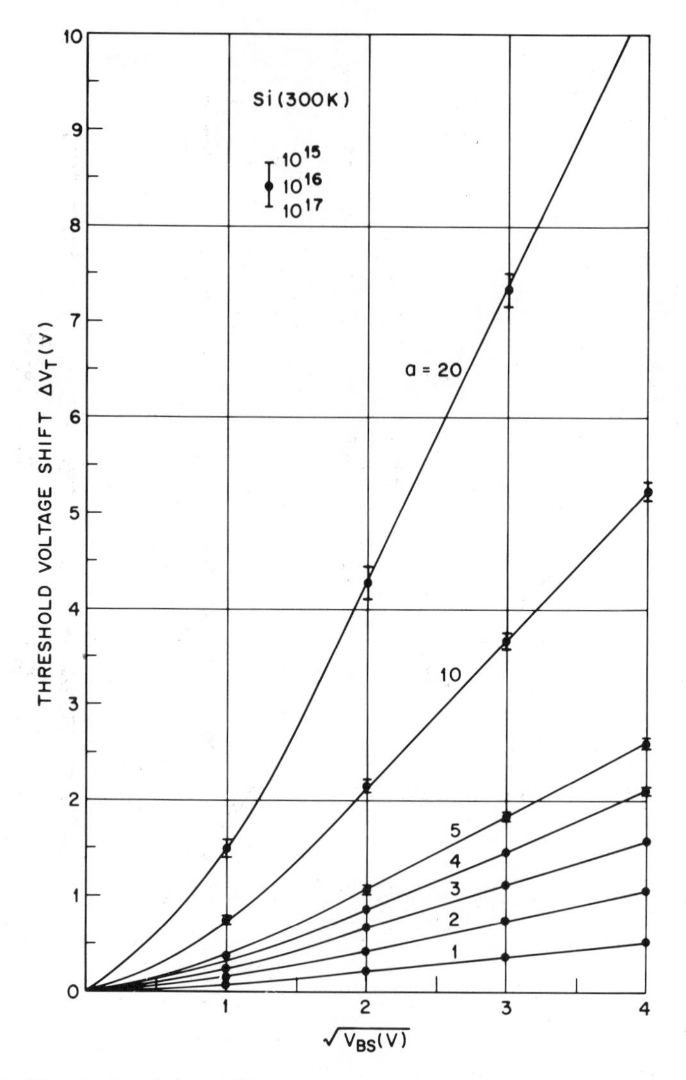

Fig. 9 Threshold voltage shift versus substrate reverse bias for various a values.

both drift and diffusion components is given by

$$J_D(x, y) = q\mu_n n\mathscr{E}_y + qD_n\nabla n$$

$$= -qD_n n(x, y)\nabla\psi_{Fn} \qquad (34)$$

where ψ_{Fn} is the electron imref measured from the bulk Fermi level. The total drain current based on the gradual-channel approximation is

$$I_D = \int_0^{x_i} J_D(x, y)Z\,dx$$

$$= \frac{1}{L}\int_0^L D_n q Z \left(\frac{\partial \psi_{Fn}}{\partial y}\right) \int_0^{x_i} n(x, y)\,dx\,dy$$

$$= \frac{Z}{L}\frac{\epsilon_s \mu_n}{L_D}\int_0^{V_D}\int_{\psi_B}^{\psi_s} \frac{e^{\beta\psi - \beta V}}{F(\beta\psi, V, n_{po}/p_{po})}\,d\psi\,dV. \tag{35}$$

The gate voltage V_G is related to the surface potential ψ_s by

$$V_G' = V_G - V_{FB} = -\frac{Q_s}{C_i} + \psi_s$$

$$= \frac{2\epsilon_s kT}{C_i q L_D} F\left(\beta\psi_s, V, \frac{n_{po}}{p_{po}}\right) + \psi_s. \tag{36}$$

Equation 35 reduces to Eq. 22 for gate voltages well above threshold. Equation 22 however, becomes inaccurate for gate voltages near threshold, and near pinch-off. For a particular device with known physical dimensions, bulk impurity concentration, and effective mobility, Eq. 35 can be calculated numerically to give accurate results for the entire range of drain voltage from the linear region to the saturation region. Figure 10 demonstrates the current saturation phenomena very well, showing a typical drain characteristic for a long-channel MOSFET.[16]

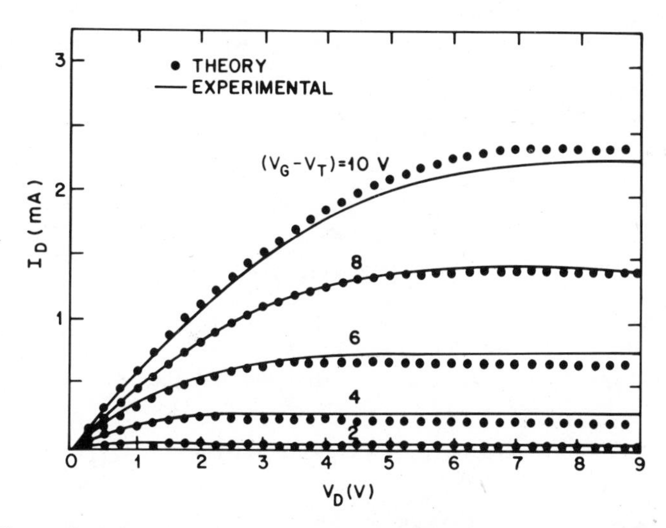

Fig. 10 Theoretical (dots) and experimental (solid lines) drain characteristics of a p-channel MOSFET having $d = 2000$ Å, $N_D = 4.6 \times 10^{14}$ cm^{-3}, and $\mu_p = 256$ cm^2/V-s. (After Pao and Sah, Ref. 16.)

8.2.3 Subthreshold Region

When gate voltage is below the threshold voltage and the semiconductor surface is in weak inversion, the corresponding drain current is called the subthreshold current.[20,21] The subthreshold region is particularly important for low-voltage, low-power applications, such as when the MOSFET is used as a switch in digital logic and memory applications, because the subthreshold region describes how the switch turns on and off.

In weak inversion, the drain current is dominated by diffusion and is derived in the same way as the collector current in a bipolar transistor with homogeneous base doping. Considering the MOSFET as an n-p-n (source–substrate–drain) bipolar transistor, we have

$$I_D = -qAD_n \frac{dn}{dy} = qAD_n \frac{n(0) - n(L)}{L} \tag{37}$$

where A is the cross section of current flow, and $n(0)$ and $n(L)$ are the electron densities in the channel at the source and the drain, respectively (Fig. 6a). These electron densities are given by

$$n(0) = n_{po} e^{\beta\psi_s} \tag{38a}$$

$$n(L) = n_{po} e^{\beta\psi_s - \beta V_D}. \tag{38b}$$

where ψ_s is the surface potential at the source. The area of current flow is given by the width Z of the device and the effective channel thickness normal to the semiconductor–insulator interface. Because of the exponential dependence of electron density on the potential ψ, the effective channel thickness corresponds to the distance in which ψ decreases by kT/q. Therefore, the effective channel thickness is $kT/q\mathscr{E}_s$ where $\mathscr{E}_s$ is the weak-inversion surface field given by

$$\mathscr{E}_s = -Q_B/\epsilon_s = \sqrt{2qN_A\psi_s/\epsilon_s}. \tag{39}$$

Substituting Eqs. 38 and 39 into 37 gives[18,22]

$$I_D = \mu_n \left(\frac{Z}{L}\right) \frac{aC_i}{2\beta^2} \left(\frac{n_i}{N_A}\right)^2 (1 - e^{-\beta V_D}) e^{\beta\psi_s} (\beta\psi_s)^{-1/2} \tag{40}$$

where we have used the relation $D_n = \mu_n kT/q$, and a is given by Eq. 33. The surface potential ψ_s is related to the gate voltage as follows:[18,19]

$$\psi_s = (V_G - V_{FB}) - \frac{a^2}{2\beta} \left\{ \left[1 + \frac{4}{a^2} (\beta V_G - \beta V_{FB} - 1) \right]^{1/2} - 1 \right\}. \tag{41}$$

Equation 40 indicates that in the subthreshold region the drain current varies exponentially with V_G, and for drain voltage V_D larger than $3kT/q$, the current becomes independent of V_D.

Equation 40 can be used to find the gate-voltage swing S, needed to reduce the current by one decade. By definition,

$$
\begin{aligned}
S &\equiv \ln 10 \cdot dV_G/d(\ln I_D) \\
&= (kT/q) \ln 10 \cdot d(\beta V_G)/d(\ln I_D) \\
&= (kT/q) \ln 10 \cdot [1 + C_D(\psi_s)/C_i]\left\{1 - \left(\frac{2}{a^2}\right)[C_D(\psi_s)/C_i]^2\right\}.
\end{aligned} \tag{42}
$$

For $a \gg (C_D/C_i)$, the subthreshold swing becomes

$$S \simeq \frac{kT}{q} \ln 10 \cdot (1 + C_D/C_i). \tag{43}$$

The term in parentheses is the capacitive divider ratio $(C_i + C_D)/C_i$.

If there is a significant interface-trap density, the capacitance C_{it} associated with the interface traps is in parallel with the depletion-layer capacitance C_D. Using Eq. 43 and substituting $(C_D + C_{it})$ for C_D, we obtain

$$S \text{ (with interface traps)} = S \text{ (no interface traps)} \times \frac{1 + (C_D + C_{it})/C_i}{1 + C_D/C_i} \tag{44}$$

where $C_{it} = qD_{it}$ and D_{it} is the interface-trap density.

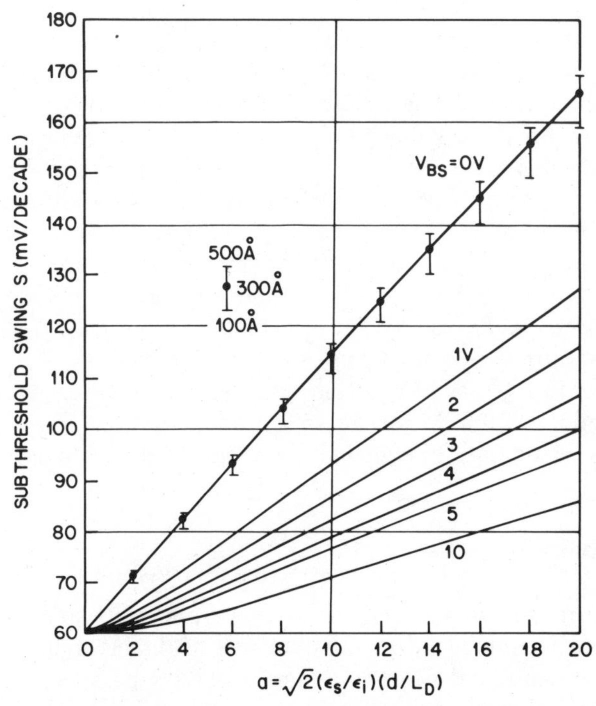

Fig. 11 Subthreshold swing versus a for various substrate reverse bias. (After Brews, Ref. 19.)

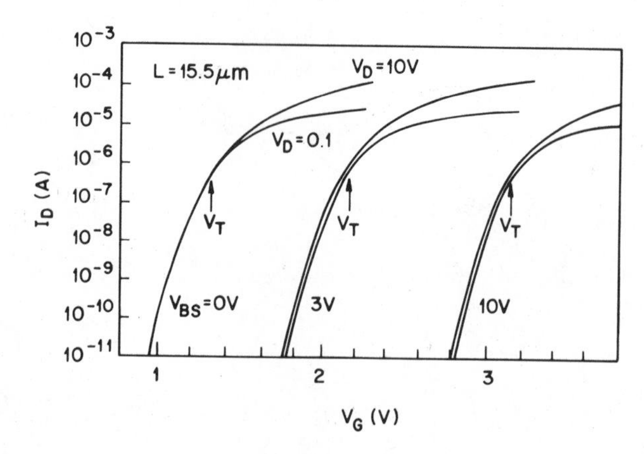

Fig. 12 Experimental subthreshold characteristics for a long-channel device (L = 15.5 μm). (After Troutman, Ref. 23.)

When a substrate bias is applied, it increases the value of ψ_s. Consequently, the depletion-layer capacitance C_D is reduced and therefore S is reduced. Figure 11 shows the calculated subthreshold swing as a function of a for different substrate-reverse biases.[19] Again the primary influence upon S is the choice of a itself. Also, the first volt of the substrate bias results in the greatest reduction of S.

Experimental subthreshold characteristics for a long-channel (15.5 μm) MOSFET are shown[23] in Fig. 12 for three values of V_{BS}. As expected, for voltages below threshold voltage (i.e., below V_T as marked on the curves), current varies exponentially with gate voltage. For a given V_{BS}, the experimental curves for drain voltages of 0.1 V and 10 V show virtually no dependence on drain voltage in the subthreshold region. This important indication of long-channel behavior is predicted by Eq. 40. The MOSFET had a gate oxide of 570 Å and a substrate doping of 5.6×10^{15} cm^{-3}. The corresponding a value is 4. The calculated subthreshold swing S is 83 mV/decade for $V_{BS} = 0$, 67 mV/decade for $V_{BS} = 3$ V, and about 63 mV/decade for $V_{BS} = 10$ V (from Fig. 11). The calculated threshold voltage shift ΔV_T is 0.75 V for $V_{BS} = 3$ V and 1.7 V for $V_{BS} = 10$ V (from Fig. 9). These results are in excellent agreement with the measured values from Fig. 12.

8.2.4 Mobility Behavior

Because current flows in the inversion layer, mobility and drift velocity are expected to be influenced by the thickness of the inversion layer. When a very small longitudinal field $\mathscr{E}_y$ is applied ($\mathscr{E}_y$ is parallel to the current flow), the drift velocity varies linearly with $\mathscr{E}_y$, and the slope is the drift mobility ($v = \mu_n \mathscr{E}_y$). Experimental measurements on ⟨100⟩ p-type Si in-

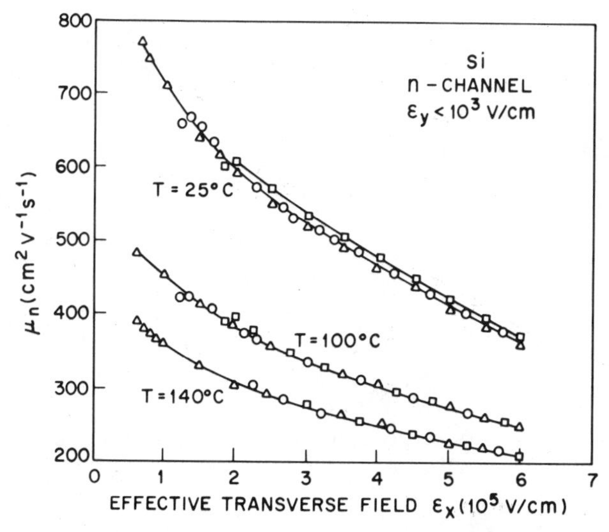

Fig. 13 Inversion layer mobility versus effective transverse field for three temperatures. (After Sabnis and Clemens, Ref. 24.)

version layers show that this mobility is a unique function of the transverse field $\mathscr{E}_x$, which is perpendicular to the current flow. This mobility is not a function of the surface processing or the doping density in the range $N_A < 10^{17}\ \text{cm}^{-3}$. The measured results are shown[24] in Fig. 13. At a given temperature, mobility decreases with increasing effective transverse field, defined as the field averaged over the electron distribution in the inversion layer, and is given by

$$(\mathscr{E}_x)_{\text{eff}} = \frac{1}{\epsilon_s}(Q_B + \tfrac{1}{2}Q_n). \tag{45}$$

When the longitudinal field increases, eventually velocity saturation occurs, similar to that of bulk silicon. The measured electron-drift velocity is shown[25,26] in Fig. 14. For a given transverse field ($\mathscr{E}_x$), the velocity is proportional to $\mathscr{E}_y$ at low longitudinal fields, and the proportionality constant is the mobility as plotted in Fig. 13. However, as $\mathscr{E}_y$ increases, the velocity tends to saturate. A general expression can be given for the drift velocity[27]

$$v_d = v_0\left[1 + \left(\frac{v_0}{v_c}\right)^2\left(\frac{v_0}{v_c} + G\right)^{-1} + \left(\frac{v_0}{v_s}\right)^2\right]^{-1/2} \tag{46}$$

where v_c, v_s, and G are fitting parameters, and

$$v_0 \equiv \mu_n(\mathscr{E}_x)\cdot\mathscr{E}_y. \tag{47}$$

That is, for a given transverse field $\mathscr{E}_x$, mobility is a unique function of $\mathscr{E}_x$.

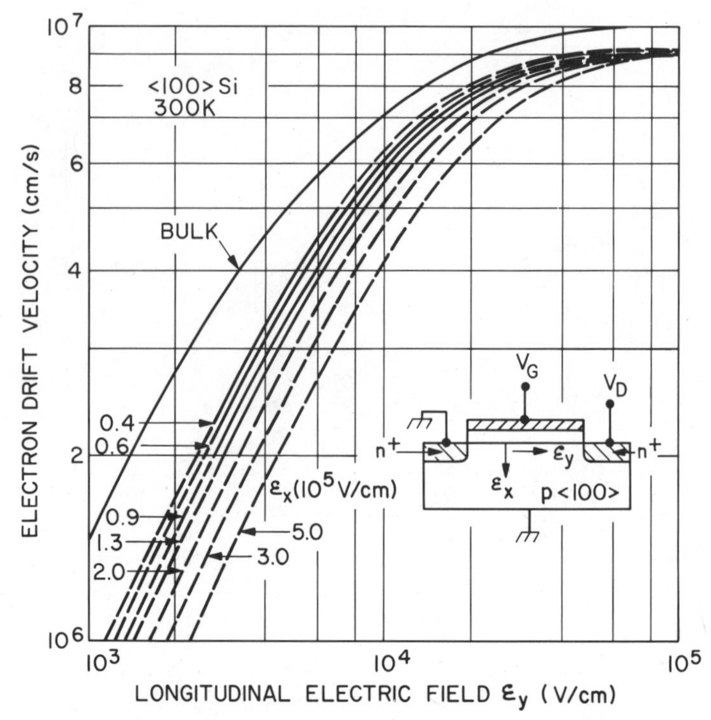

Fig. 14 Electron drift velocity versus longitudinal field for various transverse fields. The dashed curves are calculated and the solid portions indicate the regions where data were actually taken. (After Cooper and Nelson, Ref. 26).

In the limit $\mathscr{E}_y \to 0$, the drift velocity approaches $v_d = v_0 = \mu_n(\mathscr{E}_x)\mathscr{E}_y$. On the other hand, when $\mu_n \mathscr{E}_y$ is much greater than v_c and v_s, v_d is approximately equal to v_s, where v_s is a function of the transverse field.

Chapter 6 considered the effects of velocity saturation on device characteristics for the JFETs. Figure 15 shows similar results for a MOSFET. A comparison is made between the simulated current assuming a constant mobility (dashed lines) and the measured current from the same device having velocity saturation (solid lines).[28] Velocity saturation has two effects. First, saturation current is greatly reduced, especially for large gate voltages. Second, saturation current is linearly dependent on gate voltage, rather than nearly quadratic dependent as predicted by Eq. 28. Under the velocity saturation condition, the saturation current is given by

$$I_{D\,\text{sat}} = ZC_i(V_G - V_T)v_s \,. \tag{48}$$

Therefore, the transconductance g_m becomes a constant:

$$g_m = (\partial I_{D\,\text{sat}})/\partial V_G = ZC_i v_s \,. \tag{49}$$

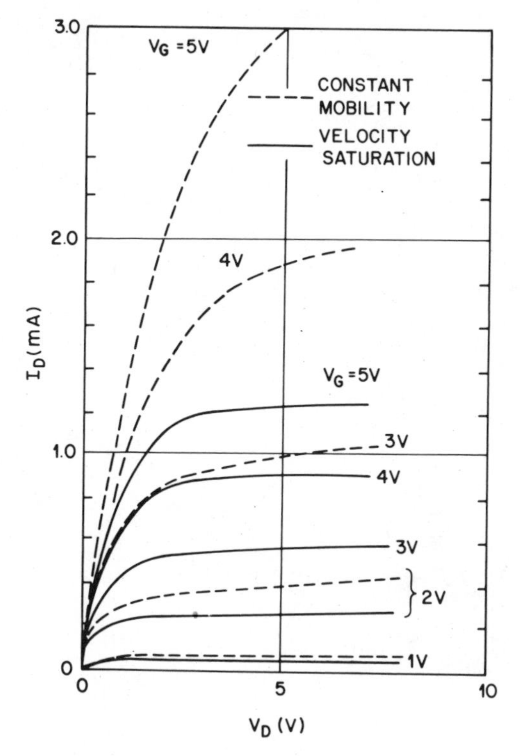

Fig. 15 Comparison of drain characteristics for constant mobility case (dashed lines) and field-dependent mobility (solid lines). (After Yamaguchi, Ref. 28.)

8.2.5 Temperature Dependence

Temperature affects device parameters and performance, especially mobility, threshold voltage, and subthreshold characteristics. The effective mobility in inversion layer has a T^{-2} power dependence on temperatures above 300 K at gate biases corresponding to strong inversion.[24]

In the linear region the threshold voltage is given by Eq. 30:

$$V_T = \phi_{ms} - \frac{Q_f}{C_i} + 2\psi_B + \frac{\sqrt{4\epsilon_s q N_A \psi_B}}{C_i}. \tag{50}$$

Because the work-function difference ϕ_{ms} and the fixed oxide charges are essentially independent of temperature, differentiating Eq. 50 with respect to temperature yields[29]

$$\frac{dV_T}{dT} = \frac{d\psi_B}{dT}\left(2 + \frac{1}{C_i}\sqrt{\frac{\epsilon_s q N_A}{\psi_B}}\right) \tag{51}$$

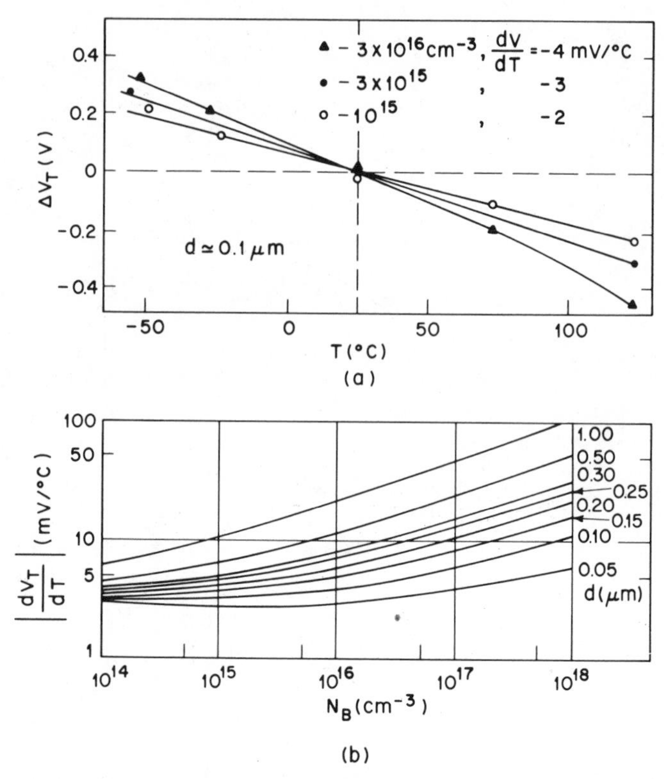

Fig. 16 (*a*) Experimental measurement of threshold voltage versus temperature. (*b*) dV_T/dT of a Si–SiO_2 system versus substrate doping with oxide thickness as a parameter. (After Vadasz and Grove, Ref. 29; Wang et al., Ref. 30.)

where

$$\frac{d\psi_B}{dT} \simeq \pm \frac{1}{T}\left[\frac{E_g(T=0)}{2q} - |\psi_B(T)|\right]. \tag{51a}$$

Figure 16*a* shows typical experimental measurements of threshold voltage near room temperature for the Si–SiO_2 systems.[29,30] The data can be represented by straight lines over this temperature range. Thus a representative figure for device behavior can be obtained by evaluating Eq. 51 at room temperature. Figure 16*b* shows the results of such calculations as a function of substrate doping for various values of oxide thickness. Also note that for a given oxide thickness, the quantity dV_T/dT generally increases with increased doping.

As temperature decreases, the MOSFET characteristics improve, especially in the subthreshold region. Figure 17 shows the transfer characteristics of a long-channel MOSFET ($L = 9\ \mu$m) with temperature as parameter.[31] Note that as temperature decreases from 296 K to 77 K, the

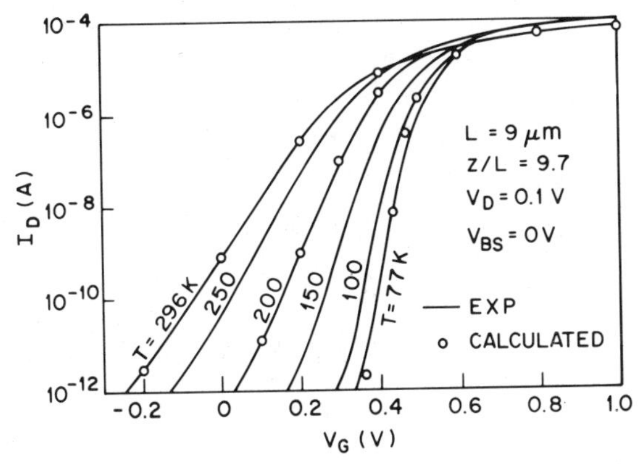

Fig. 17 Transfer characteristics for a long-channel device ($L = 9\ \mu$m) with temperature as a parameter. (After Gaensslen et al., Ref. 31.)

threshold voltage V_T increases from 0.25 V to about 0.5 V. This increase in V_T is similar to that shown in Fig. 16. The most important improvement is the reduction of the subthreshold swing S from 80 mV/decade at 296 K to 22 mV/decade at 77 K. Thus the improvement in the subthreshold swing at 77 K is about a factor of 4. This improvement comes mainly from the kT/q term in Eq. 42. Other improvements at 77 K include higher mobility, higher transconductance, higher threshold conductivity, lower power consumption, lower junction leakage current, and lower metal-line resistance. The major disadvantage is that the MOSFET must be immersed in a suitable inert coolant (e.g., liquid nitrogen) and low-temperature setup requires additional equipment.

8.2.6 Types of MOSFETs

The MOSFET is ideally a transadmittance amplifier with an infinite input resistance and a current generator at the output. In practice, however, we have other circuit parameters. An equivalent circuit is shown in Fig. 18 for the common-source connection.[32] The differential transconductance g_m was discussed previously. The input conductance G_{in} is caused by leakages through the thin gate insulator. For a thermally grown silicon dioxide layer, the leakage current between the gate and the channel is very small, of the order of 10^{-10} A/cm^2; thus the input conductance is negligible. The input capacitance C_{in} is equal to $\partial Q_M/\partial V_G$, where Q_M is the total charge on the gate.[16] In practical devices, the insulator layer and the metal gate may extend somewhat above the source and drain regions. This fringe effect will be the most important contribution to the feedback capacitance C_{fb}. The output conductance G_{out} is equal to the drain conductance. The output capacitance consists mostly of the two p-n junction capacitances connected in series through the semiconductor bulk. In the linear region, from

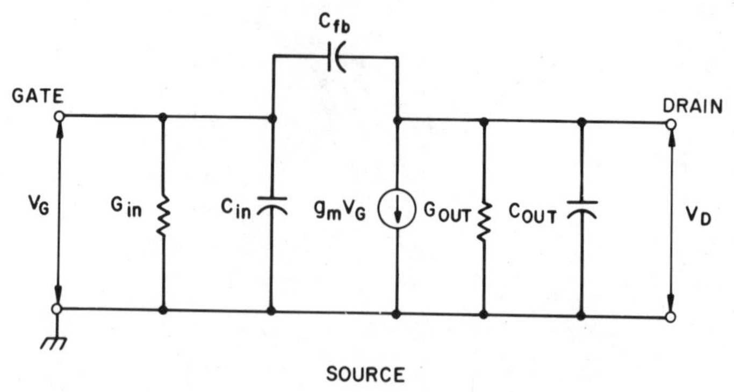

Fig. 18 Equivalent circuit of MOSFET for common-source configuration. (After Ihantola and Moll, Ref. 6.)

Eq. 26 and the fact that $C_{in} \simeq ZLC_i$, the maximum operating frequency is given by

$$f_m = \frac{\omega_m}{2\pi} = \frac{g_m}{2\pi C_{in}} \simeq \frac{\mu_n V_D}{2\pi L^2}. \tag{52}$$

In the saturation region, f_m is obtained from Eq. 49:

$$f_m \simeq \frac{v_s}{2\pi L}. \tag{53}$$

The corresponding transit time for velocity saturation is

$$\tau = \frac{L}{v_S}. \tag{54}$$

For $L = 1\ \mu m$ and $v_s = 10^7$ cm/s, the transit time is only 10 ps. However, in a typical ring oscillator with 1 μm-channel MOSFETs, the measured delay time is usually an order of magnitude longer than 10 ps. Thus the delay is mainly caused by the parasitic resistance and capacitance around the device.

There are basically four different types of MOSFET, depending on the types of inversion layer. If at zero gate bias, the channel conductance is very low, we must apply positive voltage to the gate to form the n-channel. This type is the normally-off (enhancement) n-channel MOSFET. If an n-channel exists at zero bias, we must apply a negative bias to the gate to deplete carriers in the channel to reduce channel conductance. This type is called the normally-on (depletion) n-channel MOSFET. The n-channel enhancement and depletion-mode MOSFETs are shown in Fig. 19*a*. Similarly we have the p-channel normally-off (enhancement) and normally-on (depletion) MOSFET (Fig. 19*b*).

The electrical symbol, transfer characteristics, and output characteristics of the four types are shown[33] in Fig. 20. Note that for the normally-off

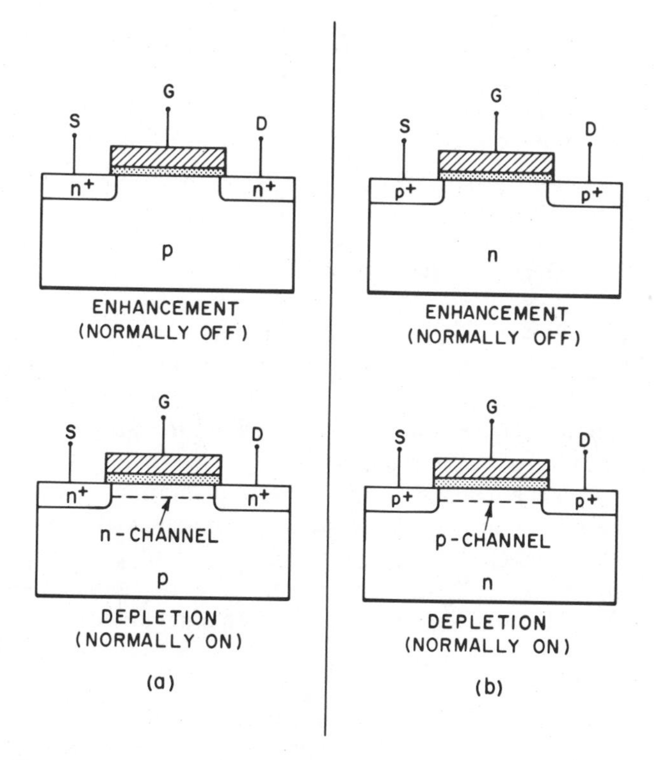

Fig. 19 Basic types of MOSFETs. (*a*) *n*-channel. (*b*) *p*-channel.

TYPE	ELECTRICAL SYMBOL	OUTPUT CHARACTERISTIC	TRANSFER CHARACTERISTIC
N - CHANNEL ENHANCEMENT (NORMALLY OFF)		I_D, +, V_T, 0, V_D →, +	I_D, −, 0, V_T, +, V_G
N - CHANNEL DEPLETION (NORMALLY ON)	G, V_G, I_D, S, D, BS	I_D, +, $V_G = 0$, −, 0, V_D →, +	I_D, −, 0, +, V_G
P - CHANNEL ENHANCEMENT (NORMALLY OFF)		− ← V_D, 0, V_T, −, I_D	V_G, −, V_T, 0, +, I_D
P - CHANNEL DEPLETION (NORMALLY ON)		− ← V_D, 0, $V_G = 0$, I_D	V_G, −, 0, +, I_D

Fig. 20 Electric symbol, transfer characteristics, and output characteristics of the four types of MOSFET. (After Gallagher and Corak, Ref. 33.)

n-channel device, a positive gate bias larger than the threshold voltage V_T must be applied before a substantial drain current flows. For the normally-on n-channel device, a large current can flow at $V_G = 0$, and the current can be increased or decreased by varying the gate voltage. The discussion above can be readily extended to p-channel devices by changing polarities.

8.3 NONUNIFORM DOPING AND BURIED-CHANNEL DEVICES

In Section 8.2 doping concentration in the channel is assumed to be constant. In practical devices, however, the doping is generally nonuniform, even for doped substrates that are initially uniform, because the thermal oxidation causes impurity redistribution. Moreover, in modern MOSFET technology, ion implantation is used extensively to improve device performance. For example, ion implantation is used for (1) a self-aligned source and drain to reduce overlap capacitances, (2) a shallow dopant at the Si–SiO_2 interface for threshold voltage adjustment, (3) a channel implant on a lightly doped substrate to reduce punch-through between source and drain, and (4) a buried-channel device by incorporating within the surface region impurities of the type opposite to that of the substrate impurities.

The impurity profiles $N(x)$ in ion-implanted devices resemble a Gaussian distribution with the maximum concentration at a projected range R_p and with a standard deviation ΔR_p:

$$N(x) = \frac{D_I}{\sqrt{2\pi}\,\Delta R_p} \exp\left[-\frac{(x-R_p)^2}{2(\Delta R_p)^2}\right] \tag{55}$$

where D_I is the ion dose per unit area (Fig. 21).[34] Both projected range and

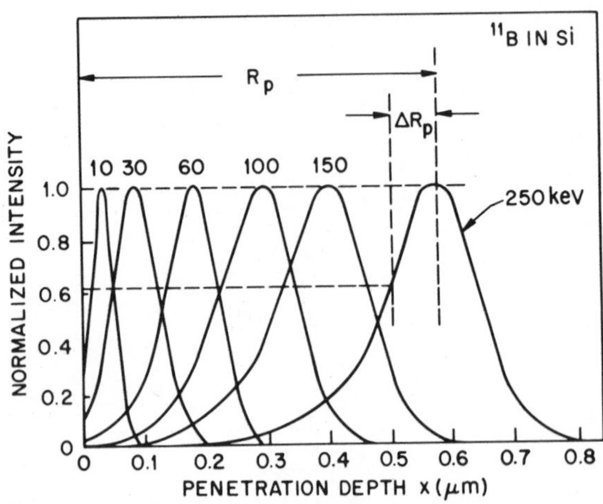

Fig. 21 Normalized range distribution of boron in silicon for different implantation energies. (After Wittmack, Maul, and Schulz, Ref. 34.)

standard deviation increase with increasing implantation energy.[34] The values of R_p and ΔR_p for boron, arsenic, and phosphorus in silicon have been given in Chapter 2.

We consider next the effect of nonuniform channel doping on device characteristics, especially on threshold voltage and subthreshold slope. We also consider the buried-channel MOSFET, which can be used as a load device (normally-on) in the basic inverter or as a normally-off high-speed device.

8.3.1 Threshold Shift

To derive the threshold voltage shift due to ion implantation, we shall first consider an idealized step-doping profile, as shown in Fig. 22 (solid line).[35] The original implant is altered after thermal annealing and the annealed profile is approximated by the step function with step depth x_s, equal to the sum of the projected range and the standard deviation of the original implant. The surface doping N_S is given by

$$(N_S - N_B)x_s = \int_0^\infty [N_A(x) - N_B]\, dx = D_I. \tag{56}$$

For the limiting case of a delta function of negative charge (e.g., ionized boron acceptors) localized at the Si–SiO_2 interface, the charge is equivalent to a reduction of the fixed oxide charge by an amount qD_I. Therefore, for

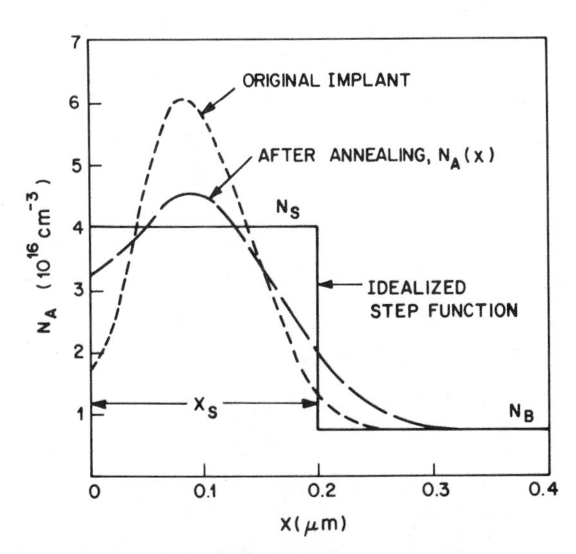

Fig. 22 Doping profile of implanted region beneath the gate oxide. The original implant is broadened by thermal annealing. A step doping is used to approximate the actual doping. (After Rideout, Gaensslen, and LeBlanc, Ref. 35.)

this case

$$V_T = V_{FB} + 2\psi_B + \frac{\sqrt{2\epsilon_s q N_B (2\psi_B + V_{BS})}}{C_i} + \frac{qD_I}{C_i} \tag{57}$$

and the threshold shift is simply

$$\Delta V_T = V_T(D_I) - V_T(D_I = 0) = \frac{qD_I}{C_i}. \tag{58}$$

For a wide x_s, that is, the maximum depletion-layer width under heavy inversion W_m is smaller than x_s, the surface region can be considered a uniformly doped region with concentration N_S. The threshold voltage is identical to that given by Eq. 31, with N_S replacing N_A and ψ_B corresponding to that for N_S.

If $W_m > x_s$, the depletion-layer width and the voltage drop across the oxide can be obtained from Poisson's equation with a high–low step-doping profile:

$$W_m = \sqrt{\frac{2\epsilon_s}{qN_B}} \left[\psi_s + V_{BS} - \frac{qx_s^2}{2\epsilon_s}(N_S - N_B) \right]^{1/2} \tag{59}$$

$$V_i = \frac{q}{C_i} \int_0^{W_m} N_A(x)\, dx$$

$$= \frac{q}{C_i} [N_B W_m + (N_S - N_B)x_s] = \frac{q}{C_i}(N_B W_m + D_I) \tag{60}$$

where ψ_s is the surface band bending under the heavy inversion and V_i is the voltage drop across the oxide. The threshold voltage is then

$$V_T = V_{FB} + \psi_s + \frac{\sqrt{2q\epsilon_s N_B}}{C_i} \left(\psi_s + V_{BS} - \frac{qx_s}{2\epsilon_s} D_I \right)^{1/2} + \frac{qD_I}{C_i}. \tag{61}$$

Equation 61 indicates that when $(V_{BS} + \psi_s)$ becomes much larger than $qx_s D_I/2\epsilon_s$, the threshold voltage for the step profile approaches that for the delta-function profile at the interface, Eq. 57.

The calculated threshold voltages as functions of substrate bias are shown in Fig. 23 using Eqs. 57 through 61. For the two uniformly doped cases, $N_B = 7.5 \times 10^{15}$ cm^{-3} and $N_B = 4 \times 10^{16}$ cm^{-3}, the corresponding a values are 3.2 and 7.4, respectively. The variations of V_T versus V_{BS}, curves (a) and (b), are similar to those shown in Fig. 9. For the delta-function dose, curve (c) is shifted by $qD_I/C_i = 1.1$ V and is parallel with curve (a). For the step doping, curve (d), when V_{BS} is small the depletion-layer width is less than 0.2 μm; therefore, V_T follows curve (b). As V_{BS} increases, the depletion-layer width W_m exceeds the depth x_s, and the variation of V_T with V_{BS} changes to that for the impulse profile, curve (c).

It is generally desirable to have low substrate sensitivity, that is, weak dependence of the threshold voltage on the substrate reverse bias. Figure 23 shows that a shallow channel implant with an appropriate ion dose can

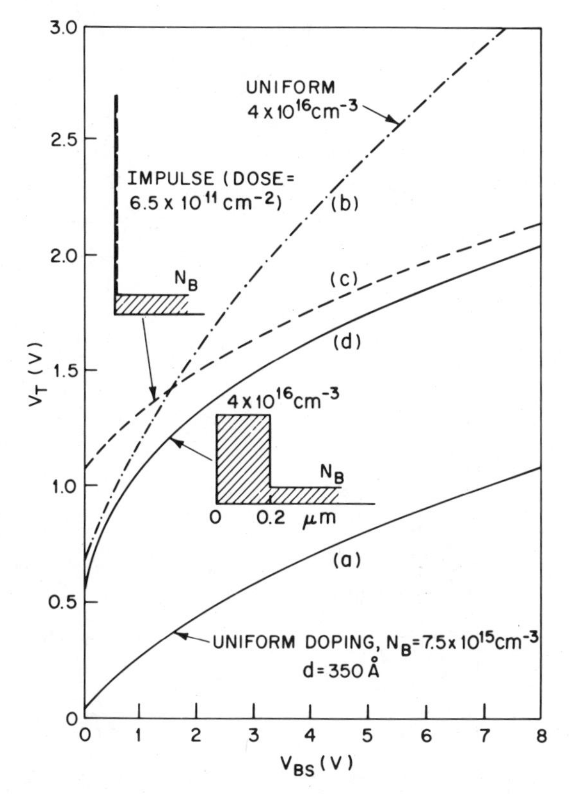

Fig. 23 Calculated substrate sensitivity for various doping profiles. (After Rideout, Gaensslen, and LeBlanc, Ref. 35.)

raise the threshold voltage while maintaining a low substrate sensitivity.

The step-profile approach described above can give first-order results for the threshold voltage. To obtain an accurate V_T, we have to consider the actual doping profile, because the step width x_s is not well defined for nonuniform doping profiles. A schematic diagram for the nonuniform doping $N_A(x)$ is shown[36] in Fig. 24*a*. The threshold voltage depends on the implanted dose D_I and the centroid of the dose x_c. Therefore, the actual implant can be replaced by a delta-function equivalent implant located at $x = x_c$ as shown in Fig. 24*b*, and

$$D_I = \int_0^{x_I} [N_A(x) - N_B]\, dx \tag{62}$$

$$x_c = \int_0^{x_I} [N_A(x) - N_B]\, x\, dx / D_I \,. \tag{63}$$

The electric field for the low–high–low doping profile is also shown in Fig. 24*b*. The area under the electric field corresponds to the surface potential

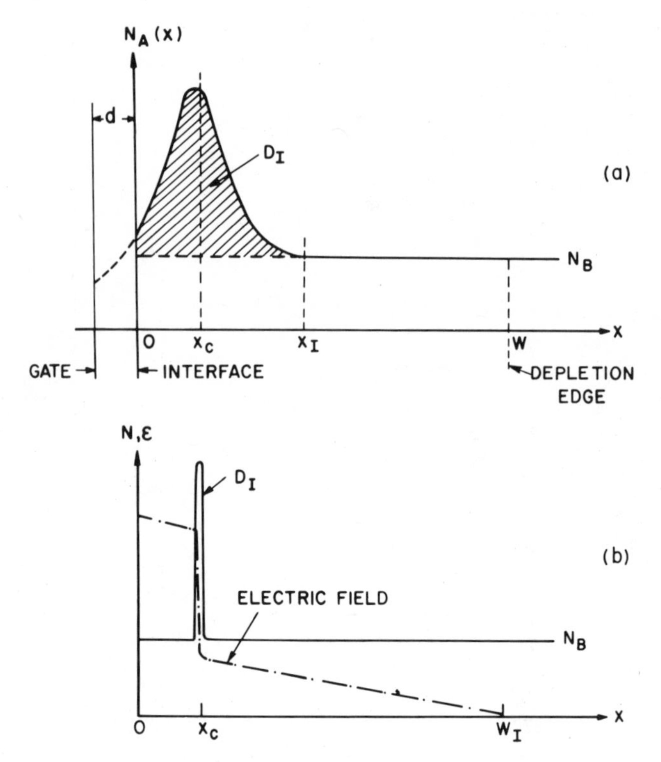

Fig. 24 (*a*) Implanted doping profile. (*b*) Delta function equivalent to (*a*). (After Brews, Ref. 36.)

ψ_s. The depletion-layer width W_I can be obtained from Poisson's equation:

$$W_I = \sqrt{2}L_D[\beta\psi_s - D_I x_c/(N_B L_D^2) - 1]^{1/2}. \tag{64}$$

The depletion-layer width for a uniformly doped device W_U is

$$W_U = \sqrt{2}L_D[\beta\psi_s - 1]^{1/2}. \tag{65}$$

Comparing Eqs. 64 and 65 shows that the depletion-layer width is reduced by the implant.

The threshold voltage shift due to an implant can be obtained once the depletion-layer width is calculated from Eq. 64. The threshold voltage shift is defined[11,36] as the difference in gate voltage needed to maintain the same inversion-layer charge density Q_n. For the uniformly doped case, with a gate bias V_{GU} and a surface-band bending ψ_{SU}, the inversion charge density is

$$Q_n = Q_s - Q_B = C_i(V_{GU} - \psi_{SU}) - qN_B W_U. \tag{66}$$

For the implanted case with gate bias V_{GI} and surface-band bending ψ_{SI}, the inversion-layer charge density is

$$Q_n = C_i(V_{GI} - \psi_{SI}) - qN_B W_I - qD_I\,. \tag{67}$$

Then the shift in gate voltage for the same Q_n is $(V_{GI} - V_{GU})$, or subtracting Eq. 66 from Eq. 67, the threshold voltage shift ΔV_T is

$$\Delta V_T = V_{GI} - V_{GU} = \psi_{SI} - \psi_{SU} - qN_B(W_U - W_I)/C_i + qD_I/C_i\,. \tag{68}$$

In practice, if Q_n is chosen in strong inversion, ψ_{SI} and ψ_{SU} differ by less than kT/q. Thus, we may make ψ_{SI} equal to ψ_{SU} in Eq. 68 and use Eqs. 64 and 65 for W_I and W_U. The threshold voltage shift is

$$\Delta V_T = \frac{qD_I}{C_i} - \frac{\sqrt{2}qL_D N_B}{C_i}\left[(\beta\psi_{SU} - 1)^{1/2} - \left(\beta\psi_{SU} - \frac{D_I x_c}{N_B L_D^2} - 1\right)^{1/2}\right]. \tag{69}$$

As discussed previously, the surface band bending ψ_{SU} under heavy inversion can be approximately given by $(2\psi_B + V_{BS})$. For a given nonuniform doping, one can evaluate D_I and x_c; the threshold voltage shift is then obtainable from Eq. 69.

Figure 25 illustrates[36] the dependence of the threshold voltage shift on the centroid x_c for different doses D_I. For $x_c = 0$, the implant is a delta function at the Si–SiO$_2$ interface. The terms in brackets in Eq. 69 equal zero and $\Delta V_T = qD_I/C_i$. As x_c increases, ΔV_T decreases. Eventually, x_c reaches the depletion-layer width W, and the depletion-layer edge becomes clamped to the implant. The dose for which x_c is equal to W can be obtained from Eq. 64:

$$D_I(x_c = W) = N_B(W_U^2 - x_c^2)/2x_c\,. \tag{70}$$

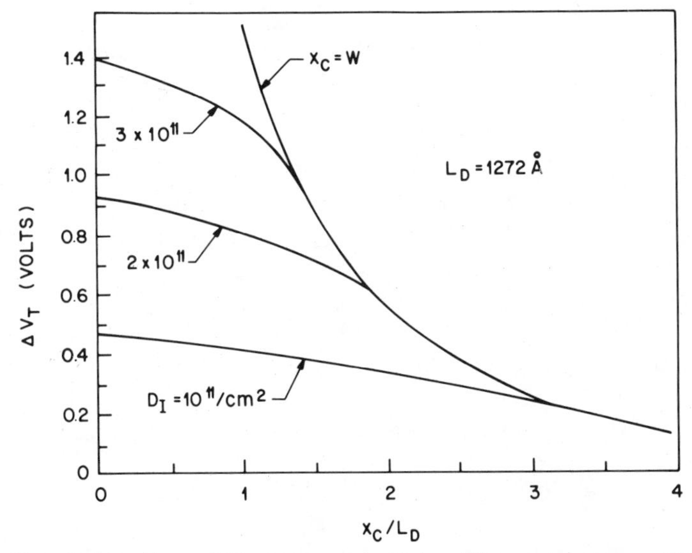

Fig. 25 Threshold voltage shift versus centroid for different ion doses. (After Brews, Ref. 36.)

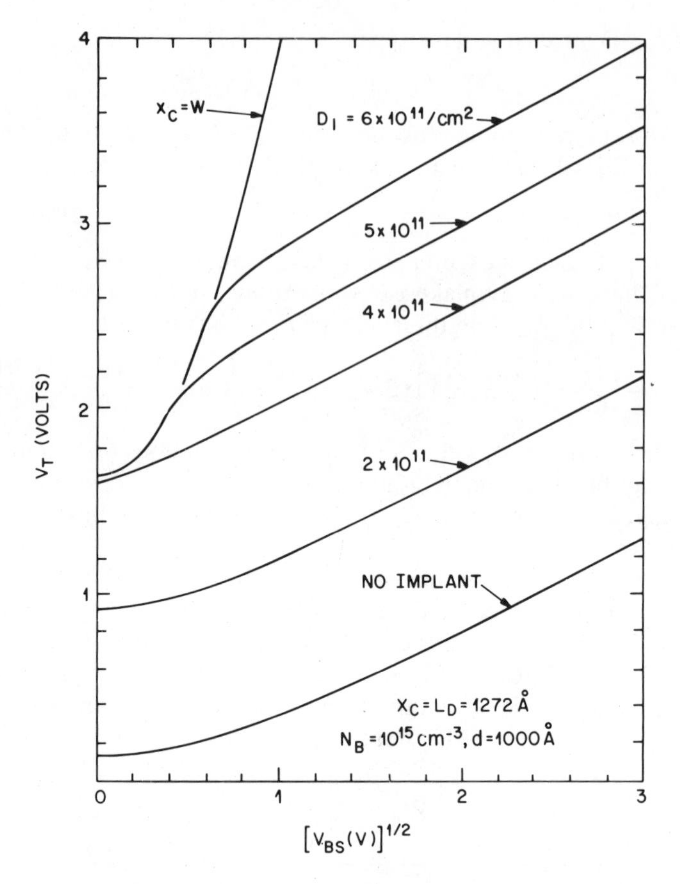

Fig. 26 Threshold voltage versus substrate bias for different ion doses. (After Brews, Ref. 36.)

Figure 26 shows the substrate sensitivity for a delta-function implant located at $x_c = L_D = 1272$ Å. For $D_I < 4 \times 10^{11}$ cm^{-2} the implant results in a parallel shift of V_T with respect to the unimplanted curve, because for small D_I, the bias dependence of the threshold voltage shift is dominated by the term qD_I/C_i in Eq. 69. For $D_I > 4 \times 10^{11}$ cm^{-2}, the depletion-layer edge becomes clamped to the implant. As V_{BS} increases, V_T follows the curve for the clamped condition $x_c = W$. For larger V_{BS}, such that $(2\psi_B + V_{BS}) > D_I x_c/(\beta N_B L_D^2)$, the term qD_I/C_i in Eq. 69 becomes dominant again, W breaks free of the implant, and V_T becomes parallel to the unimplanted curve.

8.3.2 Subthreshold Slope

In the subthreshold region, we have seen that for a uniformly doped device, ln I_D versus V_G is almost a straight line. The reciprocal slope of this line determines the gate swing needed to reduce subthreshold current to any desired level. A convenient measure of this turn-off characteristic is the gate swing S, needed to reduce the current by one decade, defined in Section 8.2.3. For a nonuniformly doped device, this parameter is given by[19]

$$S \simeq \frac{kT}{q} \ln 10 \cdot (1 + C_D/C_i) \Big/ \left[1 - \left(\frac{2}{a^2}\right)\left(\frac{C_D}{C_i}\right)^2\right] \tag{71}$$

where C_D is the depletion-layer capacitance per unit area,

$$C_D = \epsilon_s / W_I . \tag{72}$$

The subthreshold swing S is dominated by the depletion width W_I given by Eq. 64. The variation of S with implant depth for various doses is shown in Fig. 27. At $x_c = 0$, the swing is determined by the substrate doping and oxide thickness. For a given D_I, initially as the depth x_c increases, S also increases, because the implant depth reduces the depletion-layer width, increasing C_D in Eq. 71. However, once W is reduced to the point

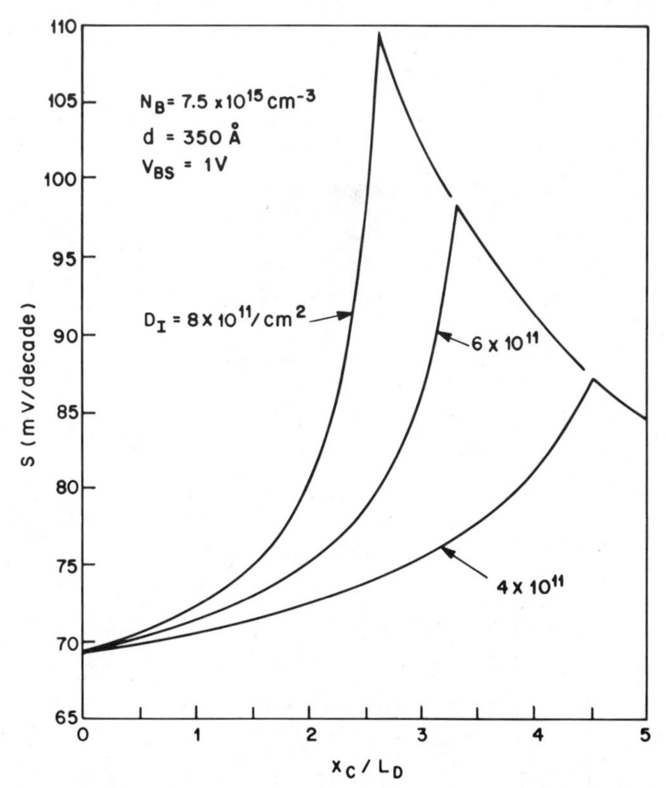

Fig. 27 Subthreshold swing versus centroid for different ion doses. (After Brews, Ref. 19.)

that $x_c = W$, the depletion-layer edge is clamped to the implant. As the implant is made deeper, the depletion-layer width is also forced deeper, maintaining the equality $x_c = W$. This deeper depletion width causes C_D to decrease and S decreases. Once the implant becomes so deep that x_c equals the value for the unimplanted device, S returns to the value for $x_c = 0$. The cusp in the curves of Fig. 27 occurs when $x_c = W$, where W is at its minimum value. This cusp will be rounded for practical implants with nonzero width. Figure 27 shows that to have a fast turnoff (i.e., small threshold voltage swing), shallow implantations should be used.

8.3.3 Buried-Channel Devices

When the type of the implanted impurities is opposite to that of the substrate (e.g., arsenic implantation into a p-type substrate), a buried channel can be formed. Figure 28a shows a cross section of a buried n-channel MOSFET, where the conducting channel is in the bulk semiconductor rather than at the Si–SiO_2 interface as in a conventional MOSFET. The actual doping profile and a step-profile approximation for the channel region are shown in Fig. 28b. A junction-depletion region is

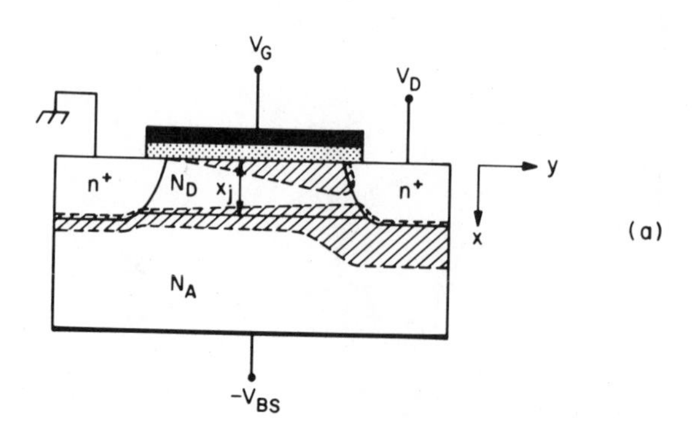

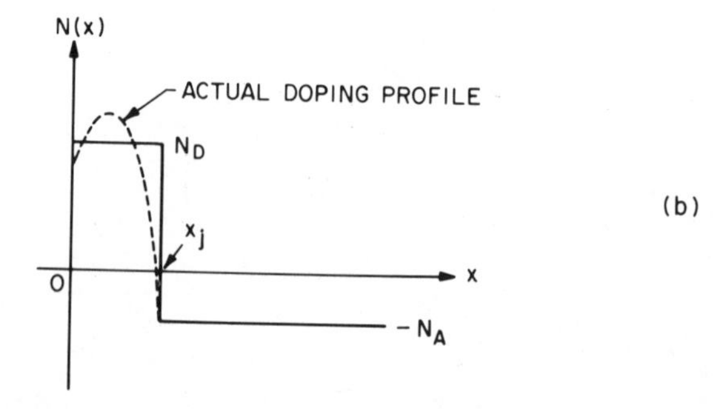

Fig. 28 (*a*) Buried-channel MOSFET. (*b*) Channel doping profile.

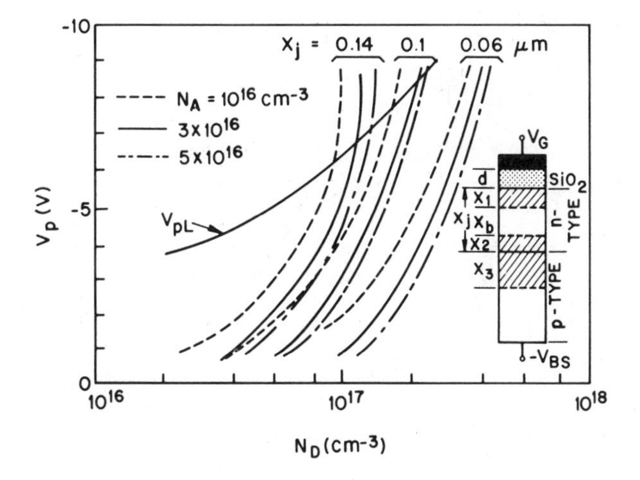

Fig. 29 Pinch-off voltage versus surface doping for different junction depths and substrate dopings. The insert shows the depletion regions near the drain. (After Merckel, Ref. 37.)

formed around the metallurgical junction at $x = x_j$; its local width depends on the applied voltages. The gate also modulates the local width of the surface depletion regions.

A buried-channel MOSFET can be made as a normally-on or normally-off device, depending on the surface doping and junction depth. Consider the normally-on case first. A cross-sectional view through the region near the source is shown[37] in the insert in Fig. 29. As the gate voltage V_G increases, the buried-channel width x_b decreases. Eventually, $x_b \to 0$ at a gate voltage called the pinch-off voltage V_P. For the step profile shown in Fig. 28*b* (assuming that $N_D \gg N_A$), the pinch-off voltage can be obtained by solving Poisson's equation and applying Gauss's law at the Si–SiO_2 interface:

$$V_P - V_{FB} = -\frac{qN_D x_j}{C_i}\left(1 + \frac{C_i}{2C_D}\right) + \frac{\sqrt{2\epsilon_s q N_A}}{C_i}\left(1 + \frac{C_i}{C_D}\right)(V_{bi} + V_{BS})^{1/2} \tag{73}$$

where $C_D = \epsilon_s / x_j$ and V_{bi} is the p-n junction built-in potential.

Figure 29 shows the variation of V_P (for $V_{BS} = 0$) as a function of the surface doping for various junction depths, x_j. For a given x_j, the magnitude of V_P increases with increased doping density. If the doping density N_D or the junction depth x_j is sufficiently large, as the magnitude of the gate voltage increases, eventually surface inversion occurs and the surface depletion layer width x_1 tends to a limiting value W_m. The corresponding limiting pinch-off voltage is given by

$$V_{PL} = V_{FB} + 2\psi_B + \sqrt{2\epsilon_s q N_D (2\psi_B)}/C_i \, . \tag{74}$$

A limiting junction depth x_{jL} can be similarly defined by

$$x_{jL} = \frac{1}{N_D}\sqrt{\frac{2\epsilon_s N_A}{q}}\left(\sqrt{2V_{bi}N_D/N_A} + \sqrt{V_{bi} + V_{BS}}\right) \tag{75}$$

such that if $x_j > x_{jL}$, it is impossible to pinch-off the device completely, because there will always be a finite thickness of the conducting channel ($x_b \neq 0$).

The drain current in the linear and saturation regions is similar to those previously derived for the JFET in Chapter 6. The drain current is given by[38]

$$I_D = Z(\mu D_I - \mu Q_s - \mu Q_B)\frac{dV}{dy} \tag{76}$$

where

$$D_I = \int_0^{x_j} [N(x) - N_A]\, dx \simeq (N_D - N_A)x_j \tag{77}$$

is the implanted net charge per unit area,

$$Q_s = -\bar{C}[V_G - V_{FB} - V(y)] \tag{78}$$

is the surface depleted or enhanced charge per unit area, $\bar{C}$ is the average gate capacitance,

$$Q_B = \sqrt{2\epsilon_s q N'_A}\,[V_{bi} + V_{BS} + V(y)]^{1/2} \tag{79}$$

is the depleted charge of the p-n junction, and $N'_A = N_D N_A/(N_D + N_A)$.

For the normal depletion mode, $(V_G - V_{FB}) < 0$, the drain current flows only in the buried channel. The bulk mobility μ_B will be used in Eq. 76. Substituting Eqs. 77 through 79 in Eq. 76 and integrating from 0 to L results in

$$I_D = \frac{\mu_B Z}{L}\{D_I V_D + \bar{C}[(V_G - V_{FB})V_D - \tfrac{1}{2}V_D^2] - \tfrac{2}{3}\sqrt{2\epsilon_s q N'_A}\,[(V_{bi} + V_{BS} + V_D)^{3/2} - (V_{bi} + V_{BS})^{3/2}]\}. \tag{80}$$

The calculated and measured results are in good agreement (Fig. 30). Note that at zero gate voltage, a large drain current is flowing, hence the name "normally-on" for this device. By varying the gate voltage, one can increase or decrease the drain current.

Figure 31*a* shows the doping profiles and depletion-layer widths prior to pinchoff in a normally-on buried-channel MOSFET. At a sufficiently large negative gate bias, the channel will be pinched off, that is, when $x_j = (x_1 + x_2)$ (Fig. 31*b*). The ionized charge density under pinch-off conditions is shown in Fig. 31*b*. The conduction below the normal pinch-off voltage is due to the presence of a region of partially depleted carriers, wherein the current is carried primarily by diffusion (Fig. 31*c*). The resulting sub-pinch-off current for a buried-channel MOSFET is thus directly analogous to the

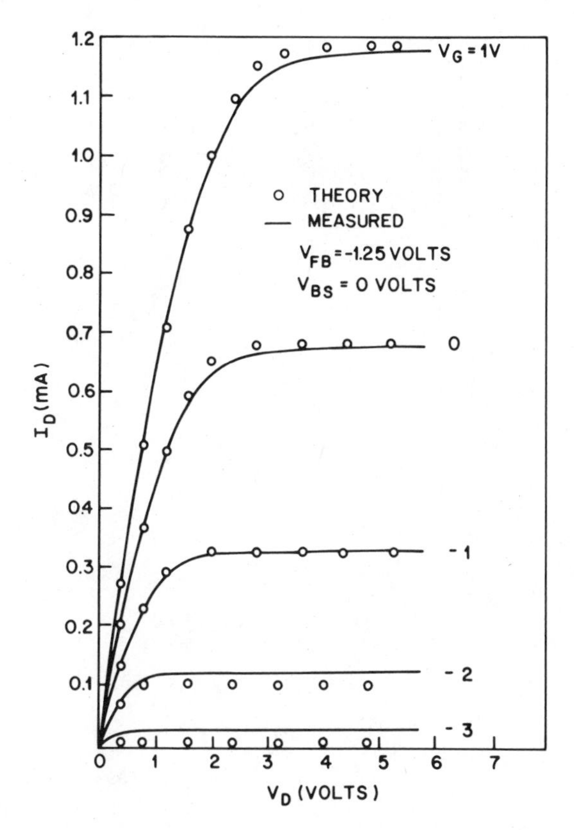

Fig. 30 Measured and calculated drain characteristics for a normally-on MOSFET. (After Huang and Taylor, Ref. 38.)

subthreshold current for an enhancement-mode MOSFET. The sub-pinch-off current will vary exponentially with the gate voltage, and the sub-pinch-off swing S is given by the capacitive divider ratio (Fig. 31*d*):[39]

$$S = \frac{kT}{q} \ln 10 \cdot [(C_1 + C_2)/C_1]$$

$$= \frac{kT}{q} \ln 10 \cdot \left[1 + \frac{\epsilon_i x_1 + \epsilon_s d}{\epsilon_i (x_2 + x_3)}\right]. \tag{81}$$

We shall next consider the normally-off buried-channel MOSFET. For such a device, at $V_G = 0$, the depletion layers from the surface and the p-n junction will pinch off the channel region, that is, $x_j = (x_1 + x_2)$. We can define a threshold voltage V_T, as the gate voltage at which the drain conductance approaches zero:[40]

$$V_{FB} - V_T = \frac{qN_D(x_j - x_1)}{C_i} + \frac{qN_D(x_j - x_1)^2}{2\epsilon_s}. \tag{82}$$

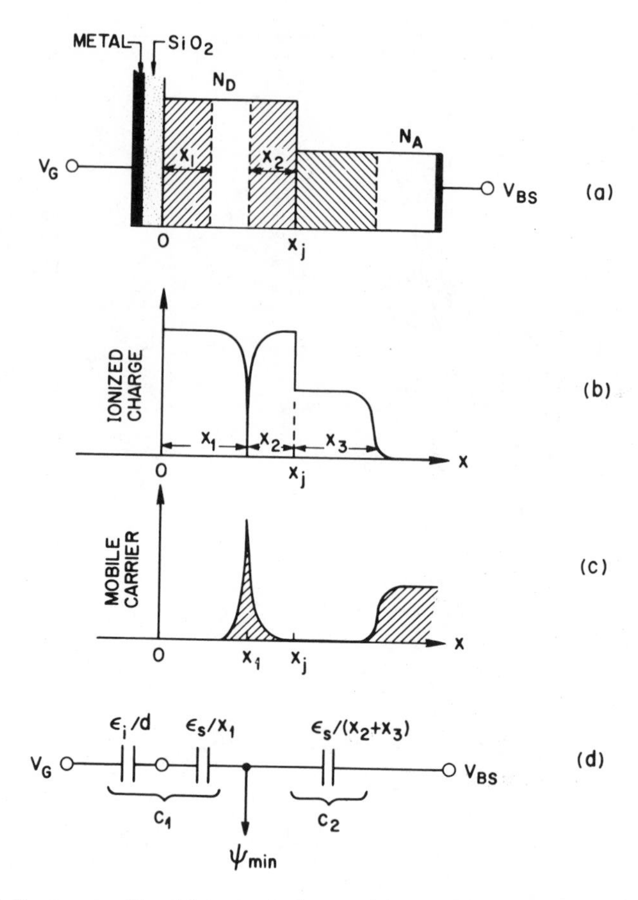

Fig. 31 (*a*) Doping profile, (*b*) ionized charge, (*c*) mobile carriers, and (*d*) equivalent capacitances in sub-pinch-off condition of a buried-channel MOSFET. (After Hendrikson, Ref. 39.)

In a practical implementation, a heavily doped boron polysilicon gate or a metal electrode with a large work function is used to give a high flat-band voltage. The device is expected to have higher mobility, as verified by Fig. 32*a*, because in bulk conduction the surface scattering can be avoided. The figure shows that for similar device parameters, the buried-channel device can have 50% higher mobility than a conventional device. Figure 32*b* shows the variation of the threshold voltage V_T with channel length. For a long-channel device, V_T is independent of channel length. The reduction of V_T as L decreases, due to the "short-channel effect," will be considered in Section 8.4. The buried-channel device is less affected by the short-channel effect than the conventional device. These characteristics make the buried-channel normally-off MOSFETs useful for high-speed applications.

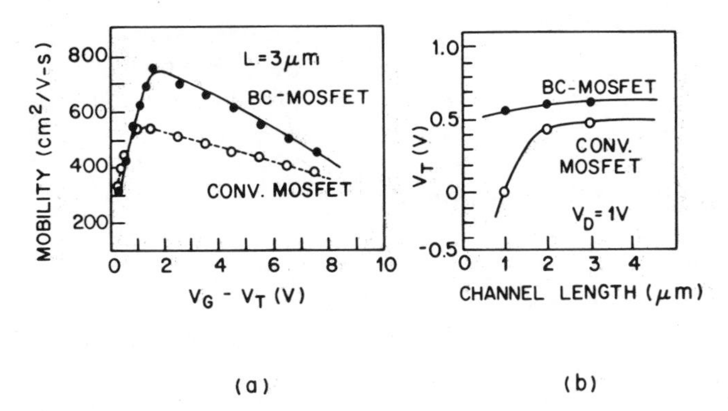

Fig. 32 Comparison of buried-channel and surface-channel MOSFET. (*a*) Mobility dependence on gate voltage. (*b*) Threshold-voltage dependence on channel length. (After Nishiuchi et al., Ref. 40.)

8.4 SHORT-CHANNEL EFFECTS

Since 1959, the beginning of the integrated-circuit era, the minimum feature length has been reduced by two orders of magnitude. We expect that the minimum dimension will continue to shrink in the foreseeable future, as illustrated in Fig. 1. As the channel length is reduced, departures from long-channel behavior, as considered in Section 8.2, may occur. These departures, the short-channel effects, arise as results of a two-dimensional potential distribution and high electric fields in the channel region.

For a given channel doping concentration, as the channel length is reduced, the depletion-layer widths of the source and drain junctions become comparable to the channel length. The potential distribution in the channel now depends on both the transverse field $\mathscr{E}_x$ (controlled by the gate voltage and the back-surface bias) and the longitudinal field $\mathscr{E}_y$ (controlled by the drain bias). In other words, the potential distribution becomes two dimensional, and the gradual-channel approximation (i.e., $\mathscr{E}_x \gg \mathscr{E}_y$) is no longer valid. This two-dimensional potential results in degradation of the subthreshold behavior, dependence of the threshold voltage on channel length and biasing voltages, and failure of current saturation due to punch-through.

As electric field is increased, the channel mobility becomes field-dependent, and eventually, velocity saturation occurs. (The mobility behavior was discussed in Section 8.2.) When the field is increased further, carrier multiplication near the drain occurs, leading to substrate current and parasitic bipolar transistor action. High fields also cause hot-carrier injection into the oxide, leading to oxide charging and subsequent threshold voltage shift and transconductance degradation.

Because short-channel effects complicate device operation and degrade device performance, these effects should be eliminated or minimized so that a physical short-channel device can preserve the "electrical" long-channel behavior.

8.4.1 Subthreshold Current

Experimental transfer characteristics (i.e., $\ln I_D$ versus V_G) are shown in Fig. 33. These MOSFETs were made using a standard n-MOS process. Substrates were ⟨100⟩-oriented, p-type silicon wafers. After growth of a

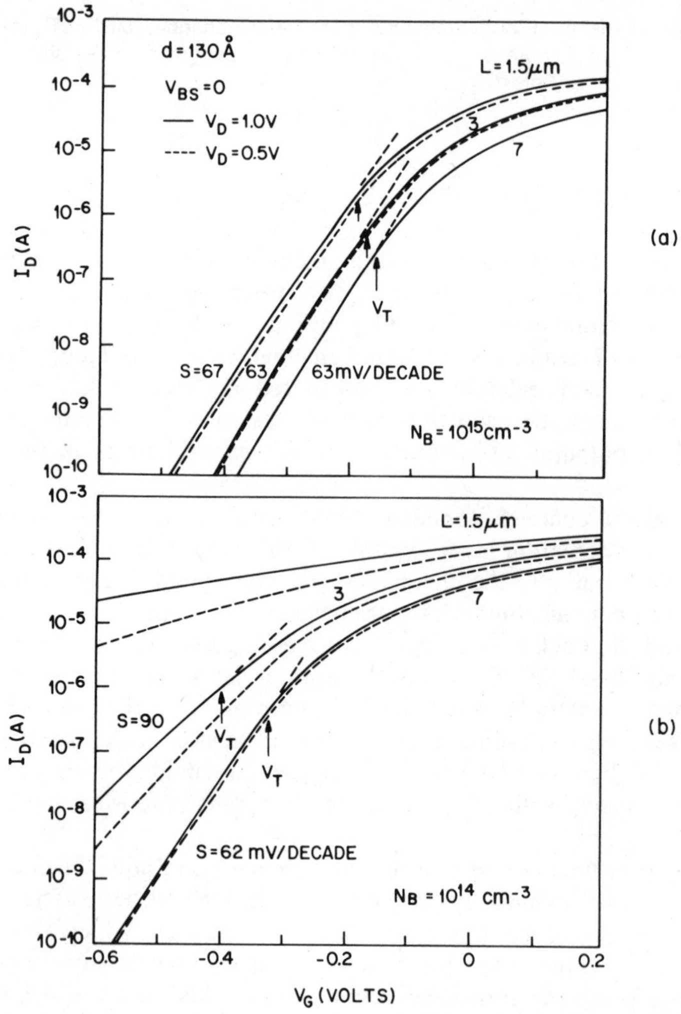

Fig. 33 Subthreshold characteristics for various channel lengths. (a) $N_A = 10^{15}\ \text{cm}^{-3}$. (b) $N_A = 10^{14}\ \text{cm}^{-3}$.

gate oxide to a given thickness, various polysilicon gate lengths ranging from about 1 to 10 μm were defined by x-ray lithography, all with a gate width of 70 μm. The source and drain were made by arsenic ion implantation and are subsequently annealed. Junction depths were varied from about 0.25 to 1.5 μm, depending on the initial implantation energy and subsequent temperature cycle. Aluminum was used for contact metallization.

In Fig. 33*a*, the gate oxide thickness is 130 Å, the substrate doping is 10^{15} cm^{-3}, and the junction depth is 0.33 μm. The device with a 7-μm channel length shows long-channel behavior, that is, the subthreshold drain current is independent of drain voltage when $V_D > 3kT/q$, Eq. 40. For $L = 3$ μm, there is a slight dependence on V_D. For an even shorter channel, $L = 1.5$ μm, the split of the curves for $V_D = 0.5$ V and $V_D = 1$ V becomes larger, with a corresponding shift of V_T (which is at the point of current departure of the I–V characteristic from the straight line). The subthreshold swing also increases, as indicated.

The situation described above becomes worse for devices with identical parameters except a lower substrate doping, $N_B = 10^{14}$ cm^{-3} (Fig. 33*b*). Note that even at 7 μm, the curves split slightly. For $L = 3$ μm, the drain current increases substantially and the subthreshold swing does also. For an even shorter channel, $L = 1.5$ μm, long-channel behavior is totally lost, and the device cannot be "turned off."

The boundary between a long-channel and a short-channel device can be defined based upon two criteria: (1) dependence of the drain current on channel length for long-channel devices, $I_D \sim 1/L$, and (2) dependence of subthreshold current on drain voltage for long-channel devices, where I_D is not a function of V_D for $V_D > 3kT/q$. Figure 34 shows plots of I_D and $\Delta I_D/I_D$ versus $1/L$, where ΔI_D is the current difference at V_T for the two drain voltages. A 10% departure from linear dependence upon $(1/L)$ for I_D or a 10% increase of $\Delta I_D/I_D$ is taken to indicate short-channel behavior.

Extensive measurements have been made on MOSFETs having gate oxide thicknesses ranging from 100 to 1000 Å, substrate dopings from 10^{14} to 10^{17} cm^{-3}, junction depths from 0.18 to 1.5 μm, and drain voltages up to 5 V. The minimum channel length L_{min} for which long-channel subthreshold behavior can be observed is found to follow a simple empirical relation:[41]

$$L_{min} = 0.4[r_j d(W_S + W_D)^2]^{1/3} \equiv 0.4(\gamma)^{1/3} \tag{83}$$

where γ is defined as the product in brackets, r_j is the junction depth in μm, d is the oxide thickness in Å, and $(W_S + W_D)$ is the sum of source and drain depletion width in a one-dimensional abrupt junction formulation:

$$W_D = \sqrt{\frac{2\epsilon_s}{qN_A}(V_D + V_{bi} + V_{BS})} \qquad \mu\text{m} \tag{84}$$

where V_{bi} is the built-in voltage of the junction and V_{BS} is the substrate bias. For V_D equal to 0, W_D equals W_S.

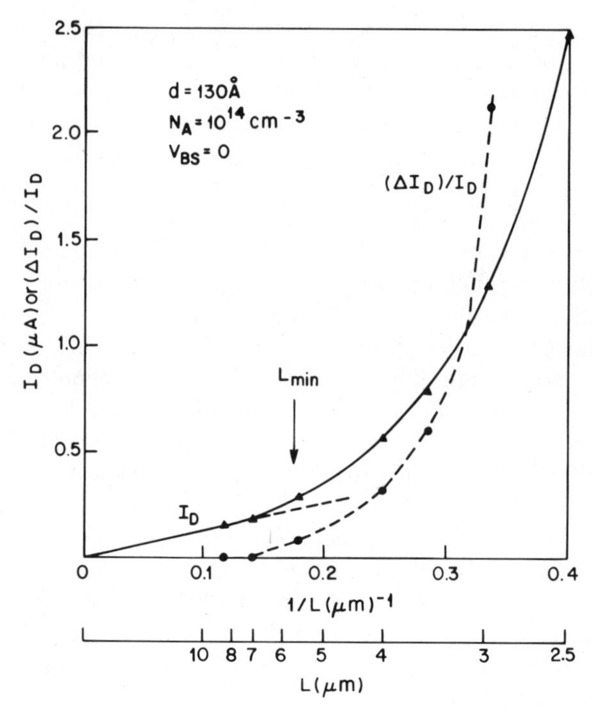

Fig. 34 Drain current and $\Delta I_D/I_D$ versus reciprocal of channel length to show the onset of short-channel effects.

Figure 35 shows Eq. 83 compared with experimental results. Also shown are L_{min} versus γ, determined from a two-dimensional computer calculation based on the afore-mentioned criteria. Equation 83 agrees with the measured and computed results within 20% in the worst cases, and is usually more accurate. This equation can therefore serve as a basic guide for MOSFET miniaturization. All devices with channel lengths that lie below the line (crosshatched area) are electrical short-channel devices, and all devices with channel lengths above the line are long-channel devices. For example, if $\gamma = 10^5$ μm^3-Å, a 10-μm channel is a short channel; but if $\gamma = 1$ μm^3-Å, a 0.5-μm channel can be considered a long channel.

Departure from long-channel behavior can be obtained by applying the charge conservation principle to the region bounded by the metal gate and bulk of the semiconductor[42] (shown in Fig. 36*a*):

$$Q'_M + Q'_O + Q'_n + Q'_B = 0 \tag{85}$$

where Q'_M is the total charge on the gate, Q'_O is the total effective oxide charges at the Si–SiO_2 interface, Q'_n is the total inversion-layer charge, and Q'_B is the total ionized impurity in the depletion region. Equation 85 may be expressed in terms of previously derived voltages,

$$V_G = V_{FB} + \psi_s + Q'_B/C_i A \tag{86}$$

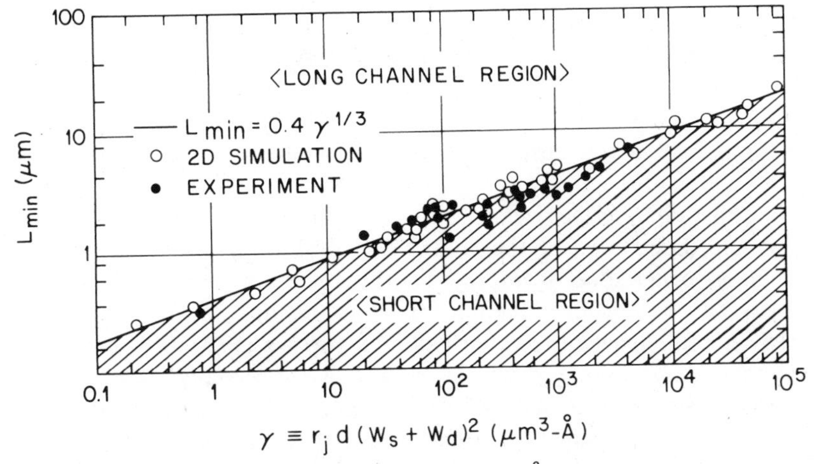

Fig. 35 Minimum channel length versus $r_j \cdot d(W_S + W_D)^2$. (After Brews et al., Ref. 41.)

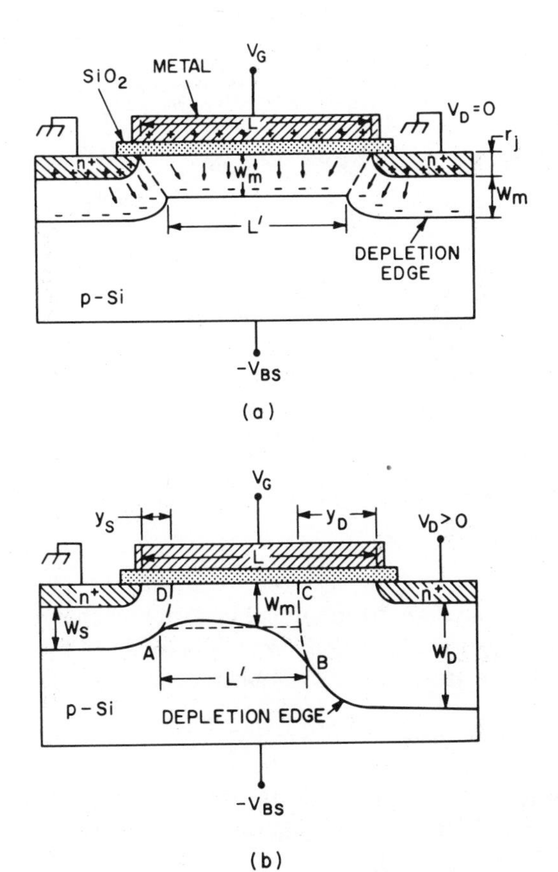

Fig. 36 Charge-conservation model. (*a*) $V_D = 0$. (*b*) $V_D > 0$. (After Yau, Ref. 42.)

where V_{FB} is the flat-band shift, ψ_s is the surface potential, and A is the area. The threshold voltage is given by setting $\psi_s \simeq 2\psi_B$,

$$V_T = V_{FB} + 2\psi_B + Q'_B/C_iA. \tag{87}$$

For long-channel devices, $Q'_B = qN_AWA$, where W is the depletion-layer width.

For short-channel devices, the full effect of Q'_B on the threshold voltage is reduced, because near the source and drain ends of the channel, some field lines originating from the source or drain terminate at the bulk charges in the channel region for $V_D = 0$ (Fig. 36*a*). For $V_D > 0$, the depletion region near the drain expands further (Fig. 36*b*). Note that the horizontal depletion-layer widths y_S and y_D are smaller than the vertical depletion-layer widths W_S and W_D, respectively, because the transverse field strongly influences the potential distribution at the surface.

Because of the reduction of the bulk charge Q'_B, surface potential for a given gate voltage increases, Eq. 86, leading to an increase of subthreshold current (from Eqs. 37 and 38). The surface potential can be found from the following expression:[43]

$$V_G - V_{FB} = \psi_s + \frac{1}{C_i}\sqrt{q\epsilon_s N_A(\psi_s + V_{BS})/2}\left(1 + \frac{L - W_D - W_S}{L - y_D - y_S}\right) \tag{88}$$

where W_D and W_S are given in Eq. 84, and

$$y_S \simeq \sqrt{\frac{2\epsilon_s}{qN_A}(V_{bi} - \psi_s)} \tag{89a}$$

$$y_D \simeq \sqrt{\frac{2\epsilon_s}{qN_A}(V_{bi} - \psi_s + V_D)}\,. \tag{89b}$$

The subthreshold current is given by

$$I_D = \mu_n\left(\frac{Z}{L - y_S - y_D}\right)\frac{aC_i}{2\beta^2}\left(\frac{n_i}{N_A}\right)^2(1 - e^{-\beta V_D})e^{\beta\psi_s}(\beta\psi_s)^{1/2}. \tag{90}$$

Equation 90 is identical to Eq. 40, except channel length is replaced by the effective channel length:

$$L_{\text{eff}} = L - y_S - y_D. \tag{90a}$$

Figure 37 shows a comparison of subthreshold behavior obtained from experimental measurements and Eq. 90. Results by two-dimensional numerical modeling are also shown. Fairly good agreement is apparent.

8.4.2 Threshold Voltage

The first-order estimation of the threshold voltage can be made using Fig. 36*a*. The total bulk charge inside the trapezoid is[42]

$$\frac{Q'_B}{Z} = qN_AW_m\left(\frac{L + L'}{2}\right). \tag{91}$$

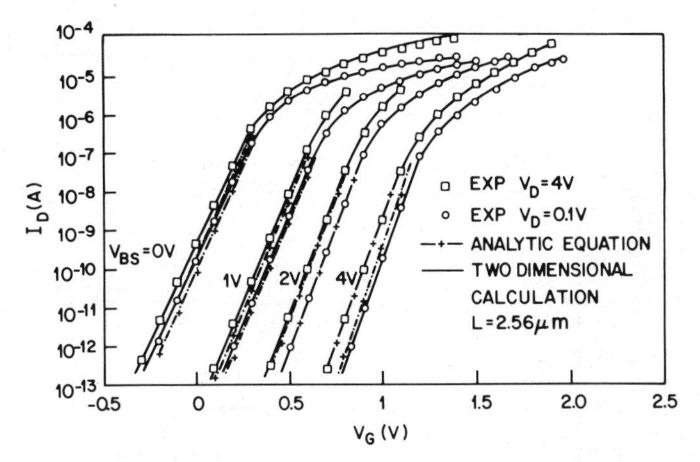

Fig. 37 Subthreshold current of a short-channel MOSFET with $d = 530$ Å, $N_A = 1.2 \times 10^{16}$ cm^{-3}, $L = 2.56$ μm, and $Z = 21.5$ μm. (After Fichtner and Potzl, Ref. 22.)

By straightforward trigonometric analysis,

$$\frac{L+L'}{2L} = 1 - \frac{r_j}{L}\left(\sqrt{1+\frac{2W_m}{r_j}} - 1\right). \tag{92}$$

The threshold voltage shift is then

$$\Delta V_T = \frac{1}{C_i}\left(\frac{Q_B'}{ZL} - qN_A W_m\right) = -\frac{qN_A W_m}{C_i}\left(1 - \frac{L+L'}{2L}\right)$$

$$= -\frac{qN_A W_m r_j}{C_i L}\left(\sqrt{1+\frac{2W_m}{r_j}} - 1\right). \tag{93}$$

To take into account the effect of the drain voltage and the substrate bias, Eq. 93 can be modified to read[22]

$$\Delta V_T = -\frac{qN_A W_m r_j}{2C_i L}\left[\left(\sqrt{1+\frac{2y_S}{r_j}} - 1\right) + \left(\sqrt{1+\frac{2y_D}{r_j}} - 1\right)\right] \tag{94}$$

where y_S and y_D are given in Eq. 89 and

$$W_m = \sqrt{2\epsilon_s(2\psi_B + V_{BS})/qN_A}\,. \tag{95}$$

Figure 38 shows the good agreement between experimental results and the calculated values from Eq. 94.

Another related threshold voltage shift is caused by the reduction of the channel width (along the z direction). This shift is related to the depletion region spreading laterally in the substrate along the channel width (see the insert in Fig. 39). Assuming that the lateral extension of the depletion region is approximately cylindrical, the total charge in the depletion region

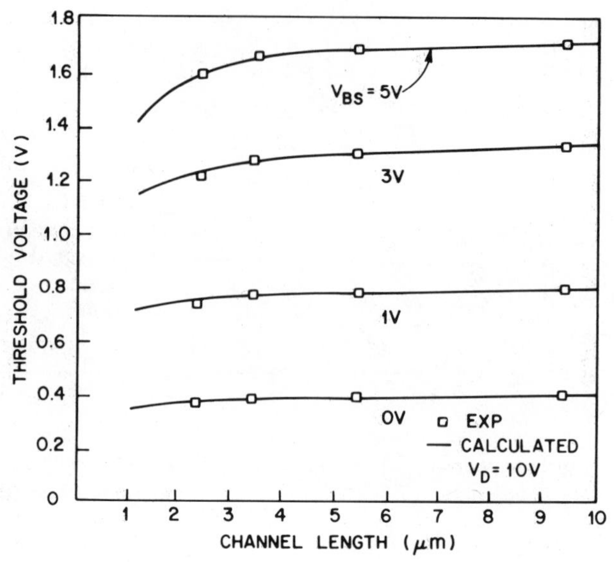

Fig. 38 Dependence of threshold voltage on channel length for a MOSFET with $d = 360$ Å, $N_A = 2.5 \times 10^{16}$ cm^{-3}, and $L = 2.4$ to $9.4\ \mu$m. (After Fichtner and Potzl, Ref. 22.)

is[44]

$$Q_{BT} = qN_AZLW\left(1 + \frac{\pi}{2}\frac{W}{Z}\right). \tag{96}$$

Equation 96 shows that the contribution in the z direction, caused by the body effect, is increased by a factor $(1 + \pi W/2Z)$. Therefore, the threshold

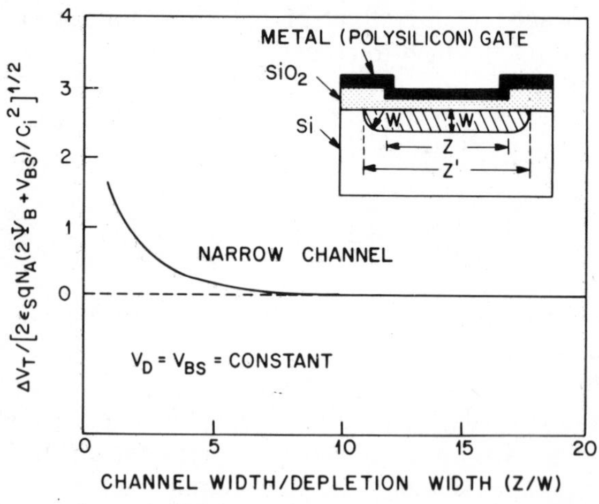

Fig. 39 Narrow-channel effect on threshold voltage. (After Merckel, Ref. 44.)

voltage is increased to

$$V_T = V_{FB} + 2\psi_B + \frac{\sqrt{2\epsilon_s q N_A(2\psi_B + V_{BS})}}{C_i}\left(1 + \frac{\pi}{2}\frac{W}{Z}\right). \tag{97}$$

Figure 39 illustrates this narrow-channel effect. The increase of threshold voltage becomes pronounced when the channel width is reduced to within an order of magnitude of the depletion-layer width. For example, for $Z = 1\ \mu m$, a substantial increase in V_T is expected when the substrate doping is less than $10^{16}\ cm^{-3}$.

8.4.3 Linear and Saturation Currents

The current–voltage characteristics in the linear and saturation regions for a device having a 0.73-μm channel length are shown[45] in Fig. 40. This device has an oxide thickness of 258 Å. The channel is implanted with 2×10^{12}-cm^{-2} boron ions at 150 keV through the oxide. The surface doping profile can be approximated by a step profile with a uniform doping of $7 \times 10^{16}\ cm^{-3}$ and a depth of 0.3 μm. When the drain voltage ranges from 0 to 5 V, the device γ, as defined by Eq. 83, varies from 3 to 10 μm^3-Å, corresponding to a minimum channel length of 0.6 to 0.9 μm. Therefore, this device is a marginal long-channel device. As a first-order approximation, the saturation current can be expressed by Eq. 28 with L replaced by an effective channel length, given by Eq. 90a. For increasing V_D, L_{eff} decreases, causing an increase in I_D.

Figure 41 shows the I–V characteristics for a device with identical parameters as the device of Fig. 40, except that channel length is reduced

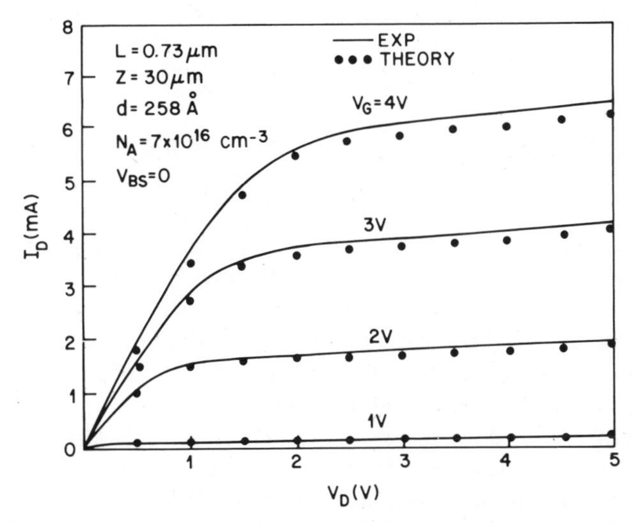

Fig. 40 Drain characteristics of a MOSFET with 0.73-μm channel length. (After Fichtner, Ref. 45.)

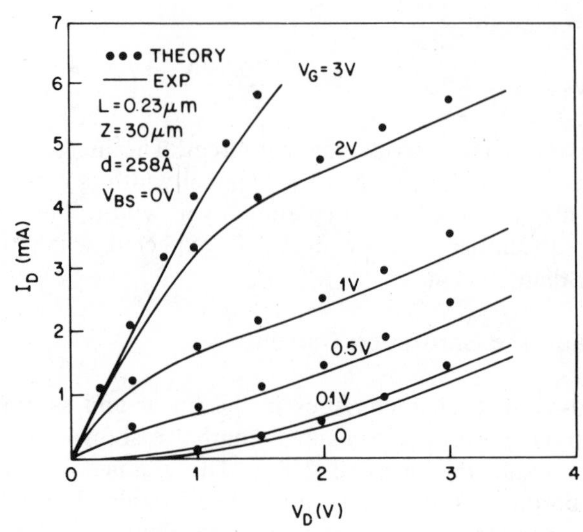

Fig. 41 Drain characteristics of a MOSFET having the same device parameters except that $L = 0.23\ \mu$m. (After Fichtner, Ref. 45.)

to 0.23 μm. This device shows severe short-channel effects. At $V_D = 0$, the sum of y_S and y_D is 0.26 μm, which is larger than the channel length. Therefore, the depletion region of the drain junction has punched through to the depletion region of the source junction. Over the drain bias range in Fig. 41, the device is operated in punch-through condition. Under such a condition, majority carriers in the source region (electrons in this case) can be injected into the depleted channel region, where they will be swept by the field and collected at the drain. The punch-through drain voltage is given as[10]

$$V_{pt} \simeq \frac{qN_A(L - y_S)^2}{2\epsilon_s} - V_{bi} . \tag{98}$$

Drain current will be dominated by the space-charge-limited current:

$$I_D \simeq 9\epsilon_s\mu_n A V_D^2 / 8L^3 \tag{99}$$

where A is the area of the n^+ regions. The space-charge-limited current increases with V_D^2 and is parallel to the inversion-layer current, which increases linearly with gate voltage. The calculated points in Figs. 40 and 41 are from a two-dimensional computer calculation incorporating the punch-through effect and field-dependent mobility effect without any adjustable parameters. The agreement with the measurements is fairly good.

Figure 42*a* and *b* show the equipotential plots for a long-channel device and the previously described 0.23-μm short-channel device, respectively. The long-channel device has identical device parameters as those of the short-channel device, except that the channel is about 10 times longer ($L = 2.3\ \mu$m). Channel lengths for both devices are about 0.27 μm shorter

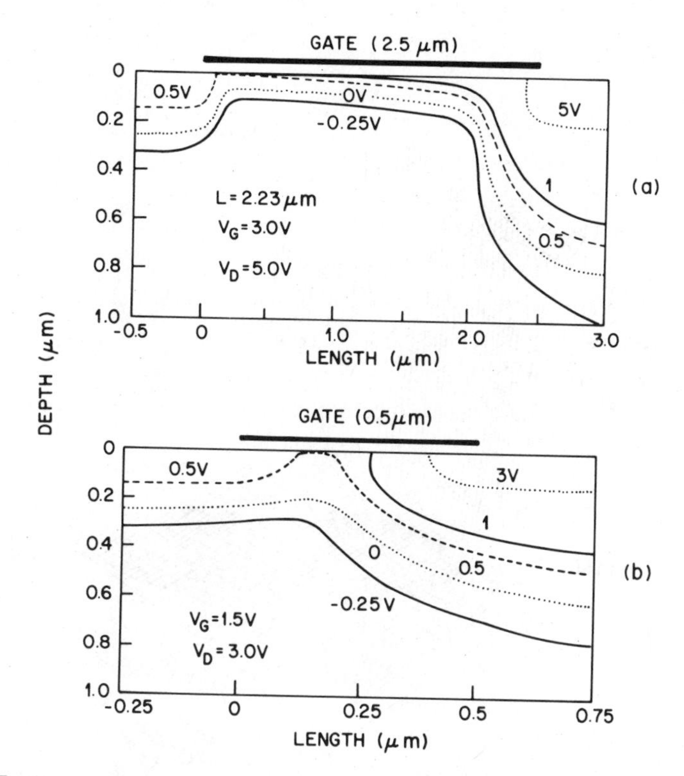

Fig. 42 Equipotential plots for the devices in Figs. 40 and 41. (*a*) 2.23-μm long-channel device. (*b*) 0.23-μm short-channel device. (After Fichtner, Ref. 45.)

than the gate lengths because of the lateral diffusion of the implanted ions. Note that the equipotential lines in the channel of Fig. 42*a* are parallel to each other and are located near the interface. In contrast, a long-channel region is not apparent in Fig. 42*b*. The equipotential lines in the channel spread deep into the bulk semiconductor, indicating a punch-through effect.

The corresponding electron densities for the foregoing two devices are shown in Figs. 43 and 44, respectively. In Fig. 43, for the long-channel device, the high electron concentrations are confined near the semiconductor surface. Because the device is operated in the saturation region, pinchoff exists near the drain. For the short-channel device, the electron density spreads deep into the bulk (Fig. 44).

Figure 45 shows constant electron density contours projected onto the *xy* plane, for the 2.23-, 0.73-, and 0.23-μm devices considered previously. As mentioned previously for the 2.5-μm-gate-length device, the drain current is confined to a depth of about 200 Å from the interface. Near the drain, pinch-off occurs, and some current flows down into the bulk of the device. For the 1.0-μm-gate-length device, the current is still confined near the interface. The pinch-off region, however, occupies a large fraction of the

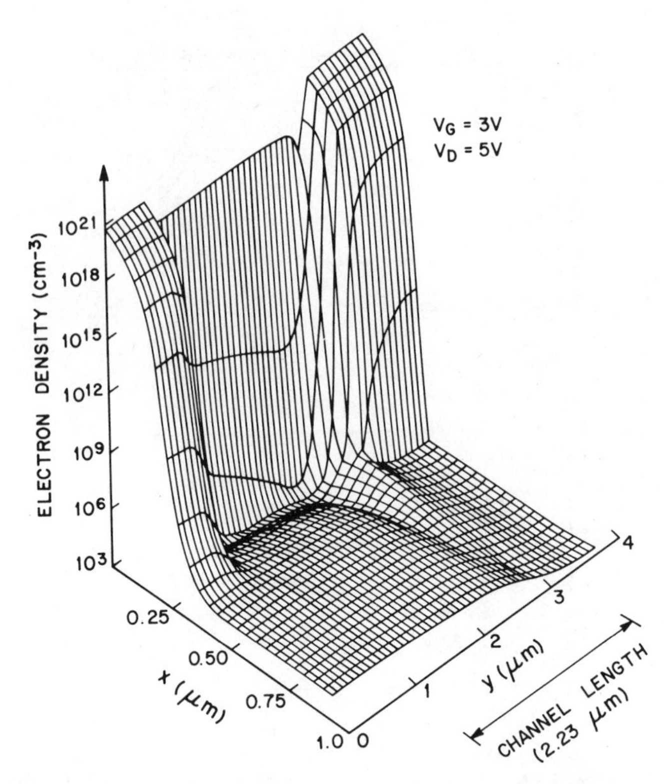

Fig. 43 Two-dimensional plot of electron density in the channel region for $L = 2.23\ \mu m$. (After Fichtner, Ref. 45.)

channel length. As we reduce the channel further to 0.23 μm (gate length 0.5 μm), the current spreads into the bulk and the long-channel behavior is lost.

Because of the complicated nature of the two-dimensional potential, it is difficult to express the current–voltage characteristics of short-channel devices in a simple form. Modifications of the basic equation, Eq. 22, have been considered by incorporating the charge-sharing concept, as illustrated in Fig. 36. With the proper selection of fitting parameters, reasonable agreement between the experimental results and analytical models has been obtained.[46–48]

8.4.4 Multiplication and Oxide Charging

In a long-channel device, when the drain voltage becomes sufficiently high, a weak avalanche occurs within the pinch-off region. From the avalanche plasma, the generated electrons enter the drain and the generated holes are collected by the substrate terminal and constitute the substrate current. The substrate current I_{BS} as a function of the gate

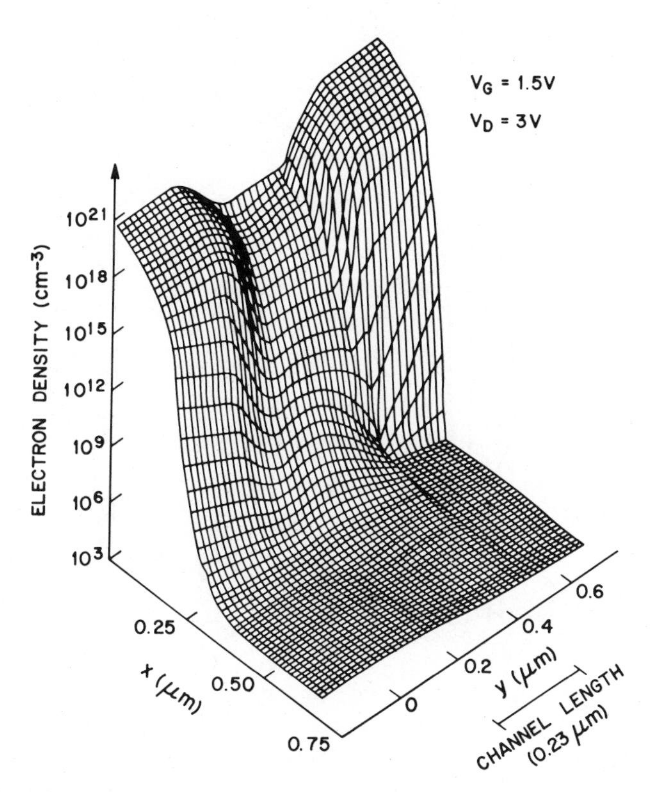

Fig. 44 Two-dimensional plot of electron density for $L = 0.23\,\mu m$. (After Fichtner, Ref. 45.)

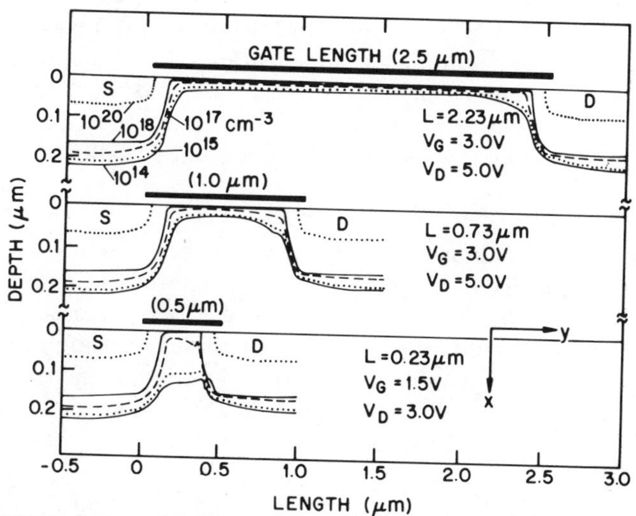

Fig. 45 Constant electron density contours for three MOSFETs with channel lengths 2.23, 0.73, and 0.23 μm. (After Fichtner, Ref. 45.)

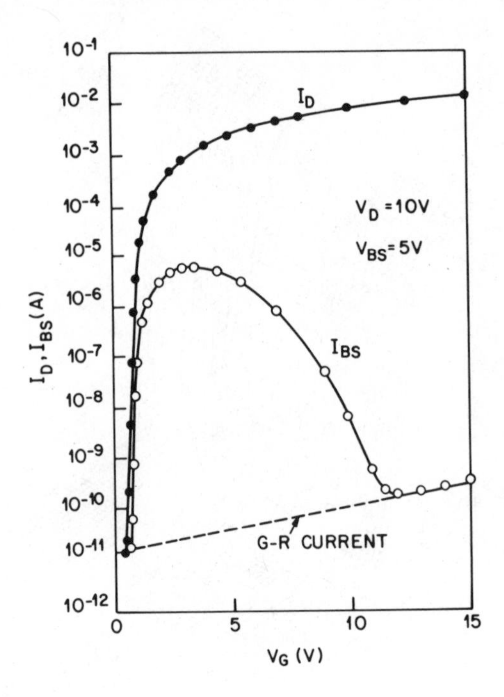

Fig. 46 Drain current and substrate current versus gate voltage of a long-channel MOSFET. (After Kamata, Tanabashi, and Kobayashi, Ref. 49.)

voltage for a long-channel device ($L = 10\ \mu$m) is shown[49] in Fig. 46. The drain current and the generation–recombination current in the depletion layers are also shown. The drain current covers the regions from subthreshold to linear to saturation. The substrate current increases first with V_G, reaches a maximum, then decreases. The maximum in I_{BS} can be explained as follows. Assuming that the impact ionization occurs uniformly in the pinch-off region, the substrate current can be written as

$$I_{BS} = I_D \alpha \Delta L \tag{100}$$

where α is the ionization coefficient, the number of electron–hole pairs generated per unit distance; and ΔL is the length of the pinch-off region. For a given V_D, as V_G increases, both I_D and $V_{D\,\text{sat}}$ increase. When $V_{D\,\text{sat}}$ increases, the lateral field $(V_D - V_{D\,\text{sat}})/L$ decreases, causing a reduction of α. Thus we have two conflicting factors. The initial increase of I_{BS} is caused by the increase of drain current with V_G, and at larger V_G, the decrease of I_{BS} is due to the decrease of α. Maximum I_{BS} occurs where the two factors balance.

For short-channel devices, an additional effect is caused by the avalanche-generated hole current. As the source–drain separation is reduced, some hole current can flow to the source (Fig. 47).[50] If the drain voltage is

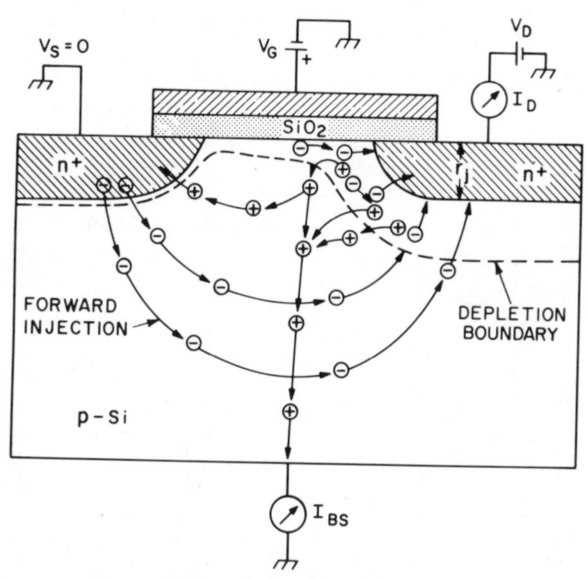

Fig. 47 Parasitic bipolar transistor action. (After Sun et al., Ref. 50.)

low, most hole current flows out the substrate terminal; the substrate current behaves similarly to that shown in Fig. 46. However, when the drain voltage is large, a substantial hole current can flow to the source, and the product of the current and substrate resistance becomes large enough (~0.6 V) to forward-bias the source–substrate junction, causing electron injection into the substrate. This injection leads to a parasitic *n-p-n* (source–substrate–drain) bipolar transistor action. The breakdown voltage of the device is then governed by the parasitic *n-p-n* transistor with a floating base, and the breakdown condition is given by

$$\alpha_{npn} M = 1. \tag{101}$$

In Eq. 101, α_{npn} is the common-base current gain, given by

$$\alpha_{npn} = \text{sech}(L_G/L_{\text{diff}}) \simeq 1 - \frac{L_G^2}{2L_{\text{diff}}} \tag{102}$$

where L_G is the effective base width (approximately equal to the gate length) and L_{diff} is the diffusion length in the substrate. The multiplication coefficient M can be expressed as

$$M = \left[1 - \left(\frac{BV_{CEO}}{V_{D\,\text{sub}}}\right)^n\right]^{-1} \tag{103}$$

where BV_{CEO} is the common-emitter breakdown voltage with open base and $V_{D\,\text{sub}}$ is the breakdown voltage of the drain-to-substrate diode. From Eqs. 101 through 103, the breakdown voltage V_B is obtained for a short-

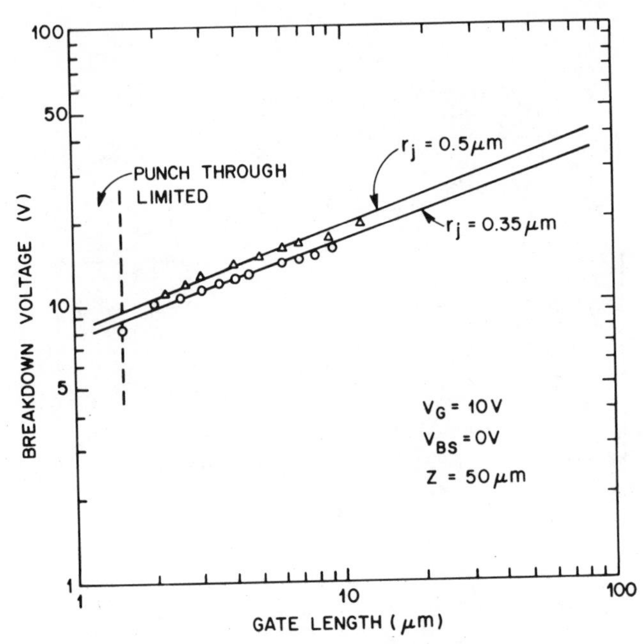

Fig. 48 Breakdown voltage versus gate length due to parasitic bipolar transistor action. (After Sun et al., Ref. 50.)

channel MOSFET as

$$V_B = BV_{CEO} \simeq \frac{V_{D\,\mathrm{sub}}}{(2)^{1/n}} \left(\frac{L_G}{L_{\mathrm{diff}}}\right)^{2/n}. \tag{104}$$

Figure 48 shows the measured V_B as a function of L_G. We can fit Eq. 104 to the data quite well, provided that the factor n is chosen to be 5.43. The difference in breakdown voltage for different r_j can be explained by the dependence of $V_{D\,\mathrm{sub}}$ on the junction curvature (as discussed in Chapter 2).

Another related high-field effect is oxide charging.[51] As the field along the channel becomes high, it is possible some electrons in the inversion layer can gain sufficient energy to surmount the Si–SiO_2 energy barrier (3.1 eV) and be injected into the gate oxide (Fig. 49, insert). Hot electrons can be injected from the avalanche plasma formed near the drain region. Thermally generated carriers can also be injected into the oxide due to a large transverse field in the bulk semiconductor. The effects of the injected hot electrons are shown in Fig. 49. Note that the threshold voltage shifts toward more positive voltage. The transconductance becomes smaller (lower slope for dI_D/dV_G) because of reduced channel mobility. The subthreshold current becomes larger because of the increased interface trap density.

Long-term operation of the device is seriously affected by oxide charg-

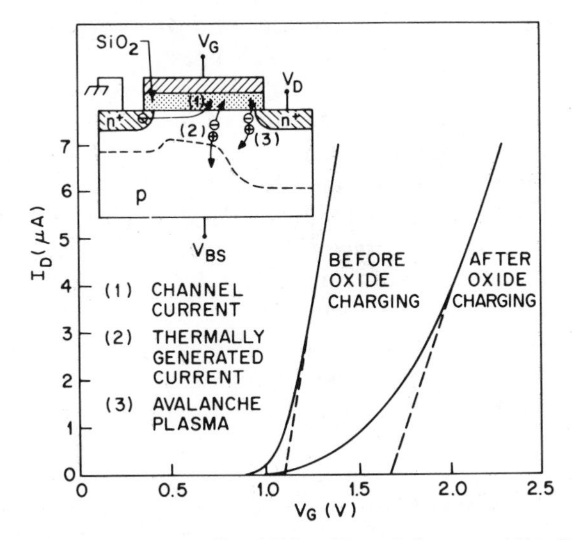

Fig. 49 Oxide charging effect. (After Ning, Osburn, and Yu, Ref. 51.)

ing, because the charging continues to increase with time during device operation. As a result of this cumulative degradation, oxide charging limits the maximum voltage levels that can be applied for a given specific device lifetime. Figure 50 shows a qualitative view of the various limitations on drain voltage.[52] As the channel length is reduced, different mechanisms limit the maximum drain voltage. These limitations are for a particular set

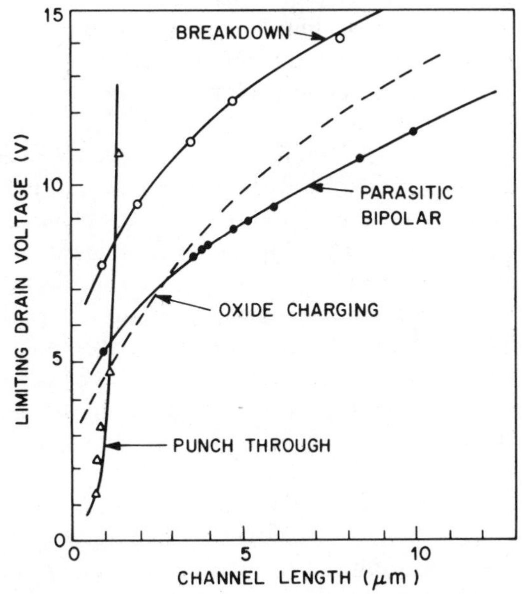

Fig. 50 Factors limiting maximum allowed drain voltage as a function of channel length. (After Matsunaga et al., Ref. 52.)

of device parameters. With other choices of device parameters, the relative importance of the various mechanisms changes.

To reduce the parasitic transistor effect, the resistance of the substrate R_{sub} can be minimized so that the product of the substrate current and R_{sub} remains smaller than 0.6 V when the drain voltage is equal to or larger than the corresponding BV_{CEO}. Then the breakdown voltage of a short-channel MOSFET will no longer be limited by BV_{CEO}; higher voltages and more reliable operation can be expected.[50] To reduce oxide charging, the density of water-related traps in the oxide should be minimized,[53] because such traps are known to capture electrons. To increase the punch-through voltage, single or double ion-implanted device structures can be made to increase the doping of the surface region. These structures will be considered in Section 8.5.

8.5 MOSFET STRUCTURES

Many device structures have been proposed to improve MOSFET performance with higher response speed, lower power consumption, more reliable operation, and higher power-handling capability. We shall now consider some representative structures.

8.5.1 Scaled-Down Device

In Section 8.4 we pointed out that short-channel effects are generally undesirable. One approach to avoid these effects is to maintain the long-channel behavior by simply scaling down all dimensions and voltages of a long-channel MOSFET, so that the internal electric fields are the same. This approach offers a conceptually simple picture for device miniaturization.

Figure 51*a* and *b* show the traditional large device and the scaled-down device,[54] respectively, in which all dimensions are shrunk by a "scaling factor," κ. This shrinking includes oxide thickness, channel length, channel width, and junction depth. The doping level is increased by κ, and all voltages are reduced by κ, leading to a reduction of the junction depletion width by about κ. Figure 51*c* compares I_D versus V_G in the linear region for the large and the scaled-down device. The threshold voltage is also reduced approximately by κ. Therefore, the number of devices per unit area increases by a factor of κ^2, the delay time due to transit across the channel, Eq. 54, decreases by κ, and the power dissipated per cell decreases by κ^2.

Note that in Fig. 51*c* the subthreshold current remains essentially the same for both devices. It remains the same because the subthreshold swing S, which is proportional to $(1 + C_D/C_i)$, remains the same as both capacitances are scaled up by the same factor κ. In addition, the junction

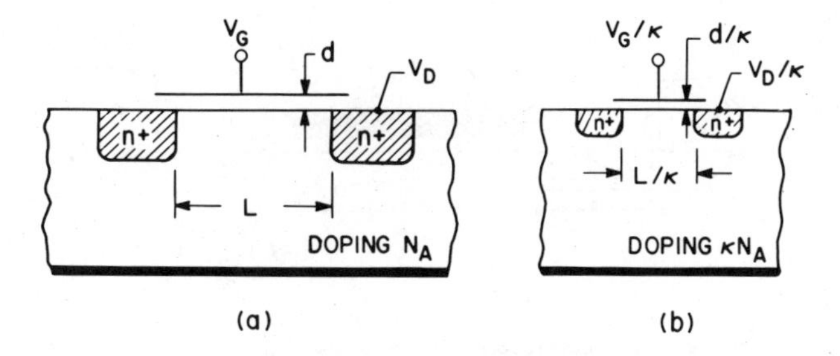

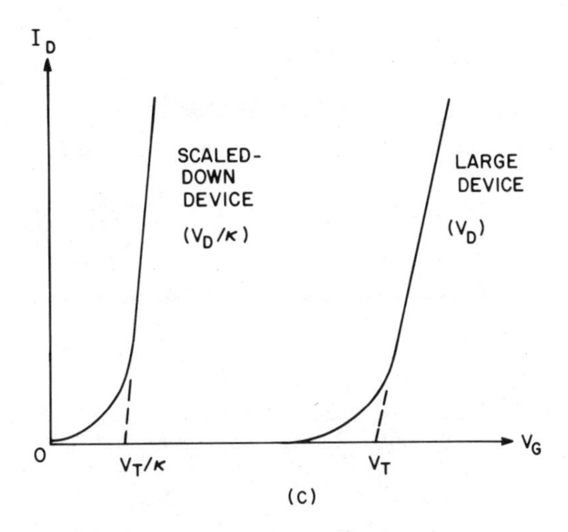

Fig. 51 Scaling approach for device miniaturization. (a) Long-channel device. (*b*) Scaled device. (*c*) Drain characteristics of these devices. (After Dennard et al., Ref. 54.)

built-in voltage and the surface potential for the onset of weak inversion do not scale (only ~10% change for 10 times increase in dopings). The range of gate voltage between depletion and heavy inversion is approximately 0.5 V. The parasitic capacitance may not scale, and the interconnect resistance increases when dimensions become smaller.

The expression for the minimum channel length, Eq. 83, can be used for a more flexible scaling approach.[41] For a given L_{min}, the value of γ is obtainable from Eq. 83, or Fig. 35, which allows the various device parameters to be adjusted independently as long as the value of γ remains the same. Therefore, all device parameters do not have to be scaled by the same factor κ. This flexibility allows one to choose new geometries that are easier to make or which optimize other aspects of device operation, rather than choosing strictly scaled geometries.

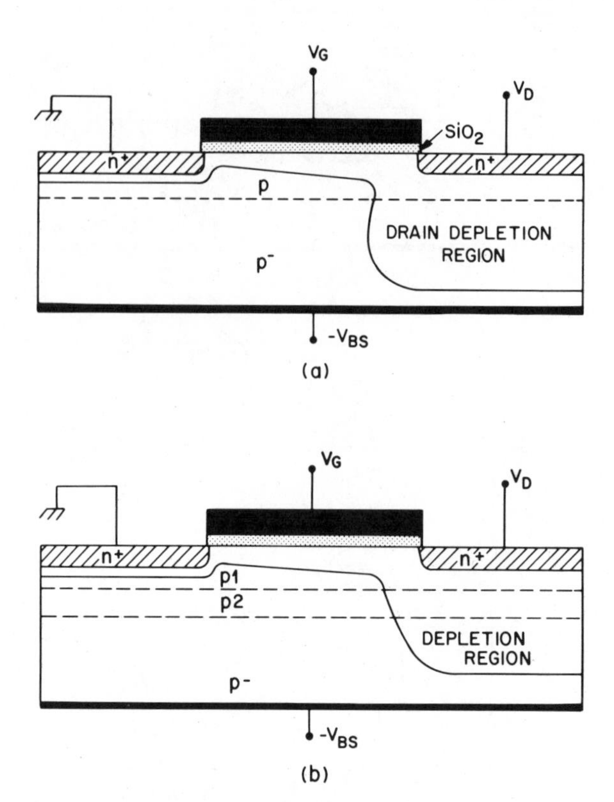

Fig. 52 HMOS structures. (*a*) Single implantation. (*b*) Double implantation. (After Shannon, Stephen, and Freeman, Ref. 55; Nihira et al., Ref. 56.)

8.5.2 HMOS

Figure 52 shows HMOS (high-performance MOS) structures. Figure 52*a* has a single ion implantation to increase the doping level at the surface region.[55] The implantation can control the threshold voltage and increase the punch-through voltage. Yet the surface region is shallow enough so that under operating conditions, the drain depletion-layer width extends into the low-doped substrate, reducing the drain capacitance. Figure 52*b* shows a double-implanted HMOS.[56] The $p1$ region contains the threshold control implant, and the $p2$ region contains the punch-through control implant. Using double implants, the HMOS with physical small-channel lengths can be tailored to minimize short-channel effects.

The implantations, however, degrade the subthreshold behavior[19] (large subthreshold swing) and can increase substrate bias sensitivity (becoming more sensitive to V_{BS}). However, various trade-offs exist and should be considered for device optimization.

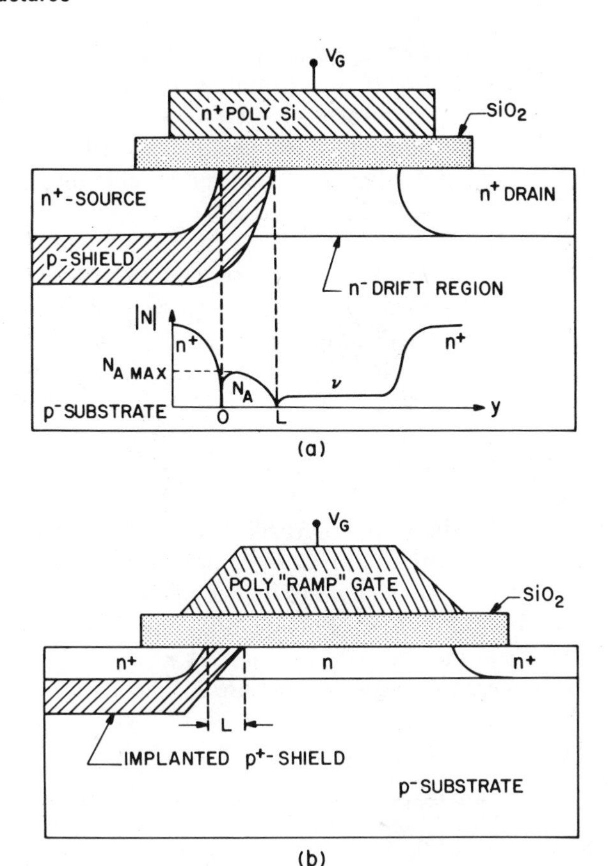

Fig. 53 (*a*) DMOS. (*b*) DIMOS structure. (After Tarui, Hayashi, and Sekigawa, Ref. 57; Tihanyi and Widmann, Ref. 58.)

8.5.3 DMOS

Figure 53*a* shows the DMOS (double-diffused MOS) structure,[57] where the channel length L is determined by the higher rate of diffusion of the p-dopant (e.g., boron), compared to the n^+-dopant (e.g., phosphorus) of the source. The channel is followed by a lightly doped drift region. Figure 53*a* also shows the doping profile along the semiconductor surface. Another version of DMOS is made by implantation. DIMOS (double-implanted MOS)[58] forms its source and drain by using a polysilicon gate as mask. The gate is tapered and the p^+-shield region is shaped by implantation through the tapered gate. The DIMOS structure improves the control in DMOS structures.

The DMOS and DIMOS structures can have very short channels and do not depend on a lithographic mask to determine channel length. Both structures have good punch-through control because of the heavily doped p-shield. The lightly doped drift region minimizes the voltage drop across the region by maintaining a uniform field ($\gtrsim 10^4$ V/cm) to achieve velocity saturation.[59] The field near the drain is the same as in the drift region, so avalanche breakdown, multiplication, and oxide charging are lessened, compared to conventional MOSFETs and HMOSs.[11]

However, the threshold voltage V_T is more difficult to control in DMOS.[60] As shown in Fig. 53*a*, V_T is determined by the maximum doping concentration $N_{A\,\max}$ along the semiconductor surface. Varying $N_{A\,\max}$ leads to variations in V_T. The localization of punch-through control to a thin p^+-shield region requires a higher doping level compared to HMOS, which leads to poorer turn-off behavior for DMOS.

8.5.4 Recessed-Channel MOSFET

The insert of Fig. 54 shows a MOSFET with a recessed channel.[61] The junction depth r_j for this structure is zero or negative. Figure 35 of Section 8.4 showed that the minimum channel length decreases as $r_j^{1/3}$. Figure 54 demonstrates that by reducing r_j, short-channel effects are minimized. For a given oxide thickness and substrate doping, as r_j decreases the onset of a large drop of V_T occurs at progressively shorter channels.

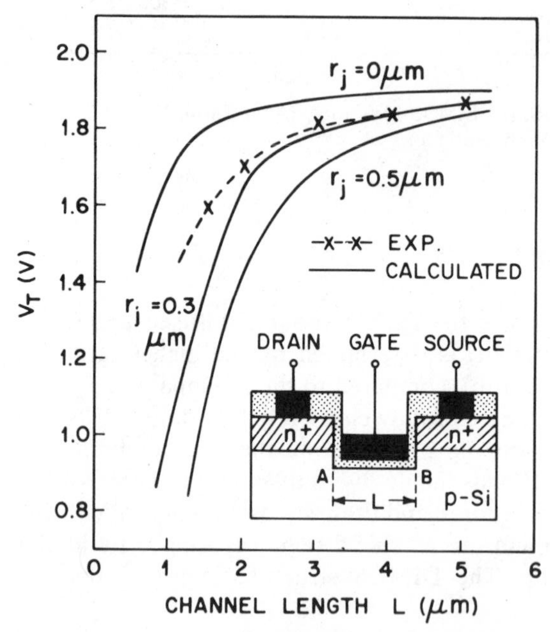

Fig. 54 Calculated and experimental V_T versus L plot for various junction depths. Insert shows a recessed-channel MOSFET. (After Nishimatsu et al., Ref. 61.)

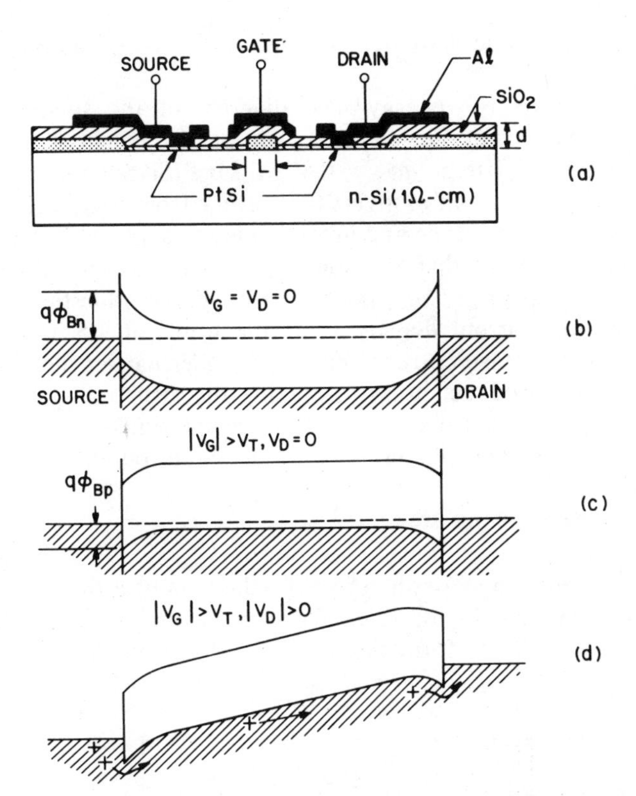

Fig. 55 MOSFET with Schottky-barrier source and drain. (*a*) Cross-sectional view of the device. (*b*), (*c*), and (*d*) Band diagrams for various biases. (After Lepselter and Sze, Ref. 62.)

The drawback of the recessed-channel structure, especially for submicron devices, is the difficulty in controlling the contour and the oxide thickness at corners A and B where the threshold voltage is determined. Also, oxide charging may be worsened, because more hot electron injection will occur.

8.5.5 Schottky-Barrier Source and Drain

Using Schottky-barrier contacts for the source and drain of a MOSFET results in performance and fabrication advantages. Figure 55*a* shows a schematic MOSFET structure with Schottky-barrier source and drain.[62] For a Schottky contact, the junction depth can effectively be made zero to minimize the short-channel effects. The high conductivity of the contact can also minimize source series resistance.

Eliminating high-temperature annealing steps can promote better quality in the oxides and better control of geometry. In addition, this structure can be made on semiconductors (such as CdS) where p-n junctions cannot be easily formed.

Figure 55*b* shows the energy-band diagram of the semiconductor at thermal equilibrium with $V_G = V_D = 0$. The barrier height of the metal to the n-substrate is $q\phi_{Bn}$ (e.g., $q\phi_{Bn} = 0.85$ eV for PtSi–Si contact). When the gate voltage is large enough to invert the surface from n-type to p-type, the barrier height between source and inversion layer is $q\phi_{Bp} = 0.25$ eV for PtSi on p-type silicon. Note that the source contact is reverse biased under operating conditions (Fig. 55*d*). For a 0.25-eV barrier, the thermionic-type reverse-saturation current density is of the order of 10^3 A/cm^2 at room temperature. To increase current density, a higher barrier metal on n-type silicon should be used. At present, making the structure on a p-type Si substrate is difficult, because metals and metal silicides that give large barrier heights on p-type silicon have not yet been found.

8.5.6 Thin-Film Transistor

The insert of Fig. 56 shows a schematic diagram of a thin-film transistor (TFT), where semiconductor (e.g., CdS) metal and insulator layers are deposited sequentially to form the device.[63] Figure 56 shows the current–

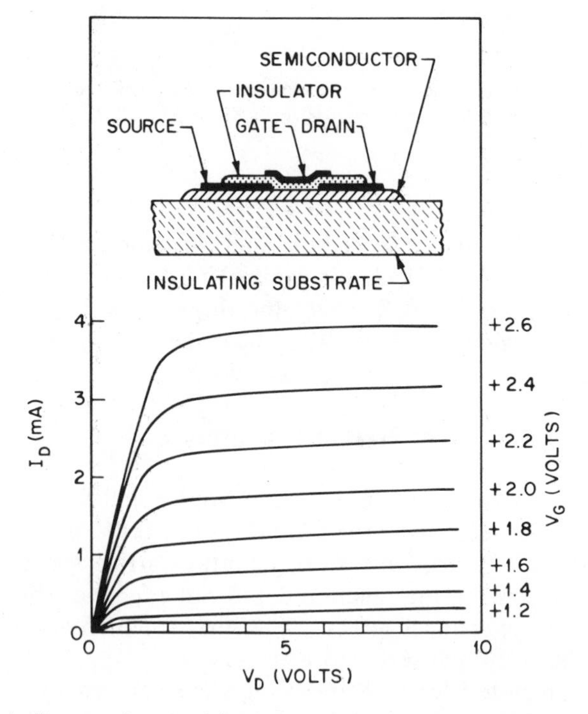

Fig. 56 Thin-film transistor and its drain characteristics. (After Weimer, Ref. 63.)

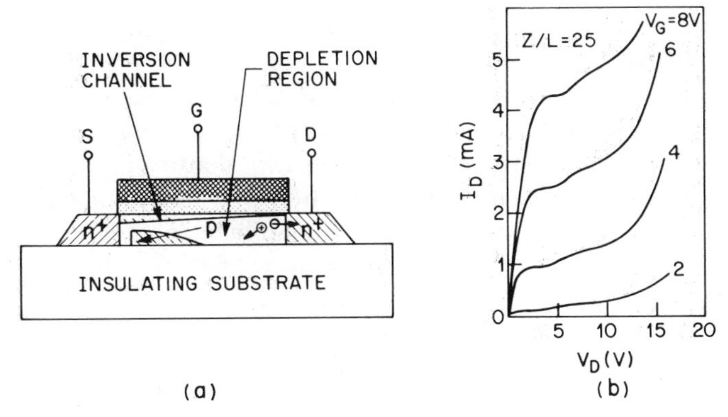

Fig. 57 (*a*) Silicon-on-insulator MOSFET. (*b*) Drain characteristics of SOS. (After Tihanyi and Schlotterer, Ref. 64.)

voltage characteristics of an enhancement TFT. These characteristics are similar to those of a MOSFET. Because the semiconductor layer is formed by deposition, more defects and crystalline imperfections occur in the layer than in the corresponding single-crystal semiconductor, resulting in more complicated transport processes in the TFT. To improve device performance, reproducibility, and reliability, the bulk and interface trap densities must be reduced.

8.5.7 SOI

Many silicon-on-insulator (SOI) devices have been proposed, including silicon-on-sapphire (SOS), silicon-on-spinel, silicon-on-nitride, and silicon-on-oxide. Figure 57*a* shows a schematic diagram of an SOI device, where single-crystal silicon is epitaxially grown on an insulator substrate (e.g., Al_2O_3 in the case of SOS).[64] The devices are made using the standard MOS process. The substrate provides the isolation between devices. Because of the reduced parasitic capacitances, the device has a higher response speed. Figure 57*b* shows a typical characteristic for an *n*-channel SOS device. The kinks in the saturation region can be explained as follows. The SOS has no substrate contact. As the drain voltage is increased, weak avalanche can occur near the drain. The generated electrons flow into the drain along with the drain current, and the generated holes flow toward the source and constitute the substrate current. The substrate current increases rapidly with drain voltage; therefore, above a certain V_D, the forward voltage of the source–substrate diode increases in a pronounced way. The device effectively has a positive substrate bias ($V_{BS} > 0$), which causes the threshold voltage to be reduced and the drain current to jump to a higher level. The threshold voltage change depends strongly on the substrate doping; higher dopings result in a larger change in V_T, and therefore a more pronounced kink.

Laser annealing techniques[65,66] have recently been applied to crystallize amorphous silicon films deposited on insulator substrates, such as Si_3N_4 and SiO_2. At present the devices show inferior performances compared to devices made on bulk single-crystal substrates, owing to high defect density in the Si film and a high interface trap density at the Si–insulator interface. With improvements in film quality, SOI devices have the potential for ultra-high-speed operation and three-dimensional device configurations, since the basic SOI layers can be stacked to form multiple-layer integrated circuits.[80]

8.5.8 VMOS

Figure 58*a* shows the VMOS (vertical or V-shaped grooved MOS) structure,[67] and Fig. 58*b* shows a modified version, the UMOS (U-shaped

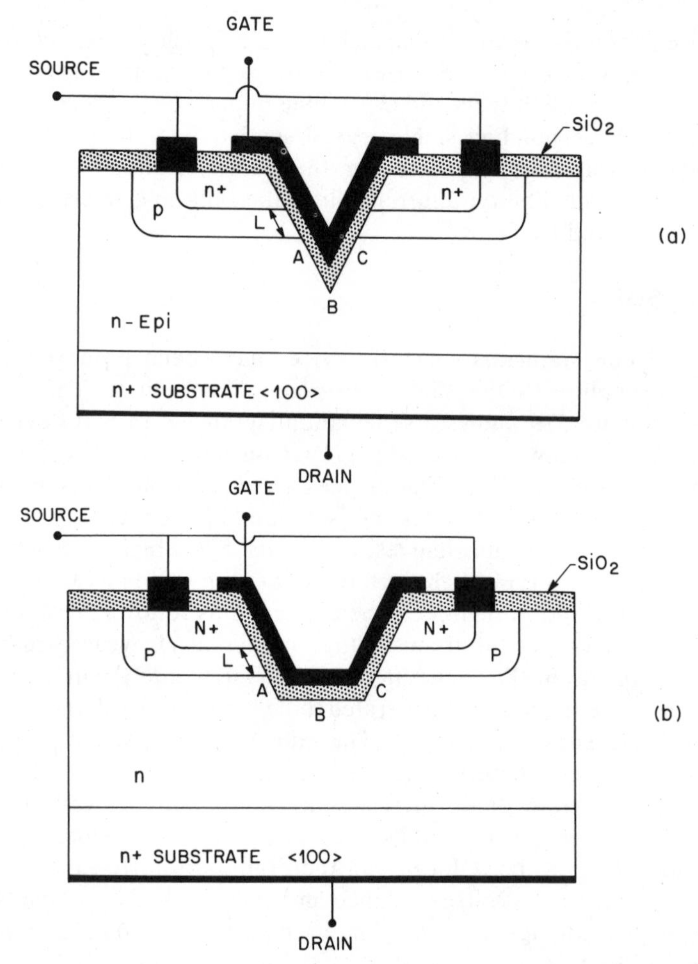

Fig. 58 (*a*) VMOS and (*b*) UMOS structures. (After Holmes and Salama, Ref. 67; Salama, Ref. 68.)

grooved MOS) structure.[68] These structures are made on ⟨100⟩-oriented silicon substrates, using a nonisotropic etch to form the notch sloping from the horizontal at 54.7°. The doping distribution shown is equivalent to a DMOS structure. The channel length is L and the two channels are parallel, one on each side of the etched groove. The device has a common drain contact at the bottom. Because many devices can be connected in parallel, these structures can handle high current and high power.[69] For planar structures with all contacts on the top surface, the n^+ contacts can be eliminated and the p region replaced by p^+; then the channel length is the ABC, and the right-hand surface contact becomes the drain electrode.

8.5.9 HEXFET

Figure 59*a* shows the HEXFET (hexagonal MOSFET) structure.[70] The operation of the HEXFET is similar to a DMOS. Each cell has an n^+ source region and a hexagonal polysilicon gate. The current flows from the source through the inverted narrow channel around the periphery of the cell and then vertically downward to the bottom drain n^+ contact. The HEXFET offers very high packing density (e.g., $\sim 10^5$ hexagonal cells per cm^2). Because of the large aspect ratio, $6NZ/L$, where N is the total number of cells per device and Z is the length of the hexagonal side, the on-resistance can be made very low.

Figure 59*b* shows device characteristics of a power HEXFET that can handle 100 A with saturation voltages comparable to bipolar transistors of

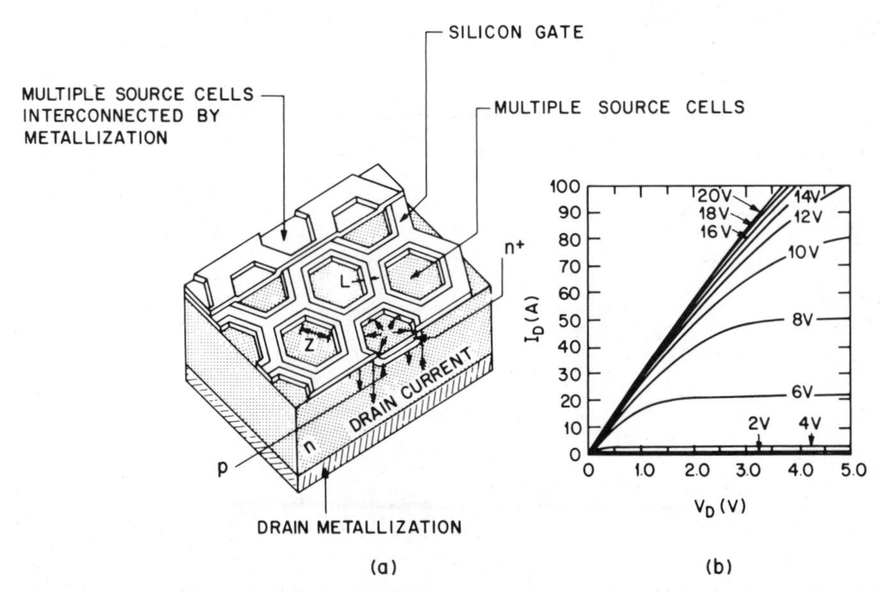

Fig. 59 (*a*) HEXFET structure. (*b*) Output characteristics. (After Collins and Pelly, Ref. 70.)

similar size. The on-resistance, which is proportional to the reciprocal of the aspect ratio $L/6NZ$, is only 0.05 Ω. By increasing the aspect ratio, lower on-resistance can be obtained.

8.6 NONVOLATILE MEMORY DEVICES

When the gate electrode of a conventional MOSFET is modified so that semipermanent charge storage inside the gate is possible, the new structure becomes a nonvolatile memory device. Since the first nonvolatile memory device proposed by Kahng and Sze[71] in 1967, various device structures have been made, and nonvolatile memory devices have been extensively used in integrated circuits[72] such as the electrically alterable read-only memory (EAROM), the erasable-programmable read-only memory (EPROM), and the nonvolatile random-access memory (NVRAM).

The two groups of nonvolatile memory devices are the floating-gate devices and the MIOS (metal–insulator–oxide–semiconductor) devices (Fig. 60). In both devices, charges are injected from the silicon across the

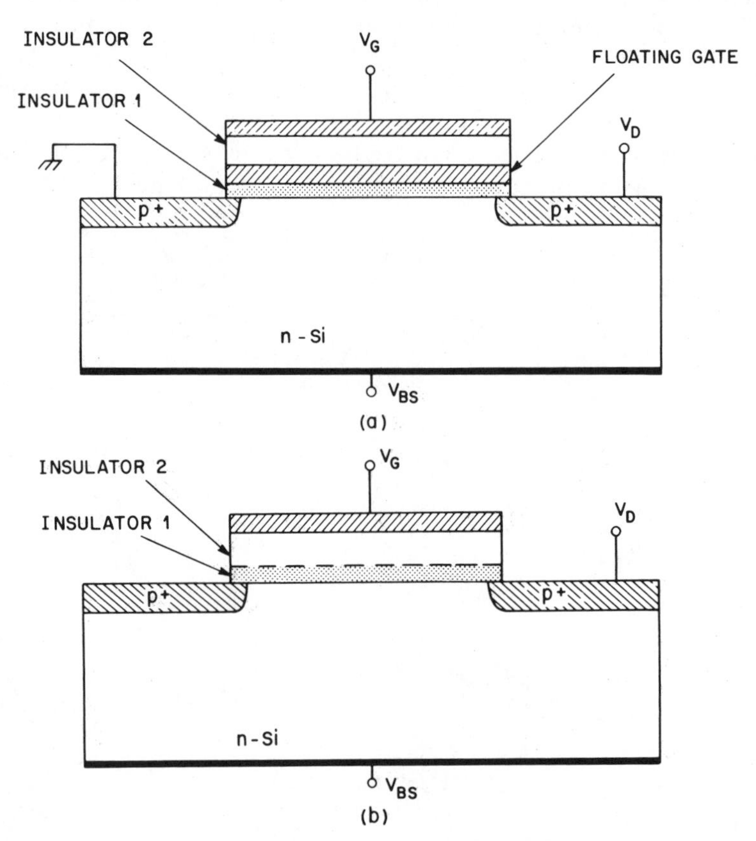

Fig. 60 (*a*) Floating-gate nonvolatile memory. (*b*) MIOS nonvolatile memory.

first insulator and stored in the floating gate or at the insulator–oxide interface of the MIOS device. The stored charge gives rise to a threshold voltage shift, and the device is at a higher-threshold voltage state. For a well-designed memory device, the charge retention time can be over 100 years. To erase the stored charge and return the device to a "lower-threshold voltage state," a gate voltage or other means (such as ultraviolet light) can be used.

8.6.1 Floating-Gate Device

The energy-band diagram of the first floating-gate device has an n-type silicon substrate, a layered gate structure of a thin oxide I(1), a floating metal gate M(1), a thick insulator I(2), and an external metal gate M(2) (Fig. 61). Upon application of a positive voltage V_G to the external gate, an electric field is established in each of the two insulators (Fig. 61a). We have, from Gauss's law, that

$$\epsilon_1 \mathscr{E}_1 = \epsilon_2 \mathscr{E}_2 + Q \qquad (105)$$

and

$$V_G = V_1 + V_2 = d_1 \mathscr{E}_1 + d_2 \mathscr{E}_2 \qquad (106)$$

where ϵ_1 and ϵ_2 are the dielectric permittivities of insulators 1 and 2; $\mathscr{E}_1$ and $\mathscr{E}_2$ are the corresponding fields, V_1 and V_2 are the voltages developed across insulators 1 and 2, and Q is the stored charge on the floating gate. From Eqs. 105 and 106 we obtain

$$\mathscr{E}_1 = \frac{V_G}{d_1 + d_2(\epsilon_1/\epsilon_2)} + \frac{Q}{\epsilon_1 + \epsilon_2(d_1/d_2)}. \qquad (107)$$

During the application of V_G, the charge on the floating gate changes with time, provided that the currents in the two insulators are not equal, that is,

$$Q(t) = \int_0^t [J_1(\mathscr{E}_1) - J_2(\mathscr{E}_2)]\, dt \qquad \text{C/cm}^2 \qquad (108)$$

where $J_1(\mathscr{E}_1)$ and $J_2(\mathscr{E}_2)$ are the current densities in insulator 1 and 2.

The current transport in insulators is generally a strong function of the electric field. When the transport is Fowler–Nordheim tunneling, the current density has the form

$$J = C_1 \mathscr{E}^2 \exp(-\mathscr{E}_0/\mathscr{E}) \qquad (109)$$

where $\mathscr{E}$ is the field, and C_1 and $\mathscr{E}_0$ are constants in terms of effective mass and barrier height. This type of current transport occurs in SiO_2 and Al_2O_3 as discussed in Chapter 7. When the transport is of the Frankel–Poole type, which occurs in Si_3N_4, the current density follows the form

$$J = C_2 \mathscr{E} \exp\left[-q\left(\phi_B - \sqrt{q\mathscr{E}/\pi\epsilon_i}\right)\Big/ kT\right] \qquad (110)$$

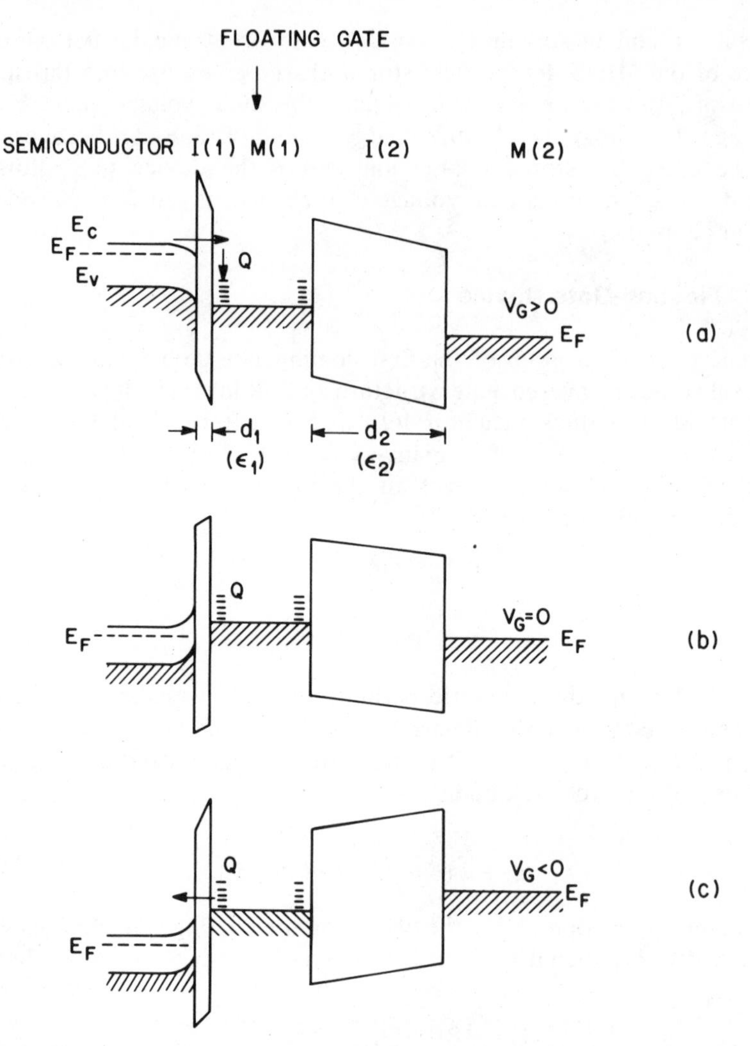

Fig. 61 Energy-band diagram of a floating-gate memory device. (a) Charging (writing mode). (*b*) Charge storage on the floating gate. (c) Discharging (erasing mode). (After Kahng and Sze, Ref. 71.)

where C_2 is a constant in terms of the trapping density in the insulator, ϕ_B the barrier height, and ϵ_i the dynamic permittivity.

After time t, the applied V_G is removed (Fig. 61*b*) and the stored charge Q causes a shift of the threshold voltage by the amount

$$\Delta V_T = -\frac{d_2}{\epsilon_2} Q. \tag{111}$$

To erase the stored charge, one can apply a negative gate voltage (Fig. 61*c*).

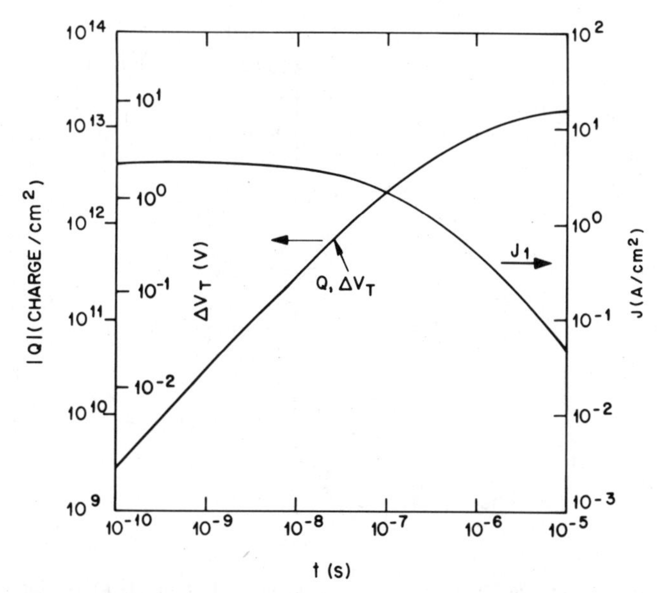

Fig. 62 Calculated charging current and stored charge as a function of charging time. (After Kahng and Sze, Ref. 71.)

Figure 62 gives the results of a theoretical computation, using Eqs. 106 through 109 with the following parameters: $d_1 = 50$ Å, $\epsilon_1 = 3.85\epsilon_0$ (for SiO_2), $d_2 = 1000$ Å, $\epsilon_2 = 30\epsilon_0$ (for ZrO_2), $V_G = 50$ V, and the current J_2 is assumed to be zero. Note that initially the stored charge increases linearly with time and then saturates. For a short time the current is almost constant and then decreases rapidly. The results above can be explained as follows: when a voltage pulse is applied at $t = 0$, the initial charge Q is zero and the initial electric field across I(1) has its maximum value $\mathscr{E}_1 = V_G/[d_1 + (\epsilon_1/\epsilon_2)d_2]$, from Eq. 107. When Q (which is negative for electrons) is sufficiently small so that $\mathscr{E}_1$ remains essentially the same, the current in turn remains the same, and Q increases linearly with time. When Q is large enough to reduce the value of $\mathscr{E}_1$ substantially, the current decreases rapidly with time and $|Q|$ begins to saturate.

Figure 62 also shows the threshold voltage shift based on Eq. 111. For this particular device, to increase the threshold voltage by 1 V, less than 0.1 μS of charging time (also called writing time) is required. Experimentally, the threshold voltage shift can be measured from the drain conductance. The change in V_T results in a change in the channel conductance g_D of the MOSFET. For small drain voltages, the channel conductance of a p-channel MOSFET is given by

$$g_D = -\frac{Z}{L}\mu_p C_i (V_G - V_T), \qquad V_G > V_T. \tag{112}$$

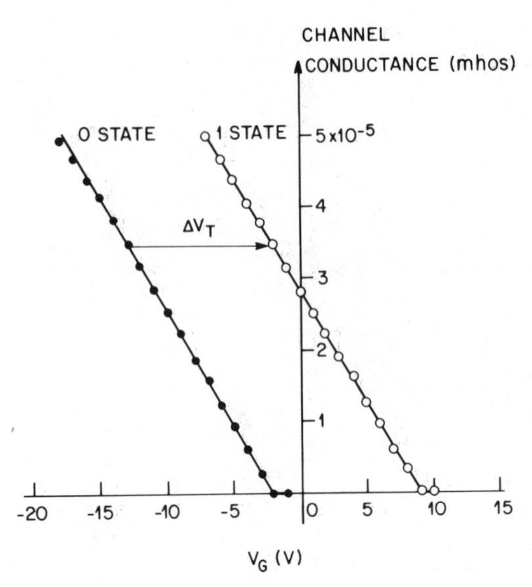

Fig. 63 Channel conductance before (0 state) and after (1 state) writing operation. (After Card and Worrall, Ref. 73.)

After altering the charge on the floating gate by Q (negative charge), the g_D versus V_G plot shifts to the right by ΔV_T (Fig. 63).[73]

The above device structure and its operation involve several essential and important concepts for the nonvolatile memory devices developed later. The tunneling-injection concept has been utilized in MIOS-type memory, and the floating-gate concept has been well developed in floating polysilicon gate memories.

The first EPROM was developed using a heavily doped polysilicon as the floating-gate material (Fig. 64*a*). The device is known[74] as a floating-gate avalanche-injection MOS memory (FAMOS). The polysilicon gate is embedded in the gate oxide and is completely isolated. The oxide thickness d_1 is of the order of 1000 Å, so that no weak spot or shorting path exists between the floating gate and the substrate. To inject charge (i.e., to write) into the floating gate, the drain junction is biased to avalanche breakdown, and electrons in the avalanche plasma are injected from the drain region into the SiO_2. These electrons are drifted by the field which is induced in the gate oxide by the capacitive coupling of the floating gate to the source and drain electrodes. To erase the FAMOS memory, ultraviolet light or x-ray is used. Electrical erasing cannot be used because the device has no external gate.

For electrical erasing, the stacked-gate avalanche-injection MOS (SAMOS) memory[75] with double-level polysilicon gates has been proposed (Fig. 64*b*). The external control gate makes electrical erasing possible and improves the writing efficiency. A similar device can be made by using

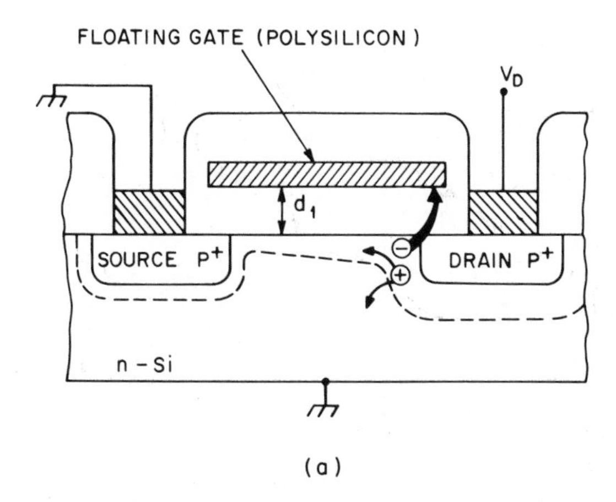

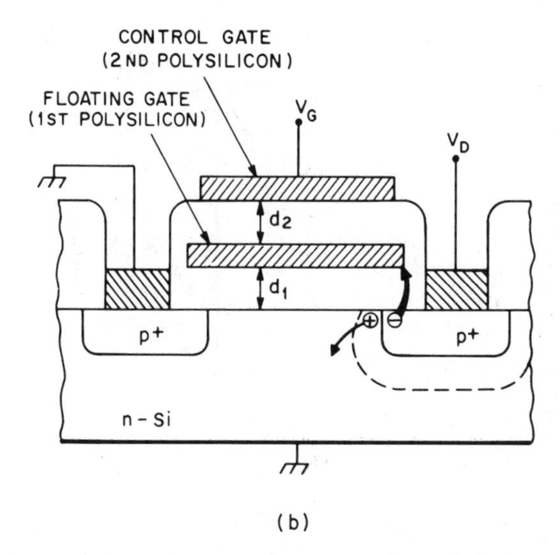

Fig. 64 (*a*) FAMOS. (After Frohman-Bentchkowsky, Ref. 74.) (*b*) SAMOS. (After Iizuka et al., Ref. 75.)

Si-rich SiO_2 between the two polysilicon gates.[81] Figure 65 shows the energy-band diagrams for the SAMOS operations. Figure 65*a* shows the equilibrium or 0 state, assuming that $V_{FB} = 0$. After avalanche injection, negative charges are stored on the floating gate causing a threshold voltage shift, and the device is in the 1 state (Fig. 65*b*). By applying a large positive voltage to the external gate, the stored charge can be removed and the device returns to the 0 state (Fig. 65*c*).

Figure 66 shows the calculated and experimental results of writing characteristics of a SAMOS.[75] For a given drain voltage, the threshold voltage increases with increasing voltage on the external gate, because the

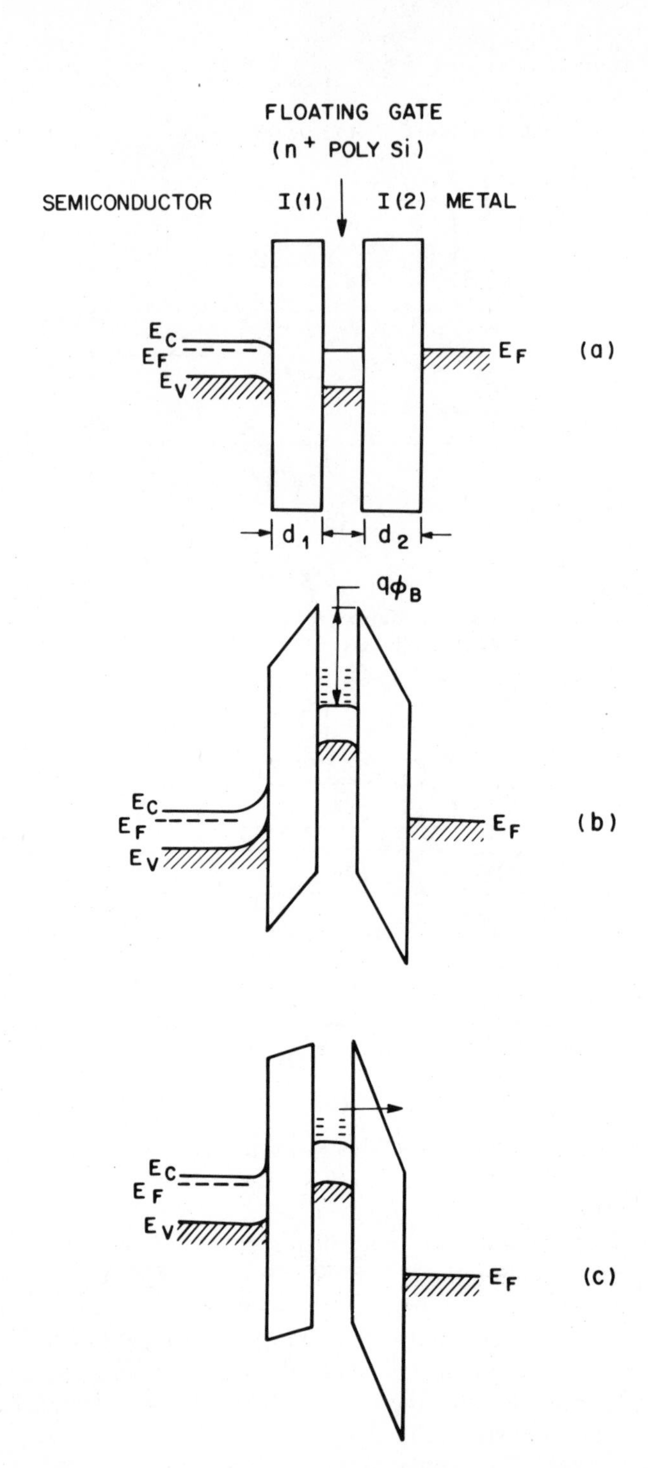

Fig. 65 Energy-band diagram for (*a*) 0 state, (*b*) 1 state, and (*c*) discharging (erasing) a SAMOS memory.

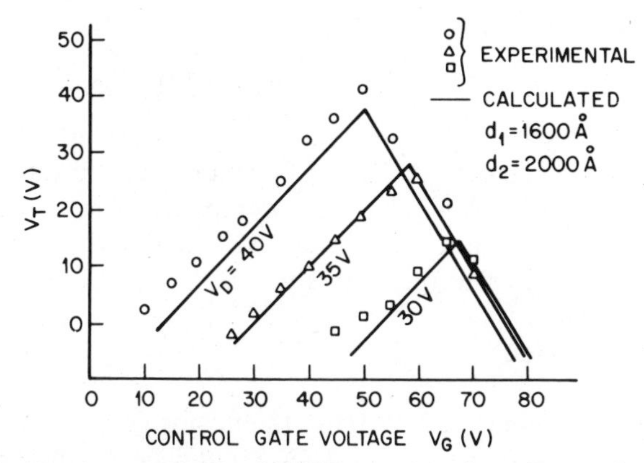

Fig. 66 Threshold voltage shift of a SAMOS as a function of control gate voltage for various drain biases. (After Iizuka et al., Ref. 75.)

gate voltage helps to transport the avalanche-generated electrons from the silicon to the floating gate (similar to that shown in Fig. 61*a*). However, as V_G increases beyond a certain point, V_T begins to decrease, because the current transport from the floating gate toward the external gate leads to a reduction of the stored charge. Figure 67 shows threshold voltage change as functions of write time and erase time, for the same device shown in Fig. 66. For a gate voltage of 20 V, a write time of about 1 μs is required to obtain a 5-V threshold shift. For a FAMOS with identical device parameters, the write time is two orders of magnitude longer. For the erasing operation, the higher the gate voltage, the faster one can remove the stored charge.

A long retention time is required for nonvolatile memory operation. The retention time is defined as the time when the stored charge decreases to

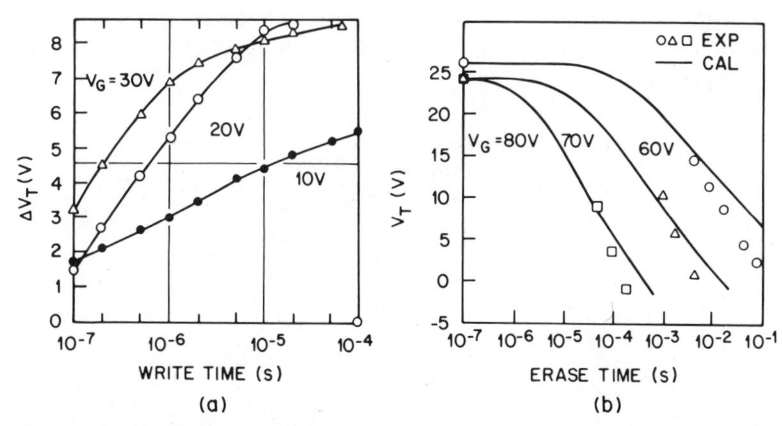

Fig. 67 Threshold voltage shift under (*a*) writing and (*b*) erasing operations of a SAMOS. (After Iizuka et al., Ref. 75.)

50% of its initial value and is given by

$$t_R = \ln 2/[\nu \exp(q\phi_B/kT)] \tag{113}$$

where ν is the dielectric relaxation frequency, and $q\phi_B$ is the barrier height indicated in Fig. 65*b*. Figure 68 shows typical calculated retention times at 125°C and 170°C with $q\phi_B = 1.7$ eV and compares them with experimental data. The values of retention time for 125°C and 170°C are found to be about 100 years and 8000 h, respectively.

8.6.2 MIOS Device

Among several kinds of MIOS (metal–insulator–SiO_2–Si) memory devices, the MNOS (metal–Si_3N_4–SiO_2–Si) device is the most popular. Other MIOS devices use different insulators to replace the silicon nitride film, such as aluminum oxide, tantalum oxide, and titanium oxide. The MIOS device has been made by using metal ions (e.g., Au) implanted into SiO_2 to alter the conduction properties of the outer oxide to form the interfacial charge storage centers.[76]

Figure 69 shows the basic band diagrams for the writing and erasing operations.[77] The current J_0 in the oxide is due to Fowler–Nordheim tunneling, and the current J_N in the silicon nitride is due to Frankel–Poole emission. The equations governing the charging behavior are identical to those described before, Eqs. 105 through 108. However, in a practical device, the traps at the oxide–nitride interface may be distributed across the nitride energy gap and extended from the interface into the bulk nitride. Various models have been considered to achieve better agreement with experimental writing characteristics. A representative result for an MNOS device with 20-Å SiO_2 and 433-Å Si_3N_4 shows that the measured and calculated results are in good agreement (Fig. 70).[78] The figure also shows that the write time to reach a given threshold voltage shift decreases rapidly with increasing V_G. For a given gate voltage the threshold voltage shift tends to saturate at a longer write time, showing the general behavior displayed in Fig. 62.

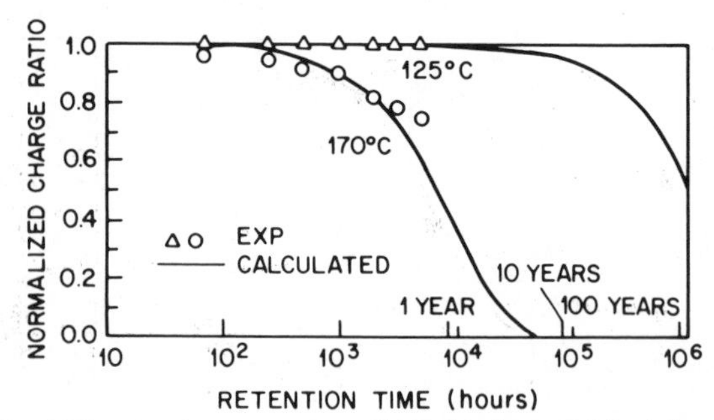

Fig. 68 Normalized charge storage versus time for two ambient temperatures. (After Nishi and Iizuka, Ref. 72.)

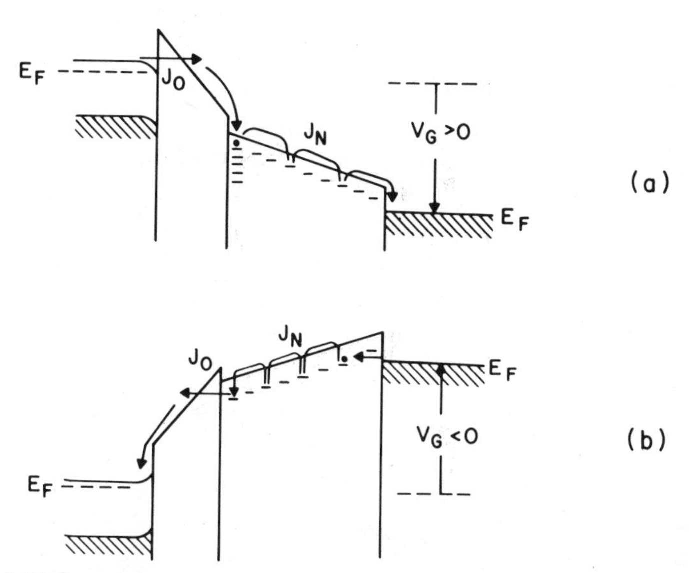

Fig. 69 (*a*) Writing and (*b*) Erasing operations of an MNOS device. (After Frohman-Bentchkowsky, Ref. 77.)

To improve MIOS device performance, a dual-dielectric charge storage device (DDC) with interfacial dopant[79] is proposed (Fig. 71*a*). The interfacial dopant (e.g., tungsten) can substantially increase the trap density at the insulator–oxide interface, thereby allowing shorter times for the write or erase operation (Fig. 71*b*). Figure 72 compares a DDC to an identical device without the interfacial dopant. The DDC is about three orders of magnitude faster than a conventional MIOS device, under the biasing conditions of $V_G = 30$ V for the writing operation and $V_G = -30$ for the erasing operation.

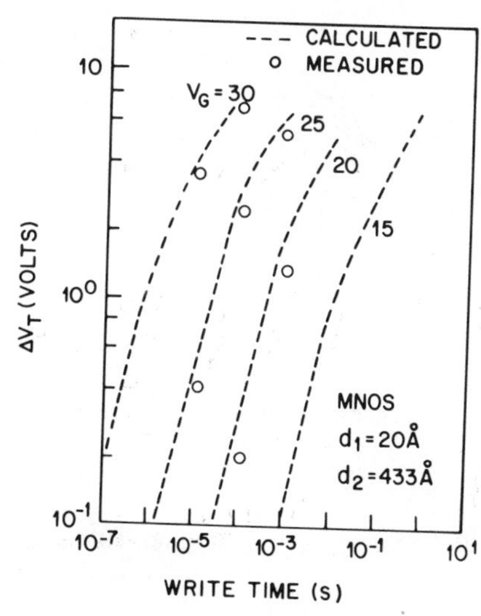

Fig. 70 Calculated and measured threshold shifts of MNOS device. (After Card and Elmasry, Ref. 78.)

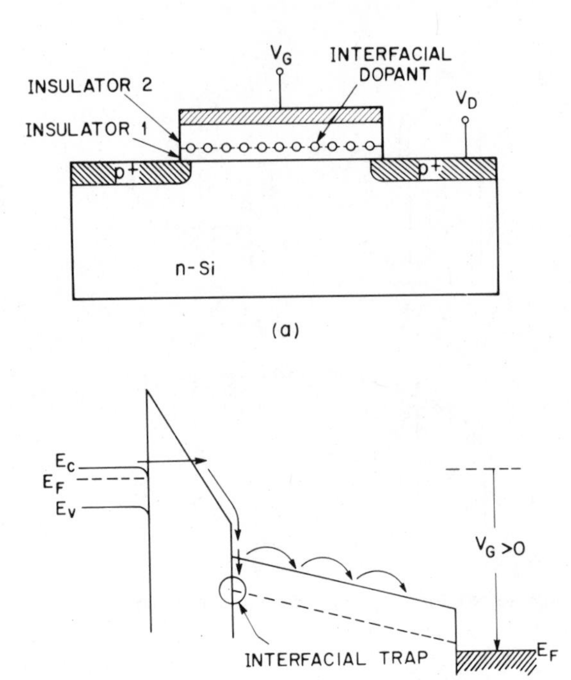

Fig. 71 Dual-dielectric storage cell. (*a*) Cross section of device. (*b*) Energy-band diagram under an applied gate voltage. (After Kahng et al., Ref. 79.)

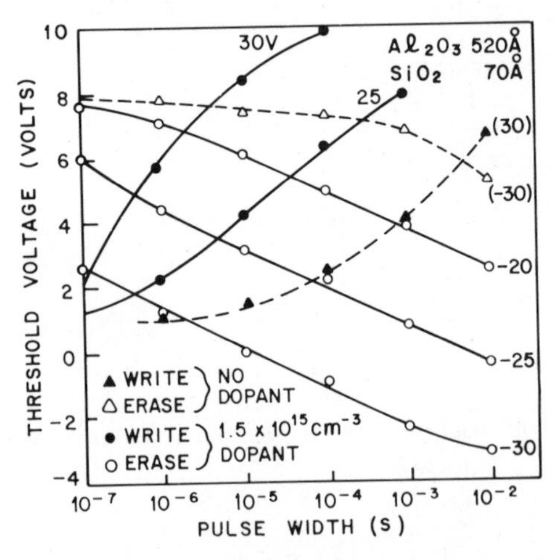

Fig. 72 Threshold voltage shift versus pulse width for writing and erasing operations of a dual-dielectric cell. (After Kahng et al., Ref. 79.)

REFERENCES

1 J. E. Lilienfeld, U.S. Patent 1,745,175 (1930).

2 O. Heil, British Patent 439,457 (1935).

3 W. Shockley and G. L. Pearson, "Modulation of Conductance of Thin Films of Semiconductors by Surface Charges," *Phys. Rev.*, **74**, 232 (1948).

4 D. Kahng and M. M. Atalla, "Silicon–Silicon Dioxide Field Induced Surface Devices," IRE Solid-State Device Res. Conf., Carnegie Institute of Technology, Pittsburgh, Pa., 1960. D. Kahng, "A Historical Perspective on the Development of MOS Transistors and Related Devices," *IEEE Trans. Electron Devices*, **ED-23**, 655 (1976).

5 H. K. J. Ihantola, "Design Theory of a Surface Field-Effect Transistor," *Stanford Electron. Lab. Tech. Rep. No. 1661-1* (1961).

6 H. K. J. Ihantola and J. L. Moll, "Design Theory of a Surface Field-Effect Transistor," *Solid State Electron.*, **7**, 423 (1964).

7 C. T. Sah, "Characteristics of the Metal–Oxide–Semiconductor Transistors," *IEEE Trans. Electron Devices*, **ED-11**, 324 (1964).

8 S. R. Hofstein and F. P. Heiman, "The Silicon Insulated-Gate Field-Effect Transistor," *Proc. IEEE*, **51**, 1190 (1963).

9 J. T. Wallmark and H. Johnson, *Field Effect Transistors, Physics, Technology, and Applications*, Prentice-Hall, Englewood Cliffs, N.J., 1966.

10 P. Richman, *MOSFET's and Integrated Circuits*, Wiley, New York, 1973.

11 J. R. Brews, "Physics of the MOS Transistor," in D. Kahng, Ed., *Applied Solid State Science*, Suppl. 2A, Academic, New York, 1981.

12 L. L. Chang and H. N. Yu, "The Germanium Insulated-Gate Field-Effect Transistor (FET)," *Proc. IEEE*, **53**, 316 (1965).

13 C. W. Wilmsen and S. Szpak, "MOS Processing for III–V Compound Semiconductors: Overview and Bibliography," *Thin Solid Films*, **46**, 17 (1977).

14 "Looking Ahead to the Year 2000, Technology," *Electronics*, **53**, (9) 530 (1980).

15 G. Moore, "VLSI: Some Fundamental Challenges," *IEEE Spectrum*, **16**, (4) 30 (1979).

16 H. C. Pao and C. T. Sah, "Effects of Diffusion Current on Characteristics of Metal–Oxide (Insulator)–Semiconductor Transistors (MOST)," *Solid State Electron*, **9**, 927 (1966).

17 A. S. Grove and D. J. Fitzgerald, "Surface Effects on *p-n* Junctions: Characteristics of Surface Space-Charge Regions under Nonequilibrium Conditions," *Solid State Electron.*, **9**, 783 (1966).

18 J. R. Brews, "A Charge-Sheet Model of the MOSFET," *Solid State Electron.*, **21**, 345 (1978).

19 J. R. Brews, "Subthreshold Behavior of Uniformly and Nonuniformly Doped Long-Channel MOSFET," *IEEE Trans. Electron Devices*, **ED-26**, 1282 (1979).

20 M. B. Barron, "Low Level Currents in Insulated Gate Field Effect Transistors," *Solid State Electron.*, **15**, 293 (1972).

21 W. M. Gosney, "Subthreshold Drain Leakage Current in MOS Field-Effect Transistors," *IEEE Trans. Electron Devices*, **ED-19**, 213 (1972).

22 W. Fichtner and H. W. Potzl, "MOS Modeling by Analytical Approximations. I. Subthreshold Current and Threshold Voltage," *Int. J. Electron.*, **46**, 33 (1979).

23 R. R. Troutman, "Subthreshold Design Considerations for IGFET's," *IEEE J. Solid State Circuits*, **SC-9**, 55 (1974).

24 A. G. Sabnis and J. T. Clemens, "Characterization of the Electron Mobility in the Inverted ⟨100⟩ Si Surface," *IEEE Tech. Dig.*, Int. Electron Device Meet., 1979, p. 18.

25 R. W. Coen and R. S. Muller, "Velocity of Surface Carriers in Inversion Layers on Silicon," *Solid State Electron.*, **23**, 35 (1980).

26 J. A. Cooper, Jr. and D. F. Nelson, "High Field Drift Velocity of Electrons at the Si-Si0_2 Interface as Determined by a Time-of-Flight Technique," J. Appl. Phys., *54*, 1445 (1983).

27 K. K. Thornber, "Relation of Drift Velocity to Low-Field Mobility and High Field Saturation Velocity," *J. Appl. Phys.*, **51**, 2127 (1980).

28 K. Yamaguchi, "Field-Dependent Mobility Model for Two-Dimensional Numerical Analysis of MOSFETs," *IEEE Trans. Electron Devices*, **ED-26**, 1068 (1979).

29 L. Vadasz and A. S. Grove, "Temperature Dependence of MOS Transistor Characteristics below Saturation," *IEEE Trans. Electron Devices*, **ED-13**, 863 (1966).

30 R. Wang, J. Dunkley, T. A. DeMassa, and L. F. Jelsma, "Threshold Voltage Variations with Temperature in MOS Transistors," *IEEE Trans. Electron Devices*, **ED-18**, 386 (1971).

31 F. Gaensslen, V. L. Rideout, E. J. Walker, and J. J. Walker, "Very Small MOSFET's for Low-Temperature Operation," *IEEE Trans. Electron Devices*, **ED-24**, 218 (1977).

32 W. Fischer, "Equivalent Circuit and Gain of MOS Field-Effect Transistors," *Solid State Electron.*, **9**, 71 (1966).

33 R. C. Gallagher and W. S. Corak, "A Metal–Oxide–Semiconductor (MOS) Hall Element," *Solid State Electron.*, **9**, 571 (1966).

34 K. Wittmack, J. Maul, and F. Schulz, *Ion Implantation in Semiconductor and Other Materials*, Plenum, New York, 1973.

35 V. L. Rideout, F. H. Gaensslen, ad A. LeBlanc, "Device Design Consideration for Ion Implanted *n*-channel MOSFETs," *IBM J. Res. Dev.*, p. 50 (Jan. 1975).

36 J. R. Brews, "Threshold Shifts Due to Nonuniform Doping Profiles in Surface Channel MOSFET's," *IEEE Trans. Electron Devices*, **ED-26**, 1696 (1979).

37 G. Merckel, "Ion Implanted MOS Transistors—Depletion Mode Devices," in F. Van de Wiele, W. L. Engle, and P. G. Jespers, Eds. *Process and Device Modeling for IC Design*, Noordhoff, Leyden, 1977.

38 J. S. T. Huang and G. W. Taylor, "Modeling of an Ion-Implanted Silicon-Gate Depletion-Mode IGFET," *IEEE Trans. Electron Devices*, **ED-22**, 995 (1975).

39 T. E. Hendrikson, "A Simplified Model for Subpinchoff Condition in Depletion Mode IGFET's," *IEEE Trans. Electron Devices*, **ED-25**, 435 (1978).

40 K. Nishiuchi, H. Oka, T. Nakamura, H. Ishikawa, and M. Shinoda, "A Normally-Off Type Buried Channel MOSFET for VLSI Circuits," *IEEE Tech. Dig.*, Int. Electron Device Meet., 1978, p. 26.

41 J. R. Brews, W. Fichtner, E. H. Nicollian, and S. M. Sze, "Generalized Guide for MOSFET Miniaturization," *IEEE Electron Devices Lett.* **EDL-1**, 2 (1980).

42 L. D. Yau, "A Simple Theory to Predict the Threshold Voltage of Short-Channel IGFET's," *Solid State Electron.*, **17**, 1059 (1974).

43 G. W. Taylor, "Subthreshold Conduction in MOSFET's," *IEEE Trans. Electron Devices*, **ED-25**, 337 (1978).

44 G. Merckel, "Short Channels—Scaled Down MOSFET's," in Ref. 37, p. 705.

45 W. Fichtner, "Scaling Calculation for MOSFET's," IEEE Solid State Circuits and Technology Workshop on Scaling and Microlithography, New York, Apr. 22, 1980.

46 G. W. Taylor, "The Effects of Two-Dimensional Charge Sharing on the Above Threshold Characteristics of Short-Channel IGFET's," *Solid State Electron.*, **22**, 701 (1979).

47 L. D. Yau, "Simple I–V Model for Short-Channel IGFET's in the Triode Region," *Electron. Lett.* **11**, 44 (1975).

48 M. Fukuma and M. Matsumura, "A Simple Model for Short Channel MOSFET's," *Proc. IEEE*, **65**, 1212 (1977).

49 T. Kamata, K. Tanabashi, and K. Kobayashi, "Substrate Current Due to Impact Ionization in MOSFET," *Jpn. J. Appl. Phys.*, **15**, 1127 (1976).

50 E. Sun, J. Moll, J. Berger, and B. Alders, "Breakdown Mechanism in Short-Channel MOS Transistors," *IEEE Tech. Dig.* Int. Electron Device Meet., Washington D.C. 1978, p. 478.

51 T. H. Ning, C. M. Osburn, and H. N. Yu, "Effect of Electron Trapping on IGFET Characteristics," *J. Electron. Mater.*, **6**, 65 (1977).

52 J. Matsunaga, M. Konaka, S. Kohyama, and H. Iizuku, 11th Int. Conf. Solid State Devices, Aug. 27, 1979, p. 45.

53 E. H. Nicollian and C. N. Berglund, "Avalanche Injection of Electrons into Insulating SiO_2 Using MOS Structures," *J. Appl. Phys.*, **41**, 3052 (1970).

54 R. H. Dennard, F. H. Gaensslen, H. Yu, V. L. Rideout, E. Bassons, and A. R. LeBlanc, "Design of Ion-Implanted MOSFET's with Very Small Physical Dimensions," *IEEE J. Solid State Circuits*, **SC-9**, 256 (1974).

55 J. M. Shannon, J. Stephen, and J. H. Freeman, "MOS Frequency Soars with Ion-Implanted Layers," *Electronics*, p. 96 (Feb. 3, 1969).

56 H. Nihira, M. Konaka, H. Iwai, and Y. Nishi, "Anomalous Drain Current in *n*-MOSFET's and Its Suppression by Deep Ion Implantation," *IEEE Tech. Dig.*, Int. Electron Device Meet., 1978, p. 487.

57 Y. Tarui, Y. Hayashi, and T. Sekigawa, "Diffusion Self-Aligned Enhance-Depletion MOS-IC," Proc. 2nd Conf. Solid State Devices, *Suppl. J. Jpn. Soc. Appl. Phys.*, **40**, 193 (1971).

58 J. Tihanyi and D. Widmann, "DIMOS—A Novel IC Technology with Submicron Effective Channel MOSFET's," *IEEE Tech. Dig.*, Int. Electron Device Meet., 1977, p. 399.

59 T. Masuhara and R. S. Muller, "Analytical Technique for the Design of DMOS Transistors," *Jpn. J. Appl. Phys.*, **16**, 173 (1976).

60 M. D. Pocha, A. G. Gonzalez, and R. W. Dutton, "Threshold Voltage Controllability in Double-Diffused MOS Transistors," *IEEE Trans. Electron Devices*, **ED-21**, 778 (1974).

61 S. Nishimatsu, Y. Kawamoto, H. Masuda, R. Hori, and O. Minato, "Grooved Gate MOSFET," *Jpn. J. Appl. Phys.*, **16**; *Suppl.* **16-1**, 179 (1977).

62 M. P. Lepselter and S. M. Sze, "SB-IGFET: An Insulated-Gate Field-Effect Transistor Using Schottky Barrier Contacts as Source and Drain," *Proc. IEEE*, **56**, 1088 (1968).

63 P. K. Weimer, "The TFT—A New Thin-Film Transistor," *Proc. IRE*, **50**, 1462 (1962).

64 J. Tihanyi and H. Schlotterer, "Influence of the Floating Substrate Potential on the Characteristics of ESFI MOS Transistors," *Solid State Electron.*, **18**, 309 (1975).

65 H. W. Lam, A. F. Tasch, T. C. Holloway, K. F. Lee, and J. F. Gibbons, "Ring Oscillators Fabricated in Laser-Annealed Silicon-on-Insulator," *IEEE Electron Devices Lett.*, **EDL-1**, 99 (1980).

66 M. W. Geis, D. C. Flanders, D. A. Antoniadis, and H. I. Smith, "Crystalline Silicon on Insulators by Graphoepitaxy," *IEEE Tech. Dig.*, Int. Electron Device Meet., 1979, p. 210.

67 F. E. Holmes and C. A. T. Salama, "VMOS—A New MOS Integrated Circuit Technology," *Solid State Electron.*, **17**, 791 (1974).

68 C. A. T. Salama, "A New Short Channel MOSFET Structure (UMOST)," *Solid State Electron.*, **20**, 1003 (1977).

69 A. Lidow, T. Herman, and H. W. Collins, "Power MOSFET Technology," *IEEE Tech. Dig.*, Int. Electron Device Meet., 1979, p. 79.

70 H. W. Collins and B. Pelly, "HEXFET, A New Power Technology, Cutts On-Resistance, Boosts Ratings," *Electron. Des.*, **17**, (12) 36 (1979).

71 D. Kahng and S. M. Sze, "A Floating Gate and Its Application to Memory Devices," *Bell Syst. Tech. J.*, **46**, 1283 (1967).

72 Y. Nishi and H. Iizuka, "Nonvolatile Memories," in D. Kahng, Ed., *Applied Solid State Science*, Suppl. 2A, Academic, New York, 1981.

73 H. C. Card and A. G. Worrall, "Reversible Floating-Gate Memory," *J. Appl. Phys.*, **44**, 2326 (1973).

74 D. Frohman-Bentchkowsky, "FAMOS—A New Semiconductor Charge Storage Device," *Solid State Electron.*, **17**, 517 (1974).

75 H. Iizuka, F. Masuoka, T. Sato, and M. Ishikawa, "Electrically Alterable Avalanche-Injection-Type MOS Read-Only Memory with Stacked-Gate Structures," *IEEE Trans. Electron Devices*, **ED-23**, 379 (1976).

76 L. I. Chen, K. A. Pickar, and S. M. Sze, "Carrier Transport and Storage Effects in Au Ion Implanted SiO_2 Structures," *Solid State Electron.*, **15**, 979 (1972).

77 D. Frohman-Bentchkowsky, "The Metal–Nitride–Oxide–Silicon (MNOS) Transistor Characteristics and Applications," *Proc. IEEE*, **58**, 1207 (1970).

78 H. C. Card and M. I. Elmasry, "Functional Modeling of Nonvolatile MOS Memory Devices," *Solid State Electron.*, **19**, 863 (1976).

79 D. Kahng, W. J. Sundburg, D. M. Boulin, and J. R. Ligenza, "Interfacial Dopants for Dual-Dielectric Charge-Storage Cells," *Bell Syst. Tech. J.*, **53**, 1723 (1974).

80 J. J. Gibbons and K. F. Lee, "One-Gate-Wide CMOS Inverter on Laser-Recrystallized Polysilicon," *IEEE Electron Devices Lett.*, **EDL-1**, 117 (1980).

81 D. J. DiMaria, K. M. DeMeyer, and D. W. Dong, "Electrically-Alterable Memory Using a Dual Electron Injector Structure," *IEEE Electron Devices Lett.*, **EDL-1**, 179 (1980).

PART IV

SPECIAL MICROWAVE DEVICES

9

Tunnel Devices

- INTRODUCTION
- TUNNEL DIODE
- BACKWARD DIODE
- MIS TUNNEL DIODE
- MIS SWITCH DIODE
- MIM TUNNEL DIODE
- TUNNEL TRANSISTOR

9.1 INTRODUCTION

In this chapter and in Chapters 10 and 11 we consider some special microwave semiconductor devices. The microwave frequencies cover the range from about 0.1 GHz (10^8 Hz) to 1000 GHz with corresponding wavelength from 300 cm to 0.3 mm. The microwave frequency range is usually grouped into different bands. The bands and the corresponding frequency ranges are listed in Table 1. For frequencies above 30 to 300 GHz, we have the millimeter-wave band because the wavelength is between 10 and 1 mm. For even higher frequencies, we have the sub-millimeter-wave band. In 1970, a new band designation for microwave frequencies has been adopted as listed in Table 2. To avoid confusion it is recommended to state both the band and the corresponding frequency range. Many semiconductor devices considered in previous chapters can be operated in the microwave region. Table 3 summarizes the representative microwave devices and their operational principles.

We consider devices associated with quantum tunneling phenomena in this chapter. The first acknowledged paper on tunnel devices discussed the

Table 1 Bands and Frequency Range[63]

Band	Waveguide Size (cm)	Frequency Range (GHz)
L	—	1.0–2.6
S	7.6 × 3.8	2.60–3.95
G	5 × 2.5	3.95–5.85
C	4.4 × 2.3	4.90–7.05
J	3.8 × 1.9	5.30–8.20
H	3.2 × 1.3	7.05–10.00
X	2.5 × 1.25	8.20–12.40
M	2.1 × 1.2	10.00–15.00
P	1.8 × 1.0	12.40–18.00
N	1.5 × 0.85	15.00–22.00
Ku	—	15.30–18.00
K	1.2 × 0.65	18.00–26.50
R	0.9 × 0.56	26.50–40.00
Millimeter		>30–300
Submillimeter		>300

tunnel diode, also referred to as the Esaki diode, and was written[1] by L. Esaki in 1958. In the course of studying the internal field emission in a degenerate germanium *p-n* junction, he discovered an "anomalous" current–voltage characteristic in the forward direction, that is, a negative-resistance region over part of the forward characteristic. Esaki explained this anomalous characteristic by the quantum tunneling concept and

Table 2 New Bands and Frequency Range[64]

Band	Frequency Range (GHz)
A	0.100–0.250
B	0.250–0.500
C	0.500–1.000
D	1.000–2.000
E	2.000–3.000
F	3.000–4.000
G	4.000–6.000
H	6.000–8.000
I	8.000–10.000
J	10.000–20.000
K	20.000–40.000
L	40.000–60.000
M	60.000–100.000
Millimeter	>30–300
Submillimeter	>300

Table 3 Microwave Semiconductor Devices

Device	Chapter	Operational Principle
Varactor diode	2	Reactance varies with bias voltage
p-i-n diode	2	Nearly constant capacitance, high breakdown voltage
Bipolar transistor	3	Electrons and holes participate in transport processes
Point-contact diode	5	Small area, small capacitance
Schottky diode	5	Majority-carrier transport, thermionic injection
JFET	6	Majority carrier, current modulated by junction-gate bias
MESFET	6	Majority carrier, current modulated by Schottky-gate bias
MOSFET	8	Minority-carrier transport in surface inversion channel
Tunnel diode	9	Tunneling in forward-biased p^+n^+ junction, negative differential resistance
Backward diode	9	Tunneling in reverse-biased junction or near zero bias, high nonlinearity
IMPATT diode	10	Avalanche and transit-time effects to generate high power
BARITT diode	10	Barrier injection and transit-time effects
TRAPATT diode	10	Trapped plasmas avalanche triggered transit diode
TED	11	Electrons transferred from low-energy high-mobility band to high-energy low-mobility band

obtained reasonable agreement between the tunneling theory and the experimental results.

The tunneling phenomenon is a majority carrier effect. In addition, the tunneling time of carriers through the potential energy barrier is not governed by the conventional transit time concept ($\tau = W/v$, where W is the barrier width and v is the carrier velocity), but rather by the quantum transition probability per unit time which is proportional to $\exp[-2\bar{k}(0)W]$, where $\bar{k}(0)$ is the average value of momentum encountered in the tunneling path corresponding to an incident carrier with zero transverse momentum and energy equal to the Fermi energy.[2] Reciprocation gives the tunneling time proportional to $\exp[2\bar{k}(0)W]$. This tunneling time is very short, permitting the use of tunnel devices well into the millimeter-wave region.

9.2 TUNNEL DIODE

Because of its mature technology and high reliability, the tunnel diode is used in special low-power microwave applications, such as local oscillator and frequency locking circuits. This section discusses the dc and microwave characteristics of tunnel diodes.

A tunnel diode consists of a simple *p-n* junction in which both *p* and *n* sides are degenerate (i.e., very heavily doped with impurities). Figure 1 shows a schematic energy diagram of a tunnel diode in thermal equilibrium. Because of the high dopings the Fermi level is located within the allowed bands themselves. The amount of degeneracy, V_p and V_n, is typically a few kT, and the depletion-layer width is of the order of 100 Å or less, which is considerably narrower than the conventional *p-n* junction.

Figure 2*a* shows a typical static current–voltage characteristic of a tunnel diode. In the reverse direction (*p* side negative with respect to *n* side) the current increases monotonically. In the forward direction the current first increases to a maximum value (peak current or I_P) at a voltage V_P, then decreases to a minimum value I_V at a voltage V_V. For a voltage larger than V_V, the current increases exponentially with the voltage. The static characteristic is the result of three current components: band-to-band tunneling current, excess current, and thermal current (Fig. 2*b*).

We first discuss qualitatively the tunneling processes at absolute zero temperature using the simplified band structures[3] as shown in Fig. 3. Note that the Fermi levels are within the bands of the semiconductor, and at thermal equilibrium (Fig. 3*b*) the Fermi level is constant across the junction. Above the Fermi level there are no filled states on either side of the junction, and below the Fermi level there are no empty states available on

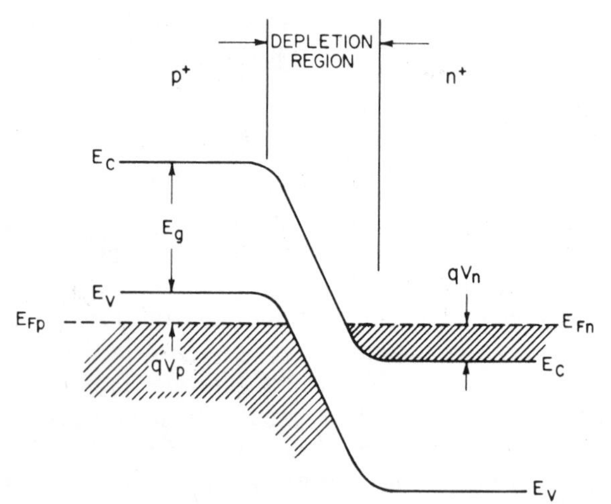

Fig. 1 Energy-band diagram of a tunnel diode in thermal equilibrium. V_p and V_n are the degeneracies on the *p*-side and *n*-side, respectively.

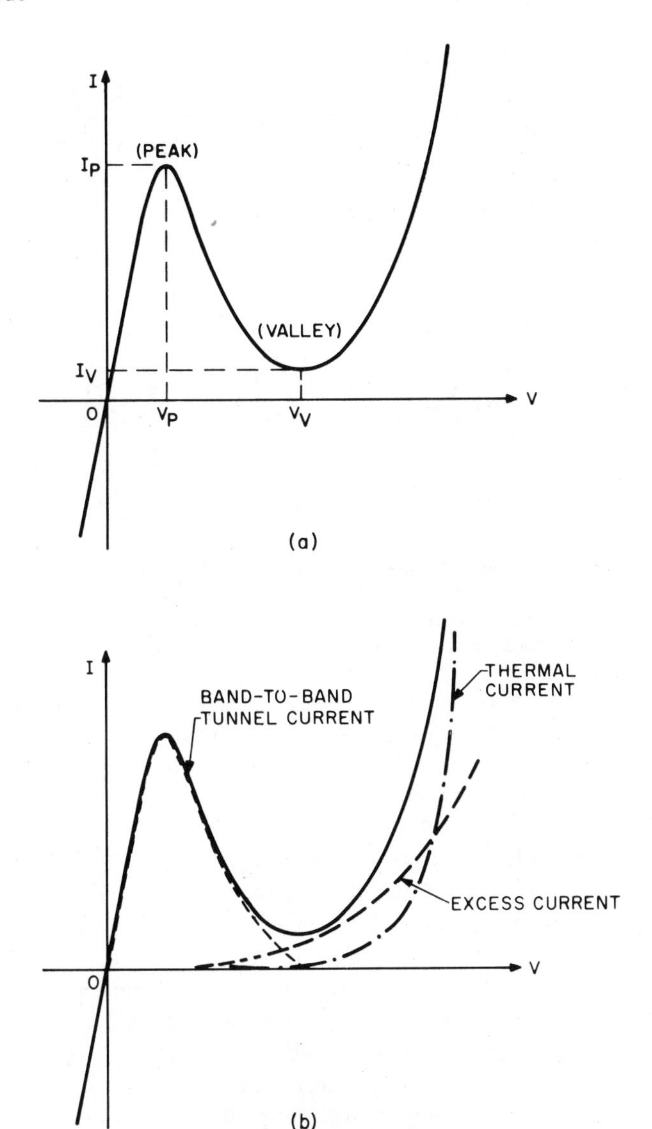

Fig. 2 (*a*) Static current–voltage characteristics of a typical tunnel diode. I_P and V_P are the peak current and peak voltage, respectively. I_V and V_V are the valley current and valley voltage, respectively. (*b*) The static characteristic is broken down into three current components.

either side of the junction. Hence tunneling currents cannot flow at zero applied voltage.

When a biasing voltage is applied, the electrons may tunnel from the valence band to the conduction band, or vice versa. The necessary conditions for tunneling are: (1) occupied energy states exist on the side from which the electron tunnels; (2) unoccupied energy states exist at the same

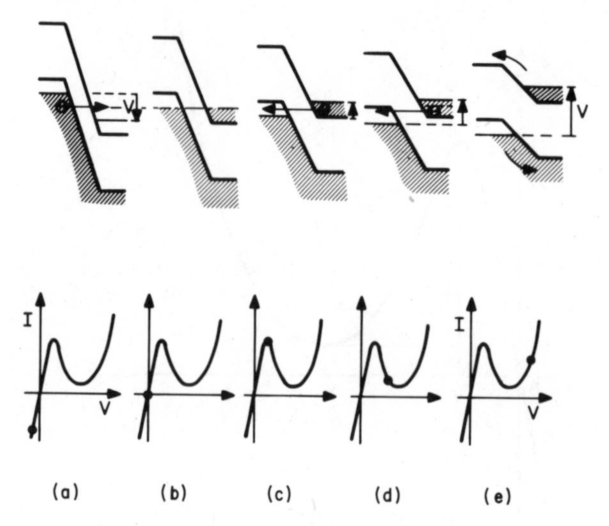

Fig. 3 Simplified energy-band diagrams of tunnel diode at (a) reverse bias; (*b*) thermal equilibrium, zero bias; (*c*) forward bias such that peak current is obtained; (*d*) forward bias such that valley current is approached; and (e) forward bias with thermal current flowing. (After Hall, Ref. 3.)

energy levels as in (1) on the side to which the electron can tunnel; (3) the tunneling potential barrier height is low and the barrier width is small enough that there is a finite tunneling probability; and (4) the momentum is conserved in the tunneling process.

Figure 3*a* shows electron tunneling from the valence band into the conduction band when a reversed bias is applied. The corresponding current is also designated by the dot on the $I-V$ curve. When a forward bias is applied (Fig. 3*c*) a band of energies exists for which there are filled states on the n side corresponding to states which are available and unoccupied on the p side. The electrons can thus tunnel from the n side to the p side. When the forward voltage is further increased, there are fewer available unoccupied states on the p side (Fig. 3*d*). If forward voltage is applied such that the band is "uncrossed," that is, the edge of the conduction band is exactly opposite the top of the valence band, there are no available states opposite filled states. Thus at this point the tunneling current can no longer flow. With still further increases of the voltage the normal thermal current will flow (Fig. 3*e*), and will increase exponentially with the applied voltage. One thus expects that as the forward voltage increases, the tunneling current increases from zero to a maximum I_P and then decreases to zero when $V = V_n + V_p$, where V is the applied forward voltage, V_n the amount of degeneracy on the n side $[V_n \equiv (E_{Fn} - E_C)/q]$, and V_p is the amount of degeneracy on the p side $[V_p \equiv (E_V - E_{Fp})/q]$, as shown in Fig. 1. The decreasing portion after the peak current gives rise to the negative resistance region.

The tunneling process can be either direct or indirect. Figure 4*a* shows

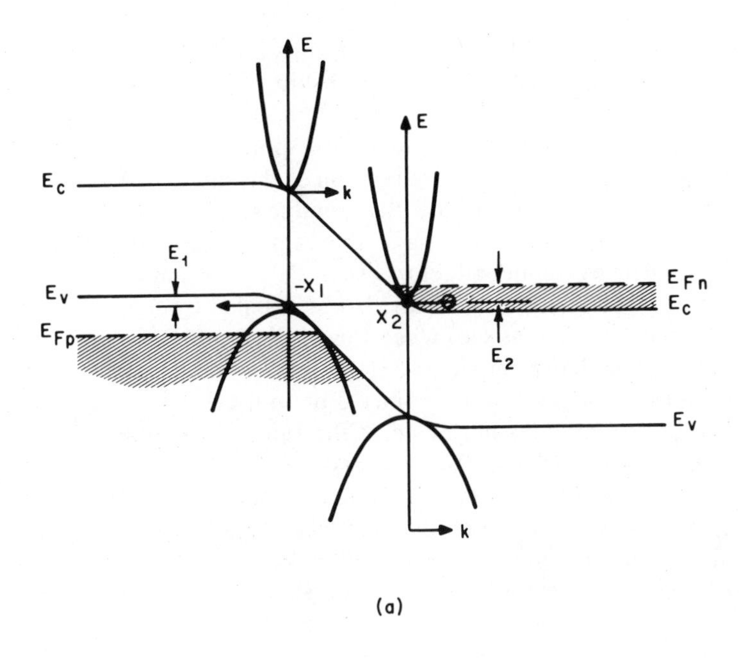

(a)

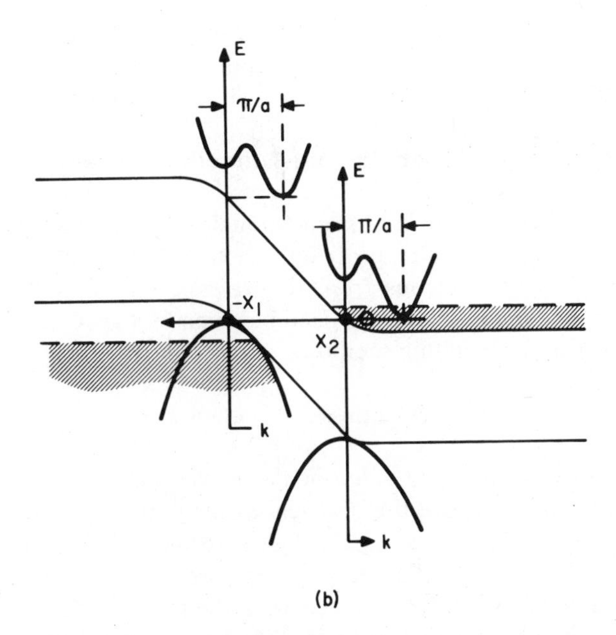

(b)

Fig. 4 (*a*) Direct tunneling process with E–k relationship ($k_{min} = k_{max}$) at the classical turning points ($-x_1$ and x_2) superimposed on the E–x relationship of the tunnel junction. (*b*) Indirect tunneling process with $k_{min} \neq k_{max}$.

direct tunneling where the E–k relationships at the classical turning points are superimposed on the E–x relationship of the tunnel junction. The electrons can tunnel from the vicinity of the minimum of the conduction-band energy-momentum surface to a corresponding value of momentum in the vicinity of the valence-band maximum of the energy-momentum surface. For direct tunneling to occur, the conduction-band minimum and the valence-band maximum must have the same momentum. This condition can be fulfilled by semiconductors, such as GaAs and GaSb, that have a direct bandgap. This condition can also be fulfilled by semiconductors with indirect bandgap (such as Ge) when the applied voltage is sufficiently large that the valence-band maximum (Γ point)[4] is in line with the indirect conduction-band minimum (Γ point). For indirect tunneling, the conduction-band minimum does not occur at the same momentum as the valence-band maximum, Fig. 4*b*. To conserve momentum, the difference in momentum between the conduction-band minimum and the valence-band maximum must be supplied by scattering agents such as phonons or impurities. For phonon-assisted tunneling, both the energy and momentum must be conserved; that is, the sum of the phonon energy and the initial electron energy tunneling from the n to the p side is equal to the final electron energy after it has tunneled to the p side, and the sum of the initial electron momentum and the phonon momentum ($\hbar k_p$) is equal to the final electron momentum after it has tunneled. In general, the probability for indirect tunneling is much lower than the probability for direct tunneling when direct tunneling is possible. Also, indirect tunneling involving several phonons has a much lower probability than that with only a single phonon.

9.2.1 Tunneling Probability and Tunneling Current

When the electric field in a semiconductor is sufficiently high (on the order of 10^6 V/cm), a finite probability exists for quantum tunneling or direct excitation of electrons from the valence band into the conduction band. The tunneling probability T_t can be given by the WKB approximation (Wentzel–Kramers–Brillouin method):[5]

$$T_t \simeq \exp\left[-2\int_{-x_1}^{x_2} |k(x)|\,dx\right] \tag{1}$$

where $|k(x)|$ is the absolute value of the wave vector of the carrier in the barrier, and $-x_1$ and x_2 are the classical turning points shown in Fig. 4.

The tunneling of an electron through a forbidden band is formally the same as a particle tunneling through a barrier. We shall consider two special potential barriers for an electron in the forbidden gap: the triangular and the parabolic barriers shown in Fig. 5*a* and *b*, respectively. For the triangular barrier the wave vector is given by

$$k(x) = \sqrt{\frac{2m^*}{\hbar^2}(PE - E)} = \sqrt{\frac{2m^*}{\hbar^2}\left(\frac{E_g}{2} - q\mathscr{E}x\right)} \tag{2}$$

where PE is the potential energy, E the incoming electron energy, E_g the bandgap of the semiconductor, and $\mathscr{E}$ the electric field. Substituting Eq. 2 into Eq. 1 yields

$$T_t \simeq \exp\left[-2\int_{-x_1}^{x_2}\sqrt{\frac{2m^*}{\hbar^2}\left(\frac{E_g}{2}-q\mathscr{E}x\right)}\,dx\right]$$
$$=\exp\left[+\frac{4}{3}\frac{\sqrt{2m^*}}{q\mathscr{E}\hbar}\left(\frac{E_g}{2}-q\mathscr{E}x\right)^{3/2}\right]\Bigg|_{-x_1}^{x_2}. \tag{3}$$

Since at

$$x = x_2, \qquad \left(\frac{E_g}{2}-q\mathscr{E}x\right)=0$$

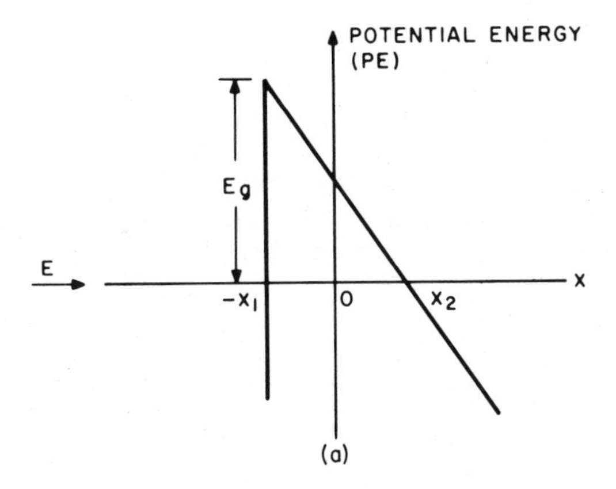

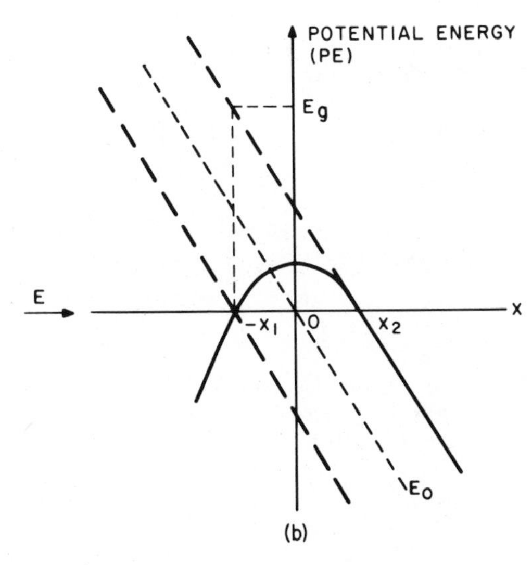

Fig. 5 (*a*) and (*b*).

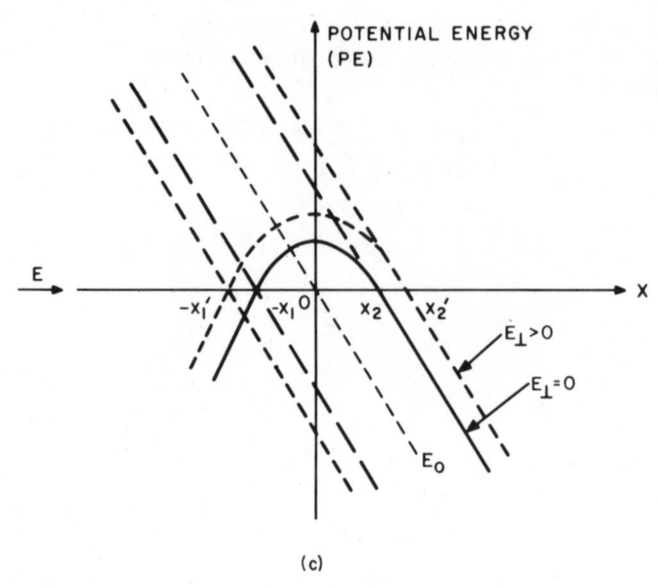

Fig. 5 (*a*) Triangular potential barrier. (*b*) Parabolic potential barrier. (*c*) Parabolic potential barrier with transverse energy components $E_\perp$. (After Moll, Ref. 10.)

and at

$$x = -x_1, \qquad \left(\frac{E_g}{2} - q\mathscr{E}x\right) = E_g$$

we have

$$T_t \simeq \exp\left(-\frac{4\sqrt{2m^*}E_g^{3/2}}{3q\hbar\mathscr{E}}\right). \tag{4}$$

For the parabolic energy barrier, E_0 is defined as the energy measured from the electron energy to the center of the band, and the form of $(PE - E)$ is

$$PE - E = \frac{(E_g/2)^2 - E_0^2}{E_g} = \frac{(E_g/2)^2 - (q\mathscr{E}x)^2}{E_g}. \tag{5}$$

This form is also the simplest algebraic function that has the correct behavior at the band edges.[6] The probability is then given by

$$\begin{aligned} T_t &\simeq \exp\left[-2\int_{-x_1}^{x_2}\sqrt{\frac{2m^*}{\hbar^2}\left(\frac{E_g^2/4 - q^2\mathscr{E}^2x^2}{E_g}\right)}\,dx\right] \\ &= \exp\left[-\frac{m^{*1/2}E_g^{3/2}}{2\sqrt{2}q\hbar\mathscr{E}}\int_{-1}^{1}(1-y^2)^{1/2}\,dy\right]\Bigg|_{y\equiv 2q\mathscr{E}x/E_g} \\ &= \exp\left(-\frac{\pi m^{*1/2}E_g^{3/2}}{2\sqrt{2}q\hbar\mathscr{E}}\right). \end{aligned} \tag{6}$$

Equation 6 is virtually identical to Eq. 4 except for the numerical constant.

Since the total momentum must be conserved in the tunneling process, the transverse momentum must be included in the calculation of the tunneling probability. Therefore, we divide the total energy into E_x and $E_\perp$, where $E_\perp$ is the energy associated with the momentum perpendicular to the direction of tunneling (or the transverse momentum) and E_x is the energy associated with momentum in the tunneling direction. Then for $E_\perp > 0$ (Fig. 5c)

$$PE - E_x = \frac{E_g^2/4 - E_0^2}{E_g} + E_\perp \tag{7}$$

and the classical turning points are at

$$-x_1', x_2' = \mp \frac{1}{q\mathscr{E}} \sqrt{E_g^2/4 + E_g E_\perp}. \tag{8}$$

We obtain from Eqs. 1, 7, and 8 the tunneling probability

$$T_t \simeq \exp\left(-\frac{\pi m^{*1/2} E_g^{3/2}}{2\sqrt{2} q\hbar\mathscr{E}}\right) \exp\left(-\frac{2E_\perp}{\bar{E}}\right) \tag{9}$$

where the first exponent is the same as in Eq. 6, corresponding to the probability of tunneling with zero transverse momentum, and $\bar{E}$ is given by

$$\bar{E} \equiv \frac{4\sqrt{2} q\hbar\mathscr{E}}{3\pi m^{*1/2} E_g^{1/2}} \tag{10}$$

which is a measure of the significant range of transverse momentum. Thus if $\bar{E}$ is very small, only the electron with small transverse momentum can tunnel. In other words, perpendicular energy further reduces the transmission by the factor $\exp(-2E_\perp/\bar{E})$. Transverse momentum must be conserved in direct tunneling. From the results above it is clear that to obtain large tunneling probability, both the effective mass and the bandgap should be small and the electric field should be large.

We next consider the tunneling current and shall present the first-order approach using the density of states in the conduction and valence bands.[1] We shall then discuss a more rigorous tunneling theory[6] in which the momentum is conserved.

At thermal equilibrium the tunneling current $I_{V\to C}$ from the valence band to the empty state of the conduction band and the current $I_{C\to V}$ from the conduction band to the empty state of the valence band should be detail-balanced. Expressions for $I_{C\to V}$ and $I_{V\to C}$ are formulated as follows:

$$I_{C\to V} = A \int_{E_C}^{E_V} F_C(E) n_C(E) T_t [1 - F_V(E)] n_V(E)\, dE \tag{11a}$$

$$I_{V\to C} = A \int_{E_C}^{E_V} F_V(E) n_V(E) T_t [1 - F_C(E)] n_C(E)\, dE \tag{11b}$$

where A is a constant, the tunneling probability T_t is assumed to be equal for both directions, $F_C(E)$ and $F_V(E)$ are the Fermi–Dirac distribution functions, and $n_C(E)$ and $n_V(E)$ are the density of states in the conduction band and valence band, respectively. When the junction is biased, the observed current I_t is given by

$$I_t = I_{C\to V} - I_{V\to C} = A\int_{E_C}^{E_V} [F_C(E) - F_V(E)]T_t n_C(E) n_V(E)\, dE. \tag{12}$$

A closed form of Eq. 12 is given by[7]

$$I_t = I_P(V/V_P)\exp(1 - V/V_P) \tag{13}$$

where I_P and V_P are the peak current and peak voltage as defined in Fig. 2. The peak voltage can be obtained by differentiating the density of states $n_C(E)$ and $n_V(E)$ with respect to E and by aligning the maximum number of electrons on the n side with the maximum number of vacancies on the p side. The result is found to be[7]

$$V_P \approx (V_n + V_p)/3 \tag{14}$$

The degeneracy on the n side, V_n, can be evaluated from the Fermi–Dirac integral and is given by[8]

$$V_n \approx \frac{kT}{q}\left[\ln\left(\frac{N_D}{N_C}\right) + 0.35\left(\frac{N_D}{N_C}\right)\right] \tag{15}$$

where N_D is the donor concentration and N_C the effective density of states in the conduction band. A similar expression is given for the degeneracy on the p side, V_p, where N_D and N_C are replaced by N_A and N_V, respectively, in Eq. 15.

Figure 6 shows the position of the peak voltage[9] as a function of the degeneracy V_n and V_p for Ge tunnel diodes. Note that the peak voltage shifts toward high values as the doping increases and the experimental values of V_p agree reasonably well with Eq. 14.

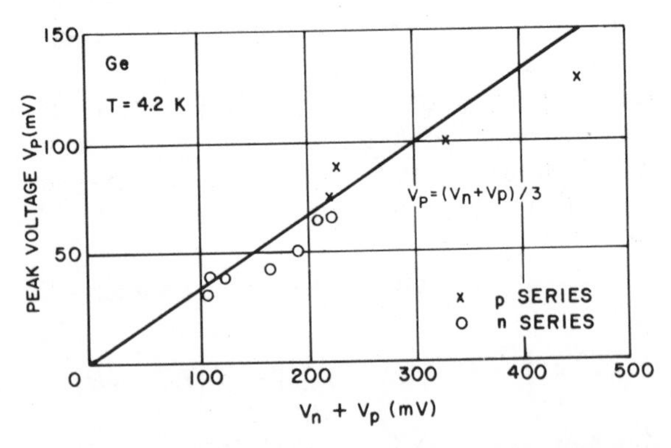

Fig. 6 Variation of peak voltage of Ge tunnel diodes as a function of the sum of V_n and V_p. (After Meyerhofer, Brown, and Sommers, Ref. 9; Demassa and Knott, Ref. 7.)

To obtain the peak current we shall first consider the direct tunneling, taking the conservation of momentum into account. Equation 9 gives the tunneling probability, where the transverse momentum is included. The incident current per unit area in the energy range $dE_x\,dE$ is given by[6, 10]

$$dJ_x = \frac{q(m_y^* m_z^*)^{1/2}}{2\pi^2\hbar^3}\, dE_x\, dE_\perp \tag{16}$$

where

$$E = E_x + E_\perp, \qquad E_x = \hbar^2 k_x^2/2m_x^*. \tag{17}$$

By taking m^* as isotropic and equal for the n and p sides, and by using Eqs. 9 and 17, we obtain J_t, the tunneling current per unit area,

$$J_t = \frac{qm^*}{2\pi^2\hbar^3}\exp\left(-\frac{\pi m^{*1/2}E_g^{3/2}}{2\sqrt{2}q\hbar\mathscr{E}}\right)\int [F_C(E) - F_V(E)]\exp(-2E_\perp/\bar{E})\, dE\, dE_\perp. \tag{18}$$

We have used Eq. 16 to give E and $E_\perp$ as the variables of integration. The limits of integration are determined by the condition $0 \le E_\perp \le E_1$, $0 \le E_\perp \le E_2$, where E_1 and E_2 are the electron energies measured from the n-band and p-band edges, respectively (Fig. 4). The limits on E are given by the band edges.

The integral over $E_\perp$ can be carried out immediately with the result

$$J_t = \frac{qm^*}{2\pi^2\hbar^2}\exp\left(-\frac{\pi m^{*1/2}E_g^{3/2}}{2\sqrt{2}\hbar q\mathscr{E}}\right)\left(\frac{\bar{E}}{2}\right)D \tag{19}$$

$$D \equiv \int [F_C(E) - F_V(E)][1 - \exp(-2E_S/\bar{E})]\, dE \tag{20}$$

$$\mathscr{E} \equiv (qV_{bi}N^*/2\epsilon_s)^{1/2} \tag{21}$$

where in Eq. 20 E_S is the smaller of E_1 and E_2, and $\bar{E}$ is given by Eq. 10. The field $\mathscr{E}$ is the average electric field of a step junction, qV_{bi} is the built-in potential energy approximately equal to the bandgap, and N^* is the effective doping concentration:

$$N^* \equiv N_A N_D/(N_A + N_D). \tag{22}$$

For a Ge tunnel diode, the appropriate effective mass in Eq. 19 is given by[11]

$$m^* = 2\left(\frac{1}{m_e^*} + \frac{1}{m_{lh}^*}\right)^{-1}$$

for tunneling from the light hole band to the ⟨000⟩ conduction band of germanium, where m_{lh}^* is the light-hole mass ($= 0.04\, m_o$) and m_e^* is the ⟨000⟩ conduction band mass ($= 0.036\, m_o$). For tunneling in the ⟨100⟩ direction to the ⟨111⟩ minima, the effective mass is given by

$$m^* = 2\left[\left(\frac{1}{3m_l^*} + \frac{2}{3m_t^*}\right) + \frac{1}{m_{lh}^*}\right]^{-1}$$

where $m_l^* = 1.6\,m_o$ and $m_t^* = 0.082\,m_o$ are the longitudinal and transverse masses of the $\langle 111 \rangle$ minima. The exponents in Eq. 19 differ, however, by only 5% in these two cases. The quantity D, Eq. 20, is an overlap integral which determines the shape of the I–V characteristic. It has the dimensions of energy (in units of eV) and depends on the temperature and the degeneracy qV_n and qV_p. At $T = 0\,\text{K}$, both F_C and F_V are step functions. Figure 7 shows the quantity D versus the forward voltage (dashed lines) for the cases $V_n = V_p$ and $V_n = 3V_p$. The maximum value of D occurs at $V = (V_n + V_p)/3$, as given in Eq. 14.

Figure 7 also shows the corresponding results of indirect tunneling (solid lines). For phonon-assisted indirect tunneling the tunneling probability is given by[6, 12]

$$T_t \simeq \exp\left[\frac{-4\sqrt{2}m_{rx}^{*\,1/2}(E_g - E_p)^{3/2}}{3q\hbar\mathscr{E}}\right] \tag{23}$$

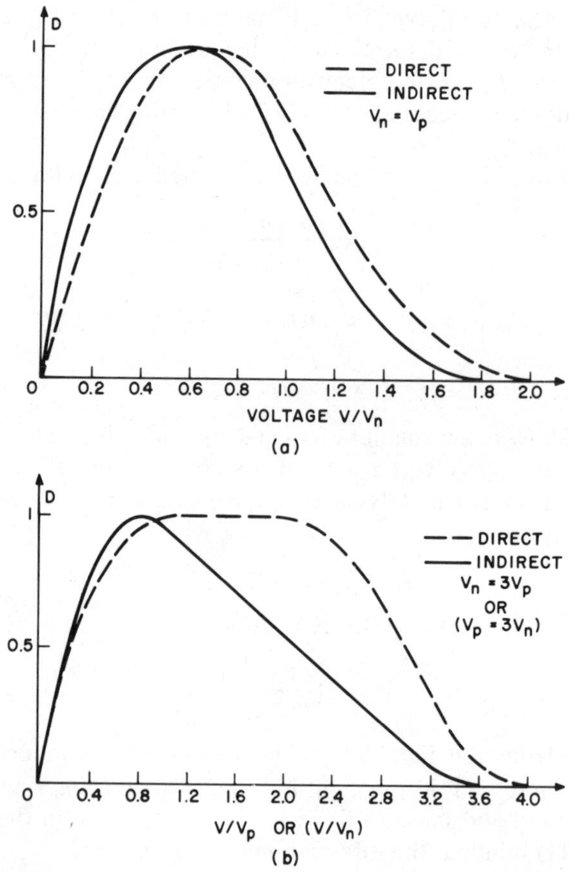

Fig. 7 Effective density of state D versus forward voltage for direct (dashed lines) and indirect tunneling (solid lines) with $\bar{E}$ very large. (a) $V_n = V_p$. (b) $V_n = 3V_p$ (or $V_p = 3V_n$). (After Kane, Ref. 6.)

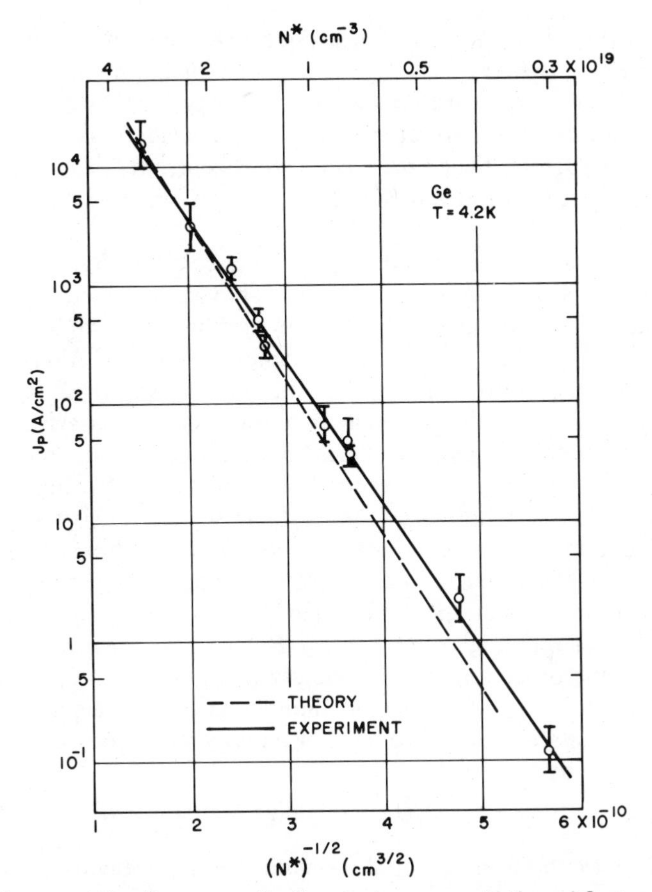

Fig. 8 Peak current density versus effective doping concentration of Ge tunnel diodes. The dashed line is calculated from Eq. 19. (After Demassa and Knott, Ref. 7; Meyerhofer, Brown, and Sommers, Ref. 9.)

where m^*_{rx} is the reduced effective mass in the tunneling direction, and E_p is the phonon energy. The expression for the tunneling current is very similar to Eq. 19, where the maximum in the forward characteristic occurs at

$$V = V_n + V_p - (V_n^2 + V_p^2)^{1/2}.$$

Figure 8 plots the peak current calculated from Eq. 19 for several Ge tunnel diodes together with experimental values.[7] The agreement is very good.

9.2.2 Current–Voltage Characteristics

As shown in Fig. 2*b*, the static *I–V* characteristic is the result of three current components: the tunneling current, the excess current, and the thermal current. For an ideal tunnel diode, the tunneling current decreases to zero at biases where $V \geq (V_n + V_p)$; for larger biases only normal diode currents caused by forward injection of minority carriers flow. In practice, however, the actual current at such biases is considerably in excess of the normal diode current, hence the term "excess current." The excess current is mainly due to carrier tunneling by way of energy states within the forbidden gap.

The excess current is derived[13] with the help of Fig. 9, where some examples of possible tunneling routes are shown. An electron starting at C in the conduction band might tunnel to an appropriate local level at A, from which it could then drop down to the valence band D. Alternatively, the electron could drop down from C to an empty level at B, from which it could tunnel to D. A third variant is a route such as $CABD$, where the electron dissipates its excess energy in a process that could be called impurity band conduction between A and B. A fourth route that should also be included is a staircase from C to D which consists of a series of tunneling transitions between local levels together with a series of vertical steps in which the electron loses energy by transferring from one level to another, a process made possible when the concentration of intermediate levels is sufficiently high. The route CBD can be regarded as the basic mechanism, the other routes being simply more complicated modifications. Let the junction be at a bias V, and consider an electron making a tunneling transition from B to D. The energy E_x through which it must tunnel is given by

$$E_x \approx E_g - qV + q(V_n + V_p) \approx q(V_{bi} - V) \tag{24}$$

where V_{bi} is the built-in potential (assuming that the electron ends up near

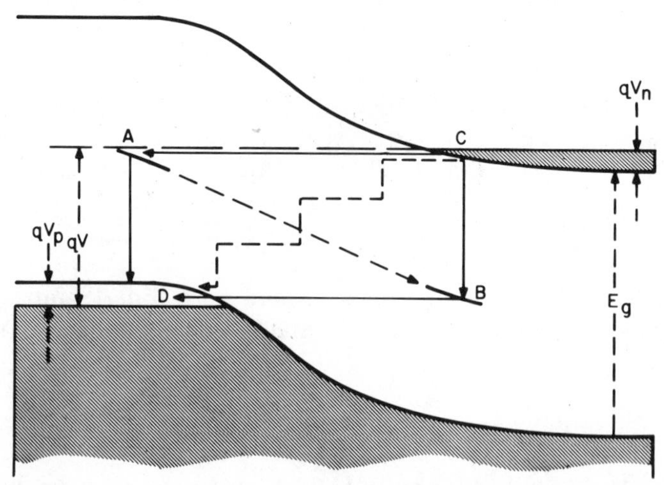

Fig. 9 Band diagram illustrating mechanisms of tunneling via states in the forbidden gap for the excess current flow. (After Chynoweth, Feldmann, and Logan, Ref. 13.)

the top of the valence band). The tunneling probability T_t for the electron on the level at B can be given by

$$T_t \simeq \exp\left(\frac{-4\sqrt{2}m_x^{*1/2}E_x^{3/2}}{3q\hbar\mathscr{E}}\right) = \exp(-\alpha_x E_x^{3/2}/\mathscr{E}). \tag{25}$$

This expression is the same as Eq. 4 except that E_g is replaced by E_x. The maximum electric field for a step junction is given by

$$\mathscr{E} = 2(V_{bi} - V)/W \tag{26}$$

where W is the depletion layer width given by

$$W = \left[\frac{2\epsilon_s}{q}\left(\frac{N_A + N_D}{N_A N_D}\right)(V_{bi} - V)\right]^{1/2}. \tag{27}$$

Let the volume density of the occupied levels at energy E_x above the top of the valence band be D_x. Then the excess current density will be given by

$$J_x \simeq A_1 D_x T_t \tag{28}$$

where A_1 is a constant and it is assumed that the excess current will vary predominantly with the parameters in the exponent of T_t rather than with those in the factor D_x. Substituting Eqs. 24 through 27 into 28 yields the expression for the excess current:[13]

$$J_x \simeq A_1 D_x \exp\{-\alpha_x'[E_g - qV + 0.6q(V_n + V_p)]\} \tag{29}$$

where α_x' is a constant. Equation 29 predicts that the excess current will increase with the volume density of bandgap levels (through D_x), and also increase exponentially with the applied voltage V (provided that $E_g \gg qV$). Equation 29 can also be written as[14]

$$J_x = J_V \exp\left[\frac{4}{3}\left(\frac{m_x^* \epsilon_s}{N^*}\right)^{1/2}(V - V_V)\right] \tag{30}$$

$$= J_V \exp[A_2(V - V_V)] \tag{30a}$$

where J_V is the valley current density at the valley voltage V_V and A_2 is the prefactor in the exponent. Experimental results of $\ln J_x$ versus V for Si and GaAs tunnel diodes exhibit linear relationships in good agreement with Eq. 30.

The thermal current is the familiar minority-carrier injection current in p-n junctions:

$$J_{th} = J_0(e^{qV/kT} - 1) \tag{31}$$

where J_0 is the saturation current density given in Chapter 2.

The complete static current–voltage characteristic is the sum of the three current components:

$$J = J_t + J_x + J_{th}$$

$$= J_P\left(\frac{V}{V_P}\right)\exp\left(1 - \frac{V}{V_P}\right) + J_V \exp[A_2(V - V_V)] + J_0 \exp\left(\frac{qV}{kT}\right). \tag{32}$$

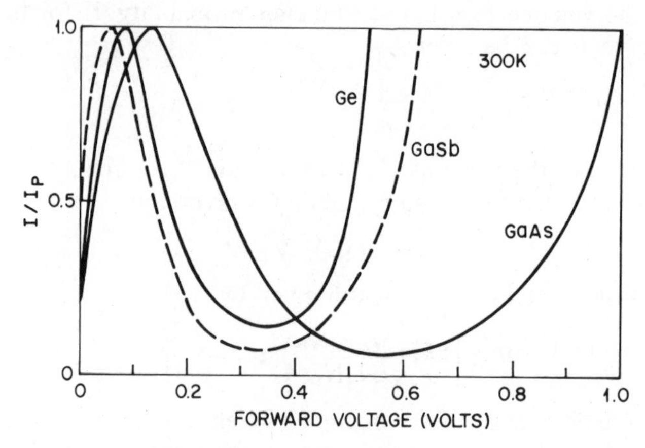

Fig. 10 Typical current–voltage characteristics of Ge, GaSb, and GaAs tunnel diodes at 300 K.

The tunneling current's contribution to the total current is significant for $V < V_V$, the excess current's contribution is significant for $V \approx V_V$, and the contribution of the thermal current is significant for $V > V_V$.

Figure 10 shows a comparison of the typical current–voltage characteristic of Ge, GaSb,[15] and GaAs[16,17] tunnel diodes at 300 K. The current ratios of I_P/I_V are 8:1 for Ge and 12:1 for GaSb and GaAs. Tunnel diodes have been made in many other semiconductors, such as Si[18] with a current ratio of about 4:1, InAs[19] with a ratio of 2:1 at room temperature and 10:1 at 4 K (because of its small bandgap), InP[20] with a ratio of 5:1 and some ternary tunnel diodes such as the $Ga_{0.7}Al_{0.3}As$ tunnel diode[21] with a ratio of 12:1. In general, the ratio for a given semiconductor can be increased by increasing the doping concentrations on both n and p sides. The ultimate limitation on the ratio depends on (1) the peak current, which depends on the effective tunneling mass and the bandgap; and (2) the valley current, which depends on the distribution and concentration of energy levels in the forbidden gap.

We shall briefly consider the I–V characteristics resulting from the effects of temperature, electron bombardment, and pressure. The temperature variation of the peak current can be explained by the change of D and E_g in Eq. 19. At high concentrations the temperature effect on D is small, and the negative value of dE_g/dT is primarily responsible for the tunneling probability change. As a result, the peak current increases with temperature. In the more lightly doped tunnel diodes, the decrease of D with temperature dominates, and the temperature coefficient is negative. For typical Ge tunnel diodes, the variation of the peak current over a temperature range[22] of −50 to 100°C is about ±10%. The valley current generally increases with increasing temperature, because the bandgap reduces with temperature, Eq. 29.

Under electron bombardment,[23] the major effect is the increase in excess current caused by increased volume density of the energy levels in the bandgap. The increased excess current can be gradually annealed out. Similar results can be observed for other radiations, such as γ rays. The stress on the $I-V$ characteristics causes the excess current in Ge and Si tunnel diodes[24] to increase. The changes are found to be reversible. This effect arises from deep-lying states associated with the strain-induced defects in the depletion region. For GaSb, however, both I_P and I_V decrease with increasing hydrostatic pressure,[25] which can be explained by an increase in the bandgap and a reduction in the degeneracy of V_n and V_p with increasing pressure.

9.2.3 Device Performance

Most tunnel diodes are made using one of the following techniques. (1) Ball alloy: A small metal alloy pellet containing the counter dopant of high-solid solubility is alloyed to the surface of a semiconductor substrate with high doping in a precisely controlled temperature–time cycle under inert or hydrogen gas (e.g., the arsenic in an arsenic-doped tin ball forms the n^+-type region on the surface of a p^+ Ge substrate). The desired peak current level I_P is obtained by an etching process. (2) Pulse bond: The contact and the junction are made simultaneously when the junction is pulse-formed between the semiconductor substrate and the metal alloy containing the counter dopant. (3) Planar processes:[26] Planar tunnel diode fabrication uses planar technology, including solution growth, diffusion, and controlled alloy. Figure 11 shows a microwave package[27] with a mounted Ge tunnel diode. The top contact to the tin ball is a soldered wire mesh. The diameter of junction is quite small, about 2.5 μm for a 6-GHz device and even less for higher-frequency diodes.

Figure 12*a* shows the symbol[28] of a tunnel diode. Figure 12*b* shows the basic equivalent circuit, which consists of four elements: the series inductance L_S, the series resistance R_S, the diode capacitance C, and the negative diode resistance $-R$.

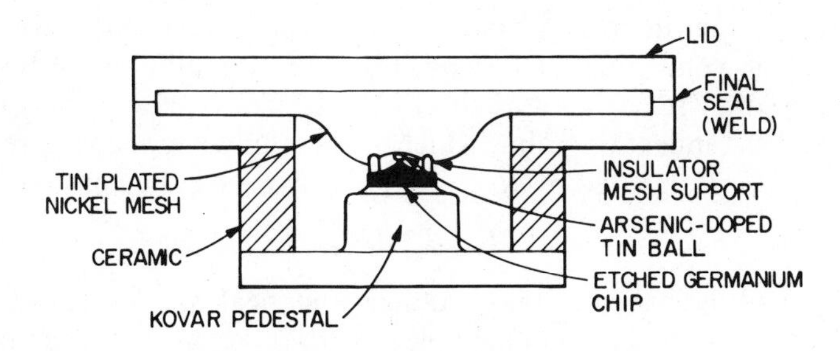

Fig. 11 A microwave package with a Ge tunnel diode mounted in it. (After Hindin, Ref. 27.)

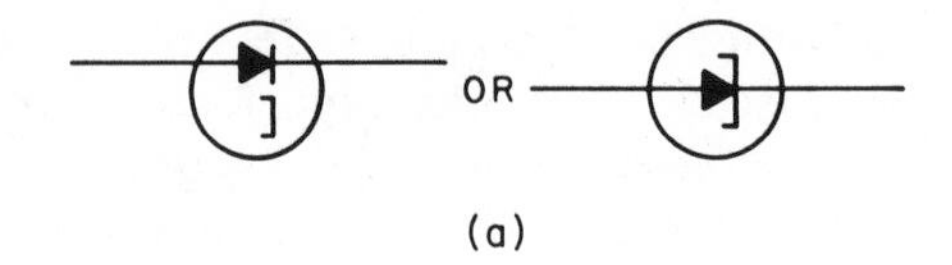

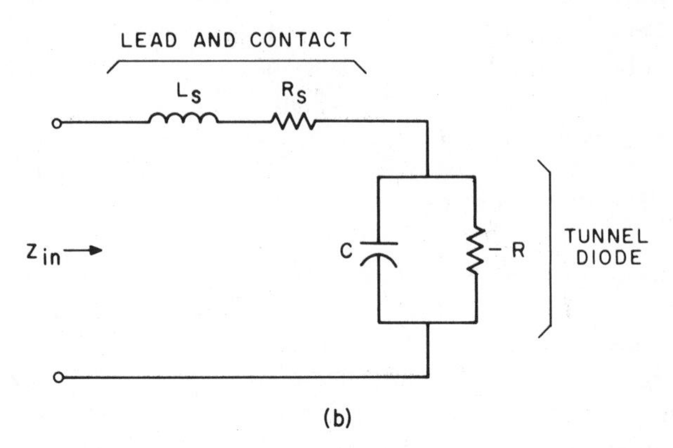

Fig. 12 (*a*) Symbol of tunnel diode. (*b*) Equivalent circuit of tunnel diode. (After Ref. 28.)

The series resistance R_S includes the lead resistance, the ohmic contacts, and the spreading resistance in the wafer, which is given by $\rho/2d$, where ρ is the resistivity of the semiconductor and d is the diameter of the diode area. The series inductance L_S in a coaxial cavity is given by[29]

$$L_S = \frac{2.303\mu_0 l}{2\pi} \ln\left(\frac{r_2}{r_1}\right) \tag{33}$$

where μ_0 is the permeability of the medium, l is the length, and r_1 and r_2 are the inner and outer radii of the coaxial line, respectively. We shall see that these parasitic elements establish important limits on the performance of the tunnel diode.

To consider the diode capacitance and negative resistance, we refer to typical current–voltage characteristic, Fig. 13*a*. The differential resistance, defined as $(dI/dV)^{-1}$, is plotted in Fig. 13*b*. The value of the negative resistance at the inflection point, which is the minimum negative resistance in the region, is designated by $R_{\min}$. This resistance can be approximated by

$$R_{\min} \approx 2V_P/I_P \tag{34}$$

where V_P and I_P are the peak voltage and peak current, respectively. Figure 13*c* shows the conductance plot (dI/dV) versus V. At the peak and valley voltages the conductance becomes zero; the diode capacitance is usually measured at the valley voltage, and is designated by C_j.

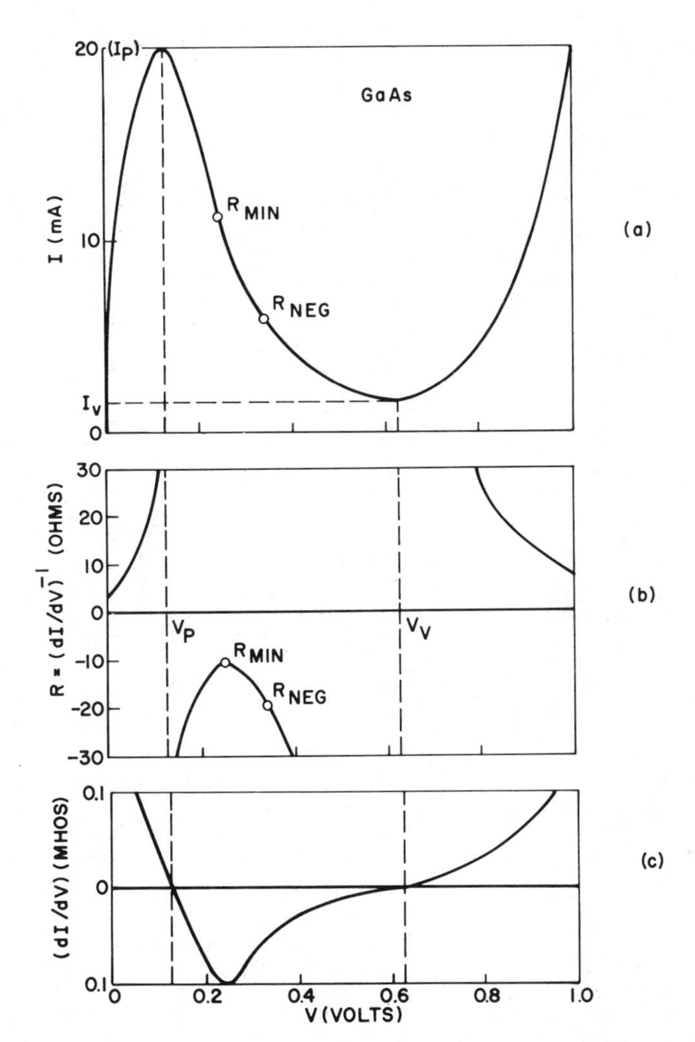

Fig. 13 (*a*) Current–voltage characteristics of a GaAs tunnel diode at 300 K. (*b*) Differential resistance $(dI/dV)^{-1}$ versus voltage, where R_{min} is the minimum resistance and R_{neg} is the resistance corresponding to the minimum noise figure. (*c*) Differential conductance $G \equiv dI/dV$ versus voltage. At peak and valley currents $G = 0$.

The input impedance Z_{in} of the equivalent circuit of Fig. 12 is given by

$$Z_{in} = \left[R_S + \frac{-R}{1+(\omega RC)^2} \right] + j\left[\omega L_S + \frac{-\omega CR^2}{1+(\omega RC)^2} \right]. \tag{35}$$

From Eq. 35 we see that the resistive (real) part of the impedance will be zero at a certain frequency, and the reactive (imaginary) part of the impedance will also be at a second frequency. We denote these frequencies by the resistive cutoff frequency f_r and the reactive cutoff frequency f_x, respectively; these frequencies are given by

$$f_r \equiv \frac{1}{2\pi RC}\sqrt{\frac{R}{R_S}-1} \tag{36}$$

$$f_x \equiv \frac{1}{2\pi}\sqrt{\frac{1}{L_S C}-\frac{1}{(RC)^2}}. \tag{37}$$

For cutoff frequencies specified at the minimum resistance and valley capacitance, we have

$$f_{r0} \equiv \frac{1}{2R_{\min}C_j}\sqrt{\frac{R_{\min}}{R_S}-1} \geq f_r \tag{38}$$

$$f_{x0} \equiv \frac{1}{2\pi}\sqrt{\frac{1}{L_S C_j}-\frac{1}{(R_{\min}C_j)^2}} \leq f_x \tag{39}$$

where f_{r0} is the maximum resistive cutoff frequency, at which the diode no longer exhibits negative resistance; and f_{x0} is the minimum reactive cutoff frequency or the self-resonant frequency, at which the diode reactance is zero and at which the diode oscillates if $f_{r0} > f_{x0}$. In most applications where the diode is operated into the negative resistance region, it is desirable to have $f_{x0} > f_{r0}$ and $f_{r0} \gg f_0$, the operating frequency. Equations 38 and 39 show that to fulfill the requirement that $f_{x0} > f_{r0}$, the series inductance L_S must be lowered. A figure of merit for tunnel diodes is the speed index, which is defined as the ratio of the peak current to the capacitance at the valley voltage, I_P/C_j. The switching speed of a tunnel diode is determined by the current available for charging the junction

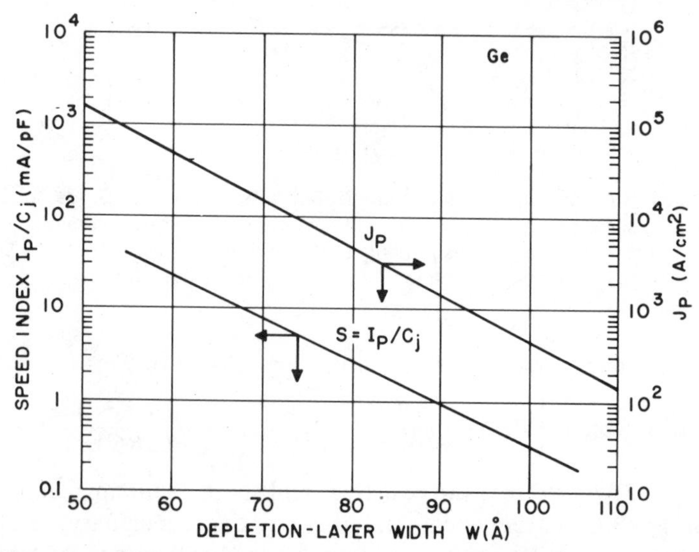

Fig. 14 Average value of peak current density and the speed index ($\equiv I_P/C_j$) versus depletion-layer width of Ge tunnel diodes at 300 K. (After Davis and Gibbons, Ref. 26.)

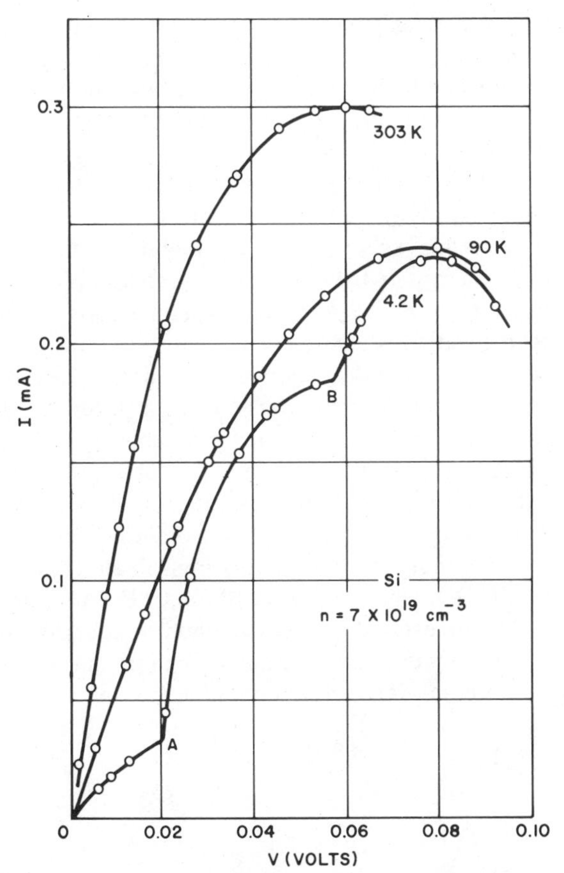

Fig. 15 Current–voltage characteristic of a Si tunnel diode at three temperatures. At 4.2 K the bending points *A* and *B* correspond to phonon-assisted tunneling processes. (After Esaki and Miyahara, Ref. 30.)

capacitance and therefore depends on the amount of current available from the power supply and the average *RC* product. Since *R*, the negative resistance, is inversely proportional to the peak current, a large speed index (or small *RC* product) is required for fast switching. Figure 14 shows the peak current and the speed index[26] versus depletion-layer width of Ge tunnel diodes at 300 K. We see that a narrow width or larger effective doping is needed to obtain a large speed index.

Another important quantity associated with the equivalent circuit is the noise figure, which is defined as

$$NF \equiv 1 + \frac{q}{2kT}|RI|_{\min} \tag{40}$$

where $|RI|_{\min}$ is the minimum value of the negative resistance-current product on the current–voltage characteristic. Figure 13 shows the corresponding value of R (designated by R_{neg}). The product $q|RI|_{min}/2kT$ is called the noise constant K and is a material constant. Typical values of K at room temperature are 1.2 for Ge, 2.4 for GaAs, and 0.9 for GaSb. The noise figure for Ge tunnel diodes is about 5 dB at 6 GHz and 6 dB at 14 GHz.

In addition to its microwave and digital applications, the tunnel diode is a useful device for the study of fundamental physical parameters. The diode can be used in tunneling spectroscopy, a technique that uses tunneling electrons of known energy distribution as a spectroscopic probe instead of photons of known frequency in optical spectroscopy. Tunneling spectroscopy has been used to study electron energy states in solids and to observe the excitation of modes such as in junction devices. Figure 15 shows an example of the $I-V$ characteristics of a Si tunnel diode.[30] As the temperature is reduced to 4.2 K, the curve reveals two bending points A and B. These bending points correspond to phonon-assisted tunneling processes, and the energies (or voltages) at A and B correspond to the acoustic and optical phonons, respectively. Similar observations are made in group III–V semiconductor junctions. Figure 16 shows the plots of the conductance (dI/dV) versus V at 4.2 K for GaP, InAs, and InSb.[31] The arrows indicate the corresponding optical phonon energies in these semiconductors. The phonon-assisted indirect tunneling can also be studied in more detail by the second-derivative technique.[32]

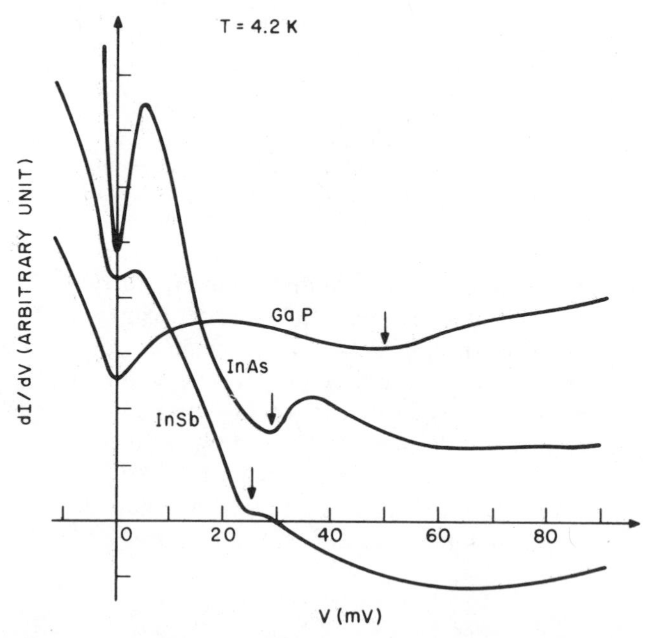

Fig. 16 Conductance (dI/dV) in arbitrary units versus voltage for tunnel diodes made from group III–V compounds. The arrows indicate positions of optical phonon energies. (After Hall, Racette, and Ehrenreich, Ref. 31.)

9.3 BACKWARD DIODE

In connection with the tunnel diode, when the doping concentrations on the p and n sides of a p-n junction are nearly or not quite degenerate, the current in the "reverse" direction for small bias, as shown in Fig. 17, is larger than the current in the "forward" direction—hence the name "backward diode." At thermal equilibrium, the Fermi level in the backward diode is very close to the band edges. When a small reverse bias (p side negative with respect to n side) is applied, the energy-band diagram is similar to Fig. 3a except that there is no degeneracy on both sides. Under reverse bias, electrons can readily tunnel from the valence band into the

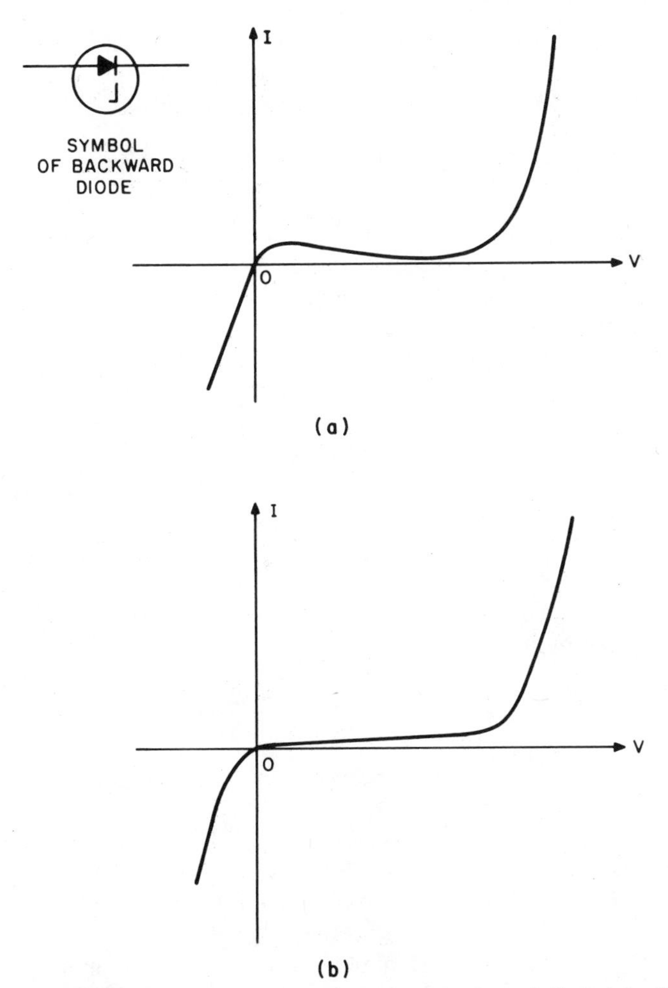

Fig. 17 Symbol and current–voltage characteristic of backward diode (a) with negative resistance and (b) without negative resistance.

conduction band and give rise to a tunneling current, given by Eq. 19, which can be written in the form

$$J \simeq B_1 \exp(+|V|/B_2) \tag{41}$$

where B_1 and B_2 are positive quantities and are slowly varying functions of the applied voltage V. Equation 41 indicates that the reverse current increases approximately exponentially with the voltage.

The backward diode can be used for rectification of small signals, microwave detection, and mixing. Similar to the tunnel diode, the backward diode has a good frequency response because there is no minority-carrier storage effect. In addition, the current–voltage characteristic is insensitive to temperature and to radiation effect, and the backward diode has very low $1/f$ noise.[33–36]

For nonlinear applications such as high-speed switching, a device figure of merit is γ, the ratio of the second derivative to the first derivative of the current–voltage characteristic. It is also referred to as the curvature coefficient:[37]

$$\gamma \equiv \frac{d^2I/dV^2}{dI/dV}. \tag{42}$$

The value of γ is a measure of the degree of nonlinearity normalized to the operating admittance level. For a forward-biased p-n junction or a Schottky barrier (refer to Chapter 5) the value of γ is simply given by q/nkT. Thus γ varies inversely with T. At room temperature, γ for an ideal p-n junction ($n = 1$) is about 40 V^{-1} independent of bias. For a reverse-biased p-n junction, however, the value of γ is very small at low voltages and increases linearly with the avalanche multiplication factor near breakdown voltage.[38] Although the ideal reverse breakdown characteristic would give a value of γ greater than 40 V^{-1}, because of the statistical distribution of impurities and the effect of space-charge resistance, much lower values of γ are expected.

For a backward diode the value of γ can be obtained from Eqs. 4, 13, 14 and 26; and is given by[39]

$$\gamma(\text{for } V = 0) = \frac{4}{V_n + V_p} + \frac{2}{\hbar}\sqrt{\frac{2\epsilon_s m^*}{N^*}} \tag{43}$$

where m^* is the average effective mass of the carriers

$$m^* \simeq m_e^* m_h^* / (m_e^* + m_h^*)$$

and N^* is the effective doping concentration given by Eq. 22. Clearly, the curvature coefficient γ depends upon the impurity concentrations on both sides of the junction and the effective masses. In contrast to Schottky barriers, the value of γ is relatively insensitive to temperature variation because the parameters in Eq. 43 are slowly varying functions of temperature.

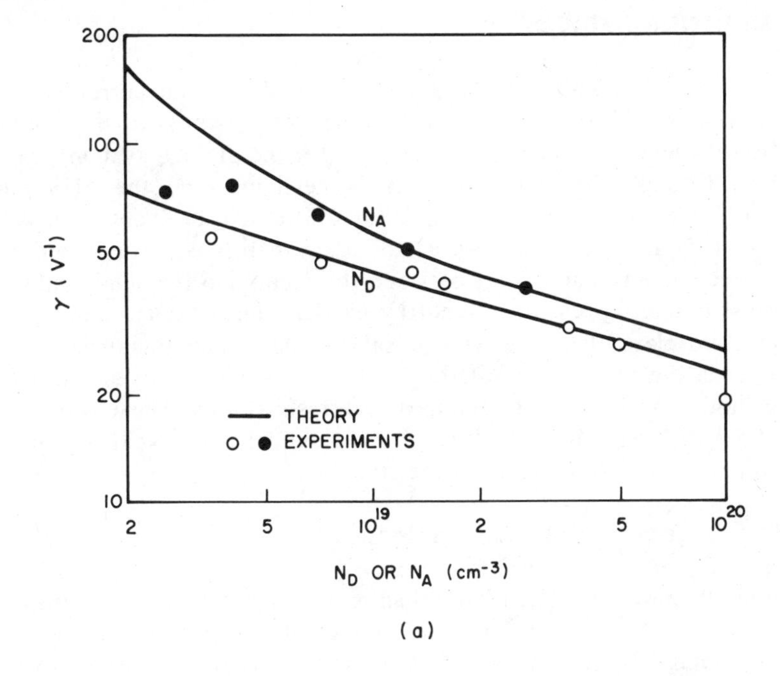

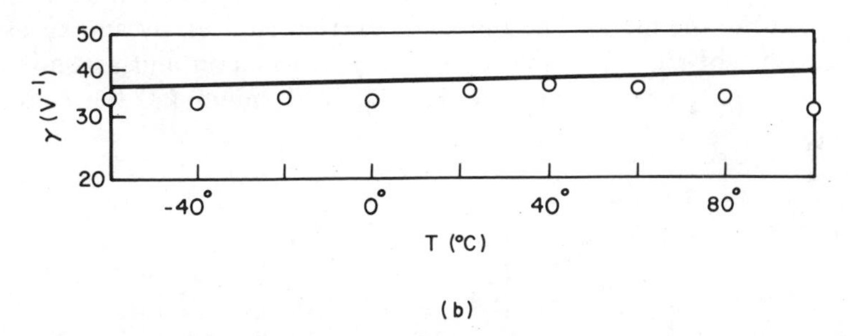

Fig. 18 (a) Curvature coefficient at 300 K for $V \simeq 0$ versus acceptor concentration in Ge (for a fixed $N_D = 2 \times 10^{19}$ cm^{-3}) or donor concentration (for a fixed $N_A = 10^{19}$ cm^{-3}). Solid lines are from computed results. Data points are from experimental measures. (*b*) Curvature coefficient versus temperature. The solid line represents computed results and the data points are measured. (After Karlovsky, Ref. 39.)

Figure 18 shows a comparison between theoretical and experimental values of γ for Ge backward diodes. The solid lines are computed from Eq. 43 using $m_e^* = 0.22m_o$ and $m_h^* = 0.39m_o$. The agreement is generally good over the doping range considered. Also note that there are two interesting features of γ for backward diodes: (1) γ can exceed 40 V^{-1}, and (2) it is insensitive to temperature variation.

9.4 MIS TUNNEL DIODE

For a metal–insulator–semiconductor (MIS) diode, the current–voltage characteristics critically depend on the insulator thickness. If the insulator layer is sufficiently thick (greater than 50 Å for the Si–SiO_2 system), carrier transport through the insulator layer is negligible and the MIS diode represents a conventional MIS capacitor (discussed in Chapter 7). Alternatively, if the insulator layer is very thin (less than 10 Å), little impediment is provided to carrier transport between the metal and the semiconductor, and the structure represents a Schottky-barrier diode (discussed in Chapter 5). The third class of device with an intermediate layer thickness (10 Å $<$ $d < 50$ Å) is the MIS tunnel diode. In this section we consider first MIS tunnel diodes on degenerate semiconductor substrates. These diodes can exhibit negative resistance similar to that of a tunnel diode. Next we consider MIS tunnel diodes on nondegenerate substrates.

9.4.1 Degenerate Semiconductor

Figure 19 shows simplified band diagrams,[40] including interface traps for MIS tunnel diodes with p^{++} and n^{++} semiconductor substrates. The band bending, image forces, and potential drops across the oxide layer at equilibrium are omitted for simplicity. Consider the p^{++} type first. Applying a positive voltage to the metal, Fig. 19*b*, causes electron tunneling from the valence band to the metal. The tunneling current is given by an expression similar to that of Eq. 12. Using a WKB approximation and assuming the conservation of energy E and transverse momentum $k_\perp$, the tunneling

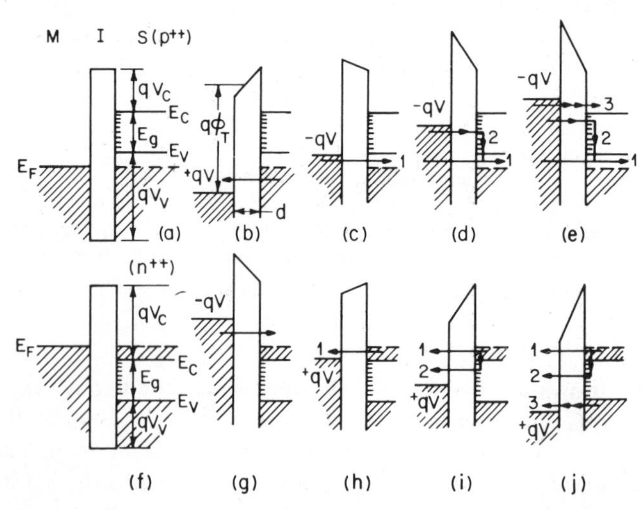

Fig. 19 Simplified band diagrams including interface traps of MIS tunnel diodes on degenerate substrates. (After Dahlke and Sze, Ref. 40.)

current density along the x direction between two conducting regions through a forbidden region can be written as[41]

$$J = \frac{q}{4\pi^2\hbar} \int\int T_t[F_1(E) - F_2(E)]dk_\perp^2\, dE \tag{44}$$

where F_1 and F_2 are the Fermi distribution functions in the two conducting regions, and T_t is the tunneling probability. For the MIS diode under consideration, the constant energy surface in k space for electrons in the semiconductor is, in general, considerably smaller than that in the metal. As a result, the tunneling of electrons from the semiconductor to the metal is always assumed to be allowable. If it is further assumed that the energy bands of the solids involved are parabolic with an isotropic electron mass m^*, Eq. 44 can be reduced to

$$J = \frac{m^* q}{2\pi^2\hbar^3} \int\int T_t\, dE_\perp\, dE \tag{45}$$

where $E_\perp$ and E are the transverse and total kinetic energy of electrons in the semiconductor. The limits of integration for $E_\perp$ are zero and E; the limits for E are simply the two Fermi levels. The tunneling probability for a rectangular barrier with an effective barrier height $q\phi_T$ and width d, Fig. 19b, can be obtained[42] from Eq. 1:

$$T_t \simeq \exp[-2(2m^*/\hbar^2)^{1/2}(q\phi_T)^{1/2}d] \tag{46}$$

$$\simeq \exp(-\alpha_T \phi_T^{1/2} d) \tag{46a}$$

where α_T approaches unity if the effective mass in the insulator equals the free electron mass, and if ϕ_T is in volts and d in Å.

This tunneling current, Fig. 19b, increases monotonically with the increasing energy range between the Fermi levels; it further increases with the decreasing insulator barrier height. Applying a small negative voltage to the metal, Fig. 19c, results in electron tunneling from the metal to the unoccupied semiconductor valence band. According to Fig. 19d, an increase of the voltage $-V$ implies an increase in the effective barrier height for electrons tunneling from the metal to the unoccupied states of the valence band, that is, a negative I–V characteristic (if $qV_C < qV_V$, as shown in Fig. 19a). However, electrons in the metal with higher energies can tunnel simultaneously into the empty interface traps and momentarily recombine with holes in the valence band, resulting in another current component. Since the insulator barrier decreases with bias, this current component has a positive I–V characteristic. Finally, an additional increase of the bias results in a third very fast-growing tunnel current component from the metal into the conduction band of the semiconductor (Fig. 19e).

Next consider the n^{++}-type semiconductors. As shown in Fig. 19f, the effective insulator barriers for the n^{++} type are expected to be smaller than those of the p^{++}-type samples; hence for a given bias, there will be larger tunnel currents. For a negative bias on the metal, electrons tunnel from the metal into the empty states of the semiconductor conduction band, resulting in a large, rapidly increasing current (Fig. 19*g*). A small positive voltage on the metal leads to increasing electron tunneling from the conduction band of the semiconductor into the metal (Fig. 19*h*). If the interface traps are filled with conduction electrons by recombination, a further increase in bias gives rise to a second current component caused by the tunneling of electrons from the interface traps into the metal. This current component increases with increasing bias since the effective insulator barrier decreases (Fig. 19*i*). For larger voltage additional tunneling from the valence band to the metal is possible, but its influence on the total $I-V$ characteristic is comparatively small because of the relatively high oxide barrier (Fig. 19*j*). Thus the band structure of the semiconductor has a much smaller influence on the tunneling characteristics of the n^{++}-type compared to p^{++}-type structures.

Figure 20 shows the measured $I-V$ characteristics at room temperature (solid lines) and liquid-nitrogen temperature (dashed lines) for three p^{++} silicon samples with oxide layers (20 Å) treated in different ways. Note that the small influence of temperature on the $I-V$ characteristics is typical for tunneling. The samples have an oxide grown in dry oxygen (Fig. 20, curve a), in steam (Fig. 20, curve b) and in steam with 30-min annealing in H_2 at 350°C (Fig. 20, curve c). The band structure of the semiconductor is most distinctly reflected by the $I-V$ characteristic for negative voltages. The current increases gradually with bias until $V \approx -1$ V, where the current starts to increase rapidly with bias. This voltage corresponds to the silicon bandgap at high doping (with some band-edge tailing effect).

The predicted negative resistance at small negative voltages is apparently masked by the tunneling of electrons from the metal into the interface traps, as discussed previously. The $I-V$ characteristics of curves (a) and (b), show in principle the same trend as curve (c) but exhibit considerably increased currents especially in the forbidden energy range (-1.1 V $< V < 0$). If these currents are assumed to be proportional to the interface-trap density, Fig. 20 leads to the conclusion that there is an increase of one or two orders of magnitude in the interface-trap density when changing from annealed to steam-grown and finally to dry-oxygen grown oxide layers. This is in qualitative agreement with an experimentally determined increase of interface traps for annealed-steam or oxygen-grown oxide layers of larger thickness ($d \approx 1000$ Å).[43] Figure 21, curve (c) shows that the effects of the semiconductor band structure and the density of interface traps on the dc tunnel characteristics are even more distinctly reflected by the conductance–voltage curves, which are obtained by differentiating the measured curves of Fig. 20. The left branches of the curves for $V < -1.1$ V represent electrons tunneling from the metal into the conduction band; the right branches, $V > 0$, represent tunneling from the valence band into the metal. The minimum

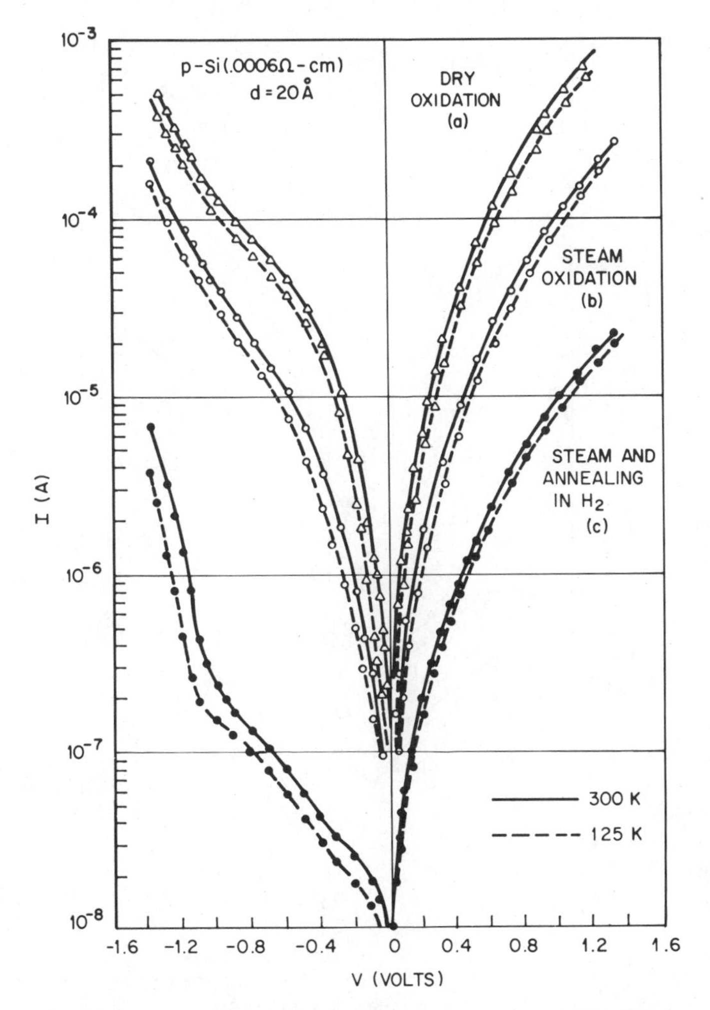

Fig. 20 Measured I–V characteristics at room temperature (solid lines) and liquid nitrogen temperature (dashed lines) for three p^{++} silicon samples with the oxide layer (20 Å) treated in different ways. (After Dahlke and Sze, Ref. 40.)

conductance at small negative voltages is a result of the superposition of two current components (Fig. 19*d*). The expected negative conductance of the first component is apparently compensated for by the larger positive conductance of the interface traps.

Figures 21 and 22 present the measured ac conductance and capacitance as functions of applied voltage and frequency, respectively. The capacitance curves of the unannealed sample are higher than the annealed samples, as expected, since the former has higher interface-trap densities. Much larger frequency dependence is observed on the ac conductance curves. For $f < 5$ kHz, the conductance curves are virtually identical to the

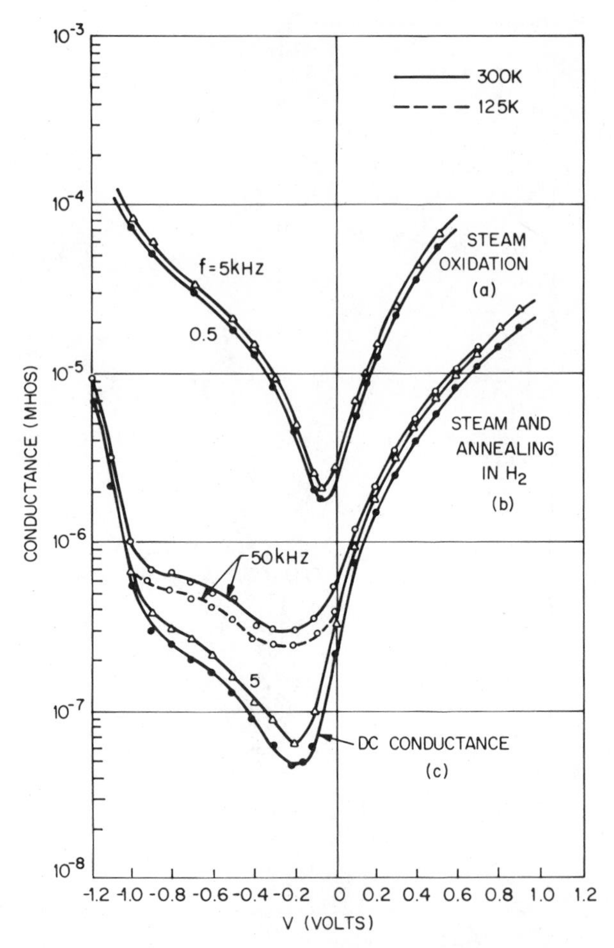

Fig. 21 Conductance–voltage curves measured at different frequencies. The dc conductance curve is obtained by differentiating the curves of Fig. 20. (After Dahlke and Sze, Ref. 40.)

dc conductance curve, The insert in Fig. 22 shows the basic equivalent circuit including the tunneling effect. The *RC* circuit to the left of the line *AA* is identical to the *RC* circuit for thicker insulator films. R_T is the equivalent resistance for the tunneling of electrons into the valence band and/or conductance band of the p^{++} semiconductor (corresponding to current components 1 and 3 in Fig. 19*c* and *e*). R_{TS} is the equivalent resistance for the tunneling of electrons into the interface traps and the recombining of them with holes in the valence band (corresponding to current component 2 in Fig. 19*d* or *e*). Both R_T and R_{TS} are functions of the applied voltage. The complete circuit can be simplified as a capacitor $C(\omega)$ in parallel with a conductor $G(\omega)$, both frequency-dependent.

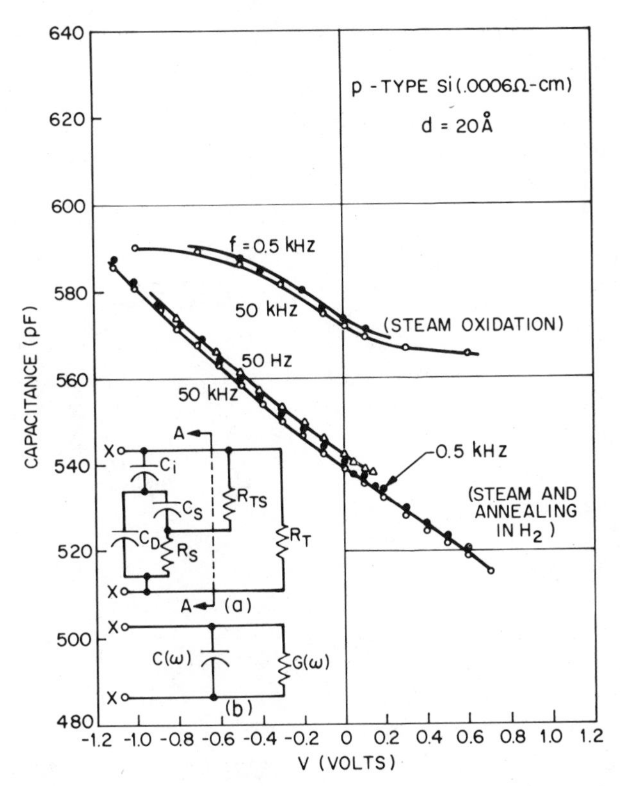

Fig. 22 Capacitance–voltage curves measured at different frequencies. The insert shows the equivalent circuits for an MIS tunnel diode. (After Dahlke and Sze, Ref. 40.)

We can show that $d[C(\omega)]/d\omega \leq 0$ and $d[G(\omega)]/d\omega \geq 0$. For thin oxide layers and/or highly doped semiconductor substrates, the capacitance increases from its high-frequency value[40]

$$C(\infty) = \frac{C_i C_D}{C_i + C_D} \tag{47a}$$

to

$$C(0) \simeq C(\infty) + AC_s \tag{47b}$$

as $\omega \to 0$, and the conductance increases from its low-frequency value

$$G(0) = \frac{1}{R_T} + \frac{1}{R_{TS} + R_s} \tag{48a}$$

to

$$G(\infty) = G(0) + \frac{A(R_{TS} + R_s)}{R_{TS} R_s} \tag{48b}$$

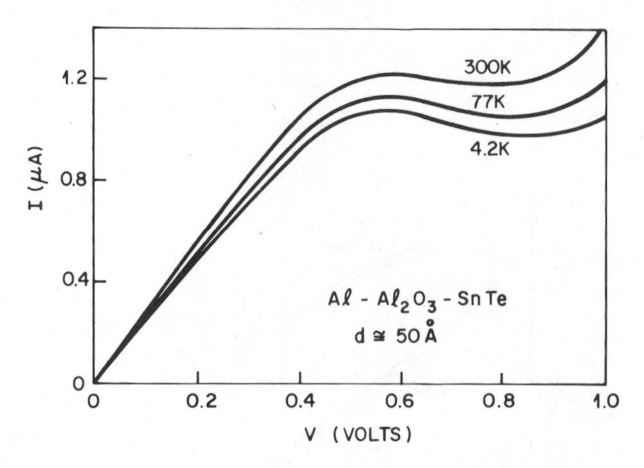

Fig. 23 MIS tunnel diode ($Al–Al_2O_3–SnTe$) *I–V* characteristics at three temperatures. (After Esaki and Stiles, Ref. 44.)

as $\omega \to \infty$, where

$$A \equiv \frac{(C_i R_{TS} - C_D R_s)^2}{(C_i + C_D)^2 (R_{TS} + R_s)^2}. \tag{49}$$

The experimental results are in good agreement with the discussion above (Figs. 21 and 22).

The negative resistance[44] has been obtained in MIS tunnel diodes of $Al–Al_2O_3–SnTe$. The SnTe is a highly doped *p* type with a concentration of 8×10^{20} cm^{-3}; the Al_2O_3 is about 50 Å thick. Figure 23 shows the measured current–voltage characteristics at three different temperatures, where the negative resistance occurs between 0.6 to 0.8 V. These results are in good agreement with theoretical prediction[41] based on Eq. 45.

9.4.2 Nondegenerate Semiconductor

For the MIS tunnel diode with a very thin insulating layer on a nondegenerate semiconductor substrate, no negative resistance is expected. The MIS tunnel diode on a nondegenerate semiconductor substrate has been used in memories as MNOS devices, in energy conversion as solar cells, and in microwave devices as the emitter–base junction of a tunnel transistor.

One of the most important parameters for this diode is the metal–insulator barrier height, which has a profound effect on the *I–V* characteristics.[45,46] Figure 24 shows schematic energy-band diagrams at thermal equilibrium for MIS tunnel diodes on *p*-type substrates with two metal-to-insulator barrier heights. For the low-barrier case ($\phi_{mi} = 3.2$ V for the $Al–SiO_2$ system) the surface of the *p*-type silicon is inverted. For the high barrier case ($\phi_{mi} = 4.2$ V for the $Au–SiO_2$ system) the surface is in

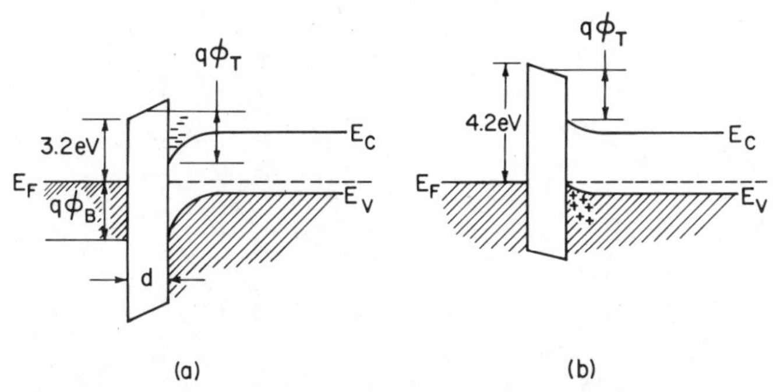

Fig. 24 Energy-band diagrams for MIS tunnel diodes on nondegenerate substrates. (*a*) Low metal–insulator barrier. (*b*) High metal–insulator barrier. (After Green, King, and Shewchun, Ref. 45.)

accumulation of holes. Two main tunnel current components exist: J_{ct}, from the conduction band to the metal, and J_{vt}, from the valence band to the metal. Both currents are given by expressions similar to Eq. 44. Figure 25 shows the theoretical I–V curves for the two diodes. For the low-barrier case, Fig. 25*a*, under small forward and reverse biases, the dominant current is the minority (electron) current J_{ct}, since the Fermi level is closer to the conduction-band edge. As the forward bias (positive voltage on semiconductor) increases, the current also increases monotonically. At a given bias, the current increases rapidly with decreasing insulator thickness. This is because the current is limited by the tunneling probability, Eq. 46, which varies exponentially with the insulator thickness. At reverse bias, the current is virtually independent of the insulator thickness for $d < 30$ Å, because the current is now limited by the rate of supply of minority carriers (electrons) through the semiconductor, and the current is similar to the saturation current in a reverse-biased *p-n* junction. Figure 25*a* also shows the experimental result for $d = 23.5$ Å. Note that there is good agreement between theory and experiment, and the current–voltage characteristics are very similar to those of a *p-n* junction.

For the high-barrier case, Fig. 25*b*, under forward bias, the dominant current is the majority (hole) tunnel current from the valence band to the metal, and the current increases exponentially with decreasing insulator thickness. Under reverse bias the current does not become independent of the insulator thickness as in Fig. 25*a*. Instead, the current increases rapidly with decreasing insulator thickness, because for majority transport the current is limited in both directions by the tunneling probability, not by the rate of carrier supply.

Substituting Eq. 46a into Eq. 45 and evaluating the integral over the energy range yields[42,47]

$$J = A^* T^2 e^{(-\alpha_T \phi_T^{1/2} d)} e^{-q\phi_B/kT} (e^{qV/kT} - 1) \tag{50}$$

where $A^* = 4\pi m_t^* q k^2 h^{-3}$ is the effective Richardson constant and ϕ_B is the

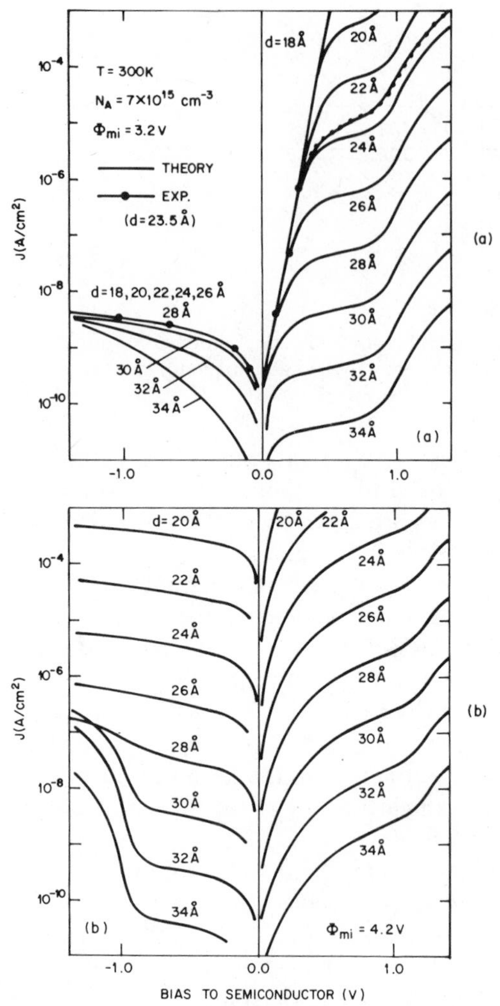

Fig. 25 Current–voltage characteristics of the MIS tunnel diodes having: (*a*) low barrier, (*b*) high barrier. (After Green, King, and Shewchun, Ref. 45.)

barrier height. Equation 50 is identical to the standard thermionic-emission equation for Schottky barriers except for the term $\exp(-\alpha_T \phi_T^{1/2} d)$, which is the tunneling probability. It is thus clear from Eq. 50 that for ϕ_T of the order of 1 V and $d > 50$ Å, the tunneling probability is about $e^{-50} \approx 10^{-22}$, and the current is indeed negligible. As d and/or ϕ_T decrease, the current increases rapidly toward the thermionic emission current level. Figure 26 shows the forward I–V characteristics[47] of four Au–SiO_2–Si tunnel diodes. For $d = 10$ Å, the current follows the standard Schottky diode behavior with an ideality factor n close to 1. As the insulator thickness increases, the current decreases rapidly and the ideality factor begins to depart from unity.

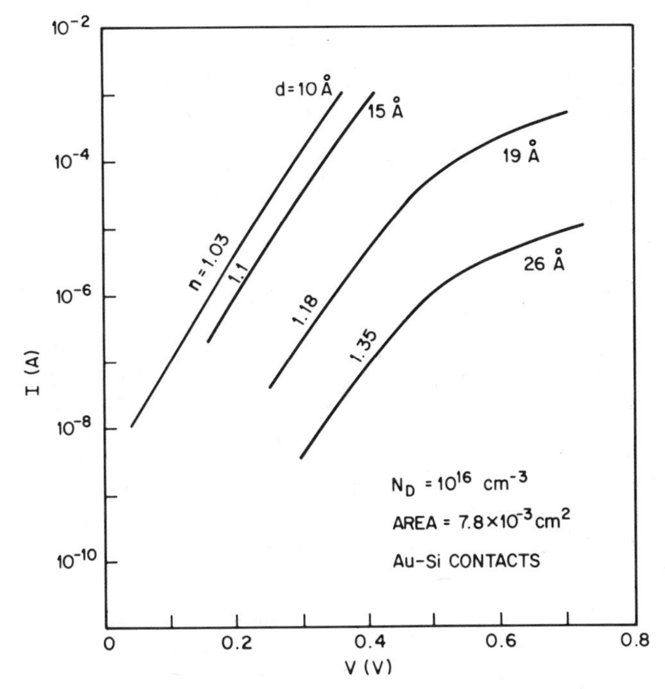

Fig. 26 Measured current–voltage characteristics of MIS tunnel diodes having different oxide thicknesses. (After Card and Rhoderick, Ref. 47.)

9.5 MIS SWITCH DIODE

The MIS switch diode is a four-layer structure as shown in Fig. 27*a*. The diode was found to display a current-controlled negative resistance, Fig. 27*b*, similar to a Shockley diode (Chapter 4).[48] When a negative bias (p^+ region is assumed to be grounded) is applied to the device, the I–V characteristic shows a high-impedance or OFF state of devices. At a sufficiently high voltage, the switching voltage V_S, the device suddenly switches to a low-voltage high-current ON state. The switching is due to either the extension of the surface-depletion region to the p^+n region (punchthrough) or by avalanching in the surface n layer.[49] Attractive features of the MIS switch diode include suitable current and voltage levels for digital circuit applications, high switching speed (1 ns or less), and high sensitivity of the switching voltage V_S to light or current injection. The initial device was built on a Si wafer and employed SiO_2 as the tunnel insulator. Later, similar behaviors were obtained from other insulators (e.g., Si_3N_4) and thick polycrystalline silicon. Current injection in the n layer via a third terminal has been shown to control the switching voltage, similar to a thyristor (Chapter 4).

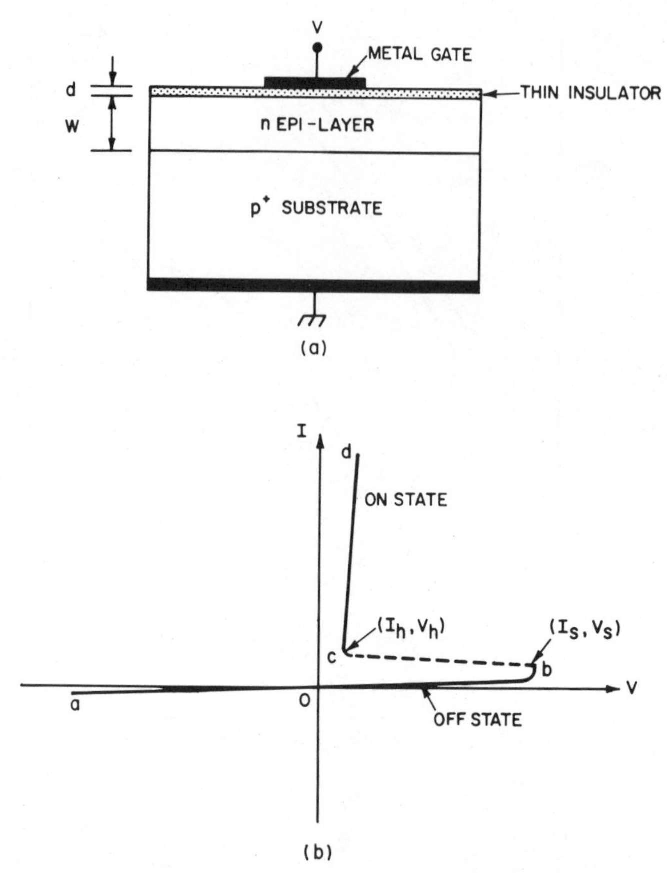

Fig. 27 (a) MIS switch diode, a four-layer structure. (*b*) Current–voltage characteristic shows current-controlled S-type negative resistance.

To understand the switching mechanism, consider the energy-band diagrams[50] in Fig. 28. Figure 28*a* shows the idealized diagram at thermal equilibrium. Figure 28*b* illustrates the diode under positive gate bias. If the conductivity of the tunnel insulator is much greater than that of the reverse-biased p^+n junction, most of the applied voltage will drop across the p^+n junction. The I–V characteristic of the device is essentially the same for the reverse-biased p^+n junction, because of generation in the depletion region:

$$J_g = \frac{qn_i W_J}{2\tau_g} \simeq \frac{n_i}{\tau_g}\left(\frac{q\epsilon_s}{2N_D}\right)(V + V_{bi})^{1/2} \tag{51}$$

where n_i is the intrinsic carrier concentration, τ_g the minority-carrier lifetime, W_J the depletion-region width, N_D the n-layer doping concentration, V_{bi} the built-in potential, and V the applied voltage. This current is illustrated by the portion $0a$ of the I–V curve in Fig. 27*b*.

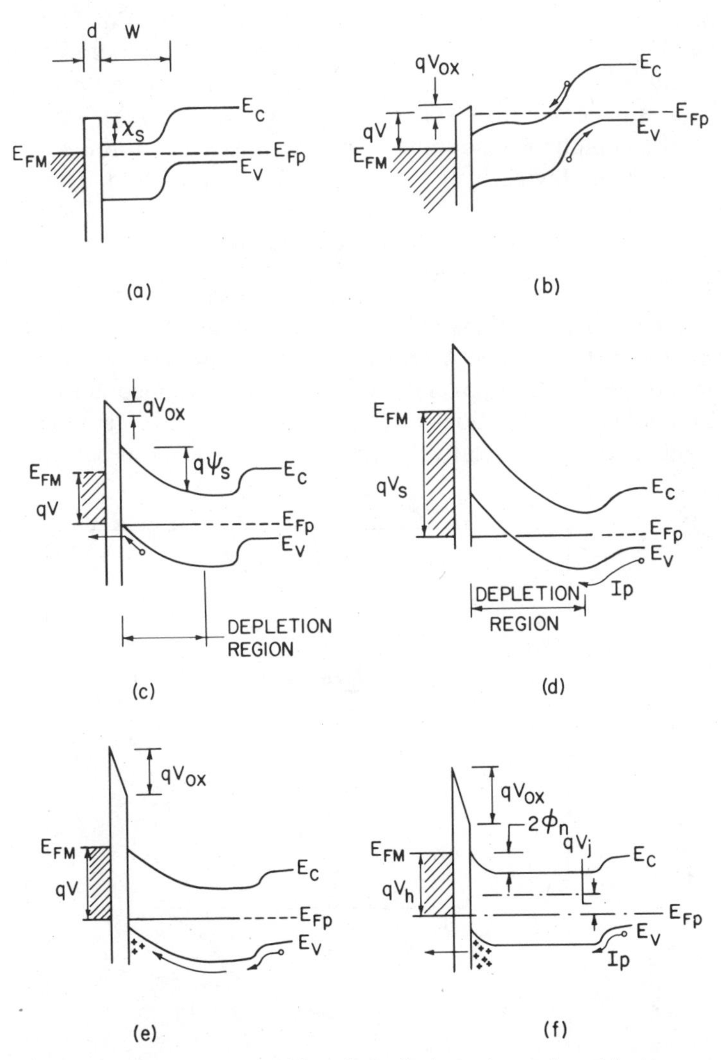

Fig. 28 Energy-band diagrams of MIS switch diode under various biasing conditions. (After Simmons and El-Badry, Ref. 50.)

When the gate voltage increases negatively, Fig. 28*c*, the free electrons are swept out of the epitaxial layer through the p^+n junction and a depletion region grows under the gate electrode. Electron–hole pairs are generated in this depletion region, the electrons being swept out of the device through the p^+n junction and the holes being swept to the SiO_2–Si interface. For thick oxides ($d \geq 50$ Å) the generated holes accumulate at the interface and give rise to inversion. If the oxide is thin enough, the holes tunnel through it; thus as the voltage increases, the surface n layer becomes depleted rather than inverted. Under these conditions, the generated current is given by an expression similar to Eq. 51 except that the term $(V + V_{bi})^{1/2}$ is replaced by $(\psi_s)^{1/2}$, where ψ_s is the surface potential. The current is again weakly dependent on voltage (curve 0*b* in Fig. 27*b*).

When a sufficiently high voltage is applied, say $V = V_S$ (the switching voltage), the depletion region under the oxide extends to the p^+n junction, Fig. 28*d*, and the device is at the punchthrough condition. This voltage is given by

$$V_S \simeq \frac{qN_D(W - W_{J0})^2}{2\epsilon_s} \tag{52}$$

where W is the n-type epitaxial layer thickness and W_{J0} is the p^+n junction depletion layer at zero bias. After punchthrough the injected holes from the p^+n junction increase exponentially with biasing voltage across the junction. Immediately after the punchthrough, the field in the oxide is insufficient to allow the relatively larger injected holes to pass through the

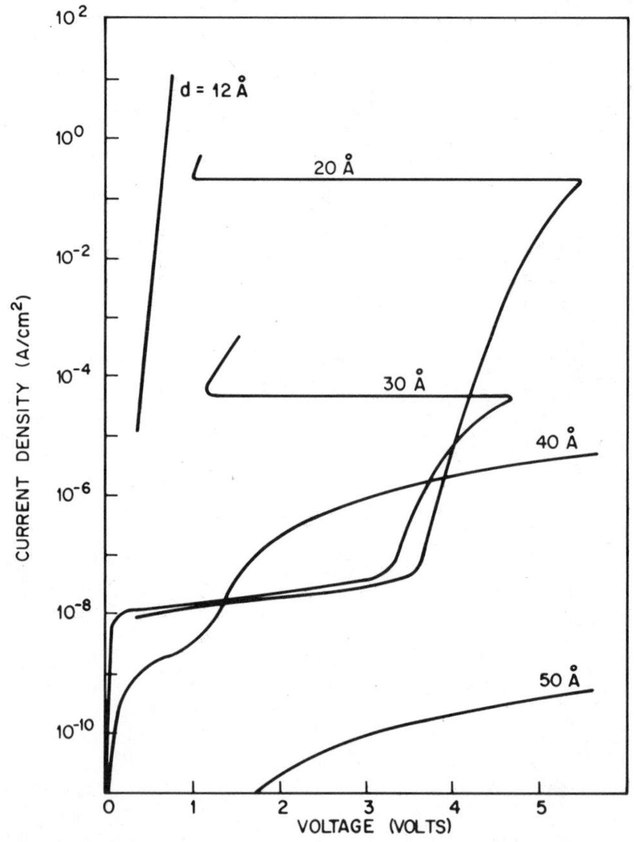

Fig. 29 Calculated *I–V* characteristic of MIS switch diodes for different values of oxide thickness. The device constants are $W = 10\ \mu m$, $N_D = 10^{14}\ cm^{-3}$, and $\tau_g = 3.5 \times 10^{-5}$ s. (After Habib and Simmons, Ref. 49.)

insulator, since the current is tunneling-limited. Consequently, an incremental voltage increase will cause holes to accumulate at the SiO_2–Si interface. The buildup of holes moves the surface region from deep depletion toward inversion, Fig. 28*e*; consequently, the ψ_s and the voltage across the device decreases. The voltage drop across the oxide increases, allowing a larger electron tunneling current to go through the p^+n junction, thus turning the junction on still further. This feedback mechanism is a regenerative one that causes the device to display a negative-resistance region, as indicated by curve *bc* in Fig. 27*b*. The holding voltage is shown in Fig. 28*f* and is given by

$$V_h = V_{ox} + \psi_s + V_J \tag{53}$$

where V_{ox} is the voltage drop across the oxide, ψ_s the surface potential, and V_J the voltage across the *p-n* junction. This consideration is for switching under the punchthrough condition. Under the avalanching condition, the same regenerative mechanism is responsible for the switching. The switching voltage in Eq. 53 will be replaced by the avalanche breakdown voltage of the *n*-type layer at the Si–SiO_2 interface.

As mentioned previously, the oxide thickness is a key parameter in the switching behavior shown in Fig. 29.[49] For thicker oxides ($d \gtrsim 50$ Å) the tunneling impedance is too high for low-voltage operation ($V \lesssim 5$ V) to meet the switching requirement. For very thin oxides ($d < 15$ Å) the p^+n junction can be turned fully on prior to the development of the deep depletion mode; thus the device displays a *p-n* junction characteristic. Switching behavior is observed only for the intermediate thicknesses (15 Å $< d <$ 40 Å).

9.6 MIM TUNNEL DIODE

A metal–insulator–metal (MIM) tunnel diode is a thin-film device in which the electrons from the first metal can tunnel into the insulator film and be collected by the second metal. The insert in Fig. 30 shows the basic energy-band diagrams of a MIM diode with similar metal electrodes. Since all of the voltage applied is dropped across the insulator, the tunnel current through the insulator is, from Eq. 44,

$$J = \frac{4\pi q m^*}{h^3} \int\int T_t[F(E) - F(E + qV)]\, dE_\perp dE. \tag{54}$$

At 0 K, Equation 54 simplifies to[51]

$$J = J_0[\bar{\phi}\exp(-A\sqrt{\bar{\phi}}) - (\bar{\phi} + V)\exp(-A\sqrt{\bar{\phi} + V})] \tag{55}$$

where

$$J_0 \equiv q^2/[2\pi h(\Delta d)^2]$$

$$A \equiv 4\pi(\Delta d)\sqrt{2mq}/h$$

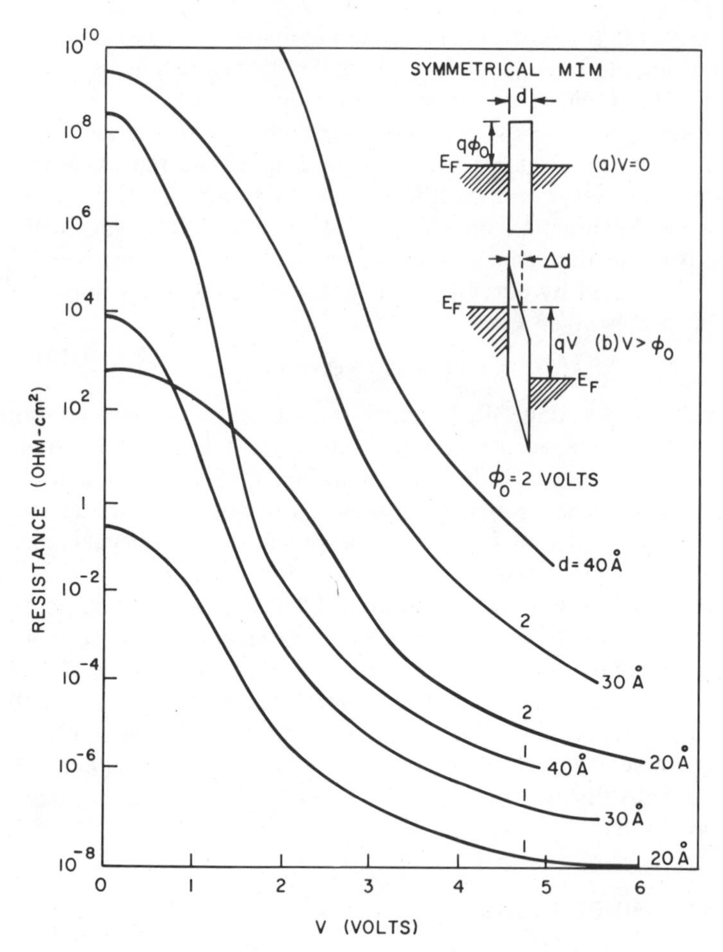

Fig. 30 Tunnel resistance of a symmetrical MIM structure. The insert shows band diagrams at $V = 0$ (zero bias) and $V > \phi_0$. (After Simmons, Ref. 51.)

and $\bar{\phi}$ is the average barrier height above the Fermi level. Equation 55 can be interpreted as a current density $J_0\bar{\phi}\exp(-A\sqrt{\bar{\phi}})$ flowing from electrode 1 to electrode 2 and a current density $J_0(\bar{\phi}+V)\exp(-A\sqrt{\bar{\phi}+V})$ flowing from electrode 2 to electrode 1.

We now apply Eq. 55 to the ideal symmetrical MIM structure (Fig. 30). By ideal we mean that the temperature effect, the image-force effect, and the field-penetration effect in metal electrodes are neglected. For $0 \le V \le \phi_0$, $\Delta d = d$, and $\bar{\phi} = \phi_0 - V/2$, the current density is given by

$$J = J_0[(\phi_0 - V/2)\exp(-A\sqrt{\phi_0 - V/2}) - (\phi_0 + V/2)\exp(-A\sqrt{\phi_0 + V/2})]. \quad (56)$$

For larger voltage, $V > \phi_0$, we have $\Delta d = d\phi_0/V$ and $\bar{\phi} = \phi_0/2$. The current density is then

$$J = \frac{q^2\mathscr{E}^2}{4\pi h\phi_0}[\exp(-\mathscr{E}_0/\mathscr{E}) - (1+2V/\phi_0)\exp(-\mathscr{E}_0\sqrt{1+2V/\phi_0}/\mathscr{E})] \quad (57)$$

where $\mathscr{E} = V/d$ is the field in the insulator and $\mathscr{E}_0 \equiv \frac{8}{3}\sqrt{\pi q}(\phi_0)^{3/2}$. For very high voltage such that $V > (\phi_0 + E_F/q)$, the second term in Eq. 57 can be neglected, and we have the well-known Fowler–Nordheim equation. Figure 30 shows the computed results for the tunnel resistance (V/J), where various barrier heights and insulator thicknesses are used. We note that the tunnel resistance decreases rapidly with increasing applied voltage.

For an ideal asymmetrical MIM structure (Fig. 31, insert) in the low-voltage range $0 < V < \phi_1$, the quantities $\Delta d = d$ and $\bar{\phi} = (\phi_1 + \phi_2 - V)/2$ are independent of the polarities. Thus the J–V characteristics are also independent of the polarity. At higher voltages, $V > \phi_2$, the average barrier height $\bar{\phi}$ and the effective tunneling distance Δd become polarity-dependent. Therefore, the currents for different polarities are different. Figure 31 illustrates the tunnel resistance as a function of V for $d = 20$, 30, and 40 Å, $\phi_1 = 1$ V, and $\phi_2 = 2$ V. Experimental results obtained from Al–Al_2O_3–Al structures with $d = 30$ Å, $\phi_1 = 1.6$ V, and $\phi_2 = 2.5$ V are in good agreement with theoretical predictions based on Eqs. 56 and 57. However, the effective area used for the experimental data is only 1% of the electrode area.[52] This is quite different from the Al–SiO_2–SiO_2 (MIS) tunnel diodes, in which the effective area is usually equal to the total Al electrode area. The small tunneling area in the MIM diode can be explained by the statistical nature of the formation of insulating films on metal substrate.[53] Only the thinnest portion in the insulator film is responsible for the tunneling current. Because of the statistical fluctuation of the thickness, the capacitance of an MIM structure is always larger than that calculated based on an average thickness of the insulator film.[54] Another factor that can affect the capacitance value is the contribution from the potential distribution in the metal electrodes. Because of the penetration of the electric field into the metal, the total capacitance is effectively equal to two capacitances in series:[55]

$$1/C = d/\epsilon_i + 2.3\lambda_m/\epsilon_m \quad (58)$$

where ϵ_i and ϵ_m are the permittivities of the insulator and the metal, respectively, and λ_m is the characteristic penetration length in the metal (~0.5 Å for the noble metals). From Eq. 58, note that the capacitance due to the second term is of the order of 10^6 pF/cm^2. The total capacitance per unit area will be much smaller than the value ϵ_i/d when d approaches a value of the order of 5 Å.

The MIM tunnel diodes have been used to study the energy–momentum relation[56] in the forbidden gap of single-crystal insulators, for example, GaSe ($E_g = 2.0$ eV, $\epsilon_s/\epsilon_0 = 8$, and a carrier density at 300 K, $p \sim 3 \times 10^{14}$ cm^{-3}). GaSe is a layered crystal structure that facilitates peeling a

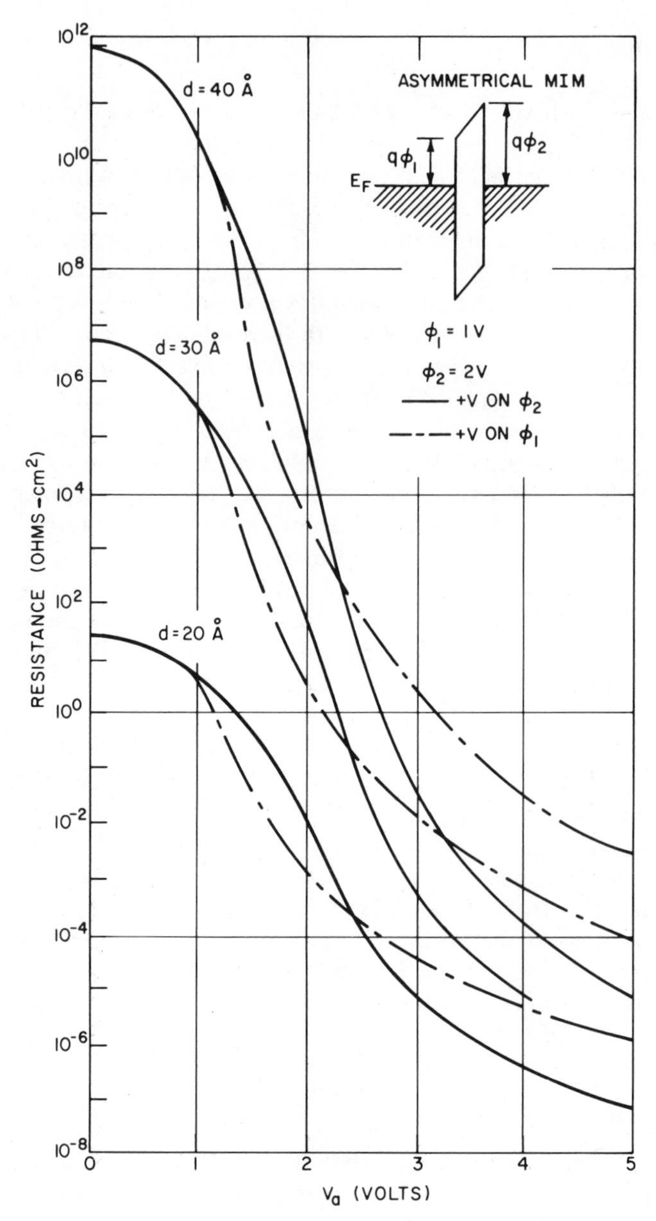

Fig. 31 Tunnel resistance as a function of applied voltage for an asymmetrical MIM structure. The insert shows the band diagram at zero bias. (After Simmons, Ref. 51.)

single-crystal specimen to the required thinness (<100 Å). An MIM tunnel structure is formed using the single-crystal specimen sandwiched between two metal electrodes, for example, Al–GaSe–Au (the Al–GaSe barrier is 1.08 V, and the Au–GaSe barrier is 0.52 V). The insert in Fig. 32 shows such a band diagram. The measured data of Al–GaSe–Au structures with different thicknesses are shown by the solid symbols in Fig. 32. Using one set of $J(V)$ curves, one can obtain the momentum–energy $k(E)$ relationship in GaSe using Eqs. 44 and 54. Once the $k(E)$ relationship is obtained, one can calculate, using no adjustable parameters, the tunneling currents for all other thicknesses. The theoretical $I-V$ curves (solid lines) of Fig. 32 have been calculated on this basis. Agreement between theory and experiment is excellent for all voltages and thicknesses.

A stable and reproducible MIM tunnel diode is the edge-MOM diode, which has uniform tunneling area.[57] Figure 33 shows a schematic diagram of the device. The overlap between the 1-μm metal (Ni) strip and an oxidized edge of a metal (Ni) strip about 100 Å thick gives rise to an overlap area of ~10^{-10} cm^2. The oxide thickness (NiO) is of the order of 10 Å. The curvature coefficient γ, Eq. 42, for the Ni–NiO–Ni edge–MOM diode is in the range 1 to 10. For signal detection experiments, the nonlinear $I-V$ characteristics of the device are invariant from audio frequencies to near infrared. The device can serve as a broad-band detector and mixer.

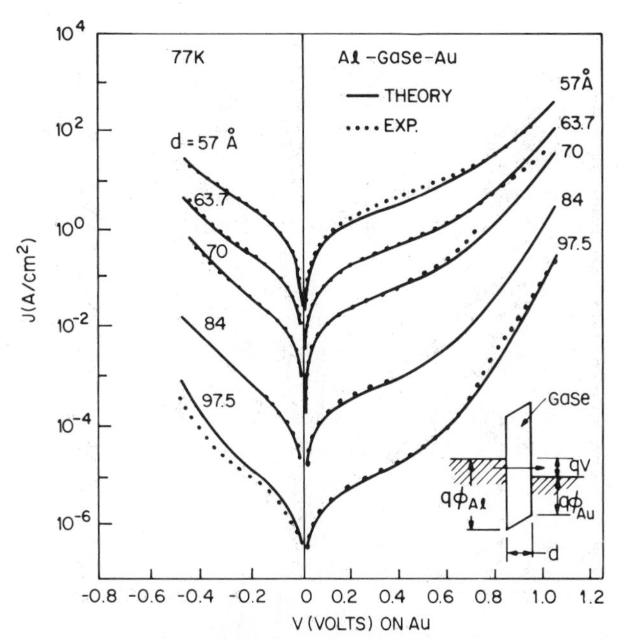

Fig. 32 Current density versus applied voltage for several Al–GaSe–Au structures. Data for both bias directions are shown as solid symbols. Solid lines are theoretical results. The insert shows the band diagram under bias. (After Kurtin, McGill, and Mead, Ref. 56.)

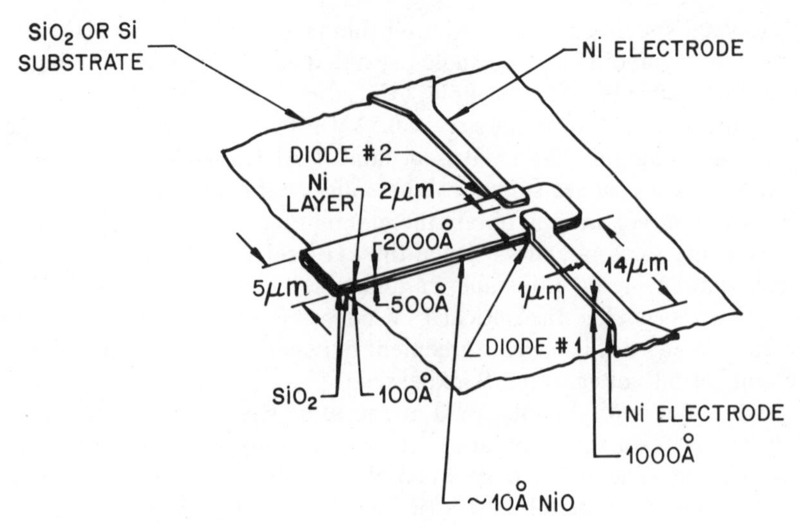

Fig. 33 Schematic diagram of Edge–MOM diode. (After Heiblum et al., Ref. 57.)

9.7 TUNNEL TRANSISTOR

Over the years, many attempts have been made to invent or discover new solid-state devices capable of achieving better performance than bipolar transistor or MOSFETs. Among the most interesting candidates are the tunnel transistors. Mead proposed the first tunnel transistor in 1960.[58] Figure 34*a* shows the device, a metal–insulator–metal–insulator–metal (MIMIM) structure in which current flow through the first MIM insulator layer occurs by tunneling. After tunneling from the emitter to the base region, the electrons have energies more than a few kT above the Fermi level in the base; these electrons are called hot electrons because they are not in thermal equilibrium with the lattice. The transistor is therefore referred to as a hot-electron transistor.[59] Spratt, Schwartz, and Kane pointed out that the current gain of such a structure could be greatly improved by replacing the collector insulator with a Schottky-barrier semiconductor layer (Fig. 34*b*).[60] The MIMS structure has a lower maximum oscillation frequency than the bipolar transistor mainly because of its longer emitter charging time (caused by larger emitter capacitance) and smaller common-base current gain (caused by hot-electron scattering in the base region).

Figure 34*c* shows the energy-band diagram of another tunnel transistor.[61] This tunnel transistor has an MIS structure in which electrons can tunnel from the emitter metal electrode into the *p*-type diffused base layer. The insert in Fig. 35 shows the cross-sectional diagram. Aluminum is used as the metal electrode on a thin SiO_2 layer, about 20 Å, thermally grown in dry oxygen (550°C for 30 min). The *p*-type base is about 1500 Å with an

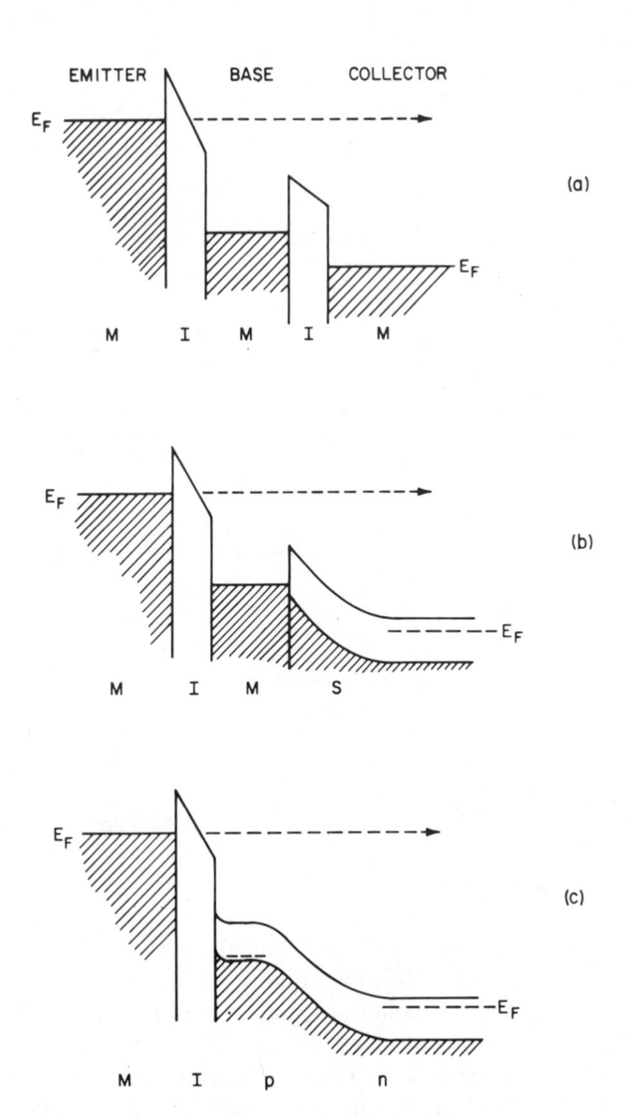

Fig. 34 (*a*) MIMIM tunnel transistor. (*b*) MIMS tunnel transistor. (*c*) MI-*p*-*n* tunnel transistor. (After Mead, Ref. 58; Spratt, Schwartz, and Kane, Ref. 60; Kisaki, Ref. 61.)

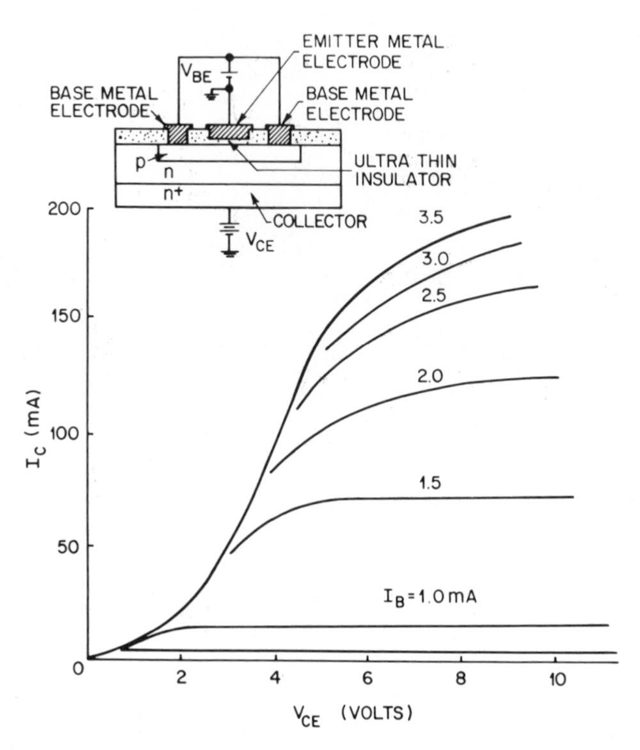

Fig. 35 Collector I–V characteristics of MI-p-n tunnel transistor. The insert shows the device cross section. (After Kisaki, Ref. 61.)

acceptor concentration of $\sim 10^{18}\ \text{cm}^{-3}$, and the n-type epitaxial layer has a donor concentration of $8 \times 10^{15}\ \text{cm}^{-3}$. Figure 35 shows the collector current–voltage characteristics. A common-emitter current gain over 100 can be obtained at low base-current levels.

Recently, a tunneling-base transistor[62] having two heterojunctions was proposed. Figure 36*a* shows the band diagram at thermal equilibrium. Three fundamental requirements have to be met for the transistor to function properly: (1) the discontinuities ΔE_C and ΔE_V must occur in the same direction on the same energy scale as shown to provide the confinement for electrons and the barrier for holes, respectively; (2) the n region must remain conductive under all conditions, and (3) the base thickness, together with the energy-band configuration, must be such that the tunneling current is dominant. To meet these requirements, one possible choice is to use the p-type $GaAs_{0.5}Sb_{0.5}$ with $N_A \lesssim 10^{16}\ \text{cm}^{-3}$ as the emitter and collector materials, and n-type $Ga_{0.5}In_{0.5}As$ with $N_D \gtrsim 10^{19}$ and $d \simeq 50$ Å as the base for which $E_{g1} \approx E_{g2} = 0.8$ eV and $\Delta E_C = \Delta E_V = 0.5$ eV. The molecular-beam epitaxy (MBE) technique can be used to grow the layers.

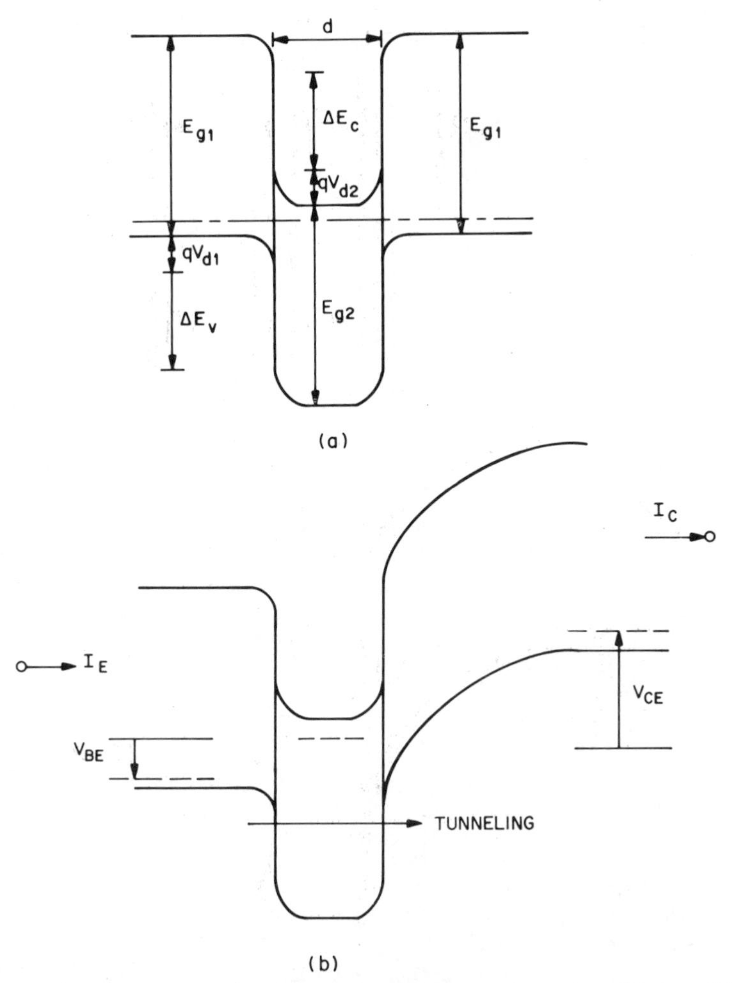

Fig. 36 Band diagrams of tunneling-base transistor (*a*) at thermal equilibrium, (*b*) under biasing condition. (After Chang and Esaki, Ref. 62.)

Figure 36*b* shows the tunnel transistor under bias conditions. The dominant current will be the tunneling current through the base, which can be expressed as $J_t \sim \exp(qV_{BE}/kT)$. Figure 37 shows, in a semiquantitative fashion, the current characteristics for the emitter (J_E) and collector (J_C) currents. The approximate equality of J_E and J_C implies a unity emitter efficiency and base transport factor and, consequently, a large current gain. This tunnel transistor has three advantages over the conventional bipolar transistor: (1) negligible base transit time because of the inherently short tunneling time, (2) small emitter capacitance because of low emitter doping, and (3) low base resistance caused by high base doping. These advantages are expected to improve the microwave performances of the tunneling-base transistor.

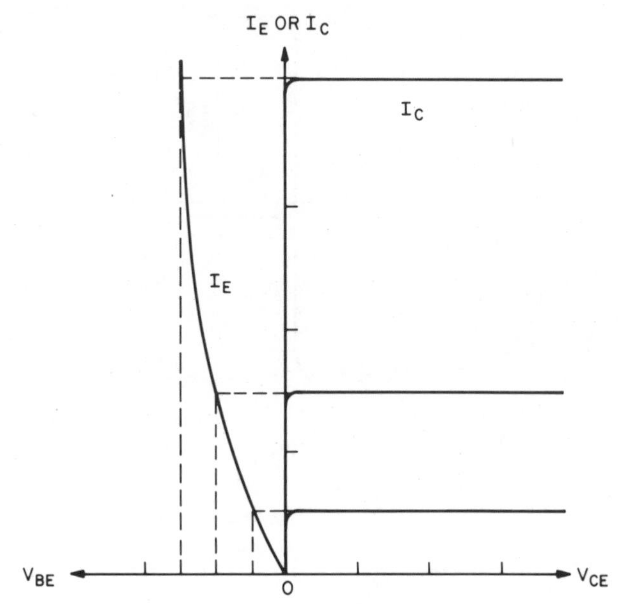

Fig. 37 Emitter and collector current as functions of base and collector voltage in a tunneling-base transistor. (After Chang and Esaki, Ref. 62.)

REFERENCES

1. L. Esaki, "New Phenomenon in Narrow Germanium *p-n* Junctions," *Phys. Rev.*, **109**, 603 (1958); "Long Journey into Tunneling," *Proc. IEEE*, **62**, 825 (1974); "Discovery of the Tunnel Diode," *IEEE Trans. Electron Devices*, **ED-23**, 644 (1976).
2. K. K. Thornber, Thomas C. McGill, and C. A. Mead, "The Tunneling Time of an Electron," *J. Appl. Phys.*, **38**, 2384 (1967).
3. R. N. Hall, "Tunnel Diodes," *IRE Trans. Electron Devices*, **ED-7**, 1 (1960).
4. J. V. Morgan and E. O. Kane, "Observation of Direct Tunneling in Germanium," *Phys. Rev. Lett.*, **3**, 466 (1959).
5. L. D. Landau and E. M. Lifshitz, *Quantum Mechanics*, Addison-Wesley, Reading, Mass., 1958, p. 174.
6. E. O. Kane, "Theory of Tunneling," *J. Appl. Phys.*, **32**, 83 (1961); "Tunneling in InSb," *Phys. Chem. Solids*, **2**, 181 (1960).
7. T. A. Demassa and D. P. Knott, "The Prediction of Tunnel Diode Voltage–Current Characteristics," *Solid State Electron.*, **13**, 131 (1970).
8. H. Kroemer, "The Einstein Relation for Degenerate Carrier Concentration," *IEEE Trans. Electron Devices*, **ED-25**, 850 (1978).
9. D. Meyerhofer, G. A. Brown, and H. S. Sommers, Jr., "Degenerate Germanium I, Tunnel, Excess, and Thermal Current in Tunnel Diodes," *Phys. Rev.*, **126**, 1329 (1962).
10. J. L. Moll, *Physics of Semiconductors*, McGraw-Hill, New York, 1964, p. 252.
11. P. N. Butcher, K. F. Hulme, and J. R. Morgan, "Dependence of Peak Current Density on Acceptor Concentration in Germanium Tunnel Diodes," *Solid State Electron.*, **3**, 358 (1962).

12 L. V. Keldysh, "Behavior of Non-Metallic Crystals in Strong Electric Fields," *Sov. J. Exp. Theor. Phys.*, **6**, 763 (1958).

13 A. G. Chynoweth, W. L. Feldmann, and R. A. Logan, "Excess Tunnel Current in Silicon Esaki Junctions," *Phys. Rev.*, **121**, 684 (1961).

14 D. K. Roy, "On the Prediction of Tunnel Diode $I-V$ Characteristics," *Solid State Electron.*, **14**, 520 (1971).

15 W. N. Carr, "Reversible Degradation Effects in GaSb Tunnel Diodes," *Solid State Electron.*, **5**, 261 (1962).

16 R. P. Nanavati and C. A. Morato De Andrade, "Excess Current in Gallium Arsenide Tunnel Diodes," *Proc. IEEE*, **52**, 869 (1964).

17 N. Holonyak, Jr., "Evidence of States in the Forbidden Gap of Degenerate GaAs and InP—Secondary Tunnel Current and Negative Resistance," *J. Appl. Phys.*, **31**, 130 (1960).

18 V. M. Franks, K. F. Hulme, and J. R. Morgan, "An Alloy Process for Making High Current Density Silicon Tunnel Diode Junction," *Solid State Electron.*, **8**, 343 (1965).

19 H. P. Kleinknecht, "Indium Arsenide Tunnel Diodes," *Solid State Electron.*, **2**, 133 (1961).

20 C. A. Burrus, "Indium Phosphide Esaki Diodes," *Solid State Electron.*, **3**, 357 (1962).

21 A. Yoshihito, M. Konagai, and Y. Sakai, "Mixed Crystal Tunnel Diode," *Jpn. J. Appl. Phys.*, **12**, 480 (1973).

22 R. M. Minton and R. Glicksman, "Theoretical and Experimental Analysis of Germanium Tunnel Diode Characteristics," *Solid State Electron.*, **7**, 491 (1964).

23 R. A. Logan, W. M. Augustyniak, and J. F. Gilber, "Electron Bombardment Damage in Silicon Esaki Diodes," *J. Appl. Phys.*, **32**, 1201 (1961).

24 W. Bernard, W. Rindner, and H. Roth, "Anisotropic Stress Effect on the Excess Current in Tunnel Diodes," *J. Appl. Phys.*, **35**, 1860 (1964).

25 V. V. Galavanov and A. Z. Panakhov, "Influence of Hydrostatic Pressure on the Tunnel Current in GaSb Diodes," *Sov. Phys. Semicond.*, **6**, 1924 (1973).

26 R. E. Davis and G. Gibbons, "Design Principles and Construction of Planar Ge Esaki Diodes," *Solid State Electron.*, **10**, 461 (1967).

27 H. J. Hindin, "Tunnel Diode Flying High," *Electronics*, **52**, 81 (1979).

28 "Standards on Definitions, Symbols, and Methods of Test for Semiconductor Tunnel (Esaki) Diodes and Backward Diodes," *IEEE Trans. Electron Devices*, **12**, 374 (1965).

29 W. B. Hauer, "Definition and Determination of the Series Inductance of Tunnel Diodes," *IRE Trans. Electron Devices*, **8**, 470 (1961).

30 L. Esaki and Y. Miyahara, "A New Device Using the Tunneling Process in Narrow p-n Junctions," *Solid State Electron.*, **1**, 13 (1960).

31 R. N. Hall, J. H. Racette, and H. Ehrenreich, "Direct Observation of Polarons and Phonons During Tunneling in Group 3–5 Semiconductor Junctions," *Phys. Rev. Lett.*, **4**, 456 (1960).

32 A. G. Chynoweth, R. A. Logan, and D. E. Thomas, "Phonon-Assisted Tunneling in Silicon and Germanium Esaki Junctions," *Phys. Rev.*, **125**, 877 (1962).

33 H. V. Shurmer, "Backward Diodes as Microwave Detectors," *Proc. Inst. Electr. Eng., Lond.*, **111**, 1511 (1964).

34 S. T. Eng, "Low-Noise Properties of Microwave Backward Diodes," *IRE Trans. Microwave Theory Tech.*, **MTT-8**, 419 (1961).

35 J. B. Hopkins, "Microwave Backward Diodes in InAs," *Solid State Electron.*, **13**, 697 (1970).

36 A. B. Bhattacharyya and S. L. Sarnot, "Switching Time Analysis of Backward Diodes," *Proc. IEEE*, **58**, 513 (1970).

37 H. C. Torrey and C. A. Whitmer, *Crystal Rectifiers*, McGraw-Hill, New York, 1948, Ch. 8.

38 S. M. Sze and R. M. Ryder, "The Nonlinearity of the Reverse Current–Voltage Characteristics of a *p-n* Junction near Avalanche Breakdown," *Bell Syst. Tech. J.*, **46**, 1135 (1967).

39 J. Karlovsky, "The Curvature Coefficient of Germanium Tunnel and Backward Diodes," *Solid State Electron.*, **10**, 1109 (1967).

40 W. E. Dahlke and S. M. Sze, "Tunneling in Metal–Oxide–Silicon Structures," *Solid State Electron.*, **10**, 865 (1967).

41 L. L. Chang, P. J. Stiles, and L. Esaki, "Electron Tunneling between a Metal and a Semiconductor: Characteristics of Al–Al_2O_3–SnTe and –GeTe Junctions," *J. Appl. Phys.*, **38**, 4440 (1967).

42 V. Kumar and W. E. Dahlke, "Characteristics of Cr–SiO_2–*n*Si Tunnel Diodes," *Solid State Electron.*, **20**, 143 (1977).

43 E. H. Nicollian and A. Goetzberger, "The Si–SiO_2 Interface Electrical Properties as Determined by the MIS Conductance Technique," *Bell Syst. Tech. J.*, **46**, 1055 (1967).

44 L. Esaki and P. J. Stiles, "New Type of Negative Resistance in Barrier Tunneling," *Phys. Rev. Lett.*, **16**, 1108 (1966).

45 M. A. Green, F. D. King, and J. Shewchun, "Minority Carrier MIS Tunnel Diodes and Their Application to Electron and Photovoltaic Energy Conversion: I. Theory," *Solid State Electron.*, **17**, 551 (1974). "II. Experiment," *Solid State Electron.*, **17**, 563 (1974).

46 V. A. K. Temple, M. A. Green, and J. Schewchun, "Equilibrium-to-Nonequilibrium Transition in MOS Tunnel Diodes," *J. Appl. Phys.*, **45**, 4934 (1974).

47 H. C. Card and E. H. Rhoderick, "Studies of Tunnel MOS Diodes I. Interface Effects in Silicon Schottky Diodes," *J. Phys. D Appl. Phys.*, **4**, 1589 (1971).

48 T. Yamamota and M. Morimoto, "Thin-MIS-Structure Si Negative Resistance Diode," *Appl. Phys. Lett.*, **20**, 269 (1972).

49 S. E.-D. Habib and J. G. Simmons, "Theory of Switching in *p-n* Insulator (Tunnel)–Metal Devices," *Solid State Electron.*, **22**, 181 (1979).

50 J. G. Simmons and A. El-Badry, "Theory of Switching Phenomena in Metal/Semi-Insulator/n-p^+ Silicon Devices," *Solid State Electron.*, **20**, 955 (1977); A. El-Badry and J. G. Simmons, "Experimental Studies of Switching in Metal Semi-Insulating np^+ Silicon Devices," *Solid State Electron.*, **20**, 963 (1977).

51 J. G. Simmons, "Generalized Formula for the Electric Tunnel Effect between Similar Electrodes Separated by a Thin Insulating Film," *J. Appl. Phys.*, **34**, 1793 (1963).

52 S. R. Pollack and C. E. Morris, "Tunneling through Gaseous Oxidized Films of Al_2O_3," *Trans. AIME*, **233**, 497 (1965).

53 Z. Hurych, "Influence of Nonuniform Thickness of Dielectric Layers on Capacitance and Tunnel Currents," *Solid State Electron.*, **9**, 967 (1966).

54 J. Pochobradsky, "On the Capacitance of Metal–Insulator–Metal Structures with Nonuniform Thickness," *Solid State Electron.*, **10**, 973 (1967).

55 H. Y. Ku and F. G. Ullman, "Capacitance of Thin Dielectric Structures," *J. Appl. Phys.*, **35**, 265 (1964).

56 S. Kurtin, T. C. McGill, and C. A. Mead, "Tunneling Currents and E–k Relation," *Phys. Rev. Lett.*, **25**, 756 (1970); "Direct Interelectrode Tunneling in GaSe," *Phys. Rev.*, **B3**, 3368 (1971).

57 M. Heiblum, S. Y. Wang, J. R. Whinnery, and T. K. Gustafson, "Characteristics of Integrated MOM Junctions at DC and at Optical Frequencies," *IEEE J. Quantum Electron.*, **QE-14**, 159 (1978).

58 C. A. Mead, "Tunnel-Emission Amplifiers," *Proc. IRE*, **48**, 359 (1960).

59 C. R. Crowell and S. M. Sze, "Hot Electron Transport and Electron Tunneling in Thin-Film Structures," in G. Haas and R. E. Thun, Eds., *Physics of Thin Films*, Vol. 4, Academic, New York, 1967.

60 J. P. Spratt, R. F. Schwartz, and W. M. Kane, "Hot Electrons in Metal Films: Injection and Collection," *Phys. Rev. Lett.*, **6**, 341 (1961).

61 H. Kisaki, "Tunnel Transistor," *Proc. IEEE*, **61**, 1053 (1973).

62 L. L. Chang and L. Esaki, "Tunnel Triode—A Tunneling Base Transistor," *Appl. Phys. Lett.*, **31**, 687 (1977).

63 Hewlett-Packard Electronic Test Instruments, Hewlett-Packard, Palo Alto, Calif., 1961.

64 S. Y. Liao, *Microwave Devices and Circuits*, Prentice-Hall, Englewood Cliffs, N.J., 1980.

10

IMPATT and Related Transit-Time Diodes

- INTRODUCTION
- STATIC CHARACTERISTICS
- DYNAMIC CHARACTERISTICS
- POWER AND EFFICIENCY
- NOISE BEHAVIOR
- DEVICE DESIGN AND PERFORMANCE
- BARITT AND DOVETT DIODES
- TRAPATT DIODE

10.1 INTRODUCTION

The word IMPATT stands for "impact ionization avalanche transit time". IMPATT diodes employ impact-ionization and transit-time properties of semiconductor structures to produce negative resistance at microwave frequencies. The negative resistance arises from two delays which cause the current to lag behind the voltage. One is the "avalanche delay" caused by finite buildup time of the avalanche current; the other is the "transit-time delay" from the finite time for the carriers to cross the "drift" region. When these two delays add up to half-cycle time, the diode electronic resistance is negative at the corresponding frequency. The experimental observation of the IMPATT oscillation was first reported by Johnston, DeLoach, and Cohen in 1965 for a silicon diode biased into reverse avalanche breakdown and mounted in a microwave cavity.[1,2]

The negative resistance arising from transit time in semiconductor diodes was first considered by Shockley in 1954 because of the potential advantages on structural simplicity of two-terminal devices compared to three-terminal transistors.[3] In 1958, Read proposed a high-frequency semiconductor diode consisting of an avalanche region at one end of a relatively high resistance region serving as the transit-time drift space for the generated charge carriers (i.e., p^+-n-i-n^+ or n^+-p-i-p^+).[4] The Read IMPATT diode oscillation was first reported by Lee et al.[5] The small-signal theory developed by Misawa and by Gilden and Hines has confirmed that a negative resistance of the IMPATT nature can be obtained from junction diode or a metal–semiconductor contact with any doping profile.[6,7]

The IMPATT diode is now one of the most powerful solid-state sources of microwave power.[8,9] At present, the IMPATT diode can generate the highest cw power output at millimeter-wave frequencies (i.e., above 30 GHz) of all solid-state devices.[10,10a] But there are two noteworthy difficulties in IMPATT circuit applications: (1) the noise is high, and sensitive to operating conditions; and (2) large electronic reactances are present, which are strongly dependent on oscillation amplitude and require unusual care in circuit design to avoid detuning or even burnout of the diode.[11]

Another transit-time device is the BARITT diode (barrier injection and transit time). The name BARITT recalls the major mechanisms of operation. The BARITT operation was first reported by Coleman and Sze in 1971 using a metal–semiconductor–metal reach-through diode.[12] Similar structures were proposed in 1968 by Ruegg based on large-signal analysis and by Wright using space-charge-limited transport mechanisms.[13,14] Because of the lack of avalanche delay time, the BARITT diode is expected to operate at lower power and lower efficiency than the IMPATT diode. On the other hand, the noise associated with carrier injection across the barrier is smaller than the avalanche noise in an IMPATT diode. The low-noise property and the stability of the device make the BARITT diode suited for low-power applications, such as local oscillators. The BARITT diodes have demonstrated superiority over IMPATT diodes and transferred-electron devices in Doppler detectors for both dc and microwave power requirements and minimum detectable signal.[15,16] A BARITT-related device is the DOVETT diode (double velocity transit time), in which the velocity of carriers near one contact is significantly less than that near the other contact. No experimental microwave results on DOVETT diodes have been reported yet. However, because of the transit-time delay in the low-velocity region, DOVETT diodes are expected to have higher efficiencies than BARITT diodes.

An IMPATT-related device is the TRAPATT diode (trapped plasma avalanche triggered transit). The TRAPATT mode of operation was discovered by Prager, Chang, and Weisbrod in 1967 from silicon avalanche diodes.[17] The operating frequency is substantially lower than the transit-time frequency and the efficiency is considerably higher. Subsequent theoretical studies established that under large-signal conditions the periodic avalanching of the diode begins at the high-field side and sweeps rapidly across the diode, leaving it substantially filled by a highly conducting plasma of holes and electrons whose space charge depresses the voltage to very low values.[18-20] Since the plasma cannot rapidly escape, this mode is called the trapped plasma mode or TRAPATT mode. The TRAPATT diode has been tried experimentally in pulsed transmitter of phased-array radar systems.[15,21]

10.2 STATIC CHARACTERISTICS

The basic members of the IMPATT diode family are the Read diode, the one-sided abrupt p-n junction, the two-sided (double-drift) diode, the hi–lo and lo–hi–lo diodes (modified Read diode), and the p-i-n diode.

We shall now consider their static characteristics, such as the field distribution, breakdown voltage, and space-charge effect. Figure 1 shows the doping profile, electric field distribution, and ionization integrand at breakdown condition for an idealized Read diode (p^+-n-i-n^+ or its dual n^+-p-i-p^+). The ionization integrand is given by

$$\langle \alpha \rangle \equiv \alpha_n \exp\left[-\int_x^W (\alpha_n - \alpha_p)\, dx' \right] \tag{1}$$

where α_n and α_p are the ionization rates of electrons and holes, respectively, and W is the depletion width.

The avalanche breakdown condition as discussed in Chapter 2 is given by

$$\int_0^W \langle \alpha \rangle\, dx = 1. \tag{2}$$

Because of the strong dependence of alphas on an electric field, we note that the "avalanche region" is highly localized, that is, most of the multiplication processes occur in a narrow region near the highest field between 0 and x_A, where x_A is defined as the width of the avalanche region (to be discussed later).

The voltage drop across the avalanche region is called V_A. It will be shown that both x_A and V_A have a profound effect on the optimum current density and the maximum efficiency of an IMPATT diode. The layer outside the avalanche region ($x_A \leq x \leq W$) is called the drift region.

There are two limiting cases of the Read doping profiles. As the N_2 region becomes zero, we have a one-sided abrupt p^+n junction. On the other hand, when the N_1 region becomes zero we have a p-i-n diode, also referred to as the Misawa diode.[6] Figure 2a describes the structure of a one-sided abrupt p-n junction. The avalanche region is highly localized. Figure 2b describes the structure of a two-sided abrupt silicon p-n junction. The avalanche region is located near the center of the depletion layer.

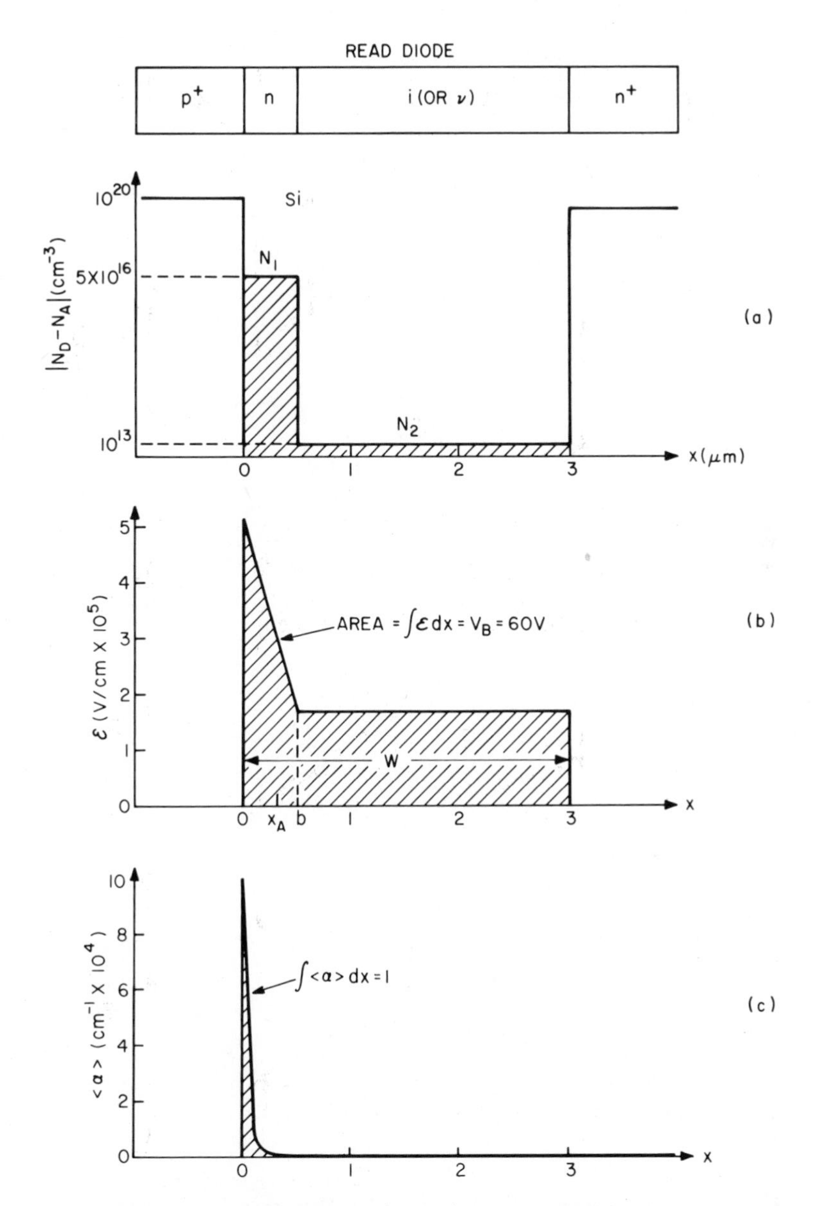

Fig. 1 Read diode (a) doping profile (p^+-n-i-n^+). (b) electric field distribution, and (c) ionization integrand at avalanche breakdown. (After Read, Ref. 4.)

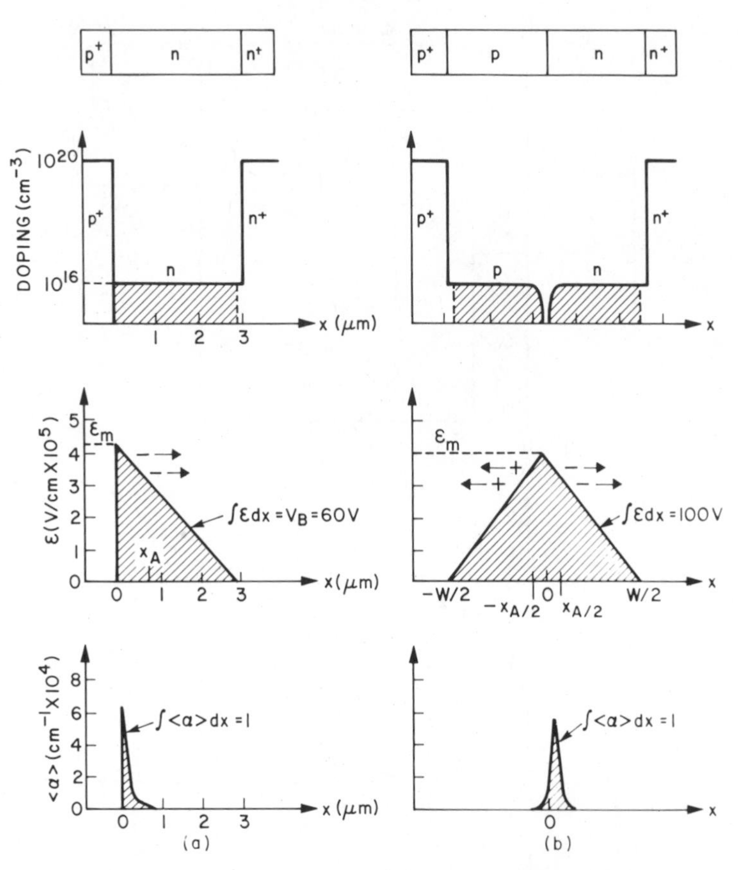

Fig. 2 Doping profile, electric field distribution, and ionization integrand for (*a*) one-sided abrupt p^+-n diode, single-drift diode; (*b*) two-sided symmetrical abrupt p^+-p-n-p^+ diode, double-drift diode.

The slight asymmetry of the integrand $\langle\alpha\rangle$ with respect to the location of the maximum field is because of the large difference between α_n and α_p in Si. If $\alpha_n \simeq \alpha_p$ as in the case of GaP, $\langle\alpha\rangle$ reduces to

$$\langle\alpha\rangle = \alpha_n = \alpha_p \tag{3}$$

and the avalanche region is symmetrical with respect to $x = 0$.

Figure 3*a* shows a modified Read diode, the hi–lo structure, in which the doping N_2 is larger than that for a Read diode.[22] Figure 3*b* shows another modified Read diode, the lo–hi–lo structure, in which a "clump" of charge Q is located at $x = b$. Since a nearly uniform high-field region exists from $x = 0$ to $x = b$, the value of the maximum field can be much lower than that for a hi–lo diode. The *p-i-n* diode has a uniform field across the intrinsic layer under low-current conditions. The avalanche region corresponds to the full intrinsic layer width.

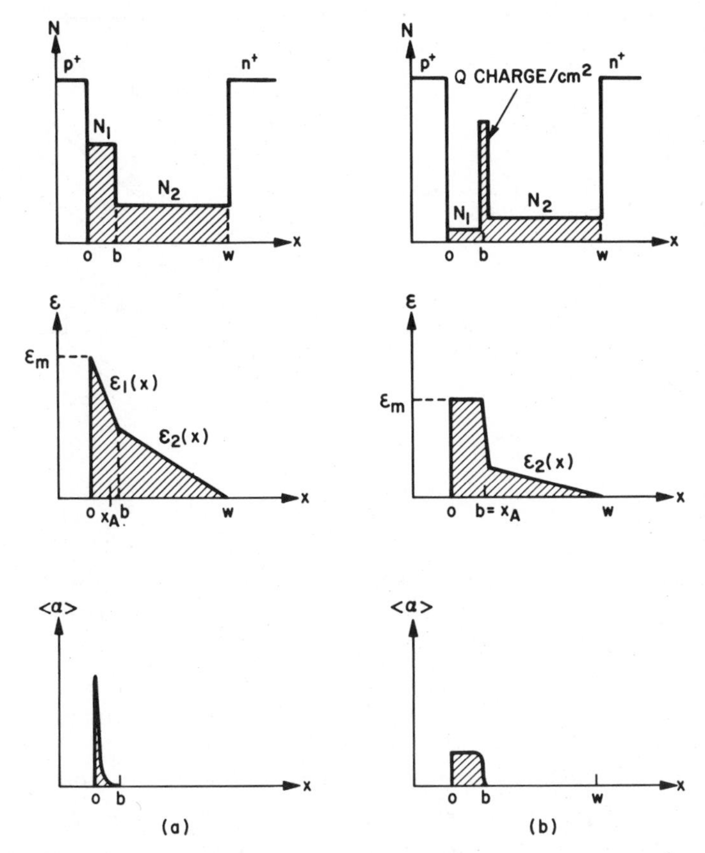

Fig. 3 Doping profile, electrical field distribution, and ionization integrand for modified Read diodes. (*a*) Hi–lo structure. (*b*) Lo–hi–lo structure.

10.2.1 Breakdown Voltage

The breakdown voltages for the one-sided abrupt junctions have been considered in Chapter 2. We can use the same method as outlined in that chapter to calculate the breakdown voltages of other diodes. For the two-sided symmetrical abrupt junctions, Fig. 2*b*, the breakdown voltage is given by

$$V_B = \frac{1}{2}\mathscr{E}_m W = \frac{\epsilon_s \mathscr{E}_m^2}{qN_B} \tag{4}$$

where $\mathscr{E}_m$ is the maximum field, which occurs at $x = 0$. The maximum fields at breakdown for Si and ⟨100⟩-oriented GaAs two-sided (symmetrical) abrupt junctions, together with the one-sided abrupt junctions, are shown[23,24] in Fig. 4. Once the doping is known, the breakdown voltage can be calculated from Eq. 4 using the maximum field value from Fig. 4. The

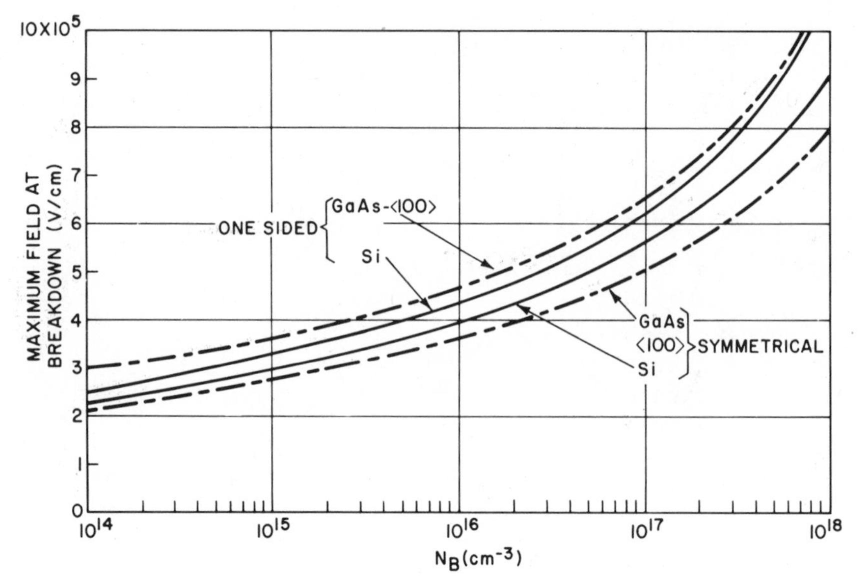

Fig. 4 Maximum electric field at breakdown versus doping for Si and GaAs one-sided and two-sided abrupt junctions. (After Sze and Gibbons, Ref. 23; Schroeder and Haddad, Ref. 24.)

applied reverse voltage at breakdown is equal to $(V_B - V_{bi})$, where V_{bi} is the built-in potential given by $2(kT/q)\ln(N_B/n_i)$ for symmetrical abrupt junctions (V_{bi} is negligible for practical IMPATT diodes).

For the Read diode and the hi–lo diode, the breakdown voltage and the total depletion width W are given by

$$V_B = \left(\mathscr{E}_m - \frac{qN_1b}{2\epsilon_s}\right)b - \frac{1}{2}\left(\mathscr{E}_m - \frac{qN_1b}{\epsilon_s}\right)(W-b) \tag{5}$$

$$W = \frac{\epsilon_s\mathscr{E}_m}{qN_2} - b\left(\frac{N_1}{N_2} - 1\right). \tag{6}$$

For a Read diode, the depletion width is limited by the thickness of the epitaxial layer, which is smaller than the width calculated from Eq. 6. The maximum field at breakdown for a Read diode or a hi–lo diode with a given N_1 is found to be essentially the same (within 1%) as the value of the one-sided abrupt junction with the same N_1, provided that the avalanche width x_A is smaller than b.[25] Therefore, the breakdown voltage can be calculated from Eqs. 5 and 6 using the maximum field value of Fig. 4.

For the lo–hi–lo diode with a very narrow clump, the breakdown voltage and the total depletion width are given by

$$V_B = \mathscr{E}_m b + \frac{1}{2}\left(\mathscr{E}_m - \frac{qQ}{\epsilon_s}\right)(W-b) \tag{7}$$

$$W = \frac{\epsilon_s}{qN_2}\left(\mathscr{E}_m - \frac{qQ}{\epsilon_s}\right) + b \tag{8}$$

where Q is the number of impurities/cm^2 in the clump. Since the maximum field is nearly constant for $0 \leq x \leq b$, $\langle \alpha \rangle$ is equal to $1/b$. The maximum field $\mathscr{E}_m$ can be calculated from the field-dependent ionization coefficient.

10.2.2 Avalanche Region and Drift Region

The avalanche region of an ideal *p-i-n* diode is the full intrinsic layer width. For the Read diode and *p-n* junctions, however, the region of carrier multiplication is restricted to a narrow region close to the metallurgical junction. The contribution to the integral in Eq. 2 decreases rapidly as x departs from the metallurgical junction. Thus a reasonable definition of the avalanche region width x_A is obtained by taking the distance over which 95% of the contribution to the integral is obtained, that is,

$$\int_0^{x_A} \langle \alpha \rangle \, dx = 0.95 \tag{9}$$

for the Read diode, hi–lo diode, and one-sided abrupt junctions, and

$$\int_{-x_A/2}^{x_A/2} \langle \alpha \rangle \, dx = 0.95 \tag{10}$$

for the two-sided junctions when the diodes are biased into breakdown.

Figure 5 shows the avalanche widths as a function of the doping for Si and GaAs diodes.[24] Also shown are the depletion widths of Si and GaAs symmetrical junctions. For a given doping, the Si n^+-p junction has a narrower avalanche width than Si p^+-n junction because of the difference in ionization rates ($\alpha_n > \alpha_p$). For a Read diode or a hi–lo diode, the avalanche region will be the same as a one-sided abrupt junction with the same doping N_1. For a lo–hi–lo diode, the avalanche region width is equal to the distance between the metallurgical junction and the charge clump $x_A = b$.

The drift region is the depletion layer excluding the avalanche region, or $x_A \leq x \leq W$. The most important parameter in the drift region is the carrier drift velocity. To obtain minimum carrier transit time across the drift region, the electric field in this region should be high enough that the generated carriers can travel at their saturation velocities v_s. For silicon the electric field should be larger than 10^4 V/cm. For GaAs, the field can be much smaller ($\sim 10^3$ V/cm).

For *p-i-n* diodes, this requirement is fulfilled automatically, since at breakdown the field (which is approximately constant over the full intrinsic width) is much larger than the required field for velocity saturation. For a Read diode the minimum field in the drift region is given by $\mathscr{E}_{\min} = \mathscr{E}_m - q[N_1 b + N_2(W - b)]/\epsilon_s$.

Clearly, from the previous discussion, a Read diode can be designed so that $\mathscr{E}_{\min}$ is sufficiently large. For abrupt junctions, some regions always have fields smaller than the minimum required field. The low-field region,

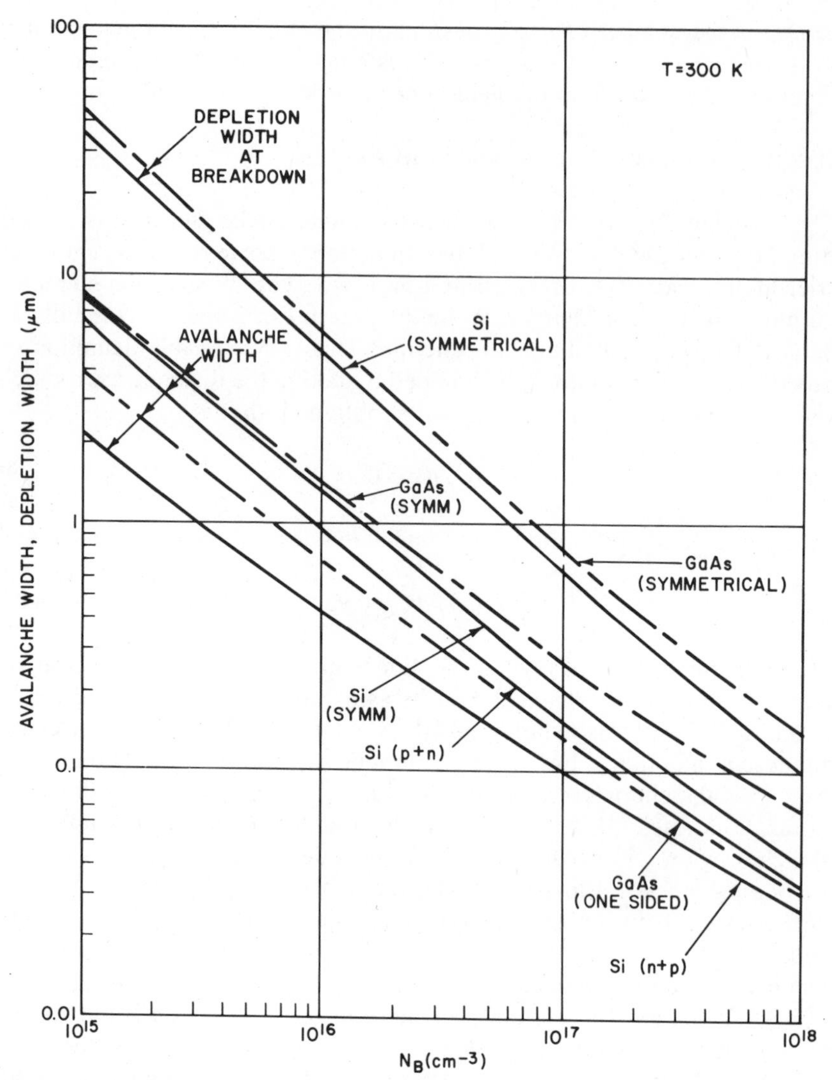

Fig. 5 Avalanche region widths for Si and GaAs junctions. Also shown are depletion widths of Si and GaAs symmetrical junctions. (After Schroeder and Haddad, Ref. 24.)

however, constitutes only a small percent of the total depletion region. For example, for a Si p^+n junction with $10^{16}\,\mathrm{cm}^{-3}$ background doping, the maximum field at breakdown is 4×10^5 V/cm. The ratio of the low-field region (for a field less than 10^4 V/cm) to the total depletion layer is $10^4/(4 \times 10^5) = 2.5\%$. For a GaAs p^+-n junction with the same doping, the low-field region is less than 0.2%. Thus the low-field region has negligible effect on the reduction of the carrier transit time across the depletion layer.

10.2.3 Temperature and Space-Charge Effects

The breakdown voltages and the maximum electric fields discussed previously are calculated for room temperature under isothermal conditions, free of space-charge effects, and in the absence of oscillation. Under operating conditions, however, the IMPATT diode is biased well into avalanche breakdown, and the current density is usually very high. This results in a considerable temperature rise in the junction and a large space-charge effect.

The ionization rates of electrons and holes decrease with increasing temperature.[26] Thus for an IMPATT diode with a given doping profile, the breakdown voltage will increase with increasing temperature. As the dc power (product of reverse voltage and reverse current) increases, both the junction temperature and the breakdown voltage increase. Eventually, the diode fails to operate, mainly because of permanent damage that results from excessive heating in localized spots. Thus the rising temperature of the junction imposes a severe limit on device operation. To prevent the rise in temperature, one must use a suitable heat sink. This will be considered in Section 10.4.

The space-charge effect is the variation of electric field in the depletion region due to generated carrier space charge. This effect gives rise to a positive dc incremental resistance for abrupt junctions and a negative dc incremental resistance for p-i-n diodes.[27]

Consider first a one-sided p^+-n-n^+ abrupt junction as shown in Fig. 6*a*. When the applied voltage V is equal to the breakdown voltage V_B, the electric field $\mathscr{E}(x)$ has its maximum absolute value $\mathscr{E}_m$ at $x = 0$. If we assume that the electrons travel at their saturation velocity v_s across the depletion region, the space-charge current I is given by

$$I = v_s \rho A \tag{11}$$

where ρ is the carrier-charge density and A the area. The disturbance $\Delta\mathscr{E}(x)$ in the electric field due to the space charge is obtained from Eq. 11 and Poisson's equation:

$$\Delta\mathscr{E}(x) \simeq \frac{Ix}{A\epsilon_s v_s}. \tag{12}$$

If we assume that all the carriers are generated within the avalanche width x_A, the disturbance in voltage caused by the carriers in the drift region $(W - x_A)$ is obtained by integrating $\Delta\mathscr{E}(x)$:

$$\Delta V_B \simeq \int_0^{W-x_A} \frac{Ix}{A\epsilon_s v_s}\,dx = I\,\frac{(W - x_A)^2}{2A\epsilon_s v_s}. \tag{13}$$

The total applied voltage is thus

$$V = V_B + \Delta V_B = V_B + IR_{SC} \tag{14}$$

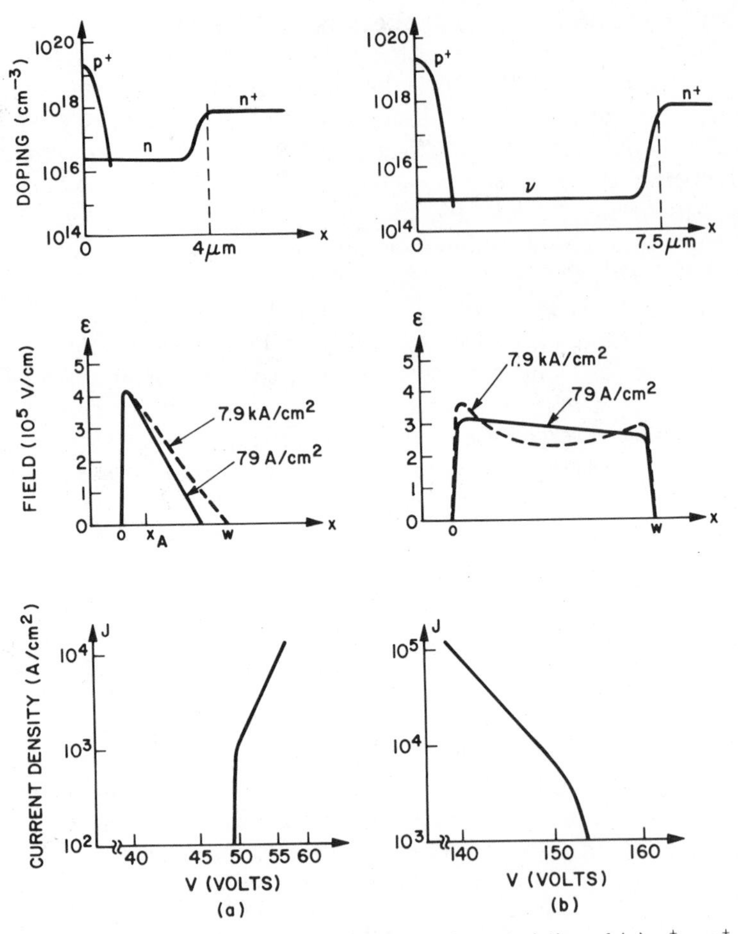

Fig. 6 Doping profile, field, and current–voltage characteristics of (a) p^+-n-n^+ and (b) p^+-ν-n^+ diodes. (After Bowers, Ref. 27.)

where R_{SC} is defined as the space-charge resistance[28] and is obtained from Eqs. 13 and 14:

$$R_{SC} \equiv \frac{\Delta V_B}{I} \simeq \frac{(W - x_A)^2}{2A\epsilon_s v_s}. \tag{15}$$

For our example shown in Fig. 6*a*, the space-charge resistance is about 20 Ω for $A = 10^{-4}$ cm^2.

For a p-i-n or a p-ν-n diode, the situation is different from that of a p^+-n junction. When the applied reverse voltage is just large enough to cause avalanche breakdown, the reverse current is small. The space-charge effect can be neglected and the electric field is essentially uniform across the depletion layer. As the current increases, more electrons are generated near the p-i boundary and more holes are generated near the n-i boundary (by impact ionization as the electric field is double-peaked, Fig. 6*b*). These space charges will cause a reduction of the field in the center of the ν region. Thus the voltage, which equals $\int_0^W \mathscr{E}\, dx$, is reduced. This reduction results in a negative incremental dc resistance for the p-ν-n diode, as shown in Fig. 6*b*.

10.3 DYNAMIC CHARACTERISTICS

10.3.1 Injection-Phase Delay and Transit-Time Effect

We consider first the injection phase delay and transit-time effect of an ideal device.[29] Assume that a conduction current pulse is injected at $x = 0$, shown in Fig. 7 with a given phase angle ϕ with respect to the total current; also assume that the applied dc voltage across the diode causes the injected carriers to travel at the saturation velocity v_s in the drift region, $0 \le x \le W$. At $x = 0$ the magnitude of the ac conduction current density $\tilde{J}_c$ equals that of the total ac current density $\tilde{J}$ with a phase delay:

$$\tilde{J}_c(x=0) = \tilde{J} \exp(-j\phi). \tag{16}$$

The total ac current anywhere in the drift region is given by the sum of the conduction current and the displacement current:

$$\tilde{J}(x) = \tilde{J}_c(x) + \tilde{J}_d(x) = \tilde{J}_c(x=0)e^{-j\omega x/v_s} + j\omega\epsilon_s\tilde{\mathscr{E}}(x) \tag{17}$$

where $\tilde{\mathscr{E}}(x)$ is the ac field. From Eqs. 16 and 17 we obtain

$$\tilde{\mathscr{E}}(x) = \frac{\tilde{J}(1 - e^{-j\omega x/v_s - j\phi})}{j\omega\epsilon_s}. \tag{18}$$

Integrating Eq. 18 gives the ac impedance

$$Z \equiv \frac{\int_0^W \tilde{\mathscr{E}}(x)\, dx}{\tilde{J}} = \frac{1}{j\omega C}\left[1 - \frac{e^{-j\phi}(1 - e^{-j\theta})}{j\theta}\right] \tag{19}$$

where C is the capacitance per unit area ϵ_s/W, and θ, the transit angle,

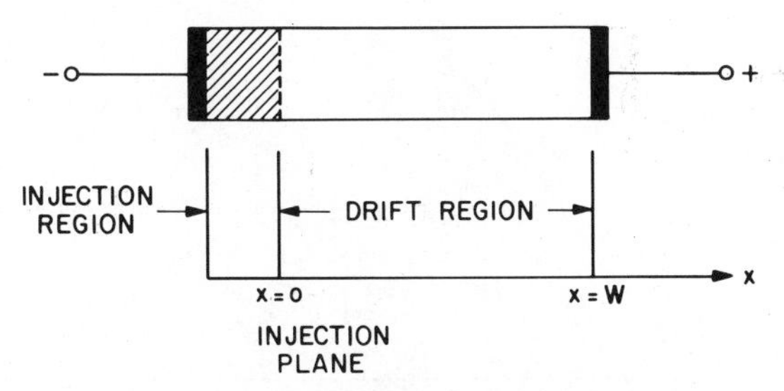

Fig. 7 Idealized diode with carrier injection at $x = 0$ and a drift region with saturation velocity.

equals $\omega W/v_s$. By taking the real and the imaginary parts of Eq. 19, we obtain

$$R = \frac{\cos\phi - \cos(\phi+\theta)}{\omega C\theta} \tag{20}$$

$$X = -\frac{1}{\omega C} + \frac{\sin(\theta+\phi) - \sin\phi}{\omega C\theta}. \tag{21}$$

We consider next the influence of the injection phase ϕ on the ac resistance R. When ϕ equals zero (no phase delay), the resistance is proportional to $(1-\cos\theta)/\theta$, which is always greater or equal to zero, as shown in Fig. 8*a*; that is, there is no negative resistance. Therefore, the transit-time effect alone cannot give rise to negative resistance. However,

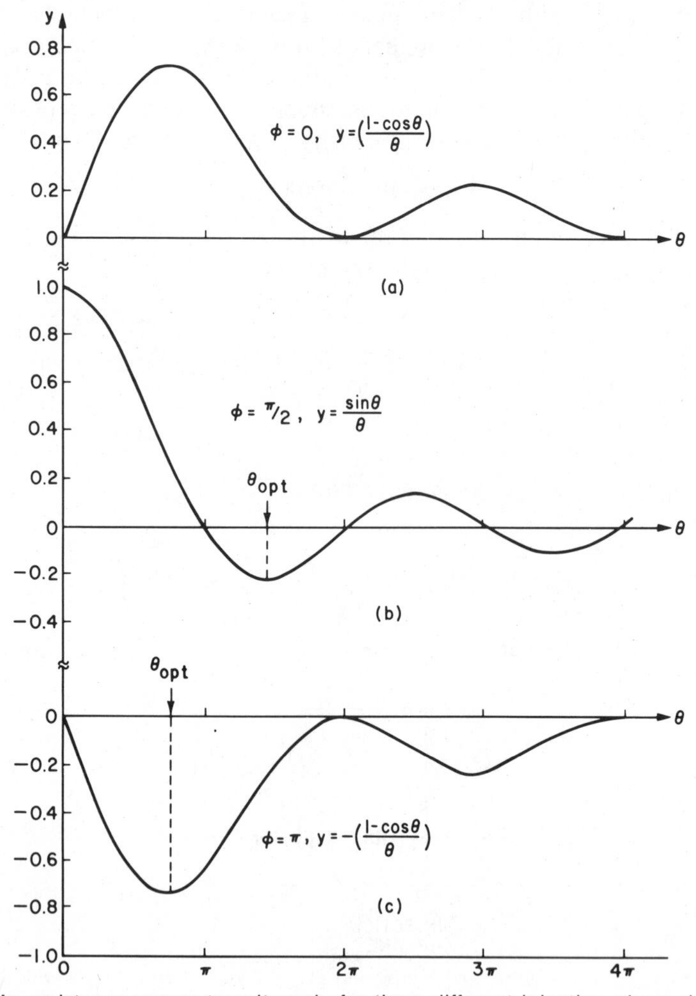

Fig. 8 Ac resistance versus transit angle for three different injection phase delays. (*a*) $\phi = 0$. (*b*) $\phi = \pi/2$. (*c*) $\phi = \pi$.

for any nonzero ϕ, the resistance is negative for certain transit angles. For example, at $\phi = \pi/2$, the largest negative resistance occurs near $\theta = 3\pi/2$, as shown in Fig. 8*b*. This is the basic operation of the BARITT diode—a phase delay of 90° is introduced by the injection of minority carriers across the barrier and a transit angle of about 270° will optimize the device operation. The detailed BARITT operation is considered in Section 10.7. For $\phi = \pi$, the largest negative resistance occurs near $\theta = \pi$, as shown in Fig. 8*c*. This corresponds to the IMPATT operation, in which the buildup of the injection current due to impact avalanche introduces a phase delay of about 180°, and the transit time in the drift region gives an additional 180° delay.

The foregoing considerations have confirmed the importance of the injection delay. The problem of finding active transit-time devices has thus been reduced to finding a means to delay the injection of conduction current into the drift region. From Fig. 8 we observe that the sum of the injection phase and the optimum transit angle, $\phi + \theta_{opt}$, is approximately equal to 2π. As ϕ increases from zero, the magnitude of the negative resistance becomes larger. Because of the larger resistance, the IMPATT diode has higher efficiency and higher power than the BARITT diode.

10.3.2 Small-Signal Analysis

The small-signal analysis was first considered by Read and developed further by Gilden and Hines.[4,7] For simplicity we assume that $\alpha_n = \alpha_p = \alpha$, and that the saturation velocities of holes and electrons are equal. Figure 9*a* shows the model of a Read diode. According to the discussion in Section 10.2, we have divided the diode into three regions: (1) the avalanche region, which is assumed to be thin so that space-charge and signal delay can be neglected; (2) the drift region, where no carriers are generated, and all carriers entering from the avalanche region travel at their saturation velocities; and (3) an inactive region that adds undesirable parasitic resistance.

The two active regions interact with one another, because the ac electric field is continuous across the boundary between them. We shall use a zero subscript to indicate dc quantities, and tilda ($\sim$) to indicate small-signal ac quantities. For quantities including both dc and ac components, no zero subscript or tilda will be added. We first define $\tilde{J}_A$ as the avalanche current density, which is the alternating conduction (particle) current density in the avalanche region, and $\tilde{J}$ as the total alternating current density. With our assumption of a thin avalanche region, $\tilde{J}_A$ is presumed to enter the drift region without delay. With the assumption of a saturation velocity v_s, the alternating conduction current density $\tilde{J}_c(x)$ in the drift region propagates as an unattenuated wave (with only phase change) at this drift velocity,

$$\begin{aligned} \tilde{J}_c(x) &= \tilde{J}_A e^{-j\omega x/v_s} \\ &\equiv \gamma \tilde{J} e^{-j\omega x/v_s} \end{aligned} \tag{22}$$

where $\gamma \equiv \tilde{J}_A/\tilde{J}$ is the complex fraction relating the avalanche current density to the total alternating current density.

At any cross section, the total alternating current density $\tilde{J}$ equals the sum of the conduction current density $\tilde{J}_c$ and the displacement current density $\tilde{J}_d$. This sum is constant, independent of position x:

$$\tilde{J} = \tilde{J}_c(x) + \tilde{J}_d(x) \neq f(x). \tag{23}$$

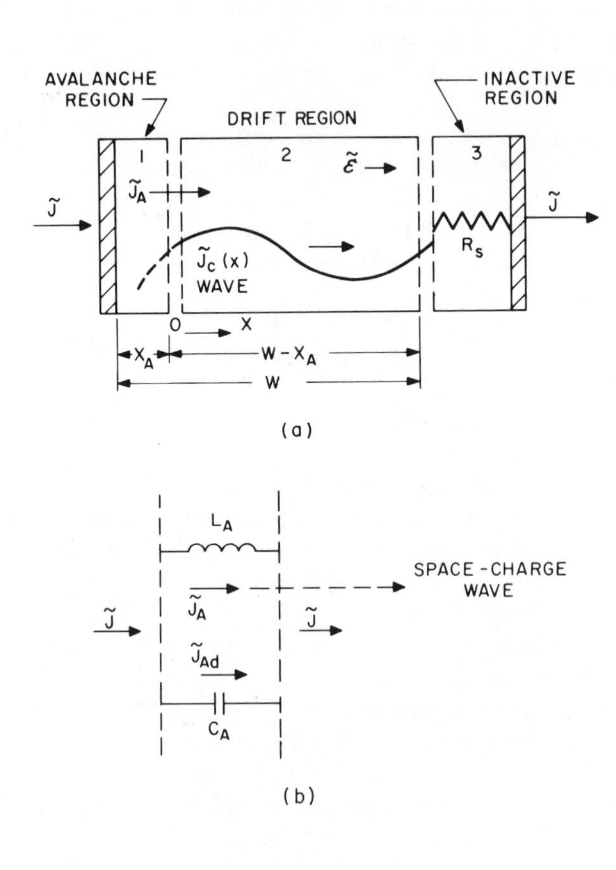

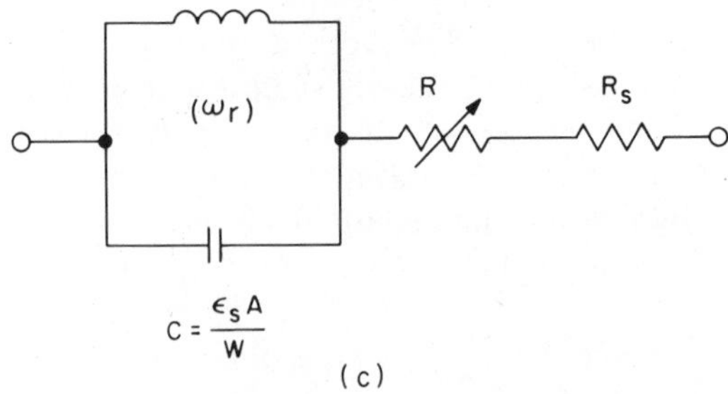

Fig. 9 (*a*), (*b*), and (*c*).

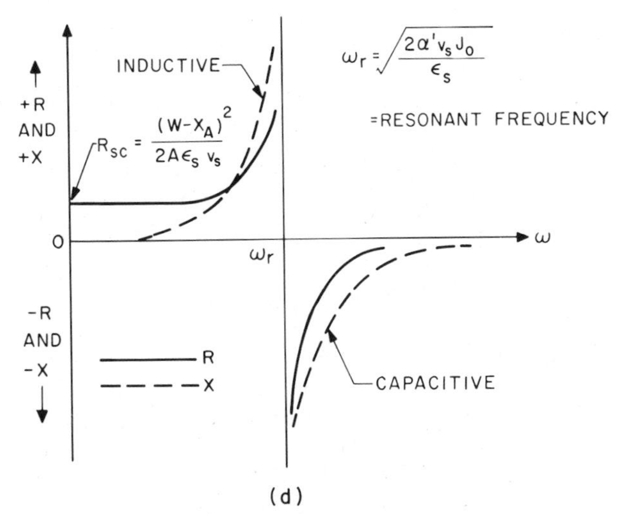

Fig. 9 (*a*) Model of Read diode with avalanche region, drift region, and inactive region. (*b*) Equivalent circuit of the avalanche region. (*c*) Equivalent circuit of Read diode for small transit angle. (*d*) Real and imaginary parts of the impedance versus resonant frequency, ω_r. (After Gilden and Hines, Ref. 7.)

The displacement current density is related to the ac field $\tilde{\mathscr{E}}(x)$ by

$$\tilde{J}_d = j\omega\epsilon_s\tilde{\mathscr{E}}(x). \tag{24}$$

Combining Eqs. 22 through 24 yields an expression for the ac electric field in the drift region as a function of x and $\tilde{J}$,

$$\tilde{\mathscr{E}}(x) = \tilde{J}\frac{1-\gamma e^{-j\omega x/v_s}}{j\omega\epsilon_s}. \tag{25}$$

Integrating $\tilde{\mathscr{E}}(x)$ gives the voltage across the drift region in terms of $\tilde{J}$. The coefficient γ is derived in the analysis that follows.

Avalanche Region Consider first the avalanche region. Under the dc condition, the direct current density $J_0(=J_{po}+J_{no})$ is related to the thermally generated reverse saturation current density $J_s(=J_{ns}+J_{ps})$ by

$$\frac{J_s}{J_0} = 1 - \int_0^W \langle\alpha\rangle\, dx. \tag{26}$$

At breakdown, J_0 approaches infinity and $\int_0^W \langle\alpha\rangle\, dx = 1$. In the dc case the ionization integral cannot be greater than unity. This is not necessarily so for a rapidly varying field. The differential equation for the current as a function of time will now be derived. Under the conditions that (1) electrons and holes have equal ionization rates and equal saturation velocities, and (2) the drift current components are much larger than the diffusion component, the basic device equations, in the one-dimensional case, can be given as follows:

$$\frac{\partial \mathscr{E}}{\partial x} = \frac{q}{\epsilon_s}(N_D^+ - N_A^- + p - n) \qquad \text{Poisson's equation} \tag{27}$$

$$\left.\begin{aligned} J_n &= qv_s n \\ J_p &= qv_s p \\ J &= J_n + J_p \end{aligned}\right\} \qquad \text{current-density equations} \tag{28}$$

$$\frac{\partial n}{\partial t} = \frac{1}{q}\frac{\partial J_n}{\partial x} + \alpha v_s(n+p) \tag{29a}$$

$$\frac{\partial p}{\partial t} = -\frac{1}{q}\frac{\partial J_p}{\partial x} + \alpha v_s(n+p) \tag{29b}$$

continuity equations.

The second terms on the right-hand side of Eq. 29 correspond to the generation rate of the electron-hole pairs by avalanche multiplication. This generation rate is so large compared to the rate of thermal generation that the latter can be neglected. Adding Eqs. 29a and 29b, using Eq. 28, and integrating from $x = 0$ to $x = x_A$ gives

$$\tau_A \frac{dJ}{dt} = -(J_p - J_n)_0^{x_A} + 2J \int_0^{x_A} \alpha\, dx \tag{30}$$

where $\tau_A = x_A/v_s$ is the transit time across the multiplication region. The boundary conditions are that the electron current at $x = 0$ consists entirely of the reverse saturation current J_{ns}. Thus at $x = 0$, $J_p - J_n = -2J_n + J = -2J_{ns} + J$. At $x = x_A$ the hole current consists of the reverse saturation current J_{ps} generated in the space-charge region, so $J_p - J_n = 2J_p - J = 2J_{ps} - J$. With these boundary conditions, Eq. 30 becomes

$$\frac{dJ}{dt} = \frac{2J}{\tau_A}\left[\int_0^{x_A} \alpha\, dx - 1\right] + \frac{2J_s}{\tau_A}. \tag{31}$$

In the dc case, J is the direct current J_0, so that Eq. 31 reduces to Eq. 26.

We will simplify Eq. 31 by substituting $\bar{\alpha}$ in place of α, where $\bar{\alpha}$ is an average value of α obtained by evaluating the integral over the extent of the avalanche region. We obtain (by neglecting the term J_s)

$$\frac{dJ}{dt} = \frac{2J}{\tau_A}(\bar{\alpha} x_A - 1). \tag{32}$$

The small-signal assumptions are now made:

$$\begin{aligned} \bar{\alpha} &= \bar{\alpha}_0 + \tilde{\alpha} e^{j\omega t} \simeq \bar{\alpha}_0 + \alpha' \tilde{\mathscr{E}}_A e^{j\omega t} \\ \bar{\alpha} x_A &= 1 + x_A \alpha' \tilde{\mathscr{E}}_A e^{j\omega t} \\ J &= J_0 + \tilde{J}_A e^{j\omega t} \\ \mathscr{E} &= \mathscr{E}_0 + \tilde{\mathscr{E}}_A e^{j\omega t} \end{aligned} \tag{33}$$

where $\alpha' \equiv \partial\alpha/\partial\mathscr{E}$ and the substitution $\tilde{\alpha} = \alpha' \tilde{\mathscr{E}}_A$ has been employed. Substituting the expressions above into Eq. 32, neglecting products of higher-order terms, leads to the expression for the ac component of the avalanche conduction current density,

$$\tilde{J}_A = \frac{2\alpha' x_A J_0 \tilde{\mathscr{E}}_A}{j\omega\tau_A}. \tag{34}$$

The displacement current in the avalanche region is given by

$$\tilde{J}_{Ad} = j\omega\epsilon_s \tilde{\mathscr{E}}_A . \tag{35}$$

These are the two components of the total circuit current in the avalanche region. For a given field, the avalanche current $\tilde{J}_A$ is reactive and varies inversely with ω as in an inductor. The other component, J_{Ad}, is also reactive and varies directly with ω as in a capacitor. Thus the avalanche region behaves as an LC parallel circuit. The equivalent circuit is shown in Fig. 9b, where the inductance and capacitance are given as (where A is the diode area)

$$L_A = \tau_A / 2J_0\alpha' A$$
$$C_A = \epsilon_s A / x_A . \tag{36}$$

The resonant frequency of this combination is given by

$$f_r \equiv \frac{\omega_r}{2\pi} = \frac{1}{2\pi}\sqrt{\frac{2\alpha' v_s J_0}{\epsilon_s}}. \tag{37}$$

The impedance of the avalanche region has the simple form

$$Z_A = \frac{x_A}{j\omega\epsilon_s A}\left(\frac{1}{1-\omega_r^2/\omega^2}\right) = \frac{1}{j\omega C_A}\left(\frac{1}{1-\omega_r^2/\omega^2}\right). \tag{38}$$

The factor γ has the form

$$\gamma \equiv \frac{\tilde{J}_A}{\tilde{J}} = \frac{1}{1-\omega^2/\omega_r^2}. \tag{39}$$

A thin avalanche region, therefore, behaves as an antiresonant circuit with a resonant frequency proportional to the square root of the direct current density J_0, Eq. 37.

Drift Region and Total Impedance Combining Eqs. 39 and 25 and integrating over the drift length $(W - x_A)$ give an expression for the ac voltage across this region,

$$\tilde{V}_d = \frac{(W - x_A)\tilde{J}}{j\omega\epsilon_s}\left[1 - \frac{1}{1-\omega^2/\omega_r^2}\left(\frac{1-e^{-j\theta_d}}{j\theta_d}\right)\right] \tag{40}$$

where θ_d is the transit angle of the drift space

$$\theta_d \equiv \frac{\omega(W - x_A)}{v_s} \equiv \omega\tau_d \tag{41}$$

and

$$\tau_d = \frac{(W - x_A)}{v_s}. \tag{41a}$$

We may also define $C_d \equiv A\epsilon_s/(W - x_A)$ as the capacitance of the drift region. From Eq. 40 we obtain the impedance for the drift region,

$$Z_d \equiv \frac{\tilde{V}_d}{\tilde{J}A} = \frac{1}{\omega C_d}\left[\frac{1}{1-\omega^2/\omega_r^2}\left(\frac{1-\cos\theta_d}{\theta_d}\right)\right] + \frac{j}{\omega C_d}\left[-1 + \frac{1}{1-\omega^2/\omega_r^2}\left(\frac{\sin\theta_d}{\theta_d}\right)\right]$$
$$= R + jX \quad \Omega \tag{42}$$

where R and X are the resistance and reactance, respectively. Obviously the real part (resistance) will be negative for all frequencies above ω_r except for nulls at $\theta_d = 2\pi \times \text{integer}$. The resistance is positive for frequencies below ω_r and approaches a finite value at zero frequency:

$$R(\omega \to 0) = \frac{\tau_d}{2C_d} = \frac{(W - x_A)^2}{2A\epsilon_s v_s} \quad \Omega.$$

The low-frequency small-signal resistance is a consequence of the space charge in the finite thickness of the drift region, and the expression above is identical to Eq. 15, derived previously.

The total impedance is the sum of the impedances of the avalanche region, drift region, and passive resistance R_s of the inactive region:

$$Z = \frac{(W - x_A)^2}{2A\epsilon_s v_s}\left(\frac{1}{1-\omega^2/\omega_r^2}\right)\frac{1-\cos\theta_d}{\theta_d^2/2} + R_S$$
$$+ \frac{j}{\omega C_d}\left[\left(\frac{\sin\theta_d}{\theta_d} - 1\right) - \left(\frac{\dfrac{\sin\theta_d}{\theta_d} + \dfrac{x_A}{W - x_A}}{1 - \omega_r^2/\omega^2}\right)\right] \quad \Omega. \tag{43}$$

Equation 43 has been cast in a form that can be simplified directly for the case of small transit angle θ_d. For $\theta_d < \pi/4$, Eq. 43 reduces to

$$Z = \frac{(W - x_A)^2}{2Av_s\epsilon_s(1 - \omega^2/\omega_r^2)} + R_S + \frac{j}{\omega C}\left[\frac{1}{(\omega_r^2/\omega^2) - 1}\right] \tag{44}$$

where $C \equiv \epsilon_s A/W$ corresponding to the total depletion capacitance. From Eq. 44, note that the first term is the active resistance, which becomes negative for $\omega > \omega_r$. The third term is reactive and corresponds to a parallel resonant circuit that includes the diode capacitance and a shunt inductor. The reactance is inductive for $\omega < \omega_r$ and capacitive for $\omega > \omega_r$. In other words, the resistance becomes negative at the frequency where the reactive component changes sign. The equivalent circuit and frequency dependence of the real and imaginary parts of the impedance are shown in Fig. 9*c* and *d*, respectively.

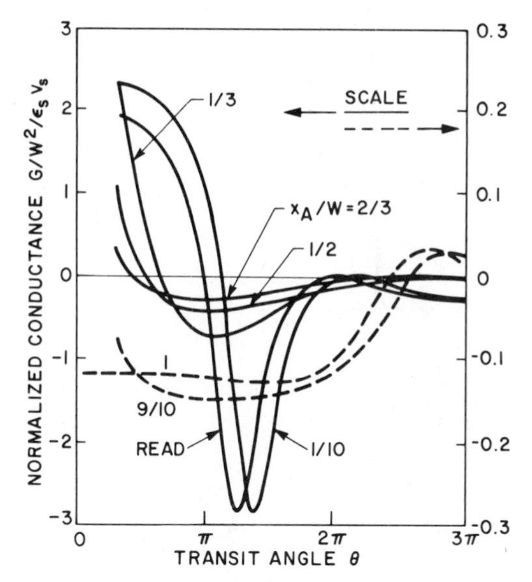

Fig. 10 Normalized conductance versus transit angle for six IMPATT diodes with different avalanche region width. (After Misawa, Ref. 30.)

The ac impedance (Z) is related to the ac admittance (Y) as follows:

$$Z \equiv \frac{\tilde{V}}{\tilde{J}A} = R + jX = \frac{1}{G + jB} = \frac{1}{Y} \tag{45}$$

where G is the conductance and B is the susceptance. The calculated small-signal conductances for six IMPATT diodes are shown[30] in Fig. 10. These diodes have identical total depletion widths but different avalanche widths ($x_A/W = \frac{1}{10}, \frac{1}{3}, \frac{1}{2}, \frac{2}{3}, \frac{9}{10}$, and 1). Note that as the avalanche width widens, the negative conductance band also widens. The Read diode has a large negative conductance in a very narrow band. On the other hand, the Misawa diode (p-i-n diode with $x_A/W = 1$) has small negative conductance, but the band is broad and flat extending to zero frequency.

10.4 POWER AND EFFICIENCY

10.4.1 Large-Signal Operation

Figure 11 shows a Read diode under large-signal operation. A high-field avalanche region exists at the p^+-n junction (Fig. 11b), where electron–hole pairs are generated; and a constant-field drift region exists in the low-doped ν region. The generated holes quickly enter the p^+ region and the generated electrons are injected into the drift region (Fig. 11c), where they do work that produces external power. As the electric field changes

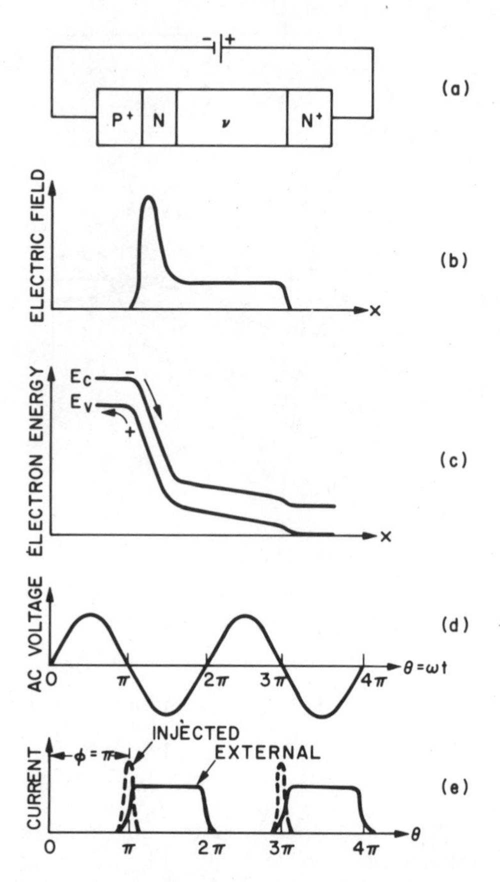

Fig. 11 Read diode (*a*) p^+-n-ν-n^+ structure, (*b*) field at avalanche breakdown, (*c*) energy-band diagram, (*d*) ac voltage, and (*e*) injected and external currents. (After Read, Ref. 4.)

periodically with time around an average value (Fig. 11*d*), the impact ionization rate per carrier follows the field change nearly instantaneously. However, the carrier density does not follow the field change in unison, because carrier generation also depends on the number of carriers already present. Even after the field has passed its maximum value, the carrier density keeps increasing because the carrier generation rate is still above the average value. The maximum carrier density is reached approximately when the field has decreased from the peak to the average value. Thus the ac variation of the injected carrier density lags the ac voltage by about 90° (this is the "avalanche delay"), even though the ionization rate is in phase with the ac field. The situation described above is illustrated as the "injected" current (Fig. 11*e*). The peak value of ac field (or voltage) occurs at $\pi/2$, but the peak of the injected carrier density occurs at π (i.e., $\phi = \pi$, as shown). The injected carriers then enter the drift region, where they traverse at saturation velocity, introducing the "transit-time delay." The induced external current is also shown in Fig. 11*e*. Comparing the ac voltage and the external current clearly shows that the diode exhibits a negative resistance at its terminals.

The detailed large-signal performance can be obtained using Eqs. 27 through 29 with approximate boundary conditions. The computed results for a Si IMPATT diode (p^+-n-ν-n^+ with $N_1 = 10^{16}\,\text{cm}^{-3}$, $b = 1\,\mu\text{m}$, $N_2 = 10^{15}\,\text{cm}^{-3}$, and $W = 6\,\mu\text{m}$) is shown in Fig. 12 for the electric field and carrier density variations at quarter-cycle intervals.[31] The electric field, hole density, and electron density are shown as functions of the distance over the depletion region. A phase plot of the terminal current and voltage of the oscillation is included. Note that (1) generation of pulses of holes and electrons begins when the voltage is maximum, and a quarter-cycle later

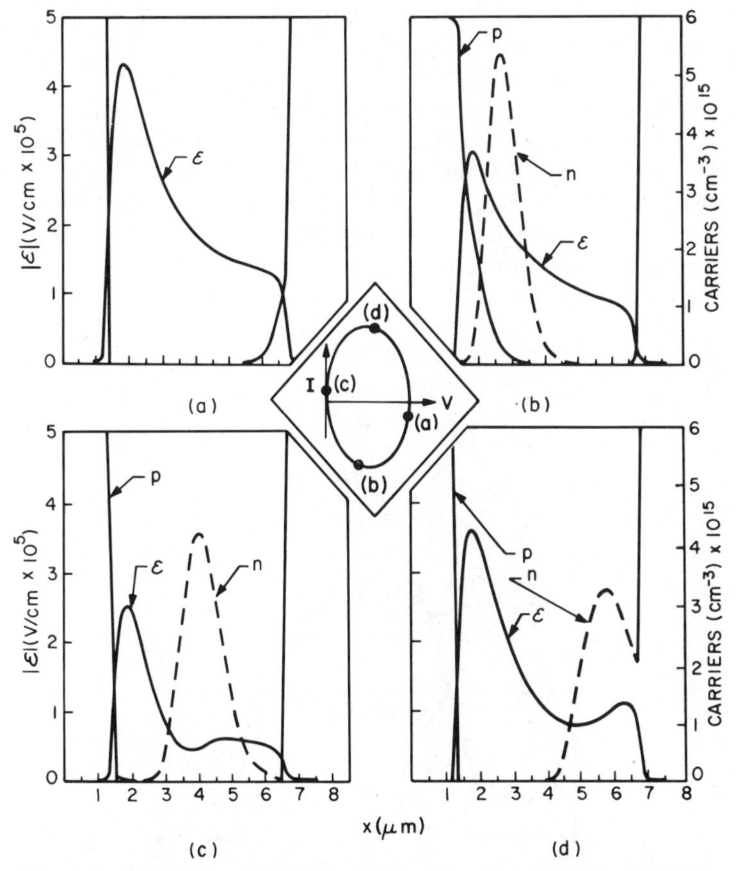

Fig. 12 Computed results for IMPATT operation. Four cases are shown at one-quarter-cycle intervals. A phase plot of the terminal current and voltage of the oscillation is included in the center insert. (After Scharfetter and Gummel, Ref. 31.)

the charge pulses are fully formed and begin drifting into their respective drift spaces; (2) holes disappear quickly from the active region while the electrons drift for approximately a half-cycle and constitute positive particle current while the ac voltage is negative; (3) for the remaining quarter-cycle the remnants of the electron charge pulse are swept out of the picture as the voltage again approaches its maximum value; (4) the displacement current is quite large and has an appreciable swing into the forward direction (positive on p^+), while the terminal voltage always remains in the reverse polarity; and (5) for sufficient modulation of diode voltage and particle current to give efficient oscillation, an extended avalanche region is required.

The terminal voltage and particle current for the diode shown in Fig. 12 are plotted in Fig. 13. The phase relations obtained between the terminal voltage and particle current are close to those of the idealized situation shown in Fig. 11.

The susceptance and negative conductance of the diode described above under large-signal operation are presented in Fig. 14 for various ac voltages. Also indicated on the plot are the calculated efficiencies. Note that the susceptance generally increases with increasing ac voltage amplitude, whereas the negative conductance decreases. The calculated result shows efficiencies as high as 18%. This efficiency is close to the highest obtained in Si IMPATT diode. However, for certain GaAs-modified Read diodes, efficiencies close to 40% have been obtained. The limitations on efficiency and power output will now be considered.

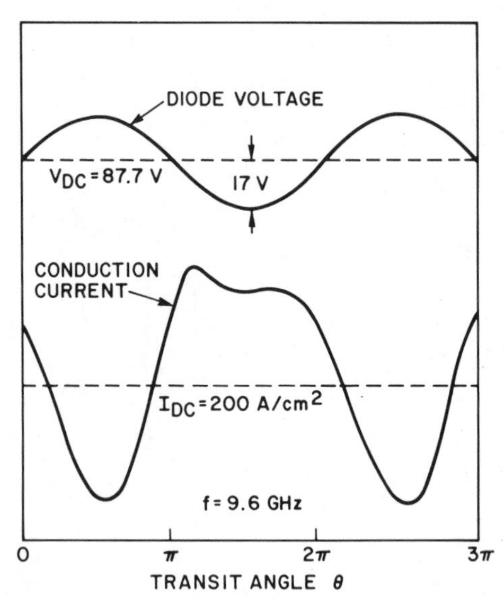

Fig. 13 Terminal voltage and particle current for the diode shown in Fig. 12 operated at 9.6 GHz. (After Scharfetter and Gummel, Ref. 31.)

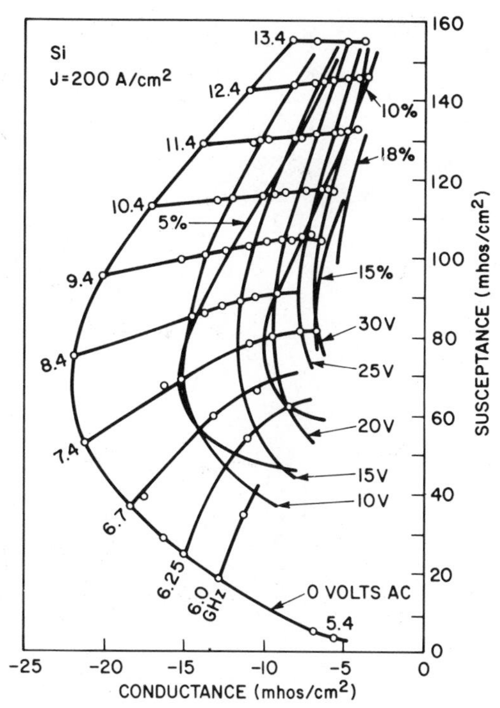

Fig. 14 Numerical calculations of negative conductance and susceptance. (After Scharfetter and Gummel, Ref. 31.)

10.4.2 Power-Frequency Limitation—Thermal

At lower frequencies, the cw performance of an IMPATT diode is limited primarily by thermal considerations, that is, by the power that can be dissipated in a semiconductor chip. A typical device mounting arrangement is shown in Fig. 15*a*. A simplified diode and heat-sink structure is shown in Fig. 15*b*. The total thermal resistance for a circular heat source of radius r at a depth d_s in the silicon is given by[32]

$$R_T = R_s + R_t + R_g + R_n + R_c$$
$$= \frac{1}{A}\left(\frac{d_s}{\kappa_s} + \frac{d_t}{\kappa_t} + \frac{d_g}{\kappa_g} + \frac{d_n}{\kappa_n}\right) + \frac{1}{4\pi\kappa_c}. \tag{46}$$

The symbols are defined in Fig. 15*b*. The last term gives the thermal spreading resistance for the infinite half-space heat sink. The various components of the thermal resistance are shown in Fig. 15*c*. The dashed curves show the thermal spreading resistance of a diamond (type II) heat sink and the corresponding total thermal resistance R_T. The thermal conductivity at 300 K is assumed to be three times that of copper, and the thermal conductivity for silicon is for the assumed maximum operating

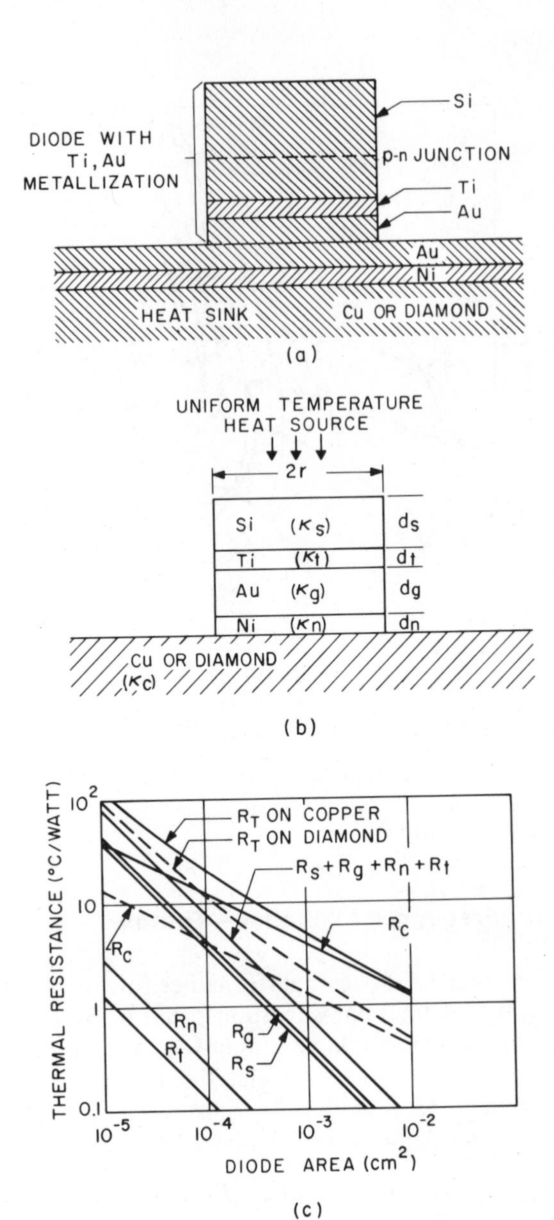

Fig. 15 (*a*) Diode mounted on copper heat sink with metallization used for gold-to-gold thermal compression bond. (*b*) Simplified diode and heat-sink structure. (*c*) Various components and the total thermal resistance versus diode area. (After Swan, Misawa, and Marinaccio, Ref. 32.)

Table 1 Thermal Conductivity and Typical Layer Thickness for Materials in a 15- to 18 GHz Diode (300 K)

Material	Thermal Conductivity[a] κ (W/cm-K)	Thickness $d(\mu m)$	d/κ $(10^{-4} \times cm^2\text{-K/W})$
Silicon	0.80	3.00	3.80
Titanium	0.16	0.02	0.13
Gold	3.00	12.50	4.20
Nickel	0.71	0.20	0.28
Copper	3.90	—	—
Diamond	20.00	—	—

[a] κ for Si is for a temperature of 500 K, which is assumed as a maximum operating temperature.

temperature of 500 K (see Table 1). Clearly, a diamond heat sink reduces the thermal resistance R_T by a factor of about 2 and R_T decreases as the diode area increases.

The power P, which can be dissipated in the diode, must equal the heat power that can be transmitted to the heat sink. Therefore, P equals $\Delta T/R_T$, where ΔT is the temperature difference between the junction and the heat sink. If the reactance $2\pi fC$ (where $C = A\epsilon_s/W$) is maintained constant, and the major contribution to the thermal resistance is from the semiconductor (assuming that $d_s \approx W$), we obtain for a given temperature increase ΔT,

$$Pf = \left(\frac{\Delta T}{R_T}\right)f \sim \left[\frac{\Delta T}{(1/A)(W/\kappa_s)}\right]\left(\frac{W}{A\epsilon_s}\right) = \frac{\kappa_s\,\Delta T}{\epsilon_s} = \text{constant}. \qquad (47)$$

Under such conditions, the cw power output will decrease as $1/f$.

10.4.3 Power-Frequency Limitation—Electronic

Because of the inherent limitations of semiconductor materials and the attainable impedance levels in microwave circuitry, the maximum output power at a given frequency of a single diode is limited. The limitations on semiconductor materials are (1) the critical electric field $\mathscr{E}_m$ at which the avalanche breakdown occurs, and (2) the saturation velocity v_s, which is the maximum attainable velocity in the semiconductor.

The maximum voltage that can be applied across a semiconductor sample is limited by the breakdown voltage, which, for a uniform avalanche, is given by $V_m = \mathscr{E}_m W$, where W is the depletion-layer width. The maximum current that can be carried by the semiconductor sample is also limited by the avalanche breakdown process, because the current in the space-charge region causes an increase in the electric field (from Poisson's equation). With the maximum field again given by $\mathscr{E}_m$, we obtain $I_m = \mathscr{E}_m \epsilon_s v_s A/W$. Therefore, the upper limit on the power input is given by the product of V_m and I_m:

$$P_m = V_m I_m = \mathscr{E}_m^2 \epsilon_s v_s A \, . \tag{48}$$

The transit-time frequency corresponding to $\theta = \pi$ transit angle is given by

$$f = \frac{v_s}{2(W - x_A)} \tag{49}$$

$$\simeq \frac{v_s}{2W} \quad \text{for} \quad W \gg x_A. \tag{49a}$$

Equation 48 can be rewritten as

$$P_m f^2 \simeq \mathscr{E}_m^2 v_s^2 / 4\pi X_c \tag{50}$$

where X_c is the reactance $(2\pi f C)^{-1}$.

Assuming that we are limited to some minimum circuit impedance, Eq. 50 predicts that the maximum power that can be given to the mobile carriers decreases as $1/f^2$. This electronic limit is expected to be dominant above millimeter-wave frequencies (>30 GHz) for Si and GaAs.

Therefore, under cw conditions, at lower frequencies we have thermal limitation ($P \sim 1/f$) and at higher frequencies we have electronic limitation ($P \sim 1/f^2$). The corner frequency for a given semiconductor depends on the maximum allowed temperature rise, the minimum attainable circuit impedance, and the product of $\mathscr{E}_m$ and v_s.

For a practical operating junction temperature of 150° to 200°C, $\mathscr{E}_m$ in Si is about 10% smaller than that in GaAs. On the other hand, v_s in Si is almost twice as large as GaAs. Therefore, in the electronic-limited range (i.e., above millimeter wave frequencies), the Si IMPATT diode is expected to have a power output about three times larger than that of a GaAs IMPATT diode operated at the same frequency.[33] In the submillimeter wave region, the uniform-field Misawa diodes are expected to be preferred, because the device has a broad negative-resistance band and the transit-time effects are not playing the dominating role in producing negative resistance as they do in Read diodes.[34]

Under pulsed conditions where thermal effects can be ignored (i.e., short pulses), the peak power capability is determined by electronic limits (i.e., $P \propto 1/f^2$) at all frequencies.

10.4.4 Limitation on Efficiency

For efficient operation of an IMPATT diode, as carriers move through the drift region, as large a charge pulse Q_m as possible must be generated in the avalanche region without reducing the electric field in the drift region below that required for velocity saturation. The motion of Q_m through the drift region results in an ac voltage amplitude mV_D, where m is the modulation factor ($m \leq 1$) and V_D is the average voltage developed across the drift region. At the optimum frequency ($\sim v_s/2W$), the motion of Q_m also results in an alternating particle current that has a phase delay of ϕ with the ac voltage across the diode. The average of the particle current is the average current J_0. The particle current swing is therefore at most from zero to $2J_0$. For a square wave of particle current and a sinusoidal variation of drift voltage, both with magnitude and phase as described above, the microwave power generating efficiency η is[31,35]

$$\eta \equiv \frac{\text{ac power output}}{\text{dc power input}} = \frac{(2J_0/\pi)(mV_D)}{J_0(V_A + V_D)} |\cos\phi|$$

$$= \frac{2m}{\pi} \frac{1}{1 + V_A/V_D} |\cos\phi| \tag{51}$$

where V_A is the dc voltage drop across the avalanche region, and the sum of V_A and V_D is the total applied dc voltage. The angle ϕ is the injection phase delay of the particle current. Under ideal conditions, ϕ is 180° and $|\cos\phi| = 1$. For double-drift diodes, the voltage V_D is replaced by $2V_D$. The ac power contribution from the avalanche region is neglected because the avalanche region voltage is inductively reactive relative to the particle current. The displacement current is capacitively reactive relative to the diode voltage and therefore contributes no average ac power.

Equation 51 clearly shows that to improve the efficiency one must increase the ac voltage modulation factor, optimize the phase delay angle toward 180°, and reduce the V_A/V_D ratio. However, V_A must be sufficiently large to initiate the avalanche process rapidly; below a certain optimum value of V_A/V_D, the efficiency falls off toward zero.[31]

If the velocity–field relation for the drifting carriers were such that the velocity were saturated at very low electric fields, m could approach unity and no falling out of velocity saturation would occur. In n-type GaAs, the velocity is effectively saturated near 10^3 V/cm, which is much smaller than the value for n-type Si, $\sim 2 \times 10^4$ V/cm. Hence much larger ac voltage swings can be expected in n-GaAs; these larger voltage swings, in turn, result in larger efficiency[36–38] in n-GaAs.

To estimate the optimum value of V_A/V_D, we shall assume that the transit-time frequency, given by Eq. 49, is about 20% larger than the resonant frequency given by Eq. 37:

$$f = \frac{v_s}{2(W - x_A)} = 1.2f_r = \frac{1.2}{2\pi}\sqrt{\frac{2\alpha' v_s J_0}{\epsilon_s}}. \tag{52}$$

The voltage V_D is given by

$$V_D = \langle \mathscr{E}_D \rangle (W - x_A) = \langle \mathscr{E}_D \rangle v_s/2f \tag{53}$$

where $\langle \mathscr{E}_D \rangle$ is the average drift field in the drift region. For 100% current modulation, $J_0 = J_{dc} = J_{ac}$, and a maximum charge $Q_m = m\epsilon_s \langle \mathscr{E}_D \rangle$ determines the current density:

$$J_0 = Q_m f = m\epsilon_s \langle \mathscr{E}_D \rangle f. \tag{54}$$

For an ionization coefficient given by $\alpha \sim \mathscr{E}^\zeta$, the value of α' can be obtained as

$$\alpha' \equiv \frac{d\alpha}{d\mathscr{E}} = \frac{\zeta\alpha}{\mathscr{E}} \simeq \frac{\zeta(W - x_A)\alpha}{V_D}. \tag{55}$$

Combining Eqs. 52 through 55 yields[35]

$$(V_A/V_D)_{\text{opt}} \simeq 4m\left(\frac{1.2}{2\pi}\right)^2 \zeta\alpha x_A. \tag{56}$$

For $f \simeq 10$ GHz, the optimum value of V_A/V_D for GaAs is 0.65 with $m \simeq 1$, while for Si the optimum value is about 1.1 with $m \simeq \frac{1}{2}$.

A plot of the efficiency versus V_A/V_D is shown in Fig. 16*a*. The maximum efficiency is obtained using the optimum values discussed above. The

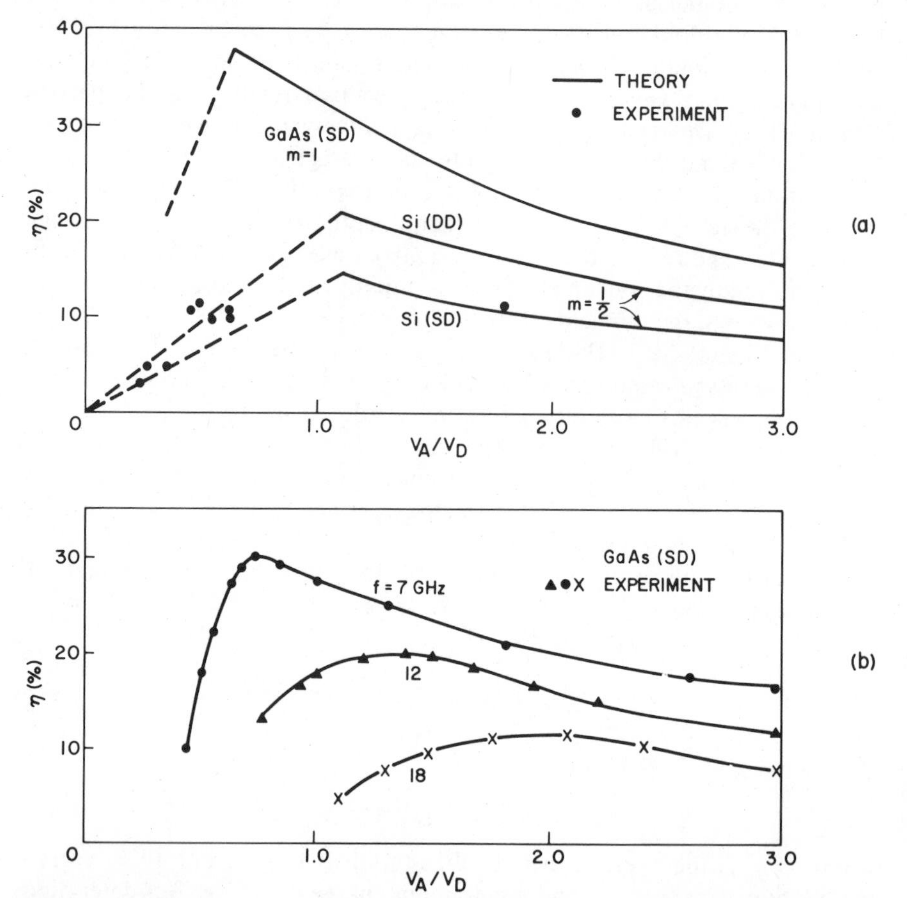

Fig. 16 (*a*) Efficiency versus V_A/V_D for Si and GaAs diodes (After Seidel, Niehaus, and Iglesias, Ref. 35.) (*b*) Measured efficiency versus V_A/V_D for GaAs hi–lo diodes. (After Nishitani et al., Ref. 39.)

expected maximum efficiency is about 15% for single-drift (SD) Si diodes, 21% for double-drift (DD) Si diodes, and 38% for single-drift n-GaAs diodes. The foregoing estimates are consistent with experimental results. At higher frequencies, the optimum ratio V_A/V_D tends to increase; this increase results in a reduction of the maximum efficiency. The experimental results for n-GaAs single-drift diodes are shown in Fig. 16*b*, which are in agreement with the foregoing discussion.[39]

In practical IMPATT diodes, many other factors reduce the efficiency. These factors include the space-charge effect, the reverse saturation current, series resistance, the skin effect, saturation of ionization rate, tunneling effect, intrinsic avalanche response time, and minority-carrier storage effects.

The space-charge effect[40] is shown in Fig. 17. The generated holes will depress the field (Fig. 17*a*). The reduction in field may turn off the avalanche process prematurely and thus reduce the 180° phase delay provided by the avalanche. As the holes drift to the right (Fig. 17*b*), the

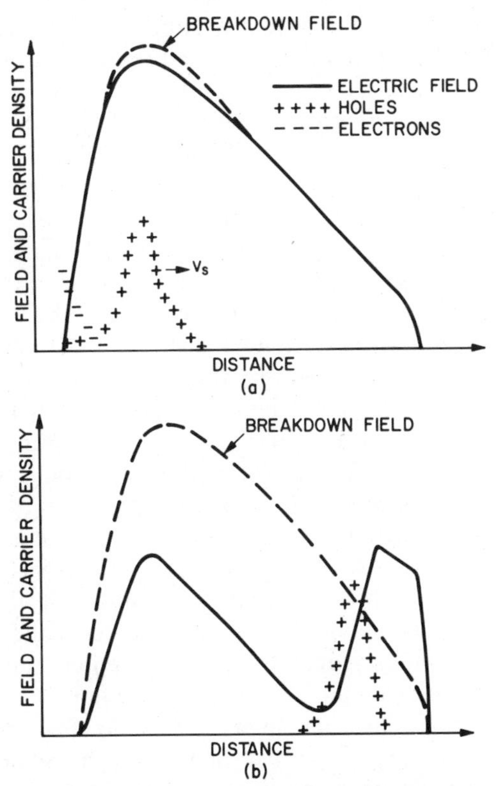

Fig. 17 Instantaneous field and charge distributions in a Read diode. (*a*) Avalanching just completed, charge beginning to move across diode. (*b*) Charge transit nearly completed. Note the strong effect of the space charge in depressing the electric field. (After Evans, Ref. 40.)

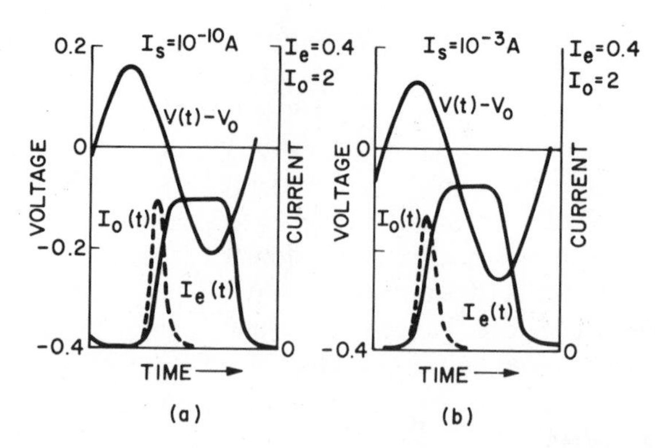

Fig. 18 Phase shifts of injection current $I_0(t)$ and terminal current $I_e(t)$, with respect to diode ac voltage. Computed results for saturation currents of (*a*) 10^{-10} A and (*b*) 10^{-3} A. (After Misawa, Ref. 41.)

space charge may also cause the field to the left of the carrier pulse to drop below that required for velocity saturation. This drop, in turn, will change the terminal current waveforms and reduce the power generated at the transit-time frequency.

The reverse-saturation-current effect[41] is shown in Fig. 18. A high saturation current causes the avalanche to build up too soon, reducing the avalanche phase delay. For $I_s = 10^{-10}$ A, $|\cos \phi|$ is close to 1, but for $I_s = 10^{-3}$ A, $|\cos \phi|$ is only 0.43 and the efficiency decreases by about a factor of 2. The minority injection from a poor ohmic contact will also increase the reverse saturation current and thus reduce the efficiency.

The effect of an unswept epitaxial layer[42,43] is shown in Fig. 19. The unswept layer gives rise to a series resistance that reduces the terminal negative resistance. Note, however, that the effect on *n*-GaAs is much smaller, because GaAs has much higher mobility at low fields.

The skin effect[44] is shown in Fig. 20. As the operating frequency of an IMPATT diode is increased into the millimeter wave region, the current will be confined to flow within a skin depth δ of the surface of the substrate. Thus the effective resistance of the substrate is increased, giving rise to a voltage drop across the radius of the diode (Fig. 20*b*). This voltage drop will cause nonuniform current distribution in the diode and an effective series resistance, both of which reduce efficiency.

For very high frequency operation, the depletion width becomes quite narrow and the field required for impact ionization becomes high. There are two major effects at such high fields. The first effect is that the ionization rate will vary slowly at high field, broadening the injected current pulse[45] and changing the terminal current waveforms so that the efficiency is reduced. The second effect is the tunneling current, which may be dominant. Since it is in phase with the field, the 180° avalanche phase delay is not provided. The tunneling mechanism has been considered

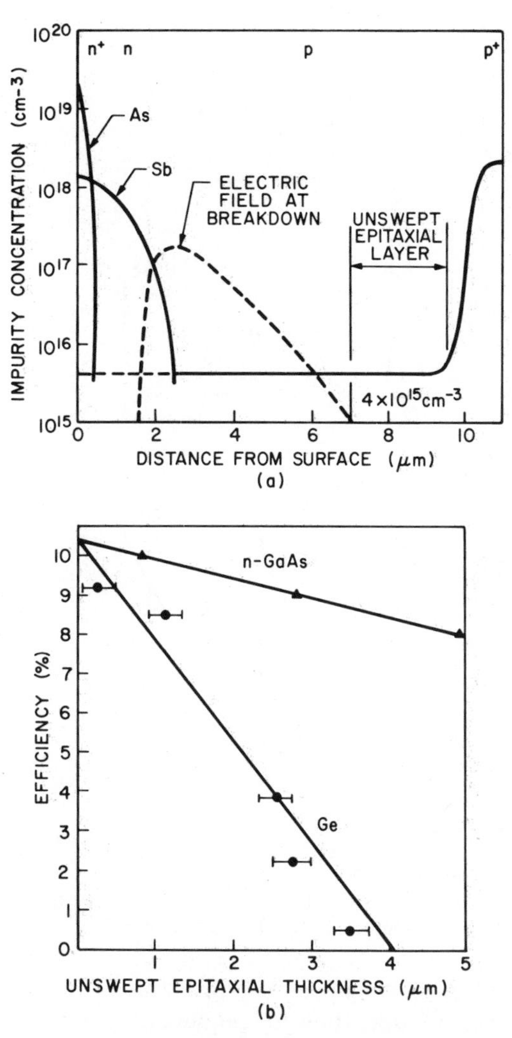

Fig. 19 (a) Electric field distribution at breakdown of a Ge IMPATT diode. (*b*) Efficiency versus unswept epitaxial layer thickness for *n*-GaAs and Ge diodes. (After Kovel and Gibbons, Ref. 42; Aono and Okuto, Ref. 43.)

for the TUNNETT (tunnel transit time) mode of operation.[4,46,47] The TUNNETT diode is expected to have lower noise than the IMPATT diode; however, the power output and efficiency will also be much lower.

Another factor that limits performance at submillimeter waves is the finite delay by which the ionization rate lags the electric field. For Si this "intrinsic avalanche response time" τ_i is less than 10^{-13} s. Since this time is very small compared to the transit time in the submillimeter wave region, the Si IMPATT diodes are expected to be efficient up to 300 GHz or higher frequencies. For GaAs, however, τ_i is found to be an order of magnitude

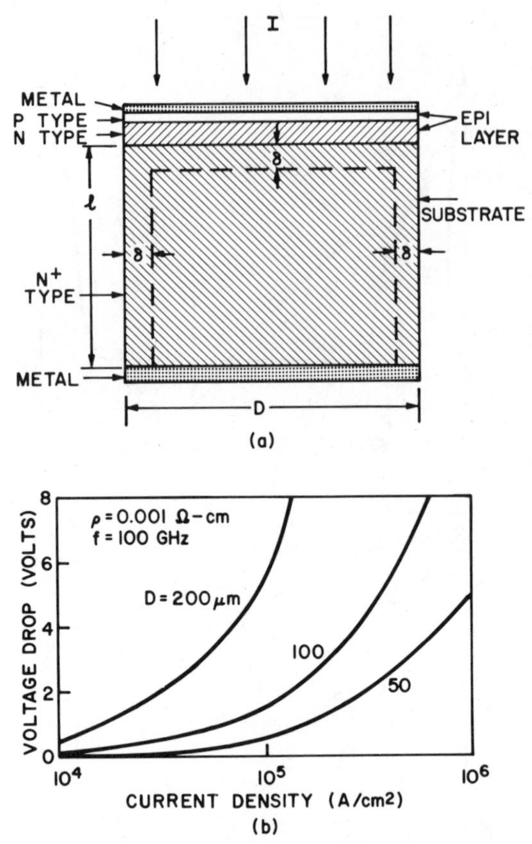

Fig. 20 Skin effect. (*a*) Current flow confined to a surface lamina of thickness δ, causing nonuniformity and resistive loss in the diode. (*b*) Calculated voltage drop in substrate at 100 GHz for several diode diameters. (After DeLoach, Ref. 44.)

longer[48] than the intrinsic avalanche response time of Si. Such a long τ_i may limit the GaAs IMPATT operation to frequencies below 100 GHz.

Minority-carrier storage effects in p^+-n (or n^+-p) diodes arising from back diffusion of the generated electrons (or holes) from the active layer into the neutral p^+ (or n^+) region can occur and will degrade the efficiency. This minority carrier will be stored in the neutral region while the remaining carriers are in transit and will diffuse back into the active region at a later time in the cycle, causing a premature avalanche which destroys the current–voltage phase relationship.

10.4.5 Burnout from Filament Formation

Burnout may occur not only if the diode is overheated, but also, more insidiously, if the carrier current fails to be uniformly distributed over the diode area and is instead concentrated into filaments of locally high

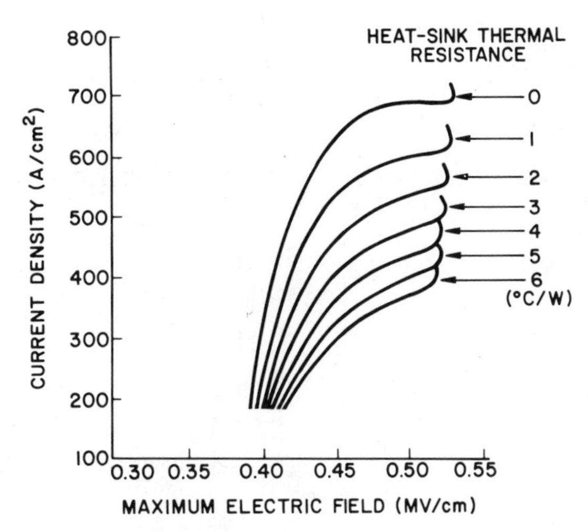

Fig. 21 Current density/electric field characteristics for Si Schottky barrier with $N_D = 6.2 \times 10^{15}\ cm^{-3}$, $\phi_{Bn} = 0.79$ V, and an ambient temperature of 22°C. (After Olson, Ref. 49.)

intensity. Such untoward behavior can often result when the diode has a dc negative conductance, because then the local region of greatest current density also has the lowest breakdown voltage. For this reason, *p-i-n* diodes are prone to easy burnout. The moving carrier space charge in the drift region tends to prevent low-frequency negative resistance and therefore helps to prevent filamentary burnout. Diodes that have positive dc resistance at low currents may develop negative dc resistance and burn out at high currents.

Figure 21 shows the current density versus electric field characteristic[49] of a GaAs Schottky-barrier diode with a doping of $6.2 \times 10^{15}\ cm^{-3}$ and a barrier height of 0.8 V. Under cw operation, as the current increases, the junction temperature also increases. Eventually, a point is reached at which the field turnaround occurs, leading to a region of differential negative resistance, and the diode will have a catastrophic failure due to high-current filament formation. Note that by providing a low heat-sink thermal resistance, one can operate the diode at higher current densities without reaching the turnaround point.

10.5 NOISE BEHAVIOR

The noise in an IMPATT diode arises mainly from the statistical nature of the generation rates of electron–hole pairs in the avalanche region. Since the noise sets a lower limit to the microwave signals to be amplified, it is important to consider the noise theory of the IMPATT diode.

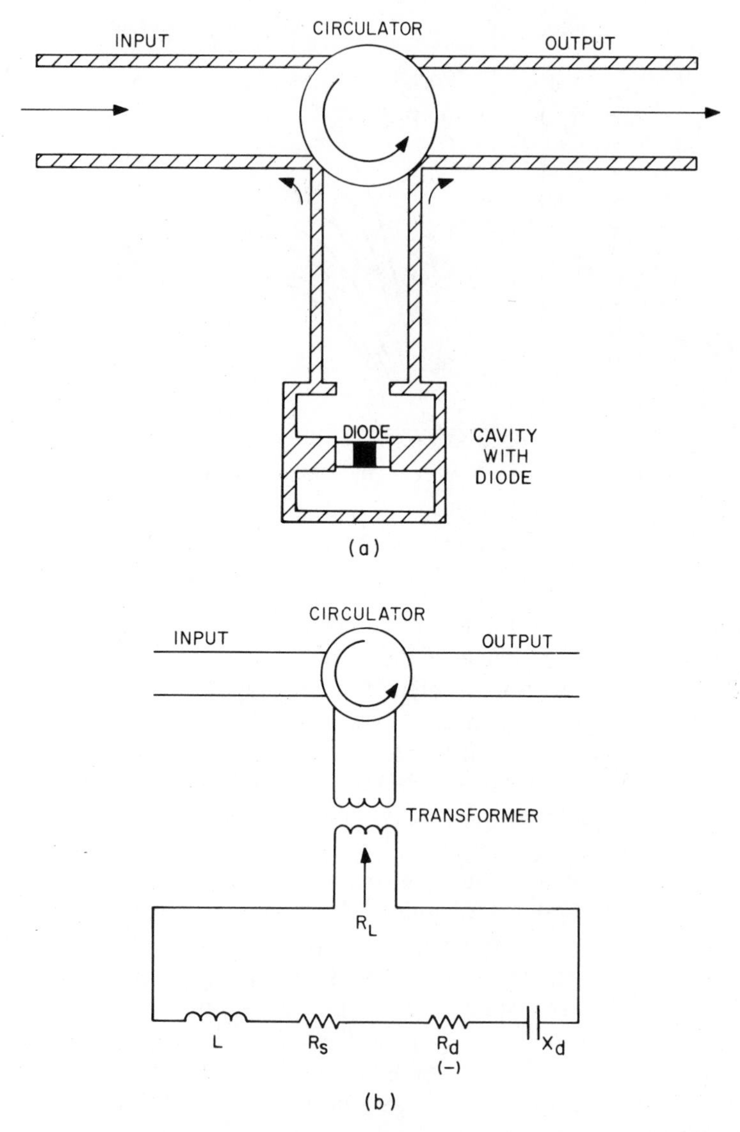

Fig. 22 (*a*) IMPATT diode inserted into a resonator. (*b*) Equivalent circuit. (After Hines, Ref. 50.)

For amplification the IMPATT diode can be inserted into a resonator that is coupled to a transmission line.[50] The line is coupled to separate input and output lines by means of a circulator, as shown in Fig. 22*a*. Figure 22*b* shows the equivalent circuit upon which the small-signal analysis is based. We shall now introduce two useful expressions for the noise performance: the noise figure and the noise measure. The noise figure *NF* is defined as

$$NF \equiv 1 + \frac{\begin{array}{c}\text{output noise power}\\ \text{arising in the amplifier}\end{array}}{(\text{power gain})(kT_0B_1)}$$

$$= 1 + \frac{\bar{I}_n^2 R_L}{P_G k T_0 B_1} \tag{57}$$

where P_G is the amplifier power gain, R_L is the load resistance, k is Boltzmann's constant, $T_0 = 290\text{ K}$, B_1 is the noise bandwidth, and $\bar{I}_n^2$ is the mean-square noise current caused by the diode and induced in the loop of Fig. 22*b*. The noise measure M is defined as

$$M \equiv \frac{\bar{I}_n^2}{4kT_0GB_1} \tag{58}$$

or

$$M \equiv \frac{\bar{V}_n^2}{4kT_0(-Z_{\text{real}})B_1} \tag{59}$$

where G is the negative conductance, $-Z_{\text{real}}$ the real part of the diode impedance, and $\bar{V}_n^2$ the mean-square noise voltage. Note that both the noise figure and the noise measure depend on the mean-square noise current (or the mean-square noise voltage). It will be shown that for frequencies above the resonant frequency f_r, the noise in the diode decreases, but so does the negative resistance. In this situation the appropriate quantity for assessing the performance of the diode as an amplifier is the noise measure, and we are interested in the minimum noise measure.

The noise figure for a high-gain amplifier is given by[50]

$$NF = 1 + \frac{qV_A/kT_0}{4\zeta\tau_A^2(\omega^2 - \omega_r^2)} \tag{60}$$

where ζ is the factor associated with the expression $\alpha \sim \mathscr{E}^\zeta$; τ_A and V_A are, respectively, the time and voltage drop across the avalanche region; and f_r is the resonant frequency given in Section 10.3. The expression above is obtained under the simplified assumptions that the avalanche region is narrow, and that the ionization coefficients of holes and electrons are equal. For $\zeta = 6$ (for Si), $\omega = 2\omega_r$, and $V_A = 3\text{ V}$, the noise figure at $f = 10\text{ GHz}$ is predicted to be 11,000 or 40.5 dB.

With realistic ionization coefficients ($\alpha_n \neq \alpha_p$ for Si) and an arbitrary doping profile, the low-frequency expression for the mean-square noise voltage is given by[51]

$$\dot{\bar{V}}_n^2 = \frac{2qB_1}{J_0A}\left(\frac{1 + W/x_A}{\alpha'}\right)^2 \sim \frac{1}{J_0} \tag{61}$$

where $\alpha' = \partial\alpha/\partial\mathscr{E}$. Figure 23 shows $\bar{V}_n^2/B_1$ as a function of frequency for a silicon IMPATT diode with $A = 10^{-4}\text{ cm}^2$, $W = 5\ \mu\text{m}$, and $x_A = 1\ \mu\text{m}$. At

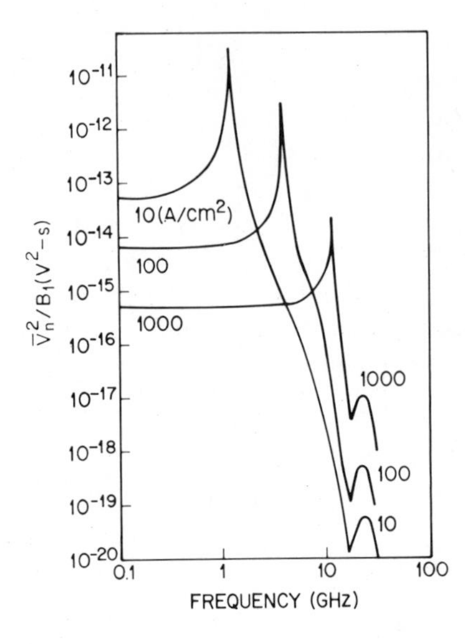

Fig. 23 Mean-square noise voltage per bandwidth versus frequency for a Si IMPATT diode. (After Gummel and Blue, Ref. 51.)

low frequencies, note that the noise voltage $\bar{V}_n^2$ is inversely proportional to the direct current density, Eq. 61. Near the resonant frequency (which varies as $\sqrt{J_0}$), $\bar{V}_n^2$ reaches a maximum and then decreases roughly as the fourth power of frequency. Noise can therefore be reduced somewhat by operating well above the avalanche frequency and keeping the current low. These conditions conflict with conditions favoring high power and efficiency, so that trade-offs are necessary to optimize for particular applications.

Figure 24 shows typical theoretical and experimental results[52] of the

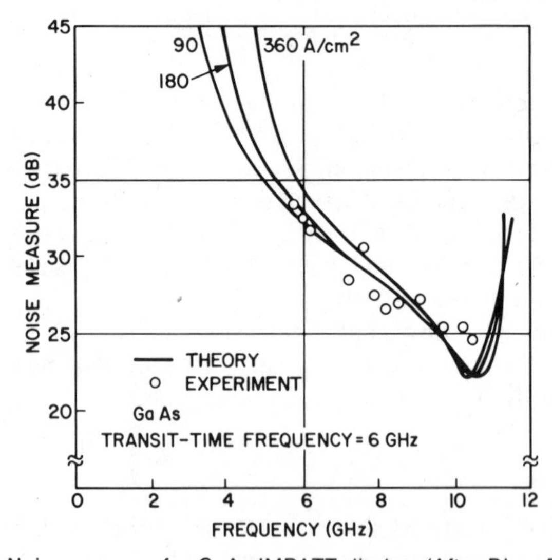

Fig. 24 Noise measure for GaAs IMPATT diodes. (After Blue, Ref. 52.)

Table 2 Noise Measure of IMPATT Diodes

Semiconductor	Ge	Si	GaAs
Small-signal noise measure (dB)	30	40	25
Large-signal oscillator noise measure (dB)	40	55	35

noise measure in a GaAs IMPATT diode. At the transit-time frequency (6 GHz), the noise measure is about 32 dB. The minimum noise measure of 22 dB, however, is obtained at about twice the transit-time frequency. One important feature of the GaAs noise measure is that it is substantially lower than that for Si IMPATT diodes. Table 2 compares the noise measures of Ge, Si, and GaAs IMPATT diodes. The amplifier and oscillator noises in the table are for a lossless circuit at a frequency that corresponds to maximum oscillator efficiency without harmonic tuning.

The main reason for the low-noise behavior in GaAs is that for a given field the electron and hole ionization rates are essentially the same in GaAs, whereas in Si they are quite different. From the avalanche multiplication integral, it can be shown that to obtain a large multiplication factor M the average distance of ionization $1/\langle\alpha\rangle$ is about equal to x_A (the avalanche width) if $\alpha_n = \alpha_p$, but is about equal to $x_A/\ln M$ if $\alpha_n \gg \alpha_p$. So, for a given x_A, considerably more ionization events must occur in Si, resulting in higher noise.

Figure 25 shows the relation between power output and FM noise measure for some Si and GaAs 6-GHz IMPATT diodes.[53] The power level

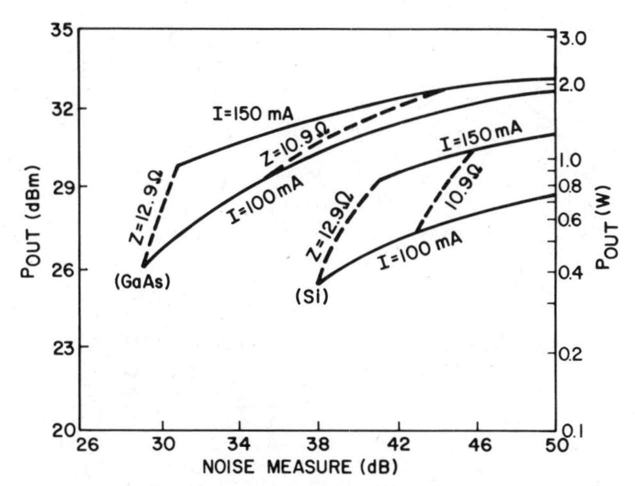

Fig. 25 Power output versus noise for a phase-locked oscillator. Locking power was held constant at 4 dBm. Contours of constant load impedance Z and constant diode current I are shown. (After Irvin et al., Ref. 53.)

is expressed with respect to a reference power of 1 mW, that is, the power is given by $10 \log(P \times 10^3)$ dBm, where P is in watts. The diodes were evaluated in a single tuned coaxial resonator circuit in which the load resistance presented to the resonator was incrementally varied by using interchangeable impedance transformers Z. At a maximum power output, the noise measure is relatively poor. A lower noise measure can be realized at the expense of a slightly reduced power output. Note again that at a given power level (say 1 W or 30 dBm), the GaAs IMPATT diode is about 10 dB quieter than a Si IMPATT diode.

10.6 DEVICE DESIGN AND PERFORMANCE

10.6.1 Diode Fabrication

Based upon the discussion presented in Section 10.4, IMPATT diodes are usually designed to maximize power output and efficiency. Figure 26 shows some typical high-power IMPATT diode designs.

The configuration of Fig. 26*a* is formed by using a double epitaxial process or by diffusion into an epitaxial layer.[54] The n^+ substrate is used to reduce series resistance. The epitaxial layer thickness should also be controlled so that at breakdown essentially no unswept epitaxial layer is left. For high-frequency operation, even the n^+ substrate should be thinned down to the order of a few micrometers to reduce losses and nonuniformity from the skin effect.

Figure 26*b* shows a Schottky-barrier IMPATT diode.[55–57] A Schottky barrier is a rectifying metal–semiconductor contact (see Chapter 5).

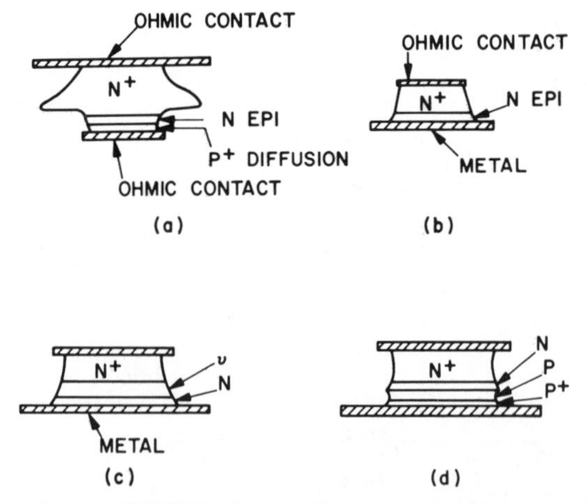

Fig. 26 Outlines of some IMPATT diode designs: (*a*) diffusion or double epitaxy, (*b*) Schottky barrier, (*c*) hi–lo diode, and (*d*) double-drift IMPATT diode using ion implantation.

Although the field distributions for Fig. 26*a* and *b* are very similar, the Schottky barrier offers some unique advantages. One advantage is that the maximum field occurs at the metal–semiconductor interface, and the heat generated can be readily conducted away from the metal contact. As shown in Fig. 26*b*, the device can be formed in a truncated cone shape so that the high-field singularity at the periphery is removed, and uniform breakdown can be obtained inside the device. Another advantage of the Schottky-barrier approach is that the device can be fabricated at relatively low temperatures, so that the original high-quality epitaxial layer can be preserved. There is also an important disadvantage, in that the metal electrode can sometimes be attacked chemically by the semiconductor under the presence of high-energy electrons and holes, thus impairing the long-term reliability of the Schottky barrier.

The Schottky-barrier approach can also be used in modified Read diode structures as shown in Fig. 26*c* by replacing the p^+ layer by the metal contact. Since the Schottky barrier is basically a majority-carrier device, the minority-carrier storage effect,[58] which occurs in the original Read structure, is eliminated and improved efficiency can be achieved.

Use of the modified Read diodes has resulted in a dramatic increase in microwave efficiency compared to the diode with constant doping profile. However, the modified Read diode demands a more stringent control of doping profiles to achieve a specific-frequency device. A self-limiting anodic etching method has been used to thin down the high-doping layer (for a hi–lo diode) or the surface low-doping layer (for a lo–hi–lo diode), so that the diode will have a controlled breakdown voltage (i.e., controlled frequency).[59] Most Schottky-barrier GaAs contact has a high barrier and low reverse saturation current. However, Pt will react with GaAs at operating temperatures to form $PtAs_2$, causing the barrier interface to move. This "chew-in" effect of Pt on GaAs will alter the breakdown voltage and degrade device performance. This effect can be controlled by the deposition of only 200 to 500 Å of platinum on the epitaxial surface and followed by deposition of a layer of tungsten or tantalum that serves to limit the Pt–GaAs reaction.[60,61]

Figure 26*d* shows a new device configuration by using an ion implantation technique.[62] For the most common dopants, for example, *B* and *P*, the depth of ion penetration is of the order of 0.5 μm/100 keV, so that $1-\mu$m depth can readily be obtained from ion implantation equipment with energy sources a few hundred keV. These energy ranges are useful in making millimeter-wave IMPATT diodes. Of particular importance is the possibility of making the double-drift structures shown in Fig. 26*d*. Power output per unit area and impedance per unit area are both approximately doubled. These structures are thus expected to have increased power output with improved efficiency.[62]

IMPATT diodes can also be made using the molecular-beam epitaxy (MBE) method.[63] Since MBE can control the doping and layer thickness

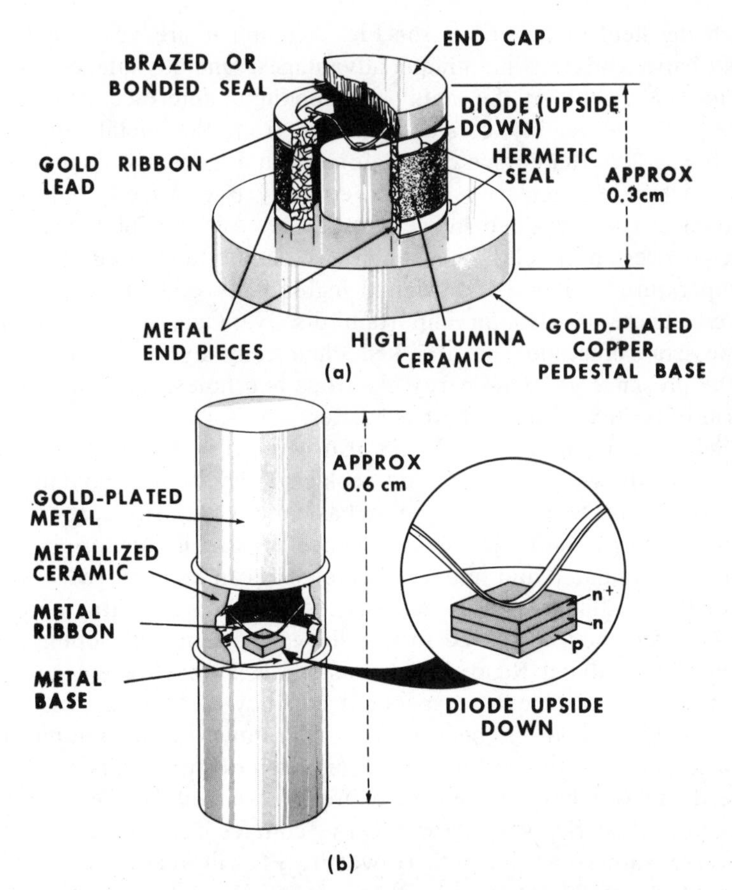

Fig. 27 Two microwave packages with IMPATT diodes mounted.

down to atomic dimensions, this method is expected to be very useful for millimeter and submillimeter IMPATT diodes.

The fabricated diode is generally mounted in a microwave package. Two typical packages are shown in Fig. 27. In both cases the diode is mounted with its diffused side or metal–electrode side in contact with a copper or diamond heat sink so that the heat generated at the junction can be readily conducted away.

10.6.2 Frequency Scaling and Microwave Performance

From the small-signal theory, we can obtain approximate relations for various device parameters as a function of operating frequency. The resistance expression in Eq. 43 can be rewritten as

$$-R \sim \frac{W^2}{2A\epsilon_s v_s}\left(\frac{1}{\omega^2/\omega_r^2-1}\right)\left(\frac{1-\cos\theta_d}{\theta_d^2/2}\right) \tag{62}$$

where θ_d is the transit angle equal to $\omega\tau$ and $\tau \simeq W/v_s$. For $-R$ to be invariant with ω/ω_r, it is required from Eq. 62 that

$$W^2/2A\epsilon_s v_s = \text{constant} \tag{63a}$$

and

$$\omega_r\tau = \text{constant}. \tag{63b}$$

Since the depletion width W is inversely proportional to the operating frequency, Eq. 49, the device area, which is proportional to W^2, is thus proportional to ω^{-2}. From the avalanche breakdown equation, Eq. 2, it can be shown that the ionization rate (α) and its field derivative (α') are inversely proportional to the depletion-layer width. Combining the relation $\alpha' \sim 1/W$ with Eqs. 37 and 63b yields the following result for the dc current density:

$$J_0 \sim \frac{\omega_r^2}{\alpha'} \sim \frac{\omega^2}{1/W} \sim \omega. \tag{64}$$

These frequency scaling relations are summarized in Table 3 and are useful as a guide for extrapolating performance and design to new frequencies.

The power-output limitations have been considered in Section 10.4. The efficiency is expected to be only weakly dependent on frequency at low frequencies. However, at millimeter-wave regions the operating current density is high ($\sim f$) and the area is small ($\sim f^{-2}$), so that the device operating temperature will be high. This high temperature, in turn, will cause the reverse-saturation-current density to increase and the efficiency to decrease. In addition, the skin effect, the tunneling effect, and other effects associated with high frequency and high field will also degrade the efficiency performance. Hence, as the frequency increases, the efficiency is expected to decrease eventually.

Figure 28 shows a useful microwave measurement circuit for the lower-frequency bands, such as the H band (6 to 8 GHz) and I band (8 to 10 GHz)

Table 3 Approximate Frequency Scaling for IMPATT Diodes

Parameter	Frequency Dependence
Junction area A	f^{-2}
Bias-current density J	f
Depletion-layer width W	f^{-1}
Breakdown voltage V_B	f^{-1}
Power output P_{out}	
Thermal limitation	f^{-1}
Electronic limitation	f^{-2}
Efficiency η	Constant

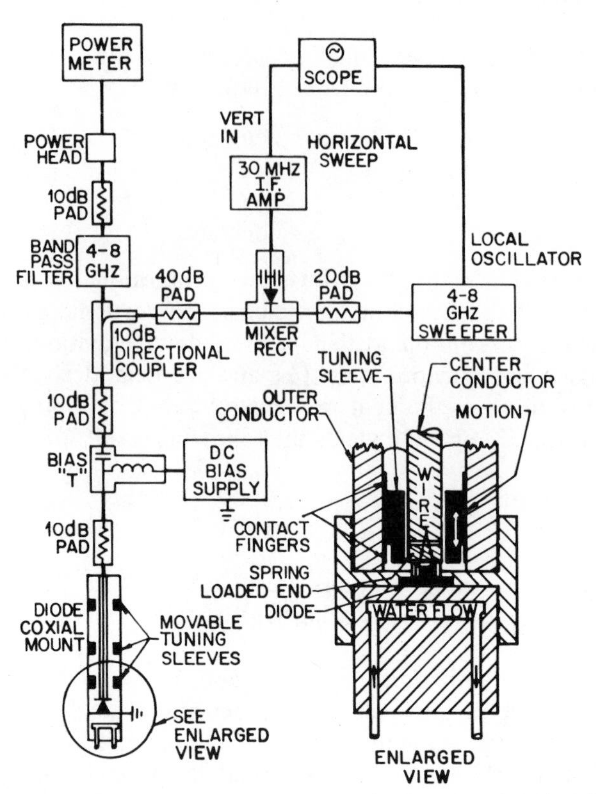

Fig. 28 Wide-range test circuit for IMPATT diodes. (After Iglesias, Ref. 65.)

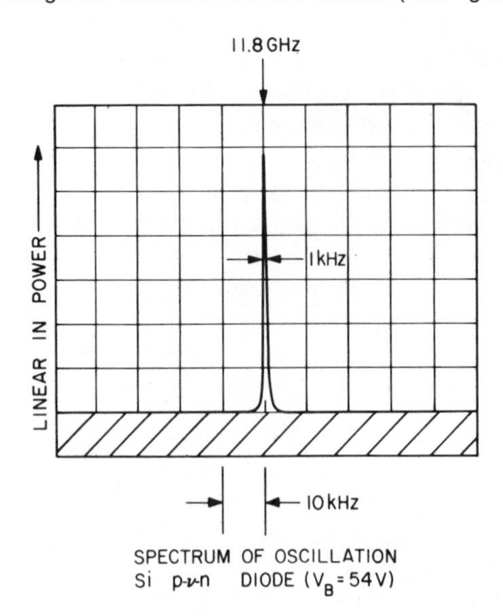

Fig. 29 Spectrum of oscillation of a *p*-*ν*-*n* diode. (After Misawa and Marinaccio, Ref. 66.)

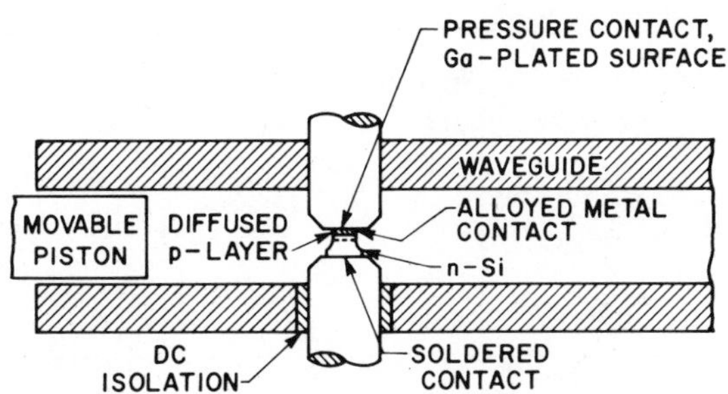

Fig. 30 Microwave cavity for millimeter-wave IMPATT diodes. (After Bowman and Burrus, Ref. 67.)

as listed[64] in Table 2 of Chapter 9. The most important features are (1) the diode mount, which provides adequate heat conduction and has movable tuning sleeves to optimize the resonant circuit; (2) the power meter, to give the microwave output power; (3) the sweeper, to detect the microwave frequencies; and (4) the oscilloscope, to display the output power signal.[65] A typical spectrum[66] of oscillation as obtained from an oscilloscope display is shown in Fig. 29 for a Si p-ν-n diode, which is operated at 11.8 GHz with a

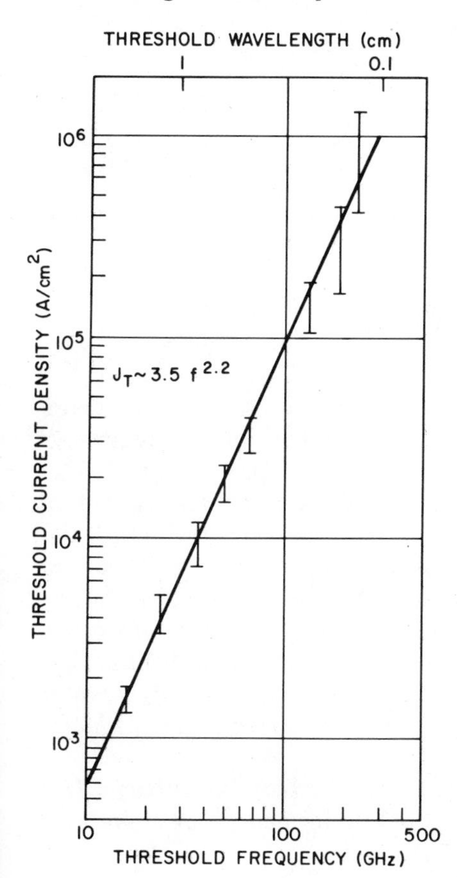

Fig. 31 Dependence of threshold frequency on direct current density. (After Bowman and Burrus, Ref. 67.)

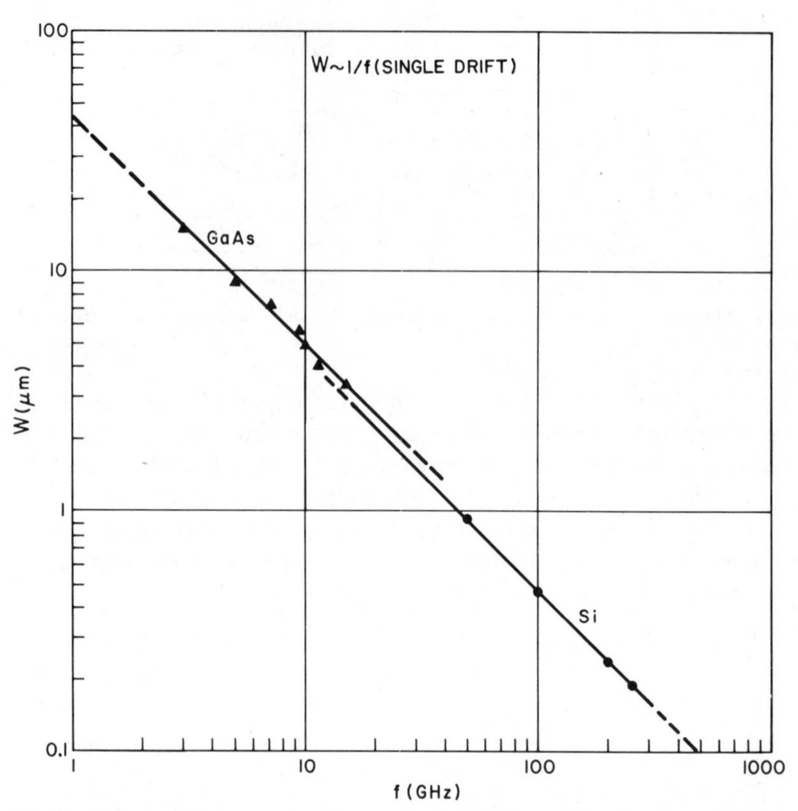

Fig. 32 Depletion width versus frequency for Si and GaAs IMPATT diodes. (After Ino, Ishibashi, and Ohmori, Ref. 68; Pribetich et al., Ref. 69.)

spectrum width less than 1 kHz. For the millimeter-wave region, both the mounting scheme and microwave circuit should be modified; a typical microwave cavity[67] is shown in Fig. 30.

Figure 31 shows the dependence of the threshold current density, that is, the minimum current density to produce oscillation, on the frequency.[67] Note that the threshold current density increases approximately as the square of the frequency, in agreement with the general behavior of the resonant frequency. To demonstrate the importance of the transit-time effect, Fig. 32 shows the optimum depletion-layer width versus the frequency for Si and GaAs IMPATT diodes.[68,69] The depletion-layer width, as expected, varies inversely with the frequency. Interestingly, at frequencies above 100 GHz, the depletion-layer width is less than 0.5 μm. This very narrow width gives some indication of the difficulty inherent in fabricating a modified Read diode or a double-drift diode at these frequencies.

At present, the highest power $\times$ (frequency)^2 product is obtained from double-drift diodes. Figure 33 compares the performance[70] of double-drift

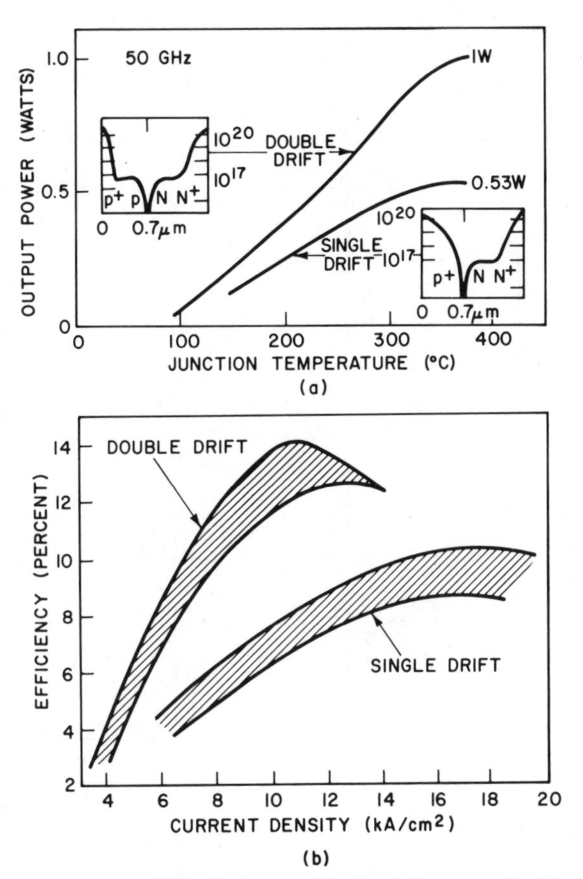

Fig. 33 (*a*) Power output and (*b*) efficiency of single-drift versus double-drift Si IMPATT diodes at 50 GHz. Range of efficiency for four diodes of each type. (After Seidel, Davis, and Iglesias, Ref. 70.)

and single-drift diodes at 50 GHz. The double-drift 50-GHz Si IMPATT diode made by ion implantation shows an output cw power over 1 W with a maximum efficiency of 14%. This result can be compared with a similar single-drift diode that delivers about 0.5 W with an efficiency of 10%. The superiority of the double-drift diodes results from the fact that both holes and electrons produced by the avalanche are allowed to do work against the radio-frequency (RF) field by traversing the drift region. In the single-drift diodes only one type of carriers is so utilized.

A summary[10,71,72] of the present state of the art of IMPATT performance is given in Fig. 34. (Also shown are results for BARITT diodes to be discussed in Section 10.7.) At lower frequencies, the power output is thermal-limited and varies as f^{-1}; at higher frequencies (>50 GHz) the power is electronic-limited and varies as f^{-2}. Around 10 GHz, efficiencies close to 40% have been obtained from lo–hi–lo Pt–GaAs diodes.[73] Figure 34

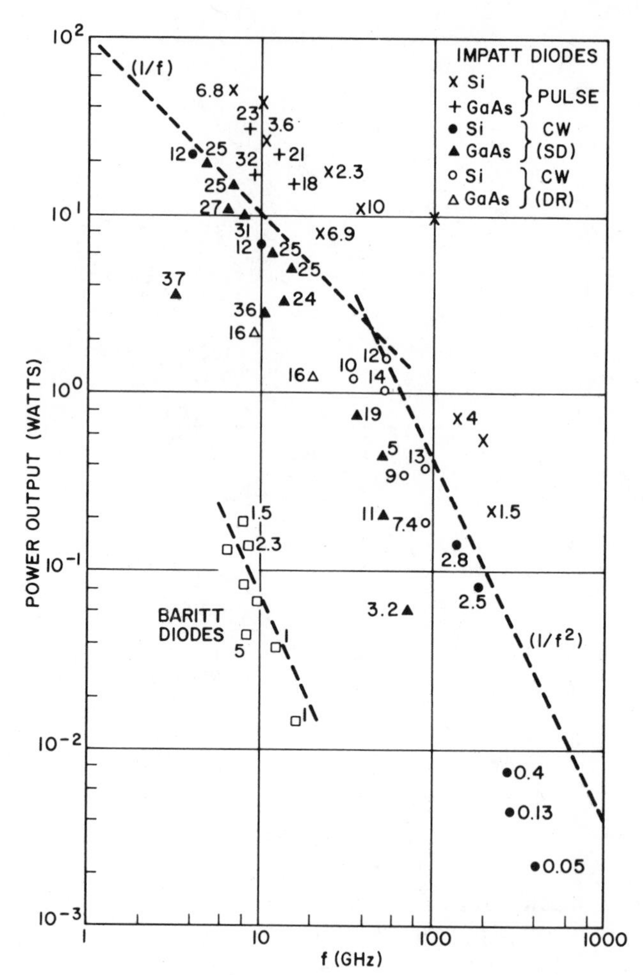

Fig. 34 State of the art of IMPATT and BARITT performances. The number against each experimental point indicates the efficiency in percent. (After Ref. 10; Ahmad and Freyer, Refs. 85 and 97.)

clearly shows that the IMPATT diode is one of the most powerful solid-state sources for the generation of microwaves. The IMPATT diodes can generate higher cw power output in the millimeter-wave frequencies than any other solid-state devices.

With improved device-fabrication technology, the IMPATT diodes are also reliable under high-temperature operations. Figure 35 shows the activation plot for the median time between failure (MTBF) of GaAs IMPATT diodes.[74] The activation energy is about 1.9 eV. At a junction temperature of 200°C, it is estimated that these diodes will have a long life ($\sim 10^7$ h), as extrapolated from short-term aging tests. Similar results have been obtained for Si IMPATT diodes.[75]

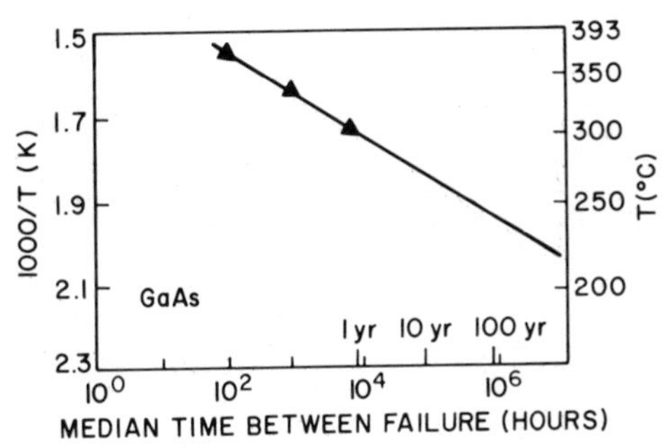

Fig. 35 Activation plot for mean time between failure of GaAs IMPATT diodes. (After Hierl et al., Ref. 74.)

10.7 BARITT AND DOVETT DIODES

The BARITT (barrier injection transit time) diode belongs to the transit-time microwave diode family.[12] The mechanisms responsible for the microwave oscillation are the thermionic injection and diffusion of minority carriers across the forward-biased barrier and a $3\pi/2$ transit angle of the injected carriers traversing the drift region. BARITT diodes tend to be low in power and efficiency but have good signal-to-noise performance, and are useful as local oscillators in microwave receivers.

The BARITT diode is basically a back-to-back pair of diodes biased into reach-through condition. We consider first the current transport of the reach-through diode. The small-signal behavior and large-signal performance will then be presented. A related device, the DOVETT diode, is considered in Section 10.7.4.

10.7.1 Current Transport

We consider first the current transport in a symmetrical metal–semiconductor–metal (MSM) structure[76] with uniformly doped n-type semiconductor (Fig. 36a). Figure 36b, c, and d show the charge distribution, electric field distribution, and energy-band diagram, respectively, for a small positive voltage applied to contact 1 with respect to contact 2 (contact 1 is forward-biased and contact 2 is reverse-biased). The depletion-layer widths are

$$W_1 = \sqrt{\frac{2\epsilon_s}{qN_D}(V_{bi} - V_1)} \tag{65}$$

and

$$W_2 = \sqrt{\frac{2\epsilon_s}{qN_D}(V_{bi} + V_2)} \tag{66}$$

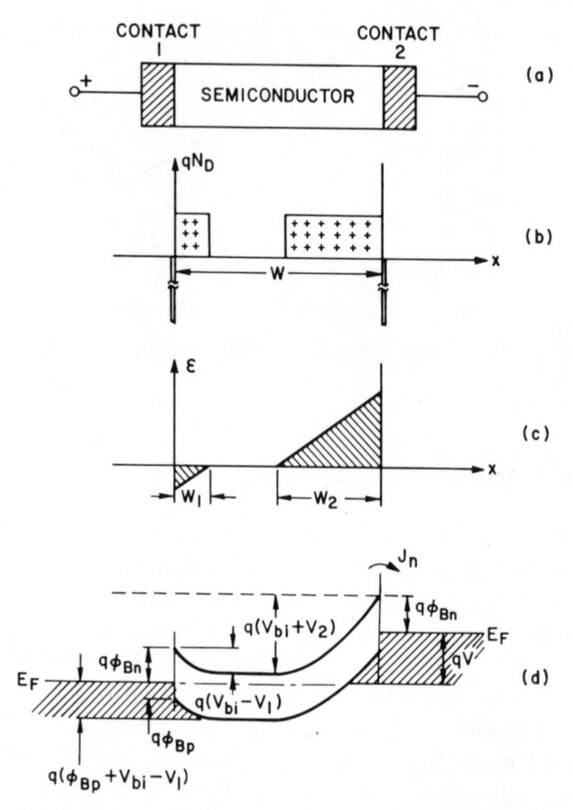

Fig. 36 Metal–semiconductor–metal (MSM) structure. (*a*) MSM with a uniformly doped *n*-type semiconductor. (*b*) Charge distribution under low bias. (*c*) Field distribution. (*d*) Energy-band diagram. (After Sze, Coleman, and Loya, Ref. 76.)

where W_1 and W_2 are the depletion widths in the n layer for the forward- and reverse-biased barriers, respectively; N_D is the ionized impurity density; and V_{bi} is the built-in voltage. Under these conditions, the current is the sum of the reverse saturation current (of a Schottky diode with a barrier height ϕ_{Bn}), generation–recombination current, and surface leakage current.

As the voltage increases, the reverse-biased depletion region will eventually reach through to the forward-biased depletion region (Fig. 37*a*). The corresponding voltage is called the reach-through voltage V_{RT}. This voltage can be obtained from the condition $W_1 + W_2 = W$, which is the length of the n region:

$$V_{RT} = \frac{qN_D}{2\epsilon_s} W^2 - W\left[\frac{2qN_D}{\epsilon_s}(V_{bi} - V_1)\right]^{1/2}$$

$$\simeq \frac{qN_D}{2\epsilon_s} W^2 - W\left(\frac{2qN_D V_{bi}}{\epsilon_s}\right)^{1/2}. \tag{67}$$

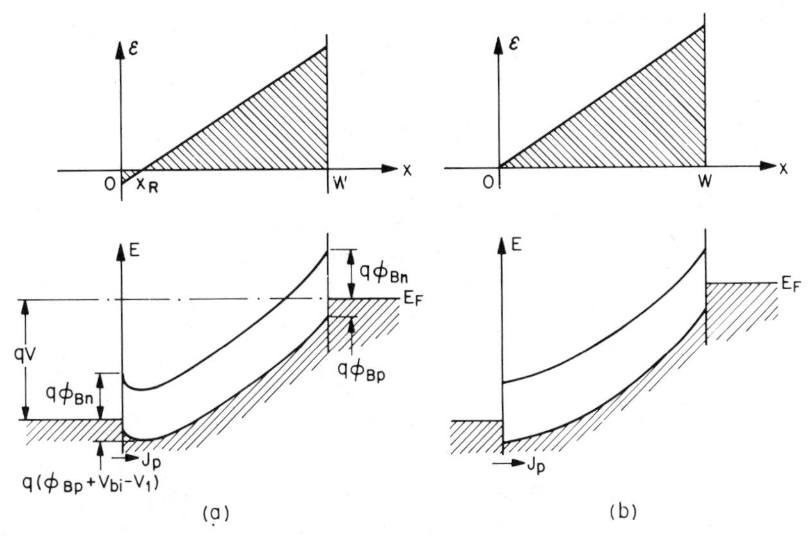

Fig. 37 (a) Field distribution and energy diagram of an MSM structure at reach-through. (*b*) Field distribution and energy diagram at flat-band condition. (After Sze, Coleman, and Loya, Ref. 76.)

If the voltage is increased further, the energy band at contact 1 can become flat. The electric field is zero at $x = 0$ when $V_{bi} = V_1$; this condition is the flat-band condition (Fig. 37*b*). The corresponding voltage is defined as the flat-band voltage V_{FB}:

$$V_{FB} \equiv \frac{qN_D W^2}{2\epsilon_s}. \tag{68}$$

The flat-band voltage versus doping for silicon is plotted in Fig. 38 for various depletion lengths. For a given length, the maximum V_{FB} is limited by the avalanche breakdown voltage.

The dc bias for a BARITT diode under microwave oscillation is generally between V_{RT} and V_{FB}. For applied voltage in this range ($V_{RT} < V < V_{FB}$), the relation between the applied voltage and the forward-biased barrier height is

$$V_{bi} - V_1 = (V_{FB} - V)^2/4V_{FB}\,. \tag{69}$$

The reach-through point x_R as shown in Fig. 37*a* is given by

$$x_R/W = (V_{FB} - V)/2V_{FB}\,. \tag{70}$$

After reach-through, the hole current thermionically emitted over the hole barrier ϕ_{Bp} becomes the dominant current:

$$J_p = A_p^* T^2 e^{-q(\phi_{Bp}+V_{bi})/kT}\left(e^{qV_1/kT} - 1\right) \tag{71}$$

where A_p^* is the effective Richardson constant (refer to Chapter 5). From

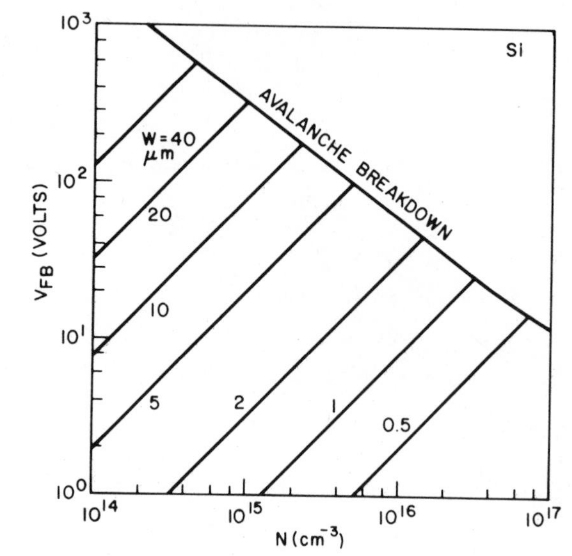

Fig. 38 Flat-band voltage versus doping concentration for Si diodes with various length. For a given length, the maximum flat-band voltage is limited by the avalanche breakdown voltage. (After Sze, Coleman, and Loya, Ref. 76.)

Eq. 69 we obtain for $V \geq V_{RT}$,

$$J_p = A_p^{*} T^2 e^{-q\phi_{Bp}} \exp\left[-\frac{q(V_{FB}-V)^2}{4kTV_{FB}}\right]. \tag{72}$$

Therefore, beyond reach-through the current will increase exponentially with applied voltage.

At current levels that are high enough for the injected carrier density to be comparable to the background ionized-impurity density, the mobile carriers will influence the field distribution in the drift region. This effect is the space-charge-limited effect. If all the mobile holes traverse the n region with the saturation velocity v_s and if $J > qv_sN_D$, the Poisson equation becomes

$$\frac{d\mathscr{E}}{dx} = \frac{\rho}{\epsilon_s} = \frac{q}{\epsilon_s}\left(N_D + \frac{J}{qv_s}\right) \simeq \frac{J}{\epsilon_s v_s}. \tag{73}$$

Integrating twice (with boundary conditions $\mathscr{E} = 0$, $V = 0$, at $x = 0$) yields[77]

$$J = \left(\frac{2\epsilon_s v_s}{W^2}\right) V = qv_sN_D\left(\frac{V}{V_{FB}}\right). \tag{74}$$

The foregoing considerations can be applied to other structures, such as the p^+-n-p^+ and p^+-i-n-π-p^+ diodes shown in Fig. 39. The expressions for the reach-through voltage and flat-band voltage of p^+-n-p^+ devices are the same as that of MSM. For the more complicated p^+-i-n-π-p^+ structures, the

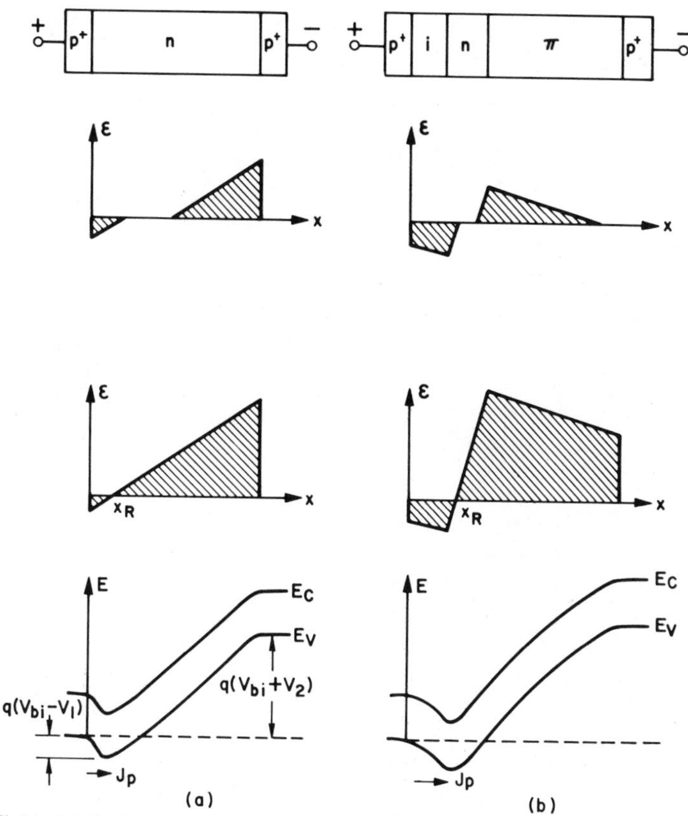

Fig. 39 Field distribution at low bias and at reach-through and the energy diagram at reach-through for (a) p^+-n-p^+ and (b) p^+-i-n-π-p^+ structures.

same procedure can be followed to obtain V_{RT}, V_{FB}, and x_R, and the relationship between the applied voltage and the voltage drop across the forward-biased junction.[78]

The current transport mechanisms of reach-through p^+-n-p^+ structure is similar to that of the MSM structure. The only difference is that in Eqs. 71 and 72, the factor $\exp(-q\phi_{Bp}/kT)$ should be absent when the carriers are injected over a forward-biased p^+-n junction,[77] that is,

$$J = A^*T^2 \exp[-q(V_{FB} - V)^2/4kTV_{FB}] = J_{FB} \exp\left[-\frac{q(V_{FB} - V)^2}{4kTV_{FB}}\right]. \tag{75}$$

For a PtSi–Si barrier, the hole barrier height $q\phi_{Bp}$ is equal to 0.2 eV. Hence at 300 K for a given voltage above reach-through, the current for the p^+-n-p^+ device will be about 3000 times larger than that for the MSM device. The value of J_{FB} ($\equiv A^*T^2$) at room temperature is about 10^7 A/cm^2. Therefore, under normal operation, the onset of the space-charge-limit effect will occur long before the current approaching J_{FB}.

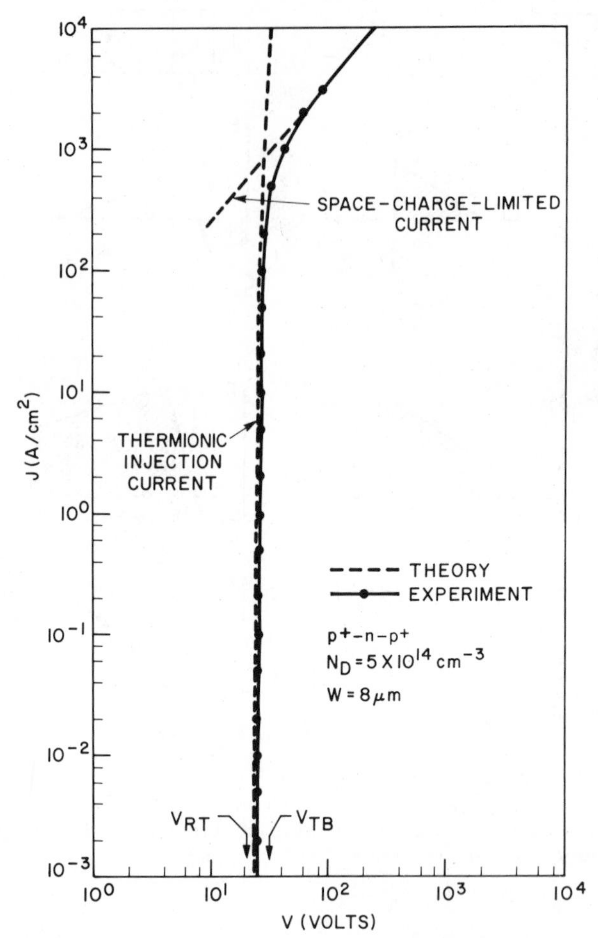

Fig. 40 Current–voltage characteristics of a Si p^+-n-p^+ reach-through diode. (After Chu, Persky, and Sze, Ref. 77.)

A typical I–V characteristic is shown in Fig. 40 for a Si p^+-n-p^+ with a background doping of $5 \times 10^{14}\ \text{cm}^{-3}$ and a thickness of 8.5 μm. The flat-band voltage is 29 V and reach-through voltage is about 21 V. We note that the current first increases exponentially and then becomes linearly dependent on voltage. The experimental results and theoretical calculation from Eqs. 74 and 75 are in good agreement.

For efficient BARITT operation, the carrier bunching must be well defined. Therefore, the current must increase very rapidly with voltage. The linear I–V relationship due to space-charge-limited effect will degrade device performance. The optimum current density is usually substantially lower than $J = qv_sN_D$.

The reach-through diode (also referred to as punch-through diode) described above is useful as a high-speed voltage limiter having a sharp current rise above reach-through voltage, little or no charge storage, and good temperature stability. Reach-through diodes with voltages as low as 1.5 V have been made with performances comparable or better than the conventional Zener diodes that operate via breakdown mechanisms (i.e., avalanche breakdown at high voltage and tunnel breakdown at low voltages).[79]

10.7.2 Small-Signal Behaviors

We will show that the BARITT diode has small-signal negative resistance; therefore, the diode can have self-starting oscillation. Consider a p^+-n-p^+ structure. When it is biased above the reach-through voltage, the electric field profile is shown in Fig. 41*a*. The point x_R corresponds to the potential maximum for hole injection, given by Eq. 70. The point a separates the low-field region from the saturation-velocity region, that is, for $\mathscr{E} > \mathscr{E}_s$, $v = v_s$, as shown in Fig. 41*b*. Under low-injection conditions,

$$a \simeq \frac{\epsilon_s \mathscr{E}_s}{qN_D} + x_R . \tag{76}$$

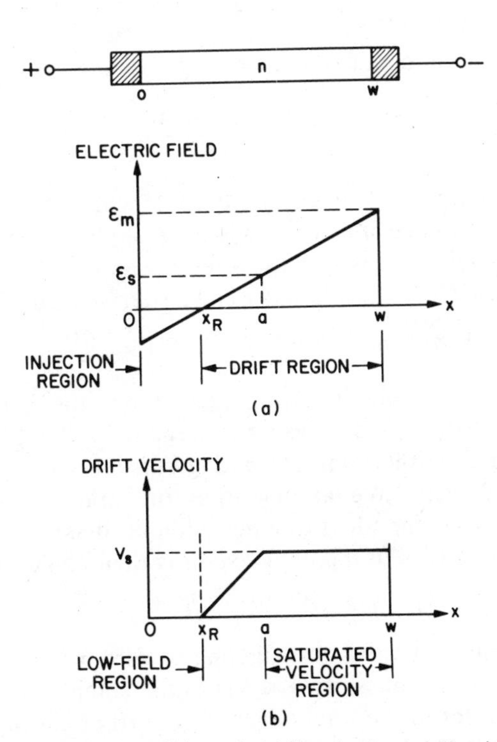

Fig. 41 (a) Field distribution and (*b*) carrier drift velocity variation in the drift region of a BARITT diode. (After Chu and Sze, Ref. 81.)

The transit time in the drift region ($x_R < x < W$) is given by:[80]

$$\tau_d = \int_{x_R}^{a} \frac{dx}{\mu_n \mathscr{E}(x)} + \int_a^W \frac{dx}{v_s}$$

$$= \int_{x_R}^{a} \frac{dx}{\mu_n q N_D x/\epsilon_s} + \frac{W-a}{v_s}$$

$$\simeq \frac{3.75\epsilon_s}{q\mu_n N_D} + \frac{W-a}{v_s}. \tag{77}$$

To derive the small-signal impedance, we shall follow an approach similar to that used in Section 10.3.2 and introduce the time-varying quantity as the sum of a time-independent term (dc component) and a small ac term:

$$J(t) = J_0 + \tilde{J}e^{j\omega t}$$

$$V(t) = V_0 + W\tilde{\mathscr{E}}e^{j\omega t}. \tag{78}$$

Substituting the foregoing expressions into Eq. 75 yields the linearized ac injected hole current density:

$$\tilde{J} = \sigma\tilde{\mathscr{E}} \tag{79}$$

where σ is the injection conductance per unit area and is given by

$$\sigma = J_0 \frac{\epsilon_s(V_{FB} - V_0)}{N_D W k T} \tag{80}$$

where J_0 is the current density given by Eq. 75, in which V is replaced by V_0. The injection conductance increases with the applied voltage, reaches a maximum, and then decreases rapidly when V_0 approaches V_{FB}. The bias voltage corresponding to the maximum σ can be derived from Eqs. 75 and 80:

$$V_0(\text{for max } \sigma) = V_{FB} - \sqrt{2kTV_{FB}/q}. \tag{81}$$

Since the ac electric field is continuous across the boundary of the injection and the drift region, these two regions will interact with one another. We define $\tilde{J}$ as the total alternating current density and $\tilde{J}_1$ as the injection current density. We assume that the injection region is thin enough that $\tilde{J}_1$ enters the drift region without delay. The alternating conduction current density in the drift region is then given by

$$\tilde{J}_c(x) = \tilde{J}_1 e^{-j\omega\tau(x)} \equiv \gamma\tilde{J}e^{-j\omega\tau(x)} \tag{82}$$

which is an unattenuated wave propagating toward $x = W$ with a transit phase delay of $\omega\tau(x)$. The quantity $\gamma \equiv \tilde{J}_1/\tilde{J}$ is the complex fraction relating the ac injection current to the total alternating current density.

At a given position in the drift region, the total alternating current density $\tilde{J}$ is equal to the sum of the conduction current $\tilde{J}_c$ and the displacement current $\tilde{J}_d$:

$$\tilde{J} = \tilde{J}_c(x) + \tilde{J}_d(x) = \text{not a function of } x. \tag{83}$$

The displacement current density is related to the ac field $\tilde{\mathscr{E}}(x)$ by

$$\tilde{J}_d(x) = j\omega\epsilon_s\tilde{\mathscr{E}}(x). \tag{84}$$

Combining Eqs. 80, 82, and 84 yields an expression for the ac electric field in the drift regions as a function of x and $\tilde{J}$,

$$\tilde{\mathscr{E}}(x) = \frac{\tilde{J}}{j\omega\epsilon_s}[1 - \gamma e^{-j\omega\tau(x)}]. \tag{85}$$

Integrating $\tilde{\mathscr{E}}(x)$ gives the ac voltage across the drift region in terms of the ac current density $\tilde{J}$. The coefficient can be expressed as

$$\gamma = \frac{\tilde{J}_1}{\tilde{J}_1 + \tilde{J}_d} = \frac{\sigma}{\sigma + j\omega\epsilon_s}. \tag{86}$$

Substituting γ into Eq. 85 and integrating over the drift length $(W - x_R)$ with the boundary conditions of $\tau = 0$ at $x = x_R$ and $\tau = \tau_d$ at $x = W$ yield the expression for the ac voltage across the drift region:

$$V_d = \frac{\tilde{J}(W - x_R)}{j\omega\epsilon_s}\left[1 - \frac{\sigma}{\sigma + j\omega\epsilon_s}\,\frac{1 - e^{j\theta_d}}{j\theta_1}\right] \tag{87}$$

where θ_d is the transit angle in the drift region,

$$\theta_d = \frac{\omega[(W - x_R) + (a - x_R)]}{v_s} = \omega\tau_d \tag{88}$$

and θ_1 is a constant given by

$$\theta_1 \equiv \omega\left(\frac{W - x_R}{v_s}\right). \tag{89}$$

We can also define $C_d = \epsilon_s/(W - x_R)$ as the capacitance of the drift region. From Eq. 87 we obtain the small-signal impedance of the structure

$$Z \equiv \frac{\tilde{V}_d}{\tilde{J}} = R_d - jX_d \qquad \Omega-\text{cm}^2 \tag{90}$$

where R_d and X_d are the small-signal resistance and reactance, respectively:

$$R_d = \frac{1}{\omega C_d}\left(\frac{\sigma}{\sigma^2 + \omega^2\epsilon_s^2}\right)\left[\frac{\sigma(1 - \cos\theta_d) + \omega\epsilon_s \sin\theta_d}{\theta_1}\right]$$

$$X_d = \frac{1}{\omega C_d} - \frac{1}{\omega C_d}\left(\frac{\sigma}{\sigma^2 + \omega^2\epsilon_s^2}\right)\left[\frac{\sigma\sin\theta_d - \omega\epsilon_s(1 - \cos\theta_d)}{\theta_1}\right].$$

Note that the real part (resistance) will be negative if the transit angle θ_d lies between the values of π and 2π, and if $|(1 - \cos\theta_d)/\sin\theta_d|$ is less than $\omega\epsilon_s/\sigma$.

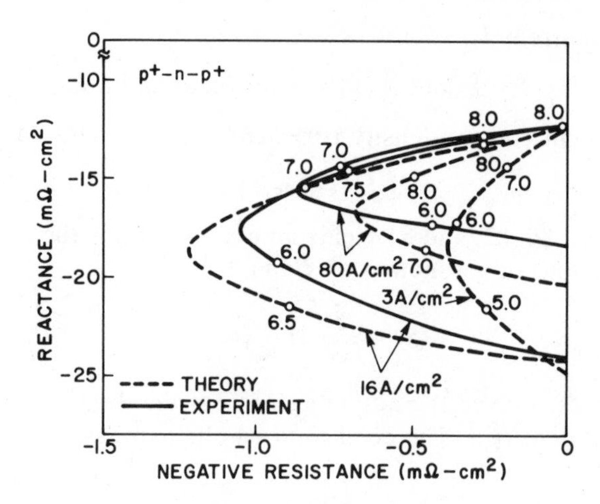

Fig. 42 Small-signal impedance for p^+-n-p^+ diode in the negative resistance region for three different bias currents. (After Chu and Sze, Ref. 81.)

Figure 42 shows the calculated[81] small-signal impedances (dashed curves) for a p^+-n-p^+ structure with $L = 8.5\ \mu m$ and $N = 5 \times 10^{14}\ cm^{-3}$. At 3 A/cm², the peak negative resistance $(-R_d)$ occurs at $f = 5.7$ GHz. As the current density increases to 16 A/cm², the magnitude of the peak $(-R_d)$ and the corresponding frequency both increase. With further increase of current to 80 A/cm², the peak value of $(-R_d)$ starts to decrease because of the reduction of the injection conductance, Eq. 80.

The measured small-signal impedance for a p^+-n-p^+ diode having the same device parameters is also shown in Fig. 42 (solid lines). The measured results and theoretical prediction (dashed lines) are in general agreement. The small-signal theory is equally applicable to MSM BARITT diodes, for which good agreements have also been obtained.

From these results, we have shown that (1) the BARITT diodes have negative small-signal resistances and therefore can have self-starting oscillation; (2) the injection over the forward-biased p^+-n junction or metal–semiconductor barrier serves as the source of carriers; and (3) the transit-time in the drift region is important for the frequency characteristics of the BARITT diodes.

The BARITT diode has been shown to be a low-noise device, with basically only two noise sources. One noise source is the shot noise of the injected carriers (injection noise). The other noise source is the random velocity fluctuation of the carriers (diffusion noise) in the drift region.

The calculated and measured results[82] in Fig. 43 show good agreement for low current densities. A noise measure of about 10 dB at 17 A/cm² is predicted for a p^+-n-p^+ diode with $N_D = 1.2 \times 10^{15}\ cm^{-3}$ and $W = 7.9\ \mu m$.

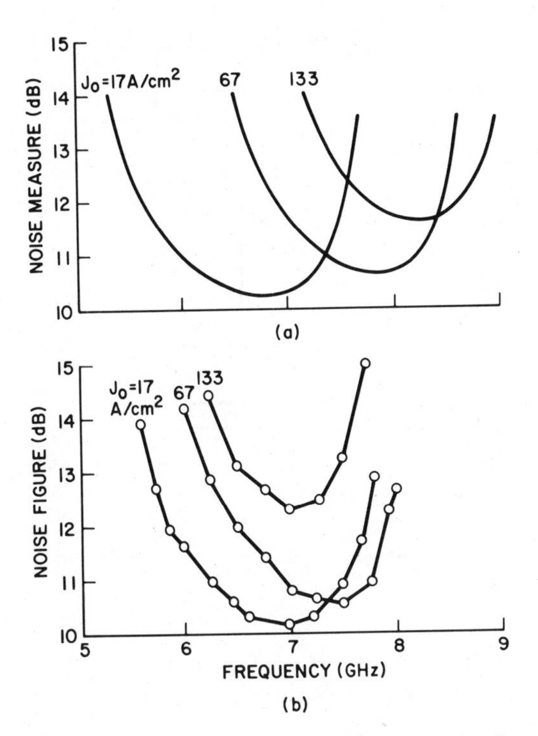

Fig. 43 (a) Calculated noise measure and (b) experimental noise figure of a Si BARITT diode with $W = 7.9\ \mu m$ and $N_D = 1.2 \times 10^{15}\ cm^{-3}$. (After Sjolund, Ref. 82.)

For higher current densities, the curves will shift to lower frequencies. Comparing these results with the noise of IMPATT diodes shows that the BARITT diodes have much lower noises.

10.7.3 Large-Signal Performance[83]

The basic large-signal BARITT operation is shown in Fig. 44. The carriers are injected as a delta function when the ac voltage reaches its peak ($\phi = \pi/2$). The induced external current travels three-fourths of a cycle to reach the negative terminal:

$$\theta_d = \omega\tau_d = 3\pi/2 \tag{91}$$

or

$$f = \frac{3}{4\tau_d} \tag{91a}$$

where θ_d is the transit angle and τ_d the carrier transit time. As a first-order approximation, the transit time is given by W/v_s. Therefore, the frequency

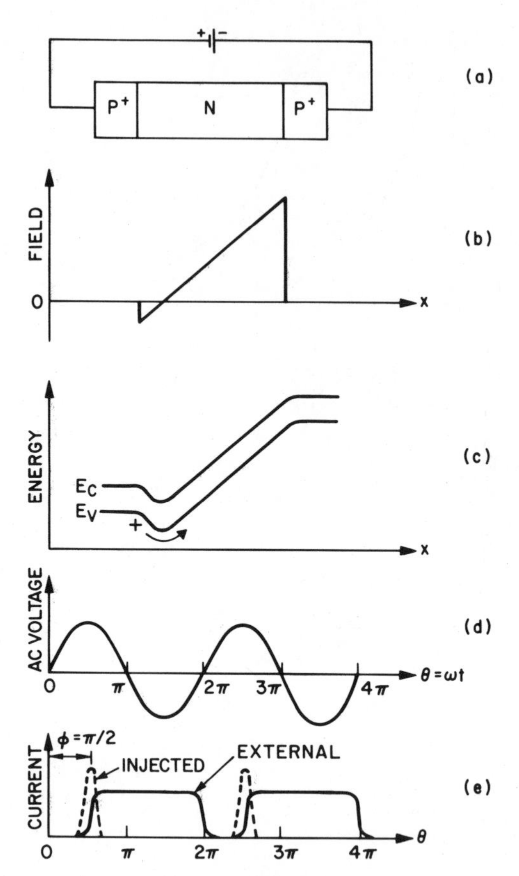

Fig. 44 Large-signal operation of BARITT diode. (*a*) p^+-n-p^+ structure. (*b*) Field distribution after reach-through. (*c*) Energy diagram. (*d*) ac voltage. (*e*) Injected and external current.

of oscillation becomes

$$f \simeq \frac{3v_s}{4W} \cdot \tag{92}$$

More accurate values for the optimum frequency can be obtained by substituting Eq. 77 into Eq. 92.

The maximum efficiency of BARITT diodes has been estimated to be of the order of 10% if the carriers are injected at $\phi = \pi/2$ (Fig. 44). However, higher efficiencies can be obtained if the carrier injection can be further delayed, that is, $\pi/2 < \phi \leq \pi$. A multilayered n^+-i-p-ν-n^+ BARITT diode,[84] which is a complementary structure of p^+-i-n-π-p^+ shown in Fig. 39*b*, has been fabricated. The n^+-i-p region serves as a retarding field to increase the injection delay time. The measured efficiency and power output are

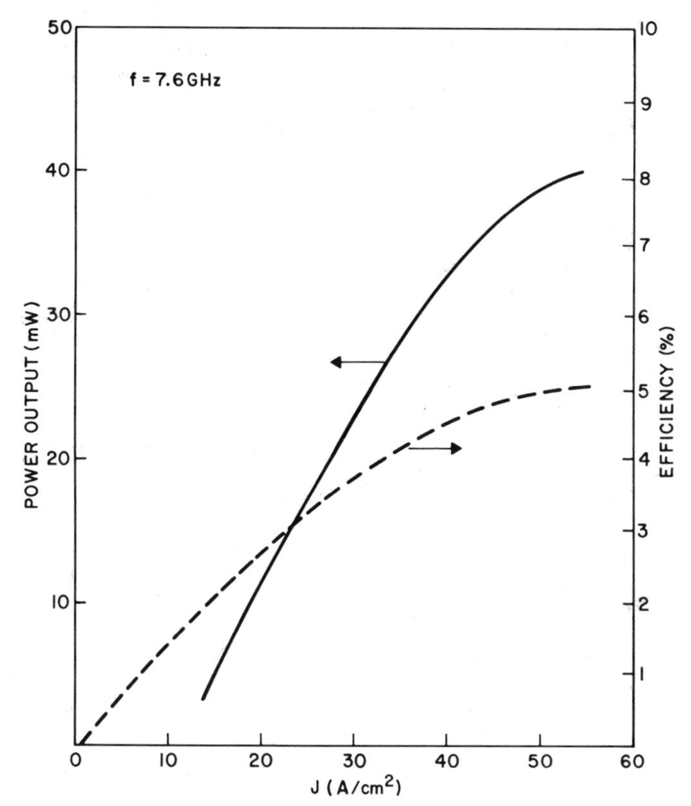

Fig. 45 Power and efficiency performance of a n^+-i-p-ν-n^+ BARITT diode. (After Eknoyan, Sze, and Yang, Ref. 84.)

shown in Fig. 45. The maximum efficiency obtained is over 5% at $J \simeq 50\ \text{A/cm}^2$. The corresponding output is 40 mW at a frequency of 7.6 GHz. Using a Pt Schottky contact on n-p^+ epitaxial layer with $N_D = 2.5 \times 10^{15}\ \text{cm}^{-3}$ and $W = 6.5\ \mu\text{m}$, an output power of 152 mW has been obtained[85] at 8.6 GHz with a peak efficiency of 2.3%. The state of the art of BARITT performance[84,85,97] is shown in Fig. 34. Although the power output near 10 GHz is about two orders of magnitude smaller than that of IMPATT diodes, so is the noise measure. By optimizing the injection delay processes, the BARITT diode is expected to realize its full potential as a low-noise microwave source with moderate power and efficiency.

10.7.4 DOVETT Diode

The DOVETT (double velocity transit time) diode is similar to the BARITT diode except that the velocity of carriers near the injection contact is significantly less than that near the collection contact.[86,87] A

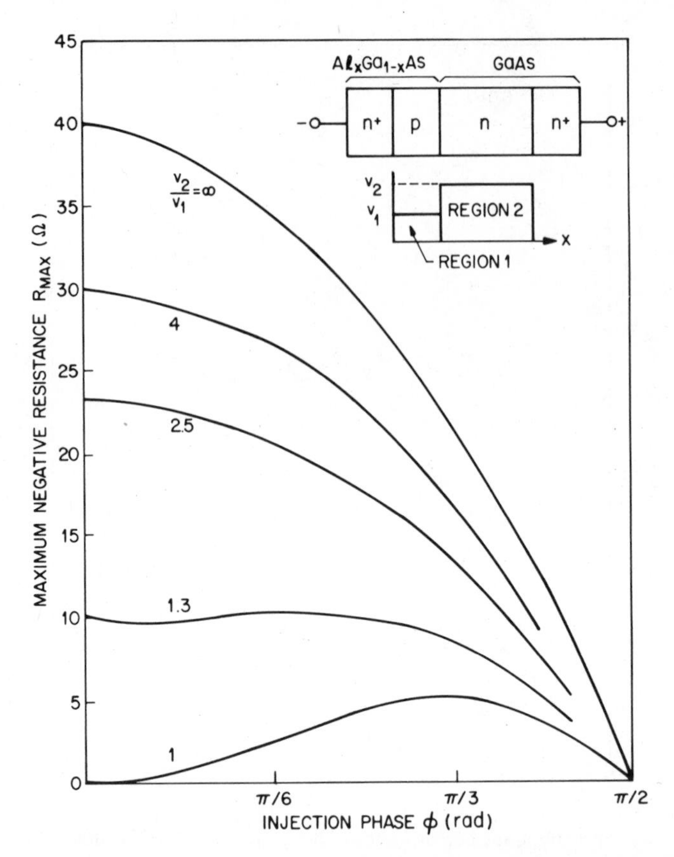

Fig. 46 Maximum negative resistance versus injection phase delay for DOVETT diodes at 10 GHz with $v_2 = 8 \times 10^6$ cm/s and $A = 10^{-4}$ cm^{-2}. The insert shows the device structure and velocity profile. (After Sitch and Robson, Ref. 87.)

heterostructure has been proposed for the DOVETT diode (Fig. 46, insert) which has an n^+-p-n-n^+ configuration where the n^+-p region is $Al_xGa_{1-x}As$ and the n-n^+ region is GaAs. Also shown in the insert is the velocity profile, where region 1 has a lower saturation velocity v_1 than the saturation velocity v_2 in region 2. The device is biased, so that the p-n anisotype heterojunction is completely punched through and large current can be injected over the forward-biased n^+-p junction.

Because of the transit-time delay in the low-velocity injection region, the negative resistance of DOVETT diodes is expected to be higher than that of BARITT diode (which is the limiting case of constant saturation velocity, that is, $v_1 = v_2$). The maximum small-signal negative resistance is given by[87]

$$R_{\max} = \frac{[v_1(1+\cos\phi) - 2v_2]\cos\phi}{A\omega^2\epsilon_s} \tag{93}$$

where ϕ is the injection phase delay and A is the device area. Figure 46 shows calculated values of R_{max} plotted against the injection angle ϕ for a range of velocity ratios. Note that in BARITT diodes ($v_2/v_1 = 1$), an injection delay of $\sim \pi/3$ radian is required for maximum negative resistance. The advantages of DOVETT diodes with a large value of velocity ratio are clearly seen from Fig. 46. The negative resistances is comparatively large, and the device can operate without any barrier injection delay ($\phi = 0$). This allows the device to be operated at higher dc current densities than BARITT diodes, where in the latter case increasing current density decreases the barrier delay until the negative resistance vanishes. An efficiency in excess of 25% has been predicted for DOVETT diodes.

10.8 TRAPATT DIODE

The TRAPATT diode is a high-power, high-efficiency device. To date the highest pulse power of 1.2 kW has been obtained at 1.1 GHz (five diodes in series[88]) and the highest efficiency[89] of 75% has been obtained at 0.6 GHz. However, the TRAPATT operation has a rather complicated manner of oscillation and requires very precise control of both device and circuit properties. In addition, the TRAPATT diode generally shows a considerably higher noise measure than the IMPATT mode; the upper operating frequency seems to be practically limited to below the millimeter-wave region.

To understand the initiating of the TRAPATT operation, we consider a current step that is applied at $t = 0$ to a n^+-p-p^+ diode (Fig. 47a). If at $t = 0$ the maximum electric field of the diode is smaller than the critical field $\mathscr{E}_m$ at breakdown, the field profile will at first move up with time (Fig. 47b). This is because[20,90]

$$J = \epsilon_s \frac{d\mathscr{E}}{dt}. \tag{94}$$

Hence the field can be expressed as

$$\mathscr{E}(x, t) = \mathscr{E}_m - \frac{qN_A}{\epsilon_s}x + \frac{Jt}{\epsilon_s} \tag{95}$$

where N_A is the doping concentration of the p region. Thus the value of t at which the electric field reaches $\mathscr{E}_m$ at a given distance x into the depletion region is obtained by setting $\mathscr{E}(x, t) = \mathscr{E}_m$, yielding

$$\frac{qN_A x}{\epsilon_s} = Jt \tag{96}$$

or upon differentiation

$$v_z \equiv \frac{dx}{dt} = \frac{J}{qN_A} \tag{97}$$

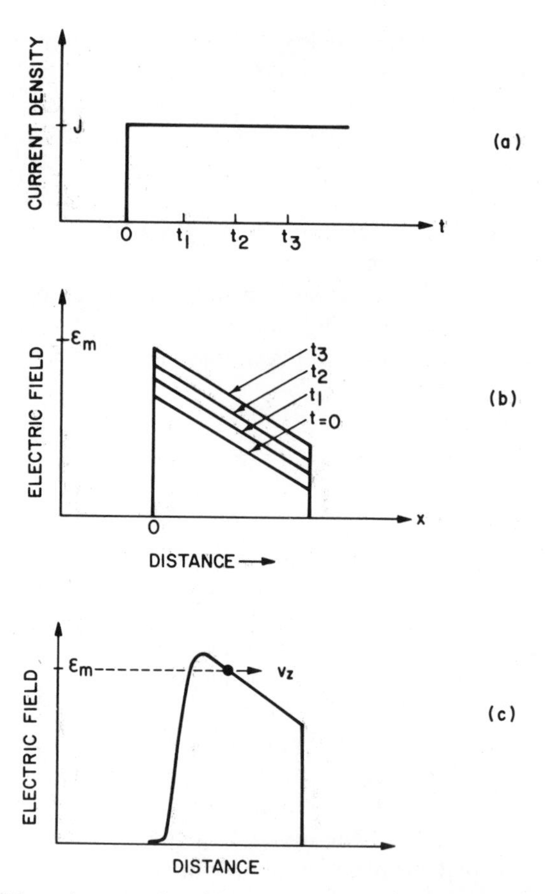

Fig. 47 TRAPATT mode operation (schematic). (*a*) Current density versus time. (*b*) Idealized electric field rising with time. (*c*) Avalanche shock front which sweeps through the diode with velocity v_z greater than the saturation velocity v_s. (After Clorfeine, Ikola, and Napoli, Ref. 20; DeLoach and Scharfetter, Ref. 90.)

where v_z is the avalanche-zone velocity and represents the velocity at which the leading edge of the avalanche region progressed through the diode (Fig. 47). For example, for $N_A = 10^{15}$ cm^{-3} and $J = 10^4$ A/cm^2, v_z has the value of 6×10^7 cm/s, which is much larger than the saturation velocity. Thus the avalanche zone[19] (or avalanche shock front) will quickly sweep across most of the diode, leaving it filled by a highly conducting plasma of holes and electrons whose space charge depresses the voltage to low values.[18,91]

The field and carrier density calculated numerically for a particular moment is shown[92] in Fig. 48*a*. The corresponding voltage and current waveforms are shown in Fig. 48*b* and *c*, respectively. (The dot indicates the situation shown on Fig. 48*a*.) Note that the voltage at the beginning of

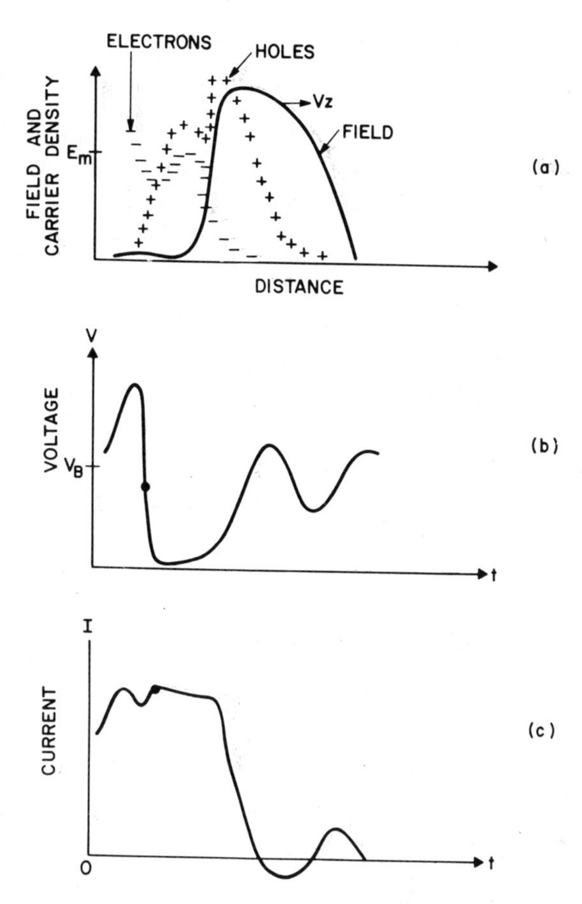

Fig. 48 TRAPATT mode operation. (*a*) Numerical calculation of instantaneous field and charge. (*b*) Voltage versus time. (*c*) Current versus time. Dots on voltage and current curves apply to the condition shown in (*a*). (After Evans, Ref. 92.)

the cycle can become considerably larger than the steady-state breakdown voltage. After the avalanche zone passes through the device, the voltage drops to low values, but soon rises again as the field polarizes the space charge. Because the drift velocity depends on field at low fields (i.e., $v = \mu\mathscr{E}$), the electrons and holes will drift at velocities determined by the low-field mobilities, and the time of transit of the carriers can become much longer than W/v_s.

Thus the TRAPATT diode can operate at comparatively low frequencies because the discharge time of the plasma, that is, the ratio Q/I of its charge to its current, can be substantially greater than the nominal transit time W/v_s of the diode at high fields. Therefore, the TRAPATT diode is still a transit-time diode in the sense that the time delay of carriers in transit, that is, the time between injection and collection, is utilized to obtain a current phase shift favorable for oscillation.

The TRAPATT diode also requires a circuit that can support harmonics of the fundamental frequency at high-voltage amplitudes. The rich harmonic content is necessary to get the requisite phase delay in the current at such low frequencies; the diode must not break down when the fundamental voltage is a maximum, but is required instead to wait until approximately one-quarter cycle later, when the avalanche is triggered by a peak in the harmonic voltage. Since the terminal voltage can be small when the terminal current is large, and vice versa, the TRAPATT diode can have high efficiency. Efficiencies substantially higher than 50% have been predicted for some silicon diodes.[93, 94]

For a TRAPATT diode, the design and performance are more complicated because of the strong device-circuit interaction that dictates most of the device performance. To reduce the overvoltage, which is the terminal voltage required to initiate the avalanche shock front, a TRAPATT diode is usually designed with a large punch-through factor. The punch-through factor F is defined in Fig. 49*a*. The larger the factor F, the closer the device resembles a *p-i-n* diode. The required depletion width and impurity concentration as a function of TRAPATT operating frequency is shown in Fig. 49*b* for three punch-through factors.

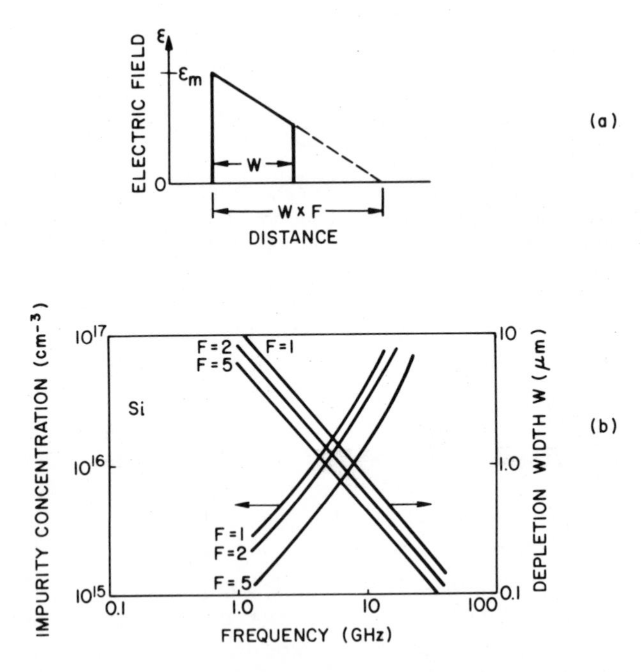

Fig. 49 Design parameter for TRAPATT diodes. The punch-through factor F is defined in (*a*). (After Evans, Ref. 40.)

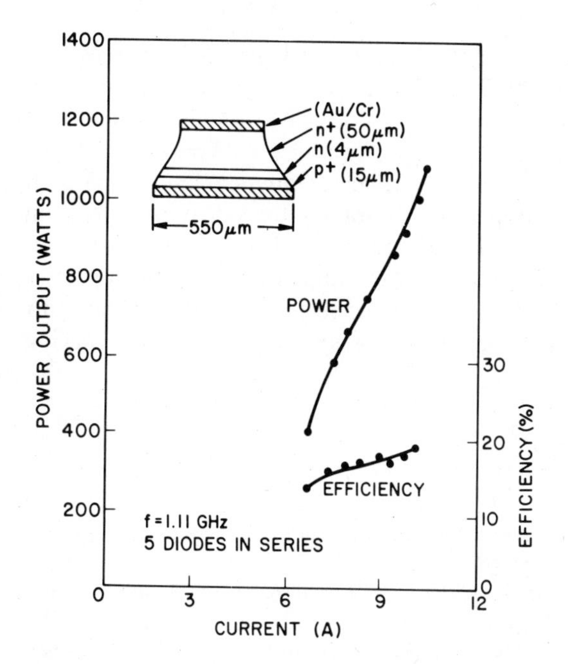

Fig. 50 Power output and efficiency versus current for a stack of five diodes at 1.11 GHz. (After Liu and Risko, Ref. 88.)

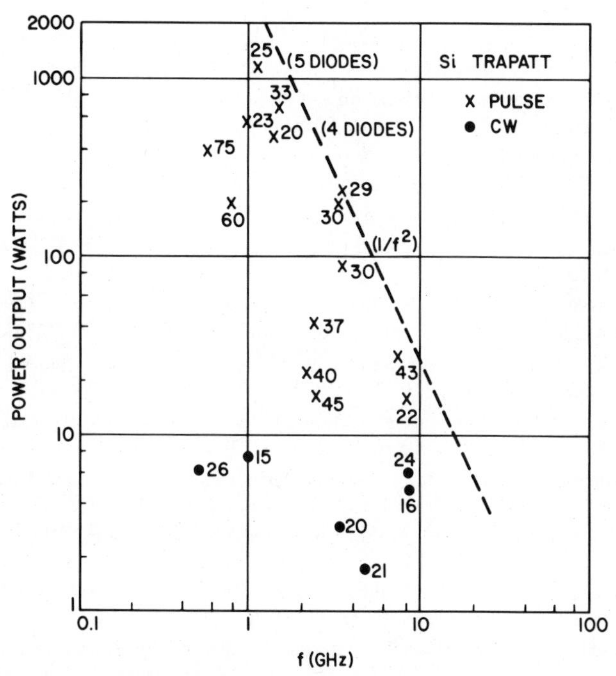

Fig. 51 State-of-the-art TRAPATT performance. The number against each experimental point indicates the efficiency in percent.

The TRAPATT diode state-of-the-art highest power performance is shown[88] in Fig. 50. The output power of a series connection of five diodes under pulse condition reaches 1.2 kW with an efficiency of 25%. A summary of power-frequency performance of TRAPATT diodes is shown[8,95,96] in Fig. 51. Note that the upper frequency limit is close to 10 GHz, and higher power outputs are all obtained under pulse operation. This remarkably high pulse power output is much larger than most other microwave semiconductor devices. But the high sensitivity to small changes in circuit, or operating conditions, or temperature, makes it difficult to maintain such performance for long times in systems subject to varying ambient conditions.

REFERENCES

1 R. L. Johnston, B. C. DeLoach, Jr., and B. G. Cohen, "A Silicon Diode Oscillator," *Bell Syst. Tech. J.*, **44**, 369 (1965).

2 B. C. DeLoach, Jr., "The IMPATT Story," *IEEE Trans. Electron Devices*, **ED-23**, 57 (1976).

3 W. Shockley, "Negative Resistance Arising from Transit Time in Semiconductor Diodes," *Bell Syst. Tech. J.*, **33**, 799 (1954).

4 W. T. Read, "A Proposed High-Frequency Negative Resistance Diode," *Bell Syst. Tech. J.*, **37**, 401 (1958).

5 C. A. Lee, R. L. Batdorf, W. Wiegman, and G. Kaminsky, "The Read Diode and Avalanche, Transit-Time, Negative-Resistance Oscillator," *Appl. Phys. Lett.*, **6**, 89 (1965).

6 T. Misawa, "Negative Resistance on *p-n* Junction under Avalanche Breakdown Conditions, Parts I and II," *IEEE Trans. Electron Devices*, **ED-13**, 137 (1966).

7 M. Gilden and M. F. Hines, "Electronic Tuning Effects in the Read Microwave Avalanche Diode," *IEEE Trans. Electron Devices*, **ED-13**, 169 (1966).

8 S. M. Sze and R. M. Ryder, "Microwave Avalanche Diodes," *Proc IEEE*, **59**, 1140 (1971).

9 W. E. Schroeder and G. I. Haddad, "Nonlinear Properties of IMPATT Devices," *Proc. IEEE*, **61**, 153 (1973).

10 Special Issue on Solid-State Microwave Millimeter-Wave Power Generation, Amplification, and Control, *IEEE Trans. Microwave Theory Tech.*, **MTT-27** (May 1979).

10a L. F. Eastman, "Microwave Semiconductor Devices: State-of-the-Art and Limiting Effects," *IEEE Tech. Dig.*, Int. Electron Device Meet., 1978, p. 364.

11 C. A. Brackett, "The Elimination of Tuning Induced Burnout and Bias Circuit Oscillation in IMPATT Oscillators," *Bell Syst. Tech. J.*, **52**, 271 (1973).

12 D. J. Coleman, Jr. and S. M. Sze, "The Baritt Diode—A New Low Noise Microwave Oscillator," IEEE Device Res. Conf., Ann Arbor, Mich., June 28, 1971; "A Low-Noise Metal–Semiconductor–Metal (MSM) Microwave Oscillator," *Bell Syst. Tech. J.*, **50**, 1695 (1971).

13 H. W. Ruegg, "A Proposed Punch-Through Microwave Negative Resistance Diode," *IEEE Trans. Electron Devices*, **ED-15**, 577 (1968).

14 G. T. Wright, "Punch-Through Transit-Time Oscillator," *Electron. Lett.*, **4**, 543 (1968).

15 H. Sobol and F. Sterzer, "Microwave Power Sources," *IEEE Spectrum*, **9**, 20 (1972).

16 J. R. East, H. Nguyen-Ba, and G. I. Haddad, "Microwave and mm Wave Baritt Doppler Detector," *Microwave J.*, **19**, 51 (1976).

17 H. J. Prager, K. K. N. Chang, and S. Weisbrod, "High Power, High Efficiency Silicon Avalanche Diodes at Ultrahigh Frequencies," *Proc. IEEE*, **55**, 586 (1967).

18 D. L. Scharfetter, D. J. Bartelink, H. K. Gummel, and R. L. Johnston, "Computer Simulation of Low-Frequency High-Efficiency Oscillation in Germanium IMPATT Diodes," *IEEE Trans. Electron Devices*, **ED-15**, 691 (1968).

19 D. J. Bartelink and D. L. Scharfetter, "Avalanche Shock Fronts in *p-n* Junctions," *Appl. Phys. Lett.*, **14**, 420 (1969).

20 A. S. Clorfeine, R. J. Ikola, and L. S. Napoli, "A Theory for the High-Efficiency Mode of Oscillation in Avalanche Diodes," *RCA Rev.*, **30**, 397 (1969).

21 K. K. N. Chang, H. Kawamoto, H. J. Prager, J. F. Reynolds, A. Rosen, and V. A. Mikenas, "TRAPATT Amplifiers for Phased-Array Radar Systems," *Microwave J.*, **16**, 27 (1973).

22 G. Salmer, H. Pribetich, A. Farrayre, and B. Kramer, "Theoretical and Experimental Study of GaAs Impatt Oscillator Efficiency," *J. Appl. Phys.*, **44**, 314 (1973).

23 S. M. Sze and G. Gibbons, "Avalanche Breakdown Voltages of Abrupt and Linearly Graded *p-n* Junctions in Ge, Si, GaAs, and GaP," *Appl. Phys. Lett.*, **8**, 111 (1966).

24 W. E. Schroeder and G. I Haddad, "Avalanche Region Width in Various Structures of IMPATT Diodes," *Proc. IEEE*, **59**, 1245 (1971).

25 G. Gibbons and S. M. Sze, "Avalanche Breakdown in Read and *p-i-n* Diodes," *Solid State Electron.*, **11**, 225 (1968).

26 C. R. Crowell and S. M. Sze, "Temperature Dependence of Avalanche Multiplication in Semiconductors," *Appl. Phys. Lett.*, **9**, 242 (1966).

27 H. C. Bowers, "Space-Charge-Limited Negative Resistance in Avalanche Diodes," *IEEE Trans. Electron Devices*, **ED-15**, 343 (1968).

28 S. M. Sze and W. Shockley, "Unit-Cube Expression for Space-Charge Resistance," *Bell Syst. Tech. J.*, **46**, 837 (1967).

29 P. Weissglas, "Avalanche and Barrier Injection Devices" in M. J. Howes and D. V. Morgan, Eds., *Microwave Devices—Device Circuit Interactions*, Wiley, New York, 1976, Chap. 3.

30 T. Misawa, "Multiple Uniform Layer Approximation in Analysis of Negative Resistance in *p-n* Junctions in Breakdown," *IEEE Trans. Electron Devices*, **ED-14**, 795 (1967).

31 D. L. Scharfetter and H. K. Gummel, "Large-Signal Analysis of a Silicon Read Diode Oscillator," *IEEE Trans. Electron Devices*, **ED-16**, 64, (1969).

32 C. B. Swan, T. Misawa, and L. Marinaccio, "Composite Avalanche Diode Structures for Increased Power Capability," *IEEE Trans. Electron Devices*, **ED-14**, 684 (1967).

33 D. L. Scharfetter, "Power-Impedance-Frequency Limitation of IMPATT Oscillators Calculated from a Scaling Approximation," *IEEE Trans. Electron Devices*, **ED-18**, 536 (1971).

34 H. W. Thim and H. W. Poetze, "Search for Higher Frequencies in Microwave Semiconductor Devices," 6th Eur. Solid State Device Res. Conf., *Inst. Phys. Conf. Ser.*, **32**, 73 (1977).

35 T. E. Seidel, W. C. Niehaus and D. E. Iglesias, "Double-Drift Silicon IMPATTs at X Band," *IEEE Trans. Electron Devices*, **ED-21**, 523 (1974).

36 R. L. Kuvas, "Carrier Transport in the Drift Region of Read-Type Diodes," *IEEE Trans. Electron Devices*, **ED-25**, 660 (1978).

37 Y. Hirachi, K. Kobayashi, K. Ogasawara, and Y. Toyama, "A New Concept for High Efficiency Operation of Hi–Lo-Type GaAs IMPATT Diodes," *IEEE Trans. Electron Devices*, **ED-25**, 666 (1978).

38 P. A. Blakey, B. Culshaw, and R. A. Giblin, "Comprehensive Models for the Analysis of High Efficiency GaAs IMPATTS," *IEEE Trans. Electron Devices*, **ED-25**, 674 (1978).

39 K. Nishitani, H. Sawano, O. Ishihara, T. Ishii, and S. Mitsui, "Optimum Design for High-Power and High Efficiency GaAs Hi–Lo IMPATT Diodes," *IEEE Trans. Electron Devices*, **ED-26**, 210 (1979).

40 W. J. Evans, "Avalanche Diode Oscillators," in W. D. Hershberger, Ed., *Solid State and Quantum Electronics*, Wiley, New York 1971.

41 T. Misawa, "Saturation Current and Large Signal Operation of a Read Diode," *Solid State Electron.*, **13**, 1363 (1970).

42 S. R. Kovel and G. Gibbons, "The Effect of Unswept Epitaxial Material on the Microwave Efficiency of IMPATT Diodes," *Proc. IEEE Lett.*, **55**, 2066, (1967).

43 Y. Aono and Y. Okuto, "Effect of Undepleted High Resistivity Region on Microwave Efficiency of GaAs IMPATT Diodes," *Proc. IEEE*, **63**, 724 (1975).

44 B. C. DeLoach, Jr., "Thin Skin IMPATTs," *IEEE Trans. Microwave Theory Tech.*, **MTT-18**, 72 (1970).

45 T. Misawa, "High Frequency Fall-Off of IMPATT Diode Efficiency," *Solid State Electron.*, **15**, 457 (1972).

46 M. E. Elta and G. I. Haddad, "Mixed Tunneling and Avalanche Mechanisms in *p-n* Junctions and Their Effects on Microwave Transit-Time Devices," *IEEE Trans. Electron Devices*, **ED-25**, 694 (1978).

47 J. I. Nishizawa, K. Motoya, and Y. Okuno, "200 GHz TUNNETT Diodes," *Jpn. J. Appl. Phys.*, **17**, *Suppl.* **17–1**, 167 (1977).

48 J. J. Berenz, J. Kinoshita, T. L. Hierl, and C. A. Lee, "Orientation Dependence of *n*-type GaAs Intrinsic Avalanche Response Time," *Electron. Lett.*, **15**, 150 (1979).

49 H. M. Olson, "A Mechanism for Catastrophic Failure of Avalanche Diodes," *IEEE Trans. Electron Devices*, **ED-22**, 842 (1975).

50 M. F. Hines, "Noise Theory for Read Type Avalanche Diode," *IEEE Trans. Electron Devices*, **ED-13**, 158 (1966).

51 H. K. Gummel and J. L. Blue, "A Small-Signal Theory of Avalanche Noise on IMPATT Diodes," *IEEE Trans. Electron Devices*, **ED-14**, 569 (1967).

52 J. L. Blue, "Preliminary Theoretical Results on Low Noise GaAs IMPATT Diodes," *IEEE Device Res. Conf.*, Seattle, Wash., June 1970.

53 J. C. Irvin, D. J. Coleman, W. A. Johnson, I. Tatsuguchi, D. R. Decker, and C. N. Dunn, "Fabrication and Noise Performance of High-Power GaAs IMPATTs," *Proc. IEEE*, **59**, 1212 (1971).

54 T. Misawa, "Microwave Si Avalanche Diode with Nearly-Abrupt-Type Junction," *IEEE Trans. Electron Devices*, **ED-14**, 580 (1967).

55 S. M. Sze, M. P. Lepselter, and R. W. MacDonald, "Metal–Semiconductor IMPATT Diode," *Solid State Electron.*, **12**, 107 (1969).

56 D. De Nobel and H. G. Kock, "A Silicon Schottky Barrier Avalanche Transit Time Diode," *Proc. IEEE*, **57**, 2088 (1969).

57 C. K. Kim and L. D. Armstrong, "GaAs Schottky-Barrier Avalanche Diode," *Solid State Electron.*, **13**, 53 (1970).

58 T. Misawa, "Minority Carrier Storage and Oscillation Efficiency in Read Diodes," *Solid State Electron.*, **13**, 1369 (1970).

59 W. C. Niehaus and B. Schwartz, "A Self-Limiting Anodic Etch-to-Voltage Technique for Fabrication of Modified Read IMPATTs," *Solid State Electron.*, **19**, 175 (1976).

60 G. E. Mahoney, "Retardation of IMPATT Diode Aging by Use of Tungsten in the Electrodes," *Appl. Phys. Lett.*, **27**, 613 (1975).

61 J. L. Heaton, R. E. Walline, and J. F. Carroll, "Low–High–Low Profile GaAs IMPATT Reliability," *IEEE Trans. Electron Devices*, **ED-26**, 96 (1979).

62 D. L. Scharfetter, W. J. Evans, and R. L. Johnston, "Double-Drift-Region (p^+pnn^+) Avalanche Diode Oscillators," *Proc. IEEE*, **58**, 1131 (1970).

63 A. Y. Cho, C. N. Dunn, R. L. Kuvas, and W. E. Schroeder, "GaAs IMPATT Diodes, Prepared by Molecular Beam Epitaxy," *Appl. Phys. Lett.*, **25**, 224 (1974).

64 S. Y. Liao, *Microwave Devices and Circuits*, Prentice-Hall, Englewood Cliffs, N.J., 1980.

65 D. E. Iglesias, "Circuit for Testing High Efficiency IMPATT Diodes," *Proc. IEEE*, **55**, 2065 (1967).

66 T. Misawa and L. P. Marinaccio, "A 1/4 Watt Si p-ν-n X-Band IMPATT Diode," Int. Electron Device Meet., Washington, D.C., Oct. 1966.

67 L. S. Bowman and C. A. Burrus, Jr., "Pulse-Driven Silicon p-n Junction Avalanche Oscillators for the 0.9 to 20 mm Band," *IEEE Trans. Electron Devices*, **ED-14**, 411 (1967).

68 M. Ino, T. Ishibashi, and M. Ohmori, "Submillimeter Wave Si p^+pn^+ IMPATT Diodes," *Jpn. J. Appl. Phys.* **16**, *Suppl.* **16–1**, 89 (1977).

69 J. Pribetich, M. Chive, E. Constant, and A. Farrayre, "Design and Performance of Maximum-Efficiency Single and Double-Drift-Region GaAs IMPATT Diodes in the 3–18 GHz Frequency Range," *J. Appl. Phys.*, **49**, 5584 (1978).

70 T. E. Seidel, R. E. Davis, and D. E. Iglesias, "Double-Drift-Region Ion-Implanted Millimeter-Wave IMPATT Diodes," *Proc. IEEE*, **59**, 1222 (1971).

71 E. D. Cohen, "Trapatts and Impatts—State of the Art and Application," *Microwave J.*, **20**, 22 (1977).

72 K. W. Grery, "Recent Advances in Microwave Devices," *Jpn. J. Appl. Phys.*, **16**, *Suppl.* **16–1**, 81 (1974).

73 C. O. Bozler, J. P. Donnelly, R. A. Murphy, R. W. Laton, and R. W. Subduny, "High-Efficiency Ion-Implanted Lo–Hi–Lo GaAs IMPATT Diodes," *Appl. Phys. Lett.*, **29**, 125 (1976).

74 T. L. Hierl, J. J. Berenz, J. Kinoshita, and I. U. Zubeck, "High Efficiency Pulsed GaAs Read IMPATT Diodes," *Electron. Lett.*, **14**, 155 (1978).

75 D. C. Potteiger, private communication.

76 S. M. Sze, D. J. Coleman, and A. Loya, "Current Transport in Metal–Semiconductor–Metal (MSM) Structures," *Solid State Electron.*, **14**, 1209 (1971).

77 J. L. Chu, G. Persky, and S. M. Sze, "Thermionic Injection and Space-Charge-Limited Current in Reach-Through p^+np^+ Structures," *J. Appl. Phys.*, **43**, 3510 (1972).

78 O. Eknoyan, E. S. Yang, and S. M. Sze, "Multilayered Ion-Implanted Baritt Diodes with Improved Efficiency," *Solid State Electron.*, **20**, 291 (1977).

79 D. De Cogan, "The Punch-Through Diode," *Microelectronics*, **8**, 20 (1976).

80 H. Nguyen-Ba and G. I. Haddad, "Effects of Doping Profile on the Performance of Baritt Devices," *IEEE Trans. Electron Devices*, **ED-24**, 1154 (1977).

81 J. L. Chu and S. M. Sze, "Microwave Oscillation in *pnp* Reach-Through Baritt Diodes," *Solid State Electron.*, **16**, 85 (1973).

82 A. Sjolund, "Small-Signal Analysis of Punch-Through Injection Microwave Devices," *Solid State Electron.*, **16**, 559 (1973).

83 S. P. Kwok and G. I. Haddad, "Power Limitation in Baritt Devices," *Solid State Electron.*, **19**, 795 (1976).

84 O. Eknoyan, S. M. Sze, and E. S. Yang, "Microwave Baritt Diode with Retarding Field—An Investigation," *Solid State Electron.*, **20**, 285 (1977).

85 S. Ahmad and J. Freyer, "High-Power Pt Schottky Baritt Diodes," *Electron. Lett.*, **12**, 238 (1976).

86 J. E. Sitch, A. Majerfeld, P. N. Robson, and F. Hasegawa, "Transit-Time-Induced Microwave Negative Resistance in GaAlAs–GaAs Heterostructure Diodes," *Electron. Lett.*, **11**, 457 (1975).

87 J. E. Sitch and P. N. Robson, "Efficiency of BARITT and DOVETT Oscillators," *Solid State Electron Devices*, **1**, 31 (1976).

88 S. G. Liu and J. J. Risko, "Fabrication and Performance of Kilowatt *L*-Band Avalanche Diodes," *RCA Rev.*, **31**, 3 (1970).

89 D. F. Kostichack, "UHF Avalanche Diode Oscillator Providing 400 Watts Peak Power and 75 Percent Efficiency," *Proc. IEEE Lett.*, **58**, 1282 (1970).

90 B. C. DeLoach, Jr. and D. L. Scharfetter, "Device Physics of TRAPATT Oscillators," *IEEE Trans. Electron Devices*, **ED-17**, 9 (1970).

91 R. L. Johnston, D. L. Scharfetter, and D. J. Bartelink, "High-Efficiency Oscillations in Germanium Avalanche Diodes Below the Transit-Time Frequency," *Proc. IEEE*, **56**, 1611 (1968).

92 W. J. Evans, "Circuits for High-Efficiency Avalanche-Diode Oscillators," *IEEE Trans. Microwave Theory Tech.*, **MTT-17**, 1060 (1969).

93 D. L. Scharfetter, "Power-Frequency Characteristics of the TRAPATT Diode Mode of High Efficiency Power Generation in Germanium and Silicon Avalanche Diodes," *Bell Syst. Tech. J.*, **49**, 799 (1970).

94 R. J. Trew, G. I. Haddad, and N. A. Masnari, "A Simplified Model of a TRAPATT Diode," *IEEE Trans. Electron Devices*, **ED-23**, 28 (1976).

95 C. O. G. Obah, E. Benko, T. A. Midford, H. C. Bowers, and P. Y. Chao, "Single-Diode 0.5 kW TRAPATT Oscillators," *Electron. Lett.*, **10**, 430 (1974).

96 R. S. Ying and T. T. Fong, "C-Band Complementary TRAPATT Diodes," *Proc. IEEE*, **62**, 287 (1974).

97 S. Ahmad and J. Freyer, "Design and Development of High-Power Microwave Silicon BARITT Diodes," *IEEE Trans. Electron Devices*, **ED-26**, 1370 (1979).

11

Transferred-Electron Devices

- INTRODUCTION
- TRANSFERRED-ELECTRON EFFECT
- MODES OF OPERATION
- DEVICE PERFORMANCES

11.1 INTRODUCTION

The transferred-electron device (TED), one of the most important microwave devices, has been extensively used as local oscillators and power amplifiers, covering the microwave frequency range from 1 to 100 GHz. The TEDs have matured to become important solid-state microwave sources used in radars, intrusion alarms, and microwave test instruments.

In 1963, Gunn discovered that coherent microwave output was generated when a dc electric field was applied across a randomly oriented, short, n-type sample of GaAs or InP that exceeded a critical threshold value of several thousand volts per centimeter.[1] The frequency of oscillation was approximately equal to the reciprocal of the carrier transit time across the length of the sample. Later, Kroemer pointed out[2] that all the observed properties of the microwave oscillation were consistent with a theory of negative differential resistance independently proposed by Ridley and Watkins[3,3a] and by Hilsum.[4,4a] The mechanism responsible for the negative differential resistance is a field-induced transfer of conduction-band electrons from a low-energy, high-mobility valley to higher-energy, low-mobility satellite valleys. The GaAs pressure experiments of Hutson et al.[5] and the $GaAs_{1-x}P_x$ alloy experiments of Allen et al.,[6] which demonstrated that the threshold electric field decreases with decreasing energy separation between the valley minima, were convincing evidence that the transferred-electron effect was indeed responsible for the Gunn oscillation.

The transferred-electron effect has also been referred to as the Ridley–Watkins–Hilsum effect or as the Gunn effect. Comprehensive reviews on TEDs have been given by Carroll,[7] Bulman, Hobson, and Taylor,[8] and Bosch and Engelmann.[9] Recently, various solid-state microwave sources have been reviewed by Thim.[10]

11.2 TRANSFERRED-ELECTRON EFFECT

11.2.1 Bulk Negative Differential Resistivity

Various physical causes give rise to bulk negative differential resistivity. One of the most important examples is the transferred-electron effect due to intervalley carrier transport. A semiconductor exhibiting bulk negative differential resistivity is inherently unstable, because a random fluctuation of carrier density at any point in the semiconductor produces a momentary space charge that grows exponentially in time. The one-dimensional continuity equation is given by*

$$\frac{\partial n}{\partial t} + \frac{1}{q}\frac{\partial J}{\partial x} = 0 . \tag{1}$$

If there is a small local fluctuation of the majority carriers from the uniform equilibrium concentration n_0, the locally created space-charge density is $(n - n_0)$. The Poisson equation and the current-density equation are

$$\frac{\partial \mathscr{E}}{\partial x} = \frac{q(n - n_0)}{\epsilon_s} \tag{2}$$

$$J = \mathscr{E}/\rho - qD\frac{\partial n}{\partial x} \tag{3}$$

where ϵ_s is the dielectric permittivity, ρ the resistivity, and D the diffusion constant. Differentiating Eq. 3 with respect to x and inserting Poisson's equation yields

$$\frac{1}{q}\frac{\partial J}{\partial x} = \frac{n - n_0}{\rho\epsilon_s} - D\frac{\partial^2 n}{\partial x^2} . \tag{4}$$

Substituting this expression into Eq. 1 gives

$$\frac{\partial n}{\partial t} + \frac{n - n_0}{\rho\epsilon_s} - D\frac{\partial^2 n}{\partial x^2} = 0. \tag{5}$$

*To avoid excessive minus signs, a positive charge will be assigned to electrons, and all operations are modified accordingly throughout this chapter.

Equation 5 can be solved by separation of variables. For the spatial response, Eq. 5 has the solution

$$n - n_0 = (n - n_0)_{x=0} \exp(-x/L_D) \tag{6}$$

where L_D is the Debye length, given by

$$L_D \equiv \sqrt{\frac{kT\epsilon_s}{q^2 n_0}} \tag{7}$$

which determines the distance over which a small unbalanced charge decays.

For the temporal response, Eq. 5 has the solution:

$$n - n_0 = (n - n_0)_{t=0} \exp(-t/\tau_R) \tag{8}$$

where τ_R is the dielectric relaxation time given by

$$\tau_R \equiv \rho\epsilon_s = \epsilon_s/q\mu n \approx \epsilon_s/q\mu n_0 \tag{9}$$

which represents the time constant for the decay of the space charge to neutrality if the differential resistivity (or the differential mobility μ) is positive. However, if the semiconductor exhibits a negative differential resistivity, any charge imbalance will grow with a time constant equal to $|\tau_R|$ instead of decay.

Bulk negative differential resistivity (NDR) devices can be classified into two groups: voltage-controlled NDR (N-shaped) and current-controlled NDR (S-shaped). The general current density versus electric-field characteristics for these two groups are shown in Fig. 1*a* and *b*, respectively.[11] Also shown are the corresponding differential resistivities ($\partial\mathscr{E}/\partial J$) from zero bias to the onset of the negative-resistivity region in Fig. 1*c* and *d*. In voltage-controlled NDR (Fig. 1*a*), the electric field can be multivalued; in current-controlled NDR (Fig. 1*b*), the electric current can be multivalued.

Figure 1 shows I–V characteristics similar to those shown in previous chapters associated with junction or contact phenomena. For example, a tunnel diode exhibits a voltage-controlled NDR, whereas a thyristor exhibits a current-controlled NDR. For bulk NDR, negative resistivity is associated with microscopic bulk semiconductor properties, such as field-enhanced trapping,[12] impact ionization of shallow impurity levels in compensated semiconductors (cryosar),[13] and electron transfer from a lower valley to higher valleys in the conduction band (transferred-electron effect). We shall consider the basic characteristics of bulk NDR devices regardless of the physical causes underlying a particular negative resistance.

Because of negative differential resistivity, the semiconductor, initially homogeneous, becomes electrically heterogeneous in an attempt to reach stability. We present next a simple argument to show that for voltage-controlled NDR devices, high-field domains (or an accumulation layer) will form, and for current-controlled NDR devices, high-current filaments will form.[11] The simple argument will be modified when we discuss the various modes of operation in later sections.

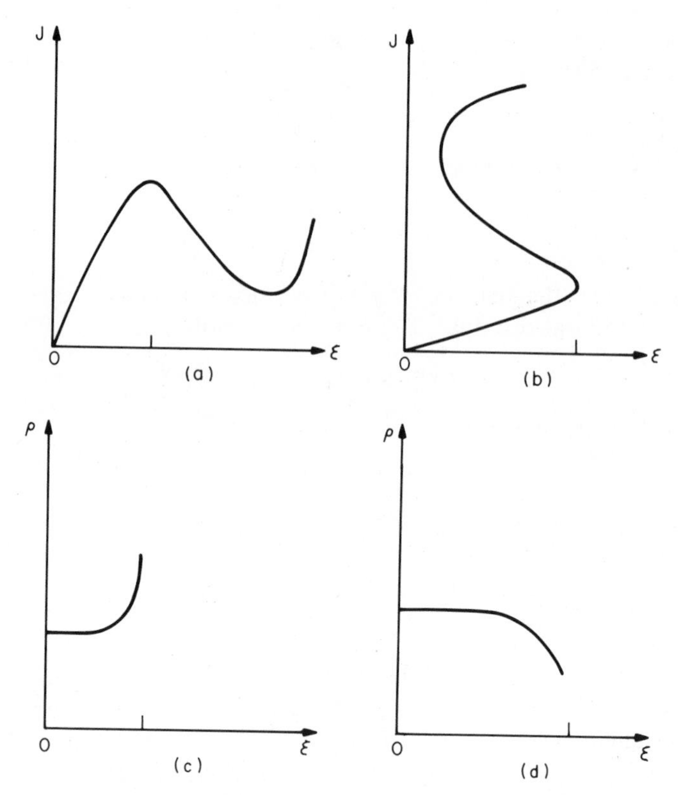

Fig. 1 Current density versus electric field characteristics. (*a*) Voltage-controlled J–$\mathscr{E}$ curve (N shape). (*b*) Current-controlled J–$\mathscr{E}$ curve (S shape). (*c*) Differential resistivities for (*a*). (*d*) Differential resistivities for (*b*). (After Ridley, Ref. 11.)

We note that in Fig. 1*c* the positive differential resistivity increases with increasing field, that is, $d\rho/d\mathscr{E} > 0$. If a region has a slightly higher field, as shown in Fig. 2*a*, the resistance there is larger. Thus less current will flow through it. This results in an elongation of the region, and a high-field domain is formed separating regions of low field. The interfaces separating low- and high-field domains lie along equipotentials, so that they are in planes perpendicular to the current direction.

For current-controlled NDR the initial positive differential resistivity decreases with increasing field; that is, $d\rho/d\mathscr{E} < 0$. If a region has a slightly higher field as shown in Fig. 2*b*, the resistance there is smaller. Thus more current will flow into it. This results in an elongation of the region along the current path, and finally in the formation of a high-current filament running along the field direction.

To consider in more detail the space-charge instability[14] of a voltage-controlled NDR, we refer to Fig. 3. Figure 3*a* shows a typical instantaneous J–$\mathscr{E}$ plot and Fig. 3*b* shows the profile of this device. Assume

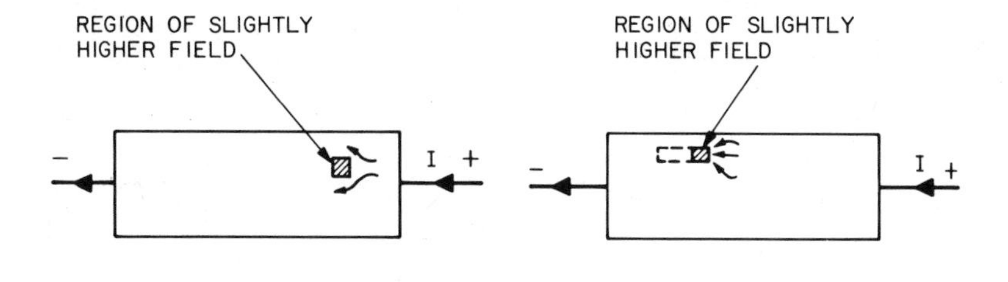

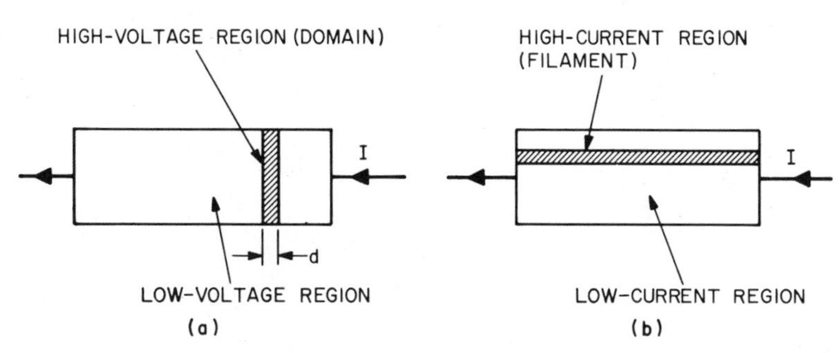

Fig. 2 (*a*) Formation of high-field domain in a voltage-controlled negative differential resistivity (NDR). (*b*) Formation of high-current filament in a current-controlled NDR.

that at point A in the device there exists an excess (or accumulation) of negative charge that could be due to a random noise fluctuation or possibly a permanent nonuniformity in doping (Fig. 3*c*). Integration once of Poisson's equation yields the electric field distribution as shown in Fig. 3*d*, where the field to the left of point A is lower than that to the right of A. If the device is biased at point $\mathscr{E}_A$ on the J–$\mathscr{E}$ curve, this condition would imply that the carriers (or current) flowing into point A are greater than those flowing out of point A, thereby increasing the excess negative space charge at A. Now the field to the left of point A is even less than it was originally, and the field to the right is greater than originally, resulting in an even greater space-charge accumulation. This process continues until the high and low fields both obtain values outside the NDR region and settle at points 1 and 2 in Fig. 3*a*, where the currents in the two field regions are equal. As a result, a traveling space-charge accumulation is formed. This process, of course, is dependent on the condition that the number of electrons inside the crystal be large enough to allow the necessary amount of space charge to build up during the transit time of the space-charge layer.[14]

The pure accumulation layer discussed above is the simplest form of space-charge instability. When there are positive and negative charges separated by a small distance, one has a dipole formation (or domain), as

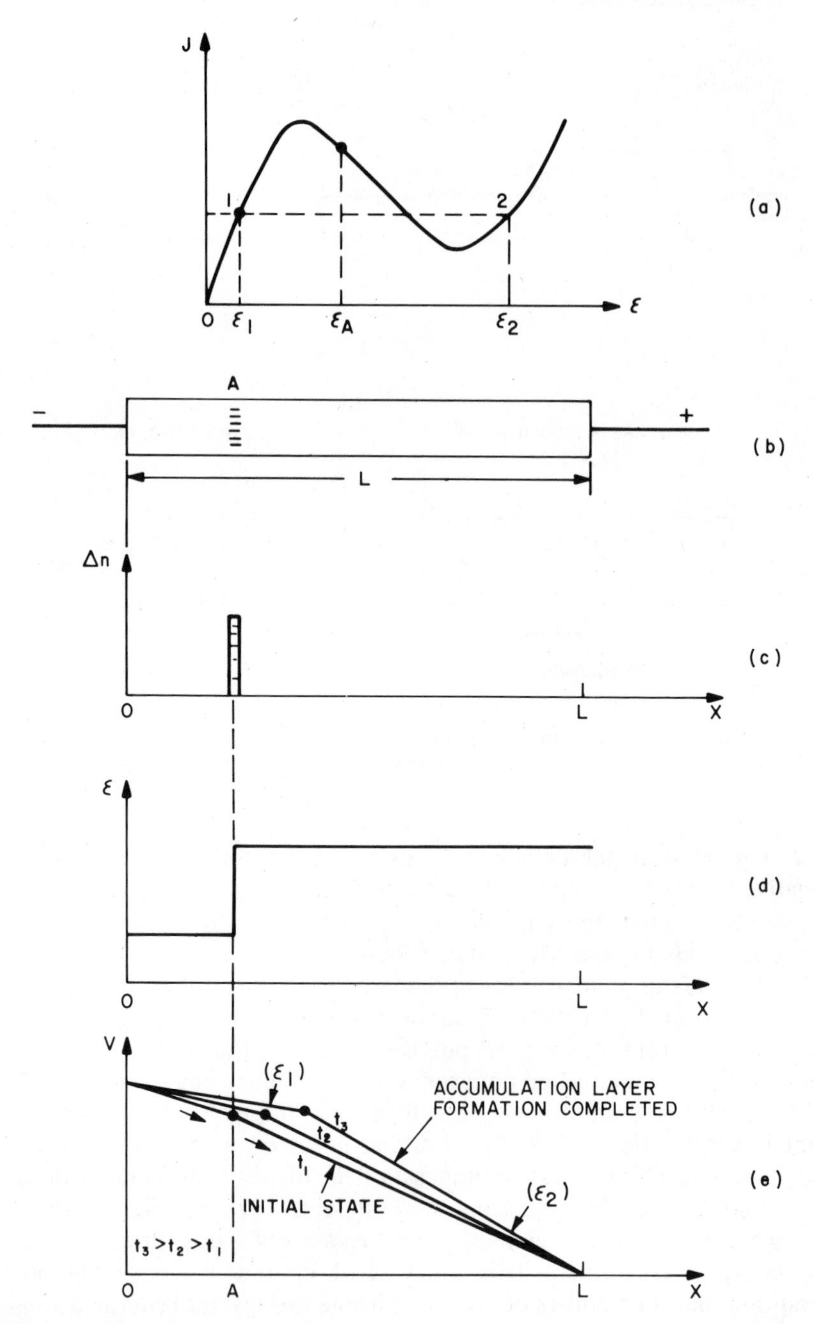

Fig. 3 Formation of an electron accumulation layer in a perturbed medium of negative resistivity. (After Kroemer, Ref. 14.)

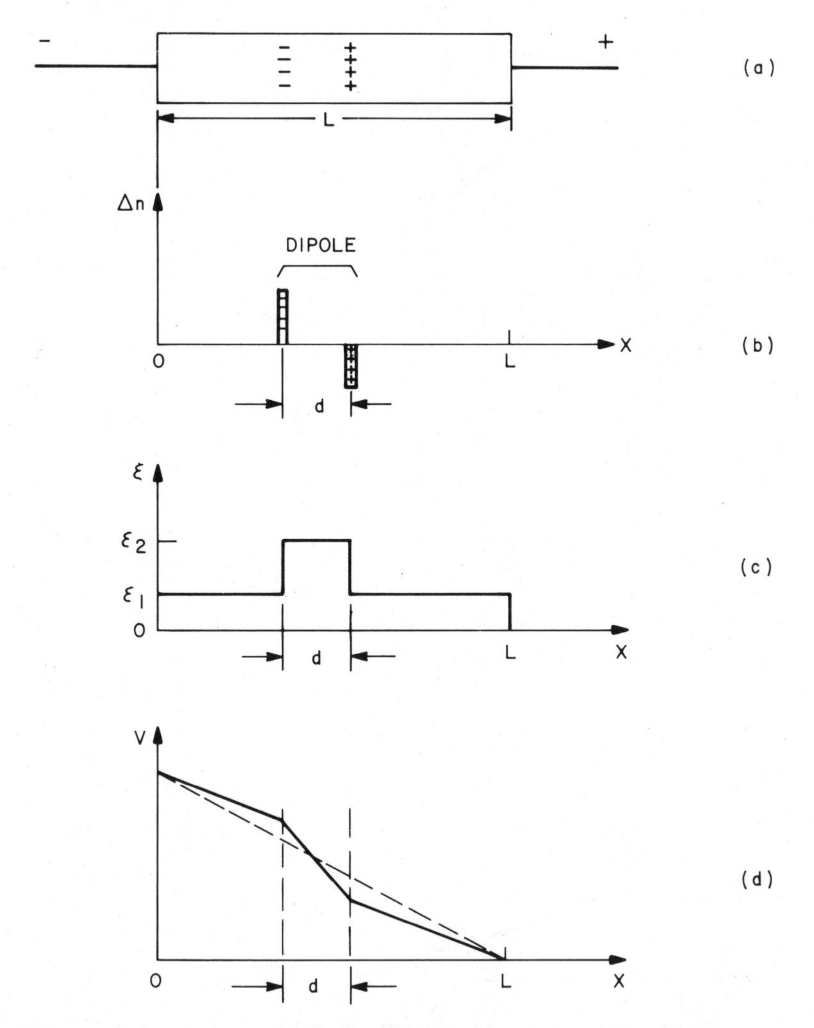

Fig. 4 Formation of an electron dipole layer in a perturbed medium of negative resistivity. (After Kroemer, Ref. 14.)

shown in Fig. 4. The electric field inside the dipole would be greater than the fields on either side of it, Fig. 4*c*. Because of the negative differential resistance, the current in the low-field region would be greater than that in the high-field region. The two field values will tend toward equilibrium levels outside the NDR region, where the high and low currents are the same, Fig. 3*a*. (Assumed here is a domain with zero wall thickness.) The dipole has now reached a stable configuration. The dipole layer moves through the crystal and disappears at the anode, at which time the field begins to rise uniformly across the sample through threshold; that is, $\mathscr{E} > \mathscr{E}_T$, as shown in Fig. 5*a*, thus forming a new dipole layer, and the process repeats itself. To

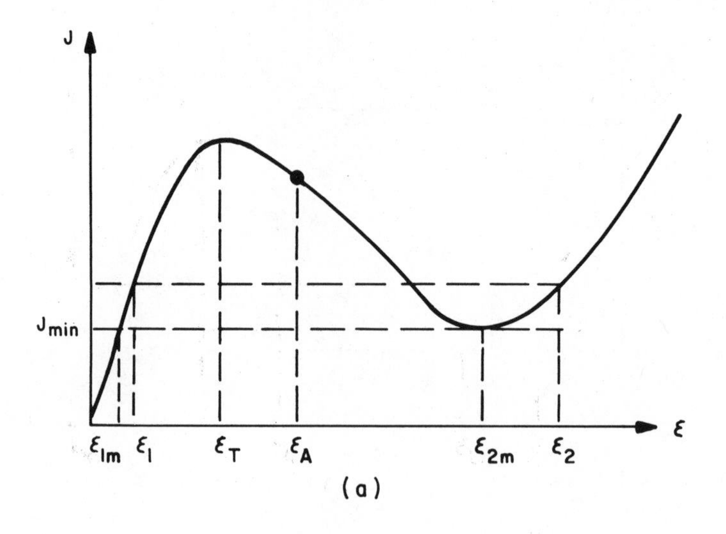

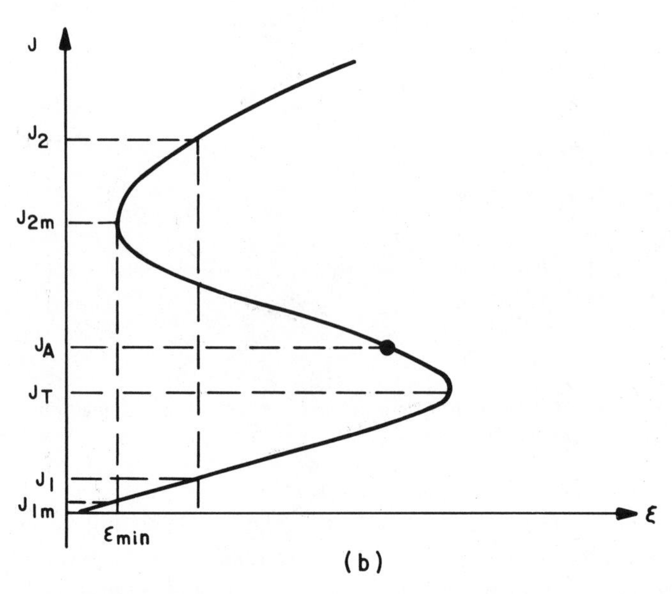

Fig. 5 (*a*) Minimum current density and corresponding field for a voltage-controlled NDR. (*b*) Minimum electric field and corresponding current density for a current-controlled NDR. (After Ridley, Ref. 11.)

estimate the dipole width d, we can use the expression

$$V = \mathscr{E}_A L = \mathscr{E}_2 d + (L - d)\mathscr{E}_1 \tag{10}$$

or

$$d = L\left(\frac{\mathscr{E}_A - \mathscr{E}_1}{\mathscr{E}_2 - \mathscr{E}_1}\right) \tag{11}$$

where L is sample length. We shall now make the assumption that the most stable situation is obtained when the input energy is a minimum. For a constant voltage, this means that the current should be a minimum. From Fig. 5a, the dipole width d is given by Eq. 11 with $\mathscr{E}_{2m}$ and $\mathscr{E}_{1m}$ replacing $\mathscr{E}_2$ and $\mathscr{E}_1$, respectively.

The case of current-controlled NDR can be treated similarly. Instead of a domain, we consider a filament of cross-sectional area a. For a given current I, under steady-state condition, we obtain from Fig. 5b

$$I = J_A A = J_2 a + (A - a)J_1 \tag{12}$$

or

$$a = A\left(\frac{J_A - J_1}{J_2 - J_1}\right) \tag{13}$$

where A is the device cross-sectional area. Under the condition of minimum energy, the voltage should be a minimum, and the area of the filament is given by Eq. 13 with J_{2m} and J_{1m} replacing J_2 and J_1, respectively.

11.2.2 Transferred-Electron Effect

The transferred-electron effect is the transfer of conduction electrons from a high-mobility energy valley to low-mobility, higher-energy satellite valleys. To understand how this effect leads to negative differential resistivity, consider the energy–momentum diagrams for GaAs and InP (Fig. 6), the two most important semiconductors for TEDs.[15,16] As can be seen, the band structures of GaAs and InP are very similar. The conduction band consists of a number of subbands. The bottom of the conduction band is located at $k = 0$ (Γ point). The first higher subband is located along the ⟨111⟩ axis (L), and the next higher subband appears along the ⟨100⟩ axis (X). Therefore, the ordering of the subbands in both semiconductors is Γ-L-X. Until Aspnes[15] made his synchrotron radiation, Schottky-barrier, electroreflectance measurements in 1976, the first subband in GaAs was generally taken as X with a separation of about 0.36 eV at room temperature. These measurements established the correct ordering of the subbands for GaAs as Γ-L-X, which is identical to that for InP (Fig. 6).

We shall now derive an approximate velocity-field characteristic based on the single-temperature model, that is, electrons in the lower valley (Γ) and the upper valleys (L) are assigned the same electron temperature T_e.[4,17]

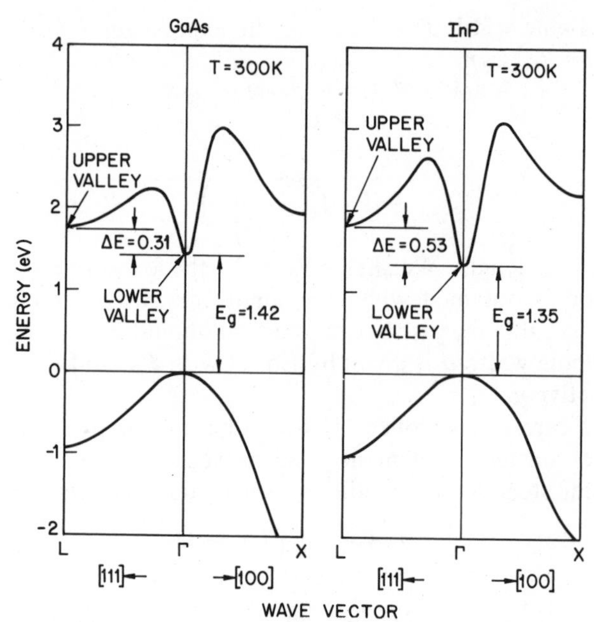

Fig. 6 Energy-band structures of GaAs and InP. The lower conduction valley is at $k = 0$ (Γ); the high valley is along the $\langle 111 \rangle$ axis (L). (After Aspnes, Ref. 15; Rees and Gray, Ref. 16.)

The energy separation between the two valleys is ΔE, which is about 0.31 eV for GaAs and 0.53 eV for InP. The lower-valley effective mass is denoted by m_1^*, and the mobility by μ_1. The same upper-valley quantities are denoted by m_2^* and μ_2, respectively. Also, the densities of electrons in the lower and upper valleys are n_1 and n_2, respectively, and the total carrier concentration is given by $n = n_1 + n_2$. The steady-state current density of the semiconductor may be written as

$$J = q(\mu_1 n_1 + \mu_2 n_2)\mathscr{E} = qnv \tag{14}$$

where the average drift velocity v is

$$v = \left(\frac{\mu_1 n_1 + \mu_2 n_2}{n_1 + n_2}\right)\mathscr{E} \simeq \frac{\mu_1 \mathscr{E}}{1 + (n_2/n_1)} \tag{15}$$

since $\mu_1 \gg \mu_2$. The population ratio between the upper and lower valleys with an energy difference of ΔE is

$$\frac{n_2}{n_1} = R \exp\left(-\frac{\Delta E}{kT_e}\right) \tag{16}$$

where R is the density-of-state ratio given by

$$R = \frac{\text{available states in all upper valley}}{\text{available states in lower valley}} = \frac{M_2}{M_1}\left(\frac{m_2^*}{m_1^*}\right)^{3/2} \tag{17}$$

The M_1 and M_2 are the number of equivalent lower and upper valleys, respectively. For GaAs, $M_1 = 1$ and there are eight upper valleys in the L direction, but they happen to be near the edge of the first Brillouin zone, and therefore $M_2 = 4$. Using the effective masses $m_1^* = 0.067m_0$ and $m_2^* = 0.55m_0$, R is found to be 94 for GaAs.

The electron temperature T_e is higher than the lattice temperature T, since the electric field accelerates the electrons and increases their kinetic energy. The electron temperature is determined through the concept of energy-relaxation time:

$$q\mathscr{E}v = \tfrac{3}{2}k(T_e - T)/\tau_e \tag{18}$$

where the energy relaxation time τ_e is assumed to be of the order of 10^{-12} s. Substituting v from Eq. 15 and n_2/n_1 from Eq. 16, we obtain from Eq. 18,

$$T_e = T + \frac{2q\tau_e\mu_1}{3k}\mathscr{E}^2\left[1 + R\exp\left(-\frac{\Delta E}{kT_e}\right)\right]^{-1}. \tag{19}$$

We can compute T_e as a function of the electric field for a given T, and from Eqs. 15 and 16 the velocity-field characteristic can be written as

$$v = \mu_1\mathscr{E}\left[1 + R\exp\left(-\frac{\Delta E}{kT_e}\right)\right]^{-1}. \tag{20}$$

The general $v - \mathscr{E}$ curves as obtained from Eqs. 19 and 20 are shown in Fig. 7 for GaAs with three lattice temperatures. Also shown is the upper-valley population fraction as a function of the field.

The simple model discussed above shows the following points: (1) there

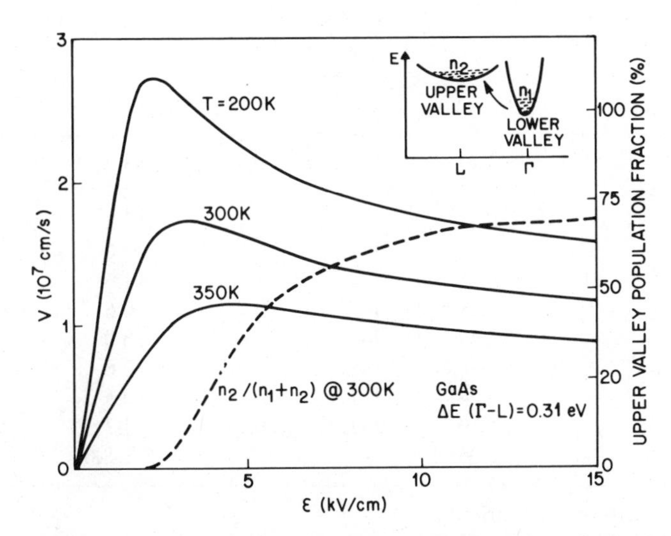

Fig. 7 Calculated velocity-field characteristics for GaAs at three lattice temperatures based on a two-valley model having a single electron temperature.

is a well-defined threshold field ($\mathscr{E}_T$) for the onset of negative differential resistivity (or negative differential mobility); (2) the threshold field increases with the lattice temperature; and (3) the negative mobility can be destroyed by having the lattice temperature too high or the energy difference ΔE too small. Therefore, certain requirements must be met for the electron transfer mechanism to give rise to bulk NDR: (1) the lattice temperature must be low enough that, in the absence of a bias electric field, most electrons are in the lower conduction-band minimum, or $kT < \Delta E$; (2) in the lower conduction-band minimum, the electrons must have high mobility, small effective mass, and low density of states, whereas in the upper satellite valleys, the electrons must have low mobility, large effective mass, and high density of states; and (3) the energy separation between the two valleys must be smaller than the semiconductor bandgap so that avalanche breakdown does not set in before electrons are transferred into the upper valleys.

Of the semiconductors satisfying these conditions, n-type GaAs and n-type InP are the most widely studied and used. The transferred-electron effect, however, has been observed in many other semiconductors, including Ge, binary, ternary, and quaternary compounds (see Table 1).[9, 18, 19] The transferred-electron effects are observed in InAs and InSb under hydrostatic pressures that are applied to decrease the energy difference ΔE, which under normal pressures is greater than the energy gap. Of particular interest is the GaInSb ternary III–V compounds for potential low-power,

Table 1 Semiconductor Materials Related to Transferred-Electron Effect at 300 K

Semiconductor	E_g (eV)	Valley Separation Between	Valley Separation ΔE (eV)	$\mathscr{E}_T$ (kV/cm)	v_p (10^7 cm/s)
GaAs	1.42	Γ-L	0.31	3.2	2.2
InP	1.35	—	0.53	10.5	2.5
Ge[a]	0.74	L-Γ	0.18	2.3	1.4
CdTe	1.50	Γ-L	0.51	11.0	1.5
InAs[b]	0.36	Γ-L	1.28	1.6	3.6
InSb[c]	0.28	Γ-L	0.41	0.6	5.0
ZnSe	2.60	Γ-L	—	38.0	1.5
$Ga_{0.5}In_{0.5}Sb$	0.36	Γ-L	0.36	0.6	2.5
$Ga_{0.3}In_{0.7}Sb$	0.24	Γ-L	—	0.6	2.9
$InAs_{0.2}P_{0.8}$	1.10	Γ-L	0.95	5.7	2.7
$Ga_{0.13}In_{0.87}As_{0.37}P_{0.63}$	1.05	—	—	5.5–8.6	1.2

[a] At 77 K, (100)- or (110)-oriented.
[b] Under 14-kbar pressure.
[c] At 77 K under 8-kbar pressure.

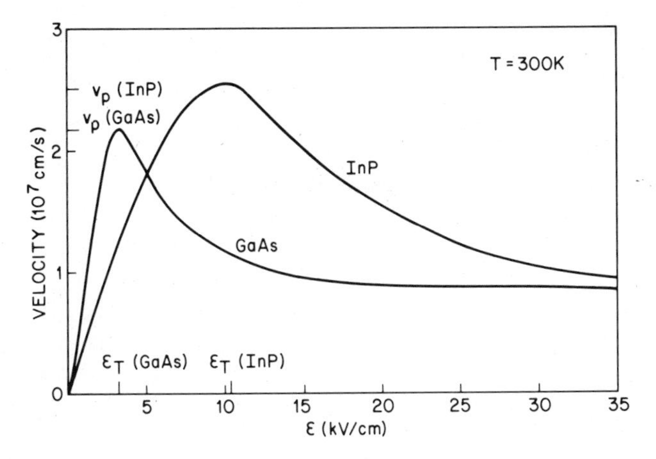

Fig. 8 Measured velocity–field characteristics for GaAs and InP. (After Ruch and Kino, Ref. 20; Rees and Gray, Ref. 16.)

high-speed applications because of their low threshold fields and high velocities. For semiconductors with large valley energy separations (e.g., $Al_{0.25}In_{0.75}As$ with $\Delta E = 1.12$ eV and $Ga_{0.6}In_{0.4}As$ with $\Delta E = 0.72$ eV) the negative resistance may become dominated by the central Γ valley:[59] Monte Carlo calculations have shown that for these semiconductors, the presence of the upper valleys is not required for a negative-resistance effect but that polar optical scattering acting in a nonparabolic central valley alone gives rise to a peak velocity and negative-resistance effect.

The measured room-temperature velocity–field characteristics for GaAs and InP are shown[16,20] in Fig. 8. The analysis based on high-field carrier transport studies are in good agreement with the experimental results.[21,22] The threshold field $\mathscr{E}_T$ defining the onset of NDR is approximately 3.2 kV/cm for GaAs and 10.5 kV/cm for InP. The peak velocity v_p is about 2.2×10^7 cm/s for high-purity GaAs and 2.5×10^7 cm/s for high-purity InP. The maximum negative differential mobility is determined to be about -2400 cm^2/V-s for GaAs and -2000 cm^2/V-s for InP. The measured relative threshold field $\mathscr{E}_T/\mathscr{E}_T$ (300 K) and the relative peak velocity v_p/v_p (300 K) in GaAs as a function of lattice temperature are shown[23] in Fig. 9. The simple model (Fig. 7) is in qualitative agreement with the experimental result.

For a known v–$\mathscr{E}$ characteristic, the current density equation, Eq. 3, can be written as

$$J = qnv(\mathscr{E}) - qD\frac{\partial n}{\partial x} \tag{21}$$

or

$$J = q(n_1\mu_1 + n_2\mu_2)\mathscr{E} - q\left(D_1\frac{\partial n_1}{\partial x} + D_2\frac{\partial n_2}{\partial x}\right) \tag{21a}$$

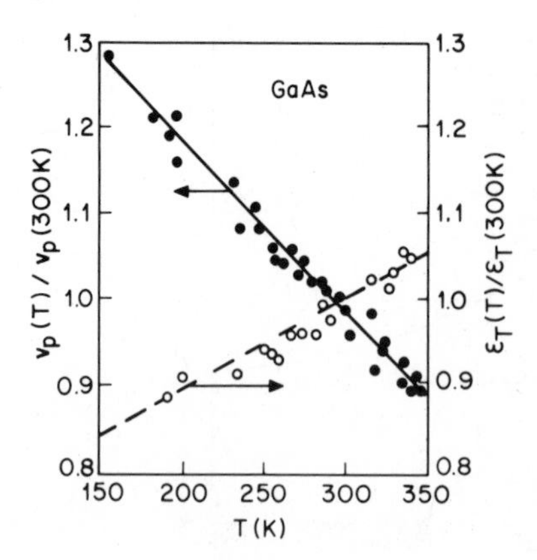

Fig. 9 Measured relative peak velocity and relative threshold field in GaAs versus temperature. (After Mojzes, Podor, and Balogh, Ref. 23.)

where the first term on the right-hand side is due to carrier drift, the second term is due to carrier diffusion, and the diffusion constant D is assumed to be independent of the field.

Under steady-state conditions and for small carrier-concentration gradients, Eq. 21 reduces to $J = qnv(\mathscr{E})$. Substituting this result in Eq. 2

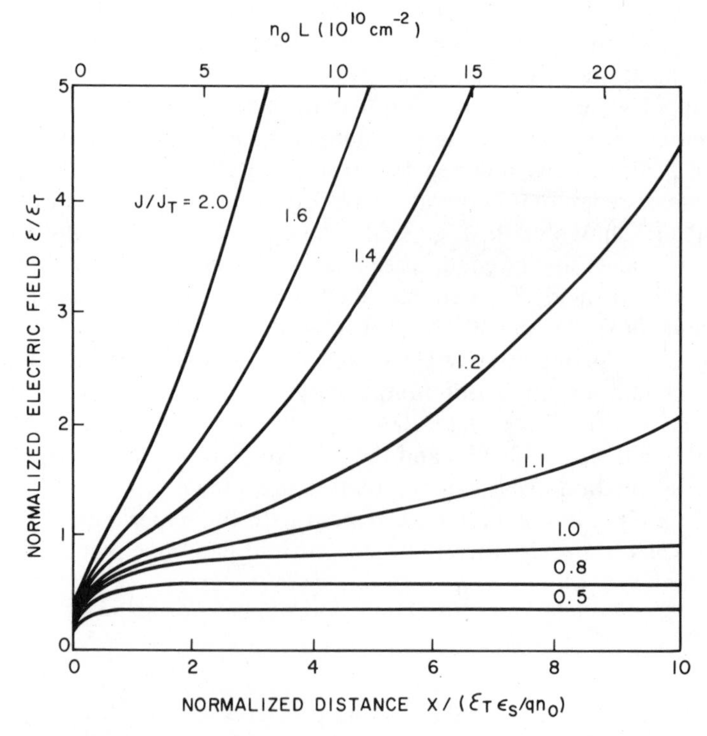

Fig. 10 Electric field versus distance x, where $\mathscr{E}(0) = 0$ at $x = 0$. (After McCumber and Chynoweth, Ref. 17.)

yields

$$\frac{\partial \mathscr{E}}{\partial x} = -\frac{q n_0}{\epsilon_s}\left[1 - \frac{J/q n_0}{v(\mathscr{E})}\right]. \tag{22}$$

This first-order nonlinear equation has the boundary condition that the field $\mathscr{E}(x)$ be continuous everywhere. Equation 22 was numerically integrated for a GaAs diode,[17] and the results are shown in Fig. 10, where J_T corresponds to $q n_0 v_p$. This figure shows that at any point x, the electric field is a monotonically increasing function of the current density. Therefore, when one properly takes into account boundary conditions, the steady-state solutions do not exhibit negative resistance. This is not surprising, however, since Shockley pointed out that bulk NDR diodes must be dc stable.[24] This is true because inside the device an internal space charge exists (i.e., excess electrons injected from the cathode) that increases with increasing bias voltage by an amount so large that the current increases despite a decrease in electron velocity due to the negative electron mobility. However, these steady-state solutions are not necessarily stable with respect to small fluctuations.

11.3 MODES OF OPERATION

Since Gunn first observed microwave oscillation in GaAs and InP TEDs in 1963, various modes of operation have been studied. Five major factors affect or determine the modes of operation:* (1) doping concentration and doping uniformity in the device, (2) length of the active region, (3) cathode contact property, (4) type of circuit used, and (5) operating bias voltage.

For a TED under nonsteady state, various space-charge layers may be formed, including accumulation layers or dipole layers (domains), as mentioned previously. Formation of a strong space-charge instability is dependent on the condition that enough charge is available in the semiconductor and the device is long enough that the necessary amount of space charge can be built up within the transit time of the electrons. These requirements set up a criterion for the various modes of operation. In Eq. 8, we have shown that for a device with negative differential mobility, the time rate of the early-stage space-charge growth is given by

$$(n - n_0) = (n - n_0)_{t=0} \exp(t/|\tau_R|)$$

where $|\tau_R| = \epsilon/q n_0 |\mu_-|$ and μ_- is negative differential mobility. If this relationship remained valid throughout the entire transit time of the space-charge layer, the maximum growth factor would be $\exp(L/v|\tau_R|)$, where

*The boundary conditions at the cathode have been applied in a somewhat questionable manner, and further investigation appears to be necessary to establish the validity of some of the statements in the literature that have been included in this chapter.

v is the average drift velocity of the space-charge layer. For large space-charge growth, this growth factor must be greater than unity, making $L/v|\tau_R| > 1$, or

$$n_0 L > \epsilon_s v/q|\mu_-|. \tag{23}$$

For n-type GaAs and InP, the right-hand side is about 10^{12} cm^{-2}. The TEDs with n_0L products smaller than 10^{12} cm^{-2} exhibit a stable field distribution (Fig. 10). Hence an important boundary separating the various modes of operation is the (carrier concentration)×(device length) product, $n_0L = 10^{12}$ cm^{-2}. We shall now consider some important operation modes, beginning with the ideal uniform-field mode.

11.3.1 Ideal Uniform-Field Mode

Under the idealized condition that no internal space charge has built up and the entire device has a uniform electric field, the current–voltage relationship for a TED can be obtained by scaling the velocity-field characteristics. The simplest voltage waveform is a square wave, as shown[25] in Fig. 11. We shall define two normalization parameters: $\alpha \equiv I_V/I_T$ and $\beta \equiv V_0/V_T$. It is apparent from the nature of the waveform assumed that the average dc current I_0 is given by

$$I_0 = (1+\alpha)I_T/2. \tag{24}$$

The dc power supplied by the device is

$$P_0 = V_0 I_0 = \tfrac{1}{2}\beta(1+\alpha)V_T I_T \tag{25}$$

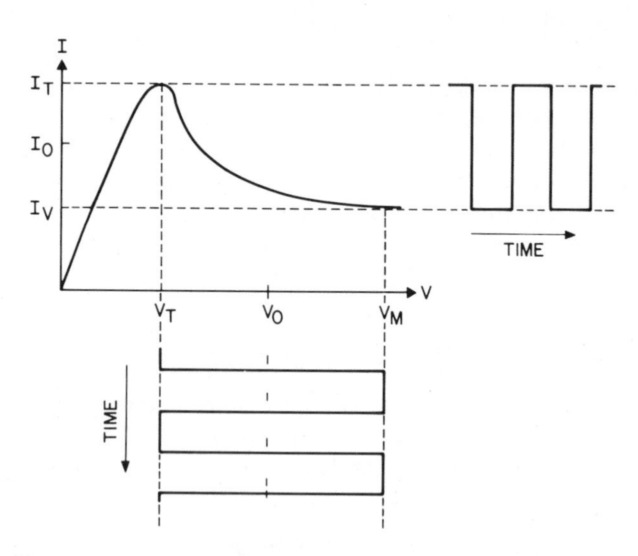

Fig. 11 Idealized square waveforms for the uniform-field mode. (After Kino and Kuru, Ref. 25.)

and the total RF power supplied to the load is

$$P_{rf} = \frac{V_M - V_T}{2}\frac{I_T - I_V}{2} = \frac{(\beta - 1)(1-\alpha)}{2} V_T I_T. \tag{26}$$

Therefore, the conversion efficiency from dc to *RF* is

$$\eta = \frac{(1-\alpha)(\beta-1)}{(1+\alpha)\beta} \tag{27}$$

and the efficiency for the fundamental frequency is

$$\eta_1 = \frac{8}{\pi^2}\eta. \tag{27a}$$

From Eq. 27, we find that maximum efficiency is obtained with a dc bias voltage as high as possible ($\beta \rightarrow \infty$) and with a peak-to-valley ratio $1/\alpha$ as large as possible. This maximum efficiency yields an ideal value of 30% for GaAs ($1/\alpha = 2.2$) and 45% for InP ($1/\alpha = 3.5$). These efficiencies should be independent of operating frequency as long as the frequency is lower than the reciprocal of the energy relaxation time and the intervalley scattering time.

Experimentally, such high efficiencies have never been obtained, and the frequency of operation is generally related to the transit-time frequency, $f_T = v/L$. The reasons are: (1) the bias voltage is limited by avalanche breakdown; (2) a space-charge layer usually forms, giving rise to a nonuniform field; and (3) the ideal current and voltage waveforms are difficult to achieve in a resonant circuit.

11.3.2 Accumulation-Layer Mode

Lightly doped or short samples ($n_0L < 10^{12}\ \text{cm}^{-2}$) exhibit a stable field distribution and positive dc resistance. Figure 10 shows the steady-state field distribution in such a device with different values of current density. When a uniform field is applied instantaneously to such a device, the accumulation-layer dynamics can be understood in a simplified manner, as Fig. 12 shows.[26] At time 1, an accumulation layer (i.e., excess electrons) is injected from the cathode so that the field distribution splits into two parts, as illustrated at time 2. The velocity on either side of the accumulation layer has been altered in the direction shown in Fig. 12*a*. Since the terminal voltage is assumed to be constant, the area under each electric-field curve of Fig. 12*b* should be equal. As the accumulation layer propagates toward the anode, this equality can only be maintained if the velocity on either side of the layer falls, as dictated by the velocity–field curve and indicated at times 3, 4, and 5. Eventually, the accumulation layer reaches the anode, time 6, and disappears there. The cathode-side field rises through the threshold, another accumulation layer is injected, and the process repeats. The current waveform is shown in Fig. 12*d*, which has a smooth form.

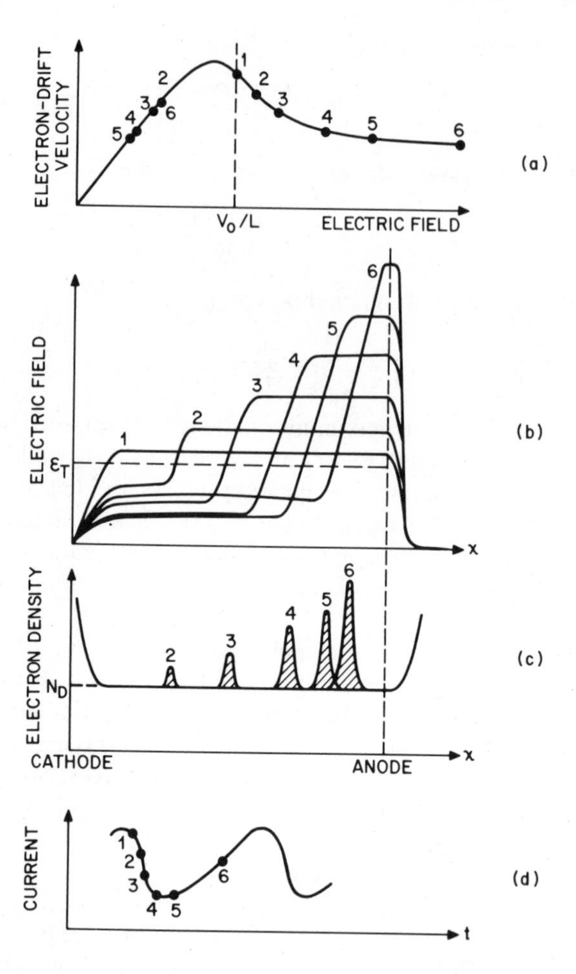

Fig. 12 Accumulation-layer transit mode under time-invariant terminal voltage. (After Hobson, Ref. 26.)

A TED with subcritical n_0L product (i.e., $n_0L < 10^{12}\ \text{cm}^{-2}$) can exhibit negative resistance in a band of frequencies near the electron transit-time frequency and its harmonics. It can be operated as a stable amplifier.[27] Figure 13 shows[28] the measured admittance as a function of frequency for n-type GaAs having $n_0 = 3 \times 10^{13}\ \text{cm}^{-3}$ and $L = 70\ \mu\text{m}$ ($n_0L = 2.1 \times 10^{11}\ \text{cm}^{-2}$). The negative conductance peaks occur at about the harmonic multiples. The experimental gain as a function of frequency is shown in Fig. 14 for a device with the same length but lower doping ($n_0 = 1.5 \times 10^{13}\ \text{cm}^{-3}$). As the electric field is increased, the peak gain occurs at progressively higher frequencies, because the negative conductance peak increases with the field and occurs at progressively higher frequencies as the field increases.

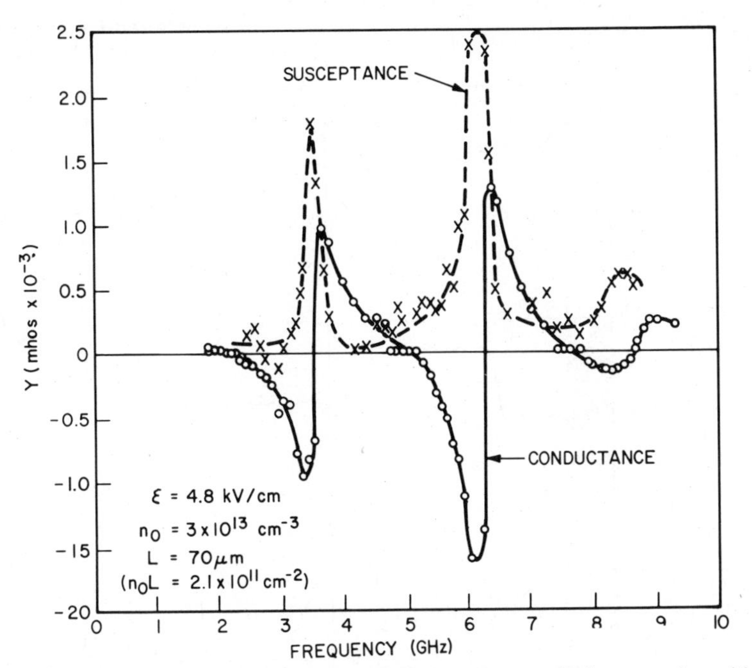

Fig. 13 Measured admittance versus frequency for a GaAs sample with $n_0 = 3 \times 10^{13}$ cm^{-3} and $L = 70$ μm. (After Hakki, Ref. 28.)

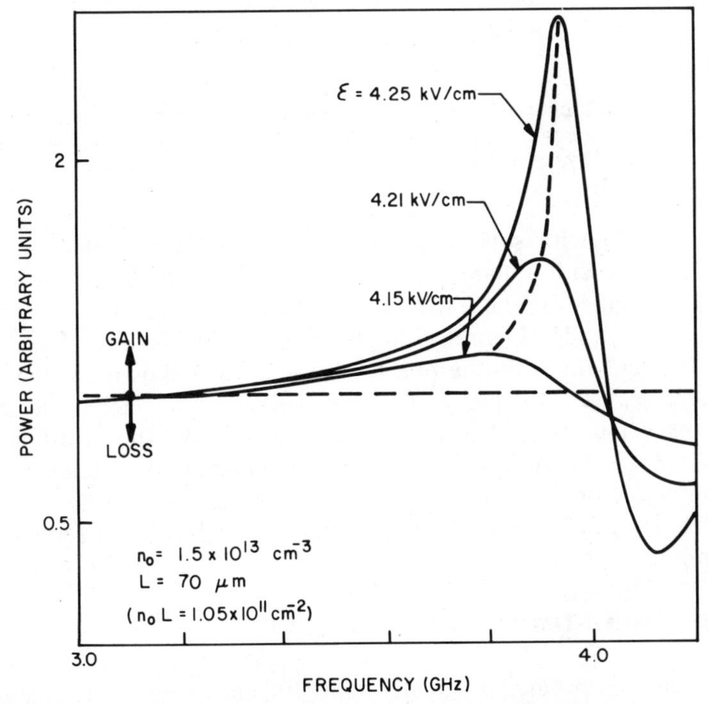

Fig. 14 Experimental gain as a function of frequency for different bias levels on a GaAs sample with $n_0 = 1.5 \times 10^{13}$ cm^{-3} and $L = 70$ μm. (After Hakki, Ref. 28.)

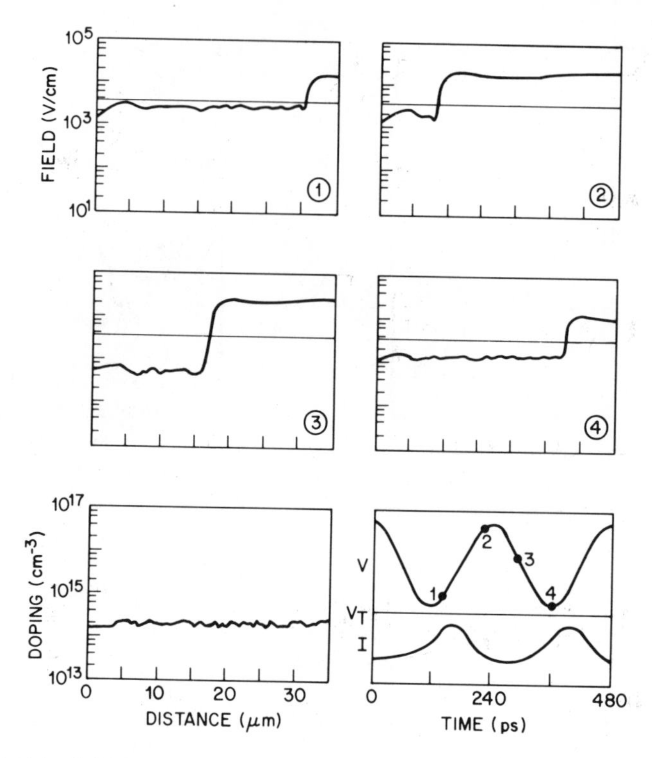

Fig. 15 Electric field versus distance at four intervals of time during one RF cycle (upper four plots). Doping profile (lower left) and voltage and current waveforms of an accumulation layer mode in a resonant circuit for GaAs TED with $fL = 1.4 \times 10^7$ cm/s and $n_0/f = 5 \times 10^4$ s-cm^{-3}. (After Thim, Ref. 29.)

When a TED with subcritical n_0L product is connected to a parallel resonant circuit with a load resistor of the order of $10R_0$, where R_0 is the low-field resistance of the TED, it will oscillate in the transit-time accumulation-layer mode. Figure 15 shows the electric field versus distance at four different times during one RF cycle.[29] Also shown are the doping fluctuation ($n_0 = 2 \times 10^{14}$ cm^{-3}) and the voltage and electronic current waveforms versus time. The voltage is always above the threshold value ($V > V_T = \mathscr{E}_T L$). These waveforms are far from ideal; the efficiency for this particular waveform is only 5%. More favorable waveforms with about 10% efficiency can be obtained if the TED is connected to a series resistor and inductor.

11.3.3 Transit-Time Dipole-Layer Mode

When the n_0L product is greater than 10^{12} cm^{-2}, the space-charge perturbations in the material increase exponentially in space and time to form mature dipole layers that propagate to the anode. The dipole is usually

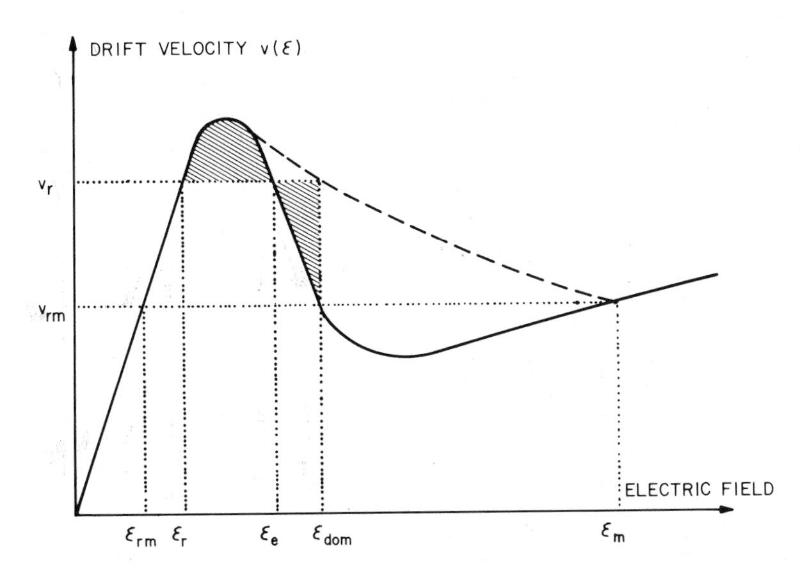

Fig. 16 Velocity versus electric field (solid line) and peak domain field versus drift velocity outside the domain (dashed line). (After Butcher, Ref. 30.)

formed near the cathode contact, since the largest doping fluctuation and space-charge perturbation exists there. The cyclic formation and subsequent disappearance at the anode of the fully developed dipole layers are what give rise to the experimentally observed Gunn oscillations. In Section 11.2, a qualitative discussion of space-charge instabilities was presented together with a short discussion of domains. In this section, the theory of dipole layers will be studied in more detail. The dipole layers are stable in the sense that they propagate with a particular velocity but do not change in any other way with time. We will assume that the electron drift velocity follows the static velocity-field characteristic shown by the solid curve in Fig. 16. The equations that determine the behavior of the electron system are Poisson's equation, Eq. 2, and the total current-density equation

$$J = qnv(\mathscr{E}) - q\frac{\partial D(\mathscr{E})n}{\partial x} + \epsilon_s \frac{\partial \mathscr{E}}{\partial t}. \tag{28}$$

This equation is similar to Eq. 21 except for the addition of the third term, which corresponds to the displacement-current component.

The type of solutions sought represent a high-field domain that propagates without change of shape with a domain velocity v_{dom}. Outside the domain, the carrier concentration and fields are at constant values given by $n = n_0$ and $\mathscr{E} = \mathscr{E}_r$, respectively. For this type of solution, both $\mathscr{E}$ and n should be functions of the single variable, $x' = x - v_{\text{dom}}t$, with the forms as shown in Fig. 17. Note that n is a double-valued function of field. The domain consists of an accumulation layer where $n > n_0$, followed by a

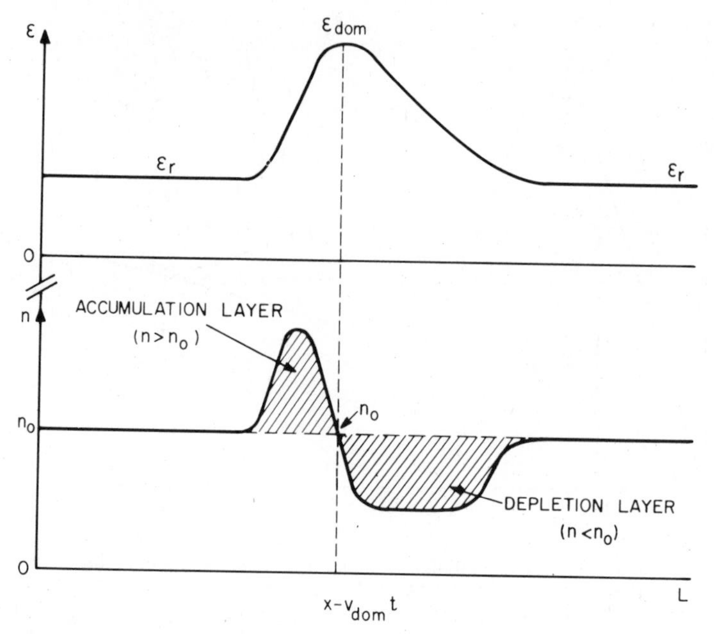

Fig. 17 Electric field and electron density distribution for a stable high-field domain propagating with velocity v_r. (After Butcher, Fawcett, and Hilsum, Ref. 31.)

depletion layer where $n < n_0$. The carrier concentration n equals n_0 at two field values, that is, $\mathscr{E} = \mathscr{E}_r$ outside the domain and at $\mathscr{E} = \mathscr{E}_{\text{dom}}$, the peak domain field.

It will be assumed that the value of the outside field $\mathscr{E}_r$ is known. (Later it will be shown that $\mathscr{E}_r$ is easily determined.) The current outside the domain consists only of conduction current and is given by $J = qn_0v_r$ where $v_r = v(\mathscr{E}_r)$. Noting that

$$\frac{\partial \mathscr{E}}{\partial x} = \frac{d\mathscr{E}}{dx'} \quad \text{and} \quad \frac{\partial \mathscr{E}}{\partial t} = -v_{\text{dom}} \frac{\partial \mathscr{E}}{\partial x'}$$

one may rewrite Eqs. 2 and 28 as

$$\frac{\partial \mathscr{E}}{\partial x'} = \frac{q}{\epsilon_s}(n - n_0) \tag{29}$$

and

$$\frac{d}{dx'}[D(\mathscr{E})n] = n[v(\mathscr{E}) - v_{\text{dom}}] - n_0(v_r - v_{\text{dom}}). \tag{30}$$

We can eliminate the variable x' from these equations by dividing Eq. 30 by 29 to obtain a differential equation for $[D(\mathscr{E})n]$ as a function of the electric field:

$$\frac{q}{\epsilon_s}\frac{d}{d\mathscr{E}}[D(\mathscr{E})n] = \{n[v(\mathscr{E}) - v_{\text{dom}}] - n_0(v_r - v_{\text{dom}})\}/(n - n_0). \tag{31}$$

In general, Eq. 31 can only be solved by numerical methods.[30–32] However, the problem may be simplified greatly by assuming that the diffusion term is independent of the electric field, $D(\mathscr{E}) = D$. Using this approximation, the solution to Eq. 31 is given by

$$\frac{n}{n_0} - \ln\left(\frac{n}{n_0}\right) - 1 = \frac{\epsilon_s}{qn_0D}\int_{\mathscr{E}_r}^{\mathscr{E}} \{[v(\mathscr{E}') - v_{\text{dom}}] - \frac{n_0}{n}(v_r - v_{\text{dom}})\}d\mathscr{E}'. \tag{32}$$

(This solution may be verified by differentiation.)

Note that, when $\mathscr{E} = \mathscr{E}_r$ or $\mathscr{E}_{\text{dom}}$, one has $n = n_0$ (Fig. 17) and the left side of Eq. 32 vanishes; therefore, the integral on the right side of the equation must vanish when $\mathscr{E} = \mathscr{E}_{\text{dom}}$. However, the integration from $\mathscr{E}$ to $\mathscr{E}_{\text{dom}}$ can represent either the integration over the depletion region when $n < n_0$ or the integration over the accumulation region where $n > n_0$. Since the first term in the integral is independent of n, whereas the contribution from the second term is different in the two cases, one must have $v_r = v_{\text{dom}}$, so that the integral vanishes for both integration over the depletion region and over the accumulation region. Then for $\mathscr{E} = \mathscr{E}_{\text{dom}}$, Eq. 32 reduces to

$$\int_{\mathscr{E}_r}^{\mathscr{E}_{\text{dom}}} [v(\mathscr{E}') - v_r]d\mathscr{E}' = 0. \tag{33}$$

This equation is satisfied by requiring that the two shaded regions in Fig. 16 have equal areas. By using this rule, the "equal-areas rule,"[30] the value of the peak domain field $\mathscr{E}_{\text{dom}}$ can be determined if the value of the outside field $\mathscr{E}_r$ is known. The dashed curve in Fig. 16 is a plot of $\mathscr{E}_{\text{dom}}$ against v_r as determined by the equal-areas rule. It begins at the peak of the velocity–field characteristic, where the field equals the threshold field, and ends at the point $(\mathscr{E}_m, v_{rm})$. For outside field values resulting in low field velocities $v(\mathscr{E}_r)$ less than v_{rm}, the equal-areas rule can no longer be satisfied and stable domain propagation cannot be supported.[31]

If the field dependence of the diffusion factor in Eq. 31 is included in the equation, one must use numerical techniques to obtain solutions. These solutions show that for a given value of outside field $\mathscr{E}_r$ there is at most one value of domain excess velocity, defined as $(v_{\text{dom}} - v_r)$, for which solutions exist. In other words, only one stable dipole domain configuration exists for each value of $\mathscr{E}_r$. Figure 18 shows[32] the excess domain velocity plotted against outside drift velocity v_r for two values of n_0. Note that for $n_0 = 10^{15}$ cm^{-3}, $v_{\text{dom}} = v_r$ for a wide range of v_r values, and in this range the equal-areas rule may be applied.

Now consider some characteristics of a high-field domain. When the domain is not in contact with either electrode, the device current is determined by the outside field $\mathscr{E}_r$ and is given as

$$J = qn_0v(\mathscr{E}_r). \tag{34}$$

Therefore, for a given carrier concentration n_0, the outside field fixes the value of J. It is convenient to define the excess voltage contained by a

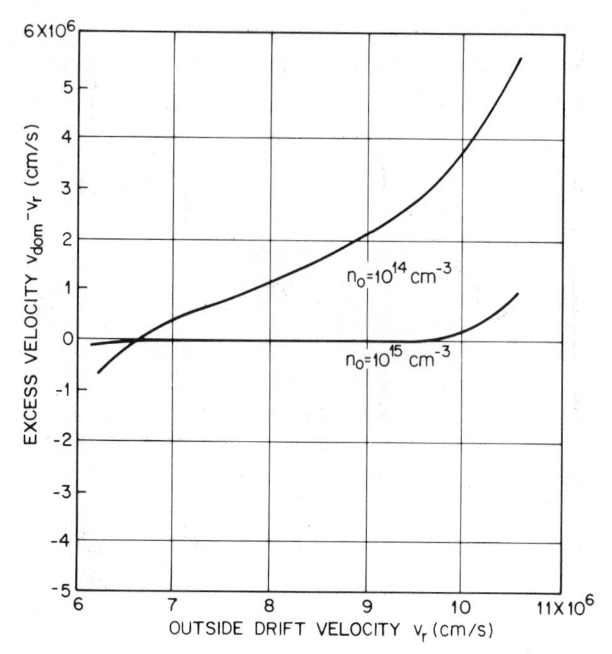

Fig. 18 The excess velocity for high-field domains versus the outside drift velocity for $n_0 = 10^{14}$ and 10^{15} cm^{-3}. (After Copeland, Ref. 32.)

high-field domain with outside field $\mathscr{E}_r$ by

$$V_{ex} = \int_{-\infty}^{\infty} [\mathscr{E}(x) - \mathscr{E}_r]dx. \tag{35}$$

The computer solutions of Eq. 35 for different values of carrier concentration and outside field are shown[32] in Fig. 19. These curves may be used to determine the outside field $\mathscr{E}_r$ in a particular diode of length L, doping concentration n_0, and bias voltage V, by noting that the following relation must hold simultaneously with Eq. 35:

$$V_{ex} = V - L\mathscr{E}_r. \tag{36}$$

The straight line defined by this equation is called the device line and is shown in Fig. 19 as a dashed line for the particular values $L = 25\ \mu$m and $V = 10$ volts. If $V/L > \mathscr{E}_T$, the threshold field, the interception of the device line, and the solutions of Eq. 35 uniquely determine $\mathscr{E}_r$, which in turn specifies the current. The slope of the device line is fixed by L; however, the intercept defining $\mathscr{E}_r$ may be varied by adjusting the bias voltage V.

When the high-field domain reaches the anode, the current in the external circuit increases and the fields in the diode readjust themselves, nucleating a new domain. Then the frequency of the current oscillations depends on, among other things, the velocity of the domain across the

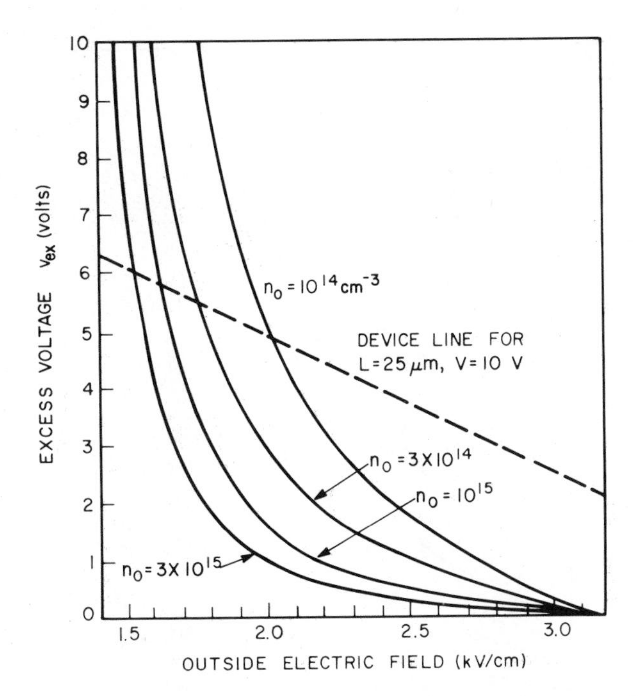

Fig. 19 Excess domain voltage versus field for various carrier concentrations. The dashed line is the device line for a sample 25 μm long and biased at 10 V. (After Copeland, Ref. 32.)

sample, v_{dom}; if v_{dom} increases, the frequency increases, and vice versa. The dependence of v_{dom} on the bias voltage can easily be determined.

Figure 20 shows a plot of the domain width versus domain excess voltage.[32] Note that for a given V_{ex}, the domain is narrower for higher doping concentrations. In the limit of zero diffusion, the domain has a triangular shape, because, when $\mathscr{E}$ in Eq. 32 lies between $\mathscr{E}_r$ and $\mathscr{E}_{\text{dom}}$, the right side of the equation approaches infinity as D approaches zero; therefore, the left-hand side must also approach infinity. This requirement implies that $n \to 0$ in the depletion region and $n \to \infty$ in the accumulation region. The electric field will vary linearly from $\mathscr{E}_{\text{dom}}$ to $\mathscr{E}_r$, and the domain width is

$$d = \frac{\epsilon_s}{qn_0}(\mathscr{E}_{\text{dom}} - \mathscr{E}_r). \tag{37}$$

The domain excess voltage is then given by

$$V_{ex} = (\mathscr{E}_{\text{dom}} - \mathscr{E}_r)d/2 = \frac{\epsilon_s(\mathscr{E}_{\text{dom}} - \mathscr{E}_r)^2}{2qn_0}. \tag{38}$$

Experimentally, only triangular domains have been obtained in GaAs and InP TEDs.

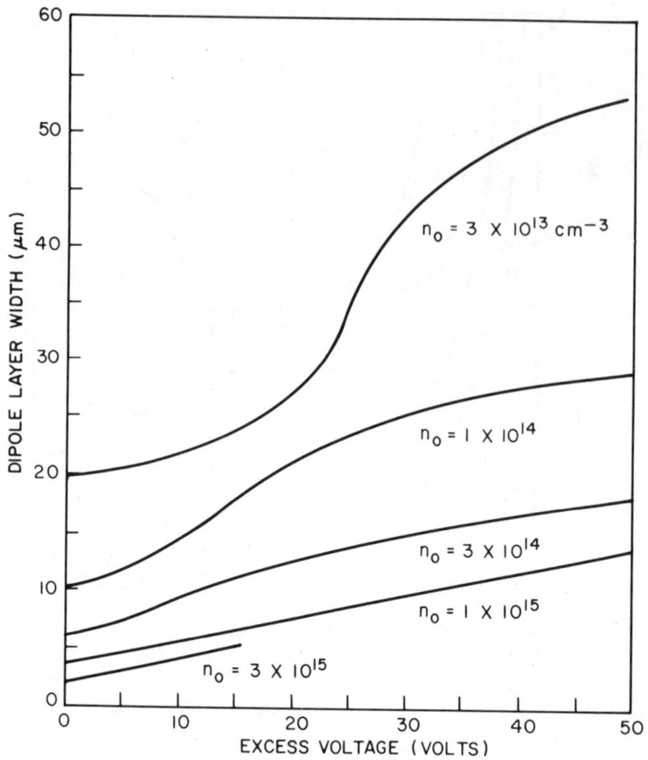

Fig. 20 Domain width versus domain excess voltage for various doping levels. (After Copeland, Ref. 32.)

When a TED with overcritical n_0L product is connected in a parallel resonant circuit, such as a high-Q microwave cavity, the transit-time dipole-layer mode can be obtained. In this mode, the high-field domain is nucleated at the cathode and travels the full length of the sample to the anode. When the dipole reaches the anode, the outside field throughout the sample begins to rise through the threshold field and a new domain is nucleated at the cathode. Figure 21 shows[33] a simulated time-dependent behavior of a domain in a GaAs device 100 μm long having a doping of 5×10^{14} cm^{-3} ($n_0L = 5 \times 10^{12}$ cm^{-2}). The time between successive vertical displays of $\mathscr{E}(x, t)$ is $16\tau_R$, where τ_R is the low-field dielectric relaxation time, Eq. 9, ($\tau_R = 1.5$ ps for this device.) Each time a domain is absorbed at the anode, the current in the external circuit increases; therefore, for samples in which the width of the domain is considerably smaller than the length of the sample, the current waveform tends to be spiked rather than the desired sinusoidal form. Figure 22 shows the experimentally obtained current waveform of a 100-μm sample[34] with an n_0L product equal to 3×10^{13} cm^{-2}. Obviously, to obtain a more nearly sinusoidal current wave-

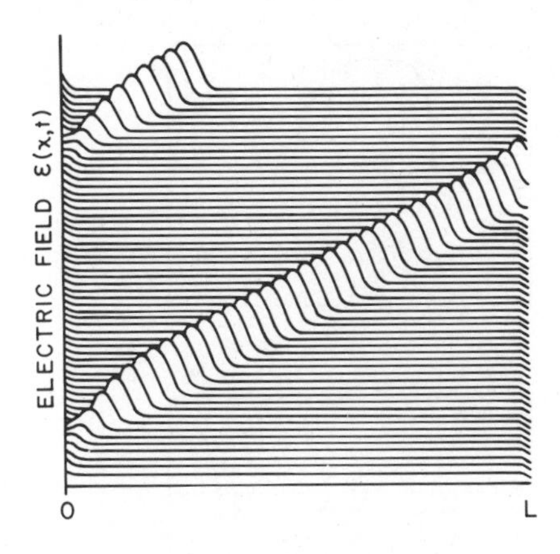

Fig. 21 Numerical simulation of the time-dependent behavior of cathode nucleated transit-time dipole-layer mode. The sample length is 100 μm with doping of 5×10^{14} cm^{-3}. Each successive instant of time is 24 ps. (After Shaw, Grubin, and Solomon, Ref. 33.)

form, one may either decrease the length of the sample (which increases the frequency in this mode) or increase the width of the domain. It can be seen from Fig. 20 that the domain width increases with decreasing doping level n_0. In general, more sinusoidal waveforms may be obtained by decreasing the n_0L product. Figure 23 shows a sequence of field distributions across a 35-μm sample during one RF cycle, together with the doping profile and the voltage and current waveforms.[29] The n_0L product is 2.1×10^{12} cm^{-2} for this device. The current waveform is much closer to sinusoidal for this device than the waveform shown in Fig. 22. Theoretical studies show that the efficiency of the transit-time mode is greatest when the n_0L product is one to several times 10^{12} cm^{-2}, so that the domain fills about half the sample and the current waveform is almost a sine wave. The maximum dc-to-RF conversion efficiency for this mode is 10%. The efficiency can be improved if the current waveform is close to a square wave. This waveform can be accomplished by adjusting the voltage to be below the threshold at the instant the dipole disappears at the anode. The

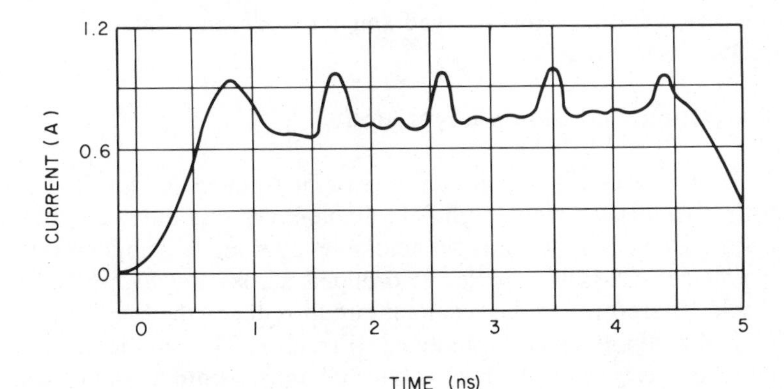

Fig. 22 Experimental current waveform of a 100-μm-long GaAs sample with a doping of 3×10^{15} cm^{-3}. (After Fukui, Ref. 34.)

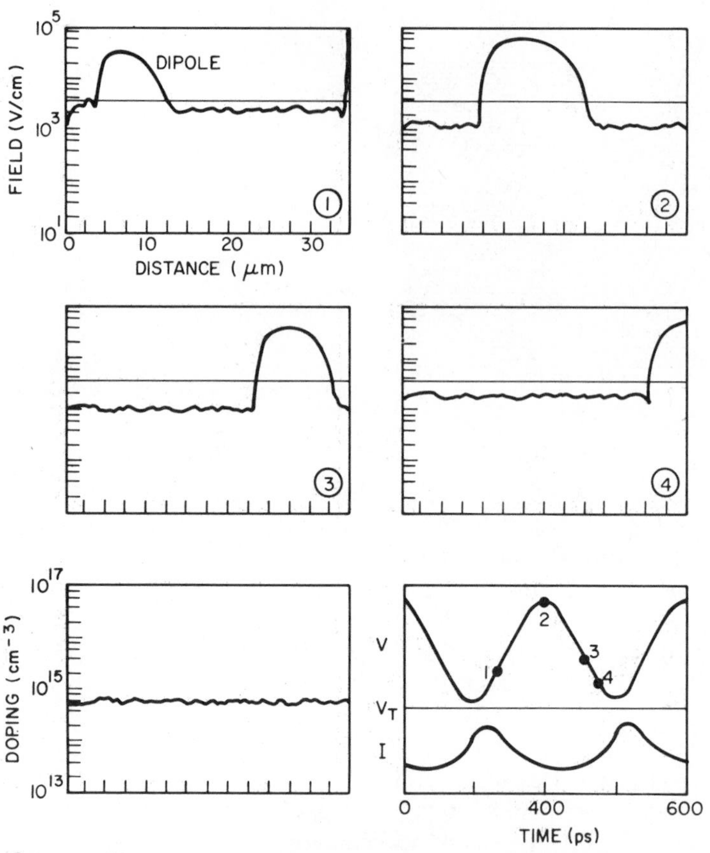

Fig. 23 Same as Fig. 15 for a transit-time dipole-layer mode. The GaAs sample has $n_0L = 2.1 \times 10^{12}\ \text{cm}^{-2}$ and $fL = 0.9 \times 10^7$ cm/s. (After Thim, Ref. 29.)

formation of a new dipole is delayed until the voltage rises above threshold. However, for this delayed domain mode, the tuning procedure is extremely complicated.

11.3.4 Quenched Dipole-Layer Mode

A TED in a resonant circuit can operate at frequencies higher than the transit-time frequency if the high-field dipole-layer was quenched before it reached the anode. In the transit-time dipole-layer mode of operation, most of the voltage across the device is dropped across the high-field dipole-layer itself. Therefore, as the resonant circuit reduces the bias voltage, the width of the dipole-layer is reduced (Fig. 20). The dipole-layer width continues to decrease as the bias voltage decreases until at some point the accumulation layer and the depletion layer neutralize each other. The bias voltage at which this occurs is V_s. Dipole-layer quenching occurs when the

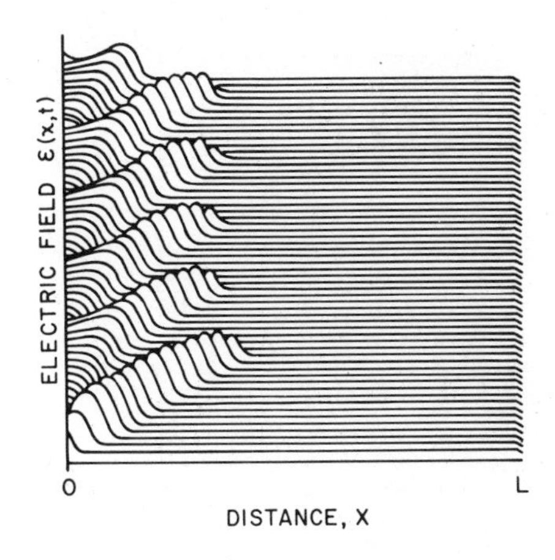

Fig. 24 Same as Fig. 21 for a quenched dipole-layer mode. (After Shaw, Grubin, and Solomon, Ref. 33.)

bias voltage across the device is reduced below V_s. When the bias voltage swings back above threshold, a new dipole-layer is nucleated, and the process repeats. Therefore, the oscillations occur at the frequency of the resonant circuit rather than at the transit-time frequency.

Figure 24 shows an example of the quenched dipole-layer mode.[33] The device has identical length and doping as that of Fig. 21. The dipole-layer is quenched at a distance of about $L/3$ from the cathode, and the operating frequency is about three times higher than the transit-time dipole-layer mode, as shown in Fig. 21.

In the quenched dipole-layer mode of operation it has been found, both theoretically[29] and experimentally,[35] that for samples in which the resonant frequency of the circuit is several times the transit-time frequency (i.e., $fL > 2\times10^7$ cm/s), and the frequency of operation is of the order of the dielectric relaxation frequency (i.e., $n_0/f \approx \epsilon_s/q|\mu_-|$, as given in Eq. 39), multiple high-field dipole-layers usually form, since one dipole does not have enough time to readjust and absorb the voltage of the other dipoles. Figure 25 shows multiple-dipole formation in a sample operated in the quenched dipole-layer mode. In this sample $n_0L = 4.2\times10^{12}$ cm^{-2}, $fL = 4.2\times10^7$ cm/s, and $n_0/f = 10^5$ s-cm^{-3}.

The upper frequency limit for this mode is determined by the speed of quenching, which in turn is determined by two time constants. The first is the positive dielectric relaxation time, whereas the second is an RC-time constant, R being the positive resistance in those regions of the diode not occupied by dipoles and C the capacitance of all the dipoles in series. The first condition gives a minimum critical n_0/f ratio of about 10^4 s-cm^{-3} for n-type GaAs and InP.[37,38]

The second time constant depends on the number of dipoles and sample length. The efficiency of quenched dipole-layer oscillators can theoretically[36] reach 13%.

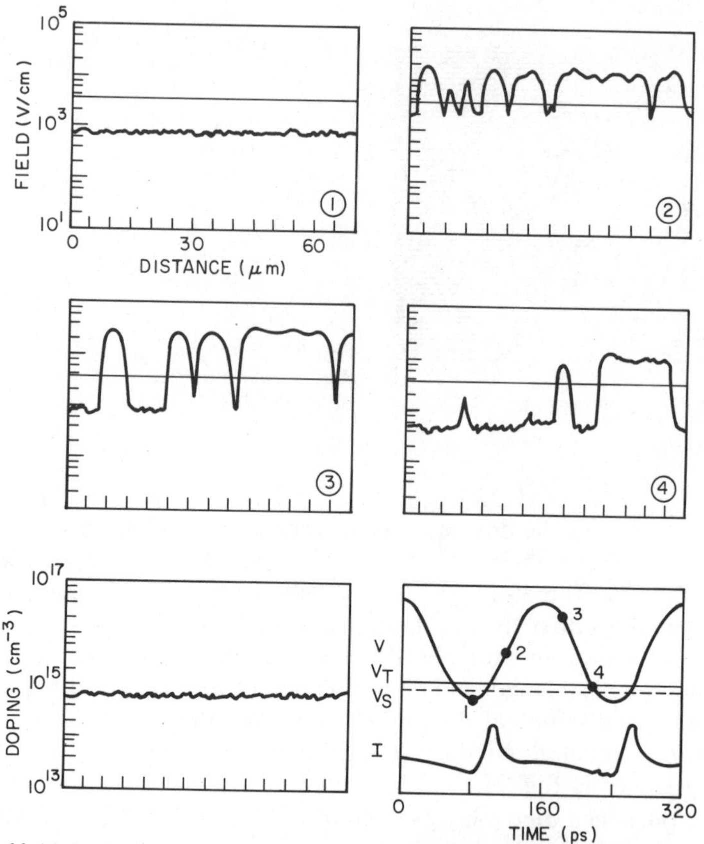

Fig. 25 Multiple-dipole formation in a sample operating in the quenched mode having $n_0L = 4.2 \times 10^{12}\ \text{cm}^{-2}$ and $fL = 4.2 \times 10^7$ cm/s. (After Thim, Ref. 29.)

11.3.5 Limited-Space-Charge Accumulation (LSA) Mode[37]

In the model of the LSA mode of operation, the electric field across the device rises from below the threshold and falls back again so quickly that it is assumed that the space-charge distribution associated with high-field dipole-layers does not have sufficient time to form. Only the primary accumulation layer forms near the cathode; the rest of the device remains fairly homogeneous, provided that doping fluctuations are sufficiently small to prevent the formation of dipole-layers. Under these conditions, a large portion of the device exhibits a uniform field, which yields efficient power generation at the circuit-controlled frequency. The higher the frequency, the shorter the distance the space-charge layer travels, leaving most of the device biased in the negative mobility range. The conditons for this mode of operation are derived from two requirements: the space charge should not have enough time to grow to an appreciable size, and the accumulation layer must be quenched completely during one RF cycle. Thus the negative τ_R, Eq. 9, should be larger than the RF period whereas the positive τ_R should be smaller. These requirements lead to the condition[38]

$$\frac{\epsilon_s}{q\mu_+} \ll \frac{n_0}{f} < \frac{\epsilon_s}{q|\mu_-|} \tag{39}$$

where μ_+ is the positive-differential mobility at low field, and μ_- an average negative-differential mobility above the threshold field. For GaAs and InP the two limiting ratios are

$$10^4 < n_0/f < 10^5 \text{ s-cm}^{-3} . \tag{40}$$

It is interesting to note that the quenched multiple dipole-layer mode also occurs in some range of n_0/f ratios if doping fluctuations are present. The LSA device is suited for generating short pulses of high-peak power, because devices with overlengths (non-transit-time mode) can be used for which heat extraction becomes difficult. However, the maximum operating frequency of LSA devices is much lower than that of transit-time devices. This lower frequency is caused by the slow energy relaxation of electrons in the lower valley, leading to increased quenching times. Computer simulations indicate that a minimum time of about 20 ps is required to stay below threshold voltage for cyclic operation in GaAs; the corresponding upper-frequency limit is about 20 GHz.[39,40] For InP a higher upper frequency is expected.

11.4 DEVICE PERFORMANCES

11.4.1 Cathode Contacts

The TEDs require extremely pure and uniform materials with a minimum of deep donor levels and traps, especially if quenching of space charge is involved in the operation. The first TEDs were fabricated from bulk GaAs and InP with alloyed ohmic contacts. Modern TEDs almost always use epitaxial layers on n^+ substrates deposited by vapor-phase epitaxy, liquid-phase epitaxy, or molecular-beam epitaxy techniques. Typical donor concentrations range from 10^{14} to 10^{16} cm^{-3}, and typical device lengths range from a few microns to several hundred microns. A typical fabrication procedure[41] using n^+-n-n^+ epitaxial GaAs is shown in Fig. 26. The TED chips are mounted in microwave packages. These packages, as well as related heat sinks, are similar to those for IMPATT diodes, discussed in Chapter 10. Some high-power TEDs are made by selective metallization and mesa etching.

To improve device performance, injection-limited cathode contacts have been used instead of the n^+ ohmic contacts.[42–44] By using an

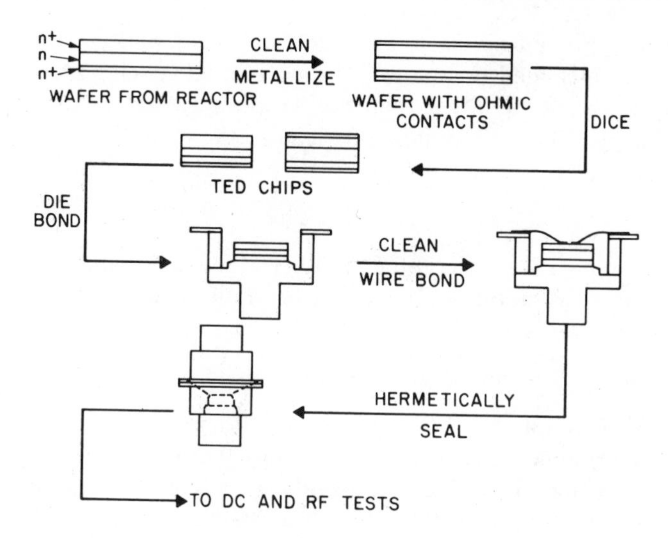

Fig. 26 Flow diagram of TED fabrication process. (After Sterzer, Ref. 41.)

injecting-limited contact, the threshold field for the cathode current can be adjusted to a value approximately equal to the threshold field $\mathscr{E}_T$ of onset of NDR. Thus a uniform electric field can result. For ohmic contacts, the accumulation or dipole-layer grows some distance from the cathode due to finite heating time of the lower-valley electrons. The "dead zone" may be as large as 1 μm, which imposes a constraint on the minimum device length and hence on the maximum operating frequency. In an injecting-limited contact, hot electrons are injected from the cathode, reducing the length of the dead zone. Since transit-time effect can be minimized, the device can exhibit a frequency-independent negative conductance shunted by its parallel-plate capacitance. If an inductance and a sufficiently large conductance are connected to this device, it can be expected to oscillate in a uniform-field mode at the resonant frequency. The theoretical efficiency will then be as derived in Section 11.3.1.

Two classes of injecting-limited contacts have been studied; one is a Schottky barrier with low-barrier height, the other is a two-zone cathode structure. Figure 27 compares these contacts with an ohmic contact. For an ohmic contact (Fig. 27*a*) there is always a low-field region near the cathode, and the field is nonuniform across the device length. For a Schottky barrier[45] under reverse bias, a reasonably uniform field can be obtained (Fig. 27*b*). The reverse current is given by (see Chapter 5)

$$J_R = A^{**}T^2 \exp(-q\phi_B/kT) \tag{41}$$

where A^{**} is the effective Richardson's constant and $q\phi_B$ is the barrier height. For current densities in the range 10^2 to 10^4 A/cm^2, the corresponding barrier height is about 0.15 to 0.3 eV. The Schottky barrier with a low barrier height is not simple to realize in III–V semiconductors, and in

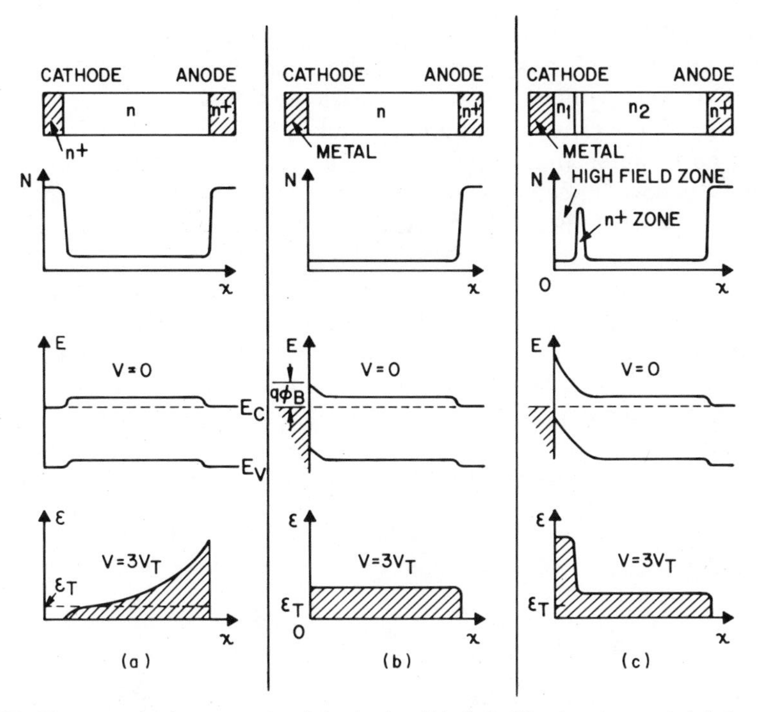

Fig. 27 Three cathode contacts: (*a*) ohmic, (*b*) Schottky barrier and (*c*) two-zone Schottky-barrier contact.

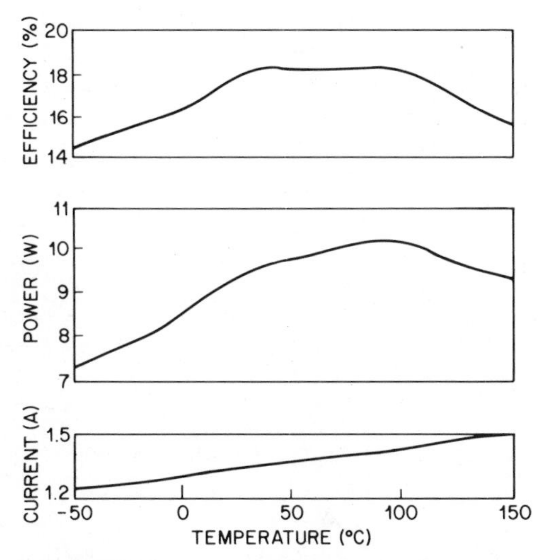

Fig. 28 Temperature dependence of efficiency, microwave power, and current for InP-pulsed TEDs with $n_0 = 2 \times 10^{15}$ cm^{-3} and $L = 8\ \mu$m having a two-zone cathode contact. (After Gray et al., Ref. 46.)

addition suffers from a rather limited temperature range, because the injected current varies exponentially with temperature (Eq. 41).

The two-zone cathode contact[46] consists of a high-field zone and an n^+ zone (Fig. 27*c*). This configuration is similar to that of a lo–hi–lo IMPATT diode (see Chapter 10). Electrons are "heated" up in the high-field zone and subsequently injected into the active region having uniform field. This structure has been successfully used over a wide temperature range. The efficiency, power output, and current of a pulsed InP TED over the temperature range −50 to +150°C are shown in Fig. 28. Note that over this temperature range, the efficiency and power only vary about ±10%. The highest efficiency obtained with an InP TED with the two-zone cathode contact is 24%. Because of Fermi-level pinning, GaAs devices with injection-limiting cathode contact have not yet been realized.

11.4.2 Power-Frequency Performance and Noise

The transfer of energy from the field to electrons and the scattering of electrons between lower valley and upper valleys take finite times. These

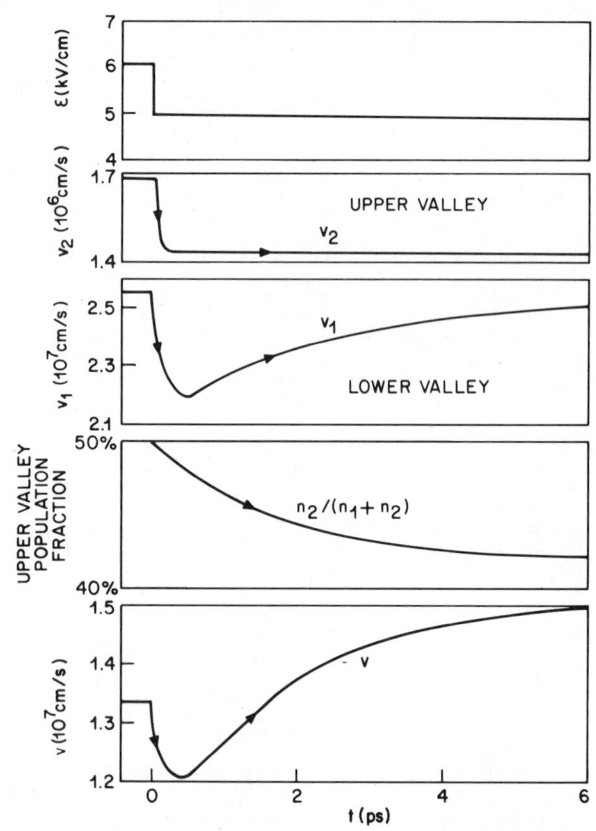

Fig. 29 Response of electrons to an electric field stepped from 6 to 5 kV/cm at $t = 0$. (After Rees, Ref. 47.)

finite times lead to an upper frequency limit corresponding to scattering and energy-relaxation frequencies. Figure 29 shows[47] the time responses of the lower- and upper-valley velocities, the average velocity, and the upper-valley population fraction when the electric field is suddenly lowered from 6 kV/cm to 5 kV/cm. Note that the velocity v_2 of the upper valley follows almost instantaneously with the field. However, the velocity v_1 of the lower valley has a slow response (with a time constant of about 5 ps). This response indicates the weak scattering of hot electrons in the lower valley. In addition, the slow decay of n_2 corresponds to the slow scattering from the upper valley to the lower valley. The response of the average velocity v_1 is thus due partly to the recovery of v_1 and partly to the intervalley transfer. Because of the finite response times, the TED have an estimated upper-frequency limit around 150 GHz.

Under transit-time conditions, the frequency of operation is inversely proportional to the device length, that is, $f = v/L$. The power–frequency relationship is given by

$$P_{rf} = V_{rf}^2/R = \mathscr{E}_{rf}^2 L^2/R = \frac{\mathscr{E}_{rf}^2 v^2}{Rf^2} \sim \frac{1}{f^2} \tag{42}$$

where V_{rf} and $\mathscr{E}_{rf}$ are the RF voltage and the corresponding field, respectively; and R is the impedance. Therefore, the output power is expected to fall as $1/f^2$. The state-of-the-art microwave power versus frequency is shown in Fig. 30 for pulsed and cw GaAs TEDs and pulsed and cw InP TEDs.[48–53,57,58] The parentheses near the data points indicate conversion efficiencies. Note that the power generally varies as $1/f^2$, as given by Eq. 42. Under pulsed condition, up to 6 kW has been obtained from GaAs overlength devices near 2 GHz. The cw power at 10 GHz is around 2 W, which is a factor of 5 lower than that of an IMPATT diode. On the other hand, the applied voltage for TEDs at a given frequency can be considerably lower than that for an IMPATT diode (by a factor of about 2 to 5).

The cw performances of InP TEDs are comparable to those of GaAs devices. However, the pulsed results of InP devices are lower, even though the theoretical efficiencies for InP are higher. These higher efficiencies mainly result from the much advanced GaAs technology.

The TEDs have two types of noise: amplitude deviations (AM noise) and frequency deviations (FM noise), both caused by thermal velocity fluctuations of the electrons. The AM noise is generally small because the amplitude is relatively stable, owing to the strong nonlinearity of the velocity–field characteristic. The FM noise has a mean-frequency deviation given by[54]

$$f_{\text{rms}} = \frac{f_0}{Q_{ex}} \sqrt{\frac{kT_{eq}(f_m)B}{P_0}} \tag{43}$$

where f_0 is the carrier frequency, Q_{ex} the external quality factor, P_0 the power output, B the measuring bandwidth. The modulation-frequency-

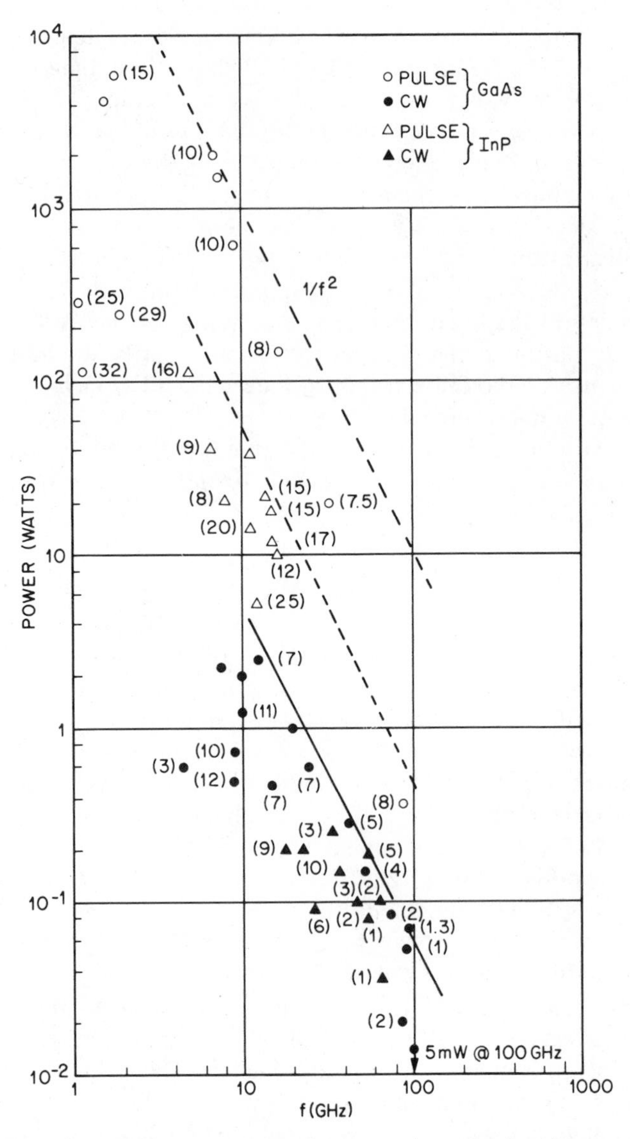

Fig. 30 Output microwave power versus frequency for pulsed and cw-operated GaAs and InP TEDs. The numbers in parentheses indicate the dc-to-rf conversion efficiencies in percentage.

dependent equivalent noise temperature T_{eq} is given by

$$T_{eq}(f_m) = qD/k|\mu_-| \tag{44}$$

where the average negative differential mobility μ_- depends on the voltage swing. Since the ratio $D/|\mu_-|$ is smaller in InP than in GaAs, the noise is expected to be lower in InP.

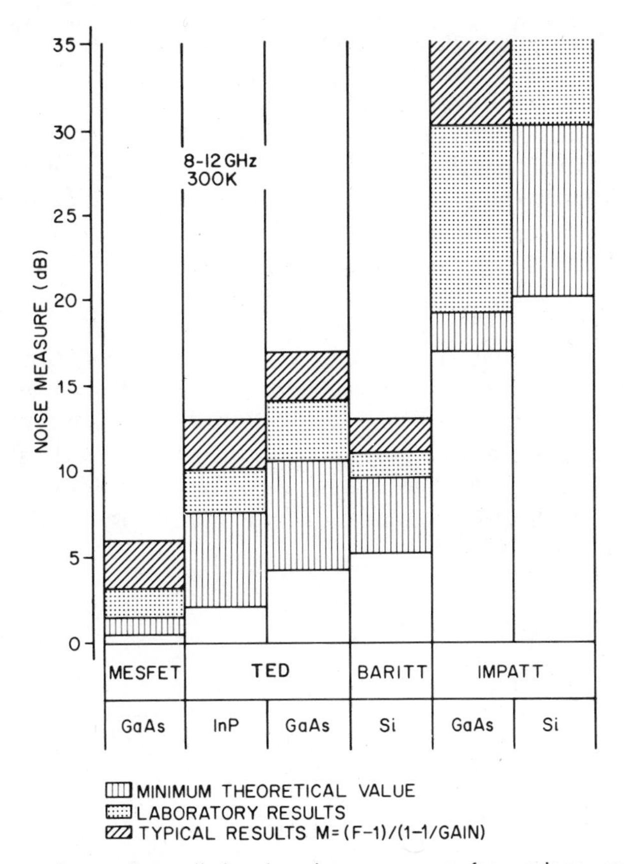

Fig. 31 Comparison of small-signal noise measures for various semiconductor microwave devices in the frequency range 8 to 12 GHz. (After Constant, Ref. 55.)

The small-signal noise measurements for various microwave devices in the 8- to 12-GHz frequency range are shown[55] in Fig. 31. The GaAs MESFET (refer to Chapter 6) has the best noise measure. The InP and GaAs TEDs follow. The Si BARRITT diode has a noise measure comparable to the TEDs. The IMPATT diode has the largest noise measures because of its avalanche processes.

11.4.3 Functional Devices

Thus far we have considered the transferred-electron effect and its application to microwave oscillators and amplifiers. The TEDs can also be used for digital and analog operations at high speed. We shall consider a TED with nonuniform cross-sectional area and a three-terminal TED.

The theory of high-field domains in one dimension can be used to analyze nonuniformly shaped oscillators if one assumes very thin high-field domains and considers phenomena in practically uniform regions in their neighborhood. These assumptions are valid if $n_0L \gg 10^{12}\ \text{cm}^{-2}$ and the variation of the cross-sectional area is gradual. Using the theory presented in the preceding section, it can be shown that there exists a value of domain excess voltage, $V_{ex} = V_{rm}$, above which the outside electric field $\mathscr{E}_r$ remains constant. The value of the outside electric field corresponding to V_{rm} is $\mathscr{E}_{rm}$, shown in Fig. 16. The current density associated with a domain with $V_{ex} = V_{rm}$ is

$$J_{rm} = qn_0v_{rm} . \tag{45}$$

Such saturated domains move in the oscillator with a constant velocity.

Let the thickness of the bulk oscillator be s and the changing width be $b(x)$, where x is measured from the cathode. If a high-field domain is nucleated from the cathode at time $t = 0$, then at time t the domain is at $x(t) = v_{rm}t$ using the foregoing assumptions. The current of the device must be constant at all cross sections, and in the vicinity of the domain is

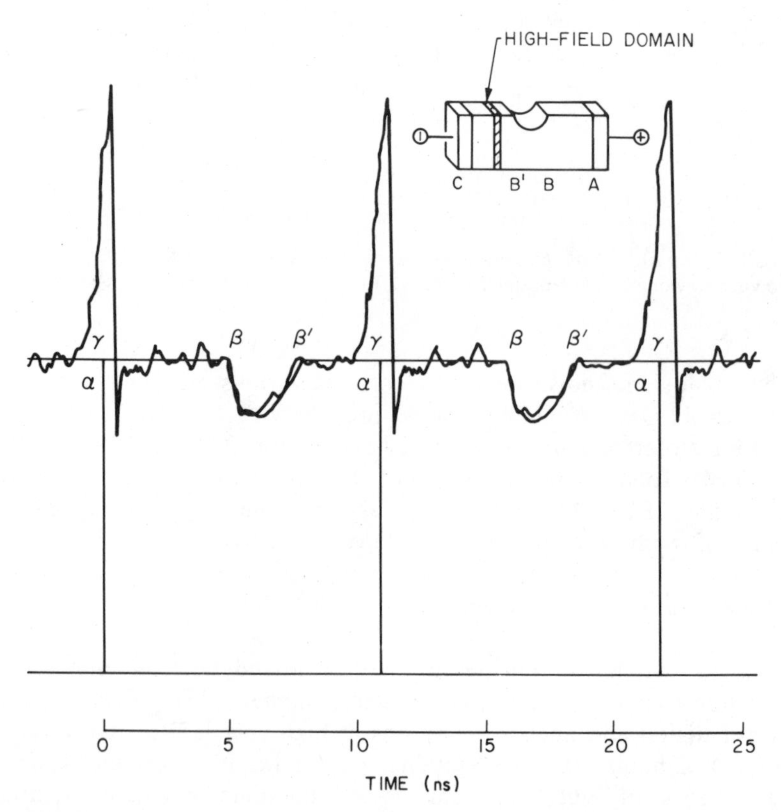

Fig. 32 Waveform generated by sample shown in the insert. (After Shoji, Ref. 56.)

given by

$$I(t) = J_{rm}sb(v_{rm}t) + I_g(t) \tag{46}$$

where $I_g(t)$ is due to the decay of the domain and is zero except at the end of the cycle. Then from Eq. 46 the current is proportional to $b(v_{rm}t)$.

Figure 32 shows the waveform of a bulk-effect oscillator for a sample shown in the insert.[56] The solid lines in this figure show the waveforms expected from this shape, and the experimental current waveform is indeed similar to the shape of the sample. The symbols α, β, β', and γ correspond to the instants of time when the domain is located at points A, B, B', and C.

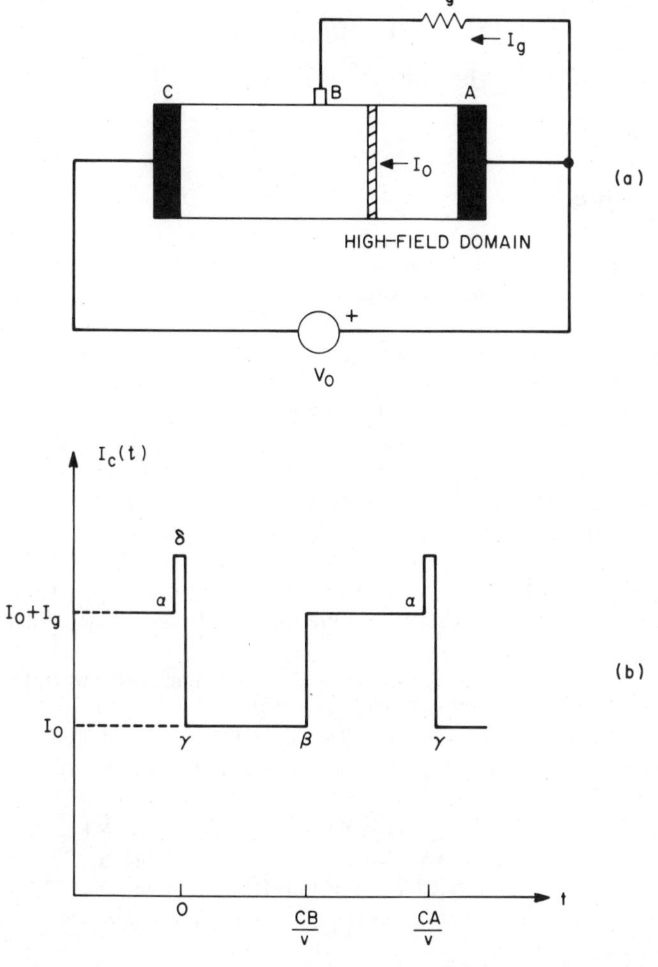

Fig. 33 (a) Controlled-current step generator. (b) Waveform of controlled-current step generator. (After Shoji, Ref. 56.)

Until now, only two terminal devices have been considered. The current waveform of a TED may be controlled by adding one or more electrode along the length of the device. Figure 33*a* shows[56] the structure of such a device with the electrode located at point *B*. The expected current waveform is shown in Fig. 33*b*. This waveform can be explained as follows. (The saturated domain theory described previously is used here again.) When the domain leaves the cathode at time $t = 0$, the cathode current $I_c(t)$ is equal to the current of the saturated domain (Aqn_0v_{rm}), and remains at this value until the domain reaches the electrode at *B*. At that time the cathode current becomes the sum of the saturated domain current and I_g, the current flowing through R_g. The current I_g is equal to the voltage sustained by the sample between *A* and *B* with a domain present, divided by R_g. The cathode current then remains at

$$I_c(t) = Aqn_0v_{rm} + I_g \tag{47}$$

until the domain is absorbed at the anode, at which time the current spikes briefly.

REFERENCES

1 J. B. Gunn, "Microwave Oscillation of Current in III–V Semiconductors," *Solid State Commun.*, **1**, 88 (1963); "Instabilities of Current in III–V Semiconductors," *IBM J. Res. Dev.*, **8**, 141 (1964).

2 H. Kroemer, "Theory of the Gunn Effect," *Proc. IEEE*, **52**, 1736 (1964).

3 B. K. Ridley and T. B. Watkins, "The Possibility of Negative Resistance Effects in Semiconductors," *Proc. Phys. Soc. Lond.*, **78**, 293 (1961).

3a B. K. Ridley, "Anatomy of the Transferred-Electron Effect in III–V Semiconductors," *J. Appl. Phys.*, **48**, 754 (1977).

4 C. Hilsum, "Transferred Electron Amplifiers and Oscillators," *Proc. IRE*, **50**, 185 (1962).

4a C. Hilsum, "Historical Background of Hot Electron Physics," *Solid State Electron.*, **21**, 5 (1978).

5 A. R. Hutson, A. Jayaraman, A. G. Chynoweth, A. S. Coriell, and W. L. Feldmann, "Mechanism of the Gunn Effect from a Pressure Experiment," *Phys. Rev. Lett.*, **14**, 639 (1965).

6 J. W. Allen, M. Shyam, Y. S. Chen, and G. L. Pearson, "Microwave Oscillations in $GaAs_{1-x}P_x$ Alloys," *Appl. Phys. Lett.*, **7**, 78 (1965).

7 J. E. Carroll, *Hot Electron Microwave Generators*, Edward Arnold, London, 1970.

8 P. J. Bulman, G. S. Hobson, and B. S. Taylor, *Transferred Electron Devices*, Academic, New York, 1972.

9 B. G. Bosch and R. W. H. Engelmann, *Gunn-Effect Electronics*, Wiley, New York, 1975.

10 H. W. Thim, "Solid State Microwave Sources," in C. Hilsum, Ed., *Handbook on Semiconductors*, Vol. 4, *Device Physics*, North-Holland, Amsterdam, 1980.

11 B. K. Ridley, "Specific Negative Resistance in Solids," *Proc. Phys. Soc. Lond.*, **82**, 954 (1963).

12 B. K. Ridley and R. G. Pratt, "A Bulk Differential Negative Resistance Due to Electron Tunneling through an Impurity Potential Barrier," *Phys. Lett.*, **4**, 300 (1963).

13 A. L. McWhorter and R. H. Rediker, "The Cryosar—A New Low-Temperature Computer Component," *Proc. IRE*, **47**, 1207 (1959).

14 H. Kroemer, "Negative Conductance in Semiconductors," *IEEE Spectrum*, **5**, 47 (1968).

15 D. E. Aspnes, "GaAs Lower Conduction Band Minimum: Ordering and Properties," *Phys. Rev.*, **14**, 5331 (1976).

16 H. D. Rees and K. W. Gray, "Indium Phosphide: A Semiconductor for Microwave Devices," *Solid State Electron Devices*, **1**, 1 (1976).

17 D. E. McCumber and A. G. Chynoweth, "Theory of Negative Conductance Application and Gunn Instabilities in 'Two-Valley' Semiconductors," *IEEE Trans. Electron Devices*, **ED-13**, 4 (1966).

18 K. Sakai, T. Ikoma, and Y. Adachi, "Velocity-Field Characteristics of $Ga_xIn_{1-x}Sb$ Calculated by the Monte Carlo Method," *Electron. Lett.*, **10**, 402 (1974).

19 R. E. Hayes and R. M. Raymond, "Observation of the Transferred-Electron Effect in GaInAsP," *Appl. Phys. Lett.*, **31**, 300 (1977).

20 J. G. Ruch and G. S. Kino, "Measurement of the Velocity-Field Characteristics of Gallium Arsenide," *Appl. Phys. Lett.*, **10**, 40 (1967).

21 P. N. Butcher and W. Fawcett, "Calculation of the Velocity-Field Characteristics for Gallium Arsenide," *Phys. Lett.*, **21**, 489 (1966).

22 M. A. Littlejohn, J. R. Hauser, and T. H. Glisson, "Velocity–Field Characteristics of GaAs with Γ-*L*-*X* Conduction-Band Ordering," *J. Appl. Phys.*, **48**, 4587 (1977).

23 I. Mojzes, B. Podor, and I. Balogh, "On the Temperature Dependence of Peak Electron Velocity and Threshold Field Measured on GaAs Gunn Diodes," *Phys. Status Solidi*, **39**, K123 (1977).

24 W. Shockley, "Negative Resistance Arising from Transit Time in Semiconductor Diodes," *Bell Syst. Tech. J.*, **33**, 799 (1954).

25 G. S. Kino and I. Kuru, "High-Efficiency Operation of a Gunn Oscillator in the Domain Mode," *IEEE Trans. Electron Devices*, **ED-16**, 735 (1969).

26 G. S. Hobson, *The Gunn Effect*, Clarendon, Oxford, 1974.

27 H. W. Thim and W. Haydl, "Microwave Amplifier Circuit Consideration," in M. J. Howes and D. V. Morgan, Eds., *Microwave Devices*, Wiley, New York, 1976, Chap. 6.

28 B. W. Hakki, "Amplification in Two-Valley Semiconductors," *J. Appl. Phys.*, **38**, 808 (1967).

29 H. W. Thim, "Computer Study of Bulk GaAs Devices with Random One-Dimensional Doping Fluctuations," *J. Appl. Phys.*, **39**, 3897 (1968).

30 P. N. Butcher, "Theory of Stable Domain Propagation in the Gunn Effect," *Phys. Lett.*, **19**, 546 (1965).

31 P. N. Butcher, W. Fawcett, and C. Hilsum, "A Simple Analysis of Stable Domain Propagation in the Gunn Effect," *Br. J. Appl. Phys.*, **17**, 841 (1966).

32 J. A. Copeland, "Electrostatic Domains in Two-Valley Semiconductors," *IEEE Trans. Electron Devices*, **ED-13**, 187 (1966).

33 M. Shaw, H. L. Grubin, and P. R. Solomon, *The Gunn–Hilsum Effect*, Academic, New York, 1979.

34 H. Fukui, "New Method of Observing Current Waveforms in Bulk GaAs," *Proc. IEEE*, **54**, 792 (1966).

35 H. W. Thim and M. R. Barber, "Observation of Multiple High-Field Domains in *n*-GaAs," *Proc. IEEE*, **56**, 110 (1968).

36 M. R. Barber, "High Power Quenched Gunn Oscillators," *Proc. IEEE*, **56**, 752 (1968).

37 J. A. Copeland, "A New Mode of Operation for Bulk Negative Resistance Oscillators," *Proc. IEEE*, **54**, 1479 (1966).

38 J. A. Copeland, "LSA Oscillator Diode Theory," *J. Appl. Phys.*, **38**, 3096 (1967).

39 D. Jones and H. D. Rees, "Electron-Relaxation Effects in Transferred-Electron Devices Revealed by New Simulation Method," *Electron. Lett.*, **8**, 363 (1972).

40 H. Kroemer, "Hot Electron Relaxation Effects in Devices," *Solid State Electron.*, **21**, 61 (1978).

41 F. Sterzer, "Transferred Electron Amplifiers and Oscillators for Microwave Application," *Proc. IEEE*, **59**, 1155 (1971).

42 H. Kroemer, "The Gunn Effect under Imperfect Cathode Boundary Condition," *IEEE Trans. Electron Devices*, **ED-15**, 819 (1968).

43 M. M. Atalla and J. L. Moll, "Emitter Controlled Negative Resistance in GaAs," *Solid State Electron.*, **12**, 619 (1969).

44 S. P. Yu, W. Tantraporn, and J. D. Young, "Transit-Time Negative Conductance in GaAs Bulk-Effect Diodes," *IEEE Trans. Electron Devices*, **ED-18**, 88 (1971).

45 D. J. Colliver, L. D. Irving, J. E. Pattison, and H. D. Rees, "High-Efficiency InP Transferred-Electron Oscillators," *Electron. Lett.*, **10**, 221 (1974).

46 K. W. Gray, J. E. Pattison, J. E. Rees, B. A. Prew, R. C. Clarke, and L. D. Irving, "InP Microwave Oscillator with 2-Zone Cathodes," *Electron. Lett.*, **11**, 402 (1975).

47 H. D. Rees, "Time Response of the High-Field Electron Distribution Function in GaAs," *IBM J. Res. Dev.*, **13**, 537 (1969).

48 G. S. Hobson, "Recent Development in Transferred Electron Devices," *J. Phys. E*, **7**, 229 (1974).

49 L. D. Irving, J. E. Pattison, P. W. Braddock, and K. W. Gray, "Improved Mean Power and Long Pulse-Width Operation of InP TEDs in J Band," *Electron. Lett.*, **14**, 116 (1978).

50 R. J. Hamilton, R. D. Fairman, S. I. Long, M. Omori, and F. B. Fank, "InP Gunn-Effect Devices for Millimeter-Wave Amplifiers and Oscillators," *IEEE Trans. Microwave Theory Tech.*, **MTT-24**, 775 (1976).

51 W. R. Day, "Gunn Oscillators," *Microwave Syst. News*, **8**, 56 (1978).

52 T. Suzuki, M. Ito, T. Ishii, and S. Mitsui, "Design and Fabrication for High Efficiency and High Output Power Gunn Diodes," *Trans. Inst. Electron. Commun. Eng., Jpn.*, **E61**, 932 (1978).

53 J. Mun, "High-Efficiency and High-Peak-Power InP Transferred-Electron Oscillators," *Electron. Lett.*, **13**, 275 (1977).

54 A. Ataman and W. Harth, "Intrinsic FM Noise of Gunn Oscillators," *IEEE Trans. Electron Devices*, **ED-20**, 12 (1973).

55 E. Constant, "Noise in Microwave, Injection, Transit Time and Transferred-Electron Devices," *Physica*, **83B**, 24 (1976).

56 M. Shoji, "Functional Bulk Semiconductor Oscillators," *IEEE Trans. Electron Devices*, **ED-14**, 535 (1967).

57 J. Ondria, "Wide-Band Mechanically Tunable W-Band (75–110 GHz) CW GaAs Gunn Diode Oscillators," Proc. 7th Bienn. Conf. Active Microwave Semicond. Devices Circuits, Cornell University, Ithaca, N.Y., 1979, p. 309.

58 J. D. Crowley, F. B. Fank, S. B. Hyder, J. J. Sowers, and D. Tringali, "Millimeter Wave InP Transferred Electron Devices," Proc. 7th Bienn. Conf. Active Microwave Semicond. Devices Circuits, Cornell University, Ithaca, N.Y., 1979, p. 331.

59 J. R. Hauser, T. H. Glisson, and M. A. Littlejohn, "Negative Resistance and Peak Velocity in the Central (000) Valley of III–V Semiconductors," *Solid State Electron.*, **22**, 487 (1979).

PART V

PHOTONIC DEVICES

12

LED and Semiconductor Lasers

12.1 INTRODUCTION

Photonic devices are those in which the basic particle of light—the photon, plays a major role. Photonic devices can be divided into three groups: (1) devices that convert electrical energy into optical radiation—the LED (light-emitting diode) and the diode laser (light amplification by stimulated emission of radiation), (2) devices that detect optical signals through electronic processes—photodetectors, and (3) devices that convert optical radiation into electrical energy—the photovoltaic device or solar cell.

The first group is considered in this chapter; photodetectors and solar cells are discussed in Chapters 13 and 14, respectively.

The electroluminescence phenomenon was discovered[1] in 1907. Electroluminescence is the generation of light by an electric current passing through a material under an applied electric field. Electroluminescent light differs from thermal radiation or incandescence in the relatively narrow range of wavelengths contained within its spectrum (for LEDs, the spectral line width is typically 100 to 500 Å). The light may even be nearly perfectly monochromatic, as in the diode laser (0.1 to 1 Å). Bergh and Dean have given a comprehensive treatment of light-emitting diodes,[2] and Gage et al.

have discussed the applications of LEDs as lamps, displays, and opto-isolators.[3]

After the invention of the maser (microwave amplification by stimulated emission of radiation) in 1954 by Townes and his collaborators[4] and the subsequent operation[5,6] of optical masers and lasers (*l* replacing *m* in maser and standing for light) in ruby, semiconductors were suggested for use as laser materials.[7-10] The theoretical calculations of Bernard and Duraffourg[11] in 1961 set forth the necessary conditions for lasing using quasi-Fermi levels. In 1962, Dumke[12] showed that laser action was indeed possible in direct bandgap semiconductors and set forth important criteria for such action. In late 1962, three groups, headed by Hall,[13] Nathan,[14] and Quist,[15] announced almost simultaneously that they had achieved lasing in semiconductors. The pulsed radiation of 0.84 μm was obtained from a liquid-nitrogen-cooled, forward-biased GaAs *p-n* junction. Shortly thereafter, Holonyak and Bevacqua[16] announced laser action in the ternary compound $GaAs_{1-x}P_x$ junction at 0.71 μm. In 1970, Hayashi et al. achieved the continuous operation of junction lasers at room temperature by the use of double heterojunctions.[17] This structure was first proposed by Kroemer[18] and Alferov and Kazarinov.[19]

Since these initial discoveries, many new laser materials have been found. The wavelength of coherent radiation has been extended from the near ultraviolet into the visible and then out to the far-infrared spectrum ($\sim$0.3 to $\sim$30 μm). Semiconductor lasers are considered the most important light sources for optical-fiber communication systems. Semiconductor lasers also have significant potential for applications in many areas of basic research and technology, such as high-resolution gas spectroscopy and atmospheric pollution monitoring. Casey and Panish[20] and Kressel and Butler[21] have published comprehensive treatments of heterostructure lasers.

12.2 RADIATIVE TRANSITIONS

The LED and semiconductor laser belong to the luminescent device family. Luminescence is the emission of optical radiation (ultraviolet, visible, or infrared) as a result of electronic excitation of a material, excluding any radiation that is purely the result of the temperature of the material (incandescence). Figure 1 shows a chart of the electromagnetic spectrum. Although different methods must be used to excite radiations of different wavelengths, all radiations are fundamentally alike. The visual range of the human eye extends only from about 0.4 to 0.7 μm. Figure 1 shows the major color bands from violet to red. The infrared region extends from 0.7 to about 1000 μm, and the ultraviolet region includes wavelengths from 0.4 to about 0.01 μm (i.e., 100 Å or 10 nm). In this and subsequent chapters, we are primarily interested in the wavelength range from near ultraviolet ($\sim$0.3 μm) to near infrared ($\sim$1.5 μm).

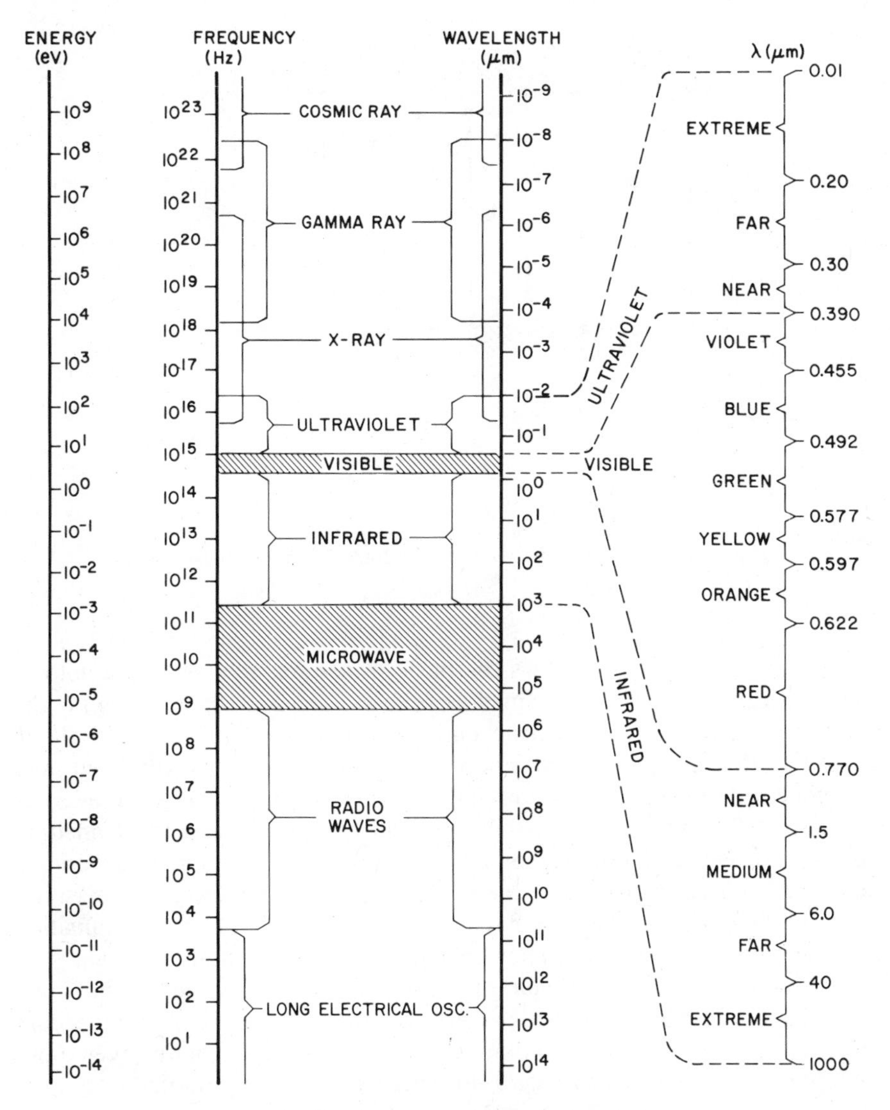

Fig. 1 Chart of electromagnetic spectrum.

Types of luminescence may be distinguished by the source of input energy:[22] (1) photoluminescence involving excitation by optical radiation, (2) cathodoluminescence by electron beam or cathode ray, (3) radioluminescence by other fast particles or high-energy radiation, and (4) electroluminescence by electric field or current. We are mainly concerned with electroluminescence, especially with injection electroluminescence, that is, optical radiation obtained by injecting minority carriers into the region of a semiconductor *p-n* junction where radiative transitions take place.

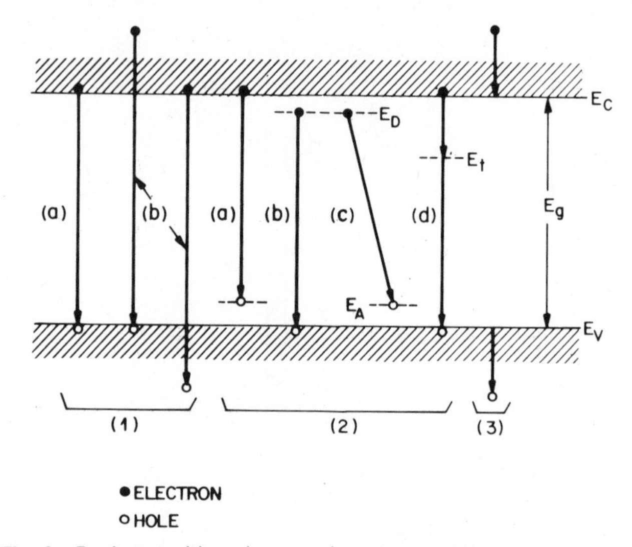

Fig. 2 Basic transitions in a semiconductor. (After Ivey, Ref. 23.)

Figure 2 schematically shows the basic transitions in a semiconductor.[23] These transitions may be classified as follows. The first classification is the interband transition: (a) intrinsic emission corresponding very closely in energy to the bandgap, where phonons or excitons may be involved, and (b) higher-energy emission involving energetic or hot carriers, sometimes related to avalanche emission. The second classification is the transition involving chemical impurities or physical defects; (a) conduction band to acceptor, (b) donor to valence band, (c) donor to acceptor (pair emission), and (d) deep levels. The third classification is the intraband transition involving hot carriers, sometimes called deceleration emission.

Not all transitions can occur in the same material or under the same conditions, and not all transitions are radiative. An efficient luminescent material is one in which radiative transitions predominate over nonradiative ones (such as the Auger nonradiative recombination).[2]

12.2.1 Emission Spectra

There are three processes for interaction between a photon and an electron in a solid. A photon may be absorbed by the transition of an electron from a filled state in the valence band to an empty state in the conduction band. In addition to being absorbed, a photon can stimulate the emission of a similar photon by the transition of an electron from a filled state in the conduction band to an empty state in the valence band. Also, an electron in the conduction band can spontaneously return to an empty state in the valence band with the emission of a photon.

The spontaneous emission rate depends on the density of filled conduction-band states and the density of empty valence-band states and may be written as[20]

$$I(h\nu) \sim \nu\langle M\rangle^2 N_C N_V F_C(E) F_V(E) \tag{1}$$

where $\langle M\rangle$ is called the transition matrix element, N_C is the density of states in the conduction band, N_V is the density of states in the valence band, $F_C(E)$ and $F_V(E)$ are, respectively, the electron and hole Fermi–Dirac distribution functions. The spontaneous emission rate generally has the form[24]

$$I(h\nu) \sim \nu^2(h\nu - E_g)^{1/2} \exp[-(h\nu - E_g)/kT] \tag{2}$$

where E_g is the bandgap.

At the impurity densities often encountered for LEDs and semiconductor lasers, the representation of localized impurity levels separated from the band edges cannot be used. Instead, the random distribution of charged impurities in the crystal results in potential fluctuations which produce tails in the conduction- and valence-band density of states (refer to Section 12.4.3).

When the donor or acceptor ionization energy goes to zero at high impurity concentrations, the carrier concentration becomes temperature-independent and all the donor or acceptor atoms (assuming no complexes) are ionized. Therefore, the free electron concentration n_0 will equal the donor concentration, and the free hole concentration p_0 will equal the acceptor concentration if the compensation is small. The best criterion for the concentration where the ionization energy goes to zero is when the ratio of the average separation r of the impurity atoms to the radius of the hydrogenic impurity a^* is about 3:

$$r/a^* \approx 3.0. \tag{3}$$

In Eq. 3,

$$r \equiv \left(\frac{3}{4\pi n}\right)^{1/3} \tag{4}$$

where n is the impurity concentration and

$$a^* = \frac{1}{2}\left(\frac{\epsilon_s}{\epsilon_0}\right)\left(\frac{m_0}{m^*}\right) \quad \text{Å} \tag{5}$$

with m^* the effective density-of-states electron mass for donors or the effective density-of-states hole mass for acceptors.

The conventional theory for optical transitions between the valence and conduction bands of direct-energy-gap materials is based on the so-called **k**-selection rule. The wave vector $\mathbf{k}_1$ of the valence-band wave function and the wave vector $\mathbf{k}_2$ of the conduction band must differ by the wave vector of the photon or the matrix element $\langle M\rangle$ is zero. Since the wave vector of the electron is much larger than the wave vector of the photon, the

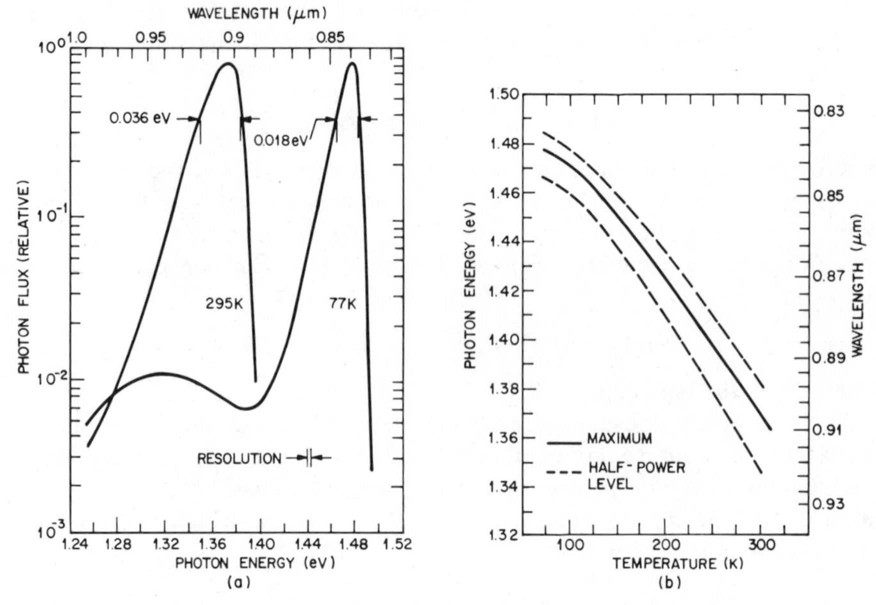

Fig. 3 (*a*) GaAs diode emission spectra at 300 and 77 K. (*b*) Dependence of emission peak and half-power width as a function of temperature. (After Carr, Ref. 25.)

k-selection rule is generally written as

$$\mathbf{k}_1 = \mathbf{k}_2. \tag{6}$$

The allowed transitions are then between initial and final states of the same wave vector and are called "direct" or "vertical" transitions. When the conduction-band minima are not at the same value of **k** as the valence band, assistance of a phonon is necessary to conserve crystal momentum, and the transition is called "indirect." When impurities are added to the semiconductor, the wave functions and matrix elements will be changed and the **k**-selection rule will not hold. Therefore, the cases generally considered are called the no-**k**-selection-rule optical transitions, and the matrix element will be energy-dependent.

Figure 3*a* shows the emission spectra for a GaAs *p-n* junction observed[25] at 77 and 300 K. The peak photon energy decreases with increasing temperature mainly because the bandgap decreases with temperature. Figure 3*b* shows a more detailed plot for the peak photon energy and the half-power points from the diode emission spectrum as a function of temperature. The width of the half-power points increases slightly with temperature expected from Eq. 2.

12.2.2 Luminescent Efficiency

For a given input excitation energy, the radiative recombination process is in direct competition with the nonradiative process. Figure 4 illustrates

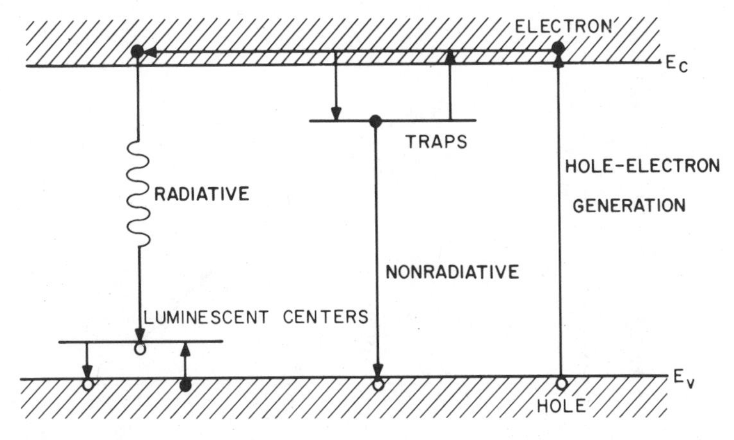

Fig. 4 Representation of radiative and nonradiative recombinations. (After Ivey, Ref. 26.)

the various competing processes.[26] The quantum efficiency n_q is the fraction of the excited carriers that combine radiatively to the total recombination and may be written in terms of the lifetimes as

$$\eta_q = R_r/R = \tau_{nr}/(\tau_{nr} + \tau_r) \tag{7}$$

where τ_{nr} is the nonradiative lifetime and τ_r is the radiative lifetime, and R_r and R are the radiative recombination rate and total recombination rate, respectively. The recombination rate and lifetime are related for p-type layers by

$$R = (n - n_0)/\tau \tag{8}$$

and for n-type layers by

$$R = (p - p_0)/\tau \tag{9}$$

where n_0 and p_0 are electron and hole concentrations in thermal equilibrium, respectively, and n and p are electron and hole concentrations under optical excitations, respectively. The minority-carrier lifetime τ is given by

$$\tau = \tau_r \tau_{nr}/(\tau_{nr} + \tau_r). \tag{10}$$

Equation 7 shows that the radiative lifetime τ_r needs to be small to give high quantum efficiency.

The insert in Fig. 5 shows a cross section of an arrangement used for efficiency measurement. The emission is measured directly as the short-circuit current of the solar cell. Figure 5 shows the measured external efficiency as a function of temperature.[25] The efficiency decreases with increasing temperature and has a value of 40, 32, or 7% at 20, 77, or 300 K, respectively. The decrease is caused by the reduction of τ/τ_r with temperature.

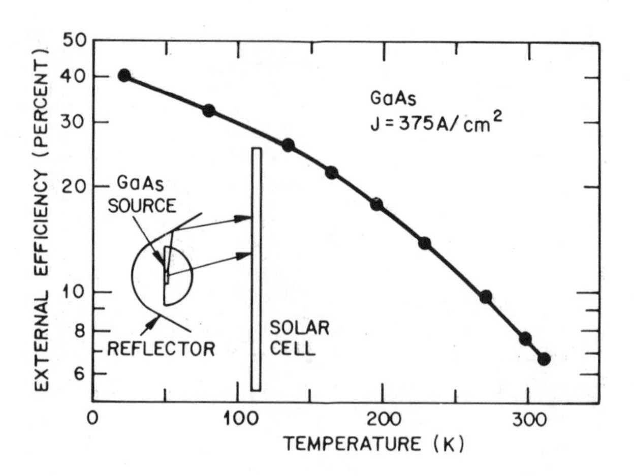

Fig. 5 External quantum efficiency versus temperature for GaAs LED. The insert shows cross section of arrangement used for efficiency measurement. (After Carr, Ref. 25.)

12.2.3 Methods of Excitation

Electroluminescence may be excited in a variety of ways, including intrinsic, avalanche, tunneling, and injection processes. For the intrinsic excitation, a powder of a semiconductor (e.g., ZnS) is embedded in a dielectric (plastic or glass) and subjected to an alternating electric field. At frequencies in the audio range, electroluminescence usually occurs. Generally, the efficiency is low ($\lesssim 1\%$). The mechanism is mainly caused by impact ionization of accelerated electrons or field emission of electrons from trapping centers.[23,27]

For the avalanche excitation, a p-n junction or a metal–semiconductor barrier is reverse-biased into avalanche breakdown. The electron–hole pairs generated by impact ionization may result in emission of either interband (avalanche emission) or intraband (deceleration emission) transitions. Electroluminescence can also result from tunneling into forward-biased or reverse-biased junctions. When a sufficiently large reverse bias is applied to a metal–semiconductor barrier (p-type degenerate), holes at the metal Fermi level can tunnel into the valence band and subsequently make a radiative recombination with electrons that have tunneled from the valence band to the conduction band.[28]

Injection electroluminescence is by far the most important method of excitation.[29] When a forward bias is applied to a p-n junction, the injection of minority carriers across the junction can give rise to efficient radiative recombination, since electric energy can be converted directly into photons. In subsequent sections we shall be concerned primarily with injection electroluminescent devices.

12.3 LIGHT-EMITTING DIODES

Light-emitting diodes (LEDs) are semiconductor p-n junctions that under proper forward-biased conditions can emit external spontaneous radiation in the ultraviolet, visible, and infrared regions of the electromagnetic spectrum. We consider first the visible LED, which has a multitude of applications in the essential information linkage between electronic instruments and their human users. We also consider the infrared LED, which is useful in opto-isolators, and is a potential light source for optical-fiber communication.

12.3.1 Visible LED

The effectiveness of light for stimulating the human eye is given by the relative eye sensitivity $V(\lambda)$, which is a function of wavelength. Figure 6 shows the relative sensitivity, as defined for a 2° viewing angle by the Commission Internationale de l'Eclairage (CIE) for photopic vision.[2] For the maximum sensitivity of the eye at 0.555 μm, $V(0.555) = 1.0$; the value of $V(\lambda)$ falls to nearly zero at the extremes of the visible spectrum at 0.39 and 0.77 μm. The luminosity of the radiant energy describes the effectiveness of the available radiant energy on vision. For normal photopic vision at the peak sensitivity of the eye (0.555 μm), 1 W of radiant energy is equivalent to 680 lm (lumen). Figure 6 also shows the six major color bands from violet to red.

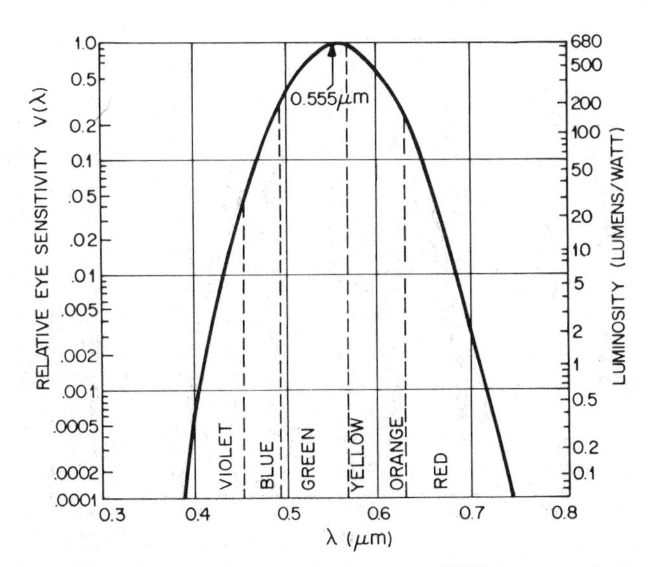

Fig. 6 Relative luminosity function as defined by the CIE for normal photopic vision. Major color bands are also indicated.

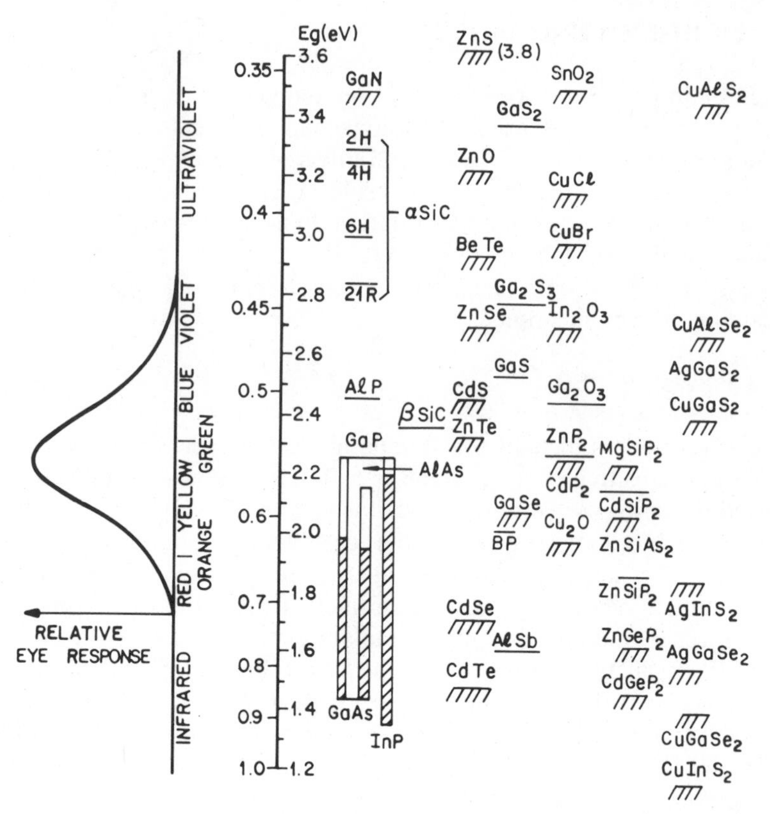

Fig. 7 Semiconductors of interest as visible LEDs, including the relative luminosity function of the human eye. (After Bergh and Dean, Ref. 2.)

Materials for Visible LED Since the eye is only sensitive to light of energy $h\nu \gtrsim 1.8\,\text{eV}$ ($\sim 0.7\,\mu\text{m}$), semiconductors of interest must have energy bandgaps larger than this limit. Figure 7 lists semiconductors that can be used as visible LEDs. Direct bandgap semiconductors, indicated by shading, are particularly important for electroluminescent devices, because the radiative recombination is a first-order transition process (no phonon involved) and the quantum efficiency is expected to be much higher than that for an indirect bandgap semiconductor, where a phonon is involved. Among all the semiconductors listed in Fig. 7, the most important is the $GaAs_{1-x}P_x$ III–V compound system.

Figure 8*a* shows the energy gap for $GaAs_{1-x}P_x$ as a function of the mole fraction x.[20] For $0 < x < 0.45$, the energy gap is direct and increasing from $E_g = 1.424\,\text{eV}$ at $x = 0$ to $E_g = 1.977$ and $x = 0.45$. For $x > 0.45$, the energy gap is indirect. Figure 8*b* shows the corresponding energy–momentum plots for selected alloy compositions.[30] As indicated, the conduction band has two minima. The one along the Γ axis is the direct minimum, whereas

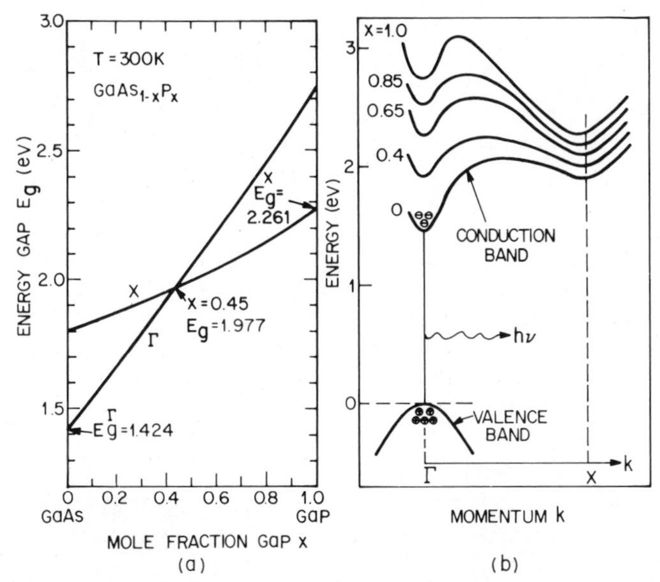

Fig. 8 (*a*) Compositional dependence of the direct and indirect energy bandgap for $GaAs_{1-x}P_x$. (After Casey and Panish, Ref. 20.) (*b*) Schematic energy–momentum diagram for $GaAs_{1-x}P_x$. The alloy compositions shown correspond to red ($x = 0.4$), orange (0.65), yellow (0.85), and green light (1.0). (After Craford, Ref. 30.)

the one along the x axis is the indirect minimum. Electrons in the direct minimum of the conduction band and holes at the top of the valence band have equal momentum; electrons in the indirect minimum have different momentum. For direct bandgap semiconductors, such as GaAs and $GaAs_{1-x}P_x$ ($x \leq 0.45$), the momentum is conserved. Interband transitions may occur with high probability. The photon energy is then approximately equal to the bandgap energy of the semiconductor. The radiative transition mechanism is predominant in direct bandgap materials. However, for $GaAs_{1-x}P_x$ with $x > 0.45$ and GaP that are indirect bandgap semiconductors, the probability for interband transitions is extremely small, since phonons or other scattering agents must participate in the process in order to conserve momentum. Therefore, for indirect bandgap semiconductors, such as GaP, special recombination centers are incorporated to enhance the radiative processes.

An efficient radiative recombination center in $GaAs_{1-x}P_x$ can be formed by incorporating specific impurities such as nitrogen.[31] When nitrogen is introduced, it replaces phosphorus atoms in the lattice sites. The outer electronic structure of nitrogen is similar to that of phosphorus (both are group V elements in the periodic table), but the electronic core structures of these atoms differ considerably. This difference gives rise to an electron trap level close to the conduction band. A recombination center produced in this way is called an isoelectronic center. The isoelectronic centers are

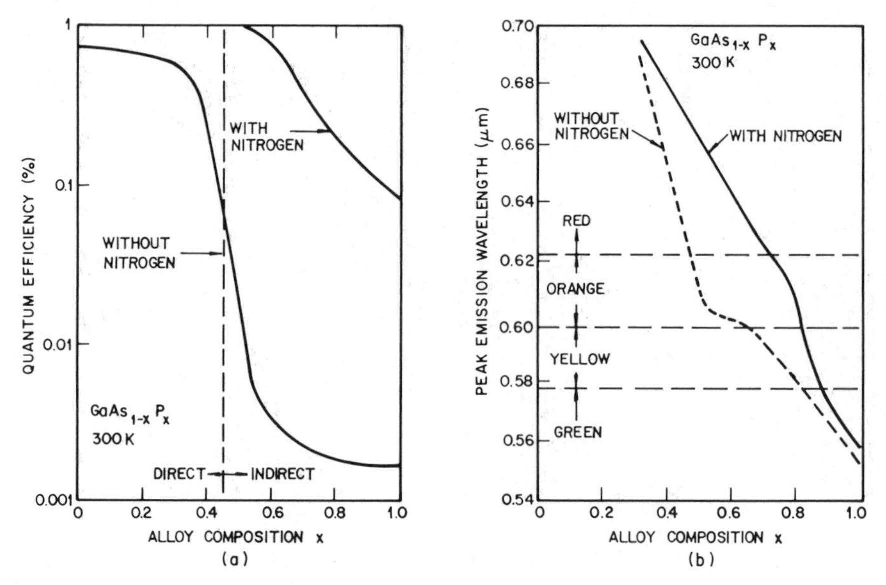

Fig. 9 (*a*) Quantum efficiency versus alloy composition. (*b*) Peak emission wavelength versus composition with and without isoelectronic impurity nitrogen. (After Groves, Herzog, and Craford, Ref. 31.)

normally neutral. In p-type material, an injected electron is first trapped at the center. The negatively charged center then captures a hole from the valence band to form the bound exciton. The subsequent annihilation of this electron–hole pair yields a photon with an energy equal to the bandgap minus an energy approximately equal to the binding energy of the center. Since the trapped electron is highly localized at the center, its momentum is diffused. Thus momentum can be conserved and the probability of direct transition is greatly enhanced. This radiative recombination mechanism is predominant in indirect bandgap materials, such as GaP.

Figure 9*a* shows the quantum efficiency versus alloy composition for $GaAs_{1-x}P_x$ with and without the isoelectronic impurity nitrogen.[31] The efficiency without nitrogen drops sharply in the composition range $0.4 < x < 0.5$ because of the proximity of the direct–indirect transition at $x = 0.45$. The efficiency with nitrogen is considerably higher for $x > 0.5$, but still decreases steadily with increasing x because of the increasing separation between the direct and indirect bandgap (Fig. 8*b*). The nitrogen-doped alloy also shows a shift of the peak emission wavelength, because of the binding energy of the isoelectronic center (Fig. 9*b*).

LED Configuration and Performance The basic LED structures are the flat-diode configurations shown[3] in Fig. 10. Generally, direct bandgap LEDs (red) are fabricated on GaAs substrates (Fig. 10*a*), and indirect bandgap LEDs (orange, yellow, and green) are fabricated on GaP sub-

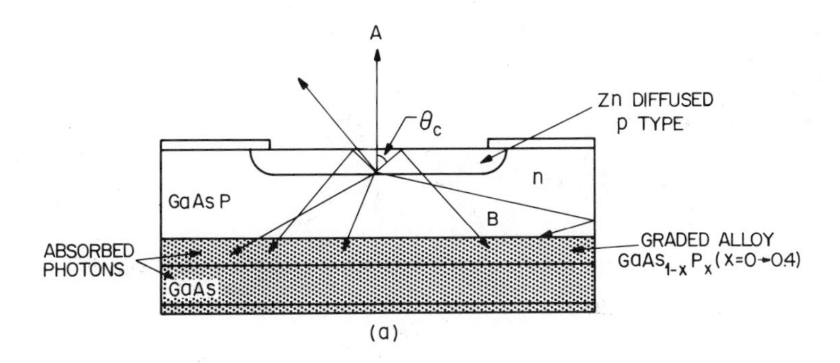

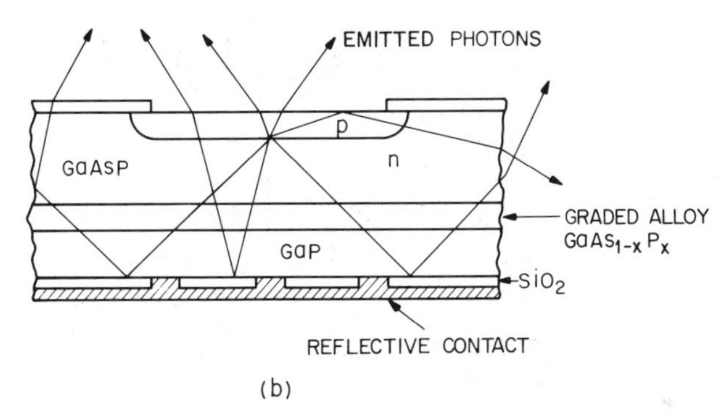

Fig. 10 Effects of (a) opaque substrate and (*b*) transparent substrate on photon emitted at the junction. (After Gage et al., Ref. 3.)

strates (Fig. 10*b*). When GaAs is used as the substrate, a graded-alloy $GaAs_{1-x}P_x$ with x varying from 0 to about 0.4 is epitaxially grown, and then followed by a layer of $GaAs_{1-x}P_x$ with a constant alloy composition (Fig. 10*a*). The graded region minimizes the nonradiative centers at the interface that result from lattice mismatch. The photons generated at the junction are emitted in all directions, but only a fraction of these photons can emerge from the surface to reach the eye of an observer.

Three loss mechanisms reduce the quantity of emitted photons: (1) absorption within the LED material, (2) fresnel loss, and (3) critical angle loss. The absorption loss for LEDs on GaAs substrates (Fig. 10*a*) is large since the substrate is opaque to light and absorbs about 85% of the photons emitted at the junction. For LEDs on GaP substrates (Fig. 10*b*), photons emitted downward can be reflected back with only about 25% absorption; the efficiency can be significantly improved. When photons pass from a medium with an index of refraction of $\bar{n}_2$ (e.g., GaAs with $\bar{n}_2 = 3.66$) to a medium of $\bar{n}_1$ (e.g., air with $\bar{n}_1 = 1.0$), a portion of the light will be reflected back to the medium interface. This loss of light is called fresnel loss, and the reflection coefficient for normal incident is

$$R = \left(\frac{\bar{n}_2 - \bar{n}_1}{\bar{n}_2 + \bar{n}_1}\right)^2. \tag{11}$$

The third loss mechanism is caused by the total internal reflection of photons incident to the surface at angles greater than the critical angle θ_c, defined by Snell's law,

$$\theta_c = \sin^{-1}(\bar{n}_1/\bar{n}_2). \tag{12}$$

For GaAs, the critical angle is about 16° and, for GaP with $\bar{n}_2 = 3.45$, the critical angle is about 17°.

For the flat-diode configuration (Fig. 10*a*), the A ray is attenuated by absorption and is partially reflected because of the fresnel loss. The B rays, which strike the semiconductor–air interface at angles $\theta \geq \theta_c$, are reflected internally. The total efficiency for electrical-to-optical conversion is given by[32]

$$\eta_F = \frac{q}{P}(1 - R)(1 - \cos\theta_c)\frac{\int \Phi(\lambda)(1 + R_1 e^{-2\alpha_1(\lambda)x_1})e^{-\alpha_2(\lambda)x_2}\,d\lambda}{\int \Phi(\lambda)\,d\lambda} \tag{13}$$

$$\approx \frac{4\bar{n}_2\bar{n}_1}{(\bar{n}_2 + \bar{n}_1)^2}(1 - \cos\theta_c) \tag{14}$$

where P is the power input, $4\bar{n}_2\bar{n}_1/(\bar{n}_2 + \bar{n}_1)^2$ is the transmission coefficient, which becomes $4\bar{n}_2/(\bar{n}_2 + 1)^2$ for light from the bulk semiconductor to air, $(1 - \cos\theta_c)$ is the solid cone, $\Phi(\lambda)$ is the photon generation rate in units of photon/s-cm^2, R_1 is the reflection coefficient at the back contact, and $\alpha(\lambda)$ and x are the absorption coefficient and thickness of the respective p- and n-type regions of the device.

Figure 11 shows the cross sections of other LEDs for (*a*) hemisphere, (*b*) truncated sphere, and (*c*) paraboloid geometries.[33] Similar expressions

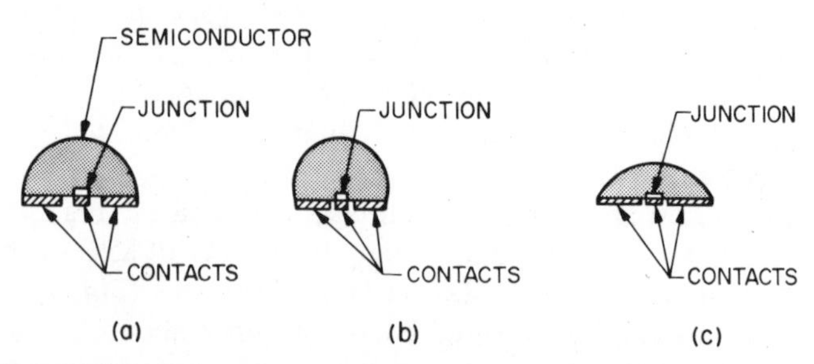

Fig. 11 Cross section of three LEDs. (*a*) Hemisphere. (*b*) Truncated sphere (Weierstrass source). (*c*) Paraboloid. (After Carr, Ref. 33.)

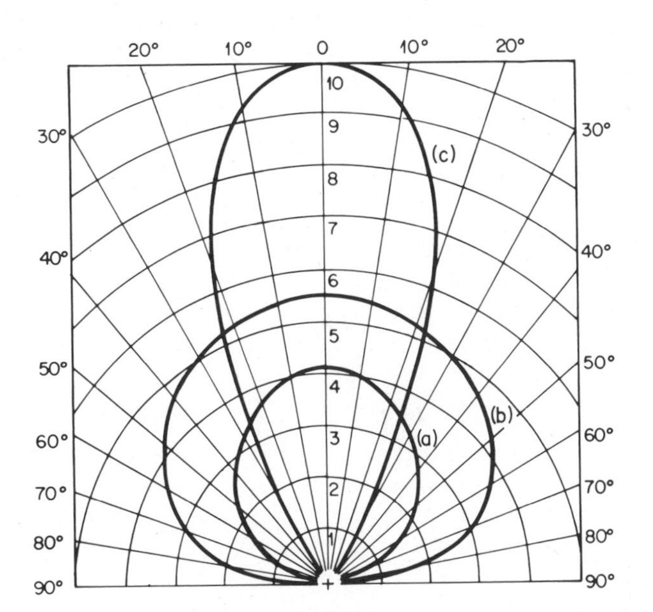

Fig. 12 Radiation patterns of LEDs: (*a*) rectangular and (*b*) hemispherical. (*c*) Parabolic geometry. (After Galginaitis, Ref. 34.)

for efficiency can be written for these geometries. The main difference between these three geometries and the flat diode geometry (Fig. 10*a*) is that the solid cone is unity. Thus the ratio is given by

$$\frac{\eta}{\eta_F} \approx \frac{1}{1-\cos\theta_c} = \frac{1}{1-\sqrt{1-1/\bar{n}_2^2}} \tag{15}$$

$$\eta/\eta_F = 2\bar{n}_2^2 \qquad \text{for} \quad \bar{n}_2 \gg 1. \tag{16}$$

For GaP with $\bar{n}_2 = 3.45$, an increase in efficiency by an order of magnitude is expected for these three LEDs (Fig. 11). Figure 12 shows the typical radiation patterns for the flat, hemispherical, and parabolic diodes.[34] It is apparent that the geometry of the device can be designed to give a desired radiation pattern.

Figure 13 shows the relative spectral responses at room temperature for some representative LEDs. The infrared LEDs of GaAs and III–V alloys will be considered in Section 12.3.2. Red LEDs can be made from the direct bandgap $GaAs_{0.6}P_{0.4}$. They can also be made from GaP doped with ZnO, where an adjacent gallium–phosphorus pair of atoms are replaced by a zinc–oxygen pair having the same total valence electrons.[2] The ZnO is another type of isoelectronic center. The resulting recombination level will lie about 0.3 eV below the GaP conduction band edge, giving rise to the 1.95-eV red light emission. LEDs for orange, yellow, and green[35] light are all made from indirect bandgap $GaAs_{1-x}P_x$ or GaP doped with isoelectronic

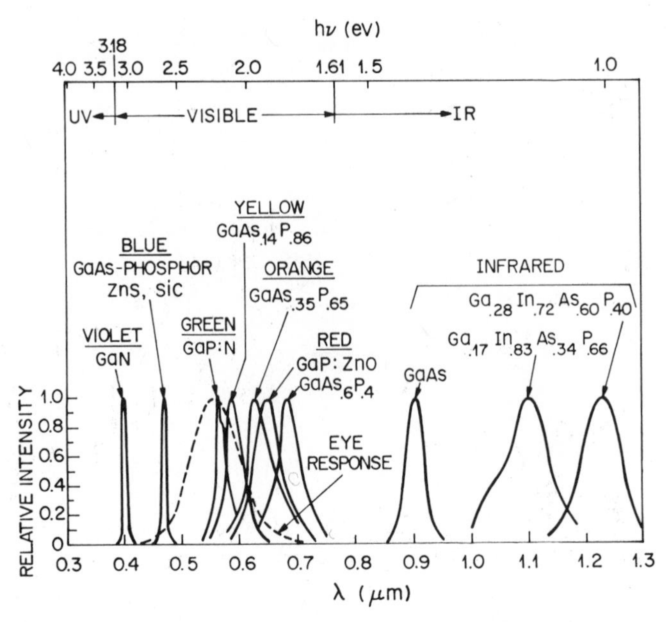

Fig. 13 Relative intensity versus wavelength for various visible and infrared LEDs.

impurity nitrogen. Blue LEDs can be obtained from ZnS, SiC,[36] and infrared-to-visible up-converters.[37] Figure 14*a* shows the basic configuration of an up-converter, where the infrared emission from a GaAs LED is absorbed by phosphor doped with rare-earth ions, such as ytterbium (Yb^{3+}) and erbium (Er^{3+}). The operation depends on the successive absorption of two photons in the infrared region followed by the emission of a single photon in the visible region (Fig. 14*b*). Figure 13 shows violet LEDs, obtained from GaN *p-n* junctions. Note that as the peak emission

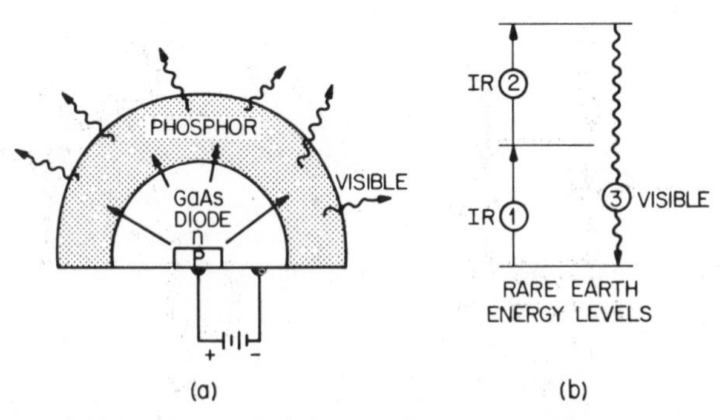

Fig. 14 (*a*) An up-converter in which GaAs emission excites light in an up-converting phosphor. (*b*) Energy level for up-conversion. (After Geusic et al., Ref. 37.)

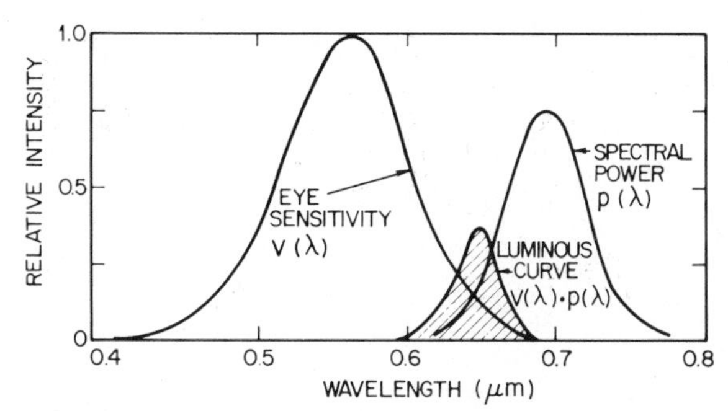

Fig. 15 Evaluation of luminous equivalent of radiation. (After Gooch, Ref. 29.)

wavelength λ_0 increases, the spectral width at half-power points also increases. This is because the spectral width for spontaneous emission is proportional to λ_0^2, as can be derived from Eq. 2.

Figure 13 also shows the eye response (dashed curve) with the peak sensitivity at 0.555 μm (green). As the wavelengths approach the red end of the visible spectrum, the sensitivity of the eye falls rapidly. When comparing the visual effects of LEDs with known emitting power, the eye response must be taken into account. For LEDs of high efficiency and broad spectral width, the luminous equivalent of the radiation is given by[29]

$$\text{luminous} = L_0 \int V(\lambda)p(\lambda)\, d\lambda \qquad \text{lm/W} \tag{17}$$

where L_0 is the peak luminosity of 680 lm/W, $V(\lambda)$ the relative sensitivity, and $p(\lambda)$ the spectrum of the radiation. For example, as shown in Fig. 15, the red emission from a GaP–ZnO diode has a peak intensity at 0.69 μm. When the spectral power curve is combined with the eye-sensitivity curve, a peak response at 0.65 μm is obtained with a luminous equivalent of 15 lm/W.

The visible LEDs can be used as indicator lamps and displays, and for opto-isolator applications.[3] Figure 16 shows schematic diagrams of various LED lamps.[2] An LED lamp contains an LED chip and a plastic lens, which is usually colored to serve as an optical filter and to enhance contrast. The lamps in Fig. 16*a* and *b* use conventional transistor and diode headers. Figure 16*c* uses a metal lead frame base containing the LED and a series resistor. A plastic lens determines the light emission pattern and the view angle of the device. Figure 16*d* is suited for a transparent semiconductor, such as GaP, which emits light through all five facets of the LED chip.

Figure 17 shows the basic formats for LED displays. The seven segments and the 3 × 5 array are usually used to display numbers from 0 to 9. For alphanumeric displays (0 to 9 and A to Z), the 14 segments and the 5 × 7 matrix array are used. The displays can be made by monolithic processes similar to those used to make silicon integrated circuits or by

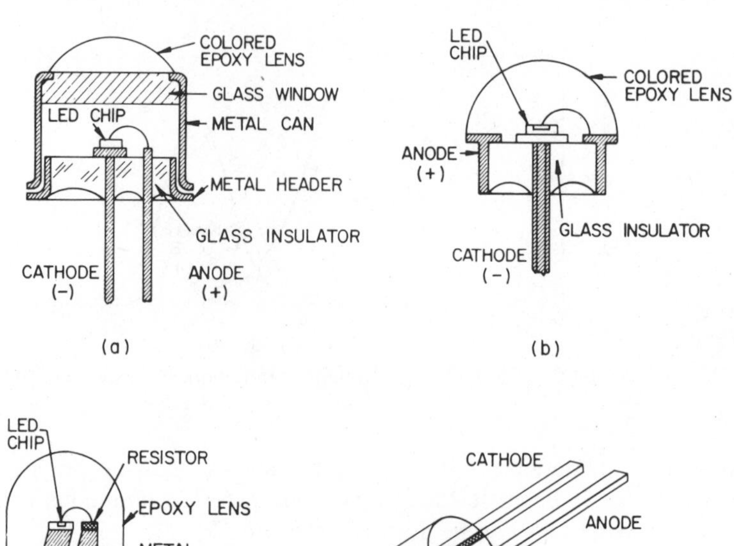

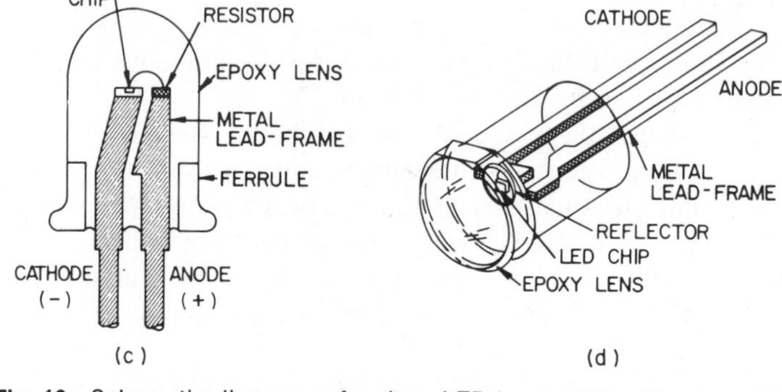

Fig. 16 Schematic diagrams of various LED lamps. (After Bergh and Dean, Ref. 2.)

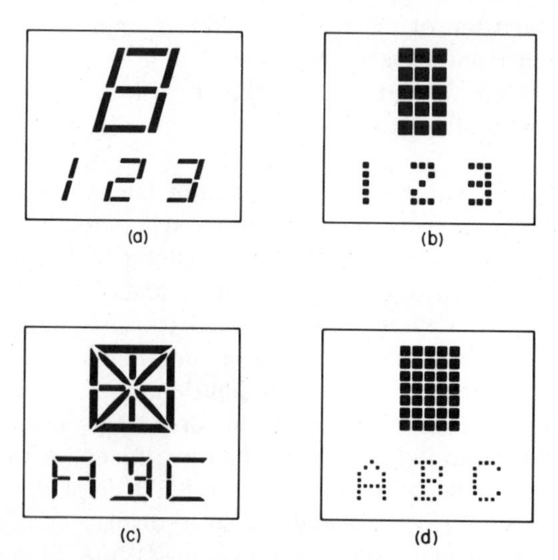

Fig. 17 LED display formats for numeric and alphanumeric. (*a*) 7 bars (numeric). (*b*) 3 × 5 array (numeric). (*c*) 14 bars (alphanumeric). (*d*) 5 × 7 array (alphanumeric).

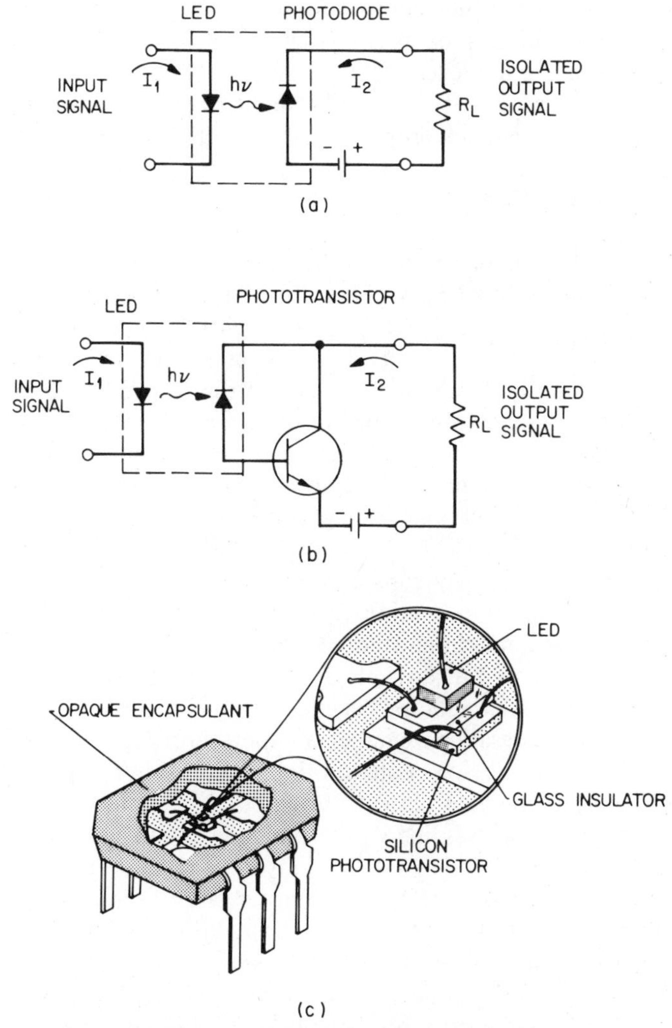

Fig. 18 (*a*) Opto-isolator. (*b*) High-gain opto-isolator. (*c*) Opto-isolator on single lead frame. (After Bergh and Dean, Ref. 2.)

using an individual LED chip mounted on a reflector to form a bar segment.

LEDs can also be used in opto-isolators where an input signal or control signal is decoupled from the output.[2,3] Figure 18*a* shows an opto-isolator having an LED as the light source and a photodiode as the detector. When an input electrical signal is applied to the LED, a light is generated and subsequently detected by the detector. The light is then converted back to an electrical signal as a current that can flow through a load resistor. The typical current transfer ratio I_2/I_1 is of the order of 10^{-3}. For a high-gain opto-isolator, Fig. 18*b*, a phototransistor is used as the detector and the current transfer ratio can be 0.1:1. Figure 18*c* shows an opto-isolator structure mounted in a single lead frame. These devices are optically coupled with the signal transmitted at the speed of light, and are electrically isolated because there is no feedback from the output to the input signal.

12.3.2 Infrared LEDs

Infrared LEDs can have various device configurations, similar to device configurations for visible LEDs (Figs. 10 and 11). Figure 13 shows the spectral response of a GaAs LED. This infrared source at the present time has the highest electroluminescent efficiency, mainly because GaAs has the most advanced material technology of all the direct bandgap semiconductors. An important application of the GaAs infrared LED is to use it as the source in an opto-isolator. Infrared LEDs have been made with many other direct bandgap semiconductors having an energy gap less than 1.5 eV. These include III–V ternary and quaternary solid solutions. Figure 13 shows examples of the infrared LEDs from a quaternary semiconductor $Ga_xIn_{1-x}As_{1-y}P_y$.[38] The energy gaps and lattice constants of these semiconductors will be discussed in Section 12.4.

Infrared LEDs are potential sources for optical-fiber communications. There are advantages and disadvantages to using LEDs as optical sources compared to using semiconductor lasers. The advantages of LEDs include higher-temperature operation, smaller temperature dependence of emitted power, simpler device construction, and simpler drive circuit. The disadvantages include lower brightness and lower modulation frequency, and wide spectral line width, typically 100 to 500 Å as compared to the narrow line width, 0.1 to 1 Å, of a laser.

The surface emitter and the edge emitter are the two basic device configurations to couple the LED light output into a small glass fiber.[39,40] For the surface emitter (Fig. 19*a*), the emitting area of the junction is confined by oxide isolation, and the contact area is usually 15 to 100 μm in diameter. The semiconductor through which the emission must be collected is made very thin, 10 to 15 μm, to minimize absorption and allow the end of the fiber to be very close to the emitting surface. The use of heterojunctions (e.g., GaAs–$Al_xGa_{1-x}As$) can increase the efficiency resulting from the carrier confinement provided by the layers of higher bandgap semiconductor (e.g., $Al_xGa_{1-x}As$) surrounding the radiative recombination region (e.g., GaAs). We will consider the carrier confinement in more detail in subsequent sections. The heterojunction can also serve as a window to the emitted radiation, because the higher bandgap confining layers do not absorb radiation from the lower bandgap emitting region.

For the edge emitter (Fig. 19*b*), the radiation can be emitted in a relatively direct beam and thus has the advantage of improved efficiency in coupling to a fiber with a small acceptance angle. The spatial distribution of

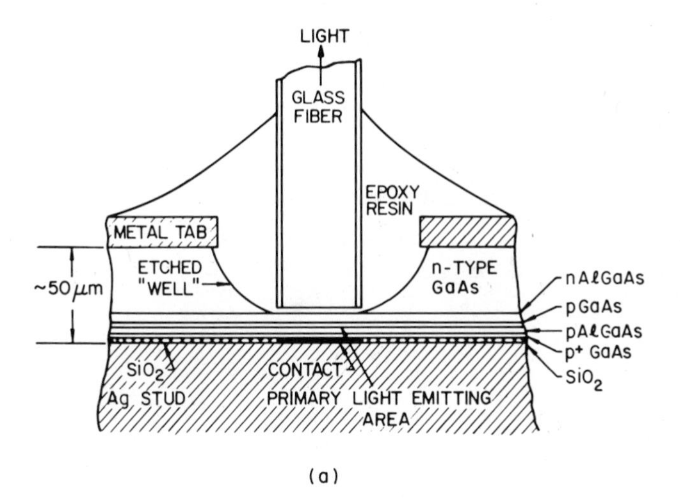

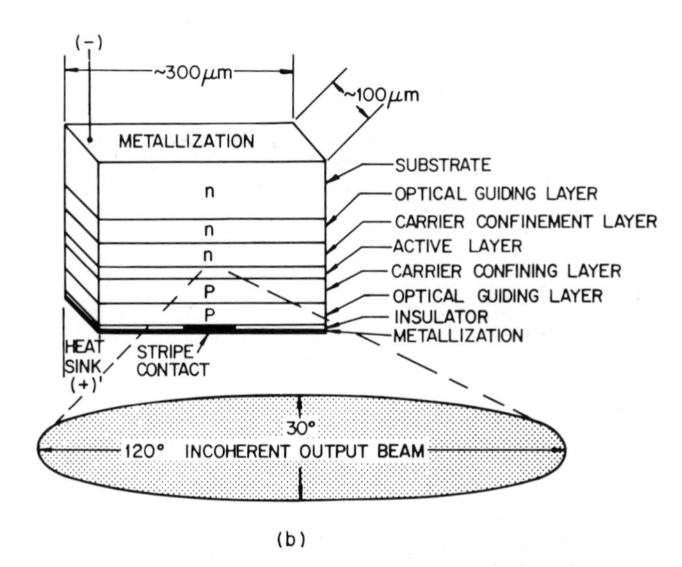

Fig. 19 (a) Small-area high-radiance AlGaAs double-heterostructure surface-emitter LED with attached fiber. (After Burrus and Miller, Ref. 39.) (*b*) Edge-emitter double-heterostructure LED. (After Ettenberg, Kressel, and Wittke, Ref. 40.)

the emitted light is similar to the distribution of a heterostructure laser, considered in Section 12.4.

The insert in Fig. 20 shows that the surface emitter can be made by confining the light-emitting area by contact resistance.[41] The diameter of the ohmic contact to the p^+ GaAs region is 30 μm. The surrounding Cr/Au metallization has high contact resistance to the p-AlGaAs (5×10^{16} cm^{-3}

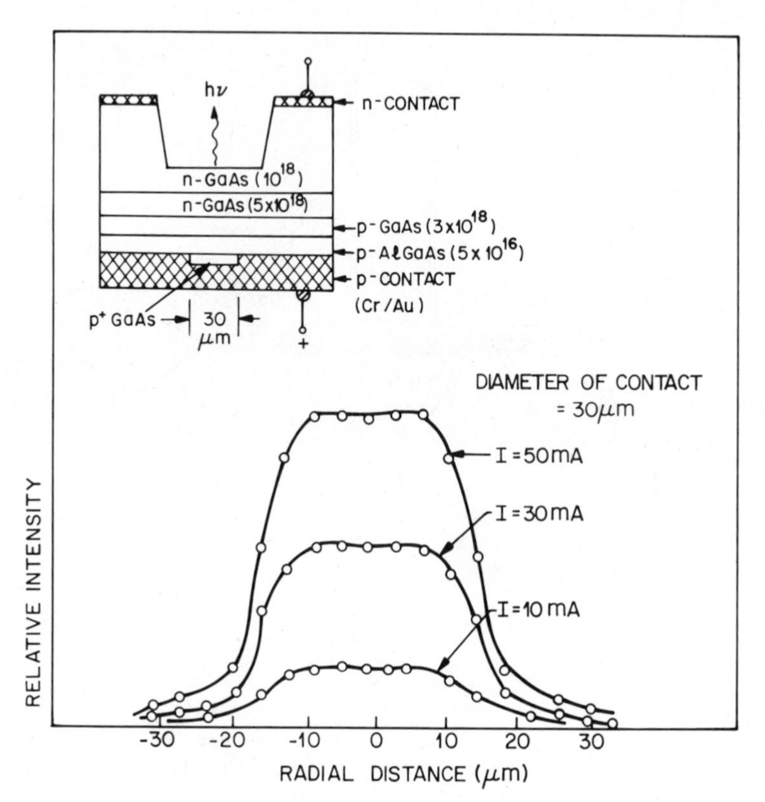

Fig. 20 Near-field intensity distribution of surface-emitter LED. The insert shows the cross section of the LED. (After Amann and Proebster, Ref. 41.)

doping). Therefore, most current will flow through the ohmic contact region. Figure 20 shows the light intensity distribution near the emitting surface. The light-emitting region is clearly confined to an area about 30 μm in diameter determined by the contact configuration.

The frequency response is another important parameter to be considered in the design of LEDs for optical communication systems. As given in Eqs. 7 and 10, the quantum efficiency and the overall lifetime are related to the radiative lifetime. With external excitation, the total radiative recombination is given by[41a]

$$R_r = Gnp/n_o p_o \tag{18}$$

where G is the total thermal generation rate. Frequently, this relation is written as

$$R_r = Bnp \tag{19}$$

where the radiative constant B is $G/n_o p_o$. When the excitation is sufficiently

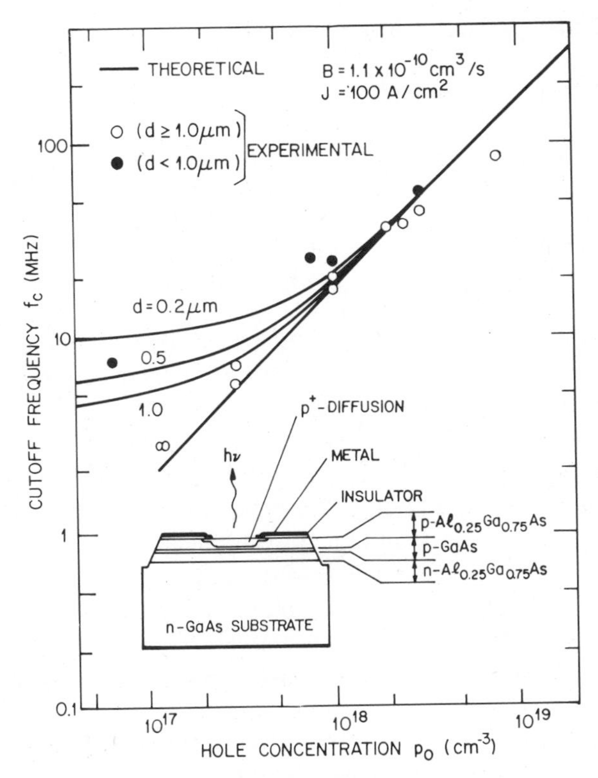

Fig. 21 Cutoff frequency versus hole concentration of active layer. (After Ikeda et al., Ref. 42.)

weak in p-type material so that $p \approx p_o$, the radiative lifetime becomes

$$\tau_r = (n - n_o)/R_r \simeq 1/Bp_o \tag{20}$$

or for n-type material with $n \approx n_o$,

$$\tau_r = (p - p_o)/R_r \simeq 1/Bn_o. \tag{21}$$

The cutoff frequency of an LED is defined as

$$f_c = \frac{1}{2\pi\tau}. \tag{22}$$

In Eq. 22, τ approaches τ_r when $\tau_r < \tau_{nr}$. Therefore, as Eqs. 20 and 21 illustrate, τ_r decreases as the doping is increased and f_c becomes larger. Figure 21 compares[42] the experimental and theoretical results for the cutoff frequency as a function of hole concentration for heterostructure surface-emitter LEDs illustrated in the insert. As can be seen, the measured cutoff frequency increases with increasing concentration p_o, in agreement with Eqs. 10 and 20. To obtain high f_c, thin recombination region and high hole concentration should be used.

12.4 SEMICONDUCTOR LASER PHYSICS

12.4.1 General Considerations

Semiconductor lasers are similar to other lasers (such as the solid-state ruby laser and He–Ne gas laser) in that the emitted radiation has spatial and temporal coherence. Laser radiation is highly monochromatic (of small bandwidth) and it produces highly directional beams of light. However, semiconductor lasers differ from other lasers in several important respects:

1 In conventional lasers, the quantum transitions occur between discrete energy levels, whereas in semiconductor lasers the transitions are associated with the band properties of materials.
2 A semiconductor laser is very compact in size (on the order of 0.1 mm long). In addition, because the active region is very narrow (on the order of 1 μm thick or less), the divergence of the laser beam is considerably larger than in a conventional laser.
3 The spatial and spectral characteristics of a semiconductor laser are strongly influenced by the properties of the junction medium (such as bandgap and refractive index variations).
4 For the *p-n* junction laser, the laser action is produced by simply passing a forward current through the diode itself. The result is a very efficient overall system that can be modulated easily by modulating the current. Since semiconductor lasers have very short photon lifetimes, modulation at high frequencies can be achieved.

Because of its compact size and capability for high-frequency modulation, the semiconductor laser is one of the most important light sources for optical-fiber communication. Figure 22 shows the loss characteristics

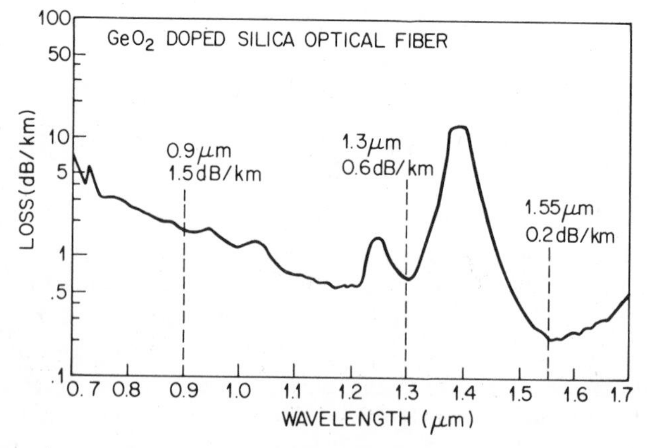

Fig. 22 Loss characteristics of a silica optical fiber. The three wavelengths of interest are also shown. (After Miya et al., Ref. 43.)

achieved in experimental fibers.[43] Three wavelengths of particular interest are also indicated on the figure. Around 0.9-μm wavelength, the GaAs–$Al_xGa_{1-x}As$ heterostructure lasers serve as the optical source, and the Si photodiode or Si avalanche photodiode is used as the optical detector. Around 1.3-μm wavelength the fiber has low loss (0.6 dB/km) and low dispersion; and around 1.55-μm wavelength, the loss reaches a minimum of 0.2 dB/km. For these two wavelengths, III–V quaternary compound lasers, such as $Ga_xIn_{1-x}As_yP_{1-y}$–InP lasers, are candidates for optical sources and Ge avalanche photodiodes as well as photodiodes in ternary or quaternary compounds are candidates for optical detectors.[44] Optical-fiber communications at longer wavelengths, such as 4 μm, have been considered.[45] To make such a system, the fiber losses should be extremely low, and efficient optical sources (e.g., $PbS_{0.1}Se_{0.9}$ LEDs)[77] and photodetectors must be developed for these long wavelengths.

Semiconductor Materials The list of semiconductor materials that have exhibited laser action has continued to grow. At present, virtually all the lasing semiconductors have direct bandgaps. This is expected since the radiative transition in a direct bandgap semiconductor is a first-order process (i.e., the momentum is automatically conserved); the transition probability is high. For indirect bandgap semiconductors, the radiative transition is a second-order process (i.e., it involves phonons or other scattering agents to conserve momentum and energy); thus the radiative transition is much weaker. Additionally, in indirect bandgap semiconductors, the free carrier loss due to injected carrier grows faster with excitation than the gain.[12]

Figure 23 shows the range of laser emission wavelengths for various

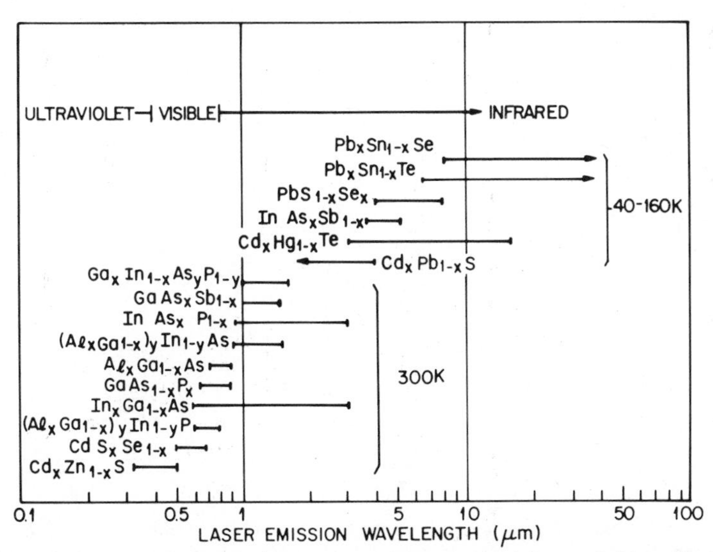

Fig. 23 Emission wavelengths either presently or potentially available with III–V and IV–VI heterostructure lasers. (After Casey and Panish, Ref. 20.)

semiconductors from near ultraviolet to far infrared.[20] GaAs was the first material to lase, and its related III–V solid solutions have been most extensively studied and developed. The IV–VI compounds, such as PbS, PbTe, PbSe, and related solid solutions, also exhibit laser action. They are direct bandgap materials with their extrema located along the ⟨111⟩ directions in the Brillouin zone,[46] in contrast to GaAs with extrema located at the zone center.

In this section, we consider basic laser physics and use examples mainly from devices of III–V compound semiconductors. When compound semiconductors are formed that have more than one group III element distributed randomly on group III lattice sites or more than one group V element distributed randomly on group V lattice sites, these compounds are crystalline solid solutions. The notation frequently used is $A_xB_{1-x}C$ for ternary and $A_xB_{1-x}C_yD_{1-y}$ for quaternary compounds, where A and B are the group III elements and C and D are the group V elements. Within each group the elements can be ordered alphabetically or according to their atomic numbers. Throughout this chapter, alphabetical order will be used within each group.

The two most important III–V compound systems are $Al_xGa_{1-x}As_ySb_{1-y}$ and $Ga_xIn_{1-x}As_yP_{1-y}$ solid solutions[20] (Fig. 24). The bandgap is plotted against the lattice constant for the III–V binary semiconductors and their intermediate ternary or quaternary compounds. To achieve heterostructures with negligible interface traps, the lattices between the two semiconductors must be closely matched. Using GaAs ($a = 5.6533$ Å) as the substrate, the ternary compound $Al_xGa_{1-x}As$ can have a lattice mismatch less than 0.1%. Similarly, using InP ($a = 5.8686$ Å) as the substrate, the quaternary compound $Ga_xIn_{1-x}As_yP_{1-y}$ can also have near perfect lattice match, as indicated by the vertical line in Fig. 24.

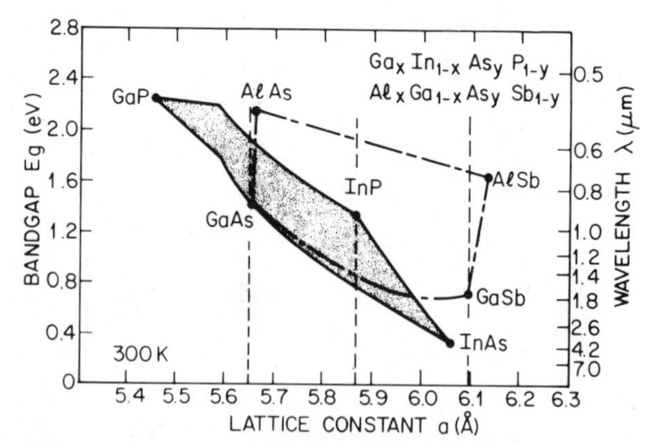

Fig. 24 Energy bandgap and lattice constant for two III–V solid solutions. (After Casey and Panish, Ref. 20.)

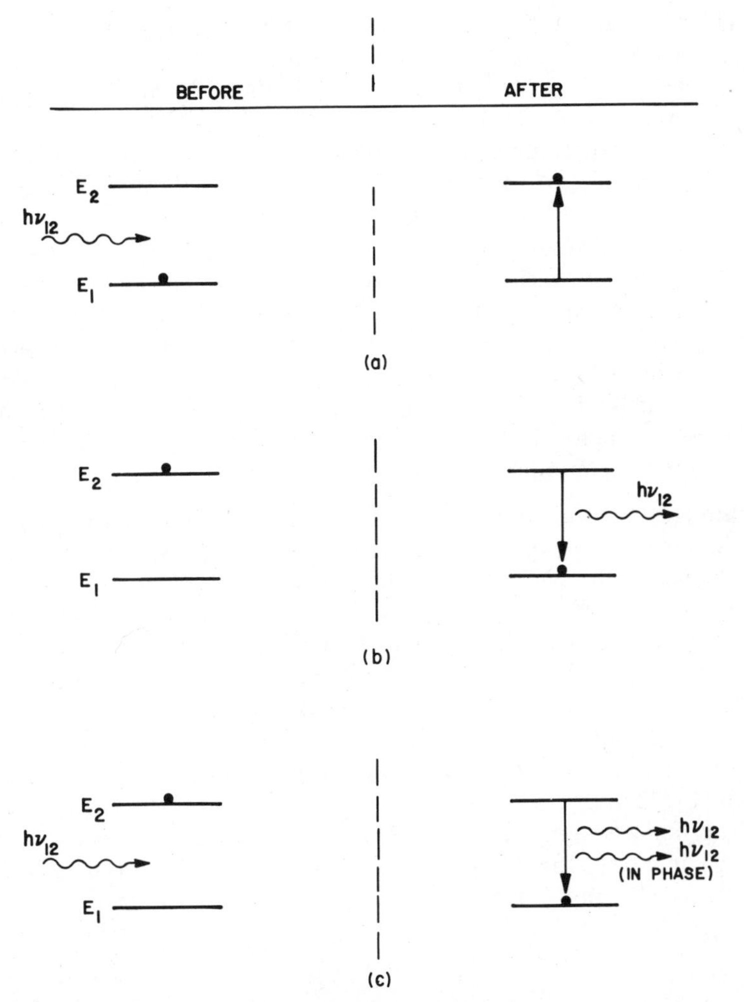

Fig. 25 The three basic transition processes between two energy levels E_1 and E_2. The black dots indicate the state of the atom. The initial state is at the left; the final state, after the process has occurred, is at the right. (*a*) Absorption. (*b*) Spontaneous emission. (*c*) Stimulated emission. (After Levine, Ref. 47.)

Stimulated Emission As mentioned previously, absorption, spontaneous emission, and stimulated emission are the three basic transition processes that relate to laser operation. We shall use a simple system to demonstrate these processes.[47] Consider two energy levels E_1 and E_2 in an atom, where E_1 is the ground state and E_2 is an excited state (Fig. 25). Any transition between these states involves the emission or absorption of a photon with frequency ν_{12} given by $h\nu_{12} = E_2 - E_1$, where h is Planck's constant. At ordinary temperatures, most of the atoms are in the ground state. This situation is disturbed when a photon of energy exactly equal to $h\nu_{12}$ impinges on the system. An atom in state E_1 absorbs the photon and thereby goes to the excited state E_2. This is the absorption process, Fig. 25*a*. The excited state of the atom is unstable and after a short time, without any external stimulus, it makes a transition to the ground state giving off a photon of energy $h\nu_{12}$. This process is called spontaneous emission, Fig. 25*b*. The lifetime for spontaneous emission (i.e., the average time of the excited state) varies considerably ranging typically from 10^{-9} to 10^{-3} s depending on various semiconductor parameters such as bandgap (direct or indirect) and density of recombination centers. An important and interesting event occurs when a photon of energy $h\nu_{12}$ impinges on an atom while it is still in the excited state. In this case, the atom is immediately stimulated to make its transition to the ground state and gives off a photon of energy $h\nu_{12}$ which is in phase with the incident radiation. This process is called stimulated emission, Fig. 25*c*.

Double Heterostructure Laser Figure 26 shows the basic structure of a *p-n* junction laser. A pair of parallel planes are cleaved or polished perpendicular to the plane of the junction. The two remaining sides of the diode are roughened to eliminate lasing in directions other than the main one. The structure is called a Fabry–Perot cavity. When a forward bias is applied to the laser diode, a current flows. Initially at low current, there is spontaneous emission in all directions. As the bias is increased, eventually a threshold current is reached at which the stimulated emission occurs and a monochromatic and highly directional beam of light is emitted from the junction (Fig. 26).

For the homostructure (e.g., GaAs *p-n* junction), the threshold current density J_{th} increases rapidly with increasing temperature. A typical value of J_{th} (obtained by pulse measurement) is about 5.0×10^4 A/cm^2 at room

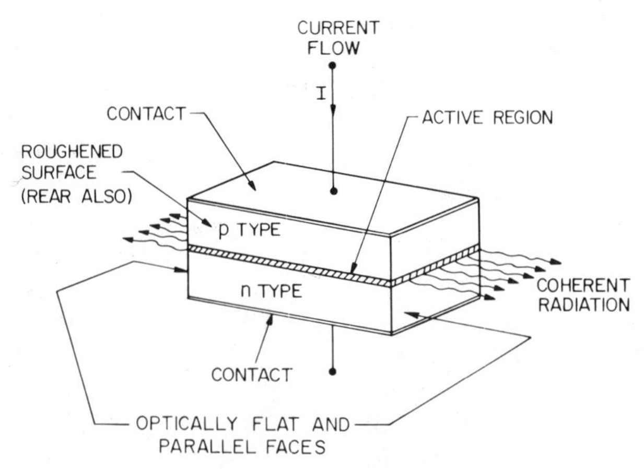

Fig. 26 Basic structure of a junction laser in the form of a Fabry–Perot cavity.

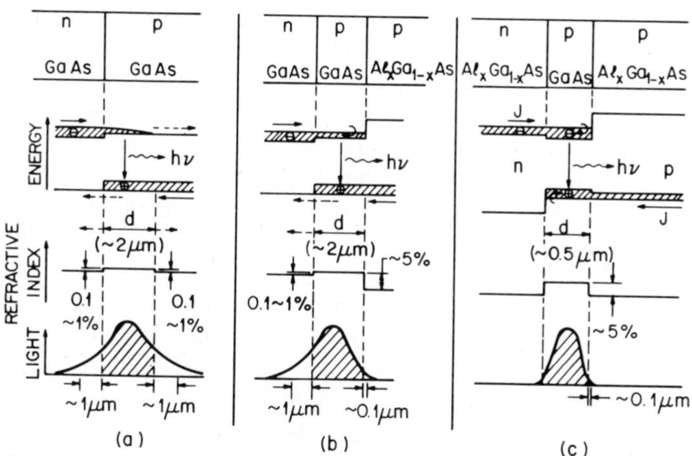

Fig. 27 Comparison of some characteristics of (*a*) homostructure, (*b*) single-heterostructure, and (*c*) double-heterostructure lasers. The top row shows energy-band diagrams under forward bias. The refractive index change for GaAs/$Al_xGa_{1-x}As$ is about 5%. The change across a homostructure is less than 1%. The confinement of light is shown in the bottom row. (After Panish, Hayashi, and Sumski, Ref. 48.)

temperature. Such a large current density imposes serious difficulties in operating the laser continuously at 300 K.

To reduce the threshold current density, heterostructure lasers have been proposed and built, using the epitaxial techniques. Figure 27 shows schematic representations of the bandgap under forward-biased conditions,[48] the refractive index change, and the optical field distribution of

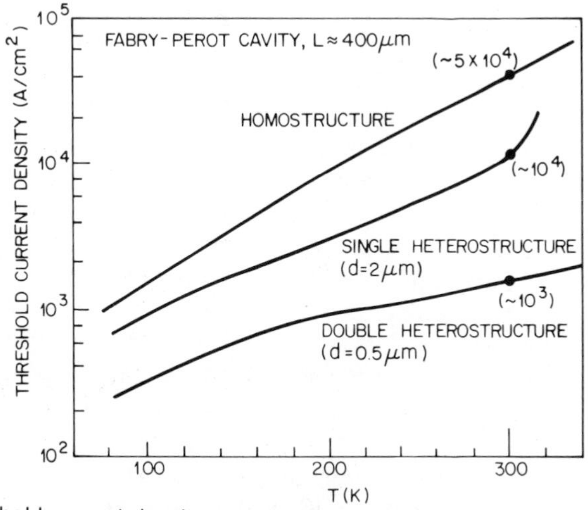

Fig. 28 Threshold current density versus temperature for three laser structures in Fig. 27. (After Panish, Hayashi, and Sumski, Ref. 48.)

light generated at the p-n junction for a homostructure, a single heterostructure, and a double heterostructure. As can be seen in the double heterostructure, the carriers are confined in the active region d by the heterojunction potential barriers on both sides, and the optical field is also confined within the active region by the abrupt reduction of the refractive index outside the active region. These confinements can enhance the stimulated emission and substantially reduce the threshold current density.

Figure 28 compares J_{th} versus operating temperature for the three structures in Fig. 27. Note that the temperature dependence is the least for the double heterostructure (DH) laser. Since J_{th} for DH lasers can be as low as 10^3 A/cm^2 or less at 300 K, continuous room-temperature operation has been achieved. This achievement has led to increased applications of semiconductor lasers in science and technology, especially for optical-fiber communication systems. In the remainder of this chapter, we will be concerned primarily with DH lasers.

12.4.2 Waveguiding

Confinement Factor In a DH laser, the light is confined and guided by the dielectric waveguide. Figure 29*a* shows a three-layer dielectric waveguide with refractive indices, $\bar{n}_1$, $\bar{n}_2$, and $\bar{n}_3$, where an active layer is sandwiched between two inactive layers. Under the condition

$$\bar{n}_2 > \bar{n}_1 \geq \bar{n}_3 \tag{23}$$

the ray angle θ_{12} at the layer 1/layer 2 interface in Fig. 29*b* exceeds the critical angle given by Eq. 12. A similar situation for θ_{23} occurs at the layer 2/layer 3 interface. Therefore, when the refractive index in the active region is larger than the index of its surrounding layers, Eq. 23, the propagation of electromagnetic radiation is guided in a direction parallel to the layer interfaces.

For the homostructure laser, the difference in the refractive index between the center waveguiding layer and the adjacent layers is only 0.1 to about 1%. For heterostructure lasers, the refractive index steps at each heterojunction can be made larger (~10%) and provide a well-defined waveguide.

To derive the detailed waveguiding properties, refer to Fig. 30, which defines an (x, y, z) Cartesian coordinate system relative to the laser. The planes $z = 0$ and $z = -L$ coincide with the two reflecting ends or the laser "mirrors," which are either cleaved or polished surfaces. The front mirror can radiate into the half-space $z > 0$. The transverse coordinates x and y coincide with the directions perpendicular and parallel to the junction plane, respectively.

Consider a symmetric three-layer dielectric waveguide with $\bar{n}_2 > \bar{n}_1 = \bar{n}_3$ (Fig. 29). For transverse electric (TE) waves polarized transversely to the direction of propagation (z direction), $\mathscr{E}_z$ equals 0. The waveguide is

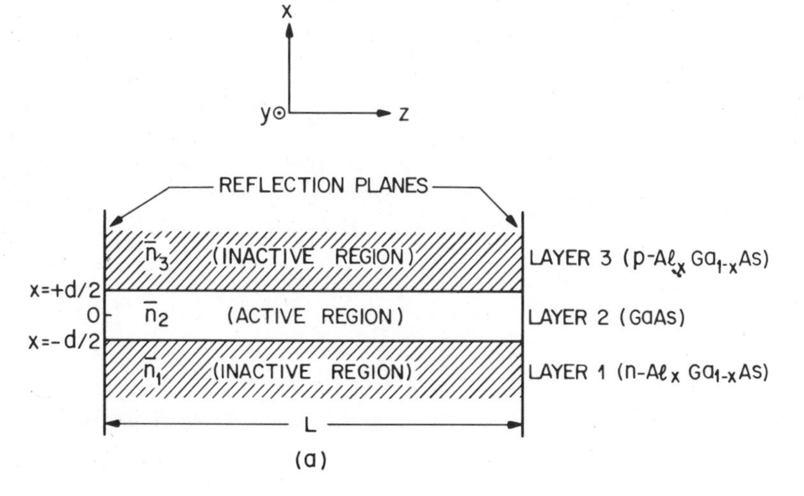

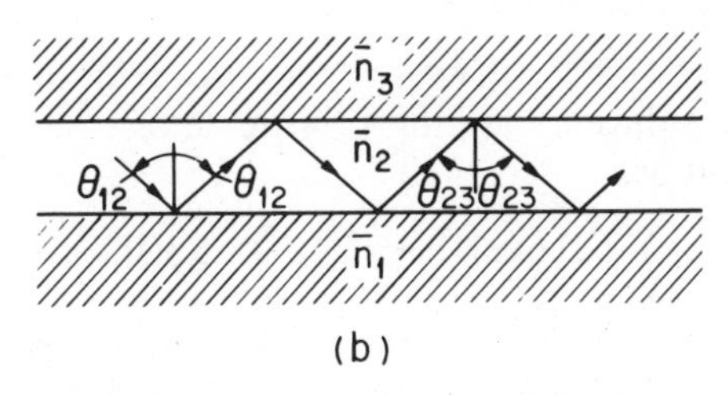

Fig. 29 (*a*) Representation of a three-layer dielectric waveguide. (*b*) Ray trajectories of the guided wave.

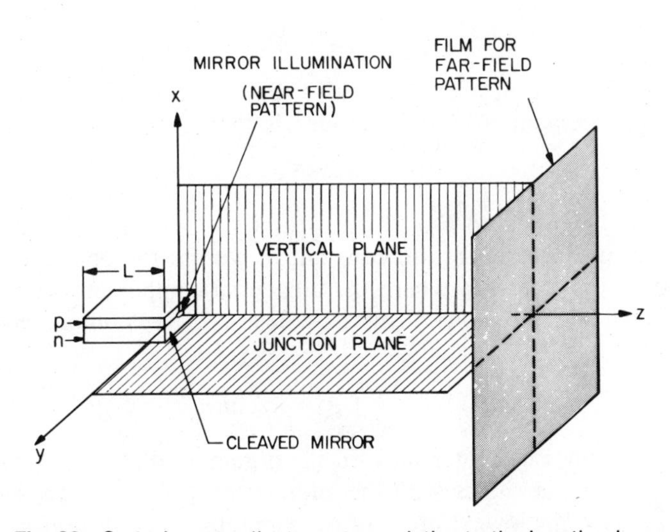

Fig. 30 Cartesian coordinate system relative to the junction laser.

considered to extend to infinity in the y direction, so that $\partial/\partial y = 0$. The wave equation is simplified to

$$\partial^2 \mathscr{E}_y/\partial x^2 + \partial^2 \mathscr{E}_y/\partial z^2 = \mu_0 \epsilon\, \partial^2 \mathscr{E}_y/\partial t^2 \tag{24}$$

where μ_0 is the permeability and ϵ is the dielectric permittivity. The solution (by separation of variables) of Eq. 24 for even TE waves within the active layer $-d/2 < x < d/2$ is given by

$$\mathscr{E}_y(x, z, t) = A_e \cos(\kappa x) \exp[j(\omega t - \beta z)] \tag{25}$$

with

$$\kappa^2 = \bar{n}_2^2 k_0^2 - \beta^2 \tag{26}$$

where $k_0 \equiv (\omega/\bar{n}_2)\sqrt{\mu_0 \epsilon}$ and β is the separation constant. The magnetic field in the z direction is given by

$$\begin{aligned} \mathscr{H}_z(x, z, t) &= (j/\omega\mu_0)/(\partial \mathscr{E}_y/\partial x) \\ &= (-j\kappa/\omega\mu_0) A_e \sin(\kappa x) \exp[j(\omega t - \beta z)]. \end{aligned} \tag{27}$$

Outside the active layer, the field must decay in order to have guided waves. For $|x| > d/2$, the solutions for the transverse electric field and the longitudinal magnetic field are

$$\mathscr{E}_y(x, z, t) = A_e \cos(\kappa d/2) \exp[-\gamma(|x| - d/2) \exp[j(\omega t - \beta z)] \tag{28}$$

and

$$\begin{aligned} \mathscr{H}_z(x, z, t) = (-x/|x|)(j\gamma/\omega\mu_0) A_e \cos(\kappa d/2) \exp[-\gamma(|x| - d/2) \\ \times \exp[j(\omega t - \beta z)] \end{aligned} \tag{29}$$

where

$$\gamma^2 \equiv \beta^2 - \bar{n}_1^2 k_0^2. \tag{30}$$

Since both κ and γ must be positive real numbers, Eqs. 26 and 30 show that the requirement for guided modes is that $\bar{n}_2 k_0^2 > \beta^2$ and $\beta^2 > \bar{n}_1^2 k_0^2$, or

$$\bar{n}_2 > \bar{n}_1. \tag{31}$$

This result is identical to Eq. 23.

To determine the separation constant β, we use the boundary condition at the dielectric interface where the tangential component of the magnetic field $\mathscr{H}_z$ must be continuous. From Eqs. 27 and 29, we obtain the eigenvalue equation

$$\tan(\kappa d/2) = \gamma/\kappa = [(\beta^2 - \bar{n}_1^2 k_0^2)/(\bar{n}_2^2 k_0^2 - \beta^2)]^{1/2}. \tag{32}$$

The solution for Eq. 32 depends on the argument of the tangent function, which has multiple values with the addition of $2\pi m$ (m is an integer). For $m = 0$, we have the lowest-order or fundamental mode. For $m = 1$, we have the first-order mode, and so on. Once the number is specified, Eq. 32 can be solved numerically or graphically. The result can then be used in Eqs. 25 through 29 for the electric and magnetic fields.

We now define a confinement factor Γ, which is the ratio of the light intensity within the active layer to the sum of light intensity both within and outside the active layer. Since the light intensity is given by the Poynting vector $\mathscr{E} \times \mathscr{H}$, which is proportional to $|\mathscr{E}_y|^2$, the confinement factor for the symmetrical three-layer dielectric waveguide can be obtained from Eqs. 25 and 28 for the even TE waves:

$$\Gamma = \frac{\int_0^{d/2} \cos^2(\kappa x)\, dx}{\int_0^{d/2} \cos^2(\kappa x)\, dx + \int_{d/2}^{\infty} \cos^2(\kappa d/2) \exp[-2\gamma(x - d/2)]\, dx}$$

$$= \left\{1 + \frac{\cos^2(\kappa d/2)}{\gamma[(d/2) + (1/\kappa)\sin(\kappa d/2)\cos(\kappa d/2)]}\right\}^{-1}. \quad (33)$$

Similar expressions may be obtained for odd TE waves as well as for the transverse magnetic (TM) waves. The confinement factor is frequently used because it represents the fraction of the energy of the propagating waves within the active layer.

To date, the most extensively studied heterostructure laser is in the GaAs–$Al_xGa_{1-x}As$ system. The energy bandgap of $Al_xGa_{1-x}As$ as a function of Al composition is shown in Fig. 31*a*. The alloy has direct bandgap up to $x \simeq 0.45$, then becomes an indirect bandgap semiconductor. For heterostructure lasers, the composition region $0 < x < 0.35$ is of most interest where the direct energy gap can be expressed as[20]

$$E_g(x) = 1.424 + 1.247x. \quad \text{(eV)} \quad (34)$$

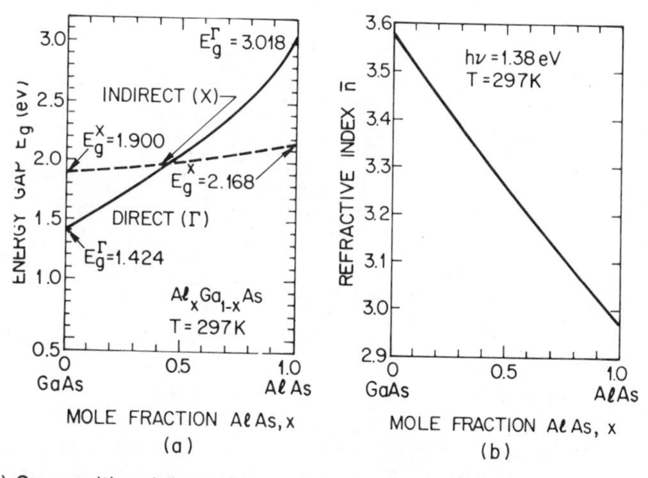

Fig. 31 (*a*) Compositional dependence of an $Al_xGa_{1-x}As$ energy gap. (*b*) Compositional dependence of the refractive index at 1.38 eV. (After Casey and Panish, Ref. 20.)

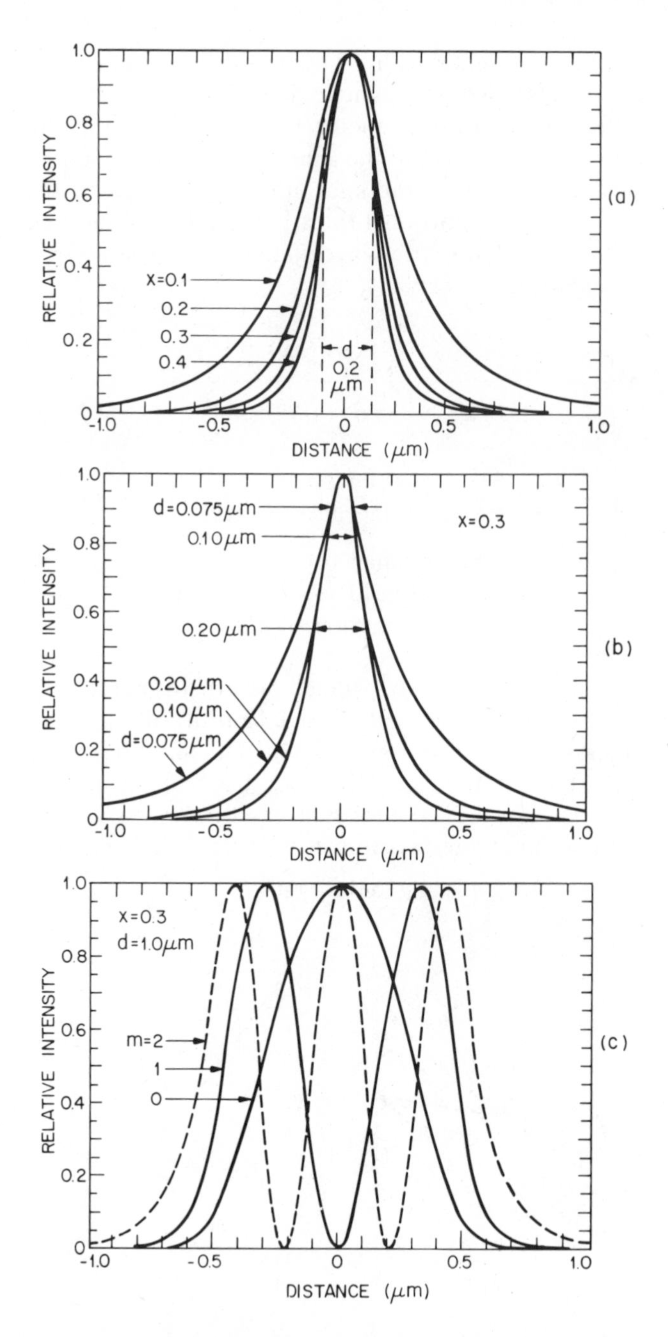

Fig. 32 Square of the electric field as a function of position within the double-heterostructure waveguide. (*a*) Variation of intensity for $d = 0.2\ \mu m$ and for different AlAs mole fractions. (*b*) Variation of intensity for $x = 0.3$ and for different d. (*c*) Intensity distribution for fundamental, first-, and second-order modes with the indicated composition and active-layer thickness. (After Casey and Panish, Ref. 20.)

Figure 31*b* shows the compositional dependence of the refractive index, which can be represented by

$$\bar{n}(x) = 3.590 - 0.710x + 0.091x^2. \tag{35}$$

For example, for $x = 0.3$ the bandgap of $Al_{0.3}Ga_{0.7}As$ is 1.798 eV, which is 0.374 eV larger than that of GaAs; its refractive index 3.385 is about 6% smaller than that of GaAs.

Figure 32*a* illustrates the influence of the composition on the optical intensity $|\mathscr{E}_y|^2$ in the direction perpendicular to the junction plane for the three-layer dielectric waveguide $Al_xGa_{1-x}As$–GaAs–$Al_xGa_{1-x}As$. The curves are calculated from Eqs. 25 and 32 for a wavelength of 0.90 μm (1.38 eV) and for the fundamental mode ($m = 0$). The active layer thickness d of 0.2 μm is held constant while the composition is varied. A significant increase in confinement occurs when x is increased from 0.1 to 0.2. Figure 32*b* shows the variation of confinement with d for $x = 0.3$. As the active layer becomes smaller, the light spreads farther into the $Al_{0.3}Ga_{0.7}As$, and less of the total intensity is within the active layer. For larger d, where higher-order modes are permitted, Fig. 32*c* shows that as the mode order increases, more of the light is outside the active region. Therefore, to improve the optical confinement, lower mode order is preferred.

Figure 33 shows the variation of the confinement factor Γ for the fundamental mode with alloy composition and d. It can be seen that Γ decreases rapidly for $d < \lambda/\bar{n}_2$ ($\simeq 0.5$ μm), where the active layer thickness becomes less than the wavelength of the radiation. Representing the fraction of the propagating mode within the active layer by Γ is an important concept for understanding the influence of the active layer thickness on the threshold current density.

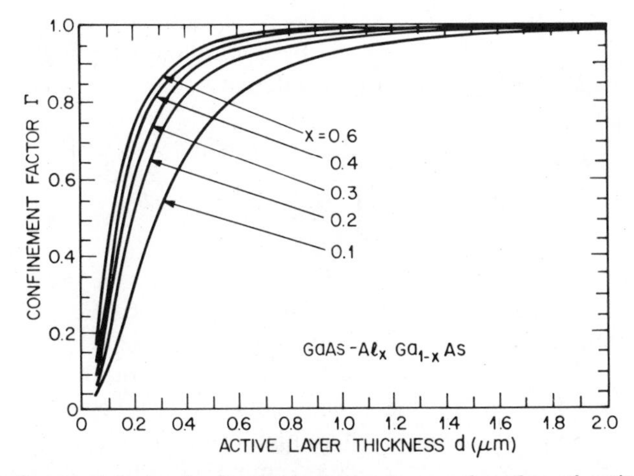

Fig. 33 Confinement factor for fundamental mode as a function of active-layer thickness and alloy composition for a GaAs–$Al_xGa_{1-x}As$ symmetric three-layer dielectric waveguide. (After Casey and Panish, Ref. 20.)

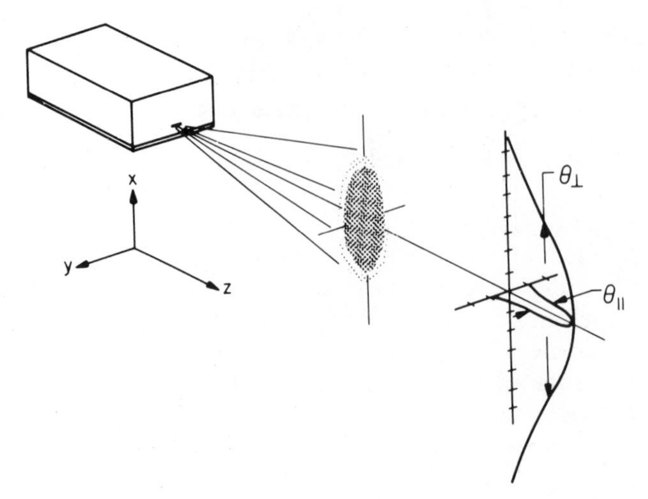

Fig. 34 Schematic representation of far-field emission of a stripe-geometry double-heterostructure laser. The full angles at half power, perpendicular to and along the junction plane, are also indicated. (After Casey and Panish, Ref. 20.)

Far-Field Pattern The far-field pattern is the intensity of the emitted radiation in free space. Figure 34 gives a schematic representation of the far-field emission of a DH laser. The full angles at half power are $\theta_\perp$ and $\theta_\parallel$ in the directions perpendicular to and along the junction plane, respectively. Typically $\theta_\parallel$ is of the order of 10°, whereas $\theta_\perp$ is considerably larger (35 to 65°), depending on the active layer thickness and alloy composition.

The far-field pattern can be obtained by first considering the TE waves in free space for $z > 0$. The wave equation is identical to Eq. 24, except that ϵ is replaced by ϵ_0 for free space. Using separation of variables and the boundary condition that $\mathscr{E}_y(x, z)$ must be continuous at $z = 0$, the far-field intensity at an angle $\theta_\perp$ relative to the intensity at $\theta_\perp = 0$ can be obtained:

$$\frac{I(\theta_\perp)}{I(0)} = \frac{\cos^2\theta_\perp \left| \int_{-\infty}^{\infty} \mathscr{E}_y(x, 0) \exp(j \sin\theta_\perp k_0 x)\, dx \right|^2}{\left| \int_{-\infty}^{\infty} \mathscr{E}_y(x, 0)\, dx \right|^2}. \tag{36}$$

For the symmetrical three-layer waveguide (DH laser), the electric-field expression of Eqs. 25 and 28 can be substituted into Eq. 36. Figure 35 shows the calculated and measured full angles at the half power $\theta_\perp$ for the far-field pattern.[49] The solid and dashed curves are the beam divergence calculated from Eq. 36 for the fundamental mode. The dashed portion represents active layer thickness where high-order modes are possible. The experimental data points are in good agreement with the calculations. For a typical active layer thickness of 0.2 μm in a GaAs–$Al_{0.3}Ga_{0.7}As$ DH laser, the angle $\theta_\perp$ is about 50°.

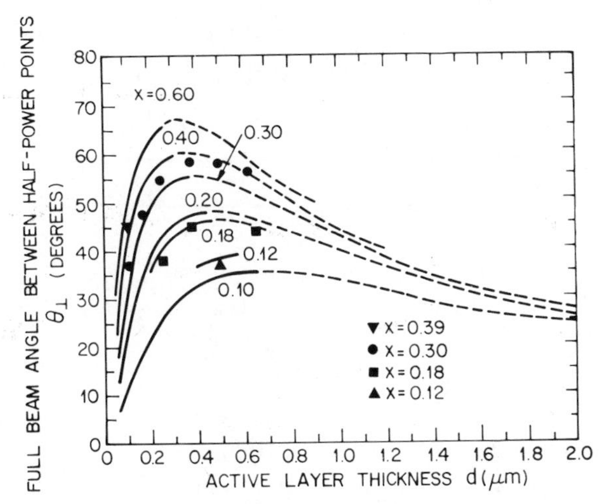

Fig. 35 Full angle at half power as a function of active-layer thickness and composition of GaAs–AlGaAs lasers. (After Casey, Panish, and Merz, Ref. 49.)

The basic laser structure in Fig. 26 is a broad-area laser since the area along the junction plane can emit radiation. Most heterostructure lasers are made in stripe geometries; Figure 36 shows two representative examples.[50] In Fig. 36*a* the oxide layer isolates all but the stripe contact, and the lasing area is restricted to a narrow region under the contact. In Fig. 36*b*, the stripe-geometry laser is fabricated by proton bombardment which produces high-resistivity regions. The lasing area is restricted to the center region which is not bombarded. The stripe widths are typically 5 to approximately 30 μm. The advantages of the stripe geometry include (1) reduction of the cross-section area, which reduces the operating current; (2) elimination of the occurrence of more than one localized high-optical intensity area (called a filament) as the biasing current increases [the stripe geometry can have single-filament operation and fundamental-mode emission along the

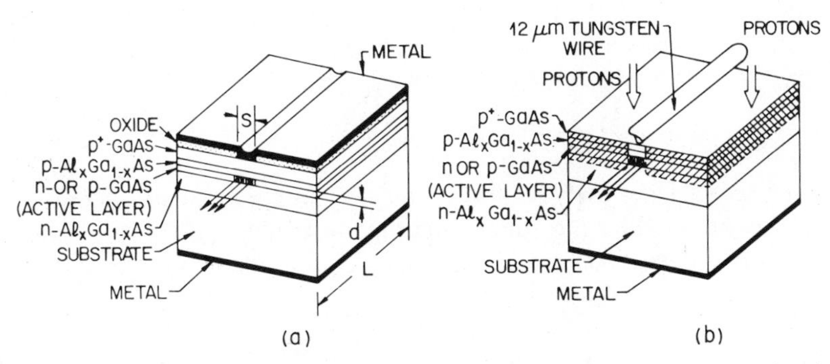

Fig. 36 Stripe-geometry DH lasers; (*a*) oxide isolation and (*b*) proton bombardment. (After D'Asaro, Ref. 50.)

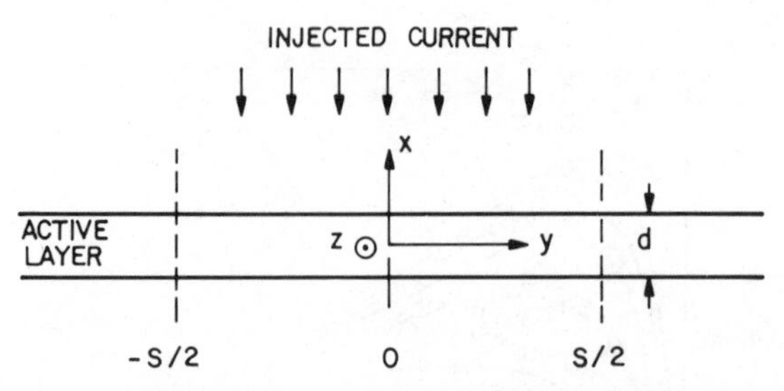

Fig. 37 Coordinate system for a stripe-geometry laser with stripe width S defined by proton bombardment.

junction plane for sufficiently narrow stripe widths ($S \lesssim 15\ \mu m$)]; (3) improved reliability by removing most of the junction perimeter from the surface; and (4) improved response time, owing to small junction capacitance.

For the stripe-geometry lasers, the electric field intensity along the direction parallel to the junction plane (y direction) is strongly influenced by the spatial variation of the dielectric permittivity. For the structure shown in Fig. 37, the wave equation with a sinusoidal time dependence given by $\exp(j\omega t)$ is[51]

$$\nabla^2 \mathscr{E}_y + (k_0^2 \epsilon/\epsilon_0)\mathscr{E}_y = 0. \tag{37}$$

In this equation, k_0 equals $2\pi/\lambda$ and ϵ/ϵ_0 is taken as two-dimensional, with the form

$$\epsilon(x, y)/\epsilon_0 = [\epsilon(0) - a^2 y^2]/\epsilon_0 \tag{38}$$

within the active layer, and

$$\epsilon(x, y)/\epsilon_0 = \epsilon_1/\epsilon_0 \tag{39}$$

in the adjacent inactive layers. In Eq. 38, $\epsilon(0)$ is the complex dielectric permittivity $\epsilon_r(0) + j\epsilon_i(0)$ at $y = 0$ in the active layer, and a is a complex constant represented by $a_r + ja_i$. An approximate solution of Eq. 37 having dielectric permittivity given by Eqs. 38 and 39 is

$$\mathscr{E}_x(x, y, z) = \mathscr{E}_y(x)\mathscr{E}_y(y)\exp(-j\beta_z z). \tag{40}$$

Since $\epsilon(x, y)$ varies slowly with y along the junction plane, $\mathscr{E}_y(x)$ is not significantly affected by the confinement along y and can be represented by the previously derived expressions, Eqs. 25 and 28. From Eq. 37 by the separation of the variable,

$$[\partial^2 \mathscr{E}_y(x)/\partial x^2] + \beta_x^2 \mathscr{E}_y(x) = 0. \tag{41}$$

Substituting Eqs. 40 and 41 into Eq. 37 and eliminating $\mathscr{E}_y(x)$ by multiplying its complex conjugate and integrating over x yields a differential equation for $\mathscr{E}_y(y)$:

$$[\partial^2\mathscr{E}_y(y)/\partial y^2] + \{k_0^2[\Gamma\epsilon(0)/\epsilon_0 + (1-\Gamma)\epsilon_1/\epsilon_0] - \beta_x^2 - \beta_z^2 - (\Gamma k_0^2 a^2 y^2/\epsilon_0)\}\mathscr{E}_y(y) = 0 \tag{42}$$

where Γ is the confinement factor.

The field distributions for $\mathscr{E}_y(y)$ represented by Eq. 42 are Hermite–Gaussian functions given by

$$\mathscr{E}_y(y) = H_p[(\Gamma^{1/2}ak_0/\epsilon_0^{1/2})^{1/2}y]\exp[-\tfrac{1}{2}(\Gamma/\epsilon_0)^{1/2}ak_0y^2] \tag{43}$$

where H_p is the Hermite polynomial of order p, which is given by

$$H_p(\xi) \equiv (-1)^p \exp(\xi^2)\partial^p \exp(-\xi^2)/\partial\xi^p \tag{44}$$

and the first three Hermite polynomials are $H_0(\xi) = 1$, $H_1(\xi) = 2\xi$, and $H_2(\xi) = 4\xi^2 - 2$. Therefore, the intensity for the fundamental mode is Gaussian and is given by

$$|\mathscr{E}_y(y)|^2 = \exp[-(\Gamma/\epsilon_0)^{1/2}a_rk_0y^2] \tag{45}$$

which demonstrates that the intensity distribution along the junction plane is influenced by a_r.

Figure 38 shows the near-field and far-field patterns along the junction

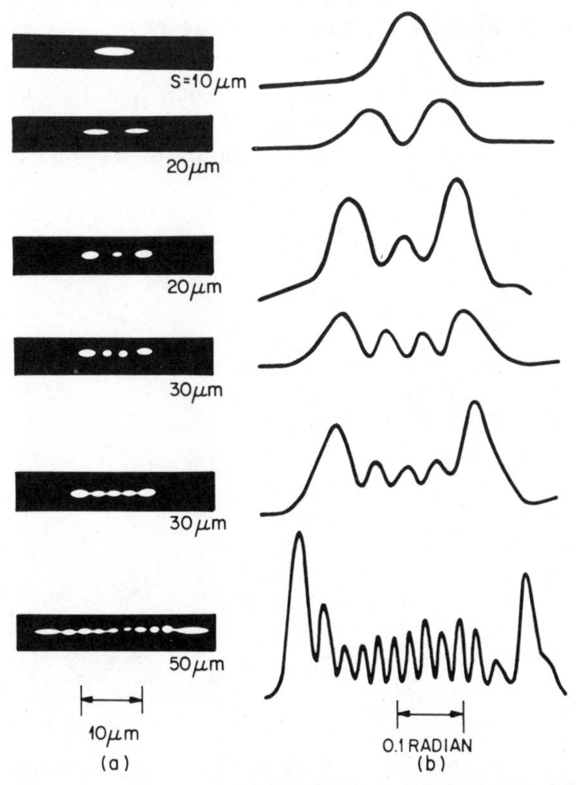

Fig. 38 Modes along the junction plane as a function of stripe width S for planar-stripe DH lasers. (a) Near-field patterns. (*b*) Far-field patterns. (After Yonezu et al., Ref. 52.)

plane for stripe-geometry lasers.[52] For a stripe width of $10\,\mu$m, a fundamental Gaussian mode distribution exists. As the stripe width increases, higher-order modes along the junction plane are observed. These modes are characteristic of the Hermite–Gaussian distribution represented by Eq. 43.

12.4.3 Threshold Current Density

Under thermal equilibrium more atoms are in the ground states than in the excited state. A population is said to be inverted if the opposite is true. If photons of energy $h\nu_{12}$ are incident on a simple system (refer to Fig. 25), where the population of level E_2 is inverted with respect to E_1, stimulated emission exceeds absorption and more photons of energy $h\nu_{12}$ leave the system than enter it. Such a phenomenon is called quantum amplification.

To consider the inversion condition for a semiconductor laser, refer to Fig. 39, which shows the energy versus density of states in a direct bandgap semiconductor.[53] Figure 39*a* shows the equilibrium condition at $T = 0$ K for an intrinsic semiconductor in which the shaded area represents the filled states. Figure 39*b* shows the situation for an inverted population at 0 K. This inversion can be achieved, for example, by photoexcitation with photon energy greater than the bandgap E_g. The valence band is empty of electrons down to an energy E_{FV}, and the conduction band filled up to E_{FC}. Now photons with energy $h\nu$ such that $E_g < h\nu < (E_{FC} - E_{FV})$ will cause downward transition and hence stimulated emission.

At finite temperatures, the carrier distributions will be "smeared" out in

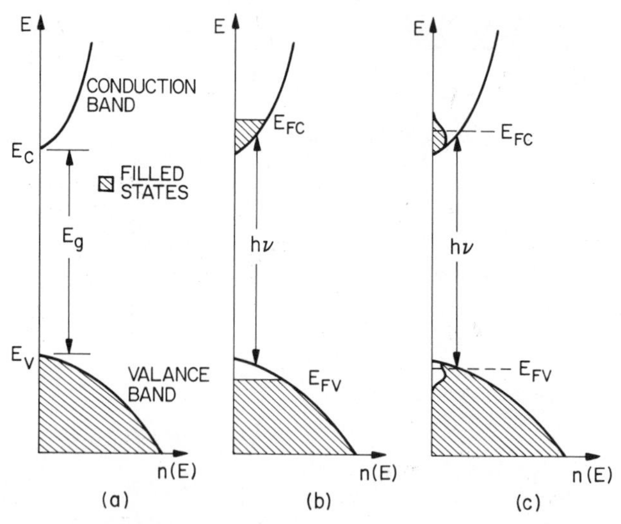

Fig. 39 Energy versus density of states in a semiconductor. (*a*) Equilibrium, $T = 0$ K. (*b*) Inverted, $T = 0$ K. (*c*) Inverted, $T > 0$ K. (After Nathan, Ref. 53.)

energy (Fig. 39*c*). Although overall thermal equilibrium does not exist, the carriers in a given energy band will be in thermal equilibrium with each other. The occupation probability of a state in the conduction band is given by the Fermi–Dirac distribution

$$F_C(E) = \frac{1}{1 + \exp\left(\dfrac{E - E_{FC}}{kT}\right)} \tag{46}$$

where E_{FC} is the imref (quasi-Fermi level) for electrons in the conduction band. A similar expression holds for the valence band.

Consider the rate of photon emission at $h\nu$ due to a transition from a group of upper states near E in the conduction band to lower states at $(E - h\nu)$ in the valence band. The rate for this emission is proportional to the product of the density of occupied upper states $n_C(E)F_C(E)$ and the density of unoccupied lower states $n_V(E - h\nu)[1 - F_V(E - h\nu)]$. The total emission rate is obtained by integrating over all energies,

$$W_{\text{spont}}(h\nu) = B \int n_C(E) n_V(E - h\nu) F_C(E)[1 - F_V(E - h\nu)] |\langle M \rangle|^2 \, dE. \tag{47}$$

In a similar manner we can write

$$W_{\text{absorption}}(h\nu) = B \int n_V(E - h\nu) n_C(E) F_V(E - h\nu)[1 - F_C(E)] |\langle M \rangle|^2 \, dE. \tag{48}$$

for the absorption rate. The coefficient B is given by

$$B = (4\pi \bar{n} q^2 h\nu / m^2 \epsilon_0 h^2 c^3) \text{ Vol} \tag{49}$$

where $\langle M \rangle$ is the matrix element and Vol is the volume of the crystal. For a net amplification we require that $W_{\text{spont}} > W_{\text{absorption}}$. From Eqs. 46, 47, and 48 with the appropriate imrefs E_{FV} and E_{FC}, we obtain[11]

$$(E_{FC} - E_{FV}) > h\nu. \tag{50}$$

Equation 50 is a necessary condition for stimulated emission to be dominant over absorption.

In a semiconductor laser, the gain g, that is, the incremental energy flux per unit length, depends on the energy-band structure and is a complicated function of doping levels, current density, temperature, and frequency. As the excitation rate is increased, the distribution functions $F_C(E)$ and $F_V(E)$ change, that is, E_{FC} increases and E_{FV} decreases. The shape of the gain curve versus photon energy also changes.

The gain can be calculated for a special distribution of density of states,[55] where both the conduction band and valence band have band tails (Fig. 40). This distribution is applicable to most GaAs DH lasers. Once the distribution of density of states is specified, for a given difference $\Delta E \equiv (E_{FC} - E_{FV})$ and a given temperature, the spontaneous function can be

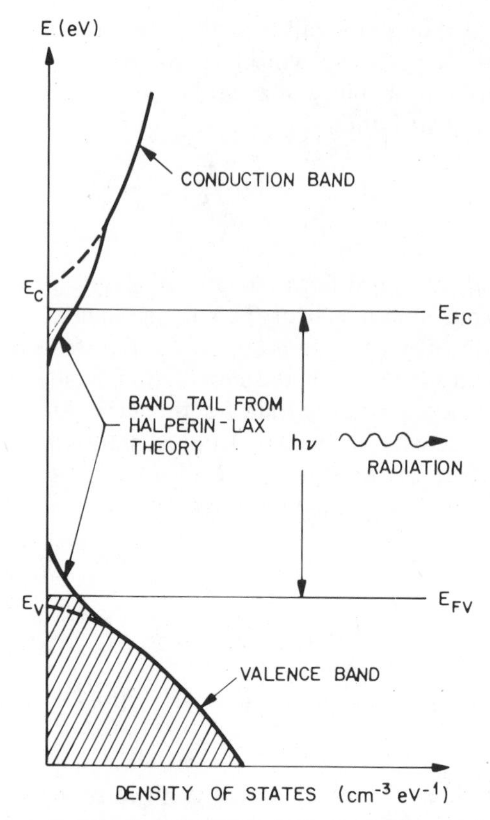

Fig. 40 Energy versus density of states where both conduction and valence bands have band tails. (After Halperin and Lax, Ref. 55.)

calculated from Eq. 47, and the gain can be obtained as a function of ΔE and $g = -\alpha$. In presenting the results, $g(h\nu)$ can also be given as a function of a nominal current density $J_{\rm nom}$, which is defined[56] for unity quantum efficiency ($\eta = 1$) as the current density required to uniformly excite a 1-μm-thick active layer. The actual current density is then given by

$$J(\mathrm{A/cm^2}) = J_{\rm nom} d/\eta. \tag{51}$$

Figure 41 shows the calculated gain for 5×10^{17}-cm^{-3} ionized acceptors with 1×10^{17}-cm^{-3} ionized donors in GaAs.[56] The gain is superlinear at low values and increases linearly with $J_{\rm nom}$ for $50 \lesssim g \lesssim 400\ \mathrm{cm^{-1}}$. The linear dashed line represents the gain which may be written as

$$g = (g_0/J_0)(J_{\rm nom} - J_0) \tag{52}$$

where $(g_0/J_0) = 5 \times 10^{-2}$ cm-μm/A and $J_0 = 4.5 \times 10^3$ A/cm^2-μm.

As discussed previously, at low current there is spontaneous emission in all directions. As the current is increased, the gain increases (Fig. 41), until

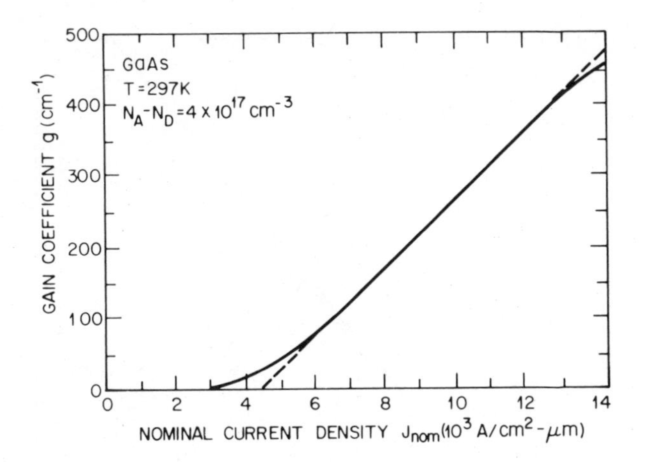

Fig. 41 Variation of the gain coefficient with the nominal current density. The dashed line represents a linear dependence. (After Stern, Ref. 56.)

the threshold for lasing is reached; that is, the gain satisfies the condition that a light wave makes a complete traversal of the cavity without attenuation:

$$R \exp[(\Gamma g - \alpha)L] = 1 \tag{53}$$

or

$$\Gamma g(\text{threshold gain}) = \alpha + \frac{1}{L} \ln\left(\frac{1}{R}\right) \tag{54}$$

where Γ is the confinement factor, α is the loss per unit length from free-carrier absorption and defect-center scattering, L is the length of the cavity, and R is the reflectance of the ends of the cavity (if the reflectances of the ends are different, $R = \sqrt{R_1 R_2}$). Equations 51, 52, and 54 may be combined to give the threshold current density as[57]

$$J_{th}(\text{A/cm}^2) = \frac{J_0 d}{\eta} + \frac{J_0 d}{g_0 \eta \Gamma}\left[\alpha + \frac{1}{L} \ln\left(\frac{1}{R}\right)\right]. \tag{55}$$

To reduce the threshold current density, one can increase η, Γ, L, and R, and reduce d and α. Figure 42 compares the calculated J_{th} from Eq. 55 to experimental results.[57] The threshold current density decreases with decreasing d reaching a minimum, and then increases again. The increase of J_{th} at very narrow active layer thickness is caused by the poor confinement factor. For a given d, J_{th} decreases with increasing AlAs composition x because of the improved confinement factor. Similar results have been obtained for InP–$Ga_xIn_{1-x}As_yP_{1-y}$–InP DH lasers.[58,59] For these

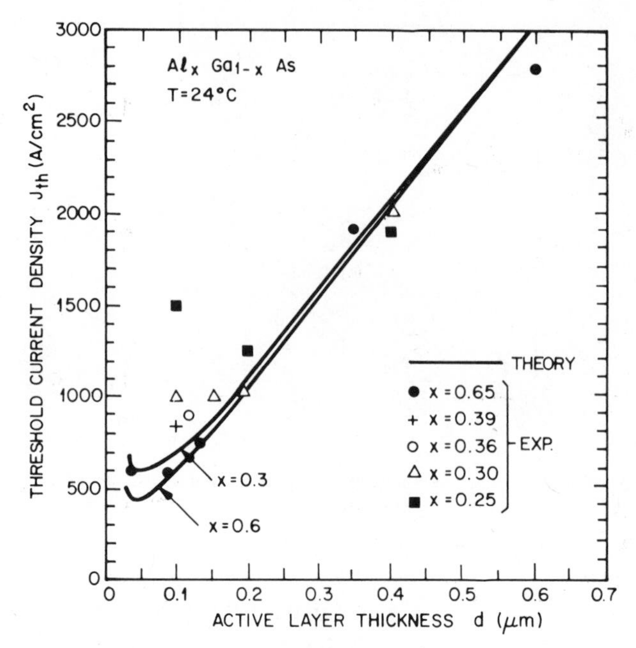

Fig. 42 Comparison of experimental J_{th} and J_{th} calculated from Eq. (55). (After Casey, Ref. 57.)

lasers with an active layer of $Ga_{0.25}In_{0.75}As_{0.54}P_{0.46}$ ($\lambda = 1.23\ \mu m$), the room-temperature J_{th} reaches a minimum value of about 1.5 kA/cm^2 at an active layer thickness near 0.2 μm.

12.5 LASER OPERATING CHARACTERISTICS

12.5.1 Device Structures

The heterostructure lasers have low threshold current density at room temperature because of (1) the carrier confinement provided by the energy barriers of higher bandgap semiconductor surrounding the active region, and (2) the optical confinement provided by the abrupt reduction of the refractive index outside the active region. In Section 12.4 we considered DH lasers with active layer thickness for carrier confinement identical to the thickness for optical confinement.

Several other configurations of heterostructure lasers exist.[60] Figure 43*a* shows the separate confinement heterostructure (SCH) laser. The energy bandgap, the refractive index, and the light intensity perpendicular to the junction plane are plotted. The step in E_g between GaAs and $Al_{0.1}Ga_{0.9}As$ is sufficient to confine the carriers within the GaAs layer d, but the step in $\bar{n}$

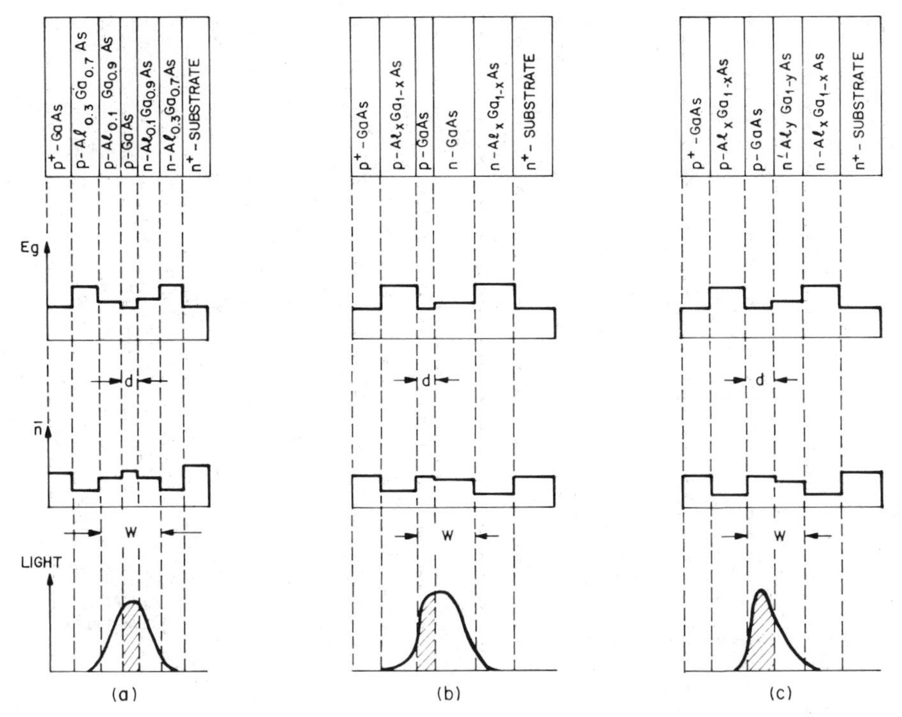

Fig. 43 Schematic representation of energy bandgap, refractive index, and light intensity for three heterostructure lasers. (*a*) Separate confinement. (*b*) Large optical cavity. (*c*) Four-layered. (After Burrus, Casey, and Li, Ref. 60.)

does not confine the light. However, the larger step in $\bar{n}$ between $Al_{0.3}Ga_{0.7}As$ and $Al_{0.1}Ga_{0.9}As$ confines the light and thereby provides the optical waveguide of width W. Low J_{th} (~ 500 A/cm^2) has been obtained from such a structure.

The large optical-cavity (LOC) heterostructure laser is similar to SCH except a p-n homojunction is sandwiched between two heterojunctions (Fig. 43b). Most of the junction current is due to the injection of electrons into the p-layer, which is the active region. The p-GaAs/p-$Al_xGa_{1-x}As$ heterojunction provides both carrier and optical confinement; the n-GaAs/n-$Al_xGa_{1-x}As$ heterojunction provides optical confinement. The advantage of LOC is the increase of optical waveguide thickness W without causing a large increase in J_{th} as for the DH laser. Since the optical waveguide cross section can be made relatively large, greater power output is available than from DH lasers.

Figure 43c shows a four-layered heterostructure laser, in which the corresponding n-GaAs layer is replaced by a slightly higher bandgap n'-$Al_yGa_{1-y}As$ layer with y approximately equal to 0.01 to 0.02. This structure can be operated in the fundamental transverse mode while

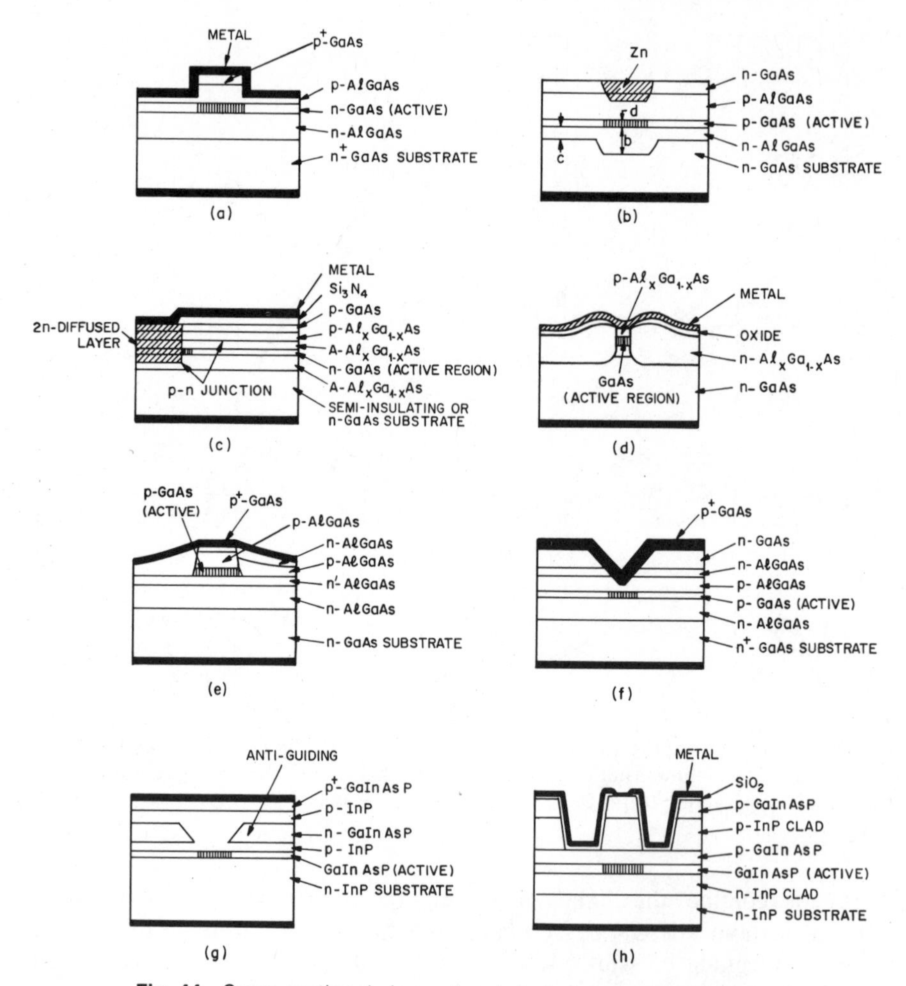

Fig. 44 Cross-sectional views of various heterostructure lasers.

maintaining a relatively low J_{th}, and it has high-power output with a relatively narrow beam divergence.

Most heterostructures have stripe geometries similar to those shown in Fig. 36. Figure 44 gives the cross sections of some other laser structures. All structures use high-quality binary semiconductors as the substrates (e.g., GaAs or InP). Successive epitaxial layers with different alloy compositions and dopings are grown by liquid-phase epitaxy, vapor-phase epitaxy, or molecular-beam epitaxy processes. The thin active layer and the small emission area are indicated in the figure for each laser structure.

Figure 44*a* shows a stripe-geometry laser with simplified processes.[61] The active stripe width is made by etching. After applying metal contacts, the device is completed by cleaving to form the Fabry–Perot cavity. Oxide isolation and proton bombardment are not required. Lateral current confinement is achieved by the high contact resistance between the metal and the low-doped *p*-AlGaAs. This laser has low J_{th}, linear light-current characteristics, and fundamental transverse mode with single longitudinal emission at 0.861 μm. (See Section 12.5.2 for longitudinal modes). Figure 44*b* shows a channeled substrate structure.[62] The channel is first formed in the substrate prior to epitaxy growth. This structure also shows good laser characteristics. Figure 44*c* shows a Zn-diffused structure,[63] called the transverse-junction-stripe (TJS) geometry laser. Since the injection in this case is from the *n*-GaAs layer to the Zn-diffused *p*-GaAs, this device is actually a very thin homojunction laser. Both single-longitudinal-mode operation and low threshold current have been obtained.

For the buried-heterostructure laser,[64] the active region is completely surrounded by higher-bandgap lower-refractive-index materials (Fig. 44*d*). This laser has an active region with a cross section as small as 1 μm^2. Threshold current as low as 15 mA has been obtained, and a nearly symmetrical far-field pattern has been observed. For the stripe-buried heterostructure,[65] the regular five-layer structure with appropriate Al concentration and doping level is grown epitaxially (Fig. 44*e*). Stripes of 10 μm width are then formed by photolithographic techniques down to the *n*′-AlGaAs layer. Then *p* and *n*-AlGaAs layers are regrown on the etched-away regions. The current is confined to the center stripe, since the regrown *p*-*n* junction adjacent to the center region is reverse-biased. The structure is a four-layer laser and shows excellent linearity in light-current characteristics and symmetry in laser output from both mirrors.

Figure 44*f* shows a novel stripe-geometry laser which uses the V-shaped groove to define the stripe width.[66] The V groove is etched into the surface of the GaAs–AlGaAs system; a Zn diffusion converts the adjacent *n*-AlGaAs layer into a *p*-type layer below the V groove. The groove profile determines the diffusion depth. Linear light-current output with a light-beam width of 30 to 40° has been obtained.

Figure 44*g* shows a self-aligned GaInAsP–InP DH laser.[67] The carrier and optical confinements are accomplished by means of the *n*-GaInAsP "antiguiding" regions, which lie outside a stripe region, buried in the *p*-InP cladding layer above the active layer. The current is injected into the narrow stripe region restricted by the reverse-biased (*p*-*n*-*p*-*n*) junctions outside the stripe width. The device shows stable operation in the fundamental transverse mode near 1.3-μm wavelength, and up to 72°C has been obtained in cw operation. Figure 44*h* shows a ridge-waveguide DH laser[68] emitting at 1.55 μm. The stripe width is determined by the width of the ridge, which is formed by the ion milling process. An advantage of this structure is that damage to the active region by the processing can be prevented. The device also operates in a fundamental transverse mode. Its 1.55-μm emission wavelength makes the laser an attractive source for low-loss, long-distance, single-mode optical communication systems.

All the laser structures described previously use cavity facets that are formed by cleaving or polishing to obtain the feedback necessary for lasing. Feedback can also be obtained by a periodic variation of the refractive index within the waveguide, which is generally produced by corrugating the interface between two dielectric layers. The insert in Fig. 45 gives an example. The periodic variation of $\bar{n}$ can give rise to con-

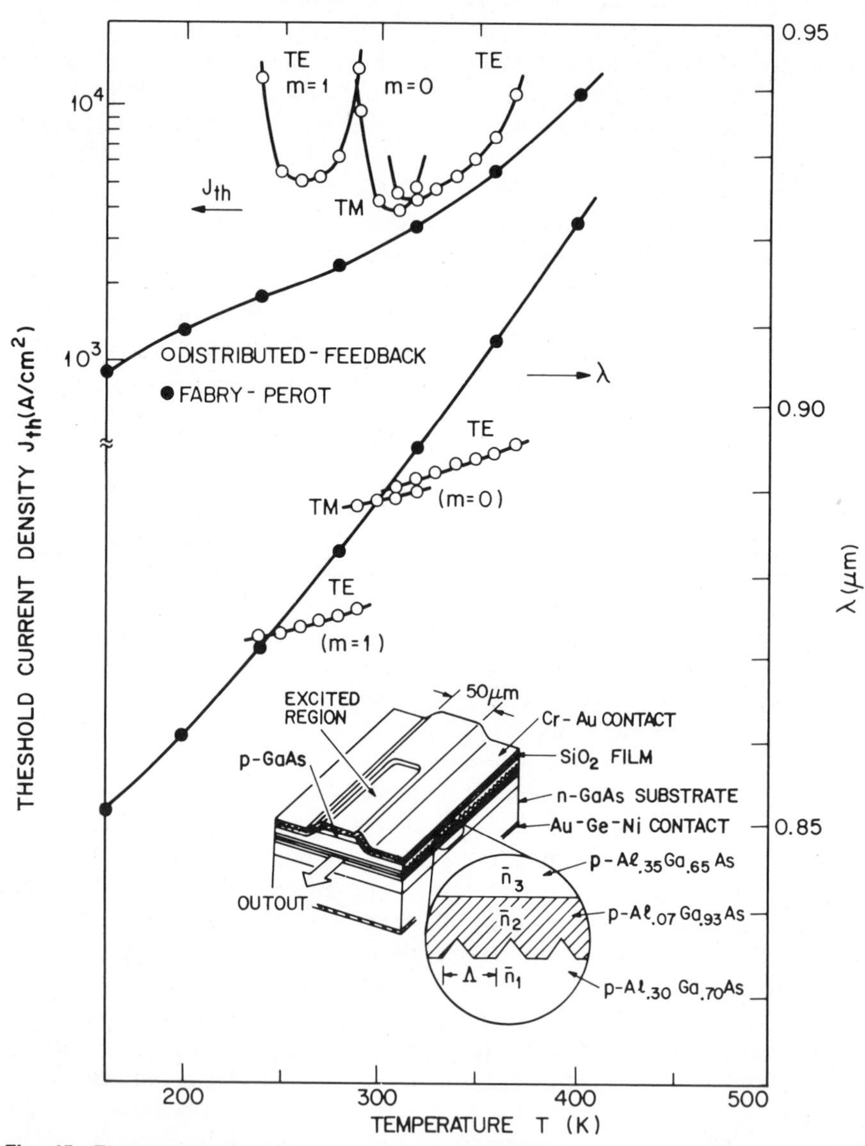

Fig. 45 Threshold current density and lasing wavelength as a function of junction temperature. The insert shows a schematic diagram of a distributed-feedback laser. (After Aiki, Nakamura, and Umeda, Ref. 70.)

structive interference. Lasers that utilize these corrugated structures are called distributed-feedback (DFB) lasers or distributed-Bragg reflector (DBR) lasers.[69] These heterostructure lasers are useful as sources in integrated optics. Figure 45 shows the emission properties of a distributed feedback laser and a Febry–Perot laser for comparison.[70] Over the temperature range 150 to 400 K, two transverse electric (TE) modes and one transverse magnetic (TM) mode were observed for the DFB laser. The laser emission of the Febry–Perot laser follows the temperature dependence of the energy gap, while the lasing wavelength of the DFB laser follows the smaller temperature dependence of the refractive index.

Semiconductor lasers have been made using superlattice structures (refer to Chapter 2). Such devices are called quantum-well heterostructure lasers.[78] When the thickness of the active layer of a DH laser is reduced to the order of carrier de Broglie wavelength ($\lambda = h/p$, where h is Planck's constant and p the momentum), two-dimensional quantization occurs and results in a series of discrete energy levels given by the bound-state energies of a finite square well. Figure 46*a* shows the quantum-well diagram for an $Al_xGa_{1-x}As$–GaAs heterostructure, where L_z is of the order of 100 Å. The energy eigenvalues are designated by E_1, E_2 for electrons; by E_{hh1}, E_{hh2}, E_{hh3} for heavy holes; and by E_{lh1}, E_{lh2} for light holes.

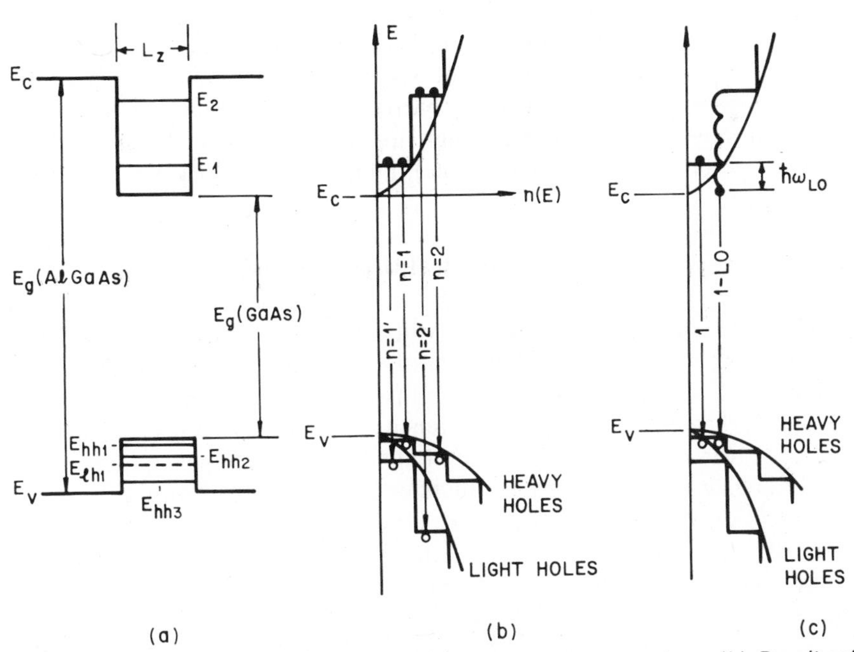

Fig. 46 (a) Square-well potential of a quantum-well heterostructure. (*b*) Density-of-states diagram. (c) Phonon scattering in a quantum-well heterostructure laser. (After Holonyak et al., Ref. 78.)

Figure 46*b* shows the corresponding density-of-state diagram. The half-parabolas that originate from the conduction-band edge E_C and valence-band edge E_V correspond to the densities of states of a bulk semiconductor. The steplike densities of states are characteristic of a quantum-well structure. Interband recombination transitions ($\Delta n = 0$ selection rule) occur from a bound state in the conduction band (say at E_1) to a bound state in the valence band (say at E_{hh1}). The energy of the transition is given by

$$\hbar\omega = E_g(\text{GaAs}) + E_1 + E_{hh1}. \tag{56}$$

Thus the recombination can proceed between two well-defined energy levels in contrast to a bulk semiconductor, where the carriers are distributed in energy over parabolically varying densities of states. Figure 46*c* shows another important feature of quantum-well heterostructures: carriers injected at higher energy can generate phonons and scatter downward in energy to ultimately a lesser density of states. In a bulk semiconductor, phonon generation is limited by the decreasing density of state, particularly at the band edge; whereas in a quantum-well system, within a constant-density-of-states region, no such limitation occurs. This process can transfer an electron to well below the confined-particle states, for example, below E_1 (Fig. 46*c*). This can lead to laser operation at energies $\hbar\omega < E_g$, instead of, as expected without phonon participation, at $\hbar\omega > E_g$.

Quantum-well heterostructure lasers in $Al_xGa_{1-x}As$–GaAs have been fabricated using metalorganic chemical vapor deposition technique. Lasers are made in stripe geometry with $Al_{0.4}Ga_{0.6}As$ confining layers and an undoped active region consisting of six $L_z \sim 120$ Å GaAs quantum wells coupled by five ~ 120 Å $Al_{0.3}Ga_{0.7}As$ barrier layers. The potential advantage of quantum-well lasers include high quantum efficiency, low threshold current (at ~ 1 mA or less is possible), and low sensitivity to temperature variations.

12.5.2 Light Output and Spectral Distribution

Figure 47 shows a typical behavior of DH lasers as the current is increased from low current densities of spontaneous emission to currents in excess of the laser threshold. The insert shows a detection scheme for measuring laser diode emission intensity. As the laser diode current is increased, the spontaneous emission (which is proportional to I_D) increases slowly at first, then I_D abruptly increases as laser oscillation begins. The extrapolation of this curve to zero I_D, as illustrated by the dashed line, gives the threshold current I_{th}. The threshold current density J_{th} is I_{th}/A, where A is the active area. From Fig. 47, the differential quantum efficiency η_D, defined as

$$\eta_D \equiv \Delta I_D / \Delta I_L \tag{57}$$

can also be obtained. For the example shown, η_D is 30%.

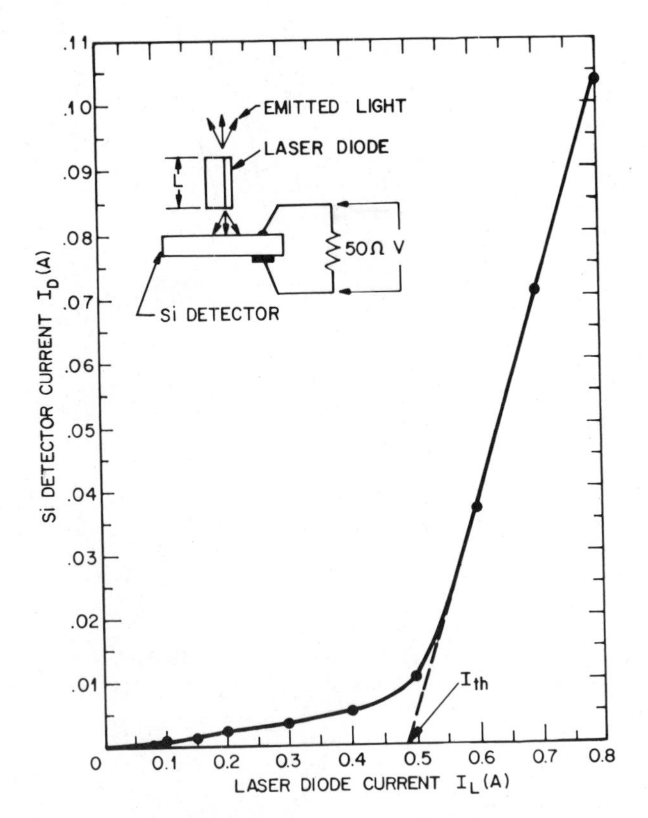

Fig. 47 Light output versus diode current for a GaAs–$Al_xGa_{1-x}As$ DH laser at room temperature. The insert shows the measurement setup. (After Casey and Panish, Ref. 20.)

Figure 48 shows the temperature dependence of a cw current threshold of a stripe-buried-heterostructure laser (Fig. 44*e*). Figure 48*a* shows cw light outputs versus injection current at various heat-sink temperatures between 25 and 115°C. Excellent linearity in light-current characteristics is observed. Figure 48*b* plots threshold currents as a function of temperature. The threshold current increases exponentially with temperature as

$$I_{th} \sim \exp(T/T_0) \tag{58}$$

where T is the heat-sink temperature in °C, and T_0 is found to be 110°C, a comparable value to that in conventional DH lasers.

At low currents the spontaneous emission has broad spectral distribution with a typical spectral width at half power 100 to 500 Å. As the current approaches the threshold, the spectral distribution becomes narrower. Figure 49 shows the emission spectra of a laser diode below, just at, and above the threshold current, indicating the narrowing of the emission when lasing is initiated.

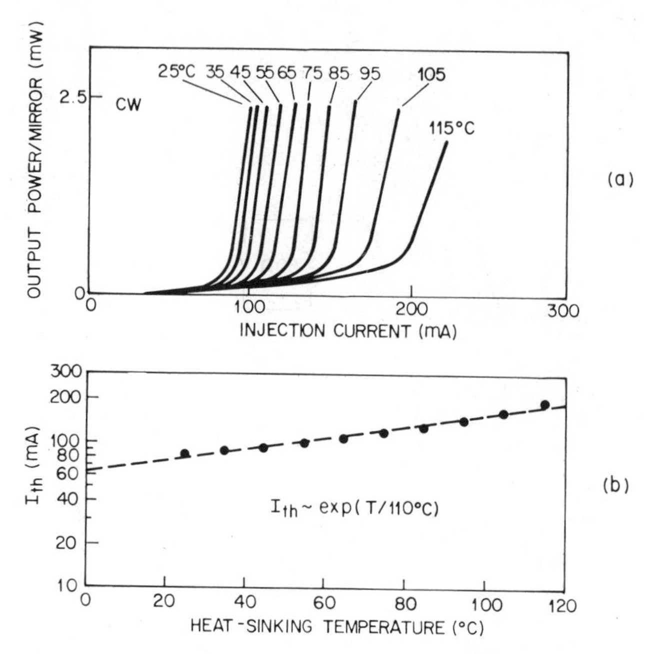

Fig. 48 (*a*) Light output versus diode current for GaAs–$Al_xGa_{1-x}As$ stripe-buried heterostructure laser. (*b*) Temperature dependence of cw current threshold. (After Tsang, Logan, and Van der Ziel, Ref. 65.)

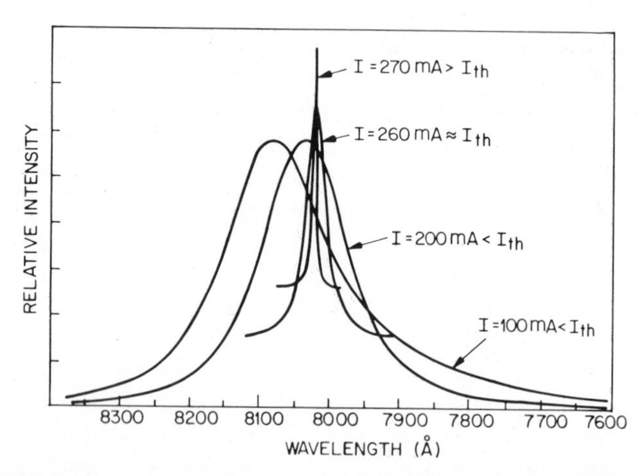

Fig. 49 Emission spectra of a diode laser below, just at, and above threshold, indicating the narrowing of the emission when lasing is initiated. (After Kressel and Butler, Ref. 21.)

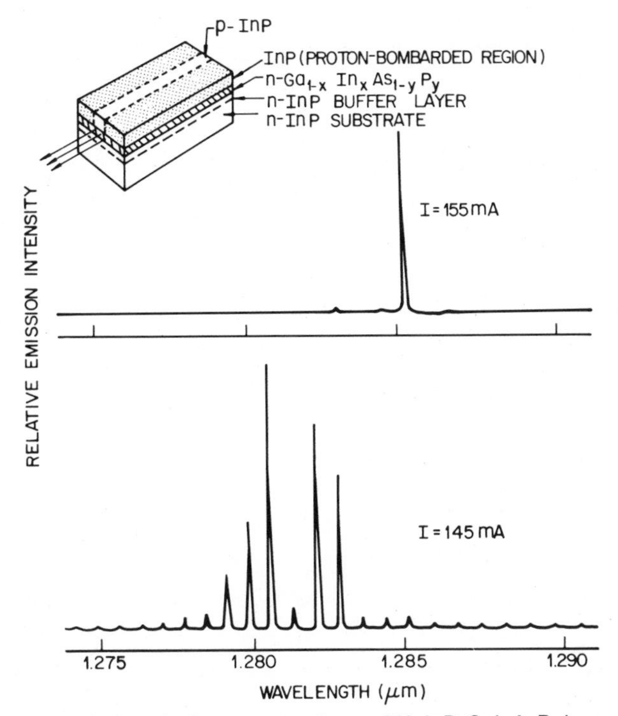

Fig. 50 High-resolution emission spectra for a DH InP–GaInAsP laser. (After Foyt, Ref. 44.)

Figure 50 shows a high-resolution spectral distribution for a proton-bombarded, stripe-geometry, InP–GaInAsP DH laser. At a current just above the threshold (145 mA) quite a few emission lines, about evenly spaced with a separation of $\Delta\lambda \simeq 7.5$ Å, exist. At a higher current ($I =$ 155 mA), there is a strong tendency for the spectrum to become single-moded with only one emission peak near 1.285 μm. These emission lines belong to the longitudinal modes that will now be derived.

With reference to Fig. 29*a*, the basic mode selection for the z direction (longitudinal direction) arises from the requirement that only an integral number m of half-wavelengths fits between the reflection planes. Thus

$$m\left(\frac{\lambda}{2\bar{n}}\right) = L \tag{59}$$

or

$$m\lambda = 2L\bar{n} \tag{60}$$

where $\bar{n}$ is the refractive index in the medium corresponding to the wavelength λ and L is the length of the semiconductor. The separation $\Delta\lambda$ between these allowed modes in the z direction is the difference in the

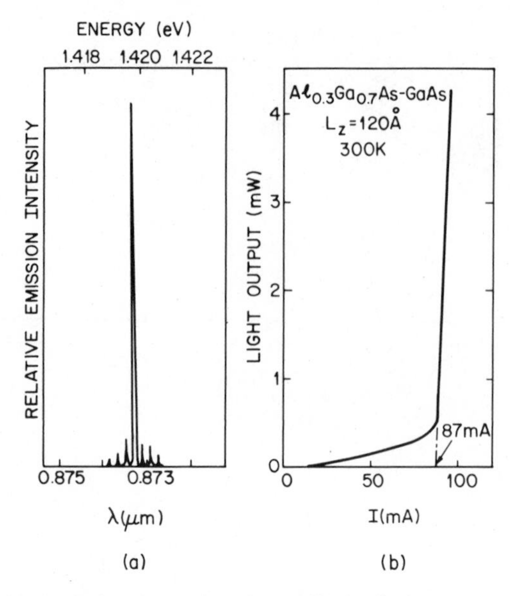

Fig. 51 (*a*) Relative emission intensity of an AlGaAs–GaAs quantum-well heterostructure laser operated continuously at 300 K. (*b*) Light output versus bias current. (After Holonyak et al., Ref. 78.)

wavelengths corresponding to m and $m+1$. Differentiating Eq. 60 with respect to λ, we obtain

$$\Delta\lambda = \frac{\lambda^2\,\Delta m}{2\bar{n}L[1-(\lambda/\bar{n})(d\bar{n}/d\lambda)]}. \tag{61}$$

for large m. The term in brackets arises from dispersion. For gas lasers, $\bar{n}$ is nearly independent of λ and the dispersion term is only a small correction. The wavelength separation between modes in the z direction is inversely proportional to the length L. Because of the short length of a semiconductor laser, the separation $\Delta\lambda$ is much larger than in a gas laser.

Figure 51*a* shows relative emission intensity of a $Al_{0.3}Ga_{0.7}As$–GaAs (five and six layers, respectively, each having $L_z = 120$ Å) quantum-well heterostructure laser operated continuously at 300 K. The peak emission line occurs at 0.8732 μm, corresponding to a transition with $\hbar\omega = (E_g + E_1 + E_{hh1} - \hbar\omega_{L0})$, where $\hbar\omega_{L0}$ is the longitudinal optical-phonon energy (Fig. 46*c*). Figure 51*b* shows the corresponding light output as a function of biasing current. The differential quantum efficiency, Eq. 57, is 85%. The temperature dependence of the threshold current can be expressed as in Eq. 58 with $T_0 = 220$°C. Therefore, the quantum-well lasers are less sensitive to temperature variations as compared to conventional DH lasers.

12.5.3 Turn-On Delay and Modulation Frequency

When a current pulse with sufficient amplitude to give stimulated emission is applied to a laser, a delay of a few nanoseconds generally occurs before the stimulated emission is observed (insert in Fig. 52). The delay time t_d is related to the minority-carrier lifetimes.

To derive the delay time, we consider the continuity equation for electrons in a p-type semiconductor. Under the conditions that the current I is uniform across the active layer d, and the injected electron concentration n is much greater than the thermal equilibrium value, the continuity equation becomes

$$dn/dt = I/qAd - n/\tau_e \tag{62}$$

where A is the area and τ_e the carrier lifetime. The first term is the uniform generation rate, and the second term is the recombination rate. A similar

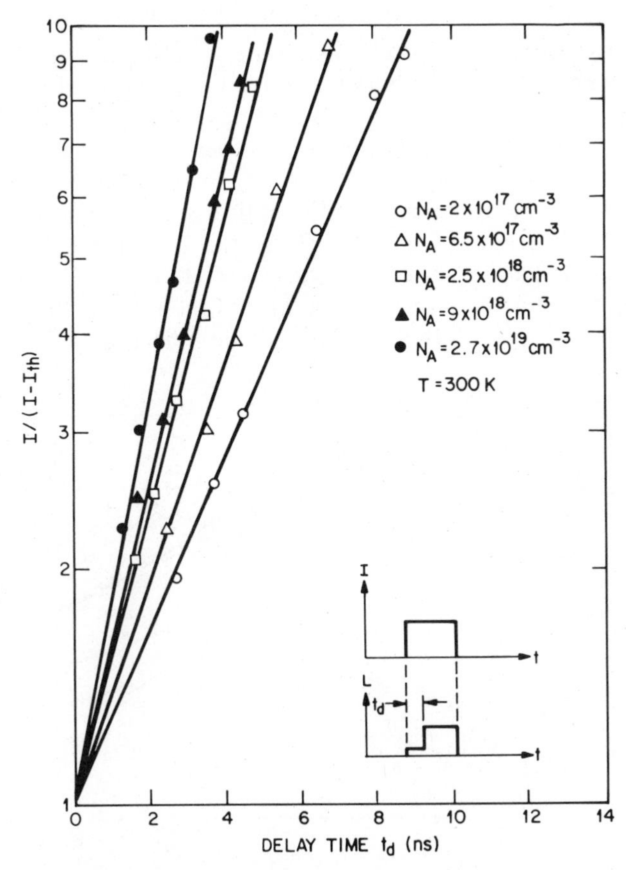

Fig. 52 Variation of laser turn-on delay with current. The delay time t_d is indicated in the insert. (After Hwang and Dyment, Ref. 71.)

expression may be written for holes in the n-type active layer. The solution of this equation for $n(0) = 0$ is

$$n(t) = (\tau_e I/qAd)[1 - \exp(-t/\tau_e)] \tag{63}$$

or

$$t = \tau_e \ln\left[\frac{I}{I - qn(t)Ad/\tau_e}\right]. \tag{64}$$

When $n(t)$ reaches the threshold value, $n(t) = n_{th}$ and $I_{th} = qn_{th}Ad/\tau_e$. Since $t = t_d$ at $n(t) = n_{th}$, the turn-on delay time is then

$$t_d = \tau_e \ln\left(\frac{I}{I - I_{th}}\right). \tag{65}$$

If the laser is prebiased to a current level $I_0 < I_{th}$, solving Eq. 62 with initial condition $n(0) = I_0\tau_e/Aqd$ gives

$$t_d = \tau_e \ln[(I - I_0)/(I - I_{th})]. \tag{66}$$

Figure 52 shows the measured results on the variation of the laser turn-on delay with current above threshold for active layers with different acceptor concentrations.[71] The delay time t_d varies logarithmically in accordance with Eq. 65; delay time decreases as N_A becomes larger (i.e., shorter carrier lifetime).

For optical fiber communications, the optical source must be able to be modulated at high frequencies. Both GaAs and GaInAsP DH lasers have good modulation characteristics well into the GHz range. Figure 53 shows a normalized modulated light output as a function of the modulation frequency for a GaInAsP–InP DH laser diode.[72] The laser diode emitting

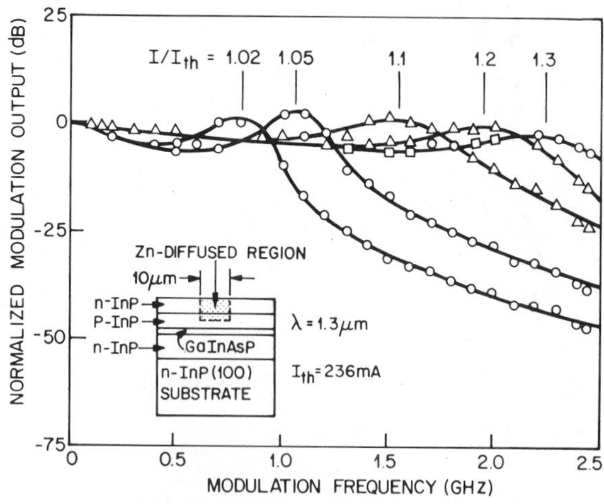

Fig. 53 Normalized modulated light output versus modulation frequency. The insert shows the laser cross section. (After Akiba, Sakai, and Yamamoto, Ref. 72.)

at 1.3 μm is directly modulated with sinusoidal current superposed on dc bias current. The light output is almost constant in the measured frequency range up to 2.5 GHz for a biasing current of 1.3 times the threshold current.

12.5.4 Laser Tuning[73]

The emission wavelength of a semiconductor laser can be varied by changing the diode current or heat-sink temperature, or by applying magnetic field or pressure. Semiconductor lasers, because of their wavelength tunability along with narrow spectral line width, high degree of stability, low input power, and structural simplicity, have significant potential for application in technology and basic research, such as molecular spectroscopy, atomic spectroscopy, high-resolution gas spectroscopy, and monitoring atmospheric pollution.

Figure 23 shows the wavelength ranges covered by compound semiconductor lasers. By choosing the appropriate composition for one of the compounds, a laser can be produced at any desired wavelength in the broad range 0.3 to over 30 μm.

Figure 54 shows a temperature tuning of a DH PbTe–$Pb_{1-x}Sn_xTe$ laser.[74]

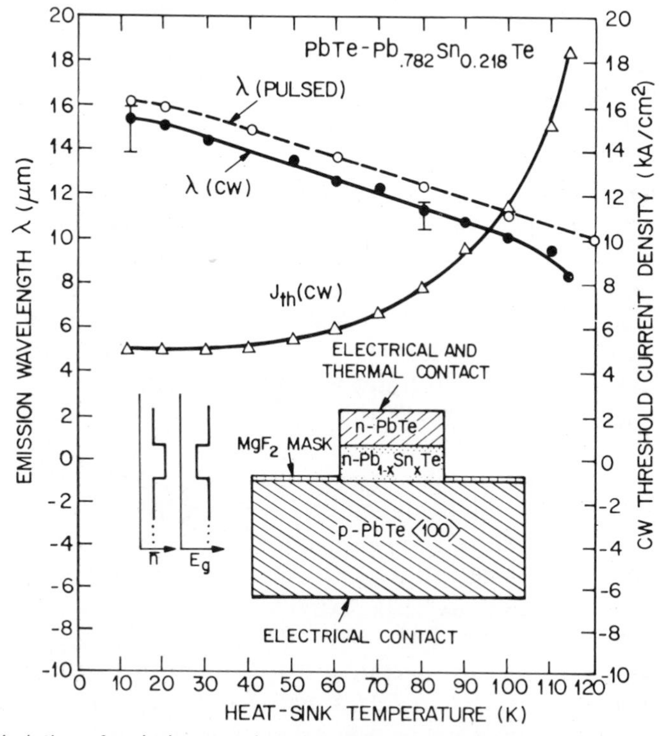

Fig. 54 Variation of emission wavelength and threshold current density as a function of temperature. The insert shows the cross section of a PbTe–PbSnTe DH laser. (After Walpole et al., Ref. 74.)

By changing the heat-sink temperature from 10 to 120 K, the emission wavelength can be varied from almost 9 to 16 μm. This variation is mainly caused by the increase of energy bandgap with increasing temperature in the IV–VI compounds.

Applying hydrostatic pressure to a diode laser can provide a very broad tuning range. The bandgap varies linearly with hydrostatic pressure for some binary compounds (e.g., InSb, PbS, and PbSe). A PbSe laser at 77 K can be tuned from 7.5 to 22 μm using hydrostatic pressure up to 14 kbars.[73]

Diode lasers can also be tuned by a magnetic field. For semiconductors with large effective mass anisotropy, the magnetic energy levels depend on the orientation of the applied magnetic field with the crystal axis. Both conduction and valence bands have their energies quantized into Landau levels. As the magnetic field increases, the energy separations between available transitions also increase, causing the emission wavelength to change. In a $Pb_{0.79}Sn_{0.21}Te$ laser at 7 K in a $\langle 100 \rangle$ magnetic field, the wavelength decreases from 15 to 14 μm when a magnetic field of 10 kG is applied.

12.5.5 Laser Degradation[20]

Injection lasers degrade by a variety of mechanisms. The three main mechanisms are (1) catastrophic degradation, (2) dark-line defect formation, and (3) gradual degradation.

For catastrophic degradation, the laser mirror under high-power operation is permanently damaged by pits or grooves forming on the mirror. Modifications of the device structures that reduce surface recombination and absorption increase the power possible at the damage limit.[75]

The dark-line defect is a network of dislocations that can form during laser operation, and it intrudes upon the optical cavity. Once started, it can grow expansively in a few hours, causing the threshold current density to increase. To reduce the probability of dark-line defect formation, quality epitaxial layers grown on substrates with low dislocation density should be used, and the laser should be carefully bonded to the heat sink to minimize strain.

By excluding instantaneous catastrophic failure and the fairly rapid degradation caused by dark-line defect formation, DH lasers have a long operating life with relatively slow degradation. DH GaAs–AlGaAs lasers under cw operation longer than 2.6×10^4 hr (3 years) at 30°C have not shown signs of degradation.[76] The extrapolated lifetime at 22°C heat-sink temperature is more than 10^6 h (over 100 years). Figure 55 shows the threshold current versus cw operating time for GaInAsP–InP DH lasers.[44] The low-threshold laser ($J_{th} = 2.9$ kA/cm^2) has unchanged characteristics after 8000 h of operation; the high-threshold laser shows evidence of only increased thermal resistance, with no sign of internal degradation. It is reasonable to assume that the lifetime obtained for GaAs DH lasers can

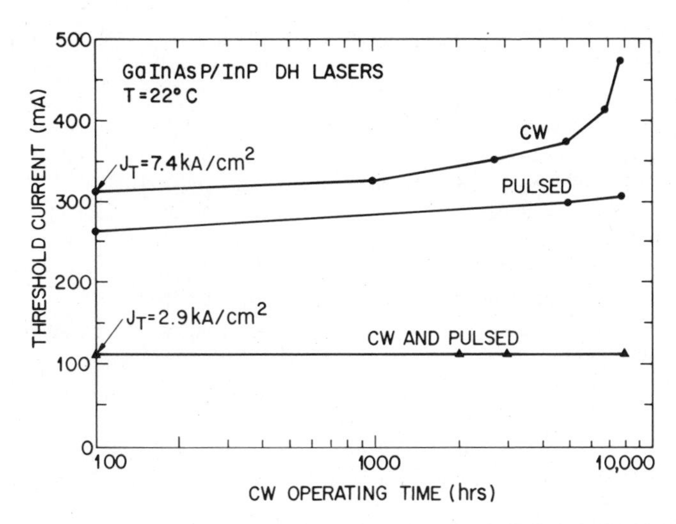

Fig. 55 Threshold current versus cw operating time for InP–GaInAsP DH lasers. (After Foyt, Ref. 44.)

also be achieved by lasers operated at longer wavelengths. The long life will meet the requirements for large-scale optical-fiber communication systems as well as satisfying the requirements for other applications.

REFERENCES

1. H. J. Round, "A Note on Carborundum," *Electron World*, **19**, 309 (1907).
2. A. A. Bergh and P. J. Dean, *Light-Emitting Diodes*, Clarendon, Oxford, 1976.
3. S. Gage, D. Evans, M. Hodapp, and H. Sorenson, *Optoelectronics Applications Manual*, McGraw-Hill, New York, 1977.
4. J. P. Gordon, H. J. Zeiger, and C. H. Townes, "Molecular Microwave Oscillator and New Hyperfine Structure in the Microwave Spectrum of NH_3," *Phys. Rev.*, **95**, 282 (1954).
5. A. L. Schawlow and C. H. Townes, "Infrared and Optical Masers," *Phys. Rev.*, **112**, 1940 (1958).
6. T. H. Maiman, "Stimulated Optical Radiation in Ruby Masers," *Nature* (*Lond.*), **187**, 493 (1960).
7. P. Aigrain (1958), as reported in Proc. Conf. Quantum Electron., Paris, 1963, p. 1762.
8. N. G. Basov, B. M. Vul, and Y. M. Popov, "Quantum-Mechanical Semiconductor Generators and Amplifiers of Electromagnetic Oscillations," *Sov. Phys. JEPT*, **10**, 416 (1960).
9. J. I. Nishizawa and Y. Watanabe, Japanese Patent (Apr. 1957).
10. W. S. Boyle and D. G. Thomas, U.S. Patent 3,059,117 (Oct. 16, 1962, filed Jan. 1960).
11. M. G. A. Bernard and G. Duraffourg, "Laser Conditions in Semiconductors," *Phys. Status Solidi*, **1**, 699 (1961).
12. W. P. Dumke, "Interband Transitions and Maser Action," *Phys. Rev.*, **127**, 1559 (1962).

13 R. N. Hall, G. E. Genner, J. D. Kingsley, T. J. Soltys, and R. O. Carslon, "Coherent Light Emission from GaAs Junctions," *Phys. Rev. Lett.*, **9**, 366 (1962).

14 M. I. Nathan, W. P. Dumke, G. Burns, F. J. Dill, Jr., and G. J. Lasher, "Stimulated Emission of Radiation from GaAs *p-n* Junction," *Appl. Phys. Lett.*, **1**, 62 (1962).

15 T. M. Quist, R. H. Rediker, R. J. Keyes, W. E. Krag, B. Lax, A. L. McWhorter, and H. J. Zeigler, "Semiconductor Maser of GaAs," *Appl. Phys. Lett.*, **1**, 91 (1962).

16 N. Holonyak, Jr., and S. F. Bevacqua, "Coherent (Visible) Light Emission from $Ga(As_{1-x}P_x)$ Junction," *Appl. Phys. Lett.*, **1**, 82 (1962).

17 I. Hayashi, M. B. Panish, P. W. Foy, and S. Sumski, "Junction Lasers which Operate Continuously at Room Temperature," *Appl. Phys. Lett.*, **17**, 109 (1970).

18 H. Kroemer, "A Proposed Class of Heterojunction Injection Lasers," *Proc. IEEE*, **51**, 1782 (1963).

19 Zn. I. Alferov and R. F. Kazarinov, Inventor's Certificate No. 181737 (1963); Zn. I. Alferov, *Fiz. Tekh. Poluprovodn.*, **1**, 436 (1967).

20 H. C. Casey, Jr. and M. B. Panish, *Heterostructure Lasers*, Academic, New York, 1978.

21 H. Kressel and J. K. Butler, *Semiconductor Lasers and Heterojunction LEDs*, Academic, New York, 1977.

22 P. Goldberg, Ed., *Luminescence of Inorganic Solids*, Academic, New York, 1966.

23 H. F. Ivey, "Electroluminescence and Semiconductor Lasers," *IEEE J. Quantum Electron.*, **QE-2**, 713 (1966).

24 A. Mooradian and H. Y. Fan, "Recombination Emission in InSb," *Phys. Rev.*, **148**, 873 (1966).

25 W. N. Carr, "Characteristics of a GaAs Spontaneous Infrared Source with 40 Percent Efficiency," *IEEE Trans. Electron Devices*, **ED-12**, 531 (1965).

26 H. F. Ivey, *Electroluminescence and Related Effects*, Suppl. 1 to L. Marton, Ed., *Advances in Electronics and Electron Physics*, Academic, New York, 1963, p. 205.

27 S. Wang, *Solid-State Electronics*, McGraw-Hill, New York, 1966.

28 P. C. Eastman, R. R. Haering, and P. A. Barnes, "Injection Electroluminescence in Metal–Semiconductor Tunnel Diodes," *Solid State Electron.*, **7**, 879 (1964).

29 C. H. Gooch, *Injection Electroluminescent Devices*, Wiley, New York, 1973.

30 M. G. Craford, "Recent Developments in LED Technology," *IEEE Trans. Electron Devices*, **ED-24**, 935 (1977).

31 W. O. Groves, A. H. Herzog, and M. G. Craford, "The Effect of Nitrogen Doping on GaAsP Electroluminescent Diodes," *Appl. Phys. Lett.*, **19**, 184 (1971).

32 W. N. Carr and G. E. Pittman, "One Watt GaAs *p-n* Junction Infrared Source," *Appl. Phys. Lett.*, **3**, 173 (1963).

33 W. N. Carr, "Photometric Figures of Merit for Semiconductor Luminescent Sources Operating in Spontaneous Mode," *Infrared Phys.*, **6**, 1 (1966).

34 S. V. Galginaitis, "Improving the External Efficiency of Electroluminescent Diodes," *J. Appl. Phys.*, **36**, 460 (1965).

35 D. R. Wight, "Green Luminescence Efficiency in GaP," *J. Phys. D*, **10**, 431 (1977).

36 W. Von Munch, "Silicon Carbide Technology for Blue Emitting Diodes," *J. Electron. Mater.*, **6**, 449 (1977).

37 J. E. Geusic, F. W. Ostermayer, H. M. Marcos, L. G. Van Uitert, and J. P. Van Der Ziel, "Efficiency of Red, Green and Blue Infrared-to-Visible Conversion Sources," *J. Appl. Phys.*, **42**, 1958 (1971).

38 A. G. Dentai, T. P. Lee, and C. A. Burrus, "Small-Area High-Radiance cw InGaAsP LEDs Emitting at 1.2 to 1.3 m," *Electron. Lett.*, **13**, 484 (1977).

39 C. A. Burrus and B. I. Miller, "Small-Area, Double Heterostructure AlGaAs Electroluminescent Diode Source for Optical-Fiber Transmission Lines," *Opt. Commun.* **4**, 307 (1971).

40 M. Ettenberg, H. Kressel, and J. P. Wittke, "Very High Radiance Edge-Emitting LED," *IEEE J. Quantum Electron.*, **QE-12**, 360 (1976).

41 M. C. Amann and W. Proebster, "Small-Area GaAs–GaAlAs Heterostructure LED with Improved Current Confinement," *Electron. Lett.*, **15**, 599 (1979).

41a R. N. Hall, *Proc. Inst. Electr. Eng. Lond.*, **B106**, 923 (1959).

42 K. Ikeda, S. Horiuchi, T. Tanaka, and W. Susaki, "Design Parameters of Frequency Response of GaAs–AlGaAs DH LED's for Optical Communications," *IEEE Trans. Electron Devices*, **ED-24**, 1001 (1977).

43 T. Miya, Y. Terunuma, T. Hosaka, and T. Miyashita, "Ultimate Low-Loss Single Mode Fiber at 1.55 μm," *Electron. Lett.*, **15**, 108 (1979).

44 A. G. Foyt, "1.0–1.6 μm Sources and Detectors for Fiber Optics Applications," IEEE Device Res. Conf., Boulder, Colo., June 25, 1979.

45 C. H. L. Goodman, "Devices and Materials for 4 μm-band Fibre-Optical Communication," *Solid State Electron Devices*, **2**, 129 (1978).

46 P. J. Lin and L. Kleinman, "Energy Bands of PbTe, PbSe and PbS," *Phys. Rev.*, **142**, 478 (1966).

47 A. K. Levine, "Lasers," *Am. Sci.*, **51**, 14 (1963).

48 M. B. Panish, I. Hayashi, and S. Sumski, "Double-Heterostructure Injection Lasers with Room Temperature Threshold As Low as 2300 A/cm^2," *Appl. Phys. Lett.*, **16**, 326 (1970).

49 H. C. Casey, Jr., M. B. Panish, and J. L. Merz, "Beam Divergence of the Emission from Double-Heterostructure Injection Lasers," *J. Appl. Phys.*, **44**, 5470 (1973).

50 L. A. D'Asaro, "Advances in GaAs Junction Lasers with Stripe Geometry," *J. Lumin.*, **7**, 310 (1973).

51 T. L. Paoli, "Waveguiding in a Stripe-Geometry Junction Laser," *IEEE J. Quantum Electron.*, **QE-13**, 662 (1977).

52 H. Yonezu, I. Sakuma, K. Kobayashi, T. Kamejima, M. Ueno, and Y. Nannichi, "A GaAs–$Al_xGa_{1-x}As$ Double Heterostructure Planar Stripe Laser," *Jpn. J. Appl. Phys.*, **12**, 1585 (1973).

53 M. I. Nathan, "Semiconductor Lasers," *Proc. IEEE*, **54**, 1276 (1966).

54 G. Lasher and F. Stern, "Spontaneous and Stimulated Recombination Radiation in Semiconductors," *Phys. Rev.*, **133**, A553 (1964).

55 B. I. Halperin and M. Lax, "Impurity-Band Tails in the High-Density Limit. I. Minimum Counting Methods," *Phys. Rev.*, **148**, 722 (1966).

56 F. Stern, "Calculated Spectral Dependence of Gain in Excited GaAs," *J. Appl. Phys.*, **47**, 5382 (1976).

57 H. C. Casey, Jr., "Room Temperature Threshold-Current Dependence of GaAs-$Al_xGa_{1-x}As$ Double Heterostructure Lasers on x and Active-Layer Thickness," *J. Appl. Phys.*, **49**, 3684 (1978).

58 R. E. Nahory and M. A. Pollack, "Threshold Dependence on Active-Layer Thickness in InGaAsP/InP DH Lasers," *Electron. Lett.*, **14**, 727 (1978).

59 M. Yana, H. Nishi, and M. Takusagawa, "Theoretical and Experimental Study of Threshold Characteristics in InGaAsP/InP DH Lasers," *IEEE J. Quantum Electron.*, **QE-15**, 571 (1979).

60 C. A. Burrus, H. C. Casey, Jr., and T. Y. Li, "Optical Sources," in S. E. Miller and A. G. Chynoweth, Eds., *Optical Fiber Communication*, Academic, New York, 1979.

61 M. C. Amann, "New Stripe-Geometry Laser with Simplified Fabrication Process," *Electron. Lett.*, **15**, 441 (1979).

62 K. Aiki, M. Nakamura, T. Kuroda, and J. Umeda, "Channeled-Substrate Planar Structure AlGaAs Injection Laser," *Appl. Phys. Lett.*, **30**, 649 (1977).

63 H. Namizaki, "Transverse-Junction-Stripe Lasers with a GaAs *p-n* Homojunction," *IEEE J. Quantum Electron.*, **QE-11**, 427 (1975).

64 T. Tsukada, "GaAs–AlGaAs Buried-Heterostructure Injection Lasers," *J. Appl. Phys.*, **45**, 4899 (1974).

65 W. T. Tsang, R. A. Logan, and J. P. Van der Ziel, "Low-Current-Threshold Stripe-Buried-Heterostructure Lasers with Self-Aligned Current Injection Stripes," *Appl. Phys. Lett.*, **34**, 644 (1979).

66 J. Gosch, "Simple Technology Irons Out Kinks in Laser Diode Output," *Electronics* **52**, (26), 59 (1979).

67 H. Nishi, M. Yano, Y. Nishitani, Y. Akita, and M. Takusagawa, "Self-Aligned Structure InGaAsP/InP DH Lasers," *Appl. Phy. Lett.*, **35**, 232 (1979).

68 I. P. Kaminow, R. E. Nahory, M. A. Pollack, L. W. Stulz, and J. C. Dewinter, "Single-Mode cw Ridge-Waveguide Laser Emitting at 1.55 m," *Electron. Lett.*, **15**, 763 (1979).

69 H. Kogelnik and C. V. Shank, "Stimulated Emission in a Periodic Structure," *Appl. Phys. Lett.*, **18**, 152 (1971); "Coupled-Wave Theory of Distributed Feedback Lasers," *J. Appl. Phys.*, **43**, 2327 (1973).

70 K. Aiki, M. Nakamura, and J. Umeda, "Lasing Characteristics of Distributed-Feedback GaAs–GaAlAs Diode Lasers with Separate Optical and Carrier Confinement," *IEEE J. Quantum Electron.*, **QE-12**, 597 (1976).

71 C. J. Hwang and J. C. Dyment, "Dependence of Threshold and Electron Lifetime on Acceptor Concentration in GaAs–$Ga_{1-x}Al_xAs$ Lasers," *J. Appl. Phys.*, **44**, 3240 (1973).

72 S. Akiba, K. Sakai, and T. Yamamoto, "Direct Modulation of InGaAsP/InP Double Heterostructure Lasers," *Electron. Lett.*, **14**, 197 (1978).

73 I. Melngailis and A. Mooradian, "Tunable Semiconductor Diode Lasers and Applications," in S. Jacobs, M. Sargent, J. F. Scott, and M. O. Scully, Eds., *Laser Applications to Optics and Spectroscopy*, Addison-Wesley, Reading, Mass., 1975.

74 J. N. Walpole, A. R. Calawa, T. C. Harman, and S. H. Groves, "Double-Heterostructure PbSnTe Lasers Grown by Molecular-Beam Epitaxy with CW Operation up to 114 K," *Appl. Phys. Lett.*, **28**, 552 (1976).

75 H. Yonezu, I. Sakuma, T. Kamojima, M. Ueno, K. Iwamoto, I. Hino, and I. Hayashi, "High Optical Power Density Emission from a Window Stripe AlGaAs DH Laser," *Appl. Phys. Lett.*, **34**, 637 (1979).

76 R. L. Hartman, N. E. Schumaker, and R. W. Dixon, "Continuously Operated AlGaAs DH Lasers with 70°C Lifetimes As Long as Two Years," *Appl. Phys. Lett.*, **31**, 756 (1977).

77 W. Lo and D. E. Swets, "Room Temperature 4.6 μm Light Emitting Diodes," *Appl. Phys. Lett.*, **36**, 450 (1980).

78 N. Holonyak, Jr., R. M. Kolbas, R. D. Dupuis, and P. D. Dapkus, "Quantum-Well Heterostructure Lasers," *IEEE J. Quantum Electron.* **QE-16**, 170 (1980).

13

Photodetectors

- INTRODUCTION
- PHOTOCONDUCTOR
- PHOTODIODE
- AVALANCHE PHOTODIODE
- PHOTOTRANSISTOR

13.1 INTRODUCTION

Photodetectors are semiconductor devices that can detect optical signals through electronic processes. The extension of coherent and incoherent light sources into the far-infrared region on one hand and the ultraviolet region on the other has increased the need for high-speed, sensitive photodetectors. A general photodetector has basically three processes: (1) carrier generation by incident light, (2) carrier transport and/or multiplication by whatever current-gain mechanism may be present, and (3) interaction of current with the external circuit to provide the output signal.

Photodetectors are important in optical-fiber communication systems operated in the near-infrared region (0.8 to 1.6 μm). They demodulate optical signals, that is, convert the optical variations into electrical variations, that are subsequently amplified and further processed. For such applications the photodetectors must satisfy stringent requirements such as high sensitivity at operating wavelengths, high response speed, and minimum noise. In addition, the photodetector should be compact in size, use low biasing voltages or current, and be reliable under operating conditions.

Anderson, DiDomenico, and Fisher[1] and Melchior[2,3] have reviewed high-speed photodetectors. A comprehensive review on infrared photodetectors has been given in *Semiconductors and Semimetals*,[4] in which a detailed treatment on avalanche photodiodes is given by Stillman and Wolfe.[5] Lee and Li have discussed photodetectors for optical-fiber communications.[6]

13.2 PHOTOCONDUCTOR

A photoconductor consists simply of a slab of semiconductor (in bulk or thin-film form) with ohmic contacts affixed to opposite ends (Fig. 1). When incident light falls on the surface of the photoconductor, carriers are generated either by band-to-band transitions (intrinsic) or by transitions involving forbidden-gap energy levels (extrinsic), resulting in an increase in conductivity. The processes of intrinsic and extrinsic photoexcitation of carriers are shown in Fig. 2.

For the intrinsic photoconductor, the conductivity is given by $\sigma = q(\mu_n n + \mu_p p)$, and the increase of conductivity under illumination is mainly due to the increase in the number of carriers. The long-wavelength cutoff for this case is given by

$$\lambda_c = \frac{hc}{E_g} = \frac{1.24}{E_g(\text{eV})} \qquad (\mu\text{m}) \tag{1}$$

where λ_c is the wavelength corresponding to the semiconductor bandgap E_g. For wavelengths shorter than λ_c, the incident radiation is absorbed by the semiconductor, and hole–electron pairs are generated. For the extrinsic case, photoexcitation may occur between a band edge and an energy level in the energy gap. Photoconductivity can take place by the absorption of photons of energy equal to or greater than the energy separation of the bandgap levels and the conduction or valence band. In this case the long-wavelength cutoff is determined by the depth of the forbidden-gap energy level.

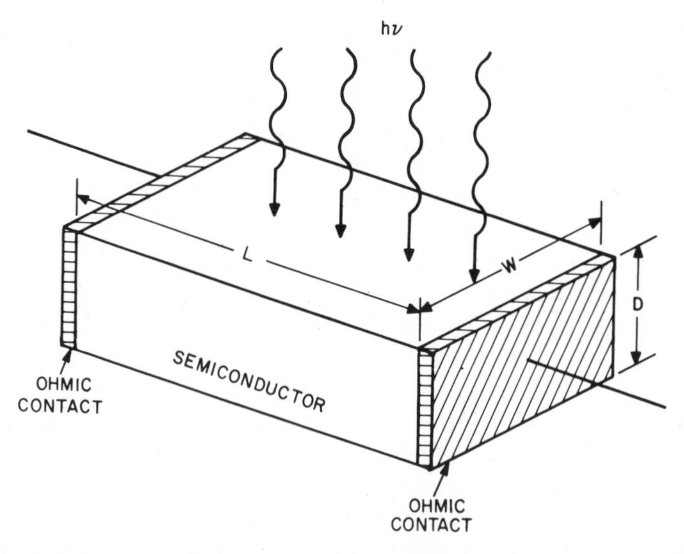

Fig. 1 Schematic diagram of a photoconductor that consists of a slab of semiconductor and two ohmic contacts at the ends.

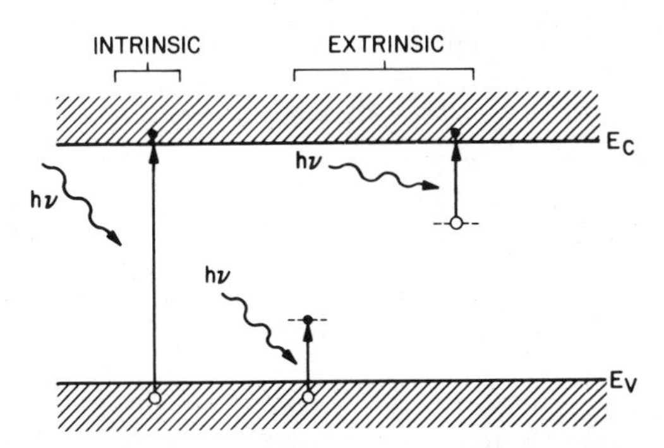

Fig. 2 Processes of intrinsic (band-to-band) and extrinsic photoexcitations.

The performance of a photodetector in general and a photoconductor in particular is measured in terms of three parameters: the quantum efficiency or gain, the response time, and the sensitivity (detectivity). First consider the principle of operation of a photoconductor under illumination (Fig. 1). At time 0, the number of carriers generated in a unit volume by a given photon flux is n_0. The number of carriers at a later time t, $n(t)$, in the same volume decays by recombination as $n = n_0 \exp(-t/\tau)$, where τ is the carrier lifetime. In other words, the recombination rate is $1/\tau$. Assuming a steady flow of photon flux impinging uniformly on the surface of a photoconductor (Fig. 1) with area $A = WL$, the total number of photons arriving at the surface is $(P_{opt}/h\nu)$ per unit time, where P_{opt} is the incident optical power and $h\nu$ is the photon energy.

At steady state, the carrier generation rate must be equal to the recombination rate. If the device thickness D is much larger than the light penetration depth $(1/\alpha)$, the total steady-state generation rate of carriers per unit volume is

$$G = \frac{n}{\tau} = \frac{\eta(P_{opt}/h\nu)}{WLD} \tag{2}$$

where η is the quantum efficiency (i.e., number of carriers generated per photon) and n is the carrier per unit volume (carrier density). The photocurrent flowing between the electrodes is

$$I_p = (\sigma\mathscr{E})WD = (q\mu_n n\mathscr{E})WD = (qnv_d)WD \tag{3}$$

where $\mathscr{E}$ is the electric field inside the photoconductor, and v_d is the drift velocity. Substituting n in Eq. 2 into Eq. 3 gives

$$I_p = q\left(\eta\frac{P_{opt}}{h\nu}\right)\left(\frac{\mu_n\tau\mathscr{E}}{L}\right). \tag{3a}$$

If we define the primary photocurrent as

$$I_{ph} \equiv q\left(\eta\frac{P_{opt}}{h\nu}\right)$$

Table 1 Typical Values of Gain and Response Time

Photodetector	Gain	Response Time (s)	Operating Temperature (K)
Photoconductor	$1 \sim 10^6$	$10^{-3} \sim 10^{-8}$	$4.2 \sim 300$
p-n junction	1	10^{-11}	300
p-i-n junction	1	$10^{-8} \sim 10^{-10}$	300
Metal–semiconductor diode	1	10^{-11}	300
Avalanche photodiode	$10^2 \sim 10^4$	10^{-10}	300
Bipolar phototransistor	10^2	10^{-8}	300
Field-effect phototransistor	10^2	10^{-7}	300

the photocurrent gain from Eq. 3a is

$$\text{gain} = \frac{I_p}{I_{ph}} = \frac{\mu_n \tau \mathscr{E}}{L} = \frac{\tau}{t_r} \tag{4}$$

where $t_r = L/v_d$ is the carrier transit time. The gain that depends upon the ratio of carrier lifetime to the transit time is a critical parameter in photoconductors. For the long-lifetime sample with short electrode spacing, the gain can be substantially greater than unity. The response time of a photoconductor is determined by the lifetime t. Some typical values[2,8] of gain and response time for various devices are given in Table 1.

Next, consider an intensity-modulated optical signal given by

$$P(\omega) = P_{\text{opt}}(1 + me^{j\omega t}) \tag{5}$$

where P_{opt} is the average optical-signal power, m the modulation index, and ω the modulation frequency. The average current I_p resulting from the optical signal is given by Eq. 3a. For the modulated optical signal, the rms optical power is $mP_{\text{opt}}/\sqrt{2}$ and the rms signal current can be written as[9]

$$i_p \approx \frac{q\eta m P_{\text{opt}}}{\sqrt{2}h\nu} \left(\frac{\tau}{t_r}\right) \frac{1}{(1 + \omega^2 \tau^2)^{1/2}}\,. \tag{6}$$

Figure 3 shows an RF equivalent circuit for a photoconductor. The conductance G consists of the dark conductance due to the dark current and the conductance induced by the average signal current and the background current. The thermal noise resulting from the conductance G is given by

$$i_G^2 = 4kTGB \tag{7}$$

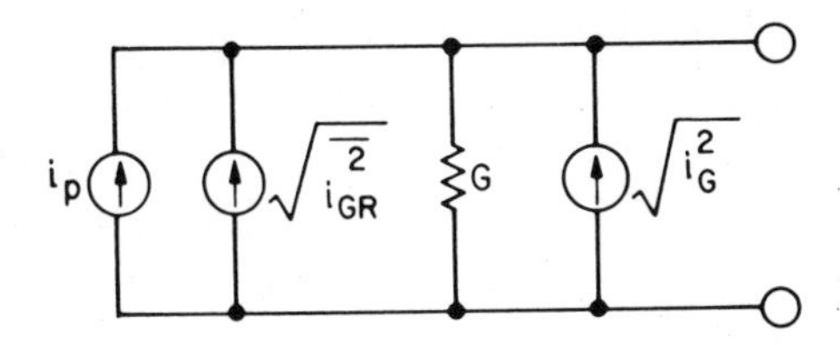

Fig. 3 Equivalent circuit of a photoconductor. (After DiDomenico and Svelto, Ref. 9.)

where k is Boltzmann's constant, T absolute temperature, and B the bandwidth. The generation–recombination noise (shot noise) is given by[10]

$$\overline{i^2_{GR}} = \frac{\tau}{t_r}\frac{4qI_0B}{1+\omega^2\tau^2} \tag{8}$$

where I_0 is the steady-state light-induced output current, which is equal to I_p $(=\tau I_{ph}/t_r)$. The signal-to-noise ratio can be obtained from Eqs. 6, 7, 8:

$$(S/N)_{\text{power}} = \frac{i_p^2}{\overline{i^2_{GR}}+\overline{i^2_G}} = \frac{\eta m^2(P_{\text{opt}}/h\nu)}{8B}\left[1+\frac{kT}{q}\frac{t_r}{\tau}(1+\omega^2\tau^2)\frac{G}{I_0}\right]^{-1} \tag{9}$$

A related figure of merit for photodetectors[11] is the noise-equivalent power (NEP) that corresponds to the incident rms optical power required to produce a signal-to-noise ratio of one in a 1-Hz bandwidth. One can obtain the NEP (i.e., $mP_{\text{opt}}/\sqrt{2}$) from Eq. 9 by setting $S/N = 1$ and $B = 1$. For infrared detectors the most used figure of merit is the detectivity D^*, which is defined as[12]

$$D^* = \frac{A^{1/2}B^{1/2}}{\text{NEP}} \qquad \text{cm(Hz)}^{1/2}/\text{W}. \tag{10}$$

To remove any ambiguity in D^*, one must state whether the radiation is from a black-body source or a monochromatic source and at what modulation frequency. It is recommended that D^* be expressed as $D^*(\lambda, f, 1)$ or $D^*(T, f, 1)$, where λ is the wavelength in μm, f the frequency of modulation in Hz, T the black-body temperature in K, and the reference bandwidth is always 1 Hz.

For a background-limited photoconductor, the ideal D^* for unit quantum efficiency is given by[12]

$$D^*(\lambda, f, 1) = \frac{c\,\exp(\zeta/2)}{2\sqrt{\pi hkT}\;\nu^2(1+2/\zeta+2/\zeta^2)^{1/2}} \qquad \text{cm(Hz)}^{1/2}/\text{W} \tag{11}$$

where c is the velocity of light and $\zeta \equiv h\nu/kT$. The ideal D^* (dashed curves) for background temperatures of 77 K and 300 K is plotted in Fig. 4. Some typical D^* values for available photoconductors such as CdS, PbS, InSb, nickle-doped germanium (Ge–Ni), and phosphorus-doped silicon (Si–P) detectors[4,7] are also shown (refer to Chapter 1 for impurity energy levels). Note that for mid-infrared to far-infrared and longer wavelengths, the photoconductors are cooled to lower temperatures (such as 77 K and

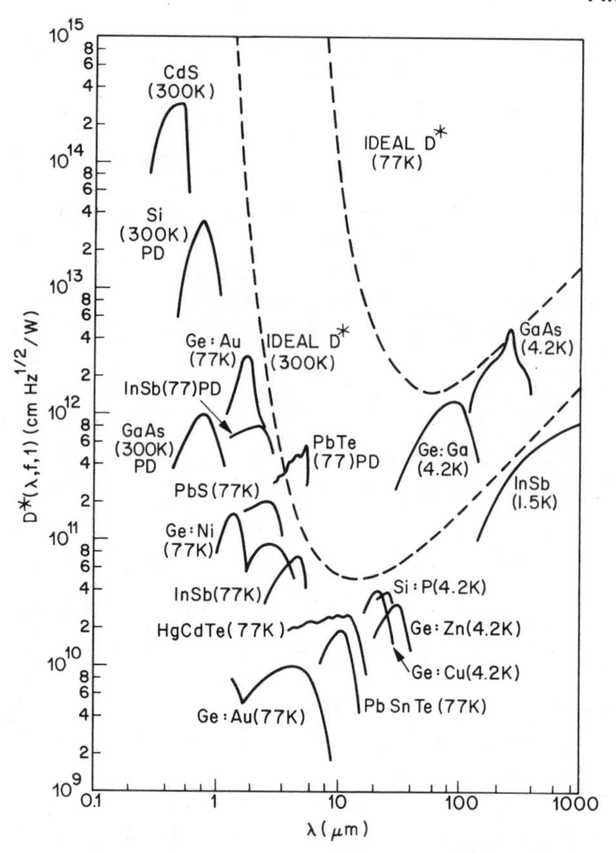

Fig. 4 Detectivity D^* as a function of wavelength for various photoconductors and photodiodes (indicated with PD). The dashed curves are the theoretical ideal D^* at 77 K and 300 K viewing an angle of 2π steradians. (After Kruse, McGlauchlin, and McQuistan, Ref. 12; Melchior, Refs. 2 and 3.)

4.2 K). The lower temperatures reduce thermal effect (which causes thermal ionization and depletes the energy levels) and increases the gain and detection efficiency. Near 0.5 μm, a CdS photoconductor gives a high sensitivity, whereas at 10 μm a HgCdTe photoconductor is preferred.[13] In the wavelength range from 100 to 400 μm, a GaAs photoconductor[14] is a better choice because of its high value of D^*. This photoconductor has high dynamic range and can give comparable performance for high-level (strong light intensity) detection. For low-level detection at microwave frequencies, however, a photodiode will provide considerably more speed and considerably higher signal-to-noise ratio. Thus photoconductors have limited use in high-frequency optical demodulators, such as in optical mixing. They have been, however, extensively used for infrared detections especially beyond a few microns, where, in spite of intensive work, satisfactory alternative detection techniques do not yet exist.

13.3 PHOTODIODE

13.3.1 General Consideration

A photodiode has a depleted semiconductor region with a high electric field that serves to separate photogenerated electron–hole pairs. For high-speed operation, the depletion region must be kept thin to reduce the transit time. On the other hand, to increase the quantum efficiency (the number of electron–hole pairs generated per incident photon), the depletion layer must be sufficiently thick to allow a large fraction of the incident light to be absorbed. Thus there is a trade-off between the speed of response and quantum efficiency.

The photodiode can be operated in a photovoltaic mode, that is, the photodiode is unbiased and connected to a load impedance similar to a solar cell (refer to Chapter 14). However, the device designs are fundamentally different. For a photodiode only a narrow wavelength range centered at the optical signal wavelength is important, whereas for a solar cell, high spectral responses over a broad solar wavelength range are required. Photodiodes are small to minimize junction capacitance, while solar cells are large-area devices. One of the most important figures of merit for photodiodes is the quantum efficiency, whereas the main concern for solar cells is the power conversion efficiency (power delivered to the load per incident solar energy).

For the visible and near-infrared range, photodiodes are usually reverse-biased with relatively large biasing voltages, because this reduces the carrier transit time and lowers the diode capacitance. The reverse voltage is, however, not large enough to cause avalanche breakdown. This biasing condition is in contrast to avalanche photodiodes, where an internal current gain is obtained as a result of the impact ionization under avalanche breakdown conditions. The photodiode family includes the *p-n* junction diode, *p-i-n* diode, metal–semiconductor diode (Schottky barrier), and heterojunction diode.

We shall now briefly consider the general characteristics of a photodiode: its quantum efficiency, response speed, and device noise.

The quantum efficiency as mentioned previously is the number of electron–hole pairs generated per incident photon:

$$\eta = (I_p/q)/(P_{opt}/h\nu) \tag{12}$$

where I_p is the photogenerated current by the absorption of incident optical power P_{opt} at a wavelength λ (corresponding to a photon energy $h\nu$).

A related figure of merit is the responsivity, which is the ratio of the photocurrent to the optical power:

$$\mathcal{R} = \frac{I_p}{P_{opt}} = \frac{\eta q}{h\nu} = \frac{\eta\lambda(\mu\text{m})}{1.24} \quad \text{A/W} \tag{13}$$

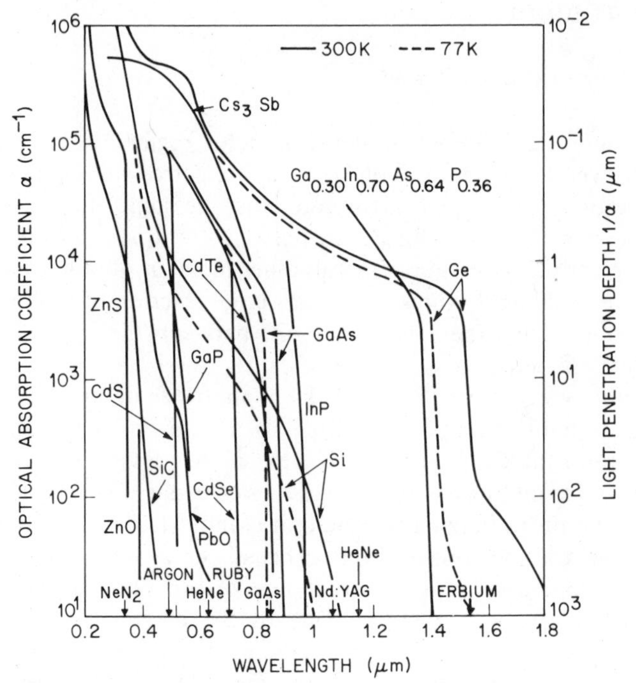

Fig. 5 Optical absorption coefficients for various photodetector materials; some laser emission wavelengths are indicated. (After Melchior, Ref. 2.)

Therefore, for a given quantum efficiency, the responsivity increases linearly with wavelength. For an ideal photodiode ($\eta = 1$), $\mathcal{R} = (\lambda/1.24)$(A/W), where λ is expressed in microns.

One of the key factors that determines the quantum efficiency is the absorption coefficient. Figures 5 and 6 show the measured intrinsic absorption coefficients α for various photodetector materials.[2] The solid curves are for 300 K and the dashed curves for 77 K. For Ge, Si, and III–V compound semiconductors, the curves shift toward longer wavelengths as the temperature increases. For some IV–VI compounds (e.g., PbSe), the opposite happens as the bandgap increases with increasing temperature. The emission wavelengths of some important lasers are also shown for reference.

Since α is a strong function of the wavelength, for a given semiconductor the wavelength range in which appreciable photocurrent can be generated is limited. The long-wavelength cutoff λ_c is established by the energy gap of the semiconductor, Eq. 1, for example, about 1.7 μm for Ge and 1.1 μm for Si. For wavelengths longer than λ_c, the values of α are too small to give appreciable absorption. The short-wavelength cutoff of the photoresponse comes about because the values of α for short wavelengths are very large ($\geqslant 10^5$ cm^{-1}), and the radiation is absorbed very near the

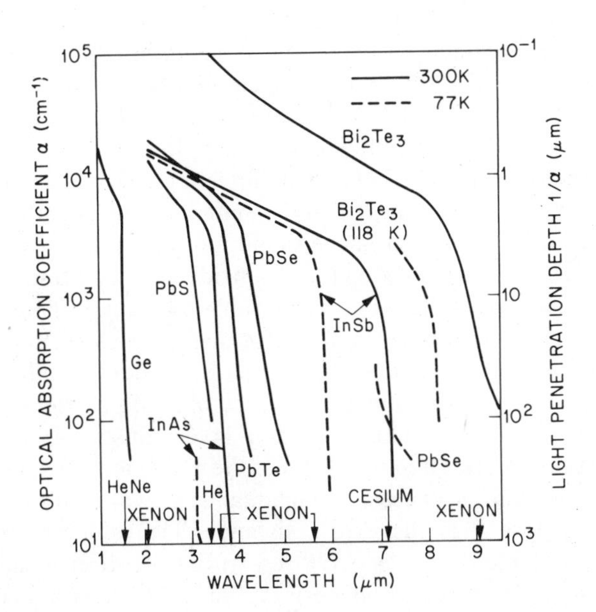

Fig. 6 Optical absorption coefficients for infrared photodetector materials. (After Melchior, Ref. 2.)

surface where the recombination time is short. The photocarriers thus can recombine before they are collected in the *p-n* junction. Figure 7 shows typical plots of quantum efficiency versus wavelength for some high-speed photodiodes. Responsivity curves are superimposed on the plot. In the ultraviolet and visible regions, metal–semiconductor photodiodes show good quantum efficiencies. In the near-infrared region, silicon photodiodes

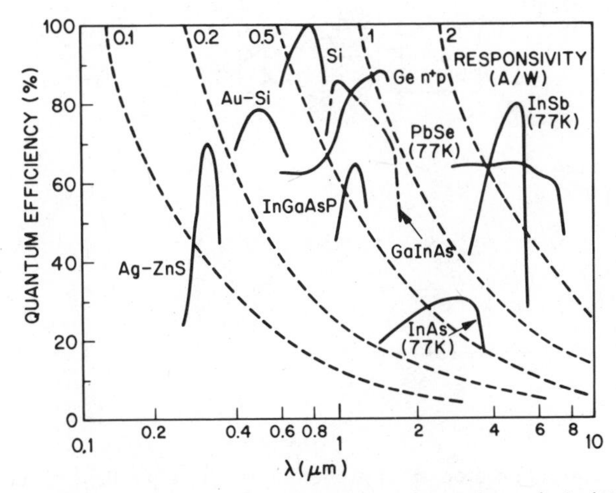

Fig. 7 Quantum efficiency and responsivity for various photodetectors.

with antireflection coating can reach 100% quantum efficiency near 0.8 to 0.9 μm. In the 1.0 to 1.6 μm region, Ge photodiodes, III–V ternary photodiodes (e.g., GaInAs), and III–V quaternary photodiodes (e.g., GaInAsP) have shown high quantum efficiencies. For longer wavelengths, photodiodes are cooled (e.g., 77 K) for high-efficiency operation.

The response speed is limited by a combination of three factors: diffusion of carriers, drift time in the depletion region, and capacitance of the depletion region. Carriers generated outside the depletion region must diffuse to the junction resulting in considerable time delay. To minimize the diffusion effect, the junction should be formed very close to the surface. Most light will be absorbed when the depletion region is sufficiently wide (of the order of $1/\alpha$); with sufficient reverse bias the carriers will drift at their saturation velocities. The depletion layer must not be too wide, however, or transit-time effects will limit the frequency response. It also should not be too thin, or excessive capacitance C will result in a large RC time constant, where R is the load resistance. The optimum compromise occurs when the depletion layer is chosen so that the transit time is of the order of one-half the modulation period. For example, for a modulation frequency of 10 GHz, the optimum depletion layer thickness in Si (with a saturation velocity of 10^7 cm/s) is about 5 μm.

To study the noise properties in a photodiode, we will consider the generalized photodetection process shown[5] in Fig. 8*a*. An optical signal and background radiation are absorbed by the photodiode, whereby electron–hole pairs are generated. These electrons and holes are then separated by the electric field and drift toward the opposite sides of the p-n junction. In the process, a displacement current is induced in the external load resister.

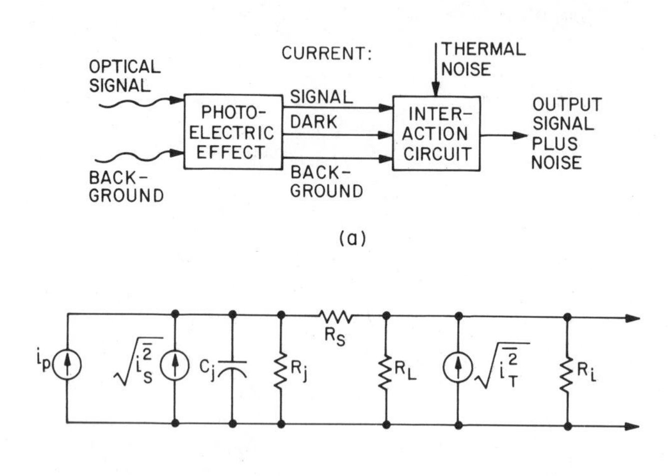

Fig. 8 (*a*) Photodetection process of a photodiode. (*b*) Equivalent circuit of a photodiode. (After Stillman and Wolfe, Ref. 5.)

To determine the current generated by this photoelectric process, we will consider an intensity-modulated optical signal given by Eq. 5. The average photocurrent due to the optical signal is

$$I_P = q\eta P_{opt}/h\nu. \tag{14}$$

For the modulated optical signal, the rms signal power is $mP_{opt}/\sqrt{2}$, and the rms signal current can be written as

$$i_p = q\eta m P_{opt}/\sqrt{2}h\nu. \tag{15}$$

The current resulting from the background radiation is I_B, and the dark current due to thermal generation of electron–hole pairs in the depletion region is I_D. Because of the randomness of the generation of all these currents, they contribute shot noise given by

$$\langle i_s^2 \rangle = 2q(I_P + I_B + I_D)B \tag{16}$$

where B is the bandwidth.

The equivalent circuit of a photodiode is shown[5] in Fig. 8*b*, where i_p, $\sqrt{i_s^2}$, C_j, R_j, and R_s are associated with the photodiode. The component C_j is the junction capacitance, R_j the junction resistance, and R_s the series resistance. The variable R_L is an external load resistor and R_i is the input resistance of the following amplifier.[55] All the resistances contribute additional thermal noise to the system. The series resistance R_s is usually much smaller than the other resistances and can be neglected. The thermal noise is given by

$$\langle i_T^2 \rangle = 4kT(1/R_{eq})B \tag{17}$$

where $1/R_{eq} = (1/R_j) + (1/R_L) + (1/R_i)$.

For a 100% modulated signal with average power P_{opt}, the signal-to-noise ratio can be written as

$$(S/N)_{\text{power}} = \frac{i_p^2 R_{eq}}{(\langle i_s^2 \rangle + \langle i_T^2 \rangle)R_{eq}} = \frac{\frac{1}{2}(q\eta P_{opt}/h\nu)^2}{2q(I_P + I_B + I_D)B + 4kTB/R_{eq}}\,. \tag{18}$$

From this equation, the minimum optical power required to obtain a given signal-to-noise ratio is

$$(P_{opt})_{\min} = \frac{2h\nu B}{\eta}\left(\frac{S}{N}\right)\left\{1 + \left[1 + \frac{I_{eq}}{qB(S/N)}\right]^{1/2}\right\} \tag{19}$$

where

$$I_{eq} = I_B + I_D + 2kT/qR_{eq}\,.$$

In the limit when $I_{eq}/qB(S/N)$ is much less than unity, the minimum optical power is determined by the quantum noise associated with the optical signal itself. In the other limit when $I_{eq}/qB(S/N)$ is much larger than unity, the background radiation and/or the thermal noise of the equivalent resistor becomes dominant. Under this condition the noise-equivalent power

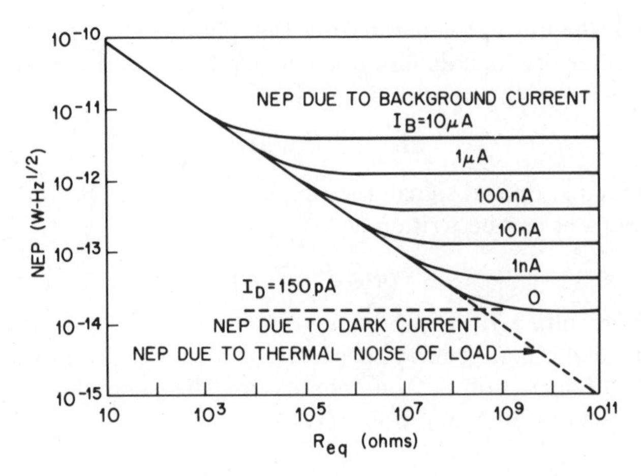

Fig. 9 Variation of NEP of *p-i-n* diode with load resistance R_{eq} for 150-pA dark current and various background current. (After Stillman and Wolfe, Ref. 5.)

(NEP) is given by

$$\begin{aligned} \text{NEP} &= \text{rms optical power } (P_{opt})_{min} \text{ with } (S/N) = 1 \text{ and } B = 1 \text{ Hz} \\ &= \sqrt{2}(h\nu/\eta)(I_{eq}/q)^{1/2} \qquad \text{W/cm}^2\text{Hz}^{1/2}. \end{aligned} \tag{20}$$

To improve the sensitivity of a photodiode, both η and R_{eq} should be increased while I_B and I_D should be decreased. The NEP calculated for a typical Si photodiode is shown[5] in Fig. 9 for $\eta = 75\%$ at $\lambda = 0.77\ \mu$m and for a dark current $I_D = 1.5 \times 10^{-10}$ A. The results show that high values of R_{eq} must be used to achieve an NEP limited by dark-current or background-current shot noise.

13.3.2 *p-i-n* Photodiode

The *p-i-n* photodiode is one of the most common photodetectors, because the depletion-region thickness (the intrinsic layer) can be tailored to optimize the quantum efficiency and frequency response. Figure 10 shows schematic representation of a *p-i-n* diode and an energy-band diagram under reverse-bias conditions together with optical absorption characteristics.[2] We shall discuss the operation of *p-i-n* photodiode in some detail with the help of Fig. 10. This discussion applies also to *p-n* junction photodiodes. Light absorption in the semiconductor produces hole–electron pairs. Pairs produced in the depletion region or within a diffusion length of it will eventually be separated by the electric field, leading to current flow in the external circuit as carriers drift across the depletion layer.

Under steady-state conditions the total current density through the

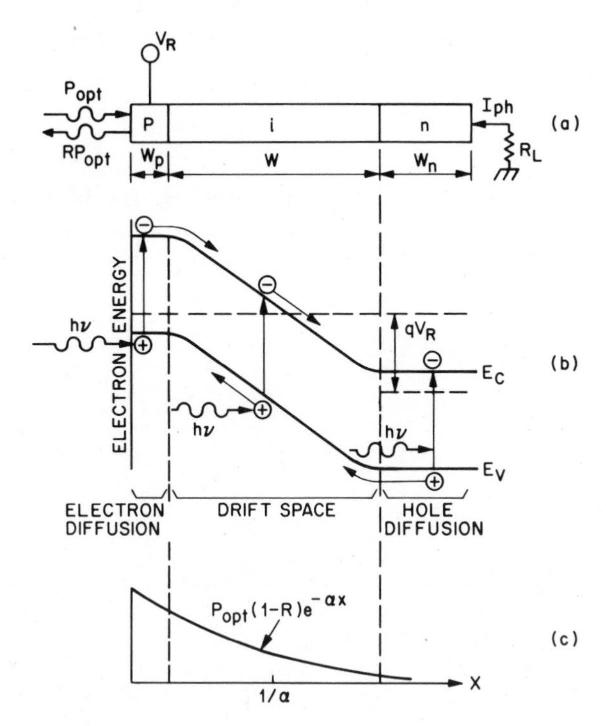

Fig. 10 Operation of photodiode. (*a*) Cross-sectional view of *p-i-n* diode. (*b*) Energy-band diagram under reverse bias. (*c*) Carrier generation characteristics. (After Melchior, Ref. 2.)

reverse-biased depletion layer is given by[15]

$$J_{\text{tot}} = J_{dr} + J_{\text{diff}} \tag{21}$$

where J_{dr} is the drift current due to carriers generated inside the depletion region and J_{diff} is the diffusion current density due to carriers generated outside the depletion layer in the bulk of the semiconductor and diffusing into the reverse-biased junction. We shall now derive the total current under the assumptions that the thermal generation current can be neglected and that the surface p layer is much thinner than $1/\alpha$. Referring to Fig. 10*c*, the hole–electron generation rate is given by

$$G(x) = \Phi_0 \alpha e^{-\alpha x} \tag{22}$$

where Φ_0 is the incident photon flux per unit area given by $P_{\text{opt}}(1-R)/Ah\nu$, where R is the reflection coefficient and A is the device area. The drift current J_{dr} is thus given by

$$J_{dr} = -q \int_0^W G(x)\,dx = q\Phi_0(1 - e^{-\alpha W}) \tag{23}$$

where W is the depletion-layer width. For $x > W$, the minority-carrier density (holes) in the bulk semiconductor is determined by the one-dimensional diffusion equation

$$D_p \frac{\partial^2 p_n}{\partial x^2} - \frac{p_n - p_{no}}{\tau_p} + G(x) = 0 \tag{24}$$

where D_p is the diffusion coefficient for holes, τ_p the lifetime of excess carriers, and p_{no} the equilibrium hole density. The solution of Eq. 24 under the boundary conditions $p_n = p_{no}$ for $x = \infty$ and $p_n = 0$ for $x = W$ is given by

$$p_n = p_{no} - (p_{no} + C_1 e^{-\alpha W}) e^{(W-x)/L_n} + C_1 e^{-\alpha x} \tag{25}$$

with $L_p = \sqrt{D_p \tau_p}$ and

$$C_1 \equiv \left(\frac{\Phi_0}{D_p}\right) \frac{\alpha L_p^2}{1 - \alpha^2 L_p^2}. \tag{26}$$

The diffusion current density is given by $J_{\text{diff}} = -qD_p(\partial p_n/\partial x)_{x=W}$,

$$J_{\text{diff}} = q\Phi_0 \frac{\alpha L_p}{1 + \alpha L_p} e^{-\alpha W} + q p_{no} \frac{D_p}{L_p} \tag{27}$$

and the total current density is obtained as

$$J_{\text{tot}} = q\Phi_0 \left(1 - \frac{e^{-\alpha W}}{1 + \alpha L_p}\right) + q p_{no} \frac{D_p}{L_p}. \tag{28}$$

Under normal operating conditions, the term involving p_{no} is much smaller so that the total photocurrent is proportional to the photon flux. The quantum efficiency can be obtained from Eqs. 12 and 28,

$$\eta = \frac{J_{\text{tot}}/q}{P_{\text{opt}}/Ah\nu} = (1 - R)\left(1 - \frac{e^{-\alpha W}}{1 + \alpha L_p}\right). \tag{29}$$

For high quantum efficiency, a low reflection coefficient with $\alpha W \gg 1$ is desirable. However, for $W \gg 1/\alpha$, the transit-time delay may be considerable. We consider next the transit-time effect.

Since the carriers require a finite time to traverse the depletion layer, a phase difference between the photon flux and the photocurrent will appear when the incident light intensity is modulated rapidly. To obtain a quantitative result for this effect the simplest case is shown in Fig. 11*a*. The applied voltage is assumed to be high enough to deplete the intrinsic region and to ensure carrier saturation velocity, v_s. For a photon flux density given by $\Phi_1 e^{j\omega t}$ (photons/s-cm^2), the conduction current density J_{cond} at point x is found to be

$$J_{\text{cond}}(x) = q\Phi_1 e^{j\omega(t - x/v_s)}. \tag{30}$$

Since $\nabla \cdot J_{\text{tot}} = 0$, we can write

$$J_{\text{tot}} = \frac{1}{W} \int_0^W \left(J_{\text{cond}} + \epsilon_s \frac{\partial \mathscr{E}}{\partial t}\right) dx \tag{31}$$

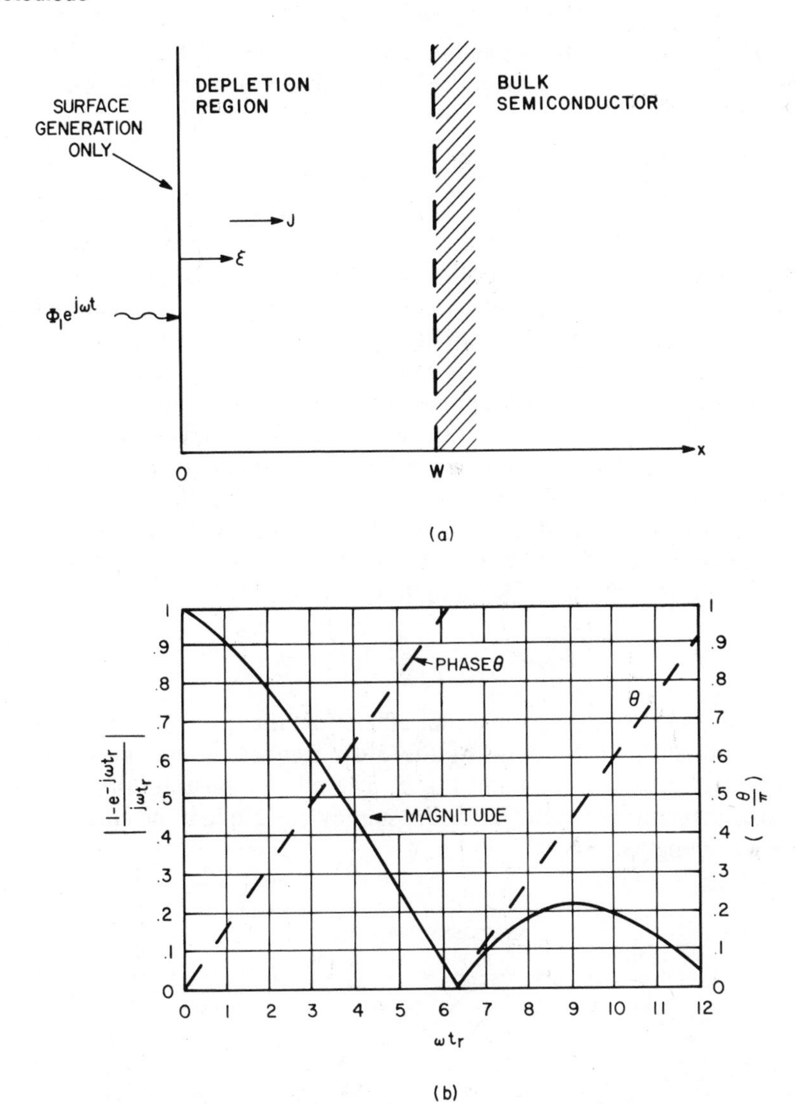

Fig. 11 (*a*) Geometry assumed for analysis of transit-time effect. (*b*) Photoresponse (normalized current or voltage) versus normalized modulation frequency of incident photon flux where $\theta \equiv \omega t_r/2$. (After Gartner, Ref. 15.)

where the second term in parentheses is the displacement-current density and ϵ_s and $\mathscr{E}$ are the permittivity and electric field, respectively. Substituting Eq. 30 into Eq. 31 yields

$$J_{\text{tot}} = \left(\frac{j\omega\epsilon_s V}{W} + q\Phi_1 \frac{1 - e^{-j\omega t_r}}{j\omega t_r}\right)e^{j\omega t} \tag{32}$$

where V is the sum of applied voltage and the built-in voltage, and $t_r \equiv W/v_s$ is the transit time of carriers through the depletion region. From Eq. 32 the short-circuit current density ($V = 0$) is given by

$$J_{sc} = \frac{q\Phi_1(1 - e^{j\omega t_r})}{j\omega t_r} e^{j\omega t}. \tag{33}$$

Figure 11*b* shows the transit-time effects at high frequencies where the amplitude and phase angle of the normalized current are plotted as functions of the normalized modulation frequency. Note that the magnitude of the ac photocurrent decreases rapidly with frequency when ωt_r exceeds unity. At $\omega t_r = 2.4$, the amplitude is reduced by $\sqrt{2}$ and is accompanied by a phase shift of $2\pi/5$. The response time of the photodetector is thus limited by the carrier transit time through the depletion layer. A reasonable compromise between high-frequency response and high quantum efficiency is obtained for an absorption region of thickness $1/\alpha$ to $2/\alpha$.

For the *p-i-n* diode the thickness of the *i* region is assumed equal to $1/\alpha$. The carrier transit time is the time required for carriers to drift through the *i* region. From Eq. 33 the 3-dB frequency is given by

$$f_{3\text{dB}} = \frac{2.4}{2\pi t_r} \simeq \frac{0.4 v_s}{W} \approx 0.4 \alpha v_s. \tag{34}$$

Figure 12 shows the internal quantum efficiency, that is, $\eta/(1 - R)$ of the Si *p-i-n* photodiode as a function of the 3-dB frequency and the depletion width calculated from Eq. 34 and Fig. 5. The curves illustrate the trade-off between the response speed (3 dB frequency) and quantum efficiency at various wavelengths.

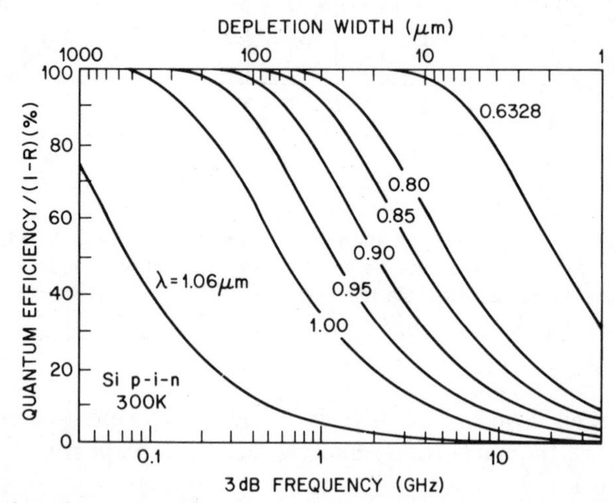

Fig. 12 Variation of quantum efficiency of Si *p-i-n* photodiode with depletion width and transit-time-limited 3-dB frequency for several wavelengths. The saturation velocity is 10^7 cm/s.

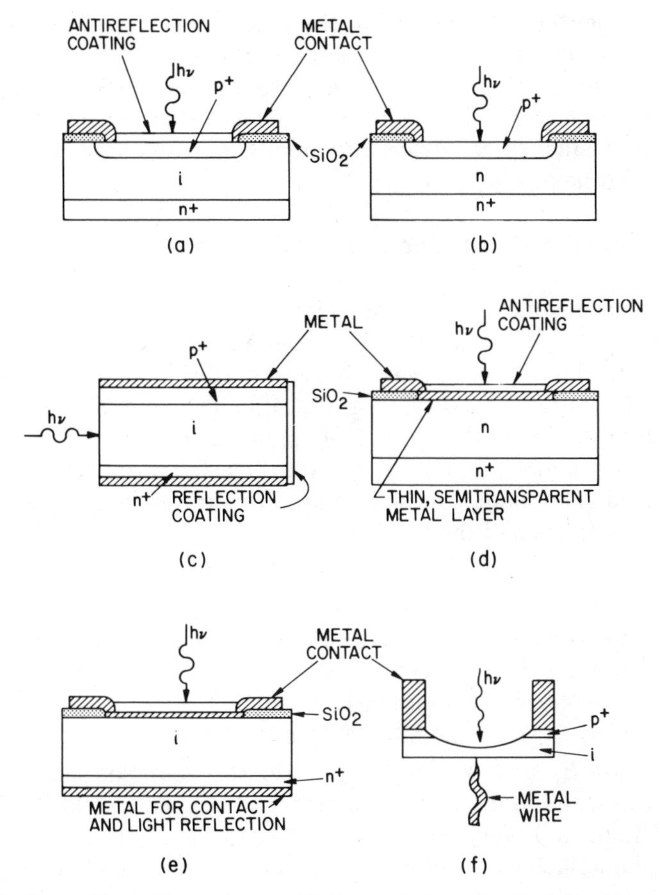

Fig. 13 Device configurations of some high-speed photodiodes. (*a*) *p-i-n* diode. (*b*) *p-n* diode. (*c*) *p-i-n* diode with illumination parallel to junction. (*d*) Metal–semiconductor diode. (*e*) Metal–*i*-*n* diode. (*f*) Semiconductor point-contact diode. (After Melchior, Ref. 2.)

The constructions of some high-speed photodiodes are shown[2] in Fig. 13. The *p-i-n* photodiode is shown in Fig. 13*a* with an antireflection coating to increase quantum efficiency. The thickness of the intrinsic region (in practice, one uses a low *n*-type doping, ν region, or a low *p*-type doping, π region) is optimized for the optical-signal wavelength and the modulation frequency. The *p-n* photodiode is a related device where the *n*-type doping is high so that the *n*-type layer is not fully depleted (Fig. 13*b*). The *p-n* photodiodes generally have a lower response speed than the *p-i-n* photodiodes because of the larger contribution of diffusion current. At a wavelength close to the long-wavelength cutoff, the required absorption depth becomes very long (for $\alpha = 10\ \text{cm}^{-1}$, $1/\alpha = 1000\ \mu\text{m}$). A compromise between quantum efficiency and response speed can be reached if the light is incident from the side, parallel to the junction as shown in Fig. 13*c*. The light can also be allowed to strike at an angle that creates multiple reflections inside the device, substantially increasing the effective absorption depth and at the same time keeping the carrier transit distance small.[16,17] The other three devices are metal–semiconductor photodiodes, to be considered next.

13.3.3 Metal–Semiconductor Photodiode

A metal–semiconductor diode can be used as a high-efficiency photodetector.[18,19] The energy-band diagram and current transport in a metal–semiconductor diode have been considered in Chapter 5, and a typical configuration is illustrated in Fig. 13*d*. To avoid large reflection and absorption losses when the diode is illuminated through the metal contact, the metal film must be very thin (~100 Å) and an antireflection coating must be used. The diode can be operated in various modes, depending on the photon energies and the biasing conditions:

1. For $E_g > h\nu > q\phi_{Bn}$ and $V < V_B$, Fig. 14*a*, where V_B is the avalanche breakdown voltage, the photoexcited electrons in the metal can surmount the barrier and be collected by the semiconductor. This process has been used extensively to determine the Schottky-barrier height and to study the hot-electron transport in metal films.[20]
2. For $h\nu > E_g$ and $V < V_B$, Fig. 14*b*, the radiation produces hole–electron pairs in the semiconductor, and the general characteristics of the diode are very similar to those of a *p-i-n* photodiode. The quantum efficiency is given by an expression identical to Eq. 29.
3. For $h\nu > E_g$ and $V \simeq V_B$ (high reverse-bias voltage), Fig. 14*c*, the diode can be operated as an avalanche photodiode (discussed in Section 13.4).

Metal–semiconductor photodiodes are particularly useful in the visible and ultraviolet regions. In these regions the absorption coefficients α in most of the common semiconductors are very high, of the order of 10^5 cm^{-1} or more, corresponding to an effective absorption length of $1/\alpha \simeq 0.1$ μm or less. It is possible to choose a proper metal and a proper antireflection coating so that a large fraction of the incident radiation will be absorbed near the surface of the semiconductor.

An interesting example shown in Fig. 15, demonstrates that low-loss transmission into the semiconductor substrate is feasible.[19] Figure 15*a* shows a particular choice of antireflection coating of 500 Å ZnS with a refractive index of 2.30. The gold film has a complex refractive index $\bar{n} = (0.28 - j0.301)$ at $\lambda = 6328$ Å (He–Ne laser wavelength), while at this wavelength the silicon substrate has a refractive index of $(3.75 - j0.018)$.

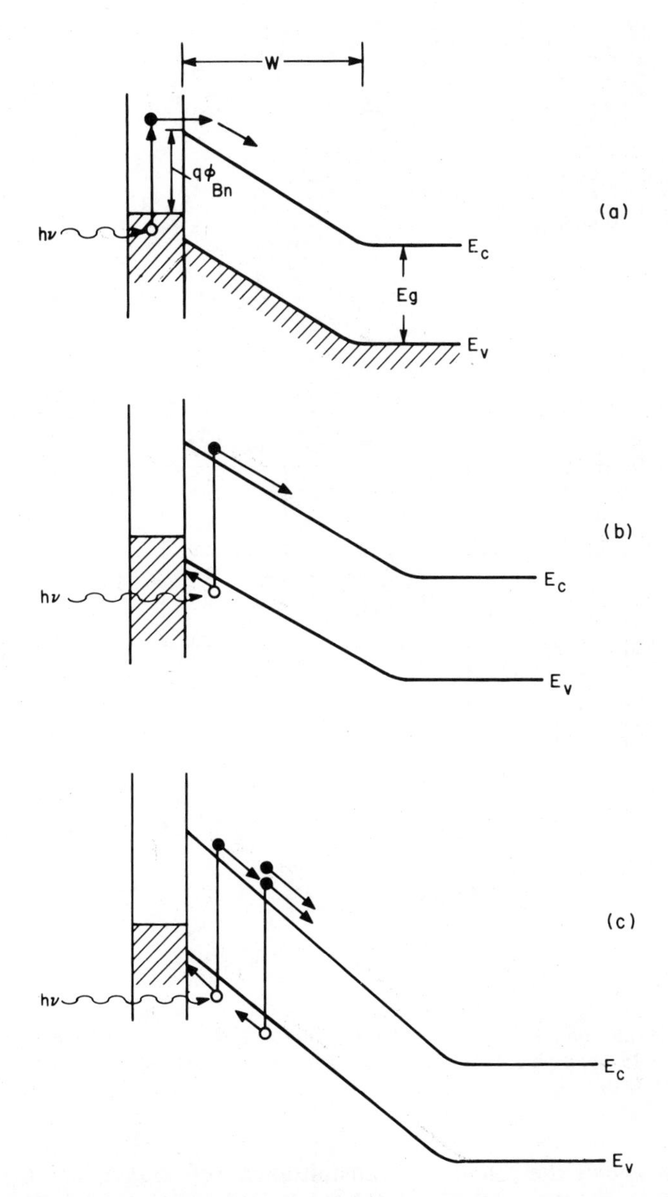

Fig. 14 (*a*) Photoelectric emission of excited electrons from metal to semiconductor ($E_g > h\nu > q\phi_{Bn}$). (*b*) Band-to-band excitation of a hole–electron pair ($h\nu > E_g$). (*c*) Hole–electron pair generation and avalanche multiplication under a large reverse bias ($h\nu > E_g$ and $V \simeq V_B$).

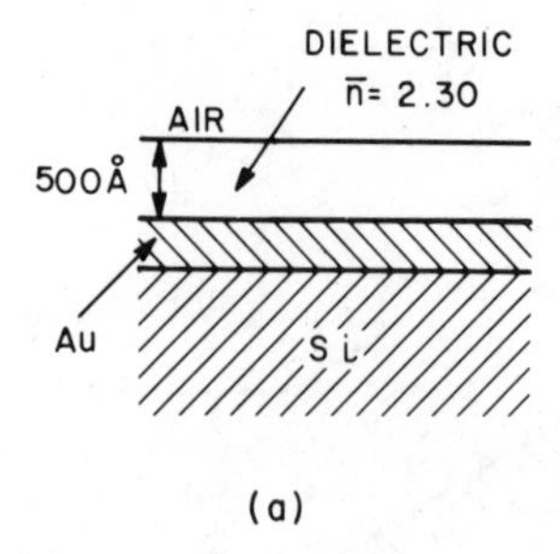

(a)

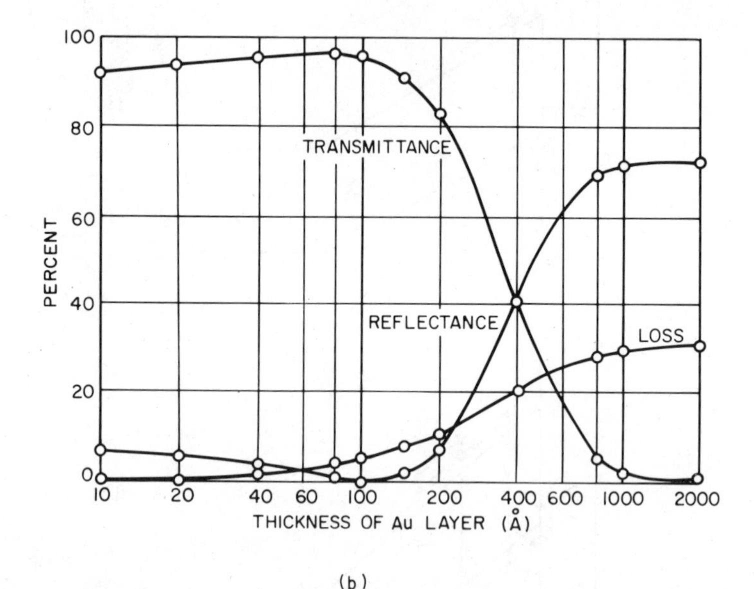

(b)

Fig. 15 (*a*) A 500-Å-thick ZnS antireflection coating. (*b*) Transmittance, reflectance, and loss in the gold films as a function of the gold layer thickness, $\lambda = 0.6328\ \mu m$. (After Schneider, Ref. 19.)

Figure 15*b* shows the calculated transmittance, reflectance, and loss in the Au films as a function of the Au layer thickness. At about 100 Å, more than 95% of the incident light will be transmitted into the silicon substrate. Experimental results with photodiodes similar to that shown in Fig. 13*d* show a net quantum efficiency of 70% at $\lambda = 6328$ Å and a very fast response of the order of 0.1 ns (1 nanosecond $= 10^{-9}$ s) pulse rise time.[17]

The quantum efficiency and the transmission coefficient of Au-*n*-Si photodiodes[21] (with 107 Å Au) are shown in Fig. 16 for photon energy from 1 to 6 eV. In the visible region (up to $h\nu = 3.1$ eV), the values of η and the transmission coefficient are practically identical. In the ultraviolet region,

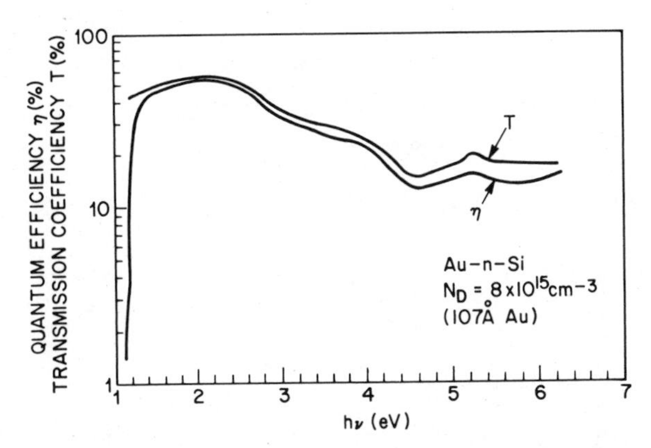

Fig. 16 Quantum efficiency and transmission coefficient of a Au–Si photodiode. (After Gutkin, Dmitriev, and Khait, Ref. 21.)

they differ by less than 25%. Over the photon energy range, the quantum efficiency is found to be practically independent of doping concentration from 2×10^{13} to 2×10^{17} cm^{-3}. This indicates that there is negligible loss due to diffusion of electrons against the electric field in the depletion region or to recombination at metal–semiconductor interface traps.[22]

By using a low-doping *i*-layer, a metal–*i*–*n* photodiode similar to a *p*-*i*-*n* diode can be made, Fig. 13*e*. A special metal–semiconductor diode is the point-contact diode[23] shown in Fig. 13*f*. The active volume is very small, and as a result both the drift time and capacitance are extremely small. It is thus suitable for very high modulation frequencies. The device is limited, however, to applications where the radiation can be focused on a spot a few microns in diameter.

13.3.4 Heterojunction Photodiode

A photodiode can be realized in a heterojunction in which the junction is formed between two semiconductors of different bandgaps (refer to Chapter 2). One advantage of a heterojunction photodiode is that the quantum efficiency does not critically depend on the distance of the junction from the surface, because a large-bandgap material can be used as a window for the transmission of optical power. In addition, the heterojunction can provide unique material combinations so that the quantum efficiency and response speed can be optimized for a given optical signal wavelength.

To obtain heterojunction with low leakage current, the lattice constants of the two semiconductors must be closely matched. Ternary III–V semiconductors $Al_xGa_{1-x}As$ (direct gap for $x < 0.4$) epitaxially grown on GaAs form heterojunctions with perfectly matched lattices[24] (with a lattice constant of 5.653 Å). These heterojunctions are important for photonic

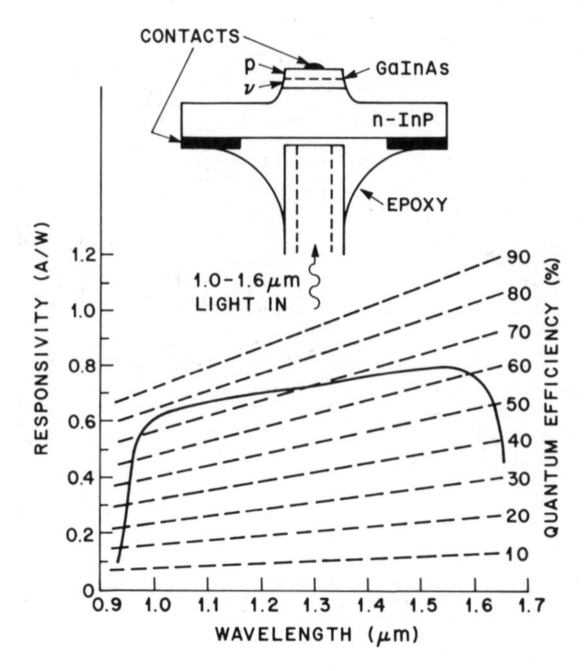

Fig. 17 Responsivity and quantum efficiency of a GaInAs *p-i-n* photodiode versus wavelength. (After Lee, Burrus, and Dentai, Ref. 28.)

devices operated in the wavelength range 0.65 to 0.85 μm. A double-heterojunction photodetector (n-$Al_{0.24}Ga_{0.76}As$/p-GaAs/p-$Al_{0.24}Ga_{0.76}As$) with an antireflection coating operated at $\lambda = 0.8075$ μm shows a quantum efficiency[25] of 92%. Similarly, a cascade structure[26] of two *p-i-n* double AlGaAs heterojunction series connected by a tunnel junction shows high efficiency at $\lambda = 0.815$ μm and has an open circuit voltage of 1.78 V.

At longer wavelengths (1 to 1.6 μm), ternary compounds such as $Ga_{0.47}In_{0.53}As$ (with $E_g = 0.73$ eV)[59] and quaternary compounds such as $Ga_xIn_{1-x}As_yP_{1-y}$ (e.g., $Ga_{0.27}In_{0.73}As_{0.63}P_{0.37}$ with $E_g = 0.95$ eV)[27] can be employed and these compounds have a perfect lattice match to a InP substrate (with lattice constant of 5.8686 Å, refer to Fig. 24 of Chapter 12). The insert in Fig. 17 shows the back-illuminated mesa structure of a p-$Ga_{0.47}In_{0.53}As$/ν-$Ga_{0.47}In_{0.53}As$/n^+–InP photodiode. The responsivity, greater than 0.6 A/W, is almost flat over the spectral range from 1.0 to 1.5 μm (Fig. 17).[28] The highest external quantum efficiency at 1.3 μm is about 70% with typical values of 55 to 60%. This device is expected to have superior performance over the Ge photodiode, because of its direct bandgap, which gives rise to a larger absorption coefficient near the intrinsic absorption edge, so that a thinner depletion width can be used to give a higher response speed.[29]

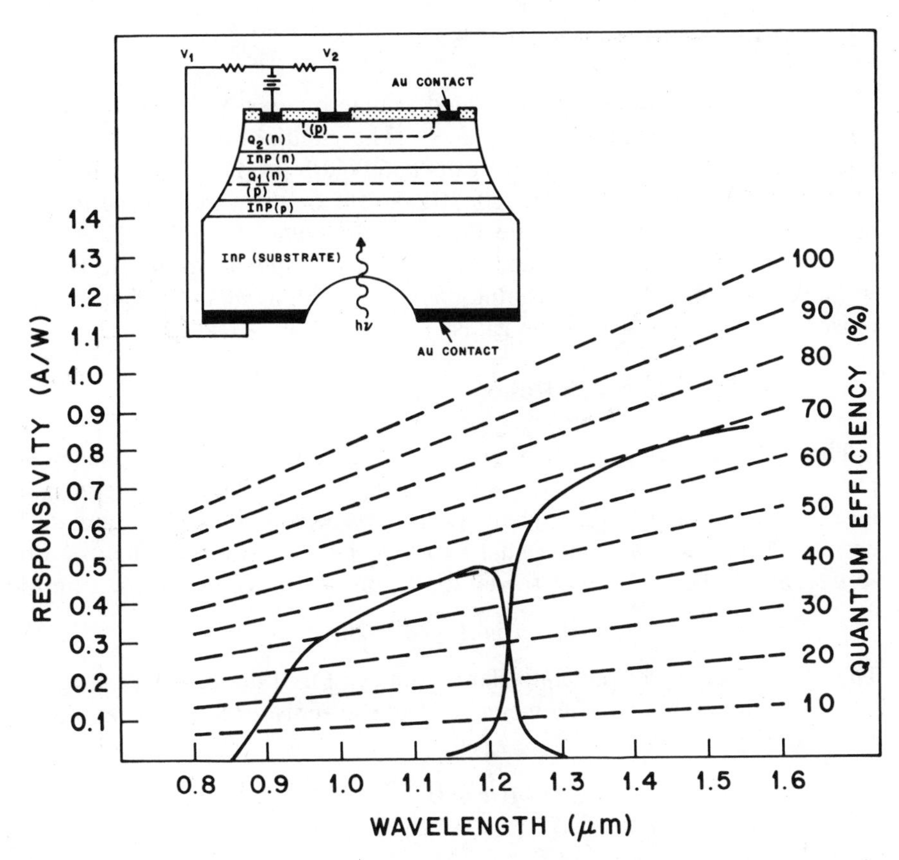

Fig. 18 Responsivity and quantum efficiency of a demultiplexing photodiode versus wavelength. The insert shows the cross section of the photodiode. (After Campbell et al., Ref. 56.)

The insert in Fig. 18 shows a heterojunction photodiode structure capable of detecting and demultiplexing two wavelength bands simultaneously.[56] The device is a multilayer structure in which wavelength discrimination is accomplished by two $Ga_xIn_{1-x}As_yP_{1-y}$ quaternary layers that are lattice-matched to InP but have different crystal composition and, consequently, different bandgaps. The bandgap of layer Q_1 (1.0 eV) is larger than that of Q_2 (0.74 eV) and light is incident through the transparent substrate. The spectral response is overlapped representing optical crosstalk between the two channels (Fig. 18). Proper utilization of the device would avoid using this overlapped region. For example, on the short-wavelength side of the overlapped region, the crosstalk is −19 dB at 1.2 μm and decreases to −30 dB at 1.15 μm; on the long-wavelength side, it is −43 dB at 1.3 μm.

13.4 AVALANCHE PHOTODIODE

Avalanche photodiodes are operated at high reverse-bias voltages where avalanche multiplication takes place.[30] The multiplication gives rise to internal current gain. The current gain–bandwidth product of an avalanche photodiode can be higher than 100 GHz, so the device can respond to light modulated at microwave frequencies.[2,31] For avalanche photodiodes, the criteria with respect to quantum efficiency and response speed are similar to those for nonavalanching photodiodes. In addition, we must consider the noise properties and avalanche gains.

13.4.1 Signal-to-Noise Ratio

The photodetection process and equivalent circuit for an avalanche photodiode are shown schematically[5] in Fig. 19*a*. The current gain mechanism multiplies the signal current, background current, and dark current. The multiplied rms signal photocurrent is identical to that of Eq. 15 except for the addition of the multiplication factor or avalanche gain M,

$$i_p = q\eta m P_{\text{opt}} M/\sqrt{2}h\nu. \tag{35}$$

The other elements of the equivalent circuit in Fig. 19*b* are the same as for the *p-i-n* photodiode. The mean-square shot-noise current after multiplication is given by

$$\langle i_s^2 \rangle = 2q(I_P + I_B + I_D)\langle M^2 \rangle B \tag{36}$$

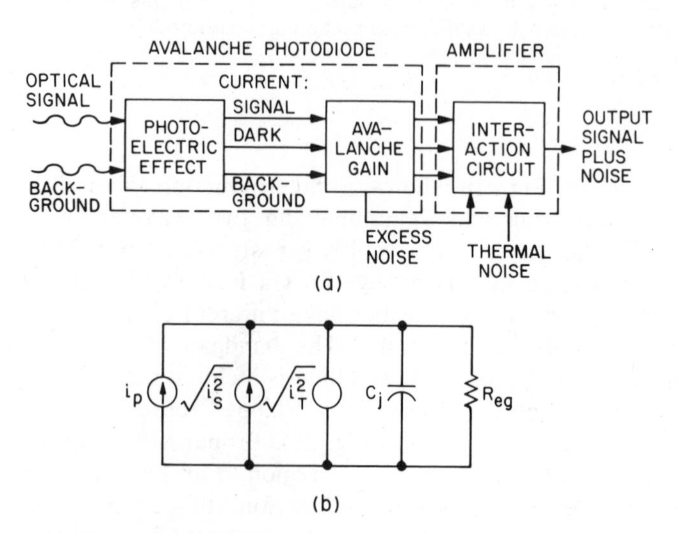

Fig. 19 (*a*) Photodetection process. (*b*) Equivalent circuit for avalanche photodiode. (After Stillman and Wolfe, Ref. 5.)

or

$$\langle i_s^2 \rangle = 2q(I_P + I_B + I_D)M^2F(M)B \tag{37}$$

where $\langle M^2 \rangle$ is the mean-square value of the internal gain, and the noise factor $F(M) = \langle M^2 \rangle / M^2$ is a measure of the increase in the shot noise compared to an ideal noiseless multiplier. The thermal noise is the same as for the *p-i-n* photodiode and is given by Eq. 17.

For a 100% modulated signal with average power P_{opt}, the signal-to-noise power ratio for the avalanche photodiode is then

$$S/N = \frac{\frac{1}{2}(q\eta P_{opt}/h\nu)^2 M^2}{2q(I_P + I_B + I_D)F(M)M^2B + 4kTB/R_{eq}}$$

$$= \frac{\frac{1}{2}(q\eta P_{opt}/h\nu)^2}{2q(I_P + I_B + I_D)F(M)B + 4kTB/R_{eq}M^2}. \tag{38}$$

From Eq. 38 we see that the avalanche gain can increase the signal-to-noise ratio by reducing the importance of the last term in the denominator. We will show that the noise factor $F(M)$ is always equal to or greater than unity and increases monotonically with multiplication except for a noiseless multiplication process. Thus there is an optimum value of M which produces the maximum signal-to-noise ratio for a given optical power. This optimum multiplication is obtained when the first term in the denominator is approximately equal to the second term.

Equation 38 can be solved for the minimum optical power P_{opt} required to produce a given S/N with avalanche gain. This power is

$$(P_{opt})_{min} = \frac{2h\nu}{\eta}\left(\frac{S}{N}\right)\left\{1 + \left[1 + \frac{I_{eq}}{qBF(M)^2(S/N)}\right]^{1/2}\right\} \tag{39}$$

where

$$I_{eq} \equiv (I_B + I_D)F(M) + 2kT/qR_{eq}M^2. \tag{40}$$

Under the condition that $I_{eq}/qBF^2(S/N)$ is negligible, the minimum detectable power is 3 dB above the quantum noise in the signal itself in a noiseless multiplication process ($F = 2$). For high frequency and large bandwidth detection, the minimum detectable power is limited by the thermal noise of the load resistance and the noise figure of the following amplifier stage. Under this condition, the term $I_{eq}/qBF(M)^2(S/N)$ is larger than unity and the noise equivalent power is given by

$$\text{NEP} = \sqrt{2}(h\nu/\eta)[I_{eq}/qF(M)^2]^{1/2}. \tag{41}$$

Since avalanche gain can substantially reduce the NEP, the avalanche photodiodes can have a significant advantage over non-avalanching photodides.

13.4.2 Avalanche Gain

The avalanche gain, also called multiplication factor, has been considered in Chapter 2. The low-frequency avalanche gain for electrons is given by

$$M = \left\{1 - \int_0^W \alpha_n \exp\left[-\int_0^x (\alpha_n - \alpha_p)\, dx'\right] dx\right\}^{-1} \tag{42}$$

where W is the depletion-layer width and α_n and α_p are the electron and hole ionization rate, respectively. For position-independent ionization coefficients, as in a p-i-n diode, the multiplication of electrons injected into the high-field region at $x = 0$ is

$$M = \frac{(1 - \alpha_p/\alpha_n) \exp[\alpha_n W(1 - \alpha_p/\alpha_n)]}{1 - (\alpha_p/\alpha_n) \exp[\alpha_n W(1 - \alpha_p/\alpha_n)]}. \tag{43}$$

For equal ionization coefficients ($\alpha_n = \alpha_p$), the multiplication takes the simple form

$$M = 1/(1 - \alpha_n W). \tag{44}$$

The breakdown voltage corresponds to the situation where $\alpha_n W = 1$. The regenerative avalanche process results in the presence of a large number of carriers in the high-field region long after the primary electrons have traversed through that region. The higher the avalanche gain (or multiplication) is, the longer the avalanche process persists, thus implying a behavior that is set by a gain–bandwidth product. For equal ionization coefficients and $M \to \infty$, the current gain–bandwidth product is found to be

$$\text{gain–bandwidth} = \frac{3}{2\pi\tau_{av}} \tag{45}$$

where $\tau_{av} = (t_n + t_p)/2$, and t_n is the electron transit time equal to W/v_n, where v_n is the electron saturation velocity. A similar expression is found for hole transit time t_p.

Figure 20 shows the calculated bandwidth[32] for an idealized p-i-n avalanche photodiode with an avalanche region of uniform electric field. The 3-dB bandwidth B, normalized to 2π times the average transit time τ_{av}, is plotted as a function of the low-frequency gain with the ratio of the ionization coefficient as a parameter. The dashed curve is for $M = \alpha_n/\alpha_p$. Above this curve, where $M < \alpha_n/\alpha_p$, the bandwidth is largely determined by the transit time of the carriers and is essentially independent of gain. Below this curve, where $M > \alpha_n/\alpha_p$, the curves are almost straight lines, indicating a constant gain–bandwidth product given by

$$M(\omega)\omega \approx \frac{1}{(W/v_n)(\alpha_p/\alpha_n)}. \tag{46}$$

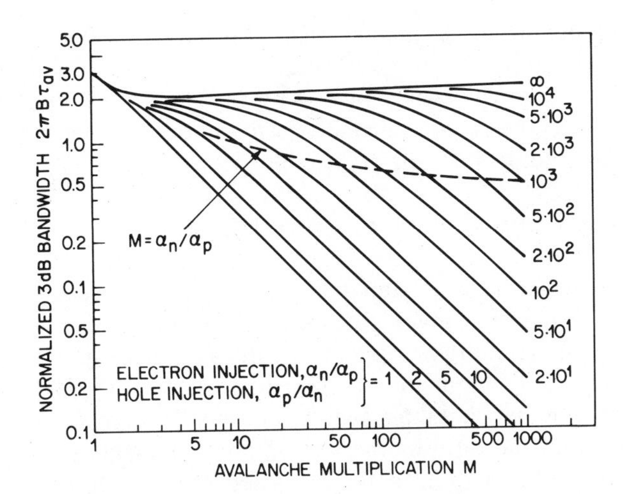

Fig. 20 Theoretical 3-dB bandwidth B times $2\pi\tau_{ac}$ (τ_{ac} = average carrier transit time) of avalanche photodiode as a function of low-frequency multiplication factor for various values of α_p/α_n for electron injection (or α_n/α_p for hole injection). (After Emmons, Ref. 32.)

Thus to obtain large gain–bandwidth products, v_n should be large and α_p/α_n and W should be small.

In a practical device, the maximum achievable dc multiplication at high light intensities is limited by the series resistance and the space-charge effect. These factors can be combined into one effective series resistance R. The multiplication for photogenerated carriers can be described by an empirical relationship as[33]

$$M_{ph} = \frac{I - I_{MD}}{I_P - I_D} = \frac{1}{\left[1 - \left(\frac{V_R - IR}{V_B}\right)^n\right]} \tag{47}$$

where I is the total multiplied current, I_p is the total primary (unmultiplied) current, and I_D and I_{MD} are the primary and multiplied dark currents, respectively. V_R is the reverse-biased voltage, V_B is the breakdown voltage, and the exponent n is a constant depending on the semiconductor material, doping profile, and radiation wavelength. For high light intensity ($I_P \gg I_D$) and $IR \ll V_B$, the maximum value of the photomultiplication is given by

$$(M_{ph})_{\max} \simeq \frac{I}{I_P} = \left.\frac{1}{1 - \left(\frac{V_R - IR}{V_B}\right)^n}\right|_{V_R \to V_B} \approx \frac{1}{nIR/V_B} \tag{48}$$

or

$$(M_{ph})_{\max} = \sqrt{V_B/nI_PR}. \tag{49}$$

When the photocurrent is smaller than the dark current, the maximum multiplication is limited by the dark current and is given by an expression similar to Eq. 49, except I_P is replaced by I_D. Thus it is important that the dark current be as low as possible so that it will not limit either the $(M_{ph})_{max}$ or the minimum detectable power, Eq. 39.

13.4.3 Avalanche Multiplication Noise

The avalanche process is statistical in nature because every electron–hole pair generated at a given distance in the depletion region does not experience the same multiplication. Since the avalanche gain fluctuates, the mean-square value of the gain is greater than the square of the mean. The excess noise can be characterized by a noise factor $F(M) = \langle M^2 \rangle / M^2$, which depends on the ratio of the ionization coefficients α_p/α_n and on the low-frequency multiplication factor M. When $\alpha_n = \alpha_p$, on the average, only three carriers, the primary and its secondary hole and electron, are present in the multiplying region for every incident photocarrier. A fluctuation that changes the number of carriers by one represents a large percentage change, and the noise factor will be large. On the other hand, if one of the ionization coefficients approaches zero (e.g., $\alpha_p \to 0$), carriers of order lnM are present in the multiplying region for every incident photocarrier; a fluctuation of one carrier is a relatively insignificant perturbation. Thus the noise factor is expected to be small if the difference between α_n and α_p is large.

For electron injection alone, the noise factor can be written as[34]

$$F = kM + (2 - 1/M)(1 - k) \tag{50}$$

where $k \equiv \alpha_p/\alpha_n$ is assumed to be constant throughout the avalanche region. For hole injection alone, the foregoing expression still applies if k is replaced by $k' \equiv \alpha_n/\alpha_p$. For the two special cases, $\alpha_p = \alpha_n$ (i.e., $k = 1$) and $\alpha_p \to 0$ (i.e., $k = 0$), Eq. 50 gives $F = M$ and $F = 2$ (at large M), respectively. The noise factor for various values of multiplication and ratios of ionization coefficients is shown in Fig. 21. We can see that a smaller value k for electron injection or a small value of k' for hole injection is desirable to minimize excess noise.

Figure 22 shows some experimental results[35] obtained from a Si avalanche photodiode with a 0.1-μA primary injection current measured at 600 kHz. The upper values (open circles) represent the noise for a hole primary photocurrent, which results from short wavelength radiation (see the insert). The lower values (closed circles) represent the noise for an electron primary photocurrent. The noise factor for electron injection is considerably lower than that for hole injection, because α_n is much larger than α_p in silicon.

The results shown in Fig. 21 can be used for the p-i-n avalanche photodiode and the lo–hi–lo type photodiode, which has a uniform electric field in the avalanche region. For a general avalanche photodiode with

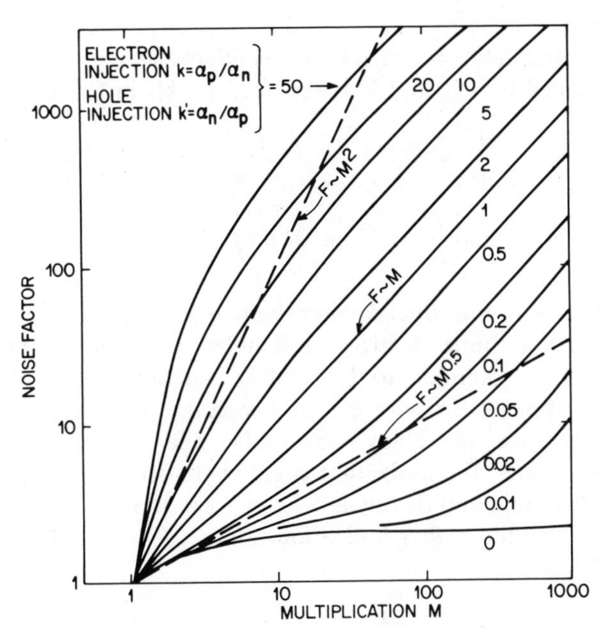

Fig. 21 Noise factors calculated for various values of multiplication and ratios of electron and hole ionization coefficients. (After McIntyre, Ref. 34.)

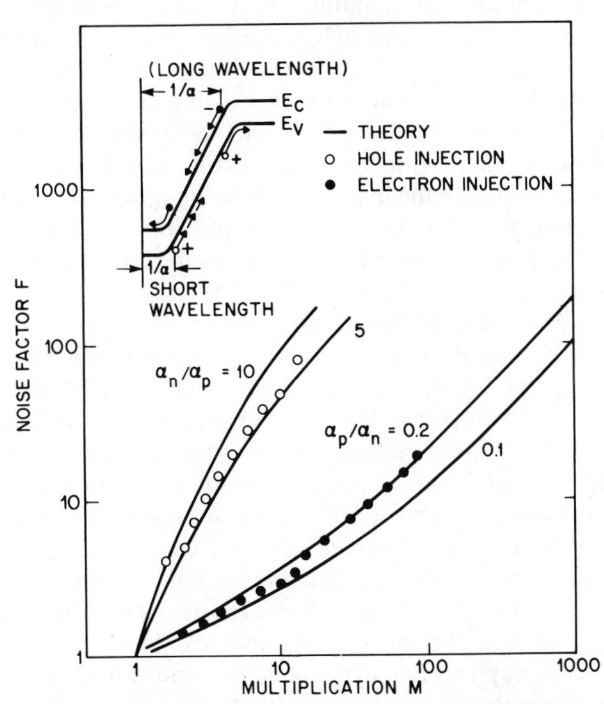

Fig. 22 Experimental results of noise factors for a silicon avalanche photodiode with a 0.1-μA primary current. The insert shows the energy-band diagram of an avalanche photodiode with an electron or hole primary current, depending on the wavelength of the incident light. (After Baertsch, Ref. 35.)

nonuniform electric field, the ionization coefficients must be weighted accordingly: k is replaced by k_{eff}, and k' is replaced by k'_{eff} in Eq. 50, where[36]

$$k_{\text{eff}} = \int_0^W \alpha_p(x)M^2(x)\,dx \Big/ \int_0^W \alpha_n(x)M^2(x)\,dx \tag{51}$$

$$k'_{\text{eff}} = k_{\text{eff}}\left[\int_0^W \alpha_p(x)M(x)\,dx \Big/ \int_0^W \alpha_n(x)M(x)\,dx\right]^{-2}. \tag{52}$$

Additional noise is introduced when light is absorbed on both sides of the junction so that both electrons and holes are injected into the avalanche region. For example, for $k_{\text{eff}} = 0.005$ and $M = 10$, the noise factor increases from about 2 for pure electron injection to 20 for 10% electron injection.[37] Therefore, to achieve low noise and wide bandwidth in an avalanche photodiode, the ionization coefficients of the carriers should be as different as possible and the avalanche process should be initiated by the carrier species with the higher ionization rate.

13.4.4 Device Performance

An avalanche photodiode requires the avalanche multiplication to be spatially uniform over the entire light-sensitive area of the diode.[38] Microplasmas, that is, small areas in which the breakdown voltage is less than that of the junction as a whole, must be eliminated. The probability of microplasmas occurring in the active area is minimized by using low dislocation materials and by designing the active area to be no larger than necessary to accommodate the incident light beam (generally from a few μm to 100 μm in diameter). The excessive leakage current along the junction edges due to the junction curvature effect[39] or high-field concentration is eliminated by using a guard-ring or surface-beveled structure.

Figure 23 shows some device configurations.[2] The guard-ring structure must have a low impurity gradient at the n-p guard-ring junction with a sufficiently large radius of curvature such that the central p^+n abrupt junction will breakdown before the guard ring does (Fig. 23*a*). A guard ring can also be used for a more complicated structure of n^+-p-π-p^+ (Fig. 23*b*). For the metal–semiconductor avalanche photodiode, a guard ring can also be used to eliminate high electric field concentration at the periphery of the contact (Fig. 23*c*). A mesa or beveled structure can have a low surface field across the junction and uniform avalanche breakdown can occur inside the device (Fig. 23*d* and *e*). To detect wavelength near the intrinsic absorption edge, a side-illuminated avalanche photodiode can be used to improve both the quantum efficiency and the signal-to-noise ratio (Fig. 23*f*).

Avalanche photodiodes have been made in various semiconductors including Ge, Si, and III–V compounds and their alloys. The key factors in selecting a particular semiconductor include the quantum efficiency at a

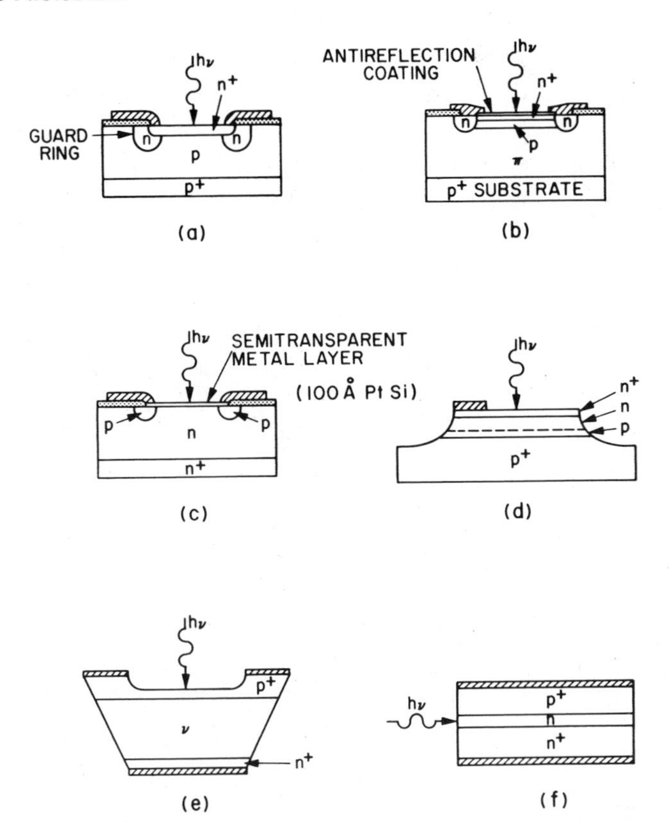

Fig. 23 Device configurations of some avalanche photodiodes. (*a*) Guard-ring structure. (*b*) Guard ring n^+-p-π-p^+ structure. (*c*) Metal–semiconductor structure. (*d*) Mesa structure. (*e*) Beveled p-i-n structure. (*f*) Side illuminated p^+-n-n^+ mesa structure. (After Melchior, Ref. 2.)

particular optical wavelength, the response speed and the noise. We shall now consider some representative device performances.

Germanium avalanche photodiodes are useful in the wavelength range from 1 to 1.6 μm because of high quantum efficiency. Since the ionization coefficients of electrons and holes are comparable in Ge, the noise factor is close to $F = M$, Eq. 50, and the mean-square shot noise current varies as M^3, Eq. 37. A high-speed Ge photodiode was fabricated using the configuration[33,40] of Fig. 23*a*. The p layer had a resistivity of 0.5 Ω-cm and a thickness of 150 μm. After diffusing antimony for a guard ring of 7.5 μm deep, arsenic was diffused to a depth of 0.4 μm to form the n^+ layer. The device had a quantum efficiency about 80% over the wavelength range 1.0 to 1.55 μm (shown in Fig. 7). At room temperature the dark current density[33] at $\frac{1}{2}V_B$ (where V_B is the breakdown voltage) was about 3×10^{-4} A/cm^2. The response waveform for a YAG laser ($\lambda = 1.06$ μm) showed a rise time of 100 ps.

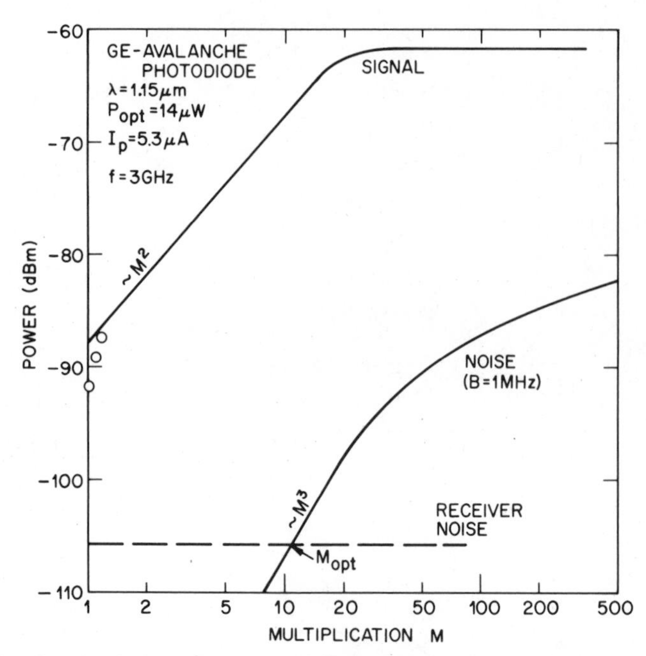

Fig. 24 Signal and noise power output of Ge avalanche photodiode in a 1-MHz band at 3 GHz. The optimum operating point for the best *S/N* ratio and sensitivity is indicated. (After Melchior, Ref. 2.)

Figure 24 shows the results[2] of the signal and noise output of a Ge avalanche photodiode for a modulation frequency of 3 GHz and an average primary current of 5.3 μA. Note that for $M < 30$, the signal power increases as M^2 and the noise power as M^3. This behavior is in good agreement with the theoretical prediction. Also note that in Fig. 24 the highest signal-to-noise ratio (~40 dB) is obtained at $M \simeq 10$, that is, where the noise contribution from the diode is about equal to the receiver noise. At higher values of M, the S/N ratio decreases because avalanche noise increases faster than the multiplied signal.

Silicon avalanche photodiodes are particularly useful in the wavelength range from 0.6 to 1.0 μm, where nearly 100% quantum efficiency has been obtained from devices having antireflection coatings. The hole-to-electron ionization coefficient ratio ($k = \alpha_p/\alpha_n$) in silicon is a strong function of the electric field; it varies from about 0.1 at 3×10^5 V/cm to 0.5 at 6×10^5 V/cm. Therefore, to minimize noise, the electric field at avalanche breakdown should be low and the ionization multiplication should be initiated by electrons.

An idealized doping profile of a p^+-π-p-π-n^+ structure is shown in Fig. 25*a*. This profile is similar to that of a lo–hi–lo IMPATT diode structure (see Chapter 10). The p^+-π-p-π-n^+ structure is called a reach-through structure,[41]

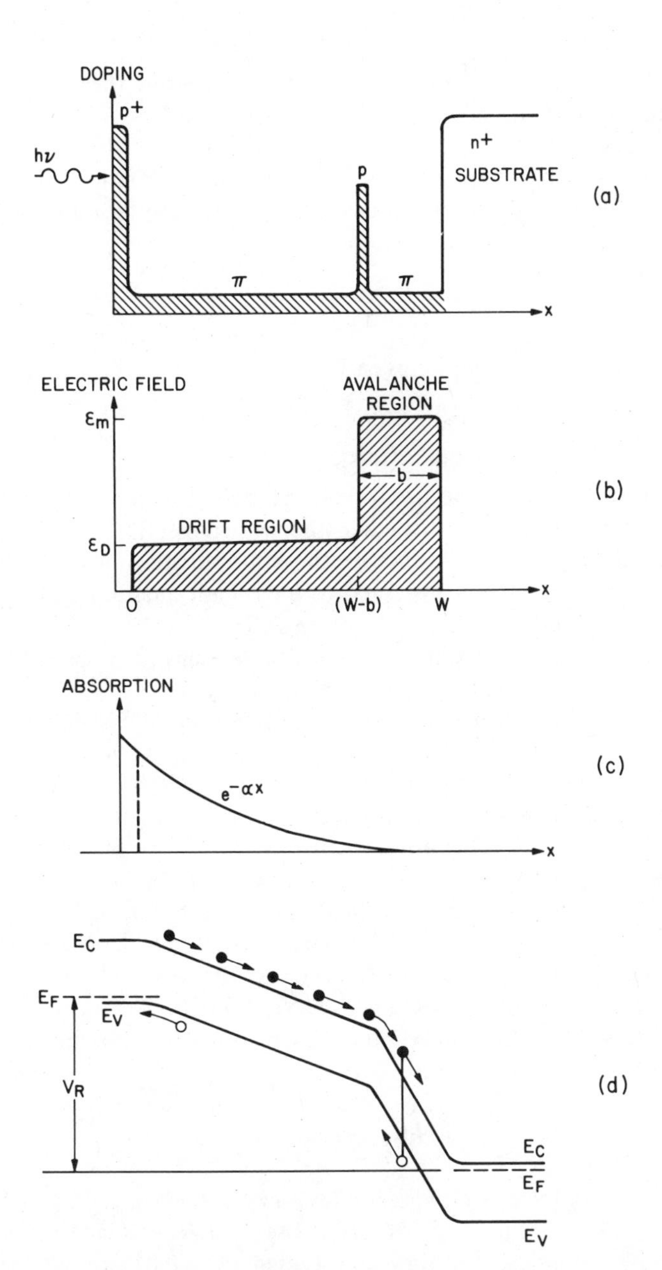

Fig. 25 Reach-through lo–hi–lo avalanche photodiode. (*a*) Doping profile. (*b*) Field distribution. (*c*) Absorption of incident light. (*d*) Energy-band diagram, showing how electrons initiate the multiplication process.

because the electric field extends all the way from the $n^+\pi$ avalanche region and reaches the p^+ layer (Fig. 25*b*). In the lower-field drift region, the carriers can travel at their saturation velocity (10^7 cm/s for $\mathscr{E}_D >$ 10^4 V/cm). In the high-field avalanche region, the maximum field $\mathscr{E}_m$ can be adjusted by adjusting the thickness b. The breakdown condition can be written as[42]

$$\alpha_n b = \frac{\ln k}{k-1} \qquad (k \equiv \alpha_p/\alpha_n). \tag{53}$$

and the breakdown voltage is given by

$$V_B \simeq \mathscr{E}_m b + \mathscr{E}_D(W - b). \tag{54}$$

For a given wavelength, we can choose a W (e.g., $W = 1/\alpha$) and then independently adjust b to optimize device performances. As seen in Fig. 25*c*, most of the light is absorbed in the π region $(W - b)$, and electrons enter the avalanche region to initiate the multiplication process (Fig. 25*d*). The p^+-π-p-π-p^+ device is expected to have high quantum efficiency, high response speed, and good signal-to-noise ratio.

In practice, it may be difficult to form the narrow p region, and the doping profile shown in Fig. 25*a* is replaced by a p^+-π-p-n^+ device, that extends the p region all the way to the n^+ substrate. (This doping profile is identical to a hi–lo IMPATT structure.) For a silicon reach-through avalanche photodiode[37] with a total depletion width of 200 μm and a π-region resistivity of 5000 Ω-cm, a quantum efficiency of 90% is obtained at 0.9 μm wavelength. The noise factor is only 4 at $M = 100$ and 20 at $M = 1000$, corresponding to a low value of $k_{\text{eff}} = 0.016$.

Under the condition that the optical signal photocurrent is much larger than the background current and the dark current, the dominant shot noise in Eq. 38 is $2qI_pM^2FB$, where F is given by Eq. 50. The optimum multiplication M_{opt} is found by setting $d(S/N)/dM = 0$. Substituting this M_{opt} into Eq. 38, we obtain a maximum signal-to-noise ratio under the large signal photocurrent condition:[43]

$$(S/N)_{\max} \sim \left(\eta/\sqrt{k_{\text{eff}}}\right)^{4/3}. \tag{55}$$

Therefore to maximize S/N, we should maximize $\eta/\sqrt{k_{\text{eff}}}$. Figure 26 shows a plot of $\eta/\sqrt{k_{\text{eff}}}$ versus b for different breakdown voltages in lo–hi–lo Si avalanche photodiodes. The optical wavelength is 0.85 μm ($\alpha = 600$ cm^{-1}), corresponding to the emission light of GaAs laser. Note that there is a maximum $\eta/\sqrt{k_{\text{eff}}}$ for a given V_B and it increases with V_B. This estimate can also be used for other impurity profiles.

Figure 27*a* shows a device configuration that is more amenable to fabrication on large-diameter silicon wafers, with good control of the doping profile through ion implantation and diffusion.[44] The cross-sectional view is shown in Fig. 23*b*. Figure 27*b* shows the field distribution,

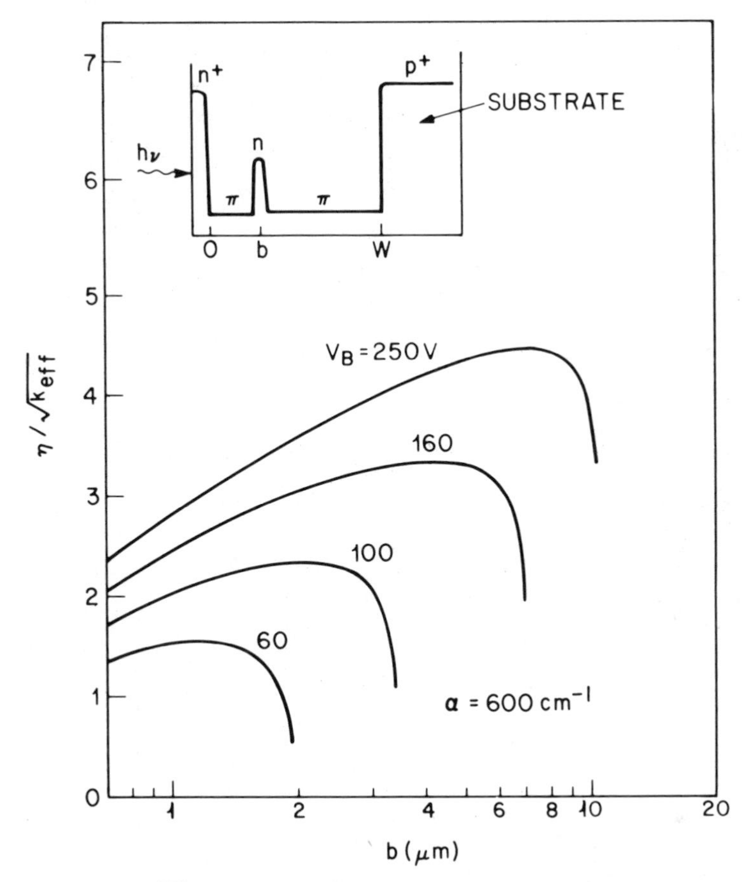

Fig. 26 A plot of $\eta/\sqrt{k_{\text{eff}}}$ versus b for silicon lo–hi–lo avalanche photodiodes. (After Kanbe and Kmura, Ref. 43.)

which has a narrow avalanche region and a long drift region. The quantum efficiency is near 100% at about 0.8-μm wavelength for the device having a SiO_2–Si_3N_4 antireflection coating (Fig. 27*c*). Because of the slight mixture of holes in initiating the multiplication process, the noise factor is higher ($k_{\text{eff}} \simeq 0.04$) than for the structure shown in Fig. 25.

Metal–semiconductor (Schottky-barrier) avalanche photodiodes are useful in the visible and ultraviolet range. The characteristics of Schottky-barrier avalanche photodiodes are similar to those of *p*-*n* junction photodiodes.[45] Schottky-barrier photodiodes have been made on a 0.5-Ω-cm *n*-type silicon substrate with a thin PtSi film (~100 Å) and a diffused guard ring[46] as shown in Fig. 23*c*. Since the edge-leakage current is eliminated by the guard ring, an ideal reverse saturation current density of 2×10^{-7} A/cm^2 can be obtained. For the Schottky-barrier photodiode, avalanche multiplication can amplify the peak value of fast photocurrent pulses of 0.8-ns

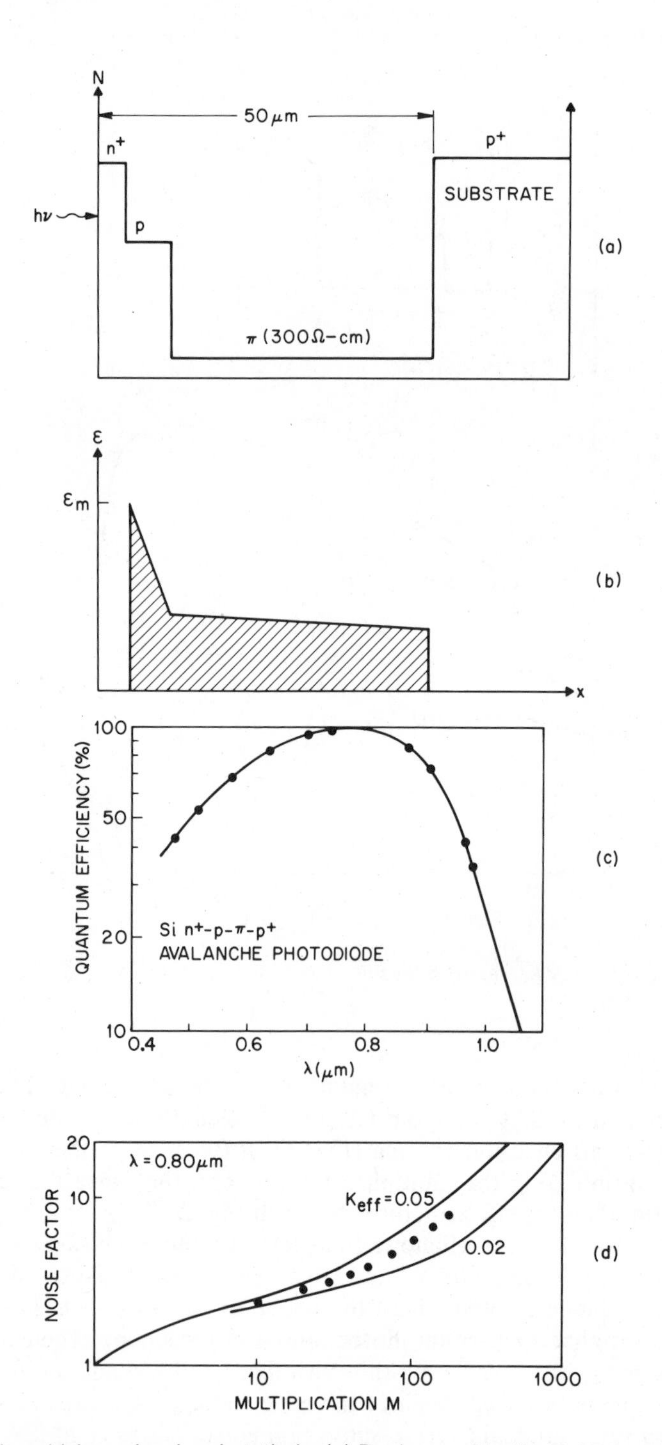

Fig. 27 Silicon hi–lo avalanche phototiode. (*a*) Doping profile. (*b*) Field distribution. (*c*) Quantum efficiency. (*d*) Noise factor. (After Melchior et al., Ref. 44.)

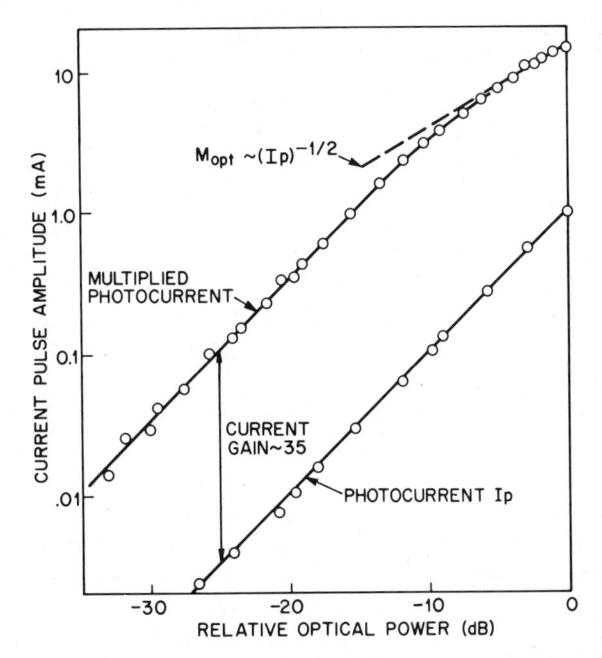

Fig. 28 Amplitudes of current pulses in a Schottky-barrier avalanche photodiode as a function of intensity of a 0.8-ns light pulse at 0.6328 μm. The lower curve shows the photocurrent at low reverse bias. The upper curve shows the highest pulse amplitudes of multiplied photocurrent reached at the breakdown voltage. (After Melchior, Lepselter, and Sze, Ref. 45.)

duration by factors of up to 35 (Fig. 28). The difference between the photocurrent shows how much current gain is possible at a certain level for pulses of certain duration (0.8 ns for the present case). The average light power from the 6328 Å laser is 0.8 mW and the peak power of the pulses corresponding to 0 dB is approximately 7 mW (Fig. 28). Noise measurements for avalanche multiplication in PtSi–Si diodes show that the noise of the multiplied photocurrent increases approximately as M^3 for light in the visible range. As the wavelength decreases, the electron primary injection photocurrent becomes dominant and the noise decreases, in agreement with noise theory.

Schottky-barrier avalanche photodiodes with *n*-type silicon substrates promise to be particularly useful as high-speed photodetectors for ultraviolet light. Ultraviolet light that is transmitted through the thin metal electrodes is absorbed within the first 100 Å of the silicon. The carrier multiplication is then mainly initiated by electrons, resulting in a low-noise and high gain–bandwidth product. Avalanche multiplication of photocurrents increases the sensitivity of detection systems consisting of Schottky-barrier photodiodes and a receiver. Amplification of high-speed photocurrent pulses is also possible.

The heterojunction avalanche photodiodes, especially in III–V alloys, have many potential advantages as alternatives to Ge and Si devices. By adjusting the alloy composition, the wavelength response of the device can be tuned. Because of the high absorption coefficients of the direct-bandgap III–V alloys, the quantum efficiency can be high, even if a narrow depletion width is used to provide high-speed response. In addition, the heterostructure window layer (larger bandgap for the surface layer) can be grown so that high-speed performance is obtained and the surface recombination loss of photogenerated carriers can be minimized.

Avalanche photodiodes have been made[47,48] in various III–V alloy systems, such as AlGaAs/GaAs, AlGaSb/GaSb, GaInAs/InP, and GaInAsP/InP. The initial stage of development shows encouraging results, especially in speed and quantum efficiency, which can be better than in existing Ge avalanche photodiodes for the wavelength range from 1 to 1.6 μm. Extensive studies are expected in this field to understand the material parameters, such as dark current, absorption coefficients, and the ratio of the ionization rates, and to improve the heterostructure materials, fabrication method, device structure, and device reliability.

On the ratio of the ionization rates, recent studies of $Al_xGa_{1-x}Sb$ junctions show that when the spin-orbit split Δ of the valence band approaches the bandgap (Fig. 29, insert) the value of k' can become very small.[57] Figure 29 shows a pronounced reduction of k' at $\Delta/E_g \approx 1$. Values of k' smaller than 0.04 have been obtained, corresponding to noise factors less than 5 at $M = 100$.

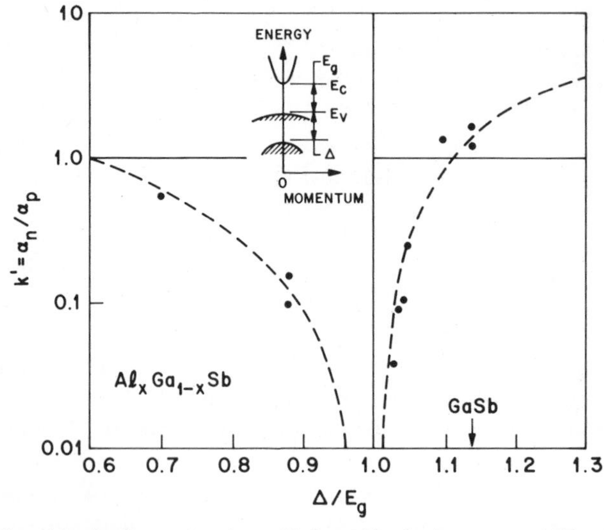

Fig. 29 Ratio of ionization rates in a $Al_xGa_{1-x}Sb$ diode versus Δ/E_g, where Δ is the spin-orbit splitting of the valence band shown in the insert. (After Hildebrand, Kuebart, and Pilkuhn, Ref. 57.)

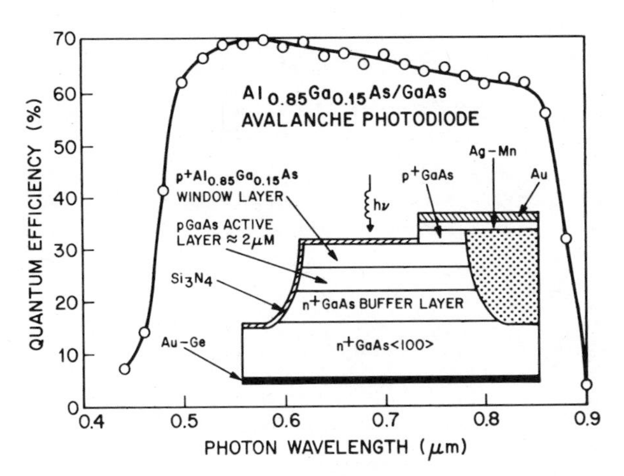

Fig. 30 Quantum efficiency versus photon wavelength for AlGaAs/GaAs heterojunction avalanche photodiode. The device configuration is shown in the insert. (After Law, Nakano, and Tomasetta, Ref. 48.)

Many heterojunction avalanche photodiodes are made using the configuration of Fig. 23*d*, where III–V semiconductor (e.g., GaAs or InP) is used as the substrate. The ternary or quaternary compound with a closely matched lattice parameter is then grown epitaxially on the substrate (e.g., by liquid or vapor phase epitaxy or the molecular-beam epitaxy method). The composition of the alloys, doping concentrations, and layer thicknesses are adjusted to optimize device performance. An $Al_{0.85}Ga_{0.15}As/GaAs$ structure is shown[48] in the insert in Fig. 30. The top layer serves as a window for the transmission of 0.5 to 0.9-μm incident light. The quantum efficiency is about 70% at 0.53 μm and can be increased to 95% with a Si_3N_4 antireflection coating (Fig. 30). The dark-current density is about 10^{-8} A/cm^2 at one-half the breakdown voltage. The rise time of the diode is 35 ps, and the avalanche gain at 273 MHz is about 100. The ratio $k(=\alpha_p/\alpha_n)$ is found to be 0.83, which is expected because the electron and hole ionization rates are comparable in ⟨100⟩-oriented GaAs. For ⟨111⟩-oriented GaAs, however, the hole ionization rate is much larger than that for electrons (refer to Chapter 1). To minimize the avalanche noise, we should use ⟨111⟩-oriented GaAs employing holes to initiate the multiplication process.

The insert of Fig. 31 shows a novel heterostructure,[49] where a planar-type *p-n* junction is formed in an InP window layer, separated from a light-absorbing lower bandgap GaInAsP layer. In this diode the depletion layer extends from *n*-InP into the *n*-GaInAsP layer at the flat-junction portion, whereas it remains inside the InP at the junction corner. Since the breakdown voltage V_B is expected to vary as $E_g^{3/2}$, V_B in InP is about 60% higher than the V_B in $Ga_{0.23}In_{0.77}As_{0.51}P_{0.49}$ when they have equal doping

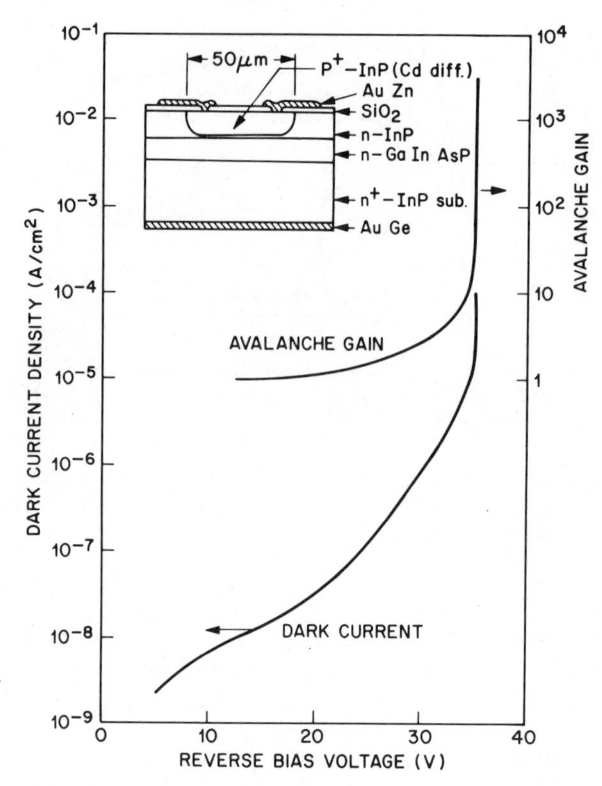

Fig. 31 Avalanche gain and dark-current dependence on reverse bias in a GaInAsP heterostructure avalanche photodiode. (After Nishida, Taguchi, and Matsumoto, Ref. 49.)

levels. This effect prevents edge breakdown in the diode structure. In addition the leakage current is reduced due to suppression of tunneling in the widegap material (e.g., InP) by the lower bandgap layer (e.g., GaInAsP). The device has a low dark-current density of 10^{-6} A/cm^2 at $\frac{1}{2}V_B$ and an avalanche gain of 3000.

Figure 32*a* shows the equilibrium energy-band diagram of a heterojunction avalanche photodiode having separate absorption and multiplication regions.[58] The p^+-n junction is formed in InP (multiplication region). The GaInAs layer grown on the n-InP is used as a light absorption region. Because the ionization rate of holes is larger by two to three times than that of electrons in InP ($k' = 0.4$), the avalanche process should be initiated by holes. In order to inject holes over the barrier into the n-InP layer, the dopings and thicknesses of n-InP and n-GaInAs layers shall be designed so that under avalanching conditions, the n-InP depletion layer is punched through into the GaInAs layer (Fig. 32*b*). The device has a quantum efficiency of 40% at 1.3 μm and 50% at 1.6 μm. Its noise factor is 3 dB lower than that of Ge avalanche photodiode operated at 1.15 μm.

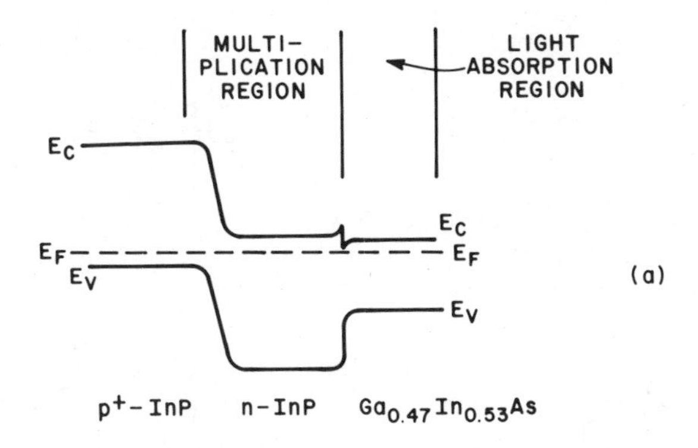

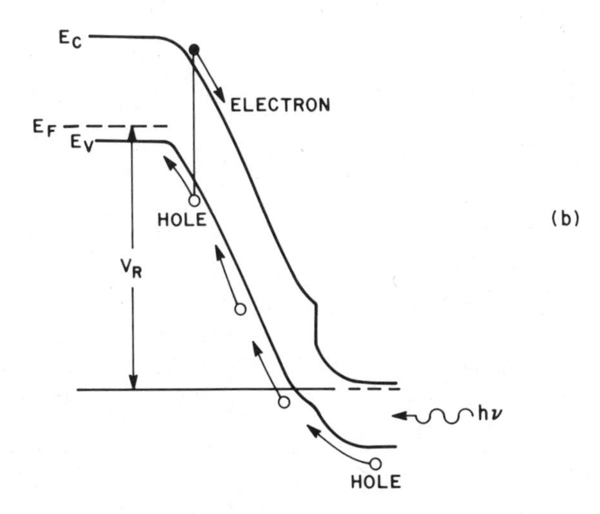

Fig. 32 (*a*) Energy-band diagram of a InP–GaInAs heterojunction at thermal equilibrium. (*b*) Energy-band diagram at avalanche breakdown. (After Susa et al., Ref. 58.)

13.5 PHOTOTRANSISTOR

All the bipolar and unipolar transistors discussed in previous chapters are potential candidates to be used as photodetectors, hence the name "phototransistor." A phototransistor[50] can have high gains through the transistor action. On the other hand, the fabrication of a phototransistor is more complicated than that of a photodiode, and the inherent larger area of a phototransistor degrades its high-frequency performances.

A bipolar phototransistor[51] is shown in Fig. 33*a*, together with its circuit model. It differs from a conventional bipolar transistor by having a large base–collector junction as the light-collecting element (represented by a parallel combination of a diode and a capacitor). This device is particularly

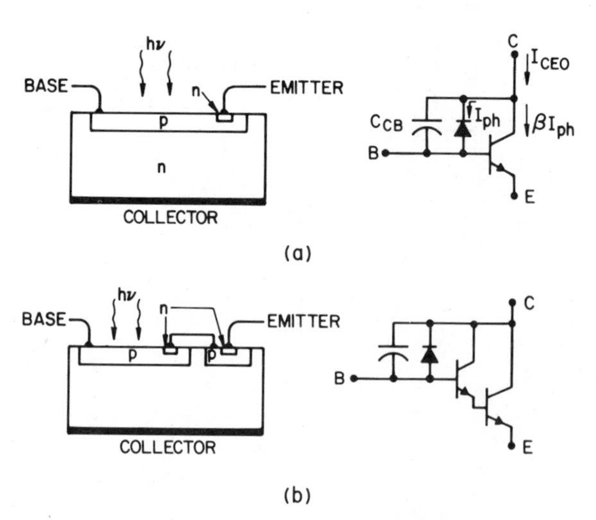

Fig. 33 (*a*) Bipolar phototransistor. (*b*) Photo-Darlington. (After Jayson and Knight, Ref. 51.)

useful in opto-isolator applications, because it offers a high current-transfer ratio (i.e., the ratio of output photodetector current to input LED current), of the order of 50% or more, as compared to a typical photodiode with a current-transfer ratio of 0.2%. When the base lead is floating, the photogenerated carriers contribute a photocurrent I_{ph} in the collector. In addition, the holes generated in the base and those swept into the base from the collector lower the base–emitter potential, so that electrons are injected across the base to the collector. The total collector current

$$I_{CEO} = I_{ph} + h_{FE}I_{ph} = (1 + h_{FE})I_{ph} \tag{56}$$

where h_{FE} is the dc common-emitter current gain, which can be much larger than unity, and the effective quantum efficiency is $(1 + h_{FE})$ times larger than that of the base–collector photodiode.

The noise equivalent power is given by an expression similar to Eq. 20 in which[52]

$$I_{eq} = I_{CEO}(1 + 2h_{fe}^2/h_{FE}) \tag{57}$$

where h_{fe} is the incremental common-emitter current gain. Therefore, there is a trade-off between low noise and high gain.

The bipolar phototransistor can be integrated with other devices. By adding a second transistor, a photo-Darlington with even higher transfer ratio can be formed (Fig. 33*b*). The frequency response of the above structures is limited by the large base–collector capacitance and is reduced further by the gain of the detector due to feedback effect. The typical response time for a photodiode is of the order of 0.01 μs, while it is about 5 μs for a phototransistor, and 50 μs for a photo-Darlington.

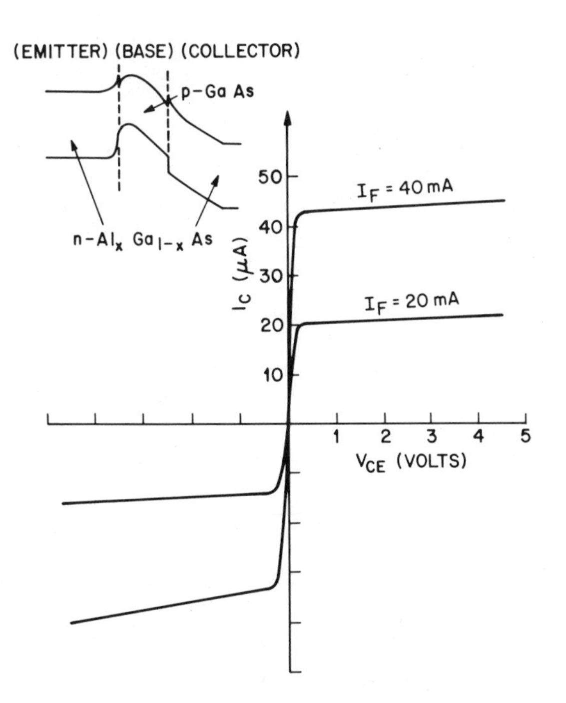

Fig. 34 Current–voltage characteristics of a bilateral heterostructure phototransistor. (After Knight et al., Ref. 53.)

Phototransistors can be made in heterostructures. The insert of Fig. 34 shows an energy-band diagram for a symmetrical *n*-AlGaAs/*p*-GaAs/*n*-AlGaAs structure.[53] The device shows high blocking voltage for both polarities of applied bias, high gain for both polarities, and linear current–voltage characteristics through the zero bias point (Fig. 34). The injection efficiency for both polarities is obtained from the difference in the bandgap between the collector–emitter layers and the narrower bandgap base. Thus the restriction of having a lightly doped base layer, as in a conventional transistor, is removed, and the depletion region at the blocking junction can be made to extend into the collector rather than into the base. This permits both high gain and high blocking voltage to be obtained in the same device. The device demonstrates a bilateral gain of 180 at blocking voltage of ± 10 V with a 2.1-μm-thick GaAs base, and a bilateral gain of 3000 with a blocking voltage of ± 2.6 V with a 0.3-μm base.

A schematic diagram of an infrared-sensing silicon MOSFET is shown[54] in the insert of Fig. 35. (For a discussion of MOSFET characteristics, refer to Chapter 8.) The *p*-type substrate is doped with a shallow acceptor, boron, and a deep acceptor, indium (0.16 eV from the valence band). The indium center can exist in a neutral state when it is occupied by a hole, or in a negatively charged state when a hole is lost to the valence band. For

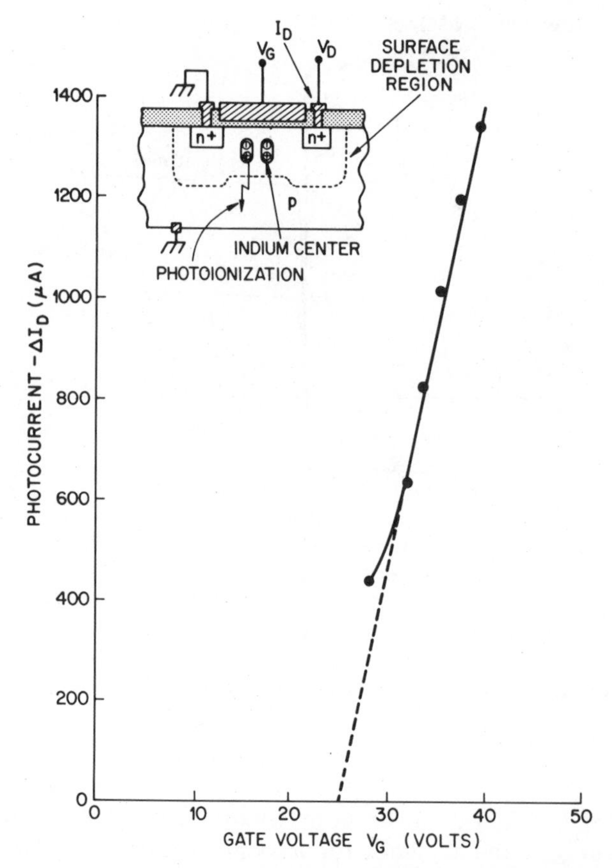

Fig. 35 Photocurrent as a function of gate voltage in the saturation region of an indium-doped MOSFET. (After Forbes, Wittmer, and Loh, Ref. 54.)

the device operation, a negative voltage is applied to the gate to cause accumulation in the channel and all the indium centers can capture a hole and thus exist in their neutral state. When infrared radiation in the 2–7 μm wavelength range illuminates the device, sufficient energy is available to photoionize the indium center by the emission of a hole to the valence band. This process increases the net acceptor charge density within the depletion region; this, in turn, modulates the conductance of the surface channel. The change in the threshold voltage ΔV_T, can be written as

$$\Delta V_T = 2\sqrt{q\epsilon_s\psi_B}\left(\sqrt{N_A + N_{In}} - \sqrt{N_A}\right)\Big/ C_i \qquad (58)$$

where ψ_B is the potential difference between the Fermi level and the intrinsic Fermi level, C_i the gate oxide capacitance per unit area, N_A the boron acceptor concentration, and N_{In} the photoionized indium concentration. The channel current in the saturation region is proportional to

$(V_G - V_T)^2$, so the change of the channel current is

$$\Delta I_D \sim \Delta V_T(V_G - V_T - \Delta V_T/2). \quad (59)$$

Figure 35 shows the photocurrent in the saturation region as a function of gate voltage. This plot closely follows the functional form of Eq. 59.

By using acceptors other than indium, one can maximize the device response for a particular wavelength range. The infrared MOSFET detector can be integrated with the storage/memory element and the output amplifier into a single device; it is useful for applications in large-scale-integrated infrared imaging arrays.

REFERENCES

1 L. K. Anderson, M. DiDomenico, Jr., and M. B. Fisher, "High-Speed Photodetectors for Microwave Demodulation of Light," in L. Young, Ed., *Advances in Microwaves*, Vol. 5, Academic, New York, 1970, pp. 1–122.

2 H. Melchior, "Demodulation and Photodetection Techniques," in F. T. Arecchi and E. O. Schulz-Dubois, Eds., *Laser Handbook*, Vol. 1, North-Holland, Amsterdam, 1972, pp. 725–835.

3 H. Melchior, "Detector for Lightwave Communication," *Phys. Today*, p. 32 (Nov. 1977).

4 R. K. Willardson and A. C. Bear, Eds. *Semiconductors and Semimetals*, Vol. 12, *Infrared Detector II*, Academic, New York, 1977.

5 G. E. Stillman and C. M. Wolfe, "Avalanche Photodiode," in Ref. 4, pp. 291–394.

6 T. P. Lee and T. Y. Li, "Photodetectors," in S. E. Miller and A. G. Chynoweth, Eds., *Optical Fiber Communications*, Academic, New York, 1979, Chap. 18.

7 P. R. Bratt, "Impurity Ge and Si Infrared Detectors," in Ref. 4, pp. 39–142.

8 R. H. Bube, "Comparison of Solid State Photoelectronic Radiation Detectors," *Trans. AIME*, **239**, 291 (1967).

9 M. DiDomenico, Jr. and O. Svelto, "Solid State Photodetection Comparison between Photodiodes and Photoconductors," *Proc. IEEE*, **52**, 136 (1964).

10 A. Van der Ziel, *Fluctuation Phenomena in Semiconductors*, Academic, New York, 1959, Chap. 6.

11 M. Ross, *Laser Receivers-Devices, Techniques, Systems*, Wiley, New York, 1966.

12 P. W. Kruse, L. D. McGlauchlin, and R. B. McQuistan, *Elements of Infrared Technology*, Wiley, New York, 1962.

13 W. L. Eisenman, J. D. Merriam, and R. F. Potter, "Operational Characteristics of Infrared Photodiode," in Ref. 4, pp. 1–38.

14 G. E. Stillman, C. M. Wolfe, and J. O. Dimmock, "Far-Infrared Photoconductivity in High Purity GaAs," in Ref. 4, pp. 169–290.

15 W. W. Gartner, "Depletion-Layer Photoeffects in Semiconductors," *Phys. Rev.*, **116**, 84 (1959).

16 H. S. Lee and S. M. Sze, "Silicon *p-i-n* Photodetector Using Internal Reflection Method," *IEEE Trans. Electron Devices*, **ED-17**, 342 (1970).

17 J. Muller, "Thin Silicon Film *p-i-n* Photodiodes with Internal Reflection," *IEEE Trans. Electron Devices*, **ED-25**, 247 (1978).

18 E. Ahlstrom and W. W. Gartner, "Silicon Surface-Barrier Photocells," *J. Appl. Phys.*, **33**, 2602 (1962).

19 M. V. Schneider, "Schottky Barrier Photodiodes with Antireflection Coating," *Bell Syst. Tech. J.*, **45**, 1611 (1966).

20 C. R. Crowell and S. M. Sze, "Hot Electron Transport and Electron Tunneling in Thin Film Structures," in R. E. Thun, Ed., *Physics of Thin Films*, Vol. 4, Academic, New York, 1967, pp. 325–371.

21 A. A. Gutkin, V. M. Dmitriev, and V. M. Khait, "Photosensitivity of Au–*n*–Si Surface-Barrier Diodes in the Spectral Range 1–6 eV," *Sov. Phys. Semicond.*, **11**, 290 (1977).

22 M. Lavagna, J. P. Pique, and Y. Marfaing, "Theoretical Analysis of Quantum Yield in Schottky Diodes," *Solid State Electron.*, **20**, 235 (1977).

23 W. M. Sharpless, "Cartridge-Type Point Contact Photodiode," *Proc. IEEE*, **52**, 207 (1964).

24 H. C. Casey, Jr. and M. B. Panish, *Heterostructure Lasers, Part B*, Academic, New York, 1978, p. 3.

25 R. C. Miller, B. Schwartz, L. A. Koszi, and W. R. Wagner, "A High-Efficiency GaAlAs Double-Heterostructure Photovoltaic Detector," *Appl. Phys. Lett.*, **33**, 721 (1978).

26 M. Ilegems, B. Schwartz, L. A. Koszi, and R. C. Miller, "Integrated Multijunction GaAs Photodetector with High Output Voltage," *Appl. Phys. Lett.*, **33**, 629 (1978).

27 C. A. Burrus, A. G. Dentai, and T. P. Lee, "InGaAsP *p-i-n* Photodiodes with Low Dark Current and Small Capacitance," *Electron. Lett.*, **15**, 655 (1979).

28 T. P. Lee, C. A. Burrus, and A. G. Dentai, "InGaAs/InP *p-i-n* Photodiodes for Lightwave Communications at 0.95 to 1.65 μm Wavelengths," *IEEE J. Quantum Electron.*, **QE-17**, 232 (1981).

29 K. Ahmad and A. W. Mabbitt, "GaInAs Photodiodes," *Solid State Electron.*, **22**, 327 (1979).

30 K. M. Johnson, "High-Speed Photodiode Signal Enhancement at Avalanche Breakdown Voltage," *IEEE Trans. Electron Devices*, **ED-12**, 55 (1965).

31 L. K. Anderson and B. J. McMurtry, "High Speed Photodetectors," *Proc. IEEE*, **54**, 1353 (1966).

32 R. B. Emmons, "Avalanche Photodiode Frequency Response," *J. Appl. Phys.*, **38**, 3705 (1967).

33 H. Melchior and W. T. Lynch, "Signal and Noise Response of High Speed Germanium Avalanche Photodiodes," *IEEE Trans. Electron Devices*, **ED-13**, 829 (1966).

34 R. J. McIntyre, "Multiplication Noise in Uniform Avalanche Diodes," *IEEE Trans. Electron Devices*, **ED-13**, 164 (1966).

35 R. D. Baertsch, "Noise and Ionization Rate Measurements in Silicon Photodiodes," *IEEE Trans. Electron Devices*, **ED-13**, 987 (1966).

36 R. J. McIntyre, "The Distribution of Gains in Uniformly Multiplying Avalanche Photodiodes: Theory," *IEEE Trans. Electron Devices*, **ED-19**, 703 (1972).

37 R. P. Webb, R. J. McIntyre, and J. Conradi, "Properties of Avalanche Photodiodes," *RCA Rev.*, **35**, 234 (1974).

38 L. K. Anderson, P. G. McMullin, L. A. D'Asaro, and A. Goetzberger, "Microwave Photodiodes Exhibiting Microplasma—Free Carrier Multiplication," *Appl. Phys. Lett.*, **6**, 62 (1965).

39 S. M. Sze and G. Gibbons, "Effect of Junction Curvature on Breakdown Voltage in Semiconductors," *Solid State Electron.*, **9**, 831 (1966).

40 H. Ando, H. Kanbe, T. Kimura, T. Yamaoka, and T. Kaneda, "Characteristics of Ge Avalanche Photodiodes in the Wavelength Region of 1–1.6 μm," *IEEE J. Quantum Electron.*, **QE-14**, 804 (1978).

41 H. W. Ruegg, "An Optimized Avalanche Photodiode," *IEEE Trans. Electron Devices*, **ED-14**, 239 (1967).

42 J. Moll, *Physics of Semiconductors*, McGraw-Hill, New York, 1964.

43 H. Kanbe and T. Kmura, "Figure of Merit for Avalanche Photodiodes," *Electron. Lett.*, **13**, 262 (1977).

44 H. Melchior, A. R. Hartman, D. P. Schinke, and T. E. Seidel, "Planar Epitaxial Silicon Avalanche Photodiode," *Bell Syst. Tech. J.*, **57**, 1791 (1978).

45 H. Melchior, M. P. Lepselter, and S. M. Sze, "Metal–Semiconductor Avalanche Photodiode," IEEE Solid-State Device Res. Conf., Boulder, Colo., June 17–19, 1968.

46 M. P. Lepselter and S. M. Sze, "Silicon Schottky Barrier Diode with Near-Ideal I–V Characteristics," *Bell Syst. Tech. J.*, **47**, 195 (1968).

47 C. E. Hurwitz and J. J. Hsieh, "GaInAsP/InP Avalanche Photodiodes," *Appl. Phys. Lett.*, **32**, 487 (1978).

48 H. D. Law, K. Nakano, and L. R. Tomasetta, "III–V Alloy Heterostructure High Speed Avalanche Photodiodes," *IEEE J. Quantum Electron.*, **QE-15**, 549 (1979).

49 K. Nishida, K. Taguchi, and Y. Matsumoto, "InGaAsP Heterostructure Avalanche Photodiodes with High Avalanche Gain," *Appl. Phys. Lett.*, **35**, 251 (1979).

50 W. Shockley, M. Sparks, and G. K. Teal, "*p-n* Junction Transistor," *Phys. Rev.*, **83**, 151 (1951).

51 J. S. Jayson and S. Knight, "Opto-Isolator," in R. Wolfe, Ed., *Applied Solid State Science*, Vol. 6, Academic, New York, 1976, pp. 119–168.

52 F. H. DeLaMoneda, E. R. Chenette, and A. Van Der Ziel, "Noise in Phototransistors," *IEEE Trans. Electron Devices*, **ED-18**, 340 (1971).

53 S. Knight, L. R. Dawson, U. G. Keramidas, and M. G. Spencer, "An Optically Triggered Double Heterostructure Linear Bilateral Phototransistor," *IEEE Tech. Dig.*, Int. Electron Device Meet., 1977, p. 472.

54 L. Forbes, L. L. Wittmer, and K. W. Loh, "Characteristics of the Indium-Doped Infrared Sensing MOSFET," *IEEE Trans. Electron Devices*, **ED-23**, 1272 (1976).

55 R. G. Smith and S. D. Personick, "Receiver Design for Optical Communication Systems," in H. Kressel, Ed., *Semiconductor Devices for Optical Communication*, Springer-Verlag, New York, 1979, Chap. 4.

56 J. C. Campbell, A. G. Dentai, T. P. Lee, and C. A. Burrus, "Improved Two-Wavelength Demultiplexing InGaAsP Photodetector," *IEEE J. Quantum Electron.*, **QE-16**, 601 (1980).

57 O. Hildebrand, W. Kuebart, and M. H. Pilkuhn, "Resonant Enhancement of Impact Ionization in $Al_xGa_{1-x}Sb$ *p-i-n* Avalanche Photodiodes," *Appl. Phys. Lett.* **37**, 801 (1980).

58 N. Susa, H. Nakagome, O. Mikami, H. Ando, and H. Kanbe, "New InGaAs/InP Avalanche Photodiode Structure for the 1–1.6 μm Wavelength Region" *IEEE J. Quantum Electron.*, **QE-16**, 864 (1980).

59 S. R. Forrest, R. F. Leheny, R. E. Nahory, and M. A. Pollack, "$In_{0.53}Ga_{0.47}As$ Photodiodes with Dark Current Limited by Generation-Recombination and Tunneling," *Appl. Phys. Lett.*, **37**, 322 (1980).

14

Solar Cells

- INTRODUCTION
- SOLAR RADIATION AND IDEAL CONVERSION EFFICIENCY
- *p-n* JUNCTION SOLAR CELLS
- HETEROJUNCTION, INTERFACE, AND THIN-FILM SOLAR CELLS
- OPTICAL CONCENTRATION

14.1 INTRODUCTION

Solar cells at present furnish the most important long-duration power supply for satellites and space vehicles. Solar cells have also been successfully employed in small-scale terrestrial applications. As worldwide energy demand increases, conventional energy resources, such as fossil fuels, will be exhausted in the not-too-distant future. Therefore, we must develop and use alternative energy resources, especially our only long-term natural resource, the sun. The solar cell is considered a major candidate for obtaining energy from the sun, since it can convert sunlight directly to electricity with high conversion efficiency, can provide nearly permanent power at low operating cost, and is virtually free of pollution. Recently, research and development of low-cost, flat-panel solar cells, thin-film devices, concentrator systems, and many innovative concepts have increased. In the near future, the costs of small solar-power modular units and solar-power plants will be economically feasible for large-scale use of solar energy.

The solar cell was first developed by Chapin, Fuller, and Pearson in 1954 using a diffused silicon *p-n* junction.[1] Subsequently, the cadmium–sulfide solar cell was developed by Raynolds et al.[2] To date, solar cells have been made in many other semiconductors, using various device configurations and employing single-crystal, polycrystal, and amorphous thin-film structures. Hovel has given a comprehensive treatment on basic solar-cell characteristics.[3] Backus has compiled a volume of classic papers on solar cells prior to 1974.[4] Pulfrey and Johnston have reviewed the photovoltaic power generations,[5,6] and Bachmann has discussed the material aspects of solar cells.[7] The most current literature can be found in the Conference Records of the Photovoltaic Specialists Conference.[8,9]

14.2 SOLAR RADIATION AND IDEAL CONVERSION EFFICIENCY

14.2.1 Solar Radiation

The radiative energy output from the sun derives from a nuclear fusion reaction. In every second about 6×10^{11} kg of H_2 is converted to He, with a net mass loss of about 4×10^3 kg, which is converted through the Einstein relation ($E = mc^2$) to 4×10^{20} J. This energy is emitted primarily as electromagnetic radiation in the ultraviolet to infrared and radio spectral regions (0.2 to 3 μm). The total mass of the sun is now about 2×10^{30} kg, and a reasonably stable life with a nearly constant radiative energy output of over 10 billion (10^{10}) years is projected.

The intensity of solar radiation in free space at the average distance of the earth from the sun is defined as the solar constant with a value[4,10] of 1353 W/m^2. The atmosphere attenuates the sunlight when it reaches the earth's surface, mainly due to water-vapor absorption in the infrared, ozone absorption in the ultraviolet, and scattering by airborne dust and aerosols. The degree to which the atmosphere affects the sunlight received at the earth's surface is defined by the "air mass." The secant of the angle between the sun and the zenith (sec θ) is called the air mass and measures the atmospheric path length relative to the minimum path length when the sun is directly overhead.

Figure 1 shows four curves related to solar spectral irradiance[10] (power per unit area per unit wavelength). The upper curve which represents the solar spectrum outside the earth's atmosphere, is the air mass zero condition (AM0). This condition can be approximated by a 5800 K black body, as shown by the dashed-dotted curve. The AM0 spectrum is the relevant one for satellite and space-vehicle applications. The AM1 spectrum represents the sunlight at the earth's surface when the sun is at zenith; the incident power is about 925 W/m^2. The AM2 spectrum is for $\theta = 60°$ and has an incident power of about 691 W/m^2.

Air mass 1.5 conditions (sun at 45° above the horizon) represent a satisfactory energy-weighted average for terrestrial applications. The number of photons in unit energy range per cm^2/s for AM1.5 is shown[11] in Fig. 2 together with that for AM0. To convert the wavelength to photon energy, we have used the relationship

$$\lambda = \frac{c}{\nu} = \frac{1.24}{h\nu\,(\text{eV})} \qquad \mu\text{m}. \tag{1}$$

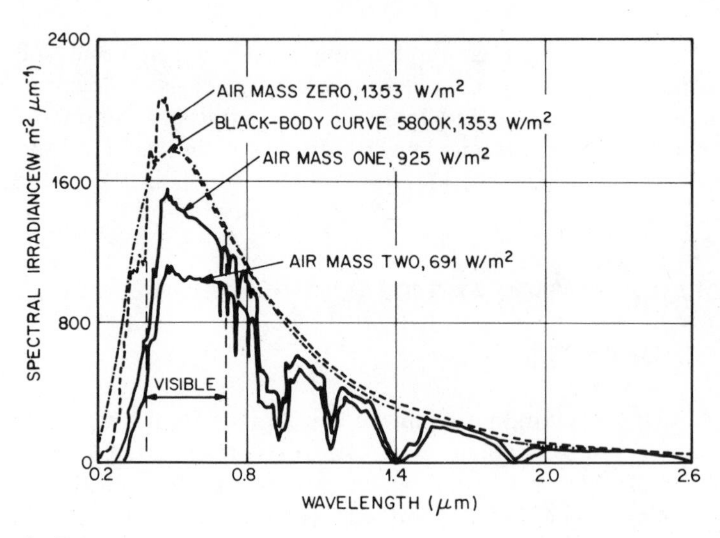

Fig. 1 Four curves related to solar spectral irradiance. (After Thekaekara, Ref. 10.)

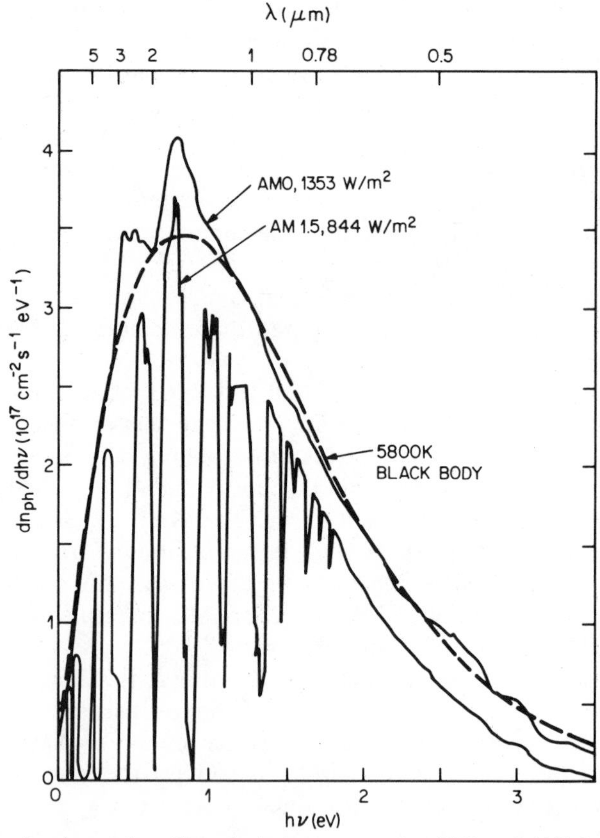

Fig. 2 Solar spectrum as a function of photon energy for AM0 and AM1.5 conditions. (After Henry, Ref. 11.)

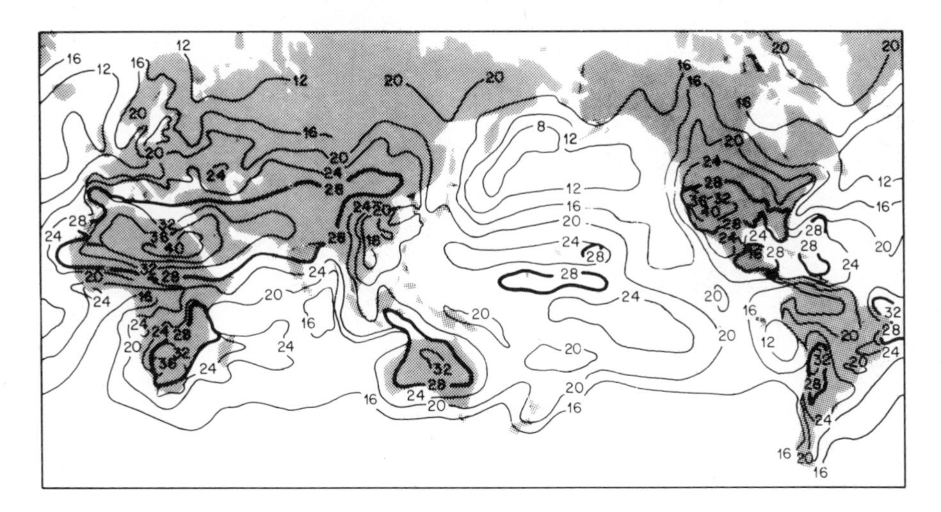

Fig. 3 Worldwide distribution of solar energy in terms of duration of sunshine. The contours refer to insolation in hundreds of hours per year. (After Pulfrey, Ref. 5.)

The total incident power for AM1.5 is 844 W/m^2.

For solar energy generation, it is necessary to know how much solar energy can be expected during the year in various locations. The worldwide distribution of solar energy in terms of sunshine duration is shown[5] in Fig. 3. The bold contour lines enclose regions that receive an average of about 8 or more hours of sunshine daily. Since all continents have sufficiently large areas covered by high average insolation, extensive worldwide utilization of solar energy can be expected in the future.

14.2.2 Ideal Conversion Efficiency

The conventional solar cell, for example, a *p-n* junction, has a single bandgap E_g. When the cell is exposed to the solar spectrum, a photon with energy less than E_g makes no contribution to the cell output (neglecting phonon-assisted absorption). A photon with energy greater than E_g contributes an energy E_g to the cell output, and the excess over E_g is wasted as heat. To derive the ideal conversion efficiency, we shall consider the energy-band diagram of a *p-n* junction under solar radiation, shown in Fig. 4*a*. The solar cell is assumed to have ideal $I-V$ characteristics. The equivalent circuit is shown in Fig. 4*b*, where a constant-current source is in parallel with the junction. The source I_L results from the excitation of excess carriers by solar radiation; I_s is the diode saturation current as derived in Chapter 2, and R_L is the load resistance.

The $I-V$ characteristics of such a device are given by

$$I = I_s(e^{qV/kT} - 1) - I_L \tag{2}$$

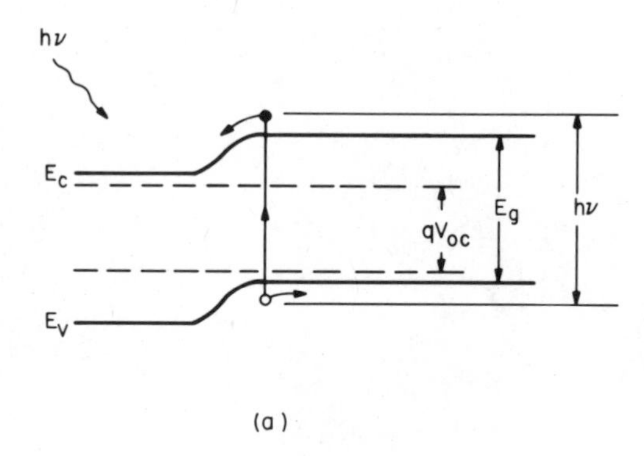

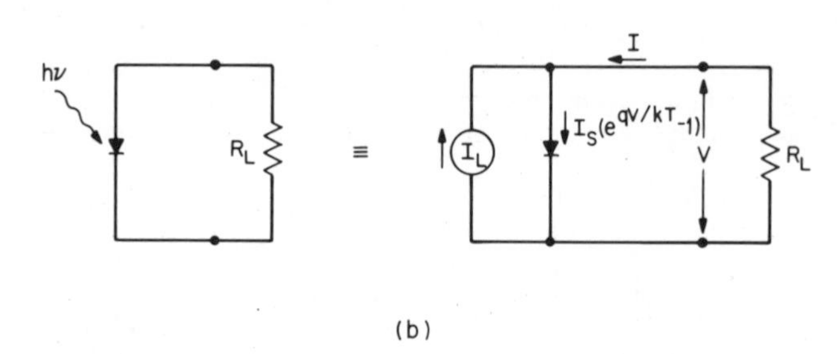

Fig. 4 (*a*) Energy-band diagram of a *p-n* junction solar cell under solar irradiation. (*b*) Idealized equivalent circuit of a solar cell.

and

$$J_s = I_s/A = qN_C N_V \left(\frac{1}{N_A} \sqrt{\frac{D_n}{\tau_n}} + \frac{1}{N_D} \sqrt{\frac{D_p}{\tau_p}} \right) e^{-E_g/kT} \tag{3}$$

where A is the device area. A plot of Eq. 2 is given in Fig. 5*a* for $I_L = 100$ mA, $I_s = 1$ nA, cell area $A = 4$ cm^2, and $T = 300$ K.[12] The curve passes through the fourth quadrant and, therefore, power can be extracted from the device. By properly choosing a load, close to 80% of the product $I_{sc}V_{oc}$ can be extracted (I_{sc} is the short-circuit current, and V_{oc} is the open-circuit voltage of the cell; the shaded area is the maximum power rectangle). The $I-V$ curve is more generally represented by Fig. 5*b*, which is an inversion of Fig. 5*a* about the voltage axis. We also define in Fig. 5*b* the quantities I_m and V_m that correspond to the current and voltage, respectively, for the maximum power output P_m ($= I_m V_m$).

From Eq. 2 we obtain for the open-circuit voltage ($I = 0$):

$$V_{oc} = \frac{kT}{q} \ln\left(\frac{I_L}{I_s} + 1\right) \approx \frac{kT}{q} \ln\left(\frac{I_L}{I_s}\right). \tag{4}$$

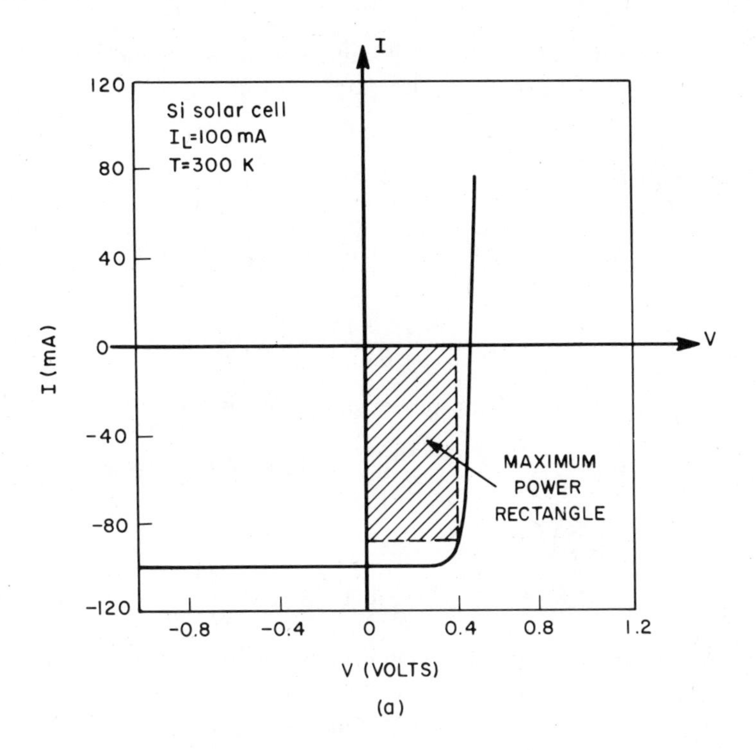

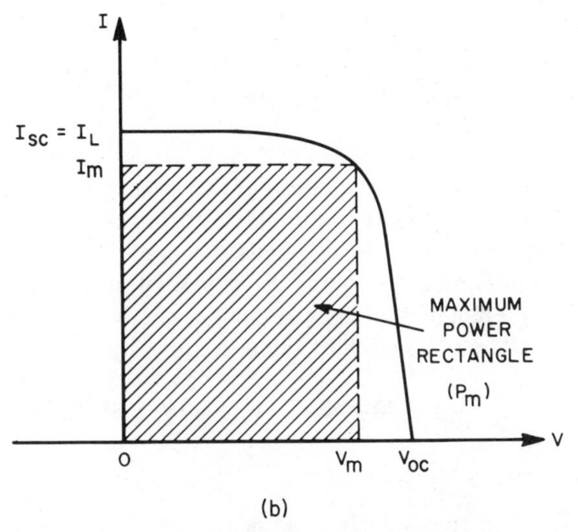

Fig. 5 (*a*) Current–voltage characteristics of a solar cell under illumination. (*b*) Inversion of (*a*) about the voltage axis. (After Prince, Ref. 12.)

Hence for a given I_L, the open-circuit voltage increases logarithmically with decreasing saturation current I_s. The output power is given by

$$P = IV = I_s V(e^{qV/kT} - 1) - I_L V. \tag{5}$$

The condition for maximum power can be obtained when $dP/dV = 0$, or

$$I_m = I_s \beta V_m e^{\beta V_m} \simeq I_L \left\{1 - \frac{1}{\beta V_m}\right) \tag{6}$$

$$V_m = \frac{1}{\beta} \ln\left(\frac{I_L/I_s + 1}{1 + \beta V_m}\right) \simeq V_{oc} - \frac{1}{\beta} \ln(1 + \beta V_m) \tag{7}$$

where $\beta \equiv q/kT$. The maximum power output P_m is then

$$P_m = I_m V_m \simeq I_L \left[V_{oc} - \frac{1}{\beta} \ln(1 + \beta V_m) - \frac{1}{\beta} \right] = I_L(E_m/q) \tag{8}$$

where

$$E_m \equiv q \left[V_{oc} - \frac{1}{\beta} \ln(1 + \beta V_m) - \frac{1}{\beta} \right]. \tag{9}$$

This energy E_m corresponds to the energy per photon delivered to the load at the maximum power point.

For a given semiconductor, the saturation current density, J_s, can be obtained from Eq. 3. At 300 K, the smallest J_s for Si is about 10^{-15} A/cm^2. Under the AM1.5 condition, the short-circuit current density J_L can be obtained graphically from Fig. 2:

$$J_L(E_g) = q \int_{h\nu = E_g}^{\infty} (dn_{ph}/dh\nu)\, d(h\nu). \tag{10}$$

The result is shown[11] in curve (1) of Fig. 6. Once J_s and J_L are known, E_m can be obtained by numerical solution of the transcendental Eqs. 4, 7, and 9. Because J_s hence E_m depends on material properties (e.g., τ, D, and doping levels), the ideal efficiency corresponds to an optimum choice of material properties that minimize J_s. Energy E_m is shown in curve (2) of Fig. 6. The ideal conversion efficiency is the ratio of the maximum power output to the incident power P_{in} and can be obtained graphically from Fig. 6.

$$\eta = \frac{P_m}{P_{in}} = \frac{I_L(E_m/q)}{P_{in}} = [V_m^2 I_s(q/kT) e^{qV_m/kT}]/P_{in} \tag{11}$$

or

$$\eta = \frac{\text{rectangular of } E_m n_{ph}}{\text{area under curve (1)}} \tag{11a}$$

where the area under curve (1) is equal to 5.2×10^{17} eV/cm^2/s. The maximum efficiency is found to be 31% for $E_g = 1.35$ eV using material parameters characteristic of III–V semiconductors.

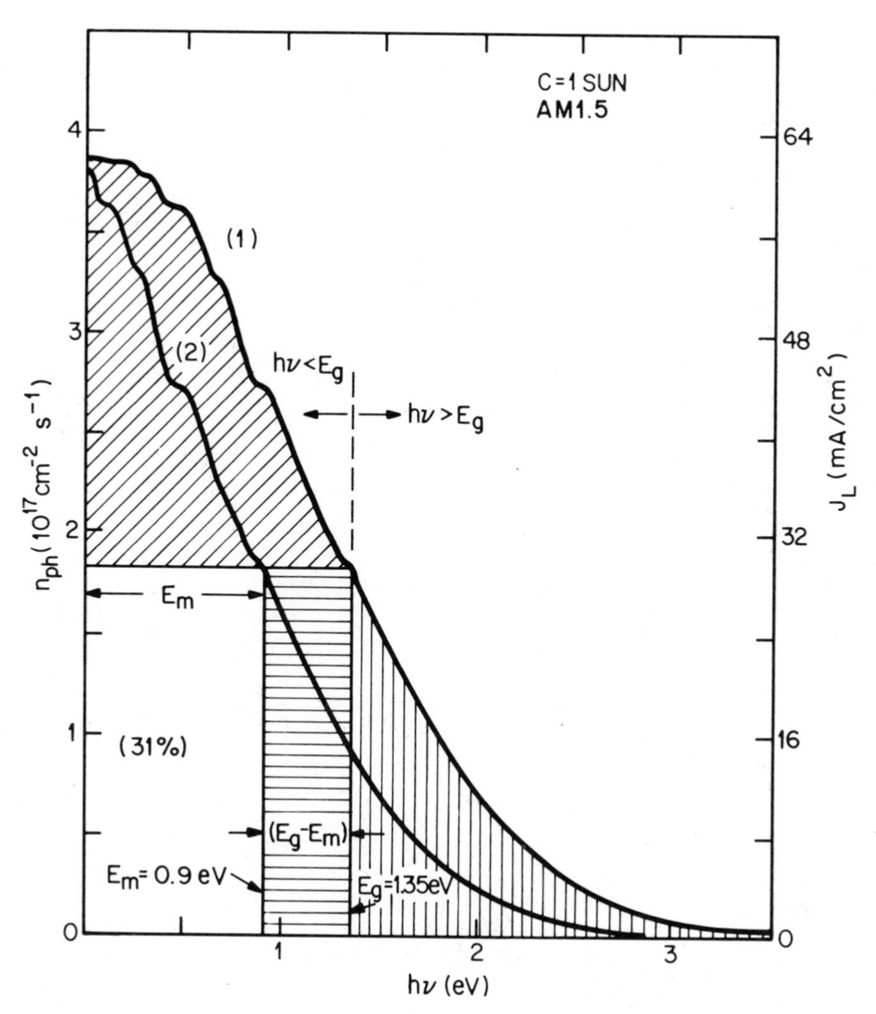

Fig. 6 Number of photons in a solar spectrum versus photon energy, and a graphical method to determine the conversion efficiency. (After Henry, Ref. 11.)

The ideal solar-cell efficiency at 300 K is shown in Fig. 7 as a function of energy bandgap.[13] The curve marked $C = 1$ is under one sun AM1.5 condition. The slight oscillations are caused by atmospheric absorption. The ideal efficiency has also been calculated using the method of detailed balance limit,[14] or assuming only radiative recombination loss;[11] the results are essentially the same as the graphical approach shown in Fig. 7. Note that the efficiency has a broad maximum and does not depend critically on E_g. Therefore, semiconductors with bandgaps between 1 and 2 eV can all be considered solar cell materials. Many factors degrade the ideal efficiency,

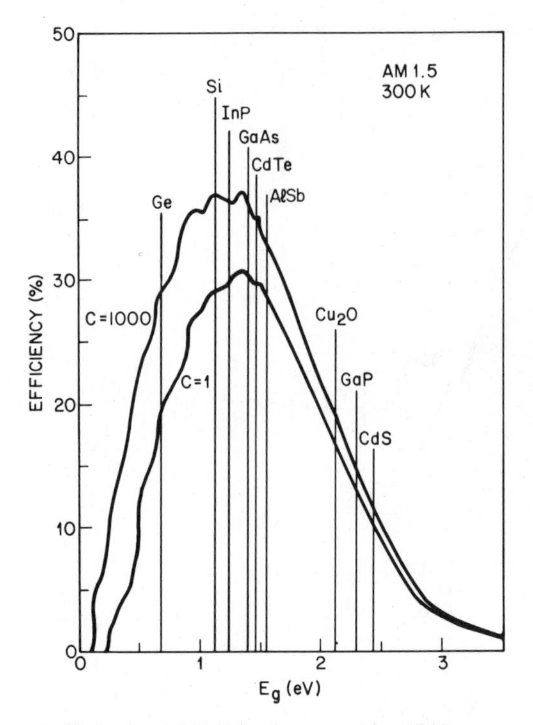

Fig. 7 Ideal solar-cell efficiency at 300 K for 1 sun and for 1000-sun concentration. (After Ref. 13.)

so that efficiencies actually achieved are lower. Practical solar cells will be considered in subsequent sections.

Figure 7 also shows the ideal efficiency at an optical concentration of 1000 suns (i.e., 844 kW/m^2). Details on optical concentration will be considered in Section 14.5. The ideal peak efficiency increases from 31% ($C = 1$) to 37% ($C = 1000$). This increase is primarily caused by the increase of V_{oc}, which in turn causes E_m to increase as indicated in Eq. 9. The short-circuit current density J_L is shown[11] in Fig. 8. Note that the horizontal distance between curves (1) and (2) is smaller than that shown in Fig. 6. Also shown in Fig. 8 is the multiple-band scheme under solar concentration. To obtain the maximum efficiency, one can use a graphical approach under the constraints that the current density across each bandgap must be the same. For two bandgaps in series, the ideal maximum efficiency is 50% with $E_{g1} = 1.56$ eV and $E_{g2} = 0.94$ eV. For three bandgaps, the ideal maximum efficiency is 56% with $E_{g1} = 1.75$ eV, $E_{g2} = 1.18$ eV, and $E_{g3} = 0.75$ eV. Beyond the three bandgaps, the efficiency increases very slowly; at 36 bandgaps, the maximum efficiency is 72%.

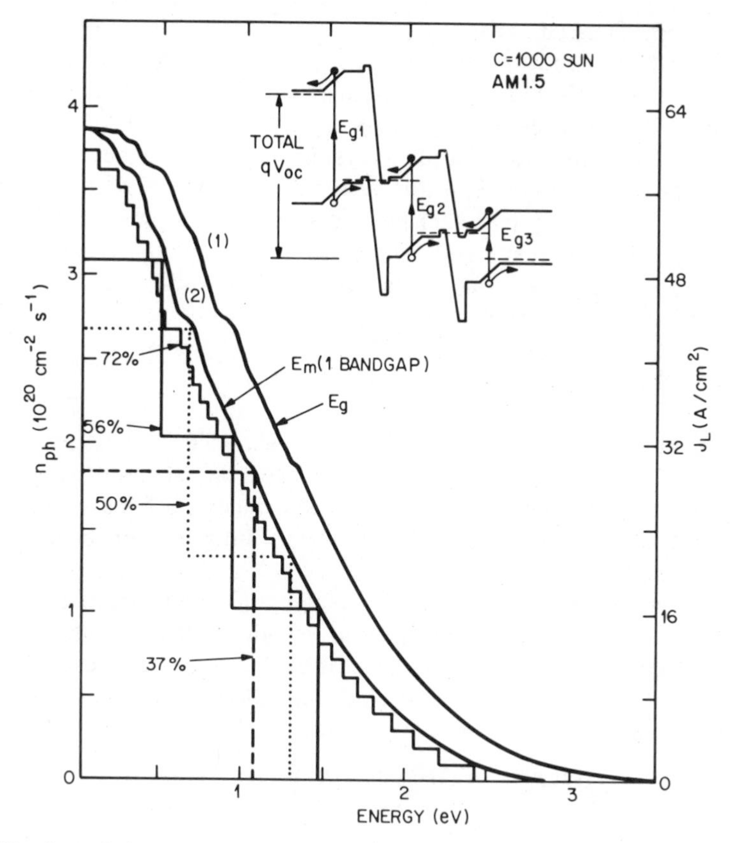

Fig. 8 Number of photons in concentrated solar spectrum, and a graphical method to determine the efficiency for multiple bandgaps in series. (After Henry, Ref. 11.)

14.3 *p-n* JUNCTION SOLAR CELLS

In this section we are concerned primarily with the silicon *p-n* junction solar cell, because it serves as a reference device for all solar cells. For satellites and space vehicles, flat-plate silicon solar cells are the most important long-duration power supply. The main concerns are cell conversion efficiency and reliability, such as degradation from high-energy particle radiation in outer space. For terrestrial applications, both flat-plate and concentrator solar cell systems have been considered. In addition to efficiency and reliability, cost is also a concern. Eventually, the terrestrial solar system will be competitive with other energy sources.

For terrestrial flat-plate systems, the basic approach attempts to reduce cell manufacturing cost as much as possible and at the same time maintain at least 10% cell conversion efficiency. This approach includes using the edge-defined film-fed growth (EFG) technique, the ribbon-to-ribbon process, and the dendrite-web process.[7] Polycrystalline silicon on ceramic or on metallurgical-grade silicon is also used to reduce substrate cost. Thin-film CdS solar cells as well as amorphous Si solar cells are also important candidates for flat-plate systems. They will be considered in Section 14.4. For the concentrator system, the basic approach attempts to improve the cell efficiency under high solar concentrations and to minimize the overall system cost. The concentrator cells will be considered in Section 14.5. In this section, the basic device characteristics of *p-n* junction solar cells will be considered.

14.3.1 Spectral Response

A typical schematic representative of a solar cell is shown in Fig. 9. It consists of a shallow *p-n* junction formed on the surface (e.g., by diffusion), a front ohmic contact stripe and fingers, a back ohmic contact that covers the entire back surface, and an antireflection coating on the front surface.

When a monochromatic light of wavelength λ is incident on the front surface, the photocurrent and spectral response, that is, the number of carriers collected per incident photon at each wavelength, can be derived as follows. The generation rate of electron–hole pairs at a distance x from the semiconductor surface is shown in Fig. 10*a* and is given by

$$G(\lambda, x) = \alpha(\lambda)F(\lambda)[1 - R(\lambda)] \exp[-\alpha(\lambda)x] \tag{12}$$

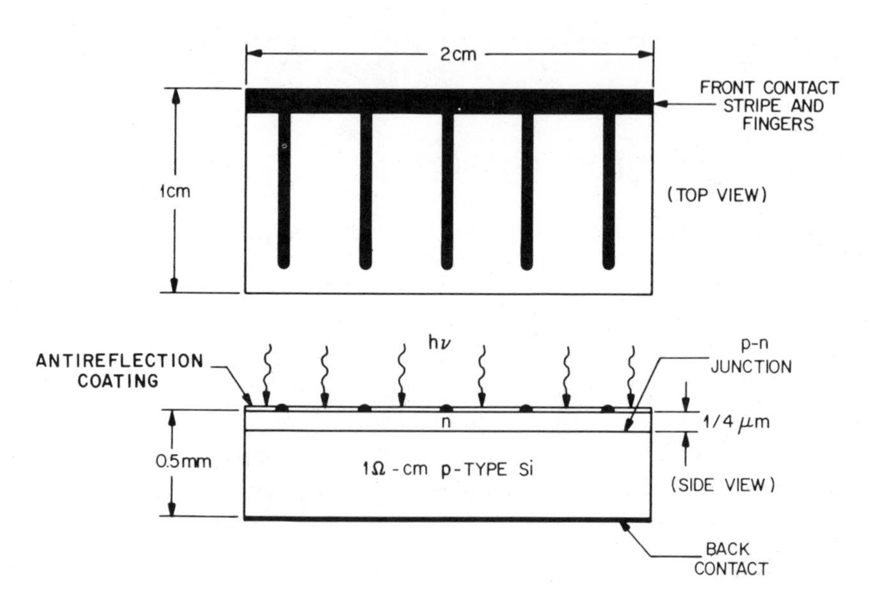

Fig. 9 Schematic representation of a silicon *p-n* junction solar cell.

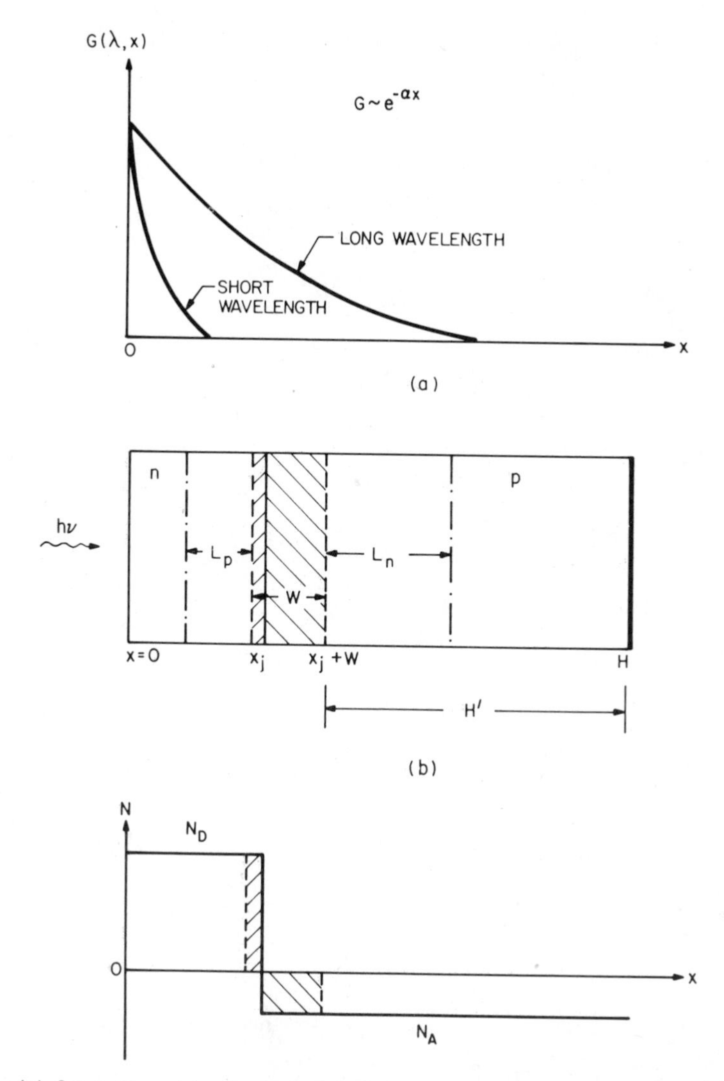

Fig. 10 (*a*) Generation rate of electron–hole pairs as a function of distance from the semiconductor surface for long- and short-wavelength light. (*b*) Solar-cell dimensions and minority-carrier diffusion lengths. (*c*) Assumed abrupt doping profile of the solar cell.

where $\alpha(\lambda)$ is the absorption coefficient, $F(\lambda)$ the number of incident photons/cm^2/s per unit bandwidth, and $R(\lambda)$ the fraction of these photons reflected from the surface.[3]

Under low-injection condition, the one-dimensional, steady-state continuity equations are

$$G_n - \frac{n_p - n_{po}}{\tau_n} + \frac{1}{q}\frac{dJ_n}{dx} = 0 \tag{13}$$

for electrons in p-type semiconductors and

$$G_p - \frac{p_n - p_{no}}{\tau_p} - \frac{1}{q}\frac{dJ_p}{dx} = 0 \tag{14}$$

for holes in n-type semiconductors. The current-density equations are

$$J_n = q\mu_n n_p \mathscr{E} + qD_n \frac{dn_p}{dx} \tag{15}$$

$$J_p = q\mu_p p_n \mathscr{E} - qD_p \frac{dp_n}{dx}. \tag{16}$$

For an abrupt p-n junction solar cell with constant doping on each side of the junction, Fig. 10b and c, there are no electric fields outside the depletion region. In the case of an n-on-p junction with an n-type front and p-type base, Eqs. 12, 14, and 16 can be combined to yield an expression for the top side of the junction:

$$D_p \frac{d^2 p_n}{dx^2} + \alpha F(1-R)\exp(-\alpha x) - \frac{p_n - p_{no}}{\tau_p} = 0. \tag{17}$$

The general solution to this equation is

$$p_n - p_{no} = A\cosh(x/L_p) + B\sinh(x/L_p) - \frac{\alpha F(1-R)\tau_p}{\alpha^2 L_p^2 - 1}\exp(-\alpha x) \tag{18}$$

where $L_p = (D_p\tau_p)^{1/2}$ is the diffusion length. There are two boundary conditions. At the surface, we have surface recombination with a recombination velocity S_p:

$$D_p \frac{d(p_n - p_{no})}{dx} = S_p(p_n - p_{no}) \qquad \text{at} \quad x = 0 \tag{19}$$

At the depletion edge, the excess carrier density is small due to the electric field in the depletion region:

$$p_n - p_{no} \simeq 0 \qquad \text{at} \quad x = x_j. \tag{20}$$

Using these boundary conditions in Eq. 18, the hole density is

$$p_n - p_{no} = [\alpha F(1-R)\tau_p/(\alpha^2 L_p^2 - 1)] \times \left[\frac{\left(\frac{S_p L_p}{D_p} + \alpha L_p\right)\sinh\left(\frac{x_j - x}{L_p}\right) + e^{-\alpha x_j}\left(\frac{S_p L_p}{D_p}\sinh\frac{x}{L_p} + \cosh\frac{x}{L_p}\right)}{(S_p L_p/D_p)\sinh(x_j/L_p) + \cosh(x_j/L_p)} - e^{-\alpha x} \right] \tag{21}$$

and the resulting hole photocurrent density at the depletion edge is

$$J_p = -qD_p\left(\frac{dp_n}{dx}\right)_{x_j} = [qF(1-R)\alpha L_p/(\alpha^2 L_p^2 - 1)] \times \left[\frac{\left(\frac{S_p L_p}{D_p} + \alpha L_p\right) - e^{-\alpha x_j}\left(\frac{S_p L_p}{D_p}\cosh\frac{x_j}{L_p} + \sinh\frac{x_j}{L_p}\right)}{(S_p L_p/D_p)\sin(x_j/L_p) + \cosh(x_j/L_p)} - \alpha L_p e^{-\alpha x_j} \right] \tag{22}$$

This photocurrent would be collected from the front side of an n-on-p junction solar cell at a given wavelength, assuming this region to be uniform in lifetime, mobility, and doping level.

To find the electron photocurrent collected from the base of the cell, Eqs. 12, 13, and 15 are used with the boundary conditions:

$$n_p - n_{po} \cong 0 \qquad \text{at} \quad x = x_j + W \tag{23}$$

$$S_n(n_p - n_{po}) = -D_n dn_p/dx \qquad \text{at} \quad x = H \tag{24}$$

where W is the depletion width and H is the width of the entire cell. Equation 23 states that the excess minority carrier density is near zero at the edge of the depletion region, while Eq. 24 states that the back surface recombination takes place at the ohmic contact.

Using these boundary conditions, the electron distribution in a uniformly doped p-type base is

$$n_p - n_{po} = \frac{\alpha F(1-R)\tau_n}{\alpha^2 L^2 - 1} \exp[-\alpha(x_j + W)]\left\{\cosh\left(\frac{x - x_j - W}{L_n}\right) - e^{-[\alpha(x-x_j-W)]} - \frac{(S_nL_n/D_n)[\cosh(H'/L_n) - \exp(-\alpha H')] + \sinh(H'/L_n) + \alpha L_n e^{-\alpha H'}}{(S_nL_n/D_n)\sinh(H'/L_n) + \cosh(H'/L_n)} \times \sinh\left(\frac{x - x_j - W}{L_n}\right)\right\} \tag{25}$$

and the photocurrent due to electrons collected at the depletion edge, $x = x_j + W$, is

$$J_n = qD_n\left(\frac{dn_p}{dx}\right)_{x_j+W} = \frac{qF(1-R)\alpha L_n}{\alpha^2 L_n^2 - 1} \exp[-\alpha(x_j + W)] \times \left\{\alpha L_n - \frac{(S_nL_n/D_n)[\cosh(H'/L_n) - \exp(-\alpha H')] + \sinh(H'/L_n) + \alpha L_n \exp(-\alpha H')}{(S_nL_n/D_n)\sinh(H'/L_n) + \cosh(H'/L_n)}\right\} \tag{26}$$

where H' as shown in Fig. 10b is the p-base neutral region. Equation 26 is derived assuming the base region to be uniform in lifetime, mobility, and doping level. If these quantities are a function of distance, numerical analysis must be used.

Some photocurrent generation takes place within the depletion region as well. The electric field in this region is generally high, the photogenerated carriers are accelerated out of the depletion region before they can recombine. The photocurrent per unit bandwidth is equal to the number of photons absorbed:

$$J_{dr} = qF(1-R)\exp(-\alpha x_j)[1 - \exp(-\alpha W)]. \tag{27}$$

The total photocurrent at a given wavelength is then the sum of Eqs. 22, 26, and 27:

$$J(\lambda) = J_p(\lambda) + J_n(\lambda) + J_{dr}(\lambda). \tag{28}$$

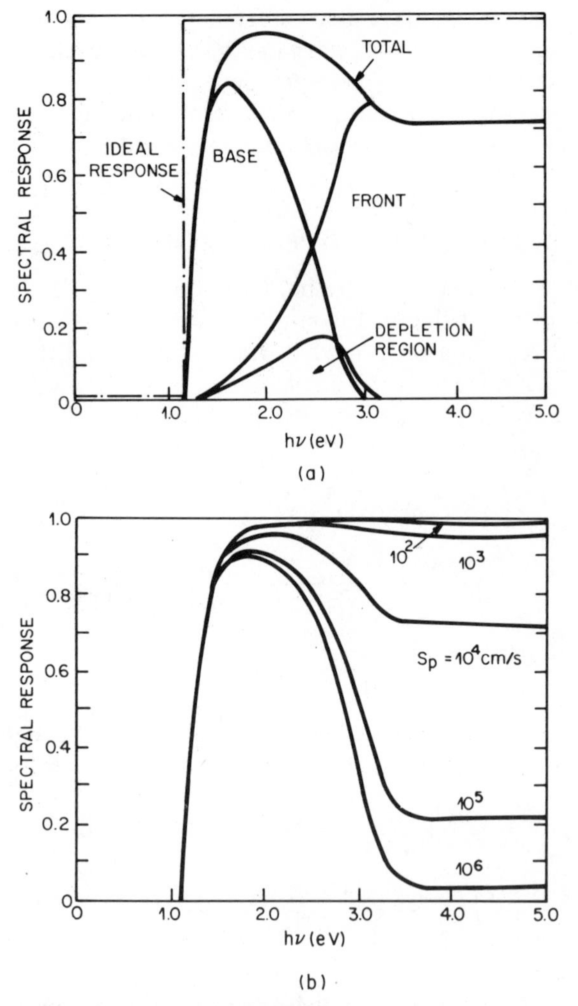

Fig. 11 (*a*) Computed internal spectral response of a Si *n*-on-*p* cell, showing the individual contributions from each of the three regions. (The dashed-dotted curve is for the ideal response.) (*b*) Computed internal spectral response of a Si *n*-on-*p* cell with different surface recombination velocities. (After Hovel, Ref. 3.)

The spectral response (SR) is equal to this sum divided by qF for externally observed response or by $qF(1-R)$ for internal spectral response:

$$\mathrm{SR}(\lambda) = \frac{1}{qF(\lambda)[1-R(\lambda)]}[J_p(\lambda) + J_n(\lambda) + J_{dr}(\lambda)]. \tag{29}$$

The ideal internal spectral response for a semiconductor with energy gap E_g is a step function that equals zero for $h\nu < E_g$ and unity for $h\nu \geq E_g$ (dashed–dotted lines in Fig. 11*a*). A realistic internal spectral response calculated for a Si n/p solar cell is shown in Fig. 11a, which departs substantially from the idealized step function at high photon energies.[3] The figure also shows the individual contributions from each of the three regions. The device parameters are $N_D = 5 \times 10^{19}\ \text{cm}^{-3}$, $N_A = 1.5 \times 10^{16}\ \text{cm}^{-3}$, $\tau_p = 0.4\ \mu\text{s}$, $\tau_n = 10\ \mu\text{s}$, $x_j = 0.5\ \mu\text{m}$, $H = 450\ \mu\text{m}$, $S_p(\text{front}) = 10^4\ \text{cm/s}$, and $S_n(\text{back}) = \infty$. At low photon energies, most carriers are generated in the base region because of the low absorption coefficient in Si. As the photon energy increases above 2.5 eV, the front region takes over. Above 3.5 eV, α becomes larger than $10^6\ \text{cm}^{-1}$, and the spectral response derives entirely from the front region. Since S_p is assumed to be quite high, the surface recombination at the front region causes large departure from the ideal response. The spectral response approaches an asymptotic value when $\alpha L_p \gg 1$ and $\alpha x_j \gg 1$ (i.e., from the front-side photocurrent, Eq. 22):

$$\text{SR} = \frac{1 + S_p/\alpha D_p}{(S_p L_p/D_p)\sinh(x_j/L_p) + \cosh(x_j/L_p)}. \tag{30}$$

The surface recombination velocity S_p has a profound effect on the spectral response especially at high photon energies. This effect is illustrated[3] in Fig. 11b for devices with the same parameters as in Fig. 11a, except that S_p varies from 10^2 to 10^6 cm/s. Note the drastic reduction in spectral response as S_p increases. Equation 30 also shows that for a given S_p, the spectral response can be improved by increasing the diffusion length L_p. Generally, to increase SR over the useful wavelength range, one should increase both L_n and L_p, and decrease both S_n and S_p.

Once the spectral response is known, the total photocurrent density obtained from the solar spectral distribution $F(\lambda)$, shown in Fig. 1, is given by

$$J_L = q\int_0^{\lambda_m} F(\lambda)[1 - R(\lambda)]\text{SR}(\lambda)\, d\lambda \tag{31}$$

where λ_m is the longest wavelength corresponding to the semiconductor bandgap. To obtain large J_L, one should minimize $R(\lambda)$ and maximize $\text{SR}(\lambda)$ over the wavelength range $0 < \lambda < \lambda_m$. The ideal photocurrent density J_L shown in Fig. 6 is for $R(\lambda) = 0$ and $\text{SR}(\lambda) = 1$.

14.3.2 *I–V* Characteristics

For a practical solar cell, the ideal equivalent circuit, Fig. 4b, will be modified to include the series resistance from ohmic loss in the front surface and the shunt resistance from leakage currents. The equivalent circuit is shown in the insert of Fig. 12. If the diode current is given by Eq. 2, the I–V characteristics are found to be[12]

$$\ln\left(\frac{I + I_L}{I_s} - \frac{V - IR_s}{I_s R_{sh}} + 1\right) = \frac{q}{kT}(V - IR_s). \tag{32}$$

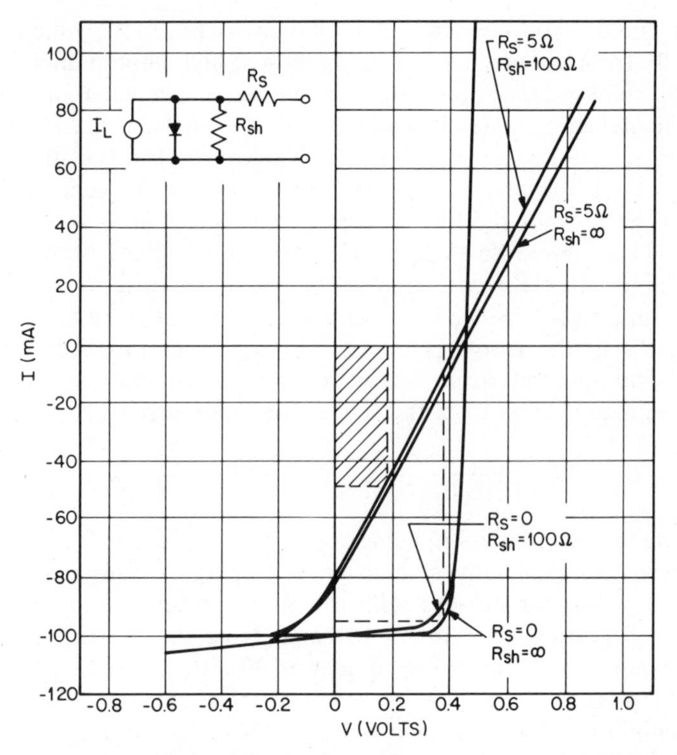

Fig. 12 Theoretical I–V characteristics for various solar cells that include series and shunt resistances. The insert shows the equivalent circuit. The parameters are identical to those shown in Fig. 5. (After Prince, Ref. 12.)

Plots of this equation, with combinations of $R_s = 0$, 5 Ω and $R_{sh} = \infty$, 100 Ω, are given in Fig. 12 with the same parameters I_s, I_L, and T as in Fig. 5. It can be seen that a shunt resistance even as low as 100 Ω does not appreciably change the power output of the device, whereas a series resistance of only 5 Ω reduces the available power to less than 30% of the optimum power with $R_s = 0$. We can thus neglect the effect of R_{sh}. The output current and output power are

$$I = I_s\left\{\exp\left[\frac{q(V - IR_s)}{kT}\right] - 1\right\} - I_L \tag{33}$$

$$P = |IV| = I\left[\frac{kT}{q}\ln\left(\frac{I + I_L}{I_s} + 1\right) + IR_s\right]. \tag{34}$$

The relative maximum available power is 1, 0.77, 0.57, 0.27, or 0.14 for R_s of 0, 1, 2, 5, or 10 Ω, respectively. The series resistance of a solar cell depends on the junction depth, impurity concentrations of p-type and n-type regions, and the arrangement of the front-surface ohmic contacts.

For a typical Si solar cell with the geometry shown in Fig. 9, the series resistance is about 0.7 Ω for n-on-p cells and about 0.4 Ω for p-on-n cells.[15] The difference in resistance is mainly the result of lower resistivity in n-type substrates.

With reference to Figs. 5 and 12, we can define a fill factor, FF:

$$FF \equiv \frac{I_m V_m}{I_L V_{oc}}. \tag{35}$$

The conversion efficiency is given by

$$\eta = \frac{I_m V_m}{P_{in}} = \frac{FF \cdot I_L V_{oc}}{P_{in}}. \tag{36}$$

To maximize the efficiency, one should maximize all three items in the numerator of Eq. 36.

For a practical solar cell, the forward current can be dominated by the recombination current in the depletion region. The efficiency is generally reduced compared with that of an ideal diode discussed in Section 14.2.2. For single-level centers, the recombination current can be expressed as[16,17]

$$I_{\text{rec}} = I'_s\left[\exp\left(\frac{qV}{2kT}\right) - 1\right] \tag{37}$$

with

$$\frac{I'_s}{A} = \frac{qn_i W}{\sqrt{\tau_p \tau_n}}. \tag{37a}$$

The energy conversion equation can again be put into closed form yielding equations similar to Eqs. 4 through 9 with the exception that I_s is replaced by I'_s, and the exponential factor is divided by 2. The efficiency for the case of recombination current is found to be much less than the ideal current case due to degradation of both V_{oc} and the fill factor. For Si solar cells at 300 K, the recombination current can cause a 25% reduction in efficiency.[18]

For solar cells having mixtures of diffusion current and recombination current, or containing many defects, the forward current may show an exponential dependence on the forward voltage as $\exp(qV/nkT)$, where n is called the ideality factor. The efficiency usually decreases with increasing values of n.

Although an individual silicon solar cell with an area of 2 cm^2 can only have 0.5 to 0.6 V of open-circuit voltage and about 30 to 60 mA of short-circuit current, solar cells can be connected in series and in parallel to form arrays, and substantially higher voltages and higher currents can be delivered to the load. An example of a solar-cell array is shown[5] in the insert in Fig. 13. This array operated at 60°C can deliver about 10 W at 11.5% efficiency under AM1 conditions (~100 mW/cm^2), as shown by the tangent of the I–V plot to the constant-power curves. At the maximum power point, the corresponding V_m and I_m for the array is 14 V and

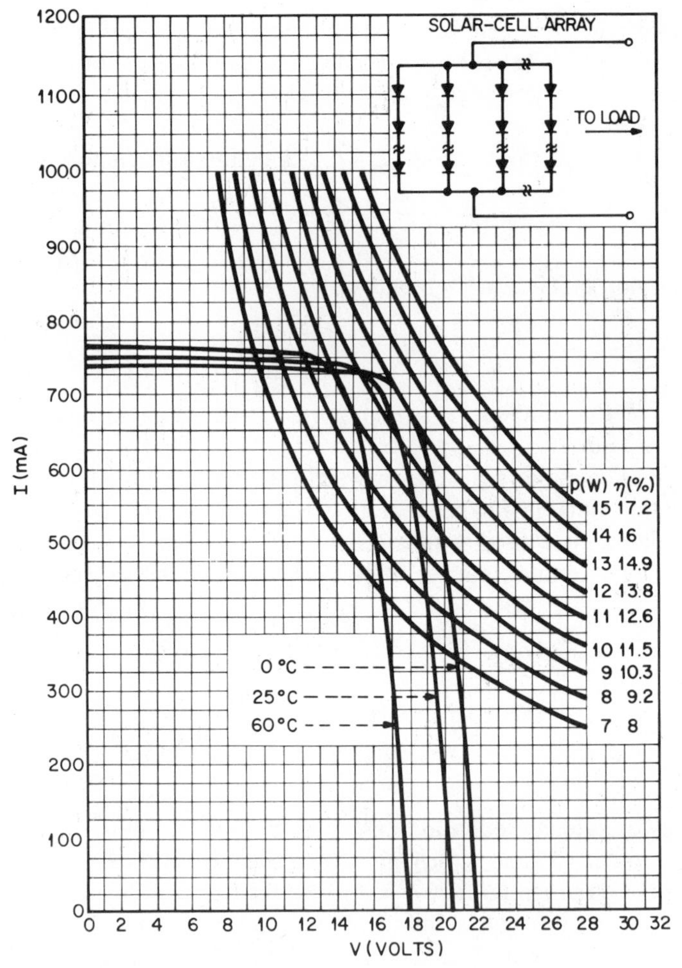

Fig. 13 Current–voltage characteristics of a solar-cell array; also shown are constant-power curves. (After Pulfrey, Ref. 5.)

720 mA, respectively. For lower-temperature operations, both power output and cell efficiency will increase.

14.3.3 Temperature and Radiation Effects

As the temperature increases, the diffusion lengths in Si and GaAs will increase, because the diffusion constant stays the same or increases with temperature, and the minority lifetime increases with temperature. The increase in minority-carrier diffusion length causes an increase in J_L. However, V_{oc} will rapidly decrease because of the exponential dependence of the saturation current on temperature. The increase in the "softness"

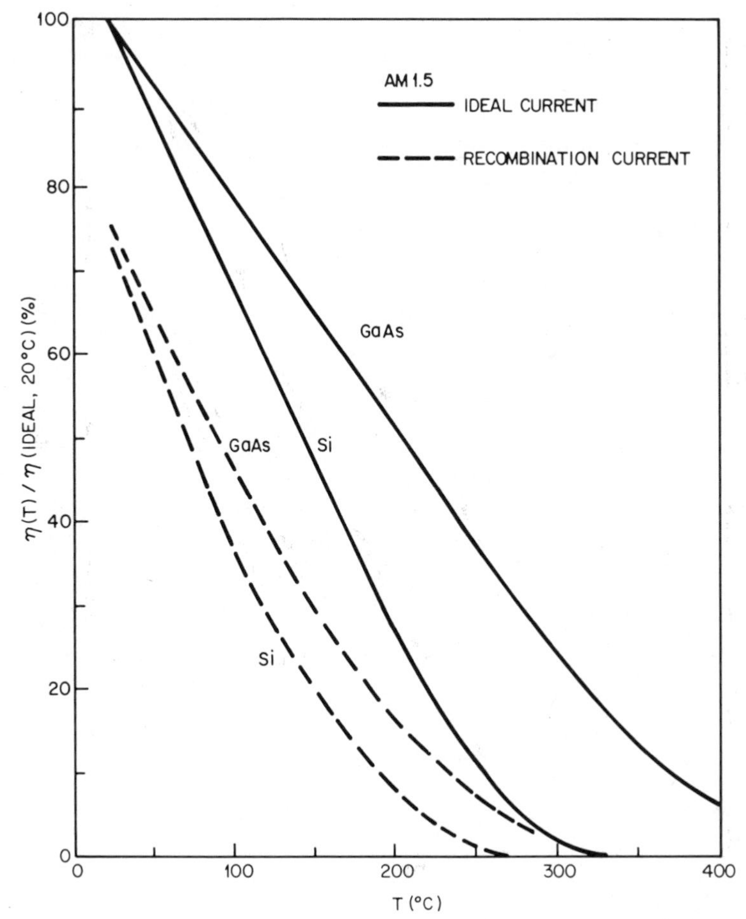

Fig. 14 Normalized efficiency for Si and GaAs *p-n* junction solar cells having ideal currents or recombination currents. (After Wysocki and Rappaport, Ref. 18.)

(roundness) in the knee of the $I-V$ curve as temperature increases will also degrade the fill factor. Therefore, the overall effect causes a reduction of efficiency as the temperature increases.

The normalized efficiency for Si and GaAs solar cells is shown[18] in Fig. 14. Under the ideal current condition, efficiency decreases linearly with temperature to about 200°C for Si and to 300°C for GaAs. For the recombination current case, the initial efficiency at 20°C is about 25% lower. The efficiency is also decreasing approximately linearly as temperature increases. Figure 14 clearly shows that excessive cell temperature will severely degrade the conversion efficiency.

For satellite applications, the high-energy particle radiation in outer space produces defects in semiconductors that cause a reduction in solar-cell power output. Assessing the expected useful life of the space solar-cell power plant is important. Its lifetime is the length of time it can deliver the electrical power necessary for successful operation of the satellite.

From the expressions of the photocurrent, Eqs. 22 and 26, note that the current will decrease with decreasing diffusion lengths, L_n and L_p. The lifetime of the excess minority carriers at any point during the bombardment of high-energy particles can be given by

$$\frac{1}{\tau} = \frac{1}{\tau_0} + K'\Phi \tag{38}$$

where τ_0 is the initial lifetime, K' is a constant, and Φ is the bombardment flux.[19] The expression above states that the recombination rate of the minority carriers is proportional to the initial number of recombination centers present plus the number introduced during bombardment, the latter being proportional to the flux. Since the diffusion length is equal to $\sqrt{D\tau}$ and D is a slowly varying function with bombardment (or doping concentration), Eq. 38 can be expressed as

$$\frac{1}{L^2} = \frac{1}{L_0^2} + K\Phi \tag{39}$$

where L_0 is the initial diffusion length and $K = K'/D$. Figure 15 shows the

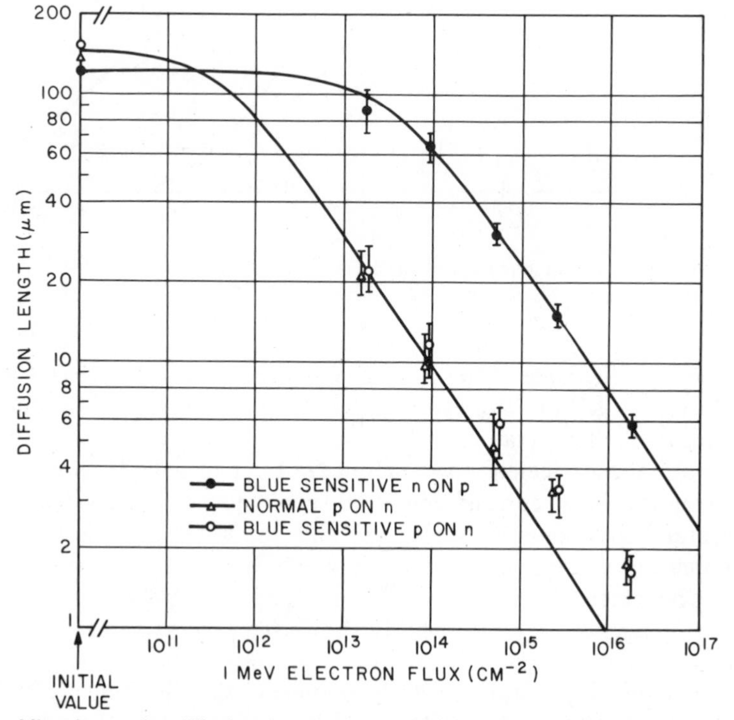

Fig. 15 Minority-carrier diffusion length versus 1-MeV electron flux. (After Rosenzweig, Gummel, and Smits, Ref. 19.)

measured substrate diffusion length as a function of a 1-MeV electron flux for three different silicon solar cells. The blue-sensitive n-on-p cell is one with n-type diffusion and with antireflection coating. The diffusion depth is adjusted to give a large spectral response near blue light (0.45 to 0.5 μm), which corresponds to the maximum of the solar energy distribution. The blue-sensitive p-on-n cell is similar to the n-on-p cell, except that the roles of p-type and n-type are interchanged. We see that the experimental results can indeed be reasonably fitted by Eq. 39. The curve passing through the points for the n-on-p cell is computed using Eq. 39 with $L_0 = 119\ \mu$m and $K = 1.7 \times 10^{-10}$. The data points for the blue-sensitive p-on-n and normal p-on-n cells can be approximately fitted by Eq. 39 with $L_0 = 146\ \mu$m and $K = 1.22 \times 10^{-8}$. It is apparent from Fig. 15 that the radiation resistance of n-on-p cells is higher than that of p-on-n cells.

To improve the radiation tolerance, lithium has been incorporated into the solar cells.[20] The Li can diffuse to and combine with radiation-induced point defects. Apparently, the Li neutralizes the defect and prevents degradation of the lifetime. In a space environment, a coverslip, such as the cerium-doped microsheet, must always be placed over the front surface to minimize the number of high-energy particles that reach the device.

14.3.4 Device Configurations

Many solar-cell configurations have been proposed for achieving higher conversion efficiency. We shall consider a few solar cells that show substantial improvements in device performances.

The "back surface field" (BSF) solar cell has shown much larger output voltage than conventional cells.[21] A schematic band diagram is shown in Fig. 16. The front surface is made in the normal way, but the back of the

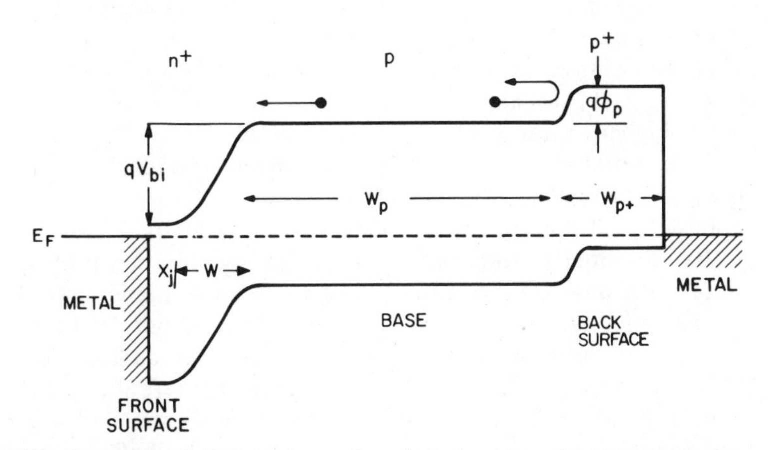

Fig. 16 Energy-band diagram for a n^+-p-p^+ back-surface field junction solar cell. (After Mandelkorn and Lamneck, Ref. 21.)

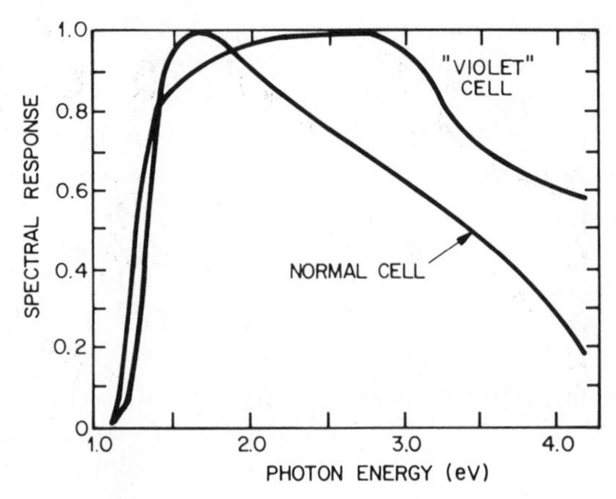

Fig. 17 Measured spectral responses of *n-p* Si solar cells. The violet cell shows considerably higher response in the violet region than a normal cell shows. (After Lindmayer and Allison, Ref. 22.)

cell, instead of containing just a metallic ohmic contact, has a very heavily doped region adjacent to the contact. The potential energy barrier $q\phi_p$ between the two base regions tends to confine minority carriers in the more lightly doped region. The BSF cell is equivalent to a normal cell of width $(x_j + W + W_p)$ having a very small recombination velocity at the back $(S_n < 100 \text{ cm/s})$. The low S_n will enhance the spectral response at low photon energies. Therefore, the short-circuit current density will increase. The open-circuit voltage is also increased due to increased short-circuit current, decreased diode recombination current at the back contact, and the added potential energy $q\phi_p$ between the p and p^+ regions.

The "violet" cell is fabricated with reduced surface doping concentration and smaller junction depth.[22] The combination of higher lifetime near the surface and narrower junction greatly improves the response at high photon energies. Figure 17 shows a comparison of the measured spectral responses between a normal cell with $x_j = 0.4\ \mu\text{m}$ and $N_D = 5 \times 10^{19} \text{ cm}^{-3}$, and a violet cell with $x_j = 0.2\ \mu\text{m}$ and a surface doping of $5 \times 10^{18} \text{ cm}^{-3}$. The violet cell has a much higher response in the violet region ($h\nu > 2.75$ eV); hence the cell name's. Its spectral response is similar to the theoretical results shown in Fig. 11*b* for front-surface $S_p \cong 10^4$ cm/s.

The "textured" cell has pyramidal surfaces produced by anisotropical etching of ⟨100⟩-oriented Si surface,[23] as shown in Fig. 18*a*. Light incident on the side of a pyramid will be reflected onto another pyramid instead of being lost. The reflectivity of bare Si is reduced from about 35% for flat surfaces to around 20% for the textured surface; the addition of an antireflection coating reduces the overall reflection to a few percent as shown in Fig. 18*b*. The reduced reflection enhances both the short-circuit current and open-circuit voltage, which in turn improves cell efficiency. An AM0 efficiency over 15% has been obtained from the textured cell.

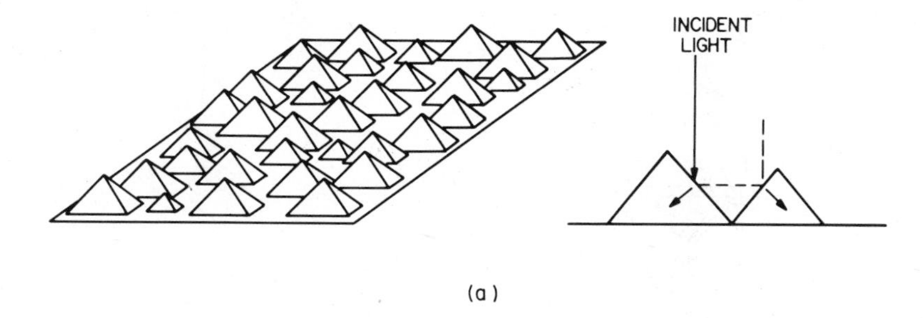

(a)

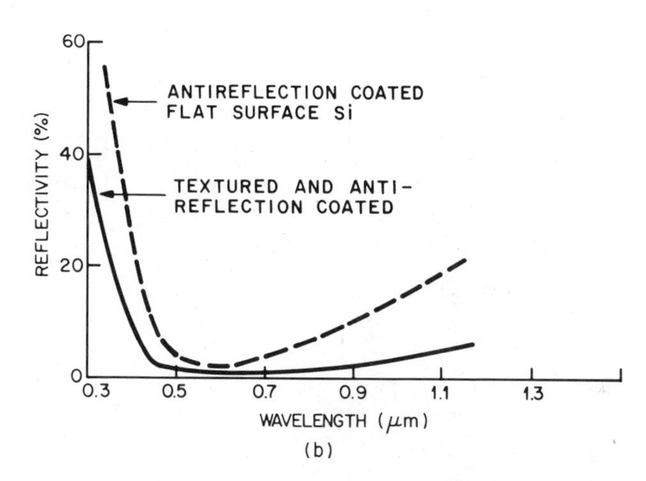

(b)

Fig. 18 (*a*) Textured cell with pyramidal surfaces. (*b*) Reflectivity versus wavelength for a flat surface cell and a textured cell. (After Arndt et al., Ref. 23.)

A new type of solar cell also using anisotropical etching of ⟨100⟩ Si has been proposed. It is called the V-groove multijunction solar cell.[24] This cell consists of many individual p^+-n-n^+ (or p^+-p-n^+) diode elements connected in series, as shown in Fig. 19*a*. The trapezoidal shape of the diode elements is defined by anisotropically etching ⟨100⟩ Si through a thermally grown silicon dioxide layer. The effective optical thickness of an element is illustrated in Fig. 19*b*. For a 50-μm-thick diode, the effective optical thickness averaged over various path lengths is over 250 μm, resulting in a fundamental collection efficiency (i.e., the ratio of the number of photons that are absorbed to the number that enter a semiconductor with $h\nu \geq E_g$) greater than 93%. Because of the higher collection efficiency, lower series resistance, and no shadowing effect on the illuminated surface (there are no

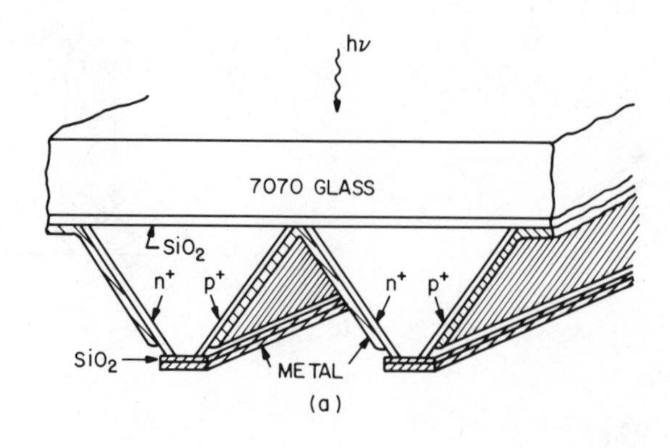

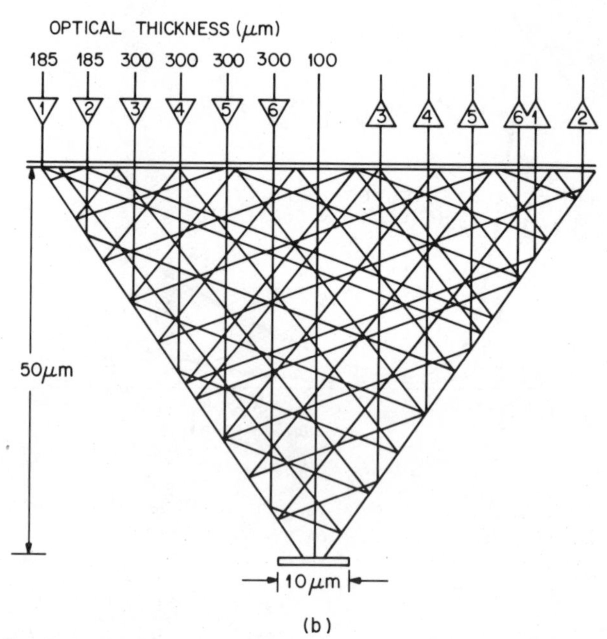

Fig. 19 (*a*) Vertical-groove multijunction solar cell. (*b*) Optical lengths for various incoming lights; the average effective optical thickness is 250 μm. (After Chappell, Ref. 24.)

collection grids), the new cell is expected to have a conversion efficiency greater than 20%.

A tandem-junction solar cell (TJC),[25] which combines the concepts of the back-surface-field cell and textured cell, is shown in Fig. 20*a*. Both n^+ and p^+ contacts are on the back side to eliminate metal shadowing and facilitate interconnection. The device behaves as a bipolar transistor with an uncontacted front n^+ emitter, shown in Fig. 20*b*. Photogenerated electrons in

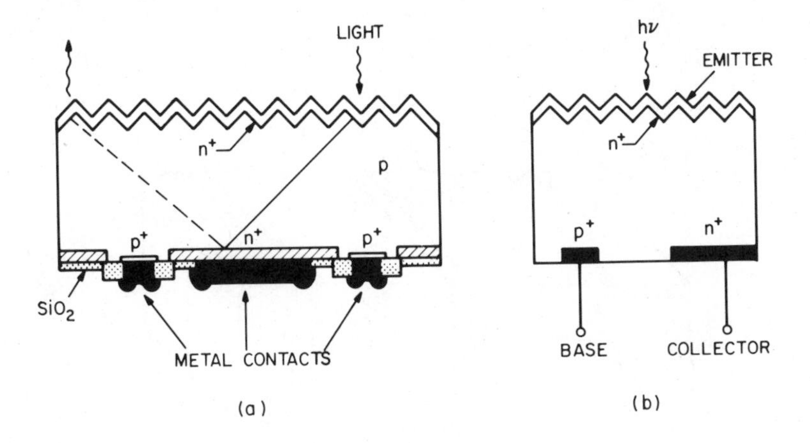

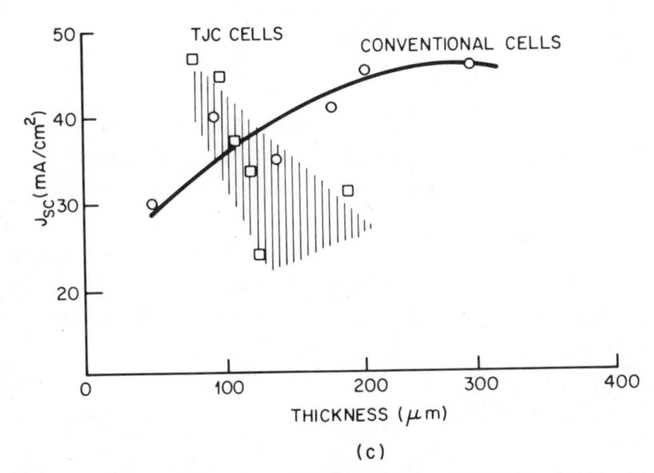

Fig. 20 (a) Cross section of a tandem-junction solar cell (TJC). (*b*) Equivalent transistor configuration. (*c*) Comparison of short-circuit current versus device thickness for TJC and conventional cells at AM0. (After Chiang, Carbajal, and Wakefield, Ref. 25.)

the emitter or in the base are collected by the n^+ collector via transistor action. The dependence of the short-circuit current on device thickness is quite different from conventional cells. A comparison is shown in Fig. 20*c*. The tandem-junction solar cell uses much thinner base material and is also expected to have an efficiency above 20%.

Another novel device configuration is the vertical-junction solar cell, which has both the junction and the metallization perpendicular to the cell surface.[26] A schematic multiple vertical-junction device is shown in Fig. 21. The diffusions and metal contacts are embedded in deeply etched grooves normal to the surface, formed by anisotropic etching of ⟨110⟩ silicon. The separation between grooves is 140 μm, which is of the same order as the

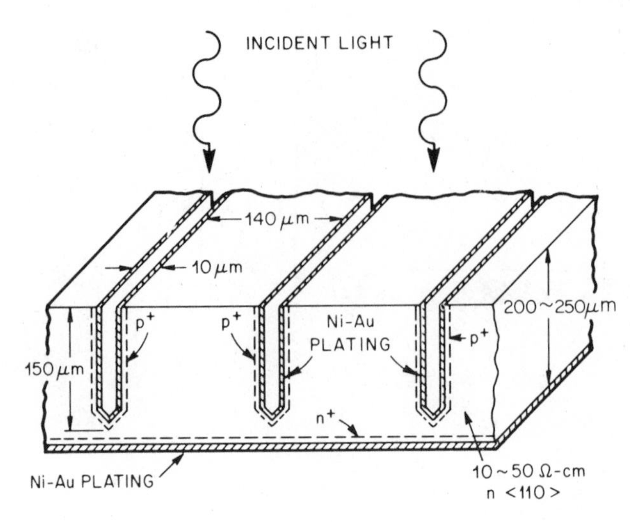

Fig. 21 Schematic diagram of low-series resistance multiple vertical-junction solar cell. (After Frank, Goodrich, and Kaplow, Ref. 26.)

minority-carrier diffusion length. There is no diffusion in the front surface, and the obscured surface area, due to the 10-μm-wide grooves, is about 7%. The contact stripes at each end and antireflection coatings are not shown.

In the vertical-junction solar cell, current flows directly through the p^+ diffusions into the metallization on the groove walls and out to the contact stripes at the edges of the cell. These contact stripes connect all the junctions electrically in parallel; therefore, the series resistance is minimized. Under one sun AM1 condition, the measured results are $V_{oc} = 0.59$ V, $J_{sc} = 33$ mA/cm^2, and $FF = 0.80$, giving an experimental efficiency of 15.6%.

Many of the aforementioned novel configurations are designed for high-intensity applications. We shall consider their performances in Section 14.5.

14.4 HETEROJUNCTION, INTERFACE, AND THIN-FILM SOLAR CELLS

14.4.1 Heterojunction Solar Cell

Heterojunctions are junctions formed between two semiconductors with different energy bandgaps (refer to Chapter 2). A typical n-on-p heterojunction band diagram in thermal equilibrium is shown in Fig. 22. Light with energy less than E_{g1} but greater than E_{g2} will pass through the first semiconductor, which acts as a window, and be absorbed by the second semiconductor. Carriers created in the depletion region and within a

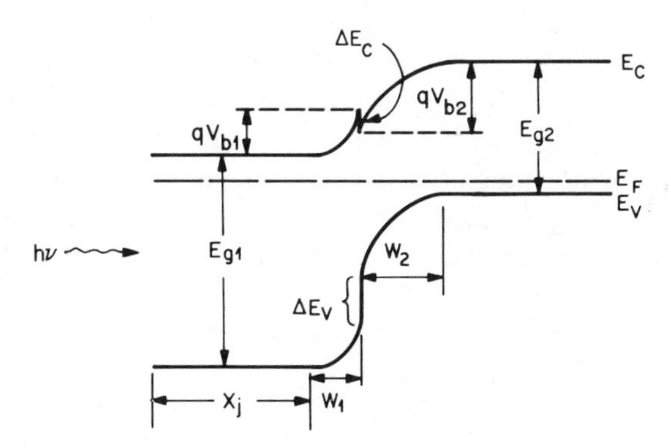

Fig. 22 Energy-band diagram of an *n*-on-*p* heterojunction in thermal equilibrium.

diffusion length of the junction will be collected similar to an *n*-on-*p* homojunction solar cell. Light with energy greater than E_{g1} will be absorbed by the first semiconductor, and carriers generated within a diffusion length from the junction or in the depletion region will also be collected.

The advantages of heterojunction solar cells over conventional *p*-*n* junction solar cells include[27] (1) enhanced short-wavelength spectral response, if E_{g1} is large enough for the high-energy photons to be absorbed inside the depletion region of the second semiconductor; (2) lower series resistance, if the first semiconductor can be heavily doped without affecting its light transmission characteristics; and (3) high radiation tolerance, if the first semiconductor is thick in addition to being high in bandgap.

The expressions for photocurrents in heterojunctions are essentially the same as in homojunctions.[3] For an *n*-on-*p* heterojunction, the hole photocurrent density in the first semiconductor is given by Eq. 22, except that α is replaced by α_1 and L_p by L_{p1}, where α_1 and L_{p1} are the absorption coefficient and diffusion length in the first semiconductor, respectively. The electron photocurrent density is given by Eq. 26, except that α is replaced by α_2, L_n by L_{n2}, and $\alpha(x_j + W)$ by $[\alpha_1(x_j + W_1) + \alpha_2 W_2]$, where α_2 and L_{n2} are the corresponding quantities for the second semiconductor, and W_1 and W_2 are the depletion widths in the two semiconductors. The photocurrent from the depletion regions is given by

$$J_{dr} = qF(1-R)[e^{-\alpha_1 x_j}(1-e^{-\alpha_1 W_1}) + e^{-\alpha_1(W_1+x_j)}(1-e^{-\alpha_2 W_2})]. \qquad (40)$$

In the expressions above, it is assumed that (1) the conduction band discontinuity ΔE_C is small (in the case of *p*-on-*n* heterojunction, ΔE_V should be small), so minority carriers in the second semiconductor will not be impeded from flowing across the junction and (2) the heterojunction has good lattice match, so there are negligible interface traps to reduce the lifetime within and around the depletion region. The spectral response for heterojunctions is given by an expression similar to Eq. 29. The long-wavelength cutoff is given by E_{g2}, and the short-wavelength response depends on the energy gap and the thickness of the first semiconductor. If E_{g1} is sufficiently larger and there are negligible recombinations at the surface and the interface, the spectral response will be similar to those shown in Fig. 11*b*, with $S_p < 10^4$ cm/s. If, on the other hand, ΔE_C and the interface trap density are large, the spectral response will be reduced.

A derivative of the heterojunction is the heteroface solar cell, which is a *p*-*n* homojunction to which a semiconductor having a larger energy gap has been added. The band structure of a *p*-$Ga_{1-x}Al_xAs$/*p*-GaAs/*n*-GaAs cell is shown[28] in the insert of Fig. 23. The wide-gap semiconductor will act as a window, admitting photons of energy less than E_{g1}. Those photons with energies between E_{g1} and E_{g2} will create carriers in the homojunction. If the absorption coefficient is high in the lower gap semiconductor, the carriers are generated in the depletion region or close to it, so the collection efficiency is high. If the heteroface material has indirect gap, as for $Ga_{1-x}Al_xAs$ with $x > 0.4$, and is thin enough, many photons with $h\nu > E_{g1}$ will pass through and

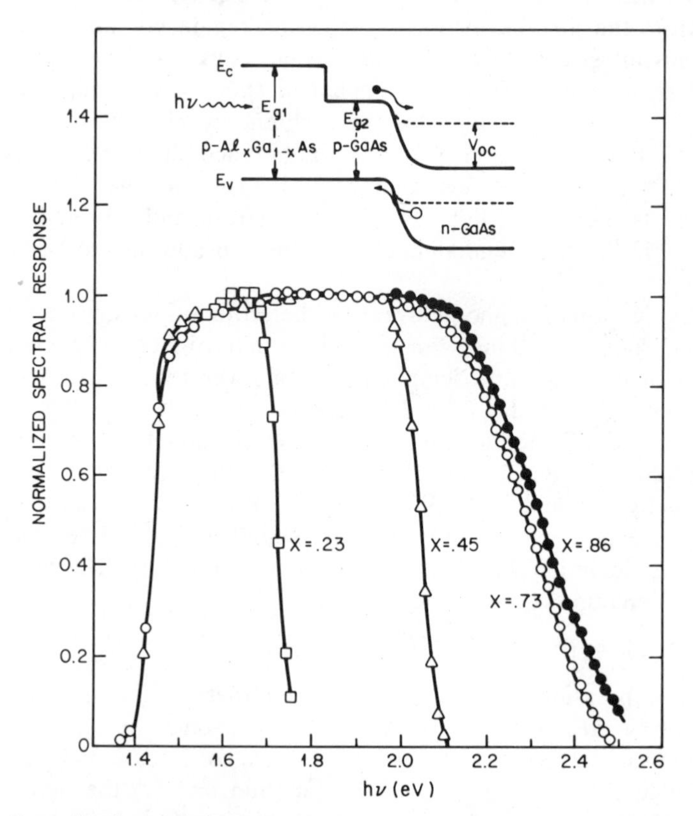

Fig. 23 Normalized spectral response for several AlGaAs/GaAs solar cells having different compositions. The insert shows the heteroface solar-cell band diagram. (After Hovel and Woodall, Ref. 28.)

be efficiently converted in the low-gap semiconductor. Figure 23 shows the normalized spectral responses of several $Ga_{1-x}Al_xAs$–GaAs solar cells, all having the same junction depths and doping levels. As the composition x increases, the bandgap E_{g1} increases; therefore, the spectral response extends to higher photon energies.

One interesting heterojunction solar cell is the conducting glass–semiconductor heterojunction. The conducting glasses include oxide semiconductors, such as indium oxide (In_2O_3, with $E_g = 3.5$ eV and electron affinity $\chi = 4.45$ eV), tin oxide (SnO_2, with $E_g = 3.5$ eV and electron affinity $\chi = 4.8$ eV), and the indium tin oxide (ITO, a mixture of In_2O_3 and SnO_2, with $E_g = 3.7$ eV and electron affinity $\chi = 4.2$ to 4.5 eV). These oxide semiconductors in thin-film form have the unique properties of good electrical conductivity and high optical transparency. They serve not only as part of the heterojunction but also as an antireflection coating.

The energy-band diagrams for an ITO/Si solar cell are shown[29] in the insert of Fig. 24. The top layer is an n-type 4000 Å ITO with 5×10^{-4} Ω–cm and the substrate is a 2 Ω–cm p-type silicon. The curves in Fig. 24 near 1 mA/cm² are all parallel to each other. The slope $d(\ln J)/dV$ is about 24 V^{-1} independent of temperature. This slope suggests a multistep tunnel process in this heterojunction. Conversion efficiencies in the 12 to 15% range

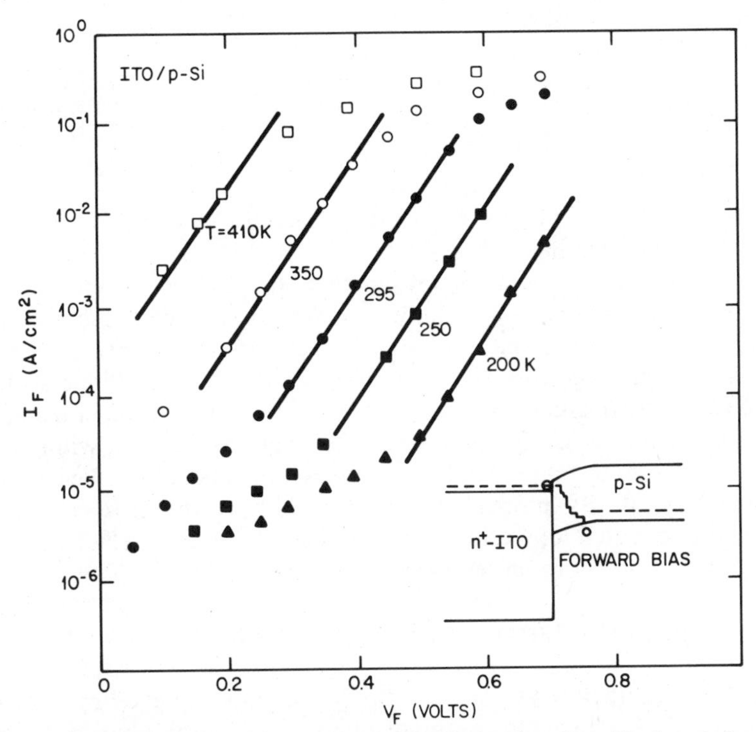

Fig. 24 Current–voltage characteristics of a ITO-Si heterojunction. The insert shows the band diagram under forward bias. (After Sites, Ref. 29.)

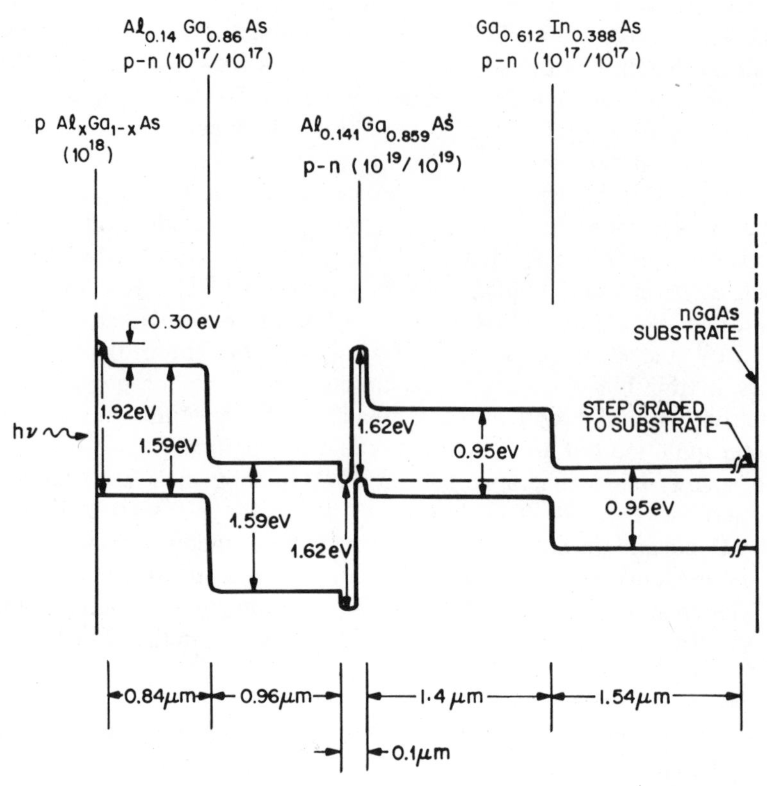

Fig. 25 An idealized cascade two-junction solar cell. (After Lamorte and Abbott, Ref. 31.)

have been obtained. For n-ITO/p-InP solar cells, efficiency of 14% under AM2 conditions has been obtained.[30]

A high-efficiency cascade heterojunction solar cell has been proposed.[31] An idealized band diagram is shown in Fig. 25. The device consists of a wide-gap cell (1.59 eV) and a narrow-gap cell (0.95 eV) joined by a heterojunction tunnel diode formed as an integral part of a monolithic structure. The design also includes a heteroface window layer to minimize surface recombination losses. Light that passes through the first cell without being absorbed will go through the ultrathin tunnel diode and be collected by the narrow-gap cell. By proper choice of bandgaps, the two cells can be designed for equal operating current. The theoretical efficiency under AM1.5 conditions at room temperature is over 30%.

14.4.2 Schottky-Barrier and MIS Solar Cells

The basic characteristics of Schottky-barrier diodes have been described in Chapter 5. A schematic energy diagram of a Schottky-barrier solar cell

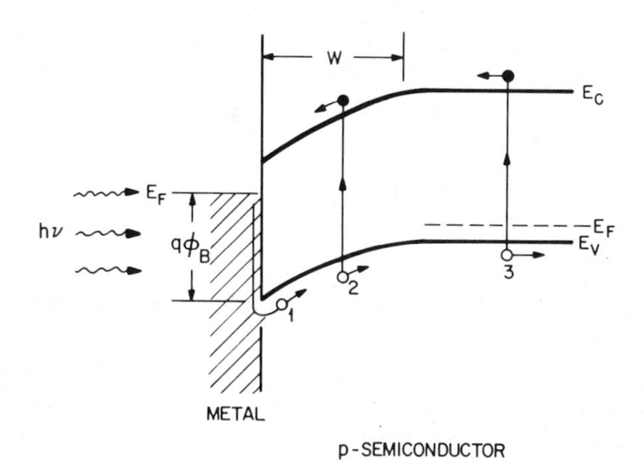

Fig. 26 Energy-band diagram of a Schottky-barrier solar cell under illumination.

under illumination is shown in Fig. 26. The metal must be thin enough to allow a substantial amount of the light to reach the semiconductor. There are three photocurrent components. Light with energy $h\nu > q\phi_B$ (the barrier height) can be absorbed in the metal and excite holes over the barrier into the semiconductor (1 in Fig. 26). This process is used in the photoelectric measurement of barrier height as described in Chapter 5. Short-wavelength light entering the semiconductor is mainly absorbed in the depletion region (2 in Fig. 26). Long-wavelength light is absorbed in the neutral region, creating electron–hole pairs just as in a p-n junction; the electrons must diffuse to the depletion edge to be collected (3 in Fig. 26). For solar-cell applications, the excitation of carriers from the metal into the semiconductor contributes less than 1% to the total photocurrent and, therefore, can be neglected.

The advantages of Schottky barriers include (1) low-temperature processing because no high-temperature diffusion is required; (2) adaptability to polycrystalline and thin-film solar cells; (3) high radiation resistance due to high electrical field near the surface; and (4) high-current output and good spectral response, because the presence of a depletion region right at the semiconductor surface can substantially reduce the effects of low lifetime and high recombination velocity near the surface.

The two major contributions to the spectral response and to the photocurrent come from the depletion region and the base neutral region. The collection from the depletion region is similar to that of a p-n junction. The high field in the depletion region will sweep the photogenerated carriers out before they can recombine, leading to a photocurrent:

$$J_{dr} = qT(\lambda)F(\lambda)[1-\exp(-\alpha W)] \tag{41}$$

where $T(\lambda)$ is the transmission coefficient of the metal for the monochromatic light of wavelength λ. The photocurrent from the base region is given by an expression identical to Eq. 26, except that $(1-R)$ is replaced by $T(\lambda)$, and $\alpha(x_j+W)$ by αW. If the back contact is ohmic and the device thickness is much greater than the diffusion length $H' \gg L_p$, the photocurrent from the base region is simplified to

$$J_n = qT(\lambda)F(\lambda)[\alpha L_n/(\alpha L_n+1)]\exp(-\alpha W). \tag{42}$$

The total photocurrent is given by the sum of Eqs. 41 and 42. To increase the photocurrent, one should increase the transmission coefficient T and diffusion length L_n. The spectral response for an ideal Schottky barrier is similar to those shown in Fig. 11*b* for $S_n < 10^4$ cm/s. The magnitude at each photon energy is lowered by reflection and absorption of light in the metal film. The transmission coefficient for gold films (10 to 100 Å) with an antireflection coating can reach 90 to 95%.

The I–V characteristics of a Schottky barrier under illumination is given by

$$I = I_s(e^{qV/nkT}-1) - I_L \tag{43}$$

and

$$I_s = AA^{**}T^2\exp(-q\phi_B/kT) \tag{44}$$

where n is the ideality factor, A the area, A^{**} the effective Richardson constant (refer to Chapter 5), and $q\phi_B$ the barrier height. The conversion efficiency is given by Eq. 11. For a given semiconductor, the efficiency can be calculated from Eqs. 11, 42, and 43 as a function of the barrier height. Figure 27 shows the ideal efficiency assuming zero reflection, unity ideality factor ($n=1$), and no resistance losses.[32] The efficiency increases with barrier height and, taking $q\phi_B(\max) = E_g$ as the limiting case, the maximum efficiency is about 25%. This figure is comparable to that of a homojunction.

To achieve high barrier heights on semiconductors, one usually uses metals with high work functions for n-type and low work functions for p-type, for example, 50 Å copper/50 Å chromium (Cr) on p-type Si, where chromium forms the Schottky barrier and copper serves as a protective coating.[33] For most metal–semiconductor systems made on uniformly doped substrates, the maximum barrier height is about $\frac{2}{3}E_g$. However, the barrier height can be increased to near the bandgap energy by inserting a heavily doped, thin semiconductor layer (100 Å with opposite doping to the substrate) between the metal and semiconductor substrate. The energy diagram for a metal–p^+-n Schottky barrier is shown[34] in the insert in Fig. 28. The effective barrier height is given by

$$q\phi_B = q\phi_m - q\chi + qV_{pm} \tag{45}$$

where $q\phi_m$ is the metal work function, χ the semiconductor electron affinity, and V_{pm} is the potential maximum given by

$$V_{pm} = \left(\frac{q}{2\epsilon_s N_A}\right)(N_A W_p - N_D W_n)^2. \tag{46}$$

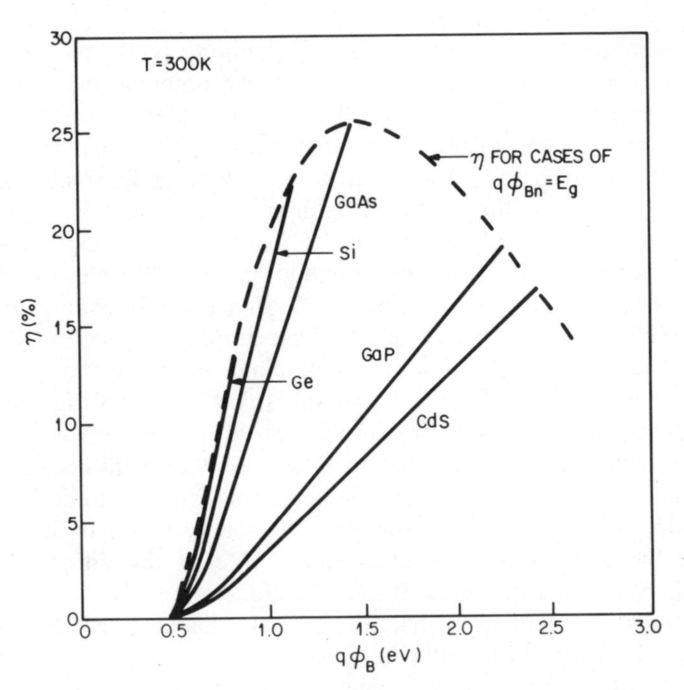

Fig. 27 Conversion efficiency versus barrier height. The envelope shows maximum efficiency calculated for $q\phi_B = E_g$. (After Pulfrey and McOuat, Ref. 32.)

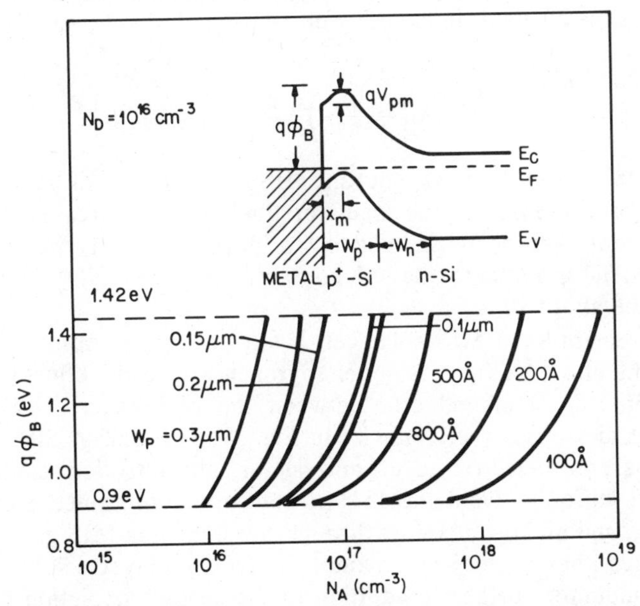

Fig. 28 Barrier height versus doping concentration and thickness of the p layer for a Au–GaAs Schottky-barrier solar cell. The insert shows the energy-band diagram of the metal–p^+-n device. (After Li, Ref. 34.)

The system described above is derived assuming $N_A W_p \gg N_D W_n$, so that the thin p region remains fully ionized and the potential maximum exists inside the p region. A complementary metal–n^+-p device can be made by interchanging n with p. The calculated GaAs barrier height for $N_D = 10^{16}\,cm^{-3}$ is shown in Fig. 28 as a function of N_A and W_p. Note that when $N_A W_p \gtrsim 2\times 10^{12}\,cm^{-2}$, the barrier height approaches the GaAs bandgap. The heavily doped surface layer can be formed by ion implantation and molecular-beam epitaxy. Another method is by a simple metallurgical reaction. The Schottky-barrier height of Al–n-type Si contact increases from 0.68 eV to about 0.9 eV under heat treatment ($<580°C$). A thin recrystalized p^+ layer (~ 100 Å) containing $\gtrsim 10^{19}\,cm^{-3}$ aluminum acceptors is formed;[35] and the band diagram is that shown in Fig. 28.

In an MIS solar cell, a thin insulating layer is formed between the metal and semiconductor substrate. The advantages of MIS solar cells include an electric field extending to the semiconductor surface in a direction that aids in collecting minority carriers generated by short-wavelength light, and the fact that the active region of the cells is free of the diffusion-induced crystal damage inherent in diffused p-n junction cells. The saturation current density is similar to that for Schottky barrier with an additional tunneling term (refer to Chapter 9):

$$J_s = A^{**}T^2 \exp(-q\phi_B/kT)\exp[-(q\phi_T)^{1/2}\delta] \tag{47}$$

where $q\phi_T$ in eV is the average barrier height presented by the insulating layer and δ in Å is the insulator thickness. Substitution of $V = V_{oc}$ and $J = 0$ in Eq. 43 yield,

$$V_{oc} = \frac{nkT}{q}\left[\ln\left(\frac{J_L}{A^{**}T^2}\right) + \frac{q\phi_B}{kT} + (q\phi_T)^{1/2}\delta\right]. \tag{48}$$

Equation 48 shows that V_{oc} of an MIS solar cell will increase with increasing δ. However, as the insulator thickness δ increases, the short-circuit current will decrease, causing a degradation of the conversion efficiency. An optimum oxide thickness for a metal–SiO_2–Si system is found[36] to be about 20 Å.

The aforementioned MIS solar cell has a uniform ultrathin metal layer covering the entire surface. A novel approach uses a thick metal MIS grid pattern with the semiconductor between the grid fingers covered by a transparent dielectric layer, shown[37] in Fig. 29. The 1000-Å SiO_2 serves as the transparent dielectric layer and as an antireflection coating. This structure is different from the ITO-Si heterojunction, since SiO_2 is non-conducting, and all currents flow through the tunneling MIS grid fingers. If positive fixed charges are in the oxide, an inversion layer will be formed at the semiconductor surface in addition to the surface depletion layer. The tunneling MIS grid with 20- to 30-Å SiO_2 will collect the photogenerated minority carriers (electrons for p-type substrate) from both the inversion and depletion layers. The inversion layer can also screen the surface

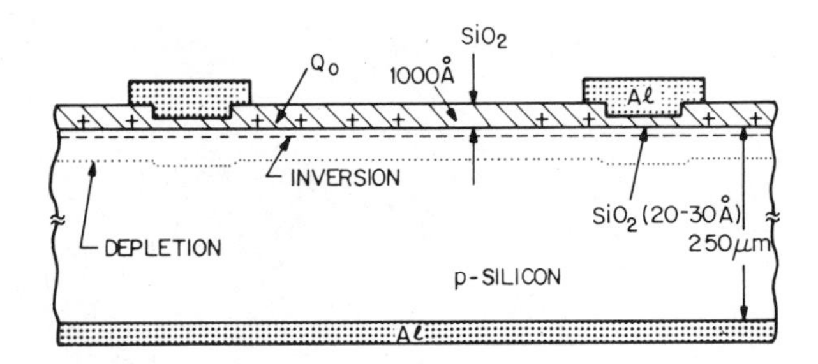

Fig. 29 Cross section of a solar cell having an MIS grid pattern. (After Van Halen et al., Ref. 37.)

recombination centers. At AM1, efficiencies of up to 18% have been obtained.[49,50] Since the oxide can be formed at low temperatures and no diffusion process is involved, this structure fabricated on polycrystalline or amorphous substrates is expected to provide a cost-effective solution for terrestrial applications.

14.4.3 Thin-Film Solar Cells

In thin-film solar cells, the active semiconductor layers are polycrystalline or disordered films that have been deposited or formed on electrically active or passive substrates, such as glass, plastic, ceramic, metal, graphite, or metallurgical silicon. A thin film of CdS, Si, GaAs, InP, CdTe, and so on, can be deposited onto the substrate by various methods, such as vapor growth, evaporation plasma, and plating. If the semiconductor thickness is larger than the inverse of the absorption coefficient, most light will be absorbed; if the diffusion length is larger than the film thickness, most photogenerated carriers can be collected.

The main advantage of thin-film solar cells is their promise of low cost, due to low-cost processing, and the use of relatively low cost materials. The main disadvantages are low efficiency and long-term instability. The low efficiency is partly caused by the grain boundary effect and partly caused by the poor quality of the semiconductor material grown on foreign substrates. The stability problem is caused by the chemical reaction of the semiconductor with ambient (such as O_2 and water vapor). Steps must be taken to ensure device reliability.

A schematic thin-film CdS solar cell is shown[38] in Fig. 30. The cells are fabricated using a substrate of electro-formed copper, coated with 0.5 μm of zinc. A layer of CdS about 20 μm thick is evaporated onto the heated substrate at 220°C. Reacting the CdS film in a cuprous ion solution forms a Cu_2S layer of 1000 Å. A transparent grid contact is deposited onto the Cu_2S and an antireflection layer applied over the Cu_2S. Figure 31 shows the energy diagram of the Cu_2S–CdS cell.[39] It is basically a heterojunction with

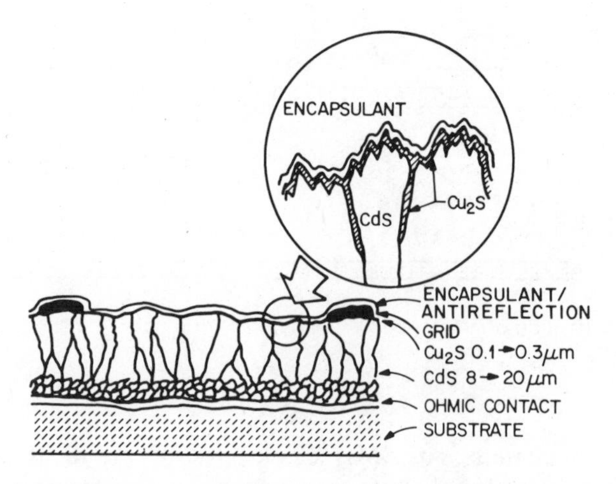

Fig. 30 Schematic thin-film CdS solar cell. (After Barnett et al., Ref. 38.)

large interface–trap density. Under front illumination, most light is absorbed in the Cu_2S. The spectral response and photocurrent are limited by high surface recombination velocity, short diffusion length, and high interface recombination. Despite these shortcomings, this cell has conversion efficiency[38] near 10%. It has been proposed that a substitution of zinc for 15 to 25% of the cadmium will increase the voltage and may lead to an efficiency of over 14%.

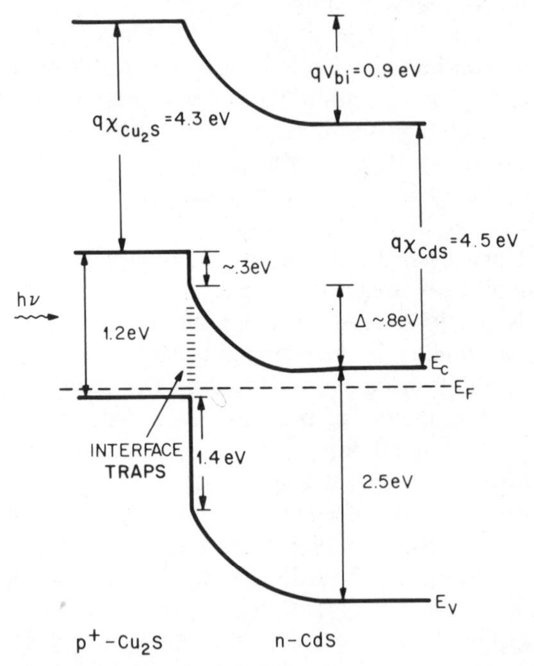

Fig. 31 Energy-band diagram of a Cu_2S–CdS solar cell. (After Burton et al., Ref. 39.)

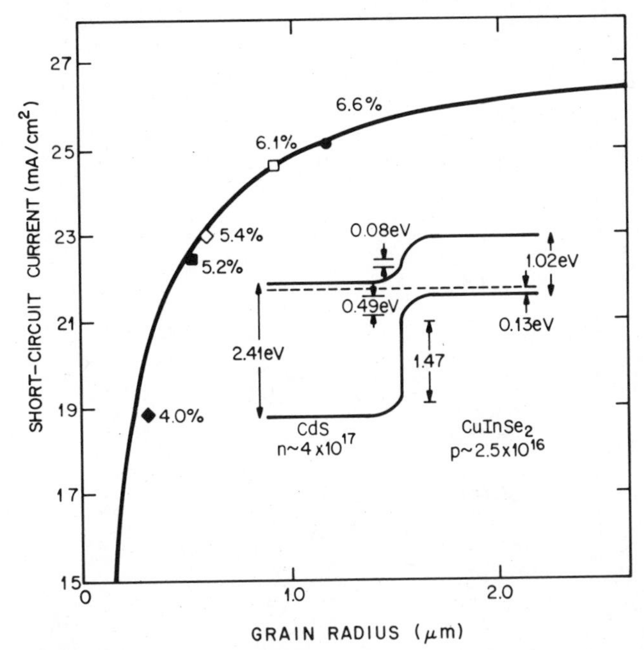

Fig. 32 Dependence of short-circuit current and efficiency upon grain radius in $CdS/CuInSe_2$ solar cells. The insert shows the energy-band diagram. (After Kazmerski et al., Ref. 40.)

Another thin-film cell is the $CuInSe_2/CdS$ heterojunction. The band diagram is shown in the insert of Fig. 32. The short-circuit current and efficiency dependent on the grain size. As the grain radius increases from 0.3 μm to above 1 μm, the efficiency increases[40] from 4 to 6.6%. Many ternary compounds are potential candidates for low-cost solar-cell applications. Figure 33 shows some of the I–III–VI_2 group and II–IV–V_2 group semiconductors,[41] which have energy gaps in the range of interest for solar photovoltaic conversion.

Amorphous silicon (a-Si) is also a material used for thin-film solar cells. Layers 1 to 3 μm thick are grown by RF glow-discharge decomposition of silane onto metal or ITO-coated glass substrates. The difference between crystalline and amorphous Si is dramatic; the former has an indirect bandgap of 1.1 eV, whereas hydrogenated a-Si has an optical absorption characteristic that resembles the characteristic expected for a crystal with a direct bandgap of 1.6 eV, shown[42] in Fig. 34. Solar cells have been prepared in *p-n* junction as well as Schottky-barrier form on thin films of hydrogenated a-Si. Some schematic diagrams of the various a-Si solar-cell structures are shown[43] in Fig. 35. Since the absorption coefficient is in the range 10^4 to 10^5 cm^{-1} across the visible portion of the solar spectrum, many carriers are photogenerated within a fraction of a micron of the illuminated

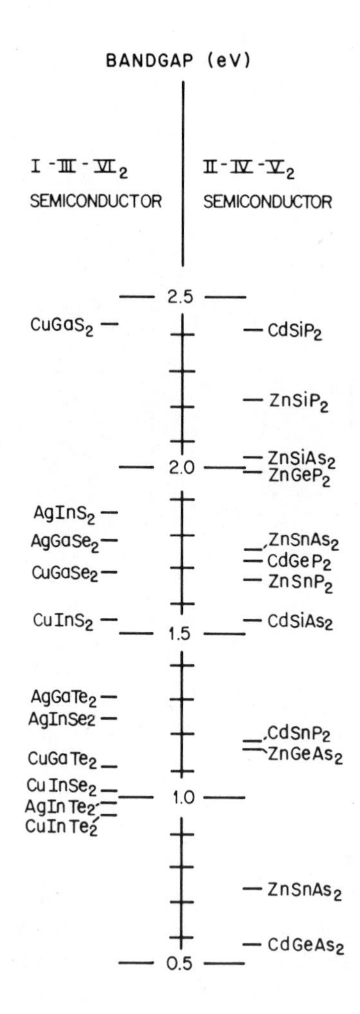

Fig. 33 Energy gaps of I–III–VI$_2$ and II–IV–V$_2$ semiconductors in the range of interest for solar photovoltaic conversion. (After Wagner and Bridenbaugh, Ref. 41.)

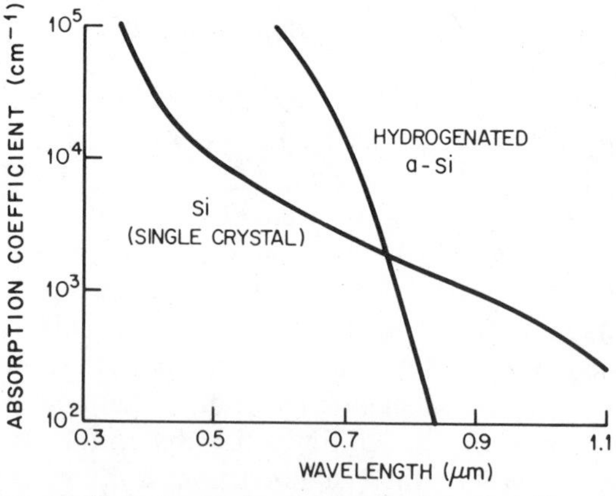

Fig. 34 Absorption coefficient versus wavelength for crystalline and hydrogenated amorphous Si. (After Gibson, LeComber, and Spear, Ref. 42.)

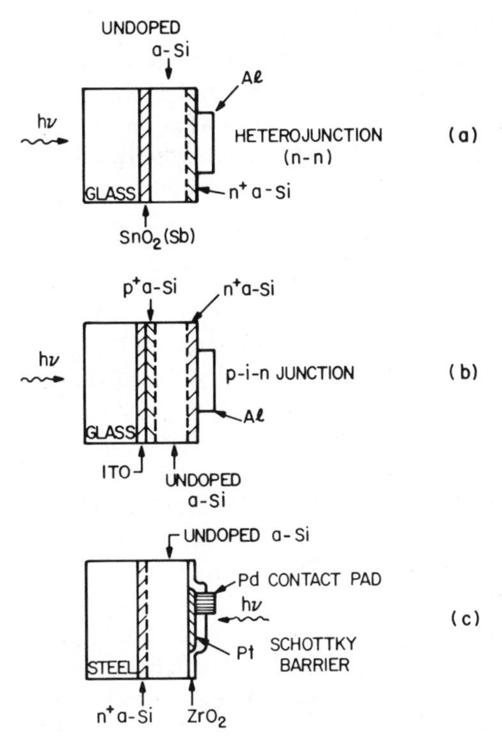

Fig. 35 Schematic diagrams of various a-Si solar cell structures. (After Carlson, Ref. 43.)

surface. The diffusion length is estimated to be about 1 μm, and the resistivity in the dark is of the order of 100 MΩ-cm, which would give a series resistance of 10^4 Ω-cm^2, even for a 1-μm-thick *i* layer in a *p-i-n* device. For this reason, the Schottky-barrier cell with a transparent metal barrier has given the best result so far ($\sim$6% efficiency).

To understand the permitted properties of disordered thin-film devices, we shall consider the idealized thin-film solar cell shown[13] in Fig. 36. Figure

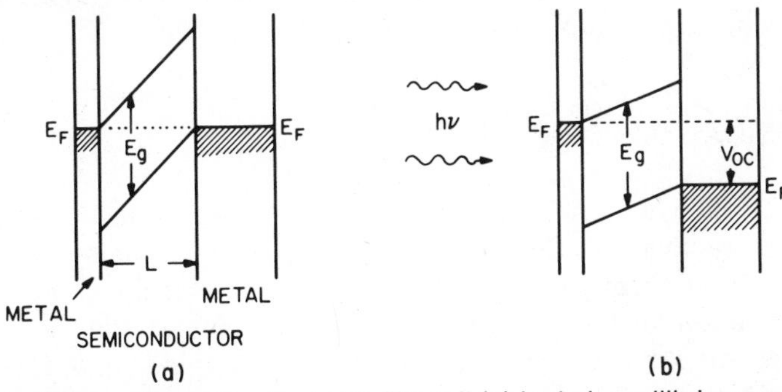

Fig. 36 Idealized configuration for a thin-film cell (*a*) in dark equilibrium and (*b*) under illumination. (After Ref. 13.)

36*a* shows the band diagram under equilibrium in the dark. It consists of a semiconductor layer of thickness L sandwiched between two metal electrodes of different barrier heights, so that their Fermi levels are pinned near the conduction and valence band edges on opposite sides of the cell. One electrode is thin enough to permit relatively unattenuated penetration of the solar spectrum. A direct-gap semiconductor with $\alpha > 10^5\ \text{cm}^{-1}$ for photons more than a few tenths of an eV beyond the absorption edge is assumed. As Fig. 36*b* suggests, the light is absorbed in a region of the strong electric field, resulting in efficient extraction of electrons and holes.

Assuming that the semiconductor contains a high concentration of traps, we can estimate at what concentration level the traps will severely degrade the device performance. In the absence of charged traps, the electric field will be uniform as shown, and is given by $\mathscr{E} = E_g/qL$. For a thickness of $1/\alpha$ ($\sim 0.1\ \mu\text{m}$) and $E_g = 1.5\ \text{eV}$, the field is $1.5 \times 10^5\ \text{V/cm}$. When traps are present with a concentration n_t, the net space charge will be n_c, where $n_c < n_t$. These charged defects will affect the field strength by $\Delta\mathscr{E} = qn_cL/\epsilon_s$. Assuming a dielectric constant of 4, we find that $\Delta\mathscr{E} \ll \mathscr{E}$ if $n_c < 10^{16}\ \text{cm}^{-3}$, implying that a total trap concentration as high as $10^{17}\ \text{cm}^{-3}$ can be tolerated without serious disturbance to the electric field in the semiconductor. A further requirement is that the space-charge-limited current must be higher than about $100\ \text{mA/cm}^2$, substantially larger than the short-circuit current density produced in one sun illumination. For a $0.1\ \mu\text{m}$ thickness, this condition also leads to a permissible trap density as high as $10^{17}\ \text{cm}^{-2}/\text{eV}$.

The electric field must also be able to extract the electrons and holes in a transit time $L/\mathscr{E}\mu$, which is short compared with the recombination lifetime $(n_t v\sigma)^{-1}$, where σ is the capture cross section ($\sim 10^{-14}\ \text{cm}^2$), and v is the thermal velocity ($\sim 10^7\ \text{cm/s}$). This condition will be satisfied if

$$\mu > \frac{n_t v\sigma L}{\mathscr{E}} = n_t v\sigma q L^2/E_g \simeq 1\ \text{cm}^2/\text{V-s} \qquad (49)$$

which is not difficult to achieve.

The considerations discussed above imply that useful solar cells can be made in semiconductors containing very high defect density if the semiconductor films are sufficiently thin and have high absorption coefficient near the band edge, coupled with requisite mobilities.

14.5 OPTICAL CONCENTRATION

Sunlight can be focused by using mirrors and lenses. Optical concentration offers an attractive and flexible approach to reducing high cell costs by substituting a concentrator area for much of the cell area. It also offers other advantages, including (1) increased cell efficiency (Fig. 7), (2) hybrid systems yielding both electrical and thermal outputs, and (3) reduced cell-temperature coefficient.

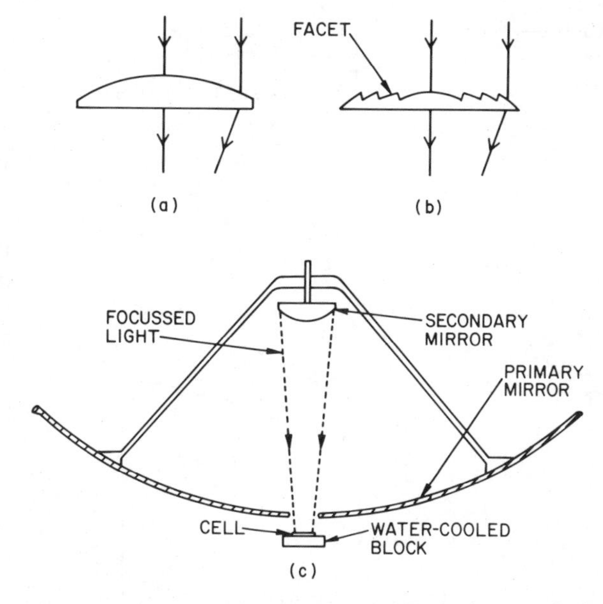

Fig. 37 (*a*) Standard lens. (*b*) Fresnel lens. (*c*) Typical concentrator module.

A standard planoconvex lens is shown in Fig. 37*a*, and an equivalent Fresnel lens is shown in Fig. 37*b*. A concentrator module for a concentration of 500 to 1000 suns is shown in Fig. 37*c*; sunlight is reflected by the primary mirror, and light from the secondary mirror is focused onto the solar cell mounted on a water-cooled block. The experimental results obtained from a silicon vertical-junction solar cell (Fig. 21) mounted in a concentrator system (Fig. 37*c*) are shown[26,44] in Fig. 38. Note that device performances improve as the concentration increases from 1 sun toward 1000 suns. The short-circuit current increases linearly with concentration. The open-circuit voltage increases at a rate of 0.1 V per decade, while the fill factor varies slightly. The efficiency, which is the product of the foregoing three factors divided by the input concentrated power, increases at a rate of about 2% per decade. With proper antireflection coatings, the projected efficiency is about 22% at 1000 suns. Therefore, one cell operated under 1000-sun concentration can produce the same power output as 1300 cells under 1 sun. Hence the optical concentration approach can potentially replace expensive solar cells with less expensive concentrator materials and a related tracking setup to minimize the overall system cost.

Under high concentrations, the carrier density approaches that of the substrate doping, and a high injection condition prevails. The current density is proportional to $\exp(qV/nkT)$, where $n = 2$. The open-circuit voltage becomes

$$V_{oc} = \frac{2kT}{q} \ln\left(\frac{J_L}{J_s} + 1\right) \tag{50}$$

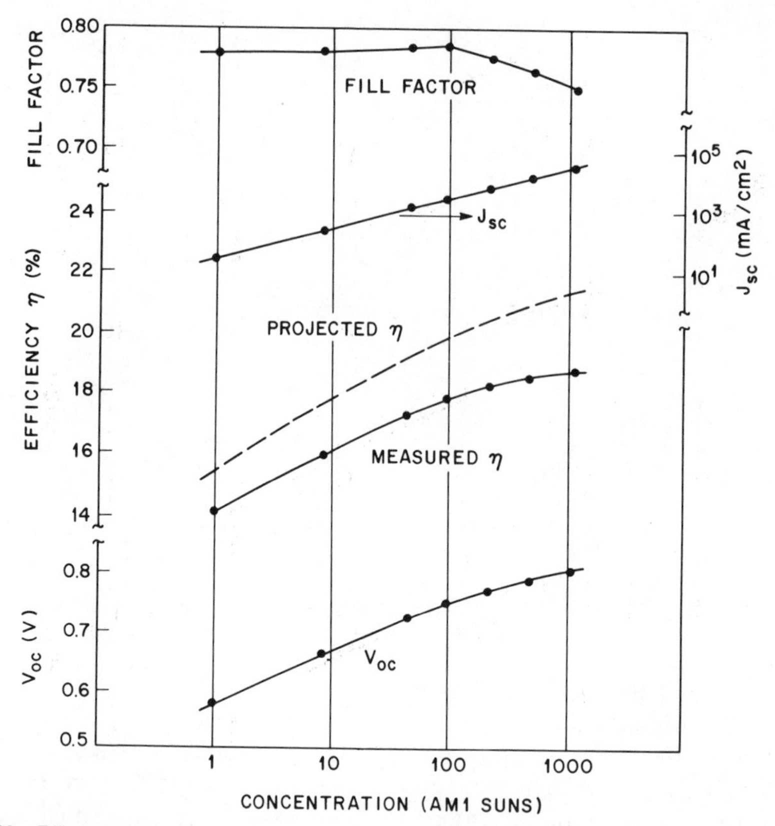

Fig. 38 Efficiency, V_{oc}, J_{sc}, and fill factor versus AM1 solar concentration for a multiple vertical junction. (After Frank, Goodrich, and Kaplow, Ref. 44.)

and J_s can be expressed as

$$J_s = \text{constant}\left(\frac{T}{T_0}\right)^{3/2} \exp\left[-\frac{E_g(T)}{2kT}\right] \tag{51}$$

where T is the operating temperature and T_0 is 300 K. The temperature dependence of V_{oc} at different concentration levels is shown[26] in Fig. 39. The temperature coefficient of V_{oc} changes from -2.07 mV/°C at 1 sun to -1.45 mV/°C at 500 suns. Thus for silicon solar cells, high solar concentration levels can decrease the efficiency losses associated with operation at elevated temperatures.

The heteroface GaAs solar cell can perform well under optical concentration. A schematic cross section of such a cell is shown[45] in the insert of Fig. 40. The measured I–V curve under 945 suns (AM1.5) is shown. The device has a fill factor of 0.79, an open-circuit voltage of 1.13 V, a maximum output power of 10.7 W, and an efficiency of 23.3% at 50°C.

The efficiency and power output can be increased further by using

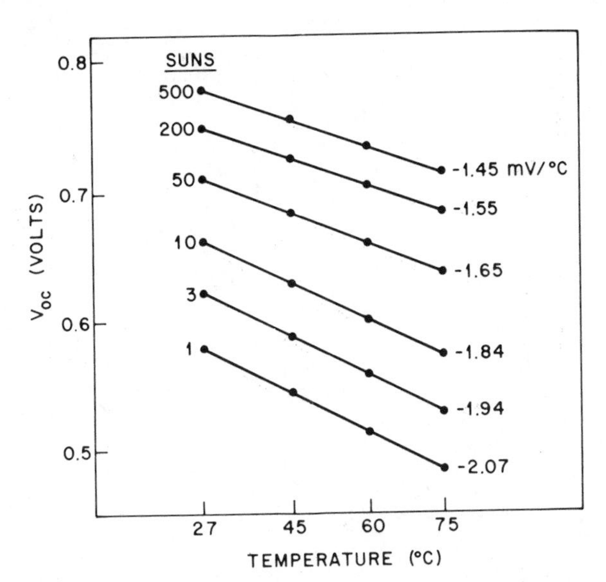

Fig. 39 Open-circuit voltage as a function of temperature for devices under various solar concentrations. (After Frank, Goodrich, and Kaplow, Ref. 26.)

multiple bandgaps (Figs. 8 and 25) and a spectral splitting arrangement, that is, by splitting the solar light flux into many narrow spectral bands and converting the flux in each band by a cell with the bandgap optimized for that band. A possible module design for spectral-splitting operation is shown[46] in Fig. 41. The dichroic mirrors split the incoming concentrated light by reflecting photons of higher energies to cell 1 and transmitting

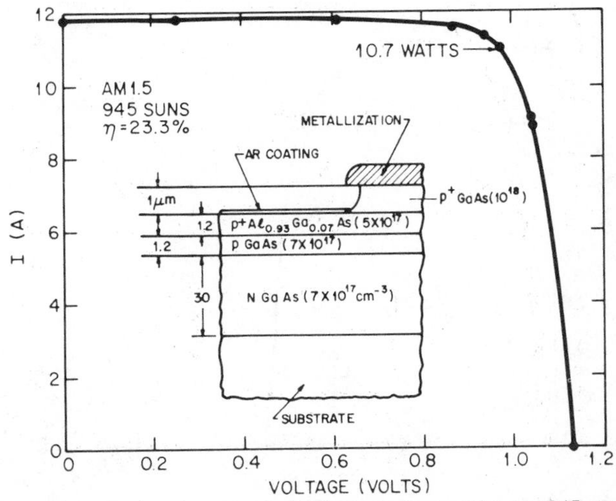

Fig. 40 Measured I–V curve for a AlGaAs/GaAs solar cell under 945 suns. The insert shows the device cross section. (After Vander Plas et al., Ref. 45.)

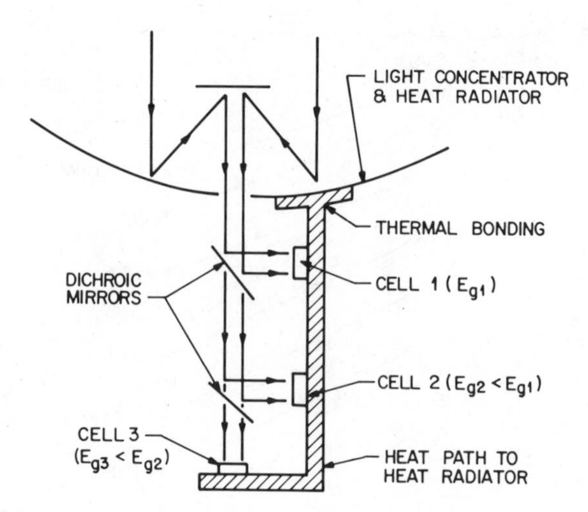

Fig. 41 Possible module design of spectral splitting for high-efficiency operation. (After Blocker, Ref. 46.)

photons of lower energies to cell 2, then to cell 3. Under 1000-sun concentration, the maximum efficiency for two spectral bands can be obtained by a graphical method as described previously (Fig. 8) and is about 60%. It is about 70% for 3 bands and about 85% for 10 bands. The ideal efficiency for n spectral-splitting bands is higher than that for n series-connected bandgaps because, in spectral splitting, the current den-

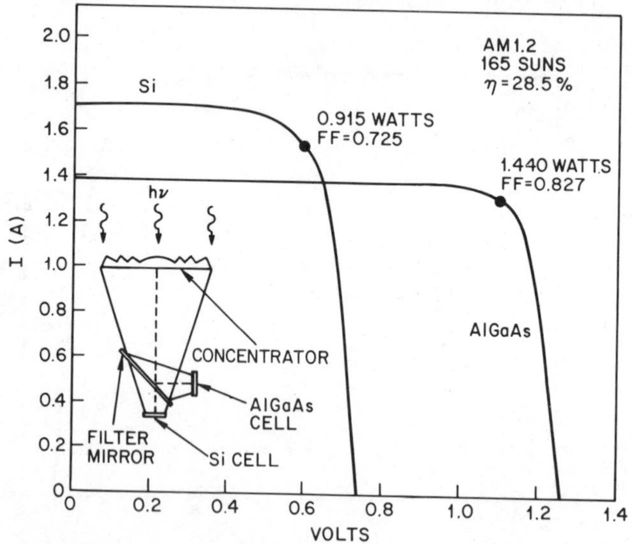

Fig. 42 Current–voltage curves for Si and AlGaAs solar cells in a spectral-splitting arrangement, operating under 165 suns. (After Moon et al., Ref. 47.)

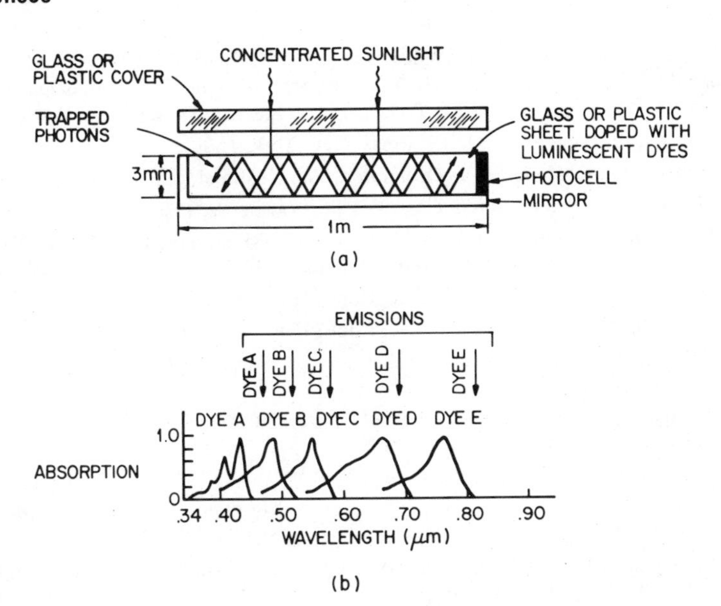

Fig. 43 Solar collector using luminescent dyes that absorb sunlight in narrow frequency bands, then reradiate photons at different wavelengths. (After Javetski, Ref. 48.)

sity for each cell can be individually adjusted to maximize the overall efficiency.

Experimentally, the $Al_xGa_{1-x}As$ cell for $x < 0.24$ and Si cell have been used in combination with a spectral-splitting filter as shown in Fig. 42. The current–voltage curves of the two cells under 165 suns (AM1.2) in the spectral-splitting arrangement are shown[47] in the figure. The maximum power outputs are 1.44 W and 0.95 W from the AlGaAs ($E_g = 1.61$ eV) and Si (1.1-eV) cell, respectively, with corresponding efficiencies of 17.4% and 11.1%. The combined efficiency is 28.5% at 30°C.

Another promising approach uses glass or plastic sheet doped with luminescent dyes that absorb sunlight in narrow frequency bands and then reradiate photons at different wavelengths in many directions (Fig. 43).[48] Some light that is reflected, becomes trapped within the sheet because the dyes cannot reabsorb it, and eventually bounces down to the sheet and to a solar cell at its edge. The system need not track the sun, because the dyes absorb light incident at any angle. The theoretical efficiency is above 50%.

REFERENCES

1 D. M. Chapin, C. S. Fuller, and G. L. Pearson, "A New Silicon *p-n* Junction Photocell for Converting Solar Radiation into Electrical Power," *J. Appl. Phys.*, **25**, 676 (1954).

2 D. C. Raynolds, G. Leies, L. L. Antes, and R. E. Marburger, "Photovoltaic Effect in Cadmium Sulfide," *Phys. Rev.*, **96**, 533 (1954).

3 H. J. Hovel, *Solar Cells*, in R. K. Willardson and A. C. Beer, Eds., *Semiconductors and Semimetals*, Vol. 11, Academic, New York, 1975; "Photovoltaic Materials and Devices for Terrestrial Applications," *IEEE Tech. Dig.* Int. Electron Device Meet., 1979, p. 3.

4 C. E. Backus, Ed., *Solar Cells*, IEEE Press, New York, 1976.

5 D. L. Pulfrey, *Photovoltaic Power Generation*, Van Nostrand Reinhold, New York, 1978.

6 W. D. Johnston, Jr., *Solar Voltaic Cells*, Dekker, New York, 1980.

7 K. J. Bachmann, "Material Aspects of Solar Cells" in E. Kaldis, Ed., *Current Topics in Material Science*, Vol. 3, North-Holland, Amsterdam, 1979.

8 Conf. Rec. 13th IEEE Photovoltaic Spec. Conf., IEEE, New York, 1978.

9 Conf. Rec. 14th IEEE Photovoltaic Spec. Conf., IEEE, New York, 1980.

10 M. P. Thekaekara, "Data on Incident Solar Energy," Suppl. Proc. 20th Annu. Meet. Inst. Environ. Sci., 1974, p. 21.

11 C. H. Henry, "Limiting Efficiency of Ideal Single and Multiple Energy Gap Terrestrial Solar Cells," *J. Appl. Phys.*, **51**, 4494 (1980).

12 M. B. Prince, "Silicon Solar Energy Converters," *J. Appl. Phys.*, **26**, 534 (1955).

13 *Principal Conclusions of the American Physical Society Study Group on Solar Photovoltaic Energy Conversion*, American Physical Society, New York, 1979.

14 W. Shockley and H. J. Queisser, "Detailed Balance Limit of Efficiency of *pn* Junction Solar Cells," *J. Appl. Phys.*, **32**, 510 (1961).

15 R. J. Handy, "Theoretical Analysis of the Series Resistance of a Solar Cell," *Solid State Electron.*, **10**, 765 (1967).

16 R. N. Hall, "Electron–Hole Recombination in Germanium," *Phy. Rev.* **87**, 367 (1952).

17 W. Shockley and W. T. Read, "Statistics of the Recombination of Holes and Electrons," *Phys. Rev.*, **87**, 835 (1952).

18 J. J. Wysocki and P. Rappaport, "Effect of Temperature on Photovoltaic Solar Energy Conversion," *J. Appl. Phys.*, **31**, 571 (1960).

19 W. Rosenzweig, H. K. Gummel, and F. M. Smits, "Solar Cell Degradation under 1 MeV Electron Bombardment," *Bell Syst. Tech. J.*, **42**, 399 (1963).

20 J. J. Wysocki, P. Rappaport, E. Davison, R. Hand, and J. J. Loferski, "Lithium-Doped Radiation-Resistant Silicon Solar Cells," *Appl. Phys. Lett.*, **9**, 44 (1966).

21 J. Mandelkorn and J. H. Lamneck, Jr., "Simplified Fabrication of Back Surface Electric Field Silicon Cells and Novel Characteristic of Such Cells," Conf. Rec. 9th IEEE Photovoltaic Spec. Conf., IEEE, New York, 1972, p. 66.

22 J. Lindmayer and J. F. Allison, "The Violet Cell: An Improved Silicon Solar Cell," Conf. Rec. 9th IEEE Photovoltaic Spec. Conf., IEEE, New York, 1972, p. 83.

23 R. A. Arndt, J. F. Allison, J. G. Haynos, and A. Meulenberg, Jr., "Optical Properties of the COMSAT Non-Reflective Cell," Conf. Rec. 11th IEEE Photovoltaic Spec. Conf., IEEE, New York, 1975, p. 40.

24 T. I. Chappell, "The V-Groove Multijunction Solar Cell" *IEEE Trans. Electron Devices*, **ED-26**, 1091 (1979).

25 S. Y. Chiang, B. G. Carbajal, and G. F. Wakefield, "Improved Performance Thin Solar Cells," *IEEE Trans. Electron Devices*, **ED-25**, 1405 (1978); W. T. Matzen, S. Y. Chiang, and B. G. Carbajal, "A Device Model for the Tandem Junction Solar Cell," *IEEE Trans. Electron Devices*, **ED-26**, 1365 (1979).

26 R. I. Frank, J. L. Goodrich, and R. Kaplow, "A Low Series Resistance Silicon Photovoltaic Cell for High Intensity Applications," Conf. Rec. 14th IEEE Photovoltaic Spec. Conf., IEEE, New York, 1980, p. 1350.

27 D. L. Feucht, "Heterojunction in Photovoltaic Devices," *J. Vac. Sci. Technol.*, **14**, 57 (1977).

28 H. J. Hovel and J. M. Woodall, "$Ga_{1-x}Al_xAs$–GaAs *p-p-n* Heterojunction Solar Cells," *J. Electrochem. Soc.*, **120**, 1246 (1973).

29 J. R. Sites, "Current Mechanisms and Barrier Height in ITO/Si Heterojunctions," *Inst. Phys. Conf. Ser.* **43**, Chap. 22 (1979).

30 K. S. S. Harsha, K. J. Bachmann, P. H. Schmidt, E. G. Spencer, and F. A. Thiel, "*n*-Indium Tin Oxide/*p*-InP Solar Cell," *Appl. Phys. Lett.*, **30**, 645 (1977).

31 M. F. Lamorte and D. Abbott, "Analysis of AlGaAs–GaInAs Cascade Solar Cell under AM0-AM5 Spectra," *Solid State Electron.*, **22**, 467 (1979).

32 D. L. Pulfrey and R. F. McOuat, "Schottky-Barrier Solar-Cell Calculations," *Appl. Phys. Lett.*, **24**, 167 (1974).

33 W. A. Anderson, S. M. Vernon, P. Mathe, and B. Lalevic, "Schottky Solar Cells on Thin Epitaxial Silicon," *Solid State Electron.*, **19**, 973 (1976).

34 S. S. Li, "A Proposed Novel MPN GaAs Schottky Barrier Solar Cell," *Jpn. J. Appl. Phys.*, **17**, *Suppl.* **17-1**, 291 (1978).

35 H. C. Card, E. S. Yang, and P. Panayoetatŏs, "Peaked Schottky-Barrier Solar Cells by Al–Si Metallurgical Reactions," *Appl. Phys. Lett.*, **30**, 643 (1977).

36 H. C. Card and E. S. Yang, "MIS-Schottky Theory under Conditions of Optical Carrier Generation in Solar Cells," *Appl. Phys. Lett.*, **29**, 51 (1976).

37 P. Van Halen, R. P. Mertens, R. J. Van Overstraeten, R. E. Thomas, and J. Van Meerbergen, "New TiO_x–MIS and SiO_2–MIS Silicon Solar Cells," *IEEE Trans. Electron Devices*, **ED-25**, 507 (1978).

38 A. M. Barnett, J. A. Bragagnolo, R. B. Hall, J. E. Phillips, and J. D. Meakin, "Achievement of 9.15% Efficiency in Thin Film CdS/Cu_2S Solar Cells," Conf. Rec. 13th IEEE Photovoltaic Spec. Conf. IEEE, New York, 1978, p. 419.

39 L. C. Burton, B. Baron, W. Devaney, T. L. Hench, S. Lorenz, and J. D. Meakin, "Studies Related to $Zn_xCd_{1-x}S$–Cu_2S Solar Cells," Conf. Rec. 12th IEEE Photovoltaic Spec. Conf., IEEE, New York, 1977, p. 526.

40 L. L. Kazmerski, P. J. Ireland, F. R. White, and R. B. Cooper, "The Performance of Copper-Ternary Based Thin-Film Solar Cells," Conf. Rec. 13th IEEE Photovoltaic Spec. Conf., IEEE, New York, 1978, p. 185.

41 S. Wagner and P. M. Bridenbaugh, "Multicomponent Tetrahedral Compounds for Solar Cells," *J. Cryst. Growth*, **39**, 151 (1977).

42 R. A. Gibson, P. G. LeComber, and W. E. Spear, "Doped Amorphous Silicon and Its Application in Photovoltaic Devices," *IEEE J. Solid State Electron Devices*, **2**, 83 (1978).

43 D. E. Carlson, "Amorphous Silicon Solar Cells," *IEEE Trans. Electron Devices*, **ED-24**, 449 (1977).

44 R. I. Frank, J. L. Goodrich, and R. Kaplow, "A Novel Silicon High-Intensity Photovoltaic Cell," GOMAC Conference, Houston, Nov. 1980.

45 H. A. Vander Plas, L. W. James, R. L. Moon, and N. J. Nelson, "Performance of AlGaAs/GaAs Terrestrial Concentrator Solar Cells," Conf. Rec. 13th IEEE Photovoltaic Spec. Conf., IEEE, New York, 1978, p. 934.

46 W. Blocker, "High Efficiency Solar Energy Conversion through Flux Concentration and Spectrum Splitting," *Proc. IEEE*, **66**, 104 (1978).

47 R. L. Moon, L. W. James, H. A. Vander Plas, T. O. Yep, G. A. Antypas, and Y. Chai, "Multigap Solar Cell Requirements and the Performance of AlGaAs and Si Cells in Concentrated Sunlight," Conf. Rec. 13th IEEE Photovoltaic Spec. Conf., IEEE, New York, 1978, p. 859.

48 J. Javetski, "A Burst of Energy in Photovoltaics," *Electronics*, **52**, 105 (1979).

49 M. A. Green, R. B. Godfrey, M. R. Willison, and A. W. Blakers, "High Efficiency Silicon minMIS Solar Cells," Conf. Rec. 14th IEEE Photovoltaic Spec. Conf., IEEE, New York, 1980, p. 684.

50 R. E. Thomas, C. F. Norman, and R. B. North, "High Efficiency MIS/Inversion Layer Silicon Solar Cells," Conf. Rec. 14th IEEE Photovoltaic Spec. Conf., IEEE, New York, 1980, p. 1350.

APPENDIXES

A. List of Symbols
B. International Systems of Units
C. Unit Prefixes
D. Greek Alphabet
E. Physical Constants
F. Lattice Constant
G. Properties of Important Semiconductors
H. Properties of Ge, Si, and GaAs at 300 K
I. Properties of SiO_2 and Si_3N_4 at 300 K

Appendix A

List of Symbols

Symbol	Description	Unit
a	Lattice constant	Å
$\mathcal{B}$	Magnetic induction	Wb/m^2
c	Speed of light in vacuum	cm/s
C	Capacitance	F
$\mathcal{D}$	Electric displacement	C/cm^2
D	Diffusion coefficient	cm^2/s
E	Energy	eV
E_C	Bottom of conduction band	eV
E_F	Fermi energy level	eV
E_g	Energy bandgap	eV
E_V	Top of valence band	eV
$\mathcal{E}$	Electric field	V/cm
$\mathcal{E}_c$	Critical field	V/cm
$\mathcal{E}_m$	Maximum field	V/cm
f	Frequency	Hz (cps)
$F(E)$	Fermi-Dirac distribution function	
h	Planck constant	J-s
$h\nu$	Photon energy	eV
I	Current	A
I_C	Collector current	A
J	Current density	A/cm^2
J_t	Threshold current density	A/cm^2
k	Boltzmann constant	J/K
kT	Thermal energy	eV
L	Length	cm or μm
m_0	Electron rest mass	kg
m^*	Effective mass	kg
$\bar{n}$	Refractive index	

Symbol	Description	Unit
n	Density of free electrons	cm^{-3}
n_i	Intrinsic density	cm^{-3}
N	Doping concentration	cm^{-3}
N_A	Acceptor impurity density	cm^{-3}
N_C	Effective density of states in conduction band	cm^{-3}
N_D	Donor impurity density	cm^{-3}
N_V	Effective density of states in valence band	cm^{-3}
p	Density of free holes	cm^{-3}
P	Pressure	N/m^2
q	Magnitude of electronic charge	C
Q_{it}	Interface–trap density	$charges/cm^2$
R	Resistance	Ω
t	Time	s
T	Absolute Temperature	K
v	Carrier velocity	cm/s
v_s	Saturation velocity	cm/s
v_{th}	Thermal velocity	cm/s
V	Voltage	V
V_{bi}	Built-in potential	V
V_{EB}	Emitter–base voltage	V
V_B	Breakdown voltage	V
W	Thickness	cm or μm
W_B	Base thickness	cm or μm
x	x direction	
∇	Differential operator	
∇T	Temperature gradient	K/cm
ϵ_0	Permittivity in vacuum	F/cm
ϵ_s	Semiconductor permittivity	F/cm
ϵ_i	Insulator permittivity	F/cm
ϵ_s/ϵ_0 or ϵ_i/ϵ_0	Dielectric constant	
τ	Lifetime or decay time	s
θ	Angle	rad
λ	Wavelength	μm or Å
ν	Frequency of light	Hz

Symbol	Description	Unit
μ_0	Permeability in vacuum	H/cm
μ_n	Electron mobility	cm^2/V-s
μ_p	Hole mobility	cm^2/V-s
ρ	Resistivity	Ω-cm
ϕ	Barrier height or imref	V
ϕ_{Bn}	Schottky barrier height on n-type semiconductor	V
ϕ_{Bp}	Schottky barrier height on p-type semiconductor	V
ϕ_m	Metal work function	V
ω	Angular frequency ($2\pi f$ or $2\pi\nu$)	Hz
Ω	Ohm	Ω

Appendix B

International System of Units

Quantity	Unit	Symbol	Dimensions
Length	meter	m	
Mass	kilogram	kg	
Time	second	s	
Temperature	kelvin	K	
Current	ampere	A	
Frequency	hertz	Hz	1/s
Force	newton	N	$kg\text{-}m/s^2$
Pressure	pascal	Pa	N/m^2
Energy	joule	J	N-m
Power	watt	W	J/s
Electric charge	coulomb	C	A-s
Potential	volt	V	J/C
Conductance	siemens	S	A/V
Resistance	ohm	Ω	V/A
Capacitance	farad	F	C/V
Magnetic flux	weber	Wb	V-s
Magnetic induction	tesla	T	Wb/m^2
Inductance	henry	H	Wb/A

Appendix C

Unit Prefixes[a]

Multiple	Prefix	Symbol	Multiple	Prefix	Symbol
10^{18}	exa	E	10^{-1}	deci	d
10^{15}	peta	P	10^{-2}	centi	c
10^{12}	tera	T	10^{-3}	milli	m
10^{9}	giga	G	10^{-6}	micro	μ
10^{6}	mega	M	10^{-9}	nano	n
10^{3}	kilo	k	10^{-12}	pico	p
10^{2}	hecto	h	10^{-15}	femto	f
10	deka	da	10^{-18}	atto	a

[a] Adopted by International Committee on Weights and Measures. (Compound prefixes should not be used; e.g., not $\mu\mu$ but p.)

Appendix D

Greek Alphabet

Letter	Lowercase	Uppercase	Letter	Lowercase	Uppercase
Alpha	α	A	Nu	ν	N
Beta	β	B	Xi	ξ	Ξ
Gamma	γ	Γ	Omicron	o	O
Delta	δ	Δ	Pi	π	Π
Epsilon	ϵ	E	Rho	ρ	P
Zeta	ζ	Z	Sigma	σ	Σ
Eta	η	H	Tau	τ	T
Theta	θ	Θ	Upsilon	υ	Υ
Iota	ι	I	Phi	ϕ	Φ
Kappa	κ	K	Chi	χ	X
Lambda	λ	Λ	Psi	ψ	Ψ
Mu	μ	M	Omega	ω	Ω

Appendix E

Physical Constants

Quantity	Symbol	Value
Angstrom unit	Å	1 Å = 10^{-4} μm = 10^{-8} cm
Avogadro constant	N_{AVO}	6.02204 × 10^{23} mol^{-1}
Bohr radius	a_B	0.52917 Å
Boltzmann constant	k	1.38066 × 10^{-23} J/K (R/N_{AVO})
Elementary charge	q	1.60218 × 10^{-19} C
Electron rest mass	m_0	0.91095 × 10^{-30} kg
Electron volt	eV	1 eV = 1.60218 × 10^{-19} J = 23.053 kcal/mol
Gas constant	R	1.98719 cal mol^{-1} K^{-1}
Permeability in vacuum	μ_0	1.25663 × 10^{-8} H/cm ($4\pi \times 10^{-9}$)
Permittivity in vacuum	ϵ_0	8.85418 × 10^{-14} F/cm ($1/\mu_0 c^2$)
Planck constant	h	6.62617 × 10^{-34} J-s
Reduced Planck constant	$\hbar$	1.05458 × 10^{-34} J-s ($h/2\pi$)
Proton rest mass	M_p	1.67264 × 10^{-27} kg
Speed of light in vacuum	c	2.99792 × 10^{10} cm/s
Standard atmosphere		1.01325 × 10^5 N/m^2
Thermal voltage at 300 K	kT/q	0.0259 V
Wavelength of 1-eV quantum	λ	1.23977 μm

Appendix F

Lattice Constants

	Element or Compound	Name	Crystal[a] Structure	Lattice Constant at 300 K (Å)
Element	C	Carbon (diamond)	D	3.56683
	Ge	Germanium	D	5.64613
	Si	Silicon	D	5.43095
	Sn	Grey Tin	D	6.48920
IV–IV	SiC	Silicon carbide	W	$a = 3.086$, $c = 15.117$
III–V	AlAs	Aluminum arsenide	Z	5.6605
	AlP	Aluminum phosphide	Z	5.4510
	AlSb	Aluminum antimonide	Z	6.1355
	BN	Boron nitride	Z	3.6150
	BP	Boron phosphide	Z	4.5380
	GaAs	Gallium arsenide	Z	5.6533
	GaN	Gallium nitride	W	$a = 3.189$, $c = 5.185$
	GaP	Gallium phosphide	Z	5.4512
	GaSb	Gallium antimonide	Z	6.0959
	InAs	Indium arsenide	Z	6.0584
	InP	Indium phosphide	Z	5.8686
	InSb	Indium antimonide	Z	6.4794
II–VI	CdS	Cadmium sulfide	Z	5.8320
	CdS	Cadmium sulfide	W	$a = 4.16$, $c = 6.756$
	CdSe	Cadmium selenide	Z	6.050
	CdTe	Cadmium telluride	Z	6.482
	ZnO	Zinc oxide	R	4.580
	ZnS	Zinc sulfide	Z	5.420
	ZnS	Zinc sulfide	W	$a = 3.82$, $c = 6.26$
IV–VI	PbS	Lead sulfide	R	5.9362
	PbTe	Lead telluride	R	6.4620

[a] D = Diamond, W = Wurtzite, Z = Zincblende, R = Rock salt.

Appendix G

Properties of Important Semiconductors

Semiconductor		Bandgap (eV)		Mobility at 300 K (cm²/V-s)[a]		Band[b]	Effective Mass m^*/m_0		ϵ_s/ϵ_0
		300 K	0 K	Elec.	Holes		Elec.	Holes	
Element	C	5.47	5.48	1800	1200	I	0.2	0.25	5.7
	Ge	0.66	0.74	3900	1900	I	1.64[c]	0.04[e]	16.0
							0.082[d]	0.28[f]	
	Si	1.12	1.17	1500	450	I	0.98[c]	0.16[e]	11.9
							0.19[d]	0.49[f]	
	Sn		0.082	1400	1200	D			
IV–IV	α-SiC	2.996	3.03	400	50	I	0.60	1.00	10.0
III–V	AlSb	1.58	1.68	200	420	I	0.12	0.98	14.4
	BN	~7.5				I			7.1
	BP	2.0							
	GaN	3.36	3.50	380			0.19	0.60	12.2
	GaSb	0.72	0.81	5000	850	D	0.042	0.40	15.7
	GaAs	1.42	1.52	8500	400	D	0.067	0.082	13.1
	GaP	2.26	2.34	110	75	I	0.82	0.60	11.1
	InSb	0.17	0.23	80000	1250	D	0.0145	0.40	17.7
	InAs	0.36	0.42	33000	460	D	0.023	0.40	14.6
	InP	1.35	1.42	4600	150	D	0.077	0.64	12.4
II–VI	CdS	2.42	2.56	340	50	D	0.21	0.80	5.4
	CdSe	1.70	1.85	800		D	0.13	0.45	10.0
	CdTe	1.56		1050	100	D			10.2
	ZnO	3.35	3.42	200	180	D	0.27		9.0
	ZnS	3.68	3.84	165	5	D	0.40		5.2
IV–VI	PbS	0.41	0.286	600	700	I	0.25	0.25	17.0
	PbTe	0.31	0.19	6000	4000	I	0.17	0.20	30.0

[a] The values are for drift mobilities obtained in the purest and most perfect materials available to date.
[b] I = indirect, D = direct.
[c] Longitudinal effective mass.
[d] Transverse effective mass.
[e] Light-hole effective mass.
[f] Heavy-hole effective mass.

Appendix H

Properties of Ge, Si, and GaAs at 300 K

Properties	Ge	Si	GaAs
Atoms/cm^3	4.42×10^{22}	5.0×10^{22}	4.42×10^{22}
Atomic weight	72.60	28.09	144.63
Breakdown field(V/cm)	$\sim 10^5$	$\sim 3 \times 10^5$	$\sim 4 \times 10^5$
Crystal structure	Diamond	Diamond	Zincblende
Density (g/cm^3)	5.3267	2.328	5.32
Dielectric constant	16.0	11.9	13.1
Effective density of states in conduction band, N_C (cm^{-3})	1.04×10^{19}	2.8×10^{19}	4.7×10^{17}
Effective density of states in valence band, N_V (cm^{-3})	6.0×10^{18}	1.04×10^{19}	7.0×10^{18}
Effective Mass, m^*/m_0			
Electrons	$m_l^* = 1.64$ $m_t^* = 0.082$	$m_l^* = 0.98$ $m_t^* = 0.19$	0.067
Holes	$m_{lh}^* = 0.044$ $m_{hh}^* = 0.28$	$m_{lh}^* = 0.16$ $m_{hh}^* = 0.49$	$m_{lh}^* = 0.082$ $m_{hh}^* = 0.45$
Electron affinity, χ(V)	4.0	4.05	4.07
Energy gap (eV) at 300 K	0.66	1.12	1.424
Intrinsic carrier concentration (cm^{-3})	2.4×10^{13}	1.45×10^{10}	1.79×10^6
Intrinsic Debye length (μm)	0.68	24	2250
Intrinsic resistivity (Ω-cm)	47	2.3×10^5	10^8
Lattice constant (Å)	5.64613	5.43095	5.6533

Properties	Ge	Si	GaAs
Linear coefficient of thermal expansion, $\Delta L/L\Delta T$ (°C^{-1})	5.8×10^{-6}	2.6×10^{-6}	6.86×10^{-6}
Melting point (°C)	937	1415	1238
Minority carrier lifetime (s)	10^{-3}	2.5×10^{-3}	$\sim 10^{-8}$
Mobility (drift) (cm^2/V-s)			
	3900	1500	8500
	1900	450	400
Optical-phonon energy (eV)	0.037	0.063	0.035
Phonon mean free path λ_0 (Å)	105	76 (electron) 55 (hole)	58
Specific heat (J/g-°C)	0.31	0.7	0.35
Thermal conductivity at 300 K (W/cm-°C)	0.6	1.5	0.46
Thermal diffusivity (cm^2/s)	0.36	0.9	0.24
Vapor pressure (Pa)	1 at 1330°C 10^{-6} at 760°C	1 at 1650°C 10^{-6} at 900°C	100 at 1050°C 1 at 900°C

Appendix I

Properties of SiO_2 and Si_3N_4 at 300 K

Insulator:	SiO_2	Si_3N_4
Structure	Amorphous	Amorphous
Melting point (°C)	~1600	—
Density (g/cm^3)	2.2	3.1
Refractive index	1.46	2.05
Dielectric constant	3.9	7.5
Dielectric strength (V/cm)	10^7	10^7
Infrared absorption band (μm)	9.3	11.5–12.0
Energy gap (eV)	9	~5.0
Thermal-expansion coefficient (°C^{-1})	5×10^{-7}	—
Thermal conductivity (W/cm-K)	0.014	—
dc resistivity (Ω-cm)		
at 25°C	10^{14}–10^{16}	$\sim10^{14}$
at 500°C	—	$\sim2\times10^{13}$
Etch rate in buffered HF[a] (Å/min)	1000	5-10

[a]Buffered HF: 34.6% (wt.) NH_4F, 6.8% (wt.) HF, 58.6% H_2O.

INDEX